SOLID WASTE LANDFILLING

SOLID WASTE LANDFILLING

Concepts, Processes, Technologies

RAFFAELLO COSSU
University of Padova, Department of Civil, Environmental and Architectural Engineering, Padova, Italy

RAINER STEGMANN
Hamburg University of Technology, Institute of Environmental Technology and Energy Economics (retired), Hamburg, Germany

ELSEVIER

Elsevier
Radarweg 29, PO Box 211, 1000 AE Amsterdam, Netherlands
The Boulevard, Langford Lane, Kidlington, Oxford OX5 1GB, United Kingdom
50 Hampshire Street, 5th Floor, Cambridge, MA 02139, United States

Library of Congress Cataloging-in-Publication Data
A catalog record for this book is available from the Library of Congress

British Library Cataloguing-in-Publication Data
A catalogue record for this book is available from the British Library

ISBN: 978-0-12-818336-6

Publisher: Candice Janco
Acquisition Editor: Candice Janco
Editorial Project Manager: Emily Thomson
Production Project Manager: Mohanapriyan Rajendran
Designer: Greg Harris

Typeset by TNQ Technologies

Credit for Visual organization: *Arcoplan Associated, Padova, Italy*

CONTENTS

14. SUSTAINABLE LANDFILLING

15. LANDFILL OPERATION

16. AFTERCARE

17. ENVIRONMENTAL IMPACT

LIST OF CONTRIBUTORS

LUCA ALIBARDI | Cranfield University, Water Science Institute, Vincent Building, College Road, Cranfield MK43 0AL, Bedfordshire, UK *l.alibardi@cranfield.ac.uk*

GIANNI ANDREOTTOLA | University of Trento, Department of Civil, Environmental and Mechanical Engineering, Via Mesiano, 77 - 38123 Trento, Italy *gianni.andreottola@unitn.it*

ANNA ARTUSO | Arcoplan Associates, Via Beato Pellegrino 23, 35137, Padova, Italy *studio@arcoplan.it*

NANJAPPA ASHWATH | Central Queensland University, Institute for Future Farming Systems, Ibis Avenue, Rockhampton North, QLD Australia 4701 *n.ashwath@cqu.edu.au*

THOMAS F. ASTRUP | Technical University of Denmark, Department of Environmental Engineering, Bygningstorvet, Lyngby, DK-2800, Denmark *thas@env.dtu.dk*

OFIRA AYALON | University of Haifa & Samuel Neaman Inst., Haifa, Israel *ofira@sni.technion.ac.il*

IOANNIS BAKAS | DTU Environment, Department of Environmental Engineering, Technical University of Denmark, 2800 Kgs. Lyngby (DK)

ALBERTO BARAUSSE | University of Padova, Department of Industrial Engineering (DII) - Via Marzolo 9, 35131 Padova, Italy *alberto.barausse@unipd.it*

EYAD S. BATARSEH | Engicon, PO Box 926963, Amman 11190, Jordan *batarsehes@cdm.com*

RICHARD BEAVEN | Waste Management Research Group, University of Southampton, School of Engineering, Highfield, Southampton, SO17 1BJ, United Kingdom. *r.p.beaven@soton.ac.uk*

PIA BENAUD | College of Life and Environmental Sciences, University of Exeter, Amory Building - Rennes Drive, Exeter EX4 4RJ, United Kingdom *P.Benaud@exeter.ac.uk*

GIOVANNI PIETRO BERETTA | Department of Earth Science "Ardito Desio", University of Milan, Via Mangiagalli 34, 20133 Milano *giovanni.beretta@unimi.it*

NICOLE D. BERGE | University of South Carolina, Department of Civil and Environmental Engineering - 300 Main Street, Columbia SC 29208, USA *berge@engr.sc.edu*

STEPHANIE C. BOLYARD | Environmental Research & Education Foundation (EREF), 3301 Benson Drive, Suite 101, Raleigh, NC 27609 *sbolyard@erefdn.org*

LINE KAI-SØRENSEN BROGAARD | DTU Environment, Department of Environmental Engineering, Technical University of Denmark, 2800 Kgs. Lyngby (DK) *gb7z@tmf.kk.dk*

LUCIANO BUTTI | Studio legale Butti & Partners, Via Leoni 4, 37121 Verona, Italy *luciano.butti@buttiandpartners.com*

GIOVANNA CAPPAI | University of Cagliari, Department of Civil - Environmental Engineering and Architecture (DICAAR), Piazza d'Armi, 1- 09123 Cagliari, Italy *gcappai@unica.it*

GIULIA CERMINARA | University of Padova, Department of Civil, Environmental and Architectural Engineering (DICEA) - Via Marzolo 9, 35131 Padova, Italy *giulia.cerminara89@gmail.com*

ELENA COSSU | Arcoplan Associates, Via Beato Pellegrino 23, 35137, Padova, Italy *elena.cossu@eurowaste.it*

RAFFAELLO COSSU | University of Padova, Department of Civil, Environmental and Architectural Engineering (DICEA) - Via Marzolo 9, 35131 Padova, Italy *raffaello.cossu@gmail.com*

ANDERS DAMGAARD | DTU Environment, Department of Environmental Engineering, Technical University of Denmark, 2800 Kgs. Lyngby (DK) *anders@brdr-damgaard.dk*

GIOVANNI DE FEO | University of Salerno, Department of Industrial Engineering, via Giovanni Paolo II 132, 84084 Fisciano, Salerno, Italy *g.defeo@unisa.it*

HANS-JÜRGEN EHRIG | University of Wuppertal, Water and Waste Management, (retired); Hohhle Gasse 106, D-53177 Bonn *hj.ehrig@t-online.de*

MARCO FAVARETTI | University of Padova, Department of Civil, Environmental and Architectural Engineering (DICEA) - Via Ognissanti 39, 35129 Padova, Italy *marco.favaretti@unipd.it*

ANDY B. FOURIE | Department of Civil, Environmental and Mining Engineering, University of Western Australia Location, 35 Stirling Highway, Perth WA 6009 Australia *andy.fourie@uwa.edu.au*

FRANCESCO GARBO | University of Padova, Department of Civil, Environmental and Architectural Engineering (DICEA) - Via Marzolo 9, 35131 Padova, Italy *francesco.garbo.2@phd.unipd.it*

HOSSEIN GHADIRI | Griffith University, 170 Kessels Road, Nathan QLD Australia 4111 *h.ghadiri@griffith.edu.au*

JEAN PIERRE GOURC | Institut de Geosciences de l'Environnement, Université Grenoble-Alpes,38058 Grenoble cedex 9, France *jeanpierre.gourc@gmail.com*

MARGARET GREENWAY | Griffith University, 170 Kessels Road, Nathan QLD Australia 4112 *m.greenway@griffith.edu.au*

VALENTINA GROSSULE | University of Padova, Department of Civil, Environmental and Architectural Engineering (DICEA) - Via Marzolo 9, 35131 Padova, Italy *valentina.grossule@phd.unipd.it*

KAI-UWE HEYER | IFAS, Ingenieurbüro für Abfallwirtschaft, Prof. R Stegmann und Partner, Schellerdamm 19 - 21, D-21079 Hamburg, Germany *heyer@ifas-hamburg.de*

OLE HJELMAR | Danish Waste Solutions ApS, Agern Allé 3, 2970 Hørsholm, Denmark *oh@danws.dk*

KARSTEN HUPE | IFAS, Ingenieurbüro für Abfallwirtschaft, Prof. R Stegmann und Partner, Schellerdamm 19 - 21, D-21079 Hamburg, Germany *hupe@ifas-hamburg.de*

MARK B. JAKSA | School of Civil, Environmental and Mining Engineering, University of Adelaide, 5005, Australia *mark.jaksa@adelaide.edu.au*

PETER KJELDSEN | Technical University of Denmark, Department of Environmental Engineering, Bygningstorvet, Lyngby, DK-2800, Denmark *pekj@env.dtu.dk*

KEITH KNOX | Knox Associates (UK) Ltd, Barnston Lodge, 50 Lucknow Avenue, Mapperley, Park, Nottingham, NG3 5BB *keith.knox@knoxuk.co.uk*

GEORGE R. KOERNER | Geosynthetic Institute, 475 Kedron Avenue Folsom, PA 19033-1208 USA *gsigeokoerner@gmail.com*

ROBERT M. KOERNER | Geosynthetic Institute, 475 Kedron Avenue Folsom, PA 19033-1208 USA *rmk27@drexel.edu*

DAVID KOSSON | Vanderbilt University, Dept. of Civil and Environmental Engineering, 400 24th Avenue South, 267 Jacobs Hall, Nashville, TN 37212 *david.kosson@vanderbilt.edu*

TIZIANA LAI | Agenzia Conservatoria delle coste - Regione Autonoma della Sardegna - Via Mameli 96, 09123 Cagliari *lai.tiziana@gmail.com*

MARIA CRISTINA LAVAGNOLO | University of Padova, Department of Civil, Environmental and Architectural Engineering (DICEA) - Via Marzolo 9, 35131 Padova, Italy *mariacristina.lavagnolo@unipd.it*

PETER LECHNER | Universität für Bodenkultur Wien, Muthgasse 107 - 1190 Wien, Austria *peter.lechner@boku.ac.at*

CATERINA LOPS | ENI, Legal Counsel, ENI SpA, Milano, Italy *caterina.lops@buttiandpartners.com*

WENJING LU | School of Environment, Division of Solid Waste Management, Sino-Italian Environment and Energy Building, Tsinghua University, Beijing 100084, China *luwenjing@tsinghua.edu.cn*

SIMONE MANFREDI | European Commission, Joint Research Centre, Directorate D "Sustainable Resources"- via Enrico Fermi 2749, 21027 Ispra, Varese *simone.manfredi@ec.europa.eu*

YASUSHI MATSUFUJI | Fukuoka University, (retired), 8-19-1 Nanakuma, Johnan-ku, Fukuoka, Japan, 814-0180 *ymatsufu@outlook.jp*

LUCA MORELLO | T&D Water and Energy Green Solutions, Via E. Fermi 6, 35020 Polverara, Padova, Italy *luca@t-and-d-italy.com*

ALDO MUNTONI | University of Cagliari, Department of Civil - Environmental Engineering and Architecture (DICAAR), Piazza d'Armi, 1- 09123 Cagliari, Italy *amuntoni@unica.it*

FEDERICO PERES | Studio legale Butti & Partners, Via leoni 4, 37121 Verona, Italy *federico.peres@buttiandpartners.com*

ALBERTO PIVATO | University of Padova, Department of Civil, Environmental and Architectural Engineering (DICEA) - Via Marzolo 9, 35131 Padova, Italy *alberto.pivato@unipd.it*

ALESSANDRA POLETTINI | University of Rome "La Sapienza", Department of Civil and Environmental Engineering, Via Eudossiana 18 – 00184 Rome, Italy *alessandra.polettini@uniroma1.it*

RAFFAELLA POMI | University of Rome "La Sapienza", Department of Civil and Environmental Engineering, Via Eudossiana 18 – 00184 Rome, Italy *raffaella.pomi@uniroma1.it*

ROBERTO RAGA | University of Padova, Department of Civil, Environmental and Architectural Engineering (DICEA) - Via Marzolo 9, 35131 Padova, Italy *roberto.raga@unipd.it*

HANS-GÜNTER RAMKE | Hochschule Ostwestfalen-Lippe, University of Applied Sciences, Campus Höxter - Faculty of Environmental Engineering and Applied Informatics, Professorship of Waste Management and Landfill Technology, An der Wilhelmshöhe 44, D-37671 Höxter, Germany *hans-guenter.ramke@hs-owl.de*

DEBRA REINHART | University of Central Florida, Civil, Environmental and Construction Engineering, 4365 Andromeda Loop N. Orlando, FL 32816 *Debra.Reinhart@ucf.edu*

GERHARD RETTENBERGER | University of Applied Science Trier and Ingenieurgruppe RUK, Auf dem Haigst 21, Stuttgart, 70597 Germany *rettenberger@ruk-online.de*

MARCO RITZKOWSKI | Hamburg University of Technology, Institute of Environmental Technology and Energy Economics, Harburger Schlossstrasse 36 - 21079 Hamburg, Germany *m.ritzkowski@tuhh.de*

TIM ROBINSON | Phoenix Engineering, Phoenix House, Scarne Mill Industrial Estate, Launceston, Cornwall, PL15 9GL, United Kingdom *Tim.Robinson@phoenix-engineers.co.uk*

MELISSA SALT | Tonkin Consulting, Lvl 2, 6 Rundle St, Kent Town SA Australia 5067 *Melissa.Salt@tonkin.com.au*

CHARLOTTE SCHEUTZ | Technical University of Denmark, Department of Environmental Engineering, Bygningstorvet, Lyngby, DK-2800, Denmark *chas@env.dtu.dk*

RAINER STEGMANN | Hamburg University of Technology, Institute of Environmental Technology and Energy Economics, (retired), Harburger Schlossstrasse 36 - 21079 Hamburg, Germany *stegmann@tuhh.de*

JIANLEI SUN | Golder Associates, 570-580 Swan St, Richmond VIC Australia 3121 *jisun@golder.com.au*

AIAKO TANAKA | Fukuoka University, 8-19-1 Nanakuma, Johnan-ku, Fukuoka, Japan, 814-0180 *ayakot@fukuoka-u.ac.jp*

DAVIDE TONINI | European Commission - Joint Research Centre, Edificio Expo, Calle Inca Garcilaso, 3-41092 Seville, Spain *davide.tonini@ec.europa.eu*

HANS A. VAN DER SLOOT | Hans van der Sloot Consultancy, Glenn Millerhof 29, 1628 TS Hoorn, The Netherlands *hans@vanderslootconsultancy.nl*

TOM VAN GERVEN | Department of Chemical Engineering - KU Leuven, Celestijnenlaan 200F, 3001 Leuven, Belgium *tom.vangerven@kuleuven.be*

ANDRE VAN ZOMEREN | Energy Research Center of the Netherlands (ECN part of TNO), Westerduinweg 3, 1755 LE Petten, The Netherlands *andre.vanzomeren@tno.nl*

VOLKMAR WILHELM | Volkmar Wilhelm, Brenzstraße 34, D-71636 Ludwigsburg *volkmar.wilhelm@online.de*

NICK D. WOODMAN | Waste Management Research Group, University of Southampton, School of Engineering, Highfield, Southampton, SO17 1BJ, United Kingdom. *n.d.woodman@soton.ac.uk*

SAM T.S. YUEN | University of Melbourne, C201 Engineering Block C, The University of Melbourne VIC Australia 3000 *stsy@unimelb.edu.au*

DIMITRIOS ZEKKOS | University of Michigan, 2350 Hayward Street, Ann Arbor 48109, USA *zekkos@geoengineer.org*

GRANT X. ZHU | Formerly at the University of Melbourne, C201 Engineering Block C, The University of Melbourne VIC Australia 3000

PREFACE

There are many important things in life that we do but are careful not to mention. Landfilling of waste is becoming one of these things!

Landfilling has gained a bad reputation, with a large part of the population maintaining that all waste problems can be solved by means of recycling (circular economy, zero waste, urban mining); indeed, landfilling (at least in Europe) is increasingly viewed as an obsolete method that should be banned. In spite of this, however, landfilling continues to represent the most widely implemented waste disposal option worldwide.

There is also a widely held belief that landfilling is an economical means of disposing of waste, which can be conveniently forgotten 30 years after landfill closure with no remaining environmental risk. These contrasting views are both equally detrimental to the safeguard of the environment.

The first tends to negatively classify the landfill itself rather than the inappropriate use of this system. Even the name "landfilling" is something to be avoided at all costs! The second view tends to completely underestimate the environmental problems deriving from the inappropriate application of landfilling, thus ultimately providing support to the first view.

Conversely, our view is the result of several decades spent in the field of landfill research, design and planning. We consider landfill as a fundamental inevitable tool for use in closing the material loop, but in the same way as all tools it, needs to be used judiciously and scientifically, bearing in mind the problems to be solved and the final targets. This has motivated us to conceive this book on sanitary landfilling.

In this volume we intend to demonstrate how landfills are the guiding theme underpinning all solid waste management strategies, including the circular economy! Indeed, primary resources are first extracted from the ground for use in industrial processes; once used, they then represent secondary resources suitable for recycling, and ultimately end up as wastes to be returned in an immobilized form to the ground the resources were originally obtained from. Indeed, a landfill is intended as a final geological deposit (sink). A world without landfilling would be at considerably higher risk of increased diffuse pollution. This approach is visualized on the front page of the book where the loop symbolizes a circular economy inevitably linked to a landfill as a sink.

Furthermore, our aim was to provide particularly detailed information relating to landfilling concepts, processes, technologies, planning, and design, on the one hand illustrating the complexity of the construction, while on the other underlining the importance of qualified operation. A significant part of the book is dedicated to the environmental sustainability of landfills in terms of analysis and control of long-term impacts.

We would like to convey our sincere thanks to all the colleagues who contributed to the book, to the Elsevier managers for their continuous support and encouragement, to Elena Cossu for the visual organization of the book (cover, internal template), to Anne Farmer for reviewing the manuscripts and to Valentina Grossule for the final riveting proof readings. And, last but by no means least, to our children and partners for their patience over the years when our devotion to the book frequently clashed with the attention and care we should have dedicated to them.

Raffaello Cossu
Rainer Stegmann

1 CHAPTER

GENERAL ASPECTS

1.1

WASTE MANAGEMENT STRATEGIES AND ROLE OF LANDFILLING

Raffaello Cossu and Rainer Stegmann

GLOBAL POLLUTION OF THE ENVIRONMENT

The increasing diffusion of contaminants in the anthroposphere undoubtedly represents a major environmental issue. Air pollution is so high in many countries that the authorities recommend staying indoors in at-risk periods. In the majority of big cities, prescribed air emission standards are not met (e.g., particles, NO_X), in many countries worldwide tap water is not suitable for drinking, huge areas of soil are contaminated and cannot be used for food production, and plants and animals accumulate toxic compounds and are not fit for human consumption. Global warming precipitates climate change generating elevated water levels, widespread storms, and flooding. Due to contamination in many areas, groundwater may not be suitable for drinking unless extensively treated.

Copious amounts of information are available relating to the pollution of snow and ice by xenobiotics, even in the polar regions. Unpolluted samples of soil and ice can only be obtained from the deep layers of ice or sediments. Plastic pollution in rivers and oceans has reached a truly frightening dimension; besides the massive accumulation of plastic that affects flora and fauna, larger particles are mechanically and/or photochemically reduced to fine plastic particles, which are dispersed throughout the environment.

Some emissions that enter the environment either "cannot" be further reduced or avoided or it would be very difficult/costly to do so, e.g., residual emissions from waste-, water-, and off-gas treatment plants, from compost applied to land, etc. Treatment plants are not able to reach an efficiency rate of 100% and emissions—although low—remain and accumulate over the years. In addition "new" waste is produced from treatment processes, including sludge from precipitation processes, loaded sorption materials, filter ashes, etc. This secondary pollution will also need to be treated and/or securely landfilled; additionally, during treatment energy is consumed, which is associated with further secondary emissions.

Another area of environmental pollution that has been scarcely investigated to date—particularly as the possibilities for reduction are very limited—is represented by the outcome of daily activities: rubber abrasion from tires, fine dust from the intensive use of agricultural land, rust and paint washed from man-made constructions, residues from clothing, house dust, erosion from asphalt covered streets, concrete and plastic from buildings, abrasion of metals from railway tracks, evaporated organics, nanoparticles, etc.

FATE OF CONTAMINANTS

In accordance with the law of Lomonosov and Lavoisier, material cannot be lost, they can only be transformed. On this basis, therefore, all emitted compounds will be distributed throughout the environment by water, air, and wind and may be biologically degraded or physically/chemically converted.

Biologically stable compounds such as aromatics, chlorinated hydrocarbons, and other xenobiotics, which may also be produced during thermal processes, are often of low water solubility; they will be adsorbed/absorbed to soil particles or other surfaces and are distributed in the environment. Some of the compounds will be degraded after longer periods of time, whereas others may remain or at times be converted into even more hazardous compounds. Little is known about the long-term behavior of nanoparticles, endocrine substances, etc., in the different media: air, water, and soil. Heavy metals remain indefinitely and may react with other elements, be adsorbed/absorbed to particles, and solubilized or chemically bound; even in a less mobile state they may be mobilized when the chemical/physical conditions change.

Environmental protection measures are heavily reliant on dilution. However, with time the concentration of pollutants in the water, air, and soil will increase. Many pollutants will accumulate over lengthy periods in the environment and create a toxic local, regional, and global environment. The latter also serves to demonstrate how dilution has never been and indeed never will represent a long-term solution. The ultimate goal is to prevent dilution as far as possible and aim to achieve the concentration and isolation of final emissions from the environment.

Prof. Rockström, together with a group of 21 scientists, proposed nine criteria that should stay within the Planetary Boundaries as a prerequisite for sustainable life on our planet: Climate Change, Ocean Acidification, Stratospheric Ozone, Global P and N Cycles, Atmospheric Aerosol Loading, Freshwater Use, Land Use Change, Biodiversity Loss, and Chemical Pollution (Fig. 1.1.1) (Rockström et al., 2009). These scientists made it very clear that these boundaries should not be crossed, to prevent an irreversible impact on the environment. This is indeed already the case for biodiversity loss and global N cycle. The group has not yet defined boundaries for Chemical Pollution and Atmospheric Aerosol Loading.

THE ROLE OF WASTE MANAGEMENT

Environmental pollution and the effects of the latter are largely determined by emissions originating from human activities as the result of emissions in a gaseous, liquid, or solid form. In the area of waste management, emissions may originate from littering, waste incineration, landfilling, biological treatment, manure, sludge and compost application on land, recycling processes, mining, production processes, waste treatment, accidents, etc. Waste research and management should also be seen as an integrated part of the efforts undertaken to remain within the planetary boundaries as long as possible.

Worldwide Waste Management Strategies

Although significant waste quantities are currently avoided and recycled, worldwide waste amounts are still on an increase up to ±1 kg/person per day. Waste composition is similar throughout all large cities

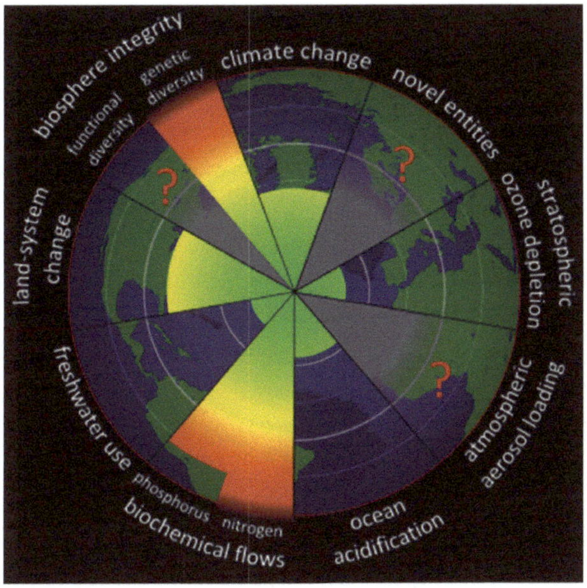

Figure 1.1.1 Visual presentation of the "planetary boundaries" by Prof. Rockström (Steffen et al., 2015).

in the world, with a tendency toward a higher degradable organic content in Asia. However, based on specific local situations in different countries, waste production and composition in rural areas may differ considerably (Chapter 1.2).

In the field of waste management, all countries have access to more or less the same tools (with the possible exception of high-technology plants): (separate) collection, landfilling, composting, anaerobic digestion, thermal treatment, and mechanical and manual sorting and separation; these tools can be used in line with specific waste production and composition, the waste management tradition, legislation, political concepts, influence of NGOs, infrastructure, climate, topography, etc.

Separate collection is carried out at varying levels in many countries. This also implies an increase in material recycling where significant amounts of packaging waste, paper, glass, and plastic materials are recovered from waste. In many regions kitchen and yard waste is separately collected, biologically treated, and converted into compost. In the United States, food waste residues are also discharged into the kitchen sink where—after grinding—the waste enters the sewage collection system.

The future of separate collection is uncertain and will depend on several factors such as:

- the specific regional and local situation with regard to waste composition, climate, space for several collection bins, etc.,
- the value of the different waste compounds recovered and available markets,
- the availability of automatic sorting systems,
- the support of the political sector,
- the fee system,
- the support of the population, etc.

Following the introduction of a large variety of bins for separate collection in several countries, these may now be reduced in the future to comprise solely a wet (organic degradable waste) and a dry waste bin for other compounds; to achieve a good quality of collected materials a third bin for residual waste may be advantageous. Furthermore, some countries may benefit from a bin for paper or other specific materials. The future of separate collection of packaging waste is, in the opinion of the author, uncertain. This will depend also on the future development of automatic mechanical separation systems (e.g., using near infrared technology). In many countries scavengers are the main players involved in the separate collection of materials of sufficient economic interest, with this informal sector contributing significantly to the recovery of materials from waste.

The installation of community centers to which that is either outside the categories collected at the kerbside or is too voluminous to be placed in the bins should be widely promoted whenever possible. Household toxic wastes, rubble, high amounts of packaging waste, garden waste, and of course household machines, electronic waste, etc. could be deposited in these centers. Although this practice might imply an increase in the amounts of residual waste, the main focus of separate collection should be on gaining high quality materials, also providing for the separate collection of the wet waste fraction.

As is already the case in many countries, cars, electrical and electronic equipment, used clothes, etc. should be separately collected and recycled. The collection of waste electric and electronic equipment (WEEE) should be addressed as a matter of priority. Huge numbers of computers are legally exported, particularly to Africa and China, as still functional used equipment. In these countries valuable materials are recovered by the informal sector by means of unacceptable primitive methods with a high risk for the public health; the high portion of nonreusable materials is discarded in an uncontrolled manner in the environment. Although car recycling is of a relatively high standard in industrialized countries, for the problem of the remaining fluff (mixture of plastic, rubber, glass, tissue materials, etc. often contaminated by car fluids) no satisfactory solution has yet been identified.

In Europe and Japan, significant amounts of municipal solid waste (MSW) are incinerated, with only relatively small amounts—especially in Japan—being treated by pyrolysis/gasification. After incineration approximately ±20% of the incoming waste is converted into bottom and fly ash; part of the bottom ash is recycled and used in road construction, for example, although the major part is landfilled. Because of the significant emission potential of the ashes (Chapters 4.2 and 20.2), there are no satisfactory sustainable solutions for the ash utilization. Due to the concern over potential negative health effects from the off-gases, numerous countries have opted not to introduce incinerators. This refusal should be overcome by increasing awareness and providing appropriate information, as well as by meeting very low emissions and ensuring the transparency of operations. Bearing this in mind, the authors are of the opinion that the way ahead in reducing the amounts and emission potential of residual waste lies in incineration. This is particularly true when the amount of residual waste to be treated is >150,000 t/a. However, thermal treatment will result in high energy recovery rates (heat is also needed in the production of electricity) and better ash quality. There is also a considerable potential for the production of high quality refuse-derived fuel (RDF) for use in cement kilns, ore melting plants, cofiring in coal power plants, etc.

A particular challenge is represented by the production of RDF with a low promoting corrosion potential. RDF production may also be an option for smaller communities to thermally treat part of their residual waste in central plants. Incineration should be the priority option for large cities.

The future of pyrolysis and gasification is hard to predict. An increased use of gasification using homogeneous waste such as tires, plastic, straw, dewatered sludge, RDF, etc. will likely be witnessed in the future.

Worldwide, landfilling remains the most frequently used method of MSW disposal (Chapters 2.1 and 2.2). In Europe it is prescribed to reduce to a minimum the amount of putrescible waste disposed of in landfills through the application of thermal or mechanical/biological pretreatment. In the latter case, the high calorific fraction is separated from the MSW and—as mentioned earlier—can be used as a fuel (RDF), whereas the residual fraction is biologically treated to low organic concentrations and subsequently landfilled (Chapter 4.1). Another pretreatment option is incineration where the bottom ashes are also landfilled. Therefore, the focus of landfilling should be on the safe disposal of nonbiodegradable residual waste. Landfills should act as final sinks to isolate nonuseable concentrated residual waste from the environment. These landfills may be combined with intermediate storage areas into which residues may be disposed and—with a high degree of certainty—may be recovered in the future. The technical standards for these storage areas should be the same as prescribed for landfills.

Hazardous and Special Wastes (HSWs) are currently forwarded for treatment to HSW incinerators and hazardous waste landfills, although codisposal with MSW continues to be applied worldwide. However, codisposal should be prohibited and industries should be forced to reduce the production of wastes, disposing of the nonuseable fraction in compliance with legally enforced standards. Loopholes allowing industries from industrialized countries to carry out their production activities in developing countries to save money by adopting lower emission and disposal standards should be accurately closed. As a general rule, the goal for landfilling for all kinds of waste is to have only one type of landfill, i.e., waste input quality standards for HSW should be the same as those applied for residual MSW. Thus, significantly concerted efforts should be made to prevent the generation and reuse of HSW and to enforce the implementation of pretreatment processes prior to landfilling (see also Chapter 2.1: the multibarrier system, waste as a barrier). The rationale underlying this strategy originates from the landfill design for MSW and HSW; the main, scarcely significant, difference between these two types of landfill is barrier thickness; a thicker barrier, however, does not necessarily imply a higher safety standard, it may only extend the time until leaking is manifested.

A ban should be implemented on the landfilling of organic HSW, and this waste should be incinerated in specifically designed high temperature furnaces; due to the high temperatures the resulting ashes are vitrified, thus resulting in a significantly lower emission potential (Chapter 4.2).

Waste Management in Economically Developing and Transient Countries

In economically developing and transient countries, waste management is frequently seen as a low priority, largely not only due to lack of financial resource but also due to a poor environmental awareness among politicians and the population (Chapter 13.3).

Waste collection is not implemented consistently throughout these countries or may be carried out in an inappropriate manner; in addition, incoming waste at landfills/open dumps is not adequately compacted. As a consequence, waste littering occurs with the effects mentioned previously, and waste dumped on the land attracts insects, birds, and mammals with a high risk of disease. The spreading of untreated manure on the fields may also be a health hazard (the spread of avian flu may be enhanced by means of an inappropriate chicken manure management).

As mentioned earlier, in developing countries (DCs) the collected waste is often disposed of in open dumps with no emission control. Scavengers then visit the dumps to collect materials for recycling, subjecting themselves to toxic fumes and dangerous waste components. Animals may stray onto the dumps in the search for food. The collection of reusable materials is often practiced by the informal sector in the communities and on the dumps. In many cases these informal systems can be quite effective and economically feasible.

The remediation of existing dumps and the design, construction, and operation of new landfills should be viewed as high priority initiatives. It is not necessary to invest large amounts of money to solve many issues resulting from the use of open dumps. Available options include adjusting of slopes, waste compaction, the collection of leachate and treatment by means of simple biological systems such as ponds, and simplified extraction and use of the gas. However, the raising of awareness in and education of the population is mandatory. By adopting these measures, not always associated with higher cost, waste avoidance and recycling may be increased, particularly in the cities. In rural areas home composting may contribute toward reducing waste and producing compost for individual use.

As indicated previously, the informal sector needs to be organized more efficiently to allow a step-by-step integration into the public sector. This would result in a decreased incidence of health risks and, in many cases, improved income potential. In Brazil, for example, where the informal sector—with the support of the community—has become increasingly organized, considerably improved working conditions have been created (Gutberlet, 2013).

Waste Management in Megacities

Waste management in megacities—i.e., cities of $\pm$>10 Mill. inhabitants—presents several highly specific issues. Limited space may impact on the placement of the necessary waste bins, with transfer stations being required to transport the waste to the treatment facilities frequently located outside the city confines. Accordingly, separate collection may only be carried out in certain areas or by using bring-based systems. The separate collection of kitchen and yard waste may be problematic due to a lack of adequate locations for biological treatment in the city; in addition there may also be a lack of areas for the implementation of compost utilization. A further challenge is represented by the development and operation of (largely automatic) anaerobic digesters suitable for integration into the cities. Compost may be used in the city gardens operated by local communities.

Megacities located in tropical areas will need to rely on a high frequency of waste collection. Due to dense traffic, waste collection will often need to be carried out at night. Bring-based systems and community centers may contribute toward increasing recycling rates and separately collecting dangerous and toxic waste from both households and the commercial sector. Where feasible, the mechanical separation of materials for recycling may be implemented at the transfer station or the treatment site. Separate collection of paper should be made compulsory in office buildings, etc. Different approaches may of course be adopted throughout the various sections of megacities (mini environment); in newly built centers the infrastructure for separate waste collection should have high priority. Alternative collection systems including subsurface containers and vacuum pipe systems may be envisaged.

Decentralized systems in which MSW (organic degradable fraction) and wastewater compounds are treated jointly may, under specific conditions, be an option (Stegmann et al., 2003).

ROLE OF LANDFILLING

As described earlier, the passage from a linear to a circular approach is progressively characterizing the modern waste management strategies.

The linear traditional approach (Take-Make-Waste) based on the extraction of raw materials, production, use, wasting, and landfilling (represented by a dotted line in Fig. 1.1.2) is obsolete and progressively abandoned.

The circular approach primarily arises from a growing need for primary raw material, as a consequence of global economic development. Attention is currently shifting from the limited and fixed stocks of raw materials to the increasing anthropogenic stocks of materials. This creates the base for

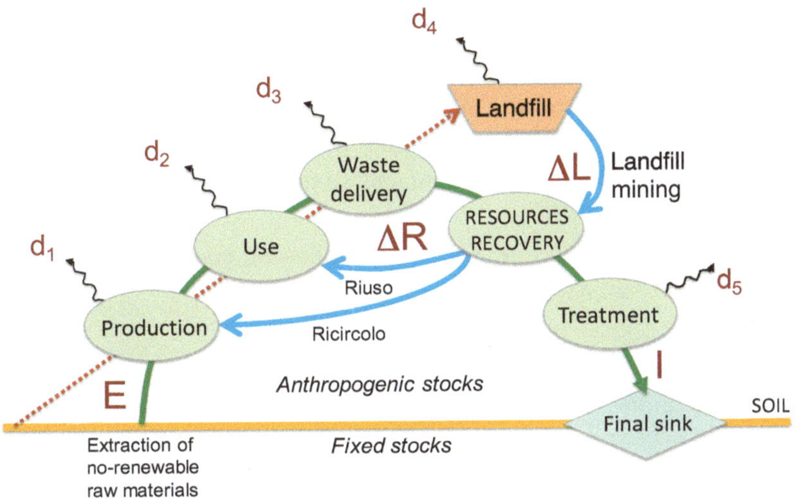

Figure 1.1.2 Role of landfilling as a final sink in circular waste management strategies in the material's life cycle. *Modified from Cossu et al. (2012).*

the development of different strategies for recovering of resources from waste (urban mining, circular economy, etc.).

As already mentioned, several waste materials, further to reusables and the traditional fractions of MSW, which are normally considered in source segregation programs (plastics, paper, cardboard, glass containers, cans, putrescibles, etc.), are being increasingly considered for recovery, reuse, or recycling as urban resources. The latter includes the following: WEEE, end-of-life vehicles, scrapped tires, construction and demolition waste, combustion residues, food waste, road sweeping waste, water treatment sludges, exhausted oils, old landfilled waste, residues from food industries, slags, and other industrial waste. These materials can alternatively be used to obtain different kinds of products, such as secondary raw materials, building materials, fuel and biofuel, composites, fertilizers, and others.

However, even using a circular approach, diffusion of contaminants, although reduced, continue to represent a problem and modern landfilling may represent, contrary to the generally held belief, a tool for use in reducing the problem.

For a better understanding of the role of landfilling, the general matter cycle, including emission control strategies and final materials sink, should be taken into consideration, as described by the scheme in Fig. 1.1.2.

Fossil materials are extracted from the soil to feed industrial production. Processing of the extracted material mobilizes the elements and compounds, which were naturally in a nonmobile form. During the production phase emissions of contaminants, even in low concentrations that respect the limits set by national regulations, are originated and may continue once the products are used (e.g., use of cars). At the end of their life, products will become residues. Handling of residues and recovery of resources (sorting, transporting, preparing for recycling) will once again produce diffuse emissions. Recovered waste resources can either be reused or recycled to the production processing. The excavation of waste from old landfills may provide additional resources (see Chapter 19.3), although giving rise to further emissions. At the end of any material recovery strategy, a final waste, which may be suitable for additional recovery, is generated. The disposal of these materials may be problematic as the toxic substances present in the recycled residues are frequently accumulated. To avoid any kind of pollution, residual wastes should be treated to immobilize the contaminants and a suitable sink should be considered to close the loop.

Considering the system described in Fig. 1.1.2, the following mass balance can then be drafted:

$$E = \Delta R + \Delta L + \Sigma di + I \quad (1)$$

where: E, extracted raw material; ΔR, recycled and reused material (secondary raw materials); ΔL, recovered material from landfill mining (secondary raw materials); di, diffuse mass emissions/loss associated with the specific steps and processes; and I, immobilized material in the final sink.

It is evident that the sum of diffuse emissions should be carefully controlled and minimized being the main cause underlying the progressive deterioration of global environmental quality.

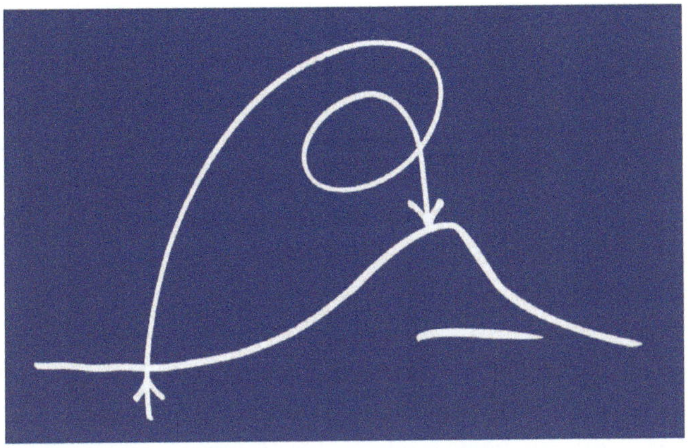

Figure 1.1.3 The COSTE (COSsu–STEgmann) logo, which describes graphically the modern role of landfilling in contributing toward reducing the global pollution problem. After production and recycling loops, materials extracted from the soil (where they were in a nonmobile form) should again be returned to the soil in a nonmobile form. Landfilling represents one of the possible final sinks.

To understand how to achieve the latter (control and minimization), a rearrangement of Eq. (1) may prove helpful:

$$\Sigma di = E - \Delta R - \Delta L - I \qquad (2)$$

Analyzing Eq. (2) it is clear what should be done to minimize diffuse emissions and consequent risks of environmental pollution:

- minimization of the extraction of raw materials
- maximization of the recovery, reuse, and recycle of residues
- maximization of the mining of old landfills
- increase of the immobilization of materials in final sink (Cossu, 2012).

The last point represents the key to allow us to fully grasp the prospect and modern role of landfilling: to act as a tool with which to close the material loop, returning to the soil, in a nonmobile form, materials that originated from the soil.

This concept has been simplified in a logo represented in Fig. 1.1.3 and which has been used to graphically characterize this book.

CONCLUSIONS

Reflecting the views relating to global pollution, it has become more and more obvious that the bar needs to be raised in waste management both in the immediate term and in the future. Environmental awareness should be raised among politicians and the population to enable them to grasp the need for further waste avoidance, collection, reuse, and recycling. To promote progress in this direction specific education should not only be provided in schools, universities, public institutions but also in the economic sector. Furthermore, new methods should be employed, making use of social media and conveying messages by modern and unconventional means, such as the binomial of waste and art (Stegmann and van der Westhuyzen, 2015). An increased involvement of the media is likewise mandatory, preferably through the introduction of approaches similar to those adopted with regard to Global Warming. If the population is made aware of why they have to reduce, reuse, and recycle waste, it will prove easier to convince them to participate and to accept the new necessary plants. However, education and information should not be ideologically driven, and it should always be made clear that while the concept of zero waste is an excellent target, it is almost impossible to achieve, and therefore waste treatment and disposal plants will always be required. Entropy alone results in the production of waste.

The NIMBY attitude likewise needs to be overcome.

Education is indeed an essential aspect, but additional measures will also be required. The major drivers promoting an increase in private, public, commercial, and industrial activities in waste avoidance, recovering, and recycling are costs, regulations, and environmental awareness. As an example, an increase in the charges in Germany for the disposal of construction and demolition (C + D waste) waste in the 1980s resulted in a shift away from landfills to material recycling; costs had reached such high levels that recycling had become economical. Similar measures would of course have to be introduced in a moderate way to avoid illegal dumping.

With regard to global pollution—and of course for many other reasons—a critical review of the use of additives in our daily products (particularly nondegradable organics, nanoparticles, metals, and other compounds that may cause health problems and negative environmental effects) should be undertaken continuously and appropriate actions should be implemented with the aim of avoiding/reducing these additives wherever feasible. Manufacturer responsibility for the product should be extended and more severely enforced. Indeed, the latter is a key issue in fostering waste avoidance and extended high quality recycling.

This therefore brings us to the conclusion that both production and treatment processes and the introduction of new products and processes need to be fully thought through from the start (Stegmann, 2016). What happens to a product once it has been used? To provide an answer to this question, the amount and quality of the residues generated should be critically reviewed (waste minimization) and their potential for recycling or final destination should be determined as part of the development process. In addition, the use of primary resources and energy should be kept to a bare minimum.

However, even if high levels of waste avoidance, reuse, and recycling are achieved, some waste materials will always need to be forwarded for disposal. To this aim, pretreatment technologies should be further developed to enhance the minimization of emission potential, thus rendering the resulting wastes increasingly immobile.

On confirming the need to minimize landfilling where possible, the fact remains that in the majority of countries landfills remain and are likely to continue to remain the backbone of waste management, being required also following the thermal treatment of waste. Landfills are sophisticated constructions when they meet all prescribed requirements. Indeed, particular emphasis should be placed on the long-term behavior of landfills and the need to implement measures to reduce the emission potential over a shorter period of time (see Chapters 2.1).

As mentioned above, all human activities undertaken will result in the production of gaseous, liquid, and solid emissions, which, in the majority of cases, will exert a detrimental effect on the environment. The only effective means of avoiding these emissions is by waste avoidance. This can be achieved in all areas of our daily life, including the context of manufacturing processes, by using fewer products for lengthier periods of time, less materials for production, etc. and by using products in a more intelligent way. Examples may include the developing of new leasing concepts for household devices, machines, electrical and electronic equipment, etc., the minimization of material and energy input in production processes, longer lasting materials, equipment and machines, etc. The products we use should be developed with a view to allowing for more efficient recycling, with educational programs promoting the use of multiple uses instead of disposable products. Many such options have already been implemented, but they represent the tip of the iceberg.

These new challenges should be viewed as an absolute necessity by all involved and undergo intensive research and application, with the social and political sector occupying a position of high responsibility. Lastly, as all disciplines are to some extent impinged on by environmental aspects, it is recommended that universities introduce a mandatory course for all students of "General Environmental Studies."

References

Cossu, R., 2012. The environmentally sustainable geological repository: the modern role of landfilling. Waste Management 32, 243–244.

Cossu, R., Salieri, V., Bisinella, V. (Eds.), 2012. Urban Mining – a Global Cycle Approach to Resource Recovery from Solid Waste. CISA Publisher. ISBN 978-88-6265-001-4.

Gutberlet, J., 2013. Recycling cooperatives addressing climate change challenges: a case study from Brasil. In: Christensen, Cossu, Stegmann (Eds.), Proceedings SARDINIA 2013 Symposium. CISA Publisher. ISBN 978-88-6265-028-1.

Rockström, J., et al., 2009. Planetary boundaries: exploring the safe operating space for humanity. Ecology and Society 14 (2), 32.

Steffen, W., Richardson, K., Rockström, J., Cornell, E., Fetzer, I., Bennett, E.M., Biggs, R., Carpenter, S.R., de Vries, W., de Witt, C., Folke, C., Gerten, D., Heinke, J., Marc, G.M., Persson, L.M., Ramanathan, V., Reyers, B., Sverker, S., 2015. Planetary boundaries: guiding human development on a changing planet. Science 347 (6223, 1259855). https://doi.org/10.1126/science.1259855.

Stegmann, R., 2016. Thinking from the end. In: Venice Symposium 2016, Conference Proceedings. CISA Publisher, Padova. ISBN 978-88-6265-009-0.

Stegmann, R., Heerenklage, J., Ritzkowski, M., Zurawski, D., 2003. Concepts for a decentralized treatment of waste and wastewater. In: Pullammanappallil, P., McComb, A., Diaz, L.F., Bidlingmaier, W. (Eds.), Proceedings of the Fourth International Conference of ORBIT Association, "Biological Processing of Organics: Advances for a Sustainable Society", Perth, Australia, Part II, April 30–May 02, 2003, pp. S.375–S.387. ISBN: 3-935974-04-03.

Stegmann, R., van der Westhuyzen, C., 2015. From waste to art and from art to waste. In: Cossu, H., Kjeldsen, M., Reinhart, Stegmann (Eds.), Proceedings Sardinia 2015 Symposium. CISA Publisher. ISBN 978-88-6265-021-2.

1.2

WASTE INPUT TO LANDFILLS

Giulia Cerminara and Raffaello Cossu

INTRODUCTION

Although there is a general similarity in type and quality of waste worldwide, there may be significant differences in the different regions, countries, cities, and communities. The knowledge about waste production and quality is essential for developing tailor-made waste management concepts and for selecting and operating adequate waste treatment facilities. Of course, only those waste materials should be landfilled that are not suitable for reuse and recycling. Different kinds of landfill concepts should be selected for different kinds of waste. Materials that are not feasible for recycling today but potentially in the future should be separately stored either in specific dedicated landfill sections or in intermediate storage facilities (e.g., soil, tires, plastic).

In general, three different kinds of landfills may be identified:

- Inert nonhazardous waste
- Municipal solid waste (MSW)
- Hazardous waste

As a fourth kind, monolandfills receiving (mass) waste with the same quality (e.g., mining waste) may be identified.

There are some general rules how to landfill waste with different kinds of properties.

In many countries, different/modified landfill standards for the different kinds of waste have to be met. In general, no liquids should be allowed to go to landfills; sludge must meet maximum water content limits to avoid mechanical stability problems of the deposited waste mass.

According to the different multibarrier concepts (see Chapter 2.1) and particularly for the use of landfills as final sinks, great attention has to be paid to the quality of the waste to be landfilled. Waste in this regard has to be seen as a barrier itself. Therefore, it might be necessary to meet limit values for specific waste categories. This means that many kinds of waste have to be treated before going to landfill.

WASTE CHARACTERIZATION

Physical Properties

Physical, mechanical, chemical, and biological characteristics of solid waste vary depending on the source and typology. The nature of the deposited waste in a landfill will affect gas and leachate production and their composition.

Moisture content, waste density, grain size distribution, field capacity, and heating value are also important as they affect the extent and rate of waste degradation processes and give an indication about the most suitable treatment and disposal solution.

Moisture Content

The moisture content of MSW is usually expressed as weight of water per unit weight of wet material.

For most MSW compounds, the moisture content can vary in a very wide range depending on the composition of the wastes, the season of the year, and weather conditions (Table 1.2.1).

Table 1.2.1 Moisture content values for different MSW fractions (Christensen, 2011)

Material Fraction	Moisture Content (%)	
	Range	Typical
Aluminum cans	2–4	3
Cardboard	4–8	5
Fines (dirt, etc.)	6–12	8
Food waste	50–80	70
Glass	1–4	2
Grass	40–80	60
Leather	8–12	10
Leaves	20–40	30
Paper	4–10	6
Plastics	1–4	2
Rubber	1–4	2
Steel cans	2–4	3
Textiles	6–15	10
Wood	15–40	20
Yard waste	30–80	60

Density

Density data are of interest when calculating the amount of total waste that can be landfilled until the prescribed height and size is reached. Table 1.2.2 reports typical densities for different MSW fractions.

In addition to the original densities, final density values after deposition, compaction, and settlement are of importance for different reasons (stability evaluations, afteruse options, long-term impacts, etc.).

Grain Size Distribution

The size and size distribution of the different waste components is of special interest when waste pretreatment is envisaged.

In landfilling particle size strongly influences the compaction degree of waste and the degradation rate of degradable fractions.

Table 1.2.2 Typical densities of different municipal solid waste fractions (Christensen, 2011)

Material Fraction	Material Density (kg/m^3)
Aluminum	2700
Steel	7700
Iron	5500
Food waste	600–750
Glass	2500
Wood	600–800
Paper	700–1150
Cardboard	700
Plastic, HDPE	960
Plastic, polypropylene	900
Plastic, polystyrene	1050
Plastic, PVC	1250

Field Capacity

The field capacity is the maximum amount of moisture that can be retained by the waste. Water in excess of the field capacity will be released as leachate. The field capacity varies with the waste typology, the degree of compaction, and the progressing of waste degradation. The field capacity of uncompacted commingled wastes from residential and commercial sources is in the range of 50%–60%.

Chemical and Physical-Chemical Properties

Important chemical properties measured for solid waste are

- Total solids (residues at 105°C for 24 h);
- Volatile matter (loss on ignition at 550°C for 4 h);
- Ash (nonvolatile solids);
- Fixed and organic carbon;
- Melting point of ash (the temperature at which the ash resulting from the burning of waste will form a solid (clinker) by fusion and agglomeration. Typical fusing temperature for the formation of clinker from solid waste ranges from 1100 to 1200°C);
- Heating value;
- Percent of carbon, hydrogen, oxygen, sulfur, and ash.

Typical values of different chemical and physical–chemical parameters characterizing different MSW fractions are reported in Table 1.2.3.

Biological Properties

Biological stability of solid waste represents the extent to which readily biodegradable organic fractions are decomposed. It is one of the main issues related to the evaluation of the long-term emission potential and the environmental impact of landfills (Cossu and Raga, 2008, Cossu et al., 2012).

The biological stability of waste material can be detected by means of respiration tests, which determine the uptake of oxygen into the waste sample and express the microbial degradation activity. The test result is usually expressed as a rate, e.g., mg O_2/kg DM/h, or as cumulative uptake over a number of days, e.g., mg O_2/kg DM during 4 days. The respiration tests can be static or dynamic, depending on the absence (static) or presence (dynamic) of continuous aeration of the biomass. The respiration index (IR_4), generally used in Germany with the acronym AT_4, may be considered static: although the oxygen consumed during the test is constantly replaced in the reactor, no airflow through the waste sample is provided, as in dynamic respiration tests (Cossu and Raga, 2008).

With the Dynamic Respiration Index (DRI) the continuous aeration is still maintained for 4 days, in a specific equipment, and the final value is calculated as average of the values measured every hour along 24h, during the highest microbial activity period (Adani et al., 2004).

Fermentation tests are an alternative way to identify the biological stabilization degree of waste material; they consist in measuring the biogas produced under anaerobic conditions for 21 days (GB_{21}), and results are expressed as normal liter of biogas per kilogram of dry matter.

The German ordinance on Environmentally Compatible Storage of Waste from Human Settlements and on Biological Waste Treatment Facilities set limit values equal to 5 mg O_2/g DM

Table 1.2.3 Typical properties of MSW fractions (Christensen, 2011)

Material Fraction	TS % Wet Weight	VS % TS	Ash % TS	Lower Heating Value (Mj/kg wet)	C (%)	H (%)	O (%)	S (%)
Vegetable food waste	23.00	96.40	5.20	2.5	47.7	6.6	39.460	0.1840
Animal food waste	42.90	94.20	8.70	9.2	56.5	7.9	18.220	0.3780
Wood	84.10	90.60	10.00	15.6	52.1	6.4	30.490	0.0836
Newsprints	87.00	92.70	8.20	14.6	44.8	5.7	44.210	0.0319
Magazines	93.80	76.70	34.00	10.6	34.2	4.2	27.450	0.0724
Advertisements	91.30	75.10	27.40	14.4	34.6	4.8	32.940	0.0784
Books and phonebooks	89.50	86.10	17.90	13.4	40.6	5.16	38.055	0.0487
Office paper	91.30	87.80	20.70	11.2	37.5	5.0	36.690	0.0643
Paper and carton containers	77.70	88.80	13.40	13.5	41.1	5.6	39.610	0.1000
Cardboard	83.50	89.00	14.00	12.2	40.9	5.4	39.480	0.0631
Plastic bottles	89.50	93.80	6.10	32.5	77.2	11.3	5.200	0.1090
Hard plastic	96.80	98.10	2.20	36.1	79.9	10.5	1.730	0.0988
Glass	88.00	0.00	100.00	0.0	0.0	0.0	0.000	0.0832
Metal containers	86.80	0.00	100.00	0.0	0.0	0.0	0.000	0.0099
Diapers and tampons	54.50	94.20	8.30	11.1	55.3	8.0	27.330	0.0718
Textiles	94.00	96.60	3.60	18.5	52.1	6.0	34.800	0.3970
Leather	93.30	89.00	12.60	22.9	61.3	7.3	13.780	0.6594

for the respiration test AT_4 and 20 NL/kg DM for the GB_{21} index (Cossu and Raga, 2008). For the Dynamic Respirometric Index a value of 1000 mg O2/kgVS/h is often proposed in Italy for the stability of MSW to be landfilled.

An alternative very promising, simple and reliable index has been proposed by Cossu et al. (2012, 2017). The index is based on the ratio BOD_5/COD measured on waste eluate. This parameter overcomes the limits of the previous mentioned indices such as high equipment costs, long lasting testing time, low representativity when inhibiting substances or inert organic impurities are present (in both cases lower values are obtained).

WASTE GENERATION

The global MSW generation in 2016 has been estimated at 2.01 billion tons and it is expected to increase to approximately 3.4 billion tons per year by 2050 (Kaza et al., 2018). Worldwide, the pro-capita waste generation averages around 0.74 kg/d but it ranges widely, from 0.11 to 4.54 kg/d (Kaza et al., 2018). The per capita waste generation rate will significantly increase in the next years (Hoornweg and Bhada-Tata, 2012).

However, global averages are broad estimates only as rates vary considerably by region, country, city, and even within cities.

MSW generation rates are influenced by economic development, the degree of industrialization, public habits, and local climate. Generally, the higher the economic development and rate of urbanization, the greater the amount of solid waste produced. Income level and urbanization are highly correlated and as incomes and living standards increase, consumption of goods and services correspondingly increases. This has a proportional effect on waste generation which can be positively attenuated by higher education levels (Ojeda Benítez et al., 2008).

Fig. 1.2.1 illustrates the average composition of MSW in several geographical areas. Middle- and lower-income countries produce more organic-rich MSW (about 60%), whereas the high-income countries produce more paper, plastics, glass, and metals.

Apart from MSW, other significant types of waste streams should be considered:

- *Construction and Demolition (C&D)* waste represents one of the largest waste stream produced in developed countries (Bournay, 2006; Behera et al., 2014), with some peaks reaching 55% of total waste generation, such as in Germany (OECD, 2008a). C&D waste can be classified as high-volume waste with relatively low impact compared with other types of waste;
- *End-of-life vehicles (ELVs)* account for about 6.5 million tons of waste in the European Union (EU) with Germany, the United Kingdom, France, Spain, and Italy responsible for approximately 70% of European car production (Eurostat, 2010). The EU Directive (2000/53/EC) aims at making dismantling and recycling of ELVs obligatory and and environmentally friendly by setting quantified targets and quality prescriptions. After the reusable parts and recyclable materials are removed from ELV, Automobile Shredder Residues (ASR) remain (Kiyotaka and Itaru, 2002). This fraction represents the 20%–25% of ELV and corresponds to an amount of approximately 2.5 million tons/year in Europe (Zorpas and Inglezakis, 2012). ASR generation in Japan is about 0.7 million t ASR/year while in the United States it reaches an approx. 5 million tons annually (EPA, 2010);

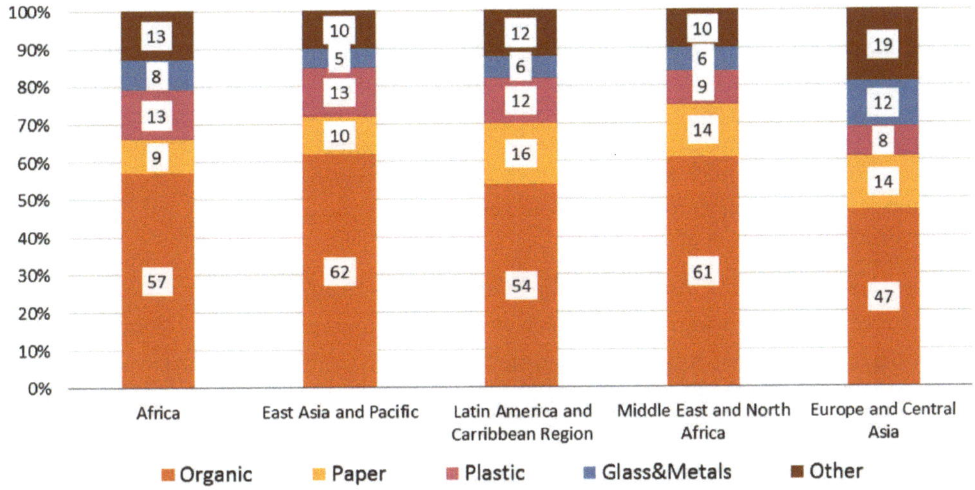

Figure 1.2.1 Municipal solid waste composition by region in the world.

- *Biomass waste* includes agricultural and forestry waste. It is estimated that globally 140 billion tons of agricultural residue are generated every year (Nakamura, 2009). Like C&D, biomass waste is a high-volume waste;
- *Waste electrical and electronic equipment (WEEE)* continues to increase dramatically due to the growing global demand for electronic and electrical goods (computers, TV-sets, fridges and cell phones, etc.). It is estimated that 315 million personal computers (PC) became obsolete in the world in 2004. Yu, et al. (2010) predicted that obsolete PCs in developing regions could amount to 400-700 million units by 2030 (compared with 200-300 million units in developed countries). In 2005 130 million mobile phones were estimated to have reached their "end of life" (UNEP 2005). The global generation of WEEE has been reported as ranging around 20-50 million t/year (Wang et al., 2014) and this figure is growing by about 2 million t/year. Only in USA WEEE generation in 2007 was worth 3.16 million tonnes (EPA 2009).
- *Scrap vehicle tires* make a significant contribution to the generation of waste. Worldwide, the amount of used automobile tires is increasing (Sienkiewicz et al., 2012). The rate of scrap tire generation in industrialized countries is approximately one passenger car tire equivalent (PTE, or 9 kg) per capita per year (Reschner, 2003). According to the European Tire and Rubber Goods industry ETRMA (2011), the global production of tires in 2011 amounted to 4.6 million tonnes. MSW incineration (MSWI) residues, sewage and industrial sludges, agricolture waste, mining waste and many others.
- *Tannery sludge.* Leather tanning is a worldwide common industry. It is known to be one of the most important industries in Mediterranean countries (Lofrano et al., 2013), but it represents an important economic field also in developing countries, as in Turkey, China, India, Pakistan, Brazil and Ethiopia. The global market of leather industry is about 215 million hides per year (Abreu and Toffoli, 2009).
- *Asbestos* is the common name applied to a group of natural, fibrous silicate minerals; it is characterized by incombustibility, high electrical and mechanical resistance, low thermal conductivity, antiseptic properties and it is highly economic (Kim and Hong, 2017; Kusiorowski et al., 2013;

Leonelli et al., 2006). An asbestos fibre is defined by the World Health Organization (WHO) as a particle having a length higher than 5 μm, and a diameter less than 3 μm (Leonelli et al., 2006). In 2013, the global asbestos production was 1.94 million tonnes; Russia, China and Brazil accounted respectively for 46.92 %, 18.96 % and 13.06 % (Li et al., 2014). Despite there is a decreasing tendency of production of asbestos since 2011, there are still about 200 million tonnes of asbestos stored worldwide, about 100 times the total production in 2013. In 2011, about 2.03 million tonnes of asbestos were consumed in the world and 61.5 % of the total was consumed in the Asia-Pacific Region (Li et al., 2014).

The waste classification in Europe is usually referred to the European Waste Catalogue (EWC). It is a classification system for waste materials and it categorizes waste based on a combination of their quality and origin. It was established on December 1993 by Commission Decision 94/3/EC and includes 645 waste types subdivided into 20 chapters.

Within each chapter, there is a list of generic waste types that are classified under the industry sector, process, or waste type. Each waste is identified by a six-digit code, which, if followed by an asterisk "*," implies that it is considered to be hazardous.

Table 1.2.4 contains the EWC in its last updated version after Commission Decision of December 18, 2014 amending Decision 2000/532/EC on the list of waste pursuant to Directive 2008/98/EC of the European Parliament and of the Council.

Table 1.2.4 The European Waste Catalogue

Code	Waste Category
01	Wastes resulting from exploration, mining, quarrying, and physical and chemical treatment of minerals
02	Wastes from agriculture, horticulture, aquaculture, forestry, hunting and fishing, and food preparation and processing
03	Wastes from wood processing and the production of panels and furniture, pulp, paper, and cardboard
04	Wastes from the leather, fur, and textile industries
05	Wastes from petroleum refining, natural gas purification and pyrolytic treatment of coal
06	Wastes from inorganic chemical processes
07	Wastes from organic chemical processes
08	Wastes from the manufacture, formulation, supply and use (MFSU) of coatings (paints, varnishes, and vitreous enamels), sealants, and printing inks

TABLE 1.2.4 The European Waste Catalogue—cont'd

Code	Waste Category
09	Wastes from photographic industry
10	Wastes from thermal processes
11	Wastes from chemical surface treatment and coating of metals and other materials; nonferrous hydrometallurgy
12	Wastes from shaping and physical and mechanical surface treatment of metals and plastics
13	Oil wastes and wastes of liquid fuels (except edible oils, 05 and 12)
14	Waste organic solvents, refrigerants and propellants (except 07 and 08)
15	Waste packaging; absorbents, wiping cloths, filter materials, and protective clothing not otherwise specified
16	Wastes not otherwise specified in the list
17	Construction and demolition wastes (including excavated soil from contaminated sites)
18	Wastes from human or animal health care and/or related research (except kitchen and restaurant wastes not arising from immediate health care)
19	Wastes from waste management facilities, off-site wastewater treatment plants and the preparation of water intended for human consumption and water for industrial use
20	Municipal wastes (household waste and similar commercial, industrial, and institutional wastes), including separately collected fractions

Source: http://eur-lex.europa.eu.

LANDFILL CLASSIFICATION AND WASTE CATEGORIES

In the European Union the national landfill regulations are inspired by the Directive 1999/31/EC, recently integrated by Directive 2018/850 (see Chapter 1.3).

Landfills, according to the cited Directive, are classified into three categories:

- landfills for hazardous waste;
- landfills for nonhazardous waste;
- landfills for inert waste.

Landfill for nonhazardous waste may be used for MSW, for any other nonhazardous waste, which fulfills the criteria for the acceptance of waste at landfill for nonhazardous waste, set out in accordance with Annex II, and for stable, nonreactive waste, with a leaching behavior equivalent to those of the nonhazardous waste.

Inert waste landfill sites shall be used only for inert waste.

According to Article 5 of 1999/31/EC Directive, "Member States shall set up a National strategy for the implementation of the reduction of biodegradable waste going to landfills [...]." Lower the amount of biodegradable organic matter landfilled, lower the hazardousness of landfill systems, and shorter the necessary aftercare period.

The following wastes may not be accepted in a landfill:

- liquid waste;
- flammable waste;
- explosive or oxidizing waste;
- hospital and other clinical waste, which is infectious;
- used tires, with certain exceptions;
- any other type of waste that does not meet the acceptance criteria laid down in Annex II.

In Table 1.2.5, a synoptic view is given of the landfilling peculiarities of some significant solid waste streams. The main features in landfill behavior, the level of mobility of potential contaminants, the landfill category they are suited for, the indicative eligible pretreatment options, the possible in situ treatment methodology, and the evidence and entity of a carbon sink effect are provided in a very general term.

In the following sections of this chapter, some selected waste categories are presented and the main characteristics important for landfilling are illustrated.

Municipal Solid Waste

MSW is composed of different fractions, as listed in Table 1.2.6.

The recycling potential of these fractions is generally very high but worldwide a consistent part of MSW is still landfilled (Fig. 1.2.2).

In some countries, particularly in Europe, MSW are mechanically and biologically pretreated (MBP) before landfilling to reduce the size of MSW components, separate fines from coarse fractions, recover valuable material, and stabilize biologically the putrescible fractions to be deposited in landfills. Detailed information on MBP technologies and landfilling of MBP waste is given, respectively, in Chapters 4.1 and 14.1.

Municipal Solid Waste Incineration Residues

Incineration of Municipal Solid Waste (MSW) plays a prominent role in several countries in Northern Europe and Japan (Allegrini et al., 2014). Incineration reduces waste mass by 70% and volume by up to 90% (Chimenos et al., 1999; Gidarakos et al., 2009; Valle-Zermeno et al., 2013) and provides a valuable source of energy. Accordingly, incineration represents an important part of the waste management system along with recycling and landfilling (Sabbas et al., 2003).

Incineration produces bottom ash (BA) as residues, typically representing 15%–30% of the input waste mass (Allegrini et al., 2014; Sivula et al., 2012), and it is approximately 80% of the total incineration residues (Chimenos et al., 1999). Bottom ashes contain iron as a main resource and to a

Table 1.2.5 Synoptic view of the main features in landfilling of some specific solid waste streams

Waste Typology	Main Feature	Mobility Level	Landfill Class	Eligible Pretreatment	In Situ Treatment	Carbon Sink
Municipal solid waste	Variable moisture and density, size and quality dishomogeneity, high compressibility, putrescible contents	high	NH	Mechanical biological treatment (MBT), Thermal treatment, washing, Sorting	Flushing, aeration	**
MBT Waste	Low mechanical strength, organics and ammonia leaching, gas production	medium	NH	Baling (with no plastic wrapping)	Aeration	**
Mixed plastic residues from recycling separately collected plastics	Low density, low organics leaching, dust emissions, increased risk of fires	low	NH/M	Washing	Flushing	***
Municipal solid waste incineration (MSWI) bottom ashes	Dust emissions, temperature rising, hydrogen production	medium	M	Carbonatization, Washing	Flushing, aeration	negligible
MSWI Fly ashes	Heavy metals leachability, no mechanical strength, small size particles	high-hazardous	H/M	Inertization, encapsulation	Roof cover, Big bag packaging	negligible
Construction and demolition (C&D) waste and excavated materials	Dishomogeneity, dust, specific case–related problems	low	I/NH	Recycling	Flushing	*
WEEE—Waste Electrical and Electronic Equipment	High content of toxic and hazardous leaching substances	high	H	Recycling	Big bag, impervious top cover	***

(Continued)

Table 1.2.5 Synoptic view of the main features in landfilling of some specific solid waste streams—cont'd

Waste Typology	Main Feature	Mobility Level	Landfill Class	Eligible Pretreatment	In Situ Treatment	Carbon Sink
ASR—Automotive Shredded Residues	Dust, nondegradable halogens	medium/ high	H/NH	Washing	Flushing, I situ aeration	**
Scraped tires	Low density, elasticity,	low	M	Recycling	Scredding	***
Tannery sludge	High moisture, residual putrescibility	medium	NH	Biostabilization, Thermal drying	Big bag packaging, Aeration	*
Asbestos	Highly health risky fibers	no	NH	Big bag packaging/ Wrapping	Soil cover	negligible
Sewage sludge (digested and dewatered)	Mechanical instability, stickiness	high	NH	Mixing with structural waste	Mixing, dedicated deposition (e.g., ditches)	**
Hazardous waste	Case-specific features	high	H	Inertization, encapsulation	Flushing	variable
Dredged material	High moisture, low organics, heavy metals, antifouling products	low/ medium	M	Dewatering, sand separation		*

H, hazardous; *I*, inert; *M*, monofill; *NH*, nonhazardous.
Number of *, represents the sinking carbon effectiveness for the individual waste typology.

Table 1.2.6 Municipal solid waste fractions

Municipal Solid Waste Fractions	
Putrescible	Food waste
	Yard waste and leaves
Cellulosic material	Newspaper
	Magazines
	Books
	Packaging material
	Cardboard
Glass	White glass
	Colored glass
Plastic material	Plastic containers
	Plastic film
	Nonrecyclable Plastic
Metals	Ferrous Metals
	Nonferrous metals
Stones and ceramics	
Leather, Wood, textiles, rubber	
Miscellaneous: Nappies, sanitary napkins, materials that do not fit in any of the above categories	
Undersieve <20 mm: Undetermined fines	

certain extent aluminum and other metals as scrap material that may be present either as scrap metals (i.e., metals in metallic form, present as individual pieces or pieces, which may subsequently be cleaned from aggregates) or as metals bound in mineral form within the aggregate matrix.

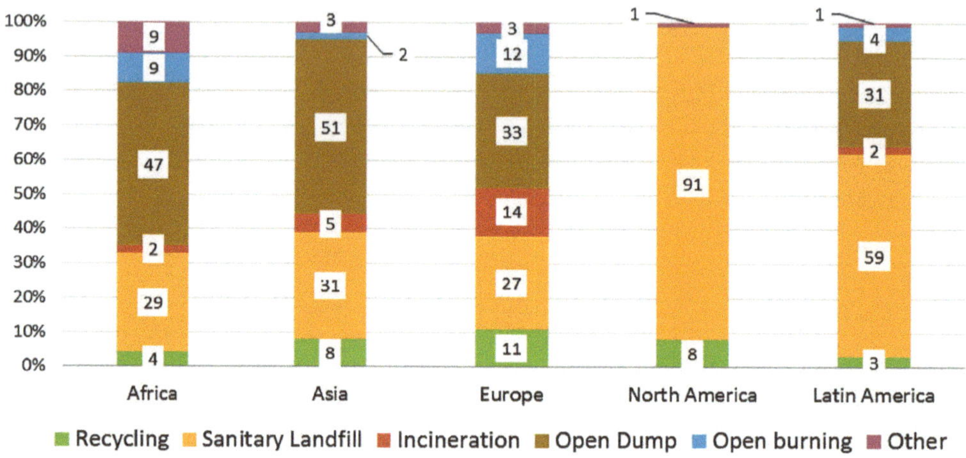

Figure 1.2.2 Municipal solid waste disposal by geographical region in the world (Hoornweg and Bhada-Tata, 2012).

After removing the metal fraction (MF), bottom ash in some countries is (partly) recycled (e.g., Germany, Denmark), in other countries they have to be landfilled, which may be in some cases also required by law (e.g., in Austria and Switzerland). Chapters 4.2 and 20.2 provide detailed information about the quality and landfilling of MSWI residues.

MSW incinerator bottom ashes may be landfilled in nonhazardous waste landfills, whereas fly ash is generally classified as a hazardous waste and has to be either treated (e.g., by vitrification) or deposited in hazardous waste landfills. In any case, bottom and fly ash should be kept separate to avoid the "contamination" of bottom ashes.

Construction and Demolition Waste

Construction and Demolition waste accounts for approximately 25%–30% of all waste generated in the EU and consists of numerous materials, including concrete, bricks, gypsum, wood, glass, metals, plastic, solvents, and excavated soil, many of which can be recycled (Rodrigues et al., 2013). So typical components in C&D waste are inert materials (e.g., concrete, bricks, etc.), which are generally believed to have low impact on the environment, but more and more plastic is used in construction, e.g., pipes, liners. However, there are also some hazardous components such as asbestos, particulate matter, insulation material, paint, asphalt, etc.

According to the generation activity C&D waste can be divided into three categories: construction waste (CW), renovation waste (RW), and demolition waste (DW) (Wu et al., 2014).

The European Union has identified C&D waste as a priority waste stream due to their high potential for recycling and reuse (Waste Framework Directive (2008/98/EC).

However, the level of recycling and reuse of C&D waste varies greatly (between less than 10% and over 90%) across the European Union. In some states, different materials have to be kept separate on site as, e.g., bricks, wood, plastic, and metals.

Waste Electrical and Electronic Equipment

WEEE is mainly generated in developed countries where the market for electronic and electrical equipment is highly saturated, as it is clearly described by the data reported in Table 1.2.7, (Ongondo et al., 2011).

The huge variety of materials and components in WEEE makes it difficult to give a generalised composition for the entire waste stream. However most studies consider five categories of materials: ferrous metals, non-ferrous metals, glass, plastics and other materials. Iron and steel are the most common materials found in EEE (by weight) and account for almost half of the total weight of WEEE. Plastics are the second largest component by weight, representing about 21% of WEEE. Non-ferrous metals, including precious metals, represent about 13% of the total weight. Over the time, metal content has remained the dominant fraction, (Ongondo et al., 2011).In addition to these materials WEEE and batteries contain numerous hazardous substances and elements such as: lead (Pb), barium (Ba), cadmium (Cd), mercury (Hg), brominated flame-retardants (BFRs) and polyvinyl chloride (PVC). In batteries, the primary hazardous compounds are Cd, Pb, and Hg, and in toners, Cd is of concern (Bigum et al., 2013). Moreover, the production of modern electronics requires the use of rare and expensive earths elements (Menad et al., 2013; Wang and Xu, 2014). As an example around 10% of total gold produced worldwide is used for EEE production. Therefore recovery and recycling of the metal fractions (MFs), is increasingly adopted due to the achievable economic benefits. Typical material fractions in waste electrical and electronic equipment are graphically presented in Fig. 1.2.3.

The non-metal fractions (NMFs), which take up a large proportion of electronic waste, are mostly treated by incineration or landfilling. Unfortunately, this fraction may still contain heavy metals, brominated flame retardants (BFRs) and other toxic and hazardous substances which may severely affect the environment if not properly managed (groundwater and air pollution risks, formation of polybrominated dibenzodioxins and dibenzofurans during uncontrolled combustion). Particularly, in developing countries, e-waste is largely deposited in open dumps or in poorly managed landfill sites. (Sabbaghi et al., 2015). In order to better promote recycling and to improve the environmental management of WEEE two directives have been implemented in the European Union.The first WEEE Directive (2002/96/EC) provided for collection strategies based on free of charge return of WEEE by the consumers.The second Directive (2012/19/EU) requires heavy metals, such as lead, mercury, cadmium, and hexavalent chromium, and flame retardants, such as polybrominated biphenyls (PBB) or polybrominated diphenyl ethers (PBDE), to be substituted by safer alternatives.

Table 1.2.7 WEEE generation, collection and treatment routes: approaches in selected countries

Country	Generation (ton/y)	Per Capita Generation (kg/Inhabitant)	Reported Discarded Items	Collection and Treatment Routes
Germany	1 100 000 (2005)	13.3	Domestic WEEE	PWMA, retailers takeback
UK	940 000 (2003)	15.8	Domestic WEEE	DTS and PCS
Switzerland	66 042 (2003)	9	Diverse range of WEEE	SWICO, S.EN.S., SLRS
China	2 212 000 (2007)	1.7	Computers, printers, refrigerators, mobile phones, TVs	Mostly informal collection and recycling
India	439 000 (2007)	0.4	Computers, printers, refrigerators, mobile phones, TVs	Informal and formal
Japan	860 000 (2005)	6.7	TVs, air conditioners, washing machines, refrigerators	Collection via retail
Nigeria	12 500 (2001–06)	–	Mobile phones charges, and batteries	Informal
Kenya	7 350 (2007)	0.2	Computers, printers, refrigerators, mobile phones, TVs	Informal
South Africa	59 650 (2007)	1.2	Computers, printers, refrigerators, mobile phones, TVs	Informal and formal
Argentina	100 000	2.5	Excludes white goods, TVs, PCs, radios, washing machines, refrigerators, and freezers	Small number of takeback schemes, municipal waste services
Brazil	679 000	3.5	Mobile and fixed phones, TVs, PCs, radios, washing machines, refrigerators, and freezers	Municipalities, recyclable waste collection
USA	2 250 000 (2007)	7.5	TVs, mobile phones, computer products	Municipal waste services; a number of voluntary schemes
Canada	86 000 (2002)	2.7	Consumer equipment, kitchen, and household appliances	A number of voluntary schemes
Australia	–	–	Computers, TVs, mobile phones, and fluorescent lamps	Proposed national recycling scheme from 2011: voluntary takeback

DTS, Distributor Takeback Scheme; *PCS*, Producer Compliance Scheme; *PWMA*, Public Waste Management Authorities; *S.EN.S.*, Swiss foundation for waste management; *SLRS*, Swiss light recycling foundation; *SWICO*, Swiss Association for information, communication and organization technology; *TV*, Television; *WEEE*, Waste Electrical and Electronic Equipment.
Source: Ongondo et al. (2011).

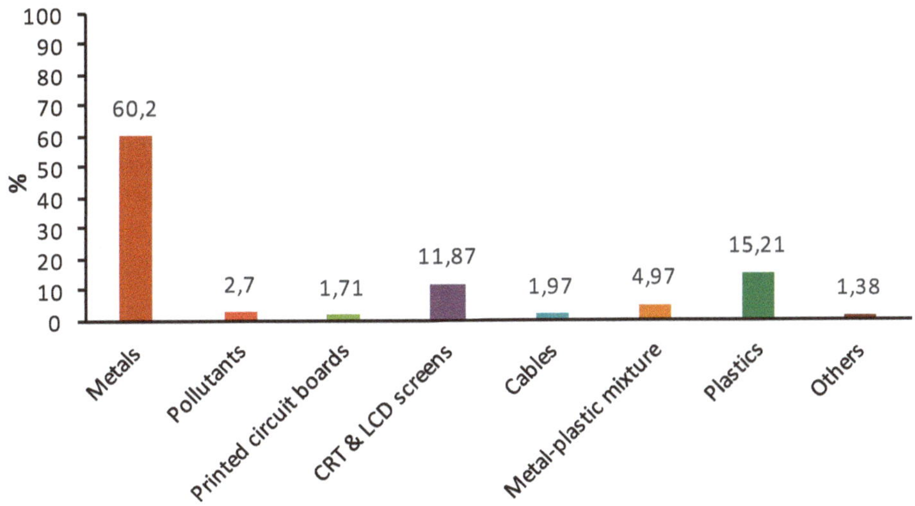

Figure 1.2.3 Typical material fractions in waste electrical and electronic equipment (Wang and Xu, 2014).

End-of-Life-Vehicles

Treatment of ELVs, aimed at recovering reusable and recyclable components, starts with a decontamination step (i.e., removal of battery, lubricants, and fuel, accounting for about 3–4 wt% of an ELV) and the dismantling of spare parts and recyclable materials (i.e., bumpers, tires, fuel tanks, glasses, accounting for about 5–10 wt% of an ELV) as preshredding operations performed in Authorized Treatment Facilities (ATFs). In the following shredding phase, the valuable metals are recovered (about 60–65 wt% of ferrous alloys and about 3–5 wt% of other metals) (Cossu et al., 2014). The residual fraction, following, the recovery of valuable materials, is named, as mentioned earlier, Automotive Shredded Residue (ASR). The composition of ASR depends on the efficiency of sorting and decontamination operations. Consequently it generally presents an heterogeneous composition (Table 1.2.8) which may render difficult treatment options (further recycling, waste to energy, etc.) other than landfilling.

The management of waste generated from ELVs is considered a significant environmental issue due to its composition, as well as to the presence of hazardous materials (PVC/chlorine, PCBs, trace elements and heavy metals) (Zorpas and Inglezakis, 2012; Sakai et al., 2014).

If ELVs are not treated appropriately, hazardous compounds may remain in the shredder residue and can influence leachability characteristics of ASR (Cossu and Lai, 2015); the insulation foam may contain fluorocarbons that could be released after disposal in a landfill, a potential risk is linked to the presence of metals (i.e., Cd, Cr, Pb, Zn, etc.) and dissolved organic carbon (DOC) content (Gonzalez-Fernandez et al., 2008; Scheutz et al., 2010; Fiore et al., 2012). ASR can be conveniently decontaminated by a washing treatment before landfilling (Cossu and Lai, 2013).

Table 1.2.8 Composition in percentage by weight of automobile shredder residues (Cossu and Lai, 2015)

Material	(%)
Textiles and foam	27–27.2
Plastics	19–20.2
Metal	1–4.6
Rubber	2.8–7
Cellulosic materials	0.2–1
Fines	45

Scrap Tires

Scrap tires are a category of waste whose disposal might be problematic due to their highly complex structure, diverse composition of the raw material, and quality of the rubber. Rubber represents the 70-80% of the tire mass, while the rest is made of steel belts and textile overlays, which during recycling have to be separated from the rubber. Tire recycling is based on the mechanical, thermal or chemical removal of the rubber fraction. Landfilling of waste tires was widely adopted in the past and it is still practised in some countries (Reschner, 2003). Waste tires may create problems because they are flexible and cannot be compacted. In addition, in case of landfill fires, they can negatively contribute to atmospheric pollution, with emissions of zinc oxide, dioxins and poly-nuclear aromatic hydrocarbons (Sharma et al., 2000). Similar environmental problems can be presented by stockpiling scrap tire in controlled or uncontrolled sites. The most noticeable problem associated with large tire storage areas is the potential fire hazard they present. Once a tire pile catches fire, it is very hard, if not impossible, to extinguish. In some instances, tire piles have been burning for several months with the black fumes being visible for many miles. Diseases (encephalitis and dengue fever) have also been reported in proximity of scrap tyre piles where, particularly under warm climate conditions, disease-carrying mosquitoes can find ideal breeding ground (Reschner, 2003). In response to the environmental problems and health hazards posed by disposal of scrap tires, most industrialized countries have put in place a legal framework to address a rational management of this waste stream. Regulations vary from country to country, but the main thrust of such legislation is to require the removal of abandoned piles, to restrict or ban landfilling and to promote reuse and recycling of tire components. Waste to energy by pyrolysis/gasification is also a feasible technical option (i.a. Conesa 2004).Recycling options for the rubber fraction may include use of granulate for reducing frictions in artificial sport fields (Van Rooij and Jongeneelen, 2010) or use as aggregate in the upper asphalt "deterioration" layer of roads (Reschner, 2003). The management of waste tires in the European Union has been regulated under the End-of-Life Vehicle Directive (2000/53/EC), which stipulates the separate collection of tyres from vehicles dismantlers and encourages the recycling of tires and their different materials. These directives have dramatically changed the tyre waste treatment routes in the EU over the last 15 years. For example, in 1996 approximately 50% of waste tires were sent to landfill, however, currently the figure is only 4% (Williams, 2013).

Tannery Sludge

The tanning process aims to transforms skins in stable and nonputrescible products namely leather, which are widely resistant against wet heat, enzymatic degradation, and thermomechanical stress. These properties may be obtained through treatments with mineral-tanning agents such as basic salts of chromium, aluminum, or zirconium and organic-tanning agents such as vegetable tannings, aldehydes, etc.

Chromium salts are the most widely used tanning substances used to produce leather (e.g., good mechanical resistance of the hides); they allow for a simplicity of operation, an extraordinary dyeing suitability, and better hydrothermal resistance in comparison with hides treated with vegetable substances (Abreu and Toffoli, 2009).

Chromium salts have a high rate of penetration into the interfibrillar spaces of the skins. In spite of that, only a fraction of the salts used in the tanning process yields the desired reaction with the skins. The rest of the salts remain as tanning residues and are subsequently sent to a treatment plant where the chromium salts end up in the remaining sludge. Chemical/physical reactions as precipitation, complexation, redox reactions, absorption, and diffusion processes influence Cr mobility in the natural environment (Chuan and Liu, 1996).

Cr(III) is stable at moderate pH and redox potential and exists mostly in cationic form; on the contrary Cr(VI) is stable under a more oxidizing environment in anionic form. Trivalent chromium form is much less toxic than the hexavalent form Cr(VI), which is very mobile and hazardous for the environment and the human health (Chuan and Liu, 1996; Kilic et al., 2011).

There are four major groups of subprocesses required to produce the final leather quality: beamhouse operations, tanyard processes, retanning, and finishing. However, for each product, the tanning process is different and the kind and amount of wastewater produced varies in a wide range (Lofrano et al., 2013).

The contaminated sludge that is produced during the tannery wastewater treatment process contains not only large quantities of chromium waste but also organic matter and salts such as chloride, sulfates, and carbonates. In addition, the sludge also contains ammonia, detergents, emulsifiers, bactericides, fungicides, coloring agents, skin proteins, hair, fats, and other components (Table 1.2.9).

Despite the environmental concerns, mainly due to the presence of chromium, landfilling has become a common practice in management of tannery sludge (Celary and Sobik-Szoltyse, 2014).

Table 1.2.9 Tannery sludge characterstics (Skrypski-Mantele and Bridle, 1995)

Parameter	Unit	Sludge
Total Solids (TS)	%	94.8
Volatile Solids (VS)	% of TS	40.8
Gross calorific value	MJ/kg	9.7
Carbon	%	22.0
Hydrogen	%	4.1
Nitrogen	%	3.9
Sulfur	%	3.2
Phosphorous	%	2.2
Chlorine	%	<0.4
Aluminum (Al)	%	1.1
Arsenic (As)	mg/kg	94
Barium (Ba)	mg/kg	360
Cadmium (Cd)	mg/kg	6
Calcium (Ca)	%	8.3
Cobalt (Co)	mg/kg	38
Chrome (Cr^{3+}/Cr^{6+})	mg/kg	8.900/<1
Copper (Cu)	mg/kg	120
Iron (Fe)	%	4.2
Mercury (Hg)	mg/kg	0.56
Nickel (Ni)	mg/kg	39
Lead (Pb)	mg/kg	200
Zinc (Zn)	%	3.3
Tin (Sn)	mg/kg	14
PCBs	μg/kg	0.3

The *Cr* release from the sludge is affected by pH; solubility decreases as pH increases, showing a minimum value in the pH range of 5.5–7.5 (Chuan and Liu, 1996).

Cr(III) may oxidize to Cr(VI) when oxidation conditions are present and appear in the leachate, representing a risk in case of uncontrolled leachate emissions (Apte et al., 2005).

Another problem that may occur in a landfill where tannery sludge has been disposed is the self-heating. It can depend on aeration, moisture content, particle size, particle porosity, specific chemicals, and biological activity (Biasin et al., 2014). Fermentation seems to be rather marginal and therefore may have little or no influence on the heating process; this is confirmed by spontaneous heating and combustion of solids that have little or no putrescible components. On the contrary, a chemical route seems to prevail (Biasin et al., 2014).

In Italy, e.g., tannery sludge is thermally dried and—after cooling off—placed in big bags that are deposited in monolandfills; air circulation may be used for keeping the temperature under control and consequently avoiding risks of internal burning (Alibardi and Cossu, 2016).

Asbestos

Asbestos has been used in about 3000 different asbestos-containing materials (ACMs) (Zaremba and Peszko, 2008); construction materials, including asbestos–cement sheets and pipes, asbestos textiles, and insulation products are the most common asbestos-containing products (Bhagia et al., 2010).

The International Agency for Research and Cancer under the WHO and the American Conference of Governmental Industrial Hygienists classifies asbestos as a carcinogenic (Kim and Hong, 2017). Such toxicity has led developed countries to either ban or restrict the use of asbestos from the early 1990s; however, there still remains a considerable amount of asbestos in use, particularly in developing countries, such as China and other countries in Southeast Asia, Africa, and South America (Kim and Hong, 2017).

Asbestos contained in the products is not degradable: it can contaminate superficial and deepwater and the surrounding atmosphere. The disposal of asbestos waste may consist of landfilling or thermal decomposition (Kusiorowski et al., 2013). Thermal decomposition is able to destroy the dangerous fibrous structure of asbestos and to form new mineral phases that can be used as secondary raw materials (Colangelo et al., 2011).

Asbestos minerals are naturally occurring hydrous silicates; therefore, during thermal treatment, at the temperature higher than 1000°C, they decompose and release chemically combined water, forming Mg–Fe silicates (Leonelli et al., 2006). The thermal decomposition of pure asbestos minerals is quite well known; on the other hand, thermal decomposition of asbestos–cement is more complex due to the multiphase reacting system. The calcium-rich cement phases and the degree of exposure to weather conditions modify the reaction paths (Kusiorowski et al., 2013).

Landfilling of asbestos is quite common and could be a very safe disposal method if the possible dispersion of airborne fibers is blocked or prevented. To this regards, the usual categorization among friable and nonfriable ACM is important. Friable asbestos even by simple hand pressure may be pulverized posing a concrete health risk.

Landfilling of asbestos waste is occurring in general under a specific permit. According to most of national regulations, asbestos waste can be landfilled in a nonhazardous waste landfill in purposely dedicated cells or trenches. Prescriptions may slightly change from different regulations.

In general the following precautions should be adopted:

- asbestos waste, particularly if friable materials are present, should be wrapped and seamed to prevent airborne fibers dispersion during handling and deposition;
- the asbestos waste should be buried immediately on their arrival at the landfill site and never stockpiled; an appropriate layer of suitable cover material should be spread on waste immediately after deposition;
- asbestos waste should not be landfilled during adverse climatic conditions, in particular when wind is blowing;
- interaction must be avoided with putrescible waste;
- asbestos waste containing hazardous components should be landfilled in hazardous waste landfills;
- capping of cells containing asbestos waste should, in general, be never removed;
- deposition of asbestos should be tracked and georeferentiated
- an emergency plan should be set.

FINAL REMARKS

Different kinds of waste, further to MSW, are landfilled worldwide. According to their natura, and amount they can be disposed of in hazardous, and non hazardous landfill sites or in mono-landfills. They may pose various kinds of problems which may need to be faced at regulatory level and to be solved during the planning, construction and operation phases. Further to the waste typologies considered in the chapter there are many more kinds of waste which are regularly landfilled. These include night soil, sewage sludge, garden and park waste, street sweeping waste, concentrates from reverse osmosis, slightly contaminated soil, light radioactive waste, etc. Information about these materials go further the general purpose of this chapter.

References

Abreu, M.A., Toffoli, S.M., 2009. Characterization of a chromium-rich tannery waste and its potential use in ceramics. Ceramics International 35, 2225–2234.

Adani, F., Confalonieri, R., Tambone, F., 2004. Dynamic respiration index as a descriptor of the biological stability of organic wastes. Journal of Environmental Quality 33 (5), 1866–1876.

Alibardi, L., Cossu, R., 2016. Pre-treatment of tannery sludge for sustainable landfilling. Waste Management 52, 202–211.

Allegrini, E., Maresca, A., Olsson, M.E., Holtze, M.S., Boldrin, A., Astrup, T.F., 2014. Quantification of the resource recovery potential of municipal solid waste incineration bottom ashes. Waste Management 34, 1627–1636.

Apaydin, O., Kurt, U., Gonullu, M.T., 2009. An investigation on tannery wastewater by electrocoagulation. Global NEST Journal 11, 546–555.

Apte, A.D., Verma, S., Tare, V., Bose, P., 2005. Oxidation of Cr(III) in tannery sludge to Cr(VI): field observations and theoretical assessment. Journal of Hazardous Materials 121, 215–222.

Ates, E., Orhon, D., Tunay, O., 1997. Characterization of tannery wastewaters for pretreatment - selected case studies. Water Science Technology 36, 217–223.

Behera, M., Bhattacharyya, S.K., Minocha, A.K., Deoliya, R., Maiti, S., 2014. Recycled aggregates from C&D waste & its use in concrete—a breakthrough towards sustainability in construction sector: a review. Construction and Building Materials 68, 501–516.

Bhagia, L.J., Vyas, J.B., Shaikh, M.I., Dodia, S.L., 2010. Chrysotile asbestos exposure in the manufacturing of thermal insulating boards. Environmental Monitoring and Assessment 167 (1–4), 559–564.

Biasin, A., Della Zassa, M., Zerlottin, M., Refosco, D., Bertani, R., Canu, P., 2014. On the understanding and control of the spontaneous heating of dried tannery wastewater sludge. Waste Management 34, 817–824.

Bigum, M., Petersen, C., Christensen, T.H., Scheutz, C., 2013. WEEE and portable batteries in residual household waste: quantification and characterization of misplaced waste. Waste Management 33, 2372–2380.

Bournay, E., 2006. Vital Waste Graphic 2, second ed., vol. 2. Basel Convention, UNEP and GRID-Arendal.

Celary, P., Sobik-Szoltyse, S., 2014. Vitrification as an alternative to landfilling of tannery sewage sludge. Waste Management 34, 2520–2527.

Chimenos, J.M., Segarra, M., Fernàndez, M.A., Espiell, F., 1999. Characterization of the bottom ash in municipal solid waste incinerator. Journal of Hazardous Materials A64, 211–222.

Christensen, T.H., 2011. Solid Waste Technology & Management. Wiley & Sons publisher, 2 volumes set. ISBN: 978-1-405-17517-3.

Chuan, M.C., Liu, J.C., 1996. Release behaviour of chromium from tannery sludge. Water Resource 30, 932–938.

Colangelo, F., Cioffi, R., Lavorgna, M., Verdolotti, L., De Stefano, L., 2011. Treatment and re cycling of asbestos-cement containing waste. Journal of Hazardous Materials 195, 391–397.

Conesa, J.A., Martín-Gullón, I., Font, R., Jauhiainen, J., 2004. Complete study of the pyrolysis and gasification of scrap tires in a pilot plant reactor. Environmental Science & Technology 38 (11), 3189–3194.

Cossu, R., Lai, T., 2013. Washing treatment of automotive shredder residue (ASR). Waste Management 33 (8), 1770–1775.

Cossu, R., Lai, T., 2015. Automotive shredder residue (ASR) management: an overview. Waste Management 45, 143–151.

Cossu, R., Lai, T., Sandon, A., 2012. Standardization of BOD5/COD ratio as a biological stability index for MSW. Waste Management 32 (8), 1503–1508.

Cossu, R., Raga, R., 2008. Test methods for assessing the biological stability of biodegradable waste. Waste Management 28 (2), 381–388.

Cossu, R., Fantinato, G., Pivato, A., Sandon, A., 2017. Further steps in the standardization of BOD5/COD ratio as a biological stability index. Waste Management 68, 16–23.

Cossu, R., Fiore, S., Lai, T., Luciano, A., Mancini, G., Ruffino, B., Viotti, P., Zanetti, M.C., 2014. Review of Italian experience on automotive shredder residue characterization and management. Waste Management 34 (10), 1752–1762.

Cotman, M., Koncan, Z., Gotvajn, A., 2004. The relationship between composition and toxicity of tannery wastewater. Water Science and Technology 49, 39–46.

Dantas, T.L.P., Josè, H.J., Moreira, R.F.P.M., 2003. Fenton and Photo-Fenton oxidation of tannery wastewater. Acta Scientiarum Technology 25, 91–95.

EPA, 2009. Municipal Solid Waste Generation, Recycling, and Disposal in the United States Detailed Tables and Figures for 2008, USEPA Office for Resource Conservation and Recovery. November 2009. http://www.epa.gov/osw/nonhaz/municipal/pubs/msw2008data.pdf.

ETRMA, 2011. End of Life Tyres: A Valuable Resource with Growing Potential. European Tyre and Rubber Manufacturers Association, Brussels, Belgium.

http://eur-lex.europa.eu/legal-content/IT/ALL/?uri=CELEX:32014D0955.

Eurostat, 2010. End-of-life vehicles (ELVs), Reuse and Recovery rate last updated on 16.04 2010. http://epp.eurostat.ec.europa.eu/portal/page/portal/waste/data/wastestreams/elvs.

Fiore, S., Ruffino, B., Zanetti, M.C., 2012. Automobile shredder residues in Italy: characterization and valorization opportunities. Waste Management 32 (8), 1548–1559.

Ganesh, R., Balaji, G., Ramanujam, R.A., 2006. Biodegradation of tannery wastewater using sequencing batch reactor—respirometric assessment. Bioresource Technology 97, 1815–1821.

Gidarakos, E., Petrantonaki, M., Anastasiadou, K., Schramm, K.-W., 2009. Characterization and hazard evaluation of bottom ash produced from incinerated hospital waste. Journal of Hazardous Materials 172, 935–942.

Gonzalez-Fernandez, O., Hidalgo, M., Margui, E., Carvalho, M.L., Queralt, I., 2008. Heavy metals' content of automotive shredder residues (ASR): evaluation of environmental risk. Environmental Pollution 153, 476–482.

Hoornweg, D., Bhada-Tata, P., 2012. What a Waste—A Global Review of Solid Waste Management. Urban Development Series, the World Bank. https://siteresources.worldbank.org/INTURBANDEVELOPMENT/Resources/336387-1334852610766/What_a_Waste2012_Final.pdf (accessed the 22nd.10.2018).

Karahan, O., Dogruel, S., Dulekgurgen, E., Orhon, D., 2008. COD fractionation of tannery wastewater—particle size distribution, biodegradability and modelling. Water Resource 42, 1083—1092.

Kaza, S., Yao, L., Bhada-Tata, P., Van Woerden, F., 2018. What a Waste 2.0. A Global snapshot of Solid Waste Management to 2050. World Bank, Washington, DC world bank. https://openknowledge.worldbank.org/handle/10986/2174.

Kilic, E., Font, J., Puig, R., Colak, S., Celik, D., 2011. Chromium recovery from tannery sludge with saponin and oxidative remediation. Journal of Hazardous Materials 185, 456—462.

Kim, Y.-C., Hong, W.-H., 2017. Optimal management program for asbestos containing building material to be available in the event of a disaster. Waste Management 64, 272—285.

Kiyotaka, K., Itaru, N., 2002. Present state of end of life vehicle recycling rates and recycling of automobile shredder residue. In: Proceedings of Japan Society of Automotive Engineers (JSAE). Annual Congress, vol. 53-02, pp. 5—8.

Kongjao, S., Damronglerd, S., Hunsom, M., 2008. Simultaneous removal of organic and inorganic pollutants in tannery wastewater using electrocoagulation technique. Korean Journal of Chemical Engineering 25, 703—709.

Koteswari, Y.N., Ramanibai, R., 2003. The effect of tannery effluent on the colonization rate of plankets: a microcosm study. Turkish Journal of Biology 27, 163—170.

Kurt, U., Apaydin, O., Gonullu, M.T., 2007. Reduction of COD in wastewater from an organized tannery industrial region by Electro-Fenton process. Journal of Hazardous Material 143, 33—40.

Kusiorowski, R., Zaremba, T., Piotrowski, J., Gerle, A., 2013. Thermal decomposition of asbestos-containing materials. Journal of Thermal Analysis and Calorimetry 113 (1), 179—188.

Lefebvre, O., Vasudevan, N., Torrijosa, M., Thanasekaran, K., Moletta, R., 2006. Anaerobic digestion of tannery soak liquor with an aerobic post-treatment. Water Resource 40, 1492—1500.

Leonelli, C., Veronesi, P., Boccaccini, D.N., Rivasi, M.R., Barbieri, L., Andreola, F., Lancellotti, I., Rabitti, D., Pellacani, G.C., 2006. Journal of Hazardous Materials B135, 149—155.

Leta, S., Assefa, F., Gumaelius, L., Dalhammar, G., 2004. Biological nitrogen and organic matter removal from tannery wastewater in pilot plant operations in Ethiopia. Applied Microbiology and Biotechnology 66, 333—339.

Lofrano, G., Belgiorno, V., Gallo, M., Raimo, A., Meriç, S., 2006. Toxicity reduction in leather tanning wastewater by improved coagulation flocculation process. Global NEST Journal 8, 151—158.

Li, J., Dong, Q., Yu, K., Liu, L., 2014. Asbestos and asbestos waste management in the Asian-Pacific region: trends, challenges and solutions. Journal of Cleaner Production 81, 218—226.

Lofrano, G., Meric, S., Zengin, G.E., Orhon, D., 2013. Chemical and biological treatment technologies for leather tannery chemicals and wastewater: a review. Science of the Total Environment 461—462, 265—281.

Mandal, T., Dasgupta, D., Mandal, S., Datta, S., 2010. Treatment of leather industry by aerobic biological Fenton oxidation process. Journal of Hazardous Material 180, 204—211.

Menad, N., Guignot, S., Houwelingen, J.A., 2013. New characterization method of electrical and electronic equipment wastes (WEEE). Waste Management 33, 706—713.

Munz, G., De Angelis, D., Gori, R., Mori, G., Casarci, M., Lubello, C., 2009. The role of tannins in conventional angogated membrane treatment of tannery wastewater. Journal of Hazardous Materials 164, 733—739.

Nakamura, T., 2009. Waste agriculture biomass convention. In: The 6th Biomass Asia Workshop in Hiroshima, 18—20 November 2009, IETC Osaka.

OECD, 2008a. Environmental Data Compendium 2006-2008. http://www.oecd.org/dataoecd/22/58/41878186.pdf.

Ojeda Benítez, S., Lozano Olvera, G., Morelos, A.R., Armijo de Vega, C., 2008. Mathematical modeling to predict residential solid waste generation. Waste management 28, 7—13.

Ongondo, F.O., Williams, I.D., Cherrett, T.J., 2011. How are WEEE doing? A global review of the management of electrical and electronic wastes. Waste Management 31, 714—730.

Orhon, D., Ates, E., Sozen, S., 2000. Experimental evaluation of the nitrification kinetics for tannery wastewaters. Water SA 26, 43—45.

Ram, B.P.K., Bajpai, Parwana HK., 1999. Kinetics of chrome-tannery effluent treatment by the activated sludge system. Process Biochem 35, 255—265.

Reschner, K., 2003. Scrap tyre recycling-Market overview and outlook. Waste Mannagement World. https://waste-management-world.com/a/scrap-tyre-recycling#author (accessed on the 20th October 2018).

Rodrigues, F., Carvalho, M.T., Evangelista, L., De Brito, J., 2013. Physical-chemical and mineralogical characterization of fine aggregates from construction and demolition waste recycling plants. Journal of Cleaner Production 52, 438–445.
Sabbaghi, M., Esmaeilian, B., Mashhadi, A.R., Behdad, S., Cade, W., 2015. An investigation of used electronic return flows: a data-driven approach to capture and predict consumers' storage and utilization behavior. Waste Management 36.
Sabbas, T., Polettini, A., Pomi, R., Astrup, T., Hjelmard, O., Mostbauera, P., Cappai, G., Magel, G., Salhofer, S., Speiser, C., Heuss-Assbichlerf, S., Klein, R., Lechner, P., 2003. Management of municipal solid waste incineration residues. Waste Management 23, 61–88.
Sakai, S.-I., Yoshida, H., Hiratsuka, J., Vandecasteele, C., Kohlmeyer, R., Rotter, V.S., Passarini, F., Santini, A., Peeler, M., Li, J., Oh, G.-J., Chi, N.K., Bastian, L., Moore, S., Kajiwara, N., Takigami, H., Itai, T., Takahashi, S., Tanabe, S., Tomoda, K., Hirakawa, T., Hirai, Y., Asari, M., Yano, J., 2014. An international comparative study of end-of-life vehicle (ELV) recycling systems. Journal of Material Cycles and Waste Management 16 (1), 1–20.
Scheutz, C., Fredenslund, A.M., Nedenskov, J., Kjeldsen, P., 2010. Release and fate of fluorocarbons in a shredder residue landfill cell: 1. Laboratory experiments. Waste Management 30 (11), 2153–2162.
Sharma, V.K., Fortuna, F., Mincarini, M., Berillo, M., Cornacchia, G., 2000. Disposal of waste tyres for energy recovery and safe environment. Applied Energy 65, 381–394.
Sienkiewicz, M., Kucinska-Lipka, J., Janik, H., Balas, A., 2012. Progress in used tyres management in the European Union: a review. Waste Management 32, 1742–1751.
Sivula, L., Oikari, A., Rintala, J., 2012. Toxicity of waste gasification bottom ash leachate. Waste Management 32, 1171–1178.
Skrypski-Mantele, S., Bridle, T.R., 1995. Environmentally sound disposal of tannery sludge. Water Resource 29, 1033–1039.
Song, Z., Williams, C.J., Edyvean, G.J., 2000. Sedimentation of tannery wastewater. Water Resource 34, 2171–2176.
Szpyrkowicz, L., Kaul, S.N., Neti Rao, N., Satyanarayan, S., 2005. Influence of anode material on electrochemical oxidation for the treatment of tannery wastewater. Water Resource 39, 1601–1613.
Thanigavel, M., 2004. In: Biodegradation of tannery effluent in fluidized bed bioreactor with low density biomass support. Annamalai University, Tamilnadu, India. M.Tech. Thesis.
UNEP, 2005. E-waste, the hidden side of IT equipment's manufacturing and use, Environmental Alert Bulletin. http://www.grid.unep.ch/product/publication/download/ew_ewaste.en.pdf.
Valle-Zermeno, R., Formosa, J., Chimenos, J.M., Martinez, M., Fernandez, A.J., 2013. Aggregate material formulated with MSWI bottom ash and APC fly ash for use as secondary building material. Waste Management 33, 621–627.
van Rooij, J.G.M., Jongeneelen, F.J., 2010. Hydroxypyrene in urine of football players after playing on artificial sports field with tire crumb infill. Int Arch Occup Environ Health 83, 105.
Wang, R., Xu, Z., 2014. Recycling of non-metallic fractions from waste electrical and electronic equipment (WEEE): a review. Waste Management 34, 1455–1469.
Williams, P.T., 2013. Pyrolysis ow waste tyres: a review. Waste Management 33, 1714–1728.
Wu, Z., Yu, A.T.W., Shen, L., Liu, G., 2014. Quantifying construction and demolition waste: an analytical review. Waste Management 34, 1683–1692.
Yu, J., Williams, E., Ju, M., Yang, Y., 2010. Forecasting Global Generation of Obsolete Personal Computers. Environmental Science & Technology 44, 3232–3237.
Zaremba, T., Peszko, M., 2008. Investigation of the thermal modification of asbestos wastes for potential use in ceramic formulation. Journal of Thermal Analysis and Calorimetry 3, 873–877.
Zorpas, A.A., Inglezakis, V.J., 2012. Automotive industry challenges in meeting EU 2015 environmental standard. Technology in Society 34 (1), 55–83.

1.3

LEGAL FRAMEWORK OF LANDFILLING IN DIFFERENT AREAS OF THE WORLD

Luciano Butti, Federico Peres and Caterina Lops

LEGAL SOURCES

Landfilling is subject to a myriad of technical regulations. However, the essential principles are usually set out in the main legislation on waste management and/or in a framework statute specifically dedicated to landfill disposal.

In the United States the "Resource Conservation and Recovery Act" (RCRA) of 1976, which amended the previous "Solid Waste Disposal Act" of 1965, sets forth the basic guidelines governing the disposal of solid and hazardous waste. The text is codified in the United States Code (USC), Title 42, Chapter 82 (§6901 et seq.). RCRA demanded to the Environmental Protection Agency (EPA) the power to enact specific implementing regulations, which are now collected in Title 40 of the Code of Federal Regulations (CFR), Parts 239–259 for nonhazardous solid wastes and Parts 260–279 for hazardous waste. Each state of the United States has its own legislation compliant with the federal one, which could be also stricter.

European Union's main act dedicated to waste management is Directive 2008/98/EC of November 19, 2008 (Waste Directive). However, landfilling is specifically addressed by Directive 1999/31/EC of April 26, 1999 on the landfill of waste (Landfill Directive), containing all the pertaining rules for waste disposal. European Union directives are framework legislation, which means that member states shall implement such directives into their national law, following the European standards, but they could also set out stricter criteria. Since 2015 all EU waste-related legislation, including the one on landfills, is under review in the framework of a transition towards a circular economy.

In Japan, the fundamental rules for managing waste are contained in the "Waste Management and Public Cleansing Law" (Law No. 137 of 1970, as amended). This basic act is completed by the "Cabinet Order of Waste Management and Public Cleansing Law" (Cabinet Order No. 300 of 1971), the "Regulations of Waste Management and Public Cleansing Law" (Ordinance No. 35 of Ministry of Health and Welfare of 1971) and the "Ordinance determining engineering standards pertaining to final disposal site for municipal solid wastes and final disposal site for industrial wastes" (Ordinance No. 1 of the Prime Minister's Office and Ministry of Health and Welfare of 1977), providing for further specifications. In this context, a mention shall be made of "The basic act for establishing a sound material-cycle society" (Act No. 110 of 2000) enacted at the beginning of the 21st century to promote a society in which products become circulative resources and proper disposal is provided just for resources not put into cyclical use.

LANDFILLING IN THE WASTE MANAGEMENT HIERARCHY

Starting from the 1970s, the development of environmental awareness in modern societies increasingly turned to consumerism, as well as the valorization of land resources, has led many countries to rethink their waste management systems, mainly based on massive landfilling, and assign the priority to alternative approaches aimed at reducing the amount of waste to be disposed. The result has been the establishment of the so-called *waste management hierarchy*, which has found proper encoding in the environmental legislation in different areas of the world.

In the US RCRA of 1976, for example, the landfill disposal is seen as the last option for managing both hazardous and nonhazardous solid waste. The Congressional findings (42 USC §6901) on environment and health, which are the premises of the act, state as follows:

> ...(7) certain classes of land disposal facilities are not capable of assuring long-term containment of certain hazardous wastes, and to avoid substantial risk to human health and the environment, reliance on land disposal should be minimized or eliminated, and land disposal, particularly landfill and surface impoundment, should be the least favored method for managing hazardous wastes and
>
> (8) alternatives to existing methods of land disposal must be developed since many of the cities in the United States will be running out of suitable solid waste disposal sites within 5 years unless immediate action is taken.

More in general, the "Pollution Prevention Act" of 1990 (Title 42, Chapter 133 of the USC, §13101) provides for an actual management hierarchy, extended to all pollution events:

> ...pollution should be prevented or reduced at the source whenever feasible; pollution that cannot be prevented should be recycled in an environmentally safe manner, whenever feasible; pollution that cannot be prevented or recycled should be treated in an environmentally safe manner whenever feasible; and disposal or other release into the environment should be employed only as a last resort and should be conducted in an environmentally safe manner.

As for the European Union, the Waste Directive contains in Article 4, titled "*Waste hierarchy,*" a clear priority order to be followed by member states.

The following waste hierarchy shall apply as a priority order in waste prevention and management legislation and policy:

(a) prevention;
(b) preparing for reuse;
(c) recycling;
(d) other recovery, e.g., energy recovery; and
(e) disposal.

The same Article 4, however, assumes that the objective of waste policies and legislation shall be in any case the "best overall environmental outcome", so "this may require specific waste streams departing from the hierarchy where this is justified by life-cycle thinking on the overall impacts of the generation and management of such waste".

In Japan, the "Basic Act for Establishing a Sound Material-Cycle Society" of 2000 illustrates its own version of the waste management hierarchy in Article 7 (titled "Basic Principles of the Cyclical Use and Disposal of Circulative Resources"):

> The cyclical use and disposal of circulative resources must be undertaken to the extent technologically and economically possible, taking it into full consideration that, for the reduction of environmental load, it is necessary to proceed under the provisions of the following items. However, consideration must be given to not proceeding under the provisions of the following items if it is deemed effective in reducing environmental load not to proceed under those provisions.
>
> (i) Regarding the entirety of, or one part of, circulative resources, that what can be reused must be reused.
>
> (ii) Regarding the entirety of, or one part of, circulative resources, that what are not reused under the preceding item (i) and can be reclaimed must be reclaimed.
>
> (iii) Regarding the entirety of, or one part of, circulative resources, that what is not reused under the above item (i) nor reclaimed under the preceding item (ii), and from which heat recovery is possible, heat recovery must be undertaken.
>
> (iv) Regarding the entirety of, or one part of, circulative resources, that what does not undergo cyclical use under the foregoing three items must be disposed of.

Indeed, the recent international trend pushes beyond the linear perspective of the *hierarchy* (where the waste that cannot be prevented is firstly reused, recycled, or recovered and, just in the last place, disposed of in landfills) toward the establishment of a zero waste *circular economy*. In December 2015 the European Commission adopted a new "Circular Economy Package" together with some revised legislative proposals on waste the target of which is to reduce landfill to maximum of 10% of all municipal waste by 2030. However, despite this programmatic intent, landfilling still remains nowadays one of the main tools for managing waste; this is certainly part of the ongoing process, but not necessarily a breach of the hierarchy, since, as specifically recalled by some of the legislative acts cited before, the choice among the managing options should look at the overall impacts, also considering the technical and economical feasibility in the given historical moment. Indeed, modern sanitary landfills are strictly regulated to minimize the impacts and maximize the efficiency; moreover, they can be also considered as a resource to be exploited by applying landfill mining principles and technologies (see Chapter 19.3).

LANDFILLS VERSUS OPEN DUMPS

Legislations variously define "landfill" as "an area of land or an excavation in which wastes are placed for permanent disposal" (US 40 CFR §257.2.) or "a waste disposal site for the deposit of the waste onto or into the land (i.e., underground)" (EU Landfill Directive, Article 2 (g)) or simply as "a site serving for landfill disposal" (Japan Cabinet Order No. 300 of 1971, Article 5).

More in general, however, a "landfill" may be referred to as a facility where waste is disposed onto or into the land following specific technical and procedural criteria established by the law to minimize or eliminate risks for the environment and the human health; this is what distinguishes it from an illegal "open dump" where waste is released into the environment without observing the basic safety legal standards.

The US RCRA of 1976, for example, is clear when prescribing that "a facility may be classified as a sanitary landfill and not an open dump only if there is no reasonable probability of adverse effects on health or the environment from disposal of solid waste at such facility" (42 USC §6944a). Similarly the EU Landfill Directive of 1999 holds that "landfill should be adequately monitored and managed to prevent or reduce potential adverse effects on the environment and risks to human health" while "it is necessary to take appropriate measures to avoid the abandonment, dumping, or uncontrolled disposal of waste" (Recitals 6–7). In Japan the Cabinet Order No. 300 of 1971 simply prohibits the dumping (Article 16, "No person shall unnecessarily dump waste").

Due to their nature of "regulated facilities" landfills are subjected to specific rules concerning permits, waste acceptance procedures, control and monitoring during the operational phase, closure and aftercare.

CLASSES OF LANDFILLS

Landfills are usually divided into different classes—with different regimes—depending on the kind of wastes they can accept.

In the European Union the Landfill Directive of 1999 (Article 4) establishes three classes of landfills: (1) landfills for hazardous waste, (2) landfills for nonhazardous waste, and (3) landfills for inert waste. The admission criteria are rather strict. It means that landfills for inert waste can only accept inert waste and landfills for hazardous waste shall only be used for hazardous waste. However, some flexibility is accorded in case of nonhazardous waste landfills that may be also used for municipal waste and certain kinds of hazardous wastes, in particular those which are stable and nonreactive, with leaching behavior equivalent to the one of nonhazardous wastes (Article 4 of the Landfill Directive specifies that "these hazardous wastes shall not be deposited in cells destined for biodegradable nonhazardous waste").

In the United States instead, landfills are divided into two main classes: (1) landfills for hazardous waste and (2) landfills for nonhazardous municipal and commercial or industrial solid waste. Landfills for hazardous waste are addressed by subtitle C of the RCRA and are regulated at federal level under the authority of the Environmental Protection Agency, while landfills for nonhazardous waste are subjected to subtitle D, which delegates to the states the relevant jurisdiction pursuant to federal guidelines.

In Japan as well, the Waste Management and Public Cleansing Law identifies two types of landfills: (1) landfills for municipal solid waste and (2) landfills for industrial waste. Each one of them is specifically addressed in a dedicated section of the law, respectively, Chapter II, Section 3 and Chapter III, Section 5.

THE PERMIT

Landfilling is not necessarily a public activity. Privates are generally allowed to build and manage landfills, given the compatibility with the urban planning, the check for the presence of other facilities and the estimation of the real needs of the nearby population.

For the managing of such activity, legislations commonly require a specific permit to be released by the public authority. The issuance of the permit is subject to a series of requirements.

In the European Union the minimum conditions for the application and for the permit are listed in articles 7, 8, and 9 of the Landfill Directive.

In the United States, as explained above, there is a distinction between solid wastes for which federal regulations (§239.6 CFR) just establish minimum permit conditions to be required by state law and hazardous wastes for which permit conditions are completely codified at federal level (§270 CFR).

The Japanese Waste Management and Public Cleansing Law also distinguishes between the permit for Municipal Solid Waste Disposal Facilities (Article 8) and the permit for Industrial Waste Disposal Facilities (Article 15).

The application usually contains information such as the following:

- identity of the applicant (could be the land owner) and the operator if different;
- description of the type of wastes to be disposed of;
- capacity of the disposal facility;
- description of the site
  - EU Landfill Directive specifies that the description shall include the hydrogeological and geological characteristics of the site; moreover, according to the Industrial Emissions Directive No. 2010/75/EU, when the facility is subjected to the integrated pollution prevention and control authorization (IPPC permit) the operator shall also submit to the competent authorities a specific "baseline report" containing information on the state of soil and groundwater contamination by relevant hazardous substances. The baseline report is intended as "a practical tool that permits, as far as possible, a quantified comparison between the state of the site described in that report and the state of the site upon definitive cessation of activities, to ascertain whether a significant increase in pollution of soil or groundwater has taken place" (Industrial Emissions Directive, Recital 24);
  - in Japan also the application shall be accompanied by drawings showing the topographical and geological features of surrounding areas (see Article 3, paragraph 5 of the Ordinance No. 35 of 1971);
  - US law requirements are particularly strict and impose the applicant to provide a wide range of detailed information and studies to prevent adverse phenomena in relation to airport safety, floodplains, wetlands, fault areas, seismic impact zones an unstable areas. The law, for example, requires information such as the location of the facility within a 100-year floodplain, with specific engineering studies on the various hydrodynamic and hydrostatic forces and on

flood protection devices designed to prevent washout; in seismic zones the applicant shall indicate the presence of faults which have had displacement in Holocene time within 3000 feet of the facility or, for underground waters, he shall identify the uppermost aquifer and aquifers hydraulically interconnected beneath the facility property, including groundwater flow direction and rate, and describe any plume of contamination that could have entered the ground at the time of the application. See CFR §258, subpart B and 257.7 through 257.13 (as recalled by §239.7) for solid waste and CFR §270.14 for hazardous waste;

- information on environmental impact assessment and the proposed methods for pollution prevention and mitigation; this usually implies the investigation of multiple parameters, such as air pollution, water pollution, noise, vibrations, bad odors, and all kind of possible effects on the surrounding living environment (see Article 3-2 of the Japanese Ordinance No. 35 of 1971; US 40 CFR §270.14(b)(8) for hazardous waste; in the European Union the environmental impact assessment is only due if required by the Council Directive 85/337/EEC on Environmental Impact Assessment);
- proposed plan for the operation and maintenance of the facility;
- proposed plan for the closure and postclosure;
- financial security
  - in the European Union, it is provided by the applicant "prior to the commencement of disposal operations to ensure that the obligations (including aftercare provisions) arising under the permit, etc. are discharged and that the closure procedures, etc. are followed" (Landfill Directive, Article 8a,iv);
  - in Japan the financial security serves as "a reserve fund for the maintenance of each final disposal site every year until the landfill disposal completes" (Waste Management and Public Cleansing Law, Article 8-5 for municipal solid waste also applying to industrial waste according to Article 15-2-3);
  - US law requires a financial assurance for closure and postclosure care (40 CFR §258.71 and §258.72 for nonhazardous municipal solid waste and 40 CFR §264.143 through §264.145 for hazardous solid waste). The operator of a hazardous solid waste landfill may choose among different financial instruments (i.e., trust fund, surety bond guaranteeing payment into a trust fund, surety bond guaranteeing performance of closure or postclosure care, letter of credit, insurance, financial test, and corporate guarantee) for which the law even provides a specific wording (40 CFR §264.151). The financial assurance could be in combination with further additional financial assurances. The operator of a nonhazardous municipal solid waste landfill shall provide a financial assurance in case he is required to undertake a corrective action program for the protection of groundwater (40 CFR §258.73), while the operator of a hazardous waste landfill shall provide a liability insurance to cover sudden and nonsudden accidental occurrences (40 CFR §264.147). The amount of the financial security is typically calculated taking into account the price to be charged by the operator for the disposal of waste in the site and the time required for the activities covered by the assurance (see EU Landfill Directive, Article 10; Japanese Waste Management and Public Cleansing Law, Article 8-5, paragraph 4 for municipal solid waste also applying to industrial waste according to Article 15-2-3). In the Unites States the financial assurance corresponds to the cost of hiring a third party to perform the assured activity (i.e., closure, postclosure care, or corrective action).

Despite the documents, studies and information furnished with the application, the issuance of the permit often depends also on some subjective requirements to be fulfilled by the operator itself and the facility personnel.

In the European Union the competent authority does not issue the permit unless it has ascertained that "the management of the landfill site will be in the hands of a natural person who is technically competent to manage the site" and that "professional and technical development and training of landfill operators and staff are provided" (Landfill Directive, Article 8a,ii).

Analogously in Japan the permit is granted only if "the capability of the applicant is sufficient for the proper and continuous conducting of the installation and operation and maintenance" of the disposal facility and if the applicant itself does not fall into specific categories (i.e., a person who declared bankrupt or a person sentenced for certain violations) as stated by the law (Japanese Waste Management and Public Cleansing Law, Article 8-2 for municipal solid waste and similarly Article 15-2 for industrial waste).

The US law requires the operator of a hazardous waste facility to submit with the application an outline of both the introductory and continuing training programs to be performed by the personnel and a brief description of how the training program is designed to meet actual job tasks (40 CFR §270.14b,12).

In certain countries the issuance of the permit is also subjected to prior publication.

In Japan, for example, the public authority provides the application and the environmental assessment document for public reference for a month and in the following two weeks every person concerned in the installation of the landfill "may submit a comment from the standpoint of the conservation of the living environment"; at the same extent, the public authority shall notify such documents to the mayors of the areas surrounding the planned facility and "listen to their opinion from the point of the conservation of the living environment" (Japanese Waste Management and Public Cleansing Law, Article 8 for municipal solid waste and similarly Article 15 for industrial waste).

In the United States the state law for nonhazardous solid waste landfills shall require that both the application and the final determination on the permit "are made available for public review and comment" and "the state shall have procedures that ensure public comments on permit determinations are considered" (40 CFR §239.6 a-b). For what concerns industrial waste the federal law specifically provides that a notice shall be published when the application is submitted and when the draft permit is issued, allowing at least 45 days for public comment, with the possibility for any interested person or the authority to require a public hearing (see, respectively, 40 CFR §124.32, §124.10, §124.11, and §124.12). At the time that the final permit decision is issued, the public authority shall also issue a response to comments and appeals may be filed within 30 days of the final decision (40 CFR §124.17 and §124.19).

Once issued the permit specifies the class of the landfill, the list of types and quantity of waste authorized to be deposited, as well as the procedures and requirements to be followed during the operational phase, closure, and aftercare. The duration of the permit could vary among countries and could be predetermined by the law or may be left to the discretion of the authorities on a case-by-case basis; in the United States, for example, the permit for hazardous waste landfills is effective for a fixed term not to exceed 10 years (40 CFR §270.50).

The commencement of the disposal operations is generally preceded by the inspection of the site by the competent authority "in order to ensure that it complies with the relevant conditions of the permit" (EU Landfill Directive, Article 8c). The Japanese Waste Management and Public Cleansing Law also provides that "the Ministry of the Environment is entitled to give necessary instructions on the inspection…in case that there is an urgent need for the conservation of living environment" (Article 8-2 for municipal solid waste).

OPERATION AND MAINTENANCE

During the operational phase the operator shall implement a specific control and monitoring program as laid out in the permit according to the procedures fixed by the law. The purpose of the monitoring is to check that the permit conditions are correctly complied with and all the processes of the landfill are executed as envisaged, with all environmental protection systems fully functioning as intended.

The EU Landfill Directive, Annex III, prescribes at least the control of leachate, relevant gas emissions and water (both surface water and groundwater) with a frequency that could vary from monthly (leachate volume and relevant gas emissions) to quarterly (leachate composition and volume and composition of surface water), every 6 months (level of groundwater) or with a site-specific frequency (groundwater composition), given the obligation for the operator to report at least annually to the competent authority. Based on the results of the monitoring, the operator shall notify to the competent authority "any significant adverse environmental effects revealed by the control and monitoring procedures and follow the decision of the competent authority on the nature and timing of the corrective measures to be taken" (Landfill Directive, Article 12).

In Japan the law requires the monitoring of multiple parameters such as retaining walls, seepages, water quality, leachate to maintain the engineering standards for final disposal set by Ordinance No. 1 of the Prime Minister's Office and Ministry of Health and Welfare of 1977 (the data to be recorded are listed in Article 4-7 of the Ordinance No. 35 of 1971).

In the United States the monitoring mostly addresses explosive gases, disease vectors (i.e., rodents, flies, mosquitoes, or other animals, including insects, capable of transmitting disease to humans), air emissions, surface water and groundwater, and includes procedures for controlling public access and prevent illegal dumping of wastes (40 CFR §258, Subpart C and E, for municipal nonhazardous solid waste and 40 CFR §257 for nonmunicipal nonhazardous solid waste).

CLOSURE AND AFTERCARE

A landfill naturally closes when the waste disposal completes and all the relevant closure conditions set out by the permit are met. However, within the limits set by the various legislations, a landfill could also start the closure procedure earlier at the request of the operator (subject to the authorization of the competent authority) or by reasoned decision of the competent authority in case of revocation of the permit (see EU Landfill Directive, Article 13(a); Japanese Waste Management and Public Cleansing Law, Article 9, paragraph 4 and Article 9-2 for municipal solid waste and Article 15-2-4 and 15-3 for industrial waste; US 40 CFR §270.42, Appendix I (a)(6) and §270.43 for hazardous waste).

Following the definite closure the operator has an obligation to carry out the aftercare of the landfill site at the conditions set by the permit and by the law. This implies that he shall continue the monitoring and control as long as required by the permit, possibly with different conditions and frequencies than during the operational phase (EU Landfill Directive, Annex III, identifies the minimum parameters to be monitored by the operator during the aftercare phase and relevant timing; Japanese Ordinance No. 35 of 1971, Article 5-5, requires the operator who notifies the completion of the landfill disposal to attach a specific "document showing the operation and maintenance method for the period from the completion of the landfill disposal to the abolition of the facility"). In the European Union the aftercare financial assurance is provided for at least 30 years, but it is specified that the obligation to monitor landfill gas, leachate, and groundwater persists "as long as the competent authority considers that a landfill is likely to cause a hazard to the environment" (Landfill Directive, Article 13d). Also for the US law the postclosure care must be conducted for 30 years, but this period may be decreased or increased by the public authority depending on the time deemed as necessary to protect human health and the environment; the postclosure care consists of at least maintaining the integrity and effectiveness of the final cover, the leachate collection system as well as the gas and groundwater monitoring systems (40 CFR §258.61(a-b) for nonhazardous waste and 40 CFR §264.117 for hazardous waste).

In general a landfill may be considered as closed or definitely terminated only when the competent authority has checked the compliance with the standards required by the law (possibly carrying out a final on-site inspection) and has communicated to the operator its approval for the closure or the definitive termination (see EU Landfill Directive, Article 13(b); Japanese Waste Management and Public Cleansing Law, Article 9, paragraph 5 for municipal solid waste also applying to industrial waste according to Article 15-2-4). US law requires the operator of a nonhazardous waste facility to notify to the authorities that a certification of completion, signed by an independent registered or approved professional engineer, has been placed in the operating record, ready for inspection (40 CFR §258.60(h) for closure and 40 CFR §258.61(e) for postclosure care), while in case of a hazardous waste facility the certification of closure shall be submitted to the authority by registered mail (40 CFR §264.115 for closure 40 CFR §264.120).

References

Council Directive 1999/31/EC, April 26, 1999. On the Landfill of Waste. Available at: http://eur-lex.europa.eu/legal-content/EN/TXT/PDF/?uri=CELEX:31999L0031&from=en.

Directive 2008/98/EC of the European Parliament and of the Council, November 19, 2008. On Waste and Repealing Certain Directives. Available at:http://eur-lex.europa.eu/LexUriServ/LexUriServ.do?uri=OJ: L:2008:312:0003:0030:en:PDF.

Japan Act No. 110, 2000. The Basic Act for Establishing a Sound Material-Cycle Society. Available at: https://www.env.go.jp/en/laws/recycle/12.pdf.

Japan Cabinet Order No. 300, September 23, 1971. Cabinet Order of Waste Management and Public Cleansing Law. Available at: https://www.env.go.jp/en/laws/recycle/02.pdf.

Japan Law No. 137, 1970. Waste Management and Public Cleansing Law. Available at: https://www.env.go.jp/en/recycle/basel_conv/files/Waste_Management_and_Public_Cleansing.pdf.

Japan Ordinance No. 35 of Ministry of Health, Welfare, 1971. Regulations of Waste Management and Public Cleansing Law. Available at: https://www.env.go.jp/en/laws/recycle/03.pdf.

Japan Prime Minister's Office, Ministry of Health and Welfare Ordinance No. 1, March 14, 1977. Ministerial Ordinance Determining Engineering Standards Pertaining to Final Disposal Site for Municipal Solid Wastes and Final Disposal Site for Industrial Wastes. Available at: https://www.env.go.jp/en/laws/recycle/04.pdf.

Pollution Prevention Act, 1990. United States Code, Title 42, Chapter 133. Available at: https://www.gpo.gov/fdsys/pkg/USCODE-2016-title42/pdf/USCODE-2016-title42-chap133.pdf.

Resource Conservation and Recovery Act, 1976. United States Code, Title 42, Chapter 82. Available at: https://www.gpo.gov/fdsys/pkg/USCODE-2016-title42/pdf/USCODE-2016-title42-chap82.pdf.

United States Code of Federal Regulations, 2017 edition, Title 40, Parts 239−279. Available at: https://www.gpo.gov/fdsys/browse/collectionCfr.action?collectionCode=CFR&searchPath=Title+40%2FChapter+I%2FSubchapter+I&oldPath=Title+40%2FChapter+I%2FSubchapter+I&isCollapsed=true&selectedYearFrom=2017&ycord=1500.

United States Environmental Protection Agency, 2005. Introduction to Permits and Interim Status (40 CFR Part 270). Available at: http://www2.epa.gov/sites/production/files/2015-09/documents/permits05.pdf.

2 CHAPTER

LANDFILL CONCEPTS

2.1

MULTIBARRIER PRINCIPLES IN LANDFILLING

Raffaello Cossu

INTRODUCTION

Independently from the quality level of the installation, any deposit of waste on ground behaves like a normal three-phase reactor (Fig. 2.1.1). Water and air enter the reactor as external input or as component of the waste mass delivered to the deposit. Depending on the kind of waste, biological, chemical, physical–chemical reactions are occurring and gases and liquids are emitted. On the base of local environmental conditions as well as on cultural, demographic, scientific, and economic development in different geographical areas, landfill emissions can create public concerns.

The major environmental concern is related to the control of emissions (leachate and gas) and particularly to the discharge of leachate into the environment.

Due to this reason a barrier system has to be installed to minimize emissions. Assuming that a barrier should be a tool for avoiding or reducing the mass transport of contaminants from the landfill into the sensitive environment, different kinds of barriers should be implemented.

All the different landfill concepts, both old and new, can then be considered as involving several kinds of barriers. The understanding of each individual barriers and the proper combination of them are the key factors for designing landfill systems according to the role of landfilling in modern waste management strategies (see Chapter 1.1).

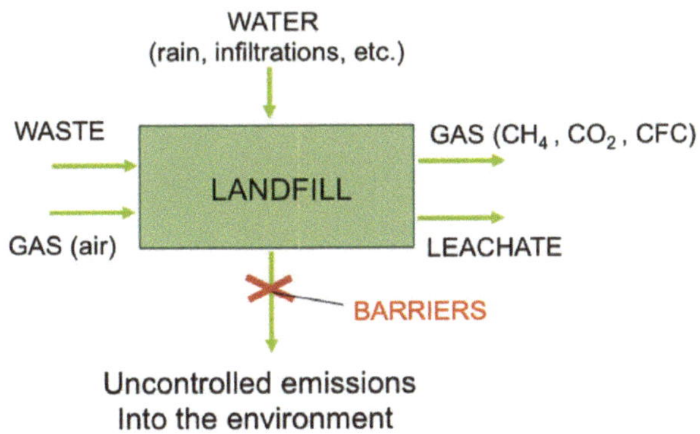

Figure 2.1.1 Material flows in a landfill.

THE MULTIBARRIER CONCEPT

Landfill can be regarded as a macrosystem, or combinations of macrosystems, to which mass transport relationships can be applied. These can be rather complicated and it is not the intent of this paper to deal with them in a rigorous way. Nevertheless the mass transfer processes are fundamental for the full understanding of the barrier concept. In the following paragraphs they will be outlined by use of elementary or simplified equations and referring, whenever possible, to a classic terminology (Bird et al., 2007).

Mass transfer terminology

Given a generic system with volume V and total surface S (Fig. 2.1.2), the system may be either homogeneous (one phase) or heterogeneous involving more than one phase. Accumulation and transformation are typical global properties involving the volume, whereas mass transport involves the surface. Transports can generally be either convective (when a flow of a species i is connected to the bulk motion of the fluid containing the species i) or diffusive (when the flow is connected to molecular motion of species i due to differences of concentration of species i).

With reference to the generic system as shown in Fig. 2.1.2, the following terms can be defined:

m_i, mass of species i in the given volume V, at time t
r_i, reaction rate for species i occurring in the considered volume V (mass of i/unit volume/unit time)
w_i, net mass flow of species i (vector quantity) transported through the surface S (mass of i/unit time). This flow will be assumed as positive when entering the system and negative when leaving it.

$$w_i = \int_S c_i v_i dS = \int_S f_i dS$$

c_i, mass concentration of species i (mass of i/volume unit of solute)
v_i, transport velocity of species i with respect to stationary coordinate axes
v, local mass average velocity

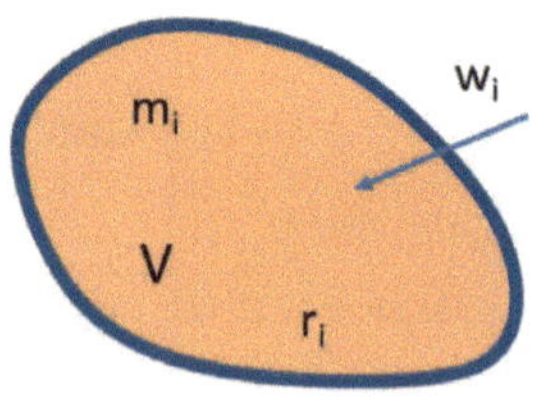

Figure 2.1.2 Scheme of a generic macroscopic flow system. Meaning of terms in the text.

$$v = \frac{\sum_{i=1}^{n} c_i v_i}{\sum_{i=1}^{n} c_i}$$

This represents the velocity of i when the transport is convective.
$n_i = c_i\, v$, convective mass flux of species i (mass of i passing through a unit area per unit time)
j_i, diffusive mass flux of species i according to Fick's first law:

$$J_i = -D\frac{\partial c_i}{\partial z}$$

$\partial c_i/\partial z$ is the concentration gradient along z
$f_i = n_i + j_i$, mass flux combining convective and diffusive transfer:

$$f_i = -D\frac{\partial c_i}{\partial z} + c_i v$$

Landfill mass balance

A landfill can be regarded as a macrosystem of volume V, with a mass content of a generic "contaminant," m, at time t, where convective and diffusive transfer of contaminant occurs and where reactions, r, involving the contaminant, take place (as "contaminant" we are taking into consideration a generic species (such as N, C, metals, etc.), which may create a nuisance to the environment in case of uncontrolled emissions). The surfaces affected by a convective mass transfer are those where a well-defined flow with a local mass average velocity can be observed. A flue pipe section is a typical surface for convective transport. The diffusive surfaces are typically those at the boundaries of phases (liquid–liquid, solid–liquid, etc.). However, the boundaries cannot always be easily determined and the kind of mass transfer assessed. In the majority of situations inside the landfill in fact both diffusive and convective flows occur. For the sake of simplicity only pure convective flow will be here considered as convective.

According to these assumptions landfill can be schematically described as in Fig. 2.1.3. S_A is the fraction of the total surface through which a total convective flow w_A enters the system, S_B is the surface fraction affected by a convective outflow w_B. Through the remaining part of the surface, S_d dispersive flow w_d (landfill emissions) occurs.

S_A connects the active core of the landfill with the external environment from which waste material and rainfall water enter the system. S_B is the surface through which process fluids are collected and transported out of the system (leachate and biogas pipelines). S_d can be considered as the boundary surface between the landfill system and the sensitive environment. The sensitive environment is intended as the part of the environment where any significant change in the quality of the compartments adversely affects the natural equilibrium presenting risks for the ecosystem. The components of the environment which are planned to play a role in attenuating landfill emissions are considered

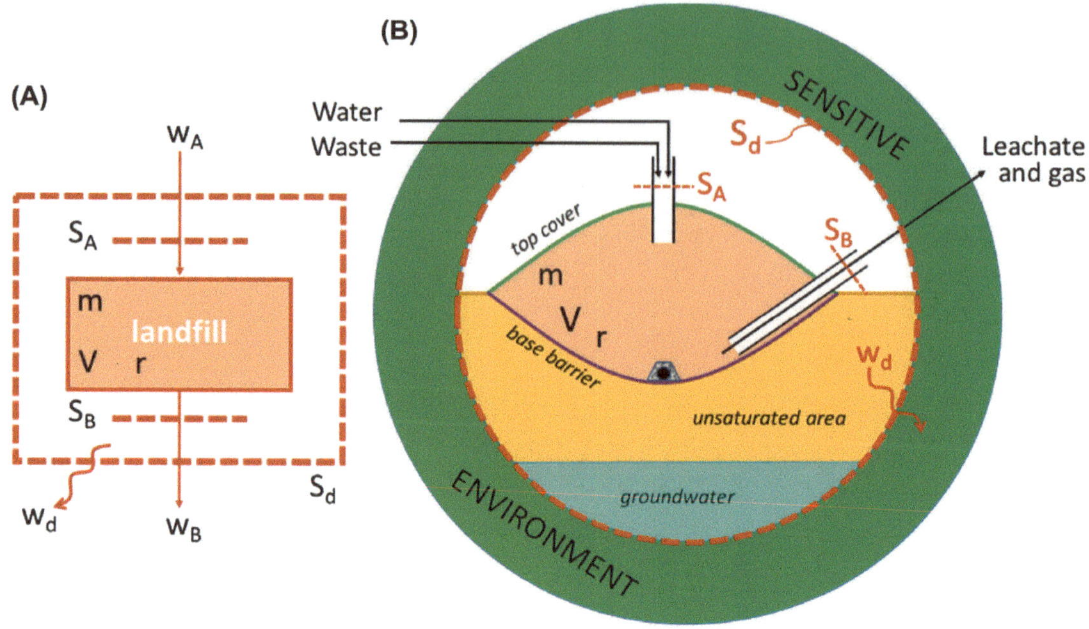

Figure 2.1.3 Scheme of the landfill system with convective flows entering the system at plane A (waste and water) and exiting at plane B (gas and leachate). Combined diffusive–convective flows (gas migration and processes, precipitation, etc.). The mass balance for the contaminant in the described landfill system.

as a part of the landfill multibarrier system (e.g., unsaturated zone underlying the landfill body, the reactive aquifer zones).

The overall reaction rate for the considered contaminant is here assumed to be including all attenuation mechanisms which may occur in the landfill system (biological degradation, sorption processes, precipitation, etc.). The mass balance for the contaminant in the described landfill system can then be written as follows:

$$\frac{dm}{dt} = w_A - w_B - w_d - rV \tag{1}$$

where dm/dt represents the accumulation of contaminants in the system.

As the difference $w_B - w_A = -\Delta w$ represents the real controlled outflow of the contaminant from the system, Eq. (1) can be rewritten as follows:

$$\frac{dm}{dt} = -\Delta w - w_d - rV \tag{2}$$

For steady-state conditions, $dm/dt = 0$.

The flux of contaminant, w_d, which reaches the sensitive environment through convective–diffusive transport, is that on which environmental impact should be determined and evaluated.

Any means or tools used in the different landfill strategies for minimizing w_d, or maximizing (which is the same) the other terms of Eq. (2), can be defined as "barriers."

Landfill barriers

According to the theoretical approach described earlier, in the following paragraphs an analytical description of the different possible barriers for controlling landfill emissions is provided.

The landfill is considered as a cascade of different kinds of barriers (Fig. 2.1.4):

- *valve barrier*, where no change in contaminant quality occurs and the flow is practically convective.
- *reactor barrier*, where reactions involving contaminants occur and transport is a combination of diffusive and convective fluxes. With regard to convective fluxes, the barrier can again be idealized as a valve.

For a good barrier efficiency, the ideal valves may either be open (as for the drainage), shut (as for the liner), or controlled (as for the water input).

Waste input barrier

The flow of contaminant contained in the waste to be landfilled, w_R, can be regarded as a convective mass flow entering the system and can be expressed as follows:

$$w_R = \sum_{R=1}^{n} c_R Q_R \tag{3}$$

where: c_R, concentration of contaminant in the waste (weight of contaminant/unit weight of waste); Q_R, amount of landfilled waste per day (weight of waste/unit time); n, number of different types of waste.

Assuming no contaminant content in the infiltrating water: $w_R = w_A$.

A reduction either of the contaminant's concentration or of the waste amount can be defined as "waste input barrier."

Water input barrier

The water that infiltrates the landfill is the result of a hydrological balance (see Chapter 6.1) and can be described as follows:

$$P_i = P + J + R^* - R - ET \pm \Delta U_s \tag{4}$$

where P, precipitation; J, irrigation or leachate recirculation; R, surface runoff; R*, runoff from external areas; ET, actual evapotranspiration; U_s, water content in top cover soil.

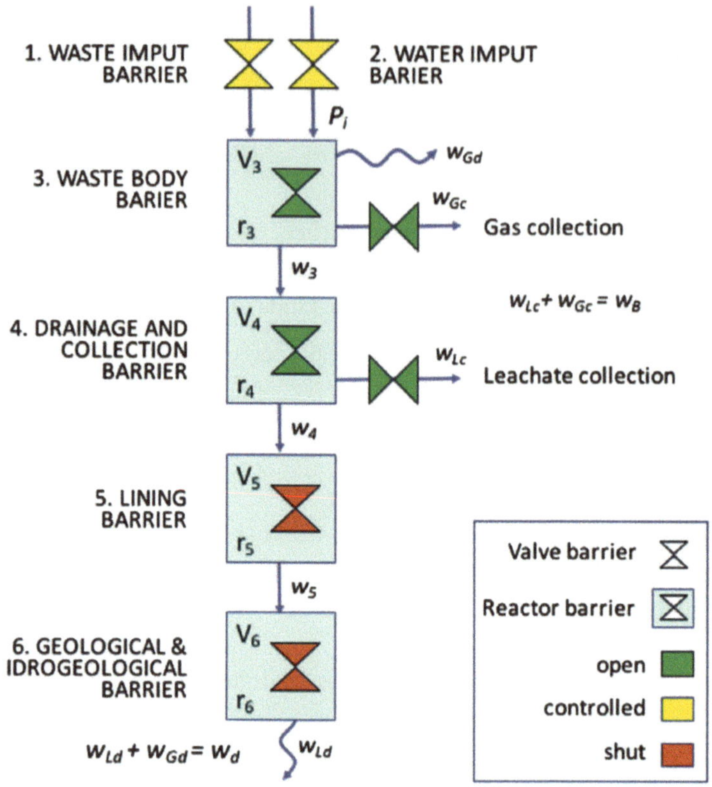

Figure 2.1.4 Scheme of different barriers and contaminant loads in a landfill system.

The infiltrating water, when excluding contribution from groundwater, surface water, runoff from external areas, and leachate recirculation, can be regarded as the rainfall water, which moves through the top cover system under the given climatic conditions. According to the quality of waste the infiltrating water may play different important roles in the processes occurring within the landfill body:

- dissolution of soluble compounds
- hydrolysis of complex biodegradable organics
- transportation and distribution of nutrients and bacteria in the waste mass.

Waste body barrier

This barrier can be analyzed as including three subbarriers:

1. the reactor barrier (transformation process of contaminants within the waste body)
2. the hydraulic waste barrier, which physically control the liquid flow through the waste
3. the daily cover barriers (control of the liquid flow and possible transformation processes within the cover layers)

1. The reactor barrier

   The anaerobic biological processes occurring in a municipal solid waste (MSW) landfill can be divided into three main steps: hydrolysis, acid fermentation, and methane fermentation (see Chapter 3.1). For each of these steps, considered as individual systems, or for the bioreactor system a mass balance can be written according to the general Eq. (2).

   The mass balance for the transport of contaminants within the landfill reactor can be expressed by the following relationship:

$$V_3\frac{dc}{dt} = \sum_{R=1}^{n} c_R Q_R - c_{3g}Q_g - c_3 Q_3 - r_3 V_3 \tag{5}$$

   The relationship can be applied to any transformation (chemical, chemicophysical, or biological), which may occur in the waste mass in relation to the waste quality.

   This relationship is fundamental for representing the evolution of modern landfilling and will be developed and discussed into details in Chapter 2.3.

   As a result of the transformation processes a flow of contaminants c_3Q_3 (contaminants concentration c_3, leachate flow rate Q_3), associated with a leachate flow rate, leaves the reactor system jointly with a contaminant flow associated with landfill gas, $c_{3g}Q_g$, if a gas is generated (c_{3g} is the contaminant concentration in the gas and Q_g is the gas flow rate). In case biodegradable organics are present in the landfilled waste, assuming steady-state conditions ($Vdc/dt = 0$), the higher the biodegradation rate, the higher the carbon in the gas phase, the lower the remaining carbon in the flowing leachate.

2. Hydraulic waste barrier

   The flux of leachate through the waste layer can be ideally described by the Darcy law:

$$v_3 = \frac{Q_3}{A} = k_R i_3 = k_R\frac{h_3}{L_3} \tag{6}$$

   where: k_R, hydraulic conductivity of waste (LT^{-1}); v_3, Darcy velocity (LT^{-1}); Q_3, leachate flow rate (L^3T^{-1}); A, surface of the waste layer (L^2); i_3, hydraulic gradient (dimensionless) in the waste layer, h_3/L_3; h_3, leachate head in the waste layer (L); L_3, thickness of the waste layer (L).

   Low permeability values of the waste material at steady flow conditions lead to an increase of the leachate head within the waste layer.

3. Daily cover barrier

   The leachate flux through the daily cover layer can follow an equation analogous to Eq. (6). L in this case is the depth of the cover material (normally 20–25 cm).

   A mass balance for the contaminant in the daily cover layer could be relevant according to the nature of the considered contaminant and cover material.

Biogas collection

The gas produced during waste degradation may be extracted and transported out of the landfill body. The concentration of the contaminants in the gas flow depends on the specific contaminant that is considered. In case of carbon, methane and carbon dioxide are the predominant forms. The ratio among them is regulated by the presence of oxygen (aerobic, anaerobic, semiaerobic environment).

The flow of contaminant associated with landfill gas emissions can be expressed as follows:

$$w_g = c_{3g} Q_g \tag{7}$$

where: w_g, contaminant mass flow (MT^{-1}); c_{3g}, gas-phase concentration of the considered contaminant (ML^{-3}); Q_g, gas flow rate ($L^{-3}T^{-1}$).

Flux through the drainage system

The granular bed of the drainage system should ideally behave as an open valve with a containment convective flux. On the contrary, several experimental studies (Ramke, 1989; Brune et al., 1994; Paksy et al., 1995) demonstrated that the granular drain behaves as a fixed bed reactor, capable of removing volatile fatty acids and other components. Biomass growth is a very serious clogging factor in drainage systems, particularly when oxygen is present.

A mass balance for the considered contaminant, similar to the general one shown in Eq. (2), can be expressed as follows:

$$V_4 \frac{dc}{dt} = c_3 Q_3 - c_4 Q_4 - c_4 Q_{Lc} - r_4 V_4 \tag{8}$$

with obvious meaning of each term (see Eq. 5 and Fig. 2.1.4).

Considering the physical design parameters of a granular drainage layer (Fig. 2.1.5), a differential equation governing steady-state drainage on a sloping barrier has been developed (McEnroe, 1994):

$$k_D y \left(\frac{dy}{dx} - s \right) + vx = 0 \tag{9}$$

where: x, horizontal distance from the upstream boundary (L); y, saturated depth over the lining layer (L); s, dimensionless slope of the lining–drainage barrier system; k_D, hydraulic conductivity of the drainage layer (LT^{-1}); v, Darcy velocity (LT^{-1}).

If the drain system is working properly, the water level in the drain layer should be below the top of the drain surface (free-drainage conditions). If the drain is undersized or clogged, y will increase.

Leachate collection

The leachate that reaches the base of the landfill can be regularly collected and transferred out of the landfill system. This permits control of the hydraulic head in the system.

Storage of leachate in the landfill body leads to increasing the head and consequently the Darcy velocity through the liners.

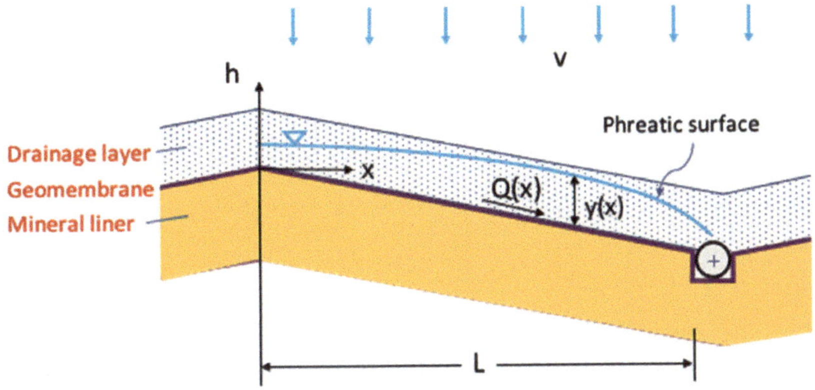

Figure 2.1.5 Definition sketch for landfill drainage system (McEnroe, 1994).

The transfer of contaminant associated with the collected leachate flow is as follows:

$$w_{Lc} = c_4 Q_{Lc} \tag{10}$$

with the meaning of the terms as can be derived from Fig. 2.1.5.

Flux through the liner system

The calculation of the mass flow through the liner system strictly depends on its design and structure. Referring to single lining systems the following structures can be considered:

- clay-only liner
- geomembrane-only liner
- geomembrane–clay composite

The transport of contaminant through a clay-only liner can be modeled by using, for the assumed one-dimensional conditions, a convective–diffusive equation as the one well known for transport of contaminants in the soil (Freeze and Cherry, 1979; Rowe, 1994):

$$n\frac{\partial c}{\partial t} = \left(nD\frac{\partial^2 c}{\partial z^2} - nv\frac{\partial c}{\partial z}\right) - \rho K_d \frac{\partial c}{\partial t} \tag{11}$$

where: c, concentration of contaminant in the leachate at depth z at time t; n, porosity (dimensionless); D, diffusion coefficient (L^2T^{-1}); ρ, dry density of the waste (ML^{-3}); K_d, partitioning coefficient ($M^{-1}L^3$); ρK_d, dimensionless measure of the carbon reduction due to sorption processes; v, Darcy velocity $= k_c\, i = k_c\, h/L$ (LT^{-1}); k_c, clay hydraulic conductivity (LT^{-1}); i, hydraulic gradient = h/L; h, leachate head (L); L, clay layer thickness (L).

In this equation the sorption process is assumed linear and K_d is the relative partitioning coefficient. For $\rho K_d = 0$ no sorption occurs and the contaminant should be considered as conservative.

Mass transport through a geomembrane-only liner can be considered as convective, occurring through the holes and cuts of the geomembrane. To this regard, it has been evaluated that the performance of a geomembrane-only liner invariably causes some damage linked to either installation (Colucci and Lavagnolo, 1995) or aging (Rollin et al., 1991). Even using a good Quality Control Programme a certain amount of ruptures should be accepted as normal design feature.

The leakage flux, v, can be calculated (Wallace and Akgun, 1994) as follows:

$$v = nq_b \tag{12}$$

where: n, number of holes per unit area (L^{-2}); $q_b = C_b a\sqrt{2gh}$, leakage through a single hole in a geomembrane (L^3T^{-1}); C_b, dimensionless coefficient dependent on the shape of the orifice; a, area of the hole (L^2); g, the gravity acceleration (LT^{-2}); h, hydraulic head (L).

For the case of geomembrane composite liner the overall convective–diffusive equation can still be used. Leakage through a hole in the geomembrane in the case of a very good geomembrane/clay contact has been calculated (Bonaparte et al., 1989; Wallace and Akgun, 1994):

$$q_b = 0.21\, a^{0.1} h^{0.9} k_c^{0.74} \tag{13}$$

where all the terms are defined as above and k_c is the hydraulic conductivity of the clay. The leakage flux, v, is also given by Eq. (12).

Flux through the unsaturated zone

The mass transport can once again be modeled using Eq. (11). Mechanical dispersive processes can be of some importance for k values above 10^{-2} m/s (Rowe, 1994).

Flux in the reactive aquifer

Different processes govern the attenuation of leachate contaminants in aquifers (Christensen, 1992). The main processes involved are dilution (mixing of leachate with unpolluted groundwater), dispersion, redox processes, biodegradation, dissolution/precipitation, complexation, ion exchange, and sorption.

The relative mass transport velocity, v_s/v_a, of the contaminant in the leachate pollution plume is as follows (Christensen, 1992):

$$\frac{v_s}{v_a} = \left(1 + \frac{\rho}{\varepsilon} K_d\right)^{-1} = R^{-1} \tag{14}$$

where: v_s, transport velocity of the contaminant; v_a, water transport velocity; ρ, bulk density of the aquifer material (ML^{-3}); e, effective porosity (dimensionless); K_d, distribution coefficient (L^3M^{-1}); R, retardation factor (dimensionless).

From Eq. (14) it is well clear how the characteristics of the aquifer, in terms of fluidodynamic (v), physical (p, e), and physical–chemical (K_d) properties, represent the "hydrogeological barrier."

DESIGN STRATEGIES AND LANDFILL CONCEPTS

Several design strategies can be applied to suit different landfill concepts, and engineering solutions can be adopted to maximize the efficiency of the different barriers as previously described.

A summarized presentation of the mass fluxes through the barriers, the possible engineering solutions, and the possibility of control is provided in Table 2.1.1.

In the following paragraphs the main landfill concepts and engineering and strategies options are discussed.

Landfill concepts are evolving continuously (see Box 2.1.1) according to the following:

- increasing knowledge of the potential of landfill emissions and their environmental impact
- development of new technologies
- cultural and social evolution
- new policies and regulations, etc.

Table 2.1.1 Summary of mass transport fluxes through the different landfill barriers, influence of different technical measures on transport parameters, evaluation of barrier efficiency, controllability level, and indication of the control period time

Flux Description	Transfer Equations	Technical Measures	Affected Parameters	Barrier Efficiency	Control Level	Control Time
Waste input	(3)	Quality of waste	c_R	H	H	PL/OP
		Waste reduction	Q_R	M/H	M	PL/OP
		Waste pretreatment:	c_R, Q_R		VH	OP
		• mechanical		L		
		• biological		H		
		• thermal		VH		
Infiltrating water	(4)	Siting	ET, P	VH	L/M	PL
		Top cover design	R, ET, P_i	H	H	OP/LT
		Waterproof landfilling	P_i	VH	H	OP

(Continued)

TABLE 2.1.1 Summary of mass transport fluxes through the different landfill barriers, influence of different technical measures on transport parameters, evaluation of barrier efficiency, controllability level, and indication of the control period time—cont'd

Flux Description	Transfer Equations	Technical Measures	Affected Parameters	Barrier Efficiency	Control Level	Control Time
Flux through the waste	(1), (5), and (6)	Leachate recirculation	k_D, R_3	VL	L	OP
		Process enhancement	r_3, w_B	M/H	L/M	OP
		Internal drainage	k_R	L/M	L	OP/AC
		Waste compaction	k_R	M	L	OP
		Waste pretreatment	k_R	H	VH	OP
Flux through the daily cover (DC)	(6)	Design and operation	k, L, r	L/M	M/H	OP
		Avoidance of DC	k	M	VH	OP
		Material quality in DC	k	L/M	M	
Flux through the drainage	(8) and (9)	Drain design	dx, s, dy, k_D	M	H	PL
		Waste pretreatment	r_4, k_D	H	M	OP
		Leachate collection	dy	H	M	OP
		Leachate recirculation	k_D	–	–	–
Flux through the liner	(11)–(13)	Lining design	h, k_c	VH	H	PL
		Direct GM/clay contact	v, q_b	H	H	PL
		CQ procedures	k_c, q_b	M/H	H	PL/OP
Transport in the unsaturated zone	(11)	Siting	K_D, v	VH	L	PL

TABLE 2.1.1 Summary of mass transport fluxes through the different landfill barriers, influence of different technical measures on transport parameters, evaluation of barrier efficiency, controllability level, and indication of the control period time—cont'd

Flux Description	Transfer Equations	Technical Measures	Affected Parameters	Barrier Efficiency	Control Level	Control Time
Transport in the aquifer	(14)	Siting	K_D, v_s, ε, ρ	H	L	PL
		Landfill below water table	v_s	L/M	VL	PL
Leachate collection	(10)	Design and operation	c_4, Q_1	H	H	PL/OP
		Leachate recirculation	Q_1	L	L	OP
		Process enhancement	c_4	M	L	OP
		Quality of waste	c_4	VH	H	PL/OP
Biogas collection	(7)	Design and operation	c_{3g}, Q_g	H	H	PL/OP
		Process enhancement	c_{3g}, Q_g	H	L	OP
		Quality of waste	c_{3g}, Q_g	VH	H	PL/OP

Transfer equation numbers and affected parameters refer to the text.
AC, aftercare; *H*, high; *L*, low; *LT*, long-term control; *M*, medium; *OP*, operational phase; *PL*, prelandfilling phase; *VH*, very high; *VL*, very low.

As stated previously, each of the different landfill concepts rely on one or more of the previously described barriers as graphically represented in Fig. 2.1.6 where indication of the main strategic barrier is applied and its relative weight is given.

Attenuate and disperse landfill

This concept is based on the waste process barrier and the geological–hydrogeological barriers and no lining barrier is considered. The consequent practical and scientific application of this concept required a particularly accurate study of the geological barrier (Robinson, 1989) and a control of the processes occurring within the waste mass. Such an option has been in the paste very popular in United Kingdom, where this landfill concept has reached the widest and highest scientific level of application.

Box 2.1.1 Technical evolution of landfilling

Gas and leachate emissions from landfilling contain substances capable of contaminating the environment. Natural phenomena or measures adopted to attenuate or prevent the uncontrolled diffusion of emissions and consequent onset of environmental and health risks are classified as barriers. The means by which these barriers are conceived has marked the technical development of landfills (Fig. 1), following the cultural development of modern society and reflecting the growing social and political awareness of environmental issues.

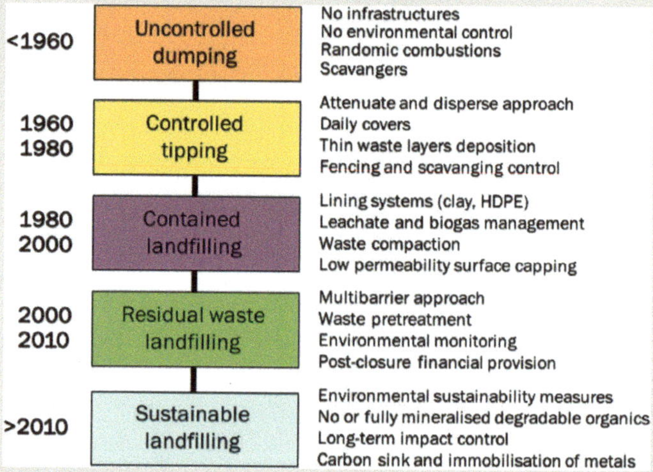

Figure 1 Chronological development of concepts, technologies, and management procedures for waste landfilling.

Not so long ago uncontrolled dumping was the main system used in the final disposal of waste worldwide. This system is still applied today as the main means of waste disposal by more than half the world population. This form of waste disposal is not supported by any means of modern infrastructure or environmental control and is invariably associated with the presence of scavengers, people who make their living from recovering marketable fractions from the deposited wastes.

In the 1960s the use of uncontrolled dumping was gradually replaced in several industrialized countries by controlled tipping, encoded by the World Health Organization in a series of guidelines used over a considerable period of time as the main technical reference point in the field.

In agreement with the principle of "dilute and disperse," the above type of landfill provided for leachate monitoring is based on attenuation throughout the unsaturated low permeability ground layers (either natural or artificial) underlying the waste. Accordingly, landfills were not sited on land characterized by highly permeable soils or at potentially vulnerable sites. Wastes were deposited in thin uncompacted layers aimed at enhancing the establishment of aerobic conditions and were covered with inert materials (preferably clay) to avoid contact between wastes and animals (dogs, birds, rodents, insects, etc.).

Provisions were laid down concerning the fencing off of landfill areas. Collection of leachate and biogas was not provided for; leachate was allowed to infiltrate into the ground, whereas biogas production was not contemplated due to the presence of aerobic conditions, which were meant (optimistically) to prevent this phenomenon. In the majority of cases things did not go exactly as planned, particularly, because little attention was paid to monitoring the unsaturated area and air circulation between the waste layers was extremely limited. Therefore, this type of landfill frequently resulted in noxious odors, leachate emissions from accumulation of leachate inside the landfill, or in nauseating puddles around the landfill.

The repeated occurrence of similar situations and an increased public awareness of environmental issues, together with technical and scientific progress in the field, have led since the 1980s to the increasing use of "Contained landfill," featuring controlled emissions, artificial lining systems, leachate drainage and collection, waste deposition by heavy compacting, anaerobic process, biogas collection, and final capping. In addition to clay as a liner, the use of synthetic membranes, initially PVC and then high-density polyethylene, became increasingly popular.

The controlled, collected emissions were naturally subjected to subsequent treatment and disposal dictating the need for specific technologies. Biogas was initially flared, although subsequently numerous recovery techniques were applied ranging from the production of thermal and electric power to upgrading of gas for domestic use and use as vehicle fuel. Increasingly complex applied treatment technologies were applied to leachate, one of the most difficult wastewaters to treat due to the inconsistency of production, high organic, and ammonia content and the wide quality variation linked to landfill aging. The simple system of aerated lagoons of the 1980s gave way to treatment options providing for high technologies of biological treatment, reverse osmosis, activated carbon absorption, chemical precipitation and oxidation, evaporation, phytoreduction, etc.

The high management costs associated with leachate treatment have resulted in a tendency to minimize, if not to totally prevent, leachate production, thereby inhibiting the inflow of rainwater to the waste body by means of an impervious surface capping of the landfill. On the one hand the latter may solve a management issue, but on the other creates an environmental problem in view of the fact that the lack of water slows down (and even totally inhibits) biological waste degradation. This may result in a sort of waste mummification and prolonged potential for long-term impact of landfills.

Throughout the years the "Contained landfill" has displayed its numerous limitations. In particular, the "blind" acceptance of the potential efficiency of new materials has led on the one hand to an excessive use (e.g., use of multiple synthetic layers without coupling them with mineral liners) and on the other to the development of landfills on vulnerable sites on which even the least shrewd mayor of the smallest village would never have thought of siting a dump or a simple landfill. Accordingly, landfills have been developed in gravel pits, in valleys with springs, and in quarries with water seepage, etc.

The limited lasting efficacy of materials and technologies, mechanical vulnerability, inappropriate management, and superficial design approach has at times resulted in severe environmental impacts that never occurred with open dumps or old simple landfills. On referring to the reclamation of old landfills one may defame the old landfills, as major reclamation interventions are quite often performed on the "modern" contained landfills built over the last 30 years. Several examples of this can be found in Italy (i.e., Vallin dell'Aquila, La Spezia; Fossano, Cuneo, etc.).

The "Residual waste landfill" was first developed in the context of a hierarchical view of waste management and rapidly became an international reference strategy. Subsequent to the various stages of Waste prevention, Material Reuse and Recycling, and Energy Recovery, residual wastes (ashes, nonrecyclables, impurities, etc.) are formed. Although landfills continue to maintain the characteristics established for a contained landfill, they rather constitute a deposit for residual wastes. The final aim of a residual landfill is to reduce the volumes of waste conferred, to minimize the production of greenhouse gases, and generally to lower environmental impacts and risks. Moreover, the majority of industrialized countries have introduced criteria for the monitoring of landfills during the postclosure stage, proposing the principle of financial provision and responsibility for landfill operators as long as the landfill continues to constitute a risk for the environment.

(Continued)

However, the evolution of the contained landfill into a residual landfill did not develop at the same pace as environmental sustainability requirements, based on avoiding leaving future generations to manage unacceptable environmental burdens.

Although present measures applied to control long-term impact (postclosure care, environmental long-term operator responsibility) are adopted by regulations, the appropriate tools for their implementation are still lacking. Indeed, termination of postclosure care is defined according to time rather than environmental performance and no criteria have been established to define acceptable conditions on which to assess operator responsibility. Furthermore, as mentioned previously, residual landfills are largely based on the technology applied for contained landfills, thus featuring the same negative characteristics ranging from the deterioration of lining and leachate drainage systems to the adoption of leachate minimization measures and related negative consequences.

The prevention of greenhouse gas emissions focuses mainly on the control of methane production by reducing the presence or biodegradable organic substances in waste. Conversely, scarce importance is given to the important role of landfills in acting as "carbon sink," permanently sequestering the nondegradable biogenic carbon present in the waste.

Lastly, the "Residual landfill," constituting a deposit for treated wastes and process residues, requires greater technical and environmental care than traditional landfills to which raw wastes are conferred. Residues may be characterized by higher concentrations of potentially polluting elements and a variation of mechanical properties that should be carefully controlled.

The modern type of landfill currently being developed on the basis of past experience, on the findings of technical and scientific research, and according to ongoing environmental challenges is the "Sustainable Landfill" incorporating the positive aspects of past landfills (e.g., control of unsaturated background, containment and treatment of emissions, waste minimization, and pretreatment) together with a coherent long-term strategy for the control of emissions and climate change issues. In particular, on the one hand the sustainable landfill aims to achieve an environmental balance over one generation (20–30 years), controlling the accumulation of mobilizable substances and uncontrolled emissions, whereas on the other hand closing the material cycle, thus constituting a final geological deposit for inorganic substances, and particularly for carbon associated with nondegradable or slowly degradable waste fractions such as plastics, rubber, and wood (when not most conveniently recycled) and with humic substances originating from the biological conversion of degradables. The role performed by the landfill as carbon sink is comparable to the role played by marine sediments (Bogner, 2005). Separate deposit of the high calorific fractions may also represent an intermediate storage for future energetic utilization of this material.

To achieve environmental sustainability in landfilling an important role is played not only by appropriate waste pretreatment but also by in situ treatment measures such as Flushing and, when in presence of a residual biological activity, aeration. In situ aeration, applied after Mechanical Biological Pretreatment or after intense anaerobic gas production, has proved to be an important tool in achieving Final Storage Quality, particularly when impervious top covers had been adopted previously.

Egyptian mummies reached the modern age unchanged as they had been carefully kept secluded from water and moisture. This is not exactly what we want from landfills.

Raffaello Cossu, Editorial—adapted from Waste management 30 (2010) 947–948.

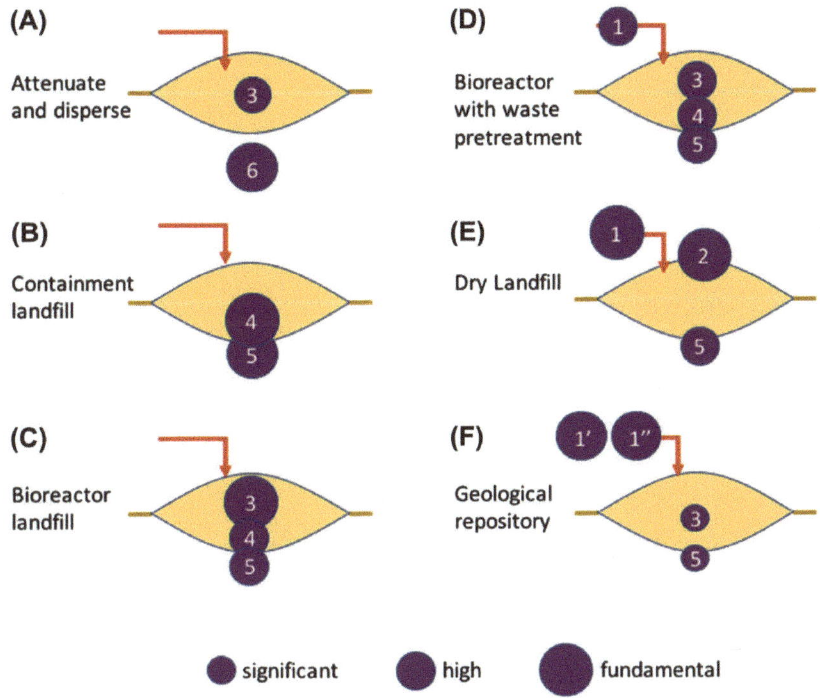

Figure 2.1.6 Schematic description of the main different landfill concepts indicating the main strategic barrier for each given concept. Of course different scenarios combining more than one concept are possible. 1, waste input barrier; 2, water inflow barrier; 3, reactor barrier; 4, drainage barrier; 5, lining barrier; 6, geological and hydrogeological barrier.

The final, w_d, mass released to the sensitive environment should have been "at a rate which gives rise to no hazard and does not cause inacceptable damage to the environment" (Knox, 1989).

Nowadays this concept has been abandoned.

Containment landfill

The concept is based on a lining, drainage, and collection barrier, which should prevent the leaching of contaminants into the subsoil. Containment has been extensively applied worldwide in connection also with the impetuous technological development of geosynthetics. But relying only on physical barrier has led to several examples of failure of this model of landfill.

Paradoxically, legally approved contained landfills may create more environmental problems than open dumps! In fact the failure of the physical artificial barriers in a site without a proper geological barrier (e.g., gravel pits) can cause severe environmental impacts (groundwater pollution, landfill gas migration), more significant than the ones which could be expected in an open dump sited in sites displaying natural protection conditions.

In general the main concern for the contained landfill is related to the long-term efficiency and behavior of the physical barriers (liners, drainage, collection systems). For MSW landfill the duration of the physical barriers already proved to be inferior to the duration of the emissions impacting the environment. And this renders that this kind of landfills is not sustainable.

Bioreactor landfill

This concept can be implemented when the wastes to be deposited contain significant amount of biodegradable organics. The bioreactor landfill relies significantly on the waste process barrier by using different methods (leachate recirculation, in situ aeration, hybrid operation aerobic—anaerobic, etc.) for enhancing the biological degradation rate. Higher amount of carbon is transferred into the gas phase (biogas) and less into the liquid phase. Consequently leachate has a lower concentration of BOD, COD, and (when nitrification is occurring) of ammonia and less odors are generated. Additionally odorous compounds are significantly reduced, facilitating the recirculation of leachate. Landfill bottom barrier (lining and drainage) is still required but the efficiency and long-term performance—particularly with regard to drainage clogging—would improve (Turk, 1995).

Bioreactor landfill with mechanical and biological pretreatment of waste

Mechanical and biological pretreatment is aimed at reducing the readily available organics and other contaminants in the waste to be landfilled (waste input barrier). Many different technologies might be involved as explained in Chapter 4.1. With this kind of landfill the waste stabilization of the landfilled waste is much quicker and the moment for the implementation of a final top cover (water infiltration barrier) can be significantly anticipated.

Dry landfill

In a dry landfill the infiltration of water is prevented by installing an impervious top cover system. The main reason for adopting a dry landfill is to keep under control the generation of leachate. The exclusion of water may be admitted only when a pretreatment allows to stabilize waste in a way that there is no need of water requested for their further transformation (degradation, leaching, etc.).

In fact in case of biodegradable waste, as already commented, the lack of water may impair the enzymatic hydrolysis, delaying significantly any biodegradation process, aerobic or anaerobic. This is why the dry landfill concept applied to MSW (the "Dry tomb") cannot be successful.

The problem is even more severe when considering predominantly inorganic and leachable waste (such as bottom ash or specific industrial waste). In this case the only factor that may reduce the emission potential of contaminants is the water addition and the consequent flushing effect. Only this effect can control the long-term emissions. Adopting a dry landfill concept for leachable waste would then result in an unsustainable system, requiring permanent maintenance of the top cover.

Rock quality storage/geological repository

As stated in Chapter 1.1, the actual role of landfilling in modern waste strategies is to be the final sink for the material cycle (Cossu, 2012). Materials extracted from the soil, where elements and compounds are in a no-mobile form, are used for production; then once products become waste and after recycling loops for recovering resources, the materials residuing in the final no-recyclable waste have to be treated and come back to the soil in a no-mobile form (final sink, geological repository). Failing to do this might increase the dispersion of materials in the environment, increasing the global pollution level. Further to several steps pretreatment, the stabilization of waste in a final sink landfill might consider a completion by in situ measures. As a backup a bottom barrier will be always installed (Fig. 2.1.6F).

CONCLUSIONS

Independently of the adopted concepts a landfill has to be designed by combining several different barriers, defining as barrier any measure or tool suitable to reduce the release of emissions to the sensitive environment.

Landfill design strategies should rely on the real, controllable efficiency of the individual barriers according to their influence on the mass transport of emissions.

Because landfills are structures which last for long time, the control of the efficiency of the barriers and the emissions should last for the same lifetime. Control has to be intended as a process including monitoring, maintenance, reaction to anomalies, and repairing. Table 2.1.1 summarizes factors and parameters related to the controllability of the different landfill barriers.

References

Bird, R.B., Steward, W.E., Lightfoot, E.N., 2007. Transport Phenomena. In: Wiley & Sons, John (Ed.), Revised 2nd Edition. ISBN: 978-0-470-11539-8.

Bogner, J.E., 2005. From waste to resources: the evolution of waste management in Europe. In: Lechner, P. (Ed.), Waste Management in the Focus of Controversial Interests. 1st BOKU Waste Conference Proceedings, Fakultas Verlag Publisher, pp. 14–22. ISBN 3-85076-721-3.

Bonaparte, R., Giroud, J.P., Gross, B.A., 1989. Rates of leakage through landfill liners. In: Geosynthetic '89 Conference Proceedings, San Diego, California, pp. 18–29.

Brune, M., Ramke, H.G., Collins, H.J., Hanert, H.H., 1994. Incrustation problems in landfill drainage systems. In: Christensen, T.H., Cossu, R., Stegmann, R. (Eds.), Landfilling of Waste: Barriers. E &FN Spon, London, pp. 569–606.

Christensen, T.H., 1992. Attenuation of leachate pollutants in groundwater. In: Christensen, T.H., Cossu, R., Stegmann, R. (Eds.), Landfilling of Waste: Leachate. Elsevier Applied Science, London, pp. 441–484.

Colucci, P., Lavagnolo, M.C., 1995. Three years field experience in electrical control of synthetic landfill liners. In: Christensen, T.H., Cossu, R., Stegmann, R. (Eds.), Proceedings Sardinia 95, Fifth International Landfill Symposium, vol. II. CISA Publisher, Cagliari, pp. 437–452.

Cossu, R., 2012. The environmentally sustainable geological repository: the modern role of landfilling. Waste Management 32, 243–244.

Freeze, A., Cherry, J., 1979. Groundwater. Prentice-Hall, Englewood Cliffs, p. 604.

Knox, K., 1989. Practice and trends in landfills in the UK. In: Christensen, T.H., Cossu, R., Stegmann, R. (Eds.), Sanitary Landfilling: Process, Technology and Environmental Impact. Academic Press, London, pp. 533–547.

McEnroe, B.M., 1994. Hydraulic of leachate collection and cover drainage. In: Christensen, T.H., Cossu, R., Stegmann, R. (Eds.), Landfilling of Waste: Barriers. E &FN Spon, London, pp. 531–541.

Paksy, A., Peeling, J.P., Robinson, J.P., Powrie, W., White, J.K., 1995. Landfill drainage as a fixed bed reactor. In: Christensen, T.H., Cossu, R., Stegmann, R. (Eds.), Proceedings Sardinia 95, Fifth International Landfill Symposium. CISA Publisher, Cagliari.

Ramke, H.G., 1989. Leachate collection systems. In: Christensen, T.H., Cossu, R., Stegmann, R. (Eds.), Sanitary Landfilling: Process, Technology and Environmental Impact. Academic Press, London, pp. 343–364.

Robinson, H., 1989. Unsaturated zone attenuation of leachate. In: Christensen, T.H., Cossu, R., Stegmann, R. (Eds.), Sanitary Landfilling: Process, Technology and Environmental Impact. Academic Press, London, pp. 453–464.

Rollin, A., Mlynarek, J., Lafleur, J., Zanescu, A., 1991. The investigation of a seven years old HDPE geomembrane used in a landfill. In: Christensen, T.H., Cossu, R., Stegmann, R. (Eds.), Proceedings. Sardinia 91, Third International Landfill Symposium. CISA Publisher, Cagliari, pp. 667–678.

Rowe, R.K., 1994. Diffusive transport of pollutants through clay liners. In: Christensen, T.H., Cossu, R., Stegmann, R. (Eds.), Landfilling of Waste: Barriers. E &FN Spon, London, pp. 219–246.

Turk, M., 1995. Maintaining proper landfill drainage by aerobic waste pretreatment. In: Christensen, T.H., Cossu, R., Stegmann, R. (Eds.), Proceedings Sardinia 95, Fifth International Landfill Symposium. CISA Publisher, Cagliari.

Wallace, R.B., Akgun, H., 1994. Leakage in double liner systems : the containment objective. In: Christensen, T.H., Cossu, R., Stegmann, R. (Eds.), Landfilling of Waste: Barriers. E &FN Spon, London, pp. 81–90.

2.2

MASS BALANCE OF CONTAMINANTS: A KEY FOR MODERN LANDFILL DESIGN

Raffaello Cossu

INTRODUCTION

Chapter 2.1 illustrates the range of different landfill barriers available. The waste input barrier and the reactor barrier (i.e., the transformation processes occurring within the landfill mass) represent the basis for the evolution of landfill concepts toward sustainability and long-term repository of contaminants (final sink).

The mass balance of a given contaminant can be measured in a landfill by calculating the initial load associated with the waste input, the load collected with gas and leachate, and the load withheld by the lining system. The difference between the initial load, the collected load, and the residual stored amount represents the load released into the environment through the uncontrolled dispersion of leachate and gas.

QUALITY OF WASTE AND MASS BALANCE IN A LANDFILL

Substances of particular environmental interest requiring constant monitoring include, among others, carbon and nitrogen, both associated with the organic compounds present in wastes. The quantities and features of wastes vary considerably according to socioeconomic and geographic aspects. As an example, in developing countries putrescible organic compounds may account for up to 80% of waste weight with scarce presence of plastics and paper, whereas in European cities putrescible organic compounds may constitute less than 20%, with a higher presence of other waste fractions. The mean amount of carbon present in a European municipal waste is approximately 20%–25% (Stegmann and Ritzkowski, 2008), distributed among the various organic categories as indicated in Table 2.2.1.

With regard to mobility in waste composition, two distinct organic fractions can be identified:

- x_S = nonmobile concentration of contaminant in the waste and
- s_S = mobile concentration of contaminant in the waste

The mobile fraction of carbon is transferred by means of leaching, biodegradation, or other reactions/conversions from the solid to the liquid (s_L = leached fraction) or gas phase (s_G = gasified fraction) or is converted, under the existing conditions, into a nonmobile solid form that contributes toward increasing the x_S fraction.

Table 2.2.1 Mean composition of a general European waste expressed as a percentage of total solids (Stegmann and Ritzkowski, 2008; Zeschmar-Lahl, 2003a, 2003b)

Component	%
ORGANIC COMPOUNDS	
Lignin	6
Hemicellulose	7
Proteins	3
Paper additives (organic + inorganic)	8
Cellulose	16
Hydrocarbons	9
Fats, resins, waxes	2
Plastic	18
INORGANIC COMPOUNDS	
Plastic additives	3
Minerals	13
Ashes	4
Hazardous substances	1
Metals	10

Mobilization may of course occur with any contaminant, including heavy metals and other nondegradable substances. The conversion processes (leaching, precipitation, complexation, etc.) are dependent on the individual substances.

Considering the landfill as black box reactor, the mass balance equation can be summarized as follows:

$$Accumulation = IN - OUT - Reacted \tag{1}$$

The term "*IN*" refers to the mass of a given contaminant in the deposited waste over a given period of time in a landfill with a volume V. It can be expressed as follows:

$$IN = \sum_{i=1}^{n} Q_i s_{Si} \tag{2}$$

where: n, number of waste streams of different typologies "*i*"; Q_i, flow rate of the individual waste typology "*i*" (t/y); s_{Si}, concentration of mobile fraction of contaminant in waste typology "*i*" (g/t).

The term "*OUT*" in Eq. (1) represents the mass released from landfill associated with emission rates of gas (q_G, Nm^3/year) and leachate (q_L, m^3/year). Taking carbon as contaminant and denoting the concentrations of carbon present, respectively, in landfill gas (CO_2, methane, etc.) and leachate (TOC) as s_G (mg/Nm^3) and s_L (mg/L), the released mass associated with carbon emissions can be calculated as follows:

$$OUT = s_L \cdot q_L + s_G \cdot q_G \tag{3}$$

By distinguishing the fractions of landfill gas and leachate collected (q_{Lc} and q_{Gc}) from the fraction dispersed in an uncontrolled manner through the landfill barrier systems (q_{Ld} and q_{Gd}), the following applies:

$$q_L = q_{Lc} + q_{Ld} \text{ and } q_G = q_{Gc} + q_{Gd} \tag{4}$$

The term "*reacted*" in Eq. (1) represents the mass of organic substance which is mineralized/stabilized and can be expressed as follows:

$$Reacted = rV \tag{5}$$

where: r, overall reaction rate ($mg/m^3 \cdot year$); V, landfill reactive volume (m^3).

The term "*accumulation*" expressed as dS/dt represents the variation of the contaminant mass (S) within the landfill versus time (t).

According to the previous equations, "*accumulation*" can be described by the following relationship:

$$\frac{ds}{dt} = \sum_{i=1}^{n} Q_i s_{si} - s_L q_L - s_G q_G - rV = \sum_{i=1}^{n} Q_i s_{si} - s_L q_{Lc} - s_L q_{Ld} - s_G q_{Gc} - s_G q_{Gd} - rV \tag{6}$$

$S(t)$ represents the residual amount of mobile carbon still present in the landfilled waste at a given time and is indicative of the mobile carbon emission potential of the system. For this reason, as long as the emission potential remains high the landfill should be protected, implying that the barrier system comprising impermeable lining and drainage systems should be efficiently active.

A graphical representation of the carbon balance terms is given in Fig. 2.2.1.

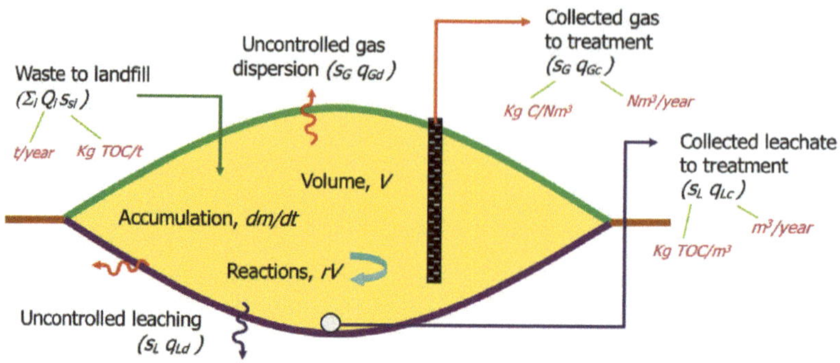

Figure 2.2.1 Graphical representation of mass balance terms, considering carbon as a contaminant, in an municipal solid waste landfill reactor.

By moving to the first part of Eq. (6), the terms expressing the uncontrolled emissions of mobile carbon, the following expression is obtained:

$$s_L q_{Ld} + s_G q_{Gd} + \frac{dS}{dt} = \sum_{i=1}^{n} Q_i s_{si} - s_L q_{Lc} - s_G q_{Gc} - rV \tag{7}$$

The terms comprised in the first part of the equations are those which should be monitored and minimized to prevent adverse environmental impacts both in the short- and the long term. By analyzing Eq. (7) it appears evident that to minimize these terms it is necessary to minimize the positive terms and maximize the negative terms on the right of the equation.

From an operational viewpoint this implies the following:

- minimizing the mobile carbon mass introduced into the landfill ($s_{Si} \cdot Q_i$), controlling both flow rates (Q_i) and quality (s_{Si}) of wastes deposited, as an example by reducing the production of waste, and applying various forms of treatment prior to landfilling;
- maximizing the mass of carbon to landfill gas and leachate collected ($s_L \cdot q_{Lc}$ and $s_G \cdot q_{Gc}$), controlling both flow and composition; and
- increasing waste stabilization (rV) by enhancing the reaction rate through the aerobization of landfilled waste (forced or natural aeration) or addition of water in the case of water shortage.

Although, as a general rule, landfill regulations and practices tend to minimize the generation of leachate by limiting or excluding the infiltration of water into the landfill body, the mass balance equation clearly indicates how water played a key role in controlling negative landfill impacts. Indeed, water controls the reaction rate r as it is a fundamental reagent required by biodegradation processes (Chapter 3.1). Without water the degradation term rV would be nullified, thus resulting in a sort of "mummification" of wastes. Moreover, water controls the transport of contaminants from solid to liquid phase. The more leachate and gas are generated, the more mobile contaminants will be removed from the system, thus preventing their uncontrolled emission into the environment.

ACCUMULATION OF MASS AND EMISSION BEHAVIOR

The accumulation over time of mobile contaminants in the landfilled waste, as described mathematically in Eq. (6), follows the trend illustrated in Fig. 2.2.2.

The accumulated mobile contaminant mass, S_t, represents the emission potential and may be expressed, for instance, in tonne of residual mobile carbon when taking into account organic degradable waste. S_t reaches a peak during landfill operation and then decreases as a consequence of degradation processes and generation of gas and leachate. Peaks and trends of the accumulation curves will depend on the adopted landfill technologies.

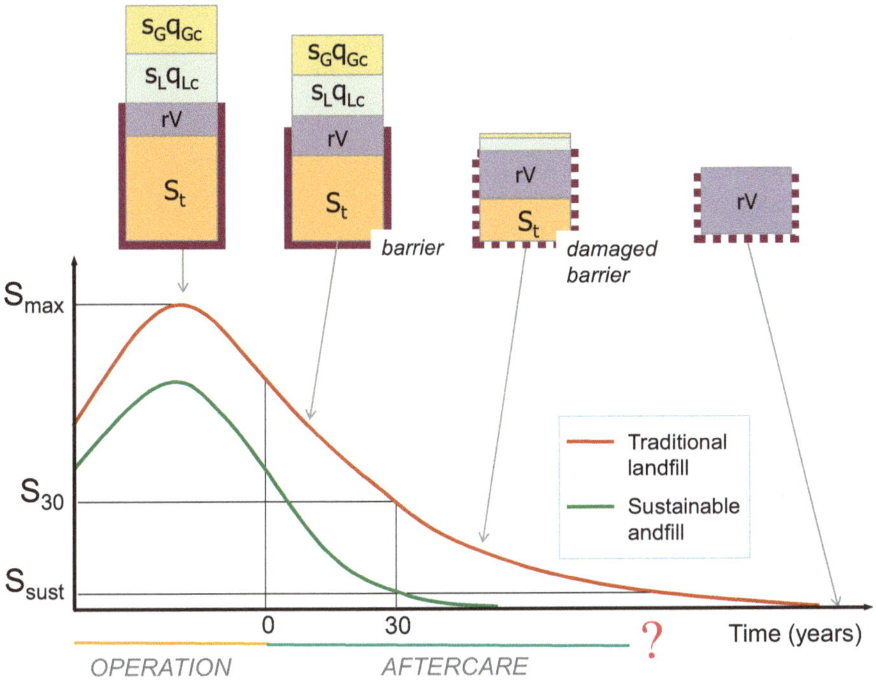

Figure 2.2.2 Curves illustrating accumulation of mobile mass of contaminants (S_t) throughout the life span of two different kinds of landfills. The figure also provides a graphical representation of mass balance terms (Eq. 6) at different times. When the barriers lose their efficacy, the lower the S_t, the lower the potential emissions of contaminant. s_L, concentration of mobile carbon (TOC) in leachate, mg/L; q_{Lc}, flow rate of collected leachate, m^3/year; s_G, concentration of mobile carbon in landfill gas (CO_2, methane, etc.), mg/Nm^3; q_{Gc}, flow rate of collected landfill gas, Nm^3/year; rV, mass of stabilized contaminant (r, reaction rate, mg/$m^3 \cdot$year, V, landfill reactive volume, m^3); S_t, amount of mobile carbon accumulated within the lining system representing the potential of residual emissions, t/year; S_{30}, amount of mobile carbon accumulated within the lining system after 30 years aftercare; S_{sust}, amount of mobile carbon accumulated within the lining system, representing final storage quality, i.e., accumulation which has reached an equilibrium with the environment.

A landfill can be deemed stable and aftercare completed once the accumulated mass has reached a value which is sustainable for the environment, S_{sust}.

The values of the term S_{sust} defines the final storage quality (FSQ) of the landfill. FSQ can be assessed based on a series of different parameters and risk analysis procedures (see Chapter 16.1 and Heyer et al., 2013).

Fig. 2.2.2 provides a graphical representation of the way in which individual mass balance terms—expressed by Eq. (6)—change over time. When the physical barriers lose efficacy due to unavoidable failures in the lining or drainage systems—such as cracking, tears, aging, clogging, etc., (see Chapter 5.1)—the residual mass of contaminants will no longer be contained and uncontrolled emissions of leachate may occur.

The quality of leachate (s_L) changes over time, reflecting the trend of residual concentration of contaminants in the waste mass. By using a first-order reaction kinetics ($ds_L/dt = -ks_L$), the concentration of a contaminant at a given time, s_{Lt}, will be as follows:

$$s_{Lt} = s_{L0}e^{-kt}$$

where: s_{L0}, initial concentration of contaminant in the leachate; t, observation time; k, reaction rate constant.

Studies aimed at predicting the long-term behavior of landfill emissions have enabled calculations to be made for a series of different contaminants in municipal solid waste (MSW) leachate with regard to the time required to reach concentrations which comply with standard limits $\left(\widehat{S}_L\right)$ set by national regulations for discharging waste waters in surface water bodies. The results are summarized in Table 2.2.2. For some parameters, such as TKN, a period of several centuries may be necessary.

The potential load of mobile contaminants associated with uncontrolled leachate emissions ($s_{L^*} \cdot q_{Ld}$) will depend on the adopted landfill technologies and, in particular, the efficiency of the multibarrier system (see Chapter 2.1). The load changes throughout the landfill life span as represented in Fig. 2.2.3. Sustainable conditions are achieved when the accumulation of mobile contaminants, S_t, reaches FSQ and uncontrolled emissions are below the target values for assessing sustainability $(s_{L^*} \cdot q_{Ld})_{sust}$.

In an open dump, uncontrolled emissions reach a peak during landfilling operations and subsequently decrease in the postclosure period.

In the case of a traditional contained landfill, uncontrolled emissions will remain below the acceptable threshold values for as long as the physical barrier systems (lining and leachate drainage) remain effective in controlling flow rate of the potential emission (q_{Ld}).

On loss of barrier efficacy uncontrolled emissions may occur, with environmental impact of the latter being dependent on the concentration of contaminants in the leachate (s_L). If the uncontrolled emissions of contaminants at a given time $(s_{L^*} \cdot q_{Ld})_t$. exceeds the target value for sustainability $(s_{L^*} \cdot q_{Ld})_{sust}$ then adverse environmental impacts may be registered.

Table 2.2.2 Estimation of the time (T_L) required for municipal solid waste leachate parameters to reach limit concentrations ($\hat{s}_L$) established by national regulations for discharging wastewater into surface water bodies in Germany (D) and Switzerland (CH), according to different experiences

P		s_{L0} (mg/L)	$\hat{s}_L$ (mg/L)	T_L (year)	References
COD	Old FRG waste	1800–8400	200 (D)	120–220	Andreas (2013)
COD	Old FRG waste	1800–8400	60 (D)	200–300	Andreas (2013)
COD	Full-scale landfill	22–22500	200 (D)	65–320	Krüempelbeck (2000)
COD	Lysimeter	500–12700	200 (D)	80–360	Heyer (2003)
TKN	Old FRG waste	1100–3600	70 (D)	120–300	Andreas (2013)
TKN	Old FRG waste	1100–3600	5 (CH)	280–580	Andreas (2013)
TKN	Full-scale landfill	1–2530	70 (D)	Decades–Centuries	Krüempelbeck (2000)
TKN	Lysimeter	200–2100	70 (D)	120–450	Heyer (2003)
Cl	Full-scale landfill	52–8700	100 (D)	25–60	Krüempelbeck (2000)
Cl	Lysimeter	340–2950	100 (D)	90–250	Heyer (2003)
AOX	Full-scale landfill	0.058–6.2	0.5 (D)	40–100	Krüempelbeck (2000)
AOX	Lysimeter	0.39–2.38	0.5 (D)	30–210	Heyer (2003)

s_{L0} is the concentration of the considered leachate at the start of the simulation time ($t = 0$). *FSL*, full-scale landfill; *LSR*, landfill simulation reactor; *O-FRG*, old waste from German landfills.

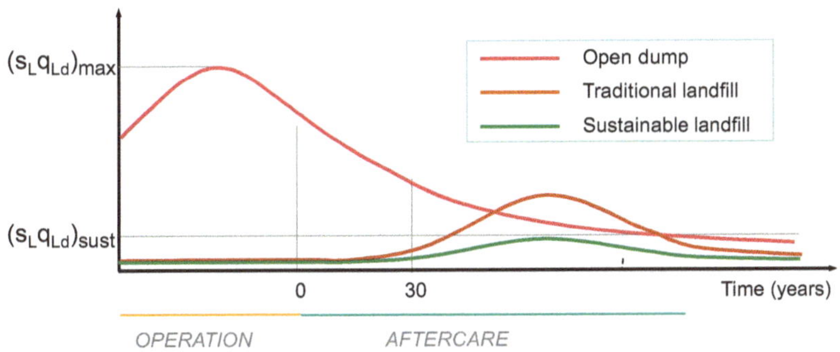

Figure 2.2.3 Curves illustrating contaminant load $s_{L*} \cdot q_{Lc}$ associated with uncontrolled emissions of leachate throughout the life span of an open dump and two different types of landfills, $(s_{L*} \cdot q_{Lc})_{sust}$, representing the sustainable target for contaminant load.

Therefore, a poor control of the accumulation of mobile contaminants in traditional landfills (no pretreatment, anaerobic conditions, scarcely permeable top covers) may produce unsustainable situations. On the contrary, a sustainable landfill designed with multibarrier concepts and aimed at optimizing mass balance terms in Eq. (7) and controlling the accumulation term *St* following loss of efficacy of the barrier system will ideally be characterized by uncontrolled emissions having low contaminant concentrations (s_L). Consequently, the mass load of contaminants $(s_{L^*} \cdot q_{Ld})_t$ could be maintained below sustainability targets.

In the light of these considerations it is clear that the expected environmental impacts from landfilling are mainly driven by the technical and administrative requirements established by national regulations.

With regard to the aftercare phase, according to the European Directive on Landfilling,[1] several national regulations, including Italian law n. 36/2003 (art. 14), have decreed a financial provision to cover aftercare costs over a period of 30 years. However, if at the end of this period the barriers have not maintained their efficacy (see Chapter 7.1) and the residual mobile mass of contaminants (S_{30})—and consequent emissions $(s_{L^*} \cdot q_{Ld})_{30}$—have not reached sustainable levels, the financial provision may not cover the costs of eventual remediation.

This aspect is critical also in view of the fact that a large majority of national regulations fail to provide for the pretreatment and in situ treatment of waste and, worse still, frequently envisage provisos that may prove detrimental in controlling the accumulation of contaminants, such as the adoption of impervious top covers, prohibition of leachate recirculation, etc.

The latter may prove to be particularly onerous for private landfill operators, in view of the liability set by the regulations for the long-term monitoring of landfills.[2] The Italian regulations additionally foresee that, in the case of negative environmental effects, the landfill operator shall be called on to comply with the remediation requirements dictated by the Authorities.[3]

[1]Council Directive 1999/31/EC, Article 10. "Cost of the landfill of waste." Member States shall take measures to ensure that all of the costs involved in the setting up and operation of a landfill site, including as far as possible the cost of the financial security or its equivalent referred to in Article 8(a) (iv), and the estimated costs of the closure and after-care of the site for a period of at least 30 years shall be covered by the price to be charged by the operator for the disposal of any type of waste in that site… (omissis).

[2]Council Directive 1999/31/EC, Article 13. "Closure and after-care procedures"… (omissis)… after a landfill has been definitely closed, the operator shall be responsible for its maintenance, monitoring and control in the after-care phase for as long as may be required by the competent authority, taking into account the time during which the landfill could present hazards. The operator shall notify the competent authority of any significant adverse environmental effects revealed by the control procedures and shall follow the decision of the competent authority on the nature and timing of the corrective measures to be taken; …(omissis)… for as long as the competent authority considers that a landfill is likely to cause a hazard to the environment and without prejudice to any Community or national legislation as regards liability of the waste holder, the operator of the site shall be responsible for monitoring and analyzing landfill gas and leachate from the site and the groundwater regime in the vicinity of the site.

[3]Legislative Decree. 36/2003, art. 13 "Operational and Post-operation management" … (omissis)… The management shall, moreover, notify the competent Authority of any significant negative environmental effects detected subsequent to the implementation of monitoring and control procedures, and shall comply with the decision taken by the competent Authority with regard to the nature of corrective actions and the terms of actualization of the same.

To conclude, landfill operators may find themselves in the paradoxical situation whereby on the one hand they are required to comply with technical requirements established by law, although these may at times extend the long-term environmental impact (e.g., sealed top covers), whereas on the other being obliged to accept responsibility for the environmental issues that may arise as a consequence!

Moreover, as underlined previously, this situation is further complicated by the total lack of specific definition of standards for environmental acceptability to be applied by the competent Authorities in assigning responsibility. In Chapter 16.1 this issue (Definition and Achievement of Final Quality Targets) is discussed in detail.

MASS BALANCE CONTROL FOR ENVIRONMENTAL SUSTAINABILITY

Implementation, from an operational point of view, of mass balance for landfill contaminants may represent the keystone to the calibration of technical interventions aimed, both during the planning, operation, and aftercare phases, at achieving environmental sustainability.

As schematically summarized in Table 2.2.3, the different terms of mass balance can be influenced by the use of a series of technologies and management methods, when referring to carbon in organic fractions.

Minimization initiatives will undoubtedly influence the quantity of waste to be landfilled.

Recycling of biodegradable organics (e.g., composting) may result in a significant reduction of waste quantities and positive variations of waste quality.

Waste treatment prior to deposition in a landfill may be undertaken using various types of processes:

- physical and physicochemical such as grinding, sieving, compacting, baling, classification, and washing (see Chapter 4.3);
- biological (anaerobic and/or anaerobic) usually combined with mechanical unit operation (see Chapter 4.1); and
- thermal (see Chapter 4.2).

The terms of mass balance may be influenced not only by pretreatment of incoming waste but also by means of in situ interventions and processes both during the operation and aftercare phase.

Possible in situ options include the following:

- enhanced anaerobic degradation;
- leachate recirculation (see Chapter 12.1);
- natural aeration (see Chapter 14.2);
- forced aeration (see Chapter 16.2);
- combination of aerobic and anaerobic sequences (hybrid landfill reactors); and
- flushing (Karnik and Parry, 1997).

Finally, a well-structured physical barrier system (liner + leachate drainage) is necessary to regulate and control storage of the residual biodegradable fraction of landfilled waste that has not yet been degraded (S_t).

Table 2.2.3 Qualitative influence of a range of technical options on mass balance terms in a municipal solid waste landfill (+++ = high, ++ = medium, + = fair)

Options	S_{Si}	Q_i	S_L	q_{Lc}	S_G	q_{Gc}	rV	S_t
Waste avoidance		+++						
Separate collection/sorting	+	++	+					
Mechanical pretreatment			+		+		+	
Biological pretreatment	++	+	+				++	
Thermal treatment	+++	++	++				+++	
Leachate recirculation			+		+		+	
In situ aeration			++		++	+	++	
Watering			+	++	+	+	+	
Physical bottom barrier								++

S_{Si}, concentration of mobile fraction of contaminant in waste typology "*I*," g/kg; Q_i, flow rate of the individual waste typology "*i*," t/year; S_L, concentration of mobile carbon (TOC) in leachate, mg/l; q_{Lc}, flow rate of collected leachate, m^3/year; S_G, concentration of mobile carbon in landfill gas (CO_2, methane, etc.), mg/Nm^3; q_{Gc}, flow rate of collected landfill gas, Nm^3/year; rV, mass of stabilized contaminant (*r*, reaction rate, mg/m^3·year; *V*, landfill reactive volume, m^3); S_t, amount of mobile carbon accumulated within the lining system, t/year.

At variance with the majority of landfill regulations which aim to control waste quality at the gate, the objective of environmental sustainability, as illustrated earlier, may be achieved by means of combined measures, including pretreatment, in situ treatment, and aftercare interventions. Therefore, assessment of sustainability targets should not be based on compliance with a specific parameter of the waste to be landfilled (taking into account only pretreatment methods) but rather by analyzing the landfill matrices (residual landfilled waste, gas, and leachate emissions) after significant periods of time to verify whether they comply with sustainability criteria (see Chapter 16.1).

Possible treatment combinations applied with a view to achieving sustainability targets are represented in Fig. 2.2.4.

These combinations are strictly linked to the characteristics of the waste. MSW may undergo mechanical treatment (Aa) by means of shredding, sieving, and sorting of recyclables, followed by an anaerobic phase in a landfill bioreactor (Ab) with intense production of gas (e.g., concentrated over an indicative period of 12–15 years) and by a posttreatment phase with forced aeration (see Chapter 14.3). This phase later on could be followed by natural aeration (semiaerobic landfilling). In the latter case, the drainage system should be implemented prior to start-up of landfill operation as the drainage

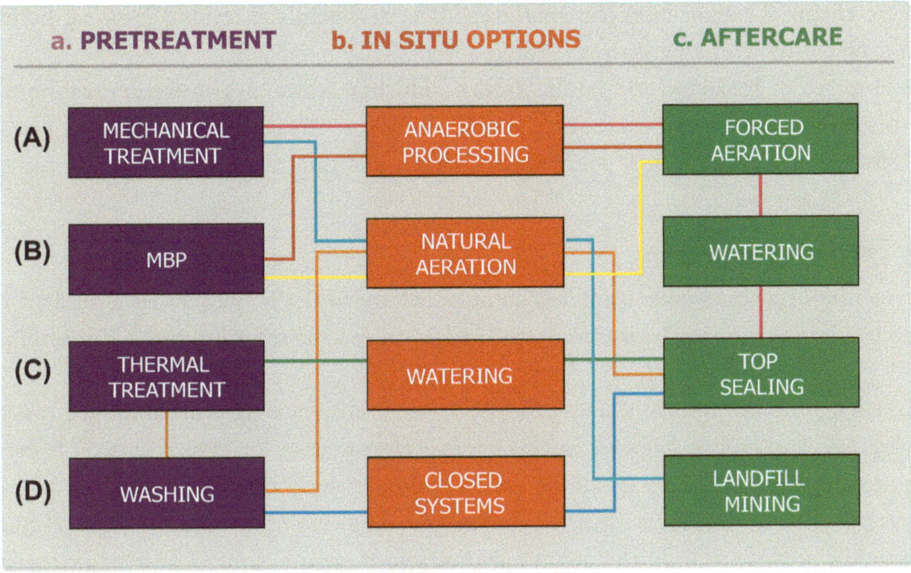

Figure 2.2.4 Combination of a series of operational alternatives capable of influencing mass balance in a landfill and promoting sustainability. The proposed combinations are theoretical and indicative, being dependent on the waste characteristics (degradability, combustibility, leachability, etc.) and not reflecting necessarily current practice.

system in semiaerobic landfills differs significantly from that adopted in anaerobic landfills (see Chapter 14.2). Further treatment for in situ washing of the waste could fully stabilize the waste allowing a sealing top cover (see Chapter 11.1).

Following mechanical pretreatment and sorting of putrescibles (Aa), MSW may alternatively undergo natural aeration (Bb) and then be stored in the medium term before being subjected to landfill mining (Dc) for the recovery of combustible fractions. This option may also be conveniently adopted with plastics, papers, and textiles which have not been, for whatever reason, recycled and when no thermal treatment is available in the area.

Subsequent to mechanical biological pretreatment, MSW may undergo either anaerobic landfilling and postcare forced aeration (Ba, Ab-Ac) or natural aeration (Bb) followed by forced aeration in the aftercare phase (Ac). Waste should be pretreated using mechanical biological systems (Ba) to significantly reduce biological activity of the organic compounds. Failure to do so may lead to in situ aeration eliciting the formation of a bacterial biomass in the drainage system capable of obstructing granular drainage beds. Also this kind of landfill may be completed by means of a posttreatment flushing (Bb), with a final sealing top cover (Cc).

Another scheme of interest is represented by the line (Ca + Da + Bb). Residual waste from thermal treatment undergoes prewashing and natural aeration during landfilling to reduce the leaching potential and promote chemical stabilization (carbonation). After assessing the residual leaching potential, the landfill may be closed with a permanent sealing cover (Cc).

In the presence of leachable and nonhazardous industrial waste, a convenient treatment solution may comprise the prewashing of waste (Da), which is subsequently landfilled under a roof or any other closed system (Db), with a top seal being applied in the aftercare phase (Cc).

ROLE OF LANDFILL AS A SINK

The construction of a sustainable landfill is complementary to the new role that a modern landfill is required to perform in closing the material cycle and, with specific regard to carbon, in acting as a *carbon sink.*

Indeed, taking into account the classification of organic substances used in calculating mass balance, as nonmobile (x_S) and mobile fractions (s_S, s_G, s_L), it is clear that an integrated system of pretreatment and in situ treatment (Fig. 2.2.5) should be aimed at progressively converting the mass of degradable organic compounds into leached or gasified carbon and achieving sustainability targets (FSQ).

Concurrently, accumulation of the nonmobile fraction of organic compounds (lignin, plastic) originally present in the waste (x_S) and the stabilization of biodegradable organics by conversion into nondegradable substances (e.g., insoluble humic substances) effectively describes the role of landfill as a carbon sink. The final quality of landfilled waste, devoid of mobile contaminants, is what is often

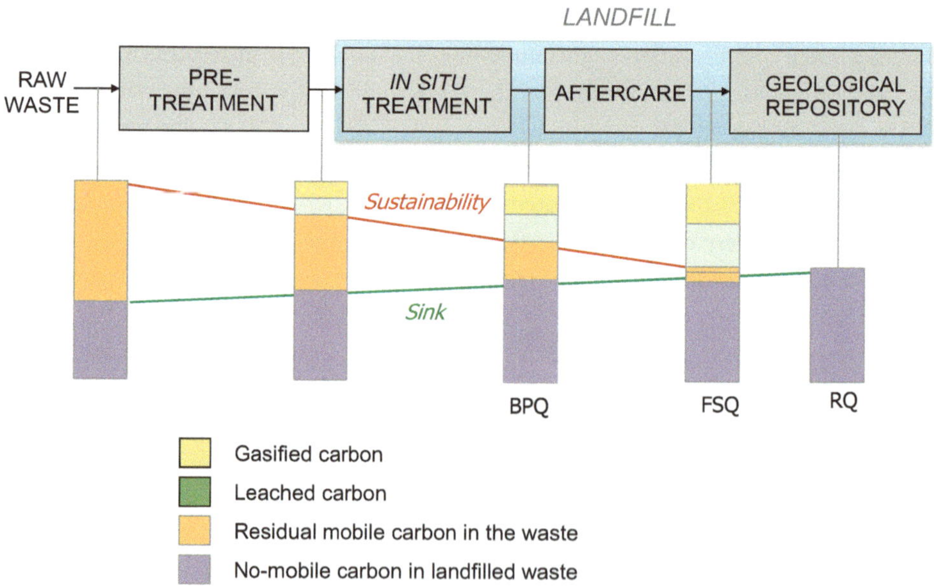

Figure 2.2.5 Time trend of the organic fractions present in landfilled wastes according to the various types of waste treatment performed, as indicated in Fig. 2.2.4G. *BPQ*, best practice quality; *FSQ*, final storage quality, *RQ*, rock quality. *Modified from Cossu (2012).*

referred to as rock quality. When this quality is reached, a landfill can be considered a geological repository (Cossu, 2012).

The role of landfill as a sink is also fundamental in closing the material loop in any circular economy strategy as discussed in Box 2.2.1.

Box 2.2.1 Back to Earth Sites: from "nasty and unsightly" landfilling to final sink and geological repository

Recycling, recovery of resources, urban mining, closing the loop, and zero waste are all terms widely used in waste management to encourage the transformation of waste into resources (Cossu and Williams, 2015).

Graphically these concepts are often represented in the form of top–down or bottom–down triangles where different waste management steps are listed in a hierarchical order. Avoidance, material recovery, and recycling are the preferred options "aimed at reducing use of natural resources and controlling contaminant potential, both on a local (water and soil contamination, odour, etc.) and a global scale (emission of greenhouse gases and other air pollutants)" (Cossu, 2009). The amount of waste sent to landfilling should be drastically reduced, with landfilling representing the least preferred option for waste management.

Transition from a linear approach (take-make-waste) to a circular approach is commonly described by a perfect cycle in which the different life stages of a product are represented: production–distribution–consumption–reuse/repair–recycling of secondary raw materials to production. The European Union has proposed an Action Plan with the introduction of measures intended to support the whole cycle (EU, 2015). What is completely missing in this strategic vision is the role of landfilling.

In the triangular hierarchic representation, landfilling is viewed as a sort of a dustbin in the system, in which to deposit all unavoidable and nonrecyclable wastes, together with residues from previous treatment. In the circular representation, landfilling is simply neglected; the landfill should no longer exist. Accordingly, more and more frequently we witness the issuing of regulations that envisage the elimination of landfills, invariably dictating rather lengthy time frame for achieving goals. As an example, the South African Polokwane Declaration, an official governmental document, sets targets of zero waste to landfills by 2022 and a 50% reduction of the same by the year 2012 (http://soer.deat.gov.za/dm_documents/polokwane_nmZiT.pdf). These declarations are frequently not based on any technical grounds but merely represent a wish and are seen by politicians currying favor with the populist as the simplest means of dealing with these issues. Other nations and regions boast of waste management systems in which landfills have been eliminated, only to later discover that this has been achieved not by means of comprehensive recycling of all wastes but rather by transferring a portion of the wastes generated elsewhere. In the same way as many of the thorny issues we encounter during a lifetime, everyone builds landfills, but most try to talk about them as little as possible.

While Europe promotes a circular economy, the nasty and unsightly landfills continue to meet one-third of the needs for waste disposal, with a particularly high concentration of facilities present in the newest Member States.

It should moreover be underlined that while landfills are indicated as the source of a wide range of well-known environmental problems (e.g., groundwater pollution, odors, greenhouses gases), in almost all industrialized countries national landfill regulations tend to provide for the construction only of "traditional" landfill sites, completely overlooking any accomplishments provided by the impressive scientific and technical developments of sustainable landfilling in recent years.

(Continued)

Taking once again the European situation as an example, the Landfill Directive dates back to 1999 (although technically speaking the document had been under consideration since 1993 but was only published in 1999 due to a difficulty in identifying a common position with the British, who were reluctant to abolish codisposal). All national legislation in European countries is based on this Directive (EU, 1997). These regulations envisage three types of landfill (for hazardous, nonhazardous, and inert waste), the control of biodegradable waste, the implementation of protective measures on the bottom and slopes of the landfill through installation of physical barriers (mineral liners, geomembranes, drainage), and the limitation of water influx. However, nowhere in the entire Directive one can find the magic words "environmental sustainability." No mention of a word or a sentence that establishes how the long-term impact of a landfill should be exhausted within the span of one generation. On the contrary, the postoperational phase, which follows the closure of a landfill, is addressed only in economic terms (art. 8 and art. 10), defining for landfill managers a financial provision relating to the covering of aftercare costs for a period of at least 30 years. Then stating in art. 13 that "… (c) after a landfill has been definitively closed, the operator shall be responsible for its maintenance, monitoring and control in the after-care phase for as long as may be required by the competent authority, taking into account the time during which the landfill could present hazards.… … (d) for as long as the competent authority considers that a landfill is likely to cause a hazard to the environment and without prejudice to any Community or national legislation as regards liability of the waste holder, the operator of the site shall be responsible for monitoring and analysing landfill gas and leachate from the site and the groundwater regime in the vicinity …."

As no environmental criteria have been established to determine when "a landfill is likely to cause hazard," the input of data during the planning and design stage relating to how to achieve the aims of FSQ is not envisaged. We all know that the long-term impact of traditional landfilling (the conceptual landfill referred to in the regulation) far exceeds the 30 years covered by the financial provision and, even worse, far exceeds the life span of the physical barriers. As a result, the designing of landfills that comply solely with the European Directive is equivalent to designing contaminated sites.

Therefore, although the regulations result in the construction of "nasty and unsightly" landfills, no steps have been taken to improve this situation. Thus, no amended legislation has been made available, no research is supported, and there is no incentive to promote the design of environmentally sustainable landfills. On the contrary, nowadays in Europe the easiest way to have a research proposal rejected is to mention, even in passing, landfills.

This leads to perverse environmental outcomes that go against the principles of the circular economy intended to eliminate these issues. During the circular management of waste (Cossu, 2013), the various stages of a product's life span generate liquid, gaseous, and solid wastes. Although the legal requirements for emission concentrations may be met, the product, because of the volume of emissions, continues to generate a load (expressed in terms of weight/time) that results in an alarming and widespread contamination. The increasing concentration of nondegradable substances in the seas, air, and glaciers at all latitudes represents one of the most pressing environmental problems. We have become so accustomed to referring to concentrations and percentages that we have forgotten that these loads are the real issue.

Taking paper as an example, the presence of problematic chemicals (e.g., MOHs (mineral oil hydrocarbons), phenols, phthalates, PCBs, toxic metals) is measurable, with these compounds tending to accumulate throughout the various circulation cycles, reaching relatively high concentrations in both the end product and the residual waste (Pivnenko et al., 2016).

The ambitious goal of closing the loop in the circular economy cannot be achieved if the issue of management of residues from production and recycling processes is not properly addressed. The generation of residual waste should not only be reduced but the sustainable management of residues that cannot be avoided should also be pursued.

By calculating a simple mass balance, starting from the moment the material resources are extracted from the ground, it becomes clear that to close the material loop, a final sink in which the residues can be safely and esthetically deposited needs to be taken into consideration. To avoid environmental issues, the deposited material should be processed to ensure that it will not result in emissions that may negatively affect the quality of the environment. Indeed, the substances present in the final residues returned to the earth should all be in a nonmobile form. This may require specific treatment procedures aimed at eliminating or significantly reducing the mobile waste fraction responsible for generating an environmental impact. The concept of Back to Earth Alternatives (BEA) is thus introduced, implying that the residues, after proper treatment, should be returned to their nonmobile state, as they were before they were extracted from the ground to be used as raw materials (Fig. 1). This would provide the actual closing of material cycles.

BEA may include the use of compost in agriculture and use of residues in the production of building materials (e.g., bricks, concrete) and construction of houses, bridges, or roads. Particularly, however, they include what we can call Back to Earth Sites (BES), i.e., permanent waste deposit sites that meet the above-described requirements.

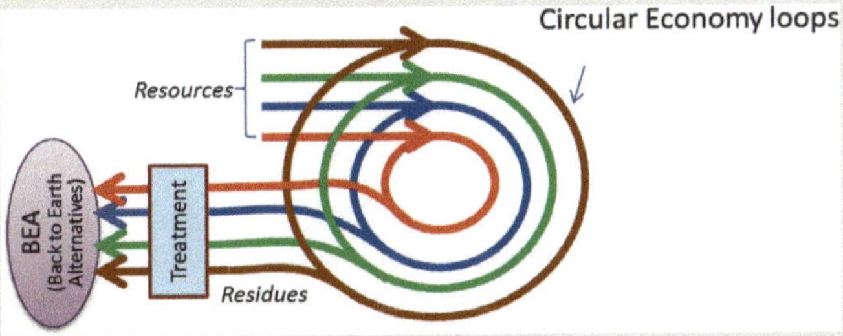

Figure 1 Circular economy conceptual scheme including the sustainable management of the residues to be sent "Back to the Earth" to close the material loop.

To implement these concepts, a proactive approach should be taken based on the following objectives:

- assessment of quantity and quality of residues in the circular economy;
- definition and demonstration of strategies for minimization of residues quantity and quality upgrading;
- treatment of the final residues (e.g., washing, mineralizing, stabilization);
- definition and demonstration of BEA for the treated residues (e.g., material sink in BES), development of new products and building materials, use in bitumen production, addition to agricultural soil;
- economic and environmental (long-term leaching) evaluation for BES and other BEA options;
- regulatory and policy framework for BES.

The policies frequently adopted to date by national and international authorities on landfilling are largely reminiscent of the three monkeys who do not see, do not hear, and do not speak; we should, however, be fully aware that a circular economy cannot exist in the absence of a sustainable closure of the material loop. In other words, we cannot have a society with no permanent deposit of residues onto or into land. With no landfilling!

Raffaello Cossu, 2016. Editorial - adapted from Waste Management, 55, 1-2.

References

Andreas, L., 2013. Long-term behavior of Municipal landfills. Emissions from Old sites in the Eastern part of Germany. In: Cossu, R., van der Sloot, H. (Eds.), Sustainable Landfilling. CISA Publisher, pp. 153–167. ISBN:978-88-6265-005-2.

Cossu, R., 2009. From triangles to cycles. Waste Management 29, 2915–2917.

Cossu, R., 2012. The environmentally sustainable geological repository: the modern role of landfilling. Waste Management 32, 243–244.

Cossu, R., 2013. The urban mining concept. Waste Management 33, 497–498.

Cossu, R., Williams, I., 2015. Urban mining: concepts, terminology, challenges. Waste Management 45, 1–3.

EU, 1997. Council Directive 1999/31/EC on the Landfill of Waste.

EU, 2015. Communication from the Commission to the European Parliament, the Council, the European Economic and Social Committee and the Committee of the Regions. Closing the Loop—an EU Action Plan for the Circular Economy, 2015 COM 614. final.

Heyer, K.-U., 2003. Emissionsreduzirung in der Deponienachsorge. Hamburger Berichte, band 21, Verlag Abfall aktuell, ISBN:3-9808180-4-7 [cit. in Heyer et al., 2013].

Heyer, K.-U., Hupe, K., Stegmann, R., 2013. Criteria for the completion of landfill aftercare. In: Cossu, R., van der Sloot, H. (Eds.), Sustainable Landfilling. CISA Publisher, pp. 611–622. ISBN:978-88-6265-005-2.

Karnik, M., Parry, C., 1997. Cost implication of operating landfills as flushing bioreactors. In: Proceedings Sardinia 97. Sixth International Landfill Symposium, vol. I. CISA, Cagliari, pp. 419–426.

Krümpelbeck, I., 2000. Untersuchungen zum Langzeitemissionsgeschehen von Siedllungsabfalldeponien. Dissertation an der Bergischen Universität Gesamthochschule Wuppertal, Bereich Siedlungwasser- und Abfallwirtschaft, Heft 3 [cit. in Heyer, 2003].

Pivnenko, K., Olsson, M.E., Götze, R., Eriksson, E., Astrup, T.F., 2016. Quantification of chemical contaminants in the paper and board fractions of municipal solid waste. Waste Management.

Stegmann, R., Ritzkowski, M., May 2008. Landfill Fundamentals. Power Point Presentation. IWWG Training Course on Sanitary Landfilling, Lima, Perù.

Zeschmar-Lahl, B., 2003a. Stoffflussanalyse an einer mechanischen Aufbereitungsanlage – Eintrag und Verbleib von Chlor und Metallen. Fachtagung Mechanisch-biologische Abfallbehandlung, Berlin, 2./3.12.2003. In: Thomé-Kozmiensky, K.J. (Ed.), Ersatzbrennstoffe, 3, pp. 231–259 [Cit. in Stegmann and Ritzkowski, 2008].

Zeschmar-Lahl, B., 2003b. Schwermetallentfrachtung am limit. Müllmagazin (4) [Cit. in Stegmann and Ritzkowski, 2008].

3 CHAPTER

PROCESSES

3.1

BIOCHEMICAL PROCESSES IN LANDFILL

Raffaello Cossu, Luca Morello and Rainer Stegmann

INTRODUCTION

Landfills are bioreactors in which physical, chemical, and biochemical processes take place. These are mainly driven and controlled by the incoming waste composition, moisture and climatic conditions, and the type of operation (e.g., slow or fast build up, high or low compaction). Municipal household solid waste consists mainly of a series of components, which are remarkably similar worldwide, although they vary in amount. Of course, many other kinds of waste enter the landfill and influence the milieu condition, such as buffer capacity; however, the main source of degradable organic substances is provided by household waste. Chapter 1.2 illustrates typical waste components found in residential waste and the elementary composition of several waste compounds.

In a landfill, organic waste is largely degraded under anaerobic conditions. Depending on the type of landfill operation, aerobic degradation processes take place in specific areas and at specific times. When the waste is unloaded and deposited, oxygen comes into contact with the waste and even after deposition may still be trapped in the pores and voids of the waste. Anaerobic processes start once all the oxygen has been consumed. The processes that take place in semiaerobic landfills and during landfill aeration in which aerobic processes dominate (see Chapters 14.2 and 16.2) are of a rather different nature.

Biological processes that take place in a landfill are not homogeneous; they occur in microenvironments and may therefore vary at short intervals (closely manifested acid and methanogenic phases). The quality and amount of gas produced and the composition of leachate reflect a typical situation of the dominant biological processes in a given period. Degradation of organics is of a longer duration in the solid phase than in the liquid phase, as the organic compounds need to migrate from the solid phase into the liquid phase by means of diffusion and dissolution. When the landfill is not water-saturated, processes take place in thin water films surrounding the particles (Fig. 3.1.1). However, if there is no movement of water through the landfill body, organic acids in the film or pores may accumulate and methanogenesis may be inhibited. In the presence of water movement, organic acids and other potentially inhibiting compounds are diluted and transported to other areas in which appropriate conditions for methanogenic processes are present. Similar situations have also been observed in areas in which leachate is stored (e.g., high water tables due to decreasing permeability in the lower parts of a landfill), where water movement is very slow.

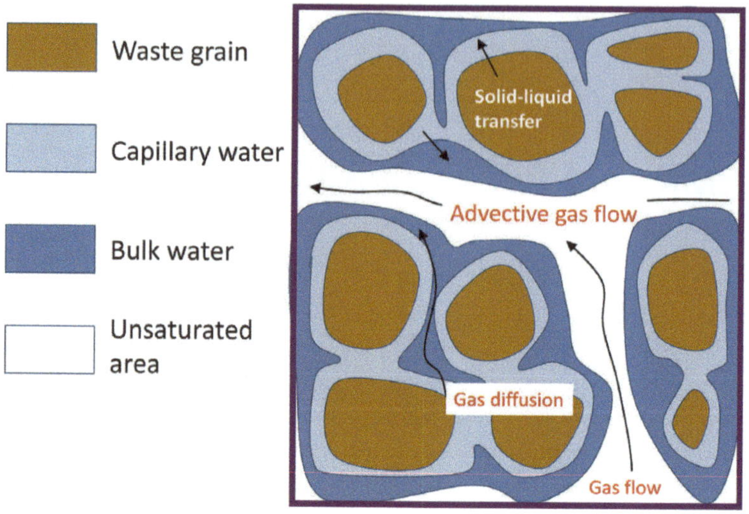

Figure 3.1.1 Different transport processes in landfills under unsaturated conditions.

Although the processes involved are highly complex, several attempts have been made to model landfill biochemistry by subdividing the waste mass into three interacting phases: gaseous, liquid, and solid. Each phase contributes toward and influences not only the biochemical processes considered in this chapter but also the physical and inorganic processes (Heimovaara et al., 2013).

In addition to hydrocarbons, a series of other, potentially toxic, compounds are introduced into the landfill, including chlorinated, bromated and fluorinated hydrocarbons, aromatic compounds, organic nitrogen, and sulfurous compounds. Due to the increasing use of a wide range of new chemicals and drugs, both microcontaminants and nanoparticles, which tend to accumulate inside the waste, may also be introduced into landfills (Pivnenko and Astrup, 2016). Some of these compounds are completely or partly degradable and may end up in the landfill gas and/or leachate (see Chapters 9.2 and 10.2). In some cases the products of degradation are more toxic than the original compounds; as an example, the anaerobic conversion of tetrachloroethylene into vinyl chloride (which is more toxic and purportedly carcinogenic), found in the gas phase, is described.

Biodegradation processes

Putrescible organic compounds are used as a source of carbon and energy by heterotrophic microorganisms (bacteria, fungi). Some heterotrophic and chemoautotrophic bacteria (nitrifying and methanogenic bacteria) use hydrocarbons solely as a source of energy and carbon dioxide as a source of carbon. Bacteria obtain energy needed for growth (anabolism) by means of redox processes (catabolism); the amount of energy yielded depends on the type of reaction involved (Table 3.1.1); aerobic respiration processes provide a much higher energy yield than methanogenesis and anaerobic processes in general. Microorganisms will invariably prefer an energetically favored reaction. Facultative bacteria are capable of modifying their catabolism to adapt to different environmental conditions.

Table 3.1.1 Energetic behavior of the main biochemical redox reactions

Biochemical Redox Reaction Name	Oxidation	Formula	Energy Yield (kcal/mol)
Aerobic respiration	O_2	$O_2 \rightarrow CO_2$	686
Denitrification	NO_3^-	$NO_3^- + H^+ \rightarrow N_2$	649
Iron reduction	Fe_3^+	$Fe_2(OH)_3 + 2H^+ \rightarrow Fe_2^+$	300
Sulfate reduction	SO_4^{2-}	$SO_4^{2-} + 2H^+ \rightarrow HS^-$	190
Methanogenesis	CO_2	$H_2 + CO_2 \rightarrow CH_4$	8.3

Modified from Anonymous (2017).

Due to the heterogeneity of the waste and the different conditions present in a landfill, all the reactions (Table 3.1.1) may take place in parallel (Berge et al., 2013). For the sake of simplicity, the biochemical processes can be subdivided into aerobic and anaerobic processes (Fig. 3.1.2).

Under anaerobic conditions biodegradable organics are hydrolyzed and finally converted mainly into methane and carbon dioxide. Generation of biological heat—although low—allows the waste body to reach and maintain mesophilic conditions (35–45°C or even higher) also due to the fact that the waste acts as an insulation material. The biomass generated is less than 5% compared to the carbon metabolized. On the one hand this implies a high reduction of carbon compounds, whereas on the other—compared to aerobic degradation—a considerably slower kinetic of metabolism. The reactions below illustrate the general formula for the aerobic (1) and anaerobic degradation (2) of glucose:

$$C_6H_{12}O_6(1\ \text{kg}) + O_2(0.64\ \text{kg}) \rightarrow CO_2(0.88\ \text{kg}) + H_2O(0.34\ \text{kg}) + \text{Biomass}(0.4\ \text{kg}) + \text{Heat}(9300\ \text{kJ}) \quad (1)$$

$$C_6H_{12}O_6(1\ \text{kg}) \rightarrow CH_4(0.25\ \text{kg}) + CO_2(0.69\ \text{kg}) + \text{Biomass}(0.056\ \text{kg}) + \text{Heat}(632\ \text{kJ}) \quad (2)$$

Due to the relatively fast growth of bacteria in the presence of oxygen, biochemical reactions are enhanced; in addition, some recalcitrant complex organic polymers may be hydrolyzed (cometabolism) and ammonia–nitrogen nitrified. Oxygen does not reach all parts of the landfill and/or may not diffuse into all solid waste particles. As a consequence, methane may still be produced locally, although it will probably be oxidized in the presence of oxygen. As methane oxidation is a slow process, methane at low concentrations may still be present in the off-gases extracted during in situ aeration (see Chapters 14.2 and 16.2). In oxygen-free zones denitrification and sulfate reduction may also occur. Under aerobic conditions a higher heat generation is achieved than during anaerobic degradation (temperatures of

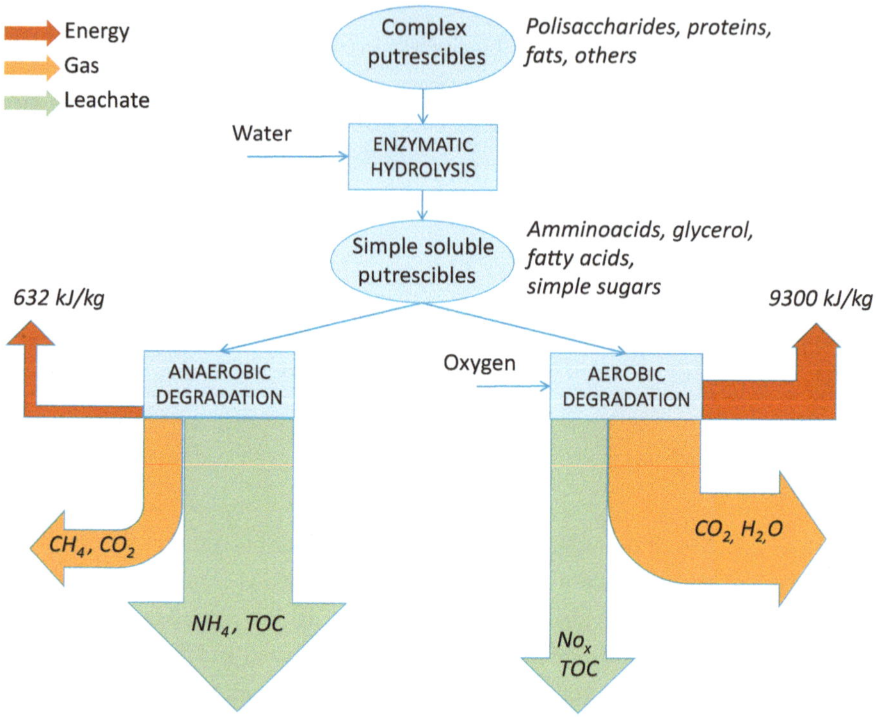

Figure 3.1.2 Overview of aerobic and anaerobic processes in a young landfill with average energy production and main catalytic products.

50–70°C can be reached in the waste body), thus positively influencing the activity of some microorganisms up to a temperature of approximately 50°C. According to the general formula for the acrobic digestion of glucose, biomass generation corresponds to approximately half the initial carbon content.

Organic substances are indeed degraded, although at the same time new large molecules may build up. The latter are represented by humic- and fulvic-like substances, consisting of large molecules with a high molecular weight. Humic substances are not clearly defined substances (see Fig. 3.1.3) but are produced in a sequence of reactions; polyphenols as precursors are released from cell walls of microorganisms, particularly fungi, and are liberated during the degradation process in substances such as lignins. By means of polyphenol oxidases these polyphenols will be oxidized to quinone, which condensate beneath each other or with various amino groups into humic substances. Subsequently, other molecules may bind to these newly formed molecules. Free radicals also play an important role in the development and cohesion of the different compounds. Molecular stability may be enhanced by means of sorption using substances such as clay minerals (Stahr et al., 2016).

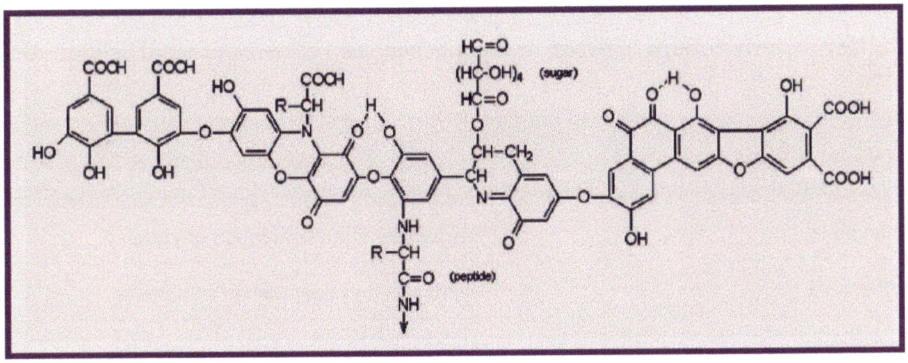

Figure 3.1.3 Scheme of a humic substance molecule (Stevenson, 1982).

Landfill microorganisms

Municipal solid waste (MSW) landfills are characterized by a wide range of active microorganisms; some bacteria are capable of catabolizing only specific substrates in a very close range of pH or temperature and do not produce spores (i.e., methanogenic bacteria). Others, however, are more flexible and can survive under a wider range of environmental conditions (i.e., facultative bacteria). Table 3.1.2 lists the key groups of bacteria (Sleat et al., 1989; Berge et al., 2013). Fecal bacteria and viruses are present in fresh landfilled waste, although their number decreases rapidly due to competition with other microbes and—when present—high temperatures; their presence in leachate is considered negligible (Andreottola and Cannas, 1992).

ANAEROBIC DEGRADATION

Anaerobic processes are highly sensitive to boundary conditions; low energy yield results in low biomass quantity and slow reaction kinetics. Inadequate pH may inhibit the metabolization process, although in a landfill this does not normally pose a problem due to the effective buffering of the system. This situation, however, may vary in a series of microenvironments. Even in the presence of small amounts of toxic substances in household waste, biological processes may be inhibited locally; temperature may also influence the process, as an example, methanogens may be psychrophilic (15–20°C), mesophilic (33–40°C), or thermophilic (55–60°C). Beyond these temperature ranges, methane production may be decreased or inhibited (McCarty, 1964). Accordingly, it should be taken into account that temperatures inside a landfill may locally vary quite significantly. In comparison with aerobic processes, the range of substrates available for metabolization is less abundant, resulting in a lower energy yield, and consequently enzymatic activity.

Table 3.1.2 Key bacterial groups involved in the biochemical processes of landfills indicating their main activity

Bacterial Group	Main Biochemical Activity
Amylolytic bacteria	Hydrolysis and degradation of starch
Proteolytic bacteria	Hydrolysis and degradation of proteins
Cellulolytic bacteria	Hydrolysis and degradation of cellulose
Hemicellulolytic bacteria	Hydrolysis and degradation of hemicellulose
Hydrogen-oxidizing methanogenic bacteria	Methane production from hydrogen and carbon dioxide
Acetoclastic methanogenic bacteria	Ethane and carbon dioxide production from acetic acid
Sulfate-reducing bacteria	Sulfate reduction to sulfite through acetic acid
Nitrosomonas	Converts ammonia nitrogen into nitrite
Nitrobacter	Converts nitrite into nitrate
Thiobacillus denitrificans	Uses inorganic sulfur to reduce nitrates in autotrophic denitrification
Planctomycetales	Anaerobic ammonium oxidation (Anammox), capable of oxidizing ammonia nitrogen in anaerobic conditions.

Anaerobic degradation and methane production

The anaerobic process is subdivided into four phases by different kinds of microorganism, although the overall process can be described by the simplified formula from Buswell and Muller (1952):

$$C_nH_aO_b + \left(n - \frac{a}{4} - \frac{b}{2}\right)H_2O \longrightarrow \left(\frac{n}{2} - \frac{a}{8} + \frac{b}{4}\right)CO_2 + \left(\frac{n}{2} + \frac{a}{8} - \frac{b}{4}\right)CH_4 \quad (3)$$

With regard to nitrogen, the main nutrient in this process, the following equation applies (Ehrig, 1984):

$$C_nH_aO_bN_d + \left(n - \frac{a}{4} - \frac{b}{2} + \frac{3d}{4}\right)H_2O \longrightarrow \left(\frac{n}{2} - \frac{a}{8} + \frac{b}{4} + \frac{3d}{8}\right)CO_2 + \left(\frac{n}{2} + \frac{a}{8} - \frac{b}{4} - \frac{3d}{8}\right)CH_4 + dNH_4 \quad (4)$$

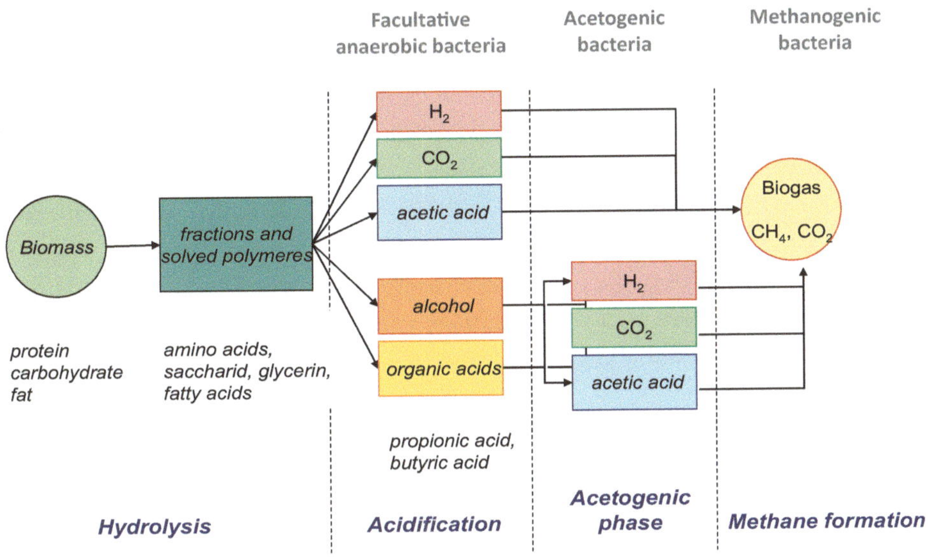

Figure 3.1.4 Scheme of the anaerobic degradation process.

These formulas allow all products to be evaluated and reagents required for the process to be calculated, based solely on knowledge of the precise composition of the initial substrate. However, this initial composition is generally unknown and is very hard to estimate with precision. Therefore, the formula will be used for theoretical applications rather than for practical use (Christensen and Kjeldsen, 1989). To estimate biogas production, mathematical models are the most widely used (see Chapter 9.1).

Anaerobic degradation can be subdivided into four different subsequential phases: Hydrolysis, Acidogenic fermentation, Acetogenic fermentation, and Methanogenesis, in addition to sulfate reduction, which may compete with the last three processes in substrate utilization (Fig. 3.1.4).

Hydrolysis

The first phase of biochemical degradation is enzymatic hydrolysis, required to reduce complex polymers into compounds that are small enough to penetrate the cellular membrane. For this purpose, some species of fermentative microorganisms produce extracellular enzymes—mainly cellulose, amylase, lipase, and protease—to catalyze the hydrolytic processes. This process is mainly relevant for the reduction of fats, carbohydrates, and proteins into fatty acids, glycerin, sugars, and amino acids (Fig. 3.1.4).

$$\text{Fats} + nH_2O \rightarrow C_3H_5(OH)_3 + R - COOH$$

$$\text{Carbohydrates} + nH_2O \rightarrow C_6H_{12}O_6$$

$$\text{Proteins} + nH_2O \rightarrow nR - CHNH_2 - COOH \longrightarrow nRCOO^- + nNH_4^+$$

Hydrolysis is the slowest process that occurs during anaerobic degradation. For this reason, the kinetics of the process regulates all subsequent reactions, thus slowing down the entire process. When there are no limitations to substrate availability and degradability, anaerobic biochemical reactions can be modeled using Monod first-order kinetic. The presence of water is fundamental for all kinds of reactions; as a consequence, an ample supply and free movement inside the waste body should be ensured. During the initial phases of landfill life, leachate is rich in hydrolytic and intermediate products, which contribute toward increasing the acidity of the system. During this phase, due to the presence of volatile fatty acids (VFAs), the formation of hydrolytic products and drop in pH, COD, and BOD_5 indices increase; in addition, ammonia nitrogen and sulfates are also produced in this phase (Fig. 3.1.5) (see Chapters 9.2 and 10.2).

Acidogenesis fermentation

During acidogenesis, the products of fermentative bacteria are VFAs, alcohols, ammonia–nitrogen, hydrogen, and carbon dioxide (Fig. 3.1.5). The most significant reactions, considering as an example glucose, are as follows:

$$C_6H_{12}O_6 + H_2O \rightarrow 2CH_3COOH + H_2 + 2CO_2 \quad \text{(acidic acid)}$$

$$C_6H_{12}O_6 \rightarrow C_3H_7COOH + 2H_2 + 2CO_2 \quad \text{(butyric acid)}$$

$$C_6H_{12}O_6 \rightarrow 2CH_3CH_2OH + 2CO_2 \quad \text{(ethanol)}$$

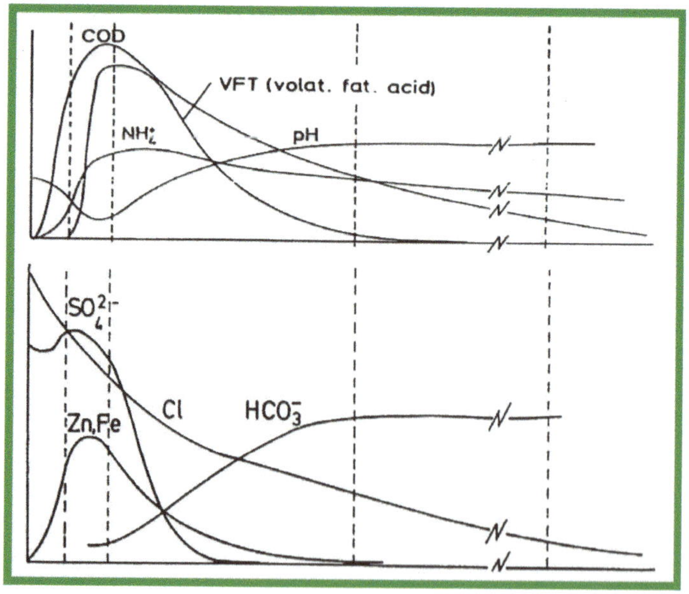

Figure 3.1.5 Variation of leachate composition versus time, during the different anaerobic degradation phases.

Numerous microorganisms are active during the acidogenic phase, with some facultative bacteria developing faster (Daverio et al., 2005). During this phase, a high presence of hydrogen-producing bacteria and hydrogen-consuming bacteria (e.g., sulfate-reducing and nitrate-reducing bacteria) is present (Senior, 1990). The carbon dioxide generated replaces residual oxygen and free nitrogen in the interstitial pores. Hydrogen gas is also produced, although, as a general rule, methanogenic bacteria immediately consume this.

Acetogenesis fermentation

During the acetogenic phase, all intermediate products resulting from acidogenesis are converted into acetic acid, hydrogen, and carbon dioxide. Acidogenic bacteria do not convert aromatic hydrocarbons into acetic acid unless they contain oxygen (Christensen and Kjeldsen, 1989).

$$C_2H_5COOH + 2H_2O \rightarrow CH_3COOH + 3H_2 + CO_2$$

$$C_3H_7COOH + 2H_2O \rightarrow 2CH_3COOH + 2H_2$$

$$CH_3CH_2OH + H_2O \rightarrow CH_3COOH + 2H_2$$

Before methanogenesis is fully developed, conversion of all hydrolytic compounds into acetic acid may elicit a decrease of pH into acid conditions and thus slow down the methanogenic processes. In addition, high concentrations of organic acids at a neutral pH (buffered system) may inhibit methanogenesis. These processes all develop throughout the diverse microenvironments present in a landfill. The gas produced consists mainly of carbon dioxide and hydrogen. Due to the presence of energized bacteria, acetogenesis can only take place at a low hydrogen concentration. When the system is in equilibrium with methanogenesis, the hydrogen produced is immediately used by methanogenic bacteria to generate methane using CO_2 as a carbon source. Accordingly, acetogenic and methanogenic bacteria live in an energetic symbiosis. When hydrogen is not consumed, it can be measured in the landfill gas and—as a consequence—higher amount of volatile acids (e.g., propionic acid) will be present in the leachate. Hydrogen is very rarely detected at significant levels in landfill gas.

Methanogenesis

The last phase of anaerobic degradation is methanogenesis, during which—using hydrogen and acetic acid as a substrate—methane and carbon dioxide are produced (Fig. 3.1.4). Bacteria performing these reactions are highly specialized and sensitive to pH and temperature; they use mainly the products from acidogenic and acetogenic reactions. While methanogens adapt to the environment in a landfill and increase over time, gas composition changes (Fig. 3.1.6). Subsequently, the process becomes more stable and biogas reaches the standard quality for an anaerobic process (50–60% CH_4, 40–50% CO_2) (see also Chapter 9.2). Two main categories of methanogens can be distinguished according to the substrate utilized: hydrogenotrophic methanogens consume hydrogen and carbon dioxide, whereas

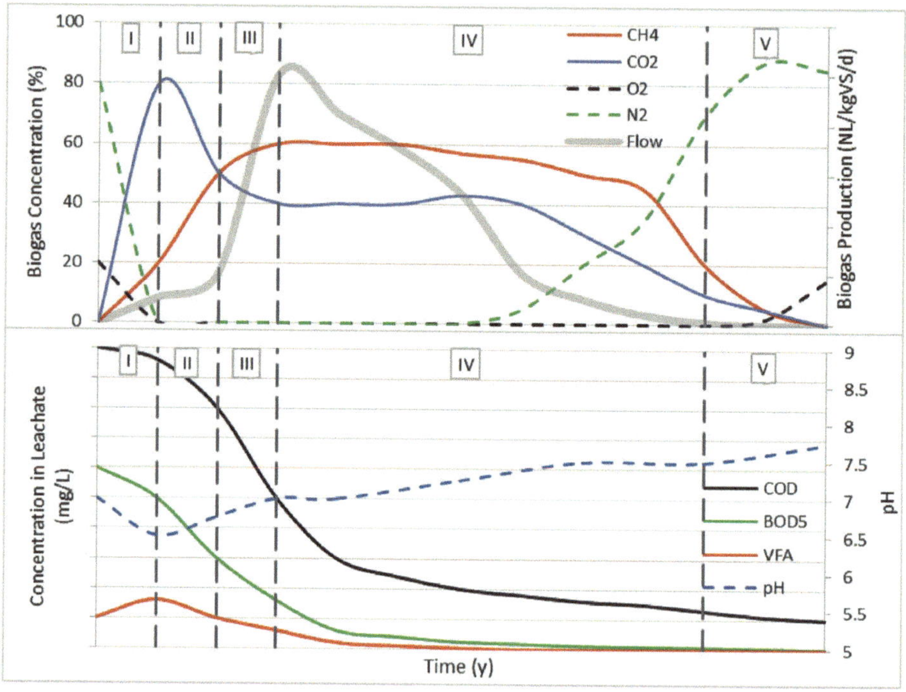

Figure 3.1.6 General biogas percentage composition, biogas flow rate, and leachate biochemical characterization during anaerobic landfill life. Behavior of concentration trend modified from Christensen and Kjeldsen (1989), Cossu et al. (2003), and Cossu et al. (2015). I: Hydrolysis, II: Acidogenesis, III: Acetogenesis, IV: Methanogenesis, V: Long-term aftercare.

acetotrophic methanogens use acetic acid. Indeed, some bacteria even use formic acid and ethanol for methane production (Christensen and Kjeldsen, 1989). The conversion of acetic acid into methane and CO_2 is the most important methanogenic reaction, and the largest amount of methane is generated by this reaction.

$$CH_3COOH \rightarrow CH_4 + CO_2$$

$$CO_2 + 4H_2 \rightarrow CH_4 + 2H_2O$$

$$HCOOH + 3H_2 \rightarrow CH_4 + 2H_2O$$

$$CH_3OH + H_2 \rightarrow CH_4 + 2H_2O$$

The above reactions transform VFAs and products originating from previous phases into biogas; as a consequence of the reduction of VFAs, COD, BOD_5, and TOC concentrations in leachate decrease

(Figs. 3.1.5 and 3.1.6). At the end of the methanogenic phase, scarcely degradable humic substances account for residual carbon concentrations. Methane and carbon dioxide concentrations in biogas remain constant or increase slightly (due to the degradation of compounds with a higher carbon content) until almost all bioavailable substrates have been degraded. Biogas production is characterized by a positive peak following the onset of stable methanogenesis (due to rapidly degradable compounds), and subsequently biogas production slowly decreases (Fig. 3.1.6).

Throughout the entire "landfill life," the composition of biogas changes (Fig. 3.1.6). Following the short initial phase in which oxygen is consumed forming high concentrations of carbon dioxide, the methanogenic phase commences and lasts for the time required to degrade more than 90% of the putrescible matter. Depending on landfill height/depth, this phase may last in excess of 30 years following landfill closure (Ritzkowski et al., 2006). After this period, air starts slowly to infiltrate into the waste mass, generating aerobic conditions; at this stage some complex organic compounds not accessible to anaerobic microorganisms will undergo degradation. This process can be enhanced by in situ aeration (Chapter 16.2).

Methane and carbon dioxide represent the main components, together accounting for more than 90% of biogas. The presence of water vapor, free nitrogen, and hydrogen are closely linked to process efficiency, whereas H_2S and $NH_{3(g)}$ concentrations depend largely on initial waste characteristics, particularly protein content.

Sulfate reduction

The sulfate reduction process occurs simultaneously with fermentation and methanogenesis, entering into competition with the latter for acetic acid and hydrogen substrates, potentially giving rise to partial inhibition of methane production (Christensen and Kjeldsen, 1989).

$$2H_2 + SO_4^{2-} + H^+ \rightarrow HS^- + 4H_2O$$

$$CH_3COOH + SO_4^{2-} \rightarrow CO_2 + HS^- + HCO_3^- + H_2O$$

$$2C_3H_7COOH + SO_4^{2-} + H^+ \rightarrow 4CH_3COOH + HS^-$$

The reactions described above are typically observed in sulfate reduction in waste/leachate (see also Figs. 3.1.5–3.1.7).

Process control parameters and nutrients

Anaerobic degradation is strongly influenced by a series of control parameters (pH, temperature, alkalinity (ALK), VFA, oxygen presence), macronutrient presence (nitrogen, phosphorus, sulfur), micronutrient availability, and concentrations of toxic compounds.

Temperature

The anaerobic degradation processes typically occur throughout the psychrophilic (15–20°C), mesophilic (33–40°C), and thermophilic (55–60°C) temperature ranges. By increasing the temperature to a higher range, different microbial groups become active, thus accelerating degradation kinetics. However, the maintenance of optimal high temperature conditions in a landfill is rather difficult and almost impossible in a full-scale anaerobic landfill. Only a combination of aerobic and anaerobic phases would yield similar results. However, temperature inside the waste body is naturally maintained at stable mesophilic conditions, thanks to the thermic characteristics of the waste mass and top cover. The energy production under anaerobic conditions is indicatively 10% of the production under aerobic conditions, but is sufficient to maintain temperatures of 35–40°C (Christensen and Kjeldsen, 1989). Due to the fact that high temperatures (≥60°C) may be reached in a landfill even under anoxic conditions (Chapter 15.1), inhibition of biological processes may occur.

Oxygen

The presence of oxygen in the waste body generally inhibits methane production. Aerobic bacteria compete heavily for the same substrate used by methanogens, with oxygen being toxic for these bacteria. Methanogens do not produce spores but can survive in the presence of localized oxygen concentration due to the development of anaerobic microenvironments (Christensen and Kjeldsen, 1989). However, in an anaerobic landfill, oxygen diffusion is limited to the upper layers and top cover and does not affect the larger mass of waste.

Moisture content

As mentioned previously, in a landfill, moisture content and water flux are fundamental for biochemical processes, supporting hydrolytic process, transporting nutrients, diluting toxins, and facilitating the mobility of microorganisms. If the top cover does not allow the introduction of fresh water into the waste body, activity will be reduced and potential emissions will remain inside the reactor. For this reason, an appropriate availability of water in the waste body is a high priority (Ritzkowski and Stegmann, 2013), to efficiently redistribute compounds and provide the necessary moisture content (Valencia et al., 2009). Moreover, to homogenize the processes occurring within the waste mass and to promote or enhance biological degradation processes, leachate recirculation may be implemented (see Chapter 12.1). Raw MSW may contain 30–60% wt of water, with moisture content in some areas reaching field capacity (50–60% of total solids). A minimum moisture content of >20–30% wt is essential to promote biological processes. In addition to a sufficiently high water content, water flux is required to facilitate/enhance biochemical processes (Ritzkowski and Stegmann, 2013). Controlled water/leachate infiltration may be implemented by installing infiltration chambers or infiltration ditches in the top cover of a landfill (Hupe et al., 2013).

pH

One of the most important parameters in an anaerobic degradation process is represented by pH, which strongly influences the efficiency of methanogenic bacteria. While fermentative bacteria work efficiently at a wide range of pH (5–9), methanogenic bacteria operate in a stricter range, indicatively between 6 and 8 (Zehnder et al., 1982; Lay et al., 1998; Chugh et al., 1998; Sekman et al., 2011; Sandip et al., 2012). Due to the production of VFA, pH may significantly decrease during the first phases of hydrolysis and fermentation. If methane production kinetics is slower than VFA production, acids accumulate and pH values may drop to values below 6, inhibiting, at least locally, the methanogenic process.

Alkalinity

The alkalinity (ALK) represents the sum of alkali salts, indirectly indicating buffer capacity, i.e., the ability to mitigate the fall in pH despite the generation of acids during fermentation processes. The main sources of ALK are the landfilled waste and products resulting from the degradation of proteins and the reduction of sulfates. No ALK is produced through the hydrolysis of carbohydrates, sugars, organic acids, and fats (McCarty, 1964).

$$NH_{3(g)} + H_2CO_3 \rightarrow NH_4^+ + HCO_3^-$$

$$4H_2 + SO_4^{2-} + CO_2 \rightarrow HS^- + HCO_3^- + 3H_2O$$

Carbonate (HCO_3^-) concentrations in leachate are a good indicator of buffer capacity. In the majority of cases, buffer capacity in the waste is sufficient to keep pH at a value in the region of 7, thus preventing inhibition of methanogens. Following an initial abundance of putrescible organic compounds and high organic acid production, mixing the waste with buffering materials is seen as a good practice; in particular, construction and demolition waste are both capable of diluting acidity and raising ALK (Christensen and Kjeldsen, 1989). Furthermore, other additives such as alkaline ashes and other similar residues rich in calcium silicates, carbonates, and bicarbonates may be used (Andreas, 2013). However, excessively high concentrations of carbonates may precipitate under specific conditions and cause clogging in drain systems (see Chapter 8.1).

Volatile fatty acids

VFAs are key contributors to the decrease of pH. VFAs are a mix of small chain acids (mainly Formic, Acetic, Propionic, Butyric, and Valeric), the concentration of which is commonly expressed in terms of acetic acid (McCarty, 1964). If the concentration of VFAs in landfill leachate is below 6000 CH_3COOH mg/L, this may not suffice to inhibit methanogens; however, this threshold value also depends on the concentration of ALK in the waste mass. The production of VFAs progressively consumes the ALK of the system before influencing pH value. For this reason, the sole monitoring of pH is not deemed sufficient to ensure the continuous monitoring of anaerobic reactors, particularly as changes are only manifested when the buffer capacity is consumed, often too late to correct VFA content. For this purpose, particularly in the context of reactor management, ALK/VFA ratio is applied. An ALK/VFA ratio of approximately 3.5 g HCO_3^-/gCH_3COOH is an optimum value for an anaerobic digester. In an anaerobic degradation system, a ratio between VFAs and ALK lower than 0.8 is required for methane production (Farquhar and Rovers, 1973). In a landfill, slightly higher values may be reached without affecting methane production (Cossu et al., 2016).

Ammonia—nitrogen

Methanogenic bacteria consume ammonia—nitrogen as a nutrient when present at concentrations of at least 49 mg/L, although concentrations exceeding 3000 mg/L are toxic for these bacteria. Ammonia ion is the main process nutrient, and its presence is fundamental in enhancing the production of ALK inside the system. At pH >8, the ammonium aqueous equilibrium promotes the production of gaseous toxic ammonia, capable of inhibiting microbial activity (McCarty, 1964).

Sulfates

Sulfates originating from the hydrolysis of proteins or from the landfilled waste are generally present in landfill leachate and may exert a negative effect on anaerobic degradation. As sulfate reduction is energetically advantageous, the reduction process will be forced to compete with methanogens for hydrogen and acetic acid. Moreover, the resulting H_2S is toxic for methanogens in an environment featuring a low pH (McCarty, 1964). As a general rule, in landfills sulfates are not present at concentrations known to produce inhibition (Fig. 3.1.7).

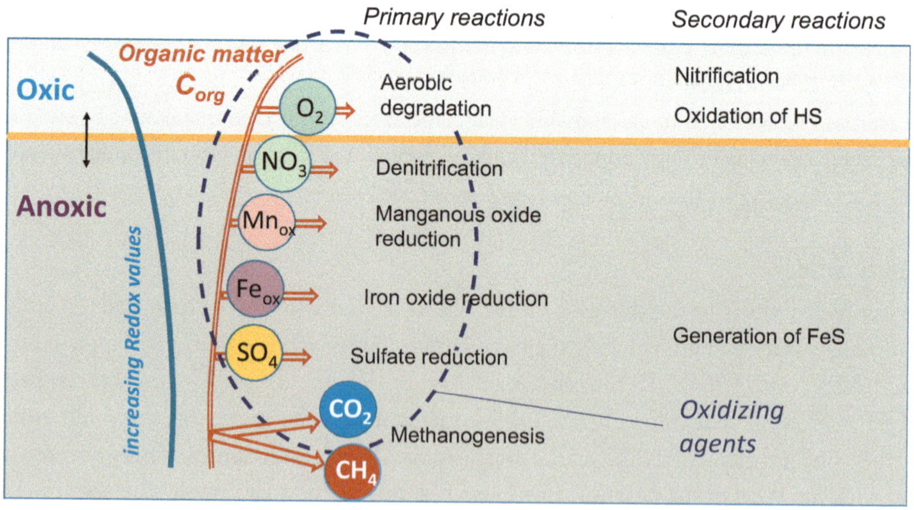

Figure 3.1.7 Redox reactions occurring during the biodegradation processes in municipal solid waste landfills. *Modified from international.eawag.ch.*

Nitrates

Nitrates and nitrites are both present in semiaerobic and aerated landfills. Due to the existence of anoxic areas in these landfills, nitrates and nitrites will be denitrified to a high extent. Denitrifiers are dependent on the presence of degradable organic compounds for metabolism and are in low competition with methanogens (Filipkowska, 2007). This effect may also occur in landfills in which aerobically treated leachate is recirculated.

Hydrogen

Hydrogen is generally converted entirely into methane during hydrogenotrophic processes or consumed by sulfate-reducing bacteria. As a consequence, hydrogen concentration in biogas is negligible (see Chapter 9.2).

Nutrients, micronutrients, and inhibitors

The main nutrients in an anaerobic degradation process are nitrogen, sulfur, and phosphorus. Nitrogen and sulfur are generally present in abundance in waste, whereas phosphorus may represent a limiting factor. The ratio between the three macronutrients is 100 C: 0.44 N: 0.08 P, highlighting that the production of new biomass is very low in percentage (Christensen and Kjeldsen, 1989; McCarty, 1964). In addition to the main nutrients, a consistent number of other elements act as fundamental micronutrients; however, due to the composition of waste, there should be no deficits of these elements. To date, no indications of toxic effects have been provided for any of the trace compounds present in landfills, although knowledge of these elements is rather lacking. In line with the concentration of these elements inside the landfill some compounds may elicit toxic effects in specific microenvironments of a landfill, although these are very difficult to quantify. This sensitivity of methanogenic bacteria to different element concentrations may create variations of kinetics inside the landfill body; further studies may contribute toward clarifying this issue. Leachate recirculation is likely fundamental in mitigating this potentially negative effect (Christensen and Kjeldsen, 1989). The effects of the different nutrients concentration ranges on anaerobic digestion processes are evident from the values reported in Table 3.1.3.

Long-term fate of organics

Depending on waste composition, reactor characteristics, and management strategies, in landfills the products of anaerobic degradation are transferred to the emissions. Generally, 90% of emissions into the environment can be ascribed to the conversion of putrescible substances into catabolite gas (CH_4 and CO_2) (Ehrig and Kruempelbeck, 2013). Biomass production affects less than 5% of available organic compounds, with carbon emissions in leachate being an order of magnitude lower than gaseous ones, even in the anaerobic degradation phase (Lornage et al., 2013; Morello et al., 2016).

The largest part of carbon deposited in a landfill will remain stored even centuries after deposition, being made up of scarcely biodegradable compounds (such as cellulose, hemicellulose, lignin, and synthetic organic polymers) (Manfredi and Christensen, 2009; De la Cruz et al., 2013). During the

Table 3.1.3 Effects of different nutrient concentration ranges on anaerobic digestion processes

Parameter	Positive Effect Range	Negative Effect Range
Mesophilic Temperature (°C)	30–40	
pH	6.6–7.6	<6 and >8.2
Alkalinity (mg$CaCO_3$/L)	1000–5000	
Volatile Fatty Acids (mgCH_3COOH/L)	<6000	>6000
Ammonia–Nitrogen (mg/L)	40–70	>3000
Phosphorus (mg/L)	>1 mg/L	
Sulfur (mg/L)	>1 mg/L	
Sodium (mg/L)	100–200	>3500
Potassium (mg/L)	200–400	>2500
Calcium (mg/L)	100–200	>2500
Magnesium (mg/L)	74–150	>1000
Iron (mg/L)	10–200	>1750
Nickel (mg/L)	0.5–30	>30
Molybdenum (mg/L)	0.1–0.35	
Tungsten (mg/L)	0.1–0.35	
Selenium (mg/L)	0.1–0.35	
Zinc (mg/L)	0.1–0.3	>400
Copper (mg/L)		>40
Chromium (mg/L)		>130
Lead (mg/L)		>340

Modified from McCarty (1964).

biochemical degradation of carbon, approximately 10% is transformed into slowly mobile forms, probably due to the formation of humic and fulvic acids, as the result of the conversion of biomass during degradation processes. They also contribute to the long-term organics repository in landfill (carbon sink). Soluble humic substances provide a relevant contribution to long-term COD emissions in leachate (Brandstätter et al., 2015).

The residual biochemical status of landfilled waste can be well described by biological stability indexes (see Box 3.1.1). Using standardized methods, these indexes allow, over a short period of time, progression of the biostabilization process to be evaluated. The indexes measure biological activity in terms of respiration (Oxygen consumption), biogas production, or BOD/COD rate in the eluate from a leaching test of the material.

AEROBIC DEGRADATION

Mandatory requirements for an aerobic process to occur are presence of oxygen, substrate and nutrients (mainly nitrogen and phosphorus) and moisture content, with the latter being fundamental for both hydrolysis and compound mobility. Under these conditions, an aerobic process, versus an anaerobic process, will afford fast reaction kinetics, thanks to its high energetic yield and to the numerous groups of available aerobic microbes in waste. Temperatures inside the waste body increase due to exothermic reactions, and pH should remain in the range between 4 and 9, generally observed in MSW landfills. Reaction kinetics is governed by enzymatic hydrolysis, which, as discussed previously, decompose the macromolecules and polymers (proteins, starch, carbohydrates) into smaller compounds available to bacteria (amino acids, sugars, fatty acids). Conversely to anaerobic process, under aerobic conditions, refractory and slowly degradable organics are also degraded (see also Chapter 16.2).

In the presence of a sufficiently high oxygen level, a nitrification process takes place, converting ammonia—nitrogen into nitrates.

Oxidation of carbon compounds

In aerobic landfills, microorganisms progressively convert the biologically available compounds into carbon dioxide, water, and new biomass. In addition, humic substances are produced.

If sufficient oxygen is provided, all other biological processes in competition for organic carbon substrate degradation are inhibited. The efficiency of the aerobic condition in a landfill has been demonstrated worldwide: reaction kinetics allows the biological stabilization time to be reduced as much as ten-fold compared with anaerobic degradation (see Chapter 2.2).

The main biodegradable polymeric constituents of organic waste can be classified as follows (Senior, 1990):

- Lignocelluloses are highly complex polymers with mainly structural function in plants.
- Cellulose and hemicellulose are partially degradable under aerobic conditions; their semirecalcitrance is mainly due to the difficult accessibility of the β-4 glucosidic bond to hydrolytic enzymes.

Box 3.1.1 Stability indexes for measuring the biological stability in solid waste sample

Biological Stability Index	Description
SRT—Static respiration test	SRTs are respiration indexes commonly used for monitoring the residual biodegradability of compost or other solid samples. SRT measures the oxygen consumed by the sample without replacing O_2 once it is consumed (Komilis and Kletas, 2012).
RI_4—Respiration index (semidynamic)	RI_4 measures the oxygen consumption of a sample in 4 days, replacing the oxygen once consumed for maintaining its concentration constant for the whole test (SEMIDYNAMIC). It is used for both solids and liquids. $RI_4 < 2.5-5$ g O_2/kgTS indicates low biodegradability (Cossu and Raga, 2008; Laner et al., 2012).
DRI—Dynamic respiration index	DRI test measures the oxygen consumption by a solid sample in a reactor crossed by a continuous flow of air. DO in the inlet and outlet is monitored (Cossu et al., 2001). The system for DRI measurement allows to use a larger amount of waste (couple of kilos), which reduce the sample heterogeneity effect but the equipment might be expensive.
BOD_5/COD ratio	BOD_5/COD is a stability index proposed by Cossu in 2001. It is based on the analysis of BOD_5 and COD on the eluate from a leaching test of the solid sample to be analyzed. Advantages of the method, when compared with other aerobic stabilization indices are as follows: use of standard equipment present in any laboratory, simple and cheap procedure, no influence by dilution effects due to the presence of impurities in the sample, possible indication of the eventual presence of toxic or inhibiting substances, sample size is not influencing the test results, and short testing time. BOD_5/COD are well correlated to the other indices and values < 0.1 indicate a stable sample (Cossu et al., 2012; Cossu et al., 2017).
GB_{21}—Gas production potential	GB_{21} index measures the anaerobic biogas production in solid waste sample in 21 days at the standard condition of 40 °C. The index results can be correlated with the respiration test ones (Cossu and Raga, 2008).
BMP—Biomethane potential	BMP test measures the total methane production of a sample, until reaching the end of biochemical process (Esposito et al., 2012). This test is longer than GB_{21} but can supply data useful for the evaluation of the reaction kinetics and the total expected methane generation.

- Lignin is another polymer the structure of which is so complex that is generally considered very low to almost nondegradable.
- Polysaccharides are the primary substrate for plants and organisms. Thanks to hydrolysis, their polymerization can easily be broken down.
- Fats are well distributed in all plants and microorganisms and are fully biodegradable compounds.
- Proteins are amino acids polymers, base components of cells. They are biodegradable and their hydrolysis will produce ammonia—nitrogen and sulfur-containing molecules.

When hydrolyzed, all these components become part of the catabolic processes that occur inside microbe cells, finally being transformed into carbon dioxide or water, or, during anabolism, new cell biomass. This process can be described by the general stoichiometric formula for the decomposition of volatile solids in the aerobic processes (Tchobanoglous et al., 1993):

$$CaHbOc + 0.5\ (ny + 2s + 0.5b - 0.5nx - c)O_2 \rightarrow nCwHxOy + sCO_2 + (0.5b - 0.5nx)H_2O$$

However, the oversimplification of the aerobic degradation process according to the stoichiometric solution of this formula may represent a potential source of error. Heat is also a reactor product, capable of increasing temperature in a landfill body up to levels of 60—70°C. For this reason, even if water is produced during aerobic processes, in the majority of cases moisture should be added to the landfill both to promote hydrolysis and to control temperature and fire. The biomass produced can be further degraded during the endogenous processes, finally producing complex nondegradable and only partially soluble compounds, such as humic acids (Brandstätter et al., 2015).

Carbon compound emissions from an aerobic landfill are made up largely of gaseous carbon dioxide, whereas organic mass emissions via leachate are more than one order of magnitude lower (Lornage et al., 2013; Cossu et al., 2015). The presence and circulation of water are of fundamental importance for the hydrolytic process, the redistribution of nutrients, and the dilution of toxics. However, aerobic processes can be equally efficient if they can count on a satisfactory moisture content inside waste (without any excess) and liquid circulation is guaranteed (Ritzkowski and Stegmann, 2013). Due to oxidation of the reduced sulfur-containing molecules, aerobic conditions may increase the sulfate content of leachate (Ritzkowski and Stegmann, 2007).

Further positive effects of aeration of landfills include the reduction of odors in biogas and avoidance of uncontrolled methane emissions, resulting in a positive effect for the global climate change issue (see Chapter 16.2).

NITROGEN BEHAVIOR

In solid waste, nitrogen is incorporated largely in organic substances and can be partially mobilized through ammonification during the hydrolysis of proteins. The ammonia—nitrogen produced can be volatilized, flushed away, or nitrified—denitrified to free nitrogen gas (Fig. 3.1.8).

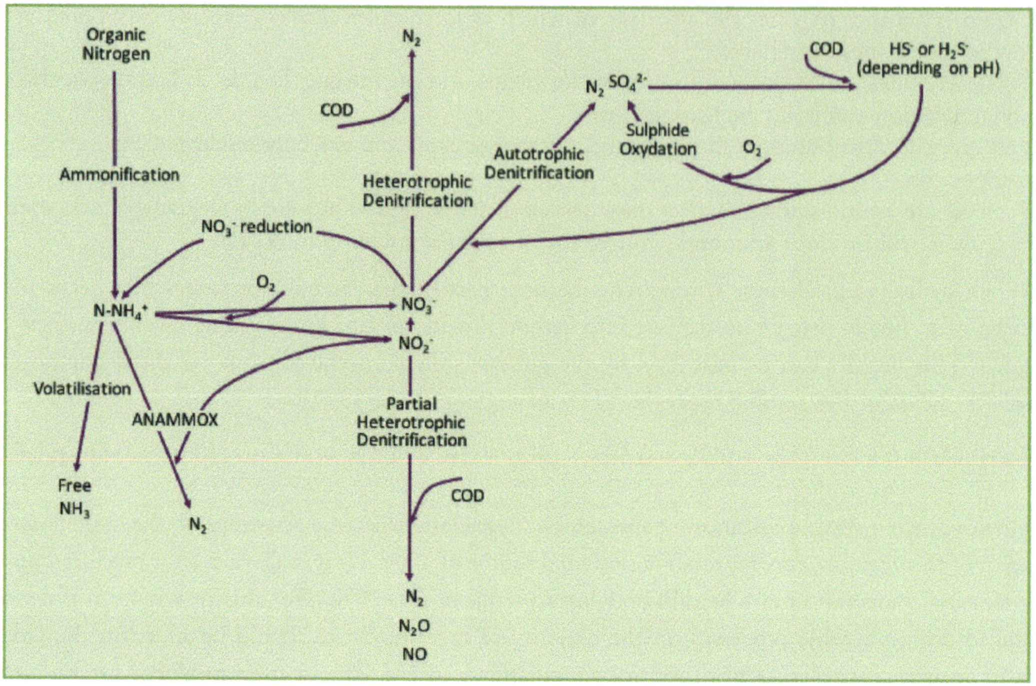

Figure 3.1.8 Principal pathways of nitrogen transformation in a landfill. ANAMMOX, anaerobic ammonium oxidation *Modified from Berge et al. (2013).*

With regard to the nitrogen mass balance, marked differences can be observed between an anaerobic and an aerated landfill. Nitrification—denitrification will occur in the presence of a satisfactory supply of air and convert ammonium ion into nitrogen gas. This is the most efficient and easiest means of reducing the nitrogen load in landfill leachate.

Ammonification

The most nitrogen-rich compound present in leachate is ammonia—nitrogen, originating mainly from the hydrolysis of proteins (yard and food wastes) during ammonification. This biological process takes place both under aerobic and anaerobic conditions and terminates when all available organic nitrogen has been consumed (Fig. 3.1.9). Ammonia nitrogen (NH_4^+) in aqueous solutions is in equilibrium with the ammonia gas form $(NH_{3(g)})$. In landfills, NH_3 formation starts at a pH higher than 8 and is also strongly affected by temperature, with higher temperature values resulting in NH_3 formation even at lower pH. Invariably, landfill pH is lower than 8.5, thus resulting in the majority of ammonia nitrogen being present in ion form. However, this may change when temperature values exceed 40°C. Alternatively, once produced, NH_4^+ may react with other organic compounds; in the presence of oxygen, be nitrified (and eventually denitrified), if the pH is high enough, it may be volatilized and, finally, progressively flushed and dissolved in the leachate. In anaerobic landfills, flushing is the only

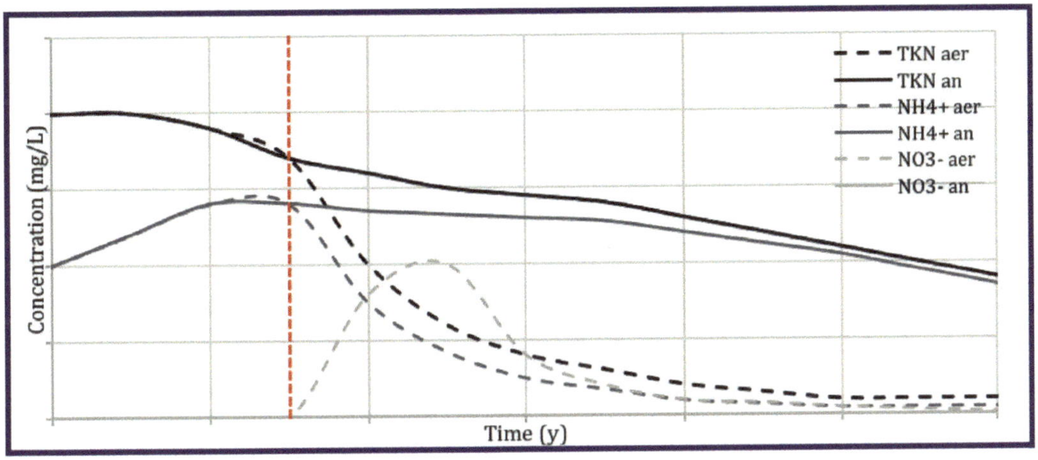

Figure 3.1.9 Ammonia and nitrate trends during an in situ aeration process in a landfill, compared with a nonaerated one. *Modified from Ritzkowski et al. (2006).*

possible means of removing ammonia nitrogen. In old traditional anaerobic landfills, ammonia nitrogen concentration represents a major problem for contamination and toxicity associated with leachate emissions, even after decades (Heyer et al., 2003). On the contrary, in aerated landfills, nitrification and NH_3 volatilization (as a consequence of the higher temperatures) are enhanced, whereas denitrification may occur in the anoxic zones.

Nitrification

By means of nitrification ammonia–nitrogen concentration may be reduced to very low values (indicatively below 50 mg/L). When nitrites and nitrates are produced simultaneously, they can be denitrified in anoxic zones of the landfill (Fig. 3.1.8).

Nitrification consists in a two-step process, in the first ammonia nitrogen is converted into nitrite by *Nitrosomonas* bacteria and in the second step nitrites are transformed into nitrates by *Nitrobacter* (Berge et al., 2013).

$$HH_4^{+} + 1.5O_2 \rightarrow NO_2^{-} + 2H^{+} + H_2O$$

$$NO_2^{-} + 0.5O_2 \rightarrow NO_3^{-}$$

Both reactions require an ample presence of oxygen and consume ALK forming nitrous acid. The first reaction is generally limiting due to the high energy demand and consequent slower bacterial growth (Tchobanoglous et al., 2003). Autotrophic bacteria involved in nitrification require a source of carbon to promote biomass growth. Temperature likewise represents an additional limiting factor and should not exceed 45°C.

Denitrification

Denitrification is an anoxic process involving the use of nitrates and nitrites to oxidize organic matter and produce energy with a yield slightly lower than aerobic respiration. The absence of oxygen and the presence of a source of biodegradable organic compounds are fundamental requirements for the process. In aerated landfills, denitrification may occur throughout all zones not directly reached by the airflow or in specific areas of waste mass specifically not aerated to guarantee anoxic conditions.

Denitrification is a heterotrophic process provided by facultative aerobes that, in the absence of oxygen, use nitrates or nitrites as electron acceptor.

$$NO_3^- + 2e^- + 2H^+ \rightarrow NO_2^- + H_2O$$

$$NO_2^- + e^- + 2H^+ \rightarrow NO + H_2O$$

$$2NO + 2e^- + 2H^+ \rightarrow N_2O + H_2O$$

$$N_2O + 2e^- + 2H^+ \rightarrow N_{2(g)} + H_2O$$

This process produces free nitrogen gas that escapes from the system. However, NO_2 and N_2O are intermediate products detectable in very low or negligible quantities.

Autographic denitrification is generally of lower relevance than the heterotrophic form, although it contributes consistently to nitrate depletion and is not limited by the requirement of readily degradable carbon substrates. To produce sulfates from HS^-, *Thiobacillus denitrificans* uses inorganic sulfur to reduce nitrates (Berge et al., 2013):

$$2NO_3^- + 1.25HS^- + 0.75H^+ \rightarrow N_{2(g)} + 1.25SO_4^{2-} + H_2O$$

This mechanism can only take place in the presence of inorganic sulfur. However, in old landfills with a low carbon content, autotrophic denitrification is favored over the heterotrophic form and may be responsible for up to 15–55% of the total denitrification process (Onay and Pohland, 2001).

ANAMMOX

Anaerobic ammonium oxidation (Anammox) is the process by which ammonia nitrogen is oxidized also under anaerobic conditions by groups of bacteria such as Planctomycetales (Berge et al., 2013):

$$NH_4^+ + 1.26NO_2^- + 0.085CO_2 + 0.02H^+ \rightarrow N_{2(g)} + 0.017H^+ + 0.24NO_3^- + 1.95H_2O$$

This process is highly complex and sensitive to environmental conditions and enters into competition with denitrifiers for nitrite conversion. Moreover, the Anammox bacterial growth rate is very low, thus requiring high retention times and a highly stable system to ensure survival. Generally, the contribution of Anammox processes to nitrogen reduction is difficult to be estimated in full-scale landfills.

Long-term nitrogen emissions

Typical highly concentrated nitrogen emissions observed in the long term are dependent on whether the landfill is aerated or remains anaerobic (Townsend et al., 2015; Ritzkowski et al., 2016). However, the majority of nitrogen initially present (60%–85%) will remain stored in the landfill body, unable to be ammonified, bound to complex nondegradable polymers and organic matter (Manfredi and Christensen, 2009).

In anaerobic landfills, during ammonification ammonia is produced in high quantities. In addition, in MSW landfills NH4 emissions due to stripping are generally negligible. As a consequence of this, ammonium ion concentrations may increase to values even higher than 3000 mgN/L, constituting one of the main problems in long-term landfill management. The progressive leaching of ammonia ions will reduce leachate concentrations, but the reaching of environmentally acceptable levels will require decades or even longer under these conditions (Ehrig and Kruempelbeck, 2013).

In aerated landfills, nitrification–denitrification processes result in a decrease of ammonia leachate concentrations allowing environmentally safe conditions to be achieved over a much shorter period of time (Ritzkowski et al., 2006) (Fig. 3.1.9).

CONCLUSIONS

To achieve an effective planning and control of landfill processes, together with minimization of emissions and cost control, it is essential to first ascertain what kind of biological processes take place in the landfill body. Due to the wide heterogeneity of waste composition and the milieu in the landfill ranging from aerobic to anaerobic conditions, the processes involved are highly complex. Furthermore, degradation causes the waste quality to change over time. To facilitate the biological processes, water content and flux are of paramount importance. The dissolution of salts into leachate affects the pH and may result in the onset of inhibition processes. Indeed, as ammonia is one of the long-lasting compounds in leachate at relatively high concentrations, knowledge of the nitrogen load is of particular importance.

References

Anonymous, 2017. http://www.esf.edu/efb/schulz/Limnology/redox.html.

Andreas, L., 2013. Long-term behaviour of municipal landfills: emissions from old sites in the Eastern part of Germany. In: Cossu, R., Van der Sloot, H. (Eds.), Sustainable Landfilling. CISA Publisher, Padova. ISBN:978-88-6265-005-2.

Andreottola, G., Cannas, P., 1992. Chemical and biological characterization of landfill leachate. In: Christensen, T.H., Cossu, R., Stegmann, R. (Eds.), Landfilling of Waste: Leachate. CRC Press. ISBN:9780419161400.

Berge, N.D., Reinhart, D.R., Townsend, T.G., 2013. The fate of nitrogen in MSW landfills. In: Cossu, R., Van der Sloot, H.A. (Eds.), Sustainable Landfilling. CISA Publisher, pp. 189–212. ISBN:978-88-6265-005-2.

Brandstätter, C., Laner, D., Fellner, J., 2015. Carbon pools and flows during lab-scale degradation of old landfilled waste under different oxygen and water regimes. Waste Management 40, 100–111.

Buswell, M.G., Muller, H.F., 1952. Mechanism of methane fermentations. Industrial and Engineering Chemistry 44, 550.

Christensen, T.H., Kjeldsen, P., 1989. Sanitary Landfilling: Process, Technology and Environmental Impact. Academic Press, Harcourt Brace Jovanovich Publishers, pp. 29–42. ISBN:978-0-12-174255-3.

Chugh, S., Clarke, W., Pullammanappallil, P., Rudolph, V., 1998. Effect of recirculated leachate volume on MSW degradation. Waste Management 564–573.

Cossu, R., Laraia, R., Adani, F., Raga, R., 2001. Test methods for the characterization of biological stability of pretreated municipal solid waste in compliance with EU directives. In: Proceedings Sardinia 2001, Eight International Waste Management and Landfill Symposium S. Margherita di Pula, Cagliari, vol. I, pp. 546–554.

Cossu, R., Raga, R., Rossetti, D., 2003. The PAF model: an integrated approach for landfill sustainability. Waste Management 23, 37–44.

Cossu, R., Raga, R., 2008. Test methods for assessing the biological stability of biodegradable waste. Waste Management 28, 381–388.

Cossu, R., Lai, T., Sandon, A., 2012. Standardization of BOD5/COD ratio as a biological stability index for MSW. Waste Management Journal 32, 1503–1508.

Cossu, R., Morello, L., Raga, R., Cerminara, G., 2016. Biogas production enhancement using semi-aerobic pre-aeration in a hybrid bioreactor landfill. Waste Management 55, 83–92.

Cossu, R., Fantinato, G., Pivato, A., Sandon, L., 2017. Further steps in the standardization of BOD5/COD ratio as a biological stability index for MSW. Waste Management. https://doi.org/10.1016/j.wasman.2017.06.035.

Daverio, E., Ligthart, J., Nieman, H., Spanjers, H., 2005. Impiego di misure calorimetriche per lo studio dell'attività batterica nei processi di digestione anaerobica. Ingegneria Ambientale XXXIV (3/4), 160–168.

De la Cruz, F.B., Chanton, J.P., Barlaz, M.A., 2013. Measurement of carbon storage in landfills from the biogenic carbon content of excavated waste samples. Waste Management 33, 2001–2005.

Ehrig, H.J., 1984. Laboratory scale tests for anaerobic degradation of municipal solid waste. In: Proceedings International Solid Wastes and Public Cleaning Association Congress, Philadelphia, 15–20 September 1984.

Ehrig, H.J., Kruempelbeck, I., 2013. Long-term behaviour of landfills: emissions from old sites in the western part of Germany. In: Cossu, R., Van der Sloot, H.A. (Eds.), Sustainable Landfilling. CISA Publisher, pp. 139–152. ISBN:978-88-6265-005-2.

Esposito, G., Frunzo, L., Liotta, F., Panico, A., Pirozzi, F., 2012. Bio-methane potential tests to measure the biogas production from the digestion and co-digestion of complex organic substrates. The Open Environmental Engineering Journal 5, 1–8.

Farquhar, G.J., Rovers, F.A., 1973. Gas production during refuse decomposition. Water, Air and Soil Pollution 2, 483–495.

Filipkowska, U., 2007. Effect of recirculation method on quality of landfill leachate and effectiveness of biogas production. Polish Journal of Environmental Study 17, 199–207.

Heimovaara, T.J., Cossu, R., Van der Sloot, H.A., 2013. State of the art and perspectives for sustainable landfills. In: Cossu, R., Van der Sloot, H.A. (Eds.), Sustainable Landfilling. CISA Publisher, pp. 19–40. ISBN:978-88-6265-005-2.

Heyer, K.U., Hupe, K., Stegmann, R., 2003. Criteria for the completion of landfill aftercare. In: Sardinia Proceedings 2003. CISA, Cagliari, Italy.

Hupe, K., Heyer, K.U., Stegmann, R., 2013. Water infiltration for enhanced in situ stabilization. In: Cossu, R., Van der Sloot, H.A. (Eds.), Sustainable Landfilling. CISA Publisher, pp. 431–443. ISBN:978-88-6265-005-2.

Komilis, D., Kletas, C., 2012. Static respiration indices to investigate compost stability: effect of sample weight and temperature and comparison with dynamic respiration indices. Bioresource Technology 121, 467–470.

Laner, D., Crest, M., Scharff, H., Morris, M.W.F., Barlaz, M.A., 2012. A review of approaches for the long term management of municipal solid waste landfills. Waste Management 32, 498–512.

Lay, J.J., Li, Y.Y., Noike, T., 1998. Developments of bacterial population and methanogenic activity in a laboratory-scale landfill bioreactor. Water Resource 32, 3673–3679.

Lornage, R., Redon, E., Lagier, T., 2013. Carbon and nitrogen mass balance for different landfill concepts, Lysimeter tests. In: Cossu, R., Van der Sloot, H.A. (Eds.), Sustainable Landfilling. CISA Publisher, pp. 95–105. ISBN:978-88-6265-005-2.

Manfredi, S., Christensen, T.H., 2009. Environmental assessment of solid waste landfilling technologies by means of LCA-modeling. Waste Management 29, 32–43.

McCarty, P.L., 1964. Anerobic Waste Treatment Fundamentals, Part One, Two and Three, Public Works.

Morello, L., Cossu, R., Raga, R., Pivato, A., Lavagnolo, M.C., 2016. Recirculation of reverse osmosis concentrate in lab-scale anaerobic and aerobic landfill simulation reactors. Waste Management 56, 262–270.

Onay, T.T., Pohland, F.G., 2001. Nitrogen and Sulfate attenuation in simulated landfill bioreactors. Wastewater Science and Technology 44, 367–372.

Pivnenko, K., Astrup, T.F., 2016. The challenge of chemicals in material lifecycles. Waste Management 56, 1–2.

Ritzkowski, M., Heyer, K.U., Stegmann, R., 2006. Fundamental processes and implications during in situ aeration of old landfills. Waste Management 26, 356–372.

Ritzkowski, M., Stegmann, R., 2007. Mechanism affecting the leachate quality in the course of landfill in situ aeration. In: Stegmann, R., Ritzkowski, M. (Eds.), Landfill Aeration. CISA Publisher, pp. 74–84. ISBN:978-88-6265-002-1.

Ritzkowski, M., Stegmann, R., 2013. Landfill aeration within the scope of post-closure care and its completion. Waste Management 33, 2074–2082.

Ritzkowski, M., Walker, B., Kuchta, K., Raga, R., Stegmann, R., 2016. Aeration of the Teuftal landfill: field scale concept and lab scale simulation. Waste Management 55, 99–107.

Sandip, M., Kanchan, K., Ashok, B., 2012. Enhancement of methane production and bio-stabilisation of municipal solid waste in anaerobic biorector landfill. Bioresource Technology 110, 10–17.

Sekman, E., Top, S., Varank, G., Bilgili, M.S., 2011. Pilot-scale investigation of aeration rate effect on leachate characteristics in landfills. Fresenius Environmental Bulletin 20.

Senior, E., 1990. Microbiology of Landfill Sites. CRC Press, Boca Raton, Florida.

Sleat, R., Harries, C., Viney, I., Rees, J.F., 1989. Activities and distribution of key microbial groups in landfills. In: Christensen, T., Cossu, R., Stegmann, R. (Eds.), Sanitary Landfilling: Processes, Technology and Environmental Impact. Academic Press, London, pp. 29–42. ISBN:978-0-12-174255-3.

Stahr, K., Kandeler, E., Herrmann, L., Streck, T., 2016. Bodenkunde und Standortlehre, 3. Auflage. Verlag Eugen Ulmer, Stuttgart.

Stevenson, F.J., 1982. Humus Chemistry Genesis, Composition, Reactions. John Wiley publisher, New York. ISBN 10:0471092991, ISBN 13:9780471092995.

Tchobanoglous, G., Theisen, H., Vigil, S., 1993. Integrated solid waste management: engineering principles and management issues. McGraw-Hill, New York, USA.

Tchobanoglous, G., Burton, F.L., Stensel, H.D., Metcalf, Eddy, 2003. Wastewater Engineering: Treatment and reuse. McGraw-Hill Education, 2003.

Townsend, T.G., Powell, J., Jain, P., Xu, Q., Tolaymat, T., Reinhart, D., 2015. Sustainable practices for landfill design and operation. In: Series: Waste Management Principles and Practice. Springer. ISBN:978-1-4939-2662-6.

Valencia, R., Van der Zon, W., Woelders, H., Lubberding, H.J., Gijzen, H.J., 2009. The effect of hydraulic conditions on waste stabilisation in bioreactor landfill simulators. Bioresource Tecnology Journal 100, 1754–1761.

Zehnder, A.B.J., Ingvorsen, K., Marti, T., 1982. Microbiology of methanogen bacteria in anaerobic digestion. In: Proceedings of the 2nd International Symposium of Anaerobic Digestion, Travemunde, 6–11 September 1981. Elsevier Biomedical Press, BV, Amsterdam, The Netherlands, pp. 45–68.

3.2

PHYSICAL/CHEMICAL REACTIONS IN LANDFILLS

Debra Reinhart and Rainer Stegmann

INTRODUCTION

This chapter focuses on the many physical/chemical processes that impact the degradation of waste, the fate and transport of waste contaminants, and the quality and quantity of leachate generated. While the chapter describes these abiotic phenomena, it is difficult to fully separate them from biotic reactions. Most of the reactions described are enabled by liquid movement through the landfill, which dissolves and then transports leached material. Dissolution of materials is the primary path in a landfill for removal of nondegradable material (e.g., metals, halides, ammonia, and dissolved recalcitrant organic matter). The rate of liquid movement is controlled by landfill characteristics such as waste heterogeneity, gas movement, extent of settling, and cover type. The hydraulic conductivity of compacted waste is relatively low, and anisotropic conditions lead to preferential movement in the horizontal direction. Channeling will occur due to waste heterogeneity; as a result some of the waste will be at or near saturation, and other regions will be fairly dry, which may limit reactions. Chapter 3.3 provides more details regarding moisture movement through landfills.

The fate and transport of contaminants (e.g., metals, trace organics) found in landfills is dependent on both specific compound properties and the surrounding waste environment. Individual compound properties (e.g., polarity, hydrophobicity) affect dissolution rates and extent. Specific conditions within the landfill also contribute such as pH, temperature, leachate quality (i.e. salt content, organic content), oxidation reduction potential, chemical/physical reactions, and physical processes (e.g. diffusion, complexation). Other factors that influence leaching of compounds include waste particle size (diffusion distance), porosity, water content, leachate flux, and temperature. Leaching tests have been used to predict the leachate quality or leaching potentials but due to the strong influence of the biological processes these have not been a good predictor for municipal solid waste (MSW) landfills. Leaching tests may help to estimate the leaching characteristic and potential of inert waste but not necessarily the leachate quality. This is due to the dependency on the liquid-to-solid ratio for the test. More realistic data can be derived from column and lysimeter tests.

METALS

The fate and transport of metals and their concentrations in leachate will be affected by liquid movement, pH, the presence of organic complexing agents (e.g., humic substances), and the presence of inorganic complexing/precipitating agents (e.g., carbonates, hydroxides, and chlorides). Studies suggest that few of the metals in leachate are present as free metal ions, but rather are primarily associated with colloidal fractions, primarily organic but also inorganic in nature (Baun and Christensen, 2004). The concentration of heavy metals in leachate does not follow any of the significant trends such as chemical oxygen demand, biochemical oxygen demand, major ions, or nutrients (Lu et al., 1985), although the concentrations in young leachates tend to be slightly higher (e.g., iron, magnesium, and zinc) than in older leachate since the lower pH during the acid formation phase enhances metal dissolution (Kjeldsen et al., 2002), as shown in Table 3.2.1. The most common metals found in leachates include iron, cadmium, copper, zinc, and nickel (Erses and Onay, 2003).

Dissolved organic matter (DOM) can play a significant role in controlling the total and free metal concentration. DOM in leachate is primarily made up of humic substances (e.g., fulvic and humic acids) and is the material responsible for the mobilization of metals in landfills. A study by Klein and Niessner (1998) found that the largest fraction of heavy metals in leachate was associated with colloidal matter, which was dominated by humic substances. These compounds contain carboxylic and phenolic functional groups (5—10 meq/g), which are primary contributors to the binding capacity of the DOM in leachate (Tipping, 1993). Metals in leachate form stable complexes with high molecular weight organic components of older leachate (Calace et al., 2001; Christensen et al., 1996). Metals will also complex with other ligands such as sulfate, carbonate, and hydroxide.

The immobilization of metals in landfills is attributed to precipitation and sorption mechanisms (e.g., adsorption, surface complexation, surface precipitation, ion exchange, and absorption). Metals complex with ligands present in leachate such as sulfides, sulfate, carbonate, hydroxide, chloride, and phosphate (Christensen et al., 1992). Carbonates and sulfides will both form precipitates with cadmium, zinc, nickel, copper, and lead. The solubility of metal carbonates is higher in comparison to metal sulfides (Christensen et al., 1994, 2001); therefore sulfide precipitation will be favored in terms of metal attenuation (Reinhart and Grosh, 1998). A study by Mårtensson et al. (1999) found that landfills do not contain enough sulfur to bind with the heavy metals at the concentrations they are normally present. Therefore free metals will precipitate with phosphates and hydroxides if conditions are favorable (Christensen et al., 2001). The pH of methanogenic leachates is typically at or above neutral, which promotes precipitation with hydroxide (Reinhart and Grosh, 1998).

Calcium carbonate ($CaCO_{3(s)}$) precipitation commonly occurs in landfills due to the high concentration of calcium and alkalinity. Fig. 3.2.1 outlines the formation of calcium carbonate in leachate. As carbon dioxide (CO_2) is produced during the methanogenic phase, a fraction of this gas will dissolve following Henry's law (Eq. 1). Once CO_2 is dissolved in the leachate, it will become part of the carbonate system, which ultimately facilitates the opportunity for $CaCO_{3(s)}$ to form. Magnesium can also precipitate with carbonate. The distribution of the species outlined in Fig. 3.2.1 is a

Table 3.2.1 Leachate composition differences between acid and methanogenic phase[a,3]

Parameter	Acidic Phase		Methanogenic Phase		Average
	Average	Range	Average	Range	
pH	6.1	4.5–7.5	8	7.5–9	
BOD_5	13,000	4000–40,000	180	20–550	
COD	22,000	6000–60,000	3000	500–4500	
BOD_5/COD	0.58		0.06		
Sulfate	500	70–1750	80	10–420	
Calcium	1200	10–2500	60	20–600	
Magnesium	470	50–1150	180	40–350	
Iron	780	20–2100	15	3–280	
Manganese	25	0.3–65	0.7	0.03–45	
Zinc	5	0.1–120	0.6	0.03–4	
Chloride					2120
Potassium					1085
Sodium					1340
Total phosphorus					6.0
Cadmium					0.005
Chromium					0.28
Cobalt					0.05
Copper					0.065
Lead					0.09
Nickel					0.17
Ammonia N					740

[a]mg/L except pH and BOD_5/COD ratio.

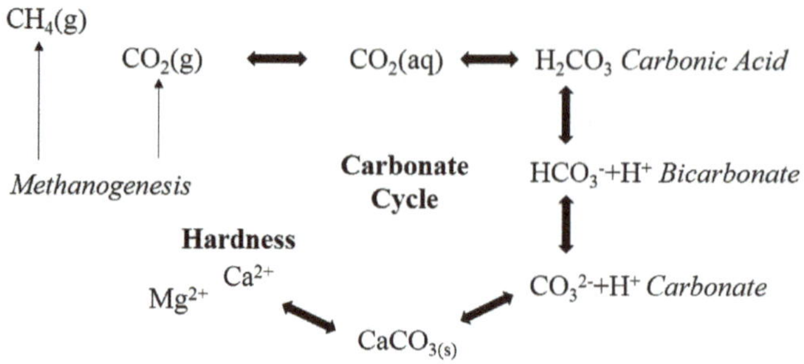

Figure 3.2.1 Carbonate cycle in a landfill: formation of calcium carbonate.

function of the overall pH of the system; bicarbonate is the dominant species at a neutral pH (Andersen, 2002).

Carbonates will form following equilibrium Eqs. (1–3). Dissociation constants shown are at 25°C and 1 bar. The formation of $CaCO_{3(s)}$ is discussed in more detail elsewhere in this text with respect to the clogging of leachate collection systems (see Chapter 8.1)

$$CO_{2(g)} + H_2O \leftrightarrow H_2CO_3 \; K_H = 10^{-1.46} \tag{1}$$

$$H_2CO_3 \leftrightarrow H^+ + HCO_3^- \; K_{a1} = 10^{-6.35} \tag{2}$$

$$HCO_3^- \leftrightarrow H^+ + CO_3^{2-} \; K_{a2} = 10^{-10.33} \tag{3}$$

TRACE ORGANIC CONTAMINANTS

A variety of trace organic contaminants are present in MSW and have been subsequently found in landfill leachate and/or gas. These contaminants, referred to as xenobiotic organic compounds (XOCs), originate from household and nonhazardous industrial wastes (Deipser and Stegmann, 1994; Slack et al., 2005; Weber et al., 2011) and are comprised of several classes of organics, including aromatic hydrocarbons, phenols, chlorinated aliphatic compounds, plasticizers, pesticides, pharmaceuticals, fire retardants, personal care products, and endocrine disrupting compounds. Trace concentrations of XOCs may also result from the biological and/or chemical transformation of landfilled waste (Allen et al., 1997; Slack et al., 2005). Specific types and concentrations of XOCs found in leachate vary among landfills and depend not only on waste composition but also on landfill operation and management practices (Slack et al., 2005). More than 1000 XOCs have been identified in leachate and/or leachate-contaminated groundwater (Christensen et al., 2001; Slack et al., 2005). Although when combined these XOCs comprise less than a few percent of total organic carbon in leachate, they are of concern because of their potential to harm human health (Christensen et al., 2001; Slack et al., 2005).

Table 3.2.2 contains an abbreviated list of representative XOCs that have been detected in landfill leachate, as well as their specific uses. More complete and detailed information associated with XOCs detected in leachate and in leachate plumes found downgradient of landfills can be found in Christensen et al. (2001), Kjeldsen et al. (2002), and Slack et al. (2005). Following release of these contaminants from their parent product, they may dissolve in leachate, sorb to the waste, and/or volatilize and

Table 3.2.2 An abbreviated list of xenobiotic organic compounds (XOCs) found in landfill leachate and their uses

XOC Category	Example Contaminant	Example Use
Halogenated hydrocarbons	1,2-Dichlorobenzene Trichloroethene	Pesticide, deodorizer, solvent Solvent, degreaser, paint removers, adhesive, cleaner
Aromatic hydrocarbons	Toluene Xylene	Solvent in paint, paint thinner, varnish Plastics, solvent in paints, nail varnish
Phenols	Phenol Bisphenol A	Disinfectant, drugs Manufacture of epoxy resins, coating on food cans
Alkylphenol	Nonylphenol Nonylphenol ethoxylate	Surfactant (also an endocrine disrupting compound) Detergents, wetting/dispersing agents, emulsifier
Pesticides	Atrazine Bentazon	Herbicide Herbicide
Phthalates	Dimethylphthalate Butylbenzyl phthalate	Plastics Plastics
Pharmaceuticals	Ibuprofen Clofibric acid	Antiinflammatory/analgesic-OTC Plant growth regulation and drug intermediate
Aliphatics	n-Tricosane	Plastics
Alcohols and ethers	Diphenylethers General alcohols	Flame retardant, plasticizer Solvents
Carboxylic acids	Benzoic acid Palmitic acid	Food preservative, perfumes, creams/drugs Food, cosmetics, and pharmaceuticals
Pyridines	Nicotine Cotinine	Insecticide, tobacco Formed from oxidation of nicotine
Terpenoids	Menthol Limonene	Flavors and fragrance Flavoring

Modified from Slack et al. (2005)

subsequently be emitted in landfill gas. The fate of these contaminants depends on waste characteristics, leachate composition, and individual contaminant characteristics.

Sorption

Attenuation of XOCs in landfills may occur via sorption. Sorption includes both adsorption (attachment to solid surfaces) and absorption (penetration into solids) processes. If sorption is irreversible, contaminants may remain immobilized within the waste mass. However, if such processes are reversible, contaminant transport will be retarded and may ultimately result in the slow release of contaminants over time (Kjeldsen et al., 2002; Reinhart et al., 1991a). The extent of organic contaminant sorption in landfills is dictated by contaminant, waste, and leachate properties, as well as landfill characteristics, such as waste density and leachate flow. Contaminant properties that greatly influence sorption processes include solubility, hydrophobicity, and polar/ionic characteristics. Generally, contaminants have an affinity for solid surfaces that share similar characteristics; charged and polar contaminants tend to sorb to charged and polar solid surfaces, while nonpolar contaminants tend to sorb to nonpolar components (e.g., organic carbon) on the solid surface (Sellers, 1999).

Large nonpolar contaminants are usually hydrophobic. Hydrophobic contaminants have low water solubility and thus have a high affinity for solids (Sellers, 1999). Contaminant hydrophobicity is often described by the octanol–water partition coefficient (K_{ow}), which quantifies the distribution of a contaminant between octanol and water under equilibrium conditions. Nonpolar organics partition to the organic carbon content of a solid (Christensen et al., 2001; Sellers, 1999) often described via a carbon-normalized partition coefficient (K_{oc}) according to the following equation:

$$K_{oc} = \frac{C_s/C_l}{f_{oc}} \tag{4}$$

where C_s is the concentration of the contaminant on the solids (mg/g solids), C_l is the concentration of the contaminant in the liquid phase (mg/L), and f_{oc} is the fraction of organic carbon in solids.

If a contaminant is polar or ionizable, sorptive behavior is dependent on the contaminant acid dissociation constant (pKa), log K_{ow}, and system pH. If leachate pH is greater than the contaminant pKa and the log K_{ow} is relatively low, sorption of these contaminants will likely be dictated by electrostatic forces and thus the surface charge of the waste components. However, if leachate pH is less than the compound pKa or the log K_{ow} is relatively large, sorption will likely be dictated by waste component hydrophobicity. Exchange reactions (e.g., replacement of positively or negatively charged surface components with positively or negatively charged organic contaminants) may also play an important role on the sorption of polar/ionic contaminants (Schwarzenbach et al., 2005). There has been little work investigating the sorptive behavior of ionic or polar contaminants on solid waste materials. Lorphensri et al. (2007) investigated the sorption of three different pharmaceuticals to different solid media and concluded that sorption of ionizable pharmaceuticals is strongly dependent on system pH and contaminant pKa and log K_{ow}. Lorphensri et al. (2007) also concluded that if the contaminant is uncharged at environmentally relevant pH values, sorption is strongly dependent on contaminant solubility and hydrophobicity. Because leachate pH values change with time, the sorptive behavior of polar or ionizable contaminants is expected to change with time in MSW landfills.

Waste properties also influence sorption of XOCs. Waste hydrophobicity is an important characteristic that generally increases with waste age; as waste ages, the more polar organic compounds preferentially degrade, leaving the more hydrophobic nonpolar compounds behind (Chen et al., 2004). Chen et al. (2004), Saquing et al. (2010), and Fang et al. (2014) measured the sorption of different hydrophobic organic contaminants to fresh and/or degraded components of MSWs. Chang et al. (2004) reported an increase in toluene sorption capacity as newspaper degraded. Wu et al. (2001) also report that the affinity for hydrophobic contaminants on solid-phase organics increases as waste degrades. Wu et al. (2001) and Saquing et al. (2010) indicated that the presence of plastics significantly influences sorption of hydrophobic organic contaminants. Plastics may serve as sinks for these organic contaminants because of their high affinity for these compounds and their subsequent slow desorption from glassy plastics (Saquing et al., 2010). Other waste properties, such as available surface area, landfill temperatures, and leachate flow, may also influence contaminant sorption. Temperature influences individual contaminant properties and thus sorptive behavior, while greater levels of leachate flow within a landfill provide additional opportunities for sorption to occur.

Using the MOCLA model to describe the distribution and fate of specific classes of XOCs, Kjeldsen and Christensen et al. (2001) predicted that many chemicals will be strongly associated with the waste, especially when the organic content of the waste is high. When modeling a typical landfill in which the fraction of solid-phase organic carbon is 0.2, 95% of most chemicals were found to sorb to the waste; however, when the fraction of organic carbon decreased to 0.02, sorption of these contaminants decreased and more appeared dissolved in the leachate stream or in the gas phase (Kjeldsen and Christensen, 2001).

Volatilization

XOCs may also volatilize. Volatilization is the transfer of liquid- or solid-phase organic contaminants to the gas phase. Trace levels of XOCs have been detected in landfill gas from field and laboratory-scale studies, such as benzene and derivatives, chlorinated hydrocarbons and aromatics, naphthalene, and carbon tetrachloride (Allen et al., 1997; Deipser and Stegmann, 1994; Slack et al., 2005). In landfills, volatilization of dissolved, free-phase, and solid-bound organic contaminants may occur, as well as contaminants adsorbed to the waste materials and/or landfilled contaminated solids (Tchobanoglous et al., 1993). Volatilization rates and extent depend on waste and contaminant properties, environmental conditions within the landfill (e.g., moisture content, temperature), leachate composition, and leachate and gas production and flow rates (Tchobanoglous et al., 1993).

Specific contaminant properties that influence volatilization processes include vapor pressure, water solubility, diffusion coefficient, and boiling point. Vapor pressure (V_p) is the measure of the tendency of a contaminant to transfer from the liquid to gas phase, with greater volatilization with higher vapor pressures. Generally, contaminants are considered volatile if the V_p is greater than 0.1 mm Hg at 20°C (Tchobanoglous et al., 1993). As contaminant water (or leachate) solubility increases, the contaminant is more hydrophilic and less likely to volatilize. The relationship between the gas and liquid phases at equilibrium is described by Henry's law:

$$C_g = C_l K_H \quad (5)$$

where C_g is the concentration of the contaminant in the gas phase (mg/L), C_l is the contaminant concentration in the liquid phase (mg/L), and K_H is the dimensionless Henry's law constant (L_{liquid}/L_{gas}). K_H can be obtained from the literature or calculated based on contaminant solubility and vapor pressure. The higher the K_H, the greater the contaminant gas phase concentration at equilibrium, and the more volatile it is. Table 3.2.3 provides guidance on specific K_H values and the tendency for volatilization (Tchobanoglous et al., 1993).

Specific in situ landfill conditions and waste properties also influence organic contaminant volatilization. Landfill conditions, such as temperature and leachate composition (e.g., pH, contaminant concentration), influence contaminant properties, ultimately influencing contaminant volatilization rates. As landfill temperature increases, contaminant solubility generally decreases and vapor pressures increase, leading to a greater rate of volatilization. The extent and rates of volatilization are also influenced by waste properties and gas extraction rates (Allen et al., 1997).

Kjeldsen and Christensen (2001) reported that for many XOCs volatilization is not as prevalent as contaminant sorption or dissolution in the leachate. Through model simulations, Kjeldsen and Christensen (2001) showed that contaminant volatilization increased as contaminant sorption, and solid-phase organic carbon content decreased. Volatilization of some alkanes (e.g., hexane and heptane), chlorinated aliphatic hydrocarbons (e.g., vinyl chloride), and freons (e.g., Freon-12, Freon-113) were significantly greater than other organic contaminants evaluated (e.g., monoaromatic hydrocarbons, chlorinated aromatic hydrocarbons, and polynuclear hydrocarbons).

Table 3.2.3 Relationship between henry's law constant and organic contaminant volatilization

Henry's Law Constant, k_H (m^3-atm/mol)[a]	Degree of Volatilization
$<10^{-7}$	Essentially nonvolatile
$10^{-7} < k_H < 10^{-5}$	Slow volatilization
$10^{-5} < k_H < 10^{-3}$	Volatilization is significant
$>10^{-3}$	Highly volatile

[a]Conversion from k_H to K_H is: $K_H = k_H/RT$, where R is the universal gas constant (m^3-atm/mol-K), T is the temperature (K), K_H is the dimensionless constant, and k_H is the constant with the units of m^3-atm/mol.
Modified from Tchobanoglous et al. (1993).

Reductive Dehalogenation of Solvents

Landfills tend to have low reduction–oxidation potential as a result of the consumption of any free or combined oxygen present during biodegradation of waste. Negative redox promotes abiotic degradation of halogenated compounds, primarily chlorinated solvents such as trichloroethene, perchloroethene, and trichloroethane that are codisposed with MSW. Dehalogenation occurs through reactions with zero-valent metals. Much of the literature describing reductive dehalogenation (Fu et al., 2014; Scheutz et al., 2011) addresses the reactions that occur during treatment of groundwater or wastewater. Although little has been written recently on reductive dehalogenation in landfills (Reinhart et al., 1991b), the presence of daughter products such as vinyl chloride, dichloroethene, and monochloroethane and the abundancy of zero-valent iron in waste suggest that this pathway exists. The process involves multiple steps: direct electron transfer from Fe^0 at the metal surface, catalyzed hydrogenolysis by H^+ or H_2, and reduction by Fe^{+2} species produced during Fe^0 corrosion (Noubactep et al., 2012). Other contaminants such as nitrate, arsenic, chromium (VI), phenol, and nitrobenzene are also susceptible to reduction by zero-valent metals. Dehalogenation and the reduction of other compounds can also occur biologically under anaerobic conditions.

ELEVATED TEMPERATURES IN LANDFILLS

Elevated temperatures have been observed in landfills, in both shallow and large, deep landfills, even under anaerobic conditions. In "hot landfills," temperatures may well exceed normal temperatures of 35–50°C, even exceeding the range that is tolerable for microbial activity (>70°C) (Fig. 3.2.2). The occurrence of elevated temperatures in the presence or absence of molecular oxygen is mediated by different mechanisms. Aerobic conditions may occur near the landfill surface, due to intrusion of air through slopes (e.g. low compacted waste exposed to wind) or drainage systems, changes in atmospheric pressure, by overpulling on landfill gas extraction wells, or by active aeration. In these cases,

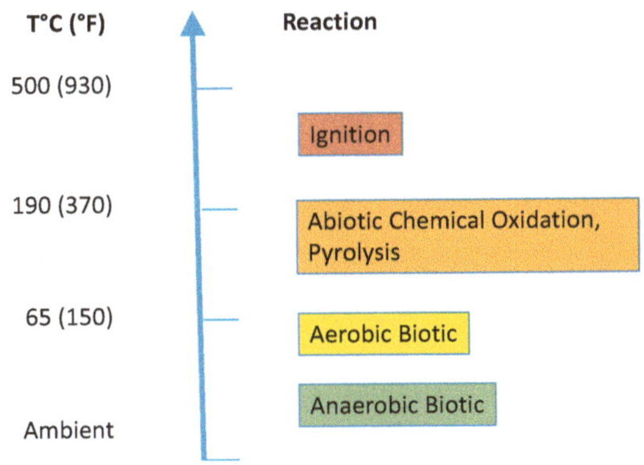

Figure 3.2.2 Temperature ranges and conditions in landfills.

elevated temperatures of up to 50–80°C may occur mainly due to the exothermic biological degradation of organic substances as described in composting processes (see also Chapter 3.1). Exothermic biotic reactions contribute to temperatures above ambient but are self-limited by the inhibition of microbes; therefore abiotic reactions appear to cause excessively high temperature conditions. In addition, exothermic chemical processes may occur such as shown in Eq. (6):

$$4FeS + 7O_2 \rightarrow 2\ Fe_2O_3 + 4SO_2 \tag{6}$$

It is unlikely that under normal landfill conditions oxygen will be present deep within a landfill (Moqbel et al., 2010). The reasons for these elevated temperatures in deep landfills, especially under anaerobic conditions, are not well understood. At high temperatures (>50–60°C), self-ignition by chemical/physical processes may occur under specific circumstances that are not completely understood, which may promote pyrolytic processes. It is further possible that the high pressures experienced at depth, due to waste overburden and/or pressure in the fluid phase, may promote abiotic exothermic processes, which may initiate pyrolytic reactions. While pyrolysis is typically endothermic, some slow pyrolytic reactions are exothermic (Moqbel et al., 2010). High salt concentrations of typical leachate may also promote chemical self-heating of waste although this needs to be verified under full-scale conditions. These reactions occur at higher temperatures, as seen in Fig. 3.2.2. Some additional examples of these exothermic chemical reactions are shown below (Moqbel, 2009):

Rust and hydrogen sulfide, oxidation of FeS:

$$2FeO(OH) + 3H_2S = 2FeS + S + 4H_20 \tag{7}$$

Scrap iron and carbonates:

$$2H_2O + 5CO_2 + 4Fe \rightarrow 4FeCO_3 + CH_4 \tag{8}$$

Aluminum dross:

$$2H_2O + 5CO_2 + Al_4C_3 + 12H_2O = 4Al(OH)_3 + 3CH_4 \tag{9}$$

$$AlN + 3H_2O = Al(OH)_3 + NH_3 \tag{10}$$

$$AlP + 3H_2O = Al(OH)_3 + PH_3\ 4Fe \rightarrow 4FeCO_3 + CH_4 \tag{11}$$

Requisite conditions for elevated temperatures include the availability of a fuel (waste), moisture, and an energy input; the latter of which can be provided by biotic oxidation in the presence of oxygen, chemical reactions, or hot loads. Fig. 3.2.3 provides an overview of the typical sequence of reactions that may lead to elevated landfill temperatures. The steps toward combustion include an increase in the temperature of the waste mass due to biotic degradation, pyrolytic decomposition of waste materials (e.g., paper), the escape of volatile compounds from the waste surface, diffusion of the pyrolyzed compounds from the solid surface into the gas phase, and gaseous and heterogeneous reactions at the waste surface. The sources of heat, then, include chemical oxidation or decomposition into simpler molecules (biotic or abiotic); aerobic or anaerobic biotic degradation of waste; oxygen adsorption, chemical

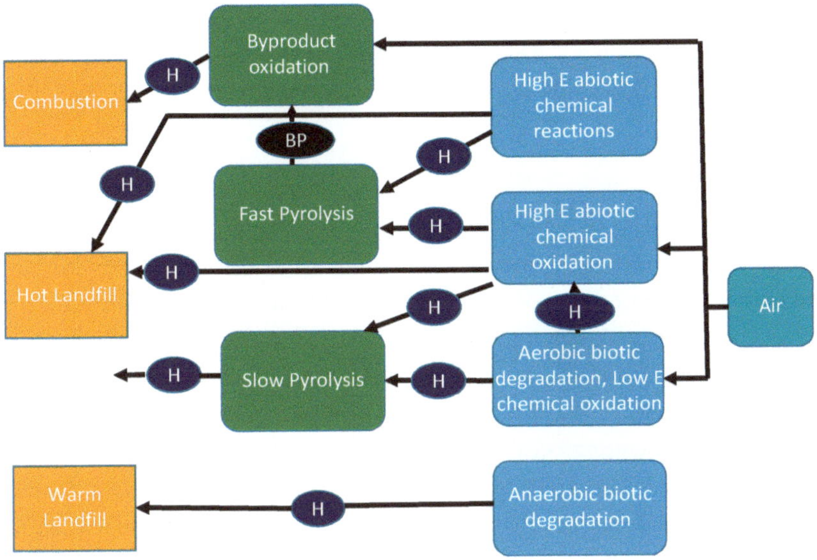

Figure 3.2.3 Contributing factors in elevated temperature landfills.

reaction, oxidative degeneration of fuel (slow pyrolysis), or oxidation of pyrolytic byproducts; and condensation of evaporated water.

Elevated temperatures are driven by the balance between heat generation and dissipation shown in Eq. (12). Dissipation of heat will prevent temperature excursions and occurs as a result of heat conduction (dry waste has a low thermal conductivity, water has a high conductivity), heat convection from waste to leachate, and water evaporation (El-Fadel, 1999). Heat is generated by both abiotic and biotic reactions:

$$\text{Energy storage} = \text{Heat conductance} + \text{Heat convection} + \text{Heat generation} - \text{Water evaporation} - \text{Gas emissions} - \text{Leachate emissions} \tag{12}$$

Elevated temperatures (>60–70°C) can have negative consequences including damage to gas collection piping and liner systems (e.g. drying out of clay barriers, softening of plastic material), odors when gas collection systems fail, inhibition of methanogenesis that leads to an accumulation of hydrogen, increased leachate strength, the potential for slope failure (if the high temperature reduces the functionality of the gas collection system, which in turn may lead to pressure accumulations), and rapid settlement. These conditions are detected by changes in leachate and gas composition, temperature, and volume. Detecting and mitigating elevated temperatures can cost tens to hundreds of millions of dollars to manage. Potential control methods for temperature control include reduction of gas extraction rates (only for cases of air intrusion), excavation and then covering hot spots with soil, introduction of liquid gases (usually N_2 or CO_2) (there are no large-scale successful examples of this), water injection in the hot spot areas, extraction of stored hot water at the landfill bottom, perimeter cut off trenches (UK Environment Agency., 2007), etc. Return of temperatures to acceptable levels can take from days to years. When temperatures are back to normal, cost for repair of the technical installations may be incurred.

EMERGING CONTAMINANTS IN LANDFILL

There has been growing concern over emerging contaminants in landfills due to the increased utilization of new and unique compounds in consumer products. The main concern with these emerging contaminants is that their potential risk to human health and the environment is largely unknown. Emerging contaminants in landfills primarily include pharmaceuticals and personal care products (PPCPs), siloxanes, fluorinated compounds, and nanoparticles. The detection and quantification of specific contaminants is challenging because of the lack of analytical techniques to detect these compounds at low levels and further complicated by the complex waste, leachate, and gas matrices. Nevertheless, there has been some research, which allows the estimation of the concentration of these compounds in landfills and their fate in the waste/leachate/gas matrix.

Pharmaceuticals and Personal Care Products in Landfill Leachate

PPCPs encompass any product that is utilized by a consumer for either cosmetic- or health-related reasons or by agribusinesses to protect the health or to improve the growth of livestock. The specific PPCPs present in landfill leachate are challenging due to the low levels and vast number of parents and daughter compounds or degradation products that could be present. A study conducted by the US Geological Survey (USGS) analyzed untreated leachate from 19 landfills across the United States (Masoner et al., 2014). This study found that 129–202 pharmaceuticals (i.e., nonprescription and prescription), household, and industrial chemicals were present in the tested samples (Masoner et al., 2014). Areas of increased precipitation were also observed to have higher concentration of PPCPs in leachate. The common chemicals detected in leachate in this study along with other literature included cotinine, bisphenol A (Joseph et al., 2011; Pan et al., 2008), lidocaine, camphor, and *N,N*-diethyltoluamide (Barnes et al., 2004). The maximum concentrations measured of these select chemicals from the USGS study are summarized in Table 3.2.4.

Siloxanes are another class of emerging contaminants that are found in the waste stream and have negative implications on landfill gas management. Siloxanes are artificial organosilicon compounds comprised of silicon, oxygen, and alkane groups (Gold, 1987; Horii and Kannan, 2008). These compounds are used in the production of personal care products such as cosmetics, soaps, food additives, water repellent windshield coatings, and deodorant (Lu et al., 2011). These products will eventually enter the waste stream due to their relatively short useful life. Once these products enter the landfill, siloxanes are converted into volatile methyl siloxanes. Volatile methyl siloxanes are present in six different forms with physical properties as shown in Table 3.2.5. The typical concentration of total organic silicon compounds ranges between 2 and 81 mg/m^3 (Surita and Tansel, 2015).

Table 3.2.4 Maximum concentration of selected chemicals detected in untreated leachate

Maximum Concentration (parts per trillion)	Chemical
7,020,000	para-Cresol (plasticizer and flame retardant, antioxidant in oils, rubber, polymers, and wood preservative)
6,380,000	Bisphenol A (used in plastics, thermal paper, and epoxy resins)
705,000	Ibuprofen (analgesic, antipyretic)
254,000	DEET (insect repellent)
147,000	Lidocaine (local anesthetic, topical antiitch treatment)
97,200	Camphor (natural compound with medicinal uses and embalming)
51,200	Cotinine (transformation product of nicotine)
2590	Carbamazepine (anticonvulsant and mood stabilizer)
168	Estrone (natural estrogenic hormone)

Modified from Masoner et al. (2014).

The presence of volatile methyl siloxanes in landfill gas limits its utilization without treatment because of the potential for detrimental effects on waste-to-energy infrastructure. The less volatile siloxanes that are combusted during power generation will be oxidized to solid silica (SiO_2) or other silicates (Si_xO_y). Silicates produced in the process are insoluble, hard, and abrasive, potentially leading to damage of moving parts, clogging of static filters, and destruction of catalytic surfaces. Silicon dioxide may deposit within the engine/combustion turbines (Fig. 3.2.4) or in the final exhaust stage, eventually damaging and shortening the life of engines and gas turbines along with requiring more frequent oil changes. Damages to spark plugs, exhaust valves, and admission valves have also been observed (Dewil et al., 2006), The current siloxane removal processes include adsorption, absorption, and deep chilling (Ajhar et al., 2010).

Fate of Fluorinated Compounds

Fluorinated compounds are typically incorporated into coatings for packaging paper, textiles, cookware, and carpets (Kissa, 2001; Sinclair et al., 2007; Stadalius et al., 2006; Washburn et al., 2005). Wastewater biosolids, which are often landfilled, have also been found to be a source of fluorinated compounds (Higgins et al., 2005). A study by Huset et al. (2011) quantified 24 fluorinated compounds in landfill leachate from four landfills in the United States. Fluorinated compounds were measured at concentration ranges of a few hundreds to 2800 ng/L (Huset et al., 2011).

Table 3.2.5 Volatile methyl siloxanes physical properties

Compound	Abbreviation	Formula	Molar Mass (g/mol)	Liquid Density at 20°C (g/L)[1]	Vapor Pressure at 20°C (Pa)[1]	Water Solubility at 25°C(mg/L)
Hexamethyldisiloxane	L2 (linear)	$C_6H_{18}OSi_2$	162	753	4456	0.93
Octamethyltrisiloxane	L3 (linear)	$C_8H_{24}O_2Si_3$	236	817	397	–
Decamethyltetrasiloxane	L4 (linear)	$C_{10}H_{30}O_3Si_4$	310	853	45.3	–
Hexamethylcyclotrisiloxane	D3 (cyclic)	$C_6H_{18}O_3Si_3$	222	950	916[2]	1.56
Octamethylcyclotetrasiloxane	D4 (cyclic)	$C_8H_{24}O_4Si_4$	297	953	88.1	0.056
Decamethylcyclopentasiloxane	D5 (cyclic)	$C_{10}H_{30}O_5Si_5$	371	958	20.8	0.017

Modified from Ajbar et al. (2010).

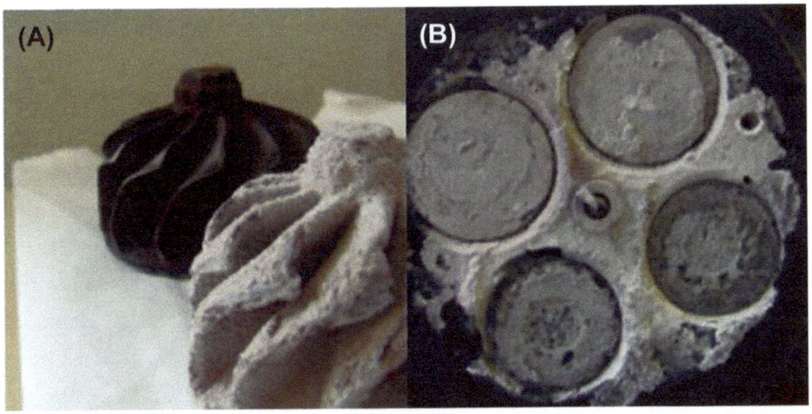

Figure 3.2.4 (A) Solid silica deposits on turbine blades (Desotec, 2015) and (B) a RICE head cylinder (Alvarez-Florez and Egusquiza, 2015).

Typically fluorinated compounds are found in the range of 0.1–150 ng/L and 0.5–1000 ng/L in surface waters (Becker et al., 2008; Huset et al., 2008; Lange et al., 2007) and wastewater (Becker et al., 2008; Huset et al., 2008), respectively. Overall these studies found that short-chained (C_4–C_7) carboxylates or sulfonates are present at greater abundance relative to longer-chain homologs ($\geq C_8$). The most abundant (on a concentration basis) fluorinated compound found in the leachates was perfluoroalkyl carboxylates (maximum concentration of 2800 ng/L) followed by perfluoroalkyl sulfonates (maximum concentration of 2300 ng/L). Sulfonamide derivatives were the third most abundant compound with methyl (C_4 and C_8) and ethyl (C_8) sulfonamide acetic acids being the most abundant (Huset et al., 2011). Fluorotelomer sulfonates were present in all leachates. Perfluoroalkyl carboxylates and perfluoroalkyl sulfonates are water soluble, which explain the affinity for these compounds to reach the leachate phase. Additionally due to these properties present in leachate, there is a potential for these contaminants to be discharged to the environment if leachate is cotreated with wastewater because traditional wastewater processes will not remove fluorinated compounds.

Fate of Nanoparticles

Over the past decade, engineered particles with nanoscale dimensions have been key to advancements in drug delivery and pharmaceuticals, cosmetics, environmental remediation, nanotechnology, biomaterials, and energy production (Bhatt and Tripathi, 2011; Colvin, 2003; Musee, 2011). Nanomaterials (NMs) have been shown to possess large surface areas and exhibit unique properties in comparison to their bulk counterparts. NMs that are found in consumer products are typically coated, functionalized, or surface-modified to achieve specific surface properties. According to an inventory completed by the Project on Emerging Nanotechnologies, 1628 consumer products containing NMs were available as of October 2013 (Project on Emerging Nanotechnologies, 2013). The primary NMs associated with these consumer products include silver, carbon, titanium, silicon/silica, zinc, and gold (Project on Emerging Nanotechnologies, 2013). In addition, NMs have been incorporated in personal care products, and clothing [such as Ag (as an antibacterial agent), SiO_2 (as a polishing and binding agent), TiO_2 (in solar cells), and ZnO (as a UV absorber in sun screen lotion)] (Reinhart et al., 2010). In addition to metallic NMs, carbon-based NMs, such as carbon nanotubes (CNTs), are also incorporated within consumer products.

The fate of NMs following disposal will most likely be a function of the product matrix and material surface characteristics. The disposal of nanowastes into landfills raises concern about the effects of the added NMs on anaerobic (waste degradation and leachate treatment) and aerobic (leachate treatment) processes associated with landfills in addition to the potential release of these NMs to the environment (e.g., mobility through liner system or leachate transport) (Bolyard et al., 2013; Reinhart et al., 2010). As an example, the next section will provide a detailed overview of an emerging area of research related to the fate of single-walled carbon nanotubes (SWNTs) in MSW landfills.

Fate of Single-walled Carbon Nanotubes in Municipal Solid Waste Landfills

CNTs are tubular cylinders of graphite (e.g., hexagon-linked carbon atoms) that have been shown to exhibit novel physiochemical properties (Popov, 2004). These engineered nanomaterials (ENMs) have large length to diameter ratios, with tube diameters typically less than 100 nm and lengths as long as several millimeters (Baughman et al., 2002; Huang et al., 2003; Popov, 2004). Because they have high Young's modulus and tensile strength, their use in composite materials is highly desirable (Baughman et al., 2002). There are two main types of CNTs: single-walled carbon nanotubes (SWNTs), which consist of a single sheet of graphite rolled in a seamless cylinder, and multiwalled carbon nanotubes (MWNTs), which consist of an array of concentric nanotubes (Baughman et al., 2002). SWNTs have been incorporated within a variety of consumer products, including sporting goods (e.g., tennis rackets, bicycle parts), compact discs, cookware, water filters, and plastics (De Volder et al., 2013; Endo et al., 2008; Project on Emerging Nanotechnologies, 2013), ultimately enhancing their characteristics.

As these NM-laden consumer products reach the end of their useful life, it is likely that they will be discarded and ultimately placed within MSW landfills. There is little known about the fate of SWNTs (or any type of CNT) in MSW landfills. Although not specifically investigated, once placed within a landfill, ENM release from these products under conditions typically found in landfills is likely (Benn et al., 2010; Petersen et al., 2011). Mechanical stress and abrasion (such as that experienced during compaction) and subsequent contact with leachate of an aggressive nature will likely aid release of nanoparticles bound in plastics/polymers/metal products (Froggett et al., 2014; Petersen et al., 2011). The form of released ENM will influence material fate and is likely product specific, depending on the parent product, NM type, release scenario (e.g., crushed/ground materials, weathering), and release environment (Froggett et al., 2014). It is possible that fractions of embedded NMs are released as either coated discrete particles, with coating composition similar to the NM coating when placed within the parent products, as coated aggregates, or incorporated within smaller components of the parent product (Froggett et al., 2014). Observations of NM released from polymer composites have been recently documented in the literature (Froggett et al., 2014; Liu et al., 2012). It should be noted that the observed release occurred under conditions less harsh than those typically found in landfills (Froggett et al., 2014).

SWNT fate once released into the leachate is complicated and dependent on several interrelated factors including leachate composition (e.g., organic content, electrolyte species and concentration), form of the released SWNT, solid waste composition, and leachate flow conditions. Two critical components of the leachate influencing SWNT transport are organic matter composition and concentration and ionic strength. Organic matter and electrolyte concentration and composition have been shown to significantly influence NM aggregation and deposition, and thus subsequent material mobility, in porous media. High molecular weight organic matter has been shown to create favorable conditions for NM stabilization, resulting in significant particle mobility (Franchi and O'Melia, 2003; Jaisi and Elimelech, 2009; Lanphere et al., 2013). High electrolyte concentrations, however, have been shown to reduce NM mobility by specific adsorption onto their surfaces (Hahn and O'Melia, 2004; Hong et al., 2009) and double layer compression. Because leachate composition, including both organics and electrolytes, changes over time as a result of physical, chemical, and biological reactions/transformations, it is likely that SWNT mobility will also change with time/waste age.

The majority of studies evaluating CNT mobility have been conducted in systems containing relatively simple porous media (e.g., sand) and narrow ranges of organic matter concentration and electrolyte concentrations (Petosa et al., 2010). Few studies have investigated the mobility of SWNTs in waste environments. Lozano and Berge (2012) conducted laboratory-scale experiments to evaluate how organics (humic acid: 20–800 mg/L), ionic strength (100–400 mM NaCl), and pH (6–8) typical of mature leachates influence SWNT behavior. Results from batch experiments suggested that the presence of high molecular weight organics, such as humic acid, acted to stabilize SWNTs present in leachate, even at high ionic strengths (>100 mM NaCl). These results also suggest that in mature landfill leachate, as long as humic acid is present, ionic strength (when represented as NaCl) will be a dominant factor influencing NM fate. It should be noted, however, that the ionic strength was represented as NaCl. In landfill, leachate multivalent ions (e.g., Ca^{+2}, Mg^{+2}, La^{+3}) will also be present and will likely be an important factor influencing NM mobility (Chen and Elimelech, 2007; Lin et al., 2009; Saleh et al., 2010). Lozano and Berge (2012) also found that pH, over the range evaluated in this study and that expected in mature leachate (from 6 to 8), had a negligible impact on NM behavior.

Khan et al. (2013) studied the laboratory-scale mobility of SWNTs dispersed in a range of synthetic leachate solutions representing both mature and young leachates through a representative solid waste environment (manufactured waste consisting of paper, metal, plastic, rabbit food, and glass). A series of 1-D saturated column experiments containing representative solid waste materials was conducted to evaluate the influence of organic matter type (e.g., humic acid and acetic acid) and concentration on SWNT transport. Results from these experiments indicated that SWNT transport may be significant in mature waste environments, with mobility decreasing with decreasing humic acid concentration. Approximately 92% of added SWNTs were retained within the waste column when in the presence of 10 mg/L humic acid, while only 30% of the added SWNTs were retained when in the presence of 400 mg/L humic acid. SWNT mobility in the presence of acetic acid was greatly inhibited (94% of introduced SWNTs were retained), suggesting their mobility in young waste environments may be small.

Solid waste composition also influences SWNT mobility. Typical MSW is heterogeneous, containing both organic and inorganic constituents and encompassing a large range of particle sizes and surface chemistries that change with waste age and decomposition. Khan et al. (2013) conducted a series of column experiments containing individual waste materials (e.g., paper, metal glass) to assess the influence of waste type on SWNT transport. Results from these experiments indicated that SWNT transport through metal, plastic, and glass was significant, with measured effluent recoveries of 77%, 88%, and 93%, respectively. The greatest SWNT retention was observed in paper, with only 20% of introduced SWNT mass recovered in the column effluent, suggesting that paper may be responsible for the majority of SWNT retention in mixed waste environments. These differences in SWNT transport suggest that the material surface characteristics, such as surface roughness and chemistry, influence transport. Flow conditions within landfills also influence NM transport. Variable flow conditions, waste heterogeneity, and preferential flow in landfills will all influence SWNT mobility. Flow velocity has been found to greatly influence the mobility of nC_{60} NMs in saturated porous media (Li et al., 2008). In addition, attachment of TiO_2 NMs to the air–water interface was observed during drainage processes (Chen et al., 2008), suggesting unsaturated conditions typical in landfills will influence SWNT mobility, possibly resulting in greater SWNT retention.

There is a distinct need for additional work evaluating and understanding SWNT (and other NM) behavior in waste environments. Although Lozano and Berge (2012) and Khan et al. (2013) have provided information associated with understanding SWNT fate within waste environments, there is a significant gap between the conditions modeled in the conducted laboratory experiments and actual conditions in MSW landfills.

CONCLUSIONS

Clearly, the complexity and changing nature of solid waste and landfill conditions lead to a large number of reactions affecting the quality of gas and leachate. While it is impossible to completely describe or fully understand the fate of landfilled materials, this chapter provides an overview of potential physical and chemical pathways. Because of the economic, health, and technical implications of these physical and chemical reactions, in particular for landfills with elevated temperatures, continued focus on these processes is important.

References

Ajhar, M., Travesset, M., Yüce, S., Melin, T., 2010. Siloxane removal from landfill and digester gas – a technology overview. Bioresource Technology 101 (9), 2913–2923. https://doi.org/10.1016/j.biortech.2009.12.018.

Allen, M.R., Braithwaite, A., Hills, C.C., 1997. Trace organic compounds in landfill gas at seven UK waste disposal sites. Environmental Science and Technology 31 (4), 1054–1061. https://doi.org/10.1021/es9605634.

Álvarez-Flórez, J., Egusquiza, E., 2015. Analysis of damage caused by siloxanes in stationary reciprocating internal combustion engines operating with landfill gas, Engineering. Failure Analysis 50, 29–38. https://doi.org/10.1016/j.engfailanal.2015.01.010.
Andersen, C.B., 2002. Understanding carbonate equilibria by measuring alkalinity in experimental and natural systems. Journal of Geoscience Education 50 (4), 389.
Barnes, K.K., Christenson, S.C., Kolpin, D.W., Focazio, M.J., Furlong, E.T., Zaugg, S.D., et al., 2004. Pharmaceuticals and other organic waste water contaminants within a leachate plume downgradient of a municipal landfill. Ground Water Monitoring and Remediation 24 (2), 119–126. https://doi.org/10.1111/j.1745-6592.2004.tb00720.x.
Baughman, R.H., Zakhidov, A.A., de Heer, W.A., 2002. Carbon nanotubes—the route toward applications. Science 297 (5582), 787–792. https://doi.org/10.1126/science.1060928.
Baun, D.L., Christensen, T.H., 2004. Speciation of heavy metals in landfill leachate: a review. Waste Management and Research 22 (1), 3–23. https://doi.org/10.1177/0734242x04042146.
Becker, A.M., Gerstmann, S., Frank, H., 2008. Perfluorooctane surfactants in waste waters, the major source of river pollution. Chemosphere 72 (1), 115–121. https://doi.org/10.1016/j.chemosphere.2008.01.009.
Benn, T., Cavanagh, B., Hristovski, K., Posner, J., Westerhoff, P., 2010. The release of nanosilver from consumer products used in the home. Journal of Environmental Quality 39, 1875–1882.
Bhatt, I., Tripathi, B.N., 2011. Interaction of engineered nanoparticles with various components of the environment and possible strategies for their risk assessment. Chemosphere 82 (3), 308–317. https://doi.org/10.1016/j.chemosphere.2010.10.011.
Bolyard, S.C., Reinhart, D.R., Santra, S., 2013. Behavior of engineered nanoparticles in landfill leachate. Environmental Science and Technology 47 (15), 8114–8122. https://doi.org/10.1021/es305175e.
Calace, N., Liberatori, A., Petronio, B.M., Pietroletti, M., 2001. Characteristics of different molecular weight fractions of organic matter in landfill leachate and their role in soil sorption of heavy metals. Environmental Pollution 113 (3), 331–339. https://doi.org/10.1016/S0269-7491(00)00186-X.
Chen, K.L., Elimelech, M., 2007. Influence of humic acid on the aggregation kinetics of fullerene (C60) nanoparticles in monovalent and divalent electrolyte solutions. Journal of Colloid and Interface Science 309 (1), 126–134. https://doi.org/10.1016/j.jcis.2007.01.074.
Chen, L., Nanny, M.A., Knappe, D.R.U., Wagner, T.B., Ratasuk, N., 2004. Chemical characterization and sorption capacity measurements of degraded newsprint from a landfill. Environmental Science and Technology 38 (13), 3542–3550. https://doi.org/10.1021/es0305914.
Chen, L., Sabatini, D.A., Kibbey, T.C.G., 2008. Role of the air—water interface in the retention of TiO_2 nanoparticles in porous media during primary drainage. Environmental Science and Technology 42 (6), 1916–1921. https://doi.org/10.1021/es071410r.
Christensen, T.H., Cossu, R., Stegmann, R., 1992. Landfilling of Waste: Leachate. Elsevier Applied Science, London, New York.
Christensen, T.H., Kjeldsen, P., Albrechtsen, H.J., Heron, G., Nielsen, P.H., Bjerg, P.L., Holm, P.E., 1994. Attenuation of landfill leachate pollutants in aquifers. Critical Reviews in Environmental Science and Technology 24 (2), 119–202. https://doi.org/10.1080/10643389409388463.
Christensen, J.B., Jensen, D.L., Christensen, T.H., 1996. Effect of dissolved organic carbon on the mobility of cadmium, nickel and zinc in leachate polluted groundwater. Water Research 30 (12), 3037–3049. https://doi.org/10.1016/S0043-1354(96)00091-7.
Christensen, T.H., Kjeldsen, P., Bjerg, P.L., Jensen, D.L., Christensen, J.B., Baun, A., et al., 2001. Biogeochemistry of landfill leachate plumes. Applied Geochemistry 16 (7–8), 659–718. https://doi.org/10.1016/S0883-2927(00)00082-2.
Colvin, V.L., 2003. The potential environmental impact of engineered nanomaterials. Nature Biotechnology 21 (10), 1166–1170.
De Volder, M.F.L., Tawfick, S.H., Baughman, R.H., Hart, A.J., 2013. Carbon nanotubes: present and future commercial applications. Science 339 (6119), 535–539. https://doi.org/10.1126/science.1222453.
Deipser, A., Stegmann, R., 1994. The origin and fate of volatile trace components in municipal solid waste landfills. Waste Management and Research 12 (2), 129–139. https://doi.org/10.1006/wmre.1994.1003.
Dewil, R., Appels, L., Baeyens, J., 2006. Energy use of biogas hampered by the presence of siloxanes. Energy Conversion and Management 47 (13–14), 1711–1722. https://doi.org/10.1016/j.enconman.2005.10.016.
El-Fadel, M., 1999. Simulating temperature variations in landfills. Journal of Solid Waste Technology and Management 26 (2), 78.
Endo, M., Strano, M., Ajayan, P., 2008. Potential applications of carbon nanotubes. In: Jorio, A., Dresselhaus, G., Dresselhaus, M. (Eds.), Carbon Nanotubes, vol. 111. Springer Berlin Heidelberg, pp. 13–62.
Erses, S.A., Onay, T.T., 2003. In situ heavy metal attenuation in landfills under methanogenic conditions. Journal of Hazardous Materials 99 (2), 159–175. https://doi.org/10.1016/S0304-3894(02)00354-0.

Fang, C-r., Long, Y-y., Shen, D-s., 2014. Sorption behavior of dibutyl phthalate and dioctyl phthalate by aged refuse. Environmental Science and Pollution Research 21 (12), 7641–7649. https://doi.org/10.1007/s11356-014-2672-7.

Franchi, A., O'Melia, C.R., 2003. Effects of natural organic matter and solution chemistry on the deposition and reentrainment of colloids in porous media. Environmental Science and Technology 37 (6), 1122–1129. https://doi.org/10.1021/es015566h.

Froggett, S., Clancy, S., Boverhof, D., Canady, R., 2014. A review and perspective of existing research on the release of nanomaterials from solid nanocomposites. Particle and Fibre Toxicology 11 (1), 17.

Fu, F., Dionysiou, D.D., Liu, H., 2014. The use of zero-valent iron for groundwater remediation and wastewater treatment: a review. Journal of Hazardous Materials 267, 194–205. https://doi.org/10.1016/j.jhazmat.2013.12.062.

Gold, V., 1987. Compendium of Chemical Terminology: IUPAC Recommendations. Blackwell Scientific Publications, Oxford (Oxfordshire), Boston.

Hahn, M.W., O'Melia, C.R., 2004. Deposition and reentrainment of brownian particles in porous media under unfavorable chemical conditions: some concepts and applications. Environmental Science and Technology 38 (1), 210–220. https://doi.org/10.1021/es030416n.

Higgins, C.P., Field, J.A., Criddle, C.S., Luthy, R.G., 2005. Quantitative determination of perfluorochemicals in sediments and domestic sludge. Environmental Science and Technology 39 (11), 3946–3956. https://doi.org/10.1021/es048245p.

Hong, Y., Honda, R.J., Myung, N.V., Walker, S.L., 2009. Transport of iron-based nanoparticles: role of magnetic properties. Environmental Science and Technology 43 (23), 8834–8839. https://doi.org/10.1021/es9015525.

Horii, Y., Kannan, K., 2008. Survey of organosilicone compounds, including cyclic and linear siloxanes, in personal-care and household products. Archives of Environmental Contamination and Toxicology 55 (4), 701–710. https://doi.org/10.1007/s00244-008-9172-z.

Huang, S., Cai, X., Liu, J., 2003. Growth of millimeter-long and horizontally aligned single-walled carbon nanotubes on flat substrates. Journal of the American Chemical Society 125 (19), 5636–5637. https://doi.org/10.1021/ja034475c.

Huset, C.A., Chiaia, A.C., Barofsky, D.F., Jonkers, N., Kohler, H.-P.E., Ort, C., et al., 2008. Occurrence and mass flows of fluorochemicals in the Glatt Valley watershed, Switzerland. Environmental Science and Technology 42 (17), 6369–6377. https://doi.org/10.1021/es703062f.

Huset, C.A., Barlaz, M.A., Barofsky, D.F., Field, J.A., 2011. Quantitative determination of fluorochemicals in municipal landfill leachates. Chemosphere 82 (10), 1380–1386. https://doi.org/10.1016/j.chemosphere.2010.11.072.

Jaisi, D.P., Elimelech, M., 2009. Single-walled carbon nanotubes exhibit limited transport in soil columns. Environmental Science and Technology 43 (24), 9161–9166. https://doi.org/10.1021/es901927y.

Joseph, L., Zaib, Q., Khan, I.A., Berge, N.D., Park, Y.-G., Saleh, N.B., Yoon, Y., 2011. Removal of bisphenol A and 17α-ethinyl estradiol from landfill leachate using single-walled carbon nanotubes. Water Research 45 (13), 4056–4068. https://doi.org/10.1016/j.watres.2011.05.015.

Khan, I.A., Berge, N.D., Sabo-Attwood, T., Ferguson, P.L., Saleh, N.B., 2013. Single-walled carbon nanotube transport in representative municipal solid waste landfill conditions. Environmental Science and Technology 47 (15), 8425–8433. https://doi.org/10.1021/es401748f.

Kissa, E., 2001. Fluorinated Surfactants and Repellents, second ed. (Rev. and Expanded).

Kjeldsen, P., Christensen, T.H., 2001. A simple model for the distribution and fate of organic chemicals in a landfill: MOCLA. Waste Management and Research 19 (3), 201–216.

Kjeldsen, P., Barlaz, M.A., Rooker, A.P., Baun, A., Ledin, A., Christensen, T.H., 2002. Present and long-term composition of MSW landfill leachate: a review. Critical Reviews in Environmental Science and Technology 32 (4), 297–336. https://doi.org/10.1080/10643380290813462.

Klein, T., Niessner, R., 1998. Characterization of heavy-metal-containing seepage water colloids by flow FFF, ultrafiltration, ELISA and AAS. Microchimica Acta 129 (1–2), 47–55. https://doi.org/10.1007/BF01246848.

Lange, F.T., Wenz, M., Schmidt, C.K., Brauch, H.J., 2007. Occurrence of perfluoroalkyl sulfonates and carboxylates in German drinking water sources compared to other countries. Water Science and Technology 56 (11), 151–158.

Lanphere, J.D., Luth, C.J., Walker, S.L., 2013. Effects of solution chemistry on the transport of graphene oxide in saturated porous media. Environmental Science and Technology 47 (9), 4255–4261. https://doi.org/10.1021/es400138c.

Li, Y., Wang, Y., Pennell, K.D., Abriola, L.M., 2008. Investigation of the transport and deposition of fullerene (C60) nanoparticles in quartz sands under varying flow conditions. Environmental Science and Technology 42 (19), 7174–7180. https://doi.org/10.1021/es801305y.

Lin, D., Liu, N., Yang, K., Zhu, L., Xu, Y., Xing, B., 2009. The effect of ionic strength and pH on the stability of tannic acid-facilitated carbon nanotube suspensions. Carbon 47 (12), 2875–2882. https://doi.org/10.1016/j.carbon.2009.06.036.

Liu, J., Katahara, J., Li, G., Coe-Sullivan, S., Hurt, R.H., 2012. Degradation products from consumer nanocomposites: a case study on quantum dot lighting. Environmental Science and Technology 46 (6), 3220–3227. https://doi.org/10.1021/es204430f.

Lorphensri, O., Sabatini, D.A., Kibbey, T.C.G., Osathaphan, K., Saiwan, C., 2007. Sorption and transport of acetaminophen, 17α-ethynyl estradiol, nalidixic acid with low organic content aquifer sand. Water Research 41, 2180–2188. https://doi.org/10.1016/j.watres.2007.01.057.

Lozano, P., Berge, N.D., 2012. Single-walled carbon nanotube behavior in representative mature leachate. Waste Management 32 (9), 1699–1711. https://doi.org/10.1016/j.wasman.2012.03.019.

Lu, J.C.S., Eichenberger, B., Stearns, R.J., 1985. Leachate from Municipal Landfills, Production and Management. Noyes Publications, Park Ridge, NJ.

Lu, Y., Yuan, T., Wang, W., Kannan, K., 2011. Concentrations and assessment of exposure to siloxanes and synthetic musks in personal care products from China. Environmental Pollution 159 (12), 3522–3528. https://doi.org/10.1016/j.envpol.2011.08.015.

Mårtensson, A.M., Aulin, C., Wahlberg, O., Ågren, S., 1999. Effect of humic substances on the mobility of toxic metals in a mature landfill. Waste Management and Research 17 (4), 296–304. https://doi.org/10.1034/j.1399-3070.1999.00053.x.

Masoner, J.R., Kolpin, D.W., Furlong, E.T., Cozzarelli, I.M., Gray, J.L., Schwab, E.A., 2014. Contaminants of emerging concern in fresh leachate from landfills in the conterminous United States. Environmental Science: Processes and Impacts 16 (10), 2335–2354. https://doi.org/10.1039/C4EM00124A.

Moqbel, S.Y., 2009. Characterizing Spontaneous Fires in Landfills (Electronic Resource). University of Central Florida, Orlando, Fla.

Moqbel, S., Reinhart, D., Chen, R.-H., 2010. Factors influencing spontaneous combustion of solid waste. Waste Management 30, 1600–1607. https://doi.org/10.1016/j.wasman.2010.01.006.

Musee, N., 2011. Nanowastes and the environment: potential new waste management paradigm. Environment International 37 (1), 112–128. https://doi.org/10.1016/j.envint.2010.08.005.

Noubactep, C., Caré, S., Crane, R., 2012. Nanoscale metallic iron for environmental remediation: prospects and limitations. Water, Air, and Soil Pollution 223 (3), 1363–1382. https://doi.org/10.1007/s11270-011-0951-1.

Pan, B., Lin, D., Mashayekhi, H., Xing, B., 2008. Adsorption and hysteresis of bisphenol A and 17α-ethinyl estradiol on carbon nanomaterials. Environmental Science and Technology 42 (15), 5480–5485. https://doi.org/10.1021/es8001184.

Petersen, E., Zhang, L., Mattison, N., O'Carroll, D., Whelton, A., Uddin, N., et al., 2011. Potential release pathways, environmental fate, and ecological risks of carbon nanotubes. Environmental Science and Technology 45, 9837–9856.

Petosa, A.R., Jaisi, D.P., Quevedo, I.R., Elimelech, M., Tufenkji, N., 2010. Aggregation and deposition of engineered nanomaterials in aquatic environments: role of physicochemical interactions. Environmental Science and Technology 44 (17), 6532–6549. https://doi.org/10.1021/es100598h.

Popov, V.N., 2004. Carbon nanotubes: properties and application. Materials Science and Engineering: R: Reports 43 (3), 61–102. https://doi.org/10.1016/j.mser.2003.10.001.

Project on Emerging Nanotechnologies, 2013. Consumer Products Inventory. From: http://www.nanotechproject.org/cpi/about/analysis/.

Reinhart, D.R., Grosh, C.J., 1998. Analysis of Florida MSW Landfill Leachate Quality.

Reinhart, D.A., Pohland, F.G., Stevens, D.K., 1991. Mathematical fate modeling of hazardous organic pollutants during codisposal with municipal refuse. Hazardous Waste and Hazardous Materials 8 (1), 85–97.

Reinhart, D.R., Pohland, F.G., Gould, J.P., Cross, W.H., 1991. The fate of selected organic pollutants codisposed with municipal refuse. Research Journal of the Water Pollution Control Federation 63 (5), 780–788. https://doi.org/10.2307/25044056.

Reinhart, D.R., Berge, N.D., Santra, S., Bolyard, S.C., 2010. Emerging contaminants: nanomaterial fate in landfills. Waste Management 30 (11), 2020–2021. https://doi.org/10.1016/j.wasman.2010.08.004.

Saleh, N.B., Pfefferle, L.D., Elimelech, M., 2010. Influence of biomacromolecules and humic acid on the aggregation kinetics of single-walled carbon nanotubes. Environmental Science and Technology 44 (7), 2412–2418. https://doi.org/10.1021/es903059t.

Saquing, J.M., Saquing, C.D., Knappe, D.R.U., Barlaz, M.A., 2010. Impact of plastics on fate and transport of organic contaminants in landfills. Environmental Science and Technology 44 (16), 6396–6402.

Scheutz, C., Durant, N.D., Hansen, M.H., Bjerg, P.L., 2011. Natural and enhanced anaerobic degradation of 1,1,1-trichloroethane and its degradation products in the subsurface – a critical review. Water Research 45 (9), 2701–2723. https://doi.org/10.1016/j.watres.2011.02.027.

Schwarzenbach, R.P., Gschwend, P.M., Imboden, D.M., 2005. General topic and overview. In: Environmental Organic Chemistry. John Wiley & Sons, Inc., pp. 3–12

Sellers, K., 1999. Fundamentals of Hazardous Waste Site Remediation. Lewis Publishers, Boca Raton.

Sinclair, E., Kim, S.K., Akinleye, H.B., Kannan, K., 2007. Quantitation of gas-phase perfluoroalkyl surfactants and fluorotelomer alcohols released from nonstick cookware and microwave popcorn bags. Environmental Science and Technology 41 (4), 1180–1185. https://doi.org/10.1021/es062377w.

Slack, R.J., Gronow, J.R., Voulvoulis, N., 2005. Household hazardous waste in municipal landfills: contaminants in leachate. Science of the Total Environment 337 (1–3), 119–137. https://doi.org/10.1016/j.scitotenv.2004.07.002.

Stadalius, M., Connolly, P., L'Empereur, K., Flaherty, J.M., Isemura, T., Kaiser, M.A., et al., 2006. A method for the low-level (ng g^{-1}) determination of perfluorooctanoate in paper and textile by liquid chromatography with tandem mass spectrometry. Journal of Chromatography A 1123 (1), 10–14. https://doi.org/10.1016/j.chroma.2006.03.037.

Surita, S.C., Tansel, B., 2015. Preliminary investigation to characterize deposits forming during combustion of biogas from anaerobic digesters and landfills. Renewable Energy 80 (0), 674–681. https://doi.org/10.1016/j.renene.2015.02.060.

Tchobanoglous, G., Theisen, H., Vigil, S., 1993. Integrated Solid Waste Management: Engineering Principles and Management Issues. McGraw-Hill.

Tipping, E., 1993. Modelling ion binding by humic acids. Colloids and Surfaces A: Physicochemical and Engineering Aspects 73 (0), 117–131. https://doi.org/10.1016/0927-7757(93)80011-3.

UK Environment Agency, 2007. Review and Investigation of Deep-seated Fires within Landfill Sites (Science Report).

Washburn, S., Bingmann, T., Braithwaite, S., Buck, R., Buxton, L., Clewell, H., Haroun, L.A., Kester, J.E., Rickard, R.W., Shipp, A., 2005. Exposure assessment and risk characterization for perfluorooctanoate (PFO) in selected consumer articles. Environmental Science and Technology 39 (11), 3904–3910.

Weber, R., Watson, A., Forter, M., Oliaei, F., 2011. Persistent organic pollutants and landfills – a review of past experiences and future challenges. Waste Management and Research 29 (1), 107–121.

Wu, B., Taylor, C.M., Knappe, D.R.U., Nanny, M.A., Barlaz, M.A., 2001. Factors controlling alkylbenzene sorption to municipal solid waste. Environmental Science and Technology 35 (22), 4569–4576.

4 CHAPTER

EX SITU PRETREATMENT

4.1

MECHANICAL BIOLOGICAL PRETREATMENT

Rainer Stegmann

INTRODUCTION

The European Union (EU) has via the Landfill Directive set targets for reducing the biologically degradable waste fraction going to landfills. This reduction shall be implemented in all European countries but has not been achieved in all cases. Based on statistical data on the waste composition of the year 1995, each member country has the following targets regarding the reduction of biodegradable municipal solid waste (MSW): 50% by 2009 and 65% by 2016, where there are some exceptions (Anonymous, 1999). Each member state can decide on its own how it will reach these target values.

From June 1, 2005 only thermally or mechanically biologically pre-treated MSW can be landfilled in Germany (Anonymus, 2000). German limit values for the quality of waste to be landfilled are presented in Table 4.1.1.

In addition, off-gas emission limit values (see Table 4.1.2) from the in-house waste treatment facilities (waste delivery, mechanical biological treatment - MBT) are set (30. BImSchV; Anonymus, 2001). The most difficult to reach value is total carbon which has to be < 55 mg total C/t of waste treated.

1) Respiration activity in 4 days (indirect method to measure the biodegradable fraction of the waste sample)
2) Gas generation potential (GB_{21}): gas production from the waste sample in 21 days
3) Total Organic Carbon (TOC) in the eluate from an elution test (1:10 solid/liquid ratio and 24 h shaking time)

In Germany, MBT is seen more as a transient solution to fill up landfills that have already installed a base liner and/or where there have been made other investments so that depreciation costs for the investment have to be paid anyway. There is not a great potential for the construction of new landfills for MBT-waste in Germany. Apart from the problems of accepting landfills by the public the costs for the new landfill in combination with pretreatment will be high (comparable with MSW incineration). The main treatment option in Germany—after separate collection and intensive recycling—is incineration (thermal pretreatment before landfilling, if the bottom ash is not recycled).

Table 4.1.1 Allocation criteria for the waste quality as a prerequisite for landfilling (class I (inert waste), class II (pretreated municipal solid waste)), German landfill ordinance, Annex 3 (2)

Parameter	Unit	Limit Values		
		MBP-Pretreated Waste (Appendix II)	All Other Wastes (Appendix I)	
		Landfill-Class 2		Landfill-Class 1
STRENGTH VALUES				
Vane shear strength	kN/m^2	$\leq$25 or axial deformation and uniaxial compression strength		
Axial deformation	%	$\leq$20		
Uniaxial compression strength	kN/m^2	$\leq$50		
ORGANIC CONTENT				
Ignition loss	mass %		$\leq$5	$\leq$3
TOC$_{solid}$	mass %	$\leq$18	$\leq$3 or ignition loss	$\leq$1 or ignition loss
Upper heating value	kJ/kg	$<$6.000 or TOC $_{solid}$		
Respiration activity	mgO$_2$/gDM	$\leq$5		
Gas formation potential (GB$_{21}$)	l/kgDM	$\leq$20		
TOC$_{Eluat}$	mg/L	$\leq$250	$\leq$100	$\leq$20
Extractable lipophilic substance	mass %	$\leq$0.8	$\leq$0.8	$\leq$0.4
ELUAT CRITERIA				
pH value		5.5–13		
Conductivity	µS/cm	$\leq$50,000	$\leq$50,000	$\leq$10,000
Phenols	mg/L	$\leq$50	$\leq$50	$\leq$0.2

TABLE 4.1.1 Allocation criteria for the waste quality as a prerequisite for landfilling (class I (inert waste), class II (pretreated municipal solid waste)), German landfill ordinance, Annex 3 (2)—cont'd

Parameter	Unit	Limit Values		
		MBP-Pretreated Waste (Appendix II)	All Other Wastes (Appendix I)	
		Landfill-Class 2		Landfill-Class 1
Arsenic	mg/L	≤0.5	≤0.5	≤0.2
Lead	mg/L	≤1	≤1	≤0.2
Cadmium	mg/L	≤0.1	≤0.1	≤0.05
Chromium-VI	mg/L	≤0.1	≤0.1	≤0.05
Copper	mg/L	≤5	≤5	≤1
Nickel	mg/L	≤1	≤1	≤0.2
Mercury	mg/L	≤0.02	≤0.02	≤0.005
Zinc	mg/L	≤5	≤5	≤2
Fluoride	mg/L	≤25	≤25	≤5
Ammonia nitrogen	mg/L	≤200	≤200	≤4
Cyanide, easy releasable	mg/L	≤0.5	≤0.5	≤0.1
AOX	mg/L	≤1.5	≤1.5	≤0.3
Water soluble proportion (evaporated residue)	mass %	≤6	≤6	≤3

MECHANICAL BIOLOGICAL TREATMENT OF MUNICIPAL SOLID WASTE

Why has the reduction of the biodegradable organic fraction in the waste before landfilling been implemented in the EU directive and what are the main achievements, when practicing mechanical biological pretreatment (MBT)?

The main emissions (leachate and biogas) are to a very high degree influenced by the organic content of the waste and the related biological processes in the landfill. When MSW is landfilled without pretreatment, emissions occur during and after the landfill operation depending on waste composition,

Table 4.1.2 Limit values for gaseous emissions 30th ordinance of the German Clean Air Act (for MBT plants)

Parameters	Target Values Clean Gas		
Sample	½ h mean	Daily mean	Monthly mean
TOC	40 mg/m^3	20 mg/m^3	55 g/t
Nitrous oxide (N_2O)	–	–	100 g/t
Ammonia (NH_3)	30 mg/m^3 (TA air)		–
Dust	30 mg/m^3	10 mg/m^3	–
Dioxins/Furans	0.1 ng/m^3		
Odor	500 OU/m^3 (single measurement)		

GE, odor units. *TA air*, Technische Anleitung (technical instructions) air

climatic conditions, etc. in the form of approximately 100–150 m^3 biogas/t wwt MSW and very roughly 5–10 m^3/ha*d highly polluted leachate. Due to biological degradation processes significant settling in the range of 20%–25% of the height of the landfill takes place over long periods (decades), which may damage, e.g., surface liners, gas extraction, and leachate collection systems. The leachate produced has to be collected and treated over many decades with great technical expenditure (Christensen et al., 1992). The biogas produced has to be extracted and used as an energy source or—if not possible—been flared.

Practicing the mechanical biological waste pretreatment (MBT), processes taking place in the landfill over long periods of time (decades) will be shortened to several months (Fig. 4.1.1). During pretreatment the emission potential of the waste will be reduced to a large extent (for the organics up to 90%) so that, compared to non-pretreated waste, significantly reduced emissions occur (Fig. 4.1.2). It is not the aim of mechanical biological pretreatment to produce "compost" to be applied in agriculture or horticulture because the content of heavy metals and other harmful or disturbing substances is in general too high (Leikam and Stegmann, 1996).

MECHANICAL BIOLOGICAL PRETREATMENT AS PART OF WASTE MANAGEMENT CONCEPTS

Although extensive waste prevention and recycling is practiced, there is still a considerable quantity of residual waste left, which is not or cannot be reused. Also if both the packaging material and the biowaste are separately collected, only a certain percentage of these compounds (±50% of the total potential) is actually recovered. As a result the waste composition may not change essentially.

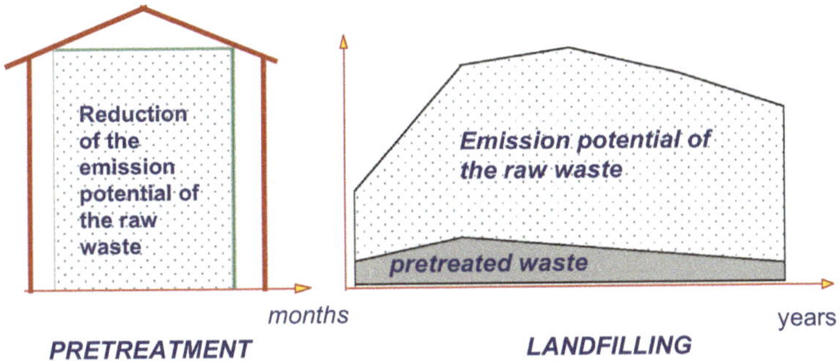

Figure 4.1.1 Shift of the emissions from landfill to pretreatment.

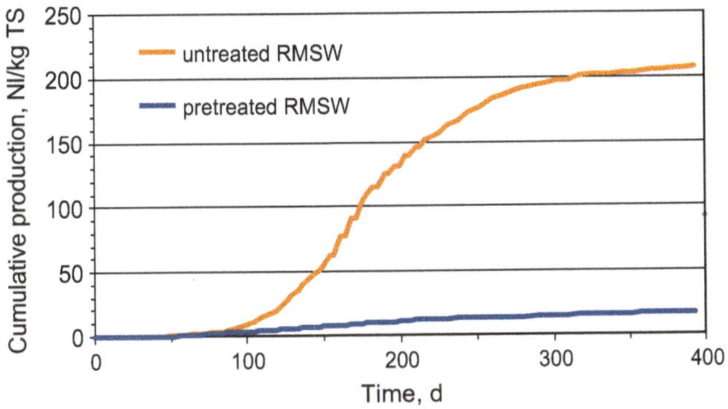

Figure 4.1.2 Cumulative gas production from untreated and aerobically pretreated residual municipal solid waste (RMSW) (4 months, laboratory scale results).

Within a waste management concept, mechanical biological pretreatment is applied to pretreat the residual MSW prior to landfilling. This includes the mechanical separation of the high calorific value fraction (refuse derived fuel, RDF) before the biological treatment (Fig. 4.1.3). RDF is used for energy production in a variety of plants (e.g., kilns, coal, power plants) (see also Thomé-Kozmiensky, 2004).

There are many different concepts for mechanical biological pretreatment (Fig. 4.1.4). Ketelsen et al. (2005) present an overview of different biological treatment technologies in use. The question which kind of system is the optimum cannot be answered since there are always advantages and disadvantages of the process. Waste composition and specific requirements have to be respected.

Mechanical Treatment

Mechanical pretreatment including shredding (see Chapter 4.3) is necessary to remove the high calorific value fraction, metals, and other materials for recycling as well as disturbing compounds that may negatively influence the technical processes. In addition it is needed to provide an adequate quality of the RDF fraction and the residual material for biological treatment and subsequent landfilling.

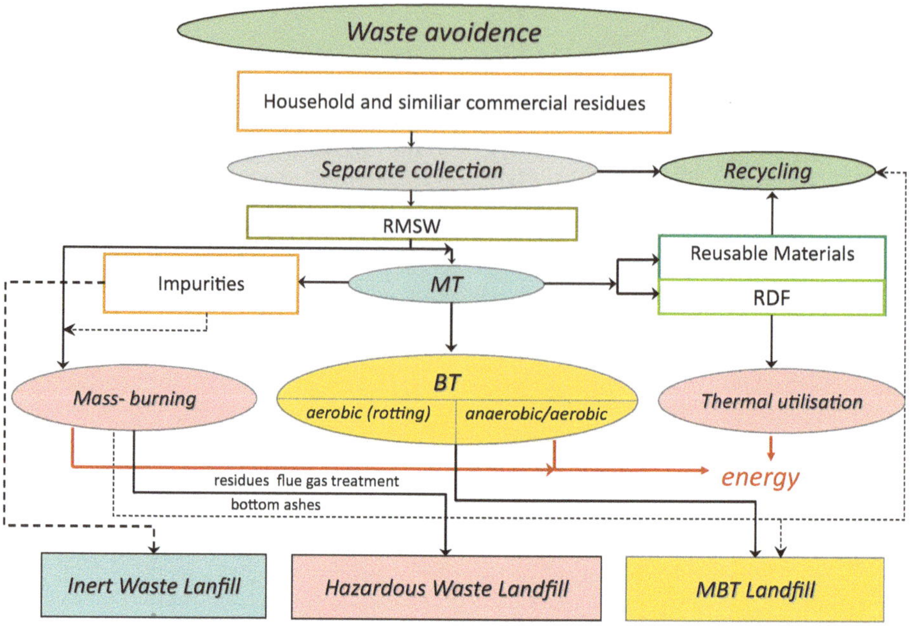

Figure 4.1.3 General concept of municipal solid waste management. *BT*, biological treatment; *MT*, mechanical treatment; *RDF*, refuse derived fuel; *RMSW*, residual municipal solid waste.

Shredders are used to partly homogenize the waste and to reduce the bulky materials to smaller pieces. Sieves, which can be rotating sieves (Fig. 4.1.5), vibrating sieves, or others are used to separate the RDF fraction from the more putriciple organic fraction (sieve diameter varies usually between 60 and 100 mm). Fig. 4.1.6 presents the grain size distribution of MSW after separate collection. Sieving is a simple method to separate specific waste components from the waste. About 95% of the vegetative compounds and 40% of the paper and plastic are in the undersize of a 100 mm rotating sieve. Bulky disturbing materials as well as metals have to be removed to protect the machines during the following treatment steps, as there are conveyer belts, turning machines, etc. Metals can be removed by means of magnets (iron) and eddy current separators (aluminum). They can be installed at different treatment steps: before and/or behind the sieve; in the latter case to remove metals from the RDF fraction; and/or in the undersize material of the sieving process before biological treatment. For the further treatment of the RDF another shredding step may be applied, and baling may be used for the storage and transport of the RDF (see Chapter 4.3).

To avoid operational problems and to reach the required target values the remaining waste (undersize) has to be prepared for the biological treatment step. For this purpose, stones, sand, remaining plastic, or other disturbing materials can be removed mainly by additional sieving steps and air classifiers (light fraction as plastic and paper).

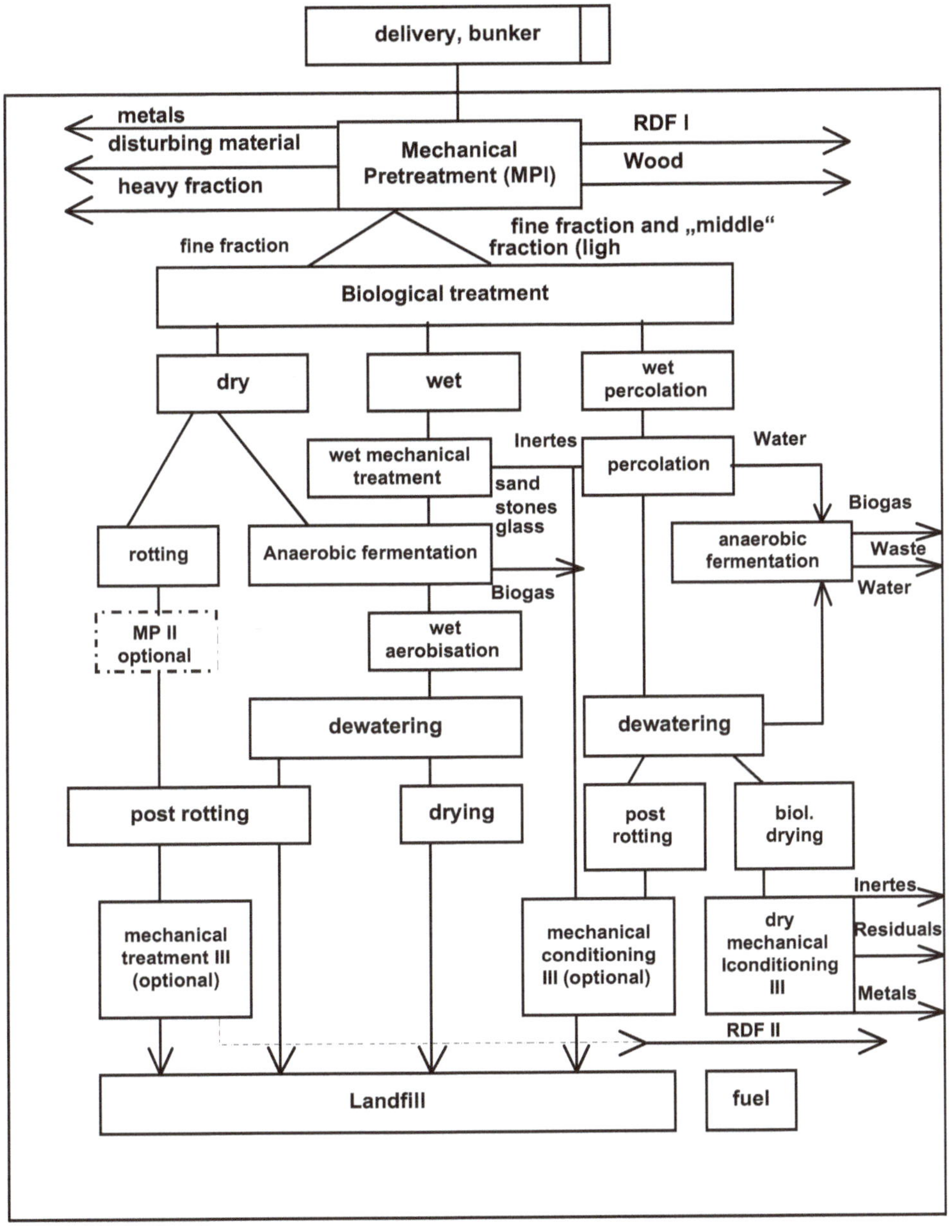

Figure 4.1.4 Overview of different technologies for mechanical biological pretreatment (Ketelsen et al., 2005).

Figure 4.1.5 Example of a rotating sieve.

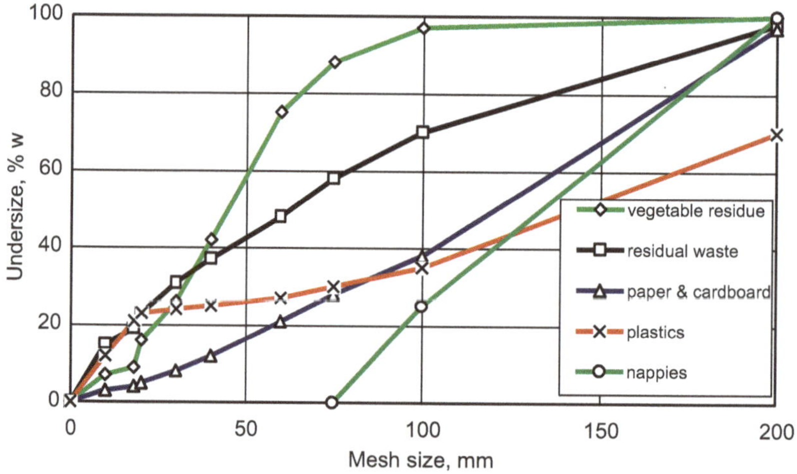

Figure 4.1.6 Grain size distribution of residual municipal solid waste (after separate collection).

Biological Pretreatment

In principle aerobic and anaerobic biological processes can be applied for the biological treatment step but they have to be adapted to the target values to be reached. Using the aerobic biological pretreatment, high odor concentrations emit especially during the initial intensive rotting phase (Doedens, 2001). As a consequence it should be practiced in enclosed systems with a forced air aeration and off-gas treatment. The postrotting or curing phase (at least 6–8 weeks) has to be operated in-house

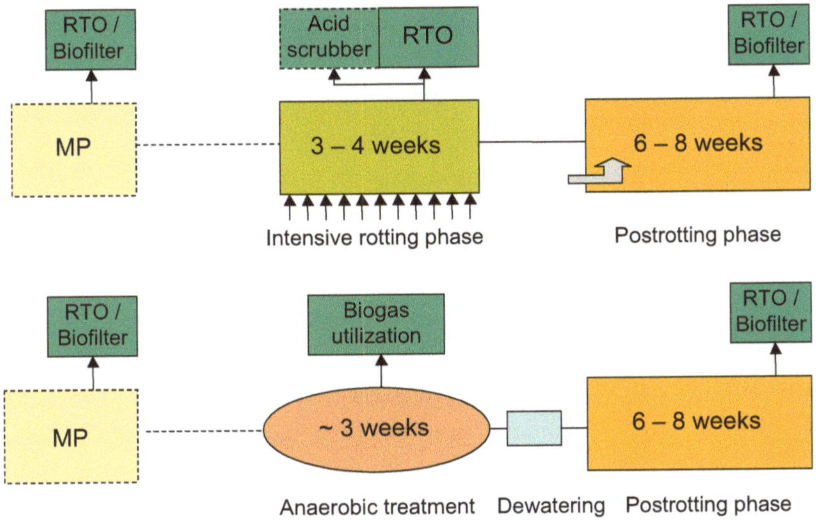

Figure 4.1.7 Mechanical biological treatment concepts for municipal solid waste: aerobic and anaerobic treatment including off-gas emission control (detention times may even be longer due to waste properties and technology used). *RTO*, regenerative thermal oxidation.

or under a roof dependent on the degree of maturation reached during the intensive rotting step. In total a rotting period of about 12–16 weeks or even longer is necessary to meet the German target values (Fig. 4.1.7). More details about the biological treatment step are presented elsewhere (Fricke et al., 2001; Leikam and Stegmann, 1997).

The organic rich fraction (undersize) is aerobically or anaerobically treated. For the aerobic processes the optimum milieu conditions have to be adjusted as, e.g., moisture content, amount of structure material allowing adequate air distribution, sufficient air supply during artificial aeration. For the aerobic treatment the technologies used for biowaste composting can also be applied. Once the material has been pretreated there are no differences between composting and aerobic pretreatment. Composting takes place mainly in big halls using adequate turning machines, in big container type composting facilities, or other systems. Important is also the odor and off-gas control, so the systems have to be gas tight and all the produced off-gases have to be captured and treated. Mechanical posttreatment is less complex or even not necessary when the aerobic processes are used. If the water content is not appropriate for landfilling, it has to be adjusted.

The anaerobic pretreatment may have several advantages compared to aerobic treatment, e.g., saving energy (no artificial aeration), gaining energy from biogas production as well as less odorous off-gas production. But the process is more complex and needs more attention. Using anaerobic treatment the capturing and removal of odors and other compounds (Table 4.1.2) is less costly compared to aerobic plants (large volume of odorous off-gas). Anaerobic fermentation should always be combined with a postrotting step, since not all organic substances (e.g., lignin-containing components) can be degraded under anaerobic milieu conditions to the degree required by (German) regulations (Table 4.1.1). In addition, the compounds are in a reduced stage and should be converted into the oxidized form. For the anaerobic treatment step, which "substitutes" the intensive rotting step, about 3 weeks and for the postrotting (curing) step at least another 6–8 weeks are necessary (see also Fig. 4.1.7).

If anaerobic treatment in the liquid phase is practiced, pretreatment has to be very extensive. In addition to the separation of the RDF, stones, sand, metal pieces, and other disturbing materials have to be removed (e.g. wood). Using wet fermentation the particle size of the substrate should be as small as possible. The pretreatment is necessary to avoid problems during transport and anaerobic fermentation of the waste in the fermentation tank (Fig. 4.1.8). For example, elevated sand content will settle in an anaerobic wet fermentation reactor and with the time this material will reach a concrete like consistence. When the wet fermentation process is used, light fractions may float on the surface of the substrate in the anaerobic tank, and together with other material may form a sludge blanket.

The pretreatment for the dry fermentation process needs not to be that extensive. There are two main anaerobic treatment concepts: the continuous and the discontinuous process. Using the discontinuous process the waste is placed into a garage type reactor with a water recirculation. To initiate the anaerobic digestion process the waste has to be seeded with already digested material (Fig. 4.1.9). After a certain period the waste is removed and subjected to aerobic posttreatment.

The continuous dry fermentation process consists, in most cases, of a long drum (AD reactor) with an internal mixing and transportation device (Fig. 4.1.10). The detention time is determined by the

Figure 4.1.8 Mechanical pretreatment plant (MBT plant Minden-Lübbecke, Germany).

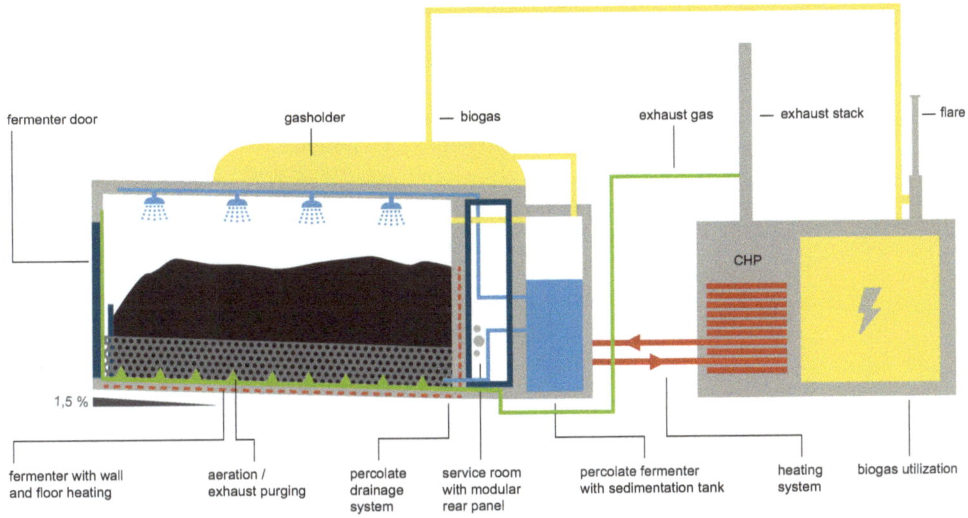

Figure 4.1.9 “Bekon” process: batch process, dry fermentation, one-stage, mesophilic, inoculation of biomass with digested residues (tm up to 34 days, biogas yield: 110 Nm^3/t wwt (Bekon Energy Technologies Co. & KG, Unterföhring, Germany).

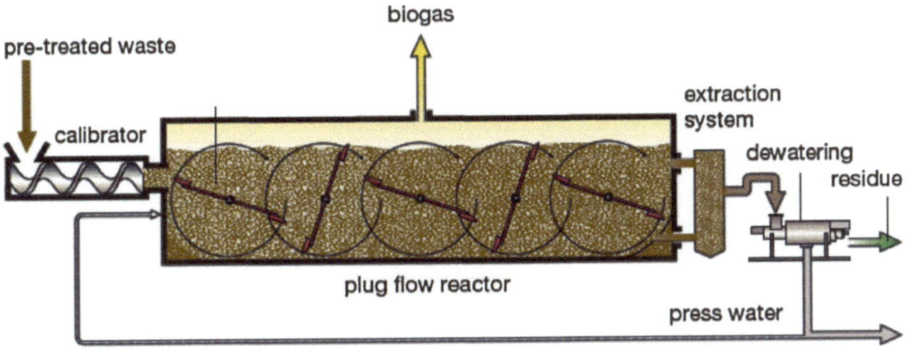

Figure 4.1.10 “Strabag” process: quasi continuous process, dry fermentation (TS = 15%–45%), one-stage, mesophilic or thermophilic, inoculation of biomass with digestion residue (Strabag Umwelttechnik GmbH, Düsseldorf).

waste throughput and the length of the reactor. These reactors are quite often used also for fermentation of separated collected biowaste.

Once the mechanical pretreatment serves for an appropriate conditioning of the substrate, then the anaerobic process is standard technology. The odorous off-gas problematic is only relevant for the pre- and posttreatment, since during anaerobic treatment only biogas is produced which is used for energy production. Fig. 4.1.11 shows as an example a mass balance of an MBT process.

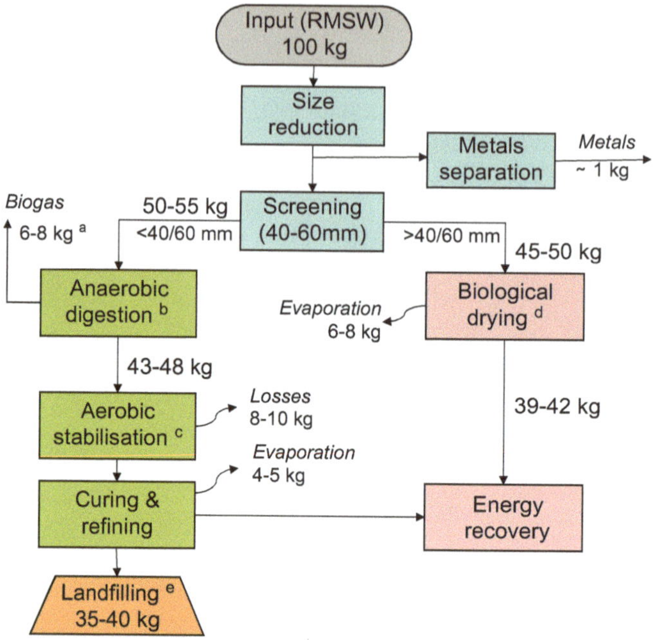

Figure 4.1.11 Example of a mass balance of an anaerobic treatment process of residual municipal solid waste (RMSW). Notes: (a) 80–100 Nm^3/t of digester input; (b) 14 days retention time in dry, single phase, thermophilic digester; (c) 4 weeks (of this at least 3 weeks in an enclosed system with forced aeration and turning); (d) 14 days biological drying process, (e) dry mass content of approximately 35% achieved through biological drying process. *Modified from Müller et al. (2001).*

When anaerobic processes are used, a dewatering and perhaps a drying step of the digested material will be in general necessary. The postrotting phase has to be practiced to meet the limit values presented in Table 4.1.1. These values cannot be met if only an anaerobic treatment takes place. If after a wet anaerobic fermentation step also a wet aerobic step follows (see Fig. 4.1.12) a dewatering and drying step may be sufficient. This is a new concept that has been proven at several plants (Heerenklage and Stegmann, 2005). The wet biological oxidation step replaces the postcomposting step.

OFF-GAS TREATMENT

The unloading as well as the mechanical biological pretreatment has to be practiced in-house in Germany. This means, that all buildings have to be ventilated and the off-gases have to be captured and treated. Due to the German off-gas emission standards (30. BImSchV; Anonymus, 2001) the treated off-gas from the whole plant has to meet target values that are presented in Table 4.1.2.

Off-gas from the unloading and mechanical pretreatment area in general can be treated in biofilters. The off-gas from the aerobic biological treatment plant, especially from the intensive rotting phase, has

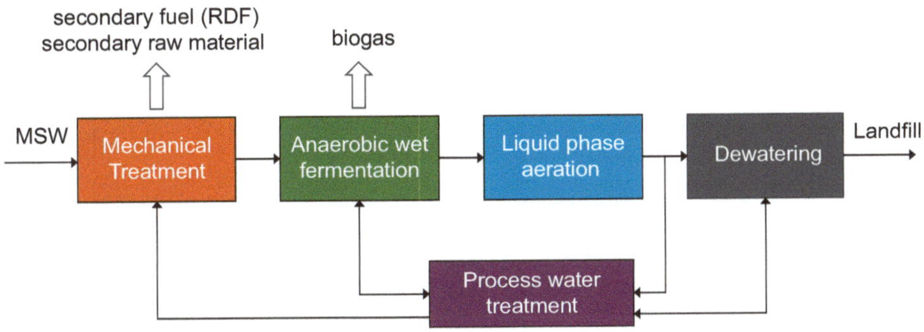

Figure 4.1.12 Waste (municipal solid waste, MSW) with subsequent aeration in the liquid phase. *RDF*, refuse derived fuel.

in general relatively high nitrogen and carbon concentrations so that a thermal treatment (e.g., regenerative thermal oxidation) for the removal of the organic compounds and an acid scrubber for the reduction of NH_3 compounds may be necessary (see also Doedens, 2001). Siloxanes and other trace compounds may cause clogging and/or corrosion. The off-gas treatment contributes significantly to the costs of the entire MSW-pretreatment process.

FINAL REMARKS

Meeting the German standards for the mechanical biological treatment of MSW prior to landfilling means that sophisticated technologies and intensive operation have to be applied. As a result this process is costly. It is a huge effort to reduce the biodegradable organic content of the residual MSW before landfilling. The advantage of this process is the production of RDF but only when it can be adequately used for energy production. Dependent on the kind of utilization, there may be different quality requirements, which would mean more efforts in the mechanical treatment. Although the biologically treated waste fraction meets the prescribed (not easy to achieve) standards, the landfill will not be free of emissions. There is still gas produced, and the leachate has to be treated prior to discharge into natural waters. There is not much experience about the amount and kind of emissions and the long-term behavior of an MBT landfill (see Chapter 14.1). The very strict air emission standards for the off-gases from the biological treatment plants (Table 4.1.2) add highly to the complexity of the plant and to the cost.

Regarding the measuring method for the determination of the target values (Table 4.1.1) for the AT4 and the GB21, there are no precise procedures available that guarantee the same results independent on the executing laboratory.

A question arises regarding the rationality of the set air emission standards that have to be met after treatment of the off-gas from the different sections of MBT-treatment plants. In Germany, these standards are not yet required for composting plants where separately collected kitchen and yard waste is treated. This may change in the future. Regarding the organic parameter (TOC < 55 g/t waste) that has to be met in the off-gas, this is only achievable after thermal treatment of the off-gases. The question is whether the high costs associated with this highly sophisticated treatment is in appropriate relation to the avoided emissions. Compared to the utilization of a bioscrubber/biofilter combination, instead of a thermal treatment mainly residual methane in the off-gas may not be reduced (see also Doedens, 2001).

There are in general two approaches for the biological treatment:

- Achieving high biological stability as set, e.g., by the German and Austrian government.
- Reduce the degree of biological stabilization only to a certain degree.

In the latter case, this pretreatment method is more a measure for the acceleration of the anaerobic processes once this partly stabilized waste is landfilled. Due to the reduced easy degradable organic content in the waste the methanogenic phase in the landfill is reached faster, so that the landfill can be run as a bioreactor landfill probably supported by means of leachate recirculation. Such a concept may be realized with a simple pretreatment (e.g., windrow composting) on the landfill surface. But also in this case a controlled operation will be necessary.

Another concept is the production of RDF from the MSW. In this case the waste is mechanically pretreated to reduce bulky materials, sand and metals and perhaps other compounds. This material is then composted—without water addition—until it has reached low water content ("dry"composting) and the biological processes have come to an end. This material can be further treated to upgrade the quality. Since the material can be stored (no biological activity), it can be easier handled. It is the goal to use this material as an RDF preferably in decentral plants for energy production. This concept has been realized in several plants in the United Kingdom and perhaps elsewhere.

References

Anonymus, 1999. Richtlinie 1999/E6 des Rates vom 26.04.1999 über Abfalldeponien. Amtsblatt der Europäischen Gemeinschaften 182.

Anonymus, 2000. Verordnung über die umweltverträgliche Ablagerung von Siedlungsabfällen und über biologische Behandlungsanlagen. http://www.bmu.de.

Anonymus, 2001. 30. Verordnung zur Durchführung des Bundesemissionsschutzgesetzes (Verordnung über Anlagen zur biologischen Behandlung von Abfällen – 30. BImSchV) vom 31.01.2001. http://www.bmu.de.

Christensen, T.H., Cossu, R., Stegmann, R. (Eds.), 1992. Landfilling of Waste: Leachate. E + FN Spon/CRC Press, ISBN 0419161406.

Doedens, H., 2001. Abgas aus Mechanisch-Biologischer Abfallbehandlung – Einleitung. In: Stegmann/Doedens/Heusel (Ed.), Abluft 2001, Hamburger Berichte 17. Verlag Abfall Aktuell, Stuttgart.

Fricke, K., Münnich, K., Ziehmann, G., Wallmann, R., 2001. Auswirkungen der neuen Abfallverordnungen auf die MBA- und Deponietechnik, (Korrespondenz Abwasser), KA-Wasser-Wirtschaft Abwasser. Abfall (48) Nr. 10.

Heerenklage, J., Stegmann, R., 2005. Mechanical biological pretreatment using the anaerobic/aerobic treatment concept in the liquid phase. In: Christensen, Cossu, Stegmann (Eds.), Proceedings from the 10th International Waste Management and Landfill Symposium, SARDINIA 2005. CISA, Padova, Italy.

Ketelsen, K., Kanning, K., Fehre, E., 2005. Vergleich von Konzepten für die biologische Stufe von MBA. In: Gallenkemper, Bidlingmaier, Doedens, Stegmann (Eds.), Münsteraner Schriften zur Abfallwirtschaft Band 8, 9. Münstereaner Abfalltage. Eigenverlag, ISBN 3-9806149-7-2.

Leikam, K., Stegmann, R., 1996. Stellenwert der mechanisch-biologischen Restabfallvorbehandlung. Abfallwirtschafts 9/96, 39–44.

Leikam, Stegmann, 1997. Mechanical-biological pretreatment of residual municipal solid waste and the landfill behaviour of pretreated waste. In: Christensen, Cossu, Stegmann (Eds.), 6th International Landfill Symposium, SARDINIA 1997, Proceedings. CISA, Italy.

Müller, W., Wellmann, R., Fricke, K., 2001. Anaerobic digestion as an integrated part of a bio-mechanical waste treatment process. In: Christensen, Cossu, Stegmann (Eds.), Proceedings from the 8th International Waste Management and Landfill Symposium, SARDINIA 2001. CISA, Padova, Italy.

Thomé-Kozmiensky, K.J., 2004. Ersatzbrennstoffe 4. TK Verlag Karl Thomé-Kozmiensky, ISBN 978-3-935317-20-7.

Further Reading

Fricke, K., Santen, H., Bahr, T., Avilar, K., Goedecke, H., Wallmann, R., Engelhard, T., 2004. Neue Verfahren zur mechanisch-biologischen Restabfallbehandlung. In: Stegmann, Rettenberger, Bidlingmaier, Ehrig, Fricke (Eds.), Deponietechnik 2004, Hamburger Berichte 22. Verlag Abfall Aktuell, Stuttgart.

Heerenklage, J., Stegmann, R., 2003. Anaerobic pretreatment of municipal solid waste with subsequent aeration in the liquid phase. In: Christensen, Cossu, Stegmann (Eds.), Proceedings from the 9th International Waste Management and Landfill Symposium, SARDINIA 2003. CISA, Padova, Italy.

Heerenklage, J., Stegmann, R., 2004. Anaerobe/Aerobe Vorbehandlung von Teststoofen in der Flüssigphase, Müll und Abfall. Heft 5, 2004.

Lechner, P., Humer, M., 2001. Methanoxidation in Deponieabdeckschichten. In: Stegmann, Doedens, Hensel (Eds.), Abluft 2001, Hamburger Berichte 17. Abfallwirtschaft, Technische Universität Hamburg-Harburg, Verlag Abfall-Aktuell, ISBN 3-9806505-9-6.

Martens, J., Stegmann, R., 2004. Die moderne Reststoff-Deponie, Konzept einer innovativen Deponierungstechnik für Reststoffe aus MBA-Anlagen, Kongress/Messe Waste to Energy. Bremen.

Stegmann, R., 1993. Design and Management of a dry landfill system (DLS). In: Christensen, Cossu, Stegmann (Eds.), Proceedings from the 4th International Landfill Symposium SARDINIA 93. CISA, Italy.

Stegmann, R., 1997. Description of a laboratory scale method to investigate anaerobic degradation processes taking place in landfills. In: Christensen, Cossu, Stegmann (Eds.), SARDINIA 97, Proceedings. CISA, Italy.

Stegmann, R., Heyer, K.-U., 2001. Landfill concept for mechanical-biologically treated residual waste. In: Christensen, Cossu, Stegmann (Eds.), SARDINIA 2001, 8th International Landfill Symposium, Proceedings. CISA, Italy.

Stegmann, R., Heyer, K.-U., 2002. Konzept für eine nachsorgearme MBV-Deponie. In: Stegmann/Rettenberger/Bidlingmaier/Ehrig (Ed.), Deponietechnik 2002, Hamburger Berichte 18. Verlag Abfall Aktuell, Stuttgart.

Stegmann, R., Heyer, K.-U., 2005. Landfilling of mechanically-biologically pretreated waste. In: Christensen, Cossu, Stegmann (Eds.), Proceedings from the 10th International Waste Management and Landfill Symposium, SARDINIA 2005. CISA, Padova, Italy.

4.2

TREATMENT AND DISPOSAL OF INCINERATION RESIDUES

Alessandra Polettini, Thomas Astrup, Giovanna Cappai, Peter Lechner, Aldo Muntoni, Raffaella Pomi, Tom Van Gerven and André van Zomeren

TYPES AND CHARACTERISTICS OF WASTE INCINERATION RESIDUES

The incineration process typically generates solid, liquid, and gaseous effluents. The solid residues (which account approximately for one-fourth of the waste mass on a wet weight basis) can be classified as follows (Sabbas et al., 2003):

- Bottom ash, consisting primarily of coarse noncombustible materials and unburned organic matter collected at the outlet of the combustion chamber in a quenching/cooling tank.
- Grate siftings, including relatively fine materials passing through the grate and collected at the bottom of the combustion chamber. Grate siftings are usually combined with bottom ash, so that in most cases it is not possible to separate the two waste streams. Together, bottom ash and grate siftings typically represent 15%–25% by mass of the waste input on a wet basis.
- Boiler and economizer ash, which represent the coarse fraction of the particulate carried over by the flue gases from the combustion chamber and collected at the heat recovery section. This stream may constitute up to 4% by mass of the waste input on a wet basis.
- Fly ash, the fine particulate matter still in the flue gases downstream of the heat recovery units, is removed before any further treatment of the gaseous effluents. The amount of fly ash produced by a municipal solid waste (MSW) incinerator is in the order of 1%–1.5% of the waste input mass on a wet basis.
- Air pollution control (APC) residues, including the particulate material captured after reagent injection in the acid gas treatment units prior to effluent gas discharge into the atmosphere. This residue may be in a solid, liquid, or sludge form, depending on whether dry, semi-dry, or wet processes are adopted for APC. The amount of APC residues generated varies largely with the type of control system adopted and may range from 0.05% to 4% of the waste input on a wet basis (Fig. 4.2.1).

Fig. 4.2.1. reports the typical solids mass flows for a grate-type MSW incinerator.

On account of the mass transfer phenomena and chemical transformations of the original waste constituents during the thermal process, as well as the modes of residues generation, they are recognized to possess dramatically different physical, chemical, mineralogical properties and environmental behavior. Fig. 4.2.2 shows typical composition ranges for bottom ash, fly ash, and APC

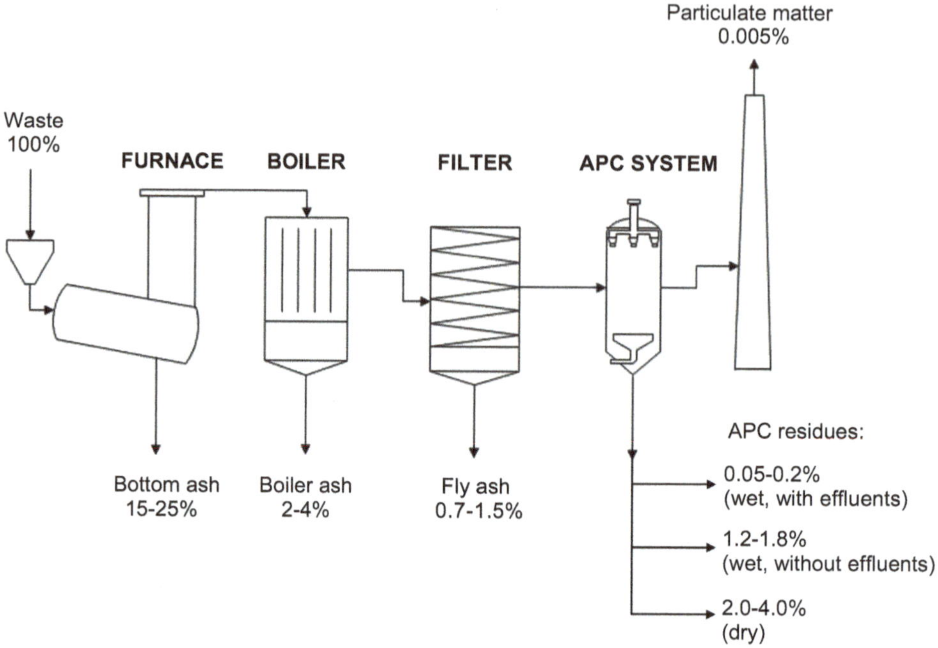

Figure 4.2.1 Typical solid mass flows for grate-type municipal solid waste incinerators. APC, air pollution control. *Adapted from European Commission (2006); Vehlow and Seifert (2012).*

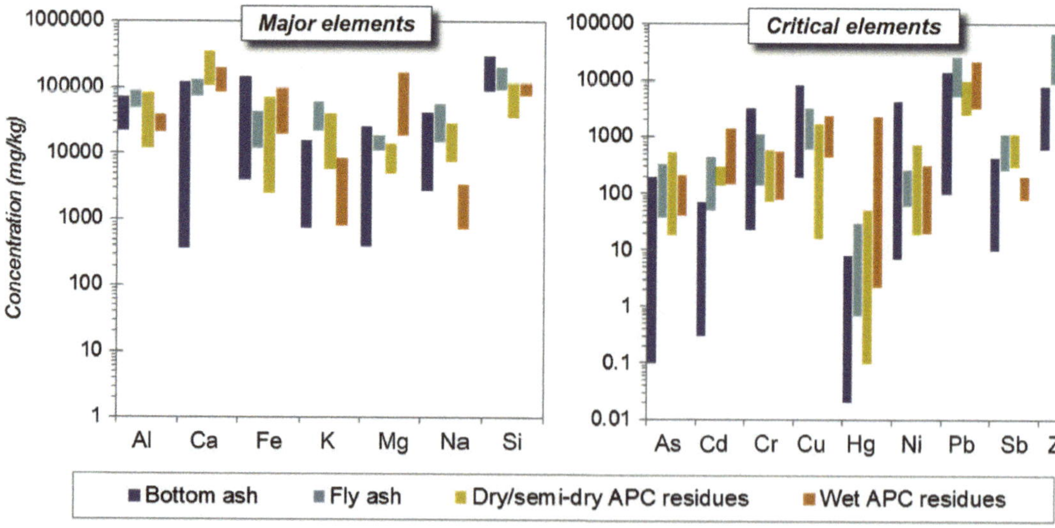

Figure 4.2.2 Typical composition ranges for different types of waste incineration residues. APC, air pollution control. *Adapted from IAWG (1997).*

residues, indicating the wide variability of residues composition for both major constituents and critical elements. Aspects such as the composition and types of the input waste, the combustion technology, the furnace operating conditions as well as the flue gas treatment system can be considered to have significant effects on the composition of the solid residues generated (e.g., Astrup et al., 2011).

ENVIRONMENTAL PROPERTIES OF WASTE INCINERATION RESIDUES

The main environmental concern in relation to incineration residues is the potential for release of contaminants by leaching. The leaching characteristics of such ashes are typically fairly more consistent than their largely variable elemental composition. Generally, fly ashes and APC residues show higher release of contaminants such as salts and trace metals than bottom ashes. In Fig. 4.2.3, the leaching of several elements from different incineration residues is given as a function of pH (using a pH-dependent leaching test). Although this figure shows the leaching over a wide range of pH values, the actual leaching of the different materials can be derived by comparing the emissions at the appropriate natural pH of the material. Leaching from incineration ashes is to a large extent controlled by the geochemical properties and by the major minerals present in the matrix (e.g., Meima and Comans, 1997; Meima and Comans, 1999; Dijkstra et al., 2002, 2006a), so that the release of contaminants is typically not affected significantly by day-to-day changes in waste input composition (Hyks and Astrup, 2009). An important aspect of incineration residues is the potential for leaching over very longtime frames. Typically, only a low portion of the metal content is leached within the first 100 years in a landfill (Astrup et al., 2006; Hyks et al., 2009). This means that most of the pollution potential remains in the residues for extended periods of time, but this also indicates that management strategies should be selected with care to be fully sustainable.

The potential environmental impacts associated to either utilization or final disposal of incineration residues generally include (1) dust emissions; (2) leachate production; (3) gas generation (see Sabbas et al., 2003 for further details).

On the basis of the anticipated effects on the environment, suitable management strategies should be sought to adopt appropriate criteria for ash treatment/utilization/disposal in an environmental sustainability perspective. To this purpose, it is clear that, in light of the mentioned differences in ash properties and related environmental behavior, the utilization and disposal strategies to be applied should reflect the intrinsic properties of the material.

MANAGEMENT STRATEGIES FOR WASTE INCINERATION RESIDUES

In a sustainability perspective, various approaches can be adopted for the management of waste incineration residues, which imply undertaking measures at different stages of their overall life cycle, namely (Sabbas et al., 2003):

- before utilization or final disposal
- during landfilling and active landfill operation
- during the passive phase of the landfill

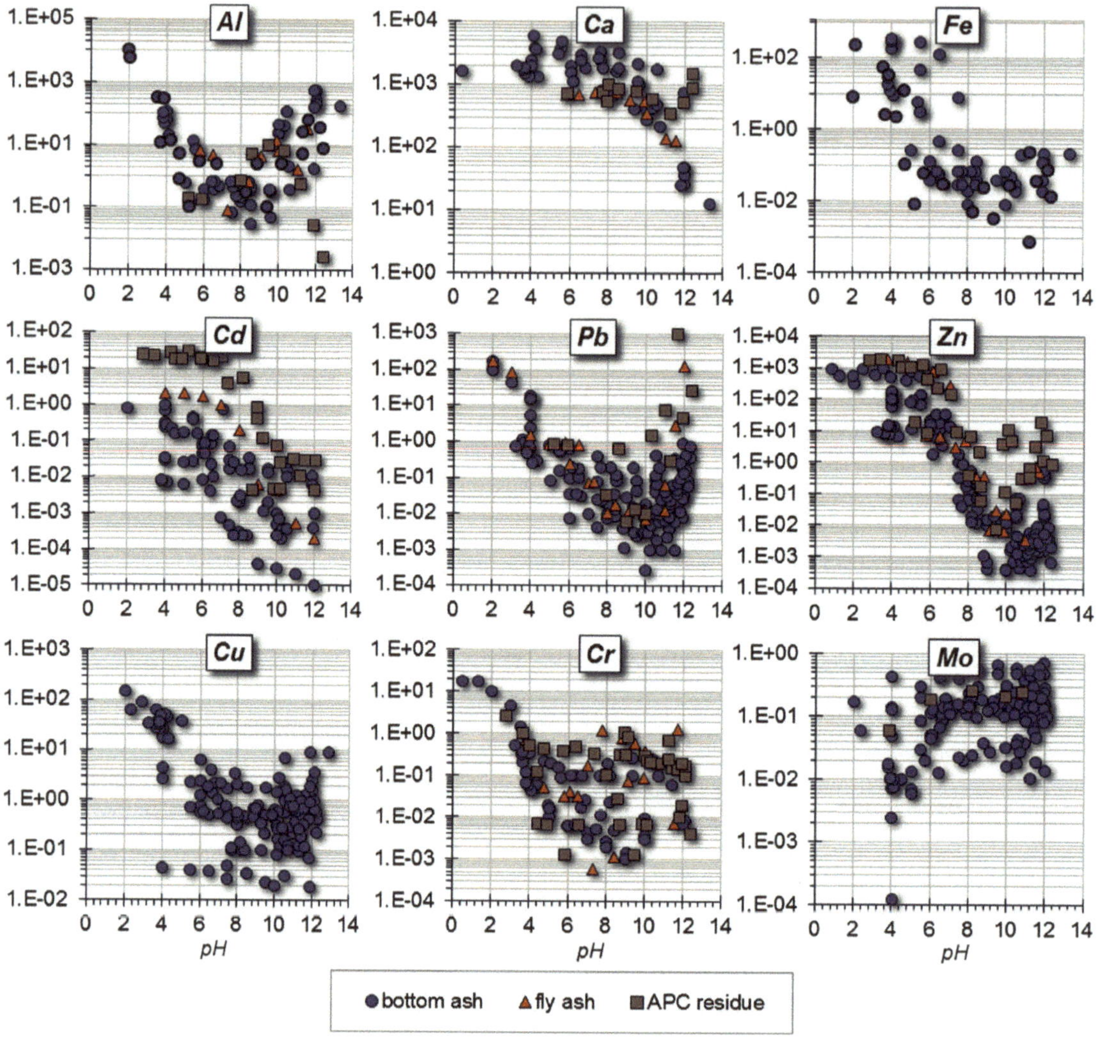

Figure 4.2.3 Leaching of major elements and metal contaminants as a function of pH from different types of waste incineration residues—all concentrations in mg/L. APC, air pollution control. *Readapted from Lechner et al. (1997).*

In general the overall target of such strategies is to produce an improvement in the environmental behavior of the residues (e.g., controlling the potential release of contaminants) and in the functional properties of the residues (e.g., enhancing mechanical strength, durability, stability). Of course, such improvements should not be pursued if the required residue treatment induces larger overall environmental impacts than the improvement of the residues warrant. This claims for a holistic assessment of the overall environmental benefits and burdens associated to each management option. While such an analysis is beyond the scopes of this chapter, the reader is referred to other studies (see, e.g., Fruergaard et al., 2010; Turconi et al., 2011) for further details.

Strategies Adopted Before Utilization or Final Disposal

Among the various type of waste incineration residues, bottom ash is recognized to be commonly well suited for use in construction works of various kinds. The other types of residues, as mentioned previously, are typically disposed in hazardous waste landfills, in several cases after appropriate pretreatment.

Due to the intrinsic properties of bottom ash, utilization is routinely practiced particularly in European countries including Denmark, France, Germany, and The Netherlands (ISWA, 2006; Astrup, 2007), or is being progressively regarded as a viable option (Belgium, Spain). Other countries such as Switzerland and the United States mainly landfill their bottom ashes (ISWA, 2006). In Table 4.2.1 an overview of bottom ash utilization practices is reported, with reference to the average amounts recycled and treatment processes typically applied prior to utilization. The differences in ash management are generally a result of traditions, legislation, and availability of natural resources, rather than being caused by differences in technical and environmental properties of the ashes. Common utilization options include metal recovery and engineering applications (road construction, building construction, embankments, noise barriers, filling materials for land reclamation, and similar options), followed by final disposal of the remnants if not suitable for utilization.

Mitigating measures may be required prior to utilization or landfilling with the purpose of improving both the environmental behavior and the technical properties of the residues to a level which of course depends on the intended final destination. Strategies adopted before reuse or disposal also include operations aimed at separating valuable materials (e.g., scrap metals) for industrial applications, at the same time rendering the quality of the remnants suitable for final disposal.

In this group of mitigating measures several pretreatment options are included, which basically can be classified as follows (IAWG, 1997; Van der Sloot et al., 2001; Sabbas et al., 2003; Polettini et al., 2007):

- physical or chemical separation
- chemical stabilization
- thermal treatment

Table 4.2.2 provides a brief overview of the pretreatment options available for incinerator residues processing. Some of the options are commonly practiced at the full scale, while others are more innovative processes, which are currently at a level of scientific development. It should be noted that while some of the listed pretreatment options are typically suited for preparation of the material to reuse, others are mainly applied in view of final disposal.

As far as bottom ash is concerned, the regulatory requirements established for utilization or landfilling in different countries are such that leaching-related limit values are not always complied by both the fresh and the weathered material. A number of components in bottom ash have been

Table 4.2.1 Overview of bottom ash utilization in different countries

Country	Bottom Ash Production (t/y)	Bottom Ash Utilization [t/y (%)]	Type of Processing			
			Screening	Ferrous Metal Recovery	Nonferrous Metal Recovery	Aging
Austria	232,000	n.a.	×	×		
Belgium	590,000	n.a.	×	×	Some	×
Denmark	645,000	630,000 (98%)	×	×	×	×
Finland	9,781	–				
France	2,995,000	2,366,000 (79%)	×	×	Often	×
Germany	3,140,000	2,025,000 (65%)	×	×	Some	×
Hungary	53,000	n.a.	n.a.	n.a.	n.a.	n.a.
Italy	750,000	150,000 (20%)	×	×	Some	
The Netherlands	1,075,000	950,000 (88%)	×	×	×	×
Norway	197,000	102,000 (52%)	n.a.	n.a.	n.a.	n.a.
Portugal	178,000	–				
Sweden	450,000	40,000 (9%)	n.a.	n.a.	n.a.	n.a.
Switzerland	640,000	–				
Spain	250,000	n.a.	n.a.	n.a.	n.a.	n.a.
United Kingdom	725,000	410,000 (56%)	×	×	Some	
United States	9,000,000	500,000 (5%)	n.a.	n.a.	n.a.	n.a.

n.a., data not available.
Readapted after ISWA (2006).

Table 4.2.2 Main processing options for incineration residues prior to utilization or final disposal

Type of Treatment	Physical separation	Chemical Separation	Chemical Stabilization	Thermal Treatment
	size separation	washing	natural and accelerated weathering	sintering
	size reduction	chemical	carbonation	vitrification
	magnetic separation	extraction	stabilization/binding with chemical agents	
	eddy current separation		blending with cement	

Readapted from Sabbas et al. (2003).

identified as critical contaminants; worldwide, the leaching of As, Cu, Mo, Pb, Sb, Zn, chloride, and sulfate is typically reported to exceed the local limit values for leaching, although the extent to which such limits are exceeded can vary from country to country depending on the required leaching test and the impact assessment method used to derive the regulatory thresholds.

When ash reuse is not allowed due to either regulatory constraints or lack of economic benefits, treatment may still be required in view of final disposal to improve its environmental behavior and to meet the limit values for the relevant landfill category (EU Landfill Directive). One of the desirable aims of the treatment may involve accelerating the chemical and mineralogical transformations that would occur in the medium to long term at the final disposal site, promoting the formation of a solid material with an environmentally acceptable contaminant release, ideally since the start of landfill disposal. This involves the concept of final storage quality (Hjelmar and Hansen, 2005), which refers to a situation where active environmental protection measures at the landfill are no longer necessary since the quantity and quality of emissions are acceptable for the environment. In other cases, where this is not feasible, the available options include extraction of contaminants from the residues prior to final disposal, or immobilization of contaminants within the waste matrix.

In the following sections the most common treatment processes will be discussed.

Physical Separation

While physical separation is rarely applied as a pretreatment of incineration residues in view of landfilling, it is often practiced for utilization purposes, aiming at isolating, mainly from bottom ash, specific materials (i.e., ferrous metals, non-ferrous metals, mineral constituents) potentially suitable for recycling, or size fractions and components that may hinder reuse in selected applications. The presence of undesirable materials/fractions in the ash may be due to two different reasons:

1. Contaminants in incineration residues are typically more abundant in the finer fractions due to their higher specific surface area.

2. The presence of fine particles in the material may have negative impacts of different nature (e.g., dust generation, high water absorption, etc.).

The separated fractions need to be treated individually to be used in civil or industrial applications, or they have to be disposed complying with the limit values established for landfilling.

Separation of the fine fractions may also positively influence technical properties such as particle size gradation, hydraulic conductivity, water demand, and ability to withstand freeze–thaw cycles. Size separation can be accomplished through both dry and wet processes. In dry separation standard drums or flat deck screens with a typical mesh size of 20–50 mm and 2–10 mm are applied to separate the oversize from the fine fraction. Wet separation can be accomplished through dense medium separation, sieving, cycloning, or high flow washing where attrition is used to facilitate fines segregation. In this respect, it should be noted that the use of high-density solutions may require the addition of solutes that may affect the quality of the separated fractions; for this reason, a careful assessment of the advantages of the size separation process should be conducted. Wet separation techniques may also combine the principle of physical separation with water extraction of soluble bottom ash constituents (see below), to an extent that depends on the residence time in the separation apparatus.

Fig. 4.2.4 reports the leaching test (EN 12457-2) results for four size fractions of an MSW incinerator bottom ash (Arickx et al., 2006):

- sludge fraction: $\Phi \in [0; 0.1]$ mm (8%),
- a sand fraction: $\Phi \in [0.1; 2]$ mm (32%)
- two gravel fractions: $\Phi \in [2; 6]$ mm (14%) and $\Phi \in [6; 50]$ mm (27%).

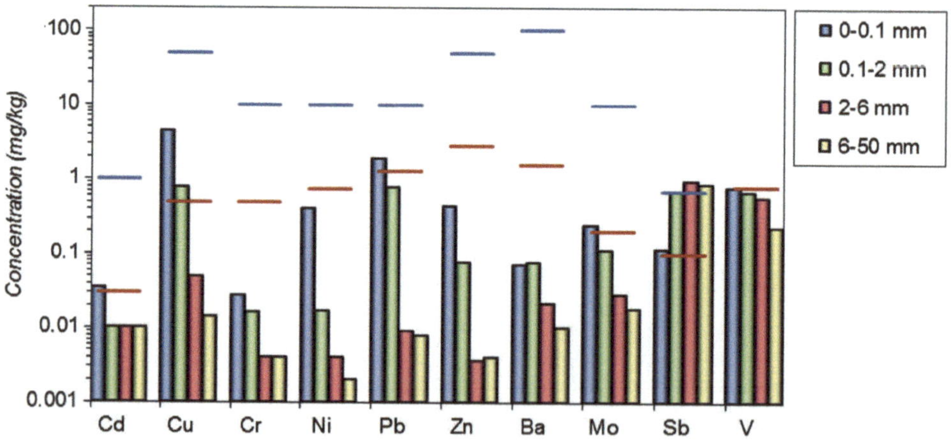

Figure 4.2.4 Effect of size separation on leaching (EN 12457-2) and comparison with regulatory limits (*red lines*: Flemish standards for formally and informally regulated metals; *blue lines*: EU limits for nonhazardous waste landfills)—As and Se were below the detection limits and are not reported here. *After Arickx et al. (2006).*

In the plots, the amounts of metals leached are compared to the limit values for utilization (formal and informal Flemish regulation) and non-hazardous waste disposal (EU Landfill Directive). The results show that leaching varied with particle size by more than two orders of magnitude. It is also noted that the fine fraction exceeded the leaching limits for utilization for several of the regulated contaminants, and also the corresponding limits for landfill disposal in a more limited number of cases (Cu and Pb).

Fig. 4.2.5 shows the results obtained from a density separation treatment using water and a $CaCl_2$ solution ($\rho = 1.35$ g/cm^3) as dense media (Van Gerven et al., 2006) to remove the (light) organic fraction from heavier particles in bottom ash. In the experiments, ~5−10 wt.% of bottom ash was recovered as the light fraction. It was found that the process was particularly effective toward Pb, reducing leaching below the limit value for utilization. Although the leached amount of Cu was significantly reduced by up to three times, the regulatory limit for reuse was still exceeded by a factor of 10. For Cr, the separation treatment was not found to have appreciable effects, so that leaching was comparable before and after the process.

Apart from the benefits associated to improved leachability and physical properties, a major drawback of size separation is that when the fine fraction accounts for a significant portion of the total ash mass, the treatment may not be valuable from an environmental point of view. The need for treatment of the wastewater stream generated by wet separation must also be accounted for when judging the overall benefits of the separation process. In addition, the introduction of substantial amounts of salts with the dense solution used for the wet separation may also be a concern for the practical feasibility of this method.

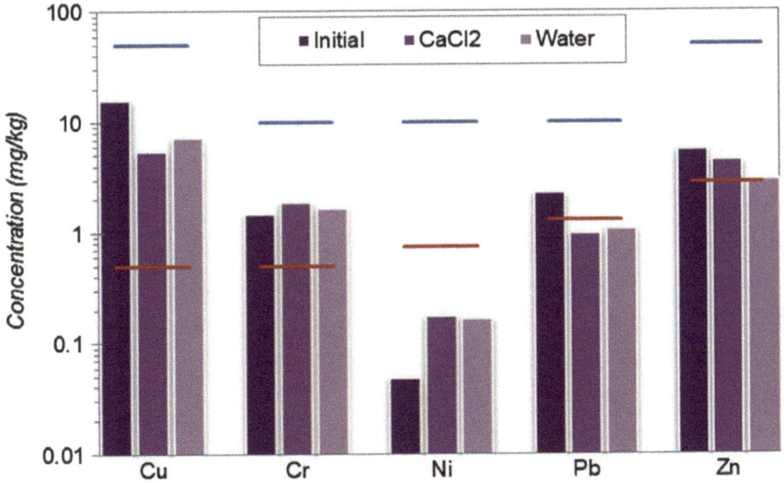

Figure 4.2.5 Effect of dense medium separation on leaching (EN 12457-2) and comparison with regulatory limits (*red lines*: Flemish standards for formally and informally regulated metals; *blue lines*: EU limits for non-hazardous waste landfills). *After Van Gerven et al. (2006).*

Ferrous and non-ferrous metals in MSWI bottom ash can account for 7%–15% and 1%–2% by weight, respectively, although this strongly depends on waste sorting strategies prior to the combustion process. The main aims of separation of ferrous and non-ferrous metals from bottom ash are recovering metal scraps in view of recycling as well as limiting the negative effects of metals such as Al, Fe, and Zn, which can cause swelling and expansion in some applications including road construction and formulation of blended cement or concrete products. Separation of ferrous and non-ferrous metals from bottom ash may be accomplished using magnets and eddy-current separators, respectively. Ferrous metal removal is a standard practice in most countries, while non-ferrous metal separation is less common although increasing in popularity due to availability of improved separation technologies, increasing economic value of metals, and increasing demand for resource recovery.

Chemical Separation

Among the chemical separation options, washing with water is the simplest process that can be applied to remove soluble constituents from the ash. Bottom ash quenching is routinely performed at the exit of the combustion chamber, however, the background concentrations in the quench water, the L/S ratio and the residence time in the quenching tank are such that this process cannot be regarded as a washing method. After quenching, bottom ash thus still contains residual soluble components which can be further extracted through a dedicated washing treatment.

Due to the alkaline nature of the material, the pH of fresh bottom ash/water suspensions typically lies between 9.5 and 12, corresponding to the range of solubility minimum for most metal species. Therefore the efficiency of trace metal removal from bottom ash through water washing is typically relatively low, and the treatment is mainly applied to remove major components that are readily soluble, such as chloride, Na, and K. Other undesired constituents such as sulfate are more hardly extracted, on account of the fact that leaching is typically solubility-controlled (at least in the L/S range adopted for washing) and the predominant mineral forms present (ettringite, gypsum, or anhydrite) are only moderately soluble in water. Enhanced sulfate dissolution yields have been reported using $NaHCO_3$ or CO_2 in the washing solution, due to the precipitation of Ca as carbonate in place of sulfate forms. However, it has also been reported that only in a few cases did assisted water washing produce a permanent effect on sulfate leaching (Astrup, 2007).

In some cases, a certain metal extraction capability may be displayed by bottom ash during water washing toward such metals as Cu, Zn, and to some extent Cr and Pb (Stegemann et al., 1995; Cossu et al., 2012; Yang et al., 2012), although it was also observed that the removal efficiency attained is typically inadequate to reduce leaching below the typical regulatory limits established by a number of countries for utilization.

For fly ash and APC residue, washing with water may also be applied as a pretreatment stage to remove soluble constituents (mainly chloride that may account for up to ~20% of the mass for APC residue) that may negatively affect the stabilization performance (in terms of leachability, water absorption, hardening, and corrosiveness) of subsequent treatments (Derie, 1996; Lundtorp et al., 1999; Mangialardi et al., 1999). Some studies (see, e.g., Derie, 1996; Nzihou and Sharrock, 2002) indicated that with an L/S ratio of 10 during water washing, about 90% of chloride salts is removed; as mentioned above, for other salts (e.g., sulfates) that are solubility controlled the efficiency of dissolution depends on the L/S ratio. The high pH of the extracting solution during water washing of, particularly, dry and semi-dry APC residues also typically results in concomitant extraction of Pb, Zn, and Cd along with soluble salts (Sabbas et al., 2003), although the amount removed from the ash is generally still too low to allow for residue detoxification (Lamers and Born, 1994).

Assisted washing may also be applied to promote the extraction of selected contaminants from waste incineration residues. The application of CO_2-based washing has been reported to allow for extraction of a number of trace metals. Fig. 4.2.6A (Polettini et al., 2005) shows the amount of metals extracted from bottom ash during washing using continuous CO_2 bubbling into deionized water (extraction times = 24 and 48 h). It was observed that, irrespective of the treatment duration, about 7–9 mg/kg were extracted in the case of Cu, Ni, and Zn, while for the other investigated metals the amount mobilized was below 1 mg/kg. Despite some removal attained for selected contaminants, assisted water washing may, however, negatively impact on the leaching characteristics of treated bottom ash. As an example of this behavior, Fig. 4.2.6B shows the leaching (EN 12457-2) of metals from the washed bottom ash. The CO_2-based washing treatment had a mobilization effect on Cr, Cu, Mo, and Ni; Pb and Zn were the only metals positively affected by the washing process.

It should be emphasized that all physical separation, leaching, and extraction processes have an associated production of residues that need either to be further treated and/or landfilled. For instance, in a washing treatment, pollutants are partly transferred from the solid waste to the liquid effluent, the quality of which should be accurately monitored to define the required treatment. Of course, the issue of generation of new effluents requires proper consideration to assess the overall environmental profile of the process.

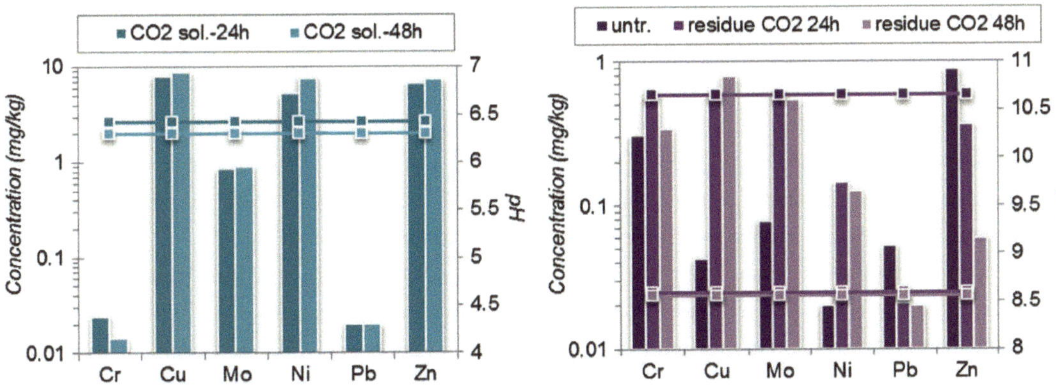

Figure 4.2.6 Effect of CO_2-water washing on (A) metal extraction from bottom ash and (B) metal leaching from the treated material. *Horizontal lines* indicate the solution pH. *After Polettini et al. (2005).*

Chemical Stabilization

The principle of chemical stabilization is to induce improvements in the physical (specific surface area, porosity, tortuosity, etc.), mechanical (durability, mechanical strength, etc.), and/or chemical properties of the material with the purpose of reducing the leachability of contaminants from the waste matrix.

For bottom ash, chemical stabilization typically involves changes that lead to the formation of low-solubility minerals with improved thermodynamic and geochemical stability. Among the treatments applied to bottom ash, natural weathering is one of the technologically simplest processes that can be applied to promote the transformation of its original constituents into more thermodynamically stable forms. Stockpiling of bottom ash under atmospheric conditions is considered as a cost-effective and simple stabilization treatment step to be applied, possibly in combination with additional processes. The treatment relies on a series of chemical and mineralogical transformations resulting from the contact between the ash and atmospheric agents (water, O_2, and CO_2) that resemble the mineralogical alteration processes documented for soils of volcanic origin. Although the natural weathering reactions require time frames in the order of hundreds to thousands of years to get to completion, several studies have shown that a certain degree of bottom ash stabilization can be attained for curing periods of some weeks to a few months. Within such durations, mineralogical alterations and pH changes may modify the leaching-controlling mechanisms altering the release of major ions and trace elements from bottom ash. Storage in open air removes mobile constituents due to washout by rainwater. It was found that the leaching of Cd, Pb, Zn, Cu, and Mo as well as chloride in some instances can be appreciably reduced after a couple of weeks to some months of outside storage (Steketee and Urlings, 1994; Meima and Comans, 1999; Chimenos et al., 2000; Gori et al., 2011). Another mechanism believed to play a role during weathering is organic matter degradation, with a reduction in DOC and a related effect on the leaching of metals (Cu above all) that are complexed by organic matter (Van Zomeren and Comans, 2004; Dijkstra et al., 2006a). However, a number of studies have also indicated stockpiling not being always adequate to meet the quality standards for bottom ash utilization, implying the need for additional treatment for specific contaminants including chloride, Cu, Cr, and Mo and possibly other metals such as Pb and Sb. An example is shown in Fig. 4.2.7, which reports the leaching test (EN 12457-2) results for the sand fraction (0.1−2 mm) of a bottom ash sample stored in contact with atmospheric air for 13 weeks (Arickx et al., 2006). A number of countries require a minimum weathering period before bottom ash utilization or disposal to promote the stabilization of the material; in Germany, for instance, natural weathering for at least 3 months is required.

Accelerated carbonation is another process that has recently deserved considerable attention due to its promising results for different types of incinerator ashes. Numerous studies (Meima et al., 2002; Fernández-Bertos et al., 2004; Cornelis et al., 2006; Dijkstra et al., 2006b; Polettini and Pomi, 2004; Polettini et al., 2005; Van Gerven et al., 2005; Costa et al., 2007; Baciocchi et al., 2010) have indicated that accelerated carbonation is capable of positively affecting some physical and chemical characteristics of bottom ash including porosity, buffering capacity, sorption properties, and leaching behavior.

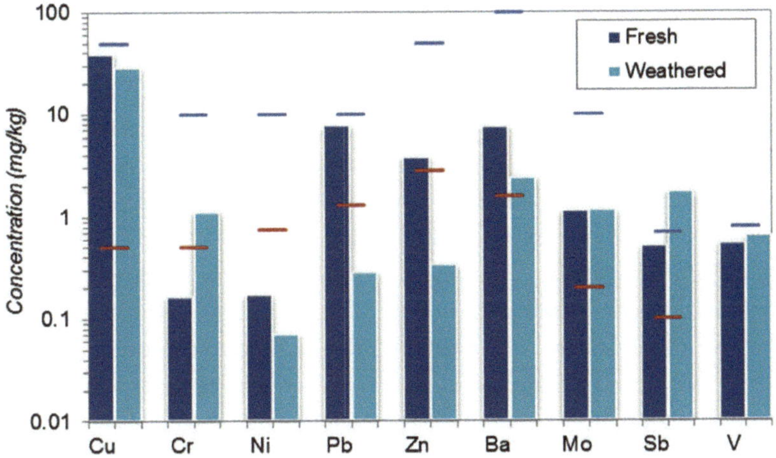

Figure 4.2.7 Effect of natural weathering of bottom ash on metal leaching (EN 12457-2) and comparison with regulatory limits (*red lines*: Flemish standards for formally and informally regulated metals; *blue lines*: EU limits for nonhazardous waste landfills). *After Arickx et al. (2006).*

In general, the overall effect of carbonation on leaching is the result of multiple mechanisms, including pH changes, precipitation of metal carbonates, and interaction with newly formed minerals, which in turn depend on both carbonation conditions and mineralogy of the starting material. The numerous processes underlying the macroscopic effects of carbonation of incinerator bottom ash make it difficult to define univocally the effect of the process on individual contaminants; however, in general terms it may be stated that Cu, Pb, and Zn are mostly immobilized by carbonation, while oxyanion-forming elements such as Cr, Mo, and Sb and also sulfate may become more mobile (Van Gerven et al., 2005; Costa et al., 2007; Baciocchi et al., 2010).

Similar considerations on the effects of carbonation on metal leaching can also be derived for APC residues (see, e.g., Costa et al., 2007; Baciocchi et al., 2009). Studies on fly ashes and APC residues have also indicated that the presence of reduction capacity in the material as Al(0) may reduce Cr(VI) to Cr(III) and thereby immobilize Cr by precipitation (Astrup et al., 2005; Hu et al., 2008). During carbonation and interaction with atmospheric O_2, the available Al(0) is also oxidized thereby removing the reduction capacity and maintaining Cr as the more mobile Cr(VI). These mechanisms may occur also in batch leaching tests indicating that leaching data regarding oxyanionic elements should be evaluated carefully (Astrup et al., 2005).

Other chemical stabilization processes used for incineration ash treatment involve the addition of species capable of increasing the sorption properties of the material, with positive effects on trace metal immobilization. The use of Al(III) and Fe(III) salts and other sorbing compounds (steel slag, apatite, allophane, bauxite) has been investigated to improve bottom ash leaching through precipitation of amorphous Al and Fe (hydr)oxides. However, the leaching of chloride and sulfate may also become a problem when Al(III) and Fe(III) salts are applied as additives. An immobilization effect has been observed for Cu and oxyanion-forming metals including Mo, Cr, and Sb, while mobilization was found for Ni, Zn, and the major cations Ca, Mg, and Na (Comans et al., 2000; Meima and Comans, 1998; Polettini et al., 2005). For fly ashes and APC residues, chemical stabilization has been carried out based on additions of Fe(II), CO_2, and phosphates, either for generation of sorption capacity or formation of sparingly soluble mineral phases (see, e.g., Lundtorp et al., 2002).

Other common chemical stabilization methods include solidification/stabilization processes (typically applied to fly ash and APC residue), which make use of hydraulic binders including cement, lime, and/or pozzolanic materials. Contaminant immobilization occurs through both physical and chemical mechanisms, either contribution depending on the specific contaminant under concern as well as on the type of binder used. The solidification/stabilization process is particularly sensitive to soluble salts, especially when hydraulic binders are used, which are negatively affected by high chloride and sulfate concentrations; furthermore, strongly amphoteric contaminants such as Pb and Zn may be more leachable from the treated matrix due to the high pore water pH, unless incorporation in the crystal lattice of the hydration products occurs or appropriate additives are used.

Although solidification/stabilization with hydraulic binders or chemical additives generally results in adequate leaching properties at relatively low costs, the increase in the volume of waste to be landfilled and the long-term performance of the process should be carefully evaluated.

For bottom ash, blending with cement or other binders can also be applied, mainly in view of recycling. Numerous applications have been considered for bottom ash use as an aggregate for production of concrete or bituminous structures and construction of road base layers. A major problem encountered is the swelling effect resulting from the oxidation of metallic aluminum (leading to voluminous aluminum hydroxide and hydrogen gas), ettringite formation, hydration of lime and magnesium oxide, and alkali–silica reactions. Other drawbacks are related to the generally weaker mechanical properties, higher water adsorption and lower resistance to freeze–thaw cycles. However, such negative effects may be minimized by appropriate pretreatment of bottom ash and careful evaluation of its content in the final product.

Thermal Treatment

Sintering and vitrification are thermal processing options that may be applied to incinerator residues to reduce volumes and obtain a strongly stable solid matrix. While vitrification involves melting the material at temperatures of 1000–1500°C producing a homogeneous liquid phase which is then rapidly cooled to generate an amorphous glassy product, sintering is conducted at temperatures below the melting point of the main constituents, leading to particle bonding and reconfiguration of chemical phases with incorporation of contaminants within the solid matrix.

When applied to bottom ash, thermal processing can result in improved characteristics in terms of mechanical strength, porosity, water absorption, chemical stability, and leaching of contaminants (see, e.g., De Gioannis et al., 2004; Hyks et al., 2011). The effect on leaching is due to both immobilization of lithophilic metals through chemical/physical mechanisms and vaporization of volatile species.

A treatment involving hydrothermal or salt-thermal processing of bottom ash producing a zeolitized material with ion exchange properties to be used for metal removal from industrial streams has provided promising results (Polettini et al., 2006).

Even low-temperature heating can improve leaching properties. A bottom ash heated for 30 min at 400°C displayed a reduction in Cu leaching, likely related to the destruction of organic matter, from 60 to 0.2 mg/kg, complying with Flemish standards for utilization (Van Gerven et al., 2006).

When applied to fly ash and APC residue, thermal processes are aimed at reducing the leachability of contaminants due to a combination of incorporation in the sintered/glassy matrix and volatilization of high-vapor-pressure compounds. Thermal treatment of incineration residues in most cases results in a more homogeneous, denser product with improved leaching properties (Sabbas et al., 2003). Vitrification also adds the benefits of physical encapsulation of contaminants in the glass matrix. However, due to the high energy consumption of thermal processes, they should only be applied if capable of producing significant improvements in ash quality as compared to other techniques. The loss of volatile species also typically requires treatment of the off-gases, an issue that needs careful evaluation when drawing the overall balance of the process. With regard to the long-term stability of the materials obtained from high-temperature processing of incinerator ashes, it is generally recognized that glasses and glass ceramics possess strong durability over time; although they may be subjected to some corrosion due to alkaline dissolution and hydrolysis of siloxane groups, the leaching of toxic metals under aggressive corrosion conditions is expected to occur to a very limited degree (see, e.g., Perret et al., 2003).

Strategies Adopted During Landfilling and Active Landfill Operation

Landfilling of incinerator residues should be regarded as a part of a proper disposal strategy, which also takes the pretreatment of the residues into account. In general, as also pointed out in other chapters of this book, landfill design should be aimed at reducing the duration of the phase requiring active environmental protection measures. At the end of the active care period, landfill emissions must reach an environmentally acceptable level.

With regard to landfill operation strategies for incineration residues, controlling the waste types to be disposed and redirecting these to appropriate landfills or landfill sections is deemed important, on account of the fact that mixing different types of wastes can negatively affect leaching. Residues of different types should be kept separate during landfilling whenever they exhibit different (in some cases even incompatible) environmental behaviors. This is, for instance, the case of co-disposal of incineration residues and organic wastes (such as MSW), which is in principle not recommended, since some metal contaminants (Cu and Pb above all) in the ash can be strongly mobilized in the presence of dissolved organic carbon. Likewise, codisposal of incineration residues of different type (e.g., bottom ash and APC residues) is also believed not to be advisable; as a matter of fact, the high content of readily soluble constituents in residues from flue gas treatment may require dedicated landfill operation strategies that would be meaningless for other types of materials. However, it should also be noted that in some instances codisposal may still be required by other constraints (lack of space for landfill siting, proximity of the landfill to waste generation sources, public acceptance, etc.) and the abovementioned disadvantages may thus not be avoided.

Short-term impacts from landfilling of waste incineration residues, including leachate production and dust generation, may also be affected by such landfill management strategies as compaction and covering operations, as well as leachate collection.

Leachate collection and treatment are of course measures required to control the liquid emissions from the landfill; however, appropriate strategies during the active care phase should be adopted for the quality of the produced leachate to reach adequate environmental standards within a reasonably short time frame. For residues with high contents of readily soluble constituents, promoting the release of contaminants by enhancing leaching through irrigation (either with water or with recycled leachate) may be beneficial with a view to attaining the final storage quality of the material in relatively short periods, thus reducing the time required for the active care phase of the landfill.

The environmental implications of leachate treatment should also be carefully evaluated when dealing with incineration residues. To this regard, given the conservative nature of most contaminants released by incineration residues, when the treatment of landfill leachate is accomplished using biological processes, control of pollutants will mostly rely on a dilution effect; conversely, if physical/chemical treatments are applied for metals and anions removal, additional residues will be generated. Since these contain elevated contaminant concentrations, they in turn require proper treatment and disposal. In both cases, release of contaminants to the environment is expected to occur. In this perspective, appropriate pretreatment of the residues also plays a crucial role in determining the environmental compatibility of the abovementioned strategies.

Another issue that deserves careful consideration is the generation of landfill gas. Although, for incineration residues monofills, minimal biological gas production rates are expected, abiotic landfill gas may still be generated in appreciable amounts (Lechner et al., 1997; Magel et al., 2001; Musselmann et al., 2000; Sabbas et al., 2003) as a result of oxidation of metallic constituents (Al, Fe, and Cu) in the presence of water. Due to its high solubility at $pH > 9.5$, lower redox potential in comparison to other elements and abundance in incineration residues, Al is regarded as the main element responsible for abiotic gas production, which is mainly related to H_2 generation according to the following reactions:

$$2Al^0 + 3H_2O \rightarrow Al_2O_3 + H_2$$
$$Al^0 + 2H_2O \rightarrow AlOOH + 1.5H_2$$
$$Al^0 + 3H_2O \rightarrow Al(OH)_3 + 1.5H_2$$

As a result of such reactions, particularly for incineration residues monofills, the potential production of abiotic landfill gas should be carefully evaluated, and dedicated gas control systems should be installed whenever required. Although separate waste collection methods aimed at separating aluminum packages certainly have an effect on the potential abiotic gas production from ash landfilling, both ferrous and nonferrous metals, as noted above, are still found in incinerator residues, indicating that gas generation will somehow still be expected. Furthermore, Al in the residues is typically associated to small-size scraps, which in fact limits the recovery rate that may be attained using eddy current separation of Al from the ashes. According to previous studies (Grosso et al., 2011), such separation techniques, which are usually applied to bottom ash only, are capable of attaining non-ferrous metal recoveries of 30% of the amount fed to the incinerator (~1% of the bottom ash mass).

A number of chemical reactions occurring in the disposal site may also be responsible for temperature development within the landfill. Examples of reactions potentially responsible for heat generation include hydration of alkaline and alkaline earth oxides, oxidation of metals, and carbonation of portlandite (Jaros and Huber, 1998). Numerous studies (see Sabbas et al., 2003; Yoshida and Rowe, 2003; Rowe and Islam, 2009, and references therein) have shown that exothermic reactions may cause temperatures as high as 90°C in MSWI residue monofills. The existence of prolonged elevated temperature conditions inside the landfill should be carefully taken into account in both the design and the operation phase, since heat-related damages of the lining and drainage systems have been reported on some instances in real landfills. Temperatures in excess of 50°C may lead to desiccation cracks in the clay liner and damages to geomembranes due to aging [involving antioxidant depletion and polymer degradation (see, e.g., Rowe, 2005; Rowe and Islam, 2009)], as well as increased diffusion coefficients and permeability of liners.

Strategies Adopted During the Passive Phase of the Landfill

During the passive phase of landfilling, protection of the environment relies on lining systems (top cover and geological barrier at the bottom) as well as on the quality attained by the waste at the end of the active care phase. Considering that artificial lining systems may not be able to guarantee adequate environmental protection for long time spans due to possible failures during the passive landfill phase (where no maintenance is performed, and material aging may play a role), it is mandatory that emissions from the landfill are reduced to an acceptable level at the time the passive phase is entered, by improving the residues quality through both waste pretreatment and appropriate management of the active phase. Thus, the stability of disposed residues at the beginning of the landfill passive phase can be considered as an additional (and likely more efficient) barrier compared to the top cover and the geological bottom liner.

Measures adopted during the active phase can also affect transformations that will occur during the passive phase. The progress of natural weathering of incineration residues during landfilling is expected to gradually lead to long-term changes in mineralogy and leaching behavior, thus affecting the potential environmental impacts of the landfill site.

OVERVIEW AND CHALLENGES

Regardless of their final destination (utilization/disposal), sustainable management of incineration residues should derive from a compromise between the following aspects:

- maximization of recovery of valuable fractions
- improvement in technical and environmental properties
- reduced footprint in terms of energy and raw materials consumption

This means that an assessment of the sustainability of management strategies should include the evaluation of the following aspects:

- ash upgrading (e.g., screening, sorting, metal recovery, maturing)
- transport of ashes and ash components
- further processing of recovered metals
- contaminant release over time by leaching
- long-term fate of the ash components
- quantity and quality of residues produced and long-term behavior, e.g., in landfills

With regard to bottom ash, metal recovery is always considered environmentally sound, as metal recycling saves production of new metal, which requires significant amounts of energy. Recovery of metals from bottom ash, therefore, saves energy and also the emissions to the environment caused by this energy consumption and the metal-making processes in general.

As pointed out above, various options for ash treatment exist aiming at improving the leaching behavior. A qualitative comparative assessment of the various treatment techniques for incinerator

Table 4.2.3 Comparative evaluation of the applicability of ash treatment methods

Principle of Treatment	Process	Bottom Ash	Fly Ash	APC Residues
Physical separation	Size separation	⇧	⇩	⇩
	Size reduction	⇧		
	Magnetic separation	⇧		
	Eddy current separation	⇧		
Chemical separation	Washing	⇧	⇧/⇔	⇧/
	Chemical extraction/mobilization	⇧	⇧/⇔	⇧/⇔
Chemical stabilization	Natural and accelerated weathering	⇧/⇔		⇧/⇔
	Carbonation	⇧	⇧	⇧
	Stabilization/binding with chemical agents	⇧	⇧	⇔
	Blending with cement	⇧	⇧	⇧/⇔
Thermal treatment	Sintering	⇧		
	Vitrification	⇧/⦸	⇧/⦸	⇩/⦸

⇧, proven efficacy; ⇔, promising results but needing further assessment/improvement/additional treatment; ⇩, ineffective or producing additional negative effects; ⦸, generally inapplicable due to costs.

residues is reported in Table 4.2.3. Finding a suitable balance in particular between the energy requirements of upgrading/treatment and the leaching properties of the final ashes at the point of utilization or landfilling, is another critical issue. In many European countries, a combination of metal recovery and natural weathering of bottom ash under ambient conditions before final utilization has provided such a balance.

The main environmental concern for incineration residues is the potential for leaching for extended periods of time. In landfills, and even in typical utilization applications such as road construction, generally only a fraction of the trace metal contents is released within a time span of, e.g., 100 years.

In conclusion, as for ash landfilling, the following guidelines can be adopted. First, proper disposal strategies are strongly dependent on the characteristics of the incineration residues to be landfilled. For residues containing only limited amounts of trace contaminants (such as bottom ash) the disposal strategy may rely, after some preliminary metal sorting and natural aging to reduce leaching to acceptable levels, on a single-stage landfill operation in which the leachate emission is controlled by passive environmental protection systems only. For residues containing significant amounts of readily soluble species in addition to trace elements (such as some kinds of APC residues), it may be advisable to adopt disposal criteria based on a two-stage operation, consisting of an active care phase with (possibly enhanced) leachate collection and treatment followed by a passive once after depletion of readily leachable constituents has occurred. The potential for gas production from incineration residue monofills should also be carefully considered—mainly due to safety reasons to prevent explosions—so that gas collection systems may be required during the early stages of landfill operation.

References

Arickx, S., Van Gerven, T., Vandecasteele, C., 2006. Accelerated carbonation for treatment of MSWI bottom ash. Journal of Hazardous Materials B137, 235–243.

Astrup, A., 2007. Pretreatment and utilization of waste incineration bottom ashes: Danish experiences. Waste Management 27, 1452–1457.

Astrup, T., Rosenblad, C., Trapp, S., Christensen, T.H., 2005. Chromium release from waste incineration air-pollution-control residues. Environmental Science and Technology 39, 3321–3329.

Astrup, T., Mosbæk, H., Christensen, T.H., 2006. Estimation of long-term leaching from waste incineration air-pollution-control residues. Waste Management 26, 803–814.

Astrup, T., Riber, C., Pedersen, A.J., 2011. Incinerator performance: effects of changes in waste input and furnace operation on air emissions and residues. Waste Management and Research 29, S57–S68.

Baciocchi, R., Costa, G., Di Bartolomeo, E., Polettini, A., Pomi, R., 2009. The effects of accelerated carbonation on CO_2 uptake and metal release from incineration APC residues. Waste Management 29, 2994–3003.

Baciocchi, R., Costa, G., Lategano, E., Marini, C., Polettini, A., Pomi, R., Postorino, P., Rocca, S., 2010. Accelerated carbonation of different size fractions of bottom ash from RDF incineration. Waste Management 30, 1310–1317.

Chimenos, J.M., Fernández, A.I., Nadal, R., Espiell, F., 2000. Short-term natural weathering of MSWI bottom ash. Journal of Hazardous Materials 79, 287–299.

Comans, R.N.J., Meima, J.A., Geelhoed, P.A., 2000. Reduction of contaminant leaching from MSWI bottom ash by addition of sorbing components. Waste Management 20, 125–133.

Cornelis, G., Van Gerven, T., Vandecasteele, C., 2006. Antimony leaching from uncarbonated and carbonated MSWI bottom ash. Journal of Hazardous Materials 137, 1284–1292.

Cossu, R., Lai, T., Pivnenko, K., 2012. Waste washing pre-treatment of municipal and special waste. Journal of Hazardous Materials 207–208, 65–72.

Costa, G., Baciocchi, R., Polettini, A., Pomi, R., Hills, C.D., Carey, P.J., 2007. Current status and perspectives of accelerated carbonation processes on municipal waste combustion residues. Environmental Monitoring and Assessment 135, 55–75.

De Gioannis, G., Muntoni, A., Polettini, A., Pomi, R., 2004. Co-treatment of different industrial wastes by means of sintering processes. In: Proc. SIDISA 2004, International Symposium on Sanitary and Environmental Engineering, Taormina, IT, 23–26 June 2004 (On CD ROM).

Derie, R., 1996. A new way to stabilise fly ash from municipal incinerators. Waste Management 16, 711–716.

Dijkstra, J.J., van der Sloot, H.A., Comans, R.N.J., 2002. Process identification and model development of contaminant transport in MSWI bottom ash. Waste Management 22, 531–541.

Dijkstra, J.J., van der Sloot, H.A., Comans, R.N.J., 2006. The leaching of major and trace elements from MSWI bottom ash as a function of pH and time. Applied Geochemistry 21, 335–351.

Dijkstra, J.J., van Zomeren, A., Meeussen, J.C.L., Comans, R.N.J., 2006. Effect of accelerated aging of MSWI bottom ash on the leaching mechanisms of copper and molybdenum. Environmental Science and Technology 40, 4481–4487.

Fernández-Bertos, M., Li, X., Simons, S.J.R., Hills, C.D., Carey, P.J., 2004. Investigation of accelerated carbonation for the stabilization of MSW incinerator ashes and the sequestration of CO_2. Green Chemistry 6, 428–436.

Fruergaard, T., Hyks, J., Astrup, T., 2010. Life-cycle assessment of waste incineration residues management. Science of the Total Environment 408, 4672–4680.

Gori, M., Bergfeldt, B., Pfrang-Stotz, G., Reichelt, J., Sirini, P., 2011. Effect of short-term natural weathering on MSWI and wood waste bottom ash leaching behaviour. Journal of Hazardous Materials 189, 435–443.

Grosso, M., Biganzoli, L., Rigamonti, L., 2011. A quantitative estimate of potential aluminium recovery from incineration bottom ashes. Resources, Conservation and Recycling 55, 1178–1184.

Hjelmar, O., Hansen, J.B., 2005. Sustainable landfill: the role of final storage quality. In: Proc. Sardinia 2005, Tenth International Waste Management and Landfill Symposium, S. Margherita di Pula (Cagliari, IT), 3–7 October 2005 (On CD-ROM).

Hu, Y., Hyks, J., Astrup, T., Christensen, T.H., 2008. Effect of drying on leaching testing of treated municipal solid waste incineration APC-residues. Waste Management and Research 26, 400–405.

Hyks, J., Astrup, T., 2009. Influence of operational conditions, waste input and ageing on leaching of contaminants from waste incineration bottom ashes. Chemosphere 76, 1178–1184.

Hyks, J., Astrup, T., Christensen, T.H., 2009. Long-term leaching from MSWI air-pollution-control residues: leaching characterization and modeling. Journal of Hazardous Materials 162, 80–91.

Hyks, J., Nesterov, I., Mogensen, E., Jensen, P.A., Astrup, T., 2011. Leaching from waste incineration bottom ashes treated in a rotary kiln. Waste Management Research 29, 995–1007.

IAWG (International Ash Working Group), Chandler, A.J., Eighmy, T.T., Hartlén, J., Hjelmar, O., Kosson, D., Sawell, S.E., van der Sloot, H.A., Vehlow, J., 1997. Municipal Solid Waste Incinerator Residues. In: Studies in Environmental Sciences, vol. 67. Elsevier Sci., Amsterdam.

ISWA, 2006. Management of Bottom Ash from WTE Plants: An Overview of Management Options and Treatment Methods (Copenhagen, Denmark).

Jaros, M., Huber, H., 1998. Development of pH and alkalinity in bottom ash – simulation of the long-term behaviour. In: Lechner, P. (Ed.), Waste Reports 08-November 1998 "Emission Behaviour of Incineration Residues". Abteilung Abfallwirtschaft/IWGA, Universität für Bodenkultur Wien.

Lamers, F.J.M., Born, J.G.P., 1994. Upgrading techniques for the quality improvement of municipal waste incineration residues. In: Environmental Aspects of Construction with Waste Materials. Studies in Environmental Sciences, vol. 60. Elsevier Sci., Amsterdam, pp. 645–654.

Lechner, P., Huber, H., Jaros, M., 1997. Assessment of the long-term behavior of MSW incinerator slag. In: Proc. Sardinia 97, Sixth International Landfill Symposium, vol. V, pp. 553–563.

Lundtorp, K., Jensen, D.L., Sørensen, M.A., Mogensen, E.P.B., Christensen, T.H., 2002. Treatment of waste incinerator air-pollution-control residues with $FeSO_4$: concept and product characterisation. Waste Management and Research 20, 69–79.

Magel, G., Heuss-Assbichler, S., Fehr, K.T., 2001. Abiotic H_2 generating reactions in a MSWI monofill. In: Proc. Sardinia 2001, Eighth International Waste Management and Landfill Symposium, vol. I, pp. 489–498.

Mangialardi, T., Paolini, A.E., Polettini, A., Sirini, P., 1999. Optimization of the solidification/stabilization process of MSW fly ash in cementitious matrices. Journal of Hazardous Materials 70, 53–70.

Meima, J.A., Comans, R.N.J., 1997. Geochemical modeling of weathering reactions in municipal solid waste incinerator bottom ash. Environmental Science and Technology 31, 1269–1276.

Meima, J.A., Comans, R.N.J., 1998. Reducing Sb-leaching from municipal solid waste incinerator bottom ash by addition of sorbent minerals. Journal of Geochemical Exploration 62, 299–304.

Meima, J.A., Comans, R.N.J., 1999. The leaching of trace elements from municipal solid waste incinerator bottom ash at different stages of weathering. Applied Geochemistry 14, 159–171.

Meima, J.A., van der Weijden, R.D., Eighmy, T.T., Comans, R.N.J., 2002. Carbonation processes in municipal solid waste incinerator bottom ash and their effect on the leaching of copper and molybdenum. Applied Geochemistry 17, 1503–1513.

Musselmann, C.N., Straub, W.A., Bidwell, J.N., Carpenter, J., 2000. Landfill gas generation at municipal waste combustor ash monofill-Franklin, New Hampshire. In: Sustainable Construction: Use of Incineration Ash.

Nzihou, A., Sharrock, P., 2002. Calcium phosphate stabilisation of fly ash with chloride extraction. Waste Management 22, 235–239.

Perret, D., Crovisier, J.L., Stille, P., Shields, G., Mäder, U., Advocat, T., Schenk, K., Chardonnens, M., 2003. Thermodynamic stability of waste glasses compared to leaching behavior. Applied Geochemistry 18, 1165–1184.

Polettini, A., Pomi, R., 2004. The leaching behaviour of incineration bottom ash as affected by accelerated ageing. Journal of Hazardous Materials B113, 209–215.

Polettini, A., Pomi, R., Ragaglia, M., 2005. Accelerated ageing as a tool for sustainable disposal of incinerator bottom ash. In: Proc. Sardinia 2005, Tenth International Waste Management and Landfill Symposium, S. Margherita di Pula (Cagliari, IT), 3–7 October 2005 (On CD ROM).

Polettini, A., Pomi, R., Lo Mastro, S., Sampieri, S., 2006. Hydrothermal zeolitization of MSWI bottom ash. In: Proc. WASCON 2006, Sixth International Conference on the Environmental and Technical Implications of Construction with Alternative Materials. Science and Engineering of Recycling for Environmental Protection, pp. 439–450.

Polettini, A., Astrup, T., Cappai, G., Lechner, P., Muntoni, A., Pomi, R., Van Gerven, T., van Zomeren, A., 2007. State-of-the-art and outlook on management of waste-to-energy bottom ashes. Part 1: Treatment. In: Proc. Sardinia 2007, Eleventh International Waste Management and Landfill Symposium, S. Margherita di Pula (Cagliari, IT), 1–5 October 2007 (On CD-ROM).

Rowe, R.K., 2005. Long-term performance of contaminant barrier systems. Geotechnique 55, 631–678.

Rowe, R.K., Islam, M.Z., 2009. Impact of landfill liner time-temperature history on the service life of HDPE geomembranes. Waste Management 29, 2689–2699.

Sabbas, T., Polettini, A., Pomi, R., Astrup, T., Hjelmar, O., Mostbauer, P., Cappai, G., Magel, G., Salhofer, S., Speiser, C., Heuss-Assbichler, S., Klein, R., Lechner, P., 2003. Management of municipal solid waste incineration residues. Waste Management 23, 61–88.

Stegemann, J.A., Schneider, J., Baetz, B.W., Murphy, K.L., 1995. Lysimeter washing of MSW incinerator bottom ash. Waste Management and Research 13, 149–165.

Steketee, J.J., Urlings, L.G.C.M., 1994. Enhanced natural stabilization of MSW bottom ash: a method for minimization of leaching. In: Environmental Aspects of Construction with Waste Materials. Studies in Environmental Science, vol. 60, pp. 233–238.

Turconi, R., Butera, S., Boldrin, A., Grosso, M., Rigamonti, L., Astrup, T., 2011. Life cycle assessment of waste incineration in Denmark and Italy using two LCA models. Waste Management and Research 29, S78–S90.

Van der Sloot, H.A., Kosson, D., Hjelmar, O., 2001. Characteristics, treatment and utilization of residues from municipal waste incineration. Waste Management 21, 753–765.

Van Gerven, T., Van Keer, E., Arickx, S., Jaspers, M., Wauters, G., Vandecasteele, C., 2005. Carbonation of MSWI-bottom ash to decrease heavy metal leaching, in view of recycling. Waste Management 25, 291–300.

Van Gerven, T., Chen, X., Evens, R., Hindrix, K., Vandecasteele, C., 2006. Upgrading MSWI bottom ash by extraction, heating or dense medium separation, in view of recycling. In: Proc. WASCON 2006, Sixth International Conference on the Environmental and Technical Implications of Construction with Alternative Materials. Science and Engineering of Recycling for Environmental Protection, pp. 507–518.

Van Zomeren, A., Comans, R.N.J., 2004. Contribution of natural organic matter to copper leaching from municipal solid waste incinerator bottom ash. Environmental Science and Technology 38, 3927–3932.

Vehlow, J., Seifert, H., 2012. Management of residues from energy recovery by thermal waste-to-energy systems and quality standards. Reports on IEA Bioenergy Task 36 Topic 5. Available online at:http://www.ieabioenergytask36.org.

Yang, R., Liao, W.-P., Wu, P.-H., 2012. Basic characteristics of leachate produced by various washing processes for MSWI ashes in Taiwan. Journal of Environmental Management 104, 67–76.

Yoshida, H., Rowe, R.K., 2003. Consideration of landfill liner temperature. In: Proc. Sardinia 2003, Ninth International Waste Management and Landfill Symposium, S. Margherita di Pula (Cagliari, IT), 6–10 October 2003 (On CD-ROM).

Further Reading

Crannell, B.S., Eighmy, T.T., Krzanowski, J.E., Eusden Jr., J.D., Shaw, E.L., Francis, C.A., 2000. Heavy metal stabilization in municipal solid waste combustion bottom ash using soluble phosphate. Waste Management 20, 135–148.

Hjelmar, O., 1996. Disposal strategies for municipal solid waste incineration residues. Journal of Hazardous Materials 47, 345–368.

European Commission, 2006. Integrated Pollution Prevention and Control (IPPC) Reference Document on the Best Available Techniques for Waste Incineration. http://eippcb.jrc.ec.europa.eu/reference/BREF/wi_bref_0806.pdf.

Lechner, P., Polettini, A., Van Gerven, T., 2006. Characterization and treatment of waste combustion residues. In: Proc. Venice 2006, Biomass and Waste to Energy Symposium, Venice (IT), 29 November–1 December 2006 (On CD ROM).

Lin, K.L., Lin, D.F., 2006. Hydration characteristics of municipal solid waste incinerator bottom ash slag as a pozzolanic material for use in cement. Cement and Concrete Composites 28, 817–823.

4.3

PHYSICAL WASTE PRETREATMENT

Raffaello Cossu and Tiziana Lai

INTRODUCTION

Physical waste treatment prior to landfilling comprises a series of operations aimed at modifying the physical properties of the waste (size, composition, density, moisture content, etc.) with the following objectives: to increase the possibility of material resource recovery, prolong landfill life span, simplify landfill operations, and reduce environmental impacts produced by landfill emissions.

To this regards, physical waste pretreatment processes may include the following:

- separation and classification,
- size reduction (shredding, milling, etc.),
- baling, and
- washing.

Size separation and classification may significantly modify waste composition and represent a crucial step in general waste management strategies. These procedures also represent an important preliminary step in biological and thermal pretreatment before landfilling. Indeed, separation and classification have already been illustrated and discussed in Chapter 4.1. This chapter will provide an insight into the other aforementioned physical pretreatment options, introducing processes, describing technologies, and discussing landfill performance of the pretreated waste.

WASTE SIZE REDUCTION

Waste size reduction or shredding is a form of mechanical pretreatment, the main result of which is waste volume reduction. After shredding, the larger fractions (e.g., >50 mm) contain mainly materials with high calorific values, whereas smaller fractions (<50 mm) are characterized by a higher content of organic and inert materials (GeoSyntec, 2003).

Several types of size reduction apparatus are used in the pretreatment of municipal solid waste (MSW), including hammer mill and shear shredder (Savage et al., 1981). The hammer mill (Fig. 4.3.1) is an impact device in which a number of hammers are fastened flexibly to an inner shaft or disk that is rotated at high speed; this apparatus is highly effective with brittle materials, while, on the contrary, it is scarcely effective with wet and ductile materials. Generally, the final product is not homogeneous with a higher fraction of fine particles. The shear shredder uses two opposing

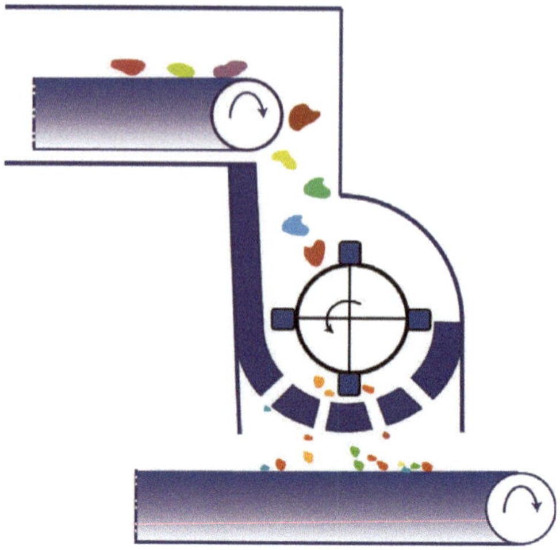

Figure 4.3.1 Hammer mill apparatus scheme.

counterrotating blades to cut or shear materials, in particular ductile and soft materials. The final product is more uniform in comparison with that obtained by hammer mill. The shear shredder is therefore more effective for the treatment of MSW (Luo et al., 2009; 2013).

The power required by a particular shredder is determined by a series of variables: internal configuration of the machine, rotational velocity of hammer tips, in addition to rate, size, composition, and moisture content of the feed. The analysis of data obtained resulted in the development of analytical relationships between comminution parameters, and the development of methods aimed at characterizing shredder performance on the basis of energy consumption and hammer wear associated with size reduction of solid waste.

Landfilling of shredded residues may provide the following advantages:

- Extension of landfill life as a result of the higher achievable compaction, which increases density by 60%–80% (Diaz et al., 2005).
- Shredded wastes are more homogeneous and manageable on side slopes (Giebelhaus, 2007).
- Increased specific surface area promoting better contact among waste, nutrients, and microorganisms and, in general, enhancing biological degradation rate. However, below a certain particle size, acid generation may increase, resulting in inhibition of methane production (Komilis et al., 1999).

In a study carried out to investigate bioreactor landfills, Warith (2002) demonstrated that waste biodegradation occurs more effectively following shredding of MSW compared with unshredded waste.

The effect of waste shredding on gas production in a field application has been reported by Giebelhaus (2007). In landfill, cells that did not receive shredded waste biogas collection ranged between 78% and 100% the expected production rate, whereas in cells containing shredded waste the biogas collected was 274% the expected volume of gas.

A cost-benefit analysis performed by Giebelhaus (2007) compared the revenue yielded by landfill management both with and without shredding pretreatment, taking into account capital costs, including cost of the shredder, operating costs, and an amortization period for equipment of 5 years. The results obtained demonstrated that, although faced with an increase in capital and operating costs, due to increased landfill capacity and biogas production, a consequent reduction in time required for long-term stability, and a reduced postclosure period resulted in an overall reduction of costs.

The use of shredding as pretreatment for MSW may be combined with treatment steps aimed at recovery of materials prior to landfilling (Hansen et al., 2007). Shredding can also be conveniently applied in combination with baling to increase waste density and with waste washing to increase the leaching effect.

BALING

The baling technique is a process comprising a series of different operational phases including sizing, arrangement, and compaction with or without wrapping of waste material with net and plastic film (TLA-Bale Tech LLC, 2009).

This technique has been widely used by farmers and the recycling industry, and in recent decades its application in the field of MSW management has become more common both for temporary storage or transportation techniques prior to other treatment process and as an alternative to traditional landfill disposal (Baldasano et al., 2003; Nammari, 2006; Wagner and Bilitewski, 2009). Fig. 4.3.2 shows wrapped and unwrapped MSW bales, respectively (Wagner and Bilitewski, 2009).

Baling–wrapping of MSW is a common storage technique applied prior to thermal treatment (Åström, 1999; Andreottola et al., 2001; Nammari et al., 2007a,b). The same technology may also be applied for waste disposal in landfills.

Figure 4.3.2 Municipal Solid Waste: wrapped bales (A) and unwrapped bales (B) (Wagner and Bilitewski, 2009).

Landfills adopting the plastic-wrapping technology include countries such as Germany, Italy, Portugal, Sweden, Korea, Lebanon, and others (Bala Press AB, 2001; El-Fadel et al., 2002; Baldasano et al., 2003; GeoSyntec, 2003; Nammari, 2006).

Bales are particularly advantageous during operations of waste loading and transport. Moreover, landfill operation proves to be less costly, due to the ease of placement of the bales using forklift tractors, and faster, creating fewer emissions such as dust, littering, and odors compared to the use of conventional landfilling methods. In addition, the use of daily cover may be reduced or completely avoided.

Two types of presses are used in the treatment of MSW, resulting in the production of different shaped cylindrical and rectangular bales (Baldasano et al., 2003; Nammari, 2006). In the cylindrical baling system, the material is fed into a cylindrical chamber, which both compresses and plastic-wraps the MSW to fix the shape of the bale and avoid the expanding of compressed material after ejection. In the rectangular baling system, two separate pieces of equipment are used in sequence: in the first phase, the press compacts waste into bales, and in the second phase, bales are wrapped in plastic (Baldasano et al., 2003; Nammari et al., 2007a).

During the compression phase varying amounts of liquid, depending on moisture content of the waste, are released, thus diminishing the potential generation of leachate (Baldasano et al., 2003).

A lower degree of compression is achieved with cylindrical than with rectangular bales (Table 4.3.1); as a consequence, the amount of liquid produced during compression is negligible (Baldasano et al., 2003).

The average achievable waste densities of bales are reported in Table 4.3.1; the range of values is mainly attributed to variation in waste composition (i.e., amount of organic fraction).

Bales are generally wrapped using low-density polyethylene (LDPE) or linear low-density polyethylene (LLDPE) films, tied with plastic or metal strips (Nammari, 2006; Markidis et al., 2013); on average 1.5 kg of LDPE per tonne of MSW is used for rectangular bales and 2.5 kg per tonne for cylindrical bales (Baldasano et al., 2003). Fig. 4.3.3 shows the disposal scheme of rectangular and cylindrical bales. Depending on the form of the bales and the emplacement technique used, a certain amount of space will remain between the bales; in the case of rectangular bales a loss of volume of 3% and for cylindrical bales a loss of about 11% can be expected (Baldasano et al., 2003).

Table 4.3.1 Comparison between characteristics of rectangular (RB) and cylindrical (CB) bales

	RB	CB	References
Density, kg/m^3	800–1100	500–800	Baldasano et al. (2003), Nammari (2006)
Area coverage index, t/m^2	2.5	4	Baldasano et al. (2003)

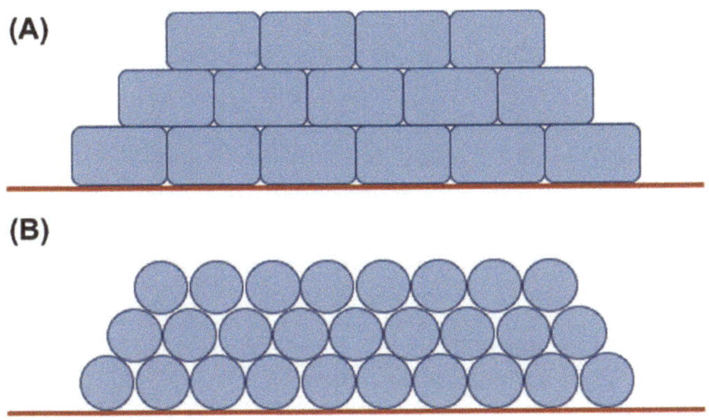

Figure 4.3.3 Disposal scheme for rectangular (A) and cylindrical (B) bales.

Compared with traditional landfill, the disposal of baled waste implies a reduction of environmental emissions and an increase in operation time of the landfill due to increased waste compaction. Furthermore, the handling of bales allows for a cleaner management and transportation, reducing inflammation hazards, and the need for daily cover soil (Fig. 4.3.4) (Baldasano et al., 2003; GeoSyntec, 2003). However, the long-term mechanical stability and environmental effects of landfilling baled wastes have not been investigated in detail (GeoSyntec, 2003). Table 4.3.2 summarizes some characteristics of landfilling of baled waste versus traditional landfilling.

MSW decomposition and associated emissions within bales depend on waste quality and type of bales. Unwrapped bales can be assimilated to an open landfill with compacted waste, whereas the wrapping of bales aims to reduce the decomposition rate and infiltration of water into the waste as in a lined closed landfill (Nammari, 2006).

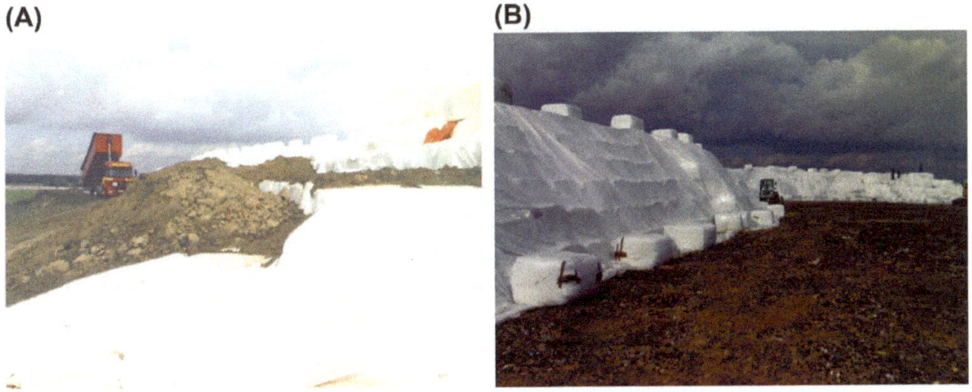

Figure 4.3.4 Covering wrapped bales with soil (A) and plastic membrane (B) (Wagner and Bilitewski, 2009).

Table 4.3.2 Characteristics of landfilling of baled waste compared with traditional landfilling (Baldasano et al., 2003; GeoSyntec, 2003)

	TL	NWB	WB
Leachate generation	+++	+++	+
Odors—emission of gases	+++	+++	+
Airborne debris and dust from waste and operational activities	+++	++	+
Pest infestation (birds, rodents, insects etc.)	+++	+++	+
Daily cover soil	+++	++	+
Risk of spontaneous ignition	+++	+++	+
Occupancy of volume	+++	+	+
Compaction	+	++	+++
Easy and clean transport	+	++	+++
Labor during operating phase	+++	++	++
Reception infrastructure	++	+++	+++
Use of moving machinery	+++	+	+
Services infrastructure	++	++	++
Waste degradation	+++	+++	+
Time required for stability	+	++	+++
Visual impact	+	++	+++
Future recycling	+	++	+++

High = +++, Medium = ++, Low = +.
TL, traditional landfill; *NWB*, not wrapped bales; *WB*, wrapped bales.

Pascual and Bajeat (1997) demonstrated how limited biodegradation occurred within the bales, all of which maintained their dimensions. Laboratory studies performed by Robles-Martínez and Gourdon (1999) showed that the main reason underlying the absence of biodegradation was the rapid acidification of waste due to accumulation of volatile fatty acids during the first days of incubation. These results have been validated during an experiment carried out on two actual waste bales

(Robles-Martínez and Gourdon, 2000): biodegradation of wrapped MSW bales was found to be very low as long as the covering layer remained in a good condition; methane was not detected in the biogas, thus confirming the absence of methanogenesis within the bales. Similar results were obtained by Hogland et al. (2001). Moreover, water content inside the bales seemed to be lower than the amount required to support biodegradation (Baldasano et al., 2003). As a consequence, as long as the wrapped bales remain intact and moisture infiltration is reduced, biodegradation is minimal (GeoSyntec, 2003).

However, El-Fadezl et al. (2002,2003), analyzing the leachate formation mechanisms and quality from presorted and baled MSW, concluded that presorting and baling of waste does not hinder waste stabilization.

More recently, Ozbay and Durmusoglu (2012, 2013) have monitored temporal variations of CO_2, CH_4, O_2, and N_2, demonstrating that aerobic decomposition takes place over the first few days, consuming the available oxygen in the bales; CO_2 production increases over the first few days, decreasing and stabilizing within a month in a range between 10% and 35%; CH_4 did not exceed 5% throughout the whole test period. Similar results had previously been obtained by Pascual and Bajeat (1997).

The temperature inside the bales is generally expected to reflect ambient temperature (Hogland et al., 2001; Baldasano et al., 2003).

The influence of shredding, waste particle size and waste density, thickness of wrapping layers, and type of wrapping film have also been investigated (Markidis et al., 2013). The results of these studies have indicated CO_2 as the prevalent gaseous product produced during decomposition, initially present at levels of approximately 80% and gradually diminishing to around 20%, whereas CH_4 concentration never exceeded an average of 17% inside the bales. Factorial analysis revealed that particle size, density, and film type (white or gray LLDPE with a UV stabilizer) significantly affected CO_2 and CH_4 generation.

Volatile organic compounds (VOCs) concentration in wastes stored in cylindrical or rectangular bales was evaluated by Nammari et al. (2007a,b). Findings obtained demonstrated that concentration of VOCs, as well as the composition of VOCs emissions, are dependent on procedures used for compacting waste into cylindrical or rectangular bales. In particular, VOCs concentration after storing was found to be higher in cylindrical than in rectangular bales. Cylindrical-baled MSW emitted larger concentrations of esters than rectangular-baled MSW; conversely, aromatic compound emissions were highest in rectangular-baled MSW, thus suggesting the presence of different degradation mechanisms for the two types of bales.

With regards to costs, the cost model applied by Baldasano et al. (2003) compared landfill costs incurred using both rectangular and cylindrical plastic-wrapped bales with the costs of conventional landfill. The results demonstrated how, due to the differences in geometry and type of machinery used (energy consumption, maintenance costs, operational characteristics, etc.), the cylindrical compacting method was less profitable than both rectangular compacting method and conventional landfill. Conversely, a landfill using rectangular plastic-wrapped bales represents an economically competitive option compared with conventional landfill, as the increase in operating costs is largely compensated by the increase in landfill capacity. However, landfilling of baled waste may affect the time required to achieve environmental stability of the landfill and, as a consequence, postclosure management costs (GeoSyntec, 2003). The use of baling methods cannot be combined with in situ treatment methods aimed at enhancing biodegradation, such as leachate recirculation. For this reason, the baling of stabilized biological waste may contribute toward overcoming these potential disadvantages.

Based on findings reported by Wagner and Bilitewski (2009), the cost of temporary storage of baled MSW ranges between 20 and 50 euro/t, depending on the environmental conditions and type of stored waste.

WASTE WASHING

The washing of waste was developed with two main aims: to achieve the wet separation of different waste fractions resulting in the recovery of cleaner recyclable materials and to improve the environmental quality of waste (e.g., residual material from thermal pretreatment processes) prior to final landfill disposal (GeoSyntec, 2003).

Waste washing prior to final disposal may represent an effective pretreatment option aimed at reducing the emission potential of specific landfilled waste. Water-soluble compounds are transferred from the solid phase into the liquid phase, from which they will subsequently need to be removed during water treatment. The remaining solid residues are assessed for material recovery and/or safe disposal subsequent to further treatment.

Waste washing with water is generally applied to reduce the readily leachable fraction and may not significantly affect the emission potentials associated with mobilization processes occurring at a slower rate from deposited waste (Cossu and Lai, 2012). Indeed, the long-term availability and mobility of pollutants, such as heavy metals, may be influenced by the diffusion of air into the landfill during the maturation phase, thus producing a rise in redox potential and consequently resulting in sulfide oxidation (Kim et al., 2005). However, other binding processes (such as carbonate) may reduce the mobility of heavy metals.

Numerous studies have demonstrated the efficacy of water washing of residues from thermal treatment plants in reducing the leachable fraction of metals, leachable salts, chlorides, and organic matters before subsequent valorization or landfill disposal (Kim et al., 2005; Ishii et al., 2009; Wang et al., 2009; Chiang and Hu, 2010; Lin et al., 2011). Washing pretreatment may also be applied for the purpose of material recovery or as pretreatment prior to the landfilling of MSW or special waste (Al-Salem et al., 2009; Jody et al., 2010; Cossu et al., 2012).

To define the most appropriate washing procedure, a series of different operative conditions should be taken into account: washing solution, liquid to solid (L/S) ratio, mixing time and speed, washing frequency, temperature, and waste size fraction. A water-washing procedure would be suitable for use in the removal of highly soluble compounds, while washing with different solutions (acid or alkaline solution, chelating agents, organic acids etc.) would be more appropriate to remove less water-soluble fractions (Nagib and Inoue, 2000; Fedje et al., 2010; Huang et al., 2011). Recently, Wang et al. (2015) investigated the influence of washing pretreatment on the total contents and bioaccessibility of heavy metals in MSW incinerator fly ash. The results of their studies revealed that both water-washing and acid-washing processes (HNO_3) resulted in the removal of a part of the soluble heavy metals present and reduced bioaccessibility. However, the use of an acid-washing procedure would better remove calcium carbonate, calcium hydroxide, and other parts of precipitates, which are poorly solubilized by the water-washing process.

The L/S ratio influences the dissolution of compounds in line with their degree of solubility. However, the mixing time may influence the kinetic of compound dissolution. In particular, a lengthy mixing time may reduce the dissolution or extraction rate. Chimenos et al. (2005) demonstrated that during washing experiments performed on air pollution control (APC) residues, maximum chloride extraction was obtained after 2 h mixing and remained steady when the time exceeded 4 h. Other elements, such as Cr, require several days to achieve an equilibrium concentration (Astrup et al., 2005).

Zhu et al. (2010) demonstrated in washing experiments how the amount of soluble chlorides decreased rapidly on a par with an increase in L/S ratio or washing frequency; in particular double washing was generally more effective in removing soluble chlorides than single washing.

The efficiency of a washing pretreatment step may be evaluated using standard batch leaching test to quantify the improvement of landfill sustainability targets calculated as an index as follows (Cossu and Lai, 2012):

$$\eta_{A,i} = (e_{r,i} - e_{w,i})/e_{r,i}$$

where $e_{r,i}$ is the concentration of the i-substance in the eluate of batch leaching test before washing; $e_{w,i}$ is the concentration of the i-substance in the eluate of batch leaching test after washing.

Waste washing as a pretreatment method applied to residues from different kinds of waste (residues from plastics sorting process, mechanical–biological treated waste, bottom ash from MSW incineration and automotive shredder residues) showed chemical oxygen demand removal rates of more than 60% for all different kinds of waste, reaching up to 85% for plastic residues. Similar removal rates were observed for dissolved organic carbon. Batch leaching tests verified the significant reduction of metals and anions released. This had been demonstrated by a decrease of electric conductivity in a range from 60% to 78% (Cossu et al., 2012).

Kim et al. (2003, 2005) evaluated, using a similar index, the effect of water washing on removal of the major elements, resulting in a decrease of leachability of trace metals such as Pb and Cd.

To set up an efficient washing treatment, water consumption should be minimized using a sufficiently high L/S ratio for each washing step but a reducing L/S ratio for the entire process (water volume used per solid mass treated). To achieve this, the same solution can be reused for several washing steps in a countercurrent or a recirculating process implemented by adding freshwater only to maintain the same L/S ratio at each stage (Chimenos et al., 2005; De Boom and Degrez, 2015).

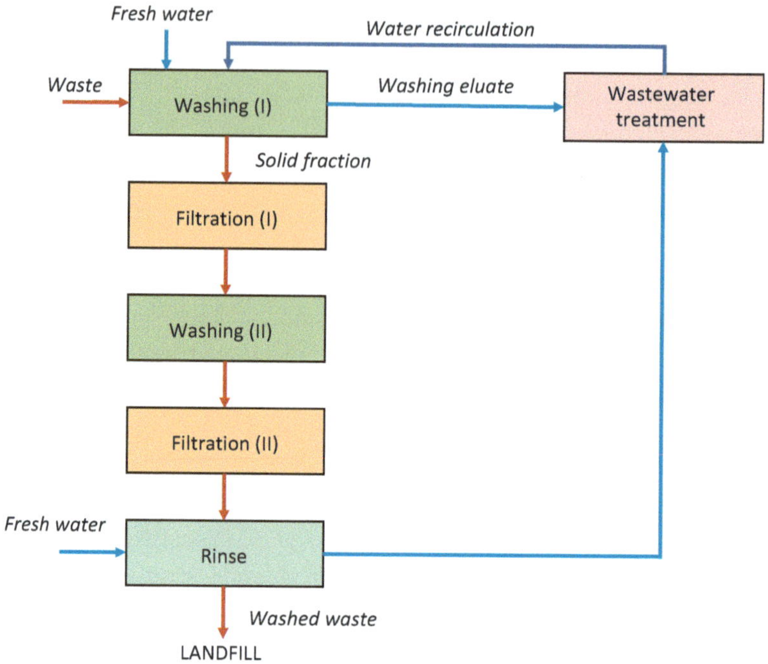

Figure 4.3.5 Washing procedure scheme, the dotted line represents phases that may occur in the presence of multiple washing steps or washing procedures with different washing solutions, i.e., rinse phase. *Modified from Chimenos et al., 2005; Cossu and Lai, 2013.*

Furthermore, the use of large amounts of water may contribute toward raising the cost of treatment and increase environmental impacts due to the subsequent need for wastewater treatment. Indeed, washing pretreatment implies a transfer of contaminants from waste to the resulting solution, which will subsequently need to be treated prior to discharge (Cossu et al., 2012; De Boom and Degrez, 2015). Physical and chemical processes, in single or multiple phases, constitute appropriate technical solutions for use in wastewater treatment (Cossu et al., 2012). The hazardous sludge generated represents an additional solid waste stream that will require further handling management (Mangialardi, 2003). A washing procedure scheme is shown in Fig. 4.3.5.

CONCLUSIONS

Physical treatment of waste before landfilling is an option that can play an important role in modern landfills where the deposit of raw waste is progressively discouraged.

In particular, residues resulting from circular economy treatment and recyclables might take advantage of baling in managing (transporting, storing, depositing) waste fractions with low putrescible contents, bulky materials, or small size residues.

Washing can guarantee the environmental sustainability by controlling the long-term emissions of soluble inorganic or soluble nonbiodegradable organic compounds.

References

Al-Salem, S.M., Lettieri, P., Baeyens, J., 2009. Recycling and recovery routes of plastic solid waste (PSW): a review. Waste Management 29 (10), 2625–2643.

Andreottola, G., Bortolon, A., Dallago, L., Ragazzi, M., 2001. A comprehensive study on MSW baling as a pre-treatment to waste-to-energy plants. In: Christensem, T.H., Cossu, R., Stegmann, R. (Eds.), Sardinia 2001: Eighth International Waste Management and Landfill Symposium. Cagliari, Italy.

Åström, J., 1999. Case Study: Lidköping Waste-to-Energy Plant. International Energy Agency. Bioenergy Task 23 Report. http://www.ieabioenergytask36.org/Publications/1998-2001%20Task%2023/Case_Studies/Lidkoping_Waste-to-Energy_Plant.PDF with the following. http://task36.ieabioenergy.com/wp-content/uploads/2016/06/Lidkoping_Waste-to-Energy_Plant-1999.pdf.

Astrup, T., Rosenblad, C., Trapp, S., Christensen, T.H., 2005. Chromium release from waste incineration air-pollution-control residues. Environmental Science & Technology 39 (9), 3321–3329.

Bala Press AB, 2001. Case Study. Jeonju City, Korea. http://www.flexus.se/data/filer/Case study Korea1065e66,69.pdf with the following link. http://www.flexus.se/en/r/28/baling-of-msw.

Baldasano, J.M., Gassó, S., Pérez, C., 2003. Environmental performance review and cost analysis of MSW landfilling by baling-wrapping technology versus conventional system. Waste Management 23, 795–806.

Chiang, K.Y., Hu, Y.H., 2010. Water washing effects on metals emission reduction during municipal solid waste incinerator (MSWI) fly ash melting process. Waste Management 30 (5), 831–838.

Chimenos, J.M., Fernandez, A.I., Cervantes, A., Miralles, L., Fernández, M.A., Espiell, F., 2005. Optimizing the APC residue washing process to minimize the release of chloride and heavy metals. Waste Management 25 (7), 686–693.

Cossu, R., Lai, T., 2012. Washing of waste prior to landfilling. Waste Management 32 (5), 869–878.

Cossu, R., Lai, T., 2013. Washing treatment of automotive shredder residue (ASR). Waste Management 33, 1770–1775.

Cossu, R., Lai, T., Pivnenko, K., 2012. Waste washing pretreatment of municipal and special waste. Journal of Hazardous Materials 207, 65–72.

De Boom, A., Degrez, M., 2015. Combining sieving and washing, a way to treat MSWI boiler fly ash. Waste Management 39, 179–188.

Diaz, L.F., Savage, G.M., Eggerth, L.L., 2005. Alternatives for the treatment and disposal of healthcare wastes in developing countries. Waste Management 25 (6), 626–637.

El-Fadel, M., Bou-Zeid, E., Chahine, W., Alayli, B., 2002. Temporal variation of leachate quality from presorted and baled municipal solid waste with high organic and moisture content. Waste Management 22, 269–282.

El-Fadel, M., Bou-Zeid, E., Chahine, W., 2003. Landfill evolution and treatability assessment of high strength leachate from MSW with high organic and moisture content. International Journal of Environmental Studies 60 (6), 603–615.

Fedje, K.K., Ekberg, C., Skarnemark, G., Steenari, B.M., 2010. Removal of hazardous metals from MSW fly ash—an evaluation of ash leaching methods. Journal of Hazardous Materials 173 (1), 310–317.

GeoSyntec Consultants, Inc, 2003. Landfill Facility Compliance Study Task 7 Report—Study of Emerging Technologies in Waste Management for MSW Landfills. Walnut Creek, California.

Giebelhaus, J., May-June. Waste shredding at the working face a landfill operator's perspective. MSW Management Forester Communications 2007, 104–114.

Hansen, T.L., la Cour Jansen, J., Davidsson, Å., Christensen, T.H., 2007. Effects of pretreatment technologies on quantity and quality of source-sorted municipal organic waste for biogas recovery. Waste Management 27 (3), 398–405.

Hogland, W., Marques, M., Nimmermark, S., Nammari, D., Moutavtchi, V., 2001. Seasonal and Long-Term Storage of Waste Fuels with Baling Technique. University of Kalmar. Report No. 112.

Huang, K., Inoue, K., Harada, H., Kawakita, H., Keisuke, O.H.T.O., 2011. Leaching behavior of heavy metals with hydrochloric acid from fly ash generated in municipal waste incineration plants. Transactions of Nonferrous Metals Society of China 21 (6), 1422–1427.

Ishii, K., Furuichi, T., Tanikawa, N., 2009. Numerical model for a watering plan to wash out organic matter from the municipal solid waste incinerator bottom ash layer in closed system disposal facilities. Waste Management 29 (2), 513–521.

Jody, B.J., Daniels, E.J., Duranceau, C.M., Pomykala, J.A., Spangenberger, J.S., September 25, 2010. End of-Life Vehicle Recycling: The State of the Art of Resource Recovery from Shredder Residue. Energy Systems Division, Argonne National Laboratory. ANL/ESD/07–8. http://www.osti.gov/scitech/biblio/1010492.

Kim, S.Y., Matsuto, T., Tanaka, N., 2003. Evaluation of pretreatment methods for landfill disposal of residues from municipal solid waste incineration. Waste Management & Research 21 (5), 416–423.

Kim, S.Y., Tanaka, N., Matsuto, T., Tojo, Y., 2005. Leaching behavior of elements and evaluation of pretreatment methods for municipal solid waste incinerator residues in column leaching tests. Waste Management & Research 23 (3), 220–229.

Komilis, D.P., Ham, R.K., Stegmann, R., 1999. The effect of municipal solid waste pretreatment on landfill behavior: a literature review. Waste Management & Research: The Journal of the International Solid Wastes and Public Cleansing Association, ISWA 17 (1), 10–19.

Lin, Y.C., Panchangam, S.C., Wu, C.H., Hong, P.K.A., Lin, C.F., 2011. Effects of water washing on removing organic residues in bottom ashes of municipal solid waste incinerators. Chemosphere 82 (4), 502–506.

Luo, S., Xiao, B., Hu, Z., Liu, S., Guo, X., 2009. An experimental study on a novel shredder for municipal solid waste (MSW). International Journal of Hydrogen Energy 34 (3), 1270–1274.

Luo, S., Yang, X., Tao, X., Luo, Y., Fu, J., 2013. Research on the optimization of a novel municipal solid waste shredder. Journal of Renewable and Sustainable Energy 5 (1), 013111.

Mangialardi, T., 2003. Disposal of MSWI fly ash through a combined washing-immobilisation process. Journal of Hazardous Materials 98 (1), 225–240.

Markidis, I., Komilis, D., Tsagas, F., Petalas, A., 2013. A fractional factorial field experiment to study the decomposition of municipal solid wastes stored in wrapped bales. Journal of Environmental Management 115, 32–41.

Nagib, S., Inoue, K., 2000. Recovery of lead and zinc from fly ash generated from municipal incineration plants by means of acid and/or alkaline leaching. Hydrometallurgy 56 (3), 269–292.

Nammari, D.R., 2006. Seasonal and Long-Term Storage of Baled Municipal Solid Waste (Ph.D. thesis). Lund University.

Nammari, D.R., Marques, M., Thörneby, L., Hogland, W., 2007a. Emissions from baled municipal solid waste: I. Methodological approach for investigation of gaseous emissions. Waste Manage Research 25, 39–48.

Nammari, D.R., Marques, M., Hogland, W., 2007b. Emissions from baled municipal solid waste: II. Effects of different treatments and baling techniques on the emission of volatile organic compounds. Waste Manage Research 25, 109–118.

Ozbay, I., Durmusoglu, E., 2012. Temporal variation of decomposition gases from baled municipal solid wastes. Bioresource Technology 112, 105–110.

Ozbay, I., Durmusoglu, E., 2013. Energy content of municipal solid waste bales. Waste Management & Research 31 (7), 674–683.

Pascual, C., Bajeat, P., 1997. Storage of Household Waste: Evaluation of the Performance of Wrapped Bale Technique. CYLERGIE & Ademe, France. Available at: https://www.fylkesmannen.no/contentassets/825b0cefe1014631bde07b85b5512d04/temporary-storage-of-household-waste.pdf with the following link.

Robles-Martínez, F., Gourdon, R., 1999. Effect of baling on the behavior of domestic wastes: laboratory study on the role of pH in biodegradation. BioresourceTechnology 69, 15–22.

Robles-Martínez, F., Gourdon, R., 2000. Long-term behaviour of baled household waste. Bioresource Technology 72, 125–130.

Savage, G.M., Diaz, L.F., Trezek, G.J., Golueke, C.G., Wiles, C., Oberacker, D., 1981. On-site evaluation of municipal solid waste shredders. Resource Recovery and Conservation 5 (4), 343–362.

TLA-Bale Tech LLC, 2009. White Paper on Baling Technology.

Wagner, J., Bilitewski, B., 2009. The temporary storage of municipal solid waste – recommendations for a safe operation of interim storage facilities. Waste Management 29, 1693–1701.

Wang, Q., Yang, J., Wang, Q., Wu, T., 2009. Effects of water-washing pretreatment on bioleaching of heavy metals from municipal solid waste incinerator fly ash. Journal of Hazardous Materials 162 (2), 812–818.

Wang, Y., Pan, Y., Zhang, L., Yue, Y., Zhou, J., Xu, Y., Qian, G., 2015. Can washing-pretreatment eliminate the health risk of municipal solid waste incineration fly ash reuse? Ecotoxicology and Environmental Safety 111, 177–184.

Warith, M., 2002. Bioreactor landfills: experimental and field results. Waste Management 22 (1), 7–17.

Zhu, F., Takaoka, M., Oshita, K., Kitajima, Y., Inada, Y., Morisawa, S., Tsuno, H., 2010. Chlorides behavior in raw fly ash washing experiments. Journal of Hazardous Materials 178 (1), 547–552.

5 CHAPTER

GEOTECHNICAL ASPECTS

5.1

WASTE MECHANICAL PROPERTIES

Jean-Pierre Gourc and Dimitrios Zekkos

INTRODUCTION

Mechanical behavior of waste is one of the most important issues in landfilling.

Considering the different causes of adverse environmental impacts and the potential hazards for the human health originating from landfilling, landfill stability is a critical one and one capable to cause massive human deaths.

A comparison between waste mechanics and soil mechanics leads to the conclusion that MSW exhibits a far more complex mechanical behavior that is due to:

- the blending of constituents in MSW, such as soil minerals and other constituents that are physically different and include flat elements such as paper and plastic, as well as fibrous elements such as textiles and wood;
- the significant evolutionary (i.e., time dependent) behavior associated to the decay of biodegradable materials.

Knowledge of the mechanical behavior of MSW is important in many aspects of landfill engineering, namely, the management of a landfill cell, the integrity of the cap barrier, the vertical expansion of a former landfill and the stability of the steep slopes.

The management of a landfill cell

From the beginning to the end of life of the landfill, proper management of the site requires a good understanding of the mechanical behavior of the waste, to address important questions such as:

- how to forecast the maximum waste capacity, i.e., the mass of waste needed to fill a new cell (Olivier et al., 2005)?
- how to monitor the biodegradation of waste, possibly by correlating settlement and biodegradation, and especially how to reliably estimate biogas production that represents a sustainable energy source (Gourc et al., 2010)?
- how to forecast the aftercare (or postclosure) period?

The integrity of the cap barrier

Landfill cap covers constitute the active barrier which could mitigate the fugitive emissions of biogas (greenhouse gas) generated due to biodegradation of waste (Staub et al., 2011). However, due to settlement of waste, and more specifically differential settlement, biogas collection pipes can be damaged and the fine soil layer of the barrier could display cracking (Camp et al., 2010; Gourc et al., 2010) increasing the amount of gas emitted to the atmosphere. Preserving, as much as possible, the integrity of the cap barrier requires a good understanding of the kinetics of settlement of the underlying waste material.

The vertical expansion of a former landfill

Given the challenges associated with identifying suitable areas to site a new sanitary landfill, there is a growing need to prioritize the vertical expansion of existing landfills (Tano et al., 2015). However, a careful consideration of the settlement of the existing waste that will be subjected to increasing loads from the additional waste is necessary to ensure the long-term continuity of the drainage and sealing functions of the new landfill's bottom liner systems.

The stability of the steep slopes

To increase the storage capacity of existing landfills that have a fixed permitted maximum elevation of the cap barrier, the only remaining solution for the owner is to increase the maximum lateral slope angle of the landfill's waste mass. This increases the risk of global instability. Conventional limit equilibrium methods of slope stability, used for soils, can be applied to waste, on the condition of a reliable assessment of the failure characteristics of the waste material (Zekkos et al., 2014).

PHYSICAL PARAMETERS FOR A RATIONAL PHYSICO-MECHANICAL APPROACH OF WASTE

Physical characterization as for soil

The composition of MSW at each landfill is a function of cultural and social waste generation habits, as well as waste management practices, and may vary significantly from region to region, as well as with time. Based on experience in developed countries, MSW includes primarily paper, plastic, textiles, and wood. Other constituents, such as metals, may be present and may represent a significant percentage of the waste mass by weight, but typically do not represent a significant percentage by volume. Other organic waste, such as food, is typically highly degradable and is rarely distinguishable after a few months.

The MSW porous medium consists of three phases: solid (*S*), liquid (*L*), and gas (*G*) (Fig. 5.1.1). In the case of MSW material, the solid phase is partly biodegradable. As a first approximation of the behavior of waste in the short term, the degradation of solid mass is not taken into account (Stoltz et al., 2010), i.e., the solid mass is assumed constant. The phases, solid, liquid, and gas, are evaluated by their mass (*M*) and volume (*V*).

$$\text{Liquid phase is characterized by density } \rho_L = \frac{M_L}{V_L} \tag{1}$$

This density is assumed to be constant with $\rho_L = 1$ Mg/m^3 for leachate, similarly to water.

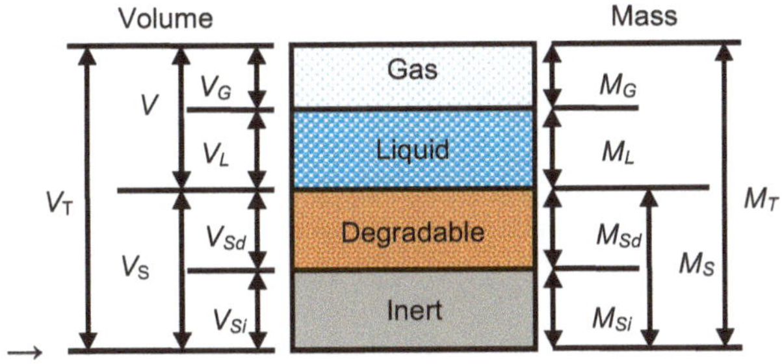

Figure 5.1.1 Volume and mass proportions of a typical municipal solid waste material.

The gas density ρ_G is not taken into account for the calculation of physical parameters defining the porous medium because the mass of gas M_G is negligible compared to the mass of liquid M_L and solid M_S. It is, however, considered in the equations of mass transfer. Concerning the MSW porous medium, three characteristic physical parameters are defined:

- Dry density ρ_d (Mg/m^3):

$$\rho_d = \frac{M_S}{V_T} \tag{2}$$

- Porosity n (m^3/m^3):

$$n = \frac{V_V}{V_T} = \frac{V_G + V_L}{V_T} \tag{3}$$

- Gravimetric liquid content w (kg/kg):

$$w = \frac{M_L}{M_S} = \frac{M_T - M_S}{M_S} \tag{4}$$

Determination of the set of parameters (ρ_d, n, and w) allows the calculation of all other physical parameters commonly used for porous medium:

- Total, or bulk, density ρ (Mg/m^3):

$$\rho = \frac{M_T}{V_T} = (1 + w)\rho_d \tag{5}$$

- Volumetric liquid content θ_L (m^3/m^3):

$$\theta_L = \frac{V_L}{V_T} = w \times \frac{\rho_d}{\rho_L} \tag{6}$$

- Volumetric gas content θ_G (m^3/m^3):

$$\theta_G = \frac{V_G}{V_T} = n - \theta_L \tag{7}$$

The density of solid fraction, also called solid or specific density, or specific gravity, is defined by

$$\rho_S = \frac{M_S}{V_S} = \frac{\rho_d}{1 - n} \tag{8}$$

In the case of soils, the solid density of silicates is well known and not much variable with a value generally close to 2.65–2.72 Mg/m^3, unless organics are present. This value is not valid for MSW, since MSW is composed of constituents of diverse nature with different solid densities. Consequently for MSW, ρ_S corresponds to an "average" and is dependent on its constituents.

PHYSICAL CHARACTERISTICS OF WASTE MATERIALS

Particles distribution

In soil mechanics, a semilog diagram is commonly used for characterizing the massive proportion of grains close to spherical shape, of different sizes, using a standard sieving process. This process is inoperative, or not relevant, for MSW that include large size constituents as well as constituents such as plastic tapes or grass fibers which are long and flexible. In addition the shape of many rigid particles is far from spherical grains, e.g., pieces of glass (Grisolia et al., 1995). In case of mechanical biological treatment (Münnich et al., 2006), after sorting, shredding, and aerobic biodegradation, the average diameter of the particles is smaller, but the diversity of shapes remains.

Solid density

From the mass percentage of every component, the mean solid density ρ_S can be theoretically assessed:

$$\rho_S = \frac{1}{\sum_i \frac{\mu_i}{\rho_{Si}}} \tag{9}$$

Taking into account the solid density ρ_{Si} of every component (i): $\rho_{Si} = \frac{M_{Si}}{V_{Si}}$ (10) and their mass percentage μ_i. Compared to silicic soils where ρ_S is very close to 2.65 Mg/m^3, for waste, it varies. For instance, it is around 2.90 Mg/m^3 for glass, 0.5 for textiles, 1.0 for plastic, 1.25 for wood and food, 1.50 for cardboard, but also varies further depending on the material type.

It turns out that this method is not feasible due to many operational difficulties that involve among other, determination of μ_i by sorting, and possible partial saturation and biodegradation of the particles. Thus, ρ_{Si} values are not considered attainable. The preferred method is based on the use of pycnometer (Stoltz et al., 2010), which allows the assessment of the volumetric gas content θ_G, and on this basis the determination of porosity (per Eq. (7)) and then the solid density (per Eq. 8).

The solid density of organic fraction is lower than that of the inert fraction. As a consequence, the decay of biodegradable fraction (that has low solid density) results in an increase of the mean solid density of the whole waste. The field test presented below confirms this trend.

During an excavation campaign on a French landfill site (Stoltz et al., 2009), a drilling operation allowed the retrieval of unique cell samples of waste that were 0.6 m in diameter and 1.2 m in height at several depths (Fig. 5.1.2) that corresponded to waste differing in age. The initial composition is assumed constant versus time, hence versus depth. For a 1.5-year-old waste at a depth between 0 and 6 m, the average $\rho_S = 1.62$ Mg/m^3, for 3-year-old waste between 6 and 12 m, $\rho_S = 1.85$ Mg/m^3, and for 7-year-old waste between 32 and 39.5 m, $\rho_S = 2.00$ Mg/m^3.

Representative elementary volume

Fig. 5.1.3 is a schematic of the assessment of the organic content of a waste material on the basis of the biogas potential, BP, that corresponds to the maximum amount of generated biogas. Due to the heterogeneity of MSW and the presence of large-size constituents, a small sample could be totally inert (BP = 0) or completely organic (BP maximum). Therefore, a minimum size of the sample is required to generate a representative value of the BP that is independent of the size of the sample (right side of the diagram).

Figure 5.1.2 Drilling operation to retrieve waste samples at different depths.

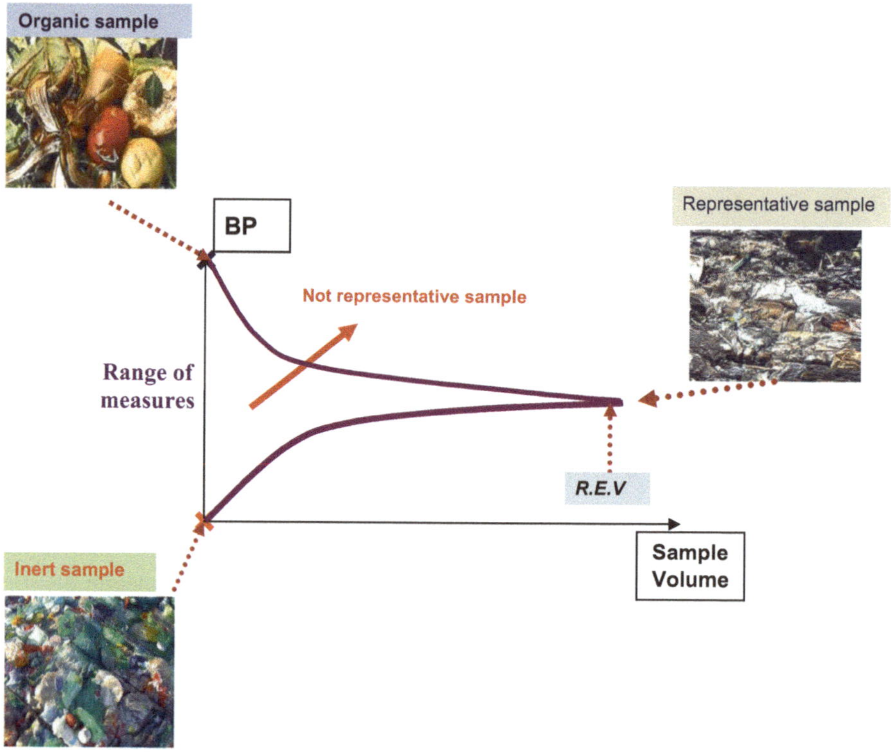

Figure 5.1.3 Representative Elementary Volume for a representative assessment of the Biogas Potential (BP).

In the same way, the assessment of the physico-mechanical properties and more generally the modeling in laboratory tests of field conditions requires a minimum size for the cells used for testing the material. In soil mechanics, it is commonly accepted that a ratio between the minimum internal dimension of the cell (L) and the maximum size of bulky particles (D_{max}), L/D_{max}, greater than or equal to around 6 is adequate. However, for fibrous and flat constituents (such as paper and plastic), a slightly lower ratio of 4 has been used, since these constituents tend to be small in thickness, easily folded, flexible and are generally placed, intentionally or unintentionally, flat (Bray et al., 2009; Athanasopoulos, 2011).

Consequently, large devices are required for testing raw MSW like the ones shown in Figs. 5.1.4 and 5.1.5 for compressibility tests with leachate recirculation in oedo-permeameters (Olivier and Gourc, 2007; Staub et al., 2013). When shredded materials or the fine fraction of sorted materials are tested, the dimensions of the experimental device can be smaller.

Compressibility

Waste is a highly compressible material. The physical parameters of the porous medium described earlier evolve with settlement. If a sample of waste placed in an oedometer cylindrical cell with rigid walls is considered, the sample, having an initial height h_0, is characterized by the set of physical

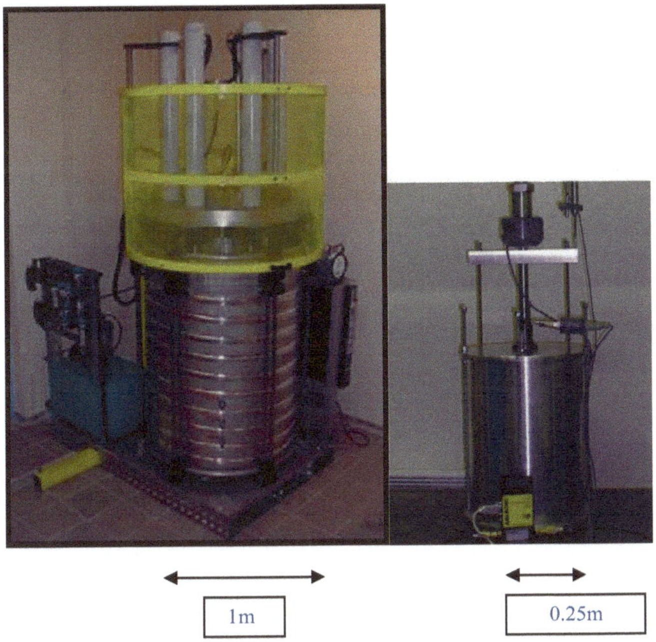

Figure 5.1.4 Examples of Oedo-permeameters of different sizes, used for municipal solid waste with control of leachate and gas flows.

Figure 5.1.5 Sample removed of the Oedo-permeameter (diameter 1 m) at the end of the compression and leachate recycling testing.

parameters (ρ_{d0}, n_0, w_0). The application of a vertical stress σ' leads to a settlement Δh (considered positive). This settlement induces an alteration of the initial physical parameters ρ_{d0}, n_0, and w_0. Taking into account that biodegradation is not being considered here, solid mass M_S (Fig. 5.1.1) does not vary during settlement. The parameter w_0 is supposed to remain constant except in the case of highly humid samples where settlement can lead to liquid drainage out of the cell. In this case, the loss of mass of liquid is noted as M_{Lout}. The corresponding change in physical parameters versus the relative settlement $\Delta h/h_0$ can be calculated with the following formula:

$$\text{Dry density}: \rho_d = \frac{\rho_{d0}}{1 - \frac{\Delta h}{h_0}} \tag{11}$$

$$\text{Porosity}: n = n_0 - \Delta n = n_0 - \frac{(1 - n_0)\frac{\Delta h}{h_0}}{\left(1 - \frac{\Delta h}{h_0}\right)} \tag{12}$$

$$\text{Gravimetric liquid content}: w = w_0 - \frac{M_{Lout}}{M_S} \tag{13}$$

The relative settlement $\Delta h/h_0$ due to vertical stress σ' can be important, since waste is a highly compressible material. Consequently, the physical parameters exhibit significant changes during compaction and landfilling operations, even without considering biodegradation.

COMPRESSIBILITY OF WASTE AND GLOBAL SETTLEMENT OF A LANDFILL

Landfilling sequence and settlement history

Waste is a highly compressible material and thus, when placed in large thicknesses, presents major technical challenges. Settlements represent a serious threat to the integrity and the performance of the cap cover and structures on top of the landfill. Hence reliable estimates of the range of expected settlements are needed.

Settlement of waste can be divided into two components (Ivanova et al., 2008; Hettiarachchi et al., 2009; Powrie et al., 2009):

- A "primary" settlement component mainly linked to the self-weight of the overlying column, quickly stabilized at constant load; this "primary" settlement includes the immediate compression considered by some authors, but isolating the immediate compression is difficult.
- A "secondary" settlement component corresponding to the progressive creep, decomposition, and raveling of the waste material during several decades.

In the literature, many studies are related to the behavior of waste under compression and many numerical models have been proposed.

In general, it is necessary to focus on secondary settlement only, because primary settlement is normally completed by the time of construction of the final cap cover. However, better understanding of primary settlement is necessary in specific cases, such as multiphase landfill cell construction, vertical expansion of landfills (Bouthot et al., 2003; Tano et al., 2015), or construction of embankments/structures on a former landfill where an increasing vertical load is applied on the waste body.

The assessment of the global settlement of a landfill cell at the level of the cap cover is conventionally based on the method of Sowers (1973); however, this is a rough approach that neglects the construction sequence of the landfill in layers. A more accurate approach (ISPM program) has been proposed by Olivier et al. (2003) and Gourc et al. (2008) and is presented below.

Based on the stacking of elementary layers of waste (height h_i of layer i) comprising a total height H of waste mass (Fig. 5.1.6), the process to evaluate the total settlement of a vertical column of waste should integrate the primary and secondary settlement of each elementary layer constituting the column. The ISPM program is used to highlight the global behavior with the elapsed time. In Fig. 5.1.7A, construction rate is assumed to be constant (same operating time (τ_i) for placement of each waste layer $h_i = h_0$) on the entire cell area. Absolute time (t) originates from the time of beginning of construction. The virtual final height of the landfill corresponds to the lifting of the n layers of constant construction thickness (n^*h_0), plus the thickness of the cap cover (h_c). However, the actual height ($H + h_c$) is lower, taking into account the surficial primary settlement $w^p{}_n$ due to the overload of every successive layer and surficial secondary settlements $w^s{}_n$ due to the secondary strain of every layer from the initial time of construction of this layer (Fig. 5.1.7B). Finally, after completion of landfilling ($t = t_c$), every layer is only subjected to secondary settlement. The right side of the diagram, for $t > t_c$ corresponds to the settlement of the cap cover. In these conditions, it is easy to understand the interest in the compressibility of waste.

In Fig. 5.1.8, the loading history of a waste sample (similar to a waste layer of initial thickness h_0) included in the landfill is simulated (Olivier and Gourc, 2007): this is the settlement of a sample of raw MSW (1 m × 1 m × initial height 0.8 m) with an initial dry density ($\rho_{d0} = 0.43$ Mg/m^3), subjected to two successive loading steps, a rapid increasing compression vertical stress (σ') until 140 kPa within 2 weeks, corresponding to waste placement in layers, inducing relative primary settlement ε_1 (with $\varepsilon = \Delta h/h_0$) and, in a second step a constant stress (σ') equal to 140 kPa, corresponding to the constant load of the waste layers above, with relative secondary settlement ε_2. In the diagram, the portion corresponding to relative secondary settlement ε_2 after 2 weeks is separated.

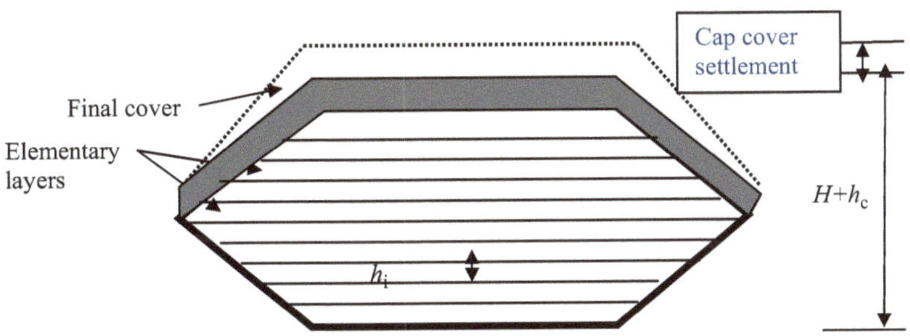

Figure 5.1.6 Typical landfill scheme, implementation of waste elementary layer by elementary layer.

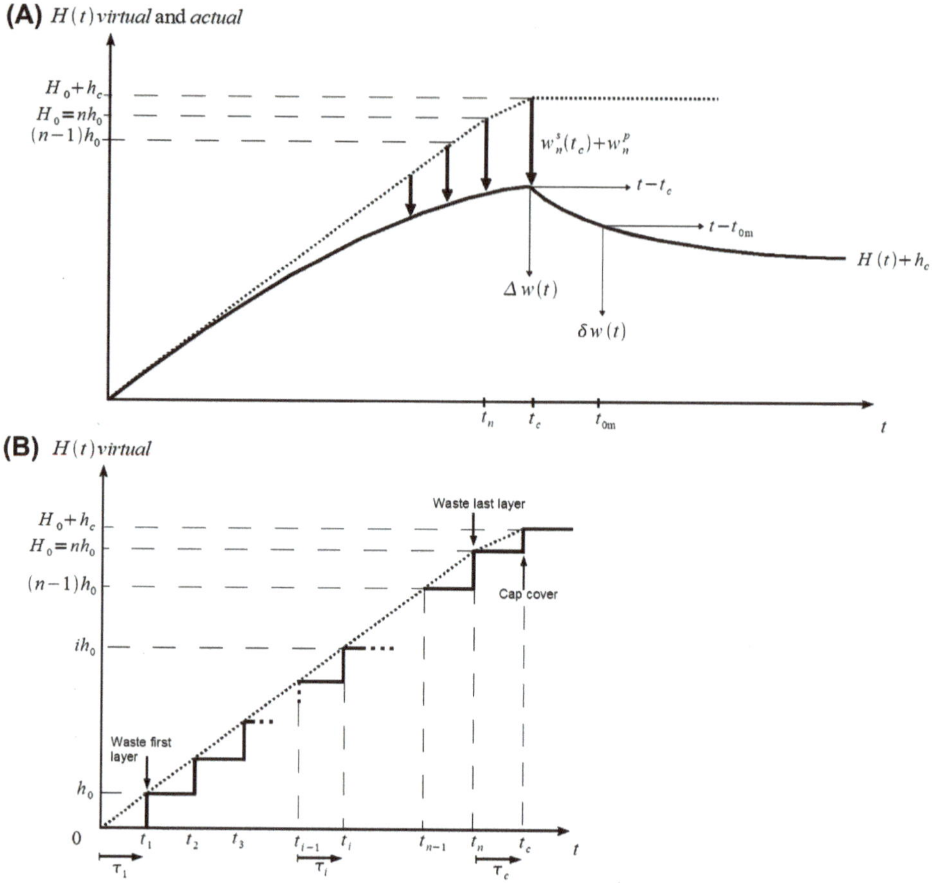

Figure 5.1.7 (A) Typical landfill construction scheme, and (B) model implementation of waste layer by layer (initial thickness h_0).

Primary settlement

Primary settlement is mainly due to compression under the load of upper waste layers. Primary settlement corresponds on site to the period of operation—compaction of waste and can be considered a purely mechanical process. It includes the immediate settlement and the stabilized settlement under the new overload. In the field, it is completed within a few weeks and within a few days in laboratory cells. It is governed by the initial arrangement of the particles (i.e., density, structure, composition, constituent sizes, shapes, and types). It is worth noting that the primary settlement of the waste sample is very high in laboratory tests (Fig. 5.1.8) and may not be representative of field conditions, due to the low initial dry density of the sample ρ_{d0}. Generally, on site, heavy compactors are used to obtain the highest density to minimize void volume. To reproduce this compaction level, significant compaction effort is needed in the laboratory. In addition, the kneading effect, due to compactor, is difficult to

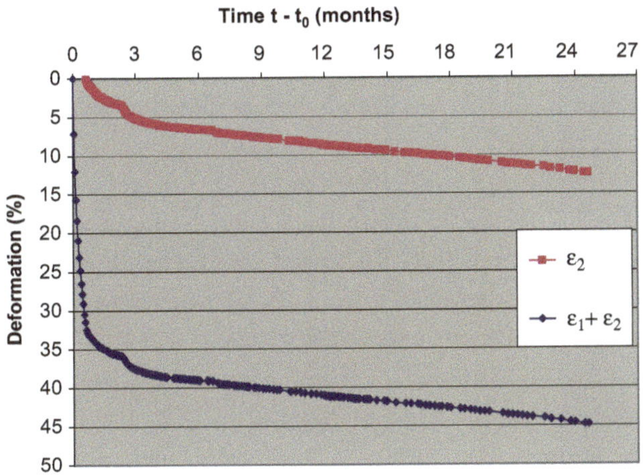

Figure 5.1.8 Typical compression test on a municipal solid waste sample.

reproduce in a laboratory cell. Dry density between 0.6 and 1.1 Mg/m^3 can be obtained on site whereas 0.45 Mg/m^3 is frequently measured in laboratory experiments when compaction effort is not significant (Kavazanjian, 2001; Zekkos et al., 2006; Hanson et al., 2010). On site, dry density is also affected by the use of daily soil covers which increases the average dry density (Zekkos et al., 2006). This method also decreases the fluid conductivity and consequently decreases the efficiency of the bioreactor process.

It is worth noting that although the assessment of waste density ρ_d is not problematic in the laboratory, the field measurement method is simple, but relatively laborious to implement (Thomas et al., 1999): it consists of excavating pits (around 2 m^3) with contours as regular as possible by means of a mechanical shovel. Excavated waste is then weighed, and the volume of the cavity is obtained by being filled with water (using a tank). This requires that the walls and the bottom of the pit are covered first by a flexible impervious membrane. Alternatively, calibrated gravel (i.e., gravel of known unit weight) can be used. The measurement of ρ_d has also been implemented in large diameter boreholes, as shown in Fig. 5.1.2, to generate an in situ unit weight profile of MSW (Stoltz et al., 2009; Zekkos et al., 2006).

The primary settlement can be described using a single settlement coefficient, the modified compression index C_R^*.Consolidation due to dissipation of pore water is not considered here, since the waste is generally not in saturated state (Bareither et al., 2012). However the classical Terzaghi relationship is commonly used.

The relative primary settlement (strain) ε_1 is assumed to be a function of the effective vertical load σ'_i.

$$\varepsilon_{1,i}(t) = \frac{h_0 - h(\sigma'_i)}{h_0} = C_R^* \cdot \log\left(\frac{\sigma'_i}{\sigma_c}\right) \tag{14}$$

where h_0 is the initial height of the waste layer before the loading phase, C_R^* the coefficient of primary compression, σ_c the preconsolidation stress. Settlement prior to the stress level $\sigma'_i = \sigma_c$ (overconsolidated phase) is assumed negligible.

Values $0.25 < C_R^* < 0.35$ were reported by the first author for laboratory tests; in the same range of values assessed by Hossain et al. (2009), but slightly higher than the values by Landva et al. (2000), or Marques et al. (2003) who reported an average value of 0.18.

The determination of primary settlement allows to infer the decrease of material porosity (n) using Eq. (12). However, this is insufficient to determine accurately the evolution of its porous structure. Tests of suction in an oedometer cell (Fig. 5.1.9) and interpretation using the Laplace relationship allowed the identification, unlike the majority of soils, of two classes of porosity: a macroporosity involved in the transfer of fluids and a microporosity (Stoltz and Tinet, 2012). The distribution of pores (diameter d_p) is strongly altered by compression. For the present specimen, under increasing compression, the dry density increased from $\rho_d = 0.55$ to $0.75\ Mg/m^3$. The diagram shows that macroporosity (large pores) is altered by compression, whereas microporosity (small pores) is insensitive in the same conditions.

Secondary settlement

Relative secondary settlement (strain) ε_2 is commonly assumed to start when primary settlement is completed ($t = t_{0M}$). Eq. (15) by Buisman for organic soils, like peats, is commonly used to determine the total secondary strain ε_2 using the secondary compression ratio $C_{\alpha\varepsilon}$:

$$\varepsilon_2 = \frac{\Delta h}{h_0} = C_{\alpha\varepsilon} \cdot \log_{10}\left(\frac{t}{t_{0M}}\right) \tag{15}$$

Based on settlement monitoring of several landfills, a relatively narrow range for the coefficient $C_{\alpha\varepsilon}$ was proposed in the literature: $0.03 < C_{\alpha\varepsilon} < 0.20$. This was possible because only conventional landfills with low and similar biodegradation rates were considered (i.e., without leachate recirculation and without pretreatment).

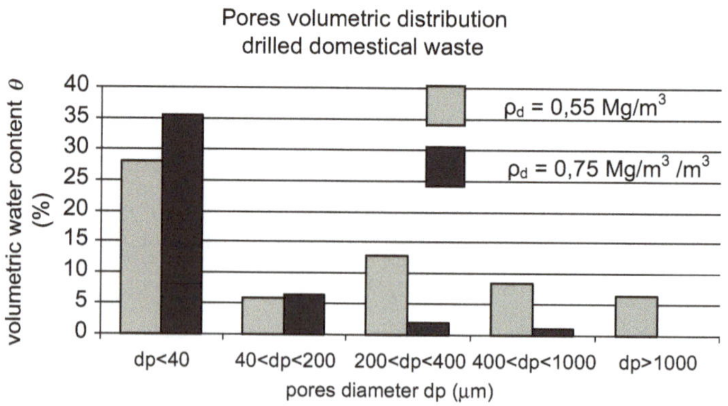

Figure 5.1.9 Volumetric distribution of pores of diameter d_p (according to the Laplace capillarity relationship) for two densities (two compression loads).

However, in modern landfills, such as bioreactor landfills, different treatments (e.g., pretreatment, leachate injection, etc.) may be applied to the waste and this can induce variable biodegradation kinetics. As the secondary settlement is partially dependent of the decay of waste, the coefficient $C_{\alpha\varepsilon}$ will change with waste treatment. Hence, mechanical and biological settlements should be separated.

A compression test in a large cell (Fig. 5.1.10) displays the influence of moisture (upper diagram) on settlement (lower diagram), and highlights the impact of leachate addition, which is a part of the bioreactor landfill technique (Olivier and Gourc, 2007). This test demonstrates interesting correlations between moisture addition and settlement: in the absence of leachate recirculation, settlement gently decreases with time, according to Eq. (15). A clear acceleration of settlement is concomitant with the process of leachate flow.

Biodegradation during the methanogenesis phase is directly correlated to biogas emission. A relation between degradation and settlement is identified, on the basis of two comparative tests (shown in Fig. 5.1.11), using the maxi-oedometer of Figs. 5.1.4 and 5.1.5. One sample is subjected to accelerated degradation by leachate injection, whereas in a similar sample degradation is inhibited (Staub et al., 2013): when the biodegradation (methanogenesis phase) starts, corresponding to a significant production of CH_4, the rate of secondary settlement is also significantly higher in comparison with the sample

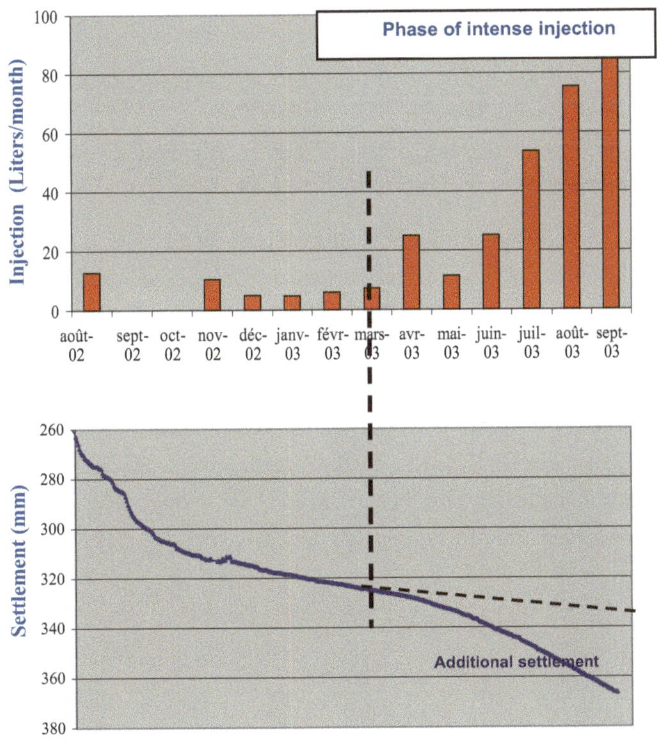

Figure 5.1.10 Effect of leachate injection on settlements in a large-scale compression cell: *dashed line* corresponds to the assumed settlement trend without leachate recirculation.

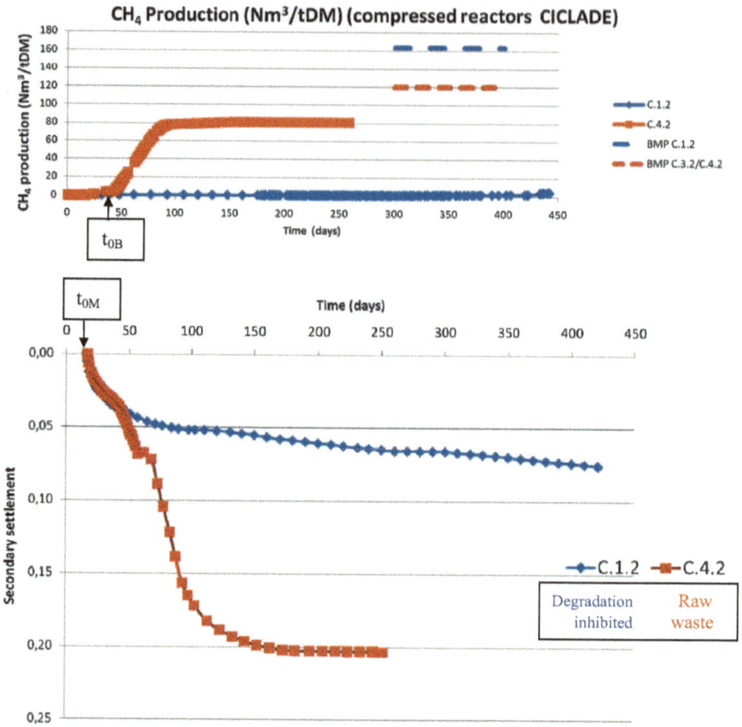

Figure 5.1.11 Correlation between methane production and settlement for two waste samples (C.4.2 raw waste and C.1.2 with biodegradation inhibited) in the oedometer–bioreactor of Figs. 5.1.4 and 5.1.5.

without biodegradation. When the production of gas is slowing down, the settlement rate is also decreasing. To model the relation between settlement and biodegradation, several different methods are proposed in the literature (McDougall, 2007), and a comparison is made by Bareither and Kwak (2015). A robust and simple method is proposed below by Gourc et al. (2010) inspired by the biomodel of Machado et al. (2008).

Secondary settlement is decoupled between a mechanical settlement ε_M corresponding to the decrease of the voids (V_G on Fig. 5.1.1), due to the mechanical rearrangement of the solid elements and the biological strain, identified as ε_B, occurring as a consequence of biodegradation (correlated to the loss of organic solid mass due to biodegradation, M_{Sd} on Fig. 5.1.1):

$$\varepsilon_2 = \frac{\Delta h_M}{h_0} + \frac{\Delta h_{sd}}{h_0} = \varepsilon_M + \varepsilon_B$$

Then, the pure mechanical settlement can be expressed by a creep relationship:

$$\varepsilon_M = \frac{\Delta h_M}{h_0} = C_{\alpha\varepsilon M} \cdot \log_{10}\left(\frac{t}{t_{0M}}\right) \tag{16}$$

where $C_{\alpha\varepsilon M}$ is the mechanical secondary compression coefficient.

Similarly to Eq. (15), mechanical secondary settlement is assumed to start at time t_{0M}. On the other hand, biodegradation is supposed to start at time t_{0B} when methanogenesis starts. t_{0B} is independent of t_{0M}, with the reasonable assumption $t_{0M} < t_{0B}$. Consequently, the results of the initial secondary settlement phase (pure mechanical settlement without biodegradation for $t_{0M} < t < t_{0B}$) can be used to obtain experimentally the best fit for parameter $C_{\alpha\varepsilon M}$. The authors propose $0.02 < C_{\alpha\varepsilon M} < 0.10$. In comparison, for organic soils like peat, it is often considered $C_{\alpha\varepsilon M} < 0.01$.

If it is assumed that degradation (and mass loss) is a direct function of recorded biogas production during the stable methanogenesis and the mass variation of the organic content is considered as a first-order degradation kinetic similar to the US EPA landfill gas model for methanogenesis (EPA, 2008), the biological settlement due to the biodegradation of the organic solid may be calculated. In this case Δh_{sd} corresponds to the loss of volume related to the loss of degradable organic mass, c is the initial gravimetric solid organic content (kg/kg), depending on the MSW composition, ρ_{d0} is the initial global dry density, ρ_{sd} is the average constitutive dry density of the degradable organic solid that is not affected during biodegradation:

$$\varepsilon_B = \frac{\Delta h_{sd}}{h_0} = \frac{\rho_d^0}{\rho_{sd}} \cdot c \cdot \left(1 - e^{-k\cdot(t - t_{0B})}\right) \tag{17}$$

$$\varepsilon_2 = \varepsilon_M + \varepsilon_B = C_{\alpha\varepsilon M} \cdot \log_{10}\left(\frac{t}{t_{0M}}\right) + \frac{\rho_d^0}{\rho_{sd}} \cdot c \cdot \left(1 - e^{-k\cdot(t - t_{0B})}\right) \tag{18}$$

It is common, as in the model above, to consider a low influence of the vertical stress on the secondary settlement process. The synthesis by Fei and Zekkos (2013), on the basis of 98 tests from 29 laboratory bioreactor simulators contributes to the assessment of the impact of the vertical loading (lower global compressibility).

References

Athanasopoulos, G.A., 2011. Laboratory testing of municipal solid waste. In: Zekkos, D. (Ed.), Geotechnical Characterization, Field Measurement, and Laboratory Testing of Municipal Solid Waste, ASCE Geotechnical Special Publication No. 209, pp. 195–205.

Bareither, C., Breitmeyer, R.J., Benson, C.H., Barlaz, M., Edil, T.B., 2012. Deer track bioreactor experiment: a field-scale evaluation of municipal solid waste bioreactor performance. Journal of Geotechnical and Geoenvironmental Engineering 138 (6), 658–670. ASCE.

Bareither, C., Kwak, S., 2015. Assessment of municipal solid waste settlement models based on field scale data analysis. Waste Management Journal 42, 101–117.

Bouthot, M., Blond, E., Quesnel, P., Fortin, A., Vermeersch, O.G., Davidson, S., Mlynarek, J., 2003. Landfill extension using the piggy-backing technique: discussion and case study in Quebec. In: Proc. Sardinia 2003, 9th International Waste Management and Landfill Symposium, Cagliari, Italy.

Bray, J.D., Zekkos, D., Kavazanjian Jr., E., Athanasopoulos, G.A., Riemer, M.F., 2009. Shear strength of municipal solid waste. Journal of Geotechnical and Geoenvironmental Engineering 135 (6), 709–722. ASCE.

Camp, S., Gourc, J.P., Plé, O., 2010. Landfill clay barrier subjected to cracking: multi-scale analysis of bending tests. Applied Clay Science 48 (3), 384–392. Elsevier.

EPA, 2008. Municipal Solid Waste Generation, Recycling, and Disposal in the United States: Facts and Figures for 2008. EPA yearly fact sheet, Cincinnati (OH), USA.
Fei, X., Zekkos, D., 2013. Factors influencing long-term settlement of municipal solid waste in laboratory bioreactor landfill simulators. Journal of Hazardous, Toxic and Radioactive Waste 17, 259–271. ASCE.
Gourc, J.P., Nousheen, K., Olivier, F., 2008. Long term settlement of domestic waste in landfill: ISPM method. Studia Geotechnica et Mechanica 30 (1–2), 31–40.
Gourc, J.-P., Staub, M.J., Conte, M., 2010. Decoupling MSW settlements into mechanical and biochemical processes – modelling and validation on large-scale setups. Waste Management 30, 1556–1568.
Grisolia, M., Napoleoni, Q., Tancredi, G., 1995. Contribution to a technical classification of MSW. In: Proceedings 5th International Landfill Symposium, Sardinia 95, vol. II, pp. 703–710.
Hanson, J.L., Yesiller, N., Von Stockhausen, A., Wong, W.W., 2010. Compaction characteristics of MSW. Journal of Geotechnical and Geoenvironmental Engineering 136 (8), 1095–1102. ASCE.
Hettiarachchi, C.H., Meegoda, J.N., Hettiarachchi, P., 2009. Effect of gas and moisture on modeling of bioreactor landfill settlement. Waste Management 29, 1018–1025.
Hossain, M.S., Gabr, M.A., Asce, F., 2009. The effect of shredding and test apparatus size on compressibility and strength parameters of degraded municipal solid waste. Waste Management 29, 2417–2424.
Ivanova, L.K., Richards, D.J., Smallman, D.J., 2008. The long-term settlement of landfill waste. Waste and Resource Management 161, 121–133.
Kavazanjian, E., 2001. Mechanical properties of MSW. In: Proceedings 8th International Waste Management and Landfill Symposium, Sardinia 2001, vol. III. CISA, pp. 415–424.
Landva, A.O., Valsangkar, A.J., Pelkey, S.G., 2000. Lateral earth pressure at rest and compressibility of municipal solid waste. Canadian Geotechnical Journal 37, 1157–1165.
Machado, S.L., Vilar, O.M., Carvalho, M.F., 2008. Constitutive model for long term municipal solid waste mechanical behaviour. Computers and Geotechnics 35, 775–790.
Marques, A.C., Filz, G.M., Vilar, O.M., 2003. Composite compressibility model for municipal solid waste. Journal of Geotechnical and Geoenvironmental Engineering 129 (4), 372–378. ASCE.
McDougall, J., 2007. A hydro-bio-mechanical model for settlement and other behaviour in landfilled waste. Computers and Geotechnics 34, 229–246.
Münnich, K., Mahler, C.F., Fricke, K., 2006. Pilot project of mechanical-biological treatment of waste in Brazil. Waste Management 26, 150–157.
Olivier, F., Gourc, J.P., Munoz, M.L., Budka, A., Denecheau, P., 2003. Validation of a multi-layer waste settlement prediction model by means of internal survey data. In: Sardinia 2003, 9th International Landfill Symposium, Cagliari, Italie, CISA Publisher, Abstract Proceedings, pp. 257–260.
Olivier, F., Gourc, J.P., Coquant, C., 2005. Predicting the storage capacity of deep landfill: Ferques bioreactor case study. In: Sardinia 2005, 10th International Landfill Symposium, Cagliari, Italy. CISA.
Olivier, F., Gourc, J.P., 2007. Hydro-mechanics of MSW subject to leachate recirculation in a large-scale compression reactor cell. Waste Management Journal 27 (1), 44–58. Elsevier.
Powrie, W., Richards, D., Velkushanova, K., 2009. Mechanisms of Settlement in Biodegradable Wastes. In: Third International Workshop "Hydro-physico-mechanics of Landfills". Braunschweig, Germany.
Sowers, G.F., 1973. Settlement of waste disposal fills. In: Proceedings of the 8th International Conference on Soil Mechanics and Foundation Engineering, Moscow, pp. 1207–2010.
Staub, M., Marcolina, G., Gourc, J.P., Simonin, R., 2011. An incremental model to assess the environmental impact of cap cover systems on MSW landfill emissions. Journal Geotextiles and Geomembranes 29, 298–312. Elsevier.
Staub, M., Gourc, J.P., Drut, N., Stoltz, G., Mansour, A., 2013. Large scale bioreactor pilots for monitoring the long term hydromechanics of MSW. Journal of Hazardous, Toxic and Radioactive Waste 17, 285–294. ASCE.
Stoltz, G., Nousheen, K., Gourc, J.P., Olivier, F., Redon, E., 2009. Bio-hydro-mechanical pro-perties from laboratory tests following a deep drilling campaign. In: Sardinia 09, 12th International Landfill Symposium, Cagliari, Italy.
Stoltz, G., Gourc, J.P., Oxarango, J.P., 2010. Characterization of the physico-mechanical parameters of MSW. Waste Management Journal 30 (8–9), 1439–1449. Elsevier.
Stoltz, G., Tinet, A.-J., Staub, M.J., Oxarango, L., Gourc, J.-P., April 2012. Moisture retention properties of municipal solid waste in relation to compression. Journal of Geotechnical and Geoenvironmental Engineering. ASCE.
Tano, F., Olivier, F., Touze-Foltz, N., Dias, D., 2015. State of the art of piggy-back landfills worldwide. In: Conference Geosynthetics 2015, Portland, USA.

Thomas, S., Aboura, A.A., Gourc, J.-P., Gotteland, P., Billard, H., Delineau, T., Gisbert, T., Ouvry, J.-F., Vuillemin, M., 1999. An in situ waste mechanical experimentation on a French landfill. In: Proc. Sardinia 1999, 7th International Landfill Symposium, S. Margherita di Pula, Italy.
Zekkos, D., Vlachakis, V.S., Athanasopoulos, G.A., 2014. The 2010 Xerolakka landfill slope instability. Environmental Geotechnics 1 (EG1), 55–65.
Zekkos, D., Bray, J.D., Kavazanjian Jr., E., Matasovic, N., Rathje, E.M., Riemer, M.F., Stokoe, K.H., October 2006. Unit weight of municipal solid waste. ASCE Journal of Geotechnical and Geoenvironmental Engineering 132 (10), 1250–1261.

5.2

MECHANICAL STABILITY AT LANDFILL SCALE

Jean-Pierre Gourc and Dimitrios Zekkos

INTRODUCTION

Prediction of the long-term settlement is useful when the postclosure and aftercare periods of the landfill are considered. In addition the kinetics of the settlement, specifically on the cap cover, could be a relevant indicator of the biodegradation level of the waste. Consequently, monitoring of at least surficial settlement is recommended. However, there is concern about the heterogeneity of the settlement distribution that requires monitoring of a significant number of surface benchmarks.

Monitoring of Landfill Settlement

Traditional surveying methods associated with a square network of landmarks are systematically used on site, but the number of benchmarks is generally not adequate to be representative. Two applications of other more efficient techniques are presented below (Gourc et al., 2001):

- *Profilometer.* The profilometer (horizontal inclinometer) was installed at a landfill horizontally just beneath the cap liner of a cell with a thickness of 22 m. This is a flexible pipe (diameter

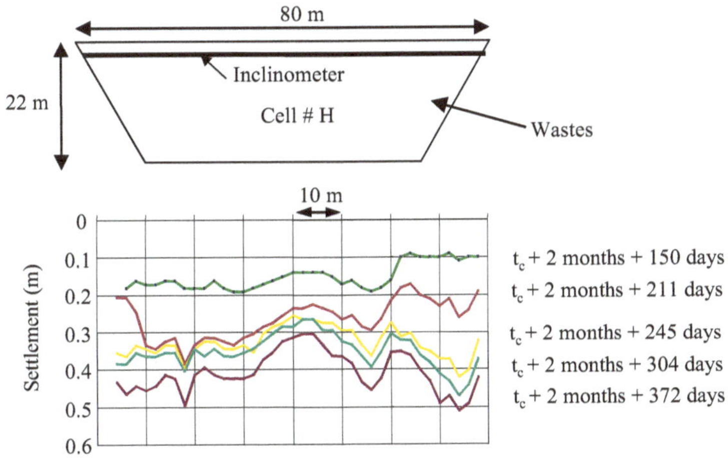

Figure 5.2.1 Evolution with time of settlement profiles generated by a horizontal inclinometer (profilometer).

63 mm) buried in a trench that makes possible to measure differential settlements at the cell surface (Fig. 5.2.1).

- *3D laser scanning*. This modern technique is appealing due to the possibility of scanning the entire surface of the landfill with a resolution that can be as low as 4 mm (Olivier et al., 2005). As shown in Fig. 5.2.2, the comparison of identical profiles at successive dates can provide the distribution of settlements over the entire surface of the landfill.

Temporary Cap Cover

Municipal solid waste (MSW) is generally highly compressible, with a settlement rate that decreases with time, as demonstrated in Fig. 5.1.8 (Chapter 5.1). Not only average total settlements, but also differential settlements (as shown in Figs. 5.2.1 and 5.2.2) are large, increasing significantly the risk of damage to the liner. The final cap cover very often includes a layer of fine soil. Large scale experimentations demonstrate that when such a fine soil layer is subjected to differential settlement, it will display cracking (Camp et al., 2010, Gourc et al. 2010) for very small extensions (strains of 1%–2%) with a high risk of landfill gas leakage (which is greenhouse gas).

To maintain the cap's integrity, despite the large settlement of the waste mass, the implementation of the "final" cover could be postponed to a later time. Recent approaches to cap lining consist of using, during the first stage of high settlement rates (2 years in Fig. 5.2.3), a geosynthetic temporary cover. In addition, experimentations demonstrate that the biogas collection efficiency of an appropriately designed geosynthetic product can reach 65% of the generated biogas.

SHEAR STRENGTH OF MUNICIPAL SOLID WASTE

The shear strength of MSW is a critical input parameter in stability analyses of landfill slopes. Primarily laboratory-testing techniques, but also field approaches have been used to assess the shear strength of MSW. A brief review of the shear strength of MSW and the approaches used to assess it is provided below.

Laboratory Approach

Main Devices

The most common laboratory device that has been used for testing MSW is the direct shear device (Landva and Clark, 1986, 1990; Zekkos et al., 2010a). Triaxial shear (Jessberger and Kockel, 1993; Grisolia et al., 1995; Bray et al., 2009), primarily in compression, and also in extension as well as compression unloading (Zekkos et al., 2012), has also been used. Limited testing has been performed in simple shear (Pelkey et al., 2001; Kavazanjian et al., 1999). In addition, unique experimental setups, such as the tension test have also been developed (Kölsch, 1995) to assess the shear resistance of the MSW, but these are not often used. Examples of these devices are shown in Fig. 5.2.4.

Specimen size: To assess the shear strength of MSW, primarily large-size testing devices, as well as smaller, conventional size, testing devices have been used. Recent studies have shown that conventional size testing devices generate test results that may be unreliable, not representative of field conditions,

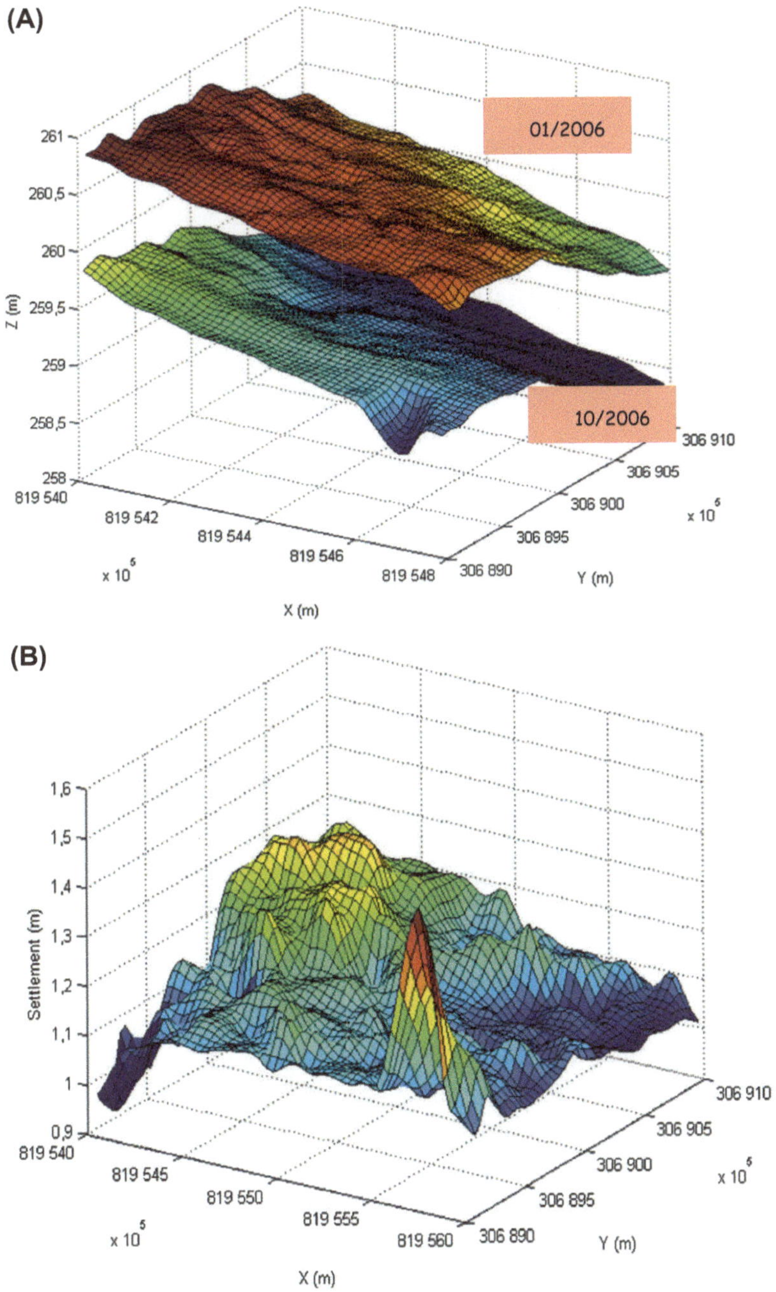

Figure 5.2.2 (A) Scan of the landfill surface elapsed time: 9 months and (B) corresponding cap cover settlement.

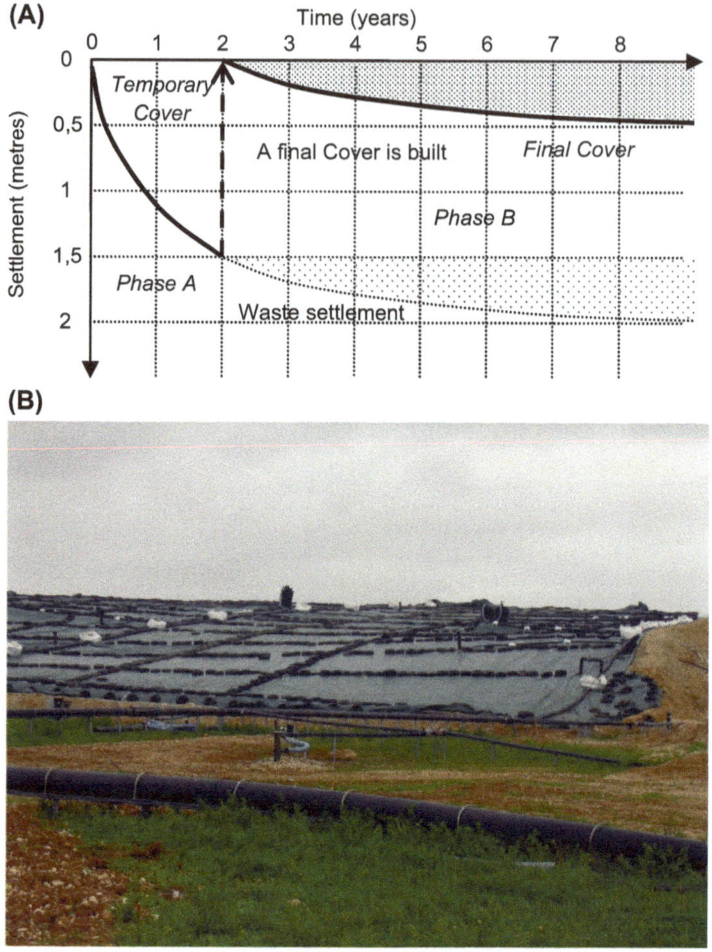

Figure 5.2.3 (A) Dual phase of the cap cover construction correlated to the waste settlement rate. (B) Example of temporary cap cover.

and excessively variable. This is primarily because of the large size of the waste constituents that is impossible to include in a conventional size testing device without processing, and thus, the specimen size is smaller than the "Representative Elementary Volume" discussed earlier. More recently, several studies have shown that a specimen that is 300 mm in diameter or larger is typically adequate (Bareither et al., 2012; Athanasopoulos, 2011; Bray et al., 2009); however, comprehensive, variable size data sets that investigate the influence of specimen size are not available.

Specimen preparation: Laboratory testing on MSW is performed on reconstituted specimens, because collection of "undisturbed" samples of MSW has proven largely unfeasible, due to a number of factors, including the type and size of waste constituents and the low "cohesion" of waste material. Reconstitution of MSW specimens is most commonly executed using variations of the moist tamping method.

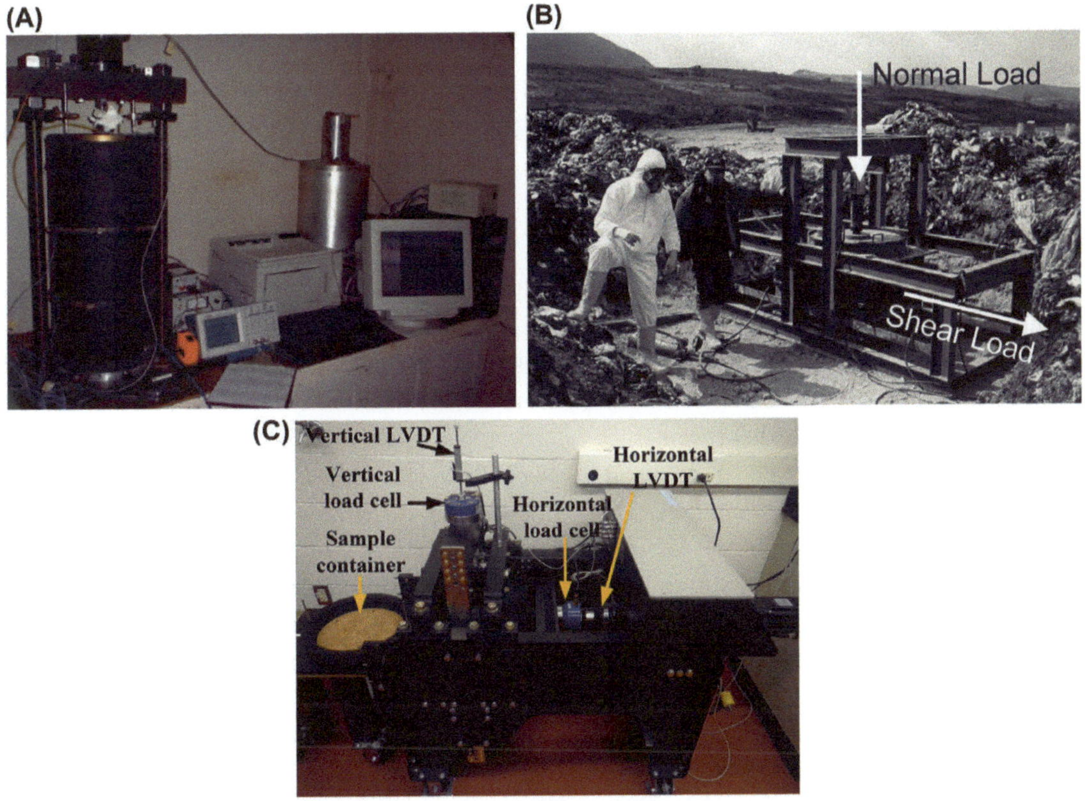

Figure 5.2.4 Examples of testing devices used in laboratory testing of MSW: (A) UC Berkeley triaxial shear; (B) in situ direct shear (Caicedo et al., 2002), and (C) University of Michigan simple shear. *MSW*, municipal solid waste; *UC*, University of California; *LVDT*, linear variable differential transformer.

A key part of the specimen preparation is the selection and processing of the waste constituents. Two different approaches have been typically used. In the waste scalping process, specimens are prepared by removing oversize-waste constituents, retaining all other waste constituents that can be accommodated and making sure that the specimen's waste composition is similar to the field composition. In the waste shredding (or milling) process, waste constituents are reduced mechanically in size. A milling process using a blender is faster, but commonly results in waste constituents that are of similar size and excessively shredded. These constituents have essentially no physical resemblance to the MSW in situ, unless the waste in the field is also preprocessed, as is the case for MBT waste that has undergone a mechanical–biological pretreatment, a common practice in countries such as Germany. In a manual laboratory shredding process, oversize particles are simply reduced to the maximum acceptable particle size. This process can only be performed manually and requires significantly longer waste specimen preparation times, but avoids excessive shredding.

Stress–Strain Relationship of Municipal Solid Waste and the Importance of Waste Composition

The stress–strain relationship of MSW in the laboratory is affected by the factors that affect soil behavior, such as confining stress and density. However, the stress–strain relationship of MSW is also significantly affected by:

1. waste composition and
2. mode of shearing, i.e., the type of testing.

Waste composition: In addition to soil materials, which are typically present in landfills due to the use of daily soil cover, a variety of waste constituents are also present, as described earlier. Procedures for their characterization for geotechnical testing purposes have been developed (Dixon and Langer, 2006; Zekkos et al., 2010b). Of the most common waste constituents found in landfills, the vast majority are 1D fibers (such as wood) and 2D sheets (such as papers, plastics, and textiles). The presence of these constituents affects the observed stress–strain behavior. Because of these differences, the stress–strain behavior of unprocessed MSW is different than that of MBT material.

Fig. 5.2.5 illustrates examples of the stress–strain or stress-displacement response of MSW. In direct shear testing, a strain-hardening relationship is typically observed (Gotteland et al., 2000), with shear stress increasing and eventually reaching a peak value as horizontal displacement increases (Mazzucato et al., 1999; Zekkos et al., 2010a). The results are generally similar in simple shear testing (Fei and Zekkos, 2015). In triaxial shear testing, most commonly, shear stress increases steadily with increasing axial strain, and in many cases, an upward curvature of the stress–strain response is observed (Gomes et al., 2002). Subsequent studies (e.g., Bray et al., 2009) have shown that this difference can be attributed to the reinforcing effect of the 1D and 2D waste constituents. During specimen preparation, these waste constituents have a tendency to become oriented in the horizontal direction, specifically due to their shape and their low density. This is also observed in-situ, even with kneading compaction (Zekkos, 2013). As a consequence, in direct shear, the shearing plane is parallel to these constituents, as shown in Fig. 5.2.6 (Bray et al., 2009), whereas in triaxial shear, the shearing plane is always oriented at some angle from the horizontal (Zekkos et al., 2012). The upward curvature observed in triaxial shear is a consequence of the progressive contribution of the 1D and 2D waste constituents to the shear resistance. Bray et al. (2009) performed a combination of direct shear and triaxial shear tests to prove the hypothesis that direct shear specimens were prepared in a split mold, and the specimens were rotated 90 degrees before being placed in the direct shear box, so that 1D and 2D constituents were generally perpendicular to the horizontal shear plane. The observed shear-displacement response exhibited an upward curvature. Triaxial testing on specimens with variable waste composition also showed that specimens without fibrous constituents reached peak shear stress conditions without an observed upward curvature. The upward curvature was observed for specimens with 1D and 2D waste constituents and was found to become more pronounced with increasing percentage of 1D and 2D waste constituents. These test results underline the importance of waste composition on the stress–strain response.

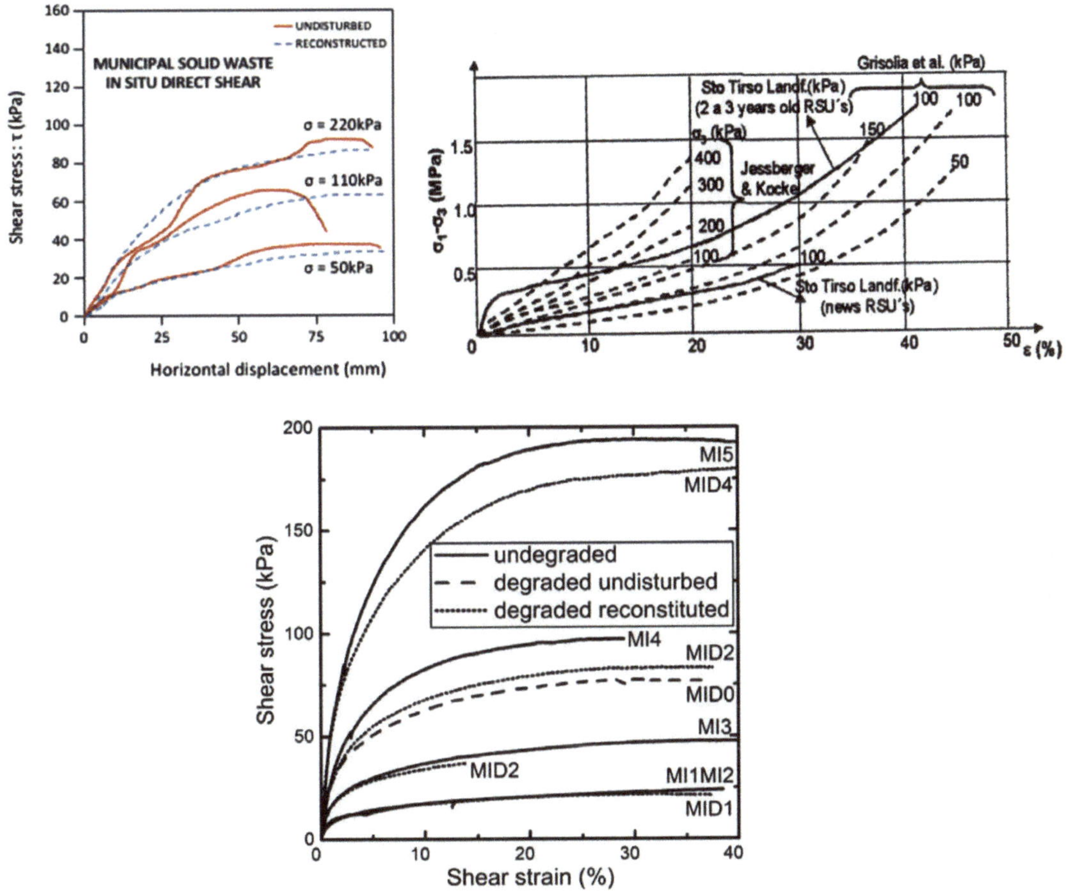

Figure 5.2.5 Examples of typical stress–strain or stress-displacement relationships for MSW in direct shear (Mazzucato et al., 1999), simple shear (Fei and Zekkos, 2015), and triaxial shear testing (Gomes et al., 2002). *MSW*, Municipal Solid Waste; *RSU*, Rifiuti Solidi Urbani (in Italian) = MSW.

In direct shear, because peak shear stress conditions are reached, the shear strength is defined being equal to the peak shear stress. Alternatively, if the specimen does not reach peak shear stress conditions, the shear resistance at maximum horizontal displacement is assumed to be equal to the shear strength. In triaxial shear, because the shear resistance is a function of the level of strain (or displacement) without typically reaching peak shear stress conditions, different definitions for shear strength have been used. Typically, shear strength is defined as the mobilized shear stress at a certain level of strain, most commonly 5% or 10% axial strain. Note that because triaxial shear testing is commonly initiated from an isotropic stress state, the strain associated with an MSW specimen reaching a K_o (in situ) stress state can be significant. Thus Bray et al. (2009) emphasized that the defined strain at failure (e.g., 5% or 10%) should be added to the strain measured to reach K_o conditions.

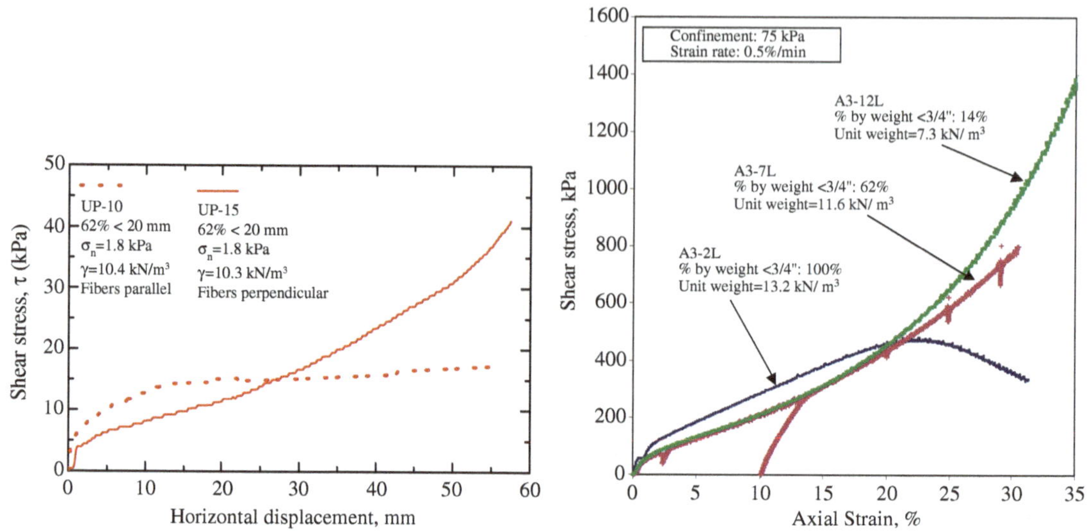

Figure 5.2.6 Examples of stress—strain and stress-displacement relationships for MSW in direct shear (Zekkos et al., 2010a), and triaxial shear testing (Bray et al., 2009) highlighting the importance of waste composition and shearing mode. *MSW*, Municipal Solid Waste.

Limitations of Laboratory Testing

Laboratory testing on MSW has limitations. Laboratory MSW specimens are always reconstituted. No definitive data exist to prove that the reconstituted specimens are representative of the field behavior of MSW. The available data, however (e.g., Mazzucato et al., 1999), indicate that reconstituted specimens are in general representative of the field conditions. This may not be surprising, since MSW is not deposited by geologic processes but is placed by humans using field procedures that can be mimicked in the laboratory. In addition, large-size specimens are needed to investigate the mechanical properties of MSW, as described earlier.

Field Approaches

In situ invasive site characterization methods, such as the standard penetration test (SPT), the cone penetration test (CPT), and the dilatometer have been used in an effort to characterize the waste material and interpret the shear strength of the material (Castelli and Maugeri, 2014; McKnight et al., 2014). These techniques test a relatively small volume of material, and as a consequence, testing results are highly variable and many times, challenging to interpret. In addition, limited in situ direct shear tests (e.g., Caicedo et al., 2002) have been performed using large shear boxes, and these test results are in general agreement with laboratory direct shear test results (Mazzucato et al., 1999).

The most common field approach used to assess the shear strength of MSW is probably the back calculation of stable landfill slopes, as well as failed landfill slopes. Back calculation of stable landfill slopes (Kavazanjian et al., 1995), requires an assumption on the factor of safety of the stable slope, and thus shear strengths back calculated from this procedure are more uncertain. Back calculation of failed slopes (Eid et al., 2000) can provide useful insights into the actual slope failure mechanisms and can be used to back calculate the shear strength of the MSW. A limitation of this approach is the uncertainty in the pore pressures (due to gases or liquids) in the waste mass. Thus, the effective stress state of the waste mass is commonly unknown.

State of the Practice on Shear Strength of Municipal Solid Waste

The shear strength (τ) of MSW is typically described by a Mohr–Coulomb failure criterion:

$$\tau = \sigma' \cdot \tan \varphi' + c' \quad (1)$$

where c' and φ' are the effective cohesion and friction angle of the MSW, and σ' is the effective normal stress.

The effective normal stress, considering the unsaturated nature of material, is given by the following equation:

$$\sigma' = (\sigma - u_G) + \chi(u_G - u_W) \quad (2)$$

where χ is a parameter that is a function of the saturation of the material; u_G and u_W are the gas pressure and water pressure respectively, and $(u_G - u_W)$ is the capillary pressure (Gourc and Staub, 2010). Most commonly, in practice, the influence of gas pressure is ignored (i.e., it is assumed that $\chi = 1$), in which case the effective normal stress is equal to the total stress minus the pore water pressure. In addition, modern MSW landfills are operated as dry tombs, in which case, there is no positive water pressure, and the total stress is equal to the effective stress.

Various recommended shear strengths of MSW are shown in Fig. 5.2.7 (Kavazanjian et al. 1995, Manassero et al. 1996, Eid et al. 2000, Bray et al. 2009). Earlier recommendations indicated that the shear strength of MSW is linearly, or bilinearly, increasing with confining (or normal) stress. Recent studies indicate that the shear strength envelope of MSW is nonlinear, i.e., the friction angle reduces with increasing normal stress. These recommended strength envelopes are consistent with direct shear testing of MSW, and thus do not account for the reinforcement effect of the fibrous waste constituents. Thus, the reinforcement effect of the fibrous waste constituents, as for example, clearly observed in triaxial testing, or direct shear testing across the fibrous constituents is generally not considered yet in design. The Bray et al. (2009) recommended shear strength captures the reduction in friction angle of MSW with increasing normal stress, an important consideration for larger size landfills.

WASTE MASS SLOPE STABILITY

A critical design component of modern landfills is the stability of the waste mass. A potential failure of the waste mass can have devastating consequences for the environment and public health, and in many instances, can even result in loss of life. The focus of this section is on deep-seated slope stability, in contrast to cover stability, which typically involves only the landfill's cover system components (in particular geosynthetic liner systems on slopes).

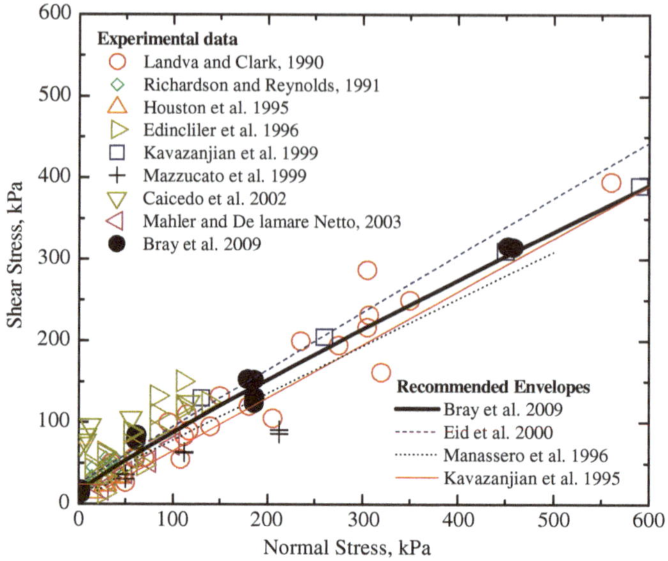

Figure 5.2.7 Recommended shear strength envelopes for Municipal Solid Waste (Modified after Bray et al., 2009).

Modes of Slope Stability Failure

There are numerous modes of deep-seated landfill slope instability. Slope instability may be entirely restricted in the waste mass and may be rotational, translational, or a combination of the two. Alternatively, a failure mode may involve the foundations soils, i.e., the failure surface may extend into the underlying foundation soils. This is particularly common when landfills are built on soft soil deposits. A waste mass may also involve a combination of a rotational failure surface through the waste with a translational component on the base containment system. The base containment system of modern landfills involves a number of soil layers and geosynthetic materials for containment as well as leachate collection. The shear resistance of the interfaces of these layers and materials may be particularly weak, and failures involving these materials are not uncommon. In fact, a deep-seated failure surface may be entirely restricted in the containment system and may not involve the waste mass, as was the case in the Kettleman Hills hazardous landfill failure (Mitchell et al., 1990). To analyze failure modes through the waste mass only, the unit weight and shear strength of the MSW are required. To analyze failure modes that involve the containment system or the foundation soils, an assessment of the shear resistance of all containment system components and their interfaces is necessary.

Static Slope Stability Methods and Example of Chart

Landfill slope stability is typically assessed using limit equilibrium methods. Modern numerical codes allow engineers to assess numerous failure surfaces and consider all modes of failure shown. The Spencer method (Spencer, 1967) or the Morgenstern–Price methods (Morgenstern and Price, 1965) should be used in executing the analyses, since these methods satisfy both force and moment equilibrium. The minimum acceptable factor of safety varies from region to region, but in general, for static conditions the following factors of safety are typically considered:

- $FS_{min} = 1.5$ for long-term or permanent deep-seated failure mechanisms;
- $FS_{min} = 1.25-1.3$ for temporary or short-term conditions, such as interim slopes, where the consequences of failure are minimal.

Seismic Slope Stability Methods

During an earthquake, ground motions amplify as the seismic waves propagate through the waste mass, primarily due to the softer nature of the waste material but also due to topographic amplification. This increased intensity of ground motions may result in instabilities in the waste mass. The stability of landfill slopes during an earthquake is of concern. In certain regions, typically with low accelerations, a calculated factor of safety greater than 1 is required in pseudostatic analyses to ensure that no failures are likely during the design earthquake. However, a factor of safety greater than 1 in pseudostatic analyses is practically impossible in seismically active regions. In these cases, the dynamic response of the landfill and its displacement needs to be analyzed. Bray (2007) provides a valuable overview of seismic slope stability methods.

Site-specific dynamic analyses can be performed to evaluate the seismic response of a landfill that is subjected to a design seismic event. These analyses may be 1D or 2D and may be coupled with sliding analyses or decoupled. The potential for ground motion amplification and the permanent seismic slope displacements of select sliding masses is calculated. Dynamic properties for the waste mass, cover, base containment, and foundation soils need to be used to execute these analyses. Dynamic properties of interest include the shear wave velocity and the shear modulus reduction, and material damping increase as a function of shearing strain, as well as the dynamic shear strength of the material. Zekkos et al. (2011) has made a comprehensive review of the dynamic properties of MSW.

Alternatively, simplified methodologies can be used. The Makdisi and Seed (1978) method that has been developed for earthfill dams, has been used in landfill practice, but improved methodologies, such as the Bray and Rathje (1998) and the Bray and Travasarou (2007) methods have also been developed and are preferable. These more recent methodologies account in an improved manner for the impact of ground motion variability and site effects. They are based on dynamic analyses, but they simplify the results in "design charts" that compare the yield coefficient (k_y) to the maximum horizontal equivalent acceleration (k_{max}) imposed by the earthquake. When the k_y/k_{max} ratio is greater than 1, seismic permanent displacements are equal to zero. As the ratio reduces, the calculated seismic permanent displacements increase. The calculated displacements are compared with the acceptable "threshold" value.

The calculated displacements are considered an index of the expected performance of the structure, especially in the case of decoupled analyses. A serviceability limit state approach is followed where, if seismic displacements are low (less than 150 mm), the consequences of shaking on stability are considered minor. If seismic displacements are between 150 mm and 600 mm, the consequences of shaking are considered moderate. Typically, in practice, a calculated displacement of 300 mm is considered a threshold between acceptable and unacceptable performance for waste masses that engage geosynthetics. Larger displacements (in the order of 1000 mm) have been accepted for failure surfaces that only engage the waste mass and do not intersect geosynthetics. The threshold value is typically set by the regulator and is also a function of other considerations such as the consequences of failure and the ability to detect and repair the damage.

Case Histories of Slope Instabilities and Lessons Learned

Instabilities Under Static Conditions

Numerous landfill stability failures have taken place in landfills under static conditions (Fig. 5.2.8). A few of the recent ones that are also well documented are listed in Table 5.2.1. These failures have been particularly instructive in understanding the main factors that have contributed to landfill slope failures. Important lessons learned include the following:

- The various modes of failure as well as the stress—strain—strength characteristics of the various landfill components that are involved in each mode of failure (e.g., cover soils, waste mass, containment system) need to be properly considered.
- The majority of failures have occurred in high moisture and/or high pressure conditions. Failures have taken place after significant rainfall events or when leachate is recirculated or pressurized into the waste mass and especially when gas collection and leachate collection systems are not available to dissipate the pressures generated. The leachate and gas pressures are affecting the effective stress of the waste mass, reducing the shear strength of the material and adversely impacting the stability of the landfill's waste mass (see Eq. 2). Note that the influence of gas pressure is suspected to play an important adverse role, because when the waste is becoming saturated in liquid, the gas conductivity becomes very low (Caicedo et al., 2002; Gourc and Staub, 2010)
- Landfill operations and especially construction of new cells, need to account for interim slope stability and the stability of adjacent cells that may be affected by these activities.

Figure 5.2.8 Examples of reported waste slides in Rumpke landfill, Ohio, in 1996, and Xerolakka landfill, Greece in 2010. *(Left) Photo: http://www.dr-koelsch.de/html/rumpke.html; (Right) Zekkos et al., 2014.*

Table 5.2.1 List of recent examples of landfill instabilities under static conditions

Landfill Name	Locations	Date of Failure	Size (m^3)	References
Rumpke	Ohio, USA	1996	1,200,000	Eid et al. (2000)
Dona Juana	Colombia	1997	1,360,000	Hendron et al. (1999), Caicedo et al. (2002)
Payatas	Philippines	2000	13,000–16,000	Merry et al. (2005), Jafari et al. (2013)
Bandung	Indonesia	2005	2,700,000	Kölsch et al. (2005)
Xerolakka	Greece	2010	12,000	Zekkos et al. (2014)

Table 5.2.2 Performance of landfills during 1994 Northridge Earthquake

Landfill	Type	Containment System	Maximum Horizontal Acceleration Rock (g)	Damage
OII	Gravel pit fill	None	0.10	Cover soil
Toyon Canyon	Canyon fill	LCRS	0.21	Cover soil; gas collection header
Sunshine Canyon	Side hill fill	None	0.46	Cover soil
Lopez Canyon	Canyon fill	None; LCRS and GLS	0.42	Cover soil; gas collection system
Chiquita Canyon	Canyon fill	None; LCRS and GLS	0.33	Cover soil; geomembrane tears
Bradley Avenue	Gravel pit fill	None; CL; LCRS and GLS	0.26	Cover soil

CL, Clay Liner; *GLS*, geosynthetic liner system; *LCRS*, leachate collection and removal system.
Modified after Augello, A.J., Matasovic, N., Bray, J.D., Kavazanjian Jr., E., Seed, R.B., 1995. Evaluation of Solid Waste Landfill Performance during the Northridge Earthquake, Earthquake Design and Performance of Solid Waste Landfills, pp. 17–50 (ASCE Geotechnical Special Publication No. 54).

Instabilities Under Seismic Conditions

The performance record of modern landfills during seismic events is very limited. In the United States, Subtitle D landfills have been strongly shaken primarily only by the 1994 Northridge earthquake. The maximum horizontal acceleration at rock (MHA) that the MSW landfills were subjected to during this earthquake did not exceed 0.36 g and, in many cases, was significantly lower. In seismically active regions, modern landfills are now designed for significantly higher levels of acceleration than the level of shaking they experienced during Northridge earthquake. A summary of the performance of the MSW landfill facilities is provided by Augello et al. (1995). Table 5.2.2 summarizes the levels of acceleration that some landfills experienced and a description of the damage observed. The current, short, performance record indicates that damage to the landfills during earthquakes is possible and may engage also the containment systems. Most common damage involves shallow failures where the motions are most amplified.

References

Athanasopoulos, G.A., 2011. Laboratory testing of municipal solid waste. In: Zekkos, D. (Ed.), Geotechnical Characterization, Field Measurement, and Laboratory Testing of Municipal Solid Waste, pp. 195–205 (ASCE Geotechnical Special Publication No. 209).

Augello, A.J., Matasovic, N., Bray, J.D., Kavazanjian Jr., E., Seed, R.B., 1995. Evaluation of Solid Waste Landfill Performance during the Northridge Earthquake. In: Earthquake Design and Performance of Solid Waste Landfills, pp. 17–50 (ASCE Geotechnical Special Publication No. 54).

Bareither, C., Benson, C., Edil, T., 2012. Effects of waste composition and decomposition on the shear strength of municipal solid waste. Journal of Geotechnical and Geoenvironmental Engineering 138 (10), 1161–1174.

Bray, J.D., Zekkos, D., Kavazanjian Jr., E., Athanasopoulos, G.A., Riemer, M.F., 2009. Shear strength of municipal solid waste. Journal of Geotechnical and Geoenvironmental Engineering, ASCE 135 (6), 709–722.

Bray, J.D., 2007. Simplified seismic slope displacement procedures. In: Invited Theme Lecture, 4th Inter. Conf. on Earthquake Geotechnical Engineering. Thessaloniki, Greece, June 26, 2007.

Bray, J.D., Travasarou, T., 2007. Simplified procedure for estimating earthquake-induced deviatoric slope displacements. Journal of Geotechnical and Geoenvironmental Engineering 133 (4), 381–392.

Bray, J.D., Rathje, E.R., 1998. Earthquake-induced displacements of solid-waste landfills. Journal of Geotechnical and Geoenvironmental Engineering, American Society of Civil Engineers 124 (3), 242–253.

Caicedo, B., Giraldo, E., Yamin, L., Soler, N., 2002. The landslide of Doña Juana landfill in Bogotá-a case study. In: Proceedings of the 4th International Congress on Environmental Geotechnics, Rio de Janeiro Brazil, 11–15 August 2002, pp. 171–175.

Camp, S., Gourc, J.P., Plé, O., 2010. Landfill clay barrier subjected to cracking: multi-scale analysis of bending tests. Applied Clay Science 48 (3), 384–392. Elsevier.

Castelli, F., Maugeri, M., 2014. Mechanical properties of municipal solid waste by SDMT. Waste Management 34 (2), 256–265.

Dixon, N., Langer, U., 2006. Development of a MSW classification system for the evaluation of mechanical properties. Waste Management 26 (3), 220–232.

Eid, H.T., Stark, T.D., Douglas, W.D., Sherry, P.E., 2000. Municipal Solid waste slope failure 1. Waste and foundation properties. Journal of Geotecnical and Geoenvironmental Engineering, ASCE 126 (5), 397–407.

Fei, X., Zekkos, D., 2015. Large-size controlled degradation and simple shear testing of municipal solid waste from Michigan. In: XVI European Conference on Soil Mechanics and Geotechnical Engineering, 13–17 September 2015, Edinburg, UK.

Gomes, C., Ernesto, A., Lopes, M.L., Moura, C., 2002. Sanitary landfill of Santo Tirso-municipal waste physical, chemical and mechanical properties. In: Proceeding of the Fourth International Congress on Environmental Geotechnics, Brazil, vol. 1, pp. 255–261.

Gotteland, P., Gourc, J.P., Aboura, A., Thomas, S., 2000. On-site determination of geomechanical characteristics of waste. In: Proceedings of Geoenvironmental Conference Geo2000-Australia.

Gourc, J.P., Camp, S., Viswanadham, B.V.S., Rajesh, S., 2010. Deformation behavior of clay cap barriers of hazardous waste containment systems: full-scale and centrifuge tests. Journal Geotextiles & Geomembranes 28 (3), 281–291. Elsevier.

Gourc, J.P., Staub, M., 2010. The bio-hydro-mechanical behavior of MSW and the improvement of landfill environmental impact. In: Invited Lecture, Proceedings 6th Int. Cong. on Environmental Geotechnics of the ISSMGE, New Delhi, India, November 2010. McGraw-Hill Publisher, pp. 24–39.

Gourc, J.P., Olivier, F., Thomas, S., Chatelet, L., Denecheau, P., Munoz, M.L., 2001. Monitoring of waste settlements on five landfills : comparison of the efficiency of different devices. In: 8th Int. Waste Management and Landfill Symposium, Sardinia 2001, vol. III, pp. 515–524.

Grisolia, M., Napoleoni, Q., Tangredi, G., 1995. The use of triaxial tests for the mechanical characterization of municipal solid waste. In: Proc. of the 5th International Landfill Symposium – Sardinia '95, vol. 2. CISA, Cagliari (Italy), pp. 761–767.

Hendron, D.M., Fernandez, G., Prommer, P.J., Giroud, J.P., Orozco, L.F., 1999. Investigation of 27th September 1997 slope failure at the Dona Juana landfill. In: Proceedings Sardinia '99, Seventh International Waste Management and Landfill Symposium, Cagliari, Italy, pp. 545–554.

Jafari, N.H., Stark, T.D., Merry, S., 2013. The July 10, 2000 Payatas landfill slope failure. International Journal of Geoengineering Case Histories 2 (3), 208–228. https://doi.org/10.4417/IJGCH-02-03-03. http://casehistories.geoengineer.org.

Jessberger, H.L., Kockel, R., 1993. Determination and assessment of the mechanical properties of waste. In: Sarsby, R.W. (Ed.), Waste Disposal by Landfill – Green '93. Balkema, Rotterdam, pp. 313–322.

Kavazanjian Jr., E., Matasovic, N., Bachus, R.C., 1999. Large diameter static and cyclic laboratory testing of municipal solid waste. In: 7th International Waste Management and Landfill Symposium, Sardinia, Italy, pp. 437–444.

Kavazanjian Jr., E., Matasovic, N., Bonaparte, R., Schmertmann, G.R., 1995. Evaluation of MSW Properties for Seismic Analysis. In: Geoenvironment 2000, pp. 1126–1141. Geotechnical Special Publication No.46.

Kölsch, F., Fricke, K., Mahler, C., Damanhuri, E., 2005. Stability of landfills, the Bandung dumpsite disaster. In: Proceedings of the 10th International Landfill Symposium.

Kölsch, F., 1995. Material values for some mechanical properties of domestic waste. In: Christensen, T.H., Cossu, R., Stegmann, R. (Eds.), Proceedings Sardinia 1995, Fifth International Waste Management and Landfill Symposium, S. Margherita di Pula, Cagliari, Italy. CISA, Environmental Sanitary Engineering Centre, Italy, pp. 711–729.

Landva, A.O., Clark, J.I., 1986. Geotechnicaltesting of wastefill. In: Proceedings, 39th Canadian Geotechnical Conference Ottawa, Ontario, pp. 371–385.

Landva, A.O., Clark, J.I., 1990. In: Landva, Knowles (Eds.), Geotechnics of Wastefill, Theory and Practice, STP 1070. ASTM, pp. 86–103.

Manassero, M., Van Impe, W.F., Bouazza, A., 1996. Waste disposal and containment. In: Proceedings, 2nd International Congress on Environmental Geotechnics, Osaka, Japan, vol. 2. Balkema, Rotterdam, The Netherlands, pp. 1425–1474.

Makdisi, F., Seed, H.D., 1978. Simplified procedure for estimation dam and embankment earthquake-induced deformations. Journal of the Geotechnical Engineering Division, ASCE 104. No, GT7.

Mazzucato, N., Simonini, P., Colombo, S., 1999. Analysis of block slide in a MSW landfill. In: Proceedings Sardinia 1999. Seventh International Waste Management and Landfill Symposium, Cagliari, Italy, 4–8 October 1999.

McKnight, T., Cho, Y., Townsend, T., Choate, A., 2014. Cone penetration testing for characterizing landfilled municipal solid waste. Journal of Geotechnical and Geoenvironmental Engineering. https://doi.org/10.1061/(ASCE)GT.1943-5606.0001243, 06014018.

Merry, S.M., Kavazanjian, E., Fritz, W., 2005. Reconnaissance of the July 10, 2000, Payatas landfill failure. Journal of Performance of Constructed Facilities, ASCE 19 (2), 100–107.

Mitchell, J., Seed, R., Seed, H.B., 1990. Kettleman Hills waste landfill slope failure. I: liner-system properties. Journal of Geotechnical Engineering 116 (4), 647–668.

Morgenstern, N.R., Price, V.E., 1965. The analysis of the stability of general slip surfaces. Géotechnique 15 (1), 79–93.

Olivier, F., Lhomme, D., Gourc, J.P., Hidra, M., 2005. The measurement of landfill settlement using 3D laser scanner imaging. In: Sardinia 2005, 10th International Landfill Symposium, Cagliari, Italy. CISA.

Pelkey, S.A., Valsangkar, A.J., Landva, A., 2001. Shear displacement dependent strength of municipal solid waste and its major constituent. Geotechnical Testing Journal 24 (4), 381–390.

Spencer, E., 1967. A method of analysis of the stability of embankments assuming parallel interslice forces. Géotechnique 17 (1), 11–26.

Zekkos, D., 2013. Experimental evidence of anisotropy in municipal solid waste. In: Proc., Coupled Phenomena in Environmental Geotechnics, Politecnico di Torino, Italy, 1–3 July 2013, pp. 69–79.
Zekkos, D., Bray, J.D., Riemer, M.F., 2008. Shear modulus and material damping of municipal solid waste based on large-scale cyclic triaxial testing. Canadian Geotechnical Journal 45 (1), 45–58.
Zekkos, D., Bray, J.D., Riemer, M.F., 2012. Drained response of municipal solid waste in large-scale triaxial shear testing. Waste Management 32 (10), 1873–1885.
Zekkos, D., Matasovic, N., El-Sherbiny, R., Athanasopoulos-Zekkos, A., Towhata, I., Maugeri, M., 2011. Chapter 4: dynamic properties of municipal solid waste. In: Geotechnical Characterization, Field Measurements, and Laboratory Testing of Municipal Solid Waste, pp. 112–134 (ASCE Geotechnical Special Publication No. 209).
Zekkos, D., Athanasopoulos, G.A., Bray, J.D., Theodoratos, A., Grizi, A., 2010a. Large-scale direct shear testing of municipal solid waste. Waste Management Journal 30, 1544–1555.
Zekkos, D., Kavazanjian Jr., E., Bray, J.D., Matasovic, N., Riemer, M.F., September 2010b. Physical characterization of municipal solid waste for geotechnical purposes. Journal of Geotechnical and Geoenvironmental Engineering, ASCE 136 (9), 1231–1241.
Zekkos, D., Vlachakis, V.S., Athanasopoulos, G.A., 2014. The 2010 Xerolakka landfill slope instability. Environmental Geotechnics 1 (1), 56–65.

6 CHAPTER

HYDRAULIC ASPECTS

6.1

LEACHATE GENERATION MODELING

Luca Alibardi and Raffaello Cossu

INTRODUCTION

Leachate is the result of two main phenomena occurring in a landfill:

- infiltration of water in the deposited waste mass;
- mass transfer of substances from waste to infiltrating water.

These two phenomena drive most of the technical and environmental issues in landfill engineering. Leachate production in fact affects both landfill design and management, including the following aspects (Fig. 6.1.1):

- water budget for biochemical reaction when biodegradable waste is landfilled;
- solubilization and removal of contaminants from the landfill environment;
- design and technology of top cover and bottom barriers;
- leachate management (recirculation, treatment, etc.);
- operation, site appearance, and landscaping.

On the one hand, leachate represents one of the main emissions from landfill sites; and current design practices are oriented at avoiding uncontrolled leachate emissions to protect surface and groundwater sources. On the other hand, leachate flows transport and circulate biodegradable organics and nutrients through landfill body, contributing to speed up biological degradation. Proper leachate extraction, treatment, and recirculation processes can therefore provide a substantial contribution to sustainable landfilling, by controlling and removing soluble pollutants to achieve, in one generation time, a state of geological storage of inorganic substances and nondegradable organic waste for the landfill (Cossu, 2012, 2009). Knowledge of leachate generation mechanisms is fundamental to properly assess and address different aspects of landfill design and management. To this regard, predictive models for leachate generation are required to define, design, and size the works, which are connected to all the above-mentioned aspects.

This chapter presents the basic understanding of the landfill hydrological balance and describes modeling approaches.

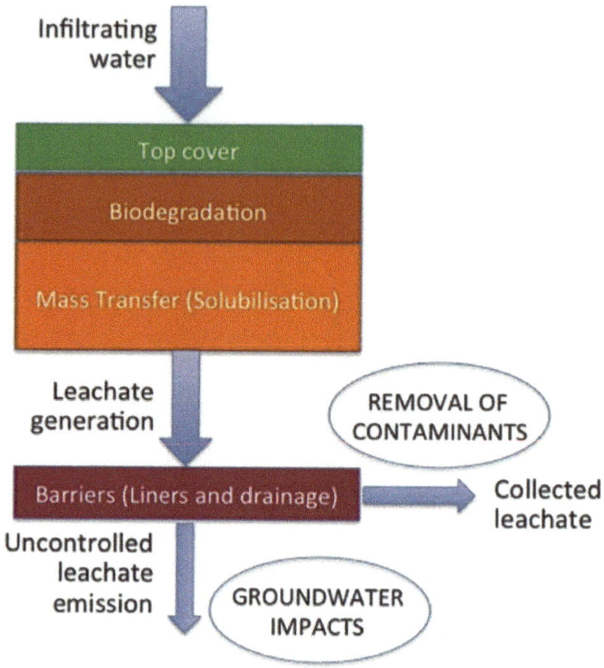

Figure 6.1.1 Landfill components that are affected by water balance and leachate generation.

WATER BALANCE IN A LANDFILL

The generation of leachate is governed by several factors that can be grouped as follows:

- climatic factors;
- landfill characteristics;
- waste characteristics; and
- processes in landfill body.

Climatic factors include rainfall frequency and intensity, temperature, and solar radiation. Rainfall is the main contributor to leachate production, and often in quick calculations leachate production is considered as a fraction of rainfall.

Landfill characteristics affecting leachate production are shape of landfill body, composition materials and slopes of top cover structures, waste deposition procedures, and waste compaction.

Waste characteristics mainly affect leachate quality but waste density, particle size, moisture content, and field capacity also affect leachate production.

Processes in landfill body affecting leachate production are the biological processes of degradation of waste that may produce or consume water, waste settling that might squeeze water out of pores, and heat generation that might evaporate water then removed by biogas extraction systems.

A general water balance for a landfill is illustrated in Fig. 6.1.2.

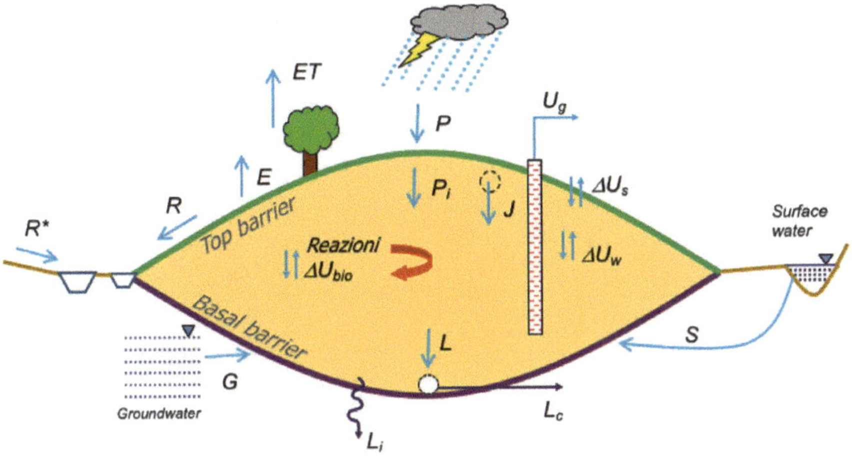

Figure 6.1.2 General water balance of a landfill. ΔU_{bio}, water production or consumption caused by chemical or biological processes; ΔU_S, variation in water content of soil; ΔU_W, variation in water content of waste; E, evaporation; ET, evapotranspiration; G, water from groundwater sources; J, irrigation or leachate recirculation; L, total leachate production; L_C, leachate collected from the drainage systems; L_i, the uncontrolled emissions through the landfill barriers; P, precipitation; P_e, effective precipitation reaching the landfill body; R^*, runoff from surrounding areas; R, surface runoff; S, water from surface water sources; U_g, water vapor contained in landfill gas.

Water balance equations can be expressed as follows. For landfill top cover:

$$P + R^* - R - E - ET = \Delta U_s$$

For landfill body:

$$P_e + J + S + G - L - U_g \pm \Delta U_{bio} = \Delta U_w$$

The general equation for leachate production considering all factors in the water balance is the following:

$$L = P + R^* + J + S + G - R - E - ET - U_g \pm \Delta U_s \pm \Delta U_w \pm \Delta U_{bio}$$

where: L: total leachate production; P: precipitation; R^*: runoff from surrounding areas; J: irrigation or leachate recirculation; S: water from surface water bodies; G: water from groundwater sources; R: surface runoff; E: evaporation; ET: evapotranspiration; U_g: water vapor contained in landfill gas; ΔU_S: variation in water content of soil; ΔU_W: variation in water content of waste; and ΔU_{bio}: water production or consumption caused by chemical or biological processes.

Factors contributing to the production of leachate are as follows: direct precipitation on landfill surface (P), surface runoff from surrounding areas (R^*), irrigation or leachate recirculation (J), and infiltrations from water sources (S, G). Factors reducing the production of leachate are as follows: evaporation or evapotranspiration (E, ET), surface runoff (R), and water vapor in landfill gas (U_b). Variation of water content in top cover soil (ΔU_S) and in waste (ΔU_W) can be positive or negative depending on incoming and outgoing water fluxes while chemical or microbial processes (ΔU_{bio}) can consume or produce water. These components of the water balance can therefore increase or reduce leachate production.

The amount of water infiltrating through the top cover layer and effectively reaching the landfill body (P_e) depends on the characteristics of rainfall (amount, frequency, and intensity) at the landfill site, the management phase of landfill (operation or postclosure), the topography, materials used, and state of the vegetation of the top cover. The effective precipitation (P_e) can be calculated as follows:

$$P_e = P + R^* - R - E - ET \pm \Delta U_s$$

The total amount of leachate that is collected from the drainage systems (L_C) is the difference between the total leachate production (L) and the uncontrolled emissions through the landfill barriers (L_i):

$$L_C = L - L_i$$

In a modern landfill, uncontrolled emissions (L_i) are avoided by the use of low permeable layers. External runoff (R^*) can be controlled by realizing ditches all around the landfill body to collect and drain away water from surrounding areas. Leachate recirculation can be used to balance humidity and nutrient availability in waste layers in an attempt to speed up biological degradation; local or national regulations might adopt different approaches in allowing or not this practice during the operational or postclosure phases of a landfill. Infiltrations of water from groundwater sources or surface water bodies (S, G) are avoided by the presence of low permeable barriers or by reducing the water table to a level lower than the bottom liner.

The equation estimating the amount of leachate that is collected from the draining system (L_C) is therefore the following:

$$L_C = P + J - R - E - ET - U_g \pm \Delta U_s \pm \Delta U_w \pm \Delta U_{bio}$$

Each part of the equation contributes differently to the leachate production. Precipitation is the main factor affecting leachate production during the entire lifetime of a landfill and during postclosure. Evaporation can have effects during waste deposition or on a nonvegetated top cover layer, whereas evapotranspiration contributes only once vegetation is established on top of the landfill. Leachate recirculation and irrigation for dust control are site-specific operations, while water content variations might be difficult to be estimated considering the potential heterogeneity inside a landfill body. A discussion of the main factors affecting leachate production is presented in next sections.

Precipitation

Precipitation is the largest contributor and the main factor governing leachate generation. Data on rainfall should be obtained from measuring apparatus on site or alternatively from the nearest meteorological station. For order of magnitude calculations, national hydrographic maps can be used.

Long periods of light rainfall contribute more than short bursts of heavy rainfall. Quick saturation of cover materials during intense storms avoids water infiltration and favors surface runoff.

Snow can contribute to leachate production during melting at wormer temperatures. No clear indications are reported in literature and a criteria of 1 cm of fresh snow equals to 1 mm of effective rainfall should be used (Canziani and Cossu, 1989).

Annual average data on precipitation can be used for the calculation, anyway the variability of rainfall within the same year or between years should be carefully evaluated. The use of average monthly precipitations can provide a more realistic representation for those countries with high variation of rainfall in different seasons.

Runoff

Surface runoff can have two contributions to leachate generation as reported in Fig. 6.1.2. Runoff from surrounding areas of landfill has a positive contribution, whereas surface runoff toward the land surrounding the landfill reduces leachate production. In both cases, the factors influencing runoff are topography of lands and top covers, morphology of soil, type of soil cover, vegetation, permeability of soil, and drainage systems (Canziani and Cossu, 1989).

Surface runoff from surrounding areas into landfill can be controlled and avoided realizing drainage ditches along landfill borders. This is particularly important for all those landfills realized under the ground level, in old quarries or along the slopes of hills or mountains. In case of "land raise" landfill, surface runoff from surroundings has on the contrary minimal impacts (Kjeldsen and Beaven, 2010).

Runoff from landfill top cover to surroundings contributes to reduce leachate generation from rainfall. The extent at which water will leave the landfill depends on slopes of top cover, soil characteristics, and vegetation. Also intensity of rainfall will affect surface runoff. During heavy rain, slow hydraulic conductivity of soil will not lead to effective infiltration, thus water will flow on landfill surface to surrounding areas.

Surface runoff may cause severe erosion effect mainly in subtropical and tropical areas in landfill with steep slopes. In colder regions, frozen soil reduces water infiltration through top covers thus leading to surface runoff of rain or melting snow (Kjeldsen and Beaven, 2010).

Surface runoff can be estimated by various methods. The simplest, most used but less reliable, method is the "Rational method," linking runoff to rainfall by a coefficient related to the characteristics of top covers (slope, type of cover material, presence, and type of vegetation). The general formula is as follows:

$$R = C \cdot P$$

Table 6.1.1 Examples of runoff coefficients

		Slope (%)		
Area type	Soil texture	0–5	5–10	>10
Bare soil	Sandy loam	0.30	0.40	0.52
	Clay	0.60	0.70	0.70–0.82
Grass	Sandy loam	0.10	0.16	0.22
	Clay	0.25–0.40	0.30–0.55	0.35–0.60

Adapted from Canziani and Cossu (1989) and Blakey (1992).

where: R = surface runoff; C = runoff coefficient; P = rainfall intensity. Both R and P are expressed in the same unit of measurement, for example, mm/day or mm/month.

Literature on hydrological, agricultural, and engineering application reports detailed values for C related to the type of soil texture, cover, and slope. Some examples of runoff coefficient data are reported in Table 6.1.1.

This approach does not consider anyway the relationship between runoff, rainfall duration, and variation of cover material compaction, leading to possible over- or underestimations. Detailed investigations on factors affecting runoff will be required for a specific landfill site if the method is to be applied with sufficient accuracy (Canziani and Cossu, 1989; Blakey, 1992).

Evaporation and Evapotranspiration

Water can be transformed into water vapor either by effect of solar radiation (evaporation) or by vegetation (transpiration). Both phenomena occur during different phases of landfill life. Evaporation (E) may occur during waste disposal as water can evaporate from waste or daily covers. After a landfill sector has been closed, water evaporates from top cover areas not yet vegetated, whereas in vegetated top covers, both evaporation and plant transpiration occur. In vegetated soils, evaporation and plant transpiration cannot be evaluated separately and evapotranspiration (ET) is used to predict their effect on water the balance.

Evaporation (E) of water from top cover occurs when vegetation is yet not available as well as from waste surface disposed in landfill. Evaporation from uncovered landfill is affected by two processes acting in opposite directions. The high infiltration capacity of waste transports rapidly water in the landfill body thus reducing the evaporation of precipitation. On the opposite direction, heat produced by aerobic biological activity on waste surface increases the energy available to water intensifying evaporation. In active landfills, infiltration seems to be the dominant process of the two. Anyway, if water content of surface areas is high due to weather conditions or by leachate recirculation, water will evaporate at the potential rate. Contribution to evaporation from biological heat is always present and it is almost constant over time. During cloudy or low sunshine periods, the role of biological heat increases and, on a whole year base, it enhances evaporation by 10% (Bendz and Bengtsson, 1996).

To evaluate water evaporation from saturated soils after heavy or prolonged rain events, evaporation rates estimated for shallow water can be used. This condition can be assumed during winter months or during rainy seasons. During dry seasons or periods of low rainfall, soil is likely to be in unsaturated conditions and evaporation can be estimated by the Turc's formula (Canziani and Cossu, 1989):

$$E = \frac{P + w}{\sqrt{\{1 + [(P + w)/L]\}}}$$

where: E = evaporation in 10 days (mm) and P = rainfall occurring in 10 days (mm).

The parameter w is the maximum quantity of water that could evaporate from the soil in 10 days in the absence of rainfall. w can vary from 1 mm for dry ground to 10 mm for moist ground and can be estimated using the following equation:

$$w = w_0 - 0.01\sqrt{t}$$

where: w_0 = 10 mm; t = time with no rain after last rainfall, in seconds, before the considered 10-day period.

L is a factor that can be calculated using the following equation:

$$L = \frac{(T + 2)/I_g}{16}$$

where: T = median air temperature in °C; I_g = effective solar radiation (cal/cm^2 per day); solar radiation can be determined by direct measurement or estimated from the theoretical I_0 using the following equations:

$I_g = I_0(0.135 + 0.68\ n/N)$ for foggy areas and for values of $n/N \leq 40$.

$I_g = I_0(0.155 + 0.69\ n/N)$ for latitudes between 54.5 and 56 degree North.

$I_g = I_0(0.16 + 0.62\ n/N)$ for latitudes between 45 and 55.5 degree North.

$I_g = I_0(0.18 + 0.662\ n/N)$ for latitudes between 35 and 45 degree North.

where: n/N = ratio between the number of effective (n) and theoretical (N) daylight hours. Values for N are reported in Table 6.1.2. I_0 = maximum possible solar radiation (Table 6.1.3).

The evaluation of evaporation using the reported procedure should be considered a rough estimation as it does not consider, for instance, the intermediate cover of landfill. Cracked soils in dry seasons in fact allow fast water infiltration that will not be removed by evaporation. On the contrary well-compacted waste layers or leveled cover reduce water infiltration enhancing its evaporation (Canziani and Cossu, 1989).

São Mateus et al. (2012) proposed the use of field tests for the evaluation of evaporation rates from cover layers in the metropolitan center landfill in Salvador, Brazil. Field evaporation tests were carried

Table 6.1.2 Mean daily duration of maximum possible sunshine (*N*, hours)

Latitude Degree North	Jan	Feb	Mar	Apr	May	Jun	Jul	Aug	Sep	Oct	Nov	Dec
60	6.7	9.0	11.7	14.5	17.1	18.6	17.9	15.5	12.9	10.1	7.5	5.9
58	7.2	9.3	11.7	14.3	16.6	17.9	17.3	15.3	12.8	10.3	7.9	6.5
56	7.6	9.5	11.7	14.1	16.2	17.4	16.9	15.0	12.7	10.4	8.3	7.0
54	7.9	9.7	11.7	13.9	15.9	16.9	16.5	14.8	12.7	10.5	8.5	7.4
52	8.3	9.9	11.8	13.8	15.6	16.5	16.1	14.6	12.7	10.6	8.8	7.8
50	8.5	10.0	11.8	13.7	15.3	16.3	15.9	14.4	12.6	10.7	9.0	8.1
48	8.8	10.2	11.8	13.6	15.2	16.0	15.6	14.3	12.6	10.9	9.3	8.3
46	9.1	10.4	11.9	13.5	14.9	15.7	15.4	14.2	12.6	10.9	9.5	8.7
44	9.3	10.5	11.9	13.4	14.7	15.4	15.2	14.0	12.6	11.0	9.7	8.9
42	9.4	10.6	11.9	13.4	14.6	15.2	14.9	13.9	12.6	11.1	9.8	9.1
40	9.6	10.7	11.9	13.3	14.4	15.0	14.7	13.7	12.5	11.2	10.0	9.3
35	10.1	11.0	11.9	13.1	14.0	14.5	14.3	13.5	12.4	11.3	10.3	9.8
30	10.4	11.1	12.0	12.9	13.6	14.0	13.9	13.2	12.4	11.5	10.6	10.2
25	10.7	11.3	12.0	12.7	13.3	13.7	13.5	13.0	12.3	11.6	10.9	10.6
20	11.0	11.5	12.0	12.6	13.1	13.3	13.2	12.8	12.3	11.7	11.2	10.9
15	11.3	11.6	12.0	12.5	12.8	13.0	12.9	12.6	12.2	11.8	11.4	11.2
10	11.6	11.8	12.0	12.3	12.6	12.7	12.6	12.4	12.1	11.8	11.6	11.5
5	11.8	11.9	12.0	12.2	12.3	12.3	12.3	12.3	12.1	12.0	11.9	11.8
0	12.0	12.0	12.0	12.0	12.0	12.0	12.0	12.0	12.0	12.0	12.0	12.0

From Shaw (1983) as reported in Canziani and Cossu (1989).

Table 6.1.3 Maximum possible solar radiation at different latitudes (cal/(cm^2 day)), assuming latent heat of water = 590 cal/g

Latitude Degree North	Jan	Feb	Mar	Apr	May	Jun	Jul	Aug	Sep	Oct	Nov	Dec
60	80	210	415	655	860	970	920	745	500	280	120	5
50	220	354	545	750	915	980	950	810	615	420	260	180
40	365	495	655	815	940	985	960	865	715	550	400	330
30	480	620	755	870	950	975	955	900	800	660	535	465
20	640	730	825	900	925	930	930	910	850	760	665	615
10	755	820	870	900	885	870	880	885	870	840	770	755
0	860	885	900	870	820	790	800	845	880	885	860	845

From Luisley et al. (1982) as reported in Canziani and Cossu (1989).

out using two reservoirs simultaneously, one filled with compacted soil and the other with water. The water reservoir was designed to reproduce potential evaporation conditions (*Ep*), whereas the soil reservoir was designed to reproduce the evaporation conditions in the soil cover (*E*) under the same weather conditions.

Experimental results of the evaporation rates in soil (*E*) and water (*Ep*) were compared and the curves *E*/*Ep* over soil volumetric water content was determined. The ratio *E*/*Ep* was used to transform the potential evaporation (*Ep*) values obtained in the weather station into soil evaporation values (*E*).

Undisturbed samples were collected to determine the dry density and water content at different points of the cover layer of the cell. The soil samples were then compacted using average field conditions of dry density and optimum water content (normal Proctor Energy).

Experimental data resulted in a liner variation with soil water content (Fig. 6.1.3) and the following equation has been reported describing the ratio between actual evaporation (*E*) and evaporation potential (*Ep*) as a function of the volumetric water content of soil cover (São Mateus et al., 2012):

$$\frac{E}{Ep} = 4.22 \cdot \theta_{soil} - 0.249$$

where: E/Ep = ratio between evaporation and evaporation potential and θ_{soil} = volumetric water content of soil cover.

Evapotranspiration (*ET*) measures the water removed from soil by the combined effect of evaporation and plant transpiration. The term potential evapotranspiration (*ETp*) is used to indicate the water

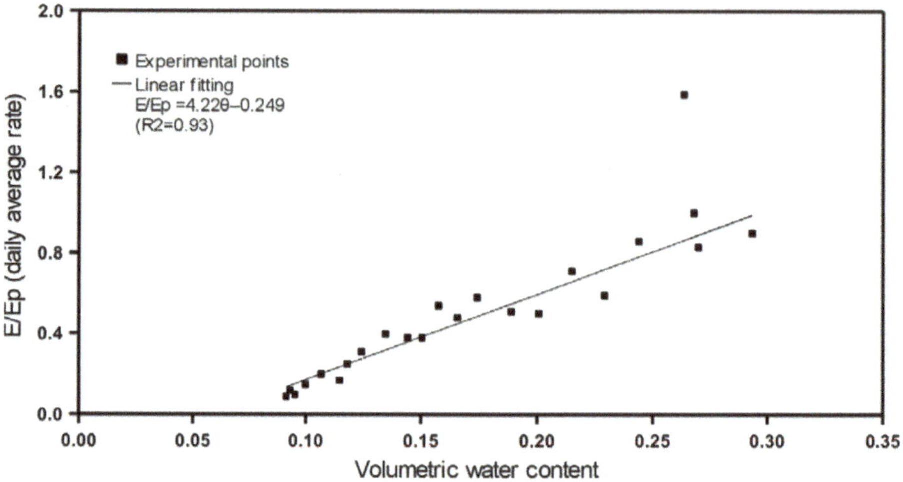

Figure 6.1.3 Variation of the ratio of daily evaporation rates (E/Ep) with soil water content. *From São Mateus et al. (2012).*

loss occurring from vegetated soils with no water deficiency, thus water is evapotranspirated at its maximum rate. Often *ET* is assumed from the Potential Evapotranspiration (*ETp*). *ETp* depends on temperature, solar input, relative humidity, wind, type, and state of crops. For a given vegetation type, the only factors affecting *ETp* are climatic parameters, therefore it can be computed from weather data (Xu and Chen, 2005; Kjeldsen and Beaven, 2010).

Canziani and Cossu (1989) proposed the formula developed by Thornthwaite for European climates. The procedure uses the mean air temperature and number of hours of daylight and it is classified as a temperature-based method.

The potential evapotranspiration is calculated using the formula:

$$ETp_i = 16\cdot\left(\frac{10\cdot T_i}{I_i}\right)^a\cdot C_i$$

where: ETp_i = potential evapotranspiration in the i-th month (mm/month); T_i = monthly average temperature (°C); $a = 6.75 \times (10 - 7 \times I_{3t}) + 7.71 \times (10 - 5 \times I_{2t}) + 1.79 \times (10 - 2 \times I_t) + 0.49239$; I_i = annual thermal indice $I_t = \sum_1^{12} i\left(\frac{T_i}{5}\right)^{1.514}$; C_i = correction coefficient relative to the month that takes account of the variation in length of day at various latitudes; it is obtained by dividing the value *N* reported in Table 6.1.2 by 10.

As previously reported, during dry periods *ET* is generally lower than *ETp* as precipitation (*P*) is not sufficient to compensate for water losses due to runoff (*R*) and potential evapotranspiration (*ETp*). In this condition, soil water content (*Us*) will reduce and therefore actual ET will be lower than *ETp*. Actual ET can be assumed as a percentage of the *ETp* and the reduction factor can be related to actual water content or to soil water deficit, defined as a difference between soil field capacity and actual water content. At low rainfall conditions and without irrigation, soil water content decreases below its field capacity and ET/ETp drops below 100% (Kjeldsen and Beaven, 2010).

Water Content of Waste and Water Accumulation Capacity

Water content of waste in landfill varies over time due to disposal procedures, landfill management processes (e.g., leachate recirculation), compaction, and degradation of waste.

The knowledge of water content of waste and its capacity to accumulate and retain water allows to estimate the time required by the leachate to be generated and reach the drainage system (Canziani and Cossu, 1989). To describe the properties of waste disposed in landfill some definitions should be addressed (Blakey, 1992):

- *initial moisture content*: water content of waste as received at the site for disposal;
- *field capacity*: field capacity represents the amount of liquid a mass of material can absorb before downward percolation occurs due to gravitational forces. This definition fits homogeneous materials such as soil. In solid waste, liquid percolates before field capacity is reached due to heterogeneity and preferential pathways. In this case the term "adsorptive capacity" better represents the amount of water retained before first leachate will be generated;
- *saturation capacity*: this represents the maximum amount of water retained in a mass of waste when pores and voids are fully saturated.

Two mechanisms of water retention act in moisture content of waste:

- *physical adsorption*: water retained by capillary forces;
- *apparent adsorption*: water retained in void spaces giving localized saturated zone.

The initial moisture content of waste will be far from the adsorptive capacity; therefore after disposal, waste may continue to adsorb rainfall or water percolating from above layers. When adsorptive capacity is reached, draining will occur and water will move downward under the influence of gravity. The lowest layers of waste will reach saturated conditions when leachate level starts to rise up.

The difference between initial moisture content of waste and adsorptive capacity (field capacity) represents the water deficit of the mass of waste and it largely affects the landfill hydrology and water drainage through landfill body.

The capacity of waste to retain water depends on several factors. Generally, at high waste densities and compaction rates, adsorptive capacity decreases. The heterogeneity of waste and low grade of compaction create preferential pathways through which water pass rapidly without being adsorbed, thus the adsorption capacity of waste will be lower than the theoretical field capacity (Blakey, 1992; Canziani and Cossu, 1989).

Water Flow and Waste Permeability

Flow of water in landfill is a complex subject, not yet completely understood. Water flow in waste can occur both in saturated and unsaturated zones. Hydraulic properties of waste may vary over time due to compaction and settlement of landfill body and waste degradation. Heterogeneity of waste varies at different scale and landfill gas production interacts with and affects water flow (Kjeldsen and Beaven, 2010).

The traditional Darcy's law can be used to describe the flow of water through top cover structures and disposed waste material. Flow (q) can be calculated imposing a hydraulic gradient (i) and a cross-sectional area (A) as for the following equation:

$$q = -KiA$$

Hydraulic conductivity (K) can be assumed from literature, obtained from lab-scale test or from in situ measurement.

Water flow in unsaturated media can also be used to predict water infiltration. Hydraulic conductivity in unsaturated condition is reduced if compared with that in saturated conditions as air is present in pores reducing the sectional area through which flow can occur, flow is limited into small pores that are less conductive and flow path through interlinked pores is difficult.

Values of hydraulic conductivity of municipal solid waste vary in the range of 10^{-5}–10^{-6} m/s. Water flow is not limited in landfill body for these values of conductivity on a long-term period, whereas short-term infiltration could be affected enhancing surface runoff (Kjeldsen and Beaven, 2010). For a detailed presentation of this topic refer to Chapter 6.2.

HYDROLOGIC EVALUATION OF LANDFILL PERFORMANCE MODEL

The Hydrologic Evaluation of Landfill Performance (HELP) model is applied for calculation of water balances for landfill cover and bottom liner systems (Berger, 2015). The HELP model was developed by the US Army Engineer Waterways Experiment Station for the US Environmental Protection Agency to assist the comparison of landfill design alternatives as judged by their water balances (Schroeder et al., 1994a).

HELP is a "quasi-two-dimensional" layer model. Several one-dimensional (1D) vertical and lateral processes are combined (vertical: precipitation, infiltration, evapotranspiration, percolation; lateral: surface runoff, lateral drainage) and two-dimensional flow is not actually modeled (Schroeder et al., 1994a; Berger, 2003).

The model accepts weather, soil, and design data and uses solution techniques that account for different components of the landfill water balance: surface storage, snowmelt, runoff, infiltration, evapotranspiration, vegetative growth, soil moisture storage, lateral subsurface drainage, leachate recirculation, unsaturated vertical drainage, and leakage through soil, geomembrane, or composite liners.

The model can simulate open or closed landfills, bottom liner systems, and cover systems, including various combinations of vegetation, cover soils, waste cells, lateral drain layers, low permeability barrier soils, and synthetic geomembrane liners (Schroeder et al., 1994a; Berger, 2015). Results are expressed as daily, monthly, annual, and long-term average water budgets (Schroeder et al., 1994a).

To perform the analysis and comparison of different design systems, HELP requires general climate and weather data from 1 to 100 years, soil characteristics, and design specifications. The climate data required by the model are growing season, average annual wind speed, average quarterly relative humidity, normal mean monthly temperatures, maximum leaf area index, evaporative zone depth, and latitude. Daily weather data include precipitation, mean temperature, and total global solar radiation.

Porosity, field capacity, wilting point, saturated hydraulic conductivity, and initial moisture storage are the necessary data for soils. The information about the different designs are slope and maximum drainage distance for lateral drainage layers, description and thickness of layers, area of the waste cells and surface characteristics, data on leachate recirculation procedures, and geomembrane characteristics. Each layer must be defined as one of four operational types: vertical percolation, lateral drainage, barrier soil liner, or geomembrane liner (Schroeder et al., 1994a).

The model has been developed in three versions, 1, 2, and 3 (Schroeder et al., 1984a,b, 1988a,b; 1994a,b). HELP version 3 was released in 1994 with improvements regarding the possibility to import weather data from external databases and as ASCII files, improvements, and simplification of submodels for evapotranspiration, surface runoff, and snowmelt. The model is also commercially available as Visual HELP or as part of the UnSat Suite Plus (Waterloo Hydrogeologic). The major features of Visual HELP are the generation of statistically reliable weather data from a database of more than 3000 international weather stations containing historical climatological data for each location and tools and design elements improving result visualization and report generation (Waterloo Hydrogeologic).

HELP model has been largely used on tests at full scale on German landfills by Berger (2000, 2002, 2003, 2015) as a result of interest from the German Federal Environmental Protection Agency in the availability of a water balance model for practical applications. The objectives of the study were to determine the limits of application of the model and to adapt HELP 3 for use in Germany (Berger, 2015). As part of the validation study and further investigations, some submodels were enhanced or replaced in new and subsequent versions (HELP 3.50 D; HELP 3.55 D; DELP 3.80 D; and HELP 3.95 D—Berger, 2015). The modifications introduced on these versions allowed improving the outputs and simulations of actual evapotranspiration, vegetative growth, and decay of runoff produced by frozen soil and of unsaturated/saturated vertical flow in vertical percolation layers. Moreover, users can change vegetation and soil properties during the modeled periods, allowing to simulate the aging of landfill profiles (Berger, 2015). A detailed presentation of validation tests has been presented by Berger (2015) showing that HELP model has been applied in several large and full-scale tests mainly in the United States and Germany. HELP model has also been used to evaluate landfill co-disposal of municipal solid waste, incineration residues, and sludge (Marques and Hogland, 2003), to evaluate waste compaction effects and top cover characteristics on leachate production for the Kimpo Metropolitan Landfill in South Korea (Jang et al., 2002), to assess the impact of waste moisture contents in leachate production in China (Yang et al., 2015). Xu et al. (2012) assessed the impacts of different liquid addition methods on HELP model outputs highlighting the importance of proper selection of waste properties (moisture content, conductivity, and field capacity) to prevent miscalculating liquid demands or design problems. HELP model has also been used for teaching modules in software applications for environmental and engineering education (Mesania and Jennings, 1998).

The main limitations of the HELP model are still on the maximum time resolution of the model and input data (daily weather data) allowing only rough estimations of surface runoff and no representation of short-term peaks. Further improvements for future versions regard the unsaturated vertical flow submodel, the structure of layers and segments, and the inclusions of other types of vegetation (Berger, 2015).

OTHER MODELS

Various alternative models have been presented in scientific literature for the prediction of leachate generation. Lobo et al. (Lobo et al., 2002a, 2002b; Lobo and Monzón, 2007) developed a dynamic simulation program named MODUELO that estimates leachate production, organic contamination, and biogas generation by using climatological information, waste production, and landfill layout data. MODUELO has been designed focusing particularly on landfills operated with the canyon/depression method. MODUELO is based on a three-dimensional (3D) discrete modeling. The types of data necessary to define the landfill are waste generation, geometric characteristics of the landfill, and climatological data. The model is composed of three parts: (1) a "present landfill layout," where active cells are defined and waste characteristics updates at any cycle (e.g., moisture, field capacity, substances dissolved in the water content); (2) a "hydrologic module" that calculates the water flow; and (3) a "contamination module" that uses moisture results to estimate organic content in the leachate and biogas production. Amount, composition, and characteristics of waste disposed in landfill are calculated by a "waste generation module." The model was calibrated and applied to the Meruelo Landfill, located in the region of Cantabria, Spain, with simulation of leachate production, also in comparison with HELP model (Lobo et al., 2003), leachate quality, and biogas production (Lobo et al., 2002b; Lobo and Monzón, 2007).

Yildiz et al. (2004) developed a mathematical model to simulate the landfill leachate behavior and distributions of moisture and leachate constituents throughout the landfill. The model includes two components: a hydraulic model and a landfill stabilization model. The hydraulic model describes leachate flow and moisture content distribution during the active life of the landfill and incorporates the effects of decreases in waste hydraulic conductivity due to overburden pressure, changes in waste age, and depth over time. The landfill stabilization model describes dissolution of organic and inorganic compounds, degradation of soluble organic matter by acidogenesis and methanogenesis, methane production, and pH changes over time. The landfill is assumed to consist of cells that are constructed progressively in time allowing to simulate the 3D development of the landfill. The model was calibrated and verified by using data from the Keele Valley Landfill, located in Ontario, Canada, with satisfactory agreement between the predicted and observed leachate flow, leachate composition, and methane production rate data but limitations were highlighted for the cases where water is not uniformly added across the entire surface of the landfill.

Komilis and Athiniotou (2014) presented a 1D monthly water budget model (MWBM) developed for application in operating landfills and continuously under development. MWBM is a simplistic water budget model that can be applied to operating landfills with a continuously changing geometrical configuration. The main processes modeled in MWBM are as follows: (1) the water entering the cells due to precipitation or from upper cells; (2) the water evaporated from the free waste surface exposed to the atmosphere; (3) the water holding capacity of a cell; and (4) the biological decomposition of the organic matter within the cell. The leachate generated from the lowest lift in the landfill represents the amount of leachate collected. The input variables in the MWBM are as follows: (1) number of lifts and the duration of each lift in months; (2) initial waste moisture content; (3) weight of intermediate soil cover on top of each cell; (4) maximum area of each lift and precipitation catchment area; (5) weight of final cover; and (6) monthly local precipitation and potential or actual evaporation. The model was calibrated and applied to the sanitary landfill in the island of Lesvos, Greece, showing adequate prediction of monthly leachate generation rates. Initial moisture content of the wastes and precipitation rates were indicated as sensible parameters during simulations, while evaporation had minimal effects on leachate generation.

Pantini et al. (2014) presented a Landfill Water Balance model (LWB) for predicting leachate production in landfills during both operative and postclosure phases. The model is based on analytical and empirical equations and uses a water balance approach accounting for the different rates of the incoming water, water losses, and water consumption. The model also includes aging and compression of waste, allowing to assess the progressive variation of hydraulic and physical properties of deposited wastes. The model accounts for the progressive variation of landfill geometry during the operative stage and temporal variation of waste hydraulic and physical—mechanical properties. The overall landfill system is discretized in single elements (cells). For each cell the model requires the number of layers, the surface area, the thickness, the progressive time of layers' disposal, the hydrological properties, and the type of the waste. Information on the top cover system (vegetation and soil cover, lateral drain layers, low permeability barriers, and geomembrane liners) is introduced to simulate the landfill postclosure phase. Results from simulation of hypothetical scenarios and comparison with HELP model showed that both waste compression and biodegradation affect leachate prediction leading to underestimation if these phenomena are neglected. The LWB model was then applied to a landfill located in the center of Italy, still in operative phase, and results indicated that water release from waste and rainfall were the main contributors to leachate generation, while evapotranspiration and runoff contributed to a lesser extent.

CONCLUSIONS

This chapter has reviewed the main components of the landfill water balance affecting leachate production and the main modeling approaches currently presented in technical and scientific literature. Leachate production is affected by climatic factors, physical, mechanical, and biochemical waste characteristics, disposal procedures and strategies, landfill management phases, and age. Intense model validations and availability of a sufficient amount of data are required to balance the influence of all these factors on modeling results and to properly assess the consistency of final simulation. Exclusion or simplification of specific factors, model assumption, or simulation limits could introduce important over- or underestimations on leachate generation. Specification of model aims, knowledge of models structure, behavior and limits, and critical review of model outputs remain essential for a proper interpretation of final results.

References

Bendz, D., Bengtsson, L., 1996. Evaporation from an active, uncovered landfill. Journal of Hydrology 182, 143–155.

Berger, K., 2000. Validation of the hydrologic evaluation of landfill performance (HELP) model for simulating the water balance of cover systems. Environmental Geology 39 (11), 1261–1274.

Berger, K., July 2002. Potential and limitations of applying HELP model for surface covers. Practice Periodical of Hazardous, Toxic, and Radioactive Waste Management 6 (3), 192–203.

Berger, K., 2003. Validation and enhancement of the HELP model to simulate to water balance of surface covers. In: SARDINIA 2003. Proceedings of the Ninth International Waste Management and Landfill Symposium, vol. 2. CD-ROM. CISA, Cagliari, Italy.

Berger, K., 2015. On the current state of the hydrologic evaluation of landfill performance (HELP) model. Waste Management 38, 201–209.

Blakey, N.C., 1992. Model prediction of landfill leachate production. In: Christensen, T.H., Cossu, R., Stegmann, R. (Eds.), Landfilling of Waste: Leachate. Elsevier Applied Science, London (Chapter 2.1).

Canziani, R., Cossu, R., 1989. Landfill hydrology and leachate production. In: Christensen, T.H., Cossu, R., Stegmann, R. (Eds.), Sanitary Landfilling: Process, Technology and Environmental Impact. Academic Press, London (Chapter 9.1).

Cossu, R., 2009. From triangles to cycles. Waste Management 29, 2915–2917.

Cossu, R., 2012. The environmentally sustainable geological repository: the modern role of landfilling. Waste Management 32, 243–244.

Waterloo Hydrogeologic. http://www.waterloohydrogeologic.com/.

Jang, Y.-S., Kim, Y.-W., Lee, S.-I., 2002. Hydraulic properties and leachate level analysis of Kimpo metropolitan landfill, Korea. Waste Management 22, 261–267.

Kjeldsen, P., Beaven, R., 2010. Landfilling: hydrology. In: Christensen, T.H. (Ed.), Solid Waste Technology & Management, vols. 1 & 2. John Wiley & Sons, Ltd, Chichester, UK.

Komilis, D., Athiniotou, A., 2014. A water budget model for operating landfills: an application in Greece. Waste Management & Research 32, 717–725.

Lobo, A.G., Monzón, T.I., 2007. MODUELO 2: a new version of an integrated simulation model for municipal solid waste landfills. Environmental Modelling & Software 22, 59–72.

Lobo, A., Herrero, J., Montero, O., Fantelli, M., Tejero, I., 2002a. Modeling for environmental assessment of municipal solid waste landfills (Part 1: hydrology). Waste Management & Research 20, 198–210.

Lobo, A., Herrero, J., Montero, O., Tejero, I., 2002b. Modelling for environmental assessment of municipal solid waste landfills (Part II: biodegradation). Waste Management & Research 20, 514–528.

Lobo, A., Muñoz, J., Sánchez, M.M., Tejero, I., 2003. Comparative analysis of three hydrological landfill models through a practical application (MODUELO 2, HELP and MODUELO 1). In: Proceedings of the Ninth International Landfill Symposium, Sardinia 2003. CISA, Cagliari, Italy.

Luisley Jr., K.R., et al., 1982. Hydrology for Engineers. McGraw Hill International Book Company, International Student Edition, Tokyo (cited in Canziani and Cossu, 1989).

Marques, M., Hogland, W., 2003. Hydrological performance of MSW incineration residues and MSW co-disposed with sludge in full-scale cells. Waste Management 23, 469–481.

Mesania, F.A., Jennings, A.A., 1998. A hydraulic barrier design teaching module based on HELP 3.04 and HELP model for Windows v2.05. Environmental Modelling & Software 13, 1–24.

Pantini, S., Verginelli, I., Lombardi, F., 2014. A new screening model for leachate production assessment at landfill sites. International Journal of Environmental Science and Technology 11, 1503–1516.

São Mateus, M.S.C., Machado, S.L., Barbosa, M.C., 2012. An attempt to perform water balance in a Brazilian municipal solid waste landfill. Waste Management 32, 471–481.
Schroeder, P.R., Morgan, J.M., Walski, T.M., Gibson, A.C., 1984a. The hydrologic evaluation of landfill performance (HELP) model. In: User's Guide for Version 1, Technical Resource Document EPA/530-SW-84-009, vol. I. US Environmental Protection Agency, Cincinnati, OH.
Schroeder, P.R., Gibson, A.C., Smolen, M.D., 1984b. The hydrologic evaluation of landfill performance (HELP) model. In: Documentation for Version 1, Technical Resource Document EPA/530- SW-84-010, vol. II. US Environmental Protection Agency, Cincinnati, OH.
Schroeder, P.R., Peyton, R.L., McEnroe, B.M., Sjostrom, J.W., 1988a. The hydrologic evaluation of landfill performance (HELP) model. In: User's Guide for Version 2, Internal Working Document EL-92-1, Report 1, vol. III. US Army Engineer Water Experiment Station, Vicksburg, MS.
Schroeder, P.R., Peyton, R.L., McEnroe, B.M., Sjostrom, J.W., 1988b. The hydrologic evaluation of landfill performance (HELP) model. In: Documentation for Version 2, Internal Working Document EL-92-1, Report 2, vol. IV. US Army Engineer Water Experiment Station, Vicksburg, MS.
Schroeder, P.R., Dozier, T.S., Zappi, P.A., McEnroe, B.M., Sjostrom, J.W., Peyton, R.L., 1994a. The hydrologic evaluation of landfill performance (HELP) model. In: User's Guide for Version 3.0. EPA/600/R-94/168a. US Environmental Protection Agency Risk Reduction Engineering Laboratory, Cincinnati, OH.
Schroeder, P.R., Dozier, T.S., Zappi, P.A., McEnroe, B.M., Sjostrom, J.W., Peyton, R.L., 1994b. The hydrologic evaluation of landfill performance (HELP) model. In: Engineering Documentation for Version 3.0. EPA/600/R-94/168b. US Environmental Protection Agency Risk Reduction Engineering Laboratory, Cincinnati, OH.
Shaw, M.E., 1983. Hydrology in Practice. Van Nostrand Reinhold Co. Ltd., UK.
Xu, C.-Y., Chen, D., 2005. Comparison of seven models for estimation of evapotranspiration and groundwater recharge using lysimeter measurement data in Germany. Hydrological Processes 19, 3717–3734.
Xu, Q., Kim, H., Jain, P., Townsend, T.G., 2012. Hydrologic evaluation of landfill performance (HELP) modeling in bioreactor landfill design and permitting. Journal of Material Cycles and Waste Management 14, 38–46.
Yang, N., Damgaard, A., Kjeldsen, P., Shao, L.-M., He, P.-J., 2015. Quantification of regional leachate variance from municipal solid waste landfills in China. Waste Management 46, 362–372.
Yildiz, E.D., Ünlü, K., Rowe, R.K., 2004. Modelling leachate quality and quantity in municipal solid waste landfills. Waste Management & Research 22, 78–92.

6.2

LANDFILL HYDRAULICS

Richard P. Beaven and Nick D. Woodman

INTRODUCTION

Leachates need to be carefully managed since components of the leachate may become contaminants if released into the natural environment in sufficient quantity. In properly engineered sanitary landfills, leachates are generally collected from within the site and treated to relevant standards prior to discharge to surface waters. Leachate levels will usually be controlled by pumping from leachate collection systems, which often comprise a network of pipes within a layer of high permeability drainage media on the base of the site. It is important to understand how pumping from a basal drainage layer will lower leachate heads and control pore water pressures in all areas of the site, not least because of the adverse impact that high pore water pressures can have on the mechanical stability of the waste. In sites where the drainage system is not performing adequately (perhaps through clogging), or in historic sites where it may be absent altogether, other leachate control systems have to be used (very often vertical wells).

Leachate (or water) movement through landfills may be through unsaturated waste, through saturated waste, or through waste, which is nominally saturated but contains bubbles of gas that exist as individual entities not linked to the gas phase in the unsaturated zone. Leachate flow (e.g., in response to pumping) is controlled to a large part by the hydrogeological properties of the waste: in broad terms the permeability (or hydraulic conductivity) of the waste will control the rate at which leachate will flow, the porosity (or more accurately the drainable fraction of the porosity) will control how much leachate levels change in response to infiltration or pumping, and collectively the "transport properties" of the waste will influence how rapidly contaminants can be transferred from the solid to liquid phase.

Water is essential to the biochemical decomposition of organic substances (e.g., Leckie et al., 1979). Hence the presence and movement of moisture in landfill also affects the biodegradation of the organic fraction of the waste, leading to the production of landfill gas and enhanced landfill settlement (e.g., Wall and Zeiss, 1995).

Municipal solid waste (MSW) landfills are known to have complex internal structure, which may change due to settlement and biodegradation. They are layered, anisotropic, and the moisture distribution may be complicated by the presence of gas. There are thought to be further heterogeneities, some of which may be at the scale of the landfill and unique to a particular site and unless specifically identified may be difficult to define or predict. Thus, landfills may exhibit various aspects of hydraulic heterogeneity at different scales. Nonetheless, under some circumstances, hydraulic properties have been observed to be consistent, and the vast majority of landfill design and analysis have adopted an assumption of a homogeneous continuum. In these cases "standard" soil and groundwater models of flow and transport have been applied extensively to landfill simulation. This important assumption has not been rigorously tested, and some authors prefer to retain empirical approaches to reflect this (Zacharof and Butler, 2004).

Where models that originated in natural media are applied to wastes, it is necessary that not only their inherent limitations are understood, but it is important to assess whether there are any special aspects relating to waste. For example, MSW typically exhibits subhorizontal alignment of (low-permeability) plastic (e.g., Hudson et al., 1999), which may be a very important factor in determining the nature of flow. Field-scale heterogeneities include the existence of daily cover layers as well as infrastructure (such as gas wells) or infrastructure remnants (such as old roadways). During the biologically active phase, landfills undergo a physical transformation in terms of substantial settlement in tandem with their biochemical degradation and reaction pathway; thus the solid itself is physically dynamic. In this light, Fellner and Brunner (2010) questioned the extent to which standard models can be applied to waste.

Empirical approaches to flow and transport suffer from providing limited physical insight, but are potentially useful tools. For example, the simulation of solute flushing in landfills has often been approximated to a continuously mixed reactor where the solute within the waste is considered to be instantaneously and fully mixed with all the liquid within the landfill (e.g., Reitzel et al., 1992). Alternatively, the residence time distribution of water in waste has been described using an empirical statistical function such as the lognormal distribution (Zacharof and Butler, 2004). Furthermore, it is frequently convenient to use empirical constitutive relationships describing parameter dependencies. For example, Powrie and Beaven (1999) described the relationship between hydraulic conductivity and effective stress as a power law. While it is acknowledged that there is a strong role for empiricism in the simulation of flow in wastes, these approaches are not considered in detail here, to focus on simple physically based approaches.

This chapter therefore examines our present understanding of landfill hydraulics—namely how water and solute pass in, out, and through the solid structure of the waste. This is done by examining the evidence for the "family" of most commonly used *physically based* equations for flow and transport. This approach is advantageous because these equations have not only been repeatedly applied to different flow situations in MSW, but provide a systematic framework for analysis.

These equations are uncoupled from the various other dynamic processes that are known to occur in waste, not least mechanical and biological processes. The extent to which such simplification is realistic or not is important for determining the usefulness of these approaches in different practical situations.

BASIC HYDROGEOLOGICAL PARAMETERS FOR WASTE

Bulk hydraulic properties of saturated wastes have been extensively characterized (e.g., see review by Beaven et al., 2011). The major properties of interest include hydraulic conductivity, water content, and porosity, all of which are strongly influenced by effective stress and the bulk density of the material.

Hydraulic Conductivity

Measurements of saturated hydraulic conductivity in both laboratory and field investigations (e.g., see review by Jain et al., 2006) have yielded values between 1×10^{-3} m/s and 1×10^{-9} m/s, although more typically values are in the range 10^{-5} to 10^{-6} m/s. The dominant influence of effective stress on compacting waste and the effect of waste density on hydraulic conductivity have been demonstrated—see Fig. 6.2.1.

Although the dry density of individual wastes increases with effective stress, the absolute value of density is dependent on waste composition and is not predictable from effective stress alone. There is much better correlation of the hydraulic conductivity of different waste materials with effective stress, with the important proviso that waste in the landfill environment has not been precompacted with site plant to a density above the equivalent effective stress.

Hydraulic conductivity is also known to be anisotropic with horizontal permeability greater than vertical permeability and is discussed in depth by Hudson (2007). The range of anisotropy estimates in the field (based on pumping test analyses) is from 2:1 to 8:1 (Landva et al., 1998; Cossu et al., 1997; Lofy, 1996) and in the laboratory from 0.5:1 to 250:1 [Landva et al. (1998), Hudson et al. (2009), Buchanan et al. (2001), and Münnich et al. (2005) who provided the upper bound from a

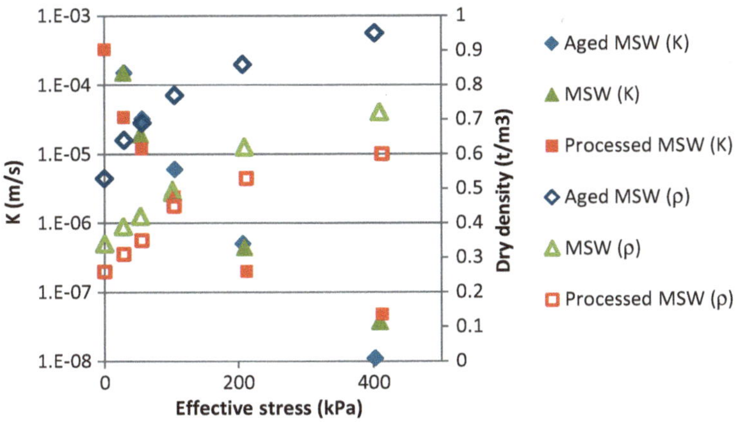

Figure 6.2.1 Dominant effect of effective stress on waste density and K. *MSW*, municipal solid waste. *After Powrie and Beaven (1999) and Beaven (2000).*

mechanically biologically treated waste at low density]. Most recently Singh et al. (2014) estimated anisotropy values as high as 100 near the surface of landfilled waste.

Porosity

For a waste that is already saturated, there are two porosity- or water content–related parameters that are of particular interest—see Box 6.2.1. The first is the total porosity (n) of the waste, which gives the total water content of the waste assuming full saturation. As indicated in the introduction, it cannot necessarily be assumed that nominally saturated waste (i.e., waste below the leachate table) in a landfill containing biodegrading waste is actually fully saturated due to the ongoing degradation of organic material and the generation of landfill gas. Typical values for the total porosity of waste are between 35%

Box 6.2.1 Water Content Definitions for Municipal Solid Waste

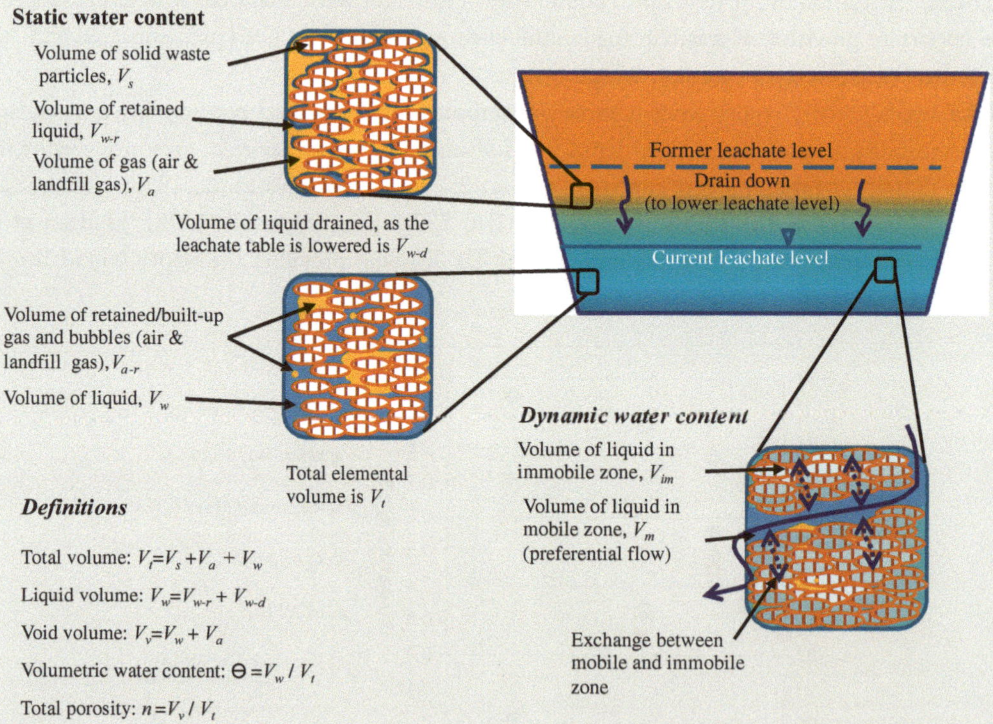

Definitions

Total volume: $V_t = V_s + V_a + V_w$

Liquid volume: $V_w = V_{w-r} + V_{w-d}$

Void volume: $V_v = V_w + V_a$

Volumetric water content: $\Theta = V_w / V_t$

Total porosity: $n = V_v / V_t$

Drainable (effective) porosity: $n_e = V_{w-d} / V_t$

Mobile Volumetric water content: $\Theta_m = V_m / V_t$

Immobile Volumetric water content: $\Theta_{im} = V_{im} / V_t$

LANDSS (2014)

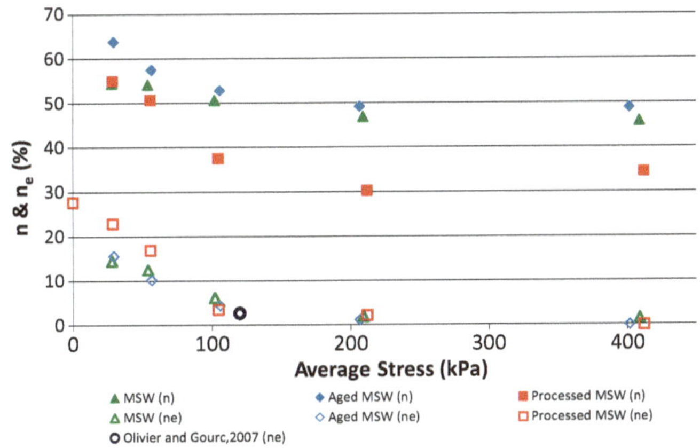

Figure 6.2.2 Variation in total and drainable porosity with effective stress. *MSW*, municipal solid waste. *After Beaven (2000).*

and 60% (expressed volumetrically). Beaven (2000) noted that changes in total porosity with increases in effective stress mainly occurred with changes in stress up to about 200 kPa. At stresses higher than ~200 kPa, there is little change in total porosity (Fig. 6.2.2).

Drainable porosity is the volume of leachate that will freely drain under gravity from the voids of a fully saturated waste material. In hydrogeology this is equivalent to the specific yield (S_y). When drainage has stopped, the waste is deemed to reach an idealized state termed *field capacity*. In practice this can take a long time to occur and is difficult to measure precisely. In crude terms the total porosity is equal to the volumetric water content at field capacity plus drainable porosity.

In a way that mirrors changes in total porosity, a very rapid reduction in drainable porosity has been demonstrated to occur as waste compacts in response to increasing the applied stress. Loosely compacted wastes, corresponding to average stresses less than 50 kPa, had drainable porosities in excess of 15%. At stresses in excess of 200 kPa, drainable porosities had fallen to ~2% or less (Fig. 6.2.2).

CONCEPTUAL MODELS

Flow in landfills may either be within the saturated or unsaturated zones, although this distinction is further complicated by the presence of landfill gas that is generated from the degradation of organic material and by the presence of perched layers.

Classical unsaturated flow as described by soil scientists usually envisages a gas phase that is connected to the atmosphere and a reduction in hydraulic conductivity with reducing degree of saturation (see Box 6.2.2). This situation probably occurs in much of the unsaturated zone of a landfill, although in many sites the gas phase will be isolated from the atmosphere (through capping) and pressure gradients are created from gas abstraction systems. Landfill gas can also occur within the nominally saturated zone of a site within discrete bubbles that may be isolated from each other.

Box 6.2.2 Relative Permeability Concepts for Municipal Solid Waste

The relative permeability of waste to liquid and gas varies between 0 and 1 with water content. The relative permeability to liquid is based on the hydraulic conductivity under fully saturated conditions, whereas the relative permeability to gas is related to gas permeability under fully drained conditions.

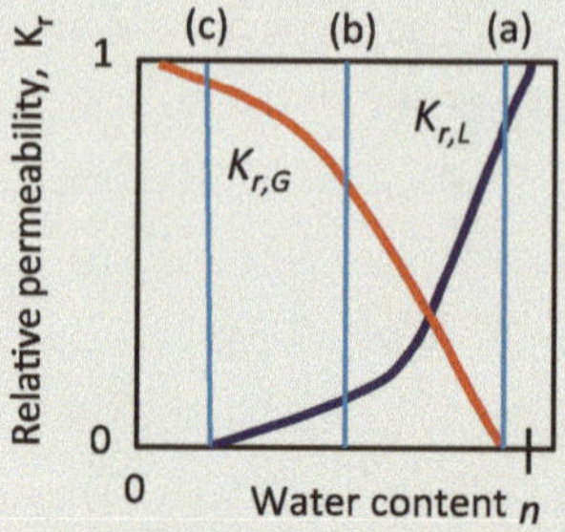

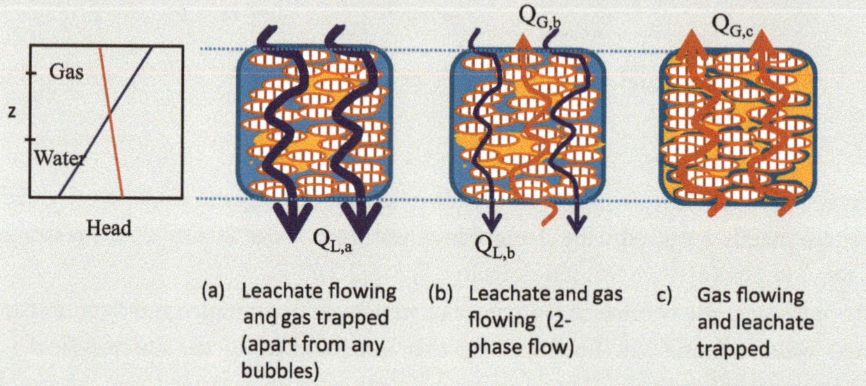

LANDSS (2014)

The volume of the bubbles respond to changes in pressure in the leachate (e.g., Hudson et al., 2001), and they will preferentially occupy larger void spaces, potentially resulting in a significant effect on leachate flow through the waste (e.g., Stoltz et al., 2010; Hudson et al., 2009).

It is often convenient to consider that landfilled wastes behave as a single continuous medium at an appropriate scale. This approach requires adoption of the concept of a "representative elementary volume" (REV). Bear (1972) defined the REV as the (smallest) macroscopic sampling volume at which a measured property does not change. The medium is considered to be a continuum with properties at a point having a value pertaining to the REV centered at that point. Continuum models in landfill assume an REV, although there is little direct evidence as to the validity of this (e.g., Woodman, 2007; Kindlein et al., 2006), nor what its actual size might be. Woodman et al. (2015) demonstrated that a 260-mm column of MSW was likely to be smaller than that required for continuum contaminant transport approaches to be effective. At the field scale, hidden large-scale structures (e.g., low-permeability cover layers, old haul roads, etc.) may invalidate homogeneity assumptions. Of course, if the location and characteristics of such structures are well constrained, then it may be possible to explicitly represent them as individual hydraulic units in a single-continuum model.

Heterogeneity may cause localized preferential flow that may invalidate a single-continuum approach. However, the waste may still be well represented by considering it to be comprised of two or more *continua* (with the potential for exchanges between them). The possibility of uniform or preferential flow in both the saturated and unsaturated zones is illustrated in Fig. 6.2.3. Preferential flow in waste might be caused by (1) macropores (i.e., larger connected macropathways that make their way through an otherwise microporous material), (2) funneling by low-permeability objects, and (3) through self-organization. Nimmo (2009) distinguished between these different types of preferential flow in unsaturated soils. In wastes these factors may affect flow in the unsaturated and saturated zones differently.

Macropore flow may occur through waste if a number of larger pathways exist that connect together. Conceivably, the more solid portion of the waste may provide a "skeleton" that keeps such a network open, although such pathways are notoriously difficult to observe and there are few reliable accounts detailing MSW macropore geometry.

It is widely recognized that the presence of plastic films and sheetlike materials in landfills can influence flow, both by creating anisotropy in vertical and horizontal hydraulic conductivity, but also by potentially shielding (locally diverting) areas of waste from water movement. Self-organization is even

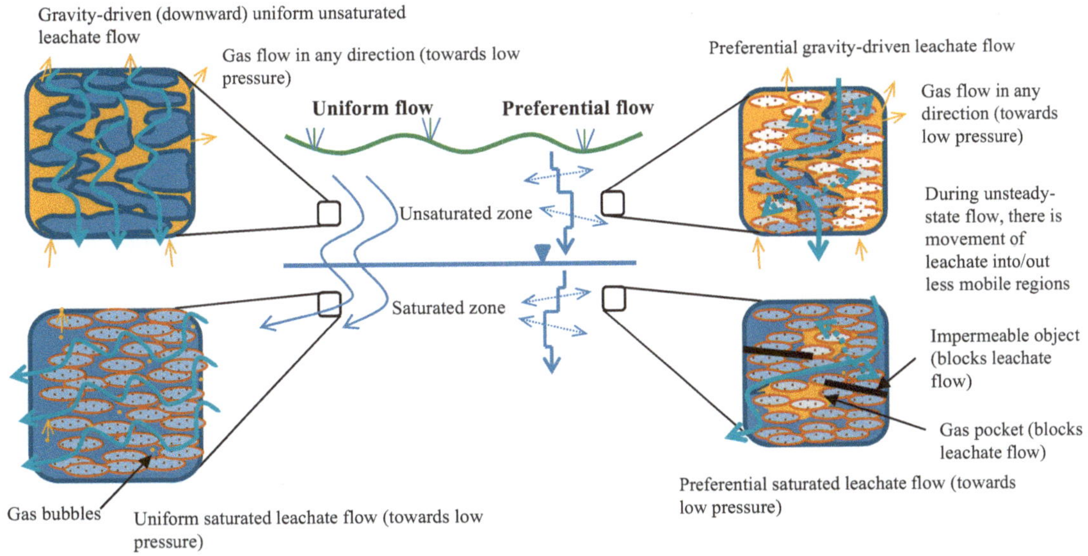

Figure 6.2.3 Conceptual sketch of different levels of flow uniformity in unsaturated and saturated municipal solid waste (LANDSS, 2014). Where flow is not steady there may be exchanges of fluid between the preferential pathways and less-mobile zones (*dashed lines*).

less well understood, although given that such phenomena occur in other geologic media it remains possible that preferential flow is either created or affected by dynamic flow effects.

Many of the existing physically based models of flow that have been applied to MSW landfill can be grouped into a small number of simple flow models based on the four broad possibilities shown in Fig. 6.2.3. Specifically (on the left side of Fig. 6.2.3) simple uniform flow in a single continuum is depicted in the saturated and unsaturated zones. This can occur in both steady-state and nonsteady-state (i.e., with pressure changing with time) conditions. The right side of Fig. 6.2.3 depicts preferential flow, again within the saturated and unsaturated zones. A simple framework of equations that relates to the scenarios for flow shown in Fig. 6.2.3 and that is discussed in this chapter is summarized in Table 6.2.1. For simplicity the equations are written for 1D vertical flow and could readily be generalized to 3D.

Saturated Flow

Single-Continuum Saturated Flow

Assuming Darcy's law to be valid in waste acting as a single-continuum, steady-state flow in landfills can be straightforwardly calculated knowledge of the fluid boundary conditions and by making assumptions about the permeability distribution. In terms of describing flow within our simple framework (Table 6.2.1), Eq. (1) gives Darcy's law in 1D. With knowledge of boundary conditions and the distribution of K, it is straightforward to estimate the direction and speed of flow within a landfill assuming Darcy's law applied to a single continuum.

Since it is often the case that saturated depths are considerably smaller than the horizontal dimensions of the landfill, it is frequently horizontal (or near-horizontal) flow that requires simulation. The flow into, out of, and between vertical wells is of interest.

Many studies into the permeability of waste materials have applied Darcy's law to calculate hydraulic conductivity from the testing of waste specimens in laboratory conditions. The majority of tests reported results from constant head experimental techniques under steady-state conditions.

Within the landfill environment the permeability of waste is normally established from pumping tests on vertical wells, which often requires application of unsteady (transient) saturated flow theory (e.g., see review by Jain et al., 2006). Unsteady uniform (single continuum) flow is described by Eq. (2) when $\Gamma_w = 0$ (Table 6.2.1). Pumping tests can also provide values of the storage coefficient, or storativity (S) of landfilled waste, which is the volume of water (or leachate) released from storage per unit surface area for each unit change in leachate head (see Box 6.2.3). If pumping results in no physical dewatering of the landfill (i.e., there is a reduction in head but no physical draining or emptying of voids), then leachate is obtained from specific storage (S_s), which essentially relates to the compressibility of both the waste (α_c) and fluid (β), as defined by Eq. (3) in Table 6.2.1. In hydrogeology, the term "confined" conditions is used to describe this situation.

Values of confined storativity determined directly from pumping tests are relatively rare in the literature, although Beaven (1996) reported a value of 0.006, from which S_s of $\sim 0.001 \text{ m}^{-1}$ can be

Table 6.2.1 Equations and timescales for flow in 1D

Condition	Steady-State Flow	Unsteady-State Pressure	Specific Storage, S_s, or Specific Moisture Capacity, $C(\psi)$	Diffusivity, D	Wave Speed, c	Characteristic Pressure Diffusion Time Over Distance, L	Characteristic Pressure Advection Time Over Distance, L
Saturated (single continuum)	$q = -K\frac{dh}{dz}$ (1)	$\frac{\partial h}{\partial t} = D\frac{\partial^2 h}{\partial z^2}$	$S_s = \rho g(\alpha_c + \theta\beta)$	$\frac{K}{S_s}$	N/A	$\frac{L^2}{D}$	N/A
Saturated (double continuum)		$\frac{\partial h}{\partial t} + \Gamma_w = D\frac{\partial^2 h}{\partial z^2}$ (2)	$S_s = \rho g(\alpha_c + \theta_m\beta)$ (3)				
Unsaturated (single continuum)	$q = -K(\psi)\left(\frac{d\psi}{dz} + 1\right)$ (4)	$\frac{\partial \psi}{\partial t} = D\frac{\partial^2 \psi}{\partial z^2} - c\frac{\partial \psi}{\partial z}$	$C(\psi) = \frac{\partial \theta}{\partial \psi}$	$\frac{K(\psi)}{C(\psi)}$	$\frac{\partial K}{\partial \theta}$	$\frac{L^2}{D}$	$\frac{L}{C}$
Unsaturated (double continuum)		$\frac{\partial \psi}{\partial t} + \Gamma_w = D\frac{\partial^2 \psi}{\partial z^2} - c\frac{\partial \psi}{\partial z}$ (5)	$C(\psi) = \frac{\partial \theta_m}{\partial \psi}$		$\frac{\partial K}{\partial \theta_m}$		

Numbers in ( ) link to equation numbers in the main text.

For saturated single-continuum flow, $q = Q/A$ is the Darcy velocity or flux, K is the saturated hydraulic conductivity, h is the head, D is the hydraulic diffusivity, S_s is the specific storage, α_c is the waste compressibility, θ is the volumetric water content, β is the fluid compressibility, c is the wave speed, and L is distance along the vertical dimension, z. Where there is a further continuum in the saturated zone, Γ_w is a nonequilibrium advective exchange term between the mobile volumetric water content θ_m and an immobile water content. For unsaturated flow, the head is $h = z + \psi$, $K(\psi)$ is the hydraulic conductivity (where ψ is the pressure head or matric potential), and $C(\psi)$ is the specific moisture capacity and $\theta(h)$ is the volumetric water content (related to the head by the "moisture retention curve"). Where there is a further continuum in the unsaturated zone, $\theta_m(h)$ is the volumetric water content of the mobile zone, which again is linked to the immobile zone by a nonequilbrium advective transfer term, Γ_w.

Box 6.2.3 Hydraulic Pumping Tests in Landfill

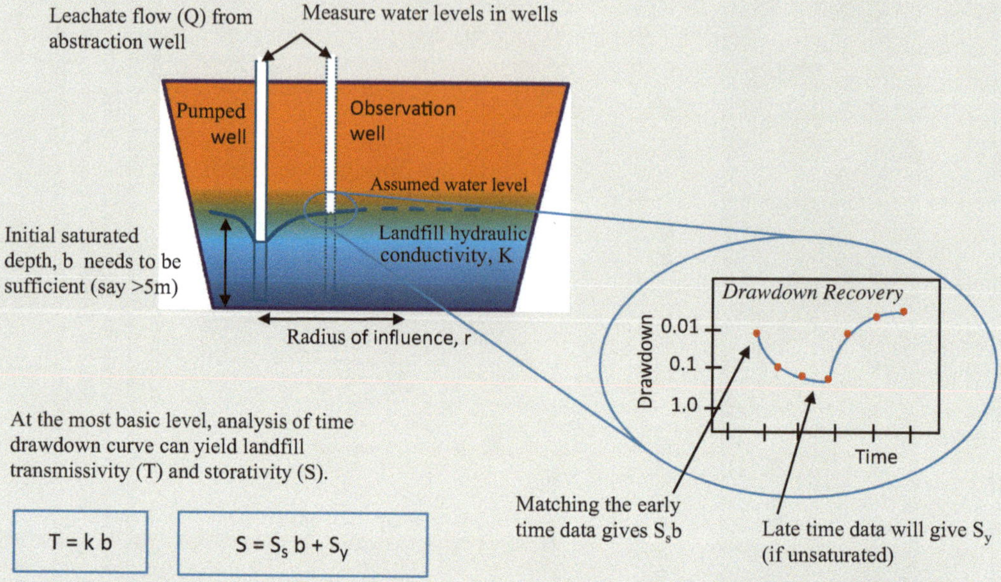

Pumping test analyses require a variety of different assumptions to be made. For example, the simplest analyses assume predominantly horizontal radial flow, isotropic conditions, and an areally extensive aquifer. There are many analytical solutions to account for different "nonideal" behaviours, many of which are applicable to the landfill environment. There is currently no consensus as to the "best" model for landfill simulation.

LANDSS (2014)

calculated. However, data are more generally available on the bulk modulus (which is the reciprocal of compressibility, α_c) of MSW with values typically between 2×10^5 and 2×10^6 Pa (e.g., McDougall, 2011). This range gives values of S_s of ~ 0.05 to $0.005\ m^{-1}$. Assuming (for example) a saturated depth of 5 m, then this range of S_s values equates to a confined storativity of between 0.25 and 0.025.

As discussed later, specific storage has particular relevance to the rate at which changes in pressure propagate through saturated systems.

More normally pumping tests in landfills will result in a physical dewatering of wastes, and the waste is considered to be "unconfined." In this case, the storage coefficient will be the sum of the specific yield (S_y) and the specific storage (S_s) and is often reported in pumping tests as a single storativity value. Calculations of storativity obtained from field pumping tests in MSW generally result in values between 1% and 15%, and these values have normally been interpreted as being equivalent to the specific yield of the waste.

Natural aquifers are normally less compressible than waste, with values of S_s typically of the order of $10^{-4}\ m^{-1}$. Thus in groundwater situations, where the analysis of a pumping test has yielded values of $S > 10^{-2}\ m^{-1}$, unconfined conditions can be inferred. In other words the high S is due to physical dewatering as the water table is lowered. However, in landfills, such an assumption cannot be made as, according to the example above, confined storativity could lead to storativity values of between 2.5% and 25% in the absence of any dewatering. This covers the range of storativity values reported in most pumping tests, and although this does not mean that dewatering *has not* occurred, care should be taken in automatically making that assumption. Hence, establishing whether MSW in landfill is confined, unconfined, or semiconfined on the basis of well pumping tests alone has added complexities. With a high anisotropy it is certainly conceivable that the waste may act in a confined/semiconfined manner.

A further source of possible error in parameter interpretation from landfill pumping tests is the effect on water levels of changing atmospheric pressure (potentially particularly important for longer-term tests).

Woodman and Rees-White (2009) have reported barometric efficiencies between 0.68 and 0.98, demonstrating that it can be important to perform barometric correction of water level data before pumping test data are analyzed, yet few studies report having done this.

Simple (standard) analysis of well pumping data from landfills may not adequately represent the complexity of the flow to the well. Complexity arises from both flow processes in the waste and the well itself. For example, the role of the unsaturated zone, liquid interaction with gas within the waste, preferential flow, and stress (depth)-dependent variation in the hydraulic properties of waste (Hudson et al. 2004) may need to be taken into account. The role of seepage faces to a pumping well (Al-Thani et al. 2004) and possible "skin" effects (reduced permeability at the edge of the well caused during drilling) may also need to be considered. Such potential complexity is probably responsible for the range in different types of pump test data observed in MSW and may obscure individual effects. Whether preferential flow provides a need for a second continuum to be simulated is discussed in the Dual- or Multicontinuum Saturated Flow section.

Dual- or Multicontinuum Saturated Flow

The diverse range of materials with very different properties and particle size distributions that make up wastes in a typical MSW landfill may cause flow, which differs to that expected by a single-continuum assumption. Voids exist in between particles, but many of the materials themselves have their own ("micro")porosity (see Box 6.2.1). Examples of this include paper, textiles, and biodegradable wastes. This secondary less-mobile porosity can be modeled as a second continuum in which the water is effectively immobile in terms of advection in the primary direction of flow. Instead, water is envisaged to only be able to move from the mobile zone in and out of this zone locally. While the nonequilibrium liquid transfer term (Γ_w) in Eq. (2) is set to zero for a single continuum, in this case it is used to simulate flow into and out of a less-mobile zone.

This approach is well established in the groundwater and soils literature (e.g., Šimunek et al., 2003) and is available in several widely known models, e.g., HYDRUS (Šimunek et al., 2005), TOUGH2 (Pruess et al., 1999), and AquiferTest Pro. The typical hydraulic response indicative of dual-porosity behavior in a pump test has an "s-shaped" response where the drawdown rate is slowed down for a period due to water coming out of the immobile zone (i.e., a delayed yield). Such curves are not universally measured in wastes and perhaps because of this, very few (if any) researchers have specifically modeled flow in saturated waste in this manner.

Nonetheless, some researchers have experienced difficulty interpreting landfill pumping test data with single-continuum models. Burrows et al. (1997) observed anomalous well water level responses which he inferred to be due to flow in heterogeneous channel. The corollary of such preferential flow would be the presence of less-mobile portions of waste, which may require a dual-porosity modeling approach. However, as has already been discussed, in addition to the possibility of highly localized and nonuniform (preferential) flow, field MSW pumping test data are likely to be influenced by a number factors, making it difficult to provide an unambiguous description of the flow pattern.

This conundrum is not restricted to pumping tests but is an example of the more generalized conceptual uncertainty that exists in terms of flow through MSW where several factors may be influencing flow and pressure measurements. To some extent this difficulty may be addressed by further flow measurement. Additionally, by detecting the presence and concentration of both naturally occurring solute and artificially added tracers, it is possible to determine more information about the pattern of flow than by observing pressures (heads) alone. In a situation where the flow is unsteady and Eq. (4) applies, solutes will be advected by the liquid flow in the mobile zone and also advected between the mobile and immobile zone. In addition, solute movement is affected by mechanical dispersion and diffusion. Describing the movement of solutes within wastes is therefore even more complicated than the distribution of pressure.

Recognizing that it is difficult to arrange in the context of horizontal pumping tests, the dynamics of solute movement can be considerably simplified if it is observed in the presence of a steady-state flow. In this instance, to adequately fit typical tracer test data in MSW under steady-state flow conditions, solute transport in the mobile saturated zone of a landfill is frequently modeled using a modified advection–dispersion equation. This includes both mechanical dispersion within an assumed mobile zone and dispersion caused by diffusion into the immobile zone that remains.

$$\frac{\partial C_m}{\partial t} + \Gamma_s = \alpha V_m \frac{\partial^2 C_m}{\partial z^2} - V_m \frac{\partial C_m}{\partial z} \tag{6}$$

where C_m is the concentration of solute in the mobile zone, Γ_s is a solute exchange term, which describes the diffusive transfer in/out of an immobile[1] zone in the event of preferential flow, α is the dispersivity, and V_m is the long-term average advective fluid velocity in the mobile zone in the z direction.

[1]An immobile zone refers to zones of waste where water is not subjected to advective flow, and transfer into and out of the zone can only occur through diffusion.

There are many different forms of exchange term, which relate to different possible geometries of the mobile pathways and immobile zones, represented as blocks by Barker (1985). In the absence of a diffusive exchange between different porosities, Γ_s is set to zero. Just as for the flow Eq. (2), the critiques of Eq. (6) are well known in natural geological media. In particular there is a substantial literature on the genesis and scaling of mechanical dispersion (e.g., Pickens and Grisak, 1981; Detwiler et al., 2000). Tracer tests in saturated waste normally involve the addition of a tracer into leachate flowing through waste and the monitoring of tracer concentrations at a point downstream. A plot of tracer concentration versus time at the outlet is referred to as a breakthrough curve (BTC), analysis of which can be matched to various transport models. Nearly all the BTCs in saturated waste known to the authors are "classical" skewed, monomodal peaks with relatively "early" breakthrough, and "long" tails indicative of significant dispersion.

For MSW there remains considerable uncertainty as to the appropriate form of exchange term. In all probability, the shape of the immobile zone (blocks) may be spatially variable. Tracer BTC data contains insufficient information to strongly discriminate between all but the most extreme block dimensions, and observations of block shapes and flow paths within the waste remain inconclusive.

Beaven et al. (2003) performed an intermittent flow test on waste in a large-scale lysimeter. The BTC showed a downward jump in concentration following flow interruption, probably indicative of diffusive exchange with less-mobile regions of waste continuing, while the advective flow had stopped (akin to observations in natural media, e.g., Brusseau et al., 1997).

Dual-porosity models have been successfully applied to BTCs in saturated MSW, and it has been concluded that relatively small mobile water contents contain the active flow (0.1%–11%) (e.g., Fellner et al., 2009; Woodman et al., 2014). This appears strongly indicative of preferential flow. However, Woodman and Beaven (2011) and Bendz and Singh (1999) caution against excessive physical interpretation of model parameters, since even a good-fitting model may be physically wrong. Alternatively, it may also be more realistic to represent the immobile zone as a "less-mobile" zone that has a slower (but nonzero) rate of advection than the mobile zone (Woodman and Beaven, 2011) via a "dual-permeability" model. These points notwithstanding solute data therefore indicate preferential flow is likely to occur commonly and can be successfully explained using dual-continuum approaches.

Even if the movement of water is as preferential as the solute data suggests this leaves a question as to whether a single continuum can adequately simulate unsteady flow. If the time taken for hydraulic equilibration between a mobile and less-mobile zone is very rapid or very slow compared to the timescale of interest, then a single continuum may remain a valid assumption (in the first instance the entire porosity comes into equilibrium and in the second only the pressure in the mobile zone is considered). To estimate such an equilibration time, as well as defining the shape and size of the immobile zone (for example, assumed to be the same as the zone effectively immobile to solute), there is a need to assume hydraulic properties of the less-mobile zone. These are not always readily available since bulk (single-continuum) hydraulic properties tend to be established most frequently. If the bulk properties may apply to both zones (for instance, in the case that the immobile zone is caused by the presence of low-permeability barriers rather than by a secondary porosity), then characteristic hydraulic equilibration times may be estimated given an assumed block size.

Eq. (5) simulates pressure as propagating according to a (modified) diffusion equation, thus a characteristic time for pressure equilibration is $L^2 S_s/K$ where L is an appropriate length scale. For pairs of K and S reported in the literature (e.g., Burrows et al., 1997; Cossu et al., 1997; Olivier and Gourc, 2007) and an immobile block size of 5 cm, the longest characteristic equilibration time is 78 s, which is very short compared to typical pumping test durations. If the immobile zone hydraulic conductivity and storativity are reduced, for example, to $K = 1 \times 10^{-9}$ m/s and $S_s = 0.001$, a characteristic time of 2500 s is predicted, which may be significant.

Unsaturated Flow

Single-Continuum Unsaturated Flow

Landfills are often managed explicitly to keep the leachate levels low, although this is not universally the case (e.g., Olivier et al., 2009). Consequently the majority of waste in a landfill will be subject to unsaturated flow processes.

Unsaturated flow dynamics may be important for a number of practical field problems. For example, it may be important in interpreting pumping responses and in understanding flow and contaminant loadings following rainfall or artificial irrigation and during recirculating flow. The propagation of drying and wetting fronts through the waste could be important in the characterization of the manner in which a landfill fills up and drains down. The rate of infiltration is important since it governs the rate of water uptake and therefore whether "ponding" or runoff is to be expected, both of which have implications for operators. In some circumstances, flow may be idealized as 1D, but in others, for example, flow to horizontal drains two or even three dimensions may be required to be modeled.

Unsaturated flow in waste has been measured in a variety of ways both in the laboratory and in the field. The response of landfills to moisture input (e.g., from rain) has long been studied. At the simplest the input and output flows have been compared (e.g., Bendz et al., 1997). Internal measurements of moisture content have been made, both in the unsaturated zone (Yuen et al., 2000) and in the position of the leachate table (Jang, 2000). Pressure and solute concentration have been monitored at the base of landfills (Knox and Shaw, 2006). Under more controlled circumstances, unsaturated flow has been measured through laboratory columns both for simulated rain and filling and drain-down conditions (e.g., Zardava, 2012). Measurements have been taken of the water retention relationship and of the permeability at different water contents.

It is useful to understand steady-state unsaturated conditions. If there is no input flow, then eventually (neglecting internal moisture generation) a stable vertical moisture profile will develop with no outflow from the waste. The moisture distribution with depth under this condition is predicted by moisture retention relationships.

Steady-state flow will arise if the inflow is constant for a sufficiently long time. Under the assumption of a single continuum and a single permeability—moisture content (or pressure) relationship, the

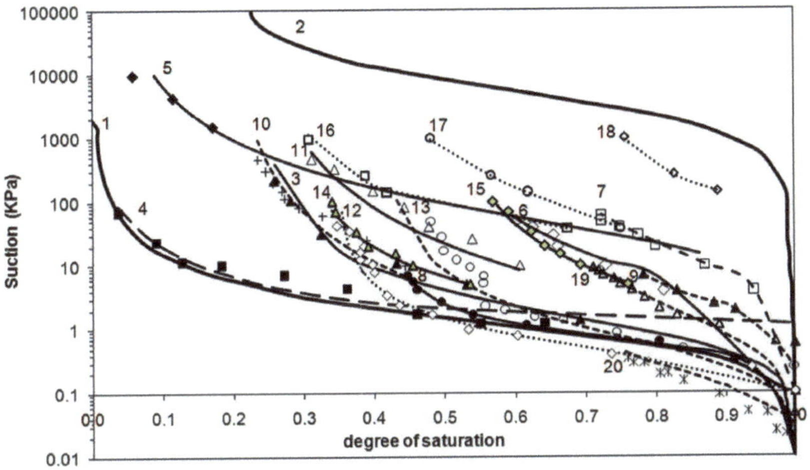

Figure 6.2.4 Comparison of moisture retention curve for municipal solid waste (MSW). Degree of saturation is defined as V_w/V_v. The data for MSW is bounded by the moisture retention curves for coarse sand (line 1) and clay (line 2). Lines 6 and 7 are for processed (mechanically biologically treated) waste and 20 for cocoa shells and peat. *Reproduced Zardava (2012).*

Buckingham–Darcy equation (Eq. 4 in Table 6.2.1) governs flow (Warrick, 2002). The suction head, ψ, increases with reducing water content, as moisture retreats into smaller and smaller pores.

The relationship between volumetric water content and suction, known as the water retention curve (WRC), and the dependence of unsaturated hydraulic conductivity on suction has been the subject of much research on waste. For transient flow, where water content is changing with time, Richards equation (Eq. 5 in Table 6.2.1, where $\Gamma_w = 0$) is commonly applied to unsaturated flow problems in waste. There are a number of ways to parameterize this equation; the form used here provides convenient comparison with the other equations, particularly Eq. (6) (note this relies on the assumption that $\frac{\partial\psi}{\partial z} \ll 1$ as per Rasmussen et al., 2000).

Fig. 6.2.4 summarizes data from a number of studies on the WRC of wastes. It includes data both from studies involving direct measurement and studies that have applied a single-continuum model based on the Richards Eq. (6) to infer the WRC. Many of these modeling applications are based on laboratory experiments, and there are fewer examples of field-scale applications.

Several authors have used the Richards equation, or an equivalent, to simulate flow through MSW. The most basic flows, namely wetting and drying fronts or an impulse of water (e.g., from a rainfall event), can be simulated using analytical solutions, under certain simplifying assumptions.

Since boundary conditions of a real landfill can be spatially and temporally varying, numerical models are best suited to simulating unsaturated flow. El-Fadel et al. (1997) reviewed a large number of such models. The Richards equation is used as the basis for the hydraulics calculations in the highly coupled landfill processes models, LDAT (White et al., 2014) and the hydro-bio-mechanical flow model of

McDougall (2007), which tend to be used for simulating complex data sets of flow in degrading and settling wastes (e.g., Beaven, 2008). The parameters that arise from fitting these models under such complicated conditions are potentially highly nonunique and are therefore potentially not as reliable as for when models are compared to data under more controlled conditions (Beven and Binley, 1992).

Haydar and Khire (2005) simulated steady-state leachate recirculation from trenches using parameters, which were set as standard in the Hydrus2D model, although the comparison to real field data was not especially well fit, due in part to the sparseness of the data. Zardava (2012) tried to fit pressure, drainage, and step outflow data from laboratory experiments with mixed results.

Yildiz et al. (2004) applied a single-continuum unsaturated model, which further assumed a unity hydraulic gradient. This achieved a fairly poor fit to the fluctuations in field-scale flow rates although a reasonable representation of the general trend was obtained. The authors attributed this to a nonuniform return of the surface runoff and acknowledged that "short-circuiting" may be occurring. McCreanor and Reinhart (2000) simulated the cumulative leachate generation at the Delaware Solid Waste Authority test cell using SUTRA. Higher permeabilities that would be expected from laboratory models would be needed to fit, and this suggested that more information is needed about field-scale unsaturated properties. They also simulated cumulative flow from the Yolo county Leachate recirculation system demonstration project. The result was a divergence between model and measurement, which increased with time and authors attributed to the models inability to simulate preferential flow.

Some support for a simple single-continuum unsaturated approach comes from the widespread successful application of the so-called "HELP" model. This is the USGS's Hydrologic Evaluation of Landfill Performance (HELP) model (Schroeder et al., 1994), which is an industry-standard model of unsaturated landfill waste. HELP can be used to predict leachate production at a site. Critically, the model neglects the capillary potential term $\left(\frac{d\psi}{dz}\right)$ in Eq. (6). Instead, HELP calculates downward flow through the waste according to Campbell's relationship for unsaturated hydraulic conductivity, which uses the Brooks—Corey moisture retention parameters. Despite this simplification, the practical usefulness of HELP is demonstrated by its widespread application. Berger (2000) performing a hydrologic validation lists several applications under which the HELP model is suitable, while also carefully analyzing the weaknesses of the approach.

In the unsaturated zone, pressure transmits not due to compressibility but by the physical flow of water. Thus, pressure fronts in the unsaturated zone contain both advective and diffusive terms. An example of a characteristic time through unsaturated waste would be the time taken for rainfall or infiltration event to be noticeable at the base of a 10-m deep landfill (for instance, marked by an increase in the flow of leachate out of the basal drain). In contrast to the saturated zone, the movement of pressure in systems conforming to Richards equation is characterized by advection and diffusion of pressure rather than only by diffusion of pressure in the case of saturated waste. Where advection dominates (e.g., White et al., 2014) the characteristic time is $\frac{L}{c}$ where c, the wave speed, can also be thought of as the speed of a wetting front moving down through the waste. c can be estimated from the unsaturated hydraulic parameters using the assumptions made by Rasmussen et al. (2000). Taking the Brooks—Corey parameters given by Breitmeyer and Benson (2011) for 795 kg/m^3 waste, assuming a prior volumetric water content of 0.35, then c is estimated as 0.2 m/day. Therefore this simple analysis with these parameters suggests that the wetting front would be expected at the base of a 10-m deep landfill after ~50 days. However, it should be noted that the wave speed is highly sensitive to input parameters. This quick analysis is only as good as the assumptions; a numerical model would give a much more complete prediction.

There is relatively little data giving hydraulic response times. Knox and Shaw (2006) observed pressure responses to rainfall using in vibrating wire piezometers installed at the base of a site. The responses were examined as the waste depth increased due to ongoing landfilling. Discernible responses were observed for a waste depth of up to 7 m, thereafter no clear response was detectible. Even at 7 m the responses took only around 2 days to appear following rainfall. The time taken for observable drainage to occur from a 1-m deep column of 100 kpa compressed MSW that was drained from full saturation has been observed to be ~2 days (Hudson, pers comm.).

Neither of these observations agrees particularly well with the prediction based on the parameters. This leaves the Richards (single-continuum, unsaturated) model under question in terms of its ability to reliably simulate the downward movement of water in landfill in all cases.

Dual- or Multicontinuum Unsaturated Flow

Dual-continuum unsaturated flow can be simply modeled using Eq. (5) in Table 6.2.1.

There is a large body of evidence suggesting that preferential flow (i.e., a proportion of the total flow through the waste passes through a subset of the total porosity) is important in unsaturated flow in MSW landfills. The possibility of preferential flow was mentioned in relatively early studies (e.g., Straub and Lynch, 1982) and has also been the topic of more recent work (Fellner et al., 2009). Evidence for preferential flow comes from a wide variety of observation. Irregular moisture content distributions have been observed (Harris, 1979; Yuen et al., 2000). Outflow has been observed from landfills prior to their reaching "field capacity" (Blight et al., 1992), and Bengtsson et al. (1994) observed different rainfall responses in young and older landfills inferring more preferential flow occurs in younger landfills. Zeiss and Major (1993) observed changing patterns of flow across a cross section with depth into a laboratory sample but not a tendency to either disperse or concentrate. Visual observations are difficult, although Stegmann and Ehrig (1989) reported they had substantial evidence of preferential flow in dug-up landfill waste.

Observations of indigenous leachate are complicated by the fact that the initial spatial distribution of each component is seldom well known. Nonetheless, observations consistent with preferential flow have been made, including evidence of reduced strength leachate following intense rainfall (Knox, 2001), seasonal variations in leachate quality within lysimeters (e.g., Ham and Bookter, 1982), and correlations between leachate concentration and flow in test cells (e.g., Akesson and Nilsson, 1997).

Han et al. (2011) applied single-porosity and dual-permeability models to laboratory step data in a synthetic waste (comprising significant amounts of paper). A dual-permeability model performed significantly better than a single-porosity model. Hudson et al. (2013) found that both the Richards equation and double-porosity model were unable to simulate data for a waste column experiment in the laboratory with intermittent flow injection. However, a double permeability model did fit successfully. Uguccioni and Zeiss (1997) concluded that to simulate their laboratory column outflow data either an explicit two-domain model is needed or a model that allows strongly varying flow parameters. Tinet et al. (2011) discussed simulation of multidomain flow behavior during infiltration in a laboratory column. They compared single- and double-porosity models with and without bimodal retention curves (based on the proposed retention curve in Ross and Smetten, 2000) with a multiporosity model. Kazimoglu (2007) came to the same conclusion.

Fellner et al. (2009) performed a field tracer test in unsaturated waste. Their analysis of the BTC suggested highly preferential flow. Fellner and Brunner (2010) anticipated preferential flow to dominate in landfill, due to heterogeneity and funneling. They cited moisture distribution data (e.g., Öman and Rosqvist, 1999; Yuen et al., 2000) as supporting the hypothesis that funneling concentrates preferential flow more toward the base of the landfill. They explicitly simulated a 1-m-wide microporous block that runs from the top of the landfill to the bottom aside a narrow macropore (channel).

Analogous to the evidence base in saturated waste, inference from tracer test data supports the wider observations of preferential flow in unsaturated waste. Rosqvist and Bendz (1999) collected tracer data which they observed had the characteristics of preferential flow. This finding was corroborated by analysis by Bendz and Singh (1999) and Rosqvist and Destouni (2000). Interestingly, Rosqvist and Destouni (2000) concluded that while a two-stream tube model performed well against laboratory data, transfer of solute between them had only minor effects on the outcome. Woodman and Beaven (2011) remodeled the data from Rosqvist and Bendz (1999) with a series of different stream tube models demonstrating robustly that the different parameters all of the models were potentially indicative of preferential flow in different ways.

The timescales for pressure equilibration in an unsaturated waste are likely to be longer than in a saturated waste since pressure transmission occurs through changes in the water content, rather than through the compressibility of water and the waste itself. Therefore this increases the chances that equilibration with a less-mobile zone is not effectively instantaneous and that the transfer needs to be simulated to make sense of typical data.

In summary, there is broad evidence for both the occurrence of preferential flow in unsaturated waste and for the necessity of simulating it. However, there is a need for the delineation of the conditions under which simpler single-continuum representations are fit for purpose. Until this is resolved, caution should be expressed over the application of single-continuum approaches to unsaturated flow in MSW.

CONCLUSIONS

This chapter has reviewed the evidence for and application of commonly used physically based models for flow in saturated and unsaturated MSW. These approaches are notable not only through their widespread use, but also in their simplicity, being decoupled from other dynamic landfill processes such as mechanical and biologically mediated compression and interaction with landfill gas.

Many of the existing models for flow through MSW can be categorized into the small family of equations elucidated in this chapter (and summarized in Table 6.2.1). This family covers saturated and unsaturated waste both in steady state and nonsteady state. For the nonsteady-state conditions, an exchange term can be introduced to allow flow to occur between a more-mobile zone and a less-mobile zone imagined to exist in a second overlapping continuum. There are examples in the landfill literature of models from all members of this family, with the exception that there are no unsteady-state saturated flow models.

There is evidence that the physical structure of waste produces a complicated porous media. However, this does not mean that more complex physically based models (such as dual- or multicontinuum models) always have to be applied. It may be practical and justifiable to use such simple models as are outlined here as long as they produce satisfactory results for the application they are being used for. However, this chapter makes clear that extensive use of these models has not necessarily provided an improved basis for their application and considerable caution is needed as to the applicability of any of these particular approaches. For a particular situation the modeler may examine whether the most applicable approach has been selected and indeed whether any of these approaches are appropriate. It is of course probable that in some situations none of these approaches will be adequate.

References

Akesson, M., Nilsson, P., 1997. Seasonal changes of leachate production and quality from test cells. Journal of Environmental Engineering 123 (9), 892–900.

Barker, J.A., 1985. Block-Geometry functions characterizing transport in densely fissured media. Journal of Hydrology 77, 263–279.

Bear, J., 1972. Dynamics of Fluids in Porous Media. American Elsevier Pub Company Inc., Toronto.

Beaven, R.P., 1996. Geotechnical and Hydrogeological Properties of Wastes. Engineering Geology of Waste Disposal. S. P. Bentley, London. The Geological Society – Special Publication No. 11.

Beaven, R.P., 2000. The Hydrogeological and Geotechnical Properties of Household Waste in Relation to Sustainable Landfilling (Ph.D. thesis). University of London, London.

Beaven, R.P., 2008. Review of responses to a landfill modelling challenge. Proceedings of the ICE – Waste and Resource Management 161 (4), 155–166. https://doi.org/10.1680/warm.2008.161.4.155.

Beaven, R.P., Barker, J.A., Hudson, A., 2003. Description of a tracer test through waste and application of a double porosity model. In: Sardinia 2003: Proceedings of the Ninth International Waste Management and Landfill Symposium, Margherita di Pula, Cagliari, Sardinia, Italy.

Beaven, R.P., Powrie, W., Zardava, K., 2011. Hydraulic properties of MSW. In: Zekkos, Dimitrios (Eds.), Geotechnical Characterization, Field Measurements and Laboratory Testing of Municipal Solid Waste. American Society of Civil Engineers, Virginia, US, pp. 1–43.

Bendz, D., Singh, V.P., 1999. Solute transport under steady and transient conditions in biodegraded municipal solid waste. Water Resources Research 35 (8), 2333–2345.

Bendz, D., Singh, V.P., Akesson, M., 1997. Accumulation of water and generation of leachate in a young landfill. Journal of Hydrology 203, 1–10.

Bengtsson, L., Bendz, D., Hogland, W., Rosqvist, H., Akesson, M., 1994. Water balance for landfills of different age. Journal of Hydrology 158, 203–217.

Berger, K., 2000. Validation of the hydrologic evaluation of landfill performance (HELP) model for simulating the water balance of cover systems. Environmental Geology 39 (11), 1261–1274.

Beven, K., Binley, A., 1992. The future of distributed models: model calibration and uncertainty prediction. Hydrological Processes 6, 279–298.

Blight, G., Ball, J., Blight, J., 1992. Moisture and suction in sanitary landfills in semiarid areas. Journal of Environmental Engineering 118 (6), 865–877.

Breitmeyer, R.J., Benson, C.J., 2011. Measurement of unsaturated hydraulic properties of municipal solid waste. Geo-frontiers, ASCE, 1433–1442. https://doi.org/10.1061/41165(397)147.

Brusseau, M.L., Hu, Q., Srivastava, R., 1997. Using flow interruption to identify factors causing nonideal contaminant transport. Journal of Contaminant Hydrology 24, 205–219.

Buchanan, D., Clark, C.F., Ferguson, N.S., Kenny, N.J., March 2001. Hydraulic characteristics of wet-pulverised municipal waste. CIWEM Journal 14–20.

Burrows, M.R., Joseph, J.B., Mather, J.D., 1997. The hydraulic properties of in-situ landfilled waste. In: Sardinia 97: Proceedings of the Sixth International Waste Management and Landfill Symposium, S. Margherita di Pula, Cagliari, Italy.

Cossu, R., Fongia, G., Muntoni, A., Mobile, A., Raga, R., 1997. Use of pumping tests for the assessment of leachate flow regime, waste hydraulic parameters and well efficiency. In: Sardinia 97: Proceedings of the Sixth International Waste Management and Landfill Symposium, S. Margherita di Pula, Cagliari, Italy.

Detwiler, R.L., Rajaram, H., Glass, R.J., 2000. Solute transport in variable-aperture fractures: an investigation of the relative importance of Taylor dispersion and macrodispersion. Water Resources Research 36 (7), 1611–1625.

El-Fadel, M., Findikakis, A.N., Leckie, J.O., 1997. Modeling leachate generation and transport in solid waste landfills. Environmental Technology 18, 669–686.

Fellner, J., Döberl, G., Allgaier, G., Brunner, P.H., 2009. Comparing field investigations with laboratory models to predict landfill leachate emissions. Waste Management 29, 1844–1851.

Fellner, J., Brunner, P.H., 2010. Modeling of leachate generation from MSW landfills by a 2-dimensional 2-domain approach. Waste Management 30, 2084–2095.

Ham, R.K., Bookter, T.J., 1982. Decomposition of solid waste in test lysimeters. Journal of the Environmental Engineering Division – Proc' ASCE EE6 108, 1147–1170.

Han, B., Scicchitano, V., Imhoff, P.T., 2011. Measuring fluid flow properties of waste and assessing alternative conceptual models of pore structure. Waste Management 31, 445–456.

Harris, M.R.R., 1979. A Study on the Behaviour of Refuse as a Landfill Material (Ph.D. thesis). Portsmouth Polytechnic.

Haydar, M.M., Khire, M.V., 2005. Leachate recirculation using horizontal trenches in bioreactor landfills. Journal of Geotechnical and Geoenvironmental Engineering 131 (7), 837–847.

Hudson, A.P., 2007. Evaluation of the Vertical and Horizontal Hydraulic Conductivities of Household Wastes (Ph.D. thesis). University of Southampton.

Hudson, A.P., Beaven, R.P., Powrie, W., 1999. Measurement of the hydraulic conductivity of household waste in a large scale compression cell. In: Sardinia 99: Seventh International Waste Management and Landfill Symposium, S. Margherita di Pula, Cagliari, Italy.

Hudson, A.P., Beaven, R.P., Powrie, W., 2001. Interaction of water and gas in saturated household waste in a large scale compression cell. In: Sardinia 2001: Eighth International Waste Management and Landfill Symposium, S. Margherita di Pula, Cagliari, Italy.

Hudson, A.P., Beaven, R.P., Powrie, W., 2009. Assessment of vertical and horizontal hydraulic conductivities of household waste in a large scale compression cell. In: Sardinia 2009: Twelfth International Waste Management and Landfill Symposium, S. Margherita di Pula, Cagliari, Sardinia, Italy.

Hudson, A.P., Beaven, R.P., White, J.K., Powrie, W., 2004. Modelling the compression behaviour of landfilled domestic waste. Waste Management 24, 259–269.

Hudson, A.P., Beaven, R.P., Woodman, N.D., 2013. Comparison of continuous and intermittent water flow for contaminant flushing of unsaturated municipal solid waste. In: MacDougall, J. (Ed.), HPM5: Hydro-Physico-Mechanics of Landfills. Napier University, Edinburgh.

Jain, P., Powell, J., Townsend, T.G., Reinhart, D.R., 2006. Estimating the hydraulic conductivity of landfilled municipal solid waste using the borehole permeameter test. Journal of Environmental Engineering 132 (6), 645–652.

Jang, Y.-S., 2000. Analysis of flow behaviour in a landfill with cover soil of a low permeability. Environmental Geology 39.

Kazimoglu, 2007. Moisture Retention and Conductivity Properties of Waste Refuse: A Laboratory Study (Ph.D. dissertation). Napier University, School of the Built Environment.
Kindlein, J., Dinkler, D., Ahrens, H., 2006. Numerical modelling of multiphase flow and transport processes in landfills. Waste Management & Research 24 (4), 376–387.
Knox, K., 2001. Development of a novel process for treatment of leachate with very high ammoniacal nitrogen concentrations. In: Proceedings Sardinia 2001, Eight Int' Waste Management and Landfill Symposium, vol. II, pp. 207–213.
Knox, K., Shaw, P.J., 2006. An Experimental Study of the Hydraulic Response of Landfills to Infiltration Events. Final Report to the Norlands Foundation. Entrust Project no 222285.003.
LANDSS, 2014. A Landfill Aftercare Decision Support System. www.southampton.ac.uk/landss.
Landva, A.O., Pelkey, S.G., Valsangkar, A.J., 1998. Coefficient of Permeability of Municipal Refuse, vols. 1–4. Environmental Geotechnics, pp. 63–67.
Leckie, J.O., Pacey, J.G., Halvadakis, C., 1979. Landfill management with moisture control. ASCE – Journal of the Environmental Engineering Division EE2 337–355.
Lofy, R.J., 1996. Zones of vacuum influence surrounding gas extraction wells. In: Cossu, R., Stegmann, R., Christensen, T.H. (Eds.), Landfilling of Waste: Biogas. E&FN Spon, London, pp. 319–394.
McCreanor, P.T., Reinhart, D.R., 2000. Mathematical modeling of leachate routing in a leachate recirculating landfill. Water Research 34 (4), 1285–1295.
McDougall, J., 2007. A hydro-bio-mechanical model for settlement and other behaviour in landfilled waste. Computers and Geotechnics 34, 229–246.
McDougall, J., 2011. Settlement – the short and the long of it. In: Geotechnical Characterization, Field Measurements and Laboratory Testing of Municipal Solid Waste. ASCE Geotechnical Special Publication #209.
Münnich, K., Bauer, J., Fricke, K., 2005. Investigation on relationship between vertical and horizontal permeabilities of MBT waste. In: Sardinia 2005: Proceedings of the 10th International Sardinia Landfill Conference, S. Margherita di Pula, Cagliari, Italy. CISA Publisher.
Nimmo, J.R., 2009. Vadose water. In: Likens (Ed.), Encyclopedia of Inland Waters, vol. 1. Elsevier, Oxford, pp. 766–777.
Olivier, F., Gourc, J.P., 2007. Hydro-Mechanical behaviour of municipal solid waste subject to leachate recirculation in a large-scale compression reactor cell. Waste Management 27, 44–58.
Olivier, F., Oxarango, L., Mugnier, V., Tinet, A.J., Marcoux, M.A., 2009. Estimating the drawdown of leachate in a saturated landfill: 3d modeling based on field pumping tests. In: Sardinia 2009: Proceedings of the 12th International Landfill Conference, S. Margherita di Pula, Cagliari, Italy. CISA Publisher.
Öman, C., Rosqvist, H., 1999. Transport fate of organic compounds with water through landfills. Water Research 33 (10), 2247–2254.
Pickens, J.F., Grisak, G.E., 1981. Modeling of scale-dependent dispersion in hydrogeologic systems. Water Resources Research 17 (6), 1701–1711.
Powrie, W., Beaven, R.P., 1999. Hydraulic properties of household waste and implications for landfills. Proceedings of the Institution of Civil Engineers-Geotechnical Engineering 137, 235–247.
Pruess, K., Oldenburg, C., Moridis, G., 1999. TOUGH2 User's Guide, Version 2.0, Rep. LBNL-43134. Lawrence Berkely National Laboratory, Berkeley.
Rasmussen, T.C., Baldwin, R.H., Dowd, J.F., Williams, A.G., 2000. Tracer vs. pressure wave velocities through unsaturated saprolite. Soil Science Society of America Journal 64, 75–85. https://doi.org/10.2136/sssaj2000.64175.
Reitzel, S., Farquahar, G., McBean, E., August 1992. Temporal characterisation of municipal solid waste leachate. Canadian Journal of Civil Engineering 668–679.
Rosqvist, H., Bendz, D., 1999. An experimental evaluation of the solute transport volume in biodegraded municipal solid waste. Hydrology and Earth System Science 3 (3), 429–438.
Rosqvist, H., Destouni, G., 2000. Solute transport through preferential pathways in municipal solid waste. Journal of Contaminant Hydrology 46, 39–60.
Ross, P., Smetten, R., 2000. A simple treatment of physical nonequilibrium water flow in soils. Soil Science Society American Journal 64, 1926–1930.
Schroeder, P.R., Aziz, N.M., Lloyd, C.M., Zappi, P.A., 1994. The Hydrologic Evaluation of Landfill Performance (HELP) Model: User's Guide for Version 3 Rep.
Šimunek, J., Jarvis, N.J., vanGenuchten, M.T., Gardenas, A., 2003. Review and comparison of models for describing non-equilibrium and preferential flow and transport in the vadose zone. Journal of Hydrology 272, 14–35.

Šimunek, J., van Genuchten, M.T., Sejna, M., 2005. The HYDRUS-1D Software Package for Simulating the Movement of Water, Heat, and Multiple Solutes in Variably Saturated Media. Version 3.0, HYDRUS Software Series 1Rep. Department of Environmental Sciences, University of California Riverside, Riverside, California, USA.

Singh, K., Kadambala, R., Jain, P., Xu, Q.Y., Townsend, T.G., 2014. Anisotropy estimation of compacted municipal solid waste using pressurized vertical well liquids injection. Waste Management & Research 32 (6), 482–491. https://doi.org/10.1177/0734242x14532003.

Stegmann, Ehrig, 1989. Leachate production and quality: results of landfill processes and operation. In: Proceedings, Paper Presented at Sardinia 89, Second International Landfill Symposium, S. Margherita di Pula, Cagliari, Sardinia, Italy.

Stoltz, G., Gourc, J.-P., Oxarango, L., 2010. Liquid and gas permeabilities of unsaturated municipal solid waste under compression. Journal of Contaminant Hydrology 118, 27–42.

Straub, W.A., Lynch, D.R., 1982. Models of landfill leaching: moisture flow and inorganic strength. Journal of Environmental Engineering, ASCE 108, 231–250.

Tinet, A.-J., Oxarango, L., Bayard, R., Benbelkacem, H., Stoltz, G., Staub, M.J., Gourc, J.-P., 2011. Experimental and theoretical assessment of the double porosity behaviour in a waste body during leachate infiltration. Waste Management 31 (8), 1797–1806.

Uguccioni, M., Zeiss, C., 1997. Comparison of two approaches to modelling moisture movement through municipal solid waste. Journal of Environmental Systems 23, 247–270.

Wall, D.K., Zeiss, C., 1995. Municipal landfill biodegradation and settlement. Journal of Environmental Engineering 121 (3), 214–224.

Warrick, A.W. (Ed.), 2002. Soil Physics Companion. CRC Press, Boca Raton.

White, J.K., Nayagum, D., Beaven, R.P., 2014. A multi-component two-phase flow algorithm for use in landfill processes modelling. Waste Management 34 (9), 1644–1656. https://doi.org/10.1016/j.wasman.2014.05.005.

Woodman, N.D., 2007. Modelling of Transport in Heterogeneous Porous Media with Application to Solute Transport in Landfills (Ph.D. thesis). University College London, London.

Woodman, N.D., Beaven, R.P., 2011. Assessment of simple preferential flow concepts for transport through municipal solid waste. In: Cossu, R., He, P., Kjeldsen, P., Matsufuji, Y., Reinhart, D., Stegmann, R. (Eds.), Sardinia 2011: Thirteenth International Waste Management and Landfill Symposium, S. Margherita di Pula, Cagliari, Sardinia, Italy.

Woodman, N.D., Rees-White, T., 2009. Barometric Efficiency of Landfill Rep., LPG/10. University of Southampton, Southampton.

Woodman, N.D., Rees-White, T.C., Stringfellow, A., Beaven, R.P., Hudson, A.P., 2014. Investigating the effect of compression on solute transport through degrading municipal solid waste. Waste Management 34 (11), 2196–2208. https://doi.org/10.1016/j.wasman.2014.06.022.

Woodman, N.D., Rees-White, T., Stringfellow, A., Beaven, R.P., Hudson, A.P., 2015. Multiple-tracer tests for contaminant transport process identification in saturated municipal solid waste. Waste Management 38 (Accepted).

Yildiz, E.D., Ünlü, K., Rowe, R.K., 2004. Modelling leachate quality and quantity in municipal solid landfills. Waste Management & Research 22 (2), 78–92.

Yuen, S.T.S., McMahon, T.A., Styles, J.R., 2000. Monitoring in situ moisture content of municipal solid waste landfills. Journal of Environmental Engineering 126 (12), 1088–1095.

Zacharof, A.I., Butler, A.P., 2004. Stochastic modelling of landfill processes incorporating waste heterogeneity and data uncertainty. Waste Management 24, 241–250.

Zardava, K., 2012. Moisture Retention and Near Saturated Flow in Mechanically Biologically Treated (MBT) Waste (Ph.D. thesis). University of Southampton.

Zeiss, C., Major, W., 1993. Moisture flow through municipal waste: patterns and characteristics. Journal of Environmental Systems 22 (3), 211–231.

7 CHAPTER

LINER SYSTEMS

7.1

PHYSICAL LANDFILL BARRIERS: PRINCIPLES AND ENGINEERING

Raffaello Cossu

INTRODUCTION

In a landfill, physical barriers represent the interface between the waste deposit and the external environment (atmosphere and subsoil). The advective and diffusive transfer of mass is controlled by these interfaces. As described in Fig. 7.1.1, two main barriers may be distinguished.

- A top barrier (also known as top cover or final capping) that regulates mass transfer between the atmosphere and the waste deposit by controlling water input (precipitation, leachate recirculation, etc.) and avoiding/minimizing the dispersion of gas emissions.

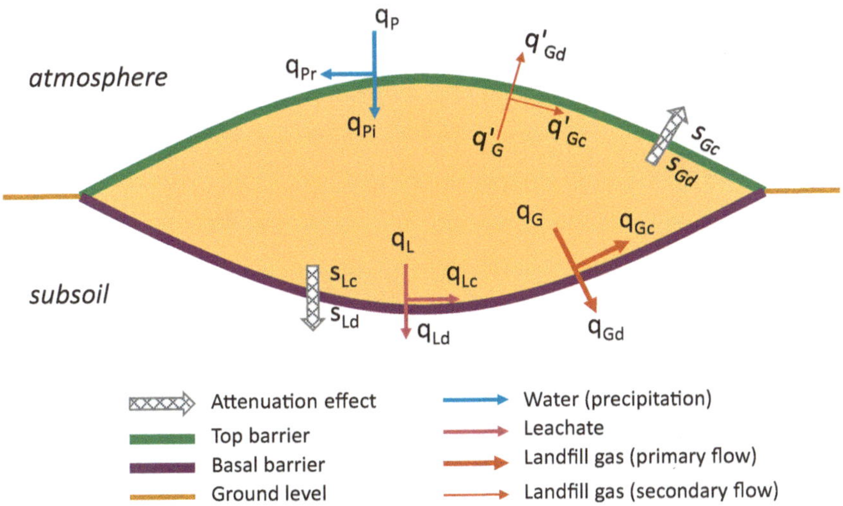

Figure 7.1.1 Scheme of the role of physical barriers in landfill design. s_{Lc}, concentration of contaminants in collected leachate; s_{Ld}, concentration of contaminants in dispersed leachate; q_{Lc}, flow rate of collected leachate; q_{Ld}, flow rate of dispersed leachate; s_{Gc}, concentration of contaminants in collected landfill gas (LFG); s_{Gd}, concentration of contaminants in dispersed LFG; q_{Gc}, flow rate of collected LFG; q_{Gd}, flow rate of dispersed LFG; q_P, precipitation flow rate; q_{Pr}, diverted precipitation water; q_{Pi}, water infiltrating into the waste mass.

- A basal barrier (also known as base barrier) that regulates mass transfer between the waste deposit and the subsoil (groundwater, aquifers, unsaturated soil, etc.) by avoiding/minimizing the dispersion of leachate and migration of gas and maximizing the collection of leachate, with the emissions containment target.

The lateral parts of physical barriers (sides, slopes, etc.) may pose specific technical problems requiring dedicated solutions, but they should be conceived as an integral part of the two main barriers, maintaining a homogeneous and coherent quality standard.

Materials used to construct the barriers are not eternal. In the long-term they will lose their properties and efficacy. For these reasons, the environmental protection of a landfill should not rely solely on physical barriers. In the context of a multibarrier system (see Chapter 2.1), their role is to provide an efficient performance by containing the emission of contaminants for the time required by other barriers to reduce the emission potential of the deposited waste.

PHYSICAL BARRIER COMPONENTS

A physical barrier is generally made up of three components:

- the liner, intended to block or minimize the advective flow of liquid and gas;
- the drainage, promoting the collection and removal of fluids blocked by the liner, thus avoiding any accumulation in the waste mass; and
- the reactive layer, intended to attenuate the load of contaminants associated with leachate and gas.

Each of these components plays a specific role in controlling the potentially uncontrolled emission of contaminants, as schematically described in Fig. 7.1.2.

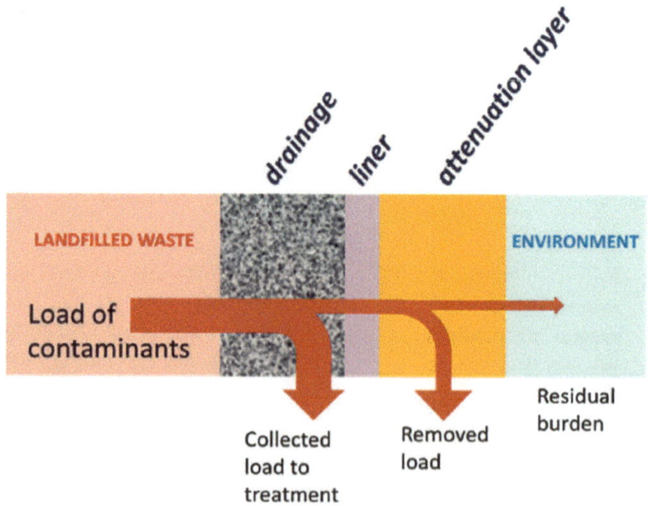

Figure 7.1.2 Scheme of the role of the three main barrier components in controlling landfill emissions (leachate and gas emissions from basal barriers, gas emissions from top barrier).

As far as possible, basal barriers should be impermeable to leachate and biogas, should include systems for leachate drainage and collection (see Chapter 8.1), and should provide—particularly along the slopes—drainage to control biogas migration. Attenuation has the function of reducing the contaminant loads associated with diffusive flow occurring through the liners. In addition, basal barriers should be structured to afford mechanical protection to the different components, mechanical support to the waste mass, and resistance to the water pressure.

Top barriers should be equipped with drainage systems both for gas, which might otherwise escape into the atmosphere, and for infiltrating water (see Chapters 8.2 and 11.1). A lining system or low-permeability layer is applied to control the rate of infiltrating water by optimizing two contrasting needs: by assuring the amount of water required for waste reactions and preventing excessive leachate generation. The reactive layer is aimed at attenuating the residual emissions of biogas, mainly by biological methane oxidation. Additionally, the top barriers should support the functional layers in line with the different afteruse options (soil for vegetation, erosion control, etc.).

In the construction of barrier systems, different kinds of materials, both natural and synthetic, may be used individually or as a component of composite fabrics, in relation to the functions required. In Fig. 7.1.3 the most typical materials used in landfill physical barrier systems are listed.

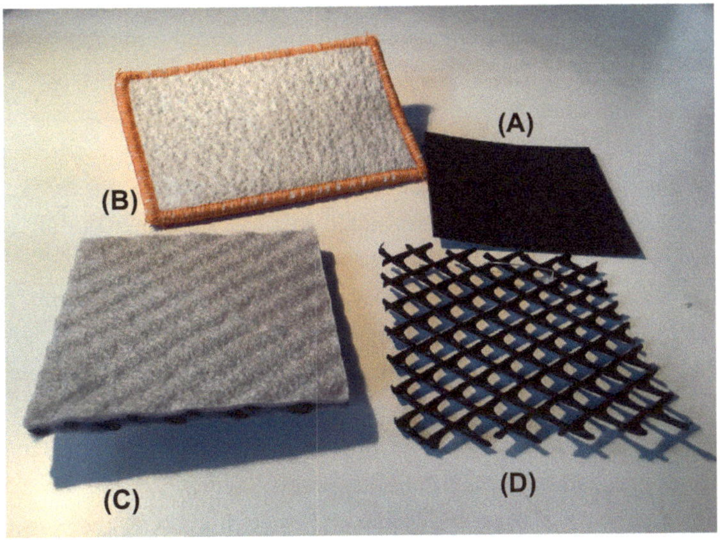

Figure 7.1.3 Picture of geosynthetics commonly used in physical landfill barriers. (A) High-density polyethylene Geomembrane 2.0 mm; (B) Geosynthetic clay liner (a sandwich of two geotextiles with a thin layer of sodium bentonite in between); (C) Geodrain (the one in the photo is a sandwich of two geotextiles with a geonet); and (D) Geonet.

MATERIALS IN PHYSICAL BARRIERS

Natural Materials

Clayey soil is the most common natural lining material. The main factors that affect the quality of clay liners are hydraulic conductivity, degree of compaction, moisture content, clay composition, field placement technique, and liner thickness. Compacted clay or in situ clay deposits should be studied and engineered in an appropriate way to ensure good performance as a liner material in accordance with design parameters. Characteristics, performances, and quality control of clay liners are discussed in Chapter 7.2.

Bentonite is a general term used to indicate clay minerals capable of swelling, when wet, up to 15–18 times their dry volume. Mixtures of bentonite and sandy soil can provide a low-permeability liner, particularly useful in areas where natural clay soil is not available.

Sand is widely used to protect synthetic liners and to increase filter stability.

Gravel is the main material for filtering and draining. The efficiency of a granular bed depends on porosity, grain shape and strength, rock quality (resistance to weathering and carbonate content), stability of the filter, drainage layer thickness, and general engineering of the system (Chapter 8.1).

Synthetic Materials

In recent years, synthetic materials have been used increasingly to enhance soil behavior properties. These materials (usually referred to as geosynthetics) are capable of performing many of the functions required of a barrier system in a landfill unit, such as lining, separation of different materials, drainage, filtration, and reinforcement (Fig. 7.1.3).

The main advantages in using geosynthetics compared to natural materials are represented by their ready availability, small volume consumption, and at times better performance. Other potential advantages such as durability, low cost, and low maintenance are, on the contrary, somewhat questionable. Those categories of geosynthetics suited to application in a landfill are as follows: geomembranes, geotextiles, geonets, geogrids, and geocomposites.

Geomembranes represent a number of different synthetic flexible membranes featuring very low permeability; these can be categorized into bituminous geomembranes and polymeric geomembranes. The former is produced by spraying asphalt or bitumen on to both woven and nonwoven textiles. Polymeric geomembranes are manufactured using several kinds of polymers and elastomers (see Chapter 7.3). The most widely used material in the production of geomembranes for landfill lining, mainly due to its chemical and biological resistance and its capability to be seamed, is High-Density Polyethylene (HDPE).

Geotextiles consist of synthetic fabrics produced either by common weaving (woven geotextiles) or by arrangement in a randomly structured mat (nonwoven geotextiles).

Geogrids are plastics formed in a wide open, regular netlike structure, purpose-developed for the reinforcement of soil.

Geonets are produced by a continuous extrusion of intersecting ribs that form a network structure, suitable for liquid drainage. The most widely used polymer in the production of geogrids and geonets is polyethylene; polyester and polypropylene are also used.

Geocomposites are products obtained by combining two or more different geosynthetics and, at times, natural materials. The most typical configurations of geocomposites are as follows: geotextile + bentonite + geotextile (bentonite geocomposite, commonly known as GCL—Geosynthetic Clay Liner) and geotextile + synthetic drainage mat + geotextile (geocomposite drain).

A detailed description of geosynthetics and discussion about design criteria is provided by Koerner (2012).

Pipes are used in leachate collection systems and in gas collecting wells. They are perforated or, more commonly, slotted. They are usually made of HDPE. Thermoplastic materials have been proven more resistant to chemical attack (Haxo and Haxo, 1994). Perforated Glass Fiber Reinforced Plastics can be conveniently used for large diameters (>60 cm) as required in particular applications such as semiaerobic landfilling (see Chapter 14.2).

FUNCTIONS OF DIFFERENT COMPONENTS IN PHYSICAL BARRIERS

The functions of physical barrier system components are as follows: lining, leachate filtration (vertical flow) and drainage (horizontal flow), reinforcement, mechanical protection, separation, erosion control, water filtration and drainage, biogas migration control, and attenuation of leachate and biogas. In Fig. 7.1.4 these functions are evaluated according to the material adopted, as also discussed herein.

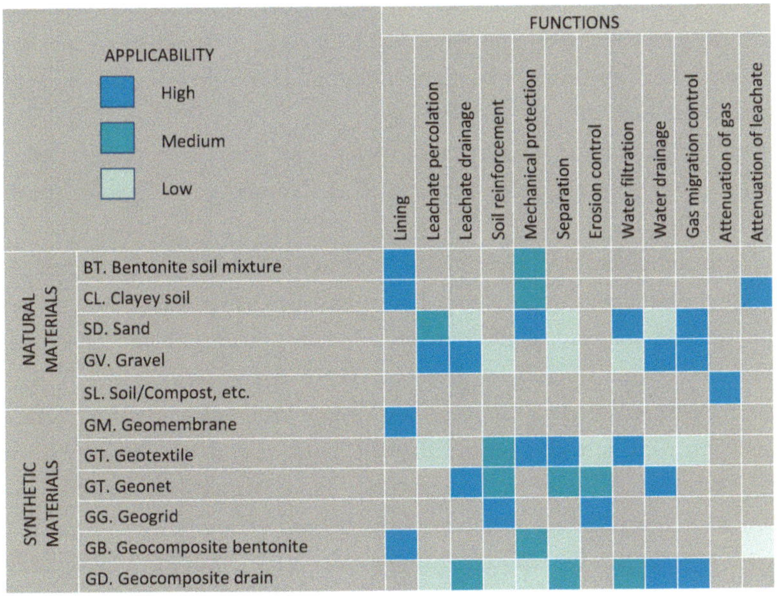

Figure 7.1.4 Level of applicability of different landfill materials with regard to the required function. The indicative evaluation is given on the base of material property, performance, and cost.

Lining is the main function in a physical barrier system and is based on the utilization of low-permeability material including synthetic membranes (geomembranes), natural soil liners (bentonite mixtures, clay), and GCL.

Leachate filtration should be homogeneous, to avoid perched water tables in the waste mass. Coarse granular material (i.e., gravel) is commonly used both in top and basal barriers. Although geotextiles have been proposed as a suitable filtering material, they may clog when in contact with leachate and accordingly their use in this context is frequently prohibited (CTD, 1997; Regione Lombardia, 2014).

Leachate drainage and collection: to avoid accumulation leachate has to be transferred directly, as fast as possible, by drainage and collection systems. For this purpose, natural granular materials are used in combination with drainage pipes. The size of the granular particles and the geometry of the drainage bed should be carefully considered to avoid clogging and damage to the drainage pipes. Geonets and geocomposite drains have been proposed for drainage and collections, although their efficiency is disputed, particularly when used as primary drainage system. Design criteria for leachate drainage and collection systems are reported in Chapter 8.1.

Soil reinforcement is fundamental in consolidating the bottom and slope of the landfill to ensure a stable foundation for the lining system. Geogrids can be conveniently used as materials for mechanical support of soil, geotextiles, geonets, and geomembranes (Koerner, 2012).

Mechanical protection: in particular, synthetic liners such as geomembranes should be mechanically protected from puncturing by sharp objects contained in the waste, by the pressure of granular draining material, and by the action of vehicles and mobile equipment in the landfill. Mechanical protection of synthetic liners can be achieved by means of layers of sand or small-sized rounded gravel. These materials can be coupled or substituted by different kind of geosynthetics such as geotextiles, geocomposites, geonets, etc.

Separation: when combining granular material of different sizes, a separation layer should be provided to avoid penetration of one material into the other. Nonwoven geotextiles are the most suitable materials for this function.

Erosion control: the completed final layer in top barriers may need to be protected against water and wind erosion. A good hydraulic design of top covers and use of special geonet mats (particularly on slopes) may afford a good solution to this problem. Geonets and mats are often shaped or coupled with other products to provide support for grass and vegetation growth.

Water filtration may be required to intercept and divert water (groundwater, surface water, precipitation water) from intrusion into the landfill through bottom and top barriers. In this case, natural granular material, woven and nonwoven geotextiles, and geocomposite drains can be successfully used. The utilization of geosynthetics may contribute toward saving space.

Water drainage and collection: the infiltrated water should be removed to avoid infiltration into the landfill, water pressure on the lining system, and water buildup. For this purpose, granular material, in combination with pipes and/or geonets may be used.

Biogas migration and emission control: landfill biogas is normally extracted and collected by means of specific wells drilled into the landfill. However, to control lateral biogas migration and surface emissions, drainage layers should be installed. These layers could be connected to the forced gas extraction and transport system, whenever present and active.

Attenuation of biogas: as described in Chapter 9.2, methane and several trace volatile organic contaminants (VOCs) are present in landfill gas and may harm the environment by contributing to the depletion of the ozone layer and to global warming. In addition, some trace elements may display significant toxicity. Residual gas in old landfills, or uncontrolled gas emissions not intercepted by existing gas extraction and transport systems, could be conveniently attenuated by active layers installed within the top cover barrier systems (Chapter 11.1).

Attenuation of leachate: attenuation by clay soils of organic contaminants, ammonium, and heavy metals present in leachate occurs within mineral liners to an extent depending on the quality of leachate and individual characteristics of the soil. The attenuation processes involved include adsorption, precipitation, ion exchange, and biodegradation (i.a. Griffin et al., 1977; Pivato and Raga, 2006: Beaven et al., 2009; Liu and Hu, 2014).

Other materials adopted to fulfill additional functions specific to top barriers will be presented and discussed in detail in Chapter 11.1.

DESIGN OF PHYSICAL BARRIERS

A landfill barrier system may be designed in line with a series of different safety levels, by using the materials presented earlier and taking into account the quality of waste and different possible technical and environmental situations.

The possible combinations and sequence of materials and products vary for the two main barriers and are dependent on the different functions requested.

Basal Barriers

Sequencing of Barrier Components

With regard to basal barriers, the sequence of potential components of the system is schematically presented in Fig. 7.1.5, in which different configurations and alternatives for the materials are reported.

Lining and attenuation functions have been represented separately to better assess the sequencing of the different components. In the case of mineral liners, these functions are performed by the same component.

Some barrier configurations proposed for use as physical landfill barriers are illustrated in Fig. 7.1.6 and discussed herein. Separation or protective layers are not represented for the sake of graphical simplification.

A. *Single mineral liner*: This configuration is the simplest and is still the most widely applied in poorly engineered landfills. It could be deemed acceptable under specific safe hydrological conditions. A minimum 1 m thickness and $k \leq 10^{-9}$ m/s should be recommended. Alternatively (A′) a GCL could be used.

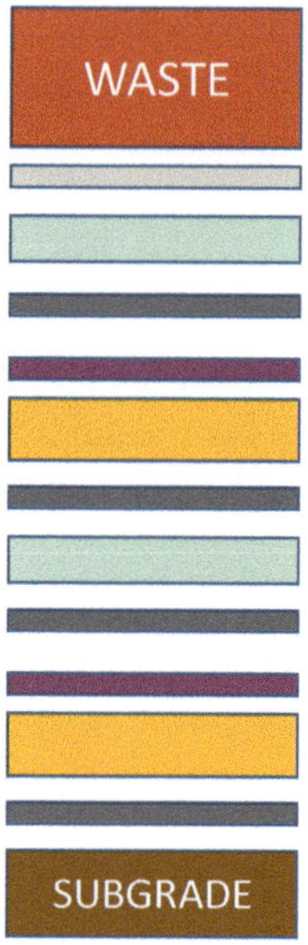

	A	A'	B	C	C'	C''	D	E	F	G	H	H'
Filter												
Drainage	GV	GV	GV	GV	GV	GV	GV	GV	GV	GV	GV	GV
Prot./sep.	GT	GT	GT	GT	GT	GT	GT	GT	GT	GT	GT	GT
Synthetic liner			GM	GM	GM	GM	GM		GM	GM	GM	GM
Mineral layer	CL	BG		CL	BG	BG	CL	CL			CL	BG
Protection				GT	GT			GT		GT	GT	
Drainage								GV	GD		GV	GD
Prot./sep.								GT			GT	
Synthetic liner							GM		GM	GM	GM	GM
Mineral layer						CL		CL		CL	CL	CL
Reinforcement												

Figure 7.1.5 Different sequences of potential components of a landfill barrier system. The various sequences are represented graphically in Figure 7.1.6. *CL*, Clay liner; *BG*, Bentonite geocomposite (GCL); *GD*, Geodrainage; *GM*, Geomembrane; *GT*, Geotextile; *GV*, Gravel.

B. *Single geomembrane liner*: This configuration may only be deemed acceptable for use in top barriers under certain circumstances (see Chapter 11.1), whereas in basal barriers it fails to provide any backup in the case of geomembrane failure.

C. *Single composite liner* (clayey soil + geomembrane): this configuration is the most widely used for basal barriers and is prescribed by several national guidelines (see Table 7.1.1). In general, a minimum clay liner thickness of 1.0 m is required with a maximum permeability of 10^{-9} m/s. Geomembranes should be installed in direct contact with the mineral liner without any kind of interposed draining material (geonets, gravel, sand, etc.). This would allow, in the case of geomembrane failure, the self-healing effect of the clayey soil to kick in. The use of GCL in basal barriers as an alternative to mineral liners is questionable for the following reasons:
- although the two materials could display similar hydraulic equivalence, the attenuation performance of GCL may differ markedly, even if it is not negligible (Rowe, 2014);
- several national regulations prescribe, without exception for the slopes, a minimum thickness of a magnitude order higher than that applied for GCL.

D. *Single sandwich composite liner* (geomembrane + clayey soil + geomembrane): this configuration increases the safety level of the single composite liner and may prevent desiccation and cracking of the mineral liner. With regard to emplacement techniques applied and stability behavior on the slopes, the installation of this system may present critical issues.

E. *Double liner of natural material*: this configuration is rarely used. The installation of separation materials (such as geotextiles) between the mineral layers and drainage layers is mandatory.

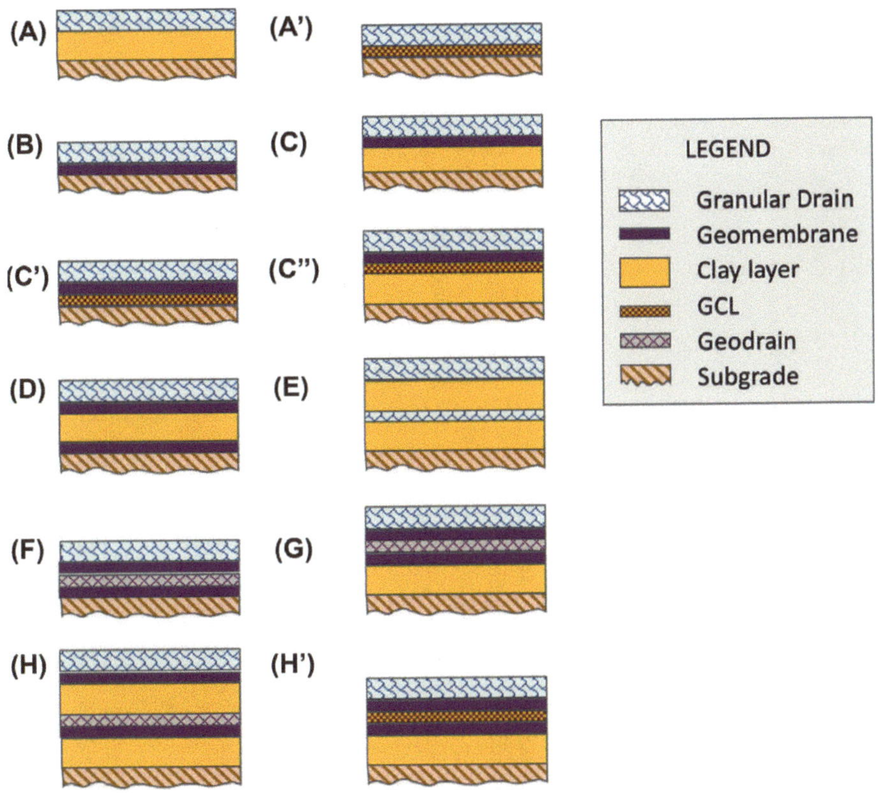

Figure 7.1.6 Possible combinations of liners in a physical barrier system. (A) single clay liner, (A') single geosynthetic clay liner (B) single geomembrane liner, (C) single composite liner, (C') single geosynthetic clay liner (C''') single geosynthetic clay liner with additional clay liner (D) single sandwich composite liner, (E) double clay liner, (F) double geomembrane liner, (G) double semicomposite liner, (H) double composite liner, (H') double layer containing geosynthetic clay liner and composite liner. Protection and separation layers are not represented. *CL*, clay layer; *GCL*, geosynthetic clay liner. *Modified from Cossu (1994).*

F. *Double liner of synthetic material*: the double synthetic liner has been proposed in the past for the lining of slopes in the presence of basal barriers. Drainage with synthetic materials should be adopted in this case. The efficacy of this configuration is severely influenced by the quality of the installation and long-term behavior of the geomembranes.

G. *Double semicomposite liner* (double synthetic top liner + single natural liner): this structure tends to combine the advantages of the single composite liner and the double synthetic liner, thus providing a further safety level. Global efficiency still depends on the long-term behavior of the geomembrane.

H. *Double composite liner*: this configuration fully exploits the possibility of control drainage for the double lining system in basal barriers. It is, of course, the most expensive solution and is

Table 7.1.1 Legal requirements for basal barriers in nonhazardous municipal solid waste (MSW) landfills in Italy, Germany, United Kingdom, United States, Japan, and China

	Italy	Germany	United Kingdom	China	Japan	United States
	Nonhazardous Waste Landfills	Landfills Class I and II[1]	Nonhazardous Waste Landfills	Nonhazardous Waste Landfills	Nonhazardous Waste Landfills	Nonhazardous Waste Landfills
Leachate drainage layer	$\geq$0.5 m	$\geq$0.5 m	$\geq$0.5 m	$\geq$0.3 m	$\geq$0.5 m	Mandatory (designed and constructed to maintain leachate level over the liner $<$ 30 cm)
Additional impermeable layer	Mandatory (geomembrane—intimate contact with compacted mineral layer)	Mandatory (geomembrane)	Mandatory (geomembrane)	Mandatory (geomembrane or geosynthetic clay liner)	Mandatory (geomembrane—intimate contact with compacted mineral layer)	Minimum 30 mil ($\approx$0.8 mm) flexible membrane liner or 60 mil ($\approx$1.5 mm) thick high-density polyethylene
Low-permeability layer	Compacted mineral layer $\geq$ 1 m, k $\leq 10^{-9}$ m/s	Compacted mineral layer $\geq$ 1 m, k $\leq 10^{-9}$ m/s	Compacted mineral layer $\geq$ 1 m,[2] k $\leq 10^{-9}$ m/s	Compacted mineral layer $\geq$ 0.75 m, k $\leq 10^{-9}$ m/s	Compacted mineral layer $\geq$ 0.5 m, k $\leq 10^{-9}$ m/s	Compacted soil $\geq$ 2 foot ($\approx$60 cm), k $\leq 10^{-9}$ m/s
Reference	D. Lgs. 36/2003. Attuazione della direttiva 1999/31/CE relativa alle discariche di rifiuti	Deponieverordnung. DepV/2009	The Landfill (England and Wales) Regulations, 1559/2002	Standards for pollution control on the landfill site of MSW (GB 16889-2008), Technical code for MSW sanitary landfill (GB 50869-2013)	Ministry of Environment No. 12, 2017	Code of Federal Regulations (Title 40)

[1]Landfills of Class I and II are landfills for nonhazardous waste (e.g., MSW). Class II landfills contain a higher amount of biodegradable waste compared with Class I.
[2]The thickness can be reduced to a minimum of 0.5 m if the permeability of the barrier is decreased (Fowmes et al., 2007).

generally adopted for hazardous waste landfills. Applications to municipal solid waste landfills are on an increase with the aim of minimizing any adverse environmental impact, particularly when the natural geological barrier is poor or lacking. The control drainage layer between the two composite liners is not always adopted. The advantages of increasing the safety degree of the barrier system (leachate leakage detection, drainage and collection of leaked leachate, reduction of pressure head on the secondary liner with consequent reduction of leaching risk) are counterbalanced by the higher construction complexity. The construction could be simplified by substituting the primary mineral liner with a GCL.

Mineral Liners

The characteristics of the hydraulic conductivity of mineral liners depend prevalently on numerous factors such as construction, methodology, leachate head buildup, material quality, transport of contaminants, etc. (see Chapter 7.2). The emplaced clay liner should be carefully engineered and controlled both in laboratory tests and in the field to provide a homogeneous performance. This aspect is often critical as national regulations frequently fail to provide precise indications on how to measure permeability parameters.

During the construction phase, particular care should be devoted to compaction and moisture control. Natural existing clay strata should be considered as a raw material, scarified, remolded, and processed to guarantee a homogeneous and controlled compliance with hydraulic requirements. Further details on construction and monitoring of mineral liners are presented in Chapter 7.2.

Synthetic Liners

The most widely used material in the manufacturing of synthetic liners is HDPE geomembrane with a thickness of 2.0–2.5 mm. HDPE geomembranes are supplied as sheet rolls and are seamed on-site. Both material and seams should be checked using a series of different tests to verify original properties and design performance (Chapter 7.3).

Mechanical support of synthetic liners should be adequate to avoid localized subsidence and/or differential settlement, capable of determining failure of the liner or significantly change the slope (see Chapter 5.2). A typical critical situation to this regard is observed when a lining system is established over an existing waste deposit; in this case, foundation reinforcement will be required. The supporting soil for the synthetic liner should be shaped to fit the design slope, well compacted and rolled. The side slope should be inferior to 85% the frictional angle between the geomembrane and the supporting soil (Jorgensen, 1987).

On installation of the geomembrane, wrinkling should be carefully avoided as this has been indicated as one of the reasons for failure (Rollin et al., 1994). Seaming should be carried out by welding of the geomembrane sheet along the direction of the leachate flux. Furthermore, seaming should be parallel to the line of the maximum slope along the side (CTD, 1997). After emplacement, the lining should be controlled according to a well-defined Quality Control Procedure (Cossu and Muntoni, 1994).

Following installation and control, geomembranes should be mechanically protected from the adjacent drainage materials (gravel, stones, etc.). To this purpose, sandy soil with a thickness of 20–30 cm may be used. As an alternative, geotextiles or geocomposites are commonly interposed between the granular and synthetic materials (Kirshner and Witte, 1994). Geotextile cushioning performance is largely dependent on operative conditions (subgrade, pressure, shape of drainage materials, etc.) as well as mass, thickness, and typology of geotextile (material, manufacturing technique, etc.). Performance tests should be employed to assess the suitability of a geotextile to provide adequate protection (Hornsey and Wishaw, 2012; Austin et al., 2014).

With regard to the leachate drainage pipeline, it is important that underlying special protective measures (GCL, geotextile, sand, soil, etc.) should be implemented. Likewise, roads used by operating vehicles should also be provided with special protective measures (Cossu, 1994).

Drainage and Collection

Design criteria for an effective leachate drainage system are presented in detail in Chapter 6.1.

In many cases, the use of drainage wells in the vertical drainage between different layers has been proposed. It is important to have flexible structures that can adapt themselves to settling and horizontal landfill movement and can resist potential damage by operating machines. These structures can be realized by using a fixed cylinder (0.8–1 m diameter) made of metallic net, filled with large size gravel, or by gradually lifting a metallic cylinder. The vertical drainage well can also be combined with biogas extraction network.

Leachate drained from the bottom of the landfill should be collected continuously to avoid leachate hydraulic head over the barrier system. Maintenance shaft and pumping stations should respect the following indications (Cossu, 1994; Lechner, 1994):

- installation should be outside the waste deposit area;
- dimensions should allow for satisfactory maintenance and working activities;
- proper safety measures and methane-detecting systems should be installed inside;
- materials should be impermeable and resistant to leachate corrosion (premanufactured polyethylene and fiberglass structures are often used);
- leachate pipes must be siphoned on arrival in manholes and water guards should be present to avoid air infiltration into the leachate drainage system and biogas presence inside the manhole. This would reduce drainage clogging and explosion risks in manholes, inspection shafts, and pumping stations.

To avoid accumulation, leachate should not be pumped, but rather collected and conveyed only by gravity. This implies therefore that landfills should be constructed on a slope or above ground level (Cossu, 1994).

Top Barriers

Top barrier configurations are strictly dependent on the following:

- typology of the top barrier in relation to the landfill life phase (daily cover, intermediate cover, temporary cover, final capping);

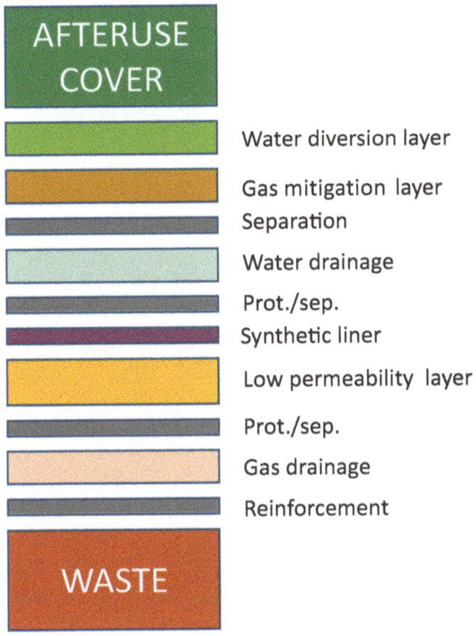

Figure 7.1.7 Sequencing of different possible components in a landfill top cover barrier.

- quality of landfilled waste;
- local situations (climatology, morphology, etc.);
- conceptual model adopted in landfill design (pretreatment, aerobic or anaerobic conditions, etc.).

A general sequencing of components for a final top barrier (final capping) is given in Fig. 7.1.7. Detailed sequencing and selection of materials, design criteria, and expected performance specific to top barriers are presented and discussed in Chapter 11.1.

LEGAL REQUIREMENTS FOR BASAL BARRIERS

All national regulations in industrialized countries prescribe minimal requirements for the construction of basal barriers. Legal requirements for nonhazardous waste in Italy, Germany, United Kingdom, and United States are reported in Table 7.1.1. In some cases, the characteristics of the layer are specified in detail (e.g., thickness and permeability value of the low-permeability layer), whereas in others the layers are mandatory but design details are not specified, leaving the designers free to implement alternative solutions.

Legal prescriptions in Italy, Germany, and United Kingdom are largely similar as national legislations are transpositions of the same European directive (Directive 1999/31/EC, also known as the "Landfill Directive"). However, it is interesting to notice that prescriptions in the United States are very similar, although implemented in a country with different characteristics (in terms of waste production and characterization) compared with Europe.

LONG-TERM PERFORMANCE AND FAILURE MECHANISMS

A physical barrier system with lining materials performing perfect containment, permeable materials performing perfect drainage, and reactive materials performing perfect emissions attenuation throughout the landfill life span simply does not exist.

The reasons for physical barrier failures, according to the barrier definition proposed by Hollnagel (1999) may be related to the following:

- Material problems (inadequate material properties for the given application, changes over time of material properties, etc.)
- Functional problems (settling, sliding, severe climatic conditions, wrong design, human mistakes during landfill operation, etc.)
- Symbol problems (poor instructions, procedures, or precautions; poor indications and communications; poor permitting, etc.)
- Immaterial problems (poor monitoring and supervision, wrong prescription by regulations, guidelines, authorities' prescriptions, etc.)

With regard to the first two problems, the main failure mechanisms manifested in relation to the individual components of a physical barrier system, both in the short- and long term, are summarized in Fig. 7.1.8, indicating the most common damage and causes of failure.

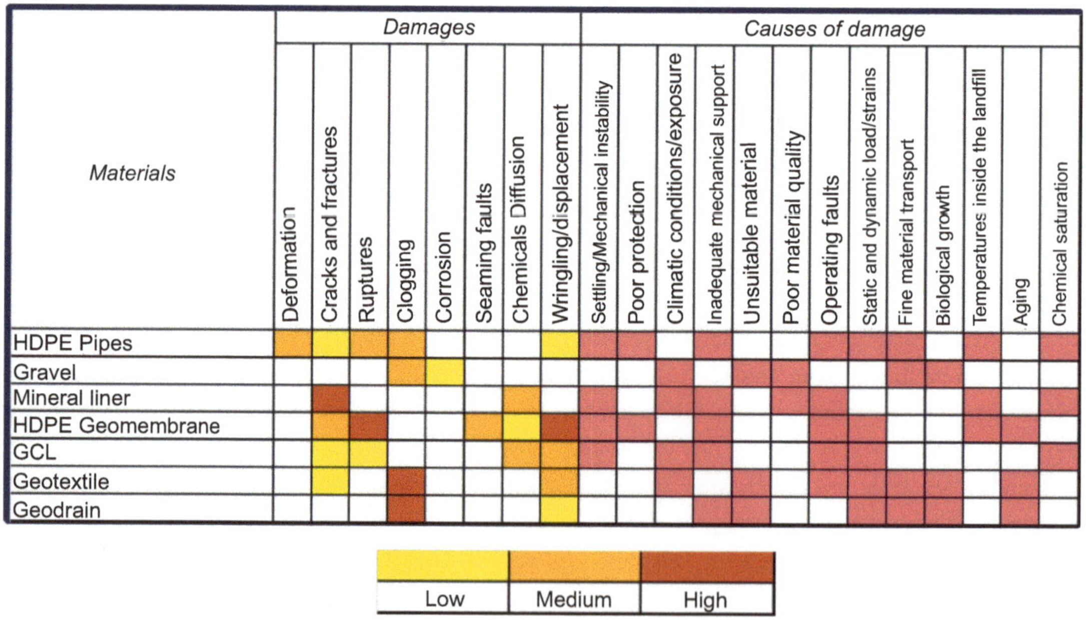

Figure 7.1.8 Potential damage and related causes for different types of materials and products indicated for use in physical landfill barriers. *GCL*, geosynthetic clay liner; *HDPE*, high-density polyethylene.

Compacted clay liners may fail to function satisfactorily as cracking may occur due to climatic conditions, temperature differences, deformation due to differential settlement, waste instability (see Chapter 5.2), poor compaction, uncontrolled groundwater infiltration, and instability of the subgrade (Pivato, 2011; Farquhar, 1994). Several studies have shown that certain substances may produce a serious effect on the permeability of clays (Quigley and Fernandez, 1994; Adar and Bilgili, 2015). Similar problems are encountered with soil—bentonite mixtures. Strong ionic concentrations may result in variations in the clay structure, influencing the swelling capacity and consequently permeability (Cancelli et al., 1994).

The main damage that could potentially occur to geomembranes are related to different kinds of ruptures (cracks, holes, brittle fractures, seam faults, etc.). The causes of the latter may be associated with poor installation, inadequate Quality Control, tensile strains, securing faults, subgrade instability, poor protection, chemical attack, and aging (Duquennoi et al., 1993; Surmann et al., 1994; Colucci and Lavagnolo, 1995; Rollin et al., 1999; Ewais et al., 2014). In addition, several VOCs are able to diffuse through the geomembrane (i.a. Rowe, 2014).

In the long term, geomembranes will inevitably degrade (Needham et al., 2006; Fowmes et al., 2007). Degradation of geomembranes due to aging is dependent on a series of factors, including depletion of antioxidants, chemical reactions, plasticizer migration, and exposure conditions, with marked differences manifested between buried and exposed geomembranes (i.a. Haxo and Haxo, 1994; Koerner et al., 2011, Chapter 7.3). In a buried HDPE 2.0 mm geomembrane exposed to waste and leachate for 7 years, "aging was detected from an increase in the yield strength, a decrease in the tensile resistance at rupture, and lowering of the elongation at break" (Rollin et al., 1994).

Geosynthetics other than geomembranes may feature similar problems to those discussed above, particularly when exposed to the atmosphere. Indeed, ultraviolet light, oxidation, and hydrolysis represent the main degradation factors.

Natural and synthetic filtering and drainage materials, particularly when subjected to biological activity, may become clogged and form incrustations (Brune et al., 1994; Cazzuffi et al., 1994). The consequent decrease displayed in drainage efficacy may result in an increased leachate head, constituting the main driving force for leakage in a basal barrier system.

CONCLUSIONS

The service life of components in a barrier system depends on the quality of materials used, sequencing of the different barrier elements, quality of the waste, quality of installation and landfill operation, and environmental conditions throughout the various landfill phases (construction, operation, postclosure).

When designing a physical barrier system, it is fundamental that all potential causes of failure of the individual barrier components should be taken into account, bearing in mind that a certain percentage of material damage, soon or later, will inevitably occur. For this reason a combined system should always be adopted to face the failure of an individual barrier component (Wallace and Akgun, 1994).

The design of physical barriers should be aimed prevalently at maintaining a containment and attenuation functions for the time required to allow the other barriers in a multibarrier system to reduce the emission potential of the waste to a sustainable level (Chapters 2.1 and 2.2, this book).

References

Adar, E., Bilgili, M.S., 2015. The Performance of four different mineral liners on the transportation of chlorinated phenolic compounds to groundwater in landfills. The Scientific World Journal. https://doi.org/10.1155/2015/171284.

Austin, R.A., Gibbs, D.T., Kendall, P.K., 2014. Geomembrane protection using cushioning geotextiles. In: Proceedings 7th International Conference on Environmental Geosynthetics, Melbourne. Accessed from: researchgate.net.

Beaven, R., Potter, H., Powrie, W., Simoes, A., Smallman, D., Stringfellow, A., 2009. Attenuation of Organic Contaminants in Leachate by Mineral Landfill Liners. Environment Agency, Science report: SC020039/SR5 Project record—ISBN: 978-1-84911-067-9.

Brune, M., Ramke, H.G., Collins, H.J., Hanert, H.H., 1994. Incrustation problems in landfill drainage systems. In: Christensen, T.H., Cossu, R., Stegmann, R. (Eds.), Landfilling of Waste: Barriers. E & FN Spon, London, pp. 569–605. ISBN 0-419-15990-8.

Cancelli, A., Cossu, R., Malpei, F., Offredi, A., 1994. Effects of leachate on the permeability of sand-bentonite mixtures. In: Christensen, T.H., Cossu, R., Stegmann, R. (Eds.), Effect of Organic Liquids on the Hydraulic Conductivity of Natural Clays, in Landfilling of Waste: Barriers. E & FN Spon, London, pp. 259–293. ISBN 0-419-15990-8.

Cazzuffi, D., Cossu, R., Lavagnolo, M.C., 1994. Efficiency of geotextiles and geocomposites in landfill drainage systems. In: Christensen, T.H., Cossu, R., Stegmann, R. (Eds.), Landfilling of Waste: Barriers. E & FN Spon, London, pp. 431–443. ISBN 0-419-15990-8.

Colucci, P., Lavagnolo, M.C., 1995. Three years field experience in electrical Control of synthetic landfill liners. In: Christensen, T.H., Cossu, R., Stegmann, R. (Eds.), Proceedings Sardinia 2001, 5th International Waste Management and Landfill Symposium, vol. 2. CISA, Environmental Sanitary Engineering, Cagliari, pp. 437–452.

Cossu, R., 1994. Principles of landfill barrier system. In: Christensen, T.H., Cossu, R., Stegmann, R. (Eds.), Landfilling of Waste: Barriers. E & FN Spon, London, pp. 3–23. ISBN 0-419-15990-8.

Cossu, R., Muntoni, A., 1994. Quality control assurance for lining and leachate collection systems. In: Christensen, T.H., Cossu, R., Stegmann, R. (Eds.), Landfilling of Waste: Barriers. E & FN Spon, London, pp. 25–34. ISBN 0-419-15990-8.

CTD—Comitato Tecnico Discariche, 1997. Linee Guida per le discariche controllate dei rifiuti solidi. CISA Publisher, Cagliari (in Italian).

Duquennoi, C., Bernhard, C., Gaumet, S., 1993. Laboratory ageing of geomembranes in landfill leachates. In: Christensen, T.H., Cossu, R., Stegmann, R. (Eds.), Proceedings Sardinia 95, Fourth International Waste Management and Landfill Symposium, vol. 2. CISA Publisher, pp. 397–404.

Ewais, A.M.R., Rowe, R.K., Brachman, R.W.I., Arnepalli, D.N., 2014. Service life of a high-density polyethylene geomembrane under simulated landfill conditions at 85°C. Journal of Geotechnical and Geoenvironmental Engineering 140, 11.

Farquhar, G.J., 1994. Experiences with Liners using natural materials. In: Christensen, T.H., Cossu, R., Stegmann, R. (Eds.), Landfilling of Waste: Barriers. E & FN Spon, London, pp. 37–53. ISBN 0-419-15990-8.

Fowmes, G.J., Dixon, N., Jones, D.R.V., 2007. Landfill stability and integrity: the UK design approach. Proceedings of the Institution of Civil Engineers Waste and Resources Management 160 (2), 51–61.

Griffin, R.A., Au, A.K., Frost, R.R., 1977. Effect of pH on adsorption of chromium from landfill-leachate by clay minerals. Journal of Environmental Science and Health. Part A: Environmental Science and Engineering 12 (8), 431–449.

Haxo, H.E., Haxo, P.D., 1994. Basic composition and properties of synthetic material in lining systems. In: Christensen, T.H., Cossu, R., Stegmann, R. (Eds.), Landfilling of Waste: Barriers. E & FN Spon, London, pp. 317–343. ISBN 0-419-15990-8.

Hollnagel, E., 1999. Accidents and barriers. In: Hoc, J.-M., Millot, P., Hollnagel, E., Cacciabue, P.C. (Eds.), Proceedings of Lez Valenciennes, 28. Presses Universitaires de Valenciennes, pp. 175–182.

Hornsey, W.P., Wishaw, D.M., 2012. Development of a methodology for the evaluation of geomembrane strain and relative performance of cushion geotextiles. Geotextiles and Geomembranes 35, 87–99.

Jorgensen, J.B., 1987. Danish code of practice for lining. In: Proceedings Sardinia 87, First International Landfill Symposium. CISA Publisher, pp. 1–15.

Kirshner, R., Witte, R., 1994. Use of geotextile for the protection of geomembranes. In: Christensen, T.H., Cossu, R., Stegmann, R. (Eds.), Landfilling of Waste: Barriers. E & FN Spon, London, pp. 475–486. ISBN 0-419-15990-8.

Koerner, R.M., 2012. Designing with Geosynthetics, sixth ed., vol. I. Xlibris. ISBN 978-1-462882885.

Koerner, R.M., Hsuan, Y.G., Koerner, G.R., 2011. Geomembrane Lifetime Prediction. Unexposed and Exposed Conditions. GRI-White Paper #6, Geosynthetic Institute, USA.

Lechner, P., 1994. Design criteria for leachate drainage and collection systems. In: Christensen, T.H., Cossu, R., Stegmann, R. (Eds.), Landfilling of Waste: Barriers. E & FN Spon, London, pp. 519–530. ISBN 0-419-15990-8.

Liu, T., Hu, L., 2014. Organic acid transport through a partially saturated liner system beneath a landfill. Geotextiles and Geomembranes 42, 428–436.

Needham, A.D., Smith, J.W.N., Gallagher, E.M.G., 2006. The service life of polyethylene geomembrane barriers. Engineering Geology 85 (1–2), 82–90.

Pivato, A., 2011. Landfill liner failure: an open question for landfill risk analysis. Journal of Environmental Protection 2, 287–297.

Pivato, A., Raga, R., 2006. Tests for the evaluation of ammonium attenuation in MSW landfill leachate by adsorption into bentonite in a landfill liner. Waste Management 26 (2), 123–132.

Quigley, R.M., Fernandez, F., 1994. Effect of organic liquids on the hydraulic conductivity of natural clays. In: Christensen, T.H., Cossu, R., Stegmann, R. (Eds.), Landfilling of Waste: Barriers. E & FN Spon, London, pp. 203–218. ISBN 0-419-15990-8.

Regione Lombardia, October 10, 2014. Guidelines for design and sustainable management of landfilling ((Linee guida per la progettazione e gestione sostenibile delle discariche)). Bollettino Ufficiale (41).

Rollin, A.L., Mlynarek, J., Lafleur, J., Zanescu, A., 1994. Performance changes in aged in-situ HDPE geomembrane. In: Christensen, T.H., Cossu, R., Stegmann, R. (Eds.), Landfilling of Waste: Barriers. E & FN Spon, London, pp. 431–443. ISBN 0-419-15990-8.

Rollin, A.L., Marcotte, M., Jacqueline, T., Chaput, L., 1999. Leak location in exposed geomembrane liners using an electrical leak detection techniques. In: Proceedings Geosynthetic 99. Industrial Fabrics Association International, Minneapolis, pp. 27–102.

Rowe, R.K., 2014. Long-term performance of landfill liners. In: Cossu, R., van der Sloot, H. (Eds.), Sustainable Landfilling. CISA Publisher, pp. 287–333. ISBN 978-88-6265-005-2.

Surmann, R., Pierson, P., Cottour, P., 1994. Geomembrane liner performance and long te rm durability. In: Christensen, T.H., Cossu, R., Stegmann, R. (Eds.), Proceedings Sardinia 95, Fourth International Waste Management and Landfill Symposium, vol. 2. CISA Publisher, pp. 405–414.

Wallace, R.B., Akgun, H., 1994. Performance changes in aged in-situ HDPE geomembrane. In: Christensen, T.H., Cossu, R., Stegmann, R. (Eds.), Landfilling of Waste: Barriers. E & FN Spon, London, pp. 81–89. ISBN 0-419-15990-8.

7.2

MINERAL LINERS

Marco Favaretti and Raffaello Cossu

INTRODUCTION

Mineral liners in landfills are a fundamental component of the physical barrier systems (Chapter 7.1). They should be purpose-designed and constructed to withstand the effect of clayey soils, compaction procedures, degree of densification, water content, and hydraulic conductivity.

This chapter describes the property of mineral liners with regard to landfill applications and presents a series of fundamental experimental methods to be performed in the laboratory and the field to ensure the correct design of mineral liners and to verify efficiency following emplacement.

Performance of clay liner in attenuating dispersed leachate emissions is finally discussed.

GEOTECHNICAL PROPERTIES OF THE MINERAL LINERS

Mineral landfill liners are in general subject to regulations imposing a geometric limit, establishing a minimum value for thickness, and a geotechnical limit setting the maximum value for hydraulic conductivity k (see Chapter 7.1).

The first limit is represented by a simple measure of length and therefore in situ control does not constitute a problem. The difficulty in determining coefficient k is associated with the precision, uncertainty and repeatability of its experimental measure, and contrasts with the prescriptive character of the regulation, fixing a specific k_{max} that cannot be overcome. Even a slight deviation from the k_{max} implies, from a purely legal point of view, nonacceptability of the used soil or the barrier. It is easy to grasp the paradox of a standard that defines with absolute precision k_{max}, without illustrating the experimental methods of determination, with possible and consequential technical and legal issues.

The mineral liner should be constructed using medium–low plasticity clayey soils (Plasticity Index, PI = 10% ÷ 40%), with a percentage of passing through ASTM 200 sieve (opening 0.075 mm) greater than 20%. The minimum value of the PI is due to the fact that below 10% it would be highly unlikely to achieve the required low hydraulic conductivity, whereas the maximum value of PI is related to the need to have a workable and compactable soil, with limited shrinkage and swelling properties. Any clods of clayey soil must not exceed 25 ÷ 50 mm. Finally, a greater than 10% presence of gravel granulometric fractions should be avoided.

The following topics will be treated in sequence:

- compaction procedures in the laboratory: interactions between water content w, saturation degree S_r, hydraulic conductivity k, and dry density ρ_d;
- main experimental procedures for the determination of hydraulic conductivity in the laboratory and in situ;
- in situ control of hydraulic conductivity and density of the mineral liners;
- short notes on transport of contaminants through mineral liners and on compatibility between clay liners and leachate.

LABORATORY COMPACTION TESTS ON CLAY SOILS

Compaction relates to the mechanical stabilization or densification of a soil, aimed at increasing its density and shear strength or decreasing hydraulic conductivity. Compaction may also be applied to influence the behavior of a clayey soil with regard to shrinkage and swelling. The properties of a compacted soil are usually studied in the laboratory by carrying out a standard (ASTM D698) or modified (ASTM D1557) version of the so-called Proctor tests. A soil sample, prepared at a specific initial water content, is inserted in successive layers inside a cylindrical steel mold and a steel hammer vertically dropped several times onto the soil (Fig. 7.2.1).

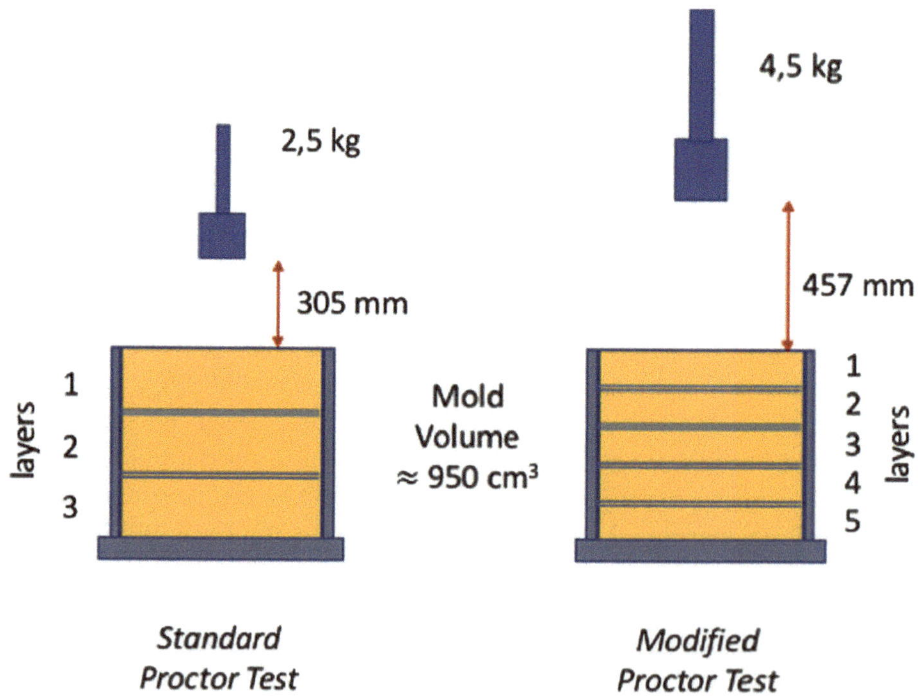

Figure 7.2.1 Schematic representation of Standard and Modified Proctor test apparatus.

Table 7.2.1 Main properties of the proctor tests apparatus

Compaction Test	Compaction Mold Size		Hammer Size		Number of Layers	Number of Drops per Layer	Height of Drop (mm)	Compaction Energy (kJ/m^3)
	ϕ (mm)	H (mm)	ϕ (mm)	Weight (N)				
Proctor standard	101.6	117	50.8	24.5	3	25	305	593
Proctor modified	101.6	117	50.8	44.5	5	25	457	2693

The size of the hammer, its height of drop, the number of drops for each soil layer, and the number of layers are standardized and reported in Table 7.2.1 according to the different Proctor procedures. The experimental results obtained are generally plotted on a "water content w versus dry density ρ_d" plane, which has a downward concavity and a maximum coordinate point defined through "optimum water content w_{opt}" and "maximum dry density $\rho_{d,max}$" (Fig. 7.2.2).

The curve w−ρ_d is uniquely defined for a specific compaction energy, as well as ideal water content w_{opt}, corresponding to the maximum degree of densification $\rho_{d,max}$ of the specimen. Dry density ρ_d is equal to the ratio between the mass of the soil particles and constant volume of the Proctor mold. An increase in ρ_d signifies that a greater amount of solid particles is contained within the mold.

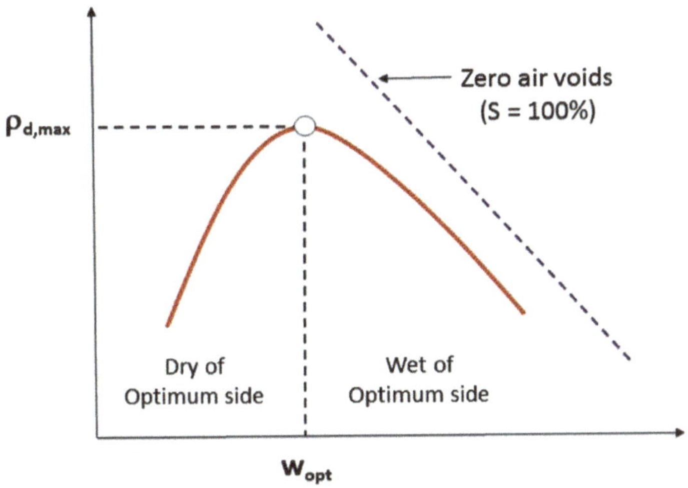

Figure 7.2.2 Typical compaction curve w versus ρ_d obtained by a Proctor test.

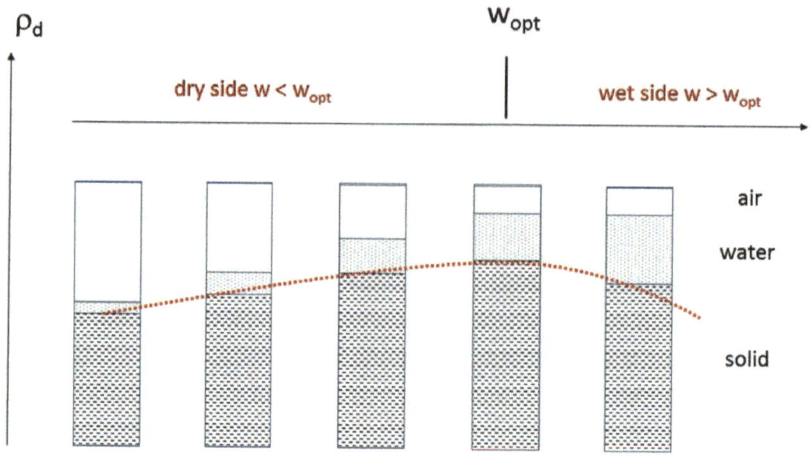

Figure 7.2.3 Variability of solid, liquid, and gas phases during a Proctor test.

On the side of the curve referred to as dry of optimum ($w < w_{opt}$), an increase in water content lubricates the particles, promoting a mutual sliding and interlocking, and giving rise to an increase in the degree of final densification ρ_d. On the wet side of optimum ($w > w_{opt}$), ρ_d decreases as the water, present in increasing quantities, not only fills the air voids (as on the dry of optimum side) but also replaces part of the volume occupied by the solid particles (Fig. 7.2.3). The two curves shown in Fig. 7.2.4 refer to the same soil, compacted to the two different compaction energies and characteristics

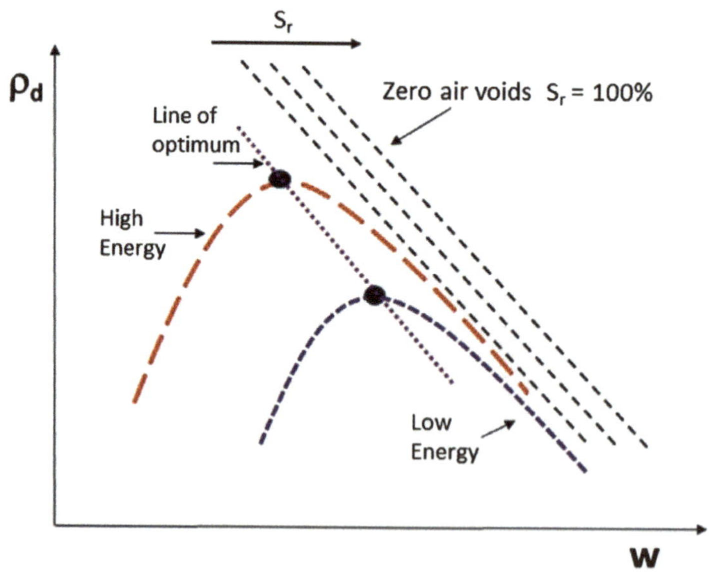

Figure 7.2.4 Typical plot w versus ρ_d of Standard and Modified Proctor test for the same soil.

of the standard and modified Proctor test procedures. The compaction energy used in the modified test is approximately 4.5-fold greater than that applied in the standard test. For this reason, the standard test is generally reserved for soils used in the construction of river embankments, earth dams, mineral liners of controlled landfills, etc., whereas the modified test is aimed at soils used in the foundations of main roads and industrial pavements. It is highly important that the most suitable Proctor test is selected because the compaction energy used in the laboratory should approximately reproduce that used in situ.

Fig. 7.2.4 shows additional theoretical (dashed) curves that represent the trend $w-\rho_d$ for constant degrees of saturation S_r (increasing to the right side), obtained through the following equation:

$$\rho_d = \frac{\rho_w \cdot S_r}{w + \frac{\rho_w}{\rho_s} \cdot S_r} = \frac{\rho_s \cdot S_r}{G \cdot w + S_r} \tag{1}$$

with ρ_s and ρ_w density of solid particles and water, respectively, and G specific gravity of soil equal to ρ_s/ρ_w. Both Proctor curves feature low and highly variable saturation levels on the dry side of optimum as the water content increases, whereas on the wet side of optimum S_r varies in a much narrower range.

Curve "$w-\rho_d$" moves to the left and upward of the diagram with increasing compaction energy. This implies that optimum water content decreases, whereas the maximum dry density increases as energy increases (Fig. 7.2.4).

Likewise, in determining hydraulic conductivity k on a large number of specimens, prepared with the same soil and the same compaction energy, we obtain a specific curve "w–k." Using N different compaction energies, we obtain as many different "w–k" curves (Fig. 7.2.5).

For all compaction energies, as the water content of the specimen increases, coefficient k drastically decreases on the dry of optimum side, even by 1 ÷ 2 orders of magnitude, whereas it increases very slowly on the wet of optimum side.

Having established a critical limit of hydraulic conductivity k_{crit}, which for municipal solid waste (MSW) landfills is generally equal to 10^{-9} m/s for the base barrier and 10^{-8} m/s for the capping barrier, it is not rare to obtain an in situ hydraulic conductivity k_{field} greater than k_{crit}. This may be due simply to the fact of having prepared the soil to an incorrect in situ water content w_{field}, much less than w_{opt}. On the dry of optimum side variability of coefficient k is, as stated above, very high. For this reason, it would be feasible to prepare the clayey material to be used in construction of the mineral liner, with a water content equal to $w_{opt} + (2\% \div 3\%)$; this approach would cause the coordinates (w; ρ_d) to fall into the wet of optimum side, less subject to changes of coefficient k.

By adopting an in situ water content w_{field} greater than the corresponding optimum water content, the barrier would be less sensitive to swelling (Lambe, 1958a,b); conversely, shrinkage would be slightly greater to that manifested on the dry of optimum side of the compaction curve (Seed and Chan, 1959).

The above, however, are only valid if the field compaction and Standard Proctor energies applied are comparable. In this case, the laboratory compaction curve $w-\rho_d$ indeed represents the actual behavior of the mineral liner.

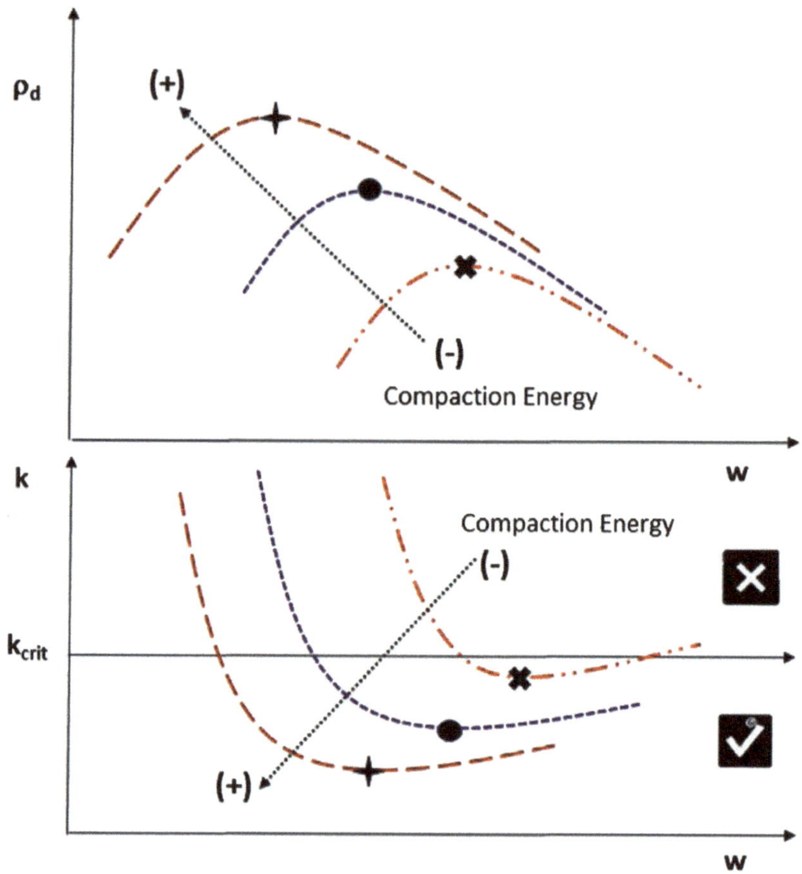

Figure 7.2.5 Plots w versus ρ_d and w versus k obtained using three different compaction energies.

What occurs, however, if the coordinate point (w_{field}; $\rho_{d,\ field}$) does not lie along the reference Standard Proctor curve? In this case, the compaction energies used in the laboratory and in situ are not comparable. Indeed, we could hypothesize that a much higher or lower energy has been applied in compacting the clayey barrier than the Standard Proctor energy. In Fig. 7.2.6 the Standard Proctor curve is indicated with a continuous line, concavity downward, and peak of coordinates (w_{opt}; $\rho_{d,\ max}$), which represents the ideal combination of water content and dry density to obtain a maximum degree of densification of the barrier. The figure shows that by varying the compaction energy, the curves move upward to the left or downward to the right (for compaction energies higher and lower, respectively, than Standard Proctor energy).

The optimum water content w_{opt} of the Standard Proctor curve corresponds to two points in the higher (point A) and lower (point B) energy curves. Point A lies on the dry of optimum side of the higher energy curve, whereas point B is located on the wet of optimum side of the lower energy curve. By being situated to the right or left of their optimum water content, important effects may be produced on structure, compressibility, permeability, shrinkage, and swelling. It is therefore clear that

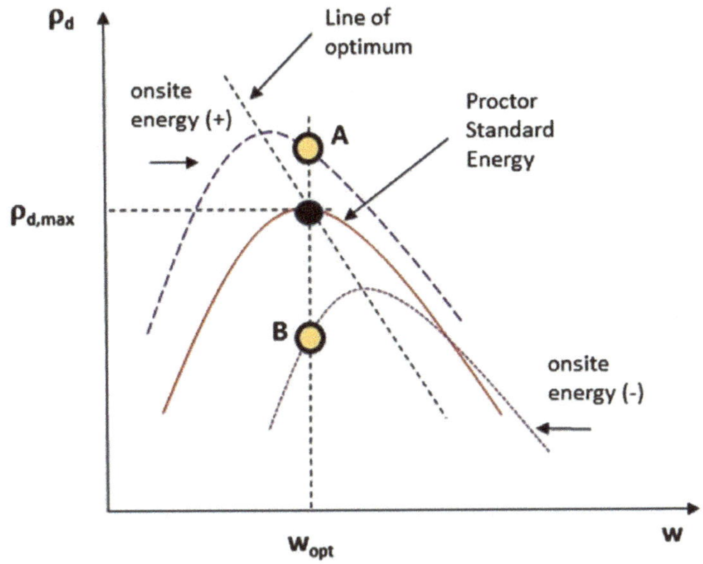

Figure 7.2.6 Plots w versus ρ_d when compaction energy used during the construction of the mineral liner differs (+ or -) from Proctor standard energy.

the control procedure currently adopted on the majority of construction sites is not always the most appropriate or reliable.

Accordingly, preliminary large-scale field tests should be carried out to define a detailed and correct compacting procedure (roller type, thickness of the layer to be compacted, number of roller passes per layer, initial water content of the soil, etc.) as similar as possible to the Standard Proctor compaction energy or to a percentage of the same. For example, on the capping of an MSW landfill, it is very difficult to achieve a degree of compaction comparable to that of the Standard Proctor test, due to the compressibility of the underlying mass.

Finally, a short note focuses on the installation and compaction of the mineral liner along the side banks of the landfill. If the slope does not exceed the value of 1H:2.5V (Fig. 7.2.7A), the installation can be completed on the banks in the same way as the lower and horizontal base. If the slope is higher, the layers should be implemented as indicated in Fig. 7.2.7B (USEPA, 1989).

HYDRAULIC CONDUCTIVITY OF THE MINERAL LINERS

Regulations relating to the design and construction of MSW landfills generally establish a hydraulic conductivity for mineral liners no greater than a critical value. However, these regulations are only prescriptive with no experimental procedure being defined to determine hydraulic conductivity in situ and

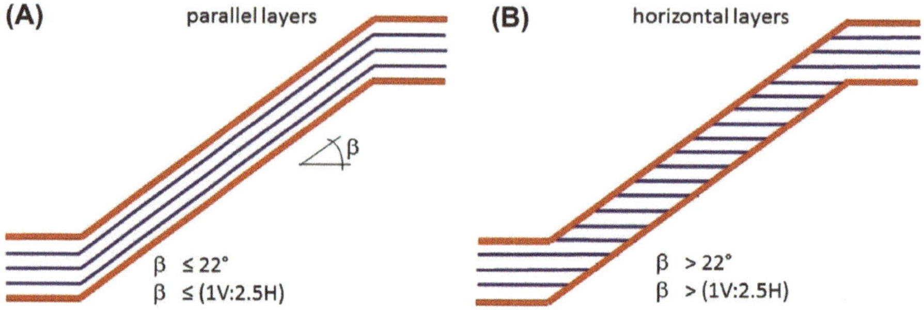

Figure 7.2.7 Layout of the mineral liner with parallel (A) and horizontal (B) lifts. *From USEPA (1989).*

in the laboratory. Therefore, a series of issues remain to be resolved and should be taken into account during the design stage:

- it is preferable to perform an in situ permeability test on the actual barrier or a laboratory permeability test on an undisturbed or reconstituted portion of the barrier?
- what should be the minimum duration adopted for the tests?
- what should be the optimal size and shape of the specimen?
- what should be the nature of the permeating fluid (distilled or tap water or real leachate)?
- what magnitude should be for the test gradient?
- is the critical hydraulic conductivity k_{crit} referred to a condition of complete saturation?
- what methods should be applied to guarantee complete saturation of the sample?

Laboratory Permeability Tests

Hydraulic conductivity k of a clayey soil can be determined in laboratory by means of various procedures using different experimental equipment. Experimental procedures include constant-head and falling-head permeability tests; inlet and outlet sections of the specimen are maintained with a constant or variable hydraulic head, respectively. The expressions used to determine coefficient k in the two cases are as follows:

Constant-head test:

$$k = \frac{Q \cdot L}{A \cdot \Delta t \cdot \Delta h} \tag{2}$$

Variable-head test:

$$k = \frac{a \cdot L}{A \cdot \Delta t} \cdot \ln\left(\frac{h_1}{h_2}\right) \tag{3}$$

where: Q: total discharge volume, m^3, in time $\Delta t = t_2 - t_1$, expressed in seconds; L: height of specimen; A: cross-sectional area of specimen; Δh: difference of the constant hydraulic head between inlet

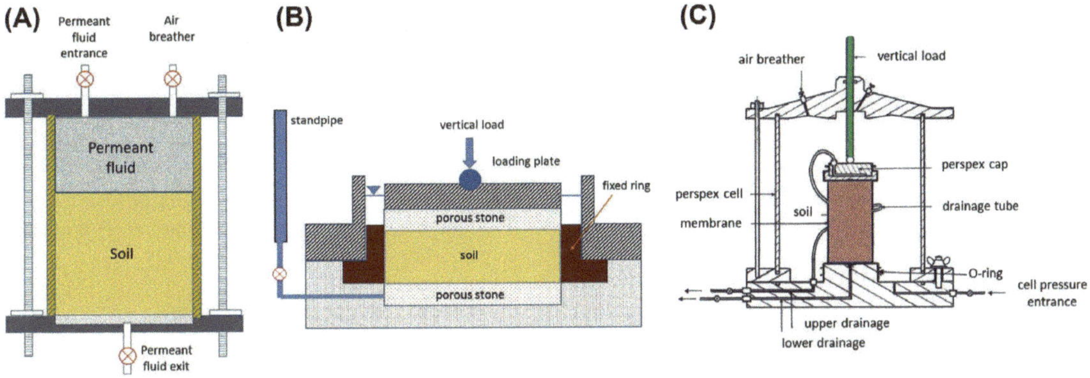

Figure 7.2.8 Laboratory permeameters: (A) compaction mold; (B) oedometric cell; (C) triaxial cell. *Modified from Daniel (1994).*

and outlet sections of the specimen; a: cross-sectional area of the graduated standpipe used in variable-head test; h_1, h_2: hydraulic heads corresponding to times t_1 and t_2 ($t_2 > t_1$) in a variable-head test.

Laboratory permeameters can be distinguished into rigid-wall (RWP) and flexible-wall (FWP) permeameters. Typical RWP are the Proctor mold (constant-head test) and the oedometric cell (variable-head test) (Fig. 7.2.8A and B), whereas triaxial cell is an FWP whereby the test is performed at constant-head (Fig. 7.2.8C).

Tests conducted using RWP are less expensive and easier to perform. If actual leachate is used as a permeating fluid, the material constituting the permeameter must be resistant to chemical aggression. The main disadvantage of RWP versus FWP is the potential sidewall leakage with consequent overestimation of coefficient k. When choosing the laboratory test to be used, the advantages and disadvantages associated with the use of an RWP or FWP (Table 7.2.2) should be specifically taken into account, together with the potential effects produced on the hydraulic conductivity by numerous factors indicated in Table 7.2.3.

Hydraulic conductivity k may be indirectly determined from the results of an oedometric test, considering a specific effective vertical stress σ'_v and the corresponding values of the coefficient of consolidation c_v and oedometric modulus M:

$$k = \frac{\gamma_w \cdot c_v}{M} \tag{4}$$

where γ_w is the unit weight of water.

Table 7.2.2 Main advantages and disadvantages of RW and FW permeameters

Permeameter	Advantages	Disadvantages
Compaction Mold (RWP)	• simple and cheap test	• specimen may be partially saturated • stress level of specimen may be not controlled • possible sidewall leakage
Oedometric Cell (RWP)	• simple and cheap test • vertical stress and strain of specimen may be controlled	• difficulty on specimen trimming • limited thickness of specimen • possible sidewall leakage
Triaxial Cell (FWP)	• stress and strain of specimen may be controlled • water volume may be controlled • sidewall leakage is limited	• complex and expensive test • difficulty on specimen trimming

Table 7.2.3 Factors influencing performance of the main laboratory permeameters

Factor	Error
Experimental high hydraulic gradient applied for a more rapid seepage through the specimen	• not representative of the in situ condition • migration of the finer grains
Migration of the finer grains due to used high hydraulic gradient	• possible underestimation of k
Degree of saturation influenced by pressurized water coming from an air/water pressure generator	• possible decrease of k due to air effects in soil pores
Specimen size and Height/Diameter ratio	• k decreases as well as H/D ratio increases
Test duration generally too short	• possible overestimation of k

In Situ Permeability Tests

The hydraulic conductivity of a mineral liner may be determined by means of one of the following in situ tests:

- Boutwell permeameter test (ASTM 6391);
- Sealed Double-Ring Infiltrometer test (SDRI) (ASTM D5093);
- Guelph permeameter (GP) (ASTM D5126).

Boutwell Permeameter Test

Each test requires a horizontal portion of a mineral liner with an area of approximately 16 m^2 and a thickness of at least 1 m. A cylinder casing should be installed in a slightly wider diameter predrilled hole. The volume between the hole wall and the side surface of the casing should be hydraulically sealed to prevent lateral water leakage. The rate of water flow into soil through the bottom of the casing is measured in one or two stages, normally using a falling-head procedure. If the soil is considered anisotropic, with different hydraulic conductivity along vertical and horizontal directions, a falling-head test may be carried out in two stages (Fig. 7.2.9A). During Stage 1 the permeant fluid can filter only across the bottom section of the borehole; when the borehole is extended below the bottom of the casing, the permeant fluid can filter both vertically and horizontally (Stage 2). The borehole is extended for Stage 2 after Stage 1 is over (Fig. 7.2.9B). A limiting hydraulic conductivity is computed from the falling-head data in both stages (K_1 and K_2). Stages 1 and 2 continue until the limiting conductivity for each stage is relatively constant.

Phase 1

Date, time, water level, and temperature in the graduated standpipe and in the Temperature Effect Gauge (TEG) should be noted throughout the test. TEG is identical to Boutwell permeameter, with a closed lower part, used to purify the experimental data from the effects of thermal variations. These effects are more relevant with smaller diameter standpipes and soils with a $k \leq 5 \times 10^{-10}$ m/s. Phase 1 is completed when a steady-flow condition is achieved.

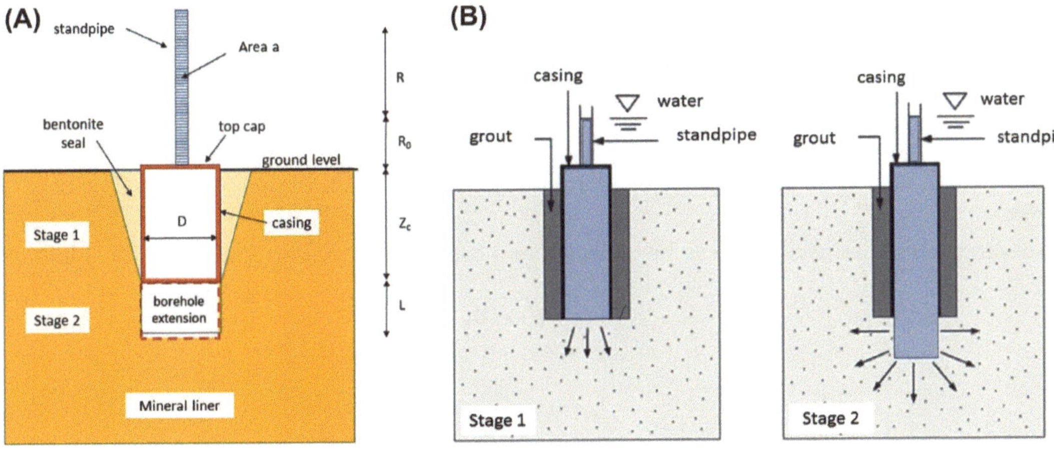

Figure 7.2.9 Boutwell permeameter test: (A) Schematic section during Stage 1 and Stage 2, (B) Seepage during Stage 1 and Stage 2.

Phase 2

Following emptying of the casing, the hole should be deepened (keeping the casing in its original position) up to one- or twofold the casing diameter. During excavation, lateral filtering side of the excavation should remain undisturbed and not smoothed. At the end of these operations, the test will develop as in the previous phase.

Methods to calculate actual vertical and horizontal hydraulic conductivities (k_v and k_h) from K_1 and K_2, determined during the Stages 1 and 2, are described as follows.

STAGE 1

$$K_1 = R_T \cdot \frac{\pi \cdot d^2}{11 \cdot D \cdot (t_2 - t_1)} \cdot \ln\left(\frac{Z_1}{Z_2}\right) \tag{5}$$

$R_T = 2.2902\ (0.9842\ T)/T^{0.1702}$ with temperature T in °C; d = internal diameter of standpipe (cm); D = internal diameter of casing (cm); $Z_1 = (Z_c + R_o + R)$ at time t_1; $Z_2 = (Z_c + R_o + R - c)$ at time t_2; c = change in TEG scale reading between times t_1 and t_2; an increase in the height of water in the TEG standpipe is positive; t_1 = time at beginning of increment(s); t_2 = time at end of increment(s).

STAGE 2

$$K_2 = R_T \cdot \frac{A}{B} \cdot \ln\left(\frac{Z_1}{Z_2}\right) \tag{6}$$

with:

$$A = d^2 \cdot \ln\left(\frac{L}{D} + \sqrt{1 + \left(\frac{L}{D}\right)^2}\right) \tag{7}$$

$$B = 8 \cdot L \cdot (t_2 - t_1) \cdot \left(1 - 0.56 \cdot e^{-1.57 \cdot L/D}\right) \tag{8}$$

and L = length of Stage 2 extension below bottom of casing (cm), D = internal diameter of Stage 2 extension.

Anisotropy of the mineral liner can be evaluated by a coefficient of anisotropy m defined as follows:

$$m = \sqrt{k_h/k_v} \tag{9}$$

$$k_v \cdot m = \frac{k_h}{m} \tag{10}$$

Once K_1, K_2, L, and D have been established, coefficient m can be determined with the following expression:

$$\frac{K_2}{K_1} = m \cdot \frac{\ln\left[\left(\frac{L}{D}\right) + \sqrt{1 + \left(\frac{L}{D}\right)^2}\right]}{\ln\left[\left(m \cdot \frac{L}{D}\right) + \sqrt{1 + \left(m \cdot \frac{L}{D}\right)^2}\right]} \tag{11}$$

And then k_v and k_H can be found by:

$$K_1 = k_v \cdot m = \frac{k_h}{m} \tag{12}$$

Sealed Double-Ring Infiltrometer Test

This test method provides a direct measurement of *infiltration rate I* (amount of permeant fluid filtering across a soil sample (m^3) per unit area (m^2) per unit time (s), expressed in m/s). Hydraulic conductivity k cannot be directly measured unless the hydraulic boundary conditions, particularly hydraulic gradient, are known.

A double-ring infiltrometer with a sealed or covered inner ring consists of an open outer and a sealed inner ring (Fig. 7.2.10).

The rings are embedded and sealed in trenches excavated in the soil. Both rings are filled with water until the inner ring is submerged. The rate of flow is measured by connecting a flexible bag filled with a known weight of water to a port on the inner ring. As water infiltrates into the ground from the inner ring, an equal amount of water flows into the inner ring from the flexible bag. After a known interval of time, the flexible bag is removed and weighed. The weight loss, converted into a volume, is equal to the

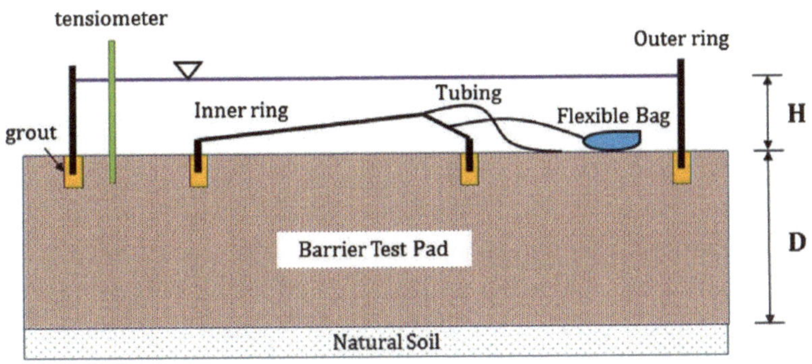

Figure 7.2.10 Section of Sealed Double-Ring Infiltrometer.

amount of water that has infiltrated into the ground. An infiltration rate is then determined from this volume of water, the area of the inner ring, and the interval of time. This process is repeated and a plot of infiltration rate versus time is constructed. The test is continued until the infiltration rate becomes steady.

The purpose of the sealed inner ring is to provide a means to measure the actual amount of flow and to eliminate evaporation losses. The purpose of the outer ring is to promote one-dimensional, vertical flow beneath the inner ring. The use of large diameter rings and large depths of embedments helps to ensure that flow is essentially one dimensional.

Changes in water temperature can introduce significant error in the volume change measurements. Temperature changes will cause water to flow in or out of the inner ring due to expansion or contraction of the inner ring and the water contained within the inner ring. The problem of temperature changes can be minimized by insulating the rings, by allowing enough flow to occur so that the amount of flow resulting from a temperature change is not significant compared to that due to infiltration, or by connecting and disconnecting the bag from the inner ring when the water in the inner ring is at the same temperature. If the soil being tested will later be subjected to increased overburden stress, then the infiltration rate can be expected to decrease as the overburden stress increases. Laboratory hydraulic conductivity tests are recommended for studies of the influence of level of stress on the hydraulic properties of the soil. The rings shall be made of metal, plastic, or fiberglass. The shape of the rings can be circular or square. The minimum width or diameter of the inner ring is 610 mm and a minimum distance of 610 mm is maintained between the inner and outer rings. Two clear flexible bags have a capacity of 1 ÷ 3 L. Representative samples of the soil to be tested shall be taken before and after the test to determine its moisture content, density, and specific gravity. The thickness of the layer being tested shall be determined as well as the approximate hydraulic conductivity of the layer beneath it.

Estimation of hydraulic conductivity k may be performed using the following expression:

$$k = \frac{I}{i} = \frac{V}{A \cdot \Delta t} \cdot \frac{1}{i} = \frac{V}{A \cdot \Delta t} \cdot \frac{L_f}{(H + L_f)} \tag{13}$$

where: I = infiltration rate; i = hydraulic gradient; A = cross-sectional area of inner ring; Δt = test duration; V = amount of water filtered during Δt; H = depth of ponded water; L_f = depth of wetting front.

Evaluation of the water front thickness may be difficult. For this reason two different procedures have been developed to determine coefficient k.

The *apparent gradient method* (Fig. 7.2.11A) is more conservative in view of the tendency to underestimate hydraulic gradient and overestimate hydraulic conductivity.

The test pad is considered fully saturated over its entire thickness. This assumption for compacted clay liners with $k < 10^{-9}$ m/s is conservative. The hydraulic gradient i is equal to:

$$i = (H + D)/D \tag{14}$$

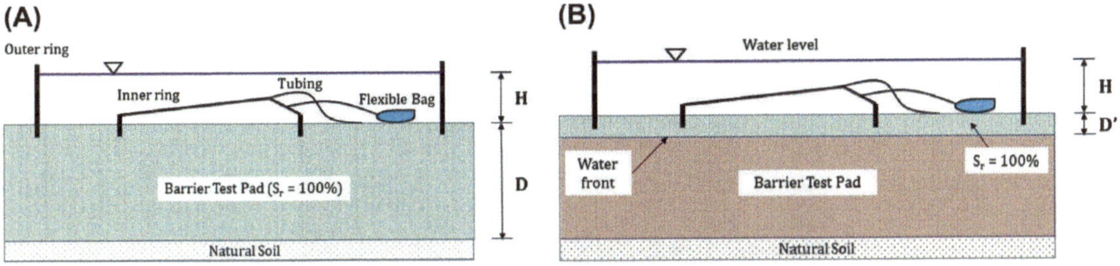

Figure 7.2.11 Different methods for permeability measurements with sealed double-ring infiltrometer: (A) apparent gradient method, (B) wetting front method.

The *wetting front method* (Fig. 7.2.11B) assumes only a partial saturation of the test pad. The water pressure at the wetting front is conservatively assumed at atmospheric pressure. Tensiometers are used to monitor the depth of wetting of the soil over time, and the variation of water content with depth is determined at the end of the test. The wetting front method is the method that is usually recommended. The hydraulic gradient i is equal to:

$$i = (H + D')/D' \tag{15}$$

Guelph Borehole Permeability Test

The GP (Fig. 7.2.12) is an instrument intended for use in rapidly determining hydraulic conductivity of a mineral liner.

The GP is easily movable and suitable for use by a single technician. Testing times range from 30′ to 120′ depending on soil nature. According to the reference standard (ASTM D5126) soil thickness of the barrier undergoing testing ranges from 0.15 to 0.75 m below ground level. A shallow cylindrical hole is excavated in the barrier and GP is inserted into the hole. The GP test employs Mariotte's bottle principle and is carried out measuring steady-state rate of water recharge into unsaturated soil from the test hole. The field hydraulic conductivity k_{fs} is estimated by means of a constant-head test procedure. k_{fs} is referred to the field-saturated bulb (Fig. 7.2.13) of soil surrounding the test hole.

Firstly, a constant hydraulic head in the hole is fixed and held at the same level as the lower part of the air tube, located at the center of the GP. When the water level in the hole starts to drop below the air inlet tip, air bubbles exit the tip and rise into the tank air space. Vacuum is then partially relieved and the tank provides water to the hole. Size of opening and geometry of the air inlet tip are suitable to control air bubble size to avoid fluctuations of well water level. When a water level is maintained constant in the hole, a bulb of saturated soil quickly develops all around, assuming shape according to soil nature, well radius, and imposed hydraulic head. When a steady-state water flow has been reached, field-saturated conductivity k_{fs} can be determined. A certain amount of air is usually entrapped in soil voids during the infiltration process, influencing the obtained values of k, which may be lower than those under conditions of complete saturation.

Figure 7.2.12 Guelph permeameter.

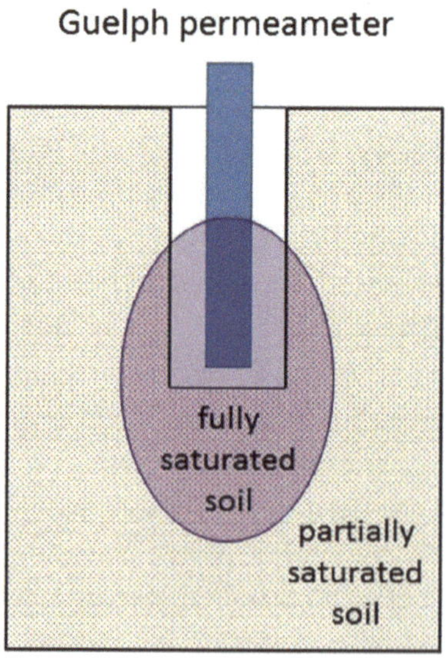

Figure 7.2.13 Bulb of saturated soil (Eijkelkamp, 2011).

Table 7.2.4 Main advantages and disadvantages of field permeameters

Permeameter	Advantages	Disadvantages
Boutwell	• cheap test equipment • easy installation • kv and kH can be evaluated • test can be performed on slopes	• test involves a limited volume of barrier • partial saturation of soil cannot be properly taken into account • testing times are long for soils having kv $\approx 10^{-10}$ m/s
Sealed double-ring infiltrometer	• moderate test equipment cost • kv can be evaluated • suitable for soils with kv $\approx 10^{-10}$ m/s • test involves a large volume of barrier • minimal lateral seepage from inner ring	• testing times are long for soils having kv $\approx 10^{-10}$ m/s • only k_v can be evaluated • wetting front suction head must be estimated • test cannot be performed on slopes
Guelph	• moderate test equipment cost • testing times are very short • uniformity of compaction can be controlled	• only an average k can be estimated • test involves a very small volume of barrier • great influence of possible discontinuities in the barrier • more suitable for soils having kv $< 10^{-8}$ m/s

From Daniel (1989).

Differences Among Testing Techniques

Boutwell permeameter and SDRI may be applied indifferently for field determination of k-coefficient. The main advantages and disadvantages deriving from use of the two devices are summarized in Table 7.2.4.

GP allows fast but relatively inaccurate tests to be performed. Its use with mineral liners in MSW landfills should therefore be limited to controlling homogeneity. In other words, by obtaining a series of fairly close k-coefficients, it may be confirmed that the barrier was constructed using homogeneous and uniformly compacted soil.

QUALITY CONTROL DURING CONSTRUCTION

Field control of compaction suitability is carried out by referring to the laboratory $w-\rho_d$ curve and determining in situ water content w_{field} and dry density ρ_d of the mineral liner.

Thus, the so-called degree of compaction D_c is obtained equal to:

$$D_c = \left[\rho_{d,field}/\rho_{d,max,lab}\right] \times 100(\%) \qquad (16)$$

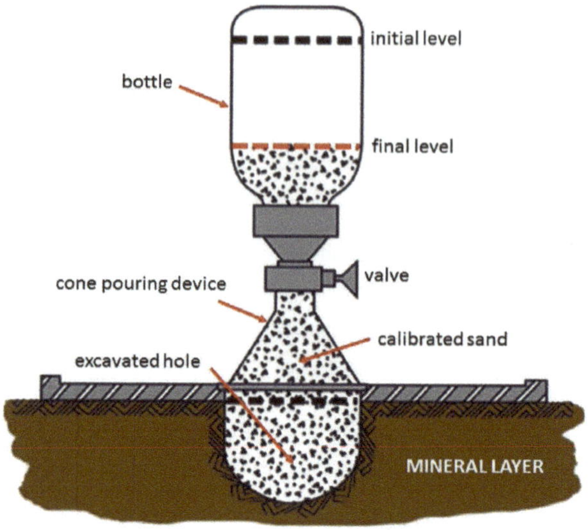

Figure 7.2.14 Sand cone test.

which should be not less than 90% ÷ 95% to confirm suitability of the mineral liner.

Field control tests may be destructive or nondestructive. In the first case, a limited amount of barrier should be excavated, while in the second measurements are made without any removal of soil. The most popular destructive test is the Sand Cone Test (ASTM D-1556-1556M). A hole of approximately 150 mm in diameter and depth is excavated at the top of the barrier. The excavated soil is stored in a plastic bag and its natural water content determined in the laboratory. The unknown volume of the excavated hole is filled with calibrated sand (dry sand with known dry density) introduced through a cone pouring device (Fig. 7.2.14). The dry mass of the excavated soil divided by the volume of calibrated sand required to fill the hole allows the dry density of the barrier to be determined. This may be compared with the Standard Proctor dry density.

The nondestructive tests include the so-called Nuclear Methods that allow the rapid and relatively reliable determination of both in situ density and water content (ASTM D6938). The instrument uses a radiative isotope (Cesium 137), which generally emits γ rays intercepted by a meter present at the lower base of the nucleodensimeter. Dense soil absorbs more radiation than loose soils (Fig. 7.2.15). The instrument is therefore able to estimate both the density of the soil and, after a few minutes, its natural moisture.

TRANSPORT OF CONTAMINANTS THROUGH CLAY LINERS

Transport of contaminants in soils may occur through three different processes:

- Advection: contaminant is transported by water flow due to a hydraulic gradient;
- Mechanical Dispersion: it consists of a mixing process occurring into an advection flow, due to different velocities among different points; and

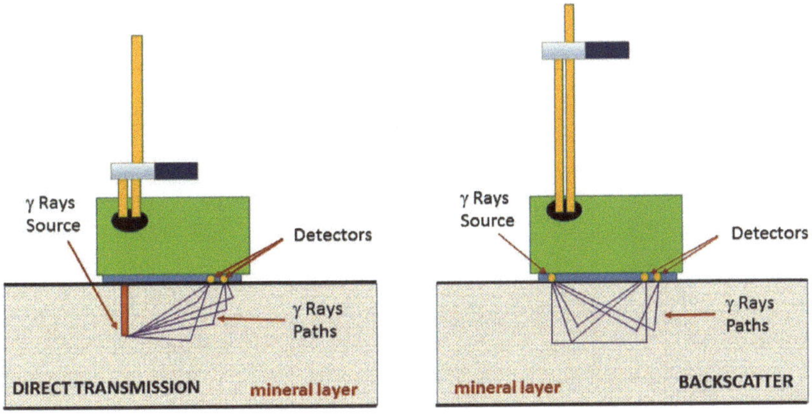

Figure 7.2.15 Nuclear density apparatus. *From www.utest.com.tr.*

- Molecular Diffusion: it is due to concentration gradients of ionic or molecular constituents that may occur also without any advection process.

The main transport process interesting clay liners is usually diffusion.

Diffusion

Mass of solute F, expressed per unit area and per unit time [$ML^{-2}T^{-1}$], solute concentration C [ML^{-3}], diffusion gradient $\partial C/\partial z$, and diffusion coefficient D [L^2T^{-1}] may be correlated, in one-dimensional process along the direction z, through the Fick's first law as follows:

$$F = -D\cdot\frac{\partial C}{\partial z} \tag{17}$$

Negative sign means that diffusion occurred toward concentration is lower (direction z). In Table 7.2.5 a set of molecular diffusion coefficients D_m of contaminants in water are reported. These coefficients are strongly depending on temperature, decreasing about 50% at 5°C.

The effective molecular diffusion D_e in saturated soils is lower than the molecular diffusion D_m of contaminants in water. This is mainly due to the tortuosity t (≤ 1, generally ranging between 0.01 ÷ 0.5) of the paths available for diffusion. The following expression can be written:

$$D = D_e = t\cdot D_m \tag{18}$$

Rate of diffusion of the solute in a porous material can be expressed by means of Fick's second law, combining Fick's first law and the continuity equation:

$$\frac{\partial C}{\partial t} = D\cdot\frac{\partial^2 C}{\partial z^2} \tag{19}$$

Table 7.2.5 Tipycal values of molecular diffusion coefficient D_m for some contaminants in water

Contaminant	D_m (m^2/s)
Na^+, K^+, Mg^{2+}, Ca^{2+}	1E-9 ÷ 2E-9
Cl^-, HCO_3^-, SO_4^{2-}	1E-9 ÷ 2E-9
Chlobenzene	8.4 E-10
Ethylbenzene	7.5 E-10
Methyl alcohol	1.5 E-9

Boundary conditions at the start of the process must be defined. In Fig. 7.2.16 a clay liner, typical of the base of an MSW landfill is drawn. At $t = 0$ the solute concentration at the top of the clay liner is C_0 (>0), whereas it is null at its bottom.

In steady condition the solute concentration C_0 may be considered constant with time. The solution of Eq. (20) can be found through the following expression:

$$C(z,t) = C_0 \cdot \mathrm{erfc}\left(\frac{z}{2 \cdot \sqrt{D \cdot t}}\right) \quad (20)$$

Attenuation

Attenuation is a process that may occur when a leachate filtrates through a mineral barrier placed at the bottom of an MSW landfill. Attenuation produces a decrease of the concentration of the leachate constituents and may develop by means of one or more among the following mechanisms:

- *Adsorption*: molecules adhere to the surface of the clay particles causing a decrease in the total dissolved solid;
- *Biological uptake*: microorganisms break down or adsorb leachate constituents attenuating leachate;
- *Cation- and Anion Exchange*: exchange of ions of one type by ions of another type without disturbing the mineral structure (isomorphous substitution);
- *Filtration*: leachate constituents are physically trapped in the random pore structure;

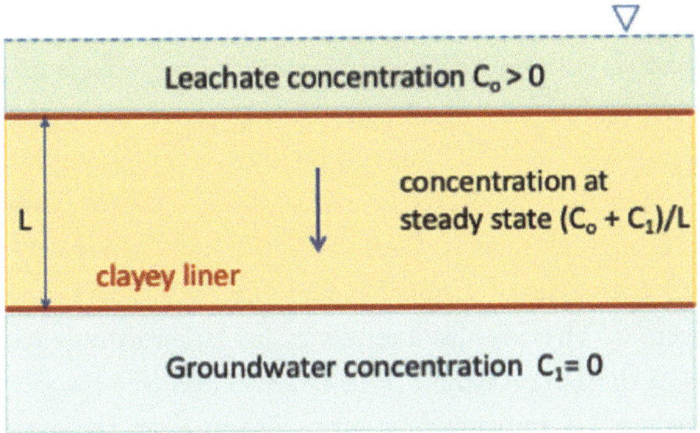

Figure 7.2.16 Clay liner system in which molecular diffusion is the main transport mechanism.

- *Precipitation*: chemical process involving a phase change in which dissolved chemical species are crystallized and deposited from a solution because their total concentration exceeds their solubility limit.

The consequences of these attenuation processes are the variation over time of both the permeability of the mineral barrier and the composition of the leachate. It is very difficult to quantify the effects of these processes in the design stage, given their duration and dependence on the specific nature of the leachate. Laboratory tests are not long enough to activate all the long-term interactions that could develop between clayey minerals and leachate. Finally, it should be emphasized that the diffusion rate of a nonattenuated contaminant, through the mineral barrier, can be very high and therefore create unexpected pollution problems.

COMPATIBILITY BETWEEN CLAY LINERS AND LEACHATE

The most common clay minerals in soils belong to three groups, kaolinite, illite, and smectite (montmorillonite). In the kaolinite crystals the unit layers are held together by hydrogen bonding. In montmorillonite the interlayers bond is due to cations exchanged for those contained in hydrating solution. Water molecules can also enter between the layers. Therefore, the van der Waals interlayer bonds are generally weak. The exchange complex of soils normally consists of a mixture of the cations Ca^{2+}, Mg^{2+}, K^+, and Na^+, in which Ca^{2+} comprises the larger part, and K^+ and Na^+ only the smaller. The ability of soils to exchange cations with those contained in the liquid solution has important effects on soil behavior. Exchangeable cations are located on particle surfaces and are attracted by their negative charges. The resulting capacity for exchangeable cations (C.E.C.) varies according to the type of clay; and it can be about 100 in kaolinite, 300 in illite, and 1000 in montmorillonite (C.E.C. expressed as milliequivalents per kg of clay). Variation occurs because of the effects of crystal size, isomorphous substitution, and pH. Oxides and organic matter also have a cation exchange capacity. The exchange of one cation for another is governed by Coulomb's law. Cations of a higher valence (higher charge) are attracted more strongly to the negatively charged clay than those of a lower valence. Hence Ca^{2+} generally replaces Na^{+} in any reaction. However, if the concentration of lower valence ions (i.e., Na^{+}) is relevant and much greater than that of higher valence (i.e., Ca^{2+}), the ion exchange phenomenon can be in an opposite direction. There is also a tendency for larger cations to enter more readily in exchange than smaller ones of the same valency. This is attributed to a greater hydration of the smaller ions that, with a larger hydrated radius, are thus more distant from the negative charges on the clay. When adsorption of water has progressed far enough for swelling, clay has then enough water for a diffuse layer of exchangeable cations to extend out from the negatively charged surface of the mineral sheets. Phenomenon is governed by the opposing effects of electrostatic attraction toward the surface and diffusion away from it in the direction of decreasing concentration. The clay surface, with its localized negative charges, and the swarm of cations of opposite charge constitute a diffuse electrical double layer. If an outer solution bathes the clay, this will affect the distribution of the cations. Also because of the charge on the clay, the concentration of anions near its surface will be lower than in the outer solution (negative adsorption of anions).

The effects of liquids may be evaluated with the Gouy–Chapman theory, which states that the thickness (t) of the diffuse double layer varies as follows:

$$t = \sqrt{\frac{D \cdot K \cdot T}{8 \cdot \pi \cdot n_0 \cdot \varepsilon^2 \cdot v}} \tag{21}$$

where D is the dielectric constant, K is the Boltzmann constant (1.38×10^{-6} erg/K), T is the temperature, n_0 is the electrolyte concentration, ε is the unit electronic charge (16×10^{-20} C), and v is the cation valence.

When the thickness of the diffused double layer decreases, the structure of the clayey soil tends to be flocculated with an increase in permeability. The process is inverse when the thickness t increases. For example, solutions containing monovalent cations tend to produce lower k than solutions with polyvalent cations.

Pore fluid composition influences not only the double-layer thickness but also the soil structure (which can alternatively become flocculated or dispersed) and the geotechnical properties of soil (Table 7.2.6). The complexity of the above-mentioned phenomena can produce large fluctuations of the geotechnical parameters of clayey soils.

Table 7.2.6 Effect of pore-fluid parameters on Diffuse Double-Layer thickness and soil structure

Parameter	Parameter related to the pore fluid	DDL Thickness	Soil Structure
Electrolyte concentration	+ −	− +	F D
Ion valence	+ −	− +	F D
Dielectric constant	+ −	+ −	D F
Temperature	+ −	+ −	D F
pH	+ −	+ −	D F

(+) increase, (-) decrease.
(F) flocculated, (D) dispersed.

CONCLUSIONS

Mineral barriers at the base of an MSW landfill must be carefully designed. The material composing them must be carefully selected and put in place using compaction procedures initially defined by test field. Clay liners can be crossed by pollutants in several ways. When hydraulic conductivity is very low, possible release of pollutants will be mainly by molecular diffusion. Attenuation processes can slow the transport of several contaminants. Leachate can attack and compromise the efficiency of the barriers, but only concentrated chemicals are of serious concern. Based on transit time calculations, clay liners can be very effective barriers and minimize release of pollutants from landfills.

References

Alberta Environment, March 1985. Design and Construction of Liners for Municipal Wastewater Stabilization Ponds. prepared by Komex Consultants. Ltd.. Alberta Environment Publication, Edmonton, Alberta, Canada.

ASTM Test Method D698: Standard Test Methods for Laboratory Compaction Characteristics of Soil Using Standard Effort.

ASTM Test Method D1557: Standard Test Methods for Laboratory Compaction Characteristics of Soil Using Modified Effort.

ASTM Test Method D1556/D1556M: Standard Test Method for Density and Unit Weight of Soil in Place by Sand-Cone Method.

ASTM Test Method 6938: Standard Test Methods for In-Place Density and Water Content of Soil and Soil-Aggregate by Nuclear Methods (Shallow Depth).

ASTM Test Method D6391: Standard Test Method for Field Measurement of Hydraulic Conductivity Using Borehole Infiltration.

ASTM Test Method D5093: Standard Test Method for Field Measurement of Infiltration rate Using Double-Ring Infiltrometer with Sealed-Inner Ring.

ASTM Test Method D5126: Standard Guide for Comparison of Field Methods for Determining Hydraulic Conductivity in Vadose Zone.

Daniel, D.E., September 1989. In situ hydraulic conductivity tests for compacted clay. Journal of Geotechnical Engineering 115 (9), 1205–1226.

Daniel, D.E., 1994. State-of-the-art: laboratory hydraulic conductivity tests for saturated soils. In: Daniel, D.E., Trautwein, S.J. (Eds.), Hydraulic Conductivity and Waste Contaminant Transport in Soil, ASTM STP 1142. ASTM, Philadelphia, PA, pp. 30–78.

Eijkelkamp Soil & Water, November 2011. Guelph Permeameter – Operating Instructions.

Evans, J.C., Fang, H.Y., Kugelman, I.J., 1985. Organic fluid effects on the permeability of soil-bentonite slurry walls. In: Proc. National Conf. On Hazardous Wastes and Environmental Emergencies, Cincinnati, OH, USA, pp. 267–271.

Lambe, T.W., 1958a. The structure of compacted clay. Journal of the Soil Mechanics and Foundation Division 84 (SM2), 1654-1–1654-34. ASCE.

Lambe, T.W., 1958b. The engineering behaviour of compacted clay. Journal of the Soil Mechanics and Foundation Division 84 (SM2), 1655-1–1655-35. ASCE.

Seed, H.B., Chan, C.K., 1959. Structure and strength characteristics of compacted clays. Journal of the Soil Mechanics and Foundation Division 85 (SM5), 87–128. ASCE.

Sharma, H.D., Lewis, S.P., 1994. Waste Containment Systems, Waste Stabilization and Landfills. John Wiley & Sons Inc., USA. ISBN:0-471-575336-4.

U.S. Environmental Protection Agency (USEPA), 1989. Requirements for Hazardous Waste Landfill Design, Construction and Closure. Seminar Publication, EPA/625-89/022, USEPA, Cincinnati, OH, 127 pp.

Further Reading

ASTM Test Method D2167: Standard Test Method for Density and Unit Weight of Soil in Place by the Rubber Balloon Method.

ASTM Test Method D6780/D6780M: Standard Test Method for Water Content and Density of Soil In situ by Time Domain Reflectometry (TDR).

Daniel, D.E., Anderson, D.C., Boynton, S.S., 1985. Fixed-walls versus flexible-wall permeameter. In: Hydraulic Barriers in Soil and Rock, ASTM STP 874. ASTM, Philadelphia, pp. 107–126. ISBN:0-08030-04117-0.

Favaretti, M., Moraci, N., 1995. Effects of leachate on the behaviour of sand-clay mixtures. In: 5th International Landfill Symposium "Sardinia 95", S. Margherita di Pula, Cagliari, Italy.

Favaretti, M., Moraci, N., 1997. Geotechnical behaviour of sand-sodium bentonite mixtures. In: 6th International Landfill Symposium "Sardinia 97", S. Margherita di Pula, Cagliari, Italy.

Holtz, R.D., Kovacs, W.D., Sheahan, T.C., 2011. An Introduction to Geotechnical Engineering, second ed. Pearson, USA. ISBN:0-13-701132-6.

Qian, X., Koerner, R.M., Gray, D.H., 2002. Geotechnical Aspects of Landfill Design and Construction. Prentice Hall, USA. ISBN:3-13-012506-7.

7.3

POLYMERIC GEOMEMBRANE COMPONENTS IN LANDFILL LINERS

George R. Koerner and Robert M. Koerner

INTRODUCTION

This chapter focuses on the various aspects of polymeric geomembrane formulation, manufacture, and fabrication as well as aspects of installation and construction quality control (CQC). The specific geomembranes discussed are those most commonly used currently as landfill barrier components.

TYPES OF GEOMEMBRANES AND FORMULATIONS

It must be recognized at the outset that all geomembranes are actually formulations of a parent resin from which they derive their generic name and varying amounts of other ingredients. The most commonly used geomembranes for solid and liquid waste containment are high-density polyethylene (HDPE), linear low-density polyethylene (LLDPE), flexible polypropylene (fPP), polyvinyl chloride (PVC), chlorosulfonated polyethylene (CSPE), and ethylene propylene diene monomer (EPDM) as shown in Table 7.3.1. They are listed according to their commonly referenced acronyms and explained in the text to follow.

It should also be recognized that the products listed in Table 7.3.1 and the references to it in the text to follow are meant to reflect on the current state of the practice. The values mentioned are not meant to be prescriptive, and future research and development may result in significant formulation changes.

Polyethylene

Polyethylene is a very common thermoplastic polymer used throughout the world. Its name originates from the monomer ethene used to create the polymer. It can be produced through radical polymerization, anionic polymerization, and cationic polymerization. This is because ethene does not have any extraneous groups that influence the stability of the propagation head of the polymer. Each of these methods results in a different type of polyethylene.

Polyethylene is classified into several different categories based mostly on its physical (mainly density) and mechanical properties. The mechanical properties of polyethylene depend significantly on variables such as the extent and type of branching, the crystal structure, and the molecular weight. The

Table 7.3.1 Commonly used polymeric geomembranes with approximate formulations by weight (Koerner, 2012)

Type	Resin	Plasticizers	Fillers	Carbon Black	Additives
HDPE	95–98	0	0	2–3	0.25–1
LLDPE	94–96	0	0	1–3	0.5–4
fPP	85–98	0	0–13	2–4	0.5–2
PVC	50–70	25–35	0–10	2–5	2–5
CSPE	40–60	0	40–50	5–10	5–15
EPDM	25–30	0	20–40	20–40	1–5

CSPE, chlorsulfonated polyethylene; *EPDM*, ethylene propylene diene terpolymer; *fPP*, flexible polypropylene; *HDPE*, high-density polyethylene; *LLDPE*, linear low-density polyethylene; *PVC*, polyvinyl chloride.

following resins are commonly used in geosynthetics and classified per ASTM D883 and ASTM F412 by density:

- HDPE greater than 0.940 g/cc
- MDPE (medium-density polyethylene) 0.940–0.926 g/cc
- LLDPE 0.925–0.919 g/cc
- LDPE (low-density polyethylene) 0.925–0.910 g/cc

HDPE has a low degree of branching and thus stronger intermolecular forces and tensile strength. The lack of branching is ensured by an appropriate choice of catalyst (e.g., Ziegler–Natta catalysts) and reaction conditions. MDPE is a branched polyethylene having a slightly lower density than HDPE. It is the most commonly used polyethylene resin in geosynthetics. It is interesting to note that only after the addition of carbon black and the additive package does the compound cross into the density threshold into HDPE. LLDPE is a substantially linear polymer, with significant numbers of short branches, commonly made by copolymerization of ethylene with longer-chain olefins. Depending on the crystallinity and molecular weight, a melting point and glass transition may or may not be observable. The temperature at which these occur varies strongly with the type of polyethylene. LDPE has a high degree of branching, which means that the chains pack into the crystal structure as well. It has therefore less strong intermolecular forces as the instantaneous-dipole induced-dipole attraction is less. This results in a lower tensile strength and increased ductility. LDPE is created by free radical polymerization.

As noted in Table 7.3.1, HDPE geomembranes are made from polyethylene resin, carbon black, and additives; the last being processing aids and long-term antioxidants. Like most geomembranes, HDPE is either smooth on both sides, smooth on one side and textured on the other, or textured on both sides.

The polyethylene resin used for geomembranes is prepared by low-pressure polymerization of ethylene as the principal monomer. The resin is usually supplied to the manufacturer in a clear, colorless pellet form which it is then mixed with master batch (carbon black and antioxidants in a carrier resin) and "let down" into the designated formulation.

It should be clearly stated that regrind or rework pellets (which have been previously processed by the same manufacturer but never been used in the field) are often added to the extruder during processing. This is in contrast to reclaimed, recycled, or postconsumer material, which should never be allowed in the formulation in any quantity.

Carbon black is added into a geomembrane formulation for general stabilization purposes, particularly for ultraviolet light stabilization. It is generally added as a preformulated concentrate in pellet form. In this "master batch" is the carbon black and possibly the antioxidants contained in a carrier resin. The carrier resin should be the same as the host resin insofar as density and melt flow are concerned.

Additives are introduced into an HDPE geomembrane formulation for the purposes of oxidation prevention, long-term durability, and as a lubricant and/or processing aid during manufacturing. It is quite difficult to write a specification for HDPE geomembranes around a particular additive, or group of additives, because they are generally proprietary. Furthermore, there is ongoing research and development in this area; thus, additives are subject to change on a regular basis.

LLDPE geomembranes are again made from polyethylene resin, carbon black, and additives for processing stabilization and long-term antioxidation. In the low-pressure polymerization of LLDPE, the random incorporation of alpha olefin comonomers produces sufficient short-chain branching to yield densities in the range of 0.915–0.930 g/cc. This results in resin properties quite different from HDPE. In particular, LLDPE has lower density, greater flexibility, less tendency toward stress cracking, greater elongation at break (both in-plane and out-of-plane), and lower modulus values at all levels of elongation.

The polyethylene resin used for LLDPE geomembranes is a linear polymer of ethylene with other alpha olefins. Carbon black is added to LLDPE geomembrane formulations for general stabilization purposes, particularly for ultraviolet light stabilization. It is generally added as a pellet preformulated into a concentrate. Additives are introduced into an LLDPE formulation for the purposes of antioxidation or processing also within the concentrate.

Flexible Polypropylene

Polypropylene is a common thermoplastic used throughout geosynthetics due to large part to its cost-effectiveness. Polypropylene is created through polymerization of propylene gas. It is obtained from high temperature cracking of petroleum hydrocarbons and propane. It is not surprising that the polypropylene and polyethylene (known collectively as polyolefins, or simply olefins) have many of the same properties. However, polypropylene differs in the following respects:

- lower density,
- service temperature is higher,
- harder and more rigid,
- more resistant to environmental stress cracking, and
- more susceptible to oxidation and chemical attack.

There are three basic structural stereostatic arrangements of polypropylene. They include isotactic, atactic, and syndiotactic. Commercially available polypropylene is 95% isotactic and is exclusively used in geosynthetics. As seen in Table 7.3.1, fPP geomembranes are made from polypropylene resin, carbon black, and additives. fPP is "a material produced by polymerization of propylene with or without other alpha olefin monomers having a 2% secant modulus of less than 300 MPa as determined by ASTM D5323." This results in properties with significantly greater flexibility than polyethylene. fPP geomembranes are either nonreinforced or reinforced with a fabric scrim between individual "plies" of the material. It is then designated as flexible polypropylene-reinforced (fPP-R). The polypropylene resin used for fPP geomembranes is a linear polymer of ethylene with other alpha olefins. The resin is generally supplied to the manufacturer in the form of pellets. Carbon black is also added to fPP geomembrane formulations for general stabilization purposes, particularly for ultraviolet light stabilization. It is generally added as a preformulated concentrate in pellet form.

Additives are introduced into an fPP formulation for the purposes of antioxidation, long-term durability, and as a lubricant and/or processing aid during manufacturing. fPP geomembranes are often manufactured with a fabric, called "reinforcing scrim," between two plys of the sheet polymer. This results in a three-ply laminated geomembrane consisting of geomembrane/scrim/geomembrane, which is laminated together under pressure, to form a unitized system. The geomembrane is said to be reinforced and then carries the designation fPP-R. Other options of multiple plys are also available. The scrim imparts dimensional stability to the material, which is important during storage, placement, and seaming. It also imparts a major increase in mechanical properties over the unreinforced type, particularly in its tensile strength, modulus of elasticity, tear, and puncture resistance of the final product.

The reinforcing scrim for fPP-R geomembranes is a woven fabric generally made from polyester yarns manufactured in a standard "basket" weave. There are many fibers (of very fine diameter) per individual yarn (e.g., 100–200 fibers per yarn depending on the desired strength). The yarns, or "strands" as they are referred to in the industry, are spaced close enough to one another to achieve the desired properties, but far apart enough to allow open space between them so that the opposing geomembrane sheet surfaces can adhere together. This is sometimes referred to as "strike-through" and is measured by a ply-adhesion test. The designation of reinforcing scrim is based on the number of yarns, or strands, per inch of woven fabric. The general range is from 6 × 6-to-20 × 20, with 10 × 10 being the most common. A 10 × 10 scrim refers to 10 strands per inch in the machine (or warp) direction and an equal number of 10 strands per inch in the cross-machine (or weft) direction.

It must also be mentioned that the polyester scrim yarns must be coated to allow for good bonding to the upper and lower fPP sheets. Various coatings, including latex, PVC, and others, have been used. The exact formulation of the coating material (or "ply enhancer") is usually proprietary.

Polyvinyl Chloride

PVC is a widely used polymer. In terms of revenue generated, it is one of the most valuable products of the chemical industry. Globally, over 50% of PVC manufactured is used in construction for house siding, piping, etc. As a building material, PVC is inexpensive, hard, and easy to assemble. In recent years, PVC has been replacing traditional building materials such as wood and concrete. Despite appearing to be an ideal building material, concerns have been raised about the environmental and human health costs of PVC.

PVC is produced from its monomer, vinyl chloride. PVC is a hard plastic that is made softer and more flexible by the addition of plasticizers, the most widely used being phthalates. When used as a geomembrane, plasticizer additions of 25%–35% are common. PVC geomembranes are made from PVC resin, plasticizer(s), fillers, and additives. PVC geomembranes can be smooth on both sides, smooth on one side and textured on the other, or textured on both sides. Texturing of PVC geomembranes is sometimes referred to as "faille."

The PVC resin used for PVC geomembranes is made by cracking ethylene dichloride into a vinyl chloride monomer. It is then polymerized to make PVC resin. The PVC resin (in the form of a porous white powder) is then compounded with other components to form a PVC compound. Plasticizers are added to PVC formulations to impart flexibility, improve handling, and modify physical and mechanical properties. When blended with the PVC resin, the plasticizer(s) must be completely mixed into the resin. Because the resin is a porous powder, and the plasticizers are liquid, mixing of the two components continues until the liquid is completely adsorbed by the powder. The result is a plasticized resin powder, which can be readily mixed with other ingredients. It is also possible to wet blend the formulation with acceptable results. There are two general categories of possible plasticizers: monomeric plasticizers and polymeric plasticizers. There are many specific types within each category. For example, monomeric plasticizers are sometimes phthalates, epoxides, and phosphates, whereas polymeric plasticizers are sometimes polyesters, ethylene copolymers, and nitrile rubber. In this regard, plasticizer formulations differ widely, e.g., between European and North American PVC geomembranes.

The filler used in a PVC formulation is a relatively small component (recall Table 7.3.1), and (if used at all) its identification can be requested. Calcium carbonate, in powder form, has been used, but other options also exist. In addition, a very small amount of carbon black is usually added to obtain the characteristic gray color of PVC geomembranes. Other additives (for the purpose of ease of manufacturing, coloring, and stabilization) are also added to the formulation in relatively small amounts. They are generally not identified but can be requested.

Chlorosulfonated Polyethylene

CSPE geomembranes consist of CSPE resin, fillers, carbon black (or colorants), and additives. The finished geomembrane is usually fabricated with a woven textile, called a "reinforcing scrim," between the individual plys of the material. It is then designated as chlorosulfonated polyethylene-reinforced (CSPE-R). There are two different types of CSPE resins used to make CSPE geomembranes. One is a completely amorphous polymer, whereas the other is a thermoplastic material containing a controlled amount of crystallinity to provide useful physical properties in the uncured state while maintaining flexibility without the need of any plasticizers. The second type is generally used to manufacture geomembranes. CSPE is made directly from branched polyethylene by adding chlorine and sulfur dioxide. The chlorosulfonic groups act as preferred cross-linking sites during the polymer aging process. In the typical commercial polymer, there is one chlorosulfonyl group for each 200 backbone carbon atoms.

CSPE resin usually arrives at the sheet manufacturing facility in containers, called Gaylords. They are somewhat pillow shaped with dimensions of 10–20 mm. They are relatively spongy in their resistance to finger pressure. Alternatively, CSPE can be premixed with carbon black in slab form, which is then referred to as a master batch. The master batch is usually made by a formulator and shipped to the manufacturing facility in a prepared form.

The amount of carbon black in CSPE geomembranes varies from 5% to 10% by weight. The carbon black functions as an ultraviolet light blocking agent and a filler, and it aids in processing. The purposes of blending fillers into the CSPE compound are to provide workability and processability. The common types of fillers are clay and calcium carbonate. Both are added in powder form and in relatively large quantities ranging from 40% to 50%. Additives, in the amount of 5%–15% by weight, are used in CSPE compounds for the purpose of stabilization, which is used to distinguish the various grades.

CSPE geomembranes are usually fabricated with a "reinforcing scrim" between two plys of the polymer sheets. This results in a three-ply laminated geomembrane consisting of geomembrane/scrim/geomembrane, which is laminated together, under pressure, to form a unitized system. The geomembrane is said to be reinforced and carries the designation CSPE-R.

Ethylene Propylene Diene Terpolymer

EPDM (ethylene propylene diene monomer) is a type of synthetic thermoset rubber, which has a wide range characteristics. Dienes currently used in the manufacture of EPDM rubbers are dicyclopentadiene, ethylidene norbornene, and vinyl norbornene. EPDM rubber, which is closely related to ethylene propylene rubber, is a copolymer of ethylene and propylene. The ethylene content is around 45%–85%. The higher the ethylene content, the higher the loading possibilities of the polymer, better mixing, and extrusion. Peroxide curing of these polymers is typical and gives a higher cross-link density compared with their amorphous counterpart. The dienes, typically comprising from 2.5% to 12% by weight of the composition, serve as cross-links when curing with sulfur and the diene functions as a coagent, which provides resistance to unwanted tackiness, creep, and flow during the end use.

The main properties of EPDM are its outstanding heat, ozone, and weather resistance. The resistance to polar substances and steam are also good. It also has excellent electrical insulating properties. EPDM can be compounded to meet specific properties and are available in a range of molecular weights, which makes them resistance to ketones, ordinary diluted acids, and alkalies to varying degrees.

Ethylene propylene diene monomer is sometimes reinforced with a fabric scrim, i.e., EPDM-R, and is often used in roofing. EPDM is a thermoset plastic, which results in some unique properties. It is also possible to blend EPDM with fPP forming a thermoplastic olefin so as to have a hybrid material capable of heat bonding when field seaming.

MANUFACTURING

Once the desired geomembrane formulation has been thoroughly mixed, it is then manufactured into a continuous sheet. The major processes used for manufacturing of the various types of sheets of geomembranes are variations of extrusion (either flat die or blown film) or calendering. Autoclaving and spread coating will also be briefly mentioned.

Blending, compounding, mixing, and/or masticating of the various components described herein is conventionally done on a weight percentage basis. However, each geomembrane's processing is somewhat unique in its equipment and procedures. Even for a particular type of geomembrane, manufacturers will use different procedures for blending or mixing. It should be noted that blending equipment must be clean and completely purged from previously produced materials of a different formulation. This might require sending a complete cycle of purging material through the system.

Flat Die Extrusion

HDPE, LLDPE, and fPP geomembranes are manufactured by taking the mixed components described earlier and feeding them into a hopper, which leads directly into a horizontal extruder as shown in Fig. 7.3.1. Many extruders are 200 mm diameter systems, which are quite large, e.g., up to 9 long. In an extruder, the components enter a feed hopper, are heated to melting temperature, and are horizontally mixed and transported via a continuous screw through a feed section, compression section, metering section, and filtering screen, and are then pressure fed into a die.

The thoroughly mixed molten polymer enters the die and is forced by pressure to flow laterally, exiting the die gap along its full width at a thickness that is precisely controlled. A typical sheet width is approximately 7.0 m. Wider sheets can be made by using two side-by-side extruders such that the melt streams meld together within the common die block. The finished dimensions, particularly thickness, can be very tightly controlled (±1.0%). The finished geomembrane sheet must be free from pinholes, surface blemishes, scratches, or other defects (e.g., nonuniform color, streaking, roughness, carbon black agglomerates, visually discernible regrind, etc.). In addition, the nominal thicknesses of the sheet should be specified.

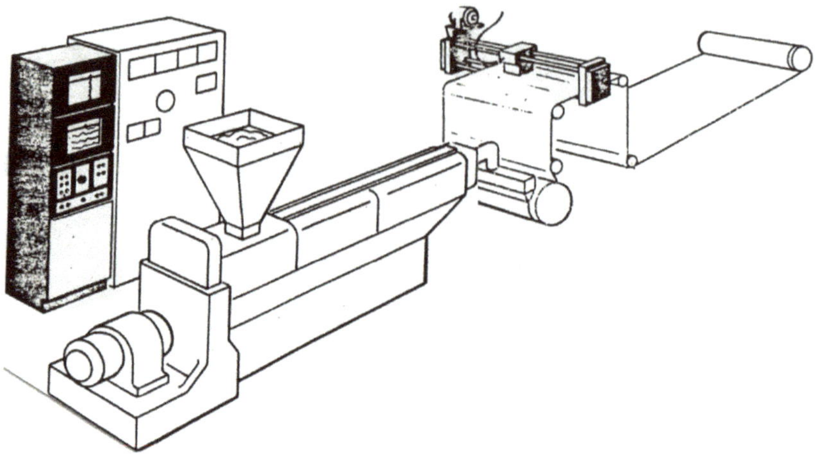

Figure 7.3.1 Schematic showing wide mouth extrusion of geomembrane (Koerner and Bove, 1984).

It is readily possible to texture sheets and create a high friction surface. Texturing can be done on one or both surfaces of the sheet. There are currently four methods used to create a textured surface on geomembrane. They are as follows:

- coextrusion via nitrogen expulsion,
- structuring via a patterned set of cooling rollers,
- hot particle impingement after heating the geomembrane, and
- applying a texture via a foaming agent.

The surface texturing for all four methods of texturing should be uniform and consistent across the roll width and length. In addition, it should be of the same type of polymer and formulation as the base sheet polymer.

Flat die extrusion can be configured to provide for different melt streams made by different extruders, a process called "coextrusion." This is accomplished by using a special feed block ahead of the die. HDPE/LLDPE/fPP sheet has been made in this manner as well as white/black sheet. It is also possible to have two extruders produce relatively thin sheets of the formulated material, one on top of the other, with a reinforcing fabric inserted between them. The three-ply material, geomembrane/scrim/geomembrane, must be passed through counterrotating rollers such that still warm polymer sheets can strike-through the scrim and adhere to one another. Several products have been produced in such a manner. However, the more common method of manufacturing scrim reinforced geomembranes is by calendering and will be described later.

Blown Film Extrusion

By using a vertically oriented circular die, the extruder can feed molten polymer in an upward orientation creating a large cylinder of polyethylene sheet as shown in Fig. 7.3.2. Because the cylinder of

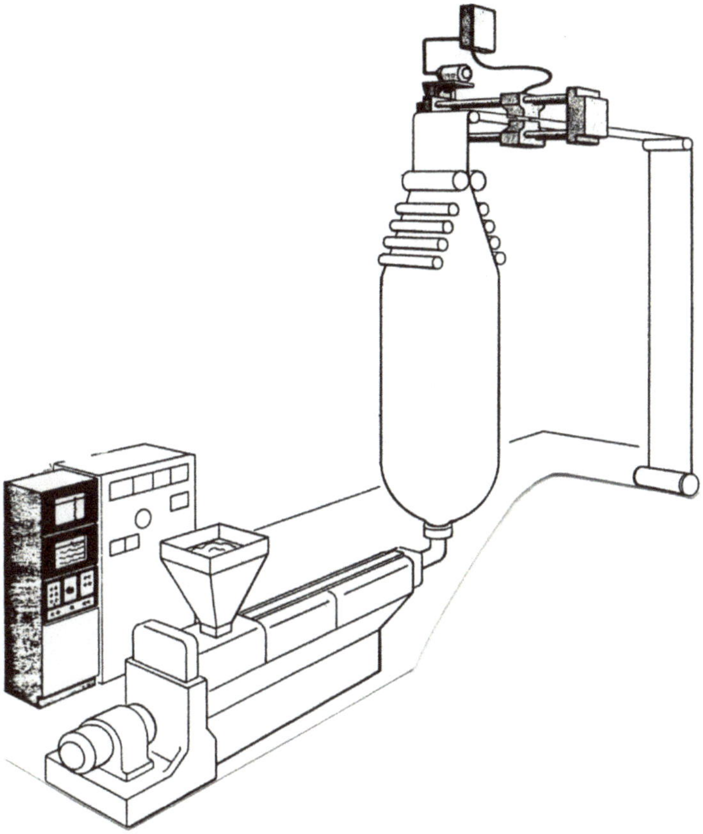

Figure 7.3.2 Schematic showing blown film extrusion of geomembrane (Koerner and Bove, 1984).

polymer is closed at the top where it passes over a set of nip rollers that advances the cylinder, a low air pressure is generally maintained within it to control its dimensional stability. Note that upward moving air is also induced outside of the cylinder to further aid in the "bubbles" stability of the gradually cooling cylinder of geomembrane. After passing through the nip rollers at the top of the system, the collapsed cylinder is cut longitudinally, opened to its full width, brought down to floor level, and rolled onto a windup core. Note that collapsing the cylinder and passing it through the nip rollers results in two subtle parallel creases. After slitting the collapsed cylinder and opening it to full width, remnants of the two creases remain at quarter distances of the sheet width from the edges.

Blown film extrusion is the most common method of manufacturing HDPE, LLDPE, and fPP geomembranes. Using this process, such geomembranes can be made smooth, one-sided textured, two-sided textured, and coextruded, using different polymers, colorants, and/or materials. A large single extruder is used to make smooth sheet by the blown film extrusion method. All thicknesses can be produced and the widths of the sheet can be as large as 10.7 m. To provide a textured surface on one or both sides of the sheet, a spiral mandrel die is used with secondary and tertiary extruders. The coextrusion texturing method utilizes a blowing agent (usually nitrogen gas) in the molten extrudate and delivers it from a small extruder immediately adjacent to the main extruder. When both sides of the sheet are to be textured, two small extruders (one internal and one external to the main extruder) are necessary. As the extrudate from these smaller extruders leaves the die and meets the cool air, the blowing agent expands, opens to the atmosphere, and creates the textured surface(s). Depending on the amount of nitrogen added, different degrees of texturing can be produced.

Calendering

PVC geomembranes are manufactured by taking proportional weights of PVC resin (a dry powder) and plasticizer (a liquid) and premixing them until the plasticizer is absorbed into the resin. Filler (in the form of a dry powder) and other additives (also usually dry powders) are then added to the plasticized resin and the total formulation is mixed in a blender. Various types of high intensity or low intensity blenders can be used. Note that PVC rework in the form of chips, rather than edge trim, can be introduced at this point.

The resulting free-flowing powder formulation is fed into a mixer, where heat is introduced, thereby initiating a reaction between the various components. These mixers can be either batch type (e.g., Banbury) or continuous type (e.g., Farrel). In these mixers, the temperature is approximately 180°C, which melts the mixture into a viscous mass. The mixed material is then removed from the discharge door or port onto a conveyor belt. From the conveyor belt the viscous material is further worked (called "masticating") in a rolling mill into a consistent, uniform color, continuous mass of 100–150 mm diameter. Finished product edge trim can also be introduced into the rolling mill. The fully mixed formulation is then fed by conveyor directly into the sizing calender, which will be described later.

CSPE geomembranes are made in the same way as PVC by mixing CSPE resin with carbon black (or their colorants) thereby making a "master batch" of these two components. Added to this mixture are fillers, additives, and lubricants in a batch-type mixer. Within the mixer, the shearing action of the rotors against the ingredients generates enough heat to cause melting and subsequent chemical reactions occur. After the mixing cycle is completed, the batch is dropped from the mixer onto a two-roll mill, then to a conveyor leading to a second two-roll mill. In moving through the roll mills, it is further mixed into a completely homogenized material having a uniform color and texture. It should be noted that edge trim is often taken from finished sheet and routed back to the roll mill for mixing and reuse.

PVC and nonreinforced CSPE formulations, irrespective of the preprocessing procedures, are manufactured into continuous geomembrane sheets by a calendering process. The viscous feed of polymer coming from the rolling mill(s) is worked and flattened between counterrotating rollers into a geomembrane sheet. Most calenders are "inverted-L" configurations, but other options are also available, see Fig. 7.3.3. These rollers are usually smooth-surfaced stainless steel cylinders and are up to 2.0 m in width. The opening distance between adjacent cylinders is set for the desired thickness of the final

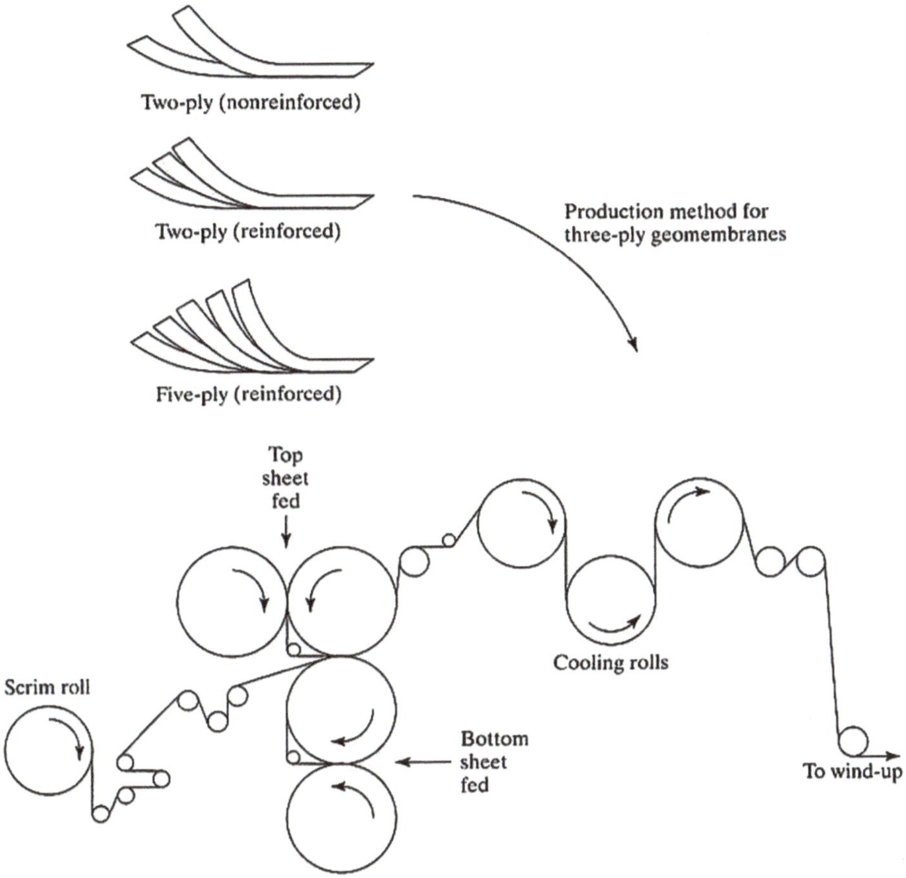

Figure 7.3.3 Processing of geomembranes by calendaring, utilizing multiple plies of material. *Compliments of Dow Chemical Co.*

sheet. A rolling bank of molten material is formed between adjacent rolls. In an inverted four roll "L" calender, several such banks are formed. They act as reservoirs for the molten material and help to fill the sheet to full width as it passes between the rolls. As the geomembrane exits from the calender, it enters an additional series of rollers for the purposes of pickoff, embossing, stripping, cooling, and cutting. At least one and perhaps two rollers are embossed, or "faille," so as to impart a slight surface texture on the geomembrane. The purpose of this embossing is to prevent the rolled geomembrane from sticking together (i.e., "blocking") during windup, storage, and transportation.

The geomembranes just described, as manufactured by the calendering process, are typically 1.0–2.0 m wide and are transported in rolls weighing up to 6.7 kN to a "panel fabrication" facility. When a specific job order is placed, the rolls are unwound and placed directly on top of one another for factory seaming into a panel. A panel will typically consist of 5–10 rolls, which are alternately seamed to one another (i.e., the left side of a particular roll is seamed to the underlying roll, whereas the right side is seamed to the overlying roll). After seaming, the completed panel is accordion folded (in a lengthwise direction) and placed on a wooden pallet. It is then covered with a protective wrapper and shipped to the job site for deployment. Note that some fabricators use other procedures for panel preparation.

Autoclaving

Thermoset polymers are also used for geomembranes, both nonreinforced and reinforced (e.g., EPDM and EPDM-R). Immediately after the rubber compound passes through the calender, the temperature of the calendered liner (with or without fabric reinforcement) is quickly reduced to room temperature when it passes over a series of cooling drums. Eventually, the rubber sheet passes through the accumulator on a wide belt conveyor and eventually goes to the splicing table. At the splicing table, even wider sheets are produced by cutting the calendered sheet (without any fabric) and splicing it to the previously cut sheet of the same dimensions. Fabric-reinforced liner is not converted to wider sheets. The two cut sheets are spliced together (in a matter of seconds) as they pass underneath a splicing beam. After the factory splice has been formed, both sides of the liner are dusted with a release agent and the dusted liner is wound onto a metal curing mandrel. Once the mandrel of rubber has been properly prepared for curing and placed on a curing rack, it is conveyed to the autoclave for curing.

The curing equipment consists of a standard autoclave, preferably with a heated jacket to reduce condensation, and a closed chamber in which a rack containing several curing mandrels is placed and steam is slowly introduced. In operating a closed chamber steam vulcanizer, the curing cycle consists of a rise to the predetermined pressure, a definite period at the required curing pressure, and a blow down to atmospheric pressure. The rack of curing mandrels is allowed to set approximately 30—45 min before preparing each mandrel for observation (inspection, cut plan, etc.) on the finishing/inspection floor. Finally, the inspected geomembranes are packaged (rolled onto a core) for distribution.

Spread Coating

An exception to the flat die or calendering methods of producing reinforced geomembranes is the spread coating process. This process is currently unique to a geomembrane type called ethylene interpolymer alloy, but it has been used to produce other specialty geomembranes in the past. It can be done in a factory or in situ in the field. The process utilizes a dense fabric substrate, commonly either a woven or nonwoven textile, and spread coats the molten polymer on its surface. Due to the dense structure of the fabric, penetration of the viscous polymer to the opposite side is usually not complete. When cooled, the sheet is turned over and the process is repeated on the opposite side or it is left in place in the field. Adherence of the polymer to the fabric is essential.

Geomembranes produced by the spread coating method are indeed multiple ply—reinforced materials, but they are produced by a method different than flat die or calendering. Manufacturing quality control (MQC) and manufacturing quality assurance (MQA) plans and specifications should be framed in a similar manner as described previously for other reinforced geomembranes.

GENERIC SPECIFICATIONS AND DELIVERY TO SITE

Generic specifications for the geomembranes described in this chapter are available. Such generic specifications cover designated test methods, limiting properties, and minimum testing frequencies. Currently, they are as follows:

GRI-GM13: HDPE: Smooth and Textured
GRI-GM17: LLDPE: Smooth and Textured
GRI-GM18: fPP: Nonreinforced and Reinforced
GRI-GM21: Ethylene Propylene Diene Terpolymer (EPDM): Nonreinforced and Reinforced
GRI-GM22: Polyethylene-Reinforced
GRI-GM25: Linear Low-Density Polyethylene-Reinforced
GRI-GM30: Coated Tape Polyethylene Barrier
ASTM D7732: PVC

The most recent versions of these generic specifications can be found on the websites of the institutes, which developed and maintain these documents (i.e., the Geosynthetic Institute and the Flexible Geomembrane Institute for PVC).

While there should be great concern and care focused on the manufacturers and installers of geomembranes, it is also incumbent that they are handled, packaged, stored, transported, restored, rehandled, and deployed in a manner so as not to cause damage. Different types of geomembranes require different types of packaging after they are manufactured. Generally, extrusion manufactured geomembranes (HDPE, LLDPE, and fPP) are wrapped around a core in roll form, whereas calendered geomembranes (PVC, CSPE-R, and EPDM-R) are accordion folded in two directions and packaged onto wooden pallets.

The extrusion-manufactured geomembranes are produced and fed directly to a windup core in full-width rolls. No external wrapping or covering is generally needed nor provided. These rolls, which weigh up to 22 kN, are either moved by forklifts using a long rod inserted into the core (called a "stinger") or they are picked up by high strength polymer slings with a crane or hoist. The slings are often dedicated to each particular roll and follow along with it until its actual deployment. The rolls are usually stored in an outdoor area. They are stacked such that one roll is nested into the valley of the two underlying rolls. In contrast, calendered geomembranes in the form of finished pallets are generally not stored longer than a few weeks at the fabrication facility or the field.

The geomembrane rolls or pallets are shipped to the job site, offloaded, and temporarily stored at a remote location on the job site in advance on spotting the material into final position and then seaming. It is the primary duty of the installation contractor [in the context of CQC or construction quality assurance (CQA)] to assure that the geomembrane supplied to the job site is the proper material that was specified in the contract, as stipulated in the plans and specifications. Clear marking should identify all rolls or pallets with the information described previously. A complete list of roll numbers should be prepared for each material type.

Prior to or on delivery of the rolls or pallets of geomembrane, the CQA Engineer should ensure that conformance test samples are obtained and sent to the proper laboratory for testing. This will generally be the laboratory of the CQA firm but may be that of the CQC firm if so designated in the CQA documents. Alternatively, conformance testing could be performed at the manufacturer's facility and, when completed, the particular lot should be marked for the particular site under investigation. This process is called MQC and can be augmented by MQA, see Daniel and Koerner (2007).

When the subgrade or subbase (either soil or some other geosynthetic) is approved as being acceptable, the rolls or pallets of the temporarily stored geomembranes are brought to their intended location, unrolled or unfolded, and accurately positioned, or "spotted" for field seaming. Before beginning to move the geomembrane rolls or pallets from their temporary storage location at the job site, the soil subgrade (or other subbase material) should be inspected and approved for its preparedness. Whenever possible, seams should be shingled downgradient and minimized. "Butt" seams are never to be allowed. If at all possible, full-roll lengths should be used on side slopes thereby avoiding seams perpendicular to side slopes. In this same regard, there should be an "as-built" plan for archiving by the owner. In addition, the installer should submit a panel layout plan for approval by the design engineer prior to placement.

SEAMING OR JOINING GEOMEMBRANES

The seaming of the deployed rolls or panels is a critical aspect of the successful functioning of any geomembrane as a barrier to liquids and gases. This section describes the various seaming methods in current use, the concept and importance of test strips (or trial seams), and destructive and nondestructive test methods.

The fundamental mechanism of seaming overlapped polymeric geomembrane sheets together is to temporarily reorganize (i.e., melt) the polymer structure of the two surfaces to be joined in a controlled manner that, after the application of pressure and a certain amount of time, results in the two sheets being bonded together. This polymer reorganization results from an input of energy that originates from either thermal or chemical methods. These methods may involve the addition of extra polymer in the bonded area.

Ideally, seaming two geomembrane sheets would result in no net loss of tensile strength or elongation across the two sheets and the joined sheets would perform as one single geomembrane sheet. However, due to geometric irregularities and possible loss of material, current seaming techniques result in minor tensile strength and/or elongation loss relative to the parent geomembrane sheet. The characteristics of the seamed geomembranes are a function of the type of geomembrane and the seaming technique deployed. These characteristics, such as required strength and elongation geomembrane type and seaming type, should be recognized by the designer when applying the appropriate design factors of safety for the overall geomembrane function and site-specific system performance.

It should be noted that the seam can be the location of the highest tensile stress in a geomembrane liner. Designers and inspectors should be aware of the importance of seeking only the highest quality geomembrane seams. The minimum seam tensile strengths and elongation (as determined by design) for various geomembranes must be predetermined by laboratory testing, knowledge of past field performance, manufacturer's literature, various journal and conference proceedings papers, or other standards setting organizations that maintain current information on seaming techniques and technologies.

The available methods of seaming thermoplastic and thermoset geomembranes discussed in this chapter are given in Tables 7.3.2 and 7.3.3.

Table 7.3.2 Various methods of joining geomembranes

Thermal Processes	Chemical Processes
Fusion: • Hot Wedge • Hot Air	*Chemical*: • Chemical • Bodied Solvent
Extrusion: • Fillet • Flat	*Adhesive*: • Chemical Adhesive • Contact Adhesive • Tape

Modified from Daniel and Koerner (1993).

Table 7.3.3 Possible field-seaming methods for various geomembranes

Type of Seaming Method	Type of Geomembrane					
	HDPE	LLDPE	fPP and fPP-R	PVC	CSPE-R	EPDM and EPDM-R
Thermal fusion (hot wedge and hot air)	A	A	A	A	A	N/A
Extrusion (fillet and flat)	A	A	A	N/A	N/A	N/A
Chemical (chemical and bodied chemical)	N/A	N/A	N/A	A	A	N/A
Adhesive (chemical and contact)	N/A	N/A	N/A	A	A	A

A, method is "applicable"; *CSPE-R*, chlorosulfonated polyethylene-reinforced; *EPDM*, ethylene propylene diene terpolymer; *EPDM-R*, ethylene propylene diene terpolymer-reinforced; *fPP*, flexible polypropylene; *fPP-R*, flexible polypropylene-reinforced; *HDPE*, high-density polyethylene; *LLDPE*, linear low-density polyethylene; *N/A*, method is "not applicable"; *PVC*, polyvinyl chloride.
Modified from Daniel and Koerner (2007).

Within the group of geomembranes that are being discussed in this chapter, there are four general categories of seaming methods: thermal fusion, extrusion welding, chemical, and adhesive. Each will be explained along with their specific variations so as to give an overview of field-seaming technology.

There are two thermal fusion methods that can be used on all thermoplastic geomembranes. In both of them, portions of the opposing surfaces are truly melted. This being the case, temperature, pressure, and seaming rate all play important roles, in that excessive melting weakens the geomembrane and inadequate melting results in low seam strength. The hot wedge, or hot shoe, method consists of an electrically heated resistance element in the shape of a wedge that travels between the two sheets to be seamed. As it melts the surfaces of the two sheets being seamed, a shear flow occurs across the upper and lower surfaces of the wedge. No grinding of surfaces is required with this method. Roller pressure is applied as the two sheets converge at the tip of the wedge to form the final seam. Hot wedge units are controllable as far as temperature, amount of pressure applied, and travel rate. A standard hot wedge creates a single uniform width seam, whereas a dual hot wedge (or "split" wedge) forms two parallel seams with a uniform unbonded space between them. This space is then used to evaluate seam quality and continuity by pressurizing the unbonded space with air and monitoring any drop in pressure that may signify a leak in the seam. In the author's opinion, dual track hot wedge seams are considered to be the premier seaming method for all types of thermoplastic geomembranes, i.e., all of the geomembranes listed in Table 7.3.3 except for EPDM, which is a thermoset polymer.

The hot air method makes use of a device consisting of a resistance heater, a blower, and temperature controls to force hot air between two sheets to melt the opposing surfaces. Immediately following the melting of the surfaces, pressure is applied to the seamed area to bond the two sheets. As with the hot wedge method, both single and dual seams can be produced. In selected situations, this technique may also be used to temporarily "tack" weld two sheets together until the final seam or weld is made and accepted. The method is not often used for production seams.

Extrusion welding is presently used on geomembranes made from polyolefins (HDPE, LLDPE, and fPP). After preparing the surfaces of the two geomembranes by light surface grinding, a ribbon of molten polymer is extruded over the edge of the two surfaces to be joined. The molten extrudate causes the prepared surfaces of the sheets to become hot and melt, after which the entire mass cools and bonds together. The extrudate must completely cover all surfaces that have been ground. The technique is called extrusion fillet seaming when the extrudate is placed over the leading edge of the seam and is called extrusion flat seaming when the extrudate is placed between the two sheets to be joined. The latter method is essentially depreciated at this time. It should be noted that extrusion fillet seaming is essentially the only practical method for seaming polyolefin geomembrane patches, for seaming in poorly accessible areas such as sump bottoms and around pipes, and for seaming of extremely short seam lengths. Temperature and seaming rate both play important roles in obtaining an acceptable bond; excessive melting weakens the geomembrane and inadequate melting results in poor extrudate flow across the seam interface and low seam strength. The polymer used for the extrudate is also important and should be the same type and formulation used to make the geomembrane sheets being joined.

The designer should specify acceptable extrusion compounds and how to evaluate them in the specifications and CQA documents.

Regarding the chemical seam types, which are used on nonpolyolefin thermoplastic geomembranes, chemical seams make use of a liquid solvent (usually methyl ethyl keytone) applied between the two geomembrane sheets to be joined. After a few seconds (required to soften the surfaces), externally applied roller pressure is applied to make complete contact and bond the sheets together. As with any of the chemical seaming processes to be described, the two adjacent materials to be bonded are transformed into a viscous phase. Care must be used to see that the proper amount of chemical is applied to achieve the desired results. Bodied chemical seams are similar to chemical seams except that 1%–20% of the parent lining resin or compound is dissolved in the chemical and then is used to make the seam. The purpose of adding the resin or compound is to increase the viscosity of the liquid for slope work and/or adjust the evaporation rate of the chemical. This viscous liquid is applied between the two opposing surfaces to be bonded. After a few seconds, roller pressure is applied to make complete contact.

For thermoset geomembranes, such as EPDM, chemical adhesive seams make use of a dissolved bonding agent (an adhesive) in the chemical or bodied chemical, which is left after the seam has been completed and cured. The adhesive thus becomes an additional element in the system. Contact adhesives are applied to the lower mating surface. After reaching the proper degree of tackiness, the upper sheet is placed on top of the lower, followed by application of roller pressure. The adhesive forms the bond and is an additional element in the system.

Whenever the plans and specifications are not written around a particular seaming method, the actual method that is used becomes a matter of choice for the installation contractor. The specific details are rarely stipulated even in the specification or CQA documents. This is to give the installation contractor the maximum latitude in selecting seaming temperatures, travel rates, mechanical roller pressures, chemical type, tack time, hand rolling pressure, etc. The role of the project's plans, specifications, and CQA documents is to adequately provide for destructive tests (on test strips and on production seams) and nondestructive tests (on production seams) to assure that the seams are fabricated to the highest quality and uniformity and are in compliance with the project's specifications.

DESTRUCTIVE SEAM TESTING

Trial seams and test strips, also called qualifying seams, are an important aspect of CQC/CQA procedures. They are meant to serve as a prequalifying experience for personnel, equipment, and procedures for making seams on the identical geomembrane material under the same climatic conditions as the actual field production seams will be made. The trial seams are usually made on two narrow pieces of excess geomembrane varying in length between 1.0 and 3.0 m, trial seams should be made in sufficient lengths, preferably as a single continuous seam for testing purposes.

The goal of these trial seams is to reproduce aspects of the actual production field-seaming activities intended to be performed in the immediately upcoming work session so as to determine equipment and operator proficiency. Ideally, trial seams can be used to estimate the quality of the production seams while minimizing damage to the installed geomembrane through destructive mechanical testing. Trial seams are typically made every 4 h (e.g., at the beginning of the morning work and after the lunch break). They are also made whenever personnel or equipment are changed, when climatic conditions reflect wide changes in geomembrane temperature, or when other conditions occur that could affect seam quality.

The destructive testing of the trial seams should be done as soon as the installation contractor feels that the strength requirements of the contract specification or CQA documents can be met. Thus, it behooves the contractor to have all aspects of the trial seam fabrication in complete working order just as would be done in the case of fabricating production field seams. For extrusion and thermal fusion seams, destructive testing can be done as soon as the seam cools. For chemical and adhesive seams this could take several days; the use of a field oven to accelerate the curing of the seam is then advisable.

Test specimens are cut from the trial seam using a 25 mm wide die. The specimens are then tested in both peel and shear using a field tensiometer. Generally, peel tests are more informative in assessing the quality of the seam. If any of the test specimens fail, a new trial seam is fabricated. The CQA personnel should observe all field trial seam tests. If the specimens pass, seaming operations can move directly to production seams. If the seam fails for a second time, corrective action needs to be taken and root cause identified, whether it may be equipment or personal in nature. Pass—fail criteria for seam tests are identified in GRI-19 seam specification.

Destructive testing of geomembrane seams is the process of actually cutting out (i.e., to sample) and removing a portion of the completed production seam and then to further die cut the sample into appropriately sized test specimens. These specimens are then tested, according to a specified procedure, to failure or to yield, depending on the type of geomembrane.

A possible procedure is to select the sampling location and cut two closely spaced 25 mm wide test specimens from the seam. The distance between these two test specimens is defined later. The individual specimens are then tested in a peel mode using a field tensiometer. If the results are acceptable, the complete seam between the two field test specimens is removed, identified, and properly distributed. If either test specimen fails, two new locations on either side of the failed specimen(s) are selected until acceptable seams are located. The seam distance between acceptable seams is usually repaired by cap stripping and suitable seaming, but other techniques are also possible. The exact procedure must be stipulated in the specifications or CQA document.

The length dimension of the field seam sample between the two test specimens just described varies according to whatever is stipulated in the plans and specifications or in accordance with the CQA documents. The options for the seam sample lengths between the two peel test specimens mentioned above that are seen in various plans, specifications, and CQA documents are as follows:

- A 36-cm sample is taken from the seam and cut into five shear and five peel specimens. ASTM D6392 gives a recommended template in this regard. The tests are conducted in the field or at a remote laboratory by, or under the direction of, the responsible CQA organization.

- A 71-cm long sample is taken from the seam and cut in half. One half is further cut into five shear and five peel test specimens, which are tested in the field or at a remote laboratory by the CQA organization (see ASTM D6392). The other half is sent to a remote laboratory for testing by the CQC organization who also does five shear and five peel tests. Alternatively, sometimes only the CQA organization does the testing and the second half of the sample is left intact and archived by the owner/operator.
- A 106-cm long sample is taken from the seam and cut into three individual 36-cm samples. Individual samples go to the CQA organization, the CQC organization, and the owner/operator. The CQA and CQC organizations each cut their respective samples into five shear and five peel test specimens and conduct the appropriate tests immediately (see ASTM D6392). The remaining sample is archived by the owner/operator.

NONDESTRUCTIVE SEAM TESTING

Although it is obviously important to conduct some destructive tests on the fabricated seams, such tests do not give adequate information on the continuity and completeness of the entire seam between sampling locations. It does little good if one section of a seam meets the specification requirements, only to have the section next to it missed completely by the field-seaming crew. Thus, continuous methods of nondestructive testing (NDT) are necessary. In each of the various methods, the goal is to validate 100% of the seams or, at least, a major percentage of them. The currently available NDT methods for evaluating the adequacy of geomembrane field seams are listed in Table 7.3.4 in the order that they will be discussed.

The dual seam method using positive air pressure was mentioned earlier in connection with the dual hot wedge or dual hot air thermal seaming methods. The air channel that results between the two bonded tracks is inflated using a hypodermic needle and is pressurized to approximately 200 kPa (30 lb/in.^{2}). There is no limit as to the length of the seam that can be tested. If the pressure drop is within an allowable amount in the designated time period (usually 5 min), the seam is acceptable; if an unacceptable drop occurs, a number of actions can be taken:

1. the seam distance can be systematically halved and the process can be repeated until the leak is located,
2. the seam length in question can be tested by some other leak detection method,
3. an extrusion fillet weld can be placed over the entire outer edge of the seam, or
4. a cap strip can be seamed over the entire seam length involved.

Details of the test can be found in ASTM D6392 and GRI-GM6. This test is well suited for long, straight-seam lengths. It is generally performed by the installation contractor, with CQA personnel viewing the procedure and documenting the results. It is considered by most people to be the premier NDT method for geomembrane seams of all that can be possibly used.

The vacuum chamber method uses a box up to 1 m (3 ft) long with a transparent top that is placed over the seam. The open bottom has a compressible gasket to provide a seal against the geomembrane seam. A vacuum of approximately 20 kPa (3 lb/in.^{2}) is then applied. When a leak is encountered, the

Table 7.3.4 Nondestructive geomembrane seam testing methods

Nondestructive Test Method	Primary User		General Comments					
	CQC	CQA	Cost of Equipment	Speed of Tests	Costs of Tests	Type of Result	Recording Method	Operator Dependency
1. Dual seam (positive pressure)	Yes	–	$500	Fast	Moderate	Yes–no	Manual	Low
2. Vacuum chamber (negative pressure)	Yes	Yes	$1200	Slow	Very high	Yes–no	Manual	High
3. Air lance	Yes	–	$200	Fast	Low	Yes–no	Manual	High
4. Mechanical point (pick) stress	Yes	–	Nil	Fast	Nil	Yes–no	Manual	Very high
5. Electric sparking	Yes	Yes	$3000	Fast	Low	Yes–no	Manual	Moderate
6. Electric wire	Yes	Yes	$500	Fast	Low	Yes–no	Manual	Moderate
7. Ultrasonic	–	Yes	$7000	Moderate	High	Yes–no	Automatic	Moderate

The leak integrity survey methods are not included in this Table. They evaluate the entire geomembrane installation and not the seams specifically, see the next section in this regard.
CQA, construction quality assurance; *CQC*, construction quality control.
Modified from Richardson and Koerner (1988).

soapy solution originally placed over the seam shows bubbles, thereby reducing the vacuum. This is due to air entering from beneath the geomembrane and passing through the unbonded zone. The test is slow to perform (a 10 s dwell time is currently recommended), and it is often difficult to make a vacuum-tight joint around the bottom of the box where it passes over the seam edges. Due to upward deformations of the liner into the vacuum box, only geomembrane thickness greater than 1.0 mm should be tested in this manner. For thinner, more flexible geomembranes, an open grid wire mesh can be used along the bottom of the box to prevent uplift. It should be recognized that 100% of the field seams cannot be inspected by the vacuum box method. The test cannot cover portions of sumps, anchor trenches, and pipe penetrations with any degree of assurance. The method is also very awkward to use on side slopes. The adequate downward pressure required to make a good seal is difficult to mobilize because it is usually done by standing on top of the box.

The air lance method uses a jet of air at approximately 350 kPa pressure coming through an orifice of 5 mm (3/16 in.) diameter. It is directed beneath the upper edge of the overlapped seam at a distance of about 100 mm (4.0 in.) from the edge of the seamed area to detect unbonded areas. When such an area is located, the air passes through the opening in the seam causing an inflation and fluttering in the localized area. A distinct change in the sound emitted can generally be heard. The method works best on relatively thin [less than 1.1 mm (45 mils)] flexible geomembranes but works only if the defect is open at the front edge of the seam where the air jet is directed. It is essentially a geomembrane installer's method to be used in a CQC manner.

The mechanical point stress or "pick" test uses a dull tool, such as a blunt screwdriver, under the leading edge of a seam. With care, an individual can detect an unbonded area, which would be easier to separate than a properly bonded area. It is a rapid test that obviously depends completely on the care and sensitivity of the person doing it. Detectability is similar to that of the air lance method, but both are very operator dependent. This test is to be performed only by the geomembrane installer as a CQC method. Design or CQA personnel should not use the pick test due to potential damage to the geomembranes.

Electric sparking is a technique used to detect pinholes in thermoplastic liners. The method uses a high-voltage (15–30 kV) current, and any leakage to ground (through an opening or hole) results in sparking. The method is used very effectively with coextruded conductive geomembranes.

The electric wire method places a copper or stainless steel wire between the overlapped geomembrane region and actually embeds it into the completed seam. After seaming, a charged probe of about 20,000 V is connected to the wire and a wand is slowly moved over the length of the seam. A seam defect between the probe and the embedded wire results in an audible alarm from the unit.

The last group of nondestructive test methods noted in can collectively be called ultrasonic methods. While they have been shown in the laboratory to be technically viable, they are rarely used in practice. The technique is basically a thickness measurement technique and is only for use with nonreinforced geomembranes. A high-frequency pulse is sent into the upper geomembrane and (in the case of good acoustic coupling and good contact between the upper and lower sheets) reflects off of the bottom of the lower one. If, however, an unbonded area is present, the reflection will occur at the unbonded interface.

LEAK INTEGRITY SURVEY

The electrical leak location survey method (or ELL method) impresses a high voltage across the geomembrane and then detects the precise locations where electrical current flows through leaks in the electrically insulating geomembrane. The development of the technology was initiated by the US EPA in 1980 (Peters et al., 1982). The first commercial leak location surveys for earth-covered and water-covered geomembranes were performed in 1985. Numerous organizations currently provide such surveys.

Techniques are available for exposed liner systems and soil-covered geomembranes. If the secondary liner system is to be investigated, it must be done before the primary liner system is placed. Most importantly, the method can be performed after the soil materials have been placed on the geomembrane, e.g., after placement of a leachate collection and removal layer. As shown, electrodes are placed in contact with conducting material above and below the geomembrane. A voltage is applied between these electrodes using an isolated power supply. Because the geomembrane is an electrical insulator, electrical current will flow only if there are holes in the geomembrane. Electrical potential data are collected using two additional moveable electrodes on the surface material in a grid or survey line pattern within the facility. The data are recorded manually or with portable data loggers for plotting and analysis. Characteristic leak signals caused by current flowing through the leaks indicate the locations of the leak.

Depending on the conductivity of the materials above and below the geomembrane, the power required can be as low as a watt. For surveys with earth cover, however, the maximum power usually needed is in the order of a few hundred watts.

The geomembrane leak location method is standardized in ASTM D6747 and D7002. These standards describe several ELL implementation methods. These are as follows:

- a method for water-covered geomembranes;
- a method using an electrically conductive layer on one surface of a specialty geomembrane;
- a method for geomembranes covered with earth materials;
- a preinstalled grid system for monitoring of leaks throughout the active life of the facility;
- a water puddle method for uncovered geomembranes; and
- use of electrically conductive sheet.

Numerous technical papers have been published describing the leak survey methods and results of surveys most importantly that of Nosko and Touze-Foltz (2000), see Tables 7.3.5 and 7.3.6 that are self-evident in their very important messages.

Table 7.3.5 Location of holes

No. of Holes	Flat Floor (1)	Corners and Edges (2)	Under Drainage Pipes (3)	Pipe Penetrations (4)	Other (5)
4194 100%	3261 77.8%	395 9.4%	165 3.9%	84 2.0%	289 6.9%

Table 7.3.6 Cause of holes versus size of holes

Size of Holes (cm^2)	Stones	%	Heavy Equipment	%	Welds	%	Cuts	%	Worker Directly	%	Total
<0.5	332	11.1	–	–	115	43.4	5	8.5	195	–	452
0.5–2.0	1720	57.6	41	6.3	105	39.6	36	61.0	105	84.4	2097
2.0–10	843	28.2	117	17.9	30	11.3	18	30.5	36	15.6	1044
>10	90	3.0	496	75.8	15	5.7	–	–	–	–	601
Amount	2985		654		265		59		231		4194
Total	71.17%		15.59%		6.32%		1.41%		5.51%		100%

COVERING, PROTECTION, AND BACKFILLING

The field deployed and seamed geomembrane must be backfilled with soil or covered with a subsequent layer of geosynthetics in a timely manner after its acceptance by the CQA personnel. If the covering layer is soil, it will generally be a drainage material such as sand or gravel depending on the required permeability of the overlying layer. Depending on the particle size, hardness, and angularity of this soil, a geotextile or other type of protection layer will often be necessary. If the covering layer is a geosynthetic, it will generally be a geonet or geocomposite drainage material, which is usually placed directly on the geomembrane. This is obviously a critical step because geomembranes are relatively thin materials with puncture and tear strengths of finite proportions. Specifications should be very clear and equivocal regarding this final step in the installation survivability of geomembranes.

It is important to recognize that most earthwork equipment used in heavy construction is generally too large and too heavy to provide safety to the underlying geomembrane. Furthermore, such equipment cannot negotiate tight areas, particularly when lined with geosynthetics. Equipment practices (e.g., speed, orientation, turning, stopping, and starting) must be greatly limited depending on the site-specific circumstances. However, under no circumstances can construction equipment drive directly on any geomembrane.

Various geosynthetic materials may be called on to cover the deployed and seamed geomembrane. Often a geotextile, a geonet, or a geocomposite will be the covering material. Sometimes, however, it will be a geogrid for covering soil reinforcement on slopes. Construction equipment cannot be allowed to operate or drive directly on any of the geosynthetics. Generators, low tire inflation all-terrain vehicles (ATVs), and other seaming-related equipment are allowed as long as they do not damage the geosynthetic. As a result, the movement of large rolls of geotextile, geonet, or geocomposite becomes very labor-intensive. Proper planning and sequencing of the operations is important for logistical control. The geosynthetic materials are laid directly on the geomembrane with no bonding of any type to the geomembrane being allowed. For example, thermally fusing of a geonet to a geomembrane should not be permitted.

The geosynthetics placed above the geomembrane will either be overlapped (as with some geotextiles), sewn (as with other geotextiles), connected with plastic ties (as with geonets), mechanically joined with rods or bars (as with geogrids), or male/female joined (as with drainage composites).

There are at least three important considerations concerning soil backfilling of geomembranes: type of soil backfill material, type of placement equipment, and considerations for waves or wrinkles in the geomembrane. Concerning the type of soil backfilling material, its particle size characteristics, hardness, and angularity are important with regard to the puncture and tear resistance of the geomembrane. In general, the maximum soil particle size is very important, with additional concerns over poorly graded soils, increased angularity, and increased hardness, being of significance. Research on the puncture resistance of geomembranes has shown that HDPE and scrim-reinforced geomembranes are more sensitive to puncture than are PVC, LLDPE, fPP, and other nonreinforced flexible geomembranes for conventional thicknesses of the respective types of geomembranes (Hullings and Koerner, 1991).

Although the truncated cone hydrostatic test is an extremely challenging index-type test, it does not reflect creep and/or stress relaxation of the geomembrane. In reviewing numerous CQA documents, it appears that the maximum backfill particle size for use with HDPE and CSPE-R geomembranes should not exceed 6 mm. LLDPE, PVC, fPP, and EPDM geomembranes appear to be able to accommodate larger soil backfill particle sizes. If the soil particle size must exceed the approximate limits given (e.g., for reasons of providing high permeability in a drainage layer), then a protection material must be placed on top of the geomembrane and beneath the soil. Nonwoven, needle punched geotextiles are generally used in this regard. They are often of a high mass per unit area, up to 1000 g/m^2, and have been made from both virgin (usually) and recycled (occasionally) fibers.

Concerning the type of placement equipment, the initial lift height of the backfill soil is very important. (Note that construction equipment should never be allowed to move directly on any deployed geomembrane.) This includes rubber tired vehicles such as automobiles and pickup trucks but does not include light-weight equipment such as ATVs. The minimum initial lift height should be determined for the type of placement equipment and soil under consideration; however, 150 mm is clearly the minimum thickness. Between this value and approximately 300 mm, low ground pressure placement equipment should be specified. Ground contact pressure equipment no greater than 40 kPa is recommended. For lift heights of greater than 300 mm, proportionately heavier placement equipment can be used.

Placement of soil backfilling should proceed from a stable working area adjacent to the deployed geomembrane and gradually progress outward. Soil is never to be dropped from dump trucks or front-end loaders directly onto the geomembrane. The soil should be pushed forward in an upward tumbling action so as not to impact directly on the geomembrane. It should be placed by a bulldozer or front-end loader and never by a motor grader, which would necessarily have its front wheels riding directly on the geomembrane.

WRINKLE MANAGEMENT

It is very difficult to achieve intimate contact of a geomembrane to the subgrade under conditions of high temperature and bright sunlight. These conditions result in expansion of all geomembrane types, with the stiffer, thicker, and black types such as HDPE resulting in fewer, but much larger waves than other types of geomembranes. The management of avoiding such waves or wrinkles is difficult. Clearly, waves that are entombed in the soil backfill covering the geomembrane will not flatten out even under extreme normal pressures (see Soong and Koerner, 1998). Thus, the geomembrane must be flat as it is backfilled.

The installer has a number of options with respect to the backfilling process. Commentary on each method follows:

1. It is not uncommon to push waves in front of the advancing backfill soil allowing them to accumulate. When too large, or no longer possible, the covering geotextile (if present) and the geomembrane are cut, overlapped, and properly seamed. This new seam must then be inspected by a nondestructive test, quite often the vacuum box method.

2. By fixing the geomembrane at its ends or intermediate points, it can be forced to lay flat so that backfilling can proceed. It must be done one roll at a time, or in discrete sections, and is slow and tedious.
3. White-surfaced geomembrane (coextruded or laminated) helps reduce waves or wrinkles by at least 50%. In this regard, covering a black geomembrane with a white geotextile is also helpful.
4. Geomembranes have been placed, seamed, and backfilled beneath a temporary tent. It obviously avoids sun but not high temperatures and results in low productivity and high costs.
5. Backfilling in the early morning hours when the geomembrane has cooled overnight and thus becomes taut is an obvious way of working with nature. In some locations it might be necessary to work at night. Both situations limit productivity somewhat but result in a flat geomembrane, which is the ultimate goal.

In summary, it is the duty of the CQA organization to see that intimate contact via flat geomembranes against the subgrade is achieved when backfilled. The method of obtaining this goal is that of the geomembrane installer and earthwork contractor. It is not up to CQA personnel to direct the field operation in this regard. CQA personnel should focus only on the result of achieving intimate contact with the subgrade via flat geomembranes.

LONG-TERM PERFORMANCE AND LIFETIME

Durability is a major issue for all polymeric materials (geomembranes), when long design lifetimes are required. Geomembranes used in critical applications have service lifetime requirements of 30—100s of years (Koerner et al., 2016). Over the past few decades, lessons have been learned from case histories and considerable research. The question of "how long will the geomembrane last?" is largely dependent on the polymeric type and its specific formulation as well as its in situ exposed environment over time. Various degradation mechanisms effecting polymeric materials can act in isolation or synergistically. They are as follows:

In-Isolation Effects

1. Ultraviolet light occurs only when the geosynthetic is exposed. Its intensity varies on the location on the globe, atmosphere, and time of year.
2. Radioactivity is not a factor unless the polymer is exposed to radioactive materials of sufficiently high intensity to cause chain scission, e.g., high-level radioactive waste materials.
3. Chemical degradation can occur in all polymers and varies from water (least aggressive) to organic solvents (most aggressive).
4. Hydrolysis degradation is the primary degradation mechanism for polyesters and polyamides. Technically, hydrolysis is a reaction with water. That is what happens when esters are hydrolyzed by waters of varying pH. Jailloux et al. (1992) suggest that the alkaline hydrolysis of esters actually involves reaction with hydroxide ions.
5. Swelling refers to a growth in bulk as a result of the uptake or adsorption of liquids.
6. Extraction refers to a release or withdrawal of components by means of a nonequilibrium driving force such as heat, pressure, diffusion, dispersion, convection, etc.
7. Delamination refers to splitting plies apart or separation into layers.

8. Oxidation occurs in all polymers and is the major degradation mechanism in polyethylene and polypropylene, Comer et al. (1998).
9. Biological degradation is generally not a factor unless biologically sensitive additives (such as low molecular weight plasticizers) are included in the polymer formulation.

Synergistic Effects

1. Temperature is an enabling variable for all of the previously mentioned mechanisms. The higher the temperature, the more rapid the degradation.
2. Applied stresses are a complicating factor, which is site specific and should be appropriately modeled in the incubation process.
3. Multiple and/or changing mechanisms over time need to be considered. It is unlikely that any of the previously described degradation mechanisms are acting alone and in-isolation from all others. When considering long service lifetimes of 100 years and beyond (Rollin, 2004), this is almost a certainty in most applications. It is, however, an extremely difficult situation to model due to the unpredictability of future events and actions.

The goal of geomembranes used in containment facilities is to manufacture the geomembrane without holes or flaws, and then to install (seam) them in a leak-free system. Additionally, the backfilling must be done so as not to compromise the installed liner system. Toward this end, references have been made to placement, seaming, destructive testing, NDT, covering, and backfilling concept as they relate to landfill applications.

As set forth by the EPA risk reduction laboratory in Cincinnati, Ohio, a system approach as just described is highly recommended for widespread implementation for geomembranes used in waste containment applications. It involves the following:

- Adequate design,
- Quality materials, which are tested for conformance to a specification,
- Best available installation, which is quality assured,
- Performance checked or monitored prior to the facility going online, and
- Careful operations and maintenance.

Availing oneself to such a quality system approach for geomembrane-lined facility will consistently upgrade the system by providing checks and balances to assure a quality end product. It will also minimize leakage rates. Fig. 7.3.4 illustrates the average primary leakage rates in nearly 200 landfills across the United States (Bonaparte et al., 2002).

Fig. 7.3.4 illustrates that a GM/GCL composite primary after final cover and in the last life cycle stage can regularly have a leakage rate under 50 lphd (5 gpad). This type of performance has further been corroborated with annual data from 34 subtitle D (municipal solid waste) double composite—lined landfills in the State of New York. The secondary flows in the detection system between the two composite liner systems have the following statistics for 2015: maximum 268 lphd, minimum 5 lphd, and mean 72 lphd.

All 34 of these New York landfills operate with hydraulic heads less than 600 mm and bottom gradient of a minimum of 2%. This is proof positive that geomembranes used in composite liner system

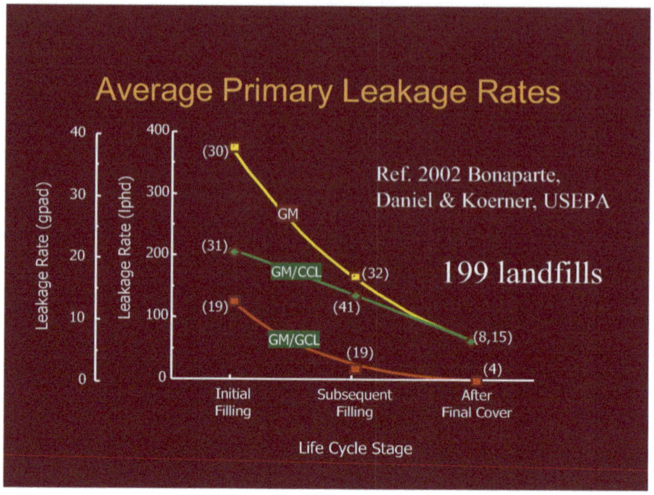

Figure 7.3.4 Graph showing average primary leakage rates of double composite liner systems in the United States (Bonaparte et al., 2002).

work well. When geomembranes are embodied in a systems approach, they may isolate waste from the surrounding environment for considerable period of time.

References

Bonaparte, R., Daniel, D.E., Koerner, R.M., 2002. Assessment and Recommendations for Improving the Performance of Waste Containment Systems. U.S. Environmental Protection Agency, Washington, DC. EPA/600/R-02/099 (NTIS PB2003-103048).

Comer, A.I., Hsuan, Y.G., Konrath, L., 1998. The Performance of Flexible Polypropylene Geomembranes in Covered and Exposed Environments. In: Proceedings of the 6th International Conference on Geosynthetics, Vol. 1. IFAI, Atlanta, Georgia, USA, pp. 359–364.

Daniel, D.E., Koerner, R.M., 1993. Geotechnical Aspects of Waste Disposal (ch. 18). In: Daniel, D.E. (Ed.), Geotechnical Practice for Waste Disposal. Chapman and Hall, London.

Daniel, D.E., Koerner, R.M., 2007. Waste Containment Facilities: Guidance for CQA and CQA of Liner and Cover Systems, second ed. ASCE Press, Reston, VA. 352 pp.

GRI-GM6, Standard Practice for "Pressurized Air Channel Test for Dual Seamed Geomembranes".

GRI-GM13, Standard Specification for "Test Properties, Testing Frequency and Recommended Warranty for High Density Polyethylene (HDPE) Smooth and Textured Geomembranes".

GRI-GM17, Standard Specification for "Test Properties, Testing Frequency and Recommended Warranty for Linear Low Density Polyethylene (LLDPE) Smooth and Textured Geomembranes".

GRI-GM18, Standard Specification for "Test Properties, Testing Frequency and Recommended Warranty for Flexible Polypropylene (fPP and fPP-R) Nonreinforced and Reinforced Geomembranes".

GRI-GM19, Standard Specification for "Seam Strength and Related Properties of Thermally Bonded Polyolefin Geomembranes".

GRI-GM21, Standard Specification for "Test Properties and Testing Frequency and Recommended Warranty for Ethylene Propylene Diene Terpolymer (EPDM and EPDM-R) Nonreinforced and Reinforced Geomembranes".

Hullings, D.E., Koerner, R.M., 1991. Puncture resistance of geomembranes using a truncated cone test. In: Proceedings, Geosynthetics '91. IFAI, pp. 273–286.

Jailloux, J.M., Nait-Ali, K.L., Freitag, F., 1992. In: Exhaustive Long-Term Study on Hydrolysis of High-Tenacity Polyester- 10 year Results, EuroGeo Conference Paper Number 212, pp. 1–6.
Koerner, R.M., 2012. Designing with Geosynthetics, sixth ed., Vol. 1 and 2. Xlibris Corporation.
Koerner, G.R., Bove, J., 1984. Geomembrane Manufacture and Installation. In: Purdue Waste Conference, West Lafayette Indiana U.S.A, pp. 231–245.
Koerner, R.M., Hsuan, Y.G., Koerner, G.R., 2016. Lifetime Prediction of Exposed Geotextiles and Geomembranes. In: Proceedings GeoAmericas 2016, Miami Beach, Florida, U.S.A, pp. 1–16.
Nosko, V., Touze-Foltz, N., 2000. Geomembrane liner failures: modeling of its influence on containment transfer. In: Proc. EuroGeo 2, Bologna, Italy, pp. 557–566.
Peters, W.R., Schultz, D.W., Duff, B.D., March 1982. Electrical resistivity techniques for locating liner leaks. In: Land Disposal of Hazardous Waste, Proceedings of the Eight Annual Research Symposium, Cincinnati, Ohio.
Richardson, G.N., Koerner, R.M., 1988. Geosynthetic Design Guidance for Hazardous Waste Landfill Cells and Surface Impoundments. EPA/600/S2-87/097.
Rollin, A.L., Marcotte, M., Jacquelin, T., Chaput, L., 2004. Leak Location Surveys in Exposed and Covered Geomembrane Liners Using an Electrical Leak Detection Technique. Geosynthetics, IFAI, Roseville, MN, pp. 615–626.
Soong, T.-Y., Koerner, R.M., 1998. Laboratory study of HDPE geomembranes waves. In: Proc. 6th Intl. Conf. on Geosynthetics. IFAI, Atlanta, pp. 301–306.

Further Reading

Darilek, G.T., Miller, L.V., 1998. Comparison of dye testing of electrical leak location testing of a solid waste liner system. In: Proc. 6th International Conference on Geosynthetics. IFAI, pp. 273–276.
GRI-GM7, Standard Practice for “Accelerated Curing of Geomembrane Test Strips Made by Chemical Fusion Methods”.
GRI-GM12, Standard Test Method for “Asperity Measurement of Textured Geomembranes Using a Depth Gage”.
GRI-GM14, Standard Guide for “Selecting Variable Intervals for Taking Geomembrane Destructive Seam Samples Using the Method of Attributes”.
GRI-GM20, Standard Guide for “Selecting Variable Intervals for Taking Geomembrane Destructive Seam Samples Using the Method of Control Charts”.
GRI-GS7, Standard Test Method for “Determining the Index Friction Properties of Geosynthetics”.
Haxo, H.E., 1988. Lining of Waste Containment and Other Impoundment Facilities. EPA/600/2-88/052, Washington, DC.
Hsuan, Y., Koerner, R.M., 1992. Stress Cracking Potential and Behavior of HDPE Geomembranes. Final Report to U.S. EPA, Contract No. CR-815692.
Hsuan, Y.G., Koerner, R.M., Kolbasuk, G., Soong, T.-Y., 2001. Potential effects of regrind on the properties of HDPE geomembranes. In: Proc. GRI-15 Conference on Hot Topics in Geosynthetics II. GSI, Folsom, PA, pp. 225–234.
Laine, D.L., Darilek, G.T., 1993. Locating leaks in geomembrane liners covered with a protection soil. In: Geosynthetics ’93, vol. 3. IFAI, Vancouver, British Columbia, Canada, pp. 1403–1412.
PGI 1104, Standard Specification for “Polyvinyl Chloride (PVC) Geomembranes”, PVC Geomembrane Institute.
Richardson, G.N., 1992. Construction Quality Management for Remedial Action and Remedial Design Waste Containment Systems. EPA/540/R-92/073. U.S. EPA, Washington, DC.
Struve, F., 1995. Extrusion of geomembranes. In: Proc. GRI-8 Conference on Geosynthetic Resins, Formulations and Manufacturing. GSI, Folsom, PA, pp. 97–115.
U. S. EPA, May 1991. Technical Guidance Document; Inspection Techniques for the Fabrication of Geomembrane Field Seams. EPA/530/SW-91/051, 174 pp.
U. S. EPA, September 1993. Technical Guidance Document; Quality Assurance and Quality Control for Waste Containment Facilities. EPA/600/R-93/182, 305 pp.

8 CHAPTER

DRAINAGE SYSTEMS

8.1

LEACHATE COLLECTION SYSTEMS

Hans-Günter Ramke

INTRODUCTION

The lining of the landfill bottoms requires the installation of a leachate collection system (LCS):

- to collect leachate
- to discharge the leachate at defined points out of the disposal area
- to avoid an accumulation of leachate above the bottom liner.

The accumulation of leachate needs to be controlled for four reasons:

- to prevent liquid levels rising to such an extent that they can spill over and cause uncontrolled pollution to ditches, drains, watercourses, etc.
- to increase the efficiency of the liners by reduction of the hydraulic head of leachate above the liner
- to reduce the contact time between waste and leachate so reducing the transport of contaminants from the solid to the liquid phase
- to ensure the stability of the waste body.

A leachate collection system generally consists of a drainage layer of inert material with high permeability, of drainage pipes to collect the leachate and discharge it out of the disposal area, of collection and inspection shafts, and of collection pipes outside the disposal area.

There are different national regulations and a number of guidance documents available on leachate drainage system design. The objective of this paper is to explain the state-of-the-art of leachate collection systems for engineering purposes, defined e.g. by the Technical Recommendation "Design Principles for Leachate Collection Systems in Bottom Liner Systems" of the Technical Committee Geotechnics of Landfills of the German Geotechnical Society (GDA E 2-14, 2011).

At the beginning of the chapter a brief overview of requirements on leachate collection systems and of design features is presented. An initial focus is the potential clogging of leachate collection systems (incrustation process), as this phenomenon presents the most severe problem for the long-life functionality of leachate collection systems. A second focus is the hydraulic calculation of drainage systems. The standard equation for sloping drainage layers is explained and exemplarily solved. Finally some more detailed recommendations for the design of leachate collection systems will be given under consideration of simple and cost effective solutions for developing countries. More comprehensive compilations are given by Ramke (1998, 2009).

GENERAL PRINCIPLES OF DESIGN

General requirements

The general requirements on LCSs in bottom liner systems can be summarized as follows:

- The LCS must ensure that the saturated thickness of leachate above the bottom liner is less than a specified level.
- The drainage layer of the LCS should extend over the entire base of a landfill and up its sloping bottom dams.
- The bottom liner has to be profiled to have sufficient gradient to promote efficient drainage to the drainage pipes. The drainage pipes should have sufficient longitudinal slope to reduce sedimentation.
- A network of perforated drainage pipes should be laid within the drainage layer with continuous gradients toward the leachate collection points and should be capable of being inspected and maintained.
- Leachate removal from the collection points should work by gravity drainage.
- All materials must be strong enough to withstand physical damage and be resistant to microbiological and chemical attacks in the environment of the landfill.
- The LCS must be able to function in spite of unavoidable incrustation processes during and after the operational life of the landfill.
- Inspection and maintenance is necessary until the system is no longer required.
- The LCS incorporates a suitable level of "redundancy" in the design to take account of failure of parts of the system and provide alternative drainage routes to collection points.

Above ground landfills on plain areas or flat slopes are more suitable than landfills in pits as they can allow free drainage of leachate by gravity. Below ground landfills in pits requires the pumping of leachate, which cannot be ensured in the long-term.

The design has to consider volume, quality, and temperature of leachate:

- Leachate volume

  The LCS is to be designed for the collection of the expected volumes of leachate generated. This will vary during the life of the site and can be estimated using a water balance calculation.
- Leachate quality

  Materials selected for use in the drainage system have to be resistant to and compatible with the expected "worst case" leachate quality. It should be borne in mind that leachate quality depends on waste composition and degradation processes and therefore may vary with time and age.
- Temperature

  The drainage system should be capable of withstanding the elevated temperatures, which can occur at the base of a landfill (between 15 and 40 °C may be expected at the bottom of landfills for municipal solid waste).

Leachate volume, leachate quality, and temperature at the base of a landfill will depend on the type of waste disposed of at the particular landfill. Landfills for construction and demolition waste, for municipal solid waste, and for industrial waste show significant different leachate properties and temperature developments, and leachate quantity can differ as well.

International guidelines and national legal requirements

There are not many international guidelines, which give technical recommendations on LCSs. One example is the UNEP publication Solid Waste Management (UNEP, 2005). Two different designs are described there: the sloped terrace design and a piped bottom collection system. The first design might be a solution on a site with a sloping bottom, whereas the second system largely corresponds to the standard system described below. Some differences in detail to the current standard (grain size, pipe diameter, use of a geotextile filter) will be explained in Recommendations on Design and Materials section.

A further Technical Guide, published by the World Bank (Rushbrook and Pugh, 1999), to Planning, Design, and Operation of Solid Waste Landfills in Middle- and Lower-Income Countries describes the abovementioned slope terrace design and an LCS, again consisting of a layer of granular material and a network of drainage pipes. The drainage material should consist of large diameter aggregates with a hydraulic conductivity of 1×10^{-4} m/s or higher. The drain pipes should have a diameter of 100 mm.

The United States have defined the requirements on LCSs in an updated version of the "Technical Manual for Solid Waste Disposal Facility Criteria" (US EPA, 1998). The drainage layer should have a minimum height of 30 cm; the hydraulic conductivity of the granular material should exceed 1×10^{-4} m/s. The minimum slope of the bottom should be 2%. In addition, geosynthetic drainage nets are mentioned as possible substitutes for the granular layers. Furthermore equations are given for the calculation of drainage pipe spacing, pipe diameters, and bearing capacity. Again a filter layer between waste, drainage layer, and drainage pipes is discussed.

The Landfill Directive of the European Union (EU, 1999) distinguishes between landfills for nonhazardous waste and hazardous waste but requires in both cases only a drainage layer with a minimum thickness of 50 cm. Furthermore, the directive states

> "In addition to the geological barrier described above a leachate collection and sealing system must be added in accordance with the following principles so as to ensure that leachate accumulation at the base of the landfills is kept to a minimum.
>
> Member states may set general or specific requirements for inert waste landfills and for the characteristics of the above mentioned technical means."

Examples for those specific national requirements are the regulations in Germany. The first national regulations were developed in the late 1980s of the last century and were set into force in the early 1990s by two general administrative regulations (TA Abfall, 1991; TA Siedlungsabfall, 1993). The basic requirements defined there need not to be changed and have proven their suitability. Due to these long and positive experiences, the German system of requirements will be explained in more detail below.

At present, the German Directive on the Landfilling of Waste (DepV, 2013), which came into force in 2002 (latest version from September 2017), defines general and specific requirements on the LCSs by a link to the engineering standard DIN 19667:2009 (Drainage of Landfills) and by a general cross

Table 8.1.1 Typical requirements on leachate collection systems in landfills

Element	Property	Specification
Drainage layer	Thickness	≥50 cm drainage layer
	Hydraulic conductivity	$\geq 1 \times 10^{-3}$ m/s (long-term)
	Material, grain size	≥30 cm coarse gravel, 16–32 mm
Drainage pipes	Inner diameter	≥250 mm
	Length	≤400 m
	Drain spacing	≤30 m
	Maximum drainage distance	≤15 m
Slopes	Cross slope	≥3%
	Longitudinal slope	≥1%

reference to the Uniform Quality Standards of the Federal States (BQS). The BQS regulate details of landfill design, suitability tests, construction, and quality management. For the LCSs, the BQS refer to Technical Recommendations of the Technical Committee Geotechnics of Landfills of the German Geotechnical Society. In particular, the Recommendation "Design Principles for Leachate Collection Systems in Bottom Liner Systems" (GDA E 2-14, 2011) will be considered here.

The requirements on LCSs in landfills following these standards are compiled in Table 8.1.1. Fig. 8.1.1 shows an example of an LCS that satisfies the current German requirements. These requirements are the same for landfills for construction and demolition waste (German landfill category I), for landfills for (pre-treated) municipal solid waste (category II), and for landfills for hazardous waste (category III).

For drainage pipes, a technical standard defines minimum requirements (DIN 4266-1:2011).

Although these specifications need not be relevant in all cases, depending on national regulations and/or site-specific conditions, they indicate the general principles and considerations for leachate collection system design.

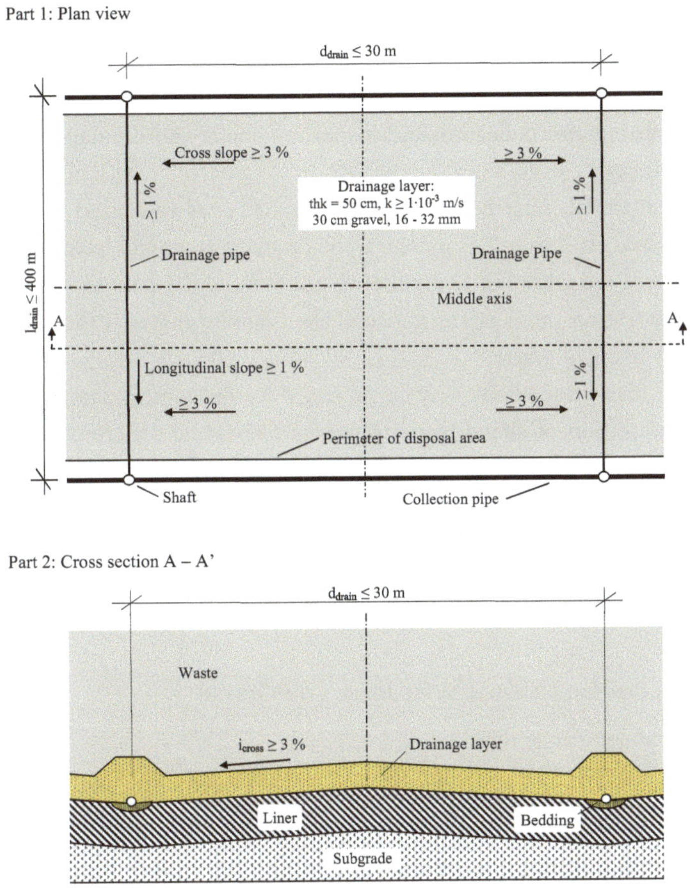

Figure 8.1.1 Plan view and cross section of a leachate collection system.

FAILURE AND CLOGGING IN LEACHATE COLLECTION DYSTEMS

Introduction

Two main reasons leading to failure of LCSs, or of their parts, can be distinguished:

- mechanical damages, especially of the drainage pipes
- clogging of the drainage layer and within the drainage pipes.

Mechanical damages of drainage elements can be prevented by careful design, material selection, and construction. However, clogging — especially by the formation of hardened, insoluble incrustations in drainage pipes and drainage layers, which is a particular issue in landfills for municipal solid waste — represents the most important problem facing the long-term effective functioning of landfill drainage systems in bottom liner systems. Incrustation material can be removed from the drainage pipes by regular flushing and if necessary by rotary cutting but cannot be removed from the pipe casing or from the drainage layer. Therefore in the following the focus will be laid on the description of intensity and causes of incrustation processes.

Mechanical damages of drainage elements

Not only drainage pipes but also collection and inspection shafts and drainage material can be susceptible to mechanical damage.

Granular drainage material must be resistant to pressure stresses caused by the waste weight and traffic load. If the material crushes under pressure, the drainage layer will have a higher content of finer particles and will show a reduction in hydraulic conductivity and be more vulnerable to incrustation processes. A suitability test on pressure resistance of the granular material can prove its long-term stability even under high loads of waste.

Also collection and inspection shafts may be subjected to mechanical damage. Typical damage may result from the decomposition of shafts made of concrete. Normal concrete is not resistant to biogas and acid leachate and cannot be used for this purpose. In general, concrete shafts outside the disposal area should be protected by an inliner made of HDPE.

Many shafts inside landfills (inside the disposal area and the waste body) have shown mechanical damages. This is caused by deformations of the waste body and by unexpected peak loads. If shafts inside the disposal area cannot be avoided, then special attention should be paid to careful design and construction.

Typical mechanical problems shown by drainage pipes include

- Depressions in the direction of flow

  Depressions occur when the bearing layer of the drainage pipes is unsuitable (cohesive soils) or insufficiently compacted. The consequence is a permanent backwater of leachate in the drainage pipe.
- Deformations of drainage pipes

  Deformation of drainage pipes without cracking (plastic buckling) is only shown by flexible drainage pipes, made of synthetics. Reasons that may result in deformation and plastic buckling are unexpected loads, insufficiently constructed pipe bearings, too high temperatures at the landfill bottom, or incorrect static calculations.
- Cracks and shards

  Cracks can be observed at flexible drainage pipes and in bending resistant drainage pipes, the formation of shards only in bending resistant drainage pipes.

  Bending resistant pipes (e.g., stoneware pipes) were used in the first decade of landfill technology in Germany. Many lines showed cracks and shards after some years. The main reason is their vulnerability to deformations of the underground. Stoneware pipes are not suitable in LCSs.

  Cracks in flexible drainage pipes are caused not only by high temperatures (e.g., on intermediate liners in the waste body) but often also by the use for pipe production of low-quality plastic material. Furthermore, stress cracks might be caused by drainage slots.

However, nearly all of these problems with drainage pipes can be avoided when the current knowledge concerning properties of synthetic materials, static calculations of pipes, and quality management during construction is considered. A compilation of the state-of-the-art is given by the quality guideline "Pipes, Shafts and Construction Elements in Landfills" (SKZ/TÜV, 2017).

Intensity of incrustation processes

The incrustation phenomenon was primarily observed in sanitary landfills for municipal solid waste in the middle of the 1980s in Germany and the United States, where control and maintenance of drainage systems were started early on as a consequence of the construction of modern landfills. A systematic survey of sanitary landfills in Germany at the end of the 1980s showed that such incrustation occurs at most landfill sites. In the drainage pipes, the degree of incrustation ranged from thin layers on the pipe wall to a significant reduction in pipe cross section (Fig. 8.1.2). Observed impairment of the drainage material ranged from local areas of incrustation to extensive solidification and can result in partial up to extensive loss of functional capability of the LCS (Fig. 8.1.3).

Figure 8.1.2 Incrustation material flushed out of a drainage pipe.

Figure 8.1.3 Detail of drainage material with incrustation of the pipe casing.

Since that time, clogging was been observed in many landfills for municipal solid waste in many countries. In the early decades of the 21st century, the problem is being seen again due to the construction of modern landfills in many emerging countries, which often show a rapid growth of height of the landfill body and high ratios of organics in the municipal solid waste.

The cause of the incrustation process was largely unknown (see Ramke, 1989), until extensive field investigations and laboratory experiments were carried out by two institutes of the Technical University of Braunschweig (Ramke and Brune, 1990 and Brune et al., 1991.) The microbiological investigations were described in detail by Brune (1991). The results of these investigations, which are still valid and have been confirmed by other research groups, will be summarized in the next section because the investigations resulted in a comprehensive understanding of the incrustation process. Based on these results, recommendations are given for construction and operation of sanitary landfill for municipal solid waste. A more detailed overview of the investigations and tests is given by Ramke (2009).

Causes of the incrustation process

The results of the chemical and microbiological field investigations and laboratory experiments brought to the conclusion that anaerobic bacteria are responsible for the formation of incrustations in drainage pipes and drainage layer material.

This conclusion is based on the following observations (compare Fig. 8.1.4):

- Anaerobic microorganisms are present in high concentrations in landfill leachate and readily colonize the surfaces of the drainage system.
- Analysis of the fine structure of the incrustation material shows it to consist of a network of aggregates of bacteria with deposits of precipitated inorganic material on their surfaces.
- On-growth experiments in which glass surfaces and gravel were exposed in the drainage pipes of two sanitary landfills showed on microscopic examination that the incrustation process is only initiated on the bacterial cells and the slime fibrils, which they excrete.

  Further precipitation centers around these seeds of incrustation until the whole biofilm becomes filled with inorganic precipitations. The aggregates of bacteria and the deposited inorganic material can accumulate to the degree already indicated.
- Chemical analysis of the incrustation material shows that its inorganic components consist of calcium, iron, magnesium, and manganese combined with carbonate and sulfur.

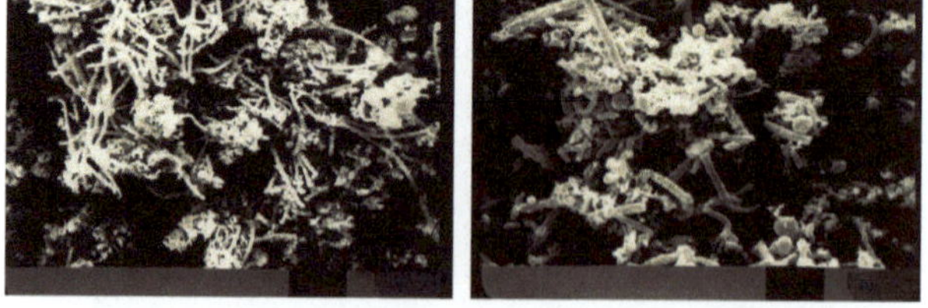

Figure 8.1.4 Typical structure of incrustation material in landfill drainage systems—progressive magnification in scanning electron microscope. *Photos: M. Brune.*

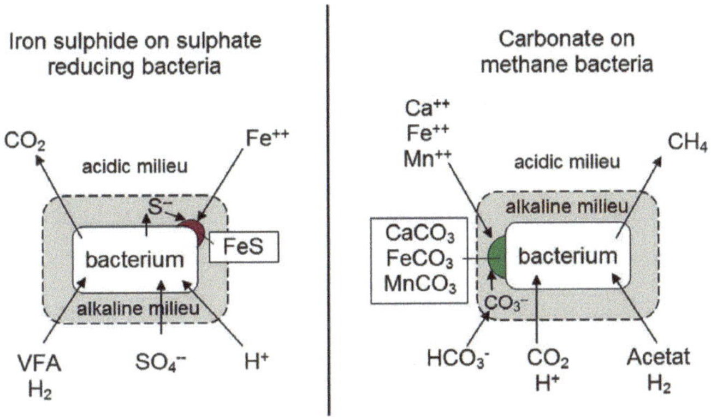

Figure 8.1.5 Schematic representation of processes leading to formation of incrustation material on bacteria (Brune et al., 1991).

Precipitation of the incrustation material is essentially caused by two processes, which are directly related to the decomposition processes in landfills for municipal solid waste (see Fig. 8.1.5):

- Bacteriogenesis of sulfidic deposits
  Iron-reducing bacteria solubilize Fe(III) by reducing it to Fe(II), whereas sulfate-reducing bacteria reduce sulfate, for example, gypsum to sulphide.
  This bioreduction of sulfate causes the milieu in the vicinity of the sulfate-reducing bacteria to become more alkaline, which results in sulfur precipitating as its insoluble metal sulfide such as FeS.
- Bacteriogenesis of carbonates
  Before calcium carbonate can be precipitated, calcium must first be mobilized from the refuse material. This is achieved by fermentative organisms producing organic acids, which lower the pH of the leachate and thus mobilize the calcium.
  Precipitation of calcium carbonate onto the surface of the methane and sulfate-reducing bacteria probably results from their metabolic consumption of hydrogen ions causing a local elevation of pH and consequently disturbing the balance between carbonate and hydrogen carbonate.

The formation of deposits in the drainage systems of sanitary landfills can thus be considered to take place in two stages:

- Fermentative bacteria together with iron- and manganese-reducing bacteria give rise to a process of mobilization, whereby a part of the organic component of the refuse is converted into volatile fatty acids being dissolved in the leachate.
  This leads to a lowering of pH value and thus causing the increasing dissolution of parts of the inorganic components of the refuse.
- In the second stage, the precipitation process, predominantly methane and sulfate-reducing bacteria in the drainage system, through their specific metabolism, causes the formation of insoluble sulfides and carbonates from metal ions dissolved in the leachate.
  This is the essential process, which leads to incrustations being formed.

Consequently, incrustations can only arise when the landfill leachate contains both easily degradable organic substances (as nutrients for the incrustation forming bacteria) and inorganic ions (calcium, iron, sulfate, hydrogen carbonate, etc.) in solution.

Recommendations for construction and operation of sanitary landfills

The main consequences to be derived from these findings for site construction and operation of landfills for municipal solid waste are as follows:

1. Incrustation of sanitary landfill drainage systems cannot be completely prevented but can be greatly reduced by appropriate methods of landfill operation.
2. The material of the drainage layer should be chosen as coarse as possible to provide a sufficient proportion of pore space and pore diameter (nothing can be said, however, about the filter stability of materials with a grain size greater than 16–32 mm).
3. Decisive for the effective life of the drainage system will be to limit the severity of incrustation through measures, involving both waste management and landfill operation, thus producing unfavorable conditions for the incrustation forming bacteria.
4. The intensity and duration of the acidic phase of biodegradation needs to be reduced. This can be achieved either by operational methods, which result in a predominantly aerobic degradation of the organic components of waste, such as
   - mechanical–biological pretreatment of municipal waste
   - slower filling of the landfill or by methods of waste management, e.g.;
   - separate collection of organic waste and
   - subsequent composting.
5. The supply of inorganic incrustation forming substances such as iron and calcium needs to be reduced. The separate landfilling of construction and demolition waste is more than advantageous.
6. Landfill operation has to consider that it is particularly critical when leachate with a high incrustation potential (heavily loaded with easily degradable organic substances and a high iron and calcium concentration) coincides with a microflora in the stable methane phase in the lower waste layers.

   For this reason, it should be avoided to establish a stable methane phase in the lower waste layers when in the following the site must filled quickly and an intensive acidic fermentation has to be expected.

These recommendations were considered in the German Technical Regulations and Directives on Landfills since 1991 and are one reason (reduced clogging potential) why the thermal or mechanical–biological pretreatment of household waste prior to disposal has been considered essential since 2005. In combination with other waste management measures, such as separate collection of different waste streams and their specific treatment, German landfills are much less afflicted by incrustation processes compared with the previous decades.

HYDRAULIC CALCULATION OF DRAINAGE SYSTEMS

Introduction

The hydraulic calculation of the complete drainage system covers the sizing of pipes, reservoirs, and the drainage layer. The hydraulic calculation of diameters of drainage and collection pipes and the volume of retention reservoirs can be undertaken with standard methods of civil engineering and is not covered here, where the drainage layer with its elements will be in the focus.

The hydraulic design of the effective drainage systems in bottom liner systems requires the determination of the following elements:

- hydraulic conductivity of the material of the drainage layer
- thickness of the drainage layer
- slope of the drainage layer
- drainage length or distance between drains.

The design criterion of drainage systems is the saturated thickness above the liner because the efficiency of the whole bottom liner system is directly related to the hydraulic pressure of leachate above the liner. In general, the saturated zone above the liner should/must not extend into the waste. Questions of leaching of waste and slope stability (landfills on sloping bottoms) have to be considered.

Flow of water in a drainage layer can be described like flow of fluids in porous media. In particular, the methods of calculating groundwater flow can be used, but the general assumptions and methods have to be adapted to these special systems. Hydraulic calculations of water flow in a sloping drainage layer normally can be performed in one dimension, using the following assumptions:

- unconfined flow, assuming the drainage layer as a phreatic aquifer
- small saturated thickness compared with length of the layer (Dupuit assumptions)
- homogeneous properties of drainage material
- uniform leakage rates
- parallel flow.

For the derivation of the basic equation describing the flow in soils above a sloping impermeable bottom or in sloping drainage layers, two different approaches are possible:

- the streamlines are parallel to the slope (first approximation of Boussinesq or extended Dupuit–Forchheimer assumption)
- the streamlines are horizontal (second approximation of Boussinesq or Dupuit–Forchheimer assumption).

For slopes less than approximately 10%, the differences between both solutions can be neglected, but for drainage layers which are components of landfill capping systems and have a lower hydraulic conductivity these differences should be considered.

In the following, the basic equation according to the second approximation of Boussinesq will be used; in Chapter 8.2 the first approximation of Boussinesq inclusive some analytical solutions for the hydraulic calculation of drainage systems in landfill capping systems will be demonstrated.

A comprehensive overview of hydraulic calculation of drainage systems of landfills under consideration of the current state of solutions is given by Ramke (2009). The following topics are covered there:

- discussion of boundary conditions
- comparison of the first and the second approximation of Boussinesq
- analytical solutions (explicit determination of height of saturated thickness)
- necessity and possibilities of unsteady and two-dimensional numerical solutions.

This section begins with a brief overview of leachate rates at landfill bottoms, then the basic equations of flow in sloping drainage layers will be explained, afterward a numerical standard solution will be presented, and finally some examples will be given.

Leachate rates at landfill bottom

The hydraulic design of bottom LCSs has to consider the leachate rate, which depends on the state of the landfill (begin, regular operation, phase of restoration).

At the beginning of landfill operation, the landfill has no or nearly no storage capacity due to the low thickness of waste. Precipitation enters the drainage system directly. The intensity and frequency of precipitation can be estimated according to the methodology of urban hydrology. Rainfalls with typical design intensities are short, and the resulting height of leachate in the drainage layer should not exceed a few centimeters. Therefore, although a hydraulic calculation of the drainage layer for this situation is unlikely to be necessary, other elements of the LCS such as drainage pipes, collection pipes, and detention reservoirs have to be designed for this loading condition. After landfill restoration, leachate generation decreases compared with the operating state of landfills with uncovered waste. Landfill covers with earth and vegetation reduce leachate generation caused by precipitation significantly, liners in landfill capping systems totally. Other components of leachate generation such as consolidation, biochemical processes, and reduction of storage capacity by decay processes are of less importance.

The operating state of landfills with uncovered waste is relevant to the hydraulic design of the drainage system. Long-term tests with lysimeters and the analysis of daily weather and discharge data of five landfills led to the classification of landfills into three groups of discharge behavior (Ramke, 1991):

I. Landfills with free storage capacity
II. Landfills with a saturated landfill body
III. Landfills with leachate recirculation.

Leachate discharge from operational landfills with a saturated landfill body and an open (active) waste surface is critical for design purposes. The following rounded leachate discharge rates, derived for Germany, might be useful under the same climatic conditions for landfills for municipal waste:

- average leachate rate (67%—value): 1 mm/d
- high leachate rate (99%—value): 10 mm/d

Hydraulic calculations can be performed with a leachate discharge rate of 10 mm/d because this value covers most of the cases of practical interest. Using this value for steady-state calculations, longer periods of higher discharges are covered too.

Under different climatic conditions and for landfills with different waste composition, regional experiences have to be considered to determine suitable leachate rates. Arid climates, e.g., in the Middle East, and tropical climates show quite different precipitation and evaporation situations compared with the temperate zone of Middle Europe. Therefore, a measuring program is recommended, which allows the application of the standard water balance equation for landfills. This will help to understand landfill reactions and to generate relevant design data.

Basic equation for sloping drainage layers

The basic equation for sloping drainage layers shall be shown for steady-state conditions under the assumptions made earlier according to the second approximation of Boussinesq because the slope of drainage layers in bottom liners is mostly less than 10%. In addition to the assumptions mentioned above, an impervious liner will be supposed. Therefore, no leakage through the liner has to be considered. The assumptions of parallel flow in a homogenous medium will lead to one-dimensional differential equations, which must be solved for the particular boundary conditions.

Fig. 8.1.6 shows the definition scheme for the basic equation on the basis of the second approximation of Boussinesq. Streamlines are horizontal, the equipotential lines are vertical. This scheme is valid for discharge on a rooflike-shaped bottom liner (left part only) or between two drains if the saturated thickness at the end of the drainage area is zero. The inflow into the drainage area over the right boundary is equal to zero. For the left boundary, a boundary condition of the first kind (predefined hydraulic head) is assumed, which corresponds to a predetermined water level in a drain or trench.

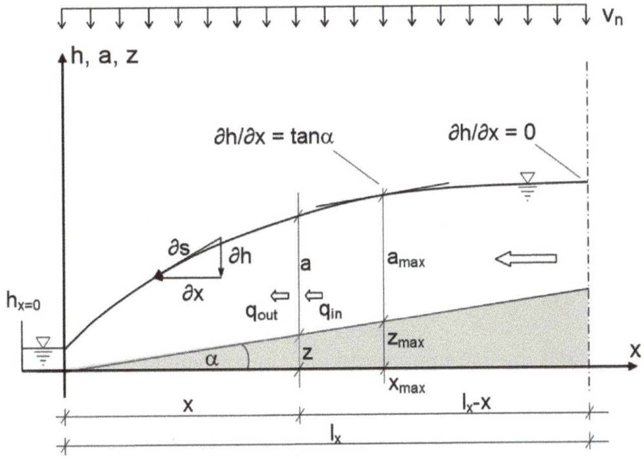

Figure 8.1.6 Definition sketch for derivation of the drainage formula on a sloping liner according to the second approximation of Boussinesq.

The basic equation for this case can be written as (see Ramke, 1991, 2009):

$$\frac{\partial h}{\partial x} = \frac{v_n}{k} \cdot \frac{(l_x - x)}{(h - x \cdot \tan \alpha)} \tag{1}$$

with $h =$ hydraulic potential, piezometric head [m]; $a =$ saturated thickness above the liner [m]; $z =$ elevation head above datum plane [m]; $x =$ horizontal axis [m]; $i =$ hydraulic gradient [-]; $l_x =$ horizontal drainage length [m]; $v_n =$ leakage rate, (rainfall) percolation rate [m/s]; $k =$ hydraulic conductivity [m/s]; $q_{in} =$ input flow into a cross section [$m^3/(m \cdot s)$]; $q_{out} =$ output flow out of a cross section [$m^3/(m \cdot s)$]; $\alpha =$ angle of slope [°].

Analytical solutions of the basic equations are possible, but the resulting formulas do not allow a direct calculation of the height of the water table or of the saturated thickness (explicit analytical solution), and the implicit solution is not very convenient. Explicit equations are only available for the calculation of the maximum saturated thickness. But even the explicit solutions for determination of the maximum saturated thickness require a careful analysis of their applicability; therefore a simple numerical solution is shown here. Ramke (2009) gives a comprehensive overview of the current state of analytical solutions.

Numerical solutions

The one-dimensional basic Eq. (1) can be solved numerically as an initial value problem. An often used approximation is the Runge-Kutta method. Ramke (1991) has described its use for this problem in detail.

The specification of the Runge-Kutta method is as follows:

$$\begin{aligned} h(x_0 + \Delta x) &= h(x_0) + \frac{1}{6} \cdot (k_1 + 2k_2 + 2k_3 + k_4) \\ k_1 &= \Delta x \cdot f(x_0, h_0) \\ k_2 &= \Delta x \cdot f(x_0 + \Delta x/2, h_0 + k_1/2) \\ k_3 &= \Delta x \cdot f(x_0 + \Delta x/2, h_0 + k_2/2) \\ k_4 &= \Delta x \cdot f(x_0 + \Delta x, h_0 + k_3) \end{aligned} \tag{2}$$

with $h(x_0) =$ value of the function at x_0; $h(x_0 + \Delta x) =$ new value of the function at "$x_0 + \Delta x$"; $\Delta x =$ increment in x-direction; $k_{1,2,3,4} =$ interim values.

The Runge-Kutta method can be implemented easily in an Excel spreadsheet. Fig. 8.1.7 gives an example (screenshot).

An increment of less than 1 cm is recommended, when leachate recharge is quite low and the hydraulic conductivity is comparative high. The initial value—the height of the water table in the drain or trench—must be adapted to the parameter set. For drainage layers consisting of coarse gravel with a very high hydraulic conductivity an initial value—predetermined water level—of 1 cm is recommended by practical considerations (see Ramke, 1991).

	A	B	C	D	E	F	G	H	I	J
1										
2	**Calculation of Saturated Thickness above a Sloping Liner**									
3	Second Approximation of Boussinesq									
4										
5										
6	Parameters	k	1.00E-04	[m/s]			Results	a_{max}	2.86E-01	[m]
7		v_n	1.16E-07	[m/s]				$a_{average}$	1.74E-01	[m]
8										
9										
10		h_o	2.00E-02	[m]						
11		tan(α)	1.00E-01	[-]						
12										
13		l_x	3.00E+01	[m]						
14		Δx	6.00E-03	[m]						
15										
16										
17	Step	x	h_0	k_1	k_2	k_3	k_4	$h_{0+\Delta x}$	a	z
18		[m]	[m]					[m]	[m]	[m]
19										
20	0	0.00	0.02	1.04E-02	8.38E-03	8.74E-03	7.42E-03	2.87E-02	2.00E-02	0.00E+00
21	1	0.01	0.03	7.43E-03	6.63E-03	6.71E-03	6.10E-03	3.54E-02	2.81E-02	6.00E-04
22	2	0.01	0.04	6.11E-03	5.65E-03	5.68E-03	5.31E-03	4.11E-02	3.42E-02	1.20E-03
23	3	0.02	0.04	5.31E-03	5.01E-03	5.03E-03	4.77E-03	4.61E-02	3.93E-02	1.80E-03
24	4	0.02	0.05	4.77E-03	4.56E-03	4.57E-03	4.38E-03	5.07E-02	4.37E-02	2.40E-03
25	5	0.03	0.05	4.38E-03	4.21E-03	4.22E-03	4.07E-03	5.49E-02	4.77E-02	3.00E-03
26	6	0.04	0.05	4.07E-03	3.93E-03	3.94E-03	3.82E-03	5.88E-02	5.13E-02	3.60E-03
27	7	0.04	0.06	3.82E-03	3.71E-03	3.71E-03	3.61E-03	6.25E-02	5.46E-02	4.20E-03

Figure 8.1.7 Example of an Excel Spreadsheet for numerical calculation of the saturated thickness in a sloping drainage layer with the Runge-Kutta method.

A lower hydraulic conductivity might require a higher initial value to avoid a "numerical backwater," caused by a predefined head too low for the resulting saturated thickness. The optimum can be found by iterations. When the average height of saturated thickness, calculated over the whole drainage length, becomes a minimum the optimal initial value has been chosen.

The calculation of LCS with a drainage length of 15 m requires 1.500 rows. Maximum and average saturated thickness can be determined directly using embedded Excel functions.

Examples

The saturated thickness in a LCS at a landfill bottom is calculated below with the Runge-Kutta method described above.

The course of the water table shall be demonstrated for a bottom liner system using typical standard parameters:

- leakage rate v_n: 10 mm/d = 1.16×10^{-7} m/s
- horizontal drainage length l_x: 15 m (rooflike shaped-landfill bottom)
- cross slope tan α: 3%

The hydraulic conductivity k is varied between $k = 10^{-2}$ down to 10^{-5} m/s. The resulting course of the water tables in the drainage layers is shown by Fig. 8.1.8. The saturated thickness is nearly zero at

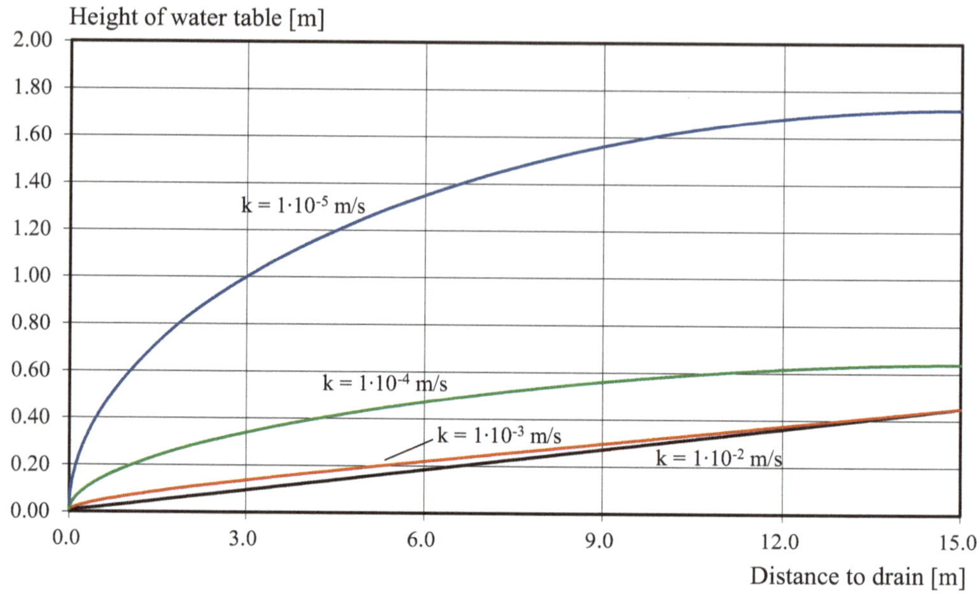

Figure 8.1.8 Course of water table in a sloping drainage layer for different hydraulic conductivities (standard leachate collection system at the bottom, $v_n = 10$ mm/d).

the end of the drainage area in the two examples with higher hydraulic conductivity. In the other two examples, there is a horizontal gradient at a distance of 15 m to the drain. The maximum saturated thickness varies between some millimeters up to 1.32 m. The standard system with a hydraulic conductivity of $k = 10^{-3}$ m/s has a maximum saturated thickness of 4.7 cm and an average saturated thickness of 2.9 cm. A hydraulic dimensioning of a drainage system in a bottom liner system can be omitted, if the standard parameters are kept, and the leakage rate of 10 mm/d is not exceeded.

A more detailed calculation of the saturated thickness with variations of cross slopes and hydraulic conductivities is shown below to demonstrate this. The saturated thickness in LCSs at landfill bottom is calculated with the Runge-Kutta method again.

The standard parameters used in the calculations were the same as used above:

- drain spacing: 30 m
- drainage length: 15 m in a rooflike-shaped bottom
- cross slope: 3%
- hydraulic conductivity: $k = 1 \times 10^{-3}$ m/s.

In Figs. 8.1.9 and 8.1.10 the maximum and the average saturated thickness are shown. Figure 8.1.9 confirms the necessity of a sufficient cross slope in the bottom drainage system. A reduction of the slope from 3% to 1% increases the saturated thickness two to three times and reduces significantly the efficiency of the whole system.

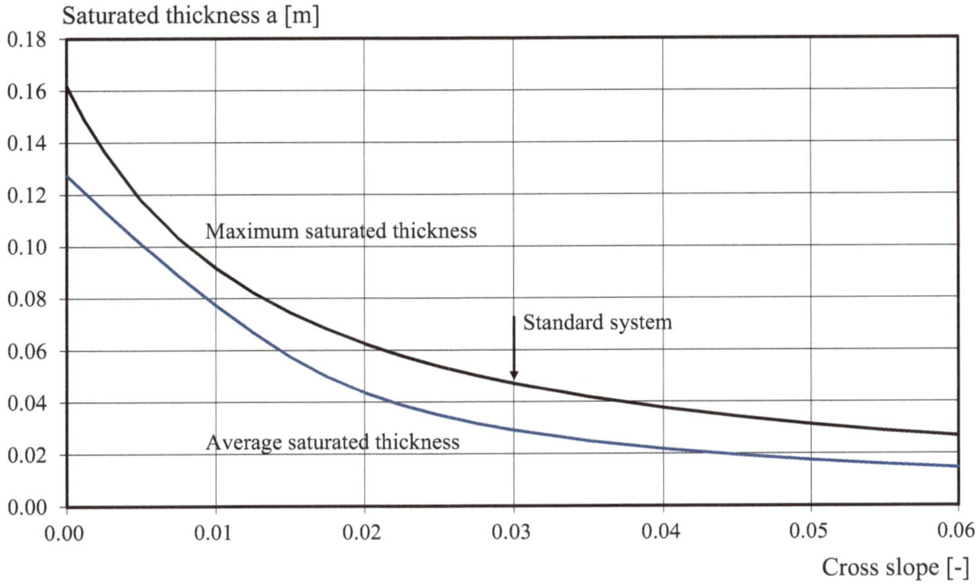

Figure 8.1.9 Saturated thickness in a drainage layer according to cross slope (drainage length 15 m, $k = 1 \times 10^{-3}$ m/s, $v_n = 10$ mm/d).

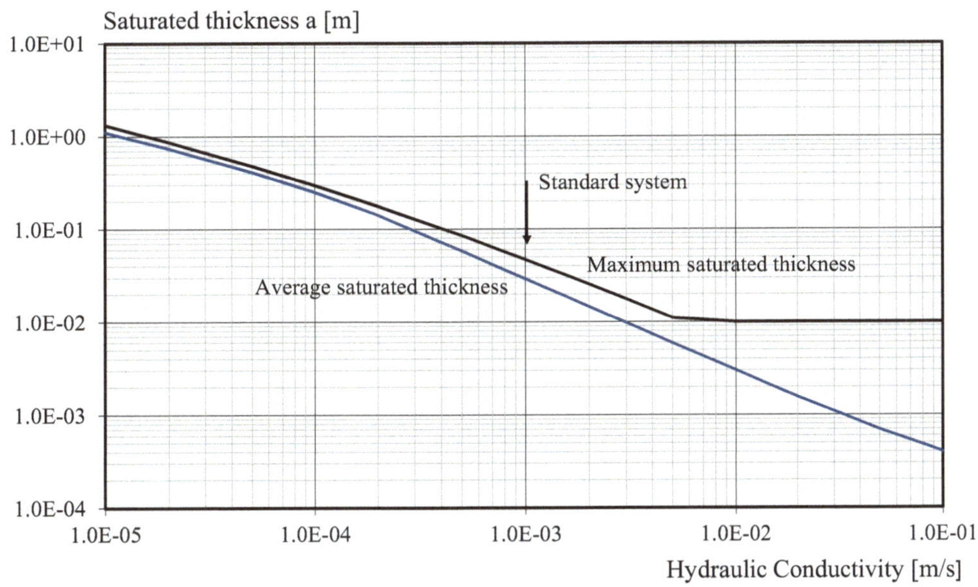

Figure 8.1.10 Saturated thickness in a drainage layer in dependence of the hydraulic conductivity (drainage length 15 m, cross slope 3%, $v_n = 10$ mm/d).

However, the long-term behavior of the hydraulic conductivity is potentially much more important than the angle of cross slope. As shown in Fig. 8.1.10, a loss of hydraulic conductivity in the drainage layer from $k = 1 \times 10^{-3}$ m/s down to $k = 1 \times 10^{-4}$ m/s – e.g., by clogging – will lead to a maximum saturated thickness of 30 cm instead of 4.7 cm. An extensive loss of hydraulic conductivity would result in an unacceptable height of leachate above the liner in a range of more than 1 m ($k = 1 \times 10^{-5}$ m/s).

This relation testifies the necessity of a minimum thickness of 30 cm of the drainage layer. Such a thickness seems not to be necessary for hydraulic reasons alone if the hydraulic conductivity is in a range of $k = 1 \times 10^{-3}$ m/s and the leachate rate is approximately 10 mm/d, but in particular clogging processes can reduce the hydraulic conductivity significantly. Coarse drainage material such as gravel 16–32 mm has a hydraulic conductivity considerable higher than $k = 1 \times 10^{-3}$ m/s, which is often the value of the hydraulic conductivity to be kept long term. But only the combination of the use of such a coarse drainage material (in LCSs), which is comparatively resistant to clogging effects and might ensure an acceptable long-term value of hydraulic conductivity, and the additional inherent safety factor, resulting from the thickness of the drainage layer and the maximum saturated thickness under "design conditions," can ensure the long-term function of the LCS.

When the hydraulic conductivity is increased to $k = 1 \times 10^{-2}$ m/s and more, which is more applicable for coarse drainage material, the average saturated thickness is lower than 3 mm. This height is much lower than the diameter of a single grain of coarse drainage material. Darcy's law, which is based on the assumption of a homogenous porous medium, is no longer applicable under these circumstances. The flow of leachate with a rate of 10 mm/d in such a drainage layer with coarse drainage material can rather be compared with a flow of a thin film of leachate between grain particles. Ramke (1991) has shown that the assumption of Darcian flow is justified for hydraulic conductivities equal and lower than $k = 1 \times 10^{-3}$ m/s, not for higher hydraulic conductivities. But due to the fact that the application of the equations based on Darcy's law results in an overestimation of the saturated thickness in coarse drainage material, it seems to be reasonable to use these equations for purposes of comparison.

RECOMMENDATIONS ON DESIGN AND MATERIALS

Standard design

The following is an example of design considerations based on satisfying the current German regulations for the landfill categories I - III (landfills for construction and demolition waste, for (pre-treated) municipal solid waste, and for hazardous waste). They relate to above ground landfills with gravity drainage and external leachate collection. Although the specific details need not be relevant to other situations, depending on national regulations and/or site specific conditions, they indicate the general principles and considerations for leachate drainage system design.

The proposals in this chapter are to a great extent identical with the design considerations of the Technical Committee Geotechnics of Landfills of the German Geotechnical Society (GDA E 2-14, 2011).

Profile and geometry

The surface of the bottom liner system has to be profiled with longitudinal and cross gradients so as to ensure gravity drainage. Therefore, it is necessary to install drainage pipes at the lowest points of the rooflike-shaped bottom liner system (see Fig. 8.1.1).

The drainage pipes should be constructed in such a manner that

- they are rectilinear to landfill edges
- they discharge to collection and inspection shafts placed outside of the disposal area
- they can be permanently inspected, maintained, and cleaned.

Pipe joints at the landfill base, which cannot be cleaned or inspected, must be avoided. To ensure maintenance, the length of the drainage pipes should not exceed 400 m, except longer pipe lengths can be controlled and flushed with local equipment.

The drainage system should satisfy the design criteria listed in Table 8.1.1, unless it can be proven otherwise. The minimum slopes must be maintained after all anticipated settlements and deformations under the applied loading from the waste body, etc. If appropriate, raised profiles should be incorporated, corresponding to the results of settlement and deformation calculations.

Protective layer

Geomembranes in bottom liner systems will be exposed to mechanical stress due to loading by the waste body and also to thermal, chemical, and biochemical effects during the construction phase, the operating phase, and the postclosure period.

In particular, the grain size and shape of the drainage material must not damage the geomembrane. Therefore, it is necessary to construct a durable and effective protective layer between the geomembrane and the drainage layer.

The protective layer can comprise geotextiles, mineral materials (sand and/or fine gravel), or a combination thereof. The permeability of the protective layer must not be considered as a drainage function. If a mineral protective layer is used, its filter stability (Terzaghi filter rule) with the drainage layer must be proved. If appropriate a geotextile with a separation function must be chosen.

In hydraulic terms, for a gravity system, the thickness of the protective layer can influence the saturated thickness of leachate above the bottom liner. Therefore, the protective layer should only be as thick as necessary to provide adequate protection of the geomembrane.

Carbonates or calcium-cemented materials should not be used. A maximum of 20% $CaCO_3$ by weight is allowed (calcareous materials can disaggregate and release calcium and carbonates, which could precipitate out and affect the drainage system).

The selected materials must be capable of withstanding all mechanical stresses and the physical/chemical and biochemical influences. The quality of material has to be proved. The adequacy of any geotextiles or sand layer, as appropriate, has to be proved in terms of its protective function.

Drainage layer

The following criteria need to be satisfied:

- thickness of drainage layer 0.5 m, including 0.3 m coarse drainage material at the bottom,
- coarse drainage material: gravel or crushed rock (double-broken rock, grain size 16–32 mm.

In selecting the grading of the drainage blanket, it is the susceptibility to incrustation, which is of primary consideration, rather than the "filter stability." German experience to date suggests that there is minimal transport of granular material in leachate from sites receiving mixed municipal waste consisting of household, commercial and construction wastes, and drained sewage sludge.

It is important to maintain a permeability value of k greater than 1×10^{-3} m/s in the drainage layer under operating conditions. Therefore, coarse-grained material should be chosen. A material grading that provides high porosity with large pore spaces should be used. For landfills for municipal solid waste (MSW), only a coarse granular material (grain size 16–32 mm) is recommendable due to the problem of incrustation processes.

The selected materials (gravel or crushed rock) must be solid. A maximum of 0.5% leachable components and a maximum of 20% grains by weight with a ratio of length: thickness >3:1 is allowed. In the case of coarse-grained gravel, the amount of crushed particles should not exceed 10%. The material qualities have to be proved and should meet the additional requirements on carbonate content according to the requirements on protective layers.

At a minimum, the lower 30 cm of the drainage layer have to consist of coarse granular material. The upper 20 cm of the total thickness may be designed as filter layer. However, a filter layer is only useful for landfills for construction and demolition debris, slags, ashes, and perhaps other types of inorganic waste. Any filter layer must have a $k \geq 1 \times 10^{-2}$ directly after construction and a long-term hydraulic conductivity of $k \geq 1 \times 10^{-3}$.

The drainage material of the drainage layer is to be distributed over the bottom liner without compaction to avoid damage of geomembranes but also crushing of the drainage material.

Often the transition between the drainage layer and the lower waste layer is discussed. It is sometimes proposed to place a geotextile or a layer of fine or pretreated waste between the drainage layer and the regular waste to prevent fine waste particles from washing out of the waste and infiltrating into the drainage layer. Both are unnecessary and potentially have deleterious effects in landfills for municipal solid waste. It is unnecessary because field investigations have shown that the stability of waste aggregates is quite high, and fines are often embedded in a cohesive matrix of organic material. Soil compatibility criteria need not to be considered for the drainage layer to ensure the retention of fines in the waste. The placement of geotextiles or finer types of waste in the transition zone may be deleterious because detailed laboratory tests have shown that in particular geotextiles are very sensitive to clogging (Ramke and Brune, 1990). Due to the very small diameters of pores in the geotextiles, incrustation processes will occur all over the surface of geotextiles and clogging material will block leachate percolation. This is to a lesser extent the same with fine municipal solid waste or mechanical–biological pretreated waste. A reduction of void diameters and an increase of specific surface provide more areas of growth for bacteria and significantly reduce the long-term permeability. Furthermore, it is critical when leachate with a high incrustation potential (heavily loaded with easily degradable organic substances and high iron and calcium concentrations) coincides with a microflora in the stable methane phase, which can occur after mechanical—biological pretreatment of waste, because this microflora is responsible for the incrustation effects.

In landfills for municipal solid waste, the lowest waste layer should have a height of approximately 2 m and consist of normal municipal solid waste without bulky components (to protect the liner) or an extraordinary content of fines (no street sweeping). Waste compaction should be started on the top of the next waste layer.

In landfills with a high content of inorganic waste (construction and demolition debris, ashes and slags, industrial residues), the situation is different. These types of waste might not have the same aggregate stability like the municipal solid waste with its high content of organics, and the filter criteria used for soils are to consider. Geotextiles must be chosen carefully to ensure the long-term permeability, and precipitation of solutes at the colder landfill bottom can occur in these types of landfills. Here the installation of a granular screening layer with finer grain size than the drainage layer, but coarser structure than the lower waste layer can be a suitable solution.

Drainage pipes

Leachate collected and accumulated within the drainage layer is collected in drainage pipes and discharged from the waste body. The drainage pipes are to be installed in the lowest points of the bottom liner system.

All drainage pipes should be capable of being inspected, maintained, and cleaned. It is recommended that the inner diameter should be 250 mm or larger, although smaller diameters may be acceptable depending on the proposed inspection and cleaning methods/equipment. To encourage leachate access, the maximum possible percentage of opening areas and opening dimensions should be chosen, commensurate with the grain size of the drainage materials. Holes or slots are located in the upper two-thirds of the circumference of the drainage pipes, the lower third is closed. Considering the load requirements of pipes, hole diameters should not be less than 12 mm, wide slots are preferred. To avoid peak stresses, oblong holes are recommended. Further information is given by DIN 19667:2009 and DIN 4266-1:2011.

It must be ensured that the drainage pipes do not damage the bottom liner system, particularly the geomembrane, under loaded conditions. Fig. 8.1.11 shows examples for bedding the drainage pipes in the liner system.

Vertical penetration of the bottom liner system (e.g., by collection shafts) is not acceptable. Horizontal penetrations (e.g., through bottom dams) have to allow inspection and control.

The pipe material chosen must be suitable to withstand all chemical and biochemical attacks by organic or inorganic leachate components. In addition, the expected postclosure physical, chemical, and biochemical stresses have to be considered. In the last two decades of modern landfill technology,

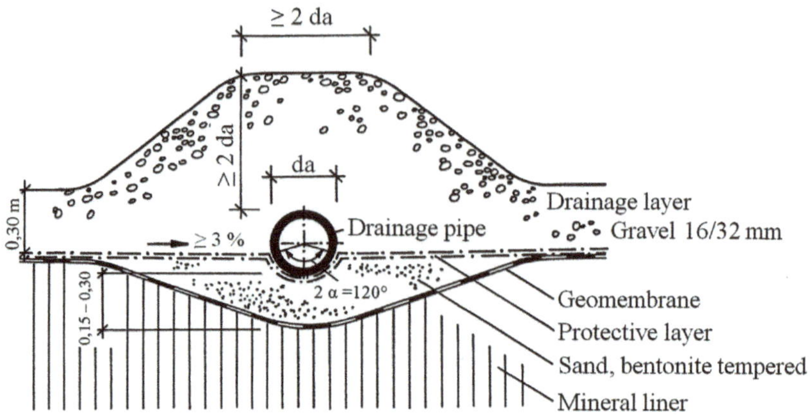

Figure 8.1.11 Example for the embedment of drainage pipes in the liner system (Ramke, 1998).

HDPE pipes have become the standard. The suitability of the drainage pipes, primarily their stability properties, temperature, and deformation behavior has to be proved. A quality guideline of the SKZ/TÜV (2017) describes more technical requirements, the related engineering standard is DIN 4266-1:2011.

Recommendations on the design of the embedment, which is essential for the long-term functionality of the drainage pipe, are given by DIN 19667:2015. Furthermore, every landfill drainage pipe needs a static calculation. For this purpose, the technical bulletin ATV-M 127-1 (1996) is very helpful.

Inspection and cleaning shafts

Where man-access is required, in principle, leachate inspection and cleaning shafts should be located outside the waste body. They must be resistant to leachate and completely waterproof. The shaft cover should be sufficiently large to allow adequate ventilation.

Furthermore, comprehensive measurement and examination of the drainage pipes should be possible from the shafts (Fig. 8.1.12). Therefore, the inner diameter of the shafts should be at least 1.5 m and the manhole should have a diameter of at least 1.0 m. Stirrups or similar aids should be placed in the shaft only if corrosion can be surely excluded, otherwise material failure and consequently danger of falling have to be expected.

In the case of purpose-designed, nonvented shafts, the whole pipe system must be sealed to avoid gas and odor emissions. Furthermore, the outlets of the drainage pipes have to be designed to prevent all gas emissions and the intrusion of air into the drainage system (e.g., by use of siphons or better by water traps). Fig. 8.1.13 shows an example of a shaft design considering these requirements. The shafts can be made of reinforced concrete (especially designed for resistance against leachate and landfill gas), synthetic materials, or a combination thereof (concrete with a synthetic inliner).

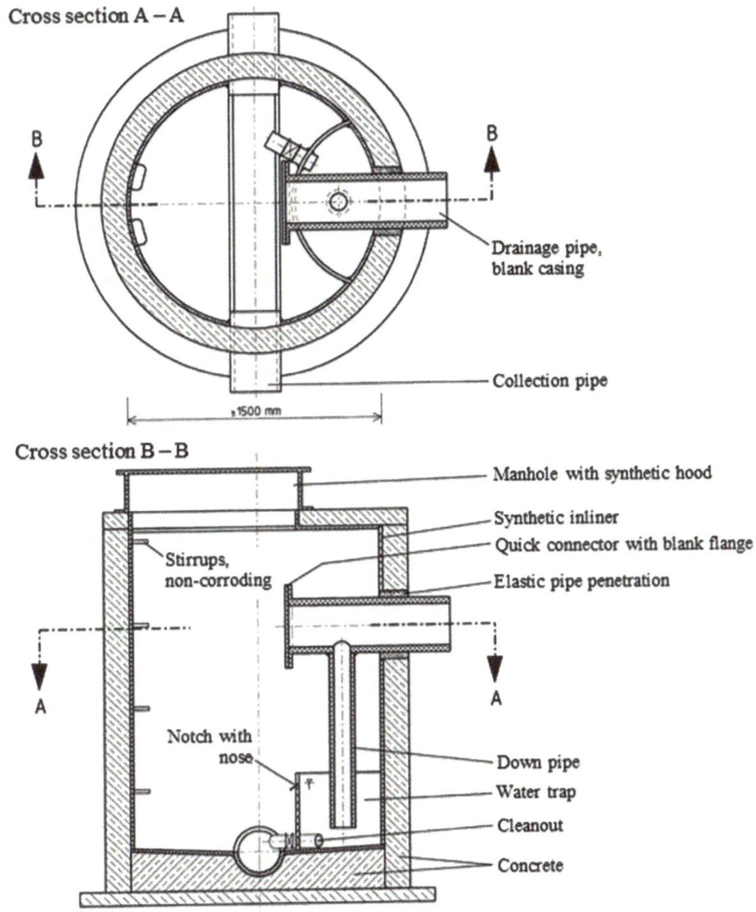

Figure 8.1.12 Example for an inspection and cleaning shaft.

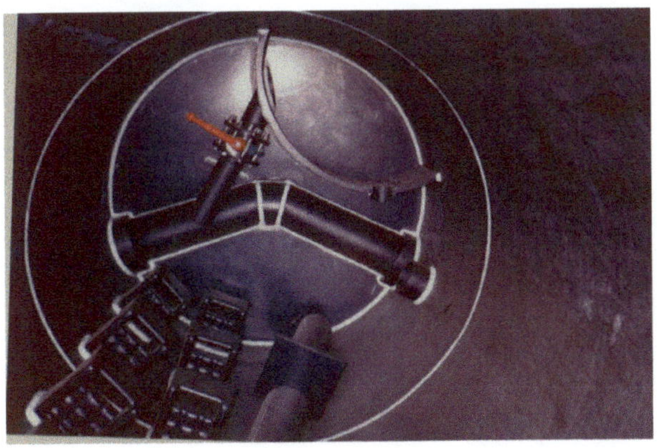

Figure 8.1.13 Photo of a shaft complying with the example. *Photo: G. Burkhardt.*

Where leachate monitoring/pumping shafts are located within the waste body, special consideration must be given to their structural stability, particularly in terms of deformation of the surrounding waste body, and excessive foundation loading on the bottom liner caused by negative skin friction.

If landfills are placed in a former clay pit, a stone quarry, or a natural hollow attempts should be made to place the monitoring and pumping shafts outside the former pit to allow gravity drainage of leachate from the drainage layer below the disposal area to the shafts outside the pit. The excavation for the shafts might require vertical large-scale drillings, whereas the pipe connection from the pit bottom to the shafts needs horizontal drillings.

Alternative design with secondary drains

In some regions, no coarse drainage material might be available, or the required coarse and washed gravel might be disproportional expansive. Under these circumstances the utilization of fine-grained drainage material or crushed rock can become necessary. The saturated thickness of leachate in a drainage layer made of finer drainage material is significant higher than in coarse drainage material, e.g., up to 30 cm in a standard system using a drainage material with a hydraulic conductivity of $k = 1 \times 10^{-4}$ m/s (see Fig. 8.1.10). Furthermore, finer drainage material is much more vulnerable to the loss of conductivity caused by incrustations. The use of crushed rock requires additional measures to protect geomembranes, and this material might not retain any fines in the waste.

To mitigate these problems, an alternative design with secondary drains, which consists of a drainage layer, made of coarse sand/fine gravel, and trenches of crushed stone, placed at 15–20 m

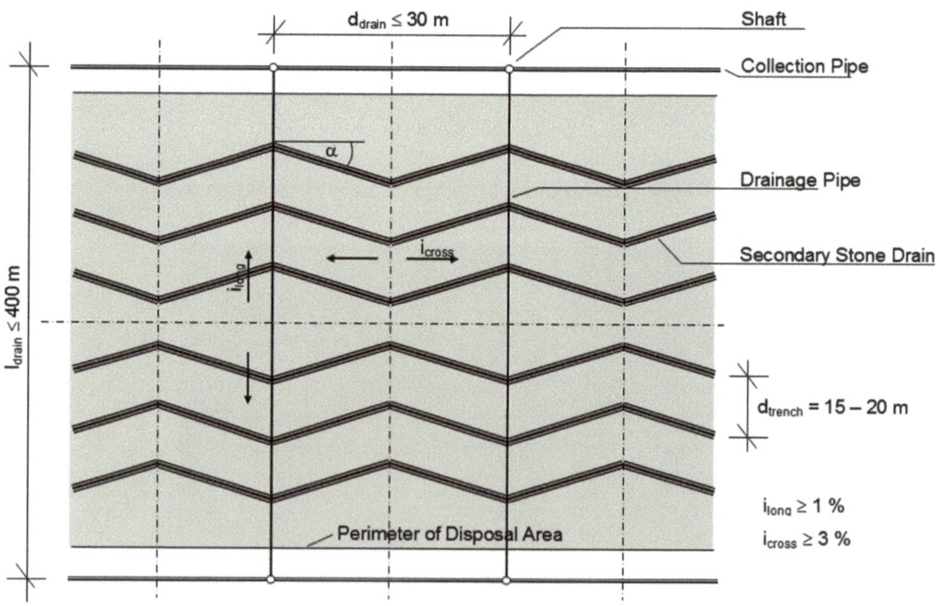

Figure 8.1.14 Leachate collection system with secondary drains.

Figure 8.1.15 Photo of a leachate collection system with secondary drains. *Photo: S. Steinkamp.*

intervals, could be acceptable (see Fig. 8.1.14). In addition to the drainage layer, the secondary high permeability stone drains can be installed on a herringbone pattern, feeding into the drainage pipes (see Fig. 8.1.15). Such a combination can improve the life span of the LCS and increase the effectiveness of the drainage system.

The secondary drains may be designed as trenches within the liner system profile. Here special consideration needs to be given to the practicality of installing and protecting the geomembrane. The protective layer above the geomembrane has to be adjusted to the peak pressures caused by sharp ends of the crushed stone.

The following dimensions and parameters are recommended for the secondary drains (see also Fig. 8.1.14):

- width of the secondary stone drain 2.5 m, height > 1.0 m
- drain spacing 15–20 m
- angle of drainage pipes to secondary drains according to the slope line (example: cross slope 3%, longitudinal slope 1%, $\alpha = 18.5$ degrees)
- drainage material: coarse-grained crushed stone, e.g., 32–150 mm.

The secondary drainage material must have a significant higher long-term hydraulic conductivity under operating conditions than the drainage layer ($k > 1 \times 10^{-2}$ m/s). For the drainage layer, a long-term hydraulic conductivity of $k = 1 \times 10^{-4}$ m/s is recommended.

The efficiency of such a system, measured as saturated thickness above the liner, is not such as good as in a standard system with a coarse drainage layer, but much better than in a system only consisting of a drainage layer with lower conductivity. This is due to the fact that the secondary stone drains are effective for drainage like the drain pipes. This results in a two-dimensional flow of leachate into both the drain pipes and the secondary drains. The water table will be dropped in two directions.

A detailed analysis performed by Ramke (1991) has shown that the average and maximum saturated thickness in a system according to Fig. 8.1.14 is in a range of approximately 50% of leachate height in a standard system according to Fig. 8.1.1 with a hydraulic conductivity of the drainage layer of $k = 1 \times 10^{-4}$ m/s (see Fig. 8.1.10).

A hydraulic calculation of such a system requires the use of groundwater models based on finite differences or finite elements because there are no analytical solutions available or simple numerical approaches such as Runge–Kutta applicable due to the two-dimensionality of flow in the drainage layer.

RECOMMENDATIONS ON MONITORING AND MAINTENANCE

Monitoring of the leachate collection system

Monitoring of LCSs begins in the phase of construction as an essential element of quality assurance.

Monitoring of the LCS comprises

- Deviation measurement
  Measurement of height and gradient of drainage pipes bottom directly after construction and yearly during landfill operation, continued in the postclosure phase until fading of settlements.
- Temperature measurement
  Measurement of temperature in the drainage pipes during landfill operation, in the postclosure phase until decrease of temperature development.
- TV inspection
  First inspection of the drainage pipes with a TV camera directly after construction after covering with gravel to detect damages, during landfill operation before and after flushing to determine intensity of incrustation and mechanical damages.

Special equipment has been developed for these purposes. An introduction is given by Kölsch (2003).

Maintenance of the leachate collection system

The maintenance of the LCS means the regular cleaning of the drainage pipes:

- Flushing
  At minimum, once a year the drainage pipes of an MSW landfill should be flushed to remove incrustation materials and fines.
  High-pressure flushing can become necessary to destroy hard incrustations at the bottom and sides of the pipes.
- Rotary cutting
  If flushing is not successful rotary cutting is required. Special attention must be given to protect the drainage pipe walls.

The results of maintenance should be recorded (intensity of incrustations, amount of incrustation material, etc.). Incrustation intensity determines flushing frequency.

References

ATV-M 127-1, 1996. Static Calculation of Leachate Drainage Pipes in Landfills (Richtlinie für die statische Berechnung von Entwässerungsleitungen für Sickerwasser aus Deponien). Deutsche Vereinigung für Wasserwirtschaft, Abwasser und Abfall (DWA), Hennef, Germany.

Brune, M., 1991. Causes of Generation of Solid and Sludgy Sediments in Leachate Collection Systems of MSW Landfills (Ursachen für die Bildung fester und schlammiger Sedimente in Entwässerungssystemen von Hausmülldeponien) (Ph.D. Thesis). Naturwissenschaftliche Fakultät, Technische Universität Braunschweig, Germany.

Brune, M., Ramke, H.-G., Collins, H.-J., Hanert, H.H., 1991. Incrustation processes in drainage systems of sanitary landfills. In: Proceedings "Sardinia 91": Third International Landfill Symposium, Cagliari. CISA—Environmental Sanitary Engineering Centre, Cagliari, Italy.

DIN 19667, 2009-10. Drainage of Landfills – Design, Construction and Operation. Deutsches Institut für Normung e.V., Beuth Verlag, Berlin, Germany.

DIN 4266—1, 2011-11. Drainage Pipes for Landfills – Part 1: Drainage Pipes Made from PE and PP. Deutsches Institut für Normung e.V., Beuth Verlag, Berlin, Germany.

European Union (EU), 1999. Council Directive 1999/31/EC of 26 April 1999 on the Landfill of Waste.

GDA E 2—14, 2011. Design Principles for Leachate Collection Systems in Bottom Liner Systems, (Entwurfsgrundsätze der Basis-Entwässerung von Deponien). Technical Recommendations of the Technical Committee Geotechnics of Landfills of the German Geotechnical Society, Essen, Germany. http://www.gdaonline.de/gda.html.

German Directive on the Landfill of Waste (DepV), 2017. Verordnung über Deponien und Langzeitlager (Deponieverordnung—DepV); 27.09.2017, BGBl. I S. 3465 (German Federal Law Gazette).

Kölsch, F., 2003. Monitoring of landfills. In: 18th International Conference on Solid Waste Technology and Management, Philadelphia, USA.

Ramke, H.-G., 1989. Leachate collection systems of sanitary landfills. In: Christensen, T., Cossu, R., Stegmann, R. (Eds.), Sanitary Landfilling: Process, Technology and Environmental Impact. Academic Press. ISBN:0-12-174255-5.

Ramke, H.-G., 1991. Hydraulic Assessment and Calculation of Leachate Collection Systems of Sanitary Landfills—Water Balances, Hydraulic Characteristics and Methods of Calculation (Hydraulische Beurteilung und Dimensionierung der Basisentwässerung von Deponien fester Siedlungsabfälle—Wasserhaushalt, hydraulische Kennwerte, Berechnungsverfahren – Dissertation) (Ph.D. thesis). Mitteilungen aus dem Leichtweiss-Institut für Wasserbau, Heft 114, Technische Universität Braunschweig, Germany.

Ramke, H.-G., 1998. Leachate collection and discharge (Sickerwassersammlung und —ableitung). In: Handbuch der Müll- und Abfallbeseitigung, Kennziffer 4545. Erich Schmidt Verlag, Berlin, Germany.

Ramke, H.-G., 2009. Leachate collection systems. In: Witt, K.J., Ramke, H.-G., Telekes, G., Imre, E. (Eds.), 1st Middle European Conference on Landfill Technology, International Conference, Organised by the Technical Committee on Geotechnics of Landfills in the DGGT and the Hungarian National Committee of the SSMGE, Budapest, February 6-8, 2008.

Ramke, H.-G., Brune, M., 1990. Integrity and Failure Mechanisms of Drainage Layers in Bottom Liner Systems, Final Research Report (Untersuchung zur Funktionsfähigkeit von Entwässerungsschichten in Deponiebasisabdichtungssystemen, Abschlussbericht). Bundesminister für Forschung und Technologie, FKZ 14504573, Germany.

Rushbrook, P., Pugh, M., 1999. Solid Waste Landfills in Middle- and Lower-Income Countries. World Bank, New York, USA. World Bank Technical Paper No. 426.

SKZ/TÜV, 2017. Pipes, Shafts and Construction Elements in Landfills (Rohre, Rohrleitungsteile, Schächte und Bauteile auf Deponien – SKZ/TÜV—LGA Güterichtlinie). TÜV Rheinland/LGA Bautechnik GmbH, Nürnberg, Germany.

TA Abfall, 1991. Technical Instruction on Storage, Chemical/Physical, Biological Treatment, Incineration and Disposal of Hazardous Wastes (Technische Anleitung zur Lagerung, chemisch/physikalischen, biologischen Behandlung, Verbrennung und Ablagerung von besonders überwachungsbedürftigen Abfällen). Second General Administrative Regulation on the Waste Law, 12.03.1991, GMB l. I S. 139 (German Federal Law Gazette).

TA Siedlungsabfall, 1993. Technical instruction on utilisation, treatment and disposal of municipal solid waste (Technische Anleitung zur Verwertung, Behandlung und sonstigen Entsorgung von Siedlungsabfällen). Third General Administrative Regulation on the Waste Law, 14. 05. 1993, BAnz. Nr. 99a (German Federal Gazette).

United Nations Environmental Program (UNEP), 2005. Solid Waste Management. UNEP, Nairobi, Kenia.

United States Environmental Protection Agency (US EPA), 1998. Technical Manual for Solid Waste Disposal Facility Criteria, Subpart D, Design Criteria, Revised April 13, 1998.

8.2

COLLECTION OF SURFACE RUNOFF AND DRAINAGE OF LANDFILL TOP COVER SYSTEMS

Hans-Günter Ramke

INTRODUCTION

After completion of disposal, a suitable top cover should be installed above the landfill body to control the interface between the waste and the atmosphere, following the concepts and design criteria described in Chapter 11.1.

Independently from the typology of the top cover, the capped landfill surface and the surrounding area require to be equipped with systems for surface water collection and drainage aimed to prevent the access of external water (watercourses, groundwater, springs, hillside flow, etc.), to collect and divert runoff from landfill slopes, and to drain and discharge the flow of percolated water preventing infiltration into the landfill body.

Ditches and interceptor drains are used for collection and discharge of external water. Surface runoff of landfill slopes is collected by swales and ditches as integrated elements of berms and the bottom dam of a landfill.

In the top cover, a drainage system (mineral granular materials or geosynthetic drainage layer, pipes, ditches, etc.) has to minimize backwater on the liner, which prevents the infiltration of water in the waste mass.

The objective of this chapter is to describe the state of the art of surface water collection and drainage. Beginning with some proposals for collection and discharge of external water, the first focus will be the hydrology of landfill top cover systems. The understanding of the water balance of the restoration layer is the precondition for derivation of design peak flows of surface runoff and design drainage rates. In both cases, the water content of the restoration layer determines, in conjunction with the current precipitation, the height and intensity of surface runoff and soil percolation. The surface landfill hydrology is comprehensively treated in Chapter 6.1.

The design and hydraulic calculation of surface runoff collection systems (SRC systems) and drainage systems is the second focus and will be explained in detail. Starting with general considerations on the integration of SCR and drainage systems in landfill top cover systems design recommendations for both components are given in accordance to some national guidelines (e.g. GDA E 2-20, 2015) and an overview report (Ramke, 2002). For each system the necessary equations for hydraulic calculations are compiled, and examples for typical situations are shown.

GENERAL PRINCIPLES OF DESIGN

General Requirements

All elements of surface water collection on landfills have to be designed in such a manner that gravity drainage is possible over a long term. Settlements of landfill surface, especially of landfills for municipal solid waste, have to be considered.

The single elements of the entire surface water collection and drainage system have to fulfill following requirements:

- Prevention of the access of external water

  The access of external water has to be prevented reliably from the landfill, in particular from all areas covered with waste (disposal areas).
- Collection and discharge of surface runoff

  All elements for collection and discharge of surface runoff from the landfill slopes are to be designed that even under storm runoff no erosion of landfill slopes occurs.
- Drainage of the landfill top cover system

  Drainage of the top cover system should ensure that percolation of the restoration layer does not lead to a backup of water on the liners higher than the thickness of the drainage layer.

The different elements of the surface water collection and drainage system require different design considerations, beginning with the hydrologic impact, continuing with the integration in the whole landfill concept, and ending with particular construction principles.

Legal Requirements

As an example for legal requirements the regulations of the Landfill Directive of the European Union (EU, 1999) shall be cited (Annex I, Chapter 2):

> "Appropriate measures shall be taken, with respect to the characteristics of the landfill and the meteorological condition, in order to
>
> - control water from precipitation entering into the landfill body
> - prevent surface water and/or groundwater from entering into the landfill body."

For both types of landfills, landfills for nonhazardous waste and landfills for hazardous waste, a drainage layer in the top cover systems is required. The only quantification concerns the thickness of the drainage layer, which is set to a height of 50 cm. Further design criteria are not given.

Detailed source of information might be found in different national regulations and guidelines, e.g., in the German Directive on the Landfilling of Waste (DepV, 2013).

PREVENTION OF ACCESS OF EXTERNAL WATER

Entering of surface water and groundwater into a landfill has to be prevented. Sources of external water can be local watercourses, hillside flow, layer groundwater, springs, or groundwater with a high groundwater table.

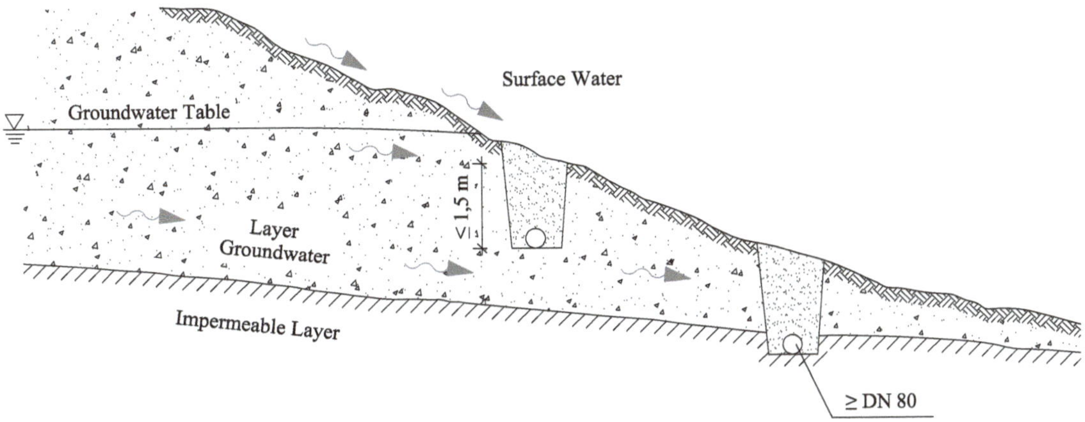

Figure 8.2.1 Design of an interceptor drains in a hilly area.

Landfill sites in the direct neighborhood or even on areas with local watercourses, leaking layer groundwater, springs, or low depth to groundwater table are strictly unsuitable. But especially in hilly regions, it might become necessary to keep off hillside flow from the landfill, which requires collection of downhill flowing precipitation water and layer groundwater.

Surface water and hillside flow have to be caught and discharged by open ditches and interceptor drains. Interceptor drains are filled with soils of high water permeability, at the bottom of the drain a drain pipe is placed, its diameter should be not less than 80 mm. Fig. 8.2.1 shows an example for the design of interceptor drains in a hilly area. Interceptor drains are perpendicular to the direction of flow and should be embedded in impermeable layers below the water bearing layers. In case of high thickness of the water bearing layer, it might be more efficient to install a series of drains instead of one deep interceptor drain.

Relevant discharge rates for hydraulic dimensioning of open ditches and drains have to be defined on the base of local climatic, hydrogeological, and morphological conditions. Local experience is a valuable source of knowledge.

HYDROLOGIC CALCULATIONS

Water Balance of Landfill Top Cover Systems

Fig. 8.2.2 defines in a schematic cross section of a landfill top cover system, consisting of a mineral liner, a drainage layer, and the restoration layer, the essential hydrologic elements.

The water balance of top cover systems is described by the following water balance equation:

$$P = ET_a + Q_{SR} + Q_{IN} + Q_{DR} + Q_{PL} \pm W_S \quad (1)$$

with P = precipitation (mm/Δt); ET_a = actual evapotranspiration (mm/Δt); Q_{SR} = surface runoff (mm/Δt); Q_{IN} = interflow (mm/Δt); Q_{DR} = lateral drainage discharge (mm/Δt); Q_{PR} = (soil)

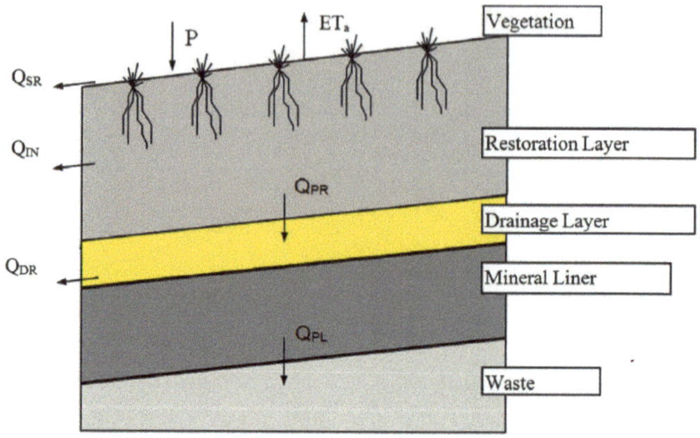

Figure 8.2.2 Definition sketch for the water balance of a landfill capping system.

percolation rate (mm/Δt), Q_{PL} = liner percolation rate (mm/Δt); W_S = change in soil water storage (mm/Δt).

For design purposes in particular, the estimation of surface runoff and lateral drainage discharge, caused by the soil percolation rate, is of importance, whereas the liner percolation rate and its relation to precipitation determine the system efficiency.

The height of the different elements depends on many variables such as climate, vegetation, types of soils in the restoration layer (topsoil and subsoil) and their water content, gradient and length of landfill slopes, and drainage layer and liner properties. Precipitation can be directly measured, and evaporation can be estimated in many areas by using empirical relations, but the prediction of surface runoff, lateral drainage discharge, and liner percolation rate requires more complex calculations.

This is due to the fact that height and intensity of surface runoff as well as the soil percolation rate (the percolation rate that passes the restoration layer, called percolation rate in the following) do not only depend on height and intensity of precipitation but also on soil water content, which changes over the year. Therefore the prediction of development of soil water content requires a long-term hydrologic modeling of the processes in the landfill top cover system to identify the critical states.

For this purpose, water balance models are useful. An example of such a model is HELP (Hydrologic Evaluation of Landfill Performance), which has been particularly developed for the hydrologic calculation of landfill capping systems. The HELP Model was developed in 1984 on behalf of the US EPA and stepwise improved in the following decades. The latest US-version dates from 1994 (Schroeder et al., 1994a,b). An intensive validation was carried out in Germany, and further enhanced versions were released (Berger, 2013). The current German version (with an English user interface) is HELP 3.95 D. The HELP Model is probably the most used model worldwide for the design and comparison of the hydrologic effectiveness of landfill capping systems. But also other model systems can be adapted to this task if they cover the essential processes. Suitable might be, e.g., models such as UNSAT-H,HYDRUS-2D, or SWAT.

Table 8.2.1 Main processes modeled in the HELP model (hydrologic evaluation of landfill performance, version 3)

System Element	Considered Processes
Vertical percolation layer (restoration layer)	Precipitation Snow accumulation, snowmelt Surface runoff Evaporation of surface water Infiltration Freezing and thawing of soil Plant growth and decay Potential and actual evapotranspiration Unsaturated/saturated vertical flow Moisture storage
Lateral drainage layer	Lateral drainage on sloping bottom
Barrier soil layer (mineral liner)	Saturated percolation according to Darcy's law
Geomembrane	Percolation through fabrication and installation defects Diffusion of water vapor through intact sections

A major task of water balance modeling of landfill top cover systems is the combination of single processes to a complete system, which reproduces the complex interactions of vegetation, potential evapotranspiration, soil water content, and real evapotranspiration. Table 8.2.1 gives an overview of the processes, which are covered by the HELP Model.

The HELP Model is a quasi-two-dimensional layer model, which simulates the processes in the vertical percolation layers, the barrier soil layer, and the geomembrane in vertical direction, and discharge in the lateral drainage layer and on the surface of the restoration layer as one-dimensional flow in lateral direction. These processes were combined and can be calculated on day-to-day basis between 1 and 100 years.

An example of water modeling, applied to real cases in Germany, is given here to demonstrate the processes affecting surface runoff and drainage discharge. For this example, a typical landfill top cover system is used:

- Vegetation: good stand of grass, leaf area index = 3.5
- Restoration layer: sandy loam, soil type No. 6, 100 cm, effective field capacity 0.105 m^3/m^3
- Drainage layer: gravel, soil type No. 21, 30 cm, hydraulic conductivity 3×10^{-3} m/s
- Mineral liner: barrier soil, soil type No. 16, 50 cm, hydraulic conductivity 5×10^{-9} m/s (modified)
- Bearing layer: sand, soil type No. 2, 30 cm, hydraulic conductivity 5.8×10^{-5} m/s

A slope with a gradient of 33% and a horizontal length of 50 m was simulated. Soil type numbers correspond to the soil type numbers in the database of the HELP Model, the hydraulic parameters were chosen according to database proposals.

This landfill top cover system is suitable in general, but the mineral liner is vulnerable to desiccation. In practice in many climates, the thickness of the restoration layer would have to be increased, but for demonstration purposes—calculation of system efficiency, surface runoff and soil percolation rate—the thickness selected is appropriate.

For the following example, three sites in Germany under different climates were selected. The weather data are embedded in the database of HELP 3.95 D. For each site a 10-year simulation was performed. The results are compiled in Table 8.2.2.

Water balance data in Table 8.2.2 are the 10 years average totals. The daily weather data were site specifically generated by the HELP weather generator, the evapotranspiration data are part of the HELP database. In all cases, it was assumed that the whole thickness of the restoration layer was accessible for evaporation.

Table 8.2.2 shows that potential and actual evapotranspiration on these three sites are quite similar, but precipitation differs significantly. The climate in Geisenheim is comparatively dry, Hamburg shows normal "coastal weather," and Luedenscheid is very wet.

These differences in precipitation rates result in different heights of rates of lateral drainage, which discharge the percolation above the liner. The absolute height of liner percolation rate ranges between

Table 8.2.2 Results of water balance modeling at three different landfill sites in germany

Term	Unit	Geisenheim	Hamburg	Luedenscheid
Precipitation	mm/a	564	802	1183
Runoff	mm/a	0	2	1
Potential evapotranspiration	mm/a	*694*	*645*	*620*
Actual evapotranspiration	mm/a	452	503	513
Lateral drainage	mm/a	73	222	563
Liner percolation rate	mm/a	47	72	102
Liner percolation ratio	%	*8.3*	*9.0*	*8.6*
Change in water storage	mm/a	−7	3	3

Italic written Numbers are not a component of water balance.

47 and 102 mm/a, but in relation to the particular precipitation the relative height of system percolation, the liner percolation ratio, is approximately 8.5% (8.3%–9.0%). The system efficiency therefore is more or less the same on the three sites evaluated.

Fig. 8.2.3 demonstrates the influence of the different precipitation at different sites on the soil water content and on the head of leachate on top of the liner. Fig. 8.2.3A describes the situation for Geisenheim in the first 5 years of simulation. In each summer, the lower limit for soil water content—the wilting point—is hit for a couple of days (water content 85 mm/m) in the restoration

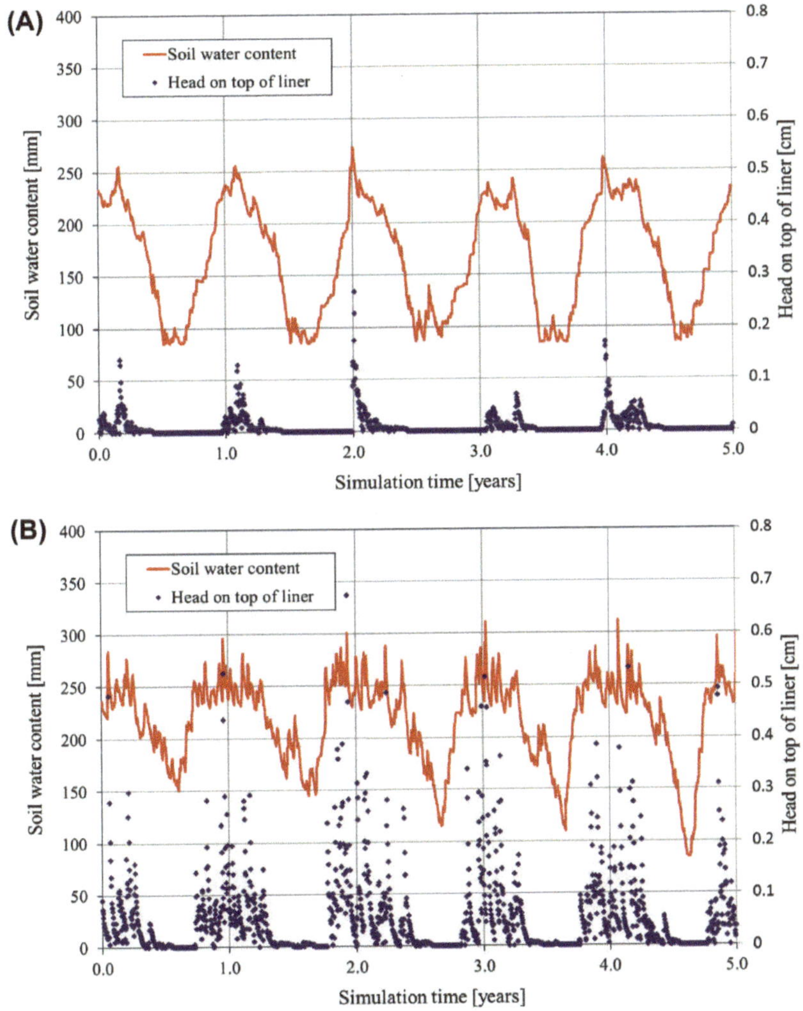

Figure 8.2.3 Time variation curves of soil water content in the restoration layer and head on top of liner for 5 years. (A): Geisenheim landfill site, (B): Luedenscheid landfill site.

layer, but in winter times the water content at field capacity is exceeded very often (190 mm/m). This results in a dam up of water above the liner in winter times and a dry drainage layer in summer times.

This yearly course of the soil water content and the height of leachate on top of the liner can be seen in the simulation for Luedenscheid too (Fig. 8.2.3B). But here the minimum of soil water content in the restoration layer is hit only once for a short period, while field capacity is exceeded much more often. In consequence of these differences in soil water content, the percolation rate is much higher, drainage discharge rates are increased, and—due to the higher and longer dam up of water in the drainage layer above the mineral liner—the liner percolation rate is significantly risen compared with Geisenheim.

It is noticeable that percolation and subsequent lateral drainage in this model only occurs when the soil water content is higher than the field capacity in the restoration layer. In practice this scheme is too simple. Preferential flow leads to vertical percolation even if field capacity is not reached in the whole layer. But the general conclusion that most of lateral drainage discharge is observed in periods with saturated restoration layers is nevertheless applicable.

These results might explain the important role of local hydrologic impacts on the water balance of a landfill top cover system. In the same manner, like described here, different system designs and different types of soils in the restoration layer can be evaluated with the HELP Model (see Berger, 2013; Ramke, 2002).

Those simulations can be a valuable base for the estimation of peak flow data of surface runoff and drainage discharge, which will be derived in the next paragraphs. But it should be mentioned here, that numerical modeling in any case needs a deep understanding of the processes and their implementation in the models.

Furthermore, particular situations under climates different from the temperate zone require additional attention. Examples are storm water periods in arid zones, snowmelt in moderate and subarctic zones, and monsoon in tropic zones.

Estimation of Surface Runoff Using the Rational Method

Introduction

Surface runoff always occurs when the access of water to the ground surface is higher than the infiltration capacity of the soil. Height and temporal development of surface runoff depends on height and intensity of rainfall and on the slope properties. Other important influencing factors are the length and the slope gradient, the type of vegetation, the type of soil and its actual infiltration capacity. Current infiltration capacity of a particular soil is determined by the soil water content.

Snowmelt is next to rain a further cause of surface runoff. But height of surface runoff caused by snowmelt depends on such a lot of particular local conditions that no general recommendations can be given. The estimation of surface runoff caused by snowmelt has to be done by usage of local experience and regional established equations.

The simulation described above has shown only very little heights of surface runoff in the average annual water balance at the three sites considered here. At first this is due to fact that Table 8.2.2 only shows the 10 years total average values and no peak events, which can be much higher. At second a detailed analysis of the simulation results disclosed that surface runoff in this simulation mostly occurs only when the soil is frozen but seldom under other weather conditions. At third the HELP Model determines surface runoff on a daily base, which might be sufficient for water balance purposes, but for design purposes shorter rain periods are essential.

For design of elements of SRC systems such as swales and ditches, the peak discharge is to be determined, whereas for the dimensioning of retention basins the knowledge of whole discharge volume is necessary, which requires the determination of the discharge hydrograph. In the following, the focus will be laid on some aspects of peak discharge determination. More advanced methods, such as the hydrograph method, are useful to apply, when all the relevant data are available.

The peak discharge can simply be estimated by the "Rational method," which is recommended for use on drainage channels draining less than 80 ha (FAO, 1998):

$$Q = \frac{10}{3600} \cdot C \cdot I_r \cdot A \tag{2}$$

with Q = peak discharge, peak rate of surface runoff (m^3/s); C = (peak) runoff coefficient (–); I_r = average rainfall intensity (mm/h); A = drainage area (ha).

Preconditions for application of the rational method (see TDT, 2014), in particular that the time of concentration is less than the duration of peak rainfall intensity, are fulfilled in most cases for landfill slopes. The minimum duration to be used for computation of rainfall intensity is 10 minutes. If the time of concentration computed or the drainage area is less than 10 minutes, then 10 minutes should be adopted for rainfall intensity computations. The next subparagraphs show the application of the rational method for landfill top cover systems. Finally an example will be given.

A disadvantage of the rational method is that the runoff coefficient does not consider the actual infiltration capacity of soils in dependence of soil water content. Therefore a more detailed approach for modeling surface runoff, based on the HELP simulations and considering the changes in soil water content will be explained in the following paragraph.

Estimation of Time of Concentration

The rational method assumes that rainfall duration is equal to the time of concentration, which is the time required for the surface runoff to flow from most distant point to the outlet. TDT (2014) explains the determination of time of concentration in detail.

The time of concentration results from the overland flow time (Kerby) and the channel flow time (Kirpich):

$$t_c = t_{ov} + t_{ch} \tag{3}$$

with t_c = time of concentration (min), t_{ov} = overland flow time (min), t_{ch} = channel flow time (min).

The Kerby equation is applicable for small watersheds, where overland flow is an important component of overall travel time:

$$t_{ov} = 1.44(C_r \cdot l_s)^{0.467} \cdot i_s^{-0.235} \tag{4}$$

with C_r = dimensionless retardance coefficient (−); l_s = length of slope (m); i_s = gradient of slope (m/m).

The retardance coefficient has a height of 0.40 for an average stand of grass, which is common for landfill slopes after restoration and vegetation. Further data are given by TDT (2014).

Kirpich has developed a simple equation for the calculation of the channel flow time:

$$t_{ch} = 0.0195 \cdot l_{ch}^{0.770} \cdot i_{ch}^{-0.385} \tag{5}$$

with l_{ch} = length of channel (m); i_{ch} = gradient of channel (m/m).

This equation should be used for predesign purposes only. When the whole system is defined channel flow time can be calculated accurately with the Manning–Strickler equation.

Estimation of Rainfall Intensity

Rainfall intensity is determined as the average rainfall rate in mm/h or mm/min for specific rainfall duration and a selected frequency. In many regions, data on rainfall intensity are available with a sufficient spatial and temporal resolution. Often these data are given as rainfall height for a defined duration. Table 8.2.3 gives an example for the relationship of height of rainfall, duration of rainfall, and annual recurrence interval. The shown design rainfall height was compiled for Hamburg.

Table 8.2.3 shows that increasing duration of rainfall and annual recurrence interval leads to increasing of height of rain. But rainfall intensity is decreasing with increasing duration of rainfall. The annual recurrence interval has to be selected due to the related risks resulting from overflow. The relationship between rainfall height, rainfall intensity, and rainfall duration is

$$I_r = \frac{h_r}{t_r} \tag{6}$$

with I_r = intensity of rainfall (mm/min); h_r = height of rainfall (mm); t_c = duration of rainfall (min).

In the north of Germany a typical rain yield factor for design purposes is in a range of 100 L/(s•ha), if a recurrence interval of 1 year and a duration of 15 minutes is selected. This is in accordance with Table 8.2.3, where the height of rainfall for this conditions is 9 mm, equivalent to a rainfall intensity of 0.6 mm/min or a rain yield factor of 100 L/(s·ha).

Runoff Coefficients

Runoff coefficients are compiled in many design guidelines. In Table 8.2.4 just a brief overview is given based on a compilation of TDT (2014).

Table 8.2.3 Height of rainfall [mm] in dependence of annual recurrence interval and duration of rainfall (Hamburg, 2003)

Duration		Annual Recurrence Interval (a)								
		0.5	1	2	5	10	20	30	50	100
5	min	3.5	5.0	6.6	8.6	10.1	11.6	12.5	13.6	15.1
10	min	5.3	7.5	9.8	12.8	15.1	17.3	18.7	20.3	22.6
15	min	6.3	9.0	11.7	15.3	18.0	20.7	22.3	24.3	27.0
30	min	8.0	11.5	15.0	19.5	23.0	26.5	28.5	31.0	34.5
60	min	9.8	14.0	18.2	23.8	28.0	32.2	34.7	37.8	42.0
90	min	11.0	15.6	20.1	26.1	30.6	35.1	37.8	41.1	45.6
120	min	12.0	16.8	21.5	27.8	32.6	37.3	40.1	43.6	48.4
3	h	13.5	18.7	23.8	30.5	35.6	40.7	43.7	47.5	52.6
6	h	16.6	22.4	28.1	35.7	41.5	47.3	50.6	54.9	60.7
8	h	18.0	24.1	30.2	38.2	44.2	50.3	53.8	58.3	64.4
12	h	20.3	26.8	33.3	41.9	48.4	54.9	58.7	63.5	70.0
18	h	21.9	28.9	35.9	45.2	52.2	59.2	63.3	68.5	75.5
24	h	23.5	31.0	38.5	48.5	56.0	63.5	67.9	73.5	81.0

Table 8.2.4 illustrates that slopes of landfills, which are generally steeper than 7%, are not covered in detail by such a standard compilation. Furthermore, the influence of vegetation and type of soils becomes obvious in the wide range of height of runoff coefficient. For practical purposes, it is not easy to decide if no practical regional experience exists, which runoff coefficient might be applicable.

Example

For the climatic conditions of Hamburg, according to the example above, a brief example of estimation of surface runoff using the rational method will be given. Design parameters of a typical landfill slope and calculation results are compiled in Table 8.2.5.

Table 8.2.4 Runoff coefficients for the rational method (TDT, 2014)

Type of Drainage Area	(Peak) Runoff Coefficient
Unimproved areas	
Sand or sandy loam soil, 0%–3%	0.15–0.20
Sand or sandy loam soil, 3%–5%	0.20–0.25
Black or loessial soil, 0%–3%	0.18–0.25
Black or loessial soil, 3%–5%	0.25–0.30
Black or loessial soil, > 5%	0.70–0.80
Deep sand area	0.05–0.15
Steep grassed slopes	0.70
Lawns	
Sandy soil, flat 2%	0.05–0.10
Sandy soil, average 2%–7%	0.10–0.15
Sandy soil, steep 7%	0.15–0.20
Heavy soil, flat 2%	0.13–0.17
Heavy soil, average 2%–7%	0.18–0.22
Heavy soil, steep 7%	0.25–0.35

The time of concentration is the sum of the time of the overland flow on the slope and the time of channel flow in the ditch or swale beside the berm at the toe of the slope. Considering a gradient of the slope of 33% and of the ditches on the berms of 8% the time of concentration is approximately 10.8 minutes. The requirement for the application of the rational method that the time of concentration is longer than 10 minutes is fulfilled.

Using Table 8.2.3 for the determination of the corresponding height of rainfall the resulting intensity of rainfall becomes approximately 0.93 mm/minute for an annual recurrence interval of 2 years. This conforms to a rainfall intensity of 56.1 mm/h.

Table 8.2.5 Example for the estimation of surface runoff using the rational method

Term	Parameter	Unit	Height
Design parameters	Horizontal width of slope	[m]	50.00
	Gradient of slope	[m/m]	0.33
	Transversal width of slope	[m]	52.65
	Horizontal length of slope	[m]	200.00
	Horizontal drainage area	[m^2]	10,000
	Gradient of berms	[m/m]	0.08
	Longitudinal length of berms	[m]	200.64
Overland flow time	Retardance coefficient	[–]	0.40
	Length of slope	[m]	52.65
	Gradient of slope	[m/m]	0.33
	Overland flow time	[min]	7.75
Channel flow time	Length of channel	[m]	200.64
	Gradient of channel	[m/m]	0.08
	Channel flow time	[min]	3.06
Time of concentration	Time of concentration	[min]	10.81
Rainfall intensity	Duration	[min]	10.81
	Annual recurrence interval	[a]	2.00
	Height of rainfall	[mm]	10.11
	Intensity of rainfall	[mm/min]	0.93
Peak discharge	Peak runoff coefficient	[–]	0.70
	Average rainfall intensity	[mm/h]	56.10
	Horizontal drainage area	[ha]	1.00
	Peak discharge	[m^3/s]	0.11

Table 8.2.4 proposes a runoff coefficient of 0.7 for steep grassed slopes, which might be suitable for landfill slopes. The resulting peak discharge is in a range of 0.11 m^3/s. But on the other hand, a value of 0.35 of the runoff coefficient, which is recommended for lawns on heavy soils with a gradient of 7% might perhaps also appropriate. This coefficient would reduce the peak discharge to the half.

It becomes obvious that the selection of a suitable runoff coefficient is of particular importance for the calculation of height of surface runoff, and that a compilation of values like Table 8.2.4 cannot cover the particular circumstances of a definite landfill. Overestimating the runoff coefficient will lead to oversized ditches and swales, whereas underestimation of the runoff coefficient holds the risks of uncontrollable discharge events.

Therefore a more site-related approach is presented in the next paragraph, which considers in particular the soil moisture.

Estimation of Surface Runoff Using the Curve Number Method

Introduction Into the Curve Number Method

In the following, the runoff coefficient will be calculated based on computations with the HELP Model, which uses the curve number method (CN method) for determination of surface runoff. Therefore a brief overview of the CN method will be given. Detailed information is compiled by the USDA (1986).

The basic equation is the relation between runoff, precipitation, and potential maximum soil water retention (infiltration capacity):

$$Q = \frac{(P - I_a)^2}{(P - I_a) + S} \qquad \text{if } P > I_a \tag{7}$$

with Q = height of runoff [mm]; P = height of precipitation (rainfall) [mm]; I_a = height of initial abstraction [mm]; S = potential maximum soil water retention [mm] after begin of runoff.

This equation describes that the discharge ratio is continuously increasing with increasing height of precipitation.

Initial abstraction includes all losses before runoff begins. The original CN method, developed of the US Soil Conservation Service, assumes the initial abstraction in a height of 20% of S ($I_a = 0.2$ S), but recent studies showed that initial losses are rather 5% of S ($I_a = 0.05$ S).

The retention capacity (S) depends on the type of soil and the land use category:

$$S = 25.4 \cdot \left[\frac{1000}{CN} - 10\right] \tag{8}$$

with CN = curve number [−].

The CN has a range of 30−100. Low numbers indicate a low runoff potential, while a CN of 100 results in a discharge, which is equal to the effective precipitation.

Table 8.2.6 Runoff curve numbers for different soil characteristics and cover types (USDA, 1986)

	Hydrologic Soil Group			
	A	B	C	D
Infiltration rates	High	Moderate	Low	Very low
Types of soil textures (no significant compaction, like soil in restoration layers)	Sand, loamy sand, or sandy loam	Silt loam or loam	Sandy clay loam	Clay loam, silty clay loam, sandy clay, silty clay, or clay
	CN_{II} values			
Cover Type	**A**	**B**	**C**	**D**
Bare soil (without vegetation)	77	86	91	94
Pasture, grassland	49	69	79	84
Meadow, continuous grass	30	58	71	78
Brush, brush–weed–grass mixture	35	56	70	77
Woods—grass combination	43	65	76	82
Woods	36	60	73	79

All CN values for fair hydrologic conditions.

CN values are tabulated for many different conditions. A quite comprehensive compilation is given by USDA (1986). A selection of CN values applicable for landfill slopes is shown in Table 8.2.6.

The CN values in Table 8.2.6 are valid for average antecedent runoff conditions (CN_{II} values) and an initial abstraction of 0.2 S. Soil moisture can be considered only by using CN values for dry (CN_I), average (CN_{II}) or wet conditions (CN_{III}).

Calculation of Runoff Coefficient With the Hydrologic Evaluation of Landfill Performance Model

The HELP Model adjusts the CN values to the soil moisture and considers furthermore lengths and gradients of slopes (Schroeder et al., 1994b).

Instead of Eq. (8), the following equation is used by the HELP Model for calculation of the maximum soil water retention S (all units are converted to SI units instead of custom units):

$$S_{SM} = \begin{cases} S_{max}\left[1 - \dfrac{SM - [(FC + WP)/2]}{UL - [(FC + WP)/2]}\right] & \text{for } SM > \ (FC + WP)/2 \\ S_{max} & \text{for } SM \leq (FC + WP)/2 \end{cases} \tag{9}$$

with S_{SM} = value of S in dependence of soil moisture [mm]; S_{max} = maximum value of S [mm]; SM = soil water storage in the vegetative or evaporative zone [mm/m]; UL = soil water storage at saturation [mm/m]; FC = soil water storage at field capacity [mm/m]; WP = soil water storage at wilting point [mm/m]. If the depth of the evaporative zone is different to the height of 1 meter, the calculated height of soil water storage SM, which is given in mm in HELP 3.95 D, has to be converted into the water content (mm/m).

The model calculates the soil moisture in seven segments of the restoration layer and weights the resulting retention parameter accordingly to the different height of the segments for the calculation of an overall retention parameter. This will not be considered here because the model does not release the internal calculations.

The maximum retention capacity is calculated in dependence of the CN value for dry conditions, but this general approach is only valid in case of a restoration layer of adequate height.

$$S_{max} = 25.4 \cdot \left(\frac{1000}{CN_I} - 10\right) \tag{10}$$

with CN_I = CN value for dry conditions [−].

CN_I is related to CN_{II} by the following polynomial (Schroeder et al., 1994b):

$$\begin{aligned} CN_I = {} & 3.751 \cdot 10^{-1} \cdot CN_{II} + 2.757 \cdot 10^{-3} \cdot CN_{II}^2 \\ & - 1.639 \cdot 10^{-5} \cdot CN_{II}^3 + 5.143 \cdot 10^{-7} \cdot CN_{II}^4 \end{aligned} \tag{11}$$

Finally the original curve number $CN_{II,o}$ is fitted to lengths and inclination of slopes:

$$CN_{II} = 100 - (100 - CN_{II,o}) \cdot \left(\frac{l_s^{*2}}{i_s^*}\right)^{CN_{II,o}^{\ -0.81}} \tag{12}$$

with $CN_{II,o}$ = original CN value for mild inclinations [−]; l_s^* = dimensionless length of slope (l_s/152.40) [−]; l_s = length of slope (m); i_s^* = dimensionless gradient of slope (i_s/0.04) [−].

The original $CN_{II,o}$ values, which correspond to the values tabulated in Table 8.2.6, are determined by the model using a regression considering soil texture and levels of vegetation.

In practice—following the approach of the HELP Model—the $CN_{II,o}$ values have to be determined, then Eq. (12) will be used to consider slope properties, and finally the maximum retention capacity can be determined for a particular restoration layer in a top cover system with Eqs. (11) and (10).

Using the output of the HELP Model, the temporal development of the actual maximum soil water retention or infiltration capacity can be computed with the daily values of soil water storage in the evaporative zone by Eq. (9). This results in the determination of the soil moisture depending curve number CN_{SM}:

$$CN_{SM} = \frac{1000}{\frac{S_{SM}}{25.4} + 10} \tag{13}$$

with CN_{SM} = CN value in dependence of soil moisture [−].

The actual maximum soil water retention S_{SM} is used for the calculation of the actual runoff coefficient in dependence of the rainfall height, which shall be indicated as CN-related runoff coefficient:

$$C_{CN} = \frac{Q}{P} = \frac{(P - I_a)^2}{P \cdot ((P - I_a) + S_{SM})} \tag{14}$$

with C_{CN} = curve number related runoff coefficient [−].

Instead of an initial abstraction of 20% of S the reduced initial abstraction in a height of 5% should be used, but it has to consider working with CN_{II} values reflecting this reduction.

Example

The previous example for a slope with a horizontal length of 50 m, a gradient of 0.33, and a good stand of grass will be used here again, but the HELP simulation is used additionally to determine the development of the actual CN-related runoff coefficient.

The water balance of four different types of restoration layers has been simulated under the climatic conditions of Hamburg. Table 8.2.7 shows the soil parameters of the variants of restoration layers and the different CN values.

Fig. 8.2.4 describes the development of soil moisture, potential soil water retention, and the CN in dependence of soil moisture for the first 2 years of simulation of variant 2 with soil No. 6 in the restoration layer, which is identical to the HELP simulation for Hamburg described above. This simulation includes year 2, which is the one with the highest precipitation and the second highest soil percolation rate. The curve of the water content in the evaporative zone (soil moisture) follows the course described above: In summer times, the water content declines, under certain circumstances down to the water content of the wilting point, increases in autumn and winter, and exceeds then the water content of field capacity frequently.

The potential soil water retention in dependence of soil moisture, calculated with Eq. (9), is in antiphase with the course of the soil water content. In summer times, the potential soil water retention comes up to its maximum according to Eq. (10). The minimum is nearly at the half of the maximum. Finally, Fig. 8.2.4 shows the development of the CN value in dependence of soil water content, calculated with Eq. (13), which changes between 41.6 (the CN_I value for dry conditions) in summer and approximately 58.5 at the day with the lowest potential soil water retention in the late second year.

Table 8.2.7 Properties of restoration layers of simulated top cover systems

Variant	Unit	I	II	III	IV
Slope geometry					
Length	[m]	50			
Gradient	[–]	0.33			
Soil parameters					
Material according to USDA		Sand	Sandy loam	Silt loam	Clay loam
Hydrologic Evaluation of Landfill Performance Model number		2	6	9	11
Thickness	[m]	1.00	1.00	1.00	1.00
Hydraulic conductivity	[m/s]	5.8×10^{-5}	7.2×10^{-6}	1.9×10^{-6}	6.4×10^{-7}
Saturation water content	[mm/m]	437	453	501	464
Field capacity water content	[mm/m]	62	190	284	310
Wilting point water content	[mm/m]	24	85	135	187
Curve number (CN) values					
$CN_{II,o}$ value, original value	[–]	45.0	60.0	75.0	81.5
CN_{II} value, slope correction	[–]	54.9	65.8	78.1	83.6
CN_I value	[–]	30.9	41.6	57.4	66.2
Retention capacity S_{max}	[mm]	568.7	356.3	188.5	129.5

All simulations of the different restoration layers are compiled in Fig. 8.2.5, which depicts the development of the CN values in dependence of soil water content for all variants. All soils show the same general course of the modified CN, but corresponding to their different CN_{II} values and—less obvious—to their different hydraulic conductivity, which influences the temporal development of soil water content.

The results of all simulations are tabulated in Table 8.2.8. The CN-related runoff coefficients, resulting from Eq. (14), were calculated with a height of initial abstraction of 5%. The height of

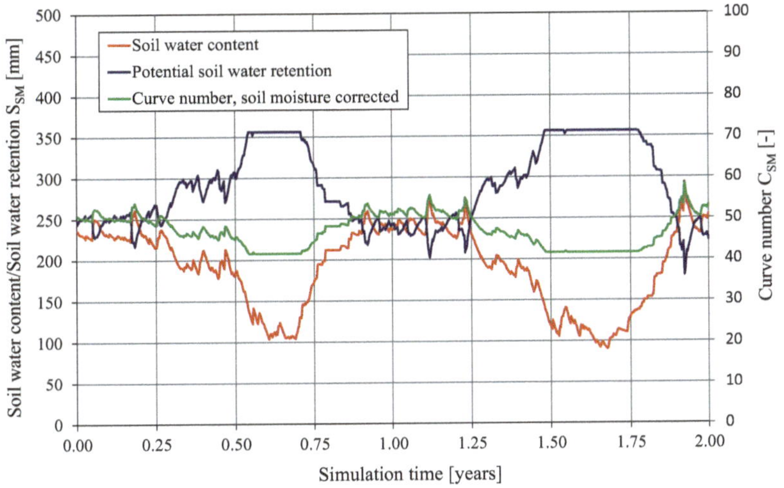

Figure 8.2.4 Time variation curves of soil water content, potential soil water retention, and the curve number in dependence of soil water content for variant 2 (Soil No. 6 in the restoration layer).

precipitation was chosen to 10 mm approximately according to the height of precipitation in the previous example and furthermore to 20 and 30 mm. Following Table 8.2.3, these data correspond in case of an annual recurrence interval of 2 years to rain durations of 10 minutes, 90 minutes, and 8 h. The resulting rainfall intensities are shown in Table 8.2.8 too.

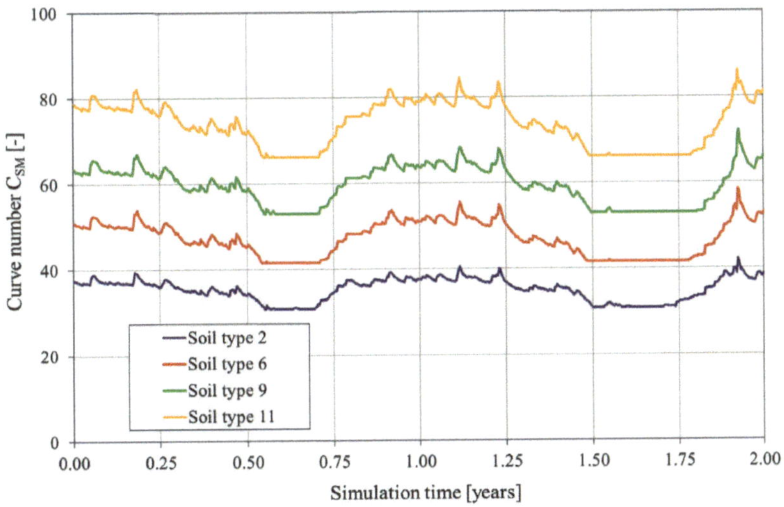

Figure 8.2.5 Time variation curve of curve numbers in dependence of soil water content for all variants.

Table 8.2.8 Calculation of curve number (CN)-related runoff coefficients

Variant		Unit	I	II	III	IV
Frequency of days with lower CN values		[%]	99.00			
CN_{SM} value, moisture dependant		[–]	39.74	54.70	67.79	82.82
Related retention capacity S_{SM}		[mm]	385.2	210.4	120.7	52.7
Initial abstraction (5% S_{SM})		[mm]	19.3	10.5	6.0	2.6
h_r	t_r	I_r	CN-related runoff coefficient			
[mm]	[min]	[mm/min]	[%]			
10	10	1.000	0.0	0.0	1.3	9.0
20	90	0.222	0.0	2.0	7.2	21.5
30	480	0.063	1.0	5.5	13.2	31.2
Frequency of days with lower CN values		[%]	99.73			
CN_{SM} value, moisture-dependent		[–]	40.40	55.71	68.92	83.84
Related retention capacity S_{SM}		[mm]	374.7	201.9	114.5	49.0
Initial abstraction (5% S_{SM})		[mm]	18.7	10.1	5.7	2.4
h_r	t_r	I_r	CN-related runoff coefficient			
[mm]	[min]	[mm/min]	[%]			
10	10	1.000	0.0	0.0	1.5	10.1
20	90	0.222	0.0	2.3	7.9	23.2
30	480	0.063	1.1	6.0	14.1	33.1

Two calculations of the CN-related runoff coefficients were performed, using moisture dependent CNs, corresponding to a frequency of 99.00% (once in 100 days) and 99.73% (once in a year) of days with lower CN_{SM} values. The differences of the resulting runoff coefficients in Table 8.2.8 are not very high between the two considered soil water contents of each variant, which is due to the fact that the moisture-dependent CN values do not differ very much.

The data in Table 8.2.8 show that on a slope with a top cover consisting of sandy soil with vegetation nearly no surface runoff occurs. The highest runoff coefficient in both cases of CN values is approximately 1% of precipitation and can be neglected. The sandy loam shows runoff only if the precipitation is higher than 20 mm, and even a 30 mm event results in runoff coefficients of only 6% at maximum. The runoff coefficients of the silt loam variant are higher and surface runoff starts with a height of rain of 10 mm. Precipitation in a height of 20 mm is discharged up to 8%, whereas 30 mm of rain show a runoff coefficient of approximately 14%. Only the clay loam, which is less suitable as topsoil, might have average runoff coefficients of 10%, 22%, and 32% for rainfall heights of 10, 20, and 30 mm.

This calculation of runoff coefficient of different variants of topsoils in the restoration layer under the same climatic conditions confirms the observation that even on slopes with cohesive soils surface runoff quite seldom occurs, when the revegetation was finished.

In opposite to the general approach of a tabulated runoff coefficient (see Table 8.2.4), which is usually much too high, a more detailed analysis will result in more realistic runoff coefficient. The HELP simulation could be a sound base for such a procedure. The calculated CN-related runoff coefficients can be used as input for the rational method after determination of recurrence interval, time of concentration, and the related average intensity of rain. But it has to be considered, that the CN-related curve numbers are average runoff coefficients and no peak runoff coefficients. A more suitable approach might be to determine on this base the effective precipitation (Q in Eq. 7) divided by the duration of rainfall (rainfall intensity I_r) as input for the calculation of peak discharge, using regional established models.

Advanced Methods of Calculation of Surface Runoff

In practice, it will be necessary to decide, which moisture-dependent CNs should be used in conjunction with which height of rainfall. Combining too infrequent situations or events will lead to an overestimation, neglecting the risk of an intensive rainfall in combination with a saturated soil with low retention capacity might result in an underestimation of the height relevant surface runoff.

Therefore a more detailed approach, which considers effects such as initial losses, change of infiltration capacity, and a discharge hydrograph, has been developed by Ramke (2002) for landfill slope purposes. Critical CN_{SM} values, generated with the HELP data like shown above, are combined with typical higher rain events in the related period, represented in the HELP database, and for each height of rain discharge hydrographs are calculated, considering variable durations of rainfall. The calculation results in a list of discharges for a period of, e.g., 10 years, and the annual recurrence interval can be related to the output—the discharge—instead to the input (height of rain and soil conditions). This extensive approach might be a suitable tool for large landfills in a region with difficult climatic conditions.

Apart these tools for estimation of surface runoff direct measurements, and local and regional experience based models are to recommend. Furthermore, regional design guidelines, e.g., for road construction or urban planning, containing empiric data of surface runoff should be considered.

Estimation of Soil Percolation Rates

The height of the soil percolation rate below the restoration layer on top of the drainage layer in a landfill top cover system depends next to the climatic conditions on the vegetation, the type of soil, and the thickness of the restoration layer. Each project requires particular hydrologic calculations to determine the design leakage rate for the hydraulic calculation of the drainage system. The frequency of days with lower soil percolation rates depends on the requirements of slope stability and the resulting acceptable height of leachate above the liner.

The following example demonstrates the procedure of estimation of soil percolation rates if no reliable data for the particular situation exist. Again the above described simulation of the same design at three different sites is used to determine the local percolation rates. The soil of the restoration layer is sandy loam. But differently from the landfill top cover system described earlier here a geomembrane is placed instead of the mineral liner to exclude percolation through the liner. Together with a gradient of the slopes of 33% and a comparatively high hydraulic conductivity of 1×10^{-2} m/s the daily lateral drainage discharge can be equated with the soil percolation rate (leakage rate) because the daily input to the drainage layer is simultaneously calculated as output at the same day.

Fig. 8.2.6 shows the course of soil water content and drainage discharge rates for the simulated standard conditions of Hamburg. It can be seen that discharge occurs only during in some months of a year, when field capacity is exceeded and infiltration in the restoration layer cannot be retained.

The resulting daily drainage discharge rates are arranged in ascending order, using a spreadsheet for all investigated sites, and shown as duration curves of percolation rates in Fig. 8.2.7. The diagram

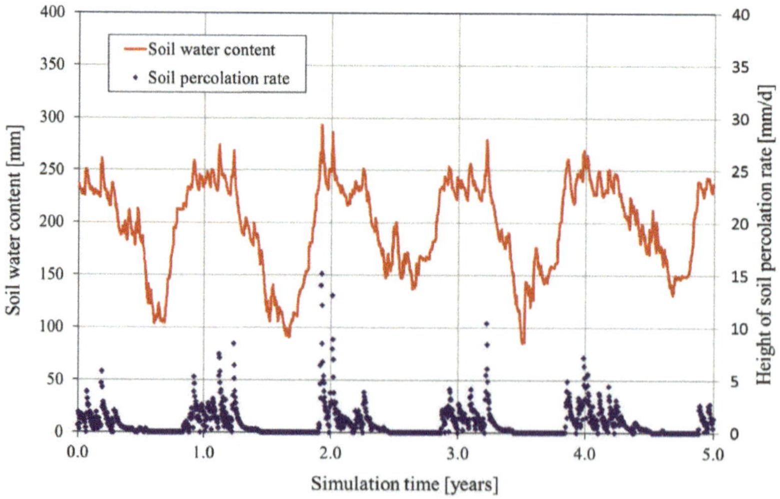

Figure 8.2.6 Time variation curve of soil water content and drainage discharge for 5 years for Hamburg.

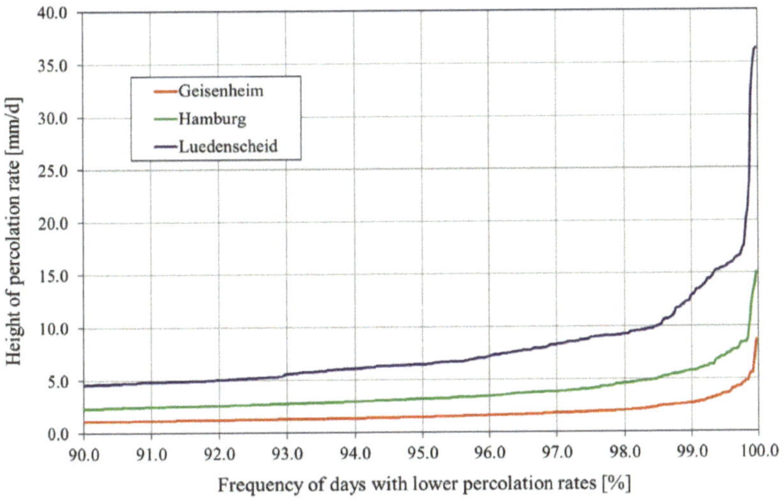

Figure 8.2.7 Duration curves of soil percolation rates.

shows the height of soil percolation rates in dependence of frequency of days with lower percolation rates. Selected was a frequency range between 90% and 100%.

It is not astonishing that the duration curve for Luedenscheid lies above the curves of Hamburg and Geisenheim, and that the curve of Geisenheim is the lowest one. Higher precipitation results in higher soil water contents for the same design if evapotranspiration is in a similar height, and higher soil water contents in combination with higher infiltration rates lead to higher soil percolation rates.

Table 8.2.9 gives the percolation rates for different frequencies of days with lower percolation rates for the different sites evaluated.

Table 8.2.9 Soil percolation rates for different sites in germany

Location			Geisenheim	Hamburg	Luedenscheid
Precipitation (average)		mm/a	564	802	1183
Frequency of days with lower percolation rates	90.0%	mm/d	1.1	2.3	4.5
	95.0%	mm/d	1.5	3.2	6.5
	99.0%	mm/d	2.7	5.8	13.0
	99.73%	mm/d	4.3	8.4	16.7
	99.97%	mm/d	8.7	15.1	36.3

A frequency of 99% of days with lower percolation rates means that the percolation rate is approximately 4 times per year higher as the quoted number. If the frequency is 99.73%, the value is exceeded once a year, and the 99.97%–value is reached or exceeded once in ten years.

Table 8.2.9 shows that in an area with low precipitation (Geisenheim) the high soil water percolation rates vary between 2.7 (99%) and 8.7 mm/d (99.73%), in a coastal zone (Hamburg) between 5.8 and 15.1, and in a very humid region (Luedenscheid) between 13.0 and 36.3 mm/d.

In addition, the percolation rates for different types of restoration layers under the climatic conditions of Hamburg were calculated using the same geometry and types of soil like for the determination of surface runoff. The results are compiled in Table 8.2.10, showing the percolation rates for 99.0% and 99.73% of frequency of days with lower percolation rates.

It can be seen that the influence of the type of soil of the restoration layer on the height of daily soil percolation rates is comparatively low. Even materials with a higher hydraulic conductivity such as sand and sandy loam (see Table 8.2.7) do not show extraordinary high soil percolation rates. Within the limits of soil materials suitable for restoration layers the local weather conditions will determine the height of soil percolation rates. A comparison of these data with data of lysimeters, test fields, and landfills has proven the range and height of the percolation rates for German climatic conditions (Ramke, 2002). But it has to be noticed that especially in the last years very high percolation rates—or better: drainage discharges—have been observed, peak values have been in a range of 25–35 mm/d (single events). This is conforming to the simulation results described here, see Fig. 8.2.7, which show the same peaks.

For the purpose of predesign of drainage layers, a percolation rate of 10 mm/d can be recommended for climatic conditions similar to Germany with an average annual precipitation rate of approximately

Table 8.2.10 Soil percolation rates for different restoration layers in hamburg

Type of Soil of Restoration Layer	Actual Evapotranspiration (Average) (mm/year)	Frequency of Days with Lower Percolation Rates (Percolation Rates (mm/d))	
		99.00%	99.73%
Sand	468	6.5	9.3
Sandy loam	503	5.8	8.4
Silt loam	519	5.7	7.9
Clay Loam	510	5.4	7.8

750 mm/a, if steady state conditions are assumed. This value will be probably exceeded only once a year and infrequent higher percolation rates can be stored in the drainage layer parallel to discharge.

The situation is different for the hydraulic dimensioning of geosynthetic drainage layers with much lower storage capacity. In this case, the decisive aspect is that daily percolation rates should be discharged daily too. Based on these considerations for the purpose of predesigns, a percolation rate of 25 mm/d is recommended by the GDA E 2-20, 2015.

The design soil percolation rates should be determined using models such as HELP or local measures on landfills in the same region with comparable design.

DESIGN AND HYDRAULIC CALCULATION OF SURFACE RUNOFF COLLECTION SYSTEMS

Integration of Surface Runoff Collection in Top Cover Systems

All elements of the SRC system are to be integrated in the overall design of the landfill body and of the landfill top cover system. Long slopes with uniform and comparative steep gradients should be avoided by reasons of better integration of the landfill in the landscape and of erosion stability. Every 10–20 m difference in height berms with longitudinal open drains should be placed for

- visual improvement of the slopes
- surface water collection and drainage (erosion control)
- maintenance purposes (firefighting, vegetation)

The berms have to be accessible for vehicles to ensure maintenance works on the slope. Fig. 8.2.8 gives an example for the integration of SRC and drainage elements in a landfill body after restoration, Fig. 8.2.9 for the design of a swale as longitudinal drainage element in a berm.

Fig. 8.2.8 visualizes that next to longitudinal open drains in berms also rough lined swales and cascades are elements of SRC systems. In addition, the placement of interceptor drains can be seen. Fig. 8.2.9 shows the direct discharge of the collected percolation from the drainage layer into the SRC system.

Directly after construction of the restoration layer revegetating should start to improve erosion stability of the slopes. When the restoration layer is covered with plants and grass intensity and ratio of discharge will be reduced significantly. If the landfill top cover system will be constructed at the end of the growing season, it should be considered to protect the topsoil with intermediate covers such as brushwood or straw mats.

Inner and outer ditches should be placed at the boundary of the landfill body and totally surround it if possible. The inner ditch collects surface and drainage water from the landfill and its slopes, the outer ditch keeps off external water. Ponds are useful for intermediate storage of collected surface water before discharging it into natural waters. The ponds can homogenize the discharge, can be used as settlement pond to remove solids from surface water, and can retain the whole volume if analyses show that surface runoff is polluted.

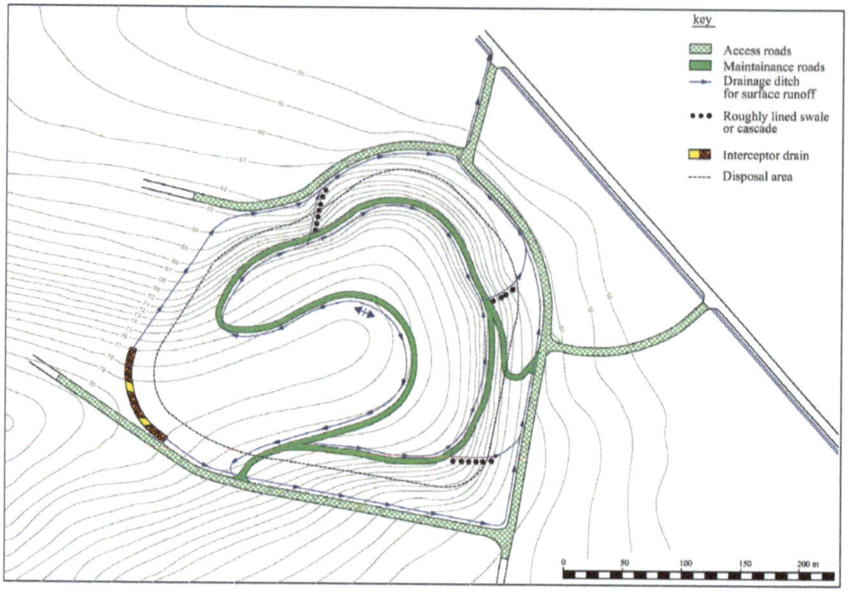

Figure 8.2.8 Example for the integration of a surface runoff collection system in a landfill body after restoration – Site plan.

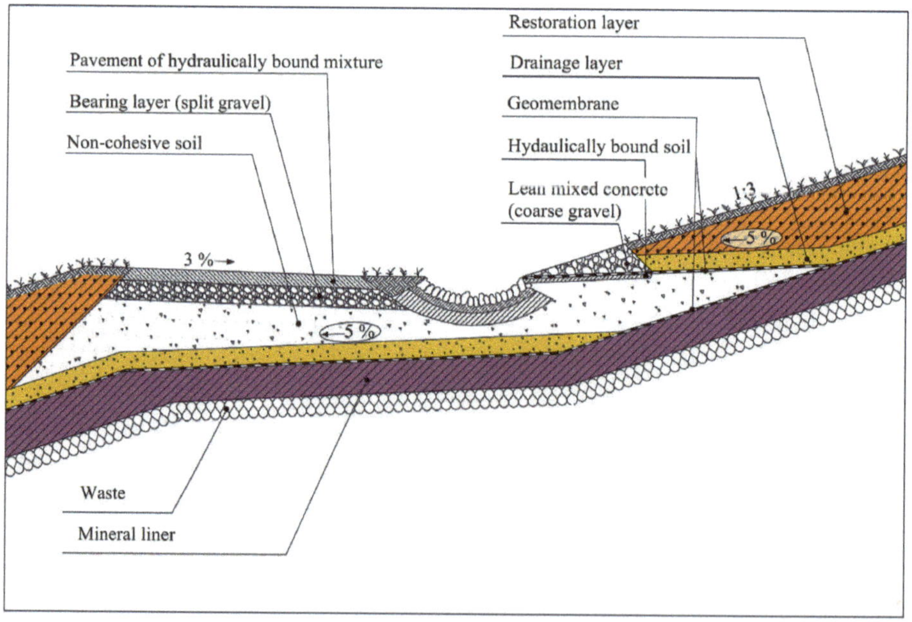

Figure 8.2.9 Example for the design of a swale as longitudinal drainage element in a berm – Direct connection of the drainage layer to the swale.

Design of Swales and Ditches

In the following, some proposals will be made for the design of swales and ditches. But it should be emphasized that regional experience is the most important source for sustainable design and construction of drainage elements because here aspects such as climatic conditions, availability of construction materials, and skills of construction companies and workers are considered.

Fig. 8.2.10 shows a cross section of a swale, Fig. 8.2.11 of a ditch. Typical for swales is the shallow cross-sectional area of flow, whereas a typical ditch is significantly deeper and has sloping sides. In Table 8.2.11, some design parameters for swales and ditches are compiled.

The lining of swales, which has to ensure sufficient discharge volume and to prevent erosion, depends on the longitudinal gradient of the bottom i_b (RAS-Ew, 2005):

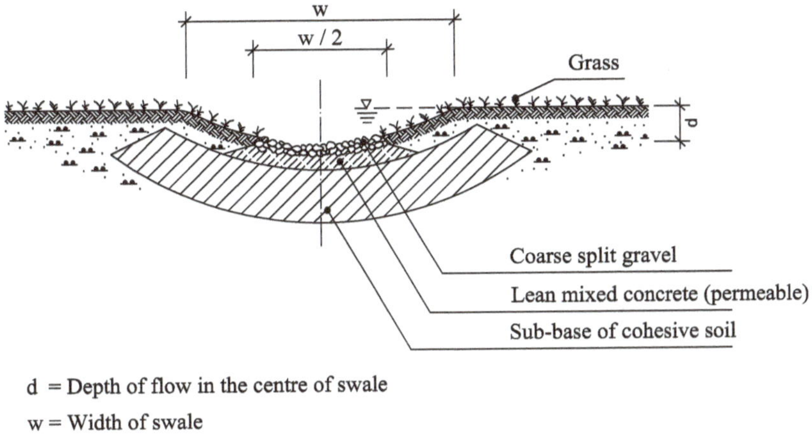

Figure 8.2.10 Example for the design of a swale with rough lining.

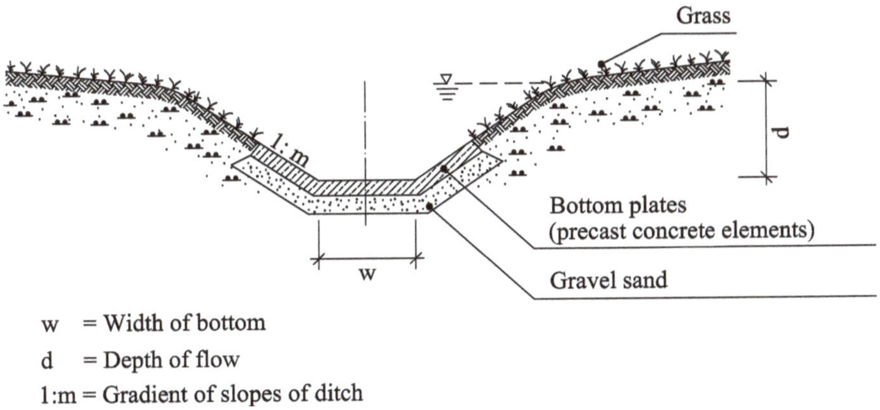

Figure 8.2.11 Example for the design of a trapezoidal ditch with soil lining (precast concrete elements).

Table 8.2.11 Parameters for designing swales and ditches (RAS-Ew, 2005)

Parameter	Unit	Type	
Swales		Vegetal lining	Rough lining
Width (w)	m	1.0–2.5	1.0–2.5
Depth (d)	m	≥0.2–w/5	-
Ditches		Vegetal lining	Soil lining
Width (w)	m	≥0.5	≥0.3
Depth (d)	m	≥0.5	0.2–0.5 or more
Gradient of slopes (banks)	1:m	1: 1.5	(1: 1.5)

- $i_b \leq 1\%$: vegetal lining (grass) or smooth lining (concrete) if necessary by hydraulic reasons
- 1% ≤ 4%: vegetal lining (grass)
- 4% ≤ 10%: rough lining if larger discharge must be expected
- ≥ 10%: rock riprap lining

Materials for smooth lining can be—apart cohesive soils such as clay or loam, which might become covered with grass and are erodible—soil cement, precast concrete elements, or geomembranes. The width of concrete lined swales can be significant smaller as shown in Table 8.2.11. Typical for rough lining is grouted riprap (see Fig. 8.2.12). Steep gradients of swales require a bottom made of coarse rock riprap.

If the space beside a slope is limited, the use of swale elements of precast concrete with a smaller bottom width and a higher depth/bottom-width ratio is recommended.

The slopes of ditches are to be grassed. Lining and erosion protection at the bottom and the lower parts of the slopes can be performed with cast-in-place concrete, precast concrete elements (see Fig. 8.2.13), paving stones and grass paving blocks, and grouted riprap.

Lining of swales and ditches has to be selected under consideration of minimization maintenance, resistance to erosive forces of flowing water, and the improvement of the hydraulic capacity.

Unpaved and unsealed swales and ditches should be used outside the direct disposal areas only because the collected water infiltrates into the underground from unsealed discharge elements in large parts. Above the disposal area the bottom of swales and ditches should be lined. In this case, the sub-base should consist of cohesive soils with low permeability (see Fig. 8.2.10) or should be lined with a

Figure 8.2.12 Example of a roughly lined swale on a steeper slope.

geomembrane. Another adequate solution is the embedding of precast concrete elements in a bed of lean-mixed concrete.

When the collected surface and drainage water is to be discharged above steeper gradients roughly lined swales and cascades become necessary. Roughly lined swales are used in general independently from driveways and should be conducted downhill in harmonic curves (see Fig. 8.2.12).

Concrete liners ensure a good hydraulic performance, but they are quite vulnerable to settlements of the landfills, even when the concrete elements will be embedded in lean-mixed concrete, and require very often intensified maintenance compared with flexible liners. Furthermore, the integration of a landfill after restoration into the surrounding landscape might be more convenient using liners with natural materials. Fig. 8.2.13 gives an example for a small ditch at the foot of a landfill slope made of precast concrete elements.

If even roughly lined swales cannot guarantee a sufficient energy conversion, cascades are to be used. Cascades are often made of very coarse rock riprap or of precast concrete elements (Fig. 8.2.14). Again the aspect of better integration in the landscape arises.

In general, it can be recommended to consider the possibility of usage of natural materials such as cohesive soils and riprap for swales and ditches. Long-term behavior, maintenance, and landscape integration should be taken into account.

Hydraulic Calculation

Hydraulic calculation of swales and ditches is a standard task of civil engineering and many guidelines demonstrate the state of the art. Helpful tools for the hydraulic calculation of swales and ditches are the FAO publication "Road Design and Construction in Sensitive Watersheds," which contains a chapter

Figure 8.2.13 Example of a ditch with concrete liner.

Figure 8.2.14 Example of a cascade constructed with precast concrete elements.

for drainage design (FAO, 1998), the "Hydraulic Design Manual" of the Texas Department of Transportation (TDT, 2014), and the classic definitive book of Chow (1959).

For dimensioning of open channels the Manning–Strickler equation can be used. The equation allows controlling the flow capacity of a single section on an open channel:

$$Q = A \cdot \frac{1}{n} \cdot r_{hy}^{2/3} \cdot i_E^{1/2} \tag{15}$$

with Q = discharge (m^3/s); A = cross-sectional area of flow (m^2); n = Manning's roughness coefficient ($s/m^{1/3}$); r_{hy} = hydraulic radius (m); i_E = gradient of the energy line (m/m); i_b = gradient of bottom of channel (m/m).

In case of uniform flow, the gradient of the energy line conforms to the gradient of the bottom of the channel. The hydraulic radius of a trapezoidal channel is calculated as (definition sketch see Fig. 8.2.11):

$$r_{hy} = \frac{A}{l_p} \tag{16}$$

with l_p = wetted perimeter (m); $A = d \cdot (w + m \cdot d)$, ($m^2$); $l_p = w + 2d \cdot (1 + m^2)^{1/2}$,(m); with d = depth of flow (m); w = width of bottom (m); 1:m = inclination of slopes of ditch (1 metre vertical in height/m metres horizontal).

For hydraulic dimensioning of swales, the next equation is recommended (see RAS-Ew, 2005), using the definitions in Fig. 8.2.10

$$Q = \frac{1}{n} \cdot d^{8/3} \cdot i_b^{1/2} \cdot \frac{w}{2d} \tag{17}$$

with Q = discharge (m^3/s); i_b = longitudinal gradient of bottom of swale (m/m); w = width of swale (m); d = depth of flow in center of swale (m).

If the capacity calculated is too low, the design parameters have to be adjusted. For a given bottom gradient either the liner of the channel can be smoothed or the bottom width can be increased. Subsequent to the hydraulic calculation the velocity of flow and the related shear stresses have to be determined. If the shear stresses are not in accordance with the permissible shear stresses of the selected liner, the design has to be adjusted and a new calculation is required (see, e.g., TDT, 2014).

For purposes of design of SRC systems on landfills, some roughness coefficients are compiled in Table 8.2.12. Further information concerning roughness coefficients, permissible shear stresses, and detailed advice on calculations are given in the above mentioned guidelines.

Fig. 8.2.15 gives an example for the discharge in two different swales in dependence of the bottom gradient, Fig. 8.2.16 for the discharge in two trapezoidal channels. The calculation was performed without consideration of freeboard and results therefore in the maximum potential discharge. Calculating the discharge, the allowed bottom gradient for different types of bottom liners is to be respected.

Table 8.2.12 Manning's roughness coefficients (TDA, 2014; RAS-Ew, 2005)

Type of Channel	Type of Liner	Manning's Roughness Coefficient n [s/m$^{1/3}$]
Earth channels	Clean, recently completed	0.016–0.020
	Earth, vegetated with short grass	0.022–0.033
	Stony bottom, weedy banks	0.025–0.040
Concrete channels	Trowel finish	0.011–0.015
	Unfinished	0.014–0.020
	Shotcrete, wavy	0.018–0.025
Swales	Concrete bottom shell	0.020–0.033
	Vegetal lining	0.033–0.050
	Rough lining	0.033–0.040

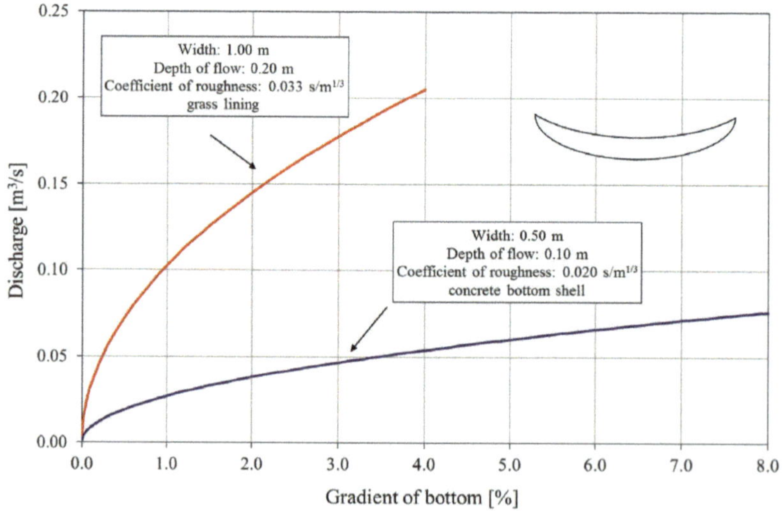

Figure 8.2.15 Discharge performance of swales.

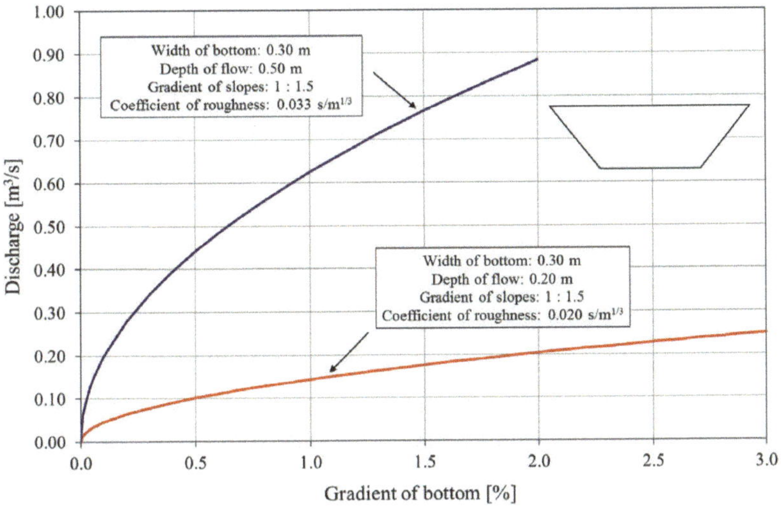

Figure 8.2.16 Discharge performance of trapezoidal ditches.

In addition to the discharge capacity also the velocity of flow and the resulting shear stresses have to be calculated and to be compared with the permissible shear stresses of the liner material selected. If the shear stresses exceed the permissible shear stresses, a more suitable liner material has to be chosen.

Finally a brief calculation of the required discharge capacity shall be given, following the basic data shown in the examples above:

- length of slope: 200 m
- width of slope: 50 m
- area of slope: 10,000 m^2 = 1 ha
- height of rainfall: 10 mm
- duration of rainfall: 10 min
- intensity of rainfall: 60 mm/h
- runoff coefficient: 0.20
- peak discharge: 0.033 m^3/s

For such a comparable low discharge a ditch is not necessary (compare Fig. 8.2.16). The best solution might be in this case to use a swale consisting of precast concrete elements. Fig. 8.2.15 demonstrates that a gradient of bottom of 2% would be sufficient to ensure free discharge.

But apart from this example in practice, it is necessary to consider not only the surface runoff of landfill slopes but also the surface runoff of berms (maintenance driveways) and the drainage discharge.

DESIGN AND HYDRAULIC CALCULATION OF DRAINAGE SYSTEMS

Integration of Drainage Systems in Top Cover Systems

Drainage layers are an integral part of landfill top cover systems. Drainage layers have to collect and to discharge the water, which percolates through the restoration layer.

The drainage layer has to be connected to the surface water collection elements such as swales and ditches.

This connection can be designed in two manners:

- direct connection of the drainage layer (Fig. 8.2.9)
- indirect connection of the drainage layer (Fig. 8.2.17)

The first alternative shows the advantages that nearly no maintenance is required, and that the whole width of the slope is available for drainage discharge, which ensures substantial reverse capacity. The second alternative requires much more maintenance, in particular, for inspection and flushing of the drainpipes. But the advantage of this solution is that an aeration of the drainage layer is inhibited, which might be necessary to prevent desiccation of a mineral liner.

Hydraulic calculations of the drainage layer are necessary as well as calculations of slope stability, and—if installed—structural analyses of drainpipes. Calculations of slope stability should consider force of flow in the drainage layer.

Design of Drainage Layers

The design of the drainage layer in the top cover system is strongly depending from the type of cover (see Chapter 11.1), the national regulations, and given local climatic conditions.

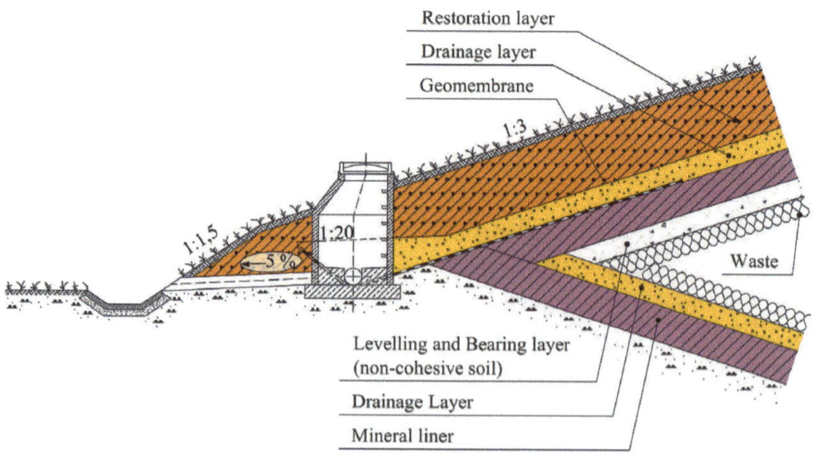

Figure 8.2.17 Example for the design of an inner ditch at landfill toe—Indirect connection of the drainage layer to the ditch.

As an example Table 8.2.13 presents the design requirements for drainage layers of the German Directive on the Landfill of Waste (DepV, 2017). These requirements are valid for all classes of landfills in Germany except Class 0 (soil deposit), which only needs a restoration layer.

Detailed design recommendations in fulfillment of these requirements are given by the German Geotechnical Society (GDA E 2-20, 2015).

The Landfill Directive of the European Union (EU, 1999) just requires a drainage layer with a thickness of >0.5 m in top cover systems of landfills for hazardous and nonhazardous wastes without further specifications.

In Europe, some national directives concretize this general request. An example is the Landfill Manual of the Irish Environmental Agency (EPA Ireland, 2000), which specifies the material of the drainage layer (granular material), its thickness (minimum thickness 0.5 m or a geosynthetic drainage layer), the hydraulic conductivity ($\geq 1 \times 10^{-4}$ m/s), and the gradient of the drainage layer (≥4%).

In the United States, the Criteria for Municipal Solid Waste Landfills (USA ECFR, 2016) determine the technical specifications to be considered for landfill design. Subpart F—Closure and Postclosure Care—requires a top cover system, including a liner, but a drainage layer is neither defined nor mentioned.

The lack of detailed specifications can also be observed in international guidelines. Rushbrook and Pugh (1999) emphasize in the World Bank Technical Paper the necessity of long-term operating SRC systems but gave no information on drainage systems in top cover systems. This is the same with the UNEP handbook on Solid Waste Management (UNEP, 2005). Components of top cover systems are discussed, but no further advice is given on details of the drainage layer. Only Savage et al. (1998) describe a top cover system inclusive a drainage layer with specifications like the Irish ones.

Due to this lack of detailed specifications in the following the German requirements listed in Table 8.2.13 shall be the base of the further discussions. The public authorities in Germany can allow

Table 8.2.13 Design parameters for drainage layers according to german regulations and recommendations (DepV, 2013; GDA, 2015)

Parameter	Unit	Dimension
Thickness	cm	≥30
Hydraulic conductivity	m/s	$\geq 1 \times 10^{-3}$
Gradient, long term	%	≥5%
Material	–	coarse sand, gravel

attenuating these requirements if the long-term operational capability can be proved, but considering the expected minimum reliability of 100 years this verification is hardly to be performed.

A thickness of the drainage layer of 30 cm in combination with a hydraulic conductivity of more than 1×10^{-3} m/s is not necessarily required, as could be demonstrated by hydraulic calculations. But facing the required long-term efficiency of the drainage system, it is useful not to reduce the thickness of the layer. Aging processes such as precipitation of dissolved components of the restoration layer, infiltration of fines, or root penetration are to be considered but hard to calculate. Therefore an inherent safety margin in terms of unused thickness and coarser pores is reasonable.

The minimum gradient of the drainage layer in the top liner system should be not less than 5% to ensure gravity discharge in all zones of the layer. Designing the top liner system, the present and future settlements of the landfill body are to be considered. Settlements are not only caused by biological degradation but also by compression and consolidation as a consequence of loads caused by the surface liner system itself.

A mineral drainage layer consists of materials such as gravel, coarse sand, crushed rock, or recycling material such as coarse slag. All these materials have to be chemical and biological resistant and must show an adequate stability. The material may contain neither easy water-soluble components nor dissoluble pollutants.

Drainage layers have to fulfill the filter criteria for granular filters compared with the adjoined layers (subsoil in the restoration layer, protective layer, or mineral liner). In particular for coarse grained mineral layers, mineral or geosynthetic filter layers become necessary in general to be designed under consideration of common filter criteria.

At the toe of a long slope, usage of special coarse drainage material is recommended, which is directly connected to the swales or ditches (see Fig. 8.2.9), to ensure free discharge. Drainage pipes might be useful for improved drainage water discharge. Design and construction drainage pipes must enable their inspection and maintenance. Minimum diameter of the drainage pipes should be 200 mm.

Hydraulic Calculation of the Drainage Layer

Basic Equations

Objective of hydraulic calculations of drainage layers is to determine the height of the water table above the liner—the saturated thickness—and to compare it with the height of the drainage layer. The saturated thickness should not exceed the height of the drainage layer to prevent slope failure caused by reduction of shear strength in the restoration layer caused by moisture penetration.

The calculations will be performed for steady-state conditions, which mean uniform and permanent recharge of the soil percolation water to the drainage layer. This results in a time-independent discharge in the drainage layer. The advantage of this approach is at first, the related equations can be solved analytical or with simple numerical models and no numerical models based on Finite Elements or Finite Differences are necessary. At second this method of calculations covers periods with higher soil percolation rates if percolation rates are used, which are seldom exceeded. This ensures long-term system safety even in critical periods.

In opposite to the leachate collection system at the bottom of the landfill, the gradient of the drainage layer in the landfill top cover system is much higher in general. Therefore the first approximation of Boussinesq is useful for the hydraulic calculation of the drainage layer, whereas the second approximation of Boussinesq is appropriate for calculating the leachate collection system (see Chapter 8.1). A comparison of both approximations and an overview of their analytical and numerical solutions are performed by Ramke (2009).

The definitions for the derivation of the basic equation according to the first approximation of Boussinesq are given by Fig. 8.2.18. The streamlines in the drainage layer are parallel to the bottom of the drainage layer—the liner; the equipotential lines are perpendicular to the slope. For the derivation of the equation a coordinate system parallel and rectangular to the slope is used.

In any cross section along the drainage length a balance of inflow and outflow must be under steady-state conditions. For the inflow we receive

$$q_{in} = v_n \cdot (l_x - (x' - \Delta x') \cdot \cos(\alpha)) \tag{18}$$

The factor "cos α" results from the conversion of the sloping x′-axis into the horizontal length. The term "Δx′" considers the portion of the water table on the shift against the x-direction:

$$\Delta x' = a' \cdot \tan \alpha \tag{19}$$

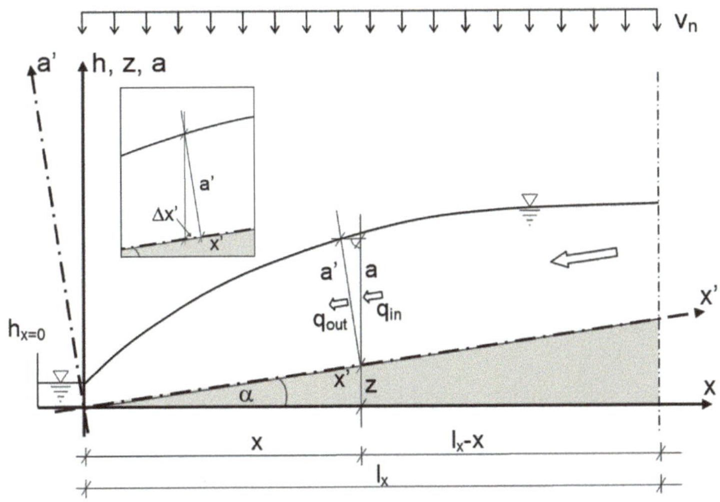

Figure 8.2.18 Definition sketch for derivation of the drainage formula on a sloping liner according to the first approximation of Boussinesq.

Details are explained in the supplement to Fig. 8.2.18. The discharge is calculated according to Darcy's law:

$$q_{out} = a' \cdot k \cdot i = a' \cdot k \cdot \frac{\partial h}{\partial x'} \tag{20}$$

The hydraulic gradient is calculated with the normal definition of the hydraulic potential, which consists of the elevation and the pressure head:

$$\begin{aligned} h &= z + a \\ h &= x' \cdot \sin\alpha + a \cdot \cos\alpha \end{aligned} \tag{21}$$

This relationship considers that the lines of equipotential are perpendicular to the sloping liner. Equating the input and output leads to the first basic equation:

$$a' \cdot k \cdot \frac{\partial(x' \cdot \sin\alpha + a' \cdot \cos\alpha)}{\partial x'} = v_n(l_x - (x' - a \cdot \tan\alpha) \cdot \cos\alpha) \tag{22}$$

After rearranging the equation and dividing by $\cos\alpha$ we receive

$$\frac{\partial a'}{\partial x'} = \frac{v_n}{k} \cdot \frac{(l_x - x' \cdot \cos\alpha + a' \cdot \sin\alpha)}{a' \cdot \cos\alpha} - \frac{\sin\alpha}{\cos\alpha} \tag{23}$$

with h = hydraulic potential, piezometric head (m); a' = saturated thickness rectangular to the liner (m); z = elevation head above datum plane (m); x' = axis parallel to the sloping drainage layer (m); i = hydraulic gradient (−); l_x = horizontal drainage length (m); v_n = leakage rate, (soil) percolation rate (m/s); k = hydraulic conductivity (m/s); q_{in} = input flow into a cross section ($m^3/(m \cdot s)$); q_{out} = output flow out of a cross section ($m^3/(m \cdot s)$); α = angle of slope (°).

Eq. (22) is the first one often used equation for the description of groundwater flow on an impermeable sloping bottom with groundwater recharge.

Analytical Solutions

Analytical solutions of the basic Eq. (22) and of Equation (1) in Chapter 8.1, which describes drainage flow according to the second assumption of Boussinesq, are possible, but the resulting formulas do not allow a direct calculation of the height of the water table (see Equation 1 in Chapter 8.1) or of the saturated thickness (Eq. 22) in dependence of the distance to the drain (x or x') for given parameters (explicit analytical solution).

Only implicit analytical solutions are possible (see McBean et al., 1982; McEnroe, 1989; Ramke, 1991). The relation between h and x or a' and x' has to be determined iteratively in this case, and the iteration scheme often shows a difficult convergence behavior (Ramke, 1991). Therefore the use of these solutions for the calculation of the course of the water table or the saturated thickness is not very recommendable compared with the easiness of the implementation of the Runge–Kutta scheme (see Chapter 8.1).

But explicit equations for the calculation of the maximum saturated thickness are available. Unfortunately an often quoted formula of Moore (1980), for the calculation of the maximum saturated thickness, mentioned in papers of the US EPA, does not work properly. Also a number of approximations were developed, but either they have a limited range of application or they are difficult in use, too. As explicit solutions, the approaches of McEnroe (1993) and Giroud and Houlihan (1995) as well as the formulas of Schmid (1993) and of Lesaffre (1987) for the determination of the maximum saturated thickness should be mentioned.

The set of equations of Schmid (1993) was developed for the boundary cases at the right end either with a horizontal gradient at the end of the drainage area or a saturated thickness about zero. The solution of Schmid (1993) complies with the solution of McEnroe (1993).

Schmid's set of equations was developed by solving the implicit formula for the condition $\partial a'/\partial x' = 0$ (gradient parallel to the x'-axis, maximum of saturated thickness). Fig. 8.2.19 explains the particular definitions for the approach of Schmid (1993).

Three cases of parameter combinations have to be distinguished

$$\textit{Case A}:\ \Delta = 4\cdot(v_n/k) - \tan^2\alpha > 0:\ a'_{max} = \sqrt{\frac{v_n}{k}}\cdot l_{x'}\cdot exp\left[\frac{\tan\alpha}{\sqrt{\Delta}}\cdot\left(arctg\frac{k\cdot\tan^2\alpha\ -\ 2\cdot v_n}{k\cdot\tan\alpha\cdot\sqrt{\Delta}}\ -\ arctg\frac{\tan\alpha}{\sqrt{\Delta}}\right)\right] \quad (24)$$

$$\text{Case B}:\ \Delta = 4\cdot(v_n/k) - \tan^2\alpha = 0: \quad (25)$$
$$a'_{max} = \sqrt{\frac{v_n}{k}}\cdot l_{x'}\cdot\frac{1}{e}\,(e = \text{Euler's Number})$$

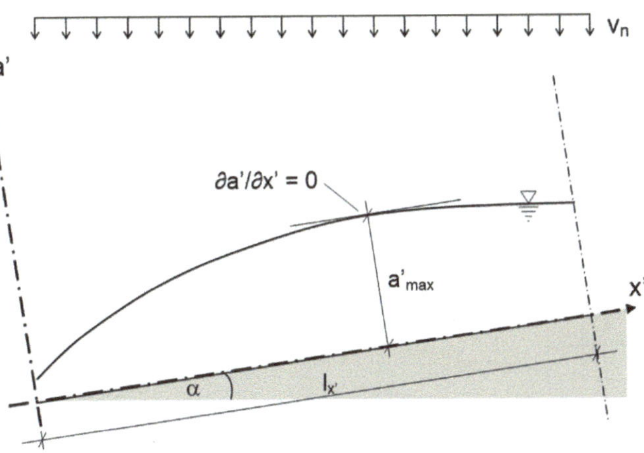

Figure 8.2.19 Definition sketch for the approach of Schmid (1993).

Case C : $\Delta = 4 \cdot (v_n/k) - \tan^2\alpha = 0$:

$$a'_{max} = \sqrt{\frac{v_n}{k} \cdot l_{x'}} \cdot \left| \frac{-2 \cdot v_n + k \cdot \tan\alpha \cdot \left(\tan\alpha - \sqrt{-\Delta}\right)}{-2 \cdot v_n + k \cdot \tan\alpha \cdot \left(\tan\alpha + \sqrt{-\Delta}\right)} \cdot \frac{\tan\alpha + \sqrt{-\Delta}}{\tan\alpha - \sqrt{-\Delta}} \right|^{\frac{\tan\alpha}{2 \cdot \sqrt{-\Delta}}} \tag{26}$$

with a'_{max} = maximum saturated thickness rectangular to the liner (m); $l_{x'}$ = length of slope parallel to the bottom (m); v_n = leakage rate, (soil) percolation rate (m/s); k = hydraulic conductivity (m/s); α = angle of slope (°).

Lesaffre (1987) has developed a very elegant approximation for discharge in a sloping drainage layer between two drains on an impermeable liner (see Fig. 8.2.20):

$$\frac{l_{s'}}{a'_{max}} = \left(\frac{4 \cdot k}{v_n} + \left(\frac{k}{v_n} - 1 \right)^2 \cdot (\tan\alpha)^2 \right)^{1/2} \tag{27}$$

with a'_{max} = maximum saturated thickness rectangular to the liner (m); $l_{s'}$ = drain spacing parallel to the slope (m); v_n = leakage rate, (soil) percolation rate (m/s); k = hydraulic conductivity (m/s); α = angle of slope (°).

Examples

Some examples will be given to demonstrate the applicability of the above explained equations for practical purposes.

Fig. 8.2.21 summarizes a comparison of the two analytical and a numerical solution according to the second approximation of Boussinesq using the Runge–Kutta method.

The diagram shows the maximum saturated thickness in dependence of the slope gradient. The calculations were performed with a length of the drainage area of 15 m, a hydraulic conductivity of

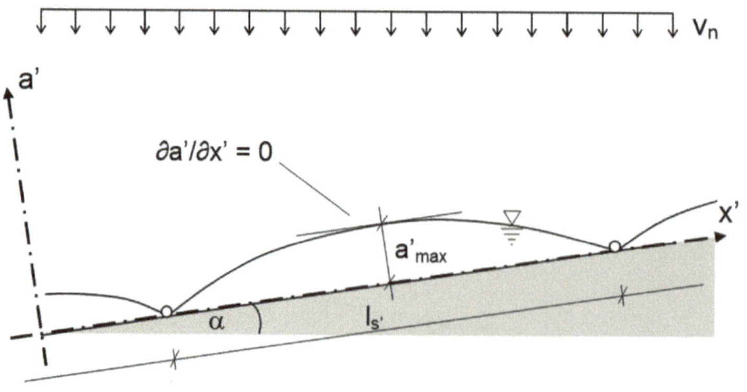

Figure 8.2.20 Definition sketch for the approach of Lesaffre (1987).

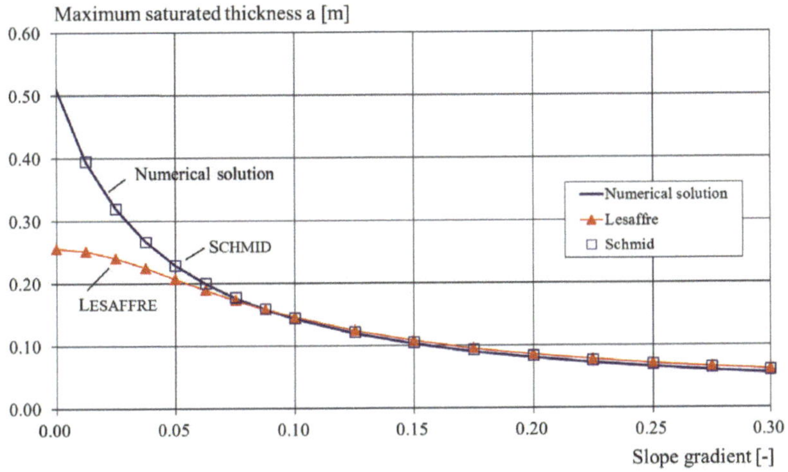

Figure 8.2.21 Maximum saturated thickness in a sloping drainage layer in dependence of slope gradient (drainage length 15 m, $k = 1 \times 10^{-4}$ m/s, $v_n = 10$ mm/d).

$k = 1 \times 10^{-4}$ m/s and a percolation rate of 10 mm/d, which is a reasonable value for climatic conditions in temperate zones (see Paragraph Estimation of Soil Percolation Rates).

It can be seen that there is an excellent correlation between the numerical solution and the analytical solution of Schmid between a slope gradient from 0% to 10%. Above the latter value, the differences between the solutions according to the first and the second approximation of Boussinesq become noticeable because of the reasons mentioned above.

The solution of Lesaffre is in conformity with the other two solutions higher than a slope gradient of 7.5%. This is due to the fact that Lesaffre considers an uphill drain, which is also dewatering the slope when the drainage layer is partly saturated at the top too. When the drainage layer near to the uphill drain is nearly unsaturated, the uphill drain has no influence.

In the first place, this example demonstrates the good applicability of the solutions presented here. In the second place, it is shown that it can be necessary also to calculate the shape of the water table and not only the maximum of saturated thickness.

Two further examples demonstrate the development of the saturated thickness in dependence of the slope gradient and the length of the slope for drainage systems in landfill top cover systems. The examples were calculated with a hydraulic conductivity of $k = 1 \times 10^{-3}$ and 1×10^{-4} m/s, the percolation rate was assumed with 5 and 10 mm/d, which covers the range of higher percolation rates.

The calculations were performed with the approaches of Lesaffre and Schmid. The results are documented in Figs. 8.2.22 and 8.2.23. The coordinate system used for the figures is parallel and perpendicular to the slope.

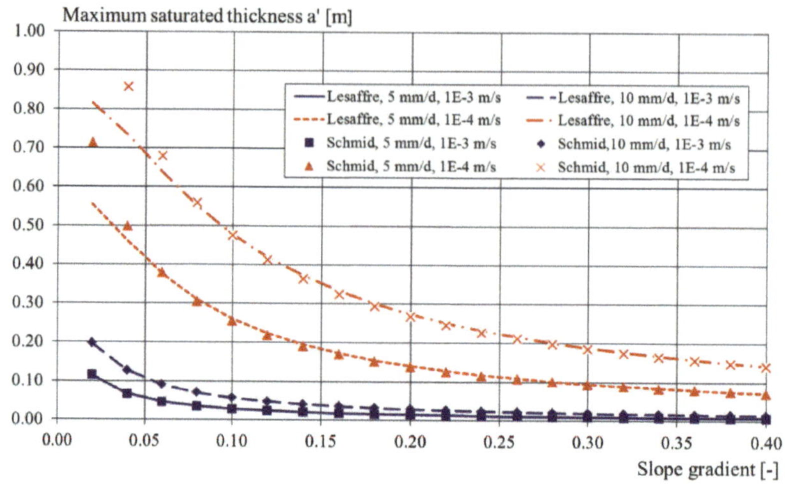

Figure 8.2.22 Maximum saturated thickness in a drainage layer in dependence of slope gradient (length of the slope 50 m, $k = 1 \times 10^{-3}$ and 1×10^{-4} m/s).

Fig. 8.2.22 shows the influence of slope gradient on the saturated thickness in a drainage layer with a length of 50 m. The maximum saturated thickness in a "standard system" with a hydraulic conductivity of $k = 1 \times 10^{-3}$ m/s and a slope gradient of 5% does not exceed much more than 10 cm if the percolation rate amounts 10 mm/d. This is much less than the often required thickness of the drainage layer of 30 cm. But if the hydraulic conductivity is reduced to $k = 1 \times 10^{-4}$ m/s, it needs a minimum slope gradient of 18% to ensure that the saturated thickness caused by this percolation rate does not exceed 30 cm.

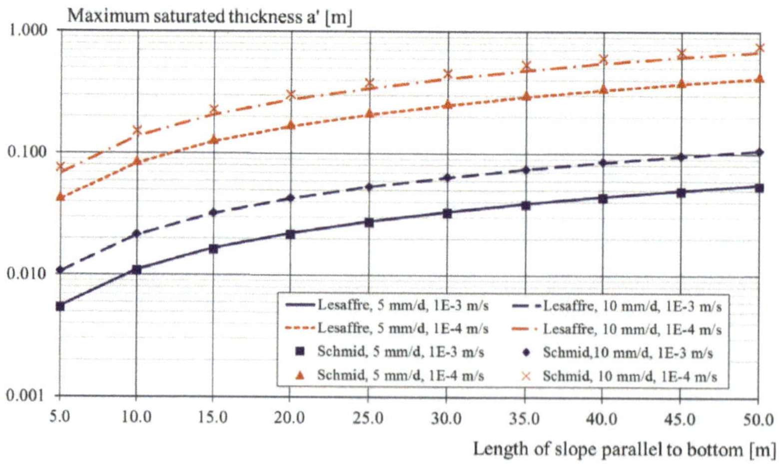

Figure 8.2.23 Maximum saturated thickness in a drainage layer in dependence of length of the slope (slope gradient 5%, $k = 1 \times 10^{-3}$ and 1×10^{-4} m/s).

Furthermore, Fig. 8.2.22 demonstrates the influence of the drain at the hillside on the saturated thickness. If both calculations show the same result, there is no flow toward the hillside drain. For a hydraulic conductivity of $k = 1 \times 10^{-3}$ m/s, there is no hillside drain necessary even if the slope gradient is just 2%. But if the conductivity is limited to $k = 1 \times 10^{-4}$ m/s flow occurs to the downhill and the uphill drains up to a slope of 6%.

The influence of the length of the slope or the drain spacing is shown by Fig. 8.2.23 (uniform slope gradient 5%). The maximum saturated thickness does not exceed 10 cm at a length of the slope of 50 m if the hydraulic conductivity is $k = 1 \times 10^{-3}$ m/s ($v_n = 10$ mm/d). For a conductivity of $k = 1 \times 10^{-4}$ m/s, the drainage area is limited to 20 m for a percolation rate of 10 mm/d and an allowable maximum saturated thickness of 30 cm.

These examples demonstrate that it is not obligatory to use drainage material with hydraulic conductivity of $k = 1 \times 10^{-3}$ m/s. But facing the abovementioned aging processes, it seems to be foresighted, to use material for the mineral drainage layer with sufficient long-term capacity.

References

Berger, K., 2013. Current and future prospects of the hydrologic evaluation of landfill performance (Help) model for simulating the water balance of landfill liner systems. In: Sardinia 2013: Proceedings of the Fourteenth International Landfill Symposium, S. Margherita di Pula. CISA—Environmental Sanitary Engineering Centre, Cagliari, Italy.

Chow, V.T., 1959. Open-Channel Hydraulics. McGraw-Hill Book Company, New York, USA.

Environmental Protection Agency Ireland (EPA Ireland), 2000. Landfill Manuals – Landfill Site Design, Wexford, Ireland.

European Union (EU), 1999. Council Directive 1999/31/EC of 26 April 1999 on the Landfill of Waste.

Food and Agriculture Organization of the United Nations (FAO), 1998. Watershed Management Field Manual—Road Design and Construction in Sensitive Watersheds; FAO Conservation Guide 13/5. FAO, Rome, Italy.

Freie und Hansestadt Hamburg (Hamburg), 2003. Design Rainfall – Rainfall Series of the City of Hamburg (Bemessungsregen – Regenreihen der Freien und Hansestadt Hamburg). Behörde für Bau und Verkehr, Amt für Bau und Betrieb, Hamburg.

GDA E 2–20, 2015. Drainage Layers in Landfill Top Cover Systems (Entwässerungsschichten in Oberflächenabdichtungssystemen). Technical Recommendations of the Technical Committee Geotechnics of Landfills of the German Geotechnical Society, Essen, Germany. http://www.gdaonline.de/gda.html.

German Directive on the Landfill of Waste (DepV), 2013. Verordnung über Deponien und Langzeitlager (Deponieverordnung—DepV) 27.09.2017, BGBl. I S. 3465 (German Federal Law Gazette).

Giroud, J.P., Houlihan, M.F., 1995. Design of leachate collection layers. In: Sardinia 95, Fifth International Landfill Symposium. CISA, Environmental Sanitary Engineering Centre, Cagliari, Italy.

Lesaffre, B., 1987. Analytical formulae for Travers drainage of sloping lands with constant rainfall. Irrigation and Drainage Systems I (2).

McBean, E.A., Poland, R., Rovers, F.A., Crutcher, A.J., 1982. Leachate collection design for containment landfills. Journal of the Environmental Engineering Division (EE 1). ASCE.

McEnroe, B., 1989. Hydraulics of leachate collection and cover drainage. In: Landfill Concepts, Environmental Aspects, Lining Technology, Leachate Management, Industrial Waste and Combustion Residues Disposal 2nd International Symposium, Porto Conte ISWA—Italian Section. Milano Academic Press, London, San Diego.

McEnroe, B., 1993. Maximum saturated depth over landfill liner. Journal of Environmental Engineering 119 (2).

Moore, C.A., 1980. Landfill and Surface Impoundment Performance Evaluation Manual. United States Environmental Protection Agency, Office of Water and Waste Management, Washington DC. EPA/560/SW-869c.

Ramke, H.-G., 1991. Hydraulic Assessment and Calculation of Leachate Collection Systems of Sanitary Landfills—Water Balances, Hydraulic Characteristics and Methods of Calculation (Hydraulische Beurteilung und Dimensionierung der Basisentwässerung von Deponien fester Siedlungsabfälle—Wasserhaushalt, hydraulische Kennwerte, Berechnungsverfahren – Dissertation). Ph.D. thesis. Mitteilungen aus dem Leichtweiss-Institut für Wasserbau, Heft 114, Technische Universität Braunschweig.

Ramke, H.-G., 2002. Collection of surface runoff and drainage of landfill capping systems (Oberflächenwassersammlung und – ableitung). In: Handbuch der Müll- und Abfallbeseitigung, Kennziffer 4542. Erich Schmidt Verlag, Berlin.

Ramke, H.-G., 2009. Leachate collection systems. In: Witt, K.J., Ramke, H.-G., Telekes, G., Imre, E. (Eds.), 1st Middle European Conference on Landfill Technology; International Conference, Organised by the Technical Committee on Geotechnics of Landfills in the DGGT and the Hungarian National Committee of the SSMGE, Budapest, February 6-8, 2008. http://www.hs-owl.de/fb8/fachgebiete/abfallwirtschaft/pdf/Budapest_2008_Internet.pdf.

RAS-Ew, 2005. Guidelines on Road Design, Part 1: Drainage (Richtlinien für die Anlage von Straßen, Teil: Entwässerung). Forschungsgesellschaft für Straßen- und Verkehrswesen (FGSV), Arbeitsgruppe "Erd- und Grundbau", Köln, Germany.

Rushbrook, P., Pugh, M., 1999. Solid Waste Landfills in Middle- and Lower-Income Countries. World Bank, New York, USA. World Bank Technical Paper No. 426.

Savage, M., Diaz, L.F., Golueke, C., Martone, C., Ham, R.K., 1998. Guidance for Landfilling Waste in Economically Developing Countries. US Environmental Protection Agency, Contract 68-C4-0022; ISWA.

Schmid, B.H., 1993. Die maximale Wassertiefe über gleichmäßig beaufschlagten, geneigten Dichtungshorizonten, Wasser und Boden, Heft 9.

Schroeder, P.R., Aziz, N.M., Lloyd, C.M., Zappi, P.A., 1994a. The hydrologic evaluation of landfill performance (HELP) model. In: User's Guide for Version 3 EPA/600/R-94/168a. US EPA, Cincinnati, OH.

Schroeder, P.R., Dozier, T.S., Zappi, P.A., McEnroe, B.M., Sjostrom, J.W., Peyton, R.L., 1994b. The hydrologic evaluation of landfill performance (HELP) model. In: Engineering Documentation for Version 3 EPA/600/R-94/168b. US EPA, Cincinnati, OH.

Texas Department of Transportation (TDT), 2014. Hydraulic Design Manual. Texas Department of Transportation, Austin, USA.

United Nations Environmental Program (UNEP), 2005. Solid Waste Management. UNEP, Nairobi, Kenia.

United States Department of Agriculture (USDA), 1986. Urban hydrology for small watersheds. In: Natural Resources Conservation Service, Conservation Engineering Division, Technical Release 55.

United States Electronic Code of Regulations (US ECFR), 2016. Part 258 – Criteria for Municipal Solid Waste Landfills, Subpart F – Closure and Post-Closure Care, ECFR-Version February 01, 2016.

9 CHAPTER

GAS MANAGEMENT

9.1

LANDFILL GAS GENERATION MODELING

Gianni Andreottola, Raffaello Cossu and Marco Ritzkowski

INTRODUCTION

Landfill gas (LFG) models are an essential tool in landfill project development to perform the following tasks:

- assess the feasibility of LFG energy recovery and its technical requirements;
- estimate of the amount of recoverable LFG that is generated over time;
- design gas extraction or disposal system (extraction wells, in situ treatment, top cover oxidation, etc.);
- due diligence evaluation of the performance of existing collection and recovery systems;
- calculate (national) greenhouse gas (GHG) inventories;
- predict residual environmental landfill impacts.

The development of LFG models started in the 1970s when several authors summarized experimental data on a rational basis (Alpern, 1973; Boyle, 1976; Ham, 1979). Qualitative models were developed by Farquhar and Rovers (1973) while quantitative models were later provided in the United States. (i.e., Palos Verdes, Scholl Canyon and Sheldon Arleta models). Nowadays the most used model for LFG prediction is Landfill Gas Emissions Model (LandGem), which is based on the approach of Monod first-order exponential kinetics.

Depending on data availability and level of scientific knowledge of the involved processes, a general classification of models is possible (statistic analysis, stocastistic, deterministic, complex models). The majority of LFG models belong to the group of deterministic models.

STRUCTURE OF LANDFILL GAS PRODUCTION MODELS

Theoretically a complete biogas model should include three submodels:

- *stoichiometric submodel:* This model provides the maximum theoretical yield of biogas from the organic waste fraction that will anaerobically decay in a landfill. Some models proposed in the literature are simply stoichiometric and provide as a result only information on LFG yields. These models are static, according to the classification previously reported.
- *kinetic submodel:* This is a dynamic model, which provides the temporal evolution of LFG generation rates. It can be either an empiric model, based on a more or less simple equation of a defined order, or a deterministic model, based on a set of equations describing the degradation of

the different biodegradable municipal solid waste (MSW) fractions, or an ecological model, which describes the dynamic of microbial populations and substrates within the landfill.

- *diffusion submodel:* This is a dynamic model, which describes the time and space variation of pressure, temperature, and gas composition within the landfill body. LFG emission rates can be obtained and the effectiveness of the gas extraction system can be verified.

In this chapter only generation models are considered, and therefore only the aspects regarding stoichiometry and kinetic of LFG will be analyzed in detail.

Estimation of maximum landfill gas yields (L_0)

A reaction representing the overall methane fermentation process for organics in solid waste can be represented by the following equation:

$$C_aH_bO_cN_d + nH_2O \rightarrow xCH_4 + yCO_2 + wNH_3 + zC_5H_7O_2N + \text{energy} \quad (1)$$

where: $C_aH_bO_cN_d$ = empirical chemical formulation for biodegradable organics in solid waste; $C_5H_7O_2N$ = chemical formulation of microbial biomass.

The biodegradable organic carbon present in MSW is transformed, during anaerobic degradation, into methane and carbon dioxide. The energy content of the organic matter is split into the free energy content of methane, the energy for bacterial synthesis, and the enthalpy of the reaction.

The fraction of the organic matter which is converted to microbial biomass, considering an infinite retention time in the system is about 4% (EMCON, 1976). Therefore, for the practical evaluation of the maximum theoretical LFG yield, cells conversion of organic matter can be neglected and Eq. (1) becomes

$$C_aH_bO_cN_d + \left(a - \frac{b}{4} - \frac{c}{2} + \frac{3d}{4}\right)H_2O \rightarrow \left(\frac{a}{2} + \frac{b}{8} - \frac{c}{4} - \frac{3d}{8}\right)CH_4 + \left(\frac{a}{2} - \frac{b}{8} + \frac{c}{4} + \frac{3d}{8}\right)CO_2 + dNH_3 \quad (2)$$

Eq. (2) has been reported several times in literature (Ham and Barlaz, 1989; Tchobanoglous et al., 1993). Once the elemental composition of the waste is known, this equation allows the evaluation of both the quantity and quality of the gas ($CH_4 + CO_2$) generation. Table 9.1.1 summarizes some typical data on elemental composition of the biodegradable fraction of MSW (Tchobanoglous et al., 1993).

Eq. (2) states that 1 mol of organic carbon is bioconverted to 1 mol of LFG. Therefore, given that 1 mol of gas at 0°C and 1 atm occupies 22.4 L,

$$1 \text{ mole C in organic matter} = 22.4\,\text{l gas}(CH_4 + CO_2) \quad (3)$$

On weight basis,

$$1 \text{ g C in organic matter} = 1.867\,\text{l gas}(CH_4 + CO_2) \quad (4)$$

Table 9.1.1 Typical data on elementary composition of municipal solid waste organic fractions expressed in percent of dry weight (Tchobanoglous et al., 1993)

	Wet Weight (%)	Dry Weight (%)	Elemental Composition (%)					
Component			C	H	O	N	S	Ash
Food wastes	11.4	4.6	4.7	4.7	4.4	13.0	10.0	4.0
Paper	42.8	55.0	50.8	53.0	61.3	18.5	60.0	55.2
Cardboard	7.5	9.8	9.2	9.4	11.1	3.7	10.0	8.0
Plastics	8.8	11.9	15.1	13.8	6.8	–	–	19.8
Textiles	2.5	3.1	3.6	3.3	2.4	14.8	–	1.4
Rubber	0.6	0.9	1.4	1.4	–	1.9	–	1.4
Leather	0.6	0.7	0.9	0.8	0.2	7.4	–	1.1
Yard wastes	23.3	11.2	11.4	10.8	10.8	40.7	20.0	8.3
Wood	2.5	2.8	2.9	2.8	3.0	–	–	0.6
Total	100	100	100	100	100	100	100	100

Eq. (2) allows to estimate the maximum theoretical yield of LFG starting from the general formula characterizing solid wastes ($C_aH_bO_cN_d$). The estimation can also be directly carried out according to either the specific formula of organic compound to be decomposed or the empiric formulas introduced to represent biodegradable fractions of MSW (Table 9.1.2). Table 9.1.2 also reports gas yield and composition deriving from the degradation of the single compounds (or classes of compounds).

The estimation of methane yield from MSW degradation can also be given in terms of chemical oxygen demand, defined as the oxygen consumption for the complete oxidation (biologically or thermally) of the organic matter (g O_2/g organic matter). This approach is typically used for wastewaters where the organic matter is present in liquid solution, but can be transferred to solid waste through the following calculations.

For the oxidation of 1 mol of CH_4, a total of 2 mol of O_2 are required:

$$CH_4 + 2O_2 \rightarrow CO_2 + 2H_2O$$
$$1\text{ mol } CH_4 \rightarrow 2\text{ mol } COD_{CH_4}$$

Table 9.1.2 Theoretical yields of methane and landfill gas ($CH_4 + CO_2$) for some typical compounds or classes of compounds

Compound (A)	Formula	COD (kg/kg)	Gas yield (Nl/kg a)	Gas Composition (% CH_4)	Reference
Cellulose	$(C_6H_{10}O_5)_x$	1.185	830	50	Gendebien et al. (1992)
Protein	$(C_4H_6ON)_y$	1.619	1100	51.5	Gendebien et al. (1992)
Lipids	$C_{55}H_{106}O_6$	2.914	1428	71.4	Gendebien et al. (1992)
Total organic carbon	C	2.67	1867 ([a])	50 ([a])	
COD	g O_2/g	1	700	50	Heidrich et al. (2011)
MSW	$C_{99}H_{149}O_{59}N$	1.489	966	54	EMCON (1976)
MSW	$(CH_2O)_n$	1.067	800	50	Hoeks (1983)
No food organic waste (assimilated to paper)	$C_{203}H_{334}O_{138}$	1.398	914	54	EMCON (1980)
Rapidly decomposable fraction	$C_{68}H_{111}O_{50}N$	1.300	875	52	Tchobanoglous et al. (1993)
Slowly decomposable fraction	$C_{20}H_{29}O_9N$	1.710	1049	55	Tchobanoglous et al. (1993)
Food waste	$C_{16}H_{27}O_8N$	1.662	993	56	EMCON (1980)

COD, chemical oxygen demand; *MSW*, municipal solid waste.
[a]based on stoichiometry and under standard conditions.

assuming

$$COD_{\text{organic matter}} = COD_{CH_4}$$

it derives

$$2\ \text{mol}\ COD_{\text{organic matter}} = 1\ \text{mol}\ CH_4$$

on a weight basis

$$1\ \text{g COD}_{\text{organic matter}} = 0.25\ \text{g CH}_4 \cdot 1$$

in terms of LFG volume

$$1\ \text{g COD}_{\text{organic matter}} = 0.35\ \text{l}_\text{N}\ \text{CH}_4$$

This approach does not allow estimating the CO_2 production and therefore it is necessary either to use Eq. (.2) or to assume a predefined CH_4/CO_2 ratio for the calculations. In any case, the amount of CO_2 given by Eq. (2) is not the real quantity that can be measured in LFG because of several reasons:

- dissolution of CO2 in leachate (on the contrary, CH4 is only slightly soluble in leachate);
- solution in equilibrium with HCO_3^- and CO_3^{2-} ions; precipitation of CO_2 as carbonates.

For these reasons, in practical calculation, the evaluation of the maximum theoretical LFG yield is often carried out applying Eq. (4) and assuming a predefined ration of $CH_4/CO_2 = 0.55-0.6$.

Furthermore, all the previous calculations have been based on the amount of organic matter in the waste, without considering its effective biodegradability. According to Andreottola and Cossu (1988), the organic carbon is roughly 50% on dry basis of MSW organic matter, and only 50% of this amount is biodegradable. The total amount of organic carbon in wet raw MSW is ranging around 200–250 kg/t (Tabasaran, 1982; Hoeks, 1983).

As far as the bioavailable fraction of organic carbon of each MSW organic fraction has to be evaluated, Andreottola and Cossu (1988) have proposed the following formula to evaluate the content of biodegradable organic carbon:

$$(OC_b)_i = OC_i \cdot (f_b)_i \cdot (1 - u_i) \cdot p_i \tag{5}$$

where: $(OC_b)_i$ = biodegradable organic carbon in the i component of waste (kg biodegradable carbon/kg wet MSW); OC_i = organic carbon content in the dry i component of waste (kg carbon/kg dry i component); $(f_b)_i$ = biodegradable fraction of OC_i (kg biodegradable carbon/kg carbon); u_i = moisture content of the i component of waste (kg water/kg wet i component); p_i = wet weight of the i component of waste (kg i component/kg MSW).

Table 9.1.3 reports values of some of the above mentioned parameters for the most significant MSW components.

The biodegradability of organic matter can also be estimated through the lignin content of MSW organic fraction, using the following relationship (Tchobanouglos et al., 1993):

$$(f_b)_i = 0.83 - 0.028 \cdot LC \tag{6}$$

where: $(f_b)_i$ = biodegradable fraction expressed on a volatile solids (VS) basis; LC = lignin content of the VS expressed as a percent of dry weight.

According to Eq. (6), wastes with high lignin content are significantly less biodegradable.

Table 9.1.3 Moisture content (u_i), organic carbon content (OC_i), and biodegradable organic fraction (f_{bi}) in different waste components (Andreottola and Cossu, 1988, modified)

Waste component	u_i (kg H_2O/kg wet component)	OC_i (kg C/kg dry component)	$(f_b)_i$ (kg biodeg. C/kg C)
Food waste	0.6	0.48	0.8
Yard waste	0.5	0.48	0.7
Paper and cardboard	0.08	0.44	0.5
Plastics and rubber	0.02	0.7	0.0
Textiles	0.1	0.55	0.2
Wood	0.2	0.5	0.5

Table 9.1.4 Biodegradability of some organic compounds present in municipal solid waste, based on lignin content (Tchobanouglos et al., 1993)

Component	Volatile solids (VS) as percent of total solids	Lignin content as percent of VS	Biodegradable fraction
Food wastes	7–15	0.4	0.82
Paper • Newsprint • Office paper • Cardboard	94.0 96.4 94.0	21.9 0.4 12.9	0.22 0.82 0.47
Yard wastes	50–90	4.1	0.72

The biodegradability of several of the organic compounds found in MSW, based on lignin content, is reported in Table 9.1.4 (Tchobanouglos et al., 1993).

According to some Authors (Tabasaran, 1982), the fraction of bioconvertible carbon to LFG depends on the temperature within the landfill:

$$(OC_b)_i = OC_i \cdot (0.014 \cdot T + 0.28) \tag{7}$$

where T is the temperature within the landfill in °C.

It seems, however, to the authors, more physically correct to recognize an influence of the temperature not for LFG yields, but for generation rates, which are strictly associated to the biological activity within the landfill. Some of the existing LFG generation models do consider the temperature impact for the determination of appropriate half-life values (e.g., Intergovernmental Pannel of Climate Change (IPCC) model).

For a better approach in the evaluation of the amount of biodegradable organic carbon, it should be subtracted to the initial content of biodegradable C, the quantity released through leaching phenomena, especially in the acid phase, when the organic content of leachate is particularly high.

Finally, according to Eq. (4) and (5), LFG specific yield (Y_{LFG}) can be evaluated as follows:

$$Y_{LFG} = 1.867 \cdot OC_i \cdot (f_b) \cdot (1 - u_i) \cdot p_i \tag{8}$$

The above reported expression represents the common theoretical basis for most of the LFG generation models.

Landfill gas generation rates

The main issue of modeling LFG production is not only to predict the amount of LFG which will be produced but also the rate and the duration of the production (Augenstein and Pacey, 1991).

The general equation which rules the biogas production is

$$\frac{dC}{dt} = f(t, C^n) \tag{9}$$

where: $t =$ time; $C =$ amount of methane or of biodegradable organics.

Eq. (9) shows the rate of substrate degradation or the rate of gas production.

In most of LFG production models, Eq. (9) is applied to a single batch of waste, which often corresponds to the amount of MSW disposed either in a single layer or in a year. As a consequence, LFG models are applied to each waste batch independently from the others. The global LFG production rate is given by the sum of the single batch contributes.

The greatest absolute exponent *n* of the dependent variable (C in Eq. (9) is the order of model (i.e., order of kinetics)).

A zero-order kinetics means that a small increment (positive or negative) of C does not influence the rate of substrate decay or biogas production. In other words, a zero-order model indicates that the rate of methane generation is independent of the amount of substrate remaining or of the amount of biogas already produced. According to some authors, many landfills have a biogas production which follows a zero-order kinetics, especially during the periods of most active gas generation; probably other factors such as moisture, nutrients etc., limit the amount of methane to be formed to a relatively constant value independent of time (Ham and Barlaz, 1989).

Most of the LFG production models follow a first-order kinetic, which means that the limiting factor is the remaining amount of substrate or the amount of biogas already produced. In this way other factors, such as moisture or nutrients availability, are supposed to be not limiting factors. Actually, in many cases the limiting factor is the water content, which plays a major role in the hydrolysis of organic matter.

Although it is well known that many factors, such as moisture, temperature, availability of nutrients, heat dissipation, moisture distribution, presence of necessary organism, influence the process of biogasification, most of the authors believe that a first-order kinetics with respect to substrate is the most suitable. This choice appears supported by the fact that the gas production gradually declines in the long term.

Table 9.1.5 (Zison, 1990) shows some examples of equations for the prediction of gas production deriving from Eq. (13) and the related curves of gas yield from a batch of waste. The rate constants (k, k_1, k_2, ..., k_i) reported in Table 9.1.5 determine the rate to which extent the specific substrate decays and—as a consequence—how much gas will be produced. When time series of field data are available, these constants are usually defined during model calibration.

Model 1 is very simple; the process of degradation of substrate is independent of the remaining amount. The gas yield is constant until all the biodegradable matter is degradated. Obviously this is a rough schematization of the process.

In Model 2 the substrate is supposed to be that high in excess that is not limiting for the gas yield rate. However, the rate declines because waste is a heterogeneous substrate, composed by different classes of biodegradable matter, each of them having its proper decay rate, so that it has to decline when the most rapidly biodegradable class is degradated.

Model 3 is the typical, simple first-order model with respect to the substrate, whose bioavailability is the limiting factor. The rate of consumption depends on the amount of the remaining substrate. It assumes that no other factors may influence the process.

Model 4, according to Zison (1990) classification, is a combination of model 2 and model 3; this model assumes that the substrate availability is the limiting factor and that the decay rate decreases with time.

Model 5 is a two-stage model. In the first stage, the rate of gas production is proportional to the amount of gas already produced, whereas in the second stage the rate is supposed to be proportional to the amount of gas to be produced. To avoid that in the first stage, the consumption of substrate is zero because of the initial amount of produced gas is zero; it assumes that stage 1 starts after 1% of the final gas yield has been produced. This model is of the kind used by EMCON Associates Inc. in the Palos Verdes landfill (EMCON, 1980).

As reported before, each biodegradable component of waste has its own degradation rate. For this reason in many models the substrate is split in several classes characterized by the relevant decay rate constant k_i.

Many authors (inter alias EMCON, 1980; Hoeks, 1983; Andreottola and Cossu, 1988) divide the substrate in two (slowly and readily biodegradable) or three (slowly, moderately and readily biodegradable) classes.

Table 9.1.5 Different equations for dpredicting landfill gas production (Zison, 1990)

	Model	Integrated form	Order	Remarks	General yield curve
1	$\frac{dC_i}{dt} = -k$	$C_2 = C_1 - k(t_2 - t_1)$	0	Constant substrate consumption; methane yield rate constant;	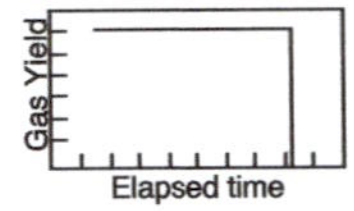
2	$\frac{dC_i}{dt} = -kt$	$C_2 = C_1 - k \cdot \ln_{t_1}^{t_2}$	0	Decay rate declines over time, declining methane production rate;	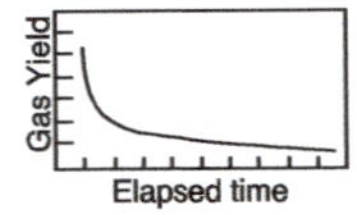
3	$\frac{dC_i}{dt} = -kC$	$C_2 = C \cdot \exp[k(t_2 - t_1)]$	1	Exponential decay of ubstrate;	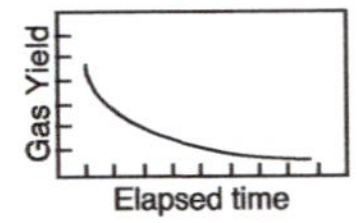
4	$\frac{dC_i}{dt} = -\frac{kC}{t}$	$C_2 = C_1 \cdot \exp\left(k_{t_1}^{t_2}\right)$	1	Combination of models 2 and 3;	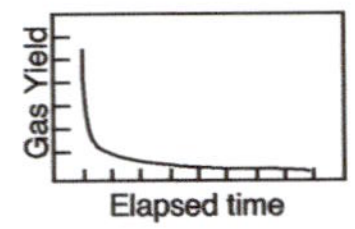
5	$\frac{dG_i}{dt} = -k_1 G$ $\frac{dL_i}{dt} = -k_2 L$	$G = \frac{L_0}{2} \cdot \exp[-k_1(t_h - t)]$ $L = \frac{L_0}{2} \cdot \exp[-k_2(t - t_h)]$	1	Twostage model, gas generation rate increases then decreases; maximal gas production rate occurs at time t_h;	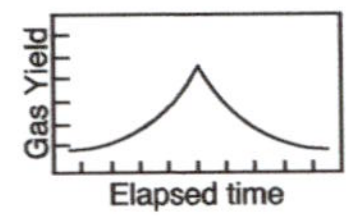

where: *C1*, concentration of substrate at earlier time t1; *C2*, concentration of substrate remaining at later time t2; C2 > 0; *Ci*, concentration of i substrate class; subscript i not shown in integrated expression for sake of clarity, should be inferred in each case; *G*, volume of gas produced prior to time t; *L*, volume of gas remaining to be produced after time t; L0 is total (ultimate) amount of gas to be produced; *th*, time for half of total gas production to occur; *k*, *k1*, *k2*, decay rate constants.

Readily biodegradable fraction is represented by food waste, moderately biodegradable fraction by yard waste, and slowly biodegradable fraction by paper, cardboard, wood, and textiles.

In many models, values of the decay rate constant depend not only on the kind of biodegradable matter but also on other factors such as moisture content, density, size of waste particles. It is possible to take into account these factors introducing appropriate corrective coefficients multiplying the constant value.

In Table 9.1.6 (Gendebien et al., 1992, modified) several values of LFG yields and generation rates are reported.

Generation time

An important parameter provided by the application of LFG production models is the period over which biogas is generated, usually indicated as generation time. It is not easy to provide a general definition of generation time. In the literature, LFG generation time between 10 and 15 and 30 years are indicated (i.a. Bridgewater and Lidgren, 1981; Ham, 1979; Andreottola and Cossu, 1988, Richards, 1989a)

Andreottola and Cossu (1988) indicate a period of gas generation equal to 30 years. An average generation time of 20 years is considered possible by Bridgewater and Lidgren (1981). Ham and Barlaz (1989) and Richards (1989a) prefer to suggest a period of 10–15 years.

Satisfactory information concerning generation time is provided by the half-life time ($T_{0.5}$), the time over which the gas generation equals half estimated yield. By definition the $T_{0.5}$ is such that the area under the production curve is equal to either side. The range of values proposed for $T_{0.5}$ is very wide. Augenstein and Pacey (1991) report several values ranging from 2 to 5 years for wet areas to 10–25 years for dry areas of the United States.

The half-life time can be also calculated in first-order kinetic models by the following expression:

$$T_{(0.5)i} = \frac{\ln 2}{k_i} \qquad (10)$$

where k_i is the decay rate constant of the *i* component of organic waste.

To evaluate the $T_{0.5}$, the experiences with lysimeters do not seem to be very useful, because the values of generation time appear an order of magnitude shorter than typical landfill values (Augenstein and Pacey, 1991).

This discrepancy could be explained taking into account that the gas production in lysimeters is due to a single batch of waste, while the gas production of the waste in a landfill is the sum of contributions of many batches disposed in different periods.

Another parameter useful to define the generation time is the lag time, which is the time passing from the placement of waste to the beginning of the significant gas production. Given a typical placement period of 1 year for batch units of waste usually considered in a gas generation model, a zero lag time corresponds to an average lag of 6 months between placement and start of generation.

Table 9.1.6 Landfill gas generation yields (L_0) for municipal solid waste reported in literature (Gendebien et al., 1992; modified)

Source	Landfill gas yields L_0 (m^3/Mg MSW)
Rhyne (1974)	437
Tabasaran (1976)	60–180
Rasch (1976)	300–500
Augenstein et al. (1976)	128
Bowermann et al. (1977)	40–50
Rettenberger (1978)	200
Stegmann (1978)	250
Ham et al. (1979)	440 120–310
Pacey (1981)	55–225
Beker (1981)	250
Hoeks and Oosthoek (1981)	200
Hoeks (1983)	320
Stegmann (1983)	120–150
Campbell (1985)	100–400
Stegmann (1986)	186–235 (USA) 120–150 (Germany)
Wise et al. (1987)	100–400
Rae (1988)	400
Richards (1989b)	135–400
range	40–500
mean	220

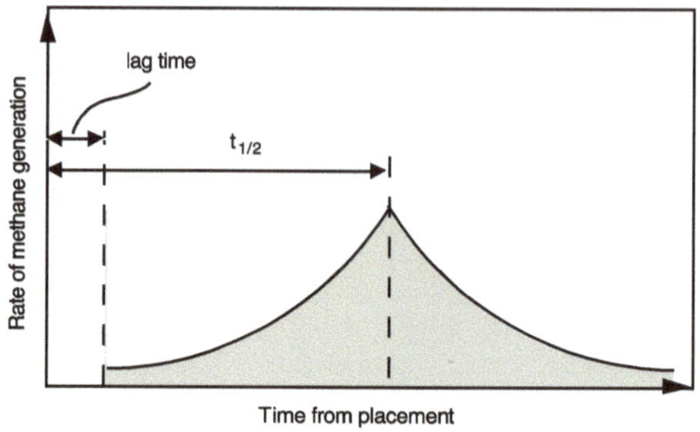

Figure 9.1.1 Generic gas generation curve with lag time and $t_{1/2}$ remarked.

Experimental data from the literature proposes a lag time ranging from a few weeks to a few months (Richards, 1989a), up to 1 year (Scheithauer, 1984; Andreottola and Cossu, 1988). The IPCC model in its latest version considers an average lag time of 6 months.

In Fig. 9.1.1, a generation curve with $T_{0.5}$ and lag time remarked is reported.

EXAMPLES OF DEVELOPED MODELS

Considering the theory described in the earlier chapters and the many variables influencing the related processes, it would be easy to understand that several LFG generation models can be studied and proposed, according to specific needs.

As anticipated in the introduction, several commercial models have been applied in the past, and they have been comprehensively presented elsewhere (Andreottola and Cossu, 1988). Nowadays, the models that are most widely applied are LanGEM, proposed by US Environmental Protection Agency (EPA), and the first-order decay (FOD) model by the IPCC.

The landfill gas emissions model

The LandGEM provides an automated estimation tool for quantifying air emissions from MSW landfills. The model was developed by the Control Technology Center of the US EPA. In 2005, the current version 3.02 has been released, and it works in Windows Excel Environment. LandGEM uses a first-order decomposition rate equation to estimate annual emissions of LFG from landfills:

$$Q_{CH_4} = \sum_{t=1}^{n} \sum_{j=0.1}^{1} kL_0 \left[\frac{M_i}{10}\right] e^{-kt_{ij}} \qquad (11)$$

where: Q_{CH4} = annual methane generation in the year of the calculation (m^3/year); i = 1 year time increment; n = (year of the calculation)−(initial year of waste acceptance); j = 0.1 year time increment; k = methane generation rate ($year^{-1}$); Lo = potential methane generation capacity

(m^3/Mg); Mi = mass of waste accepted in the ith year (t); tij = age of the section j of waste mass Mi accepted in the year i.

The model parameter k is a function of

- waste moisture content;
- availability of the nutrients for methanogens;
- pH;
- temperature.

K values obtained from the data collected in US MSW landfills range from 0.003 to 0.21 (USEPA, 1991a). These values were obtained from theoretical models using field test data and from actual field test measurements. If no user-specified K value is entered into LandGEM, a default value is used for k, equal to 0.05 $year^{-1}$ (USEPA, 1991a).

The value for the potential CH_4 generation capacity of waste (L_0) depends only on the type of waste in the landfill. The higher the cellulose content of the waste, the higher the value of L_0. The values of theoretical and obtainable L_0 range from 6.2 to 270 m^3/t of waste (USEPA, 1991b). If no user-specified L_0 value is entered into LandGEM, a default value is used for L_0, equal to 170 m^3/t of waste (USEPA, 1991a).

The Landfill Air Emission Estimation model includes both regulatory default values ($k = 0.05$ $year^{-1}$) and recommended AP-42 default values for L and k.

AP-42 is the compilation of air pollutant emission factors, published since 1972 as the primary compilation of EPA's emission factor information. It contains emission factors and process information for more than 200 air pollution source categories, including emissions from landfills (USEPA, 1995).

Recommended AP-42 defaults include a k value of 0.04 $year^{-1}$ for areas receiving 625 mm or more of rain per year. A default k of 0.02 $year^{-1}$ should be used in drier areas (<625 mm/year). An L value of 100 m^3/Mg refuse is appropriate for most landfills.

Although the recommended default k and L are based on the best fit to 21 different landfills, the predicted methane emissions range from 38 to 492% of actual.

Fig. 9.1.2 shows an example of gas production curves for a landfill with approximately 2 million tons waste in place expected at closure. The potential gas generation was modeled in two scenarios, using identical landfill parameters except that k was varied between a value for arid conditions (0.02 $year^{-1}$) and a value for wet conditions (0.065 $year^{-1}$). The graph demonstrates the significant difference in gas generation that can occur based on moisture conditions at the site.

To reduce errors deriving from the uncertainties due to the lack of reliable input data for LandGEM, the US EPA "Landfill Methane Outreach Program" developed country-specific LFG generation models, based on local tested data, of several countries (Central America, Colombia Ecuador, Mexico, China Philippines, Thailand and Ukraine) to help landfill owners and operators estimating more correctly LFG biogas production, particularly when wet conditions are occurring (USEPA, 2005).

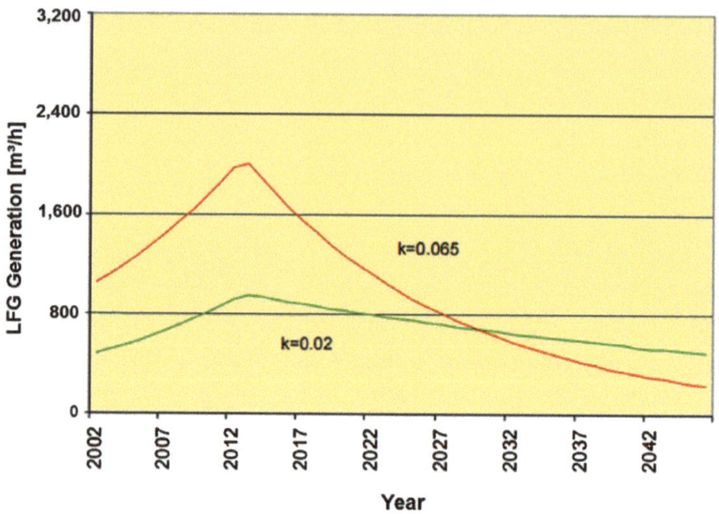

Figure 9.1.2 Landfill gas generation variance by k value.

As far as k and Lo experimental determination are concerned, Tolaymat at el. (2010) monitored for 4 years LFG composition and flow rates in a conventional landfill cell and two other cells, operated as a bioreactor. The estimated decay rate in the conventional cell (0.043 year^{-1}) was comparable to the AP-42 default value of 0.04 year^{-1}, while estimates for the two bioreactor cells were substantially higher (~0.11 year^{-1}). The data document the ability of bioreactor operation to enhance landfill methane generation, though the estimated decay rate is sensitive to the selected L_0. The values identified by Tolaymat at el. (2010) for bioreactor cells is much lower than the value proposed by LandGEM for bioreactor cells (0.7 year^{-1}) and also lower than the one derived by the applied studies of the University of Florida (0.3 year^{-1}).

It clearly appears that the present definition of bioreactor landfill is still too much affected by influencing factors such as humidity, pH, etc., giving rise to a wide range of k values.

The FOD model according to the IPCC

The FOD model (IPCC, 2006) is one of the most suitable methods available to estimate the methane generation in a landfill. Originally, the FOD model has been developed for the quantification of methane emissions from all landfills in a country in connection with the national GHG inventories. It considers the devolution of LFG generation over time based on first-order kinetics, thus resembling gas prognosis models that have been used for example by Tabasaran (1981). The FOD model estimates methane generation based on the biodegradable carbon content of waste compounds and is available as a Microsoft Excel application (IPCC, 2010). For application of the model, the amount of waste per year and a classification of the composition of the waste are required. This classification can be done either on the basis of waste origin (e.g., household waste, industrial waste, etc.) or waste composition, i.e., the fractional amounts of food waste, paper, wood, etc. By default a lag time of 6 months is considered by the model as the generation of methane does not start instantaneously after waste deposition. Moreover, the IPCC model accommodates for four different climate regions: wet boreal or temperate; dry boreal or temperate; wet tropical; and dry tropical. The selected climate conditions do also have an impact on the k value.

As for all other LFG models, the central element of the calculation is the methane generation potential (L_0) of the waste disposed in each year of landfill operation:

$$L_0 = MCF \cdot DOC \cdot DOC_F \cdot F \cdot \frac{16}{12} \quad (12)$$

where: L_0 = Methane generation potential (Mg CH_4/Mg waste); MCF = Methane correction factor, considering an eventual and partial biological stabilization of the waste under aerobic conditions; DOC = Fraction of degradable organic carbon in the waste; DOC_F = Fraction of the DOC that is converted to LFG; F = Methane concentration in the LFG (default: 0.5); 16/12 = Molecular weight ratio of methane to carbon.

The MCF may vary between 0.4 and 1.0, depending on the way the landfill is managed (IPCC, 2006). An MCF of 1.0 corresponds to a sanitary landfill where daily and intermediate covers are applied and the thin waste layers are highly compacted, thus anaerobic decay of newly deposited lifts of waste is encouraged. A lower MCF is applied for unmanaged landfills, which may allow for aerobic degradation of biodegradable organic carbon in the upper waste layers.

For DOC in the waste, the IPCC model provides table data for the estimation of this value based on waste type or waste composition. The more critical value is the DOC_F, because only some fractions of the DOC are bioavailable and will dissimilate under anaerobic conditions. Particularly lignin is the fraction of material that will resist widely anaerobic decay. For bulk MSW, the IPCC model uses a default value of $DOC_F = 0.5$. For individual waste fractions specific values for both the DOC and DOC_F are provided by the IPCC (Table 9.1.7).

In a second calculation step, the methane generation from the waste landfilled in each year is calculated and summed up according to Eq. (13):

$$CH_4(t) = \sum_{x}^{t} (e^k - 1) \cdot R(x) \cdot L_0(x) \cdot e^{-k(t-x)} \quad (13)$$

where: $CH_4(t)$ = Methane generated in year t (Mg/a); k = Methane generation constant (1/a); with $k = \ln 2/T_{0.5}$; $T_{0.5}$ = half-life time (a); t, x = years; R(x) = MSW disposed at the landfill in year x (Mg).

As opposed to the LandGEM, the IPCC model calculates the methane generated in a landfill on a mass basis, as the model adopts L_0 on a mass/mass definition. By this means, the issue of changing gas volumes under different temperatures (according to the ideal gas law) is avoided and the model outcome is directly applicable for GHG inventories where the conversion to metric tons of CO_2 equivalents is required.

Table 9.1.7 Default values for degradable organic carbon (DOC) and fraction of DOC (DOC_F) (IPCC, 2006)

Fraction	DOC (Mg C/Mg MSW)	DOC_F [-]
Wood	0.45	0.50
Paper	0.3	0.55
Textiles	0.5	0.55
Garden waste	0.16	0.55
Fats	0.2	0.77
Sludge	0.11	0.77
Demolition site waste	0.085	0.55

MSW, municipal solid waste.

Table 9.1.8 Default values for k and T0.5 (IPCC, 2006; modified)

Waste type	MAT $\leq$ 20°C; boreal		MAT > 20°C; tropical	
	Dry (MAP/PET <1)	Wet (MAP/PET >1)	Dry	Wet
Slowly degrading (e.g., textiles, cardboard)	0.04 (17)	0.06 (12)	0.045 (15)	0.07 (10)
Slowly degrading (e.g., wood, straw)	0.02 (35)	0.03 (23)	0.025 (28)	0.035 (20)
Moderately degrading (e.g., garden and park waste, nonfood putrescibles)	0.05 (14)	0.10 (7)	0.065 (11)	0.17 (4)
Rapidly degrading (e.g., food waste)	0.06 (12)	0.185 (4)	0.085 (8)	0.40 (2)

MAT, mean annual temperature; *MAP*, mean annual precipitation; *PET*, potential evapotranspiration; values in brackets refer to $T_{0.5}$.

The methane generation constant (k) is determined by a large number of factors, associated with the waste composition and specific site conditions (Table 9.1.8). As a rule, short half-life values ($T_{0.5}$ = about 2 years; k = 0.4) are associated with high moisture content and rapidly degradable substrate, such as food waste. On the other hand, slower decay rates ($T_{0.5}$ = 35 years, k = 0.02) are associated with dry conditions and slowly degradable waste such as wood or straw.

To select appropriate k values, information on the composition of waste disposed in the landfill over time and climate data are required. If these data are unavailable, a k value of 0.05 (a half-life of about 14 years) is suggested as a default value.

UNCERTAINTIES AFFECTING MODELS

The reliability of LFG generation models is in general considered as good. In most cases, the variation between the results gained by the model and real data is in the range of 10%–30%. The accuracy can be significantly enhanced if a good calibration of theoretical data with reliable field data is made. In Germany, the Federal EPA (UBA) recently reviewed the emission factors of the model approach applied for the national GHG inventory. First, results clearly demonstrate that the adjustment of factors DOC_F and/or k—based on empirical emission data—significantly reduces the discrepancy between the results of the model and real emissions (Rettenberger et al., 2014).

Uncalibrated models, if used by an experienced analyst, can estimate gas production with an accuracy within a range of 30% (Zison, 1990). There is also discussion that the prediction curves are not reliable because the model schematization is much too simple, the input data are generally inadequate, there are not enough long-term field data to calibrate the curves, and some models are based on ad hoc and empirical assumptions. Actually, the main factors influencing the model reliability are still the ones indicated by Zison almost 30 years ago (Zison, 1990):

- degree of accuracy expected;
- reliability of input data, the experience of the model users;
- similarity of the site to others, which have been previously modeled.

A consistent number of uncertainties affecting the prediction curves deal with the composition of waste, pretreatment, history of the landfill operation, landfill location, landfill structure (top cover, daily cover), leachate drainage, leachate management (continuous extraction, recycling), moisture, pH, temperature in the landfilled waste mass, efficiency of gas collection, etc. For the IPCC model, these parameters are widely reflected in the MFC. Estimates show that due to the high complexity of this factor, the associated uncertainty can be up to 15%. Moreover, inappropriate k values do impact particularly on the long-term evolution of LFG generation. The major problems are associated with the application of default values and—on a more general basis—the nonobservance of changing half-life values with increasing landfill age. Estimates show model uncertainties of up to 35% for this individual parameter.

In conclusion, it can be stated that a model aiming to describe the processes occurring in a landfill will be always affected by uncertainties, due to the intrinsic impossibility to strictly control what occurs inside the waste mass. For this reason, it is better to express the model outputs in terms of probable range instead of punctual values.

References

Alpern, R., 1973. Decomposition Rates of Garbage in Existing Los Angeles Landfills. Unpublished Master thesis. California State University, Long Beach, U.S.A.

Andreottola, G., Cossu, R., 1988. Modello matematico di produzione del biogas in uno scarico controllato. In: Christensen, Th., Cossu, R., Stegmann, R. (Eds.), Landfilling of Waste: Biogas. E&FN Spon, London. ISBN: 0 419 19400 2.

Augenstein, D.C., Wise, D.L., Wentworth, R.L., Cooney, C.L., 1976. Fuel gas recovery from controlled landfilling of municipal wastes. Resource Recovery Conservation 2, 103–117 (cit. in Gendebien et al., 1992).

Augenstein, D., Pacey, J., 1991. Modeling landfill methane generation. In: Proceedings of Sardinia 91, III International Landfill Symposium. CISA, Cagliari, Italy, pp. 115–148.

Beker, D., 1981. Development of Gas in Landfill. Rept. (60 pp.) n° SVA 3747 or IVA 48. By Inst. Afvalstoffen Onderzoek, Amersfoort, The Netherlands. (cit. in Gendebien et al., 1992).

Boyle, W.C., 1976. Energy recovery from sanitary landfills—a review. In: Schlegel, H.G., Barnea, S. (Eds.), Microbiol Energy Conversion. Pergamon Press, Oxford, UK, pp. 119–138.

Bowerman, F.R., Rohatgi, N.K., Chen, K.Y., Lockwood, R.A., 1977. A Case Study of the Los Angeles County. Palos Verdes Landfill Gas Development Project. Rept. n° EPA 600/3-77-047 for US EPA, Cincinnati, Ohio, USA, p. 114 (cit. in Gendebien et al., 1992).

Bridgewater, A., Lidgren, K., 1981. Household Waste Management in Europe. Economics and Technics. Van Nostrand-Reynold, Wokingam, UK.

Campbell, D.J.V., 1985. Landfilling. An environmentally acceptable method of waste disposal and an economic source of energy. In: Agricultural, Industrial and Municipal Waste Management in Today's Environment. Conference Proceedings, Warwick, UK. April. IMechE Conf. Publ. 1985-4 Mech. Engin. Publ, London, UK, pp. 9–14 (cit. in Gendebien et al., 1992).

EMCON, 1976. A Feasibility Study of Recovery of Methane from Parcel 1 of the Scholl Canyon Sanitary Landfill for the City of Glendale. Rept (38 pp). by EMCON Associates, San José, California, and Jacobs Engineering Co, Pasadena, California, USA (cit. in Gendebien et al., 1992).

EMCON, 1980. Methane Generation and Recovery from Landfills. Rept. by EMCON Associates, San José, California. Ann Arbor Science Publishers, Ann Arbor, Michigan, U.S.A, pp. 44–51.

Farquhar, G.J., Rovers, S.A., 1973. Gas production from landfill decomposition. Water, Soil and Air Pollution 2, 493.

Gendebien, A., Pauwels, M., Constant, M., Ledrut-Damanet, M.J., Nyns, E.J., Willumsen, H.C., Butson, J., Fabry, R., Ferrero, G.L., 1992. Landfill Gas: From Environment to Energy. Report No. EUR 14017/1 EN. Commission of the European Communities, Luxembourg.

Ham, R.K., Hekimian, K.K., Katten, S.L., Lockman, W.J., Lofy, R.J., McFaddin, D.E., Daley, E.J., 1979. Recovery, Processing, and Utilization of Gas from Sanitary Landfills. EPA-600/2-79-001.

Ham, R.K., 1979. Method for Testing a Landfill for Its Methane Potential. USA Patent N° 4 159-893. Appl.: US 857 574 (dec. 5, 1977). To Reserve Synthetic Fuels, USA. (cit. in Gendebien et al., 1992).

Ham, R.K., Barlatz, M.A., 1989. Measurement and prediction of landfill gas quality and quantity. In: Christensen, T.H., Cossu, R., Stegmann, R. (Eds.), Sanitary Landfilling: Process, Technology and Environmental Impact. Academic Press, London, United Kingdom, pp. 155–166.

Heidrich, E.S., Curtis, T.P., Dolfing, J., 2011. Determination of the internal chemical energy of wastewater. In: Environ. Sci. Technol. 2011, 45, pp. 827–832.

Hoeks, J., Oosthoek, J., 1981. Gas production from landfills. Gas (The Neth.) 101, 563–568 (cit. in Gendebien et al., 1992).

Hoeks, J., 1983. Significance of biogas production in waste tips. In: Waste Management & Research, 1, pp. 323–335 (cit. in Gendebien et al., 1992).

IPCC, 2006. Intergovernmental Panel on Climage Change. In: Pipatti, R., Svardal, P. (Eds.), Solid Waste Disposal, 2006 IPCC Guidelines for National Greenhouse Gas Inventories, Waste, Volume 5. Author, Geneva, Switzerland, pp. 1–40. – Tables 2.5 and 2.6.

IPCC, 2010. Intergovernmental Panel on Climage Change. Model can be downloaded from http://www.ipccnggip.iges.or.jp/public/2006gl/pdf/5_Volume5/IPCC_Waste_Model.xls.

Pacey, J.G., 1981. Prediction of landfill gas production and recovery. In: "Gas Research" Proc. Int. Conf. Los Angeles, California, USA, Sept.-Oct. 1981. Published by Government Institutes Inc, Rockeville, Maryland, USA, pp. 896–915. ISBN: 0 86537 094 2. (cit. in Gendebien et al., 1992).

Rae, G.W., 1988. The control of landfill gas. HM inspectorate of pollution's waste management paper. In: Landfill Gas and Anaerobic Digestion of Solid Waste, Proc. Conf. Chester, UK, Oct.. Published by Harwell Laboratories, UK, pp. 92–99 (cit. in Gendebien et al., 1992).

Rasch, R., 1976. Methane generation from waste. In: Proceedings First International Symposium "Materials and Energy from Refuse. Van Mantgen and De Does, Leiden, The Netherlands, Antwerp, Belgium, pp. 31–36. ISBN: 90 238 0834 7. (cit. in Gendebien et al., 1992).

Rettenberger, G., 1978. Origin, consequences, collection and valorization of landfill gas. In: Progress in Landfill Technology. Stuttgarter berichte zur Abfallwirtschaft, Band 9. Inst. Siedlungswasserbau, Wassergüte-une Abfallwirtschaft der Universität Stuttgart, Germany, p. 36. ISBN: 3 503 01353 9. (cit. in Gendebien et al., 1992).

Rettenberger, G., Haubrich, E., Schneider, R., 2014. Überprüfung der Emissionsfaktoren für die Berechnung der Methanemissionen aus Deponien. Studie im Auftrag des Umweltbundesamtes, Dessau, Berlin, Deutschland.

Rhyne, C.W., 1974. Landfill Gas. Rept. (21 Pp). For Office Solid Waste Management Programs. US Environmental Protection Agency (EPA), Cincinnati, Ohio, USA.

Richards, K.M., 1989a. All gas and garbage. New Scientist (June 3), 4 (cit. in Gendebien et al., 1992).

Richards, K.M., 1989b. Landfill gas: working with Gaia. Biodeterioration Abs 3, 317–331 (cit. in Gendebien et al., 1992).

Scheithauer, H., 1984. Deponiegasnutzung. Plannungen, erfahrungen, und entwicklungstendenzen (Utilization of landfill gas. Design, experiences, and development tendencies). In: Proceedings of symposium of Essen (Germany), September 1983. To be obtained from Bundesminister für forschung und tecnologie (BMFT). Bonn, Germany. (cit. in Gendebien et al., 1992).

Stegmann, R., 1978. Gases from sanitary landfills. ISWA Journal 26/27, 11–24 (cit. in Gendebien et al., 1992).

Stegmann, R., 1983. Emission at the domestic waste landfills. In: Proc. 7th "Mülltechnisches" Semin., München, Germany. Published as Ber. Wassergütewirtschaft un Gesundheitsingenieurwesen, Techn. Univ. Munchen, pp. 75–98 (cit. in Gendebien et al., 1992).

Stegmann, R., 1986. Landfill gas as an energy source."Preparing now for Tomorow's Needs". In: Proc. Conf. Chicago, Illinois, USA. Published by national Solid waste Management Assoc.. and Waste Age Magazine, Washington, DC 20036, USA, pp. 307–333 (cit. in Gendebien et al., 1992).

Tabasaran, O., 1976. Considerations on the problem landfill gas. Müll Abfall 7, 204–210 (cit. in Gendebien et al., 1992).

Tabasaran, O., 1981. Gas production from landfill. In: Bridgewater, A.V., Lidgren, K. (Eds.), Household Waste Management in Europe, Economics and Techniques. Van Nostrand Reinhold Co., New York, USA, pp. 159–175.

Tabasaran, O., 1982. Obtention et valorisation du methane a partir de dechets urbains. Tribune de Cebedeau 35, 483–488 (Belgium).

Tchobanoglous, G., Theisen, H., Vigil, S., 1993. Integrated Solid Waste Management. Engineering, Principles and Management Issues. McGraw-Hill, New York, U.S.A.

Tolaymat, T.M., Green, R.B., Hater, G.R., Barlaz, M.A., Black, P., Bronson, D., Powell, J., 2010. Evaluation of landfill gas decay constant for municipal solid waste landfills operated as bioreactors. Journal of the Air & Waste Management Association 60 (1), 91–97.

USEPA, 1991a. Air Emissions from Municipal Solid Waste Landfills. Background Information for Proposed Standards and Guidelines, EPA-450/3-90-011a (NTIS PB91-19706,). U.S.Environmental Protection Agency, Research Triangle Park, NC.

USEPA, 1991b. Regulatory Package for New Source Performance Standards and III(D) Guidelines for Municipal Solid Waste Air Emissions, Public Docket No. A-88–09 (Proposed May 1991. U.S. Environmental Protection Agency, Research Triangle Park, NC.

USEPA, 1995. AP 42. In: Compilation of Air Pollutant Emission Factors Stationary Point and Area Sources. Chp. 2: Solid Waste Disposal. (A Draft Modification Is Available since 2008), fifth ed., Volume 1.

USEPA, 2005. First-Order Kinetic Gas Generation Model Parameters for Wet Landfills. EPA-600/R-05/072.

Wise, D.L., Leuschner, A.P., Levy, P.F., Sharaf, M.A., Wentworth, R.L., 1987. Low-capital-cost fuel-gas production from combined organic residues. The Global Potential. Resource Conserv 15, 163–190 (cit. in Gendebien et al., 1992).

Zison, S., 1990. Landfill gas production curves: myths vs. reality. In: Proceedings of GRCDA/SWANA Landfill Meeting, Vancouver, Canada.

9.2

QUALITY OF LANDFILL GAS

Gerhard Rettenberger

INTRODUCTION

Landfill gas (LFG) is by definition a gas that is produced in landfills under anaerobic conditions. Therefore, LFG consists mainly of methane and carbon dioxide. In addition it contains trace constituents, which are transferred from the solid and liquid waste into the gas phase by vaporization. Also some by-products from the biological degradation process such as ammonia (NH_3) or hydrogen sulfide (H_2S) are in the LFG mostly at low concentrations. Furthermore, depending from the stage of the degradation process, hydrogen may occur. In landfills in the early stage of decomposition, 3%–5% of H_2 had been analyzed. In most cases, LFG is water vapor saturated, the moisture content depends on the temperature. At a temperature of 40°C 60 g water/m^3 LFG are in the vapor phase.

Some of the trace gases cause odors. Gas from landfills, which are in an early phase of degradation, has to be diluted with a factor of $1:10^6$ until the odor cannot be detected any more; over time this factor may decrease to $1:10^4$.

Owing to air intrusion and/or oversucking during gas extraction, often air (oxygen, nitrogen) is mixed in the landfill with LFG. LFG quality also changes with time due to the different degradation phases and waste quality. Therefore, the composition of the LFG is not constant.

The waste volumetric (σ) composition may influence the LFG quality due to the following dependencies. The resulting methane composition and the ratio of methane to carbon dioxide concentrations are as follows:

Carbohydrate	$\sigma(CH_4)$: 50%	$\sigma(CH_4)/\sigma(CO_2)$: 1
Protein	$\sigma(CH_4)$: 52%	$\sigma(CH_4)/\sigma(CO_2)$: 1.08
Fat	$\sigma(CH_4)$: 71.5%	$\sigma(CH_4)/\sigma(CO_2)$: 2.5

As those substances are decomposing with different velocities the composition of LFG changes over time.

MAIN CONSTITUENTS

Changing composition over time

The concentration of the main gas compounds in the landfill body is changing over time. A scheme illustrating the kind of progression is presented in Fig. 9.2.1.

The LFG composition trends with time, which is shown in Fig. 9.2.1, can be divided into nine phases:

Phase I–III: LFG production starts developing.
Phase IV: In the landfill a stable gas production occurs, pores and small voids in the landfill fill up with LFG. At the surface gas emissions may be measured.
Phase V: The gas production decreases but is stable. Easy degradable waste components will be biologically reduced. The LFG emissions will be lower. The composition of the gas changes because the easy degradable fraction has almost been degraded and more difficult organics are present.
Phase VI: Due to a decrease in the gas production, air can migrate into the landfill body. This process happens from the surface in to deeper layers. Aerobic processes will start. The LFG emissions are reduced and in some areas detectable only in very low concentrations.
Phase VII: As a result of aerobic processes, carbon dioxide will be generated. Residual amounts of methane will be degraded due to microbiological processes. Therefore the ratio of $\sigma(CH_4):\sigma(CO_2)$ will further decrease. There will be only very low LFG emissions.

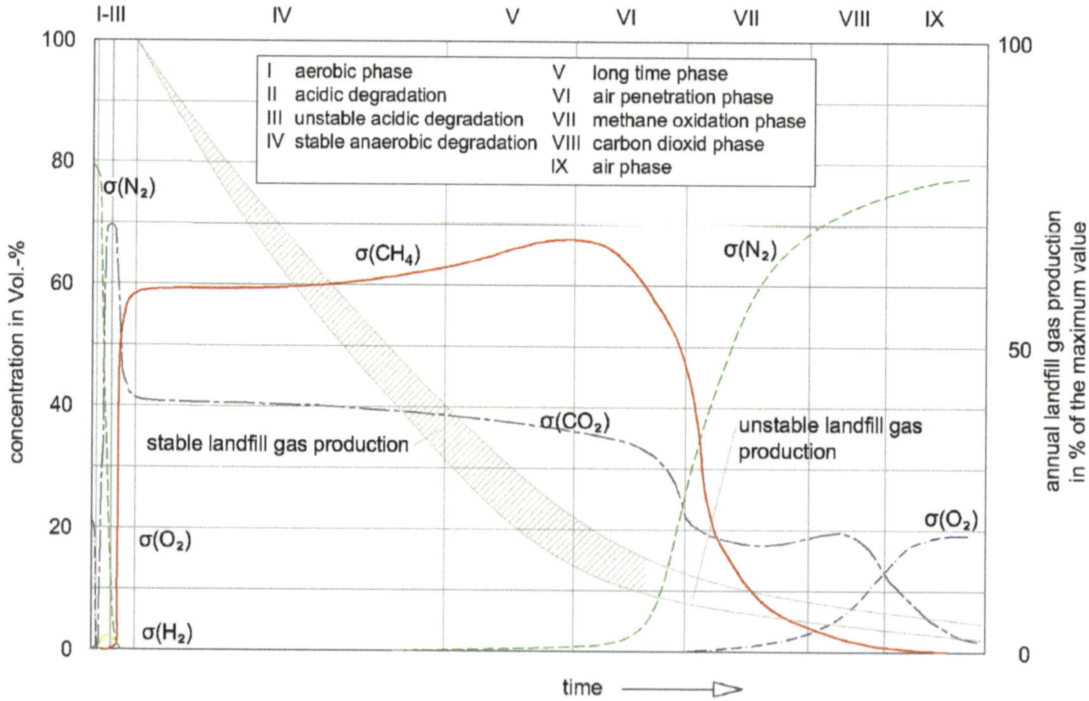

Figure 9.2.1 Gas composition pattern in the landfill body with time. (Rettenberger, 2018)

Phase VIII: The landfill body is almost in an aerobic condition. As some organic material is still left in the landfill, low concentrations of carbon dioxide will be produced.
Phase IX: The waste in the landfill is almost inert. The gas quality in the pores will be similar as in natural soil (air and some carbon dioxide).

From the results of a gas analysis the state of the landfill can be allocated to one of the above-described phases. It is very important to analyze not only methane in the LFG to receive valuable information about the ratio between the concentrations of the main gases.

Changing landfill gas composition depending from location

Taking samples out of landfills from different depths, it can be observed that the concentrations change significantly. Fig. 9.3.2 of Chapter 9.3 shows the typical gradients. These gradients are depending on the intensity of the gas production, the permeability of the waste, the wind velocity, and the kind of top cover of the landfill. When a landfill is lined with a membrane, no gradients may occur but without any cover the gradients can be observed up to a depth of several meters.

When the methane to carbon dioxide ratio reaches values of 1, methane oxidation may start.

Changing landfill gas composition during gas extraction

During gas extraction the concentration of methane and carbon dioxide decreases and the concentration of oxygen and nitrogen increase (see Chapter 9.3, Fig. 9.3.2).

Owing to increasing vacuum pressure, air is sucked in to the landfill either through the landfill surface or leaks, e.g., around gas wells. When sucking air in through the surface the ratio of oxygen to nitrogen concentration is different from the gas composition of the air: in the case of leaks the extracted gas may show the same ratio as in the air.

Gas types

To conclude from the composition of a LFG sample on the status of a landfill, it might be helpful to distinguish between different gas types:

- "Natural" changing LFG composition with time as shown in Fig. 9.2.1.
- Changing LFG composition due to methane oxidation and air intrusion in the upper layer of the landfill (see Chapter 9.3, Fig. 9.3.3).
- Typical LFG gas quality due to varying gas extraction rates and conditions (classification in six types). Table 9.2.1 shows typical mean values.

The different gas types characterize the following situations:

- Type 1: typical LFG composition in phase IV (see Fig. 9.2.1); without gas extraction or aeration, typical ratio 1.3–1.35 (Rettenberger, 2004).
- Type 2: oversucking conditions: oxygen and nitrogen concentrations show different ratios as in air; ratio of methane to carbon dioxide concentration may be around 1.3.
- Type 3: air intrusion through leaks due to oversucking: oxygen and nitrogen concentrations show the ratio of air and of methane to carbon dioxide ratio may be around 1.3.
- Type 4: mixture of type 2 and 3.

Table 9.2.1 Typical landfill gas (LFG) quality types at gas extraction plants in different conditions

LFG Type	$\sigma(CH_4)$	$\sigma(CO_2)$	$\sigma(O_2)$	$\sigma(N_2)$
1	56	43		
2	34	25	1	38
3	34	25	8	32
4	34	25	3	37
5	12	14	5	69
6	3	5	8	84

LFG types description in the text.

- Type 5: aeration by oversucking; lean gas is produced.
- Type 6: Extracted gas quality from actively aerated landfill with off-gas extraction (see Chapter 16.2), nitrogen concentrations may be higher compared to air.

Relating the results from gas samples to these types may help to conclude the status of a landfill or an LFG extraction plant.

TRACE CONSTITUENTS

LFG contains hundreds of trace constituents (Janson, 1991; Janson and Bardtke, 2005). These compounds may be organic (halogenated hydrocarbons) and/or inorganic gases (hydrogen sulfide, ammonia), which originate beyond others from solvents, cryogens, propellants, and organic silicon gases deposited in the landfill. Some of these gases are naturally generated in the landfill some of them are anthropogenic. In the early phase of gas generation, oxygen-containing gases will occur as shown in Table 9.2.2.

Some of the gases are acute toxic; some are carcinogenic, teratogenic, and/or genetically harmful. Therefore, LFG is classified as hazard.

Utilizing LFG has, in several cases, caused damages such as clogging, abrasion, or corrosion in gas engines and other facilities. The reason is mainly gases containing sulfur, chlorine, fluorine, and siloxane. As a consequence, gas treatment and/or increased maintenance may be necessary. Therefore, it is quite common and necessary to control the cumulative concentrations of these compounds. Some values are presented in Table 9.2.3. Some of the gases are acute toxic; some are carcinogenic, teratogenic, and/or genetically harmful. Therefore, LFG is classified as hazard.

In LFG, hydrogen sulfide concentrations are, in most cases, in the range of 0–70 mg/m^3 but sometimes up to >3000 mg/m^3. The sulfur-containing trace gases in LFG may have two main effects:

- they are mainly responsible for the odor of the gas;
- some of the components, such as hydrogen sulfide and mercaptans, belong to the more toxic trace gases.

Table 9.2.2 Oxygen-containing constituents in landfill gas (Rettenbeger and Stegmann, 1996)

Compound	Concentration Range (mg/m^3)
Ethanol	16–1450
Methanol	2.2–210
1-Propanol	4.1–630
2-Propanol	1.2–73
1-Butanol	2.3–73
2-Butanol	18–626
Acetone	0.27–4.1
Butanone	0.078–38
Pentanal	0.8
Hexanal	4.04
Acetic ester	2.4–263
Butyric ester	<0.9–350
Acetic butyl ester	60
Butyric propyl ester	<0.1–100
Acetic propyl ester	<0.5–50
Acetic acid	<0.06–3.4
Butyric acid	<0.02–6.8
Furan	0.01–2.4
Methylfuran	0.06–170
Tetrahydrofuran	<0.5–8.8

Table 9.2.3 Trace constituents in landfill gas in mg/m^3 CH_4 (Rettenberger, 2004; Beese, 2007)

Compound	Upper Value	Mean Value
Sulfur	2000	950
Fluorine	50	20
Chlorine	300	90
Silicon	100	42
Hydrocarbons	400	140

In Table 9.2.4, some of the sulfur-containing gases that have been detected in LFG are listed.

Anthropogenic trace gases in LFG can be differentiated into two groups: aromatic and aliphatic hydrocarbons as well as chlorinated hydrocarbons. Hydrocarbons generated in landfills had been detected in the following concentrations: e.g., limonene (3.3–269 mg/m^{3}LFG), menthene (14 mg/m^{3}LFG), fenchene (3–13 mg/m^{3}LFG). In total of these compounds up to 500 mg/m^3 LFG had been measured.

Table 9.2.5 shows as an example the following concentrations of trace gases, which dominate the composition of LFG of a landfill where municipal solid waste (MSW) and industrial/hazardous waste were codisposed.

Table 9.2.4 Sulfur-containing compounds in landfill gas (LFG) (Rettenbeger and Stegmann, 1996)

Sulfur Compounds	Concentrations (mg/m^3 LFG)
Methyl mercaptan	0.1–430
Ethyl mercaptan	0.0–120
Dimethyl sulfide	1.6–24
Dimethyl disulfide	0.02–40
Carbon disulfide	<0.5–22
Carbon oxysulfide	<0.1–1.9

Table 9.2.5 Trace constituents in landfill gas of a codisposal landfill and the ambient soil (mg/m^3 LFG) (Janson and Bardtke, 2005)

Compound	Landfill	Ambient Soil
Chloroethene (vinylchloride)	25–75	25–113
1,1-Dichloroethene	0.1–21	LOD–21
cis-1,2 Dichloroethene	1–22	1–25
Dichloromethane	18–89	3–418
Trichloromethane	0.2–13	LOD–13
Tetrachloromethane	0.004–0.04	–
1,1,1 – Trichloroethene	0.4–8	0.05–7
Trichloroethene	4–23	2–96
Tetrachloroethene	3–22	3–15
Dichlorodifluoromethane	17–70	12–1533
Trichlorofluoromethane	0.2–2	0.2–64
Benzene	4–16	6–22
Toluene	67–318	40–318
Ethylbenzene	6–35	6–43
p-/m-xylene	6–60	6–78
o-xylene	4–28	3–12
n-hexane	19–80	26–96
Cyclohexane	11–21	13–35
Hydrogen sulfide	1197–6164	0.175–12
Methanethiol	3–26	LOD–25
2-Propanethiol	4–49	LOD–33
2-Butanethiol	19–70	LOD–90
Dimethyl sulfide	4–15	LOD–4

LOD, limit of detection.

Table 9.2.6 Concentration of chlorinated hydrocarbons in landfill gas in mg/m^3 CH_4: mean value (Rettenberger, 2004), range (Rettenbeger and Stegmann, 1996)

Compound	Mean Value	Range
Chloroethene	15	LOD–520
1,1-Dichloroethene	1	1.5–320
1,2-Dichloroethene	10	8–200
Dichloromethane	3	LOD–1520
Trichloromethane	1	LOD–100
Dichloroethene	–	LOD–600
Trichloroethene	3	LOD–640
Tetradichloromethane	3	LOD–500
1,1,1-Trichloroethane	1	LOD–60
1,1-Dichloroethane	–	LOD–40
Dichlorodifluoromethane	49	10–1200
Trichlorofluoromethane	14	4–1000
1,1,2-Trichlorotrifluoroethane	2	4–60
Chlorotrifluoromethane	–	LOD–20
Chlorodifluoromethane	–	4–760
Dichlorofluoromethane	–	1–28
Chlorofluoromethane	–	0.5–55
Dichlorotetrafluoroethane	–	4–40
Dichlorobenzene	–	LOD–8

LOD, limit of detection.

Silicon-containing compounds are above all decamethyltetrasiloxane (0–15 mg/m^3 CH_4), octamethyltrisiloxane (0–2 mg/m^3 CH_4), and hexamethyldisiloxane (4–20 mg/m^3 CH_4) (Häusler and Schreier, 2005). Some of the individual trace compounds had shown the mean values (Rettenberger, 2004) and ranges (Rettenbeger and Stegmann, 1996) reported in Table 9.2.6.

These typical values originate from LFG samples from MSW landfills. In some landfills, however, the concentrations of some compounds had been considerable higher. In general over time the concentrations of these constituents decrease, in some cases, however, e.g., the hydrogen sulfide concentrations in some cases have increased extremely. Therefore, it is necessary to control these compounds at each landfill individually over long periods of time (decades).

NOTE:

- σ is used in this chapter for indicating the volume concentration of a given gas X: $\sigma(X) = V(X)/V$. It can be given as ratio or as percentage.

- m^3, if not alternatively specified, stays for cubic meter of gas volume under normal conditions (temperature: 0 °C, pressure: 10^5Pa).

References

Beese, J., 2007. Betriebsoptimierung der motorischen Gasverwertung durch den Einsatz von Gasreinigungssystemen. In: Trierer Berichte zur Abfallwirtschaft, vol. 17. Published bei Rettenberger, G., Stegmann, R., Verlag Abfall aktuell, Stuttgart.

Häusler, Th, Schreier, W., 2005. Analyse siliziumorganischer Verbindungen im Deponiegas sowie CO-Messungen zur Brandfrüherkennung. In: Trierer Berichte zur Abfallwirtschaft, vol. 16. Published bei Rettenberger, G., Stegmann, R., Verlag Abfall aktuell, Stuttgart.

Janson, O., 1991. Analytik, Bewertung und Bilanzierung gasförmiger Emissionen aus anaeroben Abbauprozessen unter besonderer Berücksichtigung der Schwefelverbindungen. In: Stuttgarter Berichte zur Abfallwirtschaft, vol. 32. Erich Schmidt Verlag GmbH & Co., Berlin.

Janson, O., Bardtke, D., 2005. Analytik der Spurenstoffe im Deponiegas in Rettenberger, G. (publisher): Neue Vrfahren und Methoden zur Sicherung und Sanierung von Altlasten am Beispiel der Deponie Gerolsheim – Kurzfassungen des Verbundvorhabens, Trierer Berichte zur Abfallwirtschaft, vol. 15. Verlag Abfall aktuell, Stuttgart.

Rettenbeger, G., Stegmann, R., 1996. Landfill gas components. In: Christensen, T.H., Cossu, R., Stegmann, R. (Eds.), Landfilling of Waste: Biogas. E&FN Spon, London, ISBN 0419194002.

Rettenberger, G., 2004. Untersuchungen zur Charakterisierung der Gasphase in Abfallablagerungen, Stuttgarter Berichte zur Abfallwirtschaft, vol. 82. Kommissionsverlag Oldenbourg Industrieverlag GmbH, München.

Rettenberger, G., 2018. Personal data set and documentation.

9.3

COLLECTION AND DISPOSAL OF LANDFILL GAS

Gerhard Rettenberger

FUNDAMENTALS

Landfill gas (LFG) extraction and utilization should be a standard procedure at all municipal solid waste landfills. By these means the emission of climate influencing gases can be significantly reduced, energy can be utilized, odors can be avoided, and explosion risks can be reduced.

There are five essential conclusions resulting from experiences with gas collection over the last decades:

1. LFG extraction systems should be installed and operated from the beginning of landfill operation.

   Fig. 9.3.1 shows the gas volumetric composition (σ) at different gas wells from the beginning of the landfill operation (1). For this test, the gas wells had been installed directly on top of the bottom liner. As can be seen in Fig. 9.3.1, the σ CH_4 to σ CO_2 ratio of 1 increases after 11–15 months after the landfill operation started. Therefore it is necessary to have the gas extraction system available during the first year of operation. Having in mind that the half-life value of the LFG production is in the early period of a landfill in the range of 3–4 years, means that without operating a gas extraction system almost 50% of the gas may be lost in this time frame.

2. Gas collection can be realized using an active or passive system. Operating an active system the gas is extracted out of the landfill by means of blowers, which generally provide vacuum pressures in the landfill body in the range of −1 to −30 hPa. Passive gas collection systems use the positive pressure in the landfill body to transport the gas out of the landfill under semicontrolled conditions. Passive systems may only be feasible, if the methane gas production is less than 0.5 L/m^2 h and a gastight top cover has been installed.

   Fig. 9.3.2 shows gradients of LFG and air components over the depth of the landfill body or a cover soil (Rettenberger, 2004). As can be seen in Fig. 9.3.2, the ratio of σ CH_4 to σ CO_2 changes with the depth. These results show that methane oxidation occurs. This process is technically used in landfill covers for biological oxidation of residual methane (see Chapter 9.4).

3. When active gas extraction is practiced, the methane concentration may change when air is sucked into the landfill.

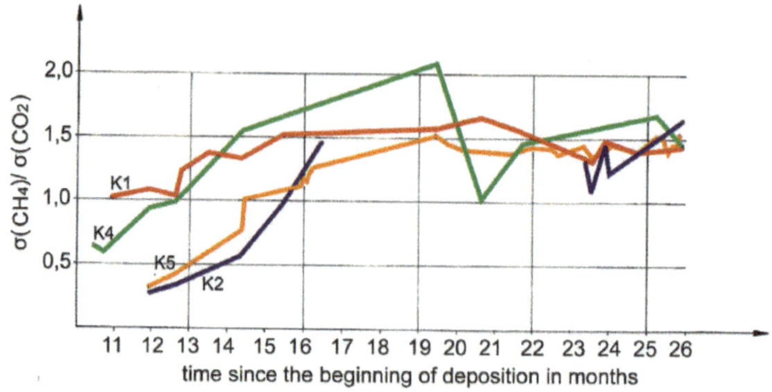

Figure 9.3.1 $\sigma(CH_4)/\sigma(CO_2)$ over time at four gas wells (K1, K2, K4, K5) since the beginning of deposition (Rettenberger, 2004).

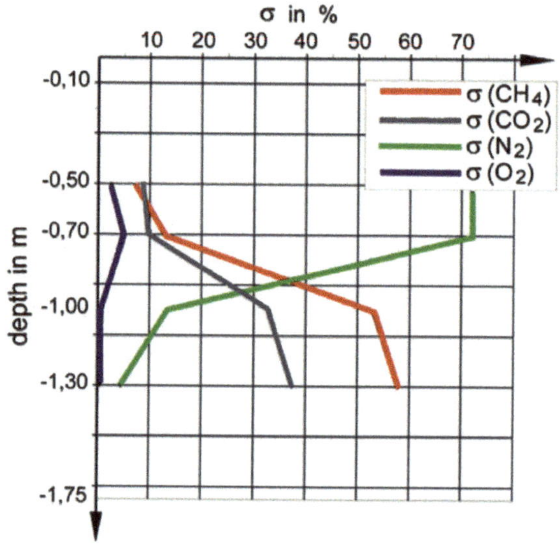

Figure 9.3.2 $\sigma(CH_4)$, $\sigma(CO_2)$, $\sigma(O_2)$, and $\sigma(N_2)$ in different depths of a pervious soil cover or an open landfill (Rettenberger, 2004).

Fig. 9.3.3 shows typical curves when oversucking occurs, which may happen when the gas extraction rate is higher than the gas production rate and/or when the gas wells are not adequately designed (Rettenberger, 2004).

The figure shows that the kind of correlation is individual for each well and is corresponding to the mathematical model, which describes the mixing of air—due to air intrusion—with the LFG (Rettenberger, 2004):

$$\sigma(CH_4) = \sigma(CH_{4,v=0})/1 + b \cdot \dot{v}^2$$

$\sigma(CH_4)$, volume concentration of methane depending from flow; $\sigma(CH_{4,v=0})$, volume concentration when flow is 0 m^3/h; b, factor calculated by statistical method; $\dot{v}$, flow.

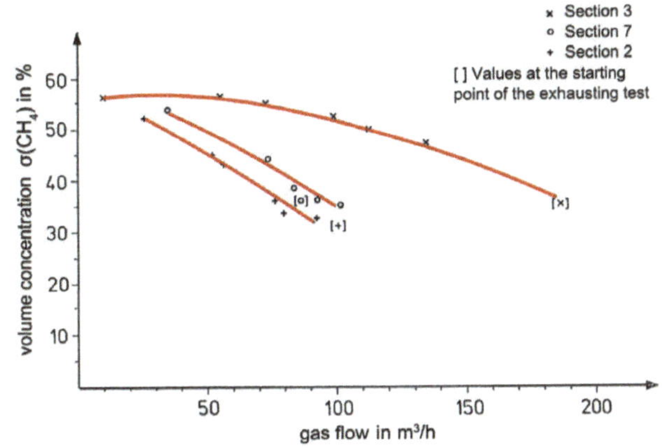

Figure 9.3.3 Methane concentrations as a function of gas flow (over sucking) during pumping tests at three wells (sections) (Rettenberger, 2004).

Together with the methane concentrations the O_2 and N_2 concentrations also vary with different gas flow rates. Fig. 9.3.4 shows the oxygen concentrations at a gas well in relation to the extracted gas volume. Significantly elevated oxygen concentrations in the landfill may be a potential for explosions, fires, and may inhibit methane production. On the other hand, there should be a certain negative pressure in the landfill body to avoid uncontrolled emissions.

As a conclusion, LFG extraction should be operated in a way that methane concentrations are in the range of 50%–45% during landfill operation and >50% after completion of the landfill. To achieve this, all wells should be adjusted individually about once per week. This is also due to the fact that the atmospheric pressure (increasing pressure change reduces and decreasing pressure change increases the gas pressure in the landfill body) and the wind (wind may cause a negative pressure inside the landfill) may have a strong influence on the gas extraction rate and/or the gas quality.

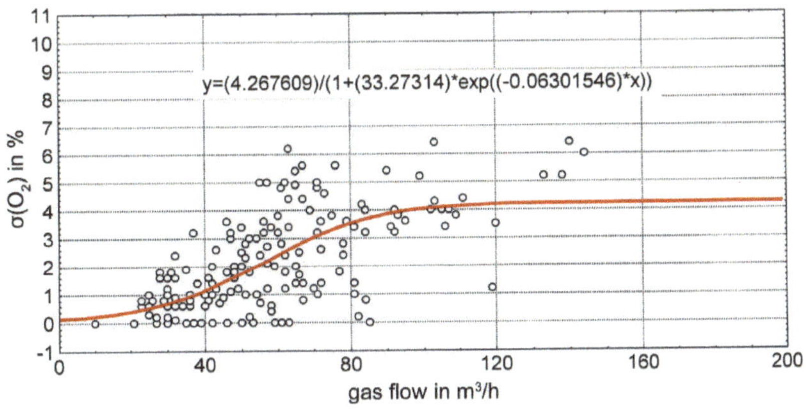

Figure 9.3.4 Volumetric concentration $\sigma(O_2)$ as a function of the gas flow at a gas well over a period of 3 months (Rettenberger, 2004).

4. Due to decreasing gas production rates in the older parts of a landfill or in the aftercare phase, more air may be sucked into the landfill that results in a so-called lean gas, i.e., methane concentrations < 20%–25%. Systematical oversucking can improve the collection efficiency rate considerably.

   When the gas flow is in the range of 2%–5% of the initial flow rate, air intrusion happens and a lean gas occurs. Owing to the oxygen content in the landfill, methane oxidation processes may occur. The gas extraction system has to be designed in a way that it can operate in a flow range of 1:20 (start of gas extraction) to 1:40 (final phase).

   As a consequence, the gas extraction system has to be continuously adapted during the operation of the plant.

5. Owing to the fact that the landfill body is very inhomogeneous and each landfill is different as well as the construction of the gas extraction system, the efficiency of a gas collection system (how much of the produced gas is captured) cannot be predicted. The design of a gas collection system is therefore not ultimate and has to be controlled either by comparing the extracted gas flow rates with the results of the gas prognosis and/or by detecting uncontrolled emissions around the wells and at the landfill surface.

Gas emissions at the landfill surface can be detected using the flame ionization detector (FID) method. Using this measuring device means to walk over the entire area of the landfill placing a little cap on top of the surface at different points and pumping small amounts of gas into the FID detector. In general the concentration range that can be measured with the detector is between 0–10,000 ppm CH_4. Using this method, most of the spots where gas is emitting can be detected. To reduce uncontrolled gas emissions, the gas extraction rates have to be adjusted, more wells may be installed, and/or the areas where the gas is emitting may be ceiled. Investigations in the last years show that most of the LFG emissions at landfills occur in a concentrated way at relatively few small areas (hot spots). Therefore, in Germany, landfill operators should identify those hot spots to take actions to avoid these emissions (Anonymous, 2016). An alternative measurement to the FID method is the control of diffusive emissions measuring methane concentrations outside of the landfill and the meteorological data to calculate the emission rate using gas distribution models. The disadvantage of this method is that the hot spots where the main emissions occur cannot be identified. More methods to measure gas emissions from a landfill are under investigation (Hrad and Huber-Humer, 2015), see also Chapter 9.5.

GAS EXTRACTION AND TRANSPORTATION

Fig. 9.3.5 shows the general system of an active gas collection system under the German regulation (Anonymous, 2001).

The gas is extracted by a blower, which can be to a high degree adjusted to different flow rates. It should provide vacuum pressures between 1 and 50 (100) hPa. The vacuum pressure should be distributed into the landfill body as homogeneous as possible. To reach this aim, a certain number of vertical

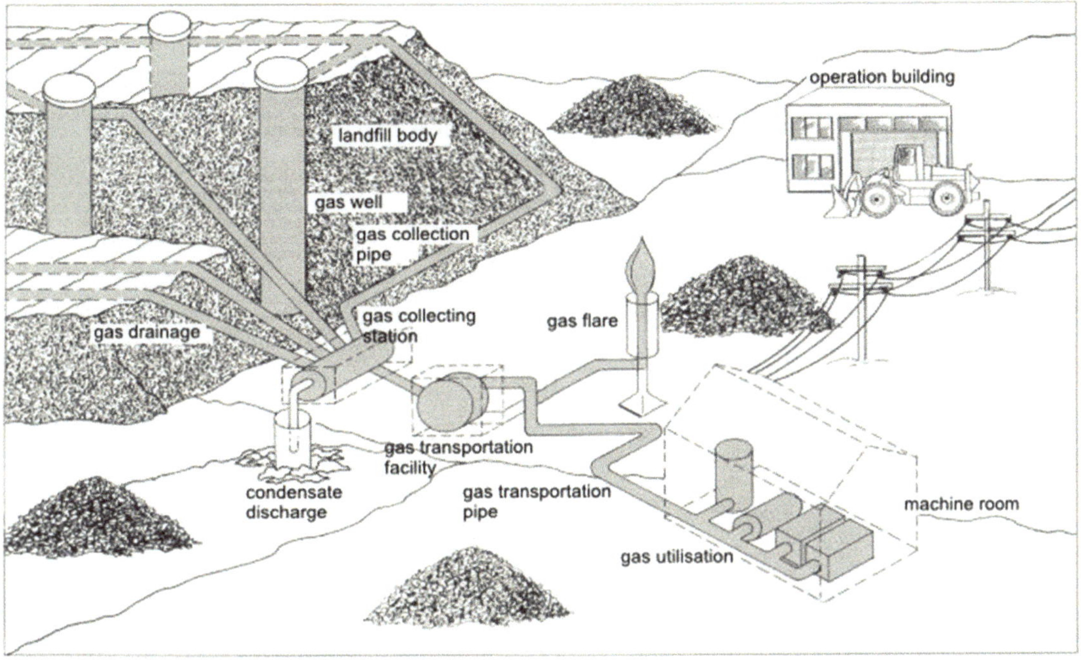

Figure 9.3.5 Scheme of an active gas collection system (Anonymous, 2001).

or horizontal gas wells have to be installed in the landfill body. There should be nearly no pressure loss inside the well, which can be achieved by choosing adequate gas pipe diameters (typical diameter $\geq$ 100 mm). If the landfill has a depth >25–30 m, two gas extraction pipes of different length (with a gas seal underneath the shorter pipe) may be installed that may be operated with different vacuum pressures (lower and higher level).

It is standard in Germany that each well is connected via a pipe to a distribution station, where the transportation pipe of several wells is connected to a main header pipe that transports the gas to the blower station. In the distribution station a valve can adjust the flow rate. In the pipe just in front the valve technical facilities for measuring in the pipe pressure, flow rate, gas composition, and temperature should be installed. In the gas distribution station, it may make sense to distinguish between lean gas and "normal" gas as it may, e.g., occur in landfills, which consist of an older and a younger part. In this case, the distribution station may be connected to two header pipes; as a result two separate gas collection systems exist. In most cases this is not necessary as long as the gas collection system can be adjusted in a way that high methane concentrations can be reached.

As the LFG is in general water saturated (at 40°C approx. 60 g/m^3, at 10°C approx 10 g/m^3), and cools down when it leaves the landfill it is necessary to remove the condensate from the pipes at their deepest points. These may be located at the gas wells, the distribution stations, or the utilization unit. Of course if necessary more knock out facilities have to be installed, e.g., in the header pipes.

In cases where there is more gas extracted as utilized, the gas has to be flared. Flaring may also be necessary when the gas utilization plant has a break down or is in maintenance. As the gas extraction rates vary with time, utilization units may not use all the extracted gas. Lean gas that is not used for energy production should also be flared until low methane concentrations. Standard flares can burn gas until methane concentrations reach 20%–25%; recently developed special flares may use gas with methane concentrations down to 8%–10% or even lower. If the gas cannot be thermally treated, it emits into the atmosphere and odor problems may occur; in addition, methane adds to the climate gas load. For all these reasons, flares should be standard at all landfills.

DESIGN AIDS FOR A GAS COLLECTION SYSTEM

Gas wells

For the design of gas wells, the following aspects should be considered:

- Inside the wells almost no pressure loss should occur.
- Landfill settlements should not cause pipe damages.
- Pipe stability should be proven.
- High temperatures (50–60°C) should not influence the pipe stability.
- Wells should be constructed in a way that air is not sucked in through the landfill surface, i.e., gas wells should be gastight against air intrusion especially at the well head and the upper part of the well.

Fig. 9.2.6 shows a typical vertical gas well; in this case with a flexible pipe in a pipe system for compensating settling, which is not used in any cases. A vertical gas well is in most cases drilled into the landfill by using an auger. The well should have the following characteristics:

- The diameter should be between 1.0 and 1.2 m, dependent on the (expected) water level in the landfill.
- They should be filled with lime-deficient course gravel (e.g., 16/32 mm).
- The central pipe should be made of PE and meet at least the pressure level of SDR 11 (Anonymous, 2005).
- The perforation should cover not more then 5% of the pipe surface. The openings can either be slots or holes. Slots and holes should be large enough to avoid clogging but smaller than the diameter of the gravel.

It should also to be considered that:

- To compensate settling the connection between the gas, extraction pipe and the horizontal pipe should be flexible.
- The upper part of the well has to be sealed (e.g., using low permeable clay)—dependent of the height of the landfill—at length from the landfill surface of at least 3 m. The gas extraction pipe—where it is slotted—should be surrounded by course gravel to avoid clogging.

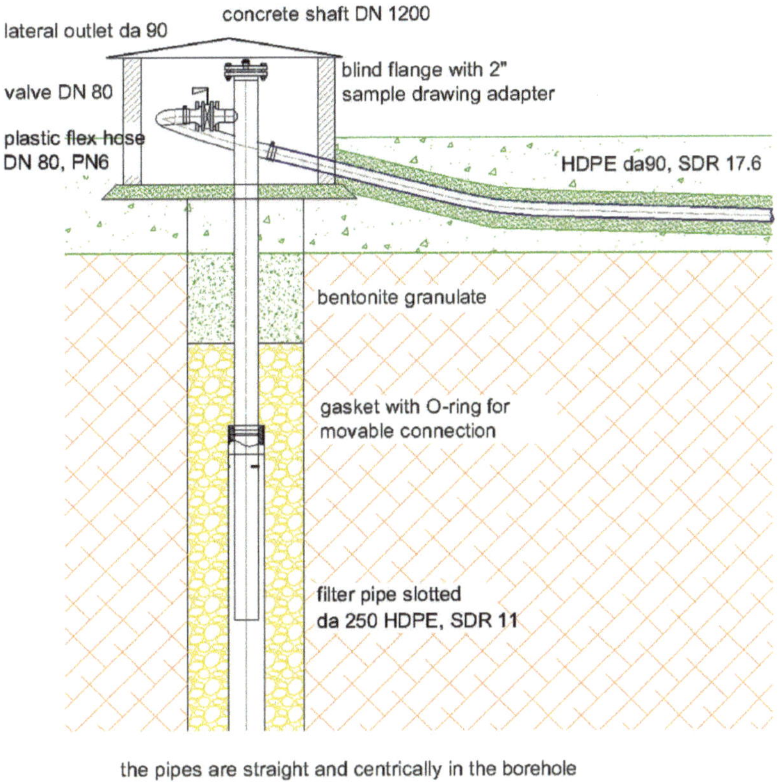

Figure 9.3.6 Typical vertical gas well. *Courtesy of Ingenieurgruppe RUK GmbH.*

The distance between the wells should be in the range of 35–50 m. At high landfills the spacing can be wider than at shallower landfills. The radius of influence can roughly be approximated using the following empirical equation (dimensions are not fitting, for details see Rettenberger, 1988):

$$R \sim 1.2\sqrt{1/p}$$

R, radius of influence, m; P, peak gas production, m^3/m^3 h (m^3 gas per m^3 landfill body per hour).

Peak gas production is the highest value in the gas prognosis curve starting from the beginning of waste disposal. If the landfill already exists, it is strongly recommended to install one or more test wells in the landfill and make some gas extraction tests. In addition, the potential water levels in the wells should be measured, and waste samples—gained during drilling—should be taken to estimate the gas production potential. The samples should be analyzed especially for AT_4, GB_{21}, $_{O}TS$, org. C, etc.

Gas extraction tests should be carried out at actual or specifically built gas extraction wells. It is recommended to make the gas extraction test using at least three wells. In the surrounding of the wells, several gas probes should be installed in different depths to measure the pressure distribution. By these means the radius of influence can be determined more exactly. The gas extraction test should last over periods of 1–3 months starting with a low extraction rate. This should be slightly increased every 2–3 days until the methane concentration is lower than 35%. As a next test phase, the methane concentration should be kept constant at 50% tonne over a period of about 2 months. After this phase, the initially practiced test procedure with the varying extraction rates and periods should be repeated.

If the gas shall be extracted from the start of landfill operation, wells are necessary that can "grow" with the landfill height, i.e., the wells can be elongated. A typical example of such a well type is presented in Fig. 9.3.7. When the landfill height increases, the steel pipe is pulled out of the landfill for some meters using an excavator or similar equipment. It is very important that the gas well stays gastight. Only during the period when the wells are enlarged in height, the gas extraction has to be interrupted.

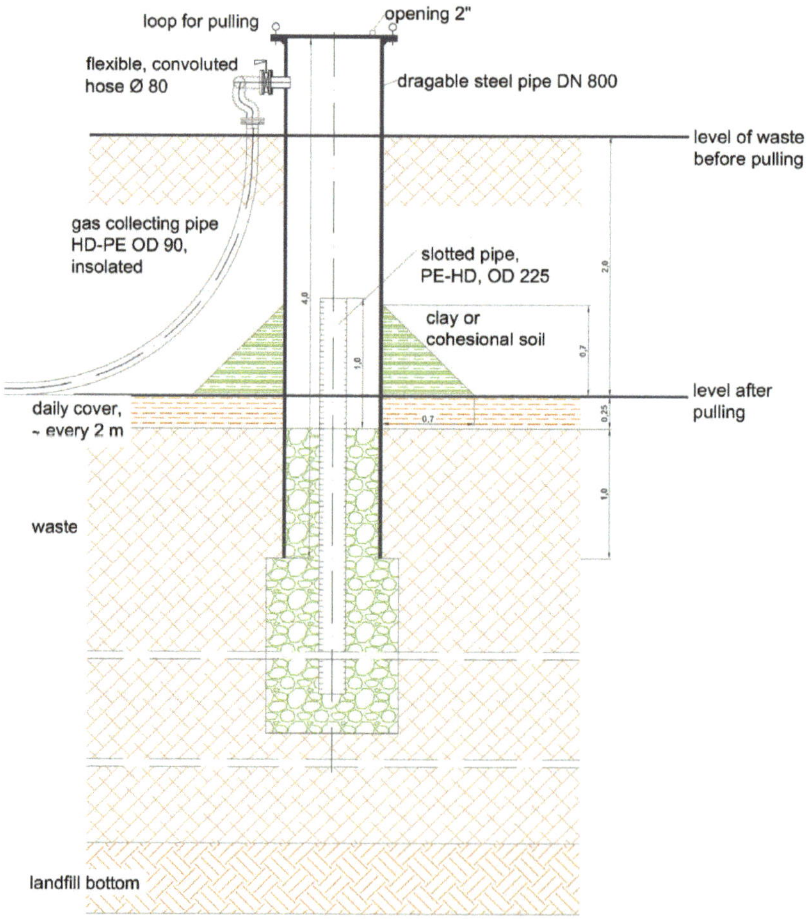

Figure 9.3.7 Extensible gas well. *Courtesy of Ingenieurgruppe RUK GmbH.*

Figure 9.3.8 Open gas wells at a landfill (inacceptable solution). *Photo by G. Rettenberger.*

Horizontal gas extraction pipes have proven their long-term functionality when they are adequately built and operated. These systems can be installed from the beginning of landfilling; they are influencing landfill operation much less than the "growing" vertical wells. Designing such systems it has to be respected that they may be partly filled up with water, which might result in clogging; in addition, settling has to be compensated. Also for these reasons PE pipes ≥250 [SDT 17,6, (4)] with slots or holes—as prescribed for the vertical wells—should be placed in, e.g., 1m2 profile trenches filled with coarse gravel or other coarse materials with low lime content and low solubility. The horizontal distances should be less then 20–40 m, and the vertical distances should be in the range of ±8 m.

An unacceptable solution of a gas extraction system is shown in Fig. 9.2.8. LFG may emit directly into the atmosphere causing high methane gas emissions. In addition, it creates a dangerous situation because there is a high potential of fires and explosions. Also due to the trace compounds and the high CO_2 concentrations in the LFG, people at the landfill may inhale toxic gases.

Header pipes and gas distribution stations

To allow condensate flow, gas transportation pipes on the landfill have to be installed with a constant slope in a depth that is not affected by frost. At the lowest points, water traps have to be installed. The diameter of the pipes should be at least d_a 90, in combination with a sufficient slope of preferably 5%. By these means a certain degree of settling may be compensated. PE is the favorable material, and the pressure strength should at least meet SDR 17.6.

Dependent on the size of the landfill, different separate collection systems may be suitable (Fig. 9.3.9): decentralized gas collection with several distribution stations and several gas flares, central ramification gas collection system with distribution stations and central flaring, central gas collection

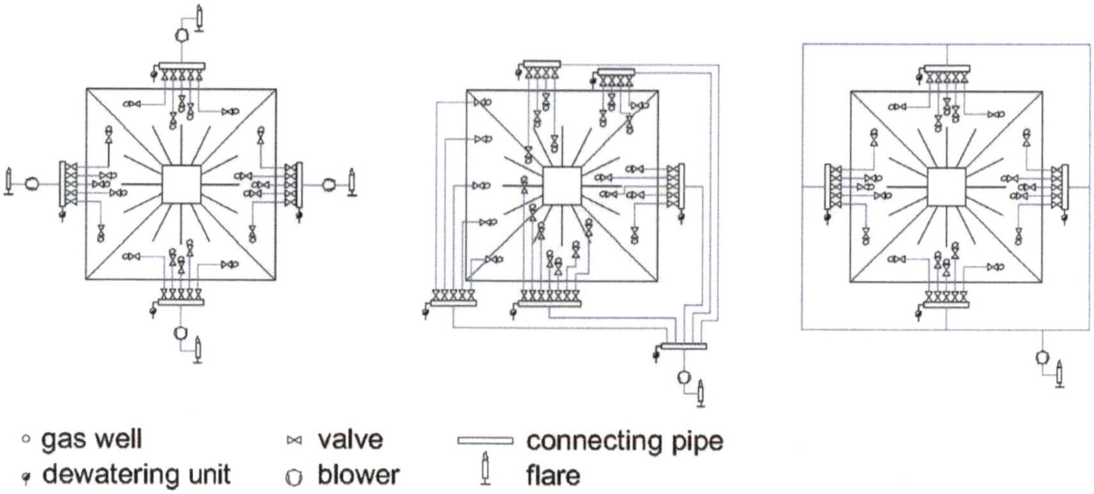

Figure 9.3.9 General layout of a gas collection system (Rettenberger, 1994).

system as ring pipeline, several distribution stations and central flaring. In the case of a separate lean gas and normal gas collection system, two collection header pipes will be necessary.

The central gas collection system has advantages concerning safety measures, quality, and flow control, and in most cases the number of different analytical equipment; in addition only one flare will be necessary, which all together will save costs.

Fig. 9.3.10 shows a distribution station for lean and "normal" gas quality pipes.

Gas extraction and flaring

As gas blowers have to face varying gas flow rates, it is necessary that they can be adequately adjusted without losing efficiency. Eccentric rotor blowers have proven successfully in practice. (see Fig. 9.3.11). Conventional blowers as, e.g., side channel blowers are only efficient if they work on the designed flow rate. In case a frequency transformer is used, the maximum flow rate can be reduced by about 50%. Meanwhile, in Germany conventional radial blowers are frequently in use.

Using gas flares, methane has to be "completely" oxidized to CO_2; in addition, the trace constituents in the LFG including the odor-producing constituents have to be destroyed. By adjusting the operation of the flare adequately, the production of dioxins will be minimized. This aim can only be achieved if the temperature in the combustion chamber of the flare is >950°C, and a retention time of >0.3 s exists. For keeping the temperature high, the combustion chamber has to be insulated.

To meet safety prescriptions, the flare has to be equipped with a flame control and an electronic ignition system. Fig. 9.3.12 shows a typical flare for LFG. It should be mentioned that in most cases LFG flares have to operate constantly over the year. Therefore permanent control and maintenance are imperative.

Figure 9.3.10 Gas distribution station with connecting pipes. *Photo by G. Rettenberger.*

Figure 9.3.11 Eccentric rotor blower. *Photo by G. Rettenberger.*

Figure 9.3.12 Gas flare with gas inlet and combustion chamber. *Photo from Lambda GmbH.*

MONITORING, CONTROL, AND CONCLUSIONS

Owing to varying conditions in the landfill body, atmospheric pressure fluctuations, increasing landfill height and age, and changing operation modes, the LFG collection system has to be permanently adjusted. Automatically operating monitoring systems may be helpful.

As an example, the following monitoring program may be followed:

Once a week:

- Measuring gas composition at each well (e.g., in the distribution station), if methane concentrations divert from the set value ±50%, the extracted flow rate should be adjusted (preferably continuous gas quality monitoring at least before the blower).
- Gas flow rate at the blower station (preferably continuous flow monitoring).

Two times a year:

- Measuring the water level in the wells, if the water level is too high and affects gas extraction, water has to be pumped out.
- Measuring the temperature in the wells.
- Analyzing the gas flow rate and gas composition including trace gases at the central blower. If the concentrations of trace gases are too high, additional gas treatment measures may be necessary; if the gas composition shows air intrusion leaks, it should be identified and repaired; if the gas flow is less then 5–7 times lower then the original flow rate, the installation has to be investigated, repaired, and if not feasible diminished and redesigned. If there is a certain tendency to

lean gases, the gas collection concept has to be revised. In this case aerobization of the landfill, passive gas venting, or lean gas treatment may be the options.

- Detecting gas emissions at the surface of a lined landfill (FID monitoring would help to identify leaks). If emissions are detected the operation mode and the gas collection system should be investigated and appropriate measures have to be taken.

Every 1–5 years:

- Controlling gas wells and pipes using camera systems, if failures are determined, actions such as cleaning or repairing have to be started.
- Safety control installations: maintenance and repairing may be necessary.

PASSIVE GAS DISPOSAL/VENTING

Passive gas venting may become relevant at old landfills where due to very low gas production rates there is no active gas extraction anymore. If a landfill is lined at the surface, a certain positive pressure inside the landfill may build up. This positive pressure may be used to control and treat gas emissions. Owing to the low pressure and flow rate, thermal treatment is not an option. At certain collection points—e.g., former gas wells—the gas may migrate out of the landfill body by passing through the landfill cover. The methane shall biologically oxidize in the soil cover before reaching the atmosphere (see also Chapter 9.5). To achieve this technically, the gas has to migrate into a kind of gas distribution layer (layer filled with sand or gravel) where the gas quality and flow is more or less equalized. From this layer, the gas migrates into the cultivation layer (top soil) where the methane shall be oxidized. Fig. 9.3.13 shows an example of a full-scale application.

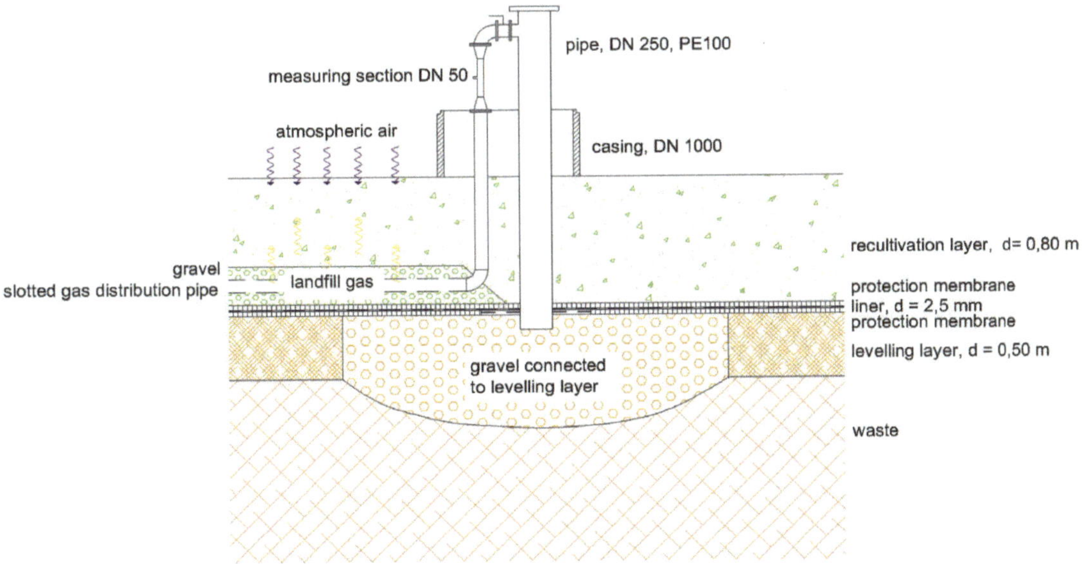

Figure 9.3.13 Passive gas venting using methane oxidation processes in the top cover. *Courtesy of Ingenieurgruppe RUK GmbH.*

References

Anonymous, 2001. DGUV Regel 114—004, 2001, Deponien, BG Bau Berufsgenossenschaft der Bauwirtschaft. Deutsche Gesetzliche Unfallversicherung e.V., Berlin.

Anonymous, 2005. GDA-Empfehlungen 2005, E 2—18: Geotechnische Belange der Deponieentgasung, Bautechnik 9.

Anonymous, 2016. VDI 3790, Blatt 2, 2016, Emissionen von Gasen, Gerüchen und Stäuben aus diffusen Quellen. Beuth Verlag, Berlin.

Hrad, M., Huber-Humer, M., 2015. Innovatives monitoring tool zur Bewertung von Methanemissionen. In: von Stegmann, R., Rettenberger, G. (Eds.), Trierer Berichte zur Abfallwirtschaft, herausgegeben, Band 22. Verlag Abfall aktuell, Stuttgart.

Rettenberger, G., 1988. Planerische, bauliche und sicherheitstechnische Gesichtspunkte bei der Erstellung von Entgasungsanlagen mit dem Ziel der Emissionsminderung. In: von Stegmann, R., Rettenberger, G. (Eds.), Hamburger Berichte, Band 1. Economica Verlag GmbH, Bonn.

Rettenberger, G., 1994. Deponie. In: Kranert, M., Cord-Landwehr, K. (Eds.), Einführung in die Abfallwirtschaft, 4. Auflage 2010. Vieweg+Teubner Verlag, Wiesbaden.

Rettenberger, G., 2004. Untersuchungen zur Charakterisierung der Gasphase in Abfallablagerungen. In: Stuttgarter Berichte zur Abfallwirtschaft, Band 82. Kommissionsverlag Oldenbourg Industrieverlag GmbH, München

NOTE- σ is used in this Chapter for indicating the volume concentration of a given gas X:: $\sigma(X) = V(X)/V$. It can be expressed as ratio or as percentage. - m^3, if not alternatively specified, stays for 1 cubic meter of gas volume under normal conditions (temperature: 0°C, pressure: 10^5Pa).

9.4

UTILIZATION OF LANDFILL GAS AND SAFETY MEASURES

Gerhard Rettenberger

INTRODUCTION

Landfill gas (LFG) might be a considerable source for energy and should be extracted and used whenever this would be convenient environmentally, technically and economically. About 60–80 m^3 of LFG per tonne of wet municipal solid waste (MSW) can be utilized over a period of approximately 15–20 years. Where LFG collection systems have been installed in the most cases, the LFG is used for energy production, but there are also examples where huge amounts of gas are flared. The reason in those cases is often that the cost of electricity in these countries is relatively low, so that the investment and operation of a gas utilization plant is not economical. Also for this reason in several cases, gas extraction and flaring is only practiced due to the financial support by a carbon credit program. This often happens in economically developing countries where there is the noncomprehensible situation that on one side energy is strongly needed and on the other side energy is wasted.

Another obstacle may also be the utility companies, which—for whichever reason—may not be interested in feeding the electricity produced from LFG into their grit. In other cases there is no grit nearby. In these situations, island solution should be considered where the energy is used on site, e.g., in form of electricity or as a fuel to substitute relative expensive diesel fuel in already existing engines. But it is especially a political task to provide the legal basis to improve this situation and support LFG utilization.

As the gas production curve is not constant—increasing during the operation phase and decreasing in the aftercare phase—it is very important that the capacity of the gas utilization plant can be relatively easy adapted to changing gas extraction rates (see also Chapter 9.1). Otherwise, too high amounts of the gas have to be flared. The kind of utilization depends very much on the specific situation. In many cases the conditions for using LFG on site or in the near vicinity are not existing because there is no energy user. Therefore in most plants the gas is used for producing electricity also due to the fact that especially in industrialized countries the public grid is usually nearby and available. In Europe, the production of electrical energy in gas engines is the most common practice and technical standard.

The upgrading of LFG natural gas quality may be due to its trace constituents and its content of oxygen and nitrogen is quite costly. This concept may be especially interesting when large amounts of LFG are produced over long periods of time. But there are also smaller treatment plants—built in containers—in use for upgrading biogas to natural gas quality, but these are not used for LFG yet.

It is also technical standard to use LFG for heat production in boilers using burners that are adapted to the—compared with natural gas—relatively low methane content. In addition, also the low temperature heat (max. 100°C) from the cooling system of the gas engine can be used. Often the heat cannot be used over the entire year so that the energy production plant is not always running at full capacity.

Only in a few cases the gas can be used in a nearby power or industrial plant where the gas is provided via a gas pipeline. Gas storage is in general costly, but there may be in the future more cases where gas engines also are operating mainly during the peak hours where higher revenues from the produced electricity can be gained. This can be achieved by storing the LFG for a couple of hours in gas storage tanks. This concept becomes more and more interesting in countries where a high amount of electricity is produced by wind power plants, which—due to changing wind conditions—do not operate constantly. Large gas storage tanks for LFG storage over longer periods (several days or weeks) may not be economical.

In general before constructing a gas utilization plant, the technical and economical conditions should be investigated very carefully. But the environmental effects also have to be discussed, because LFG emissions contribute significantly to global warming.

REQUIREMENTS

As LFG contains trace components (see Chapter 9.1), which may cause damages (e.g., corrosion) in the technical equipment such as gas engines or boilers concentrations of trace constituents should be low. The manufactures provide only a guarantee for their plants if certain limit values of trace gas concentrations will be met (see the following example from one manufacturer) (Anon, 2013):

- total chlorine $\geq$ 0–18 mg/m^3,
- total fluorine $\geq$ 0–9 mg/m^3,
- total chlorine and fluorine $\geq$ 0–18 mg/m^3,
- total sulfur $\geq$ 3–400 mg/m^3,
- total silicon $\geq$ 0–3.5 mg/m^3,
- dust $<$ 0–2 mg/m^3, and
- water vapor $<$ 50% humidity.

Other manufacturers may require different limit values. If those are not met LFG treatment may become inevitable.

In addition to the raw LFG, the concentrations of several components in the exhaust gases of flares, gas engines, and boilers are also regulated. These limit values differ from country to country. As an example, some of the threshold values that are required in Germany for reciprocating gas engines are presented (Anon, 2002).

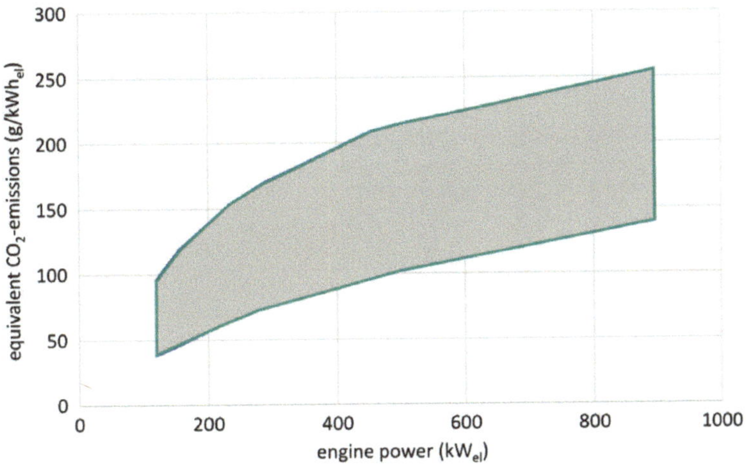

Figure 9.4.1 Unburned methane emissions as CO_2- equivalents dependent on the installed power of a gas engine (full-scale data). Courtesy of *SGS-RUK GmbH*

- CO < 1000 mg/m^3 if <3 MW (650 $mg/m^{3,}$ if >3 MW)
- NO_x < 500 mg/m^3 (lean gas Otto engine, diesel engine >3 MW); <250 mg/m^3 (normal Otto engine); <1000 mg/m^3 (diesel engine < 3 MW)
- CH_2O < 60 (40) mg/m^3
- SO_x < 350 mg/m^3
- Dust < 20 mg/m^3 (only diesel engine)

In general it should be kept in mind that the operator has the duty to minimize the emissions beyond the target values due to the state of the art.

It should be considered that there might be (diffusive) CH_4 emissions due to gas leakage; in addition, there may also be unburned methane residues in the exhaust gas of LFG engines. As may be concluded from Fig. 9.4.1 the unburned methane is not neglectable. As a consequence additional requirements to control these emissions in the future may come up.

GAS PRETREATMENT, GAS STORAGE TANKS

As indicated earlier, treatment of gas may be necessary either for emission control or damage prevention. All treatment methods of course add to the cost of the LFG utilization. In several cases, pretreatment is substituted by changing the oil in the gas engines more often because the oil acts as a sorbent.

Dewatering, drying

In most cases LFG leaves the landfill at temperatures between 35°C and 45°C. The gas is almost vapor saturated. As mentioned above, the moisture content in the gas should not exceed 50% if the gas is used in gas engines. In many cases this is achieved, e.g., by cooling the gas down to 10°C and heating it up to 20°C. In most plants in moderate or cold climate (in the latter case, freezing of the condensate in the

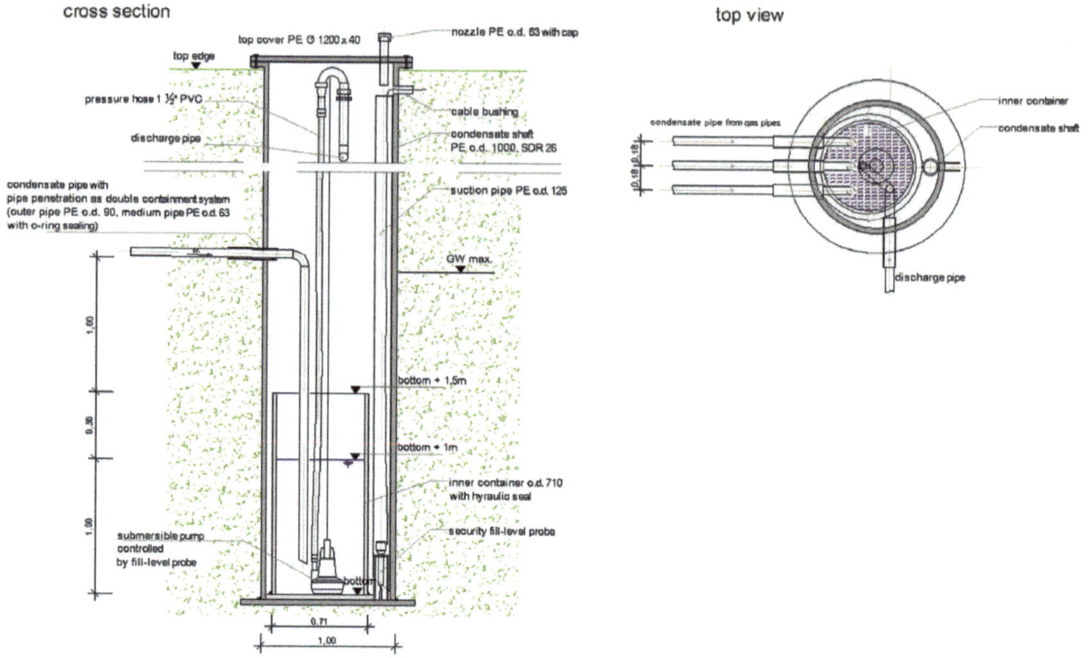

Figure 9.4.2 Siphon as a condensate knock-out in a landfill gas pipe. Courtesy of *Ingenieurgruppe RUK GmbH*

pipes or other technical devices has to be avoided) there is no separate technical moisture reduction necessary because this temperature change happens "naturally" during gas collection. The situation may be different in, e.g., tropical countries where separate cooling devices may have to be considered.

In any case, it is necessary to remove the condensate from the pipes. Therefore passive dewatering units have to be installed. As already mentioned this can be achieved, e.g., by installing a siphon at the pipe system at their lowest points. Fig. 9.4.2 shows an example of a siphon installation emplaced in the landfill body (designed for negative pressure in the pipes of up to −100 hPa). Providing adequate slopes condensate can also be redirected back into the landfill through the gas wells.

If additional dewatering is necessary, mostly cooling systems are used. In these facilities, the gas temperature is lowered down to a temperature of around 4°C.

A further dewatering action will be necessary if the gas has to be stored under high pressure in gas storage tanks made of steel. This may be necessary if the gas shall be used, e.g., in vehicles or shall be supplied into the public gas grid. In those cases, drying systems have to be used. Often adsorption filter filled with silica pellets or other appropriate sorption materials are used. These systems are operated using the pressure swing technology.

Figure 9.4.3 Activated carbon filter for landfill gas treatment placed in front of a gas engine. *Photo by G. Rettemberger.*

Removal of chlorine, fluorine, and silica compounds

To protect the gas equipment and meet the requirements of the manufacturers, the removal of chlorine, fluorine (chlorinated, fluorinated hydrocarbons), and organic silica compounds (siloxanes) from the raw gas may become necessary. Quite common is the application of adsorption processes using activated carbon (AC) pellets, which can be loaded with these compounds up to 15% of their original weight. There is a great variety of AC qualities and the contamination of the LFG is also site specific, for this reason, it might be viable to carry out tests to find out the most efficient/less costly material. After reaching the maximum loading capacity, the AC will be transported to a regeneration plant and will be substituted by unloaded AC. This might not be possible in all countries. In these cases the carbon can also be regenerated on site, which takes of course more efforts. When handling loaded AC, safety measures have to be followed. As adsorption systems show a typical breakthrough behavior, it is common to operate 2 filters in sequence. Fig. 9.4.3 shows an AC plant for LFG treatment. Other sorption materials that may be used are molecular sieves and silica gels.

Removal of sulfur

In general the concentration of hydrogen sulfide in the LFG is rather low. Therefore sulfur removal is in most cases not necessary. But there may be some exceptions especially when significant amounts of construction and demolition waste or other wastes containing gypsum have been codisposed with MSW. In these cases, sulfates will be biologically reduced to H_2S. H_2S is up to a certain concentration very odorous but cannot be smelled at higher concentrations. In addition, it is very toxic at low

Figure 9.4.4 Gas storage tank. *Photo by G. Rettemberger.*

concentrations also and very corrosive. For these reasons the codisposal of biologically degradable waste with gypsum-containing waste has to be strictly avoided.

H_2S can be biologically oxidized, but for its removal from LFG more often specially coated AC is used. These filters have a high efficiency. But biological processes also have been used in full-scale plants successfully.

Gas storage tank

As mentioned above, in only a few cases gas storage tanks have been installed at landfills. As the typical pressure range in landfills is between −10 hPa and 30 hPa, membrane gas storage tanks with or without steel casing may be used. Flexible membranes provide varying gas storage volumes. The LFG is supplied to or taken out from the storage facility by pumps. The pressure in the tank can be kept constant by means of an appropriate weight on top of the tank. Otherwise a feeding pump, which provides the necessary pressure for the user is needed. Fig. 9.4.4 shows a gas storage tank made of plastic membrane that is placed in a metal casing.

LANDFILL GAS UTILIZATION TECHNOLOGIES

Combined heat and power gas engines

LFG is mainly used in gas engines for producing electricity and heat (combined heat and power-CHP plants). The electricity can be supplied into the public grid or may be used on site as an island solution. The low temperature heat (≤100°C) can be used for heating houses or supplying heat for industrial or commercial purpose (heating green houses, fish ponds, etc.). Often the heat cannot be used constantly or at all, in these cases the cooling water of the engine has to be cooled using, in most cases, an air cooler. The advantages of CHP engines are as follows:

- Proven and well-developed technology, in general no or only minor LFG treatment necessary, operation of more than 8200 h/a possible.
- Electrical efficiency even more than 40%.
- Operation with lean LFG down to 30% methane concentration in many cases possible.
- In most cases the electricity is supplied into the public grid providing constant supply.
- Relatively cost-effective, in some countries price per kWh is—due to state regulation—financially subsidized by the utility companies.

Gas engines can be operated as stand-alone operation with asynchronous generator (island solution) or with synchronous generator for supplying electricity to the public grid.

Gas engines need constant maintenance at different levels (oil exchange, engine inspection, etc.). Under these conditions, gas engines can be operated for more than 10 years (more than 80,000 h). During this period the gas production varies to a high degree (after the operation phase of the landfill, the gas production reduces by more than 60%); the capacity of the gas engine has to be adjusted. This is necessary because the efficiency of the gas engine running at half capacity is significantly decreasing. As a consequence, in the most cases an appropriate solution may be to:

- Install several gas engines and add or remove one or more engines due to available gas volumes.
- Install gas engines completely in containers as a unit for simple exchange.
- Rent gas engines which can be exchanged with engines that have a lower capacity.

With these measures it can be avoided that lower gas volumes cannot be used and have to be flared instead (reducing the extraction rate cannot be accepted in any case).

Fig. 9.4.5 shows container units and Fig. 9.4.6 presents an installation of a CHP plant in an industrial building.

To meet the gas emission standards for exhaust gases—especially for NO_2—and to increase the electrical efficiency of the engine, in most cases turbo charged lean gas engines are used. In the future decreasing limits for formaldehyde and unburned methane in the exhaust gas can be expected and may ask for additional measures. Lean gas engines need a special gas–air mixing system to operate in a window where relatively low N_2O will be produced.

In some cases diesel engines are modified so that they can use LFG as a fuel. The disadvantage of this engine type is the necessity to inject continuously diesel of up to 15% of the total fuel consumption into the cylinder. Diesel engines are self-priming; compared with gas spark-ignited engines they can be operated at lower methane concentrations and have a high electrical efficiency. The emissions of N_2O are higher in diesel engines, which is taken care of granting higher N_2O target values in the exhaust gas. The advantages of diesel engines are more relevant for smaller plants. There are developments on the way to further improve this engine type for LFG utilization.

Figure 9.4.5 Gas engines built in container. *Photo by G. Rettemberger.*

Figure 9.4.6 Gas engine in an industrial building. *Photo by G. Rettemberger.*

Owing to the mostly abandoned locations of most of the landfills, the use of waste heat is limited. Investigations had been done using technologies to convert the heat from the engine cooling system into electrical power. Organic rankine cycle uses organic liquids that vaporize at lower temperatures than water to produce vapor that drives a steam engine, which is connected to a generator. The vapor is cooled and the organic liquid is recovered and used again. Most of the organics that have this characteristic are toxic or not environmental friendly, for this reason synthetic silicon-based organic liquids are under development. This process operated successfully at full scale and needs some further development to be more stable. It may be an option in special situations.

A proven technology is the so-called Stirling engine that may provide for smaller or older landfills electrical power of 5–8 kW_{el}, which, e.g., may be used for the electricity supply for facilities in the aftercare phase on site. One manufacturer produces only one type (5–8 kW_{el}), which can be bought from the shelf (Anon, 2016a).

Gas turbines

In a few cases in Germany (there may be more in other countries), gas turbines (20–65 kW) that use LFG have been installed. In the United States also, larger turbines (producing several hundreds kW_{el}) are in operation (Bischoff, 2008). Gas turbines need relative clean gas with very low trace compound concentrations, and the LFG has to be pressurized up to 5 bar. As an advantage that gas turbines can operate with lean gases (>25% CH_4), they need only little maintenance and meet exhaust standards quite easily. On the other hand the electrical efficiency compared with gas engines is lower (26%–32%, Bischoff, 2008). Gas turbines could close the gap between CHP engines and lean gas flares for gas utilization of landfills in the aftercare phase.

Methane supply in to a public or private gas net

For the operation of a private gas net, e.g., in Germany the national building code has to be respected. The operation itself—e.g., to supply a power unit with LFG—requires removing water/vapor from the gas and to install the pipes in the soil at a depth where freezing does not occur. Of course safety measures have to be fulfilled. Owing to the content of toxic trace compounds in the LFG, the utilization should be restrictive. These gas nets are mainly used to transport the gas to an industrial plant where it is used on site.

Owing to the content of trace compounds, it is not allowed in Germany to supply LFG into a public grid (in contrast to biogas from agricultural sources).

Internationally there are several plants where LFG is supplied into a national grid. LFG treatment technologies such as molecular sieves as a pressure swing process, the "Selexol" process or membranes have been used successfully to treat the LFG to natural gas quality (Bilitewski and Härdtle, 2013). Fig. 9.4.7 shows a "Selexol" plant, which was operated in California/United States. In general this kind of technology may only be economical at huge landfills under special economical conditions.

Burner and boiler

The direct burning of LFG to produce heat and steam is quite common and a standard technology. If the LFG gas can be used, e.g., in a plant all year around, the efficiency of the utilization of the energy content in the LFG can be more than 80%–90%, but the economical problem of these kind of plants arises when the LFG cannot be used all year around (e.g., when heat or cold is used).

In the latter cases only a certain part of the totally produced gas can be used. In addition, during the period when no or less gas is used, other disposal units (such as flares) have to be installed. Fig. 9.4.8 shows a burner/boiler unit in a garden center for heating green houses in Germany.

Figure 9.4.7 Landfill Gas treatment for supplying into the national grit using the "Selexol" process in California/USA. *Photo by G. Rettemberger.*

Figure 9.4.8 Landfill gas burner in a garden center. *Photo by G. Rettemberger.*

Using LFG in boilers has also the advantage that there are, e.g., in Germany less strict requirements regarding the raw gas quality, but of course the exhaust gas threshold values have to be met. With special burners, lean gases can also be used down to methane concentrations of 4.5% (Anon, 2016b). It is recommended to make sure that no corrosion will occur inside the boilers; it should also be checked on a routine basis if deposits—e.g., caused by siloxanes—at the inside walls of the boiler have been built up.

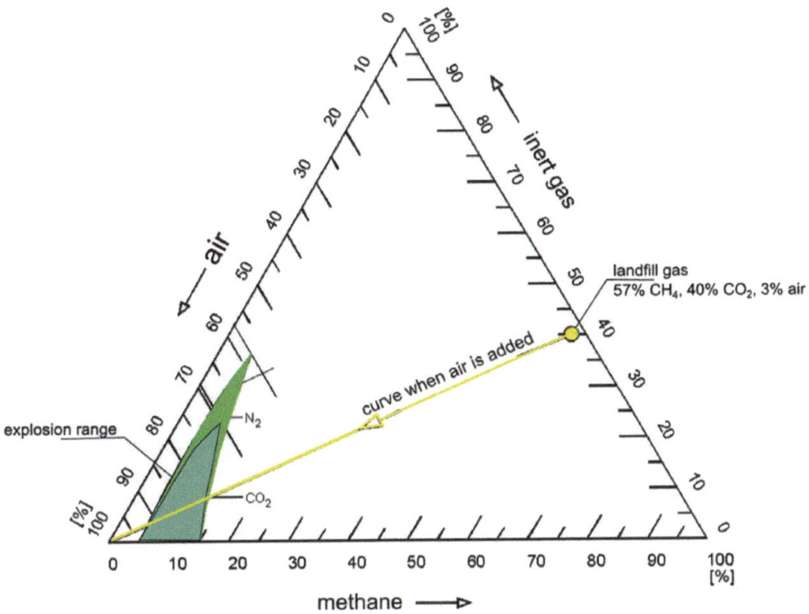

Figure 9.4.9 Diagram to assess a methane, air and inert gas mixture due to their explosion potential (Müller and Rettenberger, 1986).

SAFETY MEASURES

LFG containing oxygen or getting in contact with air can cause explosive mixtures with considerable effects to human beings as well as to the environment. If deflagration happens explosions can reach a pressure around 7 bar, but much higher pressures occur if a detonation has developed.

An explosion may occur if three premises exist as follows:

- A combustible gas with good dispersion behavior.
- A gas mixture in an explosive condition (between upper and lower explosion limits).
- An igniting source.

To find out if a gas–air mixture can be ignited the diagram presented in Fig. 9.4.9 can be used.

To ignite an explosive mixture, a source containing sufficient energy has to be available. Sources can be electrical sparks, hot surfaces, electrostatically loaded materials, or lightning.

For the safe operating of a gas extraction and utilization plant, a safety concept has to be developed. It consists of the following steps:

- Identification of any dangers regarding the actual or potential development of explosive mixtures and assessment of the risks of an explosion as well as potential effects on the complete plant (danger and risk assessment).
- Identification of potential explosion hazards to workers and the environment.
- Selection of measures against the identified dangers and risks.

- Installation of the necessary safety equipment and development of an operation and behavior mode.
- Audit the designed and installed safety concept/measures by an authorized expert.

Based on the above mentioned mentioned criteria the potential dangers and hazards of an explosion at LFG plants can be identified. Where vacuum pressure occurs (e.g., in the gas pipes from the landfill to the blower) air can be sucked into the pipe/plant and may create explosive mixture inside. In those areas where an overpressure exists (mostly in the area from the blower to the utilization facility or flare) gas may leave the pipe/plant and create explosive mixtures outside the plant. This can especially happen if the leakage—e.g., in a pipe connection—is in a building where the methane can accumulate and where there are potential ignition sources such as the blower or the electrical switch. During the risk and hazard assessment, different scenarios have to be identified where and under which conditions an explosion may occur. Based on the gained results the safety measures have to be developed.

The elaborated safety measures can be divided into three categories, which are ordered in a certain hierarchy:

Primary measures: all measures to avoid the formation of explosive mixtures (e.g., gas tight pipes, (continuous) monitoring methane and/or oxygen concentration in the gas, inertization of LFG, ventilation of buildings so that no explosive mixtures can build up, etc.).
Secondary measures: all measures that avoid that potential ignition sources may become effective (explosion proven machines, switches, etc.).
Tertiary measures: all measures that avoid damage after an explosion had happened (casings around blowers and other potential equipment that has the potential to explode).

For the development of safety measures, the first step would be to select all the primary measures that are necessary to avoid the development of explosive mixtures.

Owing to experiences, not all dangers/risks may be avoided. As additional measures, the following two aspects should be investigated:

Locating where potential danger still may occur.

- Probability of the existence of a dangerous mixture.

This is realized by characterizing areas (zones) where explosive mixtures may occur with a certain probability. A detailed definition can be found in Anon (2015a):

- zone 0: where explosive mixtures exist constantly or often,
- zone 1: where explosive mixtures exist occasionally, and
- zone 2: where explosive mixtures exist seldom and only for a short time.

The different zones can be identified using handbooks containing examples (Anon, 2001a). The zones should be marked on a map (map of zones) and posted at the plant.

When additional zones are identified during operation, additional measures have to be taken. In any case it has to be avoided that electrical equipment may become an ignition source. Only those electrical installations may be used that meet the requirements for the specific identified zone and the published

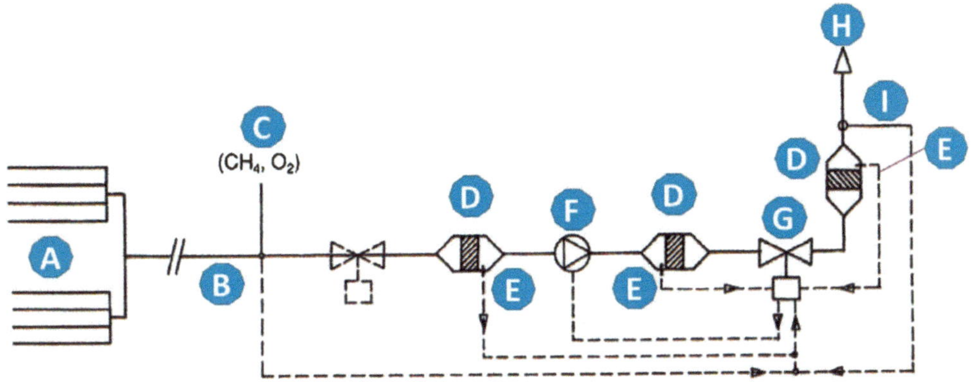

Figure 9.4.10 Tertiary explosion measures in a landfill gas collection scheme (Anon, 2001b). (A) collection pipes, (B) transportation pipe, (C) methane and oxygen concentrations control, (D) flame arrestors, (E) temperature control, (F) blower, (G) high speed valve, (H) flare, (I) flame control.

technical rules. Owing to European law, the manufacturer of a device has to give a declaration of conformity with the required standards and has to equip the product with a label showing the CE sign and the technical standards (Anon, 2015b).

If these measures are not sufficient, tertiary measures have to be taken. In that case the plant has to be constructed in a way that even if an explosion happens, the equipment will not be destroyed so that the surrounding areas will stay intact. It is important to prevent that the explosion influences other areas so that a detonation may occur. Beyond many other explosion control measures, it is important that a flame arrestor is installed in the extraction pipe—where there is a vacuum pressure—before and after the blower to prevent a backward and forward ignition. This concept is shown in Fig. 9.4.10.

The example in Fig. 9.2.10 shows the following situation: In the landfill collection pipes (A) it has to be expected that air is sucked in. Therefore zone 1 exists. This zone still exists in the transportation pipe (B) until the flare is reached (H). The blower (F) might be an ignition source. Therefore measures have to be taken. Flame arrestors (D) are installed before and after the blower and in front of the flare. The total system up to the flame arrestors is designed for explosion stability. In addition, the flame arrestor has to be temperature controlled (E), and the entire system has to be controlled with regards to methane and oxygen concentrations (C). If the upper explosion limit is reached, the plant will be shut down automatically and the closure of a high speed valve (G) will occur. The plant will also shut down in case the flame control (I) sends a signal that there is no fire.

After the installation of all necessary equipment, an audit of the plant by an authorized expert is imperative. The complete safety measures have to be documented together with a description of the plant, the assessments and the measures, the control and audit plan, the servicing and maintenance concept, and the instruction and directive activities and descriptions. All the actions have to be documented in a manual.

References

Anon, 2001a. DGUV Information 213-015, Beispielsammlung Explosionsschutzmaßnahmen bei der Arbeit auf Deponien. Deutsche Gesetzliche Unfallversicherung e.V. (DGUV), Berlin.

Anon, 2001b. DGUV Regel 114-004 Deponien. Deutsche Gesetzliche Unfallversicherung e.V. (DGUV), Berlin.

Anon, 2002. Erste Allgemeine Verwaltungsvorschrift zum Bundes-Immissionsschutzgesetz (Technische Anleitung zur Reinhaltung der Luft – TA Luft). Carl Heymanns Verlag KG, Köln.

Anon, 2013. MWM Technisches Rundschreiben 0199-99-03017/05 DE. Caterpillar Energy Solutions GmbG, Mannheim.

Anon, 2015a. Verordnung zum Schutz vor Gefahrstoffen (Gefahrstoffverordnung – GefStoffV), Bundesgesetzblatt I S 49. Bundesanzeiger Verlag, Bonn.

Anon, 2015b. Gesetz über die Bereitstellung von Produkten auf dem Markt (Produktsicherheitsgesetz – ProdSG), Bundesgesetzblatt I S 1474. Bundesanzeiger Verlag, Bonn.

Anon, 2016a. Lambda Stirling BHKW im Land Brandenburg. Lambda Gesellschaft für Gastechnik mbH, Herten. www.lambda.de.

Anon, 2016b. Flox – Brenner für die Energietechnik. e-flox GmbH, Renningen. www.e-flox.de.

Bilitewski, B., Härdtle, G., 2013. Abfallwirtschaft, Handbuch für Praxis und Lehre. Springer-Verlag, Berlin, Heidelberg.

Bischoff, V., 2008. Deponie Eichelbuck Verwertung von Deponieschwachgas in einer Mikrogasturbinenanlage. Abschlussbericht des Ingenieurbüros Roth & Partner, Karlsruhe. www4.lubw.baden-wuerttemberg.de.

Müller, K., Rettenberger, G., 1986. Gasabsauge- und Gasverwertungsanlagen an Mülldeponien – Anleitung zur Entwicklung sicherheitstechnischer Konzepte. BMFT, Bonn.

9.5

LANDFILL GAS MANAGEMENT BY METHANE OXIDATION

Peter Kjeldsen and Charlotte Scheutz

INTRODUCTION

Landfills have been found to be one of the major sources contributing to the increasing atmospheric methane (CH_4) content (Bogner et al., 2008) (see also Chapter 21.3). In Europe, landfills are the second largest (22%) anthropogenic CH_4 emission source (European Environment Agency, 2008), while in the United States landfills are the third largest anthropogenic source of CH_4 at 17% (US EPA, 2013).

The generated landfill gas (LFG) may be stored in the waste layers within the landfills or may migrate out in the surroundings and/or be emitted to the atmosphere (confer Fig. 9.5.1) through leaking leachate wells, gas extraction wells, or through areas of imperfect soil cover. Constituents in the LFG may also be degraded in the landfill cover through aerobic oxidation processes eventually in combination with anaerobic degradation processes in the deeper parts of the cover.

At some landfills with gas collection the extracted gas cannot be utilized, and the gas is either flared with a low risk of producing toxic combustion products or just emitted to atmosphere. As an alternative to gas utilization systems or as a follow-on technology when a gas utilization system gets noncost-effective, biomitigation systems may be implemented (Morris et al., 2012). They are defined here as systems based on microbial removal processes implemented at landfills to reduce emission of CH_4 (or other harmful substances). In respect to CH_4, experiments have documented that a very high CH_4 oxidation rate can be obtained in soils, compost, and other materials, high enough to significantly reduce the CH_4 emission from landfills (Scheutz et al., 2009a). Landfills may be fully covered with biologically active materials, so-called biocovers. Biomitigation systems may also imply establishment of biofilters reducing CH_4 concentration in LFG extracted from the existing gas collection system (GCS).

At old landfills with a low gas production in situ aeration may be another alternative to significantly reduce LFG emissions (see Chapter 16.2).

In this chapter an overview of the oxidation processes of LFG and the upscaling to actual biomitigation systems are presented. Different types of biomitigation systems are described and the results of a literature review of implemented biomitigation systems are reported. In addition, the methods being used to document the efficiency of biomitigation systems are described, and the existing challenges for obtaining systems with high mitigation efficiencies are highlighted.

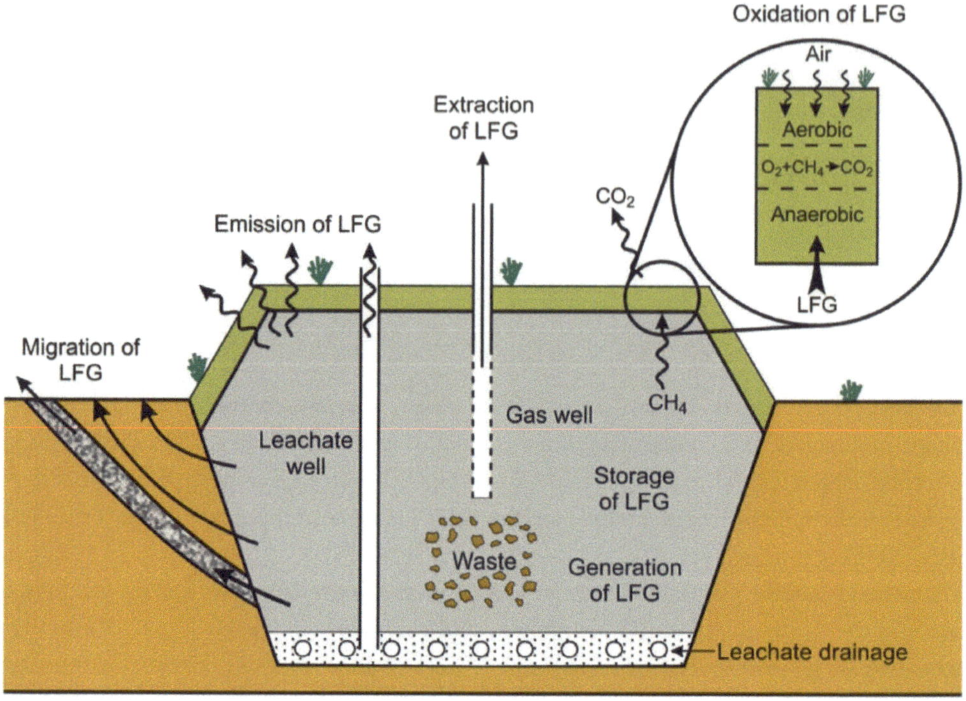

Figure 9.5.1 Conceptual model of the generation, transport, and oxidation of landfill gas (LFG).

FROM PROCESS UNDERSTANDING TO TECHNOLOGY DEVELOPMENT

The development of cost-efficient and sustainable environmental technologies is based on a detailed understanding of the governing physical, chemical, and microbial processes. This is also truly the case for engineering the microbial CH_4 oxidation process to viable full-scale biomitigation solutions. The research into the CH_4 oxidation process in landfill covers and surroundings started more than 25 years ago with the highly cited paper by Whalen et al. (1990). Now (2018) with a topical search in Web of Science combining "landfill" and "methane oxidation" more than 480 articles are found. In the early years, most research focused on characterization of the methanotrophic bacteria, laboratory experiments to determine CH_4 oxidation rates and governing environmental factors, field studies of the CH_4 oxidation in landfill soil covers, and development of simple models to simulate the dynamic behavior of CH_4 in soils. Lately (typical starting around 2005), there has been more focus on engineering the CH_4 oxidation process by implementing biomitigation technologies and also on the use of alternative materials than soils (such as matured compost, clay pellets, and other artificial materials).

The CH_4 Oxidation Process as Basis for Biomitigation—What Is Important?

In landfill covers CH_4 and O_2 may appear at the same depth due to emission of CH_4 from the waste and diffusion of O_2 from ambient air, which provides the needed conditions for sustaining methanotrophic bacteria (Scheutz et al., 2009a). Aerobic CH_4 oxidation proceeds according to the following overall reaction, producing a significant amount of heat:

$$CH_4 + 2O_2 \rightarrow CO_2 + 2H_2O + \text{heat} \quad \Delta G^\circ = -780\ \text{kJ/mol}\ CH_4 \tag{1}$$

The reaction is carried out by the so-called methanotrophic bacteria (or methanotrophs), which are unique in their ability to utilize CH_4 as a carbon and energy source (Scheutz et al., 2009a). Landfill cover soils can exhibit high capacities for CH_4 oxidation. Very high rates of CH_4 oxidation in landfill cover soils (>100 μg CH_4 g^{-1} h^{-1} and >200 g CH_4 m^{-2} day^{-1} in batch and column experiments, respectively) have been reported (Scheutz et al., 2009a).

The oxidation of CH_4 is controlled by several environmental factors such as temperature, soil moisture content, soil texture, CH_4 and O_2 supply, nutrients, etc. This implies that the climatic conditions are of huge importance for the actual CH_4 oxidation rate. In landfill soil covers or engineered biomitigation systems, temperature and soil moisture are very important parameters controlling CH_4 oxidation (Scheutz et al., 2009a). Especially temperature is an important factor in temperate and arctic climate where the ambient temperature can be limiting for the biological process during the winter season. However, several later studies of biomitigation systems using compost as bioactive material have shown significant elevated temperatures in the CH_4 oxidation layer during cold periods (Philopoulos et al., 2008; Pedersen, 2010; Dever et al., 2011; Scheutz et al., 2014; Cassini et al., 2017), which affect the biocover ability to oxidize CH_4 in a positive direction. This is supported by laboratory experiments, which investigated the influence of the temperature on CH_4 oxidation and respiration in compost samples (Scheutz et al., 2011a). Compost material was collected from the biomitigation system at Klintholm Landfill, Denmark and incubated in the laboratory at 10 different temperatures varying between 4°C and 70°C. The temperature optimum of the methanotrophic community in the biocover material was 45°C, which was much higher than temperature optima for CH_4 oxidation in landfill cover soils, which has been reported to be in the range of 15–38°C (Scheutz et al., 2009a). The results indicate that a moderately thermophilic methanotrophic community adapted to the elevated temperature conditions in the biocover had developed. The temperature optimum was comparable to maximum temperatures measured in the deeper parts of the biocover at Klintholm Landfill (Scheutz et al., 2011a).

Moisture content may also be an important environmental factor. LFG flowing through the CH_4 oxidation layer may be heated up due to the elevated temperatures. Water contained in the CH_4 oxidation layer may be evaporated. Even though that the CH_4 oxidation process produces water (confer Eq. 1) and that the LFG often is water saturated when leaving the waste layers, a desiccation of the CH_4 oxidation layer may be observed especially in hot, dry region with little infiltration of rain.

The CH_4 oxidation activity is close to zero when soil moisture content decreases below 5% (dry weight basis) (Scheutz et al., 2009a). In addition, desiccation of the CH_4 oxidation layer may lead to cracks and fissures, causing hotspots in the cover with relatively high emissions of CH_4.

Attenuation of Non-CH_4 Landfill Gas Constituents

In addition to CH_4, landfills also release numerous volatile organic compounds (VOCs), including halogenated hydrocarbons, and aromatics, of many are toxic and/or contribute to climate change or depletion of the ozone layer (Christensen et al., 1995). The trace organic compounds originate from volatilization of hazardous materials or biological degradation of organic waste disposed of at the landfill. The breakdown of different organic waste types, including food waste can result in formation of odorous compounds, often containing sulfur. Some wastes such as construction and demolition waste may contain gypsum, which in contact with water will release sulfate. Under anaerobic conditions within the landfilled mixed waste, the released sulfate may be utilized as electron donor by sulfate-reducing bacteria, and H_2S is produced. Hydrogen sulfide concentrations in the order of several thousands of ppm have been observed in LFG (Plaza et al., 2007).

Early studies carried out at Technical University of Denmark demonstrated that a number of VOCs including halogenated hydrocarbons and aromatics can be degraded in landfill cover soils in which a methanotrophic activity can be found (Kjeldsen et al., 1997; Scheutz and Kjeldsen, 2003, 2004, 2005). The research group also found that several halogenated organic compounds were dechlorinated in the anoxic part of laboratory column studies simulating landfill cover conditions whereas the generated degradation products were removed in the aerobic part of the columns as shown in Fig. 9.5.2 (Scheutz and Kjeldsen, 2004, 2005). This led the authors to conclude that both aerobic and anaerobic degradation processes in landfill covers might have a significant attenuation effect on gaseous emissions from landfill settings. Similar results of attenuation of halogenated hydrocarbons were later demonstrated for compost materials (Scheutz et al., 2009b).

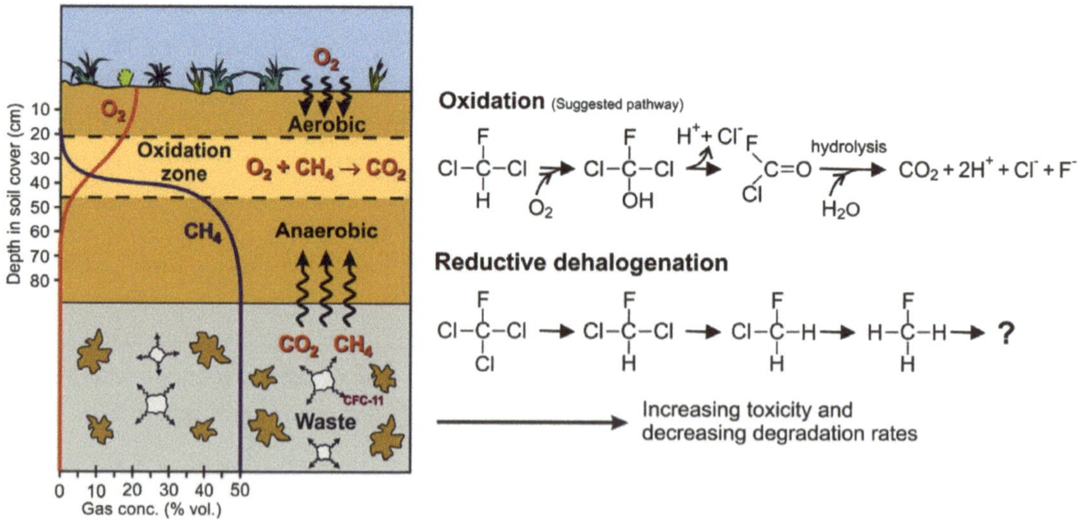

Figure 9.5.2 Combined aerobic/anaerobic degradation of CFC-11 in biocover.

Laboratory studies performed on landfill cover soils (Scheutz et al., 2004; Scheutz and Kjeldsen, 2005; Lee et al., 2010), and studies in experimental compost biocovers (Lakhouit et al., 2014; Zdeb and Pawlowska, 2013) showed high oxidation rates of several aromatic hydrocarbons such as benzene, toluene, ethyl benzene and xylenes, which are often measured in LFG. A recent study on an experimental compost biocover revealed an almost 100% reduction in odorous compounds between the raw LFG and the gas emitted from the biocover (based on olfactometry and analysis of total reduced sulfur) (Capanema et al., 2014). More details on the attenuation of trace gases in biomitigation systems are given in the review by Menard et al. (2012).

As mentioned earlier, hydrogen sulfur is an inorganic and odorous trace gas often found in high concentrations in LFG. Several studies have looked at the attenuation of H_2S in different alternative landfill cover materials. Experiments carried by Plaza et al. (2007) and Xu et al. (2010) show that a sandy soil amended with lime, layers of fine crushed concrete, compost, and yard waste could significantly reduce the H_2S emissions as a result of combined physical adsorption, chemical reaction, and biological oxidation to sulfate. Adding nitrate (e.g., by pretreated leachate recirculation) to different cover materials stimulated autotrophic denitrification where the present nitrate oxidized the H_2S to either elemental sulfur or sulfate (Sungthong and Reinhart, 2011). In another study, microbial assays show several different microbes including both sulfate-reducing and sulfur-oxidizing bacteria in an experimental landfill cover soil loaded with LFG containing H_2S in the range of 50–250 ppm (He et al., 2012).

So in summary there are several clear evidences that biomitigation systems not only are able to reduce emissions of CH_4 but also a wide range of trace gases often found in LFG.

The Role of Biomitigation Systems in Landfill Gas Management

In almost all countries of the world, municipal solid waste is still landfilled (Laner et al., 2012), and there are many landfills that have been closed within the last 10–15 years. In addition, a large number of old, uncontrolled landfills still exist in all countries producing significant volumes of LFG. In the western countries many landfills have LFG extraction systems with energy utilization or the LFG is flared. However, on a global scale, the major fraction of new as well as old landfills are not equipped with gas management systems, perhaps rather installed gas vents to avoid off-site migration and reduce explosion hazards. They are a severe CH_4 source contributing to the greenhouse effect.

The establishment of a biomitigation system may be relevant in several different cases. However, it is important to stress that LFG extraction and energy utilization systems often will be highly cost-efficient at large, newer landfills; in addition, in many countries law prescribes LFG extraction and utilization. Besides the mitigation effect by reducing emissions of CH_4, the energy production indirectly also saves CO_2 emissions (in case that the LFG-based energy replaces energy produced from fossil fuels). At high gas generation rates, biomitigation systems are also of minor relevance due to the foreseen high gas loads to the system, which may lead to inadequate retention times in the biofiltration units and a resulting low mitigation efficiency of the system.

Table 9.5.1 Scenarios where establishment of a biomitigation system at landfills could be an option

Scenario	Description
1	No gas collection system (GCS) is in place, the LFG generation is modest either due to a high landfill age or due to disposal of waste with low organic content. Installation of a GCS and a gas engine (or similar energy conversion unit) is not cost-efficient, but LFG emission is regarded as above legal limits. **1a.** Neither leachate collection system is present nor gas vents, which could be the major LFG escaping route. **1b.** A leachate collection system or gas vents are present, which may be the major LFG escaping route.
2	A GCS is in place. The gas engine (or similar energy conversion unit) is old with high running maintenance costs. A replacement of the energy conversion unit is not considered cost-efficient. **2a.** Significant fugitive or uncollected emissions from slopes, uncovered part, leachate collection system, etc. are foreseen. **2b.** No significant fugitive emissions are foreseen—may be as a result of the presence of a gas tight engineered top cover.
3	A GCS and a gas flaring system are in place. The flares have difficulties to run without the use of supporting fuel, but LFG emission is regarded as above legal limits. **3a.** Significant fugitive or uncollected emissions from slopes, uncovered part, leachate collection system, etc. are foreseen. **3b.** No significant fugitive emissions are foreseen—may be as a result of the presence of a gas tight engineered top cover.

An alternative to biocovers is the active in situ aeration where the CH_4 will be oxidized in the landfill and/or the entire landfill body becomes aerobic (see Chapter 16.2).

Table 9.5.1 gives an overview on the different scenarios where a biomitigation system may be implemented. These may lead to the establishment of different biomitigation systems depending on the specific local conditions. In the next section the different types of biomitigation systems are described.

Types of Biomitigation Systems

Table 9.5.2 defines the different biomitigation systems, which can be used for mitigation of CH_4 emissions from landfills. A *full surface biocover* is a landfill cover system that has been designed to optimize the environmental conditions for biological CH_4 consumption, so that the system functions as a vast biofilter. The cover typically consists of a basal "gas distribution layer," with high gas permeability to homogenize LFG fluxes, and an overlying "CH_4 oxidation layer," designed to support the methanotrophic populations that will consume the CH_4. Since biocovers are typically spread over an entire landfill area or landfill cells, cost becomes a critical factor in material selection, and often composted waste materials, such as garden waste, food waste, straw, stabilized dewatered sewage sludge etc., are used.

Table 9.5.2 Different types of systems for biomitigation of CH_4 emissions from landfills

Type	Abbreviations	Description
Full surface biocover	FSB	The whole landfill area is covered with a homogenous layer of bioactive coarse materials (such as a coarse soil or compost) underlain by a gas distribution layer of gravel. Gas is loaded passively to the biocover.
Biowindow system	BWS	A system incorporating the presence of an existing, low permeable soil cover. Areas of the existing cover is replaced by gas permeable, bioactive materials (such as a coarse soil or compost) underlain by a gas distribution layer of gravel. Gas is loaded passively to the biowindows.
Biofilter passive, open bed	BF-PO	A system consisting of a volume of bioactive materials where LFG is fed passively from below through a gas distribution layer. Open to the atmosphere so oxygen can diffuse into the bioactive material from above.
Biofilter passive, closed bed	BF-PC	A system consisting of a volume of bioactive materials where LFG is fed passively from below/above through a gas distribution layer. Closed to the atmosphere (for instance, in a container) so oxygen is to be part of the loading gas.
Biofilter active, open bed	BF-AO	A system consisting of a volume of bioactive materials where LFG is actively pumped from below through a gas distribution layer. The biofilter surface is open to the atmosphere so oxygen can diffuse into the bioactive material from above.
Biofilter active, closed bed	BF-AC	A system consisting of a volume of bioactive materials where LFG is actively pumped from below/above through a gas distribution layer. The biofilter is inclosed (for instance, in a container) so oxygen is to be part of the loading gas (maybe supplied by a second pump).
Bioactive intercepting trench	BIT-PO	A system consisting of a deep trench surrounding the perimeter of a landfill to collect and oxidize CH_4 in LFG migrating horizontally from the landfill. The trench may be filled with gas-distributing materials at the bottom and bioactive materials on top. Gas is routed passively to the trench.
Combined solutions	–	A system combining some of the types above; for instance, a full surface biocover to reduce fugitive emissions with a biofilter treating LFG collected from a gas extraction system.

Another critical factor is the permeability of any interim soil cover below the biocover. If the gas permeability is too low to allow the generated LFG to flow to the biocover, hot spot CH_4 releases may occur at points or areas where less tight soil cover materials have been used (e.g., landfill slopes, etc.).

A *biowindow system* accommodates the problem of an existing low permeable soil cover by construction of areas where the cover soil is replaced by biofilters, so-called biowindows (see Fig. 9.5.3). This option reduces the areas over which the LFG is escaping, leading to lower gas retention times in the filter material. A biowindow system is therefore most relevant at reduced LFG generation rates and at landfills, which are finally covered with relatively gas impermeable materials. For both the biocover and the biowindow, the gas is loaded passively to the filter.

Biofilters, like biocovers, exploit CH_4 oxidizing bacteria to mitigate low concentrated landfill CH_4 emissions. Operated as self-contained fixed bed reactors with a packing material to support and sustain a methanotrophic biofilm, biofilters can accomplish high CH_4 removal rates. Unlike biocovers, biofilters require a supply of gas, which is usually provided by a gas collection or drainage system. The supply can be either passively supported by the elevated gas pressure inside the landfill as a result of the gas generation process, or—more effectively—actively by the use of gas pumps. A biofilter can either be open bed (allowing oxygen diffusion from the atmosphere) or closed bed where the supply gas should contain all relevant gases (both CH_4 and O_2) (see Fig. 9.5.4). The use of closed bed systems may be constrained by the total gas load. At high gas loading rates, the total size of a biofilter has to be quite large, which may give rather costly solutions (Streese and Stegmann, 2003).

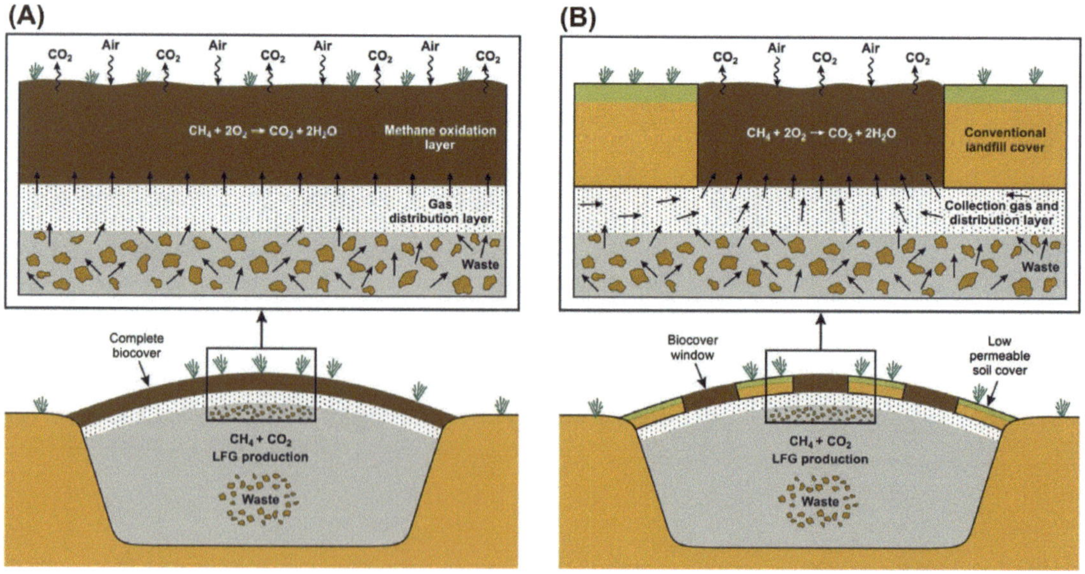

Figure 9.5.3 Illustration of the full surface biocover (A) and the biowindows concepts (B) to facilitate biological CH_4 oxidation, and thereby reduce greenhouse gas emissions from a landfill. *LFG*, landfill gas.

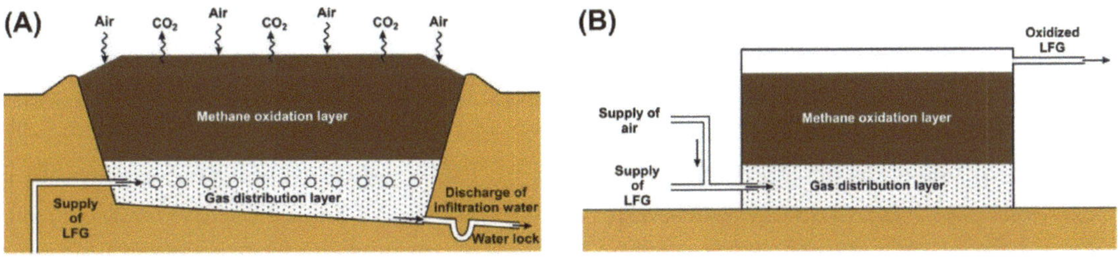

Figure 9.5.4 Two biofilter designs showing the open-bed, upflow (A) and closed-bed, upflow version (B). *LFG*, landfill gas.

Protocol for Establishing a Biomitigation System

Through the work on establishing biomitigation systems at Danish landfills a protocol framework for the establishment has been developed. The framework is shown in Fig. 9.5.5, which presents the different activities in the establishment process. A few comments are given to each of the activities steps in the following section.

Initial Characterization of a Landfill

The objective of this task is to establish the basis for implementation of the biomitigation system based on existing data (such as landfill area, total waste volume, received waste types, waste masses per year, etc.) combined with site visits and gathering of new basic data concerning the existing soil cover of the landfill. The expected gas generation from the landfill is predicted by use of an LFG generation model using the collected data on waste types, volumes, and ages.

Baseline Study of CH_4 Emissions

The objective of this task is to obtain a measurement of the baseline emissions of CH_4 from the landfill in tonnes per year. To evaluate the efficiency of an installed biomitigation system for CH_4 mitigation,

Figure 9.5.5 Description of the biocover system approach with the logical order of project activities (Scheutz et al., 2014).

the CH_4 emission after establishing the biomitigation system is to be compared with the baseline emission. The total CH_4 emission from the landfill can be determined by performance of a series of campaigns, for instance, using the tracer dispersion method (Scheutz et al., 2011b) (see Fig. 9.5.6, Mønster et al., 2014). The task should also evaluate the spatial distribution of fugitive emission from the landfill surface by screening surface CH_4 concentrations using an FID-detector or similar device eventually in combination with flux chamber measurements. Also the presence of point CH_4 source releases such as leachate wells, inspection wells, and gas vents should be evaluated and emissions measured [eventually using a small-scale version of the tracer dispersion method (Fredenslund et al., 2010)]. For scenario 2 and 3 types (Table 9.5.2), annual collected CH_4 by the GCS is determined and the significance of uncollected CH_4 (fugitive emitted and point source released) is evaluated.

Testing Available Bioactive Material

The objective of this task is to identify locally available materials for potential use in the biomitigation system, and test the materials and combinations of materials in the laboratory to determine the CH_4 oxidation capacity of available materials/combination of materials (Pedersen et al., 2011). The bioactive materials could be compost, which in many cases is produced from garden waste and sludge at the landfill. Compost is often produced in large quantities, and it can be difficult to find adequate need for the compost for normal use as soil quality improvement material. The test can be done by batch incubation tests or dynamic column tests simulating the biofiltration process (Pedersen et al., 2011).

Establishing the Full-Scale Biomitigation System

Based on the findings from the three previous tasks the type of biomitigation system is chosen (see Tables 9.5.2 and 9.5.3). Activities are made to reduce any fugitive emissions by improving the existing

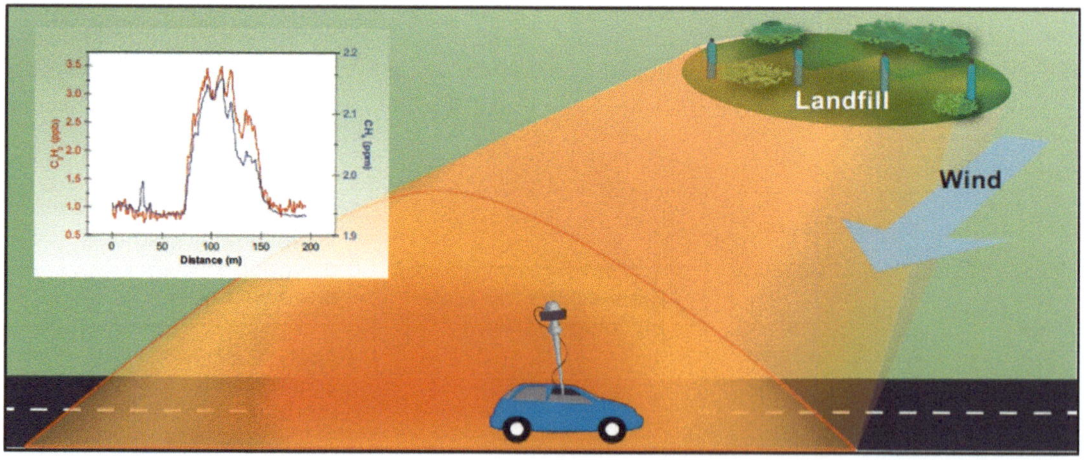

Figure 9.5.6 Schematic overview of the dynamic plume tracer dispersion method. Inserted in the figure is an example of tracer (C_2H_2) and CH_4 plume measurements at a transect downwind the landfill (Mønster et al., 2014).

Table 9.5.3 Compilation of established biomitigation systems at landfills reported in literature

Landfill Area (ha)/Waste Mass (ktonnes)/Type (Table 9.5.1)	Country	Scale[a] FS/PS	System Type[b]	Typical Gas Composition[c] ($\%CH_4/\%O_2$)	Active Material[d] Type/ A(m^2)/ D(cm)	Gas Distribution Layer[e], D(cm)	Total Efficiency Evaluated[f] GL($g/m^{-2}d^{-1}$)/ ME(%)	Measurement Methods Used[g]	Efficiency Approach[h]	Certainty in Efficiency Evaluation (+ → ++++)	References
Aikkala (3.9/200/1a)	SF	FS	FSB (With gas wells)	31–72/1–5	C/39,000/ 50	50	32–216/25–46	FC, PGP	TMMB	++	Einola et al. (2008)
Tulln	A	FS	FSB	n.r./n.r.	C/18,500/ 120–150	50	n.d./42%–96%	MSS, FC	n.r.	+	Amann et al. (2008)
St. Valentin	A	FS	FSB	n.r./n.r.	C/15,000/ 120–150	50	n.d./87%–99%	MSS, FC	n.r.	+	Amann et al. (2008)
Ameis	A	FS	FSB	n.r./n.r.	C/24,400/ 120–150	50	n.d./97%–100%	MSS, FC, PGP	n.r.	+	Amann et al. (2008)
Amstetten	A	FS	BWS	n.r./n.r.	C/1400/ n.r.	n.r.	n.r.	PGP	n.r.	-	Amann et al. (2008)
Pausendorf	A	FS	BWS	41/n.r.	C/1020/ n.r.	n.r.	n.r.	n.r.	n.r.	-	Amann et al. (2008)
Oberaich	A	FS	BWS	55/n.r.	C/182/n.r.	n.r.	n.r.	MSS	n.r.	-	Amann et al. (2008)
Fakse (12/660/1b)	DK	FS	BWS	40–65/0	C/5000/ 100	15	150/28	WLEM, MMS, FC, PGP, SCIC	TEMBA, CMB	++	Rachor et al. (2009)
Klintholm (4/480/1a)	DK	FS	BF-PO	69/0	C/4800/80	30	50/80	WLEM, MSS, FC, PGP	TEMBA	+++	Scheutz et al. (2014)
Landfill x (n.r./n.r./3)	GB	FS	BF-AO	20–40/n.r.	C and CP/ 4x150/130	(30)	530/55–99	MSS, FC, PGP, MF	TMMB, PBE	+++	Parker et al. (2013)
Outer Loop	USA	PS	FSB/BWS	n.r.	C and S/n.r./100	0/15	n.d./21–55	FC, SCIC	LMMB	+	Barlaz et al. (2004)
Leon County (n.r./n.r./1)	USA	PS	BWS	54–68/n.r.	C/2x580/ 30 or 60	15	300–700/20–60	FC, SCIC	Other	++	Bogner et al. (2010)

(*Continued*)

Table 9.5.3 Compilation of established biomitigation systems at landfills reported in literature—cont'd

Landfill Area (ha)/Waste Mass (ktonnes)/Type (Table 9.5.1)	Country	Scale[a] FS/PS	System Type[b]	Typical Gas Composition[c] (%CH_4/%O_2)	Active Material[d] Type/ A(m^2)/ D(cm)	Gas Distribution Layer[e], D(cm)	Total Efficiency Evaluated[f] GL($g/m^{-2}d^{-1}$)/ ME(%)	Measurement Methods Used[g]	Efficiency Approach[h]	Certainty in Efficiency Evaluation (+ → ++++)	References
Leduc and District Regional	CAN	PS	BWS	n.r./n.r.	C/10 or 20/150	80	40–50/20–90	FC, PGP	CMB	++	Dever (2009)
Podere Casa Rota	I	PS	BWS	20–40/<1	C/4x25/70	10	n.d./64–84	FC, PGP	CMB, PBE	+	Pecorini et al. (2013)
Horsley Park	AUS	PS	BIT-PO	49/1.9	C/15/100	250	n.d./>35	FC, PGP	CMB	++	Dever (2009)
Kelso (n.r./3000/1a)	AUS	PS	BF-PO	45–55/0.5	C/4x9/120	50	8–27/20–70	FC, PGP, MF	CMB	+++	Dever (2009), Dever et al. (2007)
St-Nicéphore	CAN	PS	BF-PO	n.r./n.r.	C/23/80	30	175/73–99	FC, PGP	TMMB	+	Capanema et al. (2013)
Landfill Y	CAN	PS	BF-PO	n.r./n.r.	C/100/90	30	n.r./n.r.	FC		-	Zeiss (2006)
Hamburg harbor sl.	DE	PS	BF-PO	n.r./n.r.	CP/17/80	20	125–600/>97%	FC, MF	TMMB	++	Gebert and Gröngröft (2006)
AV Miljø (40/2000/1b)	DK	PS	BF-AO	7/12	C/500/85	40	12/97	MSS, FC, PGP, MF		+++	Cassini et al. (2017) Scheutz et al. (2017)
Wieringermeer	NL	PS	BF-AO	n.r./n.r.	S/510/100	20	43/30–96	FC(large), PGP, MF	TMMB, PBE, CMB	+++	Geck et al. (2013)
2 French landfills	F	PS	BF-AC	2–2.5/18–20	C/17 m^3/ n.r	(30)	620/15–17, max 60	MF, FC		+++	Ducrocq et al. (2013)

[a]FS: Full Scale, PS: Pilot scale.
[b]See Table 9.5.2 for explanation.
[c]Composition of CH_4 and O_2 in LFG.
[d]Active Material Type: S (soil), C (compost matter), CP (expanded clay pellets), O (other), A: area in m^2; D: Layer thickness (cm).
[e]D: thickness of gas distribution layer (cm).
[f]GL: reported average gas load to biocover system; ME: Reported methane removal efficiency.
[g]Methods supporting performance evaluation: WLEM: whole landfill emission measurement using tracer dilution approach or other method; FC: flux chamber measurements; PGP: pore gas profiles; MSS: methane surface screen; SCIC: stable carbon isotope composition; MF: mass flow (gas flow and CH_4 content).
[h]Efficiency approach (see Evaluation of CH4 Oxidation Efficiency of the Biomitigation System): CMB: carbon mass balance; PBE: profile based efficiency; LMMB: local methane mass balance; TMMB: total methane mass balance; TEMBA: total CH_4 emission measurement—before and after.

soil cover or to reduce any hot spot CH_4 releases. For cases where a GCS is in place, uncollected point sources (such as leachate wells) are connected to the GCS and a biofilter solution may be chosen and dimensioned based on the measured CH_4 emission plus CH_4 collected, using the CH_4 oxidation capacity determined by the laboratory experiments (as described earlier) giving the needed filter volume/area.

Evaluation of CH_4 Oxidation Efficiency of the Biomitigation System

The overall objective of this task is to document the CH_4 mitigation efficiency of the established system. After establishment of the biomitigation system, an initial screening of the surface CH_4 emission and the spatial variability is performed on the entire landfill surface by a FID-detector or similar device. If areas with high CH_4 emissions are identified during the screening, the cover properties should be improved to avoid hot spot emissions. The total CH_4 emission from the landfill is determined again (using the method used in the baseline study) and results from the baseline study and the period after the biomitigation system has been established is compared and the CH_4 mitigation efficiency is determined.

Analysis of the Economic Viability of the Biocover Technology

The objective of this task is to evaluate the economic viability of the established biomitigation system. The annual reduction in emission of CO_2 equivalents is predicted for the following years based on the determined mitigation efficiency. The related costs is calculated including construction costs and running costs (for maintaining the biomitigation system), and a normalized price is determined (€/t CO_2-eq. mitigated).

Methodologies for Documentation of Biomitigation System Functionality

In general, reliable quantification of the CH_4 oxidation efficiency of a biomitigation system would require knowledge about the magnitude of the CH_4 fluxes into and out of the cover. However, in many cases such as passively operated biomitigation systems (*full surface biocovers* and *biowindows*) the CH_4 load to the cover is unknown. An exception is biofilters where the LFG is actively pumped from below/above through a gas distribution layer. In these cases the CH_4 load to the biofilter can be determined by measuring the CH_4 concentration and the flow rate in the inlet pipe to the biofilter. In addition, quantification of the CH_4 flux out of the cover is not trivial due to a number of challenges; high spatial and temporal emission variation combined with the relative large landfill area being the most important once. CH_4 surface fluxes measured within short distances (1–2 m) may vary several orders of magnitudes (Bogner et al., 1997; Rachor et al., 2009). Also changes in the atmospheric pressure can cause significant changes in the emission within short time intervals (few hours) (Fredenslund et al., 2010). Obviously obtaining an overall CH_4 flux representative for the whole biomitigation system is a challenging task. In the following different approaches for quantifying the CH_4 oxidation efficiency of a biomitigation system is presented.

Evaluation of CH_4 Oxidation in Biomitigation Systems Using Carbon Mass Balances

The CH_4 oxidation can be determined by using a carbon mass balance approach based on surface flux measurements and cover gas concentrations of CH_4 and CO_2. The method builds on the conservation of mass for carbon (Eqs. 2 and 3) and assumes that the relationship between the fluxes of CH_4 and CO_2, and the concentrations are equal at the bottom of the cover (Eq. 4). Combining Eqs. (2)–(4), results in Eq. (5) and the load of CH_4 into the cover can be calculated. The CH_4 oxidation rate can then be determined by Eq. (6). The method is based on the assumption of steady state and assimilation of carbon by the methanotrophic bacteria is not accounted for (growth equals decay).

$$J_{LFG,bottom} = J_{CO_2,surface} + J_{CH_4,surface} - J_{CO_2,respiration} \tag{2}$$

$$J_{LFG,bottom} = J_{CO_2,bottom} + J_{CH_4,bottom} \tag{3}$$

$$\frac{J_{CH_4,bottom}}{J_{CH_4,bottom} + J_{CO_2,bottom}} = \frac{C_{CH_4,bottom}}{C_{CH_4,bottom} + C_{CO_2,bottom}} \tag{4}$$

$$J_{CH_4,bottom} = \frac{C_{CH_4,bottom}}{C_{CH_4,bottom} + C_{CO_2,bottom}} \cdot (J_{CO_2,surface} + J_{CH_4,surface} - J_{CO_2,respiration}) \tag{5}$$

$$R_{CH_4,oxidation} = J_{CH_4,bottom} - J_{CH_4,surface} \tag{6}$$

$$Eff_{CH_4,oxidation} = 100 \cdot \frac{J_{CH_4,bottom} - J_{CH_4,surface}}{J_{CH_4,bottom}} \tag{7}$$

$J_{LFG,bottom}$ is the load of LFG coming into the biowindow (mole C m^{-2} day^{-1}), $J_{CO_2,respiration}$ is the CO_2 emission originating from aerobic soil respiration (mole C m^{-2} day^{-1}), $J_{CO_2,surface}$ is the CO_2 surface emission (mole C m^{-2} day^{-1}), and $J_{CH_4,surface}$ is the CH_4 surface emission (mole C m^{-2} day^{-1}). $C_{CH_4,bottom}$ and $C_{CO_2,bottom}$ are the concentrations of the CH_4 and CO_2 (% v/v) at the bottom of the biocover/window and/or the raw gas composition. R_{CH_4} oxidation is the CH_4 oxidation rate (mole C m^{-2} day^{-1}) where $J_{CH_4,bottom}$ is the load of CH_4 into the biocover/window (mole C m^{-2} day^{-1}) (Scheutz et al., 2011c). The method has been evaluated in compost column tests (Scheutz et al., 2011d). Even though the carbon mass balance approach has been used in a number of field studies (Christophersen et al., 2001; Scheutz et al., 2011c; Einola et al., 2008, 2009) there are several challenges. First of all, the approach assumes steady state, which might not be true. Also CO_2 is produced and consumed by multiple subsurface and near-surface processes in addition to CH_4 oxidation (such as soil respiration, organic matter oxidation, photosynthesis). Moreover, CO_2 is preferentially partitioned to soil moisture because it is more soluble than CH_4, which make the use of simple CO_2/CH_4 mass ratios uncertain. The surface CO_2 flux due to respiration (Eq. 2) must be determined by laboratory experiments (measurement of CO_2 generation in batch or column incubation test without CH_4 addition) or measured in field reference areas without CH_4 loadings. Neglecting CO_2 generation from respiration could be problematic for compost-based biocover systems (due the high organic content) and would result in an overestimation of the CH_4 oxidation. Finally, the method implies simultaneous measurements of surface fluxes and gas concentration measurements in the cover, which both are labor-intensive to conduct and only represents the CH_4 oxidation in a small area (<1 m^2) and a time period (<10 min). The method is thus most appropriate if applied at smaller scale biomitigation systems such as individual *biowindows* and *passively operated open or closed biofilters,* where the CH_4 load is unknown.

Evaluation of CH_4 Oxidation Efficiency From Soil Gas Profiles

A method based on gas concentrations profiles measured in the cover has been proposed to estimate CH_4 oxidation efficiency in landfill covers, biowindows, or biofilters (Gebert et al., 2011). This profile based efficiency approach assumes that the shift in the ratio of CO_2 to CH_4 in the gas profile, compared to the ratio in the raw LFG, is a result of the oxidation process and thus allows the calculation of the cumulative share of CH_4 oxidized up to a particular depth (Gebert et al., 2011). The method is based on a number of assumptions of which many are similar to those for the carbon mass balance approach. The volume of CH_4 oxidized equals the volume of CO_2 produced, and the share of oxidized CH_4, x, at a certain depth i, can be derived from Eq. (8):

$$\frac{C_{CO_2,\text{bottom}} + x}{C_{CH_4,\text{bottom}} - x} = \frac{C_{CO_2,i}}{C_{CH_4,i}} \tag{8}$$

where x is the share of oxidized CH_4 (vol.%), $C_{CH_4,\text{bottom}}$ and $C_{CO_2,\text{bottom}}$ are the concentrations of CH_4 and CO_2 (% v/v) at the bottom of the biocover/window and/or the raw gas composition, and $CH_{4,i}$ and $CO_{2,i}$ are the concentrations of CH_4 and CO_2 in depth i (% v/v).

According to Eq. (8), the concentration of CO_2 in the LFG plus the share of CO_2 produced by oxidation up to the depth i, divided by the concentration of CH_4 in the LFG minus the share of CH_4 oxidized up to this depth is equal to the ratio of concentrations between the two LFGs at the same depth i. By dividing the share of oxidized CH_4 up to depth i by the concentration of CH_4 in the LFG ($CH_{4,\text{bottom}}$), one obtains the cumulative percentage of CH_4 oxidized, i.e., the cumulated oxidation efficiency, $\text{Eff}_{CH_4,\text{oxidation}}$ (%):

$$\text{Eff}_{CH_4,\text{oxidation}} = \frac{x}{C_{CH_4,\text{bottom}}} \tag{9}$$

The method has been validated by using laboratory soil column data but has not yet been applied in the field. The method is mainly suitable in settings where the CO_2 concentrations are not significantly influenced by processes such as respiration or where CH_4 loadings and oxidation rates are high enough so that CO_2 generated from CH_4 oxidation outweighs other sources of CO_2. Both can be expected for most soil-based biofilters, biowindows, and biocovers on landfills but might not be valid for compost materials with high respiration activities. The method has the same disadvantages as the carbon mass balance method in terms of limited spatial coverage and temporary resolution, which implies that a large number of measurements are needed to obtain reliable CH_4 oxidation efficiencies for full-scale or large-scale biomitigation systems.

Evaluation of CH_4 Oxidation in Biomitigation Systems Using a CH_4 Mass Balance

The CH_4 oxidation efficiency can be determined by setting up a CH_4 mass balance for the system. For biomitigation systems, which are actively operated by feeding in CH_4 to the system such as *biofilters*, the CH_4 load to the system can be monitored by CH_4 recording the CH_4 concentration and the LFG flow to the system. The CH_4 oxidation efficiency can then be determined by comparing the CH_4 inlet load to the cover with the CH_4 surface emission from the cover. For *closed biofilters* the CH_4 out of the system can be monitored in the same way as the CH_4 load into the system. For *open biofilters*, the CH_4 surface emission can be measured using flux chambers. Due the potential risk of missing hotspots in the cover (through which a significant part of the CH_4 might be emitted), a CH_4 screening of the surface is recommended to make sure that these areas are included in the flux measurements. To obtain a reliable overall CH_4 emission of the system a rather high number of flux measurements are needed (Scheutz et al., 2017). Also it is recommended to use a geostatistical tool to interpret the flux data and calculate the total CH_4 emitted from the system.

Evaluation of CH_4 Oxidation by Comparing Before and After Total CH_4 Emission Measurements

The overall performance of a biocover or biowindow system can be done by measuring the total CH_4 emission from the biocover or biowindows system and comparing these results with previous measurements done before installation of the biocover/window system. In this way, the actual total reduction in CH_4 emissions and the oxidation efficiency of the system can be evaluated. The main challenge in measuring CH_4 emissions from landfills is the high temporal and spatial variability of the emissions. The different measurement methods available for quantifying CH_4 emissions from landfills all measure against different scales, from the top layer of a landfill to several kilometers away from the site. Measuring at the surface of the landfill has the advantage of excluding interference from surrounding sources, but it also involves the challenge of having to extrapolate the measurements to a large area with highly heterogeneous emissions. The most commonly applied method is static surface flux chambers. However, surface chambers have a limited surface area (often $<1\ m^2$), and even when taking many measurements, only a very limited area ($<1\%$) of the total landfill area will be covered. In addition, a significant part of the CH_4 emission occurs through small hotspots, and the chances of capturing one of these hotspots with a grid of chamber measurements are very small, thus an underestimation of the total landfill emission is to be expected from application of the surface flux chamber method (Scheutz et al., 2011c; Goldsmith et al., 2012; Green et al., 2010; Babilotte, 2011). One measurement method that has gained ground during the last 5 to 10 years is the mobile tracer dispersion method that combines a controlled release of tracer gas from the landfill with concentration measurements far downwind of the landfill, by using a mobile high-resolution analytical instrument (Fig. 9.5.6). The method is based on the principle that a tracer gas released at the landfill, disperses into the atmosphere likewise the CH_4 emitted from the same area. Since the ratio of their concentrations remains constant along their atmospheric dispersion, the CH_4 emission rate can be calculated when the tracer gas release rate is known (Mønster et al., 2014). The tracer dispersion method has been applied for whole site CH_4 emission quantification at a large number of Danish landfills (Mønster et al., 2015). The method has been applied by comparing before and after installation measurements for documentation of the CH_4 oxidation efficiency at two full-scale biomitigation systems in Denmark (Scheutz et al., 2011c, 2014). The method is recommended for determining the performance of *full surface biocover* and of *biowindows* systems. For a biowindows system the method will give the total CH_4 reduction but will not provide information for the individual windows CH_4 oxidation efficiencies.

OVERVIEW ON BIOMITIGATION SYSTEMS ESTABLISHED IN THE FIELD

During the last 10 years, establishment of several biomitigation systems for reduction of CH_4 emissions at landfills have been reported. The reported systems imply both full-scale systems handling the CH_4 emission from a whole landfill (or landfill cell) and pilot-scale systems only treating the CH_4 from a part of the landfill (or few landfill cells). Table 9.5.3 compiles reported systems giving details in respect to landfill setting, scale, system type, monitoring methods, mitigation efficiency, etc.

In total, 22 biomitigation systems are reported in the table, where 4 are full-scale implementations using four different system types (as defined in Table 9.5.2). Most of the biomitigation systems are using compost type materials as bioactive media in the CH_4 oxidation layer. The largest full-scale project is carried out at the Finnish landfill, Aikkala, where a *full surface biocover* is implemented on the 3.9 ha landfill surface loaded by horizontal gas distribution pipes connected to several gas collection wells. The six Austrian full-scale systems are only reported in general terms and few details are given to the methods used for performance monitoring and establish efficiency documentation. The Klintholm project is a full-scale biomitigation system established on a 4 ha large landfill cell, not equipped with a gas or leachate collection systems. The biomitigation system includes 10 passively loaded, open bed biofilters with a total area of 4800 m^2. The Klintholm system is the only project (besides the Fakse project), which bases the mitigation efficiency evaluation on the TEMBA approach (total emission measurement before and after the system establishment) using the tracer dispersion methodology (see Section *"Evaluation of methane oxidation by comparing before and after total methane emission measurements"*), and has the highest documented efficiency of the reported full-scale systems.

One of the most important challenges of the non-biofilter systems is to determine the gas load to the active biofiltration units, and it can be uncertain to determine the system efficiency in such cases. Often used approaches are the carbon mass balance approach first described in Christophersen et al. (2001)

and the profile based efficiency approach described in Gebert et al. (2011) and Section ***"Methodologies for documentation of bio-mitigation system functionality"***.

Many of the pilot-scale projects are based on either passively or actively loaded open bed biofilters with filter sizes from 10 m^2 up to over 500 m^2. The gas loading rates reported to vary a lot from 8 g CH_4/(m^2 and day) up to more than 700 g CH_4/(m^2 and day), and there seems not to be a clear correlation on gas loading rate to mitigation efficiency. It is clear that there is still a lot to learn on the most important environmental factors governing the efficiency of bio-mitigation systems.

Two of the biomitigation systems differ from the others; the Horsley Park project is of the type "Interception trench", with the main aim of reducing off-site gas migration. Another case is the last one presented in Table 9.5.3, which is a closed bed biofilter system carried out at two French landfills. The biofilters are treating very dilute LFG [CH_4 content of 2%–2.5% (vol.)] but still with high loading rates resulting in relatively low mitigation efficiencies. The efficiencies were constrained by low temperatures during winter and desiccation leading to by-passing gas flow in the biofilter material.

CHALLENGES IN BIOMITIGATION SYSTEMS AND FUTURE RESEARCH NEEDS

The experiences with research and demonstration activities in relation to biomitigation of LFG emissions have shown that there exist potentially effective biocover designs, which may be implemented in full-scale to reduce CH_4 emissions from landfills depending on the existing LFG management. The use of compost-based materials seems to be of advantage especially in regions with cold climate during winter time. In many projects, elevated temperatures to ambient temperatures have been observed in the CH_4 oxidation layer due to a self-heating process, which has resulted in high oxidation rates even during cold periods. Projects have, however, also shown that the compost is further decomposed over time due to the on-going respiration process, and longevity issues may arise due to decreasing compost permeability and diffusivity potentially deteriorating the functionality of the biofilter units.

Another challenge, which has been observed in many projects, is to obtain an even distribution of gas load to the biofilter unit whenever it is part of a full surface biocover, a biowindow system or a biofilter-based system. Research is needed to evaluate different gas distribution systems and to identify the systems, which gives the most cost-efficient solution to the gas distribution issue. For open bed systems (including full surface biocovers and biowindows) another challenge is to manage the water infiltration, which will take place as a result of precipitation. There are reports on problems with water saturated layers blocking the vertical gas transport maybe due to capillary effects in the interface between the CH_4 oxidation layer and the gas distribution layer. In the AV Miljø project (Cassini et al., 2017; Scheutz et al., 2017), a system was implemented to reduce such capillary effect, which seemed to work. However, more research is needed to avoid capillary effects.

The physical, chemical, and microbial processes, which govern the efficiency of a biomitigation system, are highly interconnected, and there is a need for developing advanced mathematical model for predicting and simulating the performance of biomitigation systems to be able to fine-tune the systems for obtaining a higher mitigation efficiency of the systems.

References

Amann, A., Bogolte, T., Santos, M.D., Hagenauer, I., Huber-Humer, M., Pauliny, W., Reichenauer, T., Watzinger, A., Wimmer, B., 2008. Leitfaden – Methanoxidationsschichten. Österreichischer verein für altlastenmanagement, Wien.

Babilotte, A., 2011. Field Comparison of Methods for Assessment of Methane Fugitive Emissions from Landfills. Report prepared for the Environmental Research & Education Foundation. http://www.erefdn.org/publications/uploads/Fugitive-Emissions_FinalReport.pdf.

Barlaz, M.A., Green, R.B., Chanton, J.P., Goldsmith, C.D., Hater, G.R., 2004. Evaluation of a biologically active cover for mitigation of landfill gas emissions. Environmental Science and Technology 38, 4891–4899.

Bogner, J., Meadows, M., Czepiel, P., 1997. Fluxes of methane between landfills and the atmosphere: natural and engineered controls. Soil Use and Management 13, 268–277.

Bogner, J., Pipatti, R., Hashimoto, S., Diaz, C., Mareckova, K., Diaz, L., Kjeldsen, P., Monni, S., Faaij, A., Qingxian, Q., Tianzhu, Z., Mohammed, A.A., Sutamihardja, R.T.M., Gregory, R., 2008. Mitigation of global greenhouse gas emissions from waste: conclusions and strategies from the intergovernmental panel on climate change (IPCC) fourth assessment report. Working group III (mitigation). Waste Management & Research 26 (1), 11–32.

Bogner, J.E., Chanton, J.P., Blake, D., Abichou, T., Powelson, D., 2010. Effectiveness of a Florida landfill biocover for reduction of CH_4 and NMHC emissions. Environmental Science & Technology 44 (4), 1197–1203.

Capanema, M.A., Ndanga, E., Lakhouit, A., Cabral, A.R., 2013. Methane oxidation efficiencies of a 6-year-old experimental landfill biocover. In: Proceedings Sardinia 203, Fourteenth International Waste Management and Landfill Symposium S. Margherita di Pula, Cagliari, Italy; 30 September – 4 October 2013. CISA, Environmental Sanitary Engineering Centre, Italy.

Capanema, M.A., Cabana, H., Cabral, A.R., 2014. Reduction of odours in pilot-scale landfill biocovers. Waste Management 34 (4), 770–779.

Cassini, F., Scheutz, C., Skov, B.H., Zishen, M., Kjeldsen, P., 2017. Mitigation of methane emissions in a pilot-scale biocover system at the AV Miljø Landfill, Denmark: 1. System design and gas distribution. Waste Management 63, 213–225.

Christensen, T.H., Kjeldsen, P., 1995. Landfill emissions and environmental impact: an introduction. In: Christensen, T.H., Cossu, R., Stegmann, R. (Eds.), SARDINIA '95, Fifth International Landfill Symposium, Proceedings, Volume III. CISA. Cagliari, Italy, pp. 3–12.

Christophersen, M., Holst, H., Chanton, J., Kjeldsen, P., 2001. Lateral gas transport in soil adjacent to an old landfill: factors governing emission and methane oxidation. Waste Management & Research: The Journal of the International Solid Wastes and Public Cleansing Association, ISWA 19, 595–612.

Dever, S.A., March 30, 2009. Passive Drainage and Biofiltration of Landfill Gas: "Behaviour and Performance in a Temperate Climate" (Dissertation 1–448). School of Civil and Environmental Engineering, University of New South Wales, Sydney, Australia.

Dever, S.A., Swarbrick, G.E., Stuetz, R.M., 2007. Passive drainage and biofiltration of landfill gas: Australian field trial. Waste Management 27, 277–286.

Dever, S.A., Swarbrick, G.E., Stuetz, R.M., 2011. Passive drainage and biofiltration of landfill gas: result of Australian field trial. Waste Management 31, 1029–1048.

Ducrocq, M., Åkerman, A., Horlaville, C., Riquier, L., 2013. Landfill gas treatment by active biofiltration: feedback from biofilter pilot trials on three French landfills. In: Proceedings Sardinia 203, Fourteenth International Waste Management and Landfill Symposium S. Margherita di Pula, Cagliari, Italy; 30 September – 4 October 2013. CISA, Environmental Sanitary Engineering Centre, Italy.

EEA (European Environment Agency), 2008. Annual European Community Greenhouse Gas Inventory 1990–2006 and Inventory Report 2008. Submission to the UNFCCC Secretariat. EEA Technical Report, No. 6/2008. ISSN: 1725-2237.

Einola, J.K.M., Sormunen, K.M., Rintala, J.A., 2008. Methane oxidation in a boreal climate in an experimental landfill cover composed from mechanically–biologically treated waste. The Science of the Total Environment 407, 67–83.

Einola, J., Sormunen, K., Lensu, A., Leiskallio, A., Ettala, M., Rintala, J., 2009. Methane oxidation at a surface-sealed boreal landfill. Waste Management 29 (7), 2105–2120.

Fredenslund, A.M., Scheutz, C., Kjeldsen, P., 2010. Tracer method to measure landfill gas emissions from leachate collection systems. Waste Management 30, 2146–2152.

Gebert, J., Gröngröft, A., 2006. Performance of a passively vented field-scale biofilter for the microbial oxidation of landfill methane. Waste Management 26, 399–407.

Gebert, J., Rower, I.U., Scharff, H., Roncato, C.D., Cabral, A.R., 2011. Can soil gas profiles be used to assess microbial CH_4 oxidation in landfill covers? Waste Management 31 (5), 987–994.

Geck, C., Gebert, J., Röwer, I.U., Scharff, H., Pfeiffer, E.M., 2013. Assessment of the efficiency of a methane oxidation biocover test field. In: Proceedings Sardinia 203, Fourteenth International Waste Management and Landfill Symposium S. Margherita di Pula, Cagliari, Italy; 30 September – 4 October 2013. CISA, Environmental Sanitary Engineering Centre, Italy.

Goldsmith, C.D., Chanton, J., Abichou, T., Swan, N., Green, R., Hater, G., 2012. Methane emissions from 20 landfills across the United States using vertical radial plume mapping. Journal of the Air & Waste Management Association 62, 183–197.

Green, R.B., Hater, G.R., Thoma, E.D., Dewees, J., Rella, C.W., Crosson, E.R., Goldsmith, C.D., Swan, N., 2010. Methane emissions measured at two California landfills by OTM-10 and an acetylene tracer method. In: Global Waste Management Symposium. San Antonio, TX, USA.

He, R., Xia, F.-F., Bai, Y., Wang, J., Shen, D.-S., 2012. Mechanism of H_2S removal during landfill stabilization in waste biocover soil, an alterative landfill cover. Journal of Hazardous Materials 217, 67–75.

Kjeldsen, P., Dalager, A., Broholm, K., 1997. Attenuation of methane and non methane organic compounds in landfill gas affected soils. Journal of the Air & Waste Management Association 47, 1268–1275.

Lakhouit, A., Schirmer, W.N., Johnson, T.R., Cabana, H., Cabral, A.R., 2014. Evaluation of the efficiency of an experimental biocover to reduce BTEX emissions from landfill biogas. Chemosphere 97, 98–101.

Laner, D., Crest, M., Scharff, H., Barlaz, M.A., 2012. A review of approaches for the long-term management of municipal solid waste landfills. Waste Management 32, 498–512.

Lee, E.H., Park, H., Cho, K.S., 2010. Characterization of methane, benzene and toluene-oxidizing consortia enriched from landfill and riparian wetland soils. Journal of Hazardous Materials 184 (1–3), 313–320.

Menard, C., Ramirez, A.A., Nikiema, J., Heitz, M., 2012. Biofiltration of methane and trace gases from landfills: a review. Environmental Reviews 20 (2), 152.

Mønster, J.G., Samuelsson, J., Kjeldsen, P., Rella, C.W., Scheutz, C., 2014. Quantifying methane emission from fugitive sources by combining tracer release and downwind measurements – a sensitivity analysis based on multiple field surveys. Waste Management 34, 1416–1428.

Mønster, J., Samuelsson, J., Kjeldsen, P., Scheutz, C., 2015. Quantification of methane emission from 15 Danish landfills using mobile tracer dispersion method. Waste Management 35, 177–186.

Morris, J.W.F., Crest, M., Barlaz, M.A., Spokas, K.A., Åkerman, A., Yan, L., 2012. Improved methodology to assess modification and completion of landfill gas management in the aftercare period. Waste Management 32, 2364–2373.

Parker, T., Childs, A., Pointer, P., Baker, D., Johnson, T., Wilson, W., 2013. Lessons learned from first full size methane oxidation biofilter in the UK. In: Proceedings Sardinia 203, Fourteenth International Waste Management and Landfill Symposium S. Margherita di Pula, Cagliari, Italy; 30 September – 4 October 2013. CISA, Environmental Sanitary Engineering Centre, Italy.

Pecorini, I., Bacchi, D., Frassinetti, L., Olivieri, F., 2013. Long term behaviour of pilot biocovers for microbial methane oxidation and odours mitigation. In: Proceedings Sardinia 203, Fourteenth International Waste Management and Landfill Symposium S. Margherita di Pula, Cagliari, Italy; 30 September – 4 October 2013. CISA, Environmental Sanitary Engineering Centre, Italy.

Pedersen, G.B., 2010. Processes in a Compost Based Landfill Biocover; Methane Emission, Transport and Oxidation (Ph.D. Thesis). Department of Environmental Engineering, Technical University of Denmark.

Pedersen, G.B., Scheutz, C., Kjeldsen, P., 2011. Availability and properties of materials for the Fakse landfill biocover. Waste Management 31 (5), 884–894.

Philopoulos, A., Felske, C., McCartney, D., 2008. Field-scale treatment of landfill gas with a passive methane oxidizing biofilter. Journal of Environmental Engineering and Science 7 (5), 531–542.

Plaza, C., Xu, Q., Townsend, T., Bitton, G., Booth, M., 2007. Evaluation of alternative landfill cover soils for attenuating hydrogen sulfide from construction and demolition (C&D) debris landfills. Journal of Environmental Management 84 (3), 314–322. https://doi.org/10.1016/j.jenvman.2006.06.001.

Rachor, I., Streese-Kleeberg, J., Gebert, J., 2009. Spatial and temporal variability of gas emissions from old landfills. In: Proceedings Sardinia 2009, Twelfth International Waste Management and Landfill Symposium, S. Margherita di Pula, Cagliari, Italy; 5–9 October 2009.

Scheutz, C., Kjeldsen, P., 2003. Capacity for biodegradation of CFCs and HCFCs in a methane oxidative counter gradient flow system simulating landfill soil covers. Environmental Science & Technology 37, 5143–5149.

Scheutz, C., Kjeldsen, P., 2004. Environmental factors influencing methane oxidation and co-oxidation of HCFCs in landfill cover soils. Journal of Environmental Quality 33, 72–79.

Scheutz, C., Kjeldsen, P., 2005. Biodegradation of trace gases in simulated landfill soil cover systems. Journal of the Air & Waste Management Association 55 (7), 878–885.

Scheutz, C., Mosbæk, H., Kjeldsen, P., 2004. Attenuation of methane and volatile organic compounds in landfill soil covers. Journal of Environmental Quality 33, 61–71.

Scheutz, C., Kjeldsen, P., Bogner, J., De Visscher, A., Gebert, J., Hilger, H., Huber-Humer, M., Spokas, K., 2009. Processes and technologies for mitigation of landfill gas emissions by microbial methane oxidation. Waste Management & Research 27 (5), 409–455.

Scheutz, C., Pedersen, G.B., Costa, G., Kjeldsen, P., 2009. Biodegradation of methane and halocarbons in simulated landfill biocover systems containing compost materials. Journal of Environmental Quality 38 (4), 1363–1371.

Scheutz, C., Meroño, A.R., Pedersen, R.B., Kjeldsen, P., 2011a. Temperatures in compost landfill covers as result of methane oxidation and compost respiration. In: Proceedings Sardinia 2011, Thirteenth International Waste Management and Landfill Symposium S. Margherita di Pula, Cagliari, Italy; 3–7 October 2011. CISA, Environmental Sanitary Engineering Centre, Italy.

Scheutz, C., Samuelsson, J., Fredenslund, A.M., Kjeldsen, P., 2011b. Quantification of multiple methane emission sources at landfills using a double tracer technique. Waste Management 31 (5), 1009–1017.

Scheutz, C., Fredenslund, A.M., Chanton, J., Pedersen, G.B., Kjeldsen, P., 2011c. Mitigation of methane emission from Fakse landfill using a biowindow system. Waste Management 31, 1018–1028.

Scheutz, C., Pedicone, A., Pedersen, G.B., Kjeldsen, P., 2011d. Evaluation of respiration in compost landfill biocovers intended for methane oxidation. Waste Management 31, 895–902.

Scheutz, C., Cassini, F., De Schoenmaeker, J., Kjeldsen, P., 2017. Mitigation of methane emissions in a pilot-scale biocover system at the AV Miljø Landfill, Denmark: 2. Methane oxidation. Waste Management 63, 203–212.

Scheutz, C., Pedersen, R.B., Petersen, P.H., Jørgensen, J.H.B., Ucendo, I.M.B., Mønster, J.G., Samuelsson, J., Kjeldsen, P., 2014. Mitigation of methane emission from an old unlined landfill in Klintholm, Denmark using a passive biocover system. Waste Management 34, 1179–1190.

Streese, J., Stegmann, R., 2003. Microbial oxidation of methane from old landfills in bio-filters. Waste Management 23 (7), 573–580.

Sungthong, D., Reinhart, D.R., 2011. Control of hydrogen sulfide emissions using autotrophic denitrification landfill biocovers: engineering applications. Frontiers of Environmental Science and Engineering in China 5 (2), 149–158.

US EPA, 2013. Inventory of U.S. Greenhouse Gas Emissions and Sinks, pp. 1990–2011. http://www.epa.gov/climatechange/Downloads/ghgemissions/US-GHG-Inventory-2013-Main-Text.pdf.

Whalen, S.C., Reeburgh, W.S., Sandbeck, K.A., 1990. Rapid methane oxidation in a landfill cover soil. Applied and Environmental Microbiology 56, 3405–3411.

Xu, Q., Townsend, T., Reinhart, D., 2010. Attenuation of hydrogen sulfide at construction and demolition debris landfills using alternative cover materials. Waste Management 30 (4), 660–666.

Zdeb, M., Pawlowska, M., 2013. Mitigation of BTEXs and p-cymene emission from municipal solid waste landfills by biofiltration-preliminary results of field experiment. Environmental Engineering IV – Proceedings of the Conference on Environmental Engineering IV 275–282.

Zeiss, C.A., 2006. Accelerated methane oxidation cover systems to reduce greenhouse gas emissions from MSW landfills in cold-semi-arid regions. Water, Air, and Soil Pollution 176, 285–306.

10 CHAPTER

LEACHATE MANAGEMENT

10.1

STRATEGIC ISSUES IN LEACHATE MANAGEMENT

Rainer Stegmann

INTRODUCTION

Sanitary landfill leachate is a strongly polluted wastewater with a variety of components. Its quality is the result of biological, chemical, and physical processes in landfills combined with the specific waste composition and the landfill water regime. With increasing leachate effluent quality standards the efforts for leachate treatment also increase. Owing to the development of a landfill in size and age, leachate amount and quality change with time, which makes it different from sewage and other kinds of wastewater.

To plan and design a leachate treatment plant, the leachate production rates and the leachate quality with time have to be known or estimated. If a landfill is planned, also the leachate treatment plant has to be designed not knowing at that stage which will be the quality and quantity. In addition a landfill "grows" with time in size and height. "fresh" waste covers already deposited waste. With increasing operation time of a landfill the "ages"—that means the different stages of biological degradation—of the landfilled waste will become more and more different. With time, also the kinds or proportions (inorganic: organic) of landfilled waste may change, because of areas in the landfill where biological degradation has already taken place to a high degree while in other landfill sections the waste has not changed at all. There is a strong relationship between the age of the landfilled waste and the leachate quality (see Chapter 3.1). The described situation makes a prediction of leachate design values very difficult. For this reason, value ranges (see Chapter 10.2) are useful. Helpful may be the investigation of leachate data from already existing landfills in this specific area.

In some cases, leachate concentrations from a young landfill may be relatively low. In this case a dilution with surface or groundwater may occur. This effect can be verified when the BOD_5:COD ratio is calculated. If the ratio is in the range of 0.4—0.5, it can be assumed that leachate from a young landfill is diluted.

Due to the variation of the leachate production rate and quality in the lifetime of a landfill, leachate treatment plants should be built in a way that this situation is taken into account. This means that the biological treatment plant may have to be enlarged with the extension of the landfill. After closure the biological treatment step may be modified for nitrogen removal or be redundant so that only physical/chemical treatment may take place. This means that the leachate treatment plant should be planned and built in a flexible way allowing enlargement/addition of the biological treatment as well as the chemical/physical step. In addition, pump connections and pipe diameter as well as the option for enlarging the buildings also for future use should be enclosed in the initial design. When parts of the plant are redundant, potential future use should be envisaged.

LANDFILL LEACHATE DISCHARGE LIMITS IN DIFFERENT COUNTRIES (EXAMPLES)

The EU Landfill Directive stipulates that leachate has to be treated properly before it is discharged to surface waters, but the directive does not provide any concrete discharge limit values. The EU member states are individually in charge to do that. As examples, some discharge limit values of the major target components are presented in Table 10.1.1.

The regulations in European countries may vary from more general regulations to site-specific considerations. The leachate discharge values for Germany are presented in Table 10.1.2.

If cotreatment of leachate in a sewage treatment plant is planned, pretreatment of the leachate will in most cases be necessary. The abovementioned 51 Anhang requires at least 75% dissolved organic carbon (DOC) degradation or a COD <400 mg/L. In addition, three different toxicity tests (fish-, daphnien-, and luminous bacteria test) have to be passed and the local authority may require additional limit values.

Leachate limit values in Italy for direct and indirect discharge are presented in Table 10.1.3. In this case, leachate is considered an industrial wastewater.

Using only biological treatment in most cases the required COD effluent concentrations for direct discharge cannot be reached. In this case an additional chemical/physical treatment step will be necessary.

In the United Kingdom the required discharge limit values are individually prescribed based on the results of a risk assessment (Table 10.1.4).

Table 10.1.1 Discharge limit values for BOD_5, COD, and ammonium in different countries (mg/L)

	Germany	France	Italy	China	Japan	Australia
			Direct			Indirect
COD	200	120	160	100	90	1500
BOD_5	20	30	40	30	60	600
TN (NH_4^+–N)	70 (NR)	30 (NR)	NR (15)	40 (25)	60 (60)	150 (100)

Modified from Wang (2013).

Table 10.1.2 Limit values for the discharge of treated leachate according to German standards (Anonymus, 1996)

Parameter	Concentrations (mg/L)
COD	200
BOD_5	20
Total nitrogen ($NH_4 + NO_2 + NO_3$)	70
Total phosphorous	3
Hydrocarbons	10
Nitrite–Nitrogen	2
AOX	0.5
Mercury	0.05
Cadmium	0.1
Chromium	0.5
Chromium (VI)	0.1
Nickel	1
Lead	0.5
Copper	0.5
Zinc	2
Cyanide, easy free settable	0.2
Sulfide	1

Table 10.1.3 Wastewater discharge limits in Italy

Parameter	Dimension	Discharge Into Water Body	Discharge Into Sewer System
BOD_5	mg/L	40	250
COD	mg/L	160	500
Suspended solids	mg/L	80	200
$N-NH_4$	mg/L	15	30
$N-NO_2$	mg/L	0.6	0.6
$N-NO_3$	mg/L	20	30
Total P	mg/L	10	10
Cadmium	mg/L	0.02	0.02
Active chlorine	mg/L	0.2	0.3
Arsenic	mg/L	0.5	0.5
Chrome VI	mg/L	0.2	0.2
Iron	mg/L	2	4

Modified from Anonymous (2006).

Table 10.1.4 Leachate benchmark values in mg/L for different types of treatment processes in UK (Anonymous, 2007)

Released Substance	Benchmark Values		
	SBR	MBR	RO
BOD_5	40	5	15
Ammoniacal–N	4	2	10
Nitrate–N	1500	150	1

The emissions to water are associated with the best available technologies: *MBR*, membrane bioreactor; *RO*, reverse osmosis; *SBR*, sequencing batch reactors. When automatic sampling systems are employed, limits may be defined such that not more than 5% of samples shall exceed the benchmark value. When spot samples are taken no spot sample shall exceed the benchmark value by more than 50%.

LEACHATE TREATMENT PLANT DESIGN AID

Leachate quality

Mainly two different phases can be identified in landfills during the anaerobic decomposition of waste: acid phase that causes a decrease of pH in the leachate but high concentrations of organic acids and inorganic ions (for example, Cl^-, SO_4^{2-}, Ca^{2+}, Mg^{2+}, Na^+) and the methanogenic phase. Heavy metal concentrations are in general comparatively low. Leachate from the acid phase is therefore characterized by high BOD_5 values (commonly > 10.000 mg/L), high BOD_5/COD ratios (commonly > 0.4) and acidic pH values (typically 5−6). Further information on the biological degradation processes in the landfill can be found in Chapter 3.1 and also elsewhere (Stegmann and Spendlin, 1989).

The stable methanogenic phase (see also Chapters 3.1 and 10.2) of anaerobic degradation is characterized by a pH range from 6 to 8. At this stage, the composition of leachate is characterized by relatively low BOD values and low ratios of BOD/COD ratios. Ammonia continues to stay at a relatively high level.

Kruse (1994) defined three characteristic periods according to the BOD_5/COD ratio:

Acid phase	$BOD_5/COD \geq 0.4$
Transient phase	$0.4 > BOD_5/COD > 0.2$
Methanogenic phase	$BOD_5/COD \leq 0.2$

Leachate quality varies due to waste composition, the degree of decomposition with time (age of the landfill), the climate (temperature, rainfall), degree of compaction, and the way of operation. A fast built up landfill section with raw waste in a moderate climate will produce a high strength leachate (high BOD and COD) while "bioreactor landfill" and landfills receiving mechanical biological pretreatment (MBT) waste will produce lower organically concentrated leachate. Table 10.1.5 shows some tendencies and estimations regarding different influences on the leachate quality.

As a design aid, the following concentration ranges may be helpful (Table 10.1.6)

In addition to the organic compounds leachate consists of a wide range of salts and other inorganic and organic compounds (some of those in trace concentrations, but due to their potential toxicity of importance, see Chapter 10.2).

Leachate quantity

Leachate production is the result of precipitation, evaporation, surface runoff, infiltration, waste storage capacity etc. (see Chapter 6.1).

Under middle European climatic conditions the measured leachate production rates are often between 15% and 25% of the annual precipitation. Values lower than 10% are mainly from very young landfills in

Table 10.1.5 Factors of influence on tendencies in leachate concentration with time based on average conditions

Landfill Type	BOD	COD	NH_4-N	Cl	Degradation of Organics in Waste
Fast build up in height, high compaction	XXX	XXX	XXX	XXX	Slow
Mechanical biological pretreatment waste	X	XX	XX	XX	Fast/Medium
Bioreactor landfill	XX	XX	XXX	XXX	Fast/Medium
Low compaction, high precipitation	XX	XX	XX	XX	Fast/Medium
High inert waste ratio (>50%)	X/XX	XX	XX	XX/XXX	Fast/Medium
High share of food waste (>50%)					
High precipitation	XXX	XXX	XXX	XXX	Fast/Medium
Low precipitation	XXX	XXX	XXX	XXX	Slow/Medium
Landfill in methanogenic stage	X	XX	XXX	XX	Slow

X, low; *XX*, medium; *XXX*, high.
Compaction: Low: <0.8 t/m^3; High: >1.0 t/m^3
Precipitation: Low: <600 mm/a; High: >1500 mm/a.
Different factors can be combined e.g., MBT waste with high or low precipitation input and/or high or low compaction, which will result in higher or lower decrease of organic concentrations.

Table 10.1.6 Design aids for leachate treatment plants (mg/L)

	BOD_5	COD	NH_4	Cl
High	>15,000	>30,000	>1,000	>2,500
Medium	±1,000	±5,000	±1,000	±2,000
Low	<200	<1,000	<1,000	<2,000

High, acidogenic phase; *low*, methanogenic phase; *medium*, transient phase.

operation with a high sorption potential. Values for leachate production above 25% (up to 50%) of the annual precipitation rate can be reached at landfills where the storage capacity has been widely used.

Of course under other climatic conditions leachate production rates are different. The "HELP model" or other models as the "Bohawald model" may be used for roughly estimating leachate production rates (see Chapters 6.1, 6.2). When using the models it should be proven if they are adapted to the specific region.

Under middle European conditions surface runoff is in general low. In countries with high precipitation rates (>1200 mm/a) surface runoff (especially after strong rainfall events) may be an issue (dependent also on the shape of the landfill). Even if the average annual evaporation rate is much higher than the precipitation rate (e.g., under arid climatic conditions), leachate production may occur especially when there are few but strong rainfall events.

If there are rainy and dry seasons (i.e., in tropical countries) leachate may be stored during the wet seasons in lagoons with some biological treatment and be recirculated to the landfill surface in the dry season. In this way also the biological processes in the landfill body are supported.

Leachate treatment options dependent on leachate quality

To reach the target values biological treatment is often combined with a chemical/physical treatment step. This is necessary because after the biological treatment there are still "non"biological degradable organics present that consist mainly of humic and fulvic like acids. These large molecules may only be further degraded after a chemical oxidation step, e.g., using ozone often in combination with UV radiation (sometimes using Fenton reagent), which by itself reduces these organic compounds to a certain extent. Otherwise these humic and fulvic like acids are reduced in the leachate by means of adsorption on activated carbon, or separation using flocculation/precipitation (not often) and/or membrane separation (nanofiltration and/or reverse osmosis) (Fig. 10.1.1).

Figure 10.1.1 Examples of raw leachate from a young landfill, biologically treated leachate, and chemically/physically treated leachate (from left to right).

In the case of biological treatment of leachate from old landfills with a low BOD the main aim is the reduction of nitrogen by nitrification/denitrification. At that stage the design values for nitrification/denitrification and not for organic substrate removal should be used (see Chapter 10.3). But in this situation denitrification can only be achieved when an external easy degradable organic substrate is added. In some cases leachate recirculation may be an option for denitrification, where the nitrates are reduced in the landfill body (see Chapters 12.1, 12.2).

If the leachate from old landfills has to be treated, chemical/physical treatment only may be an option for ammonia reduction, e.g., using membrane technology. Increasing the pH to convert NH_4 in the gaseous NH_3 form may also strip NH_4. This process may be considered at very large older landfills, but since it is not easy to control it is not often used (i.e., NENT Landfill Hong Kong, Robinson et al., 2007).

Leachate treatment can also be achieved by evaporation. In this case salts and other non-volatile inorganic and organic compounds will remain. This process is very energy intensive and also not often used.

A relatively new development is the membrane bioreactor, where the biological treatment and membrane filtration are combined. This means that the biologically treated leachate has to pass the membrane, so that also the nondegradable organics and other compounds do not appear in the effluent. By these means the sludge concentration as well as the sludge age in the bioreactor can be increased, which may result in a further reduction of not easy degradable organics. Of course, all the aspects regarding membrane filtration have to be respected (fouling etc.).

The main problem when using membrane technology is the discharge of the concentrate. In some cases, it is recirculated back to the landfill, it is solidified and landfilled, it is mixed with municipal solid waste (MSW) and incinerated in an MSW incinerator, it may be evaporated and the solids are landfilled at an HW landfill or it is otherwise treated and discharged (see also Chapter 10.4).

To reduce the main leachate pollutants not only in developing countries aerobic lagoons may be used.

PROBLEMS ASSOCIATED WITH LEACHATE TREATMENT

In general the phosphorous content in leachate is relatively low so has to be investigated if especially for the aerobic treatment the addition of phosphorous will be necessary. The relation between BOD_5:N:P should be in the range of 100:5:1. P can be added, i.e., into the influent pump shaft. Nitrogen is in an access available.

When dealing with leachate there is always a potential of the precipitation of $CaCO_3$ and Fe^{3+} that may result in clogging of pipes due to incrustations (Ramke, 1989). Crusts may also cover submersible pumps, so that they may get overheated. Clogging may also may happen on the surface of aeration aggregates (see also Chapter 8.1).

In cold climate due to long detention times, especially in lagoons, the leachate may freeze (Fig. 10.1.2). This may be taken into account by providing storage capacities to avoid the discharge of not sufficiently treated leachate.

Figure 10.1.2 Cascade-type aerated lagoon system at freezing temperatures (Plant Lingen).

If the bacteria are under stress they may produce foam. This may occur when there is a lack of phosphorous, an overloading or some toxic effects.

Odors from leachate can be avoided by leachate aeration to avoid/degrade the odorous compounds. Aeration of the leachate prior to recirculation may also be helpful to provide a controlled precipitation to avoid/reduce clogging of leachate transportation pipes.

The cost for leachate treatment are very much dependent on the target values that have to be reached, if high quality standards have to be met, the costs may be around $\pm20-40$ €/m^3.

In Chapters 6.1 and 10.1, 10.2, 10.3, 10.4, 10.5 details on leachate production, quality, and treatment are presented.

References

Anonymous, 2006. D.lgs 152/2006 (Italian Environmental Legislation).

Anonymous, February 2007. Guidance for the Treatment of Landfill Leachate in the UK. Environment Agency (Note IPPC S5.03).

Anonymus, 1996. 51.Anhang Rahmen-AbwasserVwA. Bundesministerium der Justiz und Verbraucherschutz.

Kruse, K., 1994. Langfristiges Emissionsgeschehen von Siedlungsabfalldeponien, Veröffentlichungen des Instituts für Siedlungswasserwirtschaft. TU Braunschweig (Heft 54).

Ramke, H.G., 1989. Leachate collection systems. In: Christensen, T.H., Cossu, R., Stegmann, R. (Eds.), Sanitary Landfilling, Process, Technology and Environmental Impact. Academic Press.

Robinson, H., Howard, D., Carville, M., Martin, S., 2007. The management of gas and leachate at tropical landfill sites. In: CIWEM Annual Conference 2007, Sustainable Waste Management.

Stegmann, R., Spendlin, H., 1989. Enhancement of degradation: German experiences. In: Christensen, Cossu, Stegmann (Eds.), Sanitary Landfilling, Process, Technology and Environmental Impact. Academic Press.

Wang, Y., 2013. Leachate Management in the Aftercare Period of Municipal Waste Landfills (Doctoral thesis). Aalto University.

10.2

LEACHATE QUALITY

Hans-Jürgen Ehrig and Rainer Stegmann

INTRODUCTION

Leachate quality is the result of the waste composition, water budget, and the biological, chemical, and physical conditions in the landfill body. Until now it is not possible to predict the leachate composition of a new landfill, and it is very difficult to describe the future leachate pollution of an existing landfill. As a consequence of the high organic waste content, uncontrolled anaerobic biological processes develop during the first up to 10 years of landfill operation, which—with the influence of chemical/physical processes—determine the leachate quality particularly the organic carbon content of leachate. In addition, these processes influence the pH and the mobility of a variety of other compounds as iron, calcium, etc. Reducing the period of the anaerobic phase in a landfill results in a significant reduction of the leachate pollution. After the end of landfill operation, the leachate pollution will remain at high levels over a very long period of time (decades and longer) with only a slow decrease.

Leachate is a specific wastewater with a wide range of variation in quality and quantity. The leachate production rate is a result of climatic conditions (e.g., precipitation, evaporation) and landfill specifics (e.g., infiltration, storage). Waste emplacement technology as thickness of waste layers, compaction of waste, kind of daily soil cover etc., influences the flow conditions that affect the contact between water and waste. This contact is the basis of most processes in the landfill body and the resulting leachate quality.

As a consequence of the infiltrated water and the reactive organic and inorganic waste compounds, a wide range of biological, chemical, and physical processes take place in the landfill. With increasing content of organics, the influence of biological processes on leachate composition becomes more dominant. In addition, processes such as solubility, fixation, complexation, precipitation, etc., influence the leachate composition. The interactions between the different processes are extremely complex and not predictable. As a result a nearly accurate prediction of flow rate and composition is in general very difficult.

The environmental effect of leachate varies between very low and hazardous also depending on the specific situation. Hazardous waste material in landfill does not automatically result in hazardous leachate pollution. Landfills are a very slow reacting system with the effect that the annual transfer rate of substances to the environment is relatively low. The comparatively small amounts of hazardous substances (e.g., batteries, medicine, detergents) that are part of municipal solid waste (MSW) do very often not affect the leachate concentration significantly, even over a very long period of time.

After the landfill operation has been terminated and the landfill has been closed, the concentrations of most substances in the leachate decrease with time. From that time on no more pollution potential is added to the landfill, and due to leachate and gas production the landfill emission potential will be steadily but slowly reduced. On the other hand as a result of degradation processes and changing environmental conditions in the landfill body, compounds (e.g., ammonium, reduced sulfur components) may also after landfill closure even increase over a limited period of time.

BIOLOGICAL PROCESSES AND LEACHATE QUALITY

Most MSW contains a high amount of organic degradable waste. The waste organics are degraded in the landfill by biological processes in an uncontrolled way. The kind and intensity of these processes depend on the environmental conditions such as access of oxygen, temperature, moisture content and movement, biodegradability of organics, bioavailability etc. After waste emplacement, these processes are mostly aerobic over a limited period of time. Also when the waste layers are highly compacted, air can penetrate up to 1–2 m deep into the landfill; below the top waste layers the processes change to anaerobic. The influence of aerobic processes in low compacted landfills is even higher but hardly predictable.

During the anaerobic degradation process, mainly two defined groups of microorganism are degrading the organics; they should be in an equilibrium, otherwise unwanted intermediate degradation products will be produced. The control of this equilibrium is possible in landfills only to a very limited extent. The consequences may be very high organic leachate concentrations over a limited period of time. In Fig. 10.2.1 (left side) the steps of the anaerobic biological degradation are shown. In a first phase the waste organics are hydrolyzed to smaller organic products. These intermediate products are reduced in the acidogenesis step to mostly volatile fatty acids (VFAs), alcohols, and inorganic residuals. Many products of hydrolysis and acidogenesis are in the liquid phase and leave the landfill as leachate. The effect of these processes on leachate quality is presented in Fig. 10.2.1 (right side).

The organic pollution [chemical oxygen demand (COD) and biological oxygen demand (BOD_5)] during the first years of landfill operation is in general very high. This is especially the case if the content of degradable organics is high. During this phase, COD concentrations may in some cases increase to up to 100,000 mg/L and represent mainly easy degradable organic acids. As a result the pH values are often in the range of 4.5–6.5 (Fig. 10.2.1 right side). The next steps of the anaerobic degradation process are acetogenesis and methanogenesis. Both processes must be in equilibrium: the microorganisms active during the acetogenesis metabolize the VFA and alcohols to acetic acid and H_2. These are the substrates for the methanogenic microorganism, produced as end products of fermentation biogas (40%–75% methane and 60%–25% carbon dioxide). Both processes are very sensitive, and the growth rate of the microorganisms is very low. Both groups of microorganisms may be inhibited particularly by the intermediate product VFA at higher concentrations. Hydrogen as an end product of acetogenesis

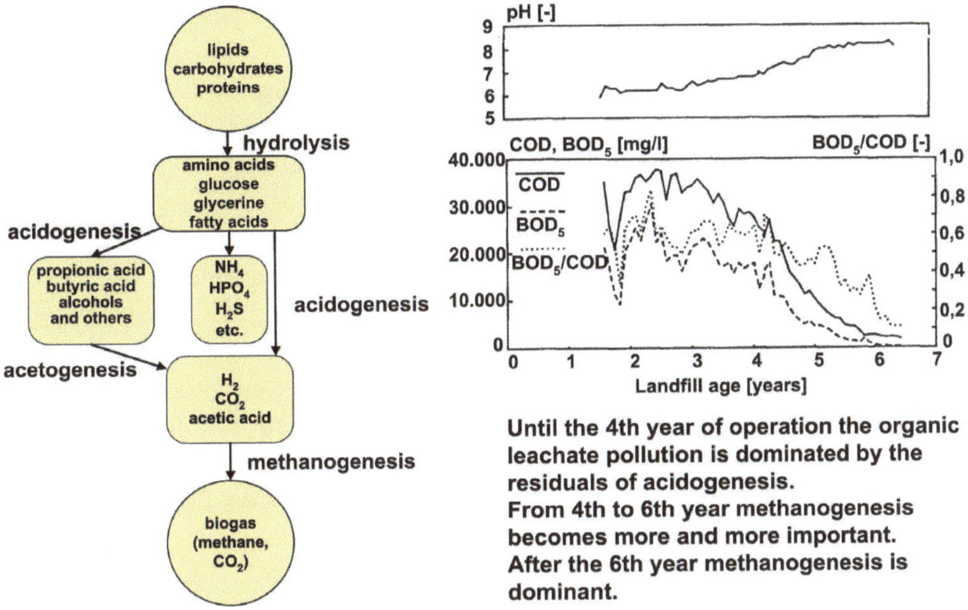

Figure 10.2.1 Sequence of anaerobic biodegradation (left side) and the effect of incomplete biodegradation on organic leachate pollution (typical older landfill with waste disposal in 2-m lifts, waste density 0.8–1.0 t/m^3 and 750 mm/year precipitation (right side) (Ehrig, 1983). *BOD*, biological oxygen demand; *COD*, chemical oxygen demand.

may inhibit the microorganisms of acetogenesis if the hydrogen is not converted to methane and by these means accumulate (see also Chapter 3.1).

After some years of landfill operation, the easy biodegradable organic waste compounds will be reduced. The methanogenic bacteria may develop in areas with lower concentrations of organic acids as in inorganic waste compartments (e.g., soil, plastic), etc., and reach out from there into areas with higher acid production. This means there will be in the same landfill areas of acid and methane production, where with time the methanogenic conditions will be dominant. As a consequence, at some stage there will be a sharp decrease of organic leachate concentrations (Fig. 10.2.1—right side). Somehow simplified the first phase with high organic concentrations is called acetic phase, and the subsequent phase with highly reduced organic degradable compounds is called methanogenic phase. The time until landfills reach the methanogenic phase differs in a wide range. Observations at different landfills in Germany have shown periods between 2 and 12 years until a stable methanogenic phase had been reached. Robinson (1995) reported that in the U.K. 33% of the observed landfills have reached the methanogenic phase after 2.5 years, 73% after 4 year, 80% after 7 years, and 93% after 10 years. Looking at the scheme of anaerobic degradation in Fig. 10.2.1, the end product—with exception of some inorganic and difficult or not degradable organic substances——is mainly biogas. Leachate from landfills in the methanogenic stage shows relatively high COD but low BOD concentrations.

The residual COD consists to a high degree of humic- and fulvic-like acids and may reach values between several hundred mg/L and 5000 mg/L. But in opposite to the organic leachate compounds produced in the acetic phase, these residual organics are less biodegradable or nonbiodegradable. These differences in biodegradability are used to characterize the anaerobic phase by means of the BOD_5/COD ratio. The acetic phase is characterized by a BOD_5/COD ratio >0.4 and the methanogenic phase by a BOD_5/COD ratio $\leq$0.1; a ratio between 0.4 and 0.1 characterizes the transitional phase. This phase can often be observed at landfills because the collected leachate represents a mix of the different leachate qualities from the different landfill sections that are of different age. Leachate samples from boreholes in the upper part of a landfill in the methanogenic phase in some cases show high organic leachate concentrations typical for the acetic phase. In such cases the fully established anaerobic microorganism community of the lower layers reduces the intermediate products from above layers.

The change of the leachate quality as a result of different biological phases in the landfill is a great challenge when planning a leachate treatment plant. Owing to this fact, it needs the treatment of two different wastewaters. This problem leads to considerations to reduce the time period of the acetic phase or even to prevent it. In the following, some options are presented:

- Recirculating the leachate to a landfill part with fully established methanogenic phase (prevents overloading, this part acts as an anaerobic filter) (see Chapters 10.3, 12).
- Installation of a 2-m layer of composted waste with a coarse structure as the first lift at the bottom of the landfill that will act as a biological filter.
- Degradation of the easy biodegradable organic waste prior to landfilling, i.e., mechanical biological treatment (MBT), or on site composting before compaction.
- Separate collection and treatment of the food and yard waste.
- Improving the environmental conditions for microorganisms with increased moisture content (controlled leachate recirculation).
- Slow built-up of the landfill to final height.

Fig. 10.2.2 (left side) shows the COD leachate concentration pattern of six different landfills. These are operated in three different modes where always two landfills were operated in the same way. In the past, in Europe, waste was mainly emplaced in 2-m layers; this has been changed to placement and high compaction of waste in thin layers ($\pm$50 cm). As a consequence of the old operation mode, in many cases high organic leachate concentrations over several years have been observed.

At landfills where the waste is compacted in thin layers, the density will be higher. In addition—as the results in Fig. 10.2.2 show—this operation mode has a positive environmental effect. Compared to the emplacement in 2-m lifts the COD concentrations in the leachate are lower and the duration of the acetic phase is shorter. The waste input of both landfill examples is relative low, which results in a relatively slow built-up of the landfill to final height.

In Fig. 10.2.2 (right side) an estimated uncontrolled biological degradation of organic matter in the surface layer is shown. Measurements at landfills show that the air penetrates about 1–2 m into the landfill. Having installed 2-m layers, some parts of each layer have only limited access to air. But if there is a 0.3- to 0.5-m layer—even with a higher density—the entire lift has contact with air as

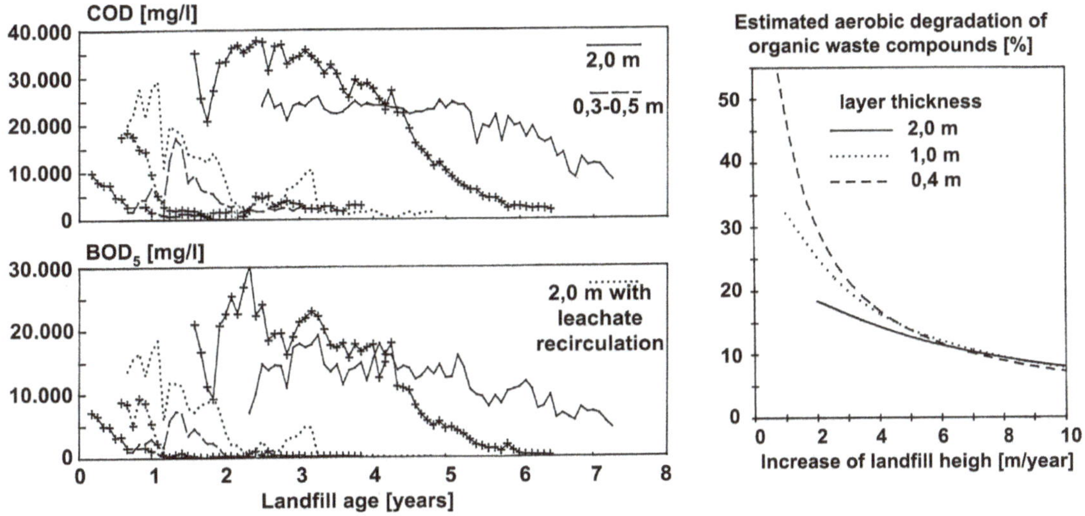

Figure 10.2.2 Effect of waste lift height and leachate recirculation on organic leachate quality—two landfills of each type (left side). Estimated uncontrolled aerobic biodegradation in surface landfill layers dependent on the increase of landfill height (right side) (Ehrig, 1989). *BOD*, biological oxygen demand; *COD*, chemical oxygen demand.

long as it is exposed to the atmosphere. The lower the increase of landfill height with time, the longer is the exposure of the waste to air with the consequence of a relatively fast aerobic degradation of the easily degradable organics. After covering such a layer with new waste, the environmental conditions change to an anaerobic milieu. But since the organic matter is partly aerobically degraded, the growth of acidogenic microorganisms is reduced and less organic acids are produced. As a consequence, the environmental conditions (e.g., pH, organic acid concentrations) for the development of methanogenic microorganisms will improve. If daily cover is used (which is not practiced in Germany at most cases), the soil should have a relatively high permeability.

At several landfills, leachate is/has been recirculated back to the landfill. In the early days, landfill operators practiced it for economic reasons because they expected as long as the total amount of leachate was recirculated there are no leachate treatment costs. But experience shows that on the longer run this practice caused a lot of problems due to elevated leachate production, water built up in the landfill, and seepage of leachate through the slopes. As a consequence the amount of recirculated leachate should be controlled, and the process has to be monitored. In addition, only treated leachate with low organic acid concentrations should be used (see Chapter 12). On the other hand, Fig. 10.2.2 (left side) shows that the intensity and time of the acetic phase will be reduced during a controlled recirculation. This is due to the fact that the water flux through the landfill body (i.e., products of microbial activity are transported faster out of the landfill body) and also the moisture content of waste increase. Owing to the partly reduction of VFA in the recirculated leachate, the pH increases with additional positive effects on anaerobic degradation in lower layers of the landfill body. At very fast growing landfills ($\geq$10 m/year), leachate recirculation has shown no effect because the concentration of organic acids is high, and the increase of the water addition is not sufficient to leach out significant amounts of organic acids. In addition the aerobic phase in the surface layer is too short. At high precipitation rates (e.g., >1200 mm/a), recirculation is not effective any more (more information on leachate recirculation is given in Chapter 12).

At several landfills the leachate from the acetic phase section of the landfill (young part) is recirculated to older sections of the landfill where the methanogenic phase is already established. Some indications on the treatment capacity of the latter section are presented in Chapter 10.5.

Another possibility to reduce or prevent the production of high-strength acetic leachate is the emplacement of a layer of composted waste as a first lift at the bottom of the landfill. Owing to the reduced organic content of the composted waste (it should have a coarse structure) and its positive environmental conditions for methanogenic microorganisms (e.g., pH), the high organic concentrations of the leachate from the upper layers can be partly anaerobically degraded. It should be considered that composted waste is not a uniform material compared to a fixed film reactor with exact given flow path. Also for this reason a 2-m layer of composted waste should be realized. At very fast growing landfills of 10 m/year and more, the thickness of composted waste should be increased up to 4 m.

LEACHATE QUALITY

As a consequence of the waste composition, the environmental conditions and the landfill operation leachate quality can vary in a wide range. Tables 10.2.1–10.2.4 present leachate quality data from landfills in several countries. Box 10.2.1 shows median, 75% quantile, and maximum leachate data from 55 MSW landfills in North Rhine Westphalia (Germany).

The most important leachate compounds are organic substances represented by the COD and BOD_5 and nitrogen compounds measured as ammonium and organic nitrogen. At first sight the frequently high COD and BOD_5 values during the acetic phase seem to be the most important challenge. But these organics can be relatively easily biologically degraded. High COD leachate concentrations during the methanogenic phase require specific treatment for elimination. In addition to the COD the organic quality of wastewater is often measured as TOC (total organic carbon). To compare both values a relationship of the COD/TOC ratio amounts to approximately 2.7–3.3 and may be used to convert COD into TOC or vice versa. The COD/TOC ratio describes the oxidation stage of organic substances and it will be higher for non oxidised compounds (e.g. hydrocarbons).

Inorganic nitrogen is a degradation product of different organic materials as there are garden and yard waste, textiles, paper, and food waste. In contrast to the organic pollution the nitrogen concentrations do not decrease during phase changes. In many cases a slight increase with landfill age can be observed. One reason for the different behavior of both components is the way they leave the landfill and their aggregate state. The products of biodegraded organics are in the liquid stage as acids and alcohols or in the gaseous stage as CH_4 and CO_2. The products of the biodegradation of organic

Table 10.2.1 Leachate analysis of German landfills

Parameter		(1) Aver.	(1) Min.	(1) Max.	(2) Aver.	(2) Min.	(2) Max.	(3) 1–5 years	(3) 6–10 years	(3) 11–20 years	(3) 21–30 years	(4) Min.	(4) Max.
pH		6.1	4.5	7.5	8	7.5	9	7.3	7.5	7.6	7.7	5.4	9
BOD_5	mg/L	13,000	4,000	40,000	180	20	550	2,285	800	275	185	6	16,000
COD	mg/L	22,000	6,000	60,000	3,000	500	4,500	3,810	2,485	1,585	1,160	22	22,700
NH_4	mgN/L		(5)		750	30	3,000	405	600	555	445	0,4	7000
NO_3	mgN/L		(5)		3	0.1	50	3.6	7.6	12	9		200
NO_2	mgN/L		(5)		0.5	0	25	0.06	0.63	0.5	0.8		11.7
Organic N	mgN/L		(5)		600	10	4,250						
Total P	mg/L		(5)		6	0.1	30						
Alkalinity (as $CaCO_3$)	mg/L		(5)		6,700	300	11,500						
AOX	μg/L		(5)		2,000	320	3,500	2,765	1,930	1505	1,130	20	7,500
Cl	mg/L		(5)		2,100	100	5,000	1,300	2,135	1,760	1,025	13	28,000
SO_4	mg/L	500	70	1,750	80	10	420	98	146	93	83	1.1	1,810
Na	mg/L		(5)		1,350	50	4,000	815	1,125	905	645	13	4,700
K	mg/L		(5)		1,100	10	2,500	1,220	910	695	595	25	2,200

(Continued)

Table 10.2.1 Leachate analysis of German landfills—cont'd

Parameter		(1) Aver.	(1) Min.	(1) Max.	(2) Aver.	(2) Min.	(2) Max.	(3) 1–5 years	(3) 6–10 years	(3) 11–20 years	(3) 21–30 years	(4) Min.	(4) Max.
Mg	mg/L	470	50	1,150	180	40	350	290	205	145	115	15	1,167
Ca	mg/L	1,200	10	2,500	60	20	600	375	465	325	155	12	10,000
Zn	mg/L	5	0.1	120	0.6	0.03	4	1.1	1.5	0.53	0.54		
Mn	mg/L	25	0.3	65	0.7	0.03	45	3.9	1.5	1.1	0.9	0.05	43
Fe	mg/L	780	20	2,100	15	3	280	50	15	9.9	8.7	0.08	550
As	µg/L	(5)			160	5	1,600	15	21	42	14	1	370
Cd	µg/L	(5)			6	0.5	140	11	4	4	3	0.13	70
Cr	µg/L	(5)			300	30	1,600	156	224	164	177	5	2,570
Cu	µg/L	(5)			80	4	1,400	711	73	62	36	2	40,000
Hg	µg/L	(5)			10	0.2	50						
Ni	µg/L	(5)			200	20	2,050	199	154	135	115	3	1,930
Pb	µg/L	(5)			90	8	1,020	156	56	67	34	5	1,300

(1) and (2) Weekly analysis at 15 landfills over 6 years (Ehrig, 1980; Ehrig and Hagedorn, 1994), (3) and (4) leachate analysis of >50 landfills over the entire landfill age (Ehrig, 2001; Krümpelbeck, 2000).

Aver., average; *Max.*, maximum; *Min.*, minimum; (1) acetic phase; (2) methanogenic phase; (3) average leachate analysis at landfills with an age of 1–5 years, 6–10 years, 11–20 years, and 21–30 years; (4) minimum and maximum values of all landfills (3); (5) the same values as during methanogenic phase (2).

Table 10.2.2 Leachate analysis from different countries: (1) acetic phase and (2) methanogenic phases leachate analysis from UK landfills (Robinson, 1995), (3) French landfill (Amokkrane et al., 1997), (4) Italian landfill (Lopez et al., 2004), (5) and (6) Greek landfills (Tatsi et al., 2003; Loizidou et al., 1992), and (7) Algerian landfill (Salem et al., 2008)

Parameter		(1) Min.	(1) Max.	(1) Aver.	(2) Min.	(2) Max.	(2) Aver.	(3) Aver.	(4)	(5) Aver.	(6) Aver.	(7)
pH		5.12	7.8	6.73	6.8	8.2	7.52	8.2	8.2	6.2	7.9	8.27
BOD_5	mg/L	2,000	68,000	18,632	97	1,770	374	200	2300	70,900	1,050	980
COD	mg/L	2,740	152,000	36,817	622	8,000	2,307	4,100	10,540	26,800	5,350	3,792
TOC	mg/L	1,010	29,000	12,217	184	2270	733	1,430	3,900			
NH_4	mgN/L	194	3,610	922	283	2,040	889	1,040	5,210	3,100	940	85.8
NO_3	mgN/L	<0.2	18	1.8	0.2	2.1	0.86	124		150	55	14.6
NO_2	mgN/L	0.01	1.4	0.2	<0.01	1.3	0.17				1.75	
TKN	mgN/L									3,400	1,100	
Phosphate	mgP/L	0.6	22.6	5	0.3	18.4	4.3	8.4	32	167	8.8	58.2
Alkalinity (as $CaCO_3$)	mg/L	2,720	15,870	7,251	3,000	9,130	5,376		21,470	12,880	4,950	
Cl	mg/L	659	4,670	1,805	570	4,710	2,074	5,420	4,900	3,260	4,120	4,569
SO_4	mg/L	<5	1,560	676	<5	322	67	550			210	3,056
Na	mg/L	474	2,400	1,371	474	3,650	1,480	3,000	3,970	(6)		

(Continued)

Table 10.2.2 Leachate analysis from different countries: (1) acetic phase and (2) methanogenic phases leachate analysis from UK landfills (Robinson, 1995), (3) French landfill (Amokkrane et al., 1997), (4) Italian landfill (Lopez et al., 2004), (5) and (6) Greek landfills (Tatsi et al., 2003; Loizidou et al., 1992), and (7) Algerian landfill (Salem et al., 2008)—cont'd

Parameter		(1) Min.	(1) Max.	(1) Aver.	(2) Min.	(2) Max.	(2) Aver.	(3) Aver.	(4)	(5) Aver.	(6) Aver.	(7)
K	mg/L	350	3,100	1,143	100	1,580	854	880	3,460	Min.	Max.	
Mg	mg/L	25	820	384	40	1,580	250	110	24.1	85.2	140	
Ca	mg/L	270	6,240	2,241	23	501	151	68	15.7			
Mn	mg/L	1.4	164	32.9	0.04	3.59	0.46		0.04			0.41
Fe	mg/L	48.3	2,300	654	1.6	160	27.4	0.91	2.7	5	16.2	8.23
Zn	mg/L	0.09	140	17.4	0.03	6.7	1.14	0.73	0.16			1.43
As	µg/L	<1	148	24	<1	485	34					
Cd	µg/L	<10	100	20	<10	80	15	100	<20	<100		<30
Cr	µg/L	30	300	130	<30	560	90		2,210	700	1,910	200
Cu	µg/L	20	1,100	130	<20	620	130	390		90	280	390
Hg	µg/L	<0.1	1.5	0.4	<0.1	0.8	0.2					
Ni	µg/L	<30	1870	420	<30	600	170	810	310	670	1350	370
Pb	µg/L	<40	650	280	<40	1,900	200	460	<30	<100		3,490

Aver., average; *Max.*, maximum; *Min.*, minimum.

Table 10.2.3 Leachate analysis from different countries (Robinson, 2007; Robinson and Carville, 2007) (1) and (2) South Africa, (3) Hong Kong, (4) Thailand, (5) Indonesia, (6) New Zealand, and (7) Korea

Parameter		(1)	(2)	(3)	(3)	(4)	(4)	(4)	(5)	(6)	(6)	(7) Min.	(7) Max.
pH		7.5	8.2	8.6	7.8	7.6	7.2	7.0	8.4	7.2	7	7.5	8.3
BOD_5	mg/L	170	550	167	117				TOC (mg/L) 968	76	737	342	1058
COD	mg/L	760	4,560	2,580	873	2,700	1,560	1,980		1,181	1,969	1,840	4,601
NH_4	mgN/L	435	1,554	2,563	1,156	3,032	1,424	1,350	2,000			860	1,740
NO_3	mgN/L	<0.1	9.2	2.5	<0.1	<1	<1	<1	<1			<1	3
NO_2	mgN/L		0.4	<0.1	<0.1	0.7	<0.4	0.3	<1			<1	120
Phosphate	mgP/L	1.4	13	27.6	22.2	1.7	15	14	12			8	14
Alkalinity (as $CaCO_3$)	mg/L	2,422	9,652	11,500	4,940	23,910	12,505	15,970	7,840			3,713	10,113
Cl	mg/L	1,690	4,626	2,740	821	3,802	2,498	3,650	2,330	859	973	667	2,156
SO_4	mg/L					15	6.4	1.6	159	1	1	6	80
Na	mg/L	590	2,825	2,100	217	2,453	1,460	2,179	1,130	669	429	550	1,707
K	mg/L		1,615	1,000	375	1,932	1,010	1,819	1,600	471	649	436	1,123

(Continued)

Table 10.2.3 Leachate analysis from different countries (Robinson, 2007; Robinson and Carville, 2007) (1) and (2) South Africa, (3) Hong Kong, (4) Thailand, (5) Indonesia, (6) New Zealand, and (7) Korea—cont'd

Parameter		(1)	(2)	(3)	(3)	(4)	(4)	(4)	(5)	(6)	(6)	(7) Min.	(7) Max.
Mg	mg/L	80	195	31	18	121	132	182	56	95	160	79	178
Ca	mg/L	105	198	19	22	55	126	199	86			41	297
Mn	mg/L	0.86				0.24	0.6	1.65	0.47	0.4	6.56		
Fe	mg/L	12/ 18.8	9.35	5.5	7.8	2.77	1.57	3.08	6.23	0.2	0.89		
Zn	mg/L	0.17/ 0.12		2	0.9	0.15	0.61	0.24	0.46	1.24	1.65		
As	μg/L	62								32	12		
Cd	μg/L	<1				<50	<50	<50	<20	10	20		
Cr	μg/L	80				780	<500	160	250	60	70		
Cu	μg/L	<10	64			<50	<50	<50	386	50	50		
Hg	μg/L	<50											
Ni	μg/L	120				<1,000	<1,000	<1,000	380	110	100		
Pb	μg/L	<4/7.3	20			<1,000	<1,000	<1,000	<300	70	150		

Max., maximum; *Min.*, minimum.

Table 10.2.4 Leachate analysis from Chinese landfills (Yue, 2016)

Parameter		(1) Min.	(1) Max.	(2)	(3)	(4)	(5) Min.	(5) Max.	(6) Min.	(6) Max.
		Beijing		Taiwan	Shanghai	Chongqing	Shenzhen		Changchun	
pH		6.8	8.5	7.9	7.4–7.8	8.21	6.2	8.0	7.2	7.5
BOD_5	mg/L	90	18,200	164	1,619	1,011	1,000	36,000	465	820
COD	mg/L	2,070	33,300	2,483	8,592	4,867	3,000	60,000	2,027	3,212
NH_4	mgN/L	1,100	2,930		534	2,701	400	1,500	1,013	1,044
SS	mg/L	324	4,720	156	584	737	100	6,000		
TS	mg/L			9,619						
DS	mg/L	10,200	24,000	9,463						
Conductivity	mS/cm	18	29	24						
Volatile fatty acid	mg/L					672				

Max., maximum; *Min.*, minimum.

(1) Li et al. (2008): Determined in six landfills of Beijing: Asuwei, Anding, Beishenshu, Liulitun, Gaoantun, and Jiaojiapo landfill.

(2) Fan et al. (2006): The landfill has been operated for about 10 years with total capacity of 3.4 million m^3. The waste consists of municipal solid waste (MSW) (20%–60%) and incinerator bottom ash (40%–80%).

(3) Zou et al. (2011): Laogang landfill in Shanghai started operation in 1989, receiving about 5000 t of refuse per day. 61% of the refuse is construction wastes, 38% agricultural and domestic waste.

(4) Liu (2007): Chang Sheng Qiao landfill in Chongqing has operated for 3.5 years, mainly receiving MSW. Because of the relative high annual average temperature, the biological stabilization rate of waste was high.

(5) Jiang et al. (2002): Waste composition in Shenzhen: 60% food waste, 7% ceramics textiles, 10% paper, 13% plastics, 4% glass, 1% metal.

(6) Guan (2005): The leachate sampled from Peijia landfill in Changchun was typical middle or old age leachate.

Box 10.2.1 Leachate quality from municipal solid waste landfills (German class II) in North Rhine Westphalia (NRW), Germany (values based on 55 class II landfills and 2200–5800 analysis; dependent on parameter) (Anonymus, 2010)

Parameter		Median	75% Quantile	Maximum
SO_4	mg/L	77	431	11,000
Cl	mg/L	1,153	1,998	10,668
Na	mg/L	912	1,509	10,100
K	mg/L	518	840	4,900
Ca	mg/L	127	271	5,490
Mg	mg/L	92	155	1,121
Conductivity	mS/m	1,091	3,845	42,500
COD	mg/L	1,300	3,809	49,250
TOC	mg/L	512	1,305	16,320
NH_4-N	mg/L	615	1,149	8,160
AOX	µg/L	680	1,461	22,900
As	µg/L	20	111	18,000
Cd	µg/L	1.5	30	1,000
Cr	µg/L	80	322	4,120
Cu	µg/L	40	224	4,100
Hg	µg/L	0.5	1.9	40
Ni	µg/L	116	283	3,560
Pb	µg/L	13	246	5,240
Zn	µg/L	110	996	23,300

nitrogen compounds such as proteins are ammonium and organic acids, acetate as well as CO_2 and H_2 (Sleat et al., 1989). To migrate through the landfill, the ammonium has to dissolve in the leachate. Due to the organic content of the waste, the mobilized ammonia may partly adsorb to organic material. In addition, the solubility of some proteins is not high. Nitrogen compounds of several proteins may be bound in not easily degradable organics. All together and with the high potential of nitrogen in the waste and the limited transport may be the reason of the long-term release of N in the leachate.

The concentrations of most organic trace compounds in leachate are mostly low, because the solubility of many trace compounds in leachate is low and their amount in the emplaced waste is comparably low too. At several old landfills, organic halogens are/have been measured in concentrations of some milligrams per liter. Organic halogens with a high vapor pressure leave the landfill with the gas flow. But some halogens with a lower vapor pressure are in the liquid stage and may be dissolved in the leachate.

Some inorganic compounds such as sulfate, iron, calcium etc., show the same tendency between acetic and methanogenic phase as COD and BOD_5. Many organic substances and demolition material contain sulfate. The biological reduction of sulfate to sulfide has a lower energy gain for microorganisms than hydrolysis and acidogenesis but a higher energy gain than methanogenesis. For this reason, sulfate is reduced in nearly the same period of time as the biological phase changes. The sulfate content of demolition waste can be released over a long period of time, which may result in an increase of sulfide concentrations in leachate at old and closed landfills unless they are not precipitated as metal sulfides.

The concentrations of calcium, iron, manganese etc., in the leachate are influenced by their content in the emplaced waste (e.g., demolition waste). In addition the pH influences the solubility conditions. The relatively low pH values during the acetic phase result in an increase of the abovementioned compound concentrations, and with the change to the methanogenic phase where relatively higher pH values can be measured, these concentrations decrease.

The concentrations of heavy metals in leachate vary by species and with time as well as between different landfills and sometimes between different parts of the same landfill. The solubility of heavy metals depends mainly on the pH value and the redox potential. Metals may adsorb at organic waste compounds or may react with inorganic substances. A typical metal fixation is the formation of metal sulfides, which precipitate out due to their relatively low solubility in leachate. Under aerobic conditions, the sulfides may be oxidized to sulfates and metals may be solubilized again but will mainly coprecipitate with the newly formed hydroxides. Owing to high CO_2 concentrations in the landfill, some heavy metals such as zinc and cadmium may also precipitate as carbonates. For these reasons the heavy metal content in the leachate is in general relatively low either under aerobic as well as anaerobic conditions.

At most landfills, mainly the biodegradable organics, nonbiodegradable organics, and nitrogen compounds in the leachate are in focus. Nitrogen is measured as ammonium and organic nitrogen. The differentiation between biodegradable and nonbiodegradable organics is relatively difficult. As a representative parameter of these two groups of organics with different behavior, the COD is measured. In addition to the oxidation of the organics, the oxygen consumption of some inorganic compounds also is included in the COD. The parameter BOD_5 describes the biodegradable fraction in 5 days (in some countries also BOD_7 BSB 7). For this reason the not easily degradable organics in the leachate, e.g., from the methanogenic phase are not or only partly measured. A more meaningful value would be, e.g., a BOD_{20}. The measurement of lower BOD values is often difficult. Low BOD values and high ammonium concentrations in the leachate often go together in methanogenic phase. The toxicity effect of ammonium on BOD measurement requires a high dilution of the leachate for the BOD analysis. With increasing sample dilution, the potential error of the BOD analysis may be higher. Of course also the potential nitrification in the BOD bottle has to be suppressed. The solution may be a laboratory batch treatment test using a fixed film reactor where the COD of the constant influent and the constant effluent are measured. In this case the difference of the COD concentrations would represent the biodegradable fraction.

LONG-TERM ASPECTS OF LEACHATE QUALITY

At the end of landfill operation, leachate will still be produced unless the landfill surface is completely covered with an impermeable liner. As long as water infiltrates into the landfill, substances will be dissolved and transported out of the landfill. Especially for the parameters COD and nitrogen, it is necessary to estimate how long leachate treatment may be necessary. At landfills with a long-term leachate monitoring, leachate emissions can be estimated; but monitoring phases of 20–30 years are in most cases too short for long-term prediction because in many cases the leachate concentrations are still high because the landfill has not reached the final stable phase. Investigating pilot-scale MSW test cells it was possible to study the biological waste stabilization over 1–2 years; as a result, at the end of the experiments very low concentrations had been reached. Using pilot-scale data for long-term prediction of the leachate quality and quantity of full-scale landfills, it is necessary to find a parameter to adjust both timescales. Instead of a timescale the water infiltration to dry solids ratio (w/s ratio) shows a relative good agreement of the different data sets (Ehrig, 2001; Ehrig and Krümpelbeck, 2001; Krümpelbeck, 2000; Krümpelbeck and Ehrig, 2000). A typical description of such transport processes is an exponential equation (Fig. 10.2.3). Input data for the regression analysis are the leachate concentration at different times (C_t) and the time (t); the parameter k as a constant value includes the average annual infiltration/dry solids rate (mm/year/kg dry solids of a 1 m^2 waste column). The result of the regression analysis is the concentration (C_0) at the end of landfill operation (year = 0) and a coefficient F. The parameter F describes the curve of this function and depends on the behavior of the specific pollution potential as there are solubility, transfer conditions etc., for each parameter. The coefficient differs between different parameters [average value: F (COD) = 1; F (ammonium) = 0.4]. Examples with different concentrations of COD and ammonium at the end of landfill operation and different w/s ratios are presented in Fig. 10.2.3. Considering the inhomogeneity of waste and landfills and the high fluctuation of leachate concentrations, the parameter F should be used for long-term prediction in a range of ±25%.

A similar approach has also been published by Wang et al. (2013). Considering the inhomogeneous nature of the waste in the landfills, a comparison between both estimations shows a relative good agreement. Other authors such as Weber (2002) used also an exponential model for the long-term leachate quality prediction but did not consider the effect of the same infiltration rate at landfills with different heights (e.g., 10 and 30 m).

Fig. 10.2.4 shows the practical application of the equation $C_t = C_0 * e^{k*t}$ ("calculated" in Fig. 10.2.4). Average annual COD and ammonium concentrations—starting at the end of the operation—were monitored in the leachate of 16 relative small German (state of Bavaria) landfills and are presented. An exponential regression was calculated to statistically describe the data set (see "Expon" in Fig. 10.2.4); in addition the exponential leachate prediction model was applied (Fig. 10.2.3). Fig. 10.2.4 shows on one side a good agreement of such estimations and on the other side their limits. Two years after the end of operation the COD concentrations drop sharply mainly due to the fact that after the end of operation most landfills in Germany had been covered with a

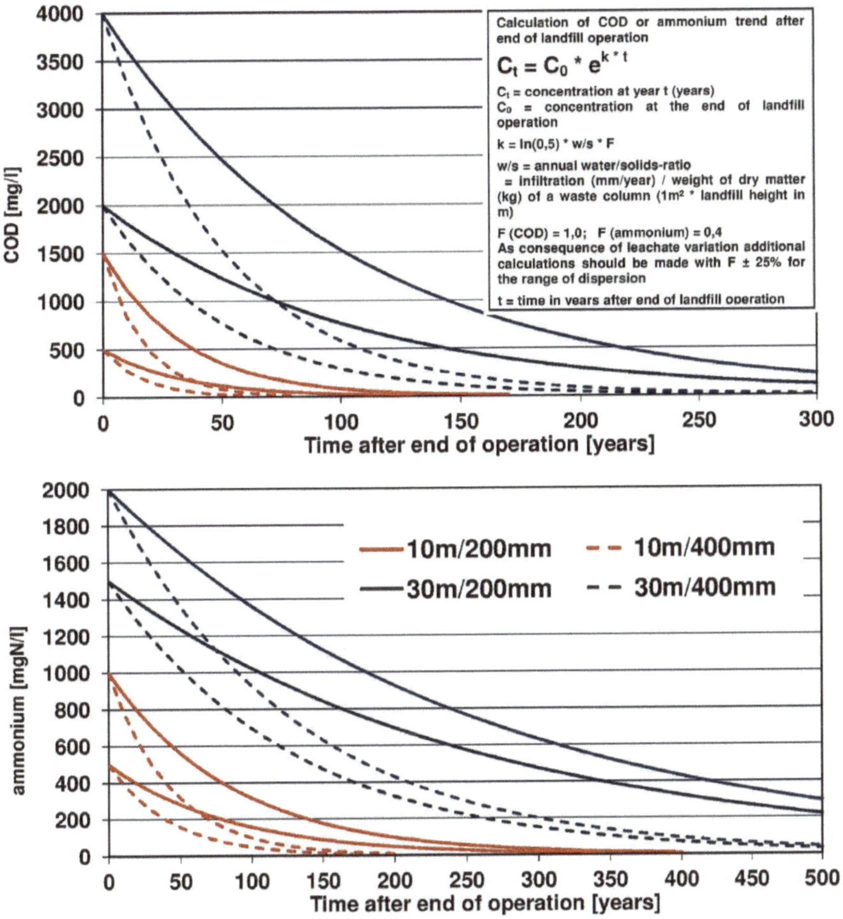

Figure 10.2.3 Calculated long-term chemical oxygen demand (COD) and ammonium leachate concentrations dependent on landfill height (m) and annual infiltration rate (mm) (two examples for each case: typical low and high concentrations at the end of operation—explanation see Long-Term Aspects of Leachate Quality section) (Krümpelbeck, 2000; Ehrig, 2001).

top liner. The prediction of the leachate quality at this early stage using a statistical evaluation or the application of the exponential model would give very erratic results. Fig. 10.2.4 shows a very good agreement of calculated and measured COD values after these initial 2 years (relative stable conditions); this is not the case for the long-term behavior of ammonium. At several landfills the ammonium concentrations after some decades of operation or after the end of the operation may be relatively constant or may even increase for a certain period. Some thoughts about potential reasons for the high NH_4 fluctuations are discussed above. An estimation of future tendencies would be more feasible after the end of the period of increasing or constant concentrations.

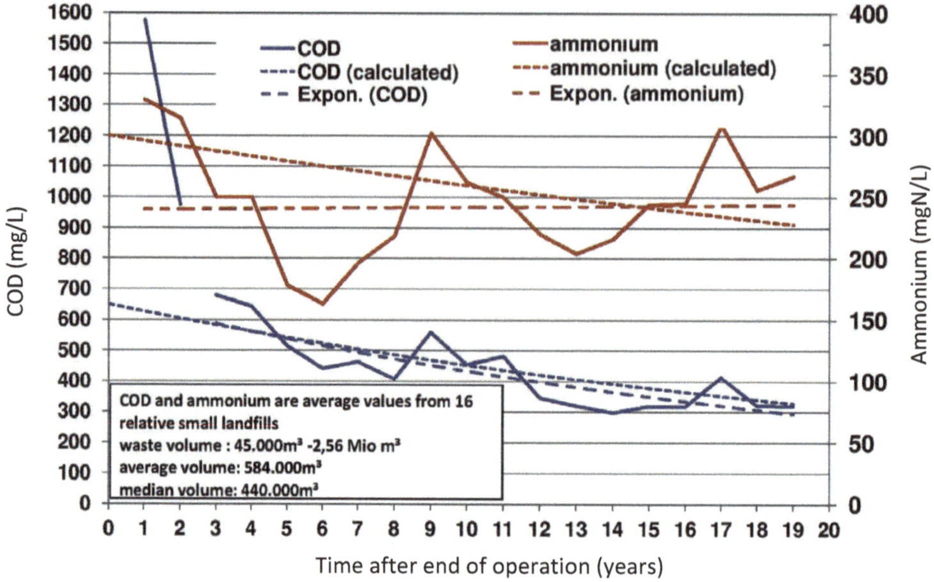

Figure 10.2.4 Comparison of average annual COD and NH_4-N concentrations in leachate of 16 Bavarian landfills after the end of landfill operation—*solid line*: average values; *dashed line*: exponential regression using the measured values; *dotted line*: calculated with $C_t = C_0 * e^{k*t}$ (see Long-Term Aspects of Leachate Quality section). *COD and ammonium data from Huber and Schatz (2007).*

LEACHATE QUALITY FROM LANDFILLS FOR MECHANICALLY AND BIOLOGICALLY TREATED WASTE (MECHANICAL BIOLOGICAL TREATMENT LANDFILLS)

As already mentioned the most critical parameters of leachate pollution are COD and ammonium; these parameters represent the high long-term emission potential. As a consequence high costs for leachate treatment over decades have to be encountered. Both parameters are predominantly influenced by the biodegradable organic waste fraction and the intensity of the biodegradation processes in the landfill. This situation and the uncontrolled production of the greenhouse gas methane during operation were the basis for the decision to ban biodegradable waste from landfilling. Several states in the EU allow only biologically or thermally treated MSW to be landfilled (see also Chapters 1.1, 1.2, 2.1, and 4.1).

Several aerobic and anaerobic biological treatment processes and their combinations are used to biologically stabilize MSW to prescribed limit values. Owing to different mechanical and biological treatment processes and the degree of MSW stabilization, the leachate pollution can vary to a certain extent.

There are several MBT landfills in operation but the period of operation is relatively short and most landfills are combinations of the closed raw waste and MBT landfills. For this reason the leachate are often mixtures of both types of landfills. Robinson et al. (2004) collected and analyzed leachate samples at different German landfills. The results are presented in Table 10.2.5. At these landfills more than

Table 10.2.5 Leachate analysis from different landfills containing more than 90% mechanical biological treatment waste (Robinson et al., 2004)

Parameter		(1)	(2)	(3)	(4)	(5)
pH		8.3	7.9	7.5	8.4	8.5
BOD_5	mg/L	202	35	1-55	6	3
COD	mg/L	4670	1620	700-2500	869	1020
TOC	mg/L	1480	543	300–950	308	340
NH_4	mgN/L	1130	197	0–27	34.2	1.8
NO_3	mgN/L	<0.3	<0.3	15–66	6.3	4.7
NO_2	mgN/L	<0.1	0.1	0.1–1.7	1.0	0.3
Phosphate	mgP/L	12.4	2.8		0.4	0.3
Alkalinity (as $CaCO_3$)	mg/L	6120	2010		1670	895
Cl	mg/L	2270	2290		901	1090
SO_4	mg/L	117	449		414	878
Na	mg/L	1520	1250		622	789
K	mg/L	728	777		393	387
Mg	mg/L	88	104		64	67
Ca	mg/L	176	329		232	255
Mn	mg/L	1.38	2.94		1.45	1.61
Fe	mg/L	19.5	13.9		2.59	1.31
Zn	mg/L	0.67	0.2		0.19	0.53
As	μg/L	<1	6		<1	4
Cd	μg/L	6	3		2	3
Cr	μg/L	494	96		42	90
Cu	μg/L	180	150		80	250

(Continued)

Table 10.2.5 Leachate analysis from different landfills containing more than 90% mechanical biological treatment waste (Robinson et al., 2004)—cont'd

Parameter		(1)	(2)	(3)	(4)	(5)
Hg	μg/L	<0.1	<0.1		<0.1	<0.1
Ni	μg/L	260	82		51	97
Pb	μg/L	130	40		10	20
AOX	μg/L	790	1500	100–900	180	370

Pretreatment:
(1) Passive windrows.
(2) Turned windrows.
(3) Turned windrows, active aeration.
(4) Container.
(5) Container + windrows.
Landfill age:
(1) 9 years.
(2)–(5) 2–5 years.

90% of the landfilled waste was MBT waste, which had been aerobically treated. The intensity of the aerobic treatment was different and increases in principle from column 1 to 5 (Table 10.2.5). With the exception of one landfill, the landfill age was relatively short amounting to 2–5 years. As a consequence the presented data may give only an indication on leachate quality from MBT landfills. But also due to this limitation it becomes obvious that with increasing biological stabilization the concentrations of nitrogen (ammonium and nitrate) as well as BOD_5 and COD decrease. Comparing leachate quality data from old/very old MSW landfills (decades or longer) particularly with the NH_4 concentrations from MBT landfill shows that the NH_4 concentrations are significantly lower in the leachate from MBT landfills; in addition, the BOD_5 values are also very strongly reduced. But the remaining nonbiodegradable organics measured as COD and TOC are still relatively high in the leachate from MBT landfills. Considering that these MBT waste landfills are only some years old with a relatively small waste volume, the COD concentrations are in the same range as in leachate from smaller old MSW landfills. This implies that leachate from MBT landfills also has to be collected and treated for a long nonpredictable time period. The higher density, different structure and composition of the MBT waste results in a higher water adsorption capacity (see Chapters 4.1, 14.1). Therefore at least over a period of several years low leachate production rates can be expected.

The data in Table 10.2.5 do not show significant differences in the inorganic leachate quality from MSW and MBT landfills. During mechanical treatment the high calorific value (coarse fraction) is separated from the fraction <10 mm or smaller. The fine fraction consists of biologically degradable waste and inorganic fines as dust etc. Owing to the comparatively higher concentrations of the inorganic compounds in the MBT waste higher inorganic leachate concentrations also could be expected; but the presented data here do not give any indications in this regard.

The data in Table 10.2.5 include only MBT waste landfills with aerobic treatment. Theoretically there might be some differences in leachate quality whether the treatment happened under aerobic or anaerobic conditions. Dependent on the time of aerobic posttreatment after the anaerobic treatment step elevated COD and ammonium leachate concentrations may occur especially when the postcomposting step is relatively short. In case of an anaerobic wet treatment process, some compounds may be washed out. But this has to be proven by data from the field.

INFLUENCE OF LANDFILL IN SITU AERATION ON LEACHATE QUALITY

Many laboratory-scale investigations show that organic leachate concentrations decrease relatively fast with an increasing degree of biological stabilization. While the BOD_5 values decrease to very low values (<25 mg/L), the COD will still remain elevated (>100 mg/L). Leachate quality data from laboratory-scale plants also show a decrease of the NH_4-N concentrations to low values, which may be <10 mg/L. But these tests were operated at constant temperatures of 35°C, i.e., also for this reason the leachate quality data cannot be transferred to full-scale landfills, where during aeration temperatures up to ±50°C will be reached in most cases. In addition a higher water flux was installed. But these data show the potential that leachate quality may be reached due to landfill aeration. The elevated temperatures may result in faster carbon degradation; organic nitrogen ammonification may be enhanced that may result in an elevated pH. At temperatures >45°C nitrification will be inhibited, and due to the higher pH, NH_3 may be produced; in how far this may cause (local) inhibition processes in the landfill body is difficult to predict. After the temperatures decrease <45°C, the reduction of the ammonium can be expected. At the bottom of landfills, often high water content/levels can be found. Owing to the fact that these areas are difficult to aerate, this situation may influence the leachate quality negatively and may result in an increase of the relative low concentrations—reached in the aerated part—of some leachate parameter (see also Ritzkowski, 2011).

LEACHATE QUALITY FROM BOTTOM ASH LANDFILLS

In many countries, incineration of MSW is a common practice. In Germany it is regarded as a treatment step before landfilling. Bottom ash is in Germany, Denmark, and some other countries partly used and partly landfilled either in Class I (inert waste landfills) or Class II (MSW landfill (Germany for treated waste)); fly ash is landfilled in hazardous waste landfills (German class III landfill). In many cases, ferrous material and sometimes aluminum are separated from bottom ash prior to landfilling. In other countries, bottom and fly ashes are still codisposed in mono landfills. During incineration, almost all organics are oxidized. Compared with bottom ash the leachable amount of heavy metals from fly ash is considerably higher. The high temperatures during incineration destroy chemical bonds of several inorganic components. Exothermic reactions of new chemical bonds can be observed in fresh bottom ash over a period of about 3 months. If fresh bottom ash is landfilled, increasing temperatures can be measured. During this phase the leachability of several inorganic compounds is relatively high (see also Chapter 20).

The quality of incineration and the kind of treatment of the residuals may have an effect on the quality of the leachate, but it is very difficult to verify this. Tables 10.2.6 and 10.2.7 show the variation of data from different bottom ash landfills. Some of these landfills contain fly ash or noncombustible inorganic waste also.

Leachate of bottom ash landfills is often highly polluted with chloride, sulfate, sodium, and potassium. The heavy metal concentrations are in the same range as in leachate from MSW landfills. At several landfills, the COD respectively TOC concentrations are relatively high, but the amount of biodegradable compounds is very low. The data presented by Johnson et al., 1999 (Table 10.2.6) show that by separating out bulky materials from the bottom ash COD and TOC values in the leachate may be reduced. In this case also ferrous material had been separated and the bottom ash was stored over periods of 3–6 month (until the exothermic reactions have come to an end and the temperature drops down to nearly ambient values). The effect on chloride, sulfate etc., in leachate seems to be relatively limited. Notable are the low heavy metal concentrations in the leachate of this landfill.

Table 10.2.6 Leachate quality of bottom ash landfills

Parameter		Aver.(1)	Min.(1)	Max.(1)	Min.(2)	Max.(2)	Aver.(3)	Min.(3)	Max.(3)	Aver.(4)
pH		7.7	7.4	8	7.2	10.5	10.3	8.7	11.3	10.5
TOC	mg/L	126	6	400			10.3	4.4	44.8	8
COD	mg/L	42	22	62	75	1100				
BOD_5	mg/L	859	17	1700	<2	130				
Na	mg/L	5050	300	9800	1600	7300	1024	225	2098	1746
K	mg/L	1650	100	3200	600	4300	460	181	944	737
Ca	mg/L	521	401	641	32	1000	329	124	644	536
Mg	mg/L				0.09	66	15.2	2.5	26.2	18.2
Cl	mg/L	2784	290	9300	160	14,300	1672	391	3657	3042
NH_4	mgN/L	72	0.3	190	1	87				

Table 10.2.6 Leachate quality of bottom ash landfills—cont'd

Parameter		Aver.(1)	Min.(1)	Max.(1)	Min.(2)	Max.(2)	Aver.(3)	Min.(3)	Max.(3)	Aver.(4)
NO_3	mgN/L	58	25	90						
SO_4	mg/L	1562	985	2900	640	7200	1190	529	2141	1786
AOX	mg/L	1.8	0.01	4.54						
Fe	mg/L	0.75	0.5	1	<0.01	18.5				
Mn	mg/L	0.055	0.05	0.06			0.0047	0.0015	0.018	
Pb	μg/L	19	0.3	75	<0.5	40	2.7	0.5	7	
Cu	μg/L	123	40	220	<0.5	210	101	6.9	479	33
Zn	μg/L	66	2	250	<10	910	5.7	2.9	10.5	5.6
Cd	μg/L	2	0.9	5	<0.1	2	1.3	0.38	5.4	1.7
Ni	μg/L	20	1	70	<2	42				
Cr	μg/L	17	13	25	<1	80	10.9	8.5	16.1	11.6
Hg	μg/L	0.1			<0.05	3				
As	μg/L				5	25				
Mo	μg/L						522	145	1315	602
W	μg/L						113	34	234	63
Sb	μg/L						32	11	57	19
V	μg/L						22	11	48	35

Aver., average; *Min.*, minimum; *Max.*, maximum.
(1) ATV-Fachausschuss 3.6 (1997).
Unknown number of bottom ash landfills (Germany).
(2) Hjelmar et al. (1995).
Bottom ash landfill (Denmark).
(3) Johnson et al. (1999).
Bottom ash landfill (Switzerland).
The ash was stored in piles over 3–6 month before dumping. Ferrous material and bulky unburned organic material were removed.
Landfill age during sampling time: 3–4 years.
(4) average values during dry weather periods.

Table 10.2.7 Leachate quality of bottom ash landfills

Parameter		Average(1) 1973–75	Average(1) 2001–03	Average(2) 1990–93	Average(2) 1997–99	Average(2) 2004–06	Average(3) and Standard Deviation
pH		10.1–9.1	8.9–8.6	10	9.3	9	7.8 ± 0.5
TOC	mg/L			387	47	30	
COD	mg/L			1540	140	103	2400 ± 1100
K	mg/L	3600	210	1142	492	545	
Ca	mg/L	610	14	252	212	195	
Mg	mg/L	0.27	17	6	5	7	
Cl	mg/L	9200	340	3476	4033	4150	1900 ± 380
NH_4	mgN/L	34	0.35	160	48	18	
NO_3	mgN/L			0.5	5	3	
SO_4	mg/L	3100	1500	1040	3000	4300	554 ± 193
AOX	mg/L			3.5	0.56	0.85	
Fe	mg/L			0.04	0.1	0.14	1.3 ± 1.4
Pb	μg/L	<20	30	70	10	<10	19 ± 31
Cu	μg/L	<30	86	40	20	<5	290 ± 270
Zn	μg/L	33	940	70	20	20	58 ± 36
Cd	μg/L	<5	1.2	10	5	<1	1.2 ± 2.8
Ni	μg/L	50	72	200	17	8	64 ± 32
Cr	μg/L	33	<2	10	10	7	500 ± 310
Hg	μg/L	<1	<0.005	3	10		0.26 ± 0.25
As	μg/L	14	6.2	10	20	50	

(1) Hjelmar and Hansen (2005). Bottom ash (85%) and fly ash landfill (15%) (Denmark). Landfill operation 1973–1976.
(2) Morf and Kuhn, 2010. Bottom ash landfill (Switzerland). Dumping on this landfill part was mostly finished in 1996.
(3) Øygard et al. (2005). Landfill for bottom ash, construction, and inorganic industrial waste (Norway). Landfill operation starts 1999. Analysis of the landfill leachate throughout 2002 and 2003.

LEACHATE FROM OTHER LANDFILL TYPES

As an example of leachate quality, data (median, 75% quantile, and maximum) from different types of landfills monitored in North Rhein Westphalia (state in Germany) over certain periods of time are presented in Boxes 10.2.2–10.2.4.

Box 10.2.2 Leachate quality from inert waste landfills (German class I) in North Rhine Westphalia (NRW), Germany (values based on 25 class I landfills and 334–1031; analysis dependent on parameter) (Anonymus, 2010)

Parameter		Median	75% Quantile	Maximum
SO_4	mg/L	398	1,761	6,600
Cl	mg/L	230	529	2,712
Na	mg/L	330	1,043	3,350
K	mg/L	106	305	1,540
Ca	mg/L	140	308	1,230
Mg	mg/L	18	151	1,100
Conductivity	mS/m	292	774	7,120
COD	mg/L	59	572	6,500
TOC	mg/L	35	215	2,335
NH_4-N	mg/L	10	195	1,294
AOX	µg/L	50	503	8,400
As	µg/L	5	45	590
Cd	µg/L	0.7	4	34
Cr	µg/L	10	29	270
Cu	µg/L	10	68	1,370
Hg	µg/L	0.2	0.9	5.5
Ni	µg/L	20	74	1,110
Pb	µg/L	5	97	2,800
Zn	µg/L	49	964	18,900

Box 10.2.3 Leachate quality from hazardous waste landfills (German class III) in North Rhine Westphalia (NRW), Germany (values based on 14 class III landfills and 640–2250 analysis; dependent on parameter) (Anonymous, 2010)

Parameter		Median	75% Quantile	Maximum
SO_4	mg/L	1,083	4,577	44,800
Cl	mg/L	6,200	27,485	189,600
Na	mg/L	2,302	19,906	89,500
K	mg/L	500	8,240	43,000
Ca	mg/L	304	610	10,200
Mg	mg/L	123	220	1,700
Conductivity	mS/m	2,100	7,680	96,300
COD	mg/L	658	3,781	35,000
TOC	mg/L	102	1,192	8,700
NH_4-N	mg/L	258	1,497	9,845
AOX	µg/L	1,000	5,412	34,000
As	µg/L	150	4,376	38,000
Cd	µg/L	2	4,386	87,000
Cr	µg/L	56	2,954	50,000
Cu	µg/L	40	1,960	21,000
Hg	µg/L	0.4	9	274
Ni	µg/L	320	2,590	55,000
Pb	µg/L	8	481	8,200
Zn	µg/L	50	18,498	590,000

Box 10.2.4 Leachate quality from sewage sludge landfills (mostly anaerobic digested sewage sludge) in North Rhine Westphalia (NRW), Germany (values based on five sewage sludge landfills and 121–222 analysis; dependent on parameter) (Anonymous, 2010)

Parameter		Median	75% Quantile	Maximum
SO_4	mg/L	210	569	2200
Cl	mg/L	190	479	1900
Na	mg/L	67	171	650
K	mg/L	23	57	300
Ca	mg/L	165	364	786
Mg	mg/L	30	86	250
Conductivity	mS/m	256	1022	4980
COD	mg/L	96	831	1850
TOC	mg/L	24	152	1000
NH_4-N	mg/L	10	153	820
AOX	μg/L	50	114	530
As	μg/L	5	28	109
Cd	μg/L	0.1	2	11
Cr	μg/L	10	25	150
Cu	μg/L	12	495	7600
Hg	μg/L	0.1	0.2	1
Ni	μg/L	16	586	4700
Pb	μg/L	10	32	200
Zn	μg/L	20	162	1300

CONCLUSIONS

Leachate production rates and quality have to be known to design a leachate treatment plant. Since the plant has to be built in the course of the landfill construction, leachate loads have to be predicted. The data presented here are examples mainly from landfills in Europe, but the quality characteristics are similar all over the world. On the other hand it can be seen from the wide range of concentrations that each landfill is different and leachate quality has to be monitored from the beginning of landfill operation. Important are the waste input quality and the way the landfill is operated. Tendencies are summarized in Table 10.2.8. Operating the landfill in a way that leachate pollution is reduced will result in a significant cost savings; in addition, with an improved way of operation, additional positive effects can be achieved (see Chapters 1.1, 2.1, 2.2, 3, 14.1, 16.2).

Table 10.2.8 Factors of influence on tendencies in leachate concentrations with time based on average conditions (different factors can be combined e.g. MBT waste with high or low precipitation input and/or high or low compaction, which will result in higher or lower decrease of organic concentrations).

Landfill type	BOD	COD	NH_4-N	Cl	Time until BOD/COD ratio <0,1			
					<3 years	3-6 years	6-9 years	>9 years
Fast build up in heigh, high compaction	●●●●●	●●●●●	●●●●	●●●●				↔
MBT waste	● / ●●	●●/●●●	● / ●● 1)	●●/●●●	↔			
Bio- reactor landfill	●● / ●●●	●●● / ●●●●	●●●●	●●●●	↔	↔		
Low compaction * (< 0,6 t wwt/m³)	●● / ●●●	●●●	●●●	●●●	↔	↔		
High share of inert waste * (> 50%)	● / ●●	●● / ●●●	●●	●●●	↔			
High share of food waste * (> 50%)	●●●●●	●●●●●	●●●●●	●●●●		↔	↔	
High precipitation * (> 1500mm/year)	●●●●●	●●●●●	●●●●●	●●●●		↔	↔	
Low precipitation * (< 600mm/year)	●●●●	●●●●	●●●	●●●			↔	↔
Landfill in methanogenic stage *	●●	●●●	●●●●	●●●●	↔			
compost layer > 2m at landfill bottom *	● / ●●	●● / ●●●	●●●●	●●●●	↔			
Average landfill *	●●●●	●●●●	●●●●	●●●●		↔	↔	↔
● : very low / ●●●●● : very high	↔	Typical interval in years reaching BOD/COD>0,1 after start of operation						
1) sometimes very low NH_4-N and increased NO_X could be possible								
* Average landfill in Europe: 2m lifts, compaction 0,8-1,0 t wwt/m³, precipitation: 750mm/year								

References

Anonymous, 2010. Beschaffenheit von Deponiesickerwasser in Nordrhein-Westfalen LANUV-Fachbericht 24. Landesamt für Natur, Umwelt und Verbraucherschutz, Nordrhein-Westfalen Recklinghausen.

Amokkrane, A., Comel, C., Veron, J., 1997. Landfill leachate pretreatment by coagulation-flocculation. Water Research 31 (11), 2775.

ATV-Fachausschuss 3.6, 1997. Arbeitsbericht "Ablagerung von Schlacken aus Hausmüllverbrennungsanlagen vor dem Hintergrund der TA Siedlungsabfall" (Work report "landfilling of MSW incinerator bottom ash for the background of the technical guidance MSW"). Korrespondenz Abwasser 44, 122.

Ehrig, H.J., 1980. Beitrag zum quantitativen und qualitativen Wasserhaushalt von Müll-de-ponien: Quantitative and qualitative water inventory of landfills, vol. 26. Veröffentlichungen des Instituts für Stadtbauwesen, University Braun-schweig.

Ehrig, H.J., 1983. Quality and quantity of sanitary landfill leachate. Waste Management & Research 54.

Ehrig, H.J., 1989. Leachate quality. In: Christensen, T.H., Cossu, R., Stegmann, R. (Eds.), Sanitary Landfilling: Process, Technology and Environmental Impact, p. 213.

Ehrig, H.J., Hagedorn, S., 1994. Berichte der Abwassertechnischen Vereinigung e.V.. Behandlung von Deponiesickerwasser: Treatment of landfill leachate, vol. 44, p. 497.

Ehrig, H.J., 2001. Sickerwasser aus Abfallablagerungen (leachate from landfills). In: ATV-Handbuch Industrieabwasser, Dienstleistungs- und Veredelungsindustrie (Eds.), ATV-dvwk (German Association for Water, Wastewater and Waste). Verlag Ernst & Sohn, ISBN 3-433-01468-X, p. 347.

Ehrig, H.J., Krümpelbeck, I., 2001. The emission behavior of old landfills in the aftercare phase. In: Eight International Waste Management and Landfill Symposium, Sardinia, 1-5 October 2001.

Fan, H.J., Shu, H.Y., Yang, H.S., et al., 2006. Characteristics of landfill leachates in central Taiwan. Science of the Total Environment 361, 25—37.

Guan, L., 2005. Treatment to Organic Pollutant in the Municipal Landfill Leachate by the Submerged Ultrafiltration Membrane Bioreactor (Master thesis). Jilin University (in Chinese).

Hjelmar, O., Johannessen, L.M., Knox, K., Ehrig, H.J., Flyvbjerg, J., Winther, P., Christensen, T.H., 1995. Management and Composition of Leachate from Landfills. EU commission, Management of waste, DGXI, C7 unpublished report.

Hjelmar, O., Hansen, J.B., 2005. Sustainable landfill: the role of final storage quality. In: Tenth International Waste Management and Landfill Symposium, Sardinia, 3-7, October 2005.

Huber, W., Schatz, S., 2007. Entwicklung der Sickerwasserzusammensetzung in der Nachsorgephase – Auswertung von bayerischen Altdeponien (Development of leachate composition in the aftercare phase – analysis at old Bavarian landfills). In: 18. Nürnberger Deponieseminar 2007.

Jiang, H.T., Zhou, G.M., Gao, T.Y., 2002. The characteristics of MSW landfill leachate. Environmental Protection Science 3, 11–13 (in Chinese).

Johnson, C.A., Kaeppeli, M., Brandenberger, S., Ulrich, A., Baumann, W., 1999. Hydrological and geochemical factors affecting leachate composition in MSW incinerator bottom ash. Part II: the geochemistry of leachate from landfill Lostorf, Switzerland. Journal Contaminant Hydrology 40, 239.

Krümpelbeck, I., 2000. Veröffentlichungen des Lehrstuhls für Abfall- und Siedlungswasserwirtschaft der Bergischen Universität Wuppertal. Untersuchungen zum langfristigen Verhalten von Siedlungsabfalldeponien: Investigations on long term behaviour of landfills, vol. 3.

Krümpelbeck, I., Ehrig, H.J., 2000. Emissionsverhalten von Altdeponien (Emission behaviour of old landfills). Deponietechnik, Hamburg.

Li, C.P., Li, G.X., Luo, Y.M., Li, Y.F., 2008. Fuzzy mathematics-based ground water quality evaluation of six MSW landfills in Beijing. Environmental Science 10, 2729–2735 (in Chinese).

Liu, G.Q., 2007. Study on Characteristic and Removal Efficiency of DOM in Leachate (Master thesis). Chongqing University (in Chinese).

Loizidou, M., Vithoulkas, N., Kapitanios, E., 1992. Physical chemical treatment of leachate from landfill. Journal Environmental Science and Health A 27 (4), 1059.

Lopez, A., Pagano, M., Volpe, A., di Pinto, A.C., 2004. Fenton's pretreatment of mature landfill leachate. Chemosphere 54, 1005.

Morf, L.S., Kuhn, E., 2010. Qualität von Sickerwasser aus Zürcher Schlackenkompatimenten (Leachate quality from landfill parts – Zürich). AWEL Amt für Abfall, Wasser, Energie und Luft, Kanton Zürich.

Øygard, J.D., Gjengedal, E., Måge, A., 2005. Mass-balance estimation of heavy metals and selected anions at a landfill receiving MSWI bottom ash and mixed construction wastes. Journal of Hazardous Materials a 123, 70.

Ritzkowski, M., 2011. How does landfill aeration impact on leachate composition? In: Cossu, Diaz, Stegmann (Eds.), Sardinia 2011: Proceedings Thirteenth International Waste Management and Landfill Symposium.

Robinson, H., 1995. A Review of the Composition of Leachates from Domestic Wastes in Landfill Sites. Environment Agency, UK. CWM 072/95.

Robinson, H.D., Knox, K., Bone, B.D., 2004. Improved Definition of Leachate Source Term from Landfills – Phase 1: Review of Data from European Landfills, Science Report P1–494/SR1. Environment Agency, UK, ISBN 1844323269.

Robinson, H., June 2007. The Composition of Leachates from Very Large Landfills. An International Review, vol. 81. CWRM, p. 19.

Robinson, H.D., Carville, M., June 2007. The design, commissioning and operation of leachate treatment plants at large landfills in tropical regions. In: Paper Presented to "Waste a Global Resource", the 2007 Annual Exhibition and Conference of the UK Chartered Institution of Wastes Management. Torbay, UK.

Salem, Z., Hamouri, K., Djemnaa, R., Allia, K., 2008. Evaluation of landfill leachate pollution and treatment. Desalination 220, 108.

Sleat, R., Harries, C., Viney, I., Rees, J.F., 1989. Activities and distribution of key microbial groups in landfill. In: Christensen, Cossu, Stegmann (Eds.), Sanitary Landfilling. Academic Press.

Tatsi, A.A., Zouboulis, A.I., Matis, K.A., Samaras, P., 2003. Coagulation-flocculation pretreatment of sanitary landfill leachate. Chemosphere 53, 737.

Wang, Y., Pelkonen, M., Kaila, J., 2013. Optimization of landfill leachate management in the aftercare period. Waste Management & Research 21, 789.

Weber, B., 2002. Wie lange muss die Sickerwasserbehandlung dauern? (How long must be leachate treated?). In: Tagung Deponienachsorge, 10./11.9.2002, Hamburg.

Yue, D., 2016. Personal Communication.

Zou, L.S., Tang, J., Ye, K.Z., 2011. Choice of landfill leachate treatment in Shanghai old port. Journal of Yunnan University of Nationalities (Natural Sciences Edition) 20, 23–26 (in Chinese).

10.3

BIOLOGICAL LEACHATE TREATMENT

Hans-Jürgen Ehrig, Rainer Stegmann and Tim Robinson

INTRODUCTION

Biological treatment is worldwide the most common practice for leachate treatment, because at most municipal solid waste landfills the main leachate compounds are organic substances and nitrogen. In most cases biological methods are also the most economical technologies. But not all organic substances are biodegradable with the consequence of higher chemical oxygen demand (COD) effluent values as the discharge requirements allow. Additional chemical and/or physical treatment will be necessary to meet the discharge values. The main nitrogen component in landfill leachate is ammonium. If ammonium concentrations exceed some thousand milligrams per liter, other methods as biological treatment (e.g., stripping) may be more economic. The advantage of such methods is the possibility of reusing the nitrogen, which may become a more important aspect in the future.

For biological leachate treatment, different processes are used. Some of them are presented in this chapter:

Anaerobic biological treatment

- Parts of the landfill body used as bioreactor
- Technical scale reactors (filter, upflow sludge bed reactor, mixed digester)

Aerobic biological treatment

- Aerated lagoons
- Activated sludge plants (conventionally plant, membrane bioreactor, sequencing batch reactor)
- Fixed film reactors [rotating biological contactors (RBCs)]

GENERAL REMARKS

As the result of local climatic conditions, uneven distribution of rainfall and other influences, in most cases, daily leachate flow rates show wide variations. To reduce the influence of such variations on the treatment process, storage capacity for untreated leachate will in general be necessary. Most treatment processes, especially those involving biological treatment, generally operate more stable and more reliably at relatively constant loading rates. As a consequence a very careful estimation of the maximum load of the biological treatment plant over several days is necessary.

Biological processes react very sensitive to low temperatures. Overall the activity of microorganism increases with increasing temperature; the activity may decrease at temperatures above 39–41°C. As a consequence of this temperature dependency, the necessary plant volume must increase with decreasing temperature. In many cases it is very difficult to get stable effluent quality values from biological plants operating at temperatures below 5°C. Therefore it is helpful to reduce/avoid that leachate gets cooler between the landfill and the treatment plant. Some possibility may be the reduction of cold air access into the transportation pipe. The storage tank should be covered with floating cover, e.g., with plastic balls. Heating with waste heat from landfill gas engines may sometimes be an option.

The organic pollution of water is mostly described by the parameters such as total organic carbon (TOC), chemical oxygen demand (COD), and biological oxygen demand (BOD, measured over a period of 5 or 7 days as BOD_5 or BOD_7). Analyzing TOC and COD almost the total amount of organics is measured. The relationship between biodegradable and total organics is nearly the same as in domestic sewage. But for leachate in many cases the BOD_5 underestimates the part of the biodegradable organics. Only highly polluted leachate from landfills in the acetic phase shows the same relationship as sewage with BOD_5/COD ratios of approximately 0.5–0.6. With increasing landfill age and transition to the methanogenic phase the BOD_5/COD ratio in leachate decreases below 0.1. Such decrease is mostly described as a transfer to more nonbiodegradable organic compounds. As shown in Fig. 10.3.1 also at BOD_5/COD ratios <0.1 a significant part of the COD is still biodegradable.

To get more information on the biodegradability of leachate from the methanogenic phase, it may be helpful to measure the BOD over a longer period of time (20–30 days). This may cause some problems due to the fact that at low BOD concentrations for analysis no or only minor dilution of the samples is necessary. To avoid toxic effects due to high ammonia concentrations in the leachate, high

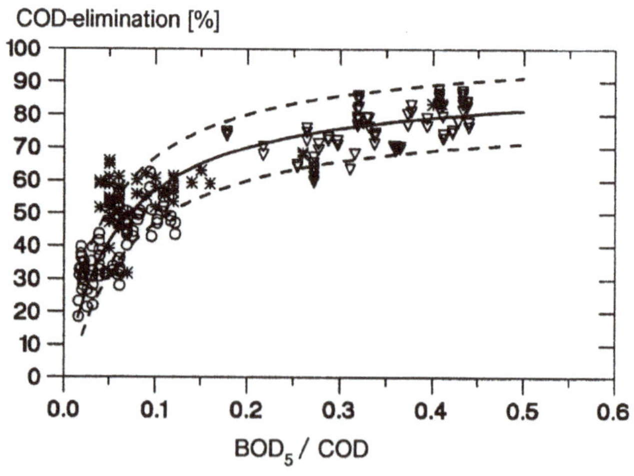

Figure 10.3.1 Relationship between BOD_5/COD ratio and biological COD removal (Chang, 1993). *BOD*, biological oxygen demand; *COD*, chemical oxygen demand.

dilution factors are required. Using undiluted leachate for the BOD measurement, the results may not be reliable. A solution to determine the biologically degradable part of a leachate sample may be a laboratory-scale biological degradation test using the difference of the COD of raw and treated leachate as the degradable COD. Fig. 10.3.1 shows results of such tests.

ANAEROBIC TREATMENT

Anaerobic biological degradation is often used for the treatment of wastewater and sludge with high organic concentrations. The anaerobic process consists of different degradation steps (Chapter 10.2) and can be summarized as follows:

$$\text{Organic substances} + H_2O \rightarrow CO_2 + CH_4 + \text{energy (heat)} + \text{biomass}$$

In particular the acetogenesis and methanogenesis are very sensitive processes where specific groups of microorganism are active. Both processes relate on each other and must be in balance, otherwise the anaerobic degradation ends in the acid phase with a remaining high organic pollution and intense odor. Therefore it is necessary to stabilize and control intensively pH, temperature, acid concentration, loading rates, etc. (Box 10.3.1).

In general after anaerobic treatment, aerobic posttreatment will be necessary to meet the discharge requirements and to eliminate remaining odor problems.

The leachate quality is the result of biological and leaching processes in the landfill, which is an uncontrolled anaerobic bioreactor (see Chapter 10.2). In the first phase after starting landfill operation, the different process steps are not in balance with resulting high organic leachate pollution. Leachate that is produced during this early landfill phase can be very effectively treated with anaerobic technology. This so-called acid phase may last over a period of several years (2–12 years). During the following methanogenic phase, an anaerobic treatment system is not viable because of a relatively low amount of biodegradable organics in the leachate ("Biological processes and leachate quality" in Chapter 10.2).

A relative simple technology for the anaerobic treatment of high-polluted acid phase leachate is the use of landfill parts that are in the methanogenic stage (Box 10.3.1). The highly polluted leachate from the part of the landfill in the acid phase has to be recirculated in an adequate way where the leachate has to be injected subsurface (see Chapters 12.1 and 12.2). In addition a gas collection system has to be in place. With the required low loading rates of the landfill, the treatment efficiency may in many cases be higher than using high-technology treatment facilities, which are in general operated at significant higher loading rates. Due to the heterogeneity of the landfill body, the range of leachate quality variation may be considerably higher.

A few publications (Cossu, 1981; Frascari et al., 2004) present data of leachate treatment in anaerobic lagoons (Box 10.3.1). Due to the production of methane and odorous substances, the environmental impact of anaerobic lagoons may be significant.

Anaerobic filter and upflow anaerobic sludge bed reactors (UASBs) are discussed as technical systems for anaerobic biological leachate treatment. Only few experiences with both systems exist.

Box 10.3.1 Conditions and loading rates of anaerobic biological treatment systems

Anaerobic process (conditions, influences, and products):

- Optimal pH value: 7.0–7.3.
- Optimal temperature: 30–33°C.
- Inhibition effects of increasing ammonia concentrations (with increasing pH a higher part of ammonium is converted to free ammonia).
- Only 5%–10% of organic carbon is converted into biomass.
- As result of low biomass production the consumption of nutrients as P and N is low.
- The energy production as heat is low.
- An important end product is the produced methane that should be used.

Example: Rough comparison of potential of electrical energy production and need during complete anaerobic and aerobic degradation:
Leachate concentration: 20,000 mg COD/L.
Anaerobic: energy produced as methane approximately 167 MJ/m^3 (can be converted to approximately 14 kW h/m^3 electric power).
Aerobic: Energy requirement for stoichiometric oxygen supply approximately 30 kW h/m^3.
Methanogenic landfill section acting as an anaerobic reactor:

- Required landfill height >10 m.
- The landfill section needs a gas collection system and should be sealed at the surface to reduce methane emissions.
- The leachate should be spread on the area as equal as possible.
- Maximum load 100–500 kg COD per 10.000 m^2 and m effective landfill height (effective landfill height = total height − 2 m at surface − 2 m at bottom).
- Higher loading rates can produce serious clogging problems with reduced permeability of the landfill body.
- Considering that landfills are very heterogeneous, operation should start with lower loading rates; areas with low treatment capacity should be excluded.
- Possible effluent values: 1000–4000 mg COD/L and 100–400 mg BOD_5/L.

Anaerobic lagoons:

- Loading rate: 1–2 kg BOD_5/m^{3*}day, influent ≈ 27,000 mg COD/L and 13,000 mg BOD_5/L; 58%–60% elimination rate (Cossu, 1981).
- Loading rate: 0.031–0.073 kg BOD_5/m^{3*}day, influent ≈ 4000–7000 mg COD/L and ≈1000 –3000 mg BOD_5/L; 25%–60% COD and 30% –90% BOD_5 reduction (Frascari et al., 2004).

Anaerobic filter (AF) and upflow anaerobic sludge bed reactor (UASB) (Mennerich, 1988):

- COD influent: 20,000–80,000 mg/L
- AF: COD effluent < 4.000 mg/L with COD load ≤ 4 kg COD/m^{3*}day
- UASB: COD effluent < 10.000 mg/L with COD load ≤ 4–5 kg COD/m^{3*}day (sludge growth and sludge coating with inorganic precipitates increase nearly parallel).
- Inorganic precipitates reduce treatment efficiency to very low elimination rates.

Blakey et al. (1992) made an investigation in laboratory scale on anaerobic lagoons and UASB reactors. The COD concentrations exceeded >10,000 mg/L. The results show degradation rates of about 70%–80% COD in 200 days at temperatures of 10°C and 90% degradation at 25 degrees in about 60–80 days (10% at 4°C). Mennerich (1988) published several results of similar experiments where he also described the problems when operating anaerobic treatment over a longer period of time.

pH values of the highly polluted leachate in the acetic phase are in general between 5.5 and 7. During anaerobic treatment the pH increases to values >7, and calcium and other components precipitate. The precipitates plug the pores of anaerobic filters (also coarse pores) and cover the microorganism on the filter surface. In UASB reactors the precipitates also covers the microorganism of the sludge bed. The result in both cases is a sharp decrease of the treatment efficiency. Methods for prereduction of calcium are used for other wastewaters but seem to be too costly for leachate treatment over only a limited period of time.

Considering the operational problems and the limited operation time only, nonsophisticated technology should be used for anaerobic pretreatment. Examples may be anaerobic filter with cheap and coarse filter material or completely mixed reactors as they are used for sewage sludge digestion. Mainly for very large landfills with long operation periods, anaerobic leachate treatment may be an option. In any case a second aerobic treatment step will be obligatory to further reduce organics and remove ammonia.

AEROBIC TREATMENT PROCESSES

General description

Aerobic biological treatment is mostly used for leachate treatment to reduce biodegradable organic substances and nitrogen components. During the acetic phase, volatile fatty acids such as acetic acid, propionic acid, butyric acid etc., are the main components representing often more than 70% of COD. With the change to the methanogenic phase the organic substances consist more of larger molecules (humic- and fulvic-like acids), which are of low or "no" biological degradability. This can be described by the BOD_5/COD ratio that decreases from >0.5 to <0.1.

A large number of heterotrophic microorganisms degrade the organic substances (organic carbon) in several steps (see Box 10.3.2). The aerobic degradation process of organic substances can be started in a plant in the relative short time of 1–2 weeks. Robinson and Knox 2001 and Robinson et al. (2003a) published data that aerobic treatment processes could also remove trace organics and heavy metals from leachate. Heavy metals may be adsorbed at the surface of accumulated microorganism and inorganic substances. Trace organics may also be adsorbed or degraded in cotreatment with easy biodegradable organics.

Besides organic leachate pollution, nitrogen components such as ammonium and organic nitrogen are the other main compounds that have to be removed during leachate treatment (both components are often analyzed together as Total Kjeldahl nitrogen—TKN). Nitrogen can be effectively removed by biological treatment and it may be more costly than the removal of the organics.

Nitrogen concentrations can vary in a wide range from some hundred milligrams per liter up to 2000–3000 mgN/L. In contrast to the organic leachate concentrations, the nitrogen values stay stable and even may increase with landfill age (see Chapter 10.2). The main nitrogen parameter is typically ammonium, but the content of organic nitrogen could reach up to 30% of the total nitrogen. Particularly during the methanogenic phase the organic nitrogen may be partly incorporated into humic

Box 10.3.2 Aerobic biological processes: general features of organic carbon elimination

Generalized equation of aerobic biological degradation of organic carbon

$$Corg + O_2 \Rightarrow H_2O + CO_2 + \text{energy (heat)} + \text{biomass}$$

- The biodegraded COD represents the minimum oxygen requirement of this process. High energy input is necessary to provide oxygen.
- The produced heat cannot be used because of losses into the atmosphere during aeration.
- 20%–30% of the organic carbon is converted into biomass (with COD > 15,000 mg/L, this value can rise up to 50%). The produced excess sludge must be removed and treated/disposed.
- Results of sludge analysis at different leachate treatment plants do not indicate any/low pollution by organic micropollutants or heavy metals (has to be checked).
- The production of new biomass requires nitrogen of about 2.5%–3% of degraded COD (phosphorous requirement about 0.3–0.5% of degraded COD). Most leachate have a deficiency in phosphorous.
- At aerobic conditions in the bioreactor, several inorganic substances (most important iron and calcium) may precipitate and produce inorganic sludge.
- The combination of organic and inorganic sludge can produce serious clogging problems at all mechanical devices and pipes.
- Foaming can be a serious problem at biological leachate treatment plants; it can be reduced by the addition of antifoaming agents. Most antifoaming agents are biodegradable oily substances and have an extremely high COD. At high dosing rates they can comprise a significant proportion of the total organic load of the treatment plant.

matrix, which would make it difficult or even impossible to remove organic N by means of biological processes.

Inorganic nitrogen may be in the status of ammonium (NH_4) or ammonia (NH_3). The environmental effect of each parameter is quite different. While ammonium in the above presented concentration range in leachate, has no toxic effect on organism, ammonia is very toxic to many organism, particularly to water organism. The balance between both parameters shifts to the ammonia side with increasing temperature and pH value. That could be a serious problem for anaerobic processes (See section on "Anaerobic treatment") but also for aerobic processes at partly high ammonium concentrations. The high fish and water organism toxicity of ammonia and the high oxygen requirement for ammonium oxidation in natural watercourses leads to very low required effluent values.

The reduction of ammonium can be achieved with a two-step biological treatment process: nitrification and denitrification. As shown in Box 10.3.3 the nitrification process is a relative sensitive process and should be carefully designed and operated. If adequately designed—the dimensioning of the plant should be based on the nitrification process—the organic degradation and the nitrification processes take place simultaneously.

Box 10.3.3 Aerobic biological processes: general features of nitrogen elimination

Organic nitrogen

- Hydrolyzed or degraded under anaerobic or aerobic conditions to ammonium
- Production of 3.57 kg alkalinity as $CaCO_3$ per kg N

Ammonium

- Oxidation of ammonium in two steps—nitrification (following equations give a general description)
- **First step: Oxidation by species of Nitrosomonas to nitrite:**
- $NH_4 + 1.5\ O_2 \Rightarrow NO_2 + 2H + H_2O +$ biomass
- **Second step: Oxidation of nitrite by species of Nitrobacter to nitrate:**
- $NO_2 + 0.5\ O_2 \Rightarrow NO_3 +$ biomass
- **First and second step combined**
- $NH_4 + 2O_2 \Rightarrow NO_3 + 2H + H_2O +$ biomass
- Approximately 4.6 kg O_2 is used to oxidize 1 kg N (variation ±20%)
- 7.14 kg alkalinity as $CaCO_3$ is removed per kilogram N nitrification (production of 2 mol H). Depending on different operation conditions as aeration, mixing etc., a variation ±30% is possible
- Approximately 0.2 kg biomass is produced per kilogram N nitrification

Influences on nitrification process:

- pH values should be above 6.5. Sharp decrease of nitrification $\leq$ pH 5.8–6. Risk of ammonia toxicity at pH > 7.5–8. Optimum pH = 7–7.5.
- The reduction of alkalinity reduces pH. If no alkalinity is remaining at neutral pH, the process will also be inhibited (minimum remaining alkalinity approximately 400 mg $CaCO_3$/L). Alkalinity could be added as caustic soda or lime.
- Both groups of microorganism have a slow growth rate and need a high sludge age.
- The oxygen concentration in the nitrification reactor should be >2 mg O_2/L.
- Several organic and inorganic substances could be toxic for one or both groups of microorganism.
- At low pH values nitrite could inhibit the process.

Nitrate

- Reduction of nitrate: denitrification
- Organic $C + NO_3 \Rightarrow CO_2 + H_2O + N_2 + HCO_3 +$ biomass + energy (heat)
- 3.57 kg alkalinity as $CaCO_3$ is produced per kg N denitrification
- Approximately 4.5 kg of biodegradable COD is required per kg N denitrification (±20%–30%)
- Using internal organic carbon 2.9 kg O_2 per kg N denitrification could be recovered

Influences on denitrification process:

- Dissolved oxygen of >1–2 mg O_2/L could inhibit the process

Nitritation and denitritation

- Reduced sludge age and low oxygen content during nitritation
- 3.4 kg O_2 used for nitritation, 2.7–3 kg degradable COD is required for denitrification, using internal organic carbon 1.7 kg O_2 could be recovered

Partial nitritation and Anammox

- Reduced sludge age and low oxygen content during nitritation. Only 55% of ammonium should be oxidized (only 1.8 kg O_2 per kg nitrogen required)

Anammox (anaerobic ammonium oxidation)

- $NH_4 + 1.32NO_2 + 0{,}066HCO_3 + 0.13H \Rightarrow 1.02N_2 + 0.26NO_3 + 2.03H_2O +$ biomass
- 11% of nitrogen oxidized to nitrate. Increase of pH values.
- Growth rate of microorganism is extreme slow.
- Increased nitrite concentrations and dissolved oxygen inhibit the microorganism.
- Temperatures above 20°C are necessary.

If sufficient organic degradable carbon is available, nitrogen may be used during the aerobic carbon degradation for biomass production; this will reduce the necessity for nitrification significantly.

In most cases the discharge concentrations of nitrate are limited. Just like ammonium nitrate also causes eutrophication in the receiving waters. Also for this reason, NO_3 should be reduced to gaseous nitrogen (N_2) by means of denitrification (Box 10.3.3). Comparing the equation for denitrification with the equation of aerobic biological degradation of organic carbon in Box 10.3.2, it becomes obvious that both are quite similar. In principle, denitrification is the anoxic degradation of organic carbon where the oxygen is substituted by NO_3. A large number of heterotrophic microorganisms can use nitrate instead of O_2. Their energy yield is somewhat lower than using O_2. For this reason the supply of oxygen must be prevented as far as possible.

To reduce nitrate, sufficient organic degradable carbon must be available. Under ideal conditions the organic degradable C should be available in the stoichiometric amount to realize complete consumption of nitrate. During the acidic phase of the landfill the carbon supply is in general no problem, but during the long-term methanogenic phase the amount of the degradable organic content in the leachate is low. As a consequence an external organic carbon source has to be supplied. This carbon should have a COD concentration of at least 600,000–700,000 mg/L; it has to be easy biodegradable and should not increase the COD values in the effluent significantly. Of course no other contaminants (heavy metals, hazardous organics) should be added. Since it is very difficult to find a waste or by-product of industrial processes with such properties in most cases methanol, acetic acid, ethandiol, etc., are used as an external carbon source. Using methanol, inhibition effects shall be prevented by means allowing microbial adapting phases with slow increase of methanol addition. This process is quite cost intensive.

To reduce the cost for nitrogen elimination of low BOD leachate, another process may be used, where the ammonium is oxidized only to nitrite (NO_2). The nitrite is subsequently reduced during anoxic organic carbon degradation (nitritation and denitritation). Using this process combination a reduction of energy for oxygen supply (no oxidation of NO_2 to NO_3) and for the external carbon dosing is possible (only reduction of NO_2 instead of NO_3).

For the nitrogen removal from leachate of landfills in the methanogenic phase, the so-called Anammox process may be important. As described in the previous paragraph, ammonium may be oxidized only to nitrite. But only approximately 55% of ammonium should be oxidized to nitrite. In the absence of dissolved oxygen in a separate tank or in a next phase, using the nitrite oxidizes the remaining ammonium. Besides gaseous nitrogen, only 11% of nitrogen is oxidized to nitrate. For establishing the complete process the necessity of intensive operation control increases.

Aerated lagoons

Biological treatment of leachate may also be achieved in lagoons, which are not equipped with a specific system (e.g., settling tank) to separate the microorganism (sludge) before the effluent will be discharged. The construction of lagoons can differ in a wide range: from nearly natural ponds to concrete tanks. Considering the high degree of leachate pollution, lagoons should be sealed at the basis. Dependent on the depth, lagoons can be operated without or with technical aeration.

To reach a sufficient oxygen supply for the aerobic degradation, lagoons with natural aeration by air diffusion into the water body need a large area and a low depth. With depth of more than 1–1.2 m, the lower part of the lagoon becomes anaerobic. If the oxygen content in the upper part is high enough, odor components may be biologically degraded. Lagoons without technical aeration are only useful for low polluted leachate with BOD_5 values of some hundred milligrams per liter. With higher concentrated leachate the reactor will become widely anaerobic. During cold seasons the biological activity is significantly reduced up to very low values.

In Germany first experiences in leachate treatment were gained during the 1970s and 1980s using as a very simple technology artificially aerated lagoons. Fig. 10.3.2 shows the scheme of an aerated lagoon using diffused air aeration and surface aeration.

Aerated lagoons are self-regulated systems. In such a system the elimination rate is determined by the combination of flow rate, lagoon volume, and water temperature. The produced biomass leaves the system with the effluent. As a consequence the solids retention time (SRT) has to be equal to the hydraulic retention time (HRT), which results in relatively long detention times with the consequence of large volumes. With the exception of reducing the incoming leachate, the system cannot be influenced. Most aerated lagoons are operated without any prestorage or equalization pond. To meet the required effluent values, it is necessary to remove the solids from the effluent stream. One solution is a subsequent nonaerated lagoon. The settled and mineralized sludge could be removed in intervals of several years.

Also with low loading rates the variation of effluent concentrations may be high (Fig. 10.3.3). In many areas during winter the water temperature may reach almost 0°C. The effect of reduced temperatures on effluent BOD_5 values is presented in Fig. 10.3.3. If the temperatures in the lagoon reach values below 10°C over a longer period of time, nitrification processes will decrease—in some cases even to zero—and at increasing temperatures it may require several weeks until nitrification is completely recovered. During high loading phases (>100–200 g BOD_5/m^3 day) of the lagoon (Fig. 10.3.3), the reduction of the nitrification rate starts already at temperatures around 15°C. The reason is that at such temperatures the HRT is shorter than the generation time of nitrifying microorganism. With increasing temperatures, nitrification starts with a lag phase and intense intermediate

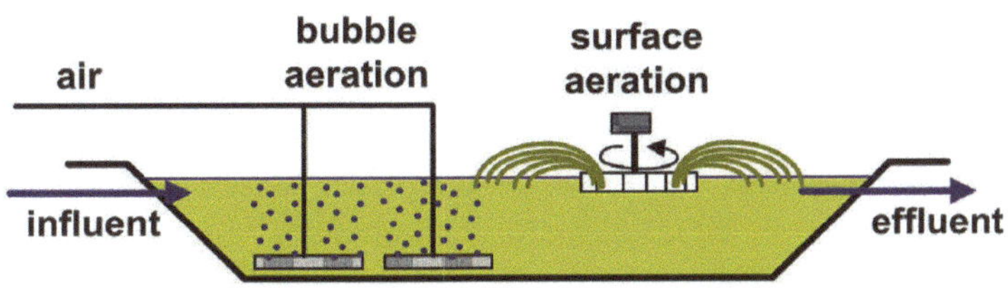

Figure 10.3.2 Aerated lagoon with bubble aeration (left side) and surface aerator (right side).

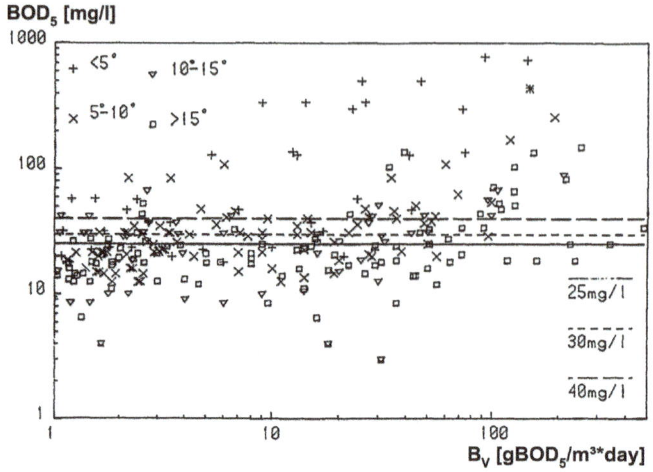

Figure 10.3.3 Aerated lagoon: volumetric loading rate B_V (kg BOD_5/m^3 day) and BOD_5 effluent concentrations depending on the temperature in the lagoon. *BOD*, biological oxygen demand.

production of nitrite (NO_2) until full nitrification is reached. Only at average temperatures of 15°C and higher, with volumetric loading rates <20 g BOD_5/m^3 day, BOD_5 values mostly below 25–30 mg/L and full nitrification may be expected (Albers, 1985). As a result of the relative low growth rate of the microorganism in leachate with BOD_5/COD ratios <0.1, the BOD_5 load should be reduced to 5 g BOD_5/m^3 day. Similar results are published by Maehlum and Haarstad (1997) showing good treatment rates at temperatures >10°C, but poor nitrification at lower temperatures. The variation of the effluent data during these periods can only partly be explained by varying influent loading rates. With all year-round temperatures of 20°C and higher, lagoons may be operated at loading rates of 100–200 g BOD_5/m^3 day. Fisher and Fell (1999) published values from aerated lagoons, which operated as sequencing batch reactors (SBRs) (temperature 20–25°C)—where the nitrogen loading rates were between 60 and 80 gN/m^3 day.

At the above given range of loading rates the oxygen requirement (OV) was calculated as follows: OV [gO_2/m^3 day] = 10[1,4 + 0,46*log(B_V)] where Bv is the volumetric loading).

As a result of the endogenous respiration, the oxygen requirement is relative high. The technical aeration may be much higher to achieve sufficient turbulence and prevent the settling of microorganism. Using mechanical stirrer can reduce the high-energy input for aeration.

Without technical installations a denitrification is in most cases not possible. Chen et al. (1997) reported results from a very large lagoon leachate treatment plant in Hong Kong, which operated as an SBR. It was designed on the basis of an extensive investigation (Robinson et al., 1995). This plant not only achieved a high reduction in COD values and ammonium (from levels of up to 150,000 mg COD/L and 8400 mgN/L, respectively), but also significant reduction of nitrate (Table 10.3.1). One reason for this effect is the high production of biomass during the degradation of the very high

Table 10.3.1 Performance values of a lagoon leachate treatment plant (Chen et al., 1997)

Parameter	Influent (Average)	Influent (Peak Values)	Effluent
pH	7	4.8–7.8	7.7–8.5
COD	40,000	153,300	600–1000
BOD_5	25,000	na	na
$N\text{-}NH_4$	3500	8400	<2
Total N			<1500

COD values. The produced sludge flocks have an anaerobic/anoxic core for denitrification and an aerobic surface. Due to the high organic leachate concentrations, an increasing part of nitrogen is used to build up new biomass (approximately 2.6%–3% of degraded COD). As a result these simultaneous happening processes were able to achieve an overall nitrogen reduction of approximately 55%.

Fig. 10.3.3 shows that also at very low loading rates (1–2 $gBOD_5/m^3$ day) effluent BOD_5 values above 25–30 mg/L may be measured. This variation of effluent concentrations had been observed at all loading rates at varying temperatures. This fluctuation could be equalized in a subsequent reed bed. In addition a slight reduction of low biodegradable COD and nitrogen concentrations during cold weather periods could be observed.

Activated sludge treatment plant

The most common biological treatment process for leachate is the activated sludge process (Fig. 10.3.4).

The plant consists of an aeration tank and a sludge separation unit (clarifier). This sludge separation unit in combination with a transport system to return the produced biomass back to the activated sludge tank is the characteristic of an activated sludge system. With this procedure the SRT is

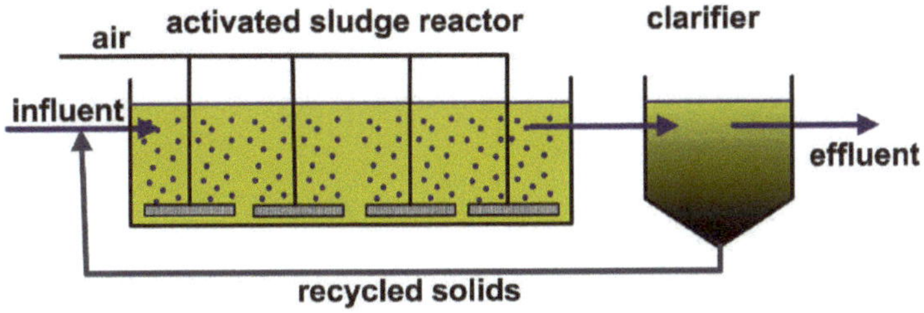

Figure 10.3.4 Activated sludge plant with clarifier for sludge separation.

uncoupled from the HRT. With the return sludge the process may be externally controlled, e.g., by increasing the biomass content in the activated sludge reactor and higher elimination rates per volume may be achieved. The limit of sludge concentration in the activated sludge tank is given by the capability of the sludge separation unit and the ability of the aeration unit to keep the sludge in suspension. If this limit is reached, biomass must be withdrawn as excess sludge.

With settling tanks as shown in Fig. 10.3.4, sludge concentrations [mixed liquor suspended solids (MLSS)] of maximum 10–12 kg/m^3 (average 6 kg/m^3) can be realized. This limit was the reason to develop new separation devices, which allow higher MLSS concentrations. The most common solution particularly for activated sludge plants with relative small flow rates is the ultrafiltration of the activated sludge tank effluent. Ultrafiltration is the filtration of the effluent through membranes with a pore size of 0.005–0.2 μm (molecular to macromolecular range). With such pore size all solids including single bacteria are retained. Such systems can handle MLSS concentrations up to 35 kg/m^3. But in most cases activated sludge plants for leachate treatment operate with values up to 12 and 20 kg/m^3. One reason for limiting MLSS is the problem to filter sludge in leachate treatment plants. Another important point is the required energy input into the activated sludge tank to reach sufficient turbulence to hold the sludge in suspense. Fig. 10.3.5 shows an ultrafiltration system at the effluent side of a conventionally

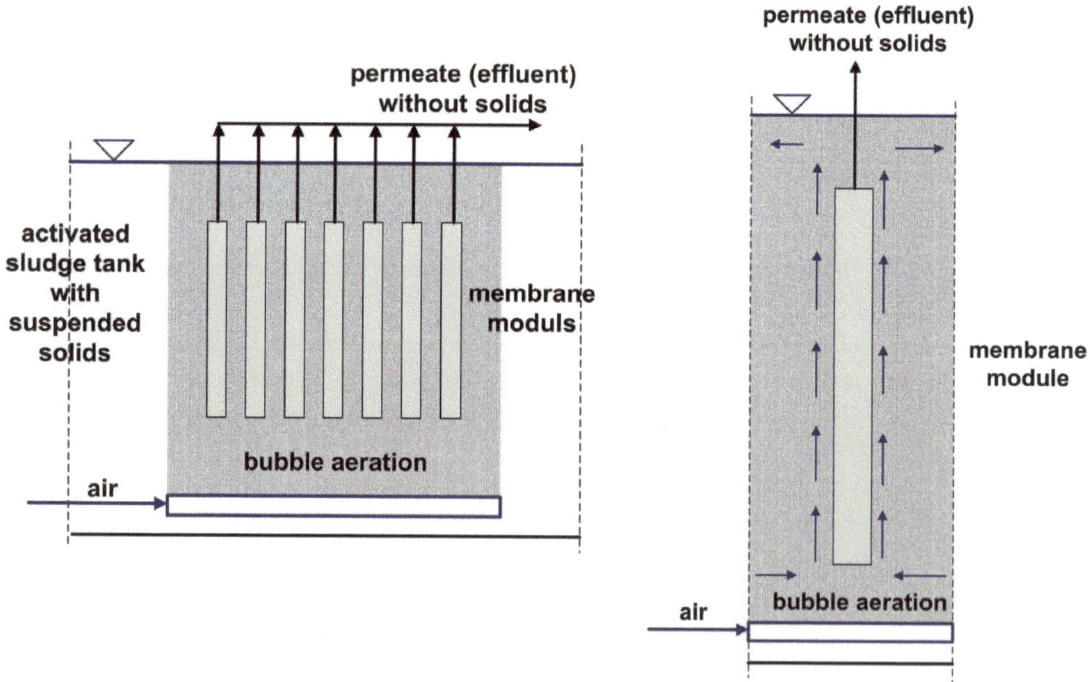

Figure 10.3.5 Sludge separation/retention in an activated sludge plant with partial vacuum ultrafiltration.

activated sludge tank. Normally the ultrafiltration consists of several membrane modules. The water is sucked by partial vacuum through the membrane. To prevent blocking of the membrane surface the water–solids mixture should flow parallel to the membrane and the water is sucked—orthogonal to the main flow—through the membrane. The parallel flow is initiated by very intensive aeration below the ultrafiltration membrane packages. This parallel stream should clean the membrane continuously.

Due to the higher sludge content by using an ultrafiltration system and due to thermal insulation of the activated sludge reactor to prevent cooling, the required volume of the activated sludge reactor can be reduced. As a consequence, in Germany, activated sludge plants at several medium size landfills are completely enclosed. The activated sludge plant consists of several prefabricated tanks in one or two parallel rows located in a large hall. In these cases a pressure-driven ultrafiltration unit is installed outside of the activated sludge tanks (Fig. 10.3.6).

To clean the membrane continuously the flow velocity through the ultrafiltration unit is quite high amounting to some meters per second parallel to the ultrafiltration membrane. Such a filtration is called cross-flow filtration.

Activated sludge plants are designed to remove biodegradable organic substances and nitrogen. For nitrification an oxidation supply is necessary with O_2 concentrations >2 mg O_2/L and denitrification is only possible during the absence of oxygen. As a consequence, two tanks are required for this process or the processes must react in one tank in sequence. For leachate treatment the activated sludge plant is often subdivided in different tanks (Figs. 10.3.6 and 10.3.7). Fig. 10.3.7 shows two activated sludge plants with different positions of the nitrification and denitrification tank. The influent of a predenitrification tank contains only organic carbon for the denitrification and ammonium. The nitrate must be pumped back from the nitrified effluent of the activated sludge tank. The influent of the postdenitrification tank contains nitrate. The organic carbon will be biodegraded in the aerated nitrification

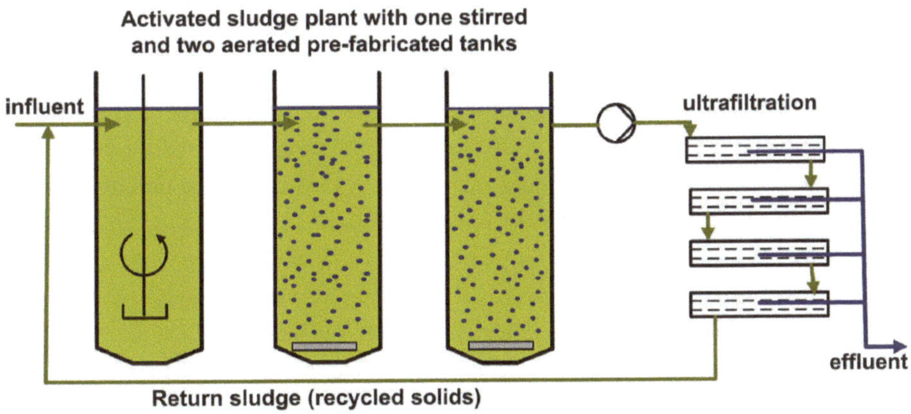

Figure 10.3.6 Activated sludge plant (prefabricated tanks) with pre-denitrification in a stirred reactor and separate pressure ultrafiltration for sludge separation.

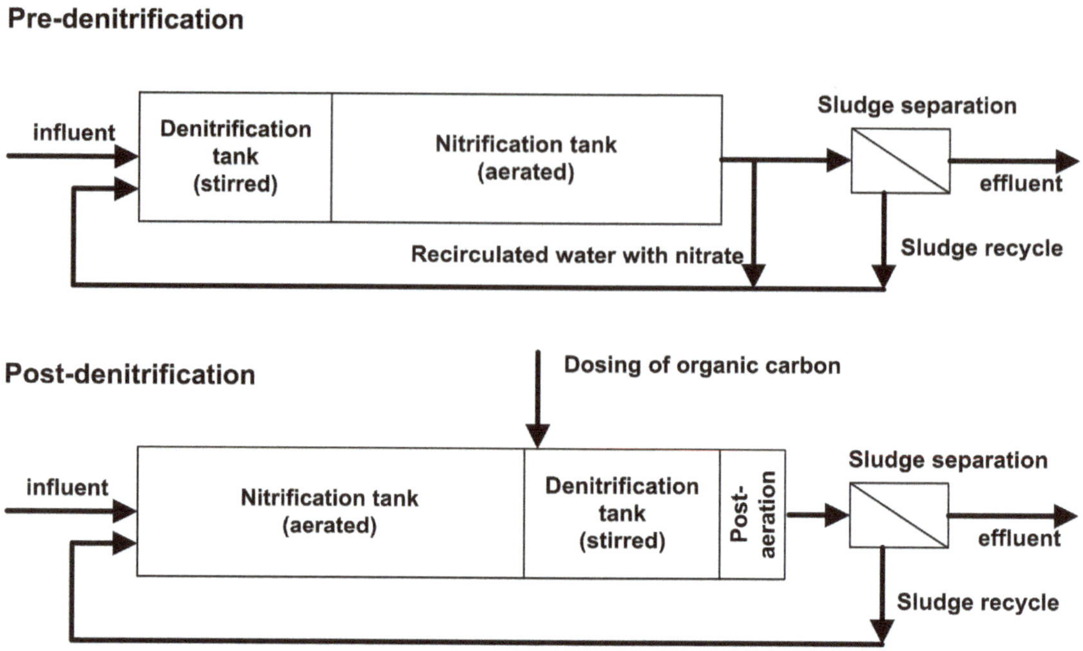

Figure 10.3.7 Activated sludge plant with pre- and postdenitrification.

tank. External carbon must be added to achieve denitrification. The oxygen requirement of the microorganism respiration is too low to reduce nitrate in a limited period of time.

Predenitrification is mostly used if the organic carbon of the raw leachate could be used. That is often possible during the acidic phase of a landfill or, e.g., from other parts of the landfill. This system needs a return-circuit water system for the transport of nitrogen-rich sludge into the denitrification tank (Fig. 10.3.6, Fig. 10.3.7 and Box 10.3.4). The postdenitrification should be the preferred option if external carbon must be added. After the postdenitrification tank a postaeration chamber is necessary to remove eventually remaining carbon substances. For operating denitrification in one reactor, it should be considered that only the easier biodegradable components will be used by the bacteria. If partly organics compounds in the leachate (endogenous respiration) and partly an external carbon source shall be used, a combination of predenitrification and postdenitrification could be installed.

Chang (1998,1999) has published a procedure to dimension an activated sludge plant for leachate treatment with denitrification and nitrification. In addition to the calculation of the tank volume for denitrification, a main point is the calculation of the necessary aerobic sludge age for nitrification. The calculation of the volume of the nitrification tank with a given sludge age is difficult because an equation with three unknown values (nitrification volume V_N, heterotrophic active biomass x_H, and autotrophic active biomass x_A) must be solved. A practical procedure is an iterative calculation approach using a spreadsheet. To simplify this procedure Box 10.3.5 includes loading rates for denitrification and nitrification processes with the addition of an external carbon source using the initial values from

Box 10.3.4 General features of predenitrification

- The volume of recirculated water (Q_{RW}) determines the maximum denitrification rate and thereby the NO_3 concentration in the effluent. Example: $Q_{RW} = 90\%$ of $Q_{influent}$, i.e., 90% of the nitrate load returns to the denitrification tank and 10% leaves the plant via the effluent. In general, the maximum denitrification rate is limited by the hydraulic conditions. Using the recirculation rate $RR = Q_{RW}/Q_{influent}$ the maximum elimination rate in % is $100 - 100/(1 + RR)$
- The recirculation rate should be limited to $RR = 10-12 * Q_i$; otherwise at higher rates an increasing amount of oxygen would be transported from the nitrification tank into the denitrification tank. Higher recirculation rates would decrease the hydraulic retention time in the denitrification tank, which might hinder the adaption of microorganism to the changing conditions from oxygen to nitrate. High recirculation rates may also result in short-circuiting in the denitrification tank with the possibility of direct transport of nitrate to then effluent of the tank.
- To establish constant O_2 values of approximately 2 mg/L and to reduce increased oxygen transfer from the nitrification tank into the denitrification tank an automatic oxygen control is necessary. Such a control system combined with a flexible oxygen supply system is very important especially during phases of low loading.
- The positioning of the influent and effluent pipe in the denitrification tank should prevent short-circuiting (also at lower recirculation rates).

Chang (1998,1999) for leachate from the methanogenic phase. The loading rates for nitrification are in the same range as in the data presented by Strachan et al. (2007) with 0.006–0.040 kg N/kg MLSS*day, Robinson et al. (2007) (0.021 kg N/kg MLVSS*day; MLVSS, mixed liquor volatile suspended solids), and Pelkonen et al. (1999) (0.012 and 0.03 kg N/kg MLVSS at 7–10°C). The same values as in Box 10.3.5 may be used if the carbon of the leachate is to be used for denitrification. The maximum relation between the biodegradable COD and the biodegradable nitrogen should not exceed the factor 5–6.

For the dimensioning of the denitrification and nitrification tank volume, the above described calculations should be used; the volumes are sufficient also for the carbon degradation. In case the biodegradable COD is six times higher than the biodegradable nitrogen, the COD load should be considered for dimensioning. In this case the COD load may be the more important value. If the nitrogen concentrations are lower than 2.6%–3% of biodegradable COD, all nitrogen will be used for biomass production and only the COD load is relevant for the dimensioning of the tank volumes. At temperatures around 15°C a sludge loading rate of 0.2 kg biodegradable COD per kilogram MLVSS and day may be used (temperature correction $1.1^{(T - 15)}$ with T in °C). With COD concentrations higher than 20,000 mg/L a 50% increase of the loading rate may be possible. The oxygen requirement for the COD and nitrogen removal is presented in Box 10.3.6.

As shown before the activity of microorganism and the resulting elimination rates depend significantly on the temperature in the reactor. With increasing temperatures the reaction rate increases so that smaller reactor volumes may be used. As a consequence it is helpful to reduce the cooling of

Box 10.3.5 Design procedure of activated sludge plants with nitrification and denitrification for methanogenic leachate

Basic data:
Flow rate: Q (m³/day)
Influent concentration of NH_4 + organic nitrogen or TKN in mgN/L
The biodegradable nitrogen concentration N_{bd} can be calculated to NH_4 + 50% of organic nitrogen or (NH_4 + TKN)/2
Lowest operation temperature in °C
Estimated suspended solids concentration in the activated sludge reactor (MLSS)
Mixed liquor volatile suspended solids (MLVSS) approximately 80% of MLSS
Nitrogen amount of approximately 2.6% of total COD (biodegradable COD) is used for biomass buildup and can be reduced before calculation (leachate and added COD)
Necessary COD: approximately 4.5 * denitrified nitrogen (only COD components with the highest and nearly similar K_{dn} values are used for denitrification. For methanogenic leachate the total amount of 4.5*N must be added. Depending on operation conditions the factor 4.5 can vary between 3.5 and >6 in praxis. Using partly acidic phase leachate COD, the factor should be increased to 5–6.)
Denitrification volume:

$$V_D = Q\,(m^3/day) * N\,(kg/m^3)/[K_{dn}\,(kgN/kgMLVSS * day) * MLVSS\,(kg/m^3)]$$

Denitrification rate K_{dn} (kgN/(kg MLVSS * day) at 15°C

	K_{dn} (kg MLVSS * day)
Endogenous	0.015
Leachate COD	0.020
Acetic acid	0.150
Ethandiol	0.150
Methanol[a]	0.090

[a]value could increase up to 0.150 after a long adaption phase.

Temperature correction $K_{dn} * 1.1^{[T(°C)-15]}$
Nitrification volume:

$$V_N = Q\,(m^3/day) * N\,(kg/m^3)\;/\;[K_N\,(kgN/kgMLVSS * day) * MLVSS\,(kg/m^3)]$$

Nitrification rate K_N (kgN/kgMLVSS * day) depending on COD and biodegradable nitrogen of influent leachate at 15°C

Nitrogen	**COD**		
	1000 mg/L	**2000 mg/L**	**3000 mg/L**
250 mgN/L	0.025	0.020	0.015
500 mgN/L	0.050	0.045	0.040
750 mgN/L	0.065	0.060	0.055
1000 mgN/L	0.075	0.070	0.065
1500 mgN/L	0.090	0.085	0.080

Temperature correction $K_N * 1.1^{[T(°C)-15]}$
COD = only leachate COD

After Chang (1998, 1999).

the reactor volume. This can be done using a cover on top of denitrification and nitrification tanks. Alternatively the whole plant can be enclosed. But some enclosed plants in Germany got temperature problems during summer time. The heat input from aeration and the heat of intensive pumping through the ultrafiltration system increased the water temperature up to around 40°C. But in the range >38–40°C most biological processes show a sharp decrease of activity. At 45°C nitrification processes

Box 10.3.6 Oxygen requirement of activated sludge plants and aeration systems

Oxygen requirement of organic carbon degradation.
The oxidation of ammonium nitrogen to nitrate nitrogen requires 4.6 kg O_2 per kg nitrogen.
Denitrification recovers 2.9 kg O_2 per kg denitrified nitrogen.
For the whole process of nitrification and denitrification 4.6–2.9 = 1.7 kg O_2 per kg nitrogen is required.
$V (m^3) = V_{dn} (m^3) + V_N (m^3)$ (Vdn: volume denitrification, Vn: volume nitrification)
Oxygen requirement for carbon reduction:
O_2-C (kg O_2/day) = 0.7 * Q (m^3/day) * COD' (kg/m^3) + 0.02 * V (m^3) * MLSS (kg/m^3) COD' (see Box 10.3.5)
Oxygen requirement for nitrification:
N_e (kg/m^3) = concentration of nitrate nitrogen in the effluent
N_{dn} (kg/m^3) = concentration of denitrified nitrate nitrogen
O_2-N (kgO_2/day) = Q (m^3/day) * [4.6 * N_e + 1.7 * N_{dn}]
Total oxygen requirement depends on oxygen saturation [C_S (mg O_2/L)] and oxygen concentration [C_X (mg O_2/L)] in the activated sludge tank
O_2 (kg/day) = (O_2-C + O_2-N) * C_S/(C_S – C_X)
C_S = 9.1 mg O_2/L at 20°C; C_X should be approximately 2 mg O_2/L

- The oxygen content in the nitrification tank should be limited to values near 2 mg O_2/L. Higher oxygen values lead to COD oxidation in the denitrification tank.
- Most common aeration systems are bubble aeration and surface aerators. Some enclosed plants with high prefabricated tanks are equipped with injector aerators.
- The compressed air from bubble aerators is heated up by the blowers and may increase the water temperature.
- Inorganic components may precipitate during aeration and clog particularly the aeration system. Ceramic aerator pipes for bubble aeration often blocked after a very short period of time. Aerator pipes consisting of flexible plastic membranes as pipes or plates show almost no clogging effect if a thin incrustation layer is peeled off with every switch of airflow. Such effect is possible when the openings for aeration are shaped like a wedge with the larger opening to the inside.
- Surface aerators allow an intensive contact of water and air, which may result during cold periods in lower water temperatures.
- At some enclosed activated sludge plants, reactors are operated with increased air pressure of typically 300–500 kPa. An advantage is the higher solubility of oxygen in water, but the required higher pressure requires more energy.
- For energy effectiveness it is necessary to adjust the oxygen supply to the actual oxygen requirement. It is necessary to estimate a possible variation of oxygen demand. For short-term overloading as a reserve an aeration capacity of approximately 10% should be available. Without further information the oxygen supply should be able to be adjusted down to 50% of calculated oxygen requirement.

decrease to almost zero. As a consequence the activated sludge may be cooled or pure oxygen may be used to reduce heat input from aeration.

The composition of leachate may vary in a wide range with the consequence that the kinetic parameters of biological processes can be strongly influenced. Pilot scale tests are helpful to verify published design parameters for the specific situation. The values presented in Box 10.3.5 should be used with caution at very low reactor temperatures (<5°C) and for high sludge concentrations (>12–15 kg/m^3).

Landfill leachate is a very complex wastewater with a lot of known and unknown substances. Particularly during changes of the milieu conditions (contact with oxygen, turbulence, etc.) inorganic substances as, e.g., carbonates may precipitate that may result in clogging of facilities that in contact with the activated sludge, e.g., pipes, pumps, and particularly aeration system.

Chloride concentrations of more than 20 g/L can particularly influence the nitrification process. Longer adaption phases can be required. Above 40 g Cl/L a significant reduction of nitrification rates has been observed (Ehrig and Albers, 1987).

Organic leachate compounds can create extreme foaming in the aeration tank. Foaming may also be the result of bacterial stress. Reasons may be overloading the system, low temperatures, lack of phosphorous, etc. In any way the P content in leachate is in general relatively low, as a result in all biological treatment plants in most cases P has to be added continuously/periodically to the activated sludge.

To prevent problems with foaming, addition of antifoaming agent may be necessary. To restrict foam production a daily addition of some liters of antifoaming agent can be necessary. Most antifoaming agents are biodegradable but the COD is extremely high. With addition of only some liters, the organic loading of the activated sludge plant can be considerably increased. The end product of the biologically degraded antifoaming agent may be in some cases the reason for an increase of the COD effluent concentration.

As mentioned above ultrafiltration can be very effective to separate sludge and effluent water. Some operation parameters of an ultrafiltration sludge separation systems are shown in Box 10.3.7. But ultrafiltration components (particularly the membranes) are influenced by the produced sludge of the activated sludge plant and by other organic and inorganic leachate compounds. Ultrafiltration systems need additional cleaning procedures of the ultrafiltration membranes to prevent blocking of the fine pore system. The blocking processes called fouling and scaling. Scaling is the crystallization of dissolved substances. The membranes are a nearly ideal place for the growth of microorganism. The produced biofilm reduces the flux through the membrane and is called biofouling. In addition also inorganic substances can react with other inorganic or organic substances with the result of fouling. To remove such coating, chemical cleaning is necessary.

In opposite to ultrafiltration systems the conventional sludge separation with clarifier needs no or very little energy input. Such a secondary settling tank can be operated with a load of 0.5 m^3/m^2 h. The increased inorganic content of sludge leads to settling conditions that are generally good, but may result also in sludge settling on the inclined walls (slope of 60 degrees) of the clarifier. Vertical walls and the use of rotating sludge scrapers are a common solution to avoid these problems even though sedimentation tanks have only a diameter of 4–5 m.

Design and operation data of several leachate-activated sludge plants in Germany are presented in Table 10.3.2. At all of these plants the sludge is separated with settling tanks. These plants were built before 1996. The limited knowledge of leachate volumes, pollution, and degradation kinetics lead to relative large activated sludge tank volumes. The advantage of these plants is the large buffer capacity against fluctuation in flow rates and concentrations.

Box 10.3.7 General features of sludge separation by ultrafiltration

Partial vacuum ultrafiltration:

- Vacuum up to 50 kPa
- Flux: 8–12 L/m^2 h

Ultrafiltration:

- Pressure: 200–500 kPa
- Flux: 60–100 L/m^2 h

In general the flux rate decreases with time, as a result regular cleaning is necessary.

- The traditional cross-flow ultrafiltration (Fig. 10.3.6) is cleaned in situ using specific chemicals.
- Modules in activated sludge tanks are relative often cleaned by change of pressure and reverse flow of the permeate. In longer time intervals a chemical cleaning is necessary. Due to the toxicity of the chemicals for the microorganism, this procedure shall not be applied in activated sludge tanks. As a consequence the filtration modules have to be removed from the activated sludge tank and cleaned in a separate tank. With several modules and only cleaning one module per time, it is possible to operate the activated sludge plant in a continuously. At some plants for domestic wastewater treatment the ultrafiltration modules are arranged in a small downstream chamber of the activated sludge plant. For cleaning the activated sludge from this chamber has to be removed and the filtration modules can be cleaned in the empty chamber. After cleaning the remaining cleaning solution has to be removed and the chamber is again opened to the activated sludge tank. In such a case the treatment process will be interrupted.
- For increasing the operation time of membranes (without or with membrane changing) the reduction of sludge concentration may be helpful.
- Ultrafiltration destroys the biomass flocs, which are important for the settling process. Very small flocs with poor settling properties have a larger surface area, allowing an increased contact between leachate compounds and oxygen. Both effects have the potential to increase the treatment efficiency. These effects are very important for the nitrification process.
- Inorganic leachate substances can impair the membranes.

Corresponding data of activated sludge plants with sludge separation by ultrafiltration are presented in Table 10.3.3. Note that the volumetric loading rates are significantly higher than ones in Table 10.3.2.

Denecke et al. (2007) and Walter et al. (2007) published a very interesting modification of an activated sludge plant for nitrification and denitrification. The plant consists of four tanks in a sequence (each volume 165 m^3). One tank was designed for denitrification. The three other tanks are used for nitrification. As a consequence of relatively low COD values, the addition of methanol was necessary to achieve denitrification. The operation of this plant was modified over a period of several years to partial nitritation and anaerobic ammonium oxidation (Anammox). The oxygen content in the aeration tanks was reduced slowly to very low values <1 mg O_2/L. As a result an increasing part of ammonium was not oxidized to nitrate but only to nitrite. In the denitrification tank the dosage of methanol could be reduced also with the effect of a reduced sludge production. The lower sludge production results in longer SRT (higher sludge age), which enhances the development of microorganism for the anaerobic ammonium oxidation. This development increases the anaerobic ammonium oxidation and reduces the

Table 10.3.2 Performance values of activated sludge plants with settling tank (ATV, 1996; Mester, 1993)

Plant No.	Volume (m^3)	Inflow (m^3/day)	COD (mg/L)		BOD_5 (mg/L)		NH_4 (mgN/L)	
			Influent	Effluent	Influent	Effluent	Influent	Effluent
1	2087	96					528	4.6
2	3880	93	1571	614			633	0.81
2	*Results of two different years*		1506	709			579	0.26
3	1780	55	320–5800	238–854			125–1350	<1.2
3	Average		1745	516			681	<1.2
4	4650	120	1200–4000	300–1000		<20	800–1900	0.1–9
4	Average		2600	900			1100	0.3
5	2380	87–103	379–1929	72–401		3–5	289–823	0.4

Table 10.3.3 Performance values of activated sludge plants using ultrafiltration for sludge separation (Ehrig and Robinson, 2011)

Plant No.	Volume (m^3)	Inflow (m^3/day)	Membrane Area (m^2)	COD (mg/L)		BOD_5 (mg/L)		NH_4 (mgN/L)	
				Influent	Effluent	Influent	Effluent	Influent	Effluent
1	98	65–120	84	1500	400			900	100
2	240	264	226	2000	<400	150	<5	600	<5
3	114	70	160	3000	<200			1250	<150
4	129	80	43.6	1000 –12000	500 –800	1500 –6000	<5	400 –800	<10

requirement of methanol addition. Over a period of 3 years the methanol consumption was reduced from over 1.5 kg methanol per kg $N_{eliminated}$ to 0.2–0.3 kg methanol per kilogram $N_{eliminated}$. The energy consumption for the aeration was reduced by approximately 60%, and the excess sludge production was reduced in the same range. It must be considered that such a modified operation requires intensive process control which results in associated costs. In addition a secure temperature level of more than 20°C is necessary. This approach of modifying a nitrification–denitrification plant over a longer period of time into a nitritation–Anammox plant seems to be very reasonable. The risk of problems with the direct installation of the very sensitive nitritation–Anammox plant could in this case be avoided.

Sequencing batch reactors

The SBR is a very interesting alternative for biological leachate treatment and quite often used. Activated sludge plants need separate tanks for each treatment step while SBR plant consists on the first view only of one tank. All above described processes take place in this one tank in sequences (Fig. 10.3.8).

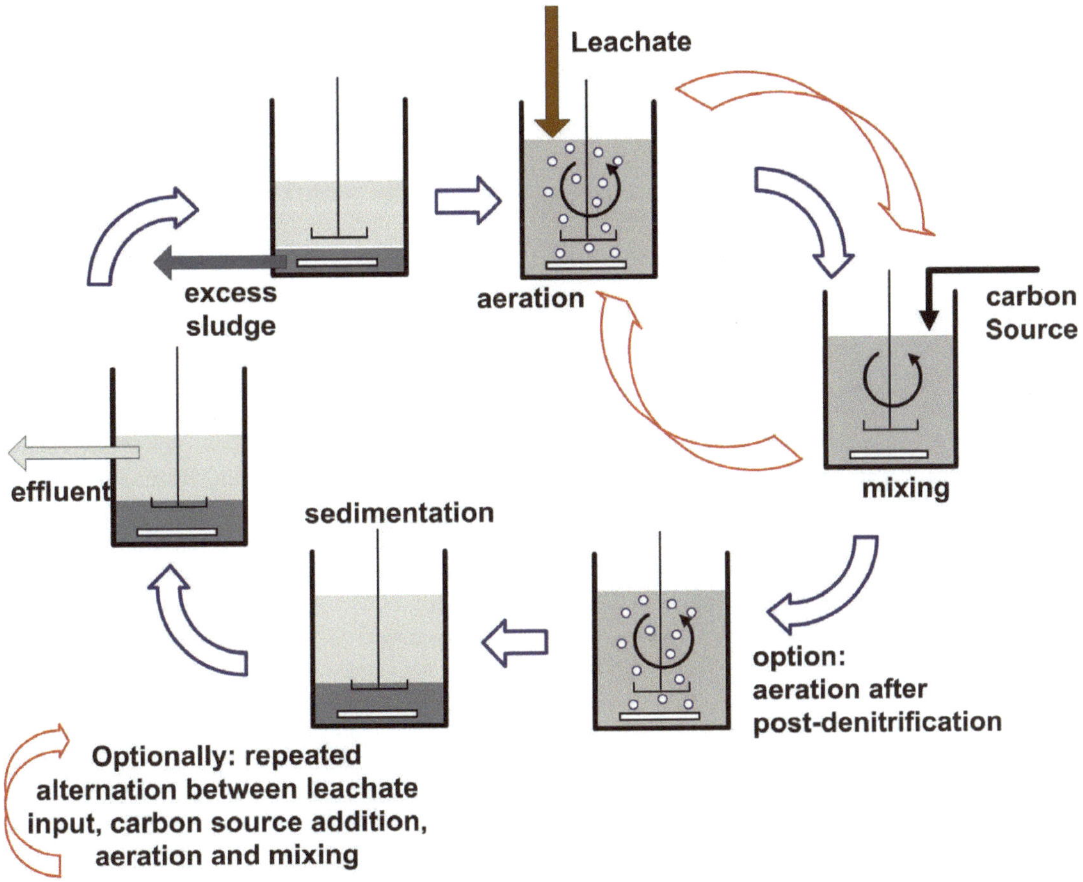

Figure 10.3.8 Example of a sequential batch reactor operation cycle.

This system offers a wide range of process variability. At conventional activated sludge plants it is very difficult to change the function of the installed tanks and the tank volume is difficult to change. But in SBR plants it is no problem to change phases and their duration. Fig. 10.3.8 shows a possible cycle of leachate treatment in an SBR with postdenitrification. The cycle includes the following phases:

- Aeration for nitrification and leachate influent (feeding over a short or longer period of time)
- Mixing with addition of an external carbon source for denitrification
- Postaeration to remove remaining external carbon
- Sedimentation
- Discharge of treated effluent
- Discharge of excess sludge (eventually combined with the removal of effluent and not in every cycle)

Not only mixing and aeration phases can be modified; in principle, one cycle could include several mixing and aeration phases. Further information on SBR operation is given in Box 10.3.8.

Box 10.3.8 General features of sequencing batch reactor plant operation

- The complete treatment process takes place in one reactor [two parallel sequencing batch reactor (SBRs) can be reasonable].
- Influent and effluent storage tanks are required.
- Common cycle periods are 24, 12, 8, and 6 h. For leachate treatment mostly 24 and 12 h used. It should be considered that each cycle includes phases of settling and discharge (phases without biological activity). These phases cause an increase in the required volume.
- Aerators and mixing devices operate in a cycle only during certain periods of time requiring a higher energy demand. The volume of an SBR tank has nearly the same size as a tank volume of a conventional activated sludge plant. Effective aeration systems and mixing devices are necessary to reach sufficient turbulence and a uniform and fast oxygen supply; energy input should be as low as possible.
- The total SBR tank is during certain periods used as sedimentation tank. Different from sedimentation tanks in traditional activated sludge plants, there is no turbulence (e.g., from leachate in- and outflow) that may disturb the sedimentation process; this results in a more effective sedimentation of the sludge. The effluent has to be removed during a relative short period of time; the relatively high flow rate could lead to a higher release of small sludge particles. This phenomenon should be respected when the discharge system is designed and operated.
- Solid concentrations in the effluent may be too high for following chemical–physical posttreatment steps. Since discharge phases are too short for treatment, e.g., in an ultrafiltration unit a storage tank for the supply of the filtration step will be necessary.
- To avoid cold water temperatures SBR plants including pre- and poststorage tanks should be covered.
- As a consequence of the relative short aeration phase, the oxygen supply is high and the risk of intensive foaming increases.
- Precipitated inorganic substances can be very problematic; during settling phases the material may accumulate and compact at the bottom of the reactor.

As shown in Fig. 10.3.8, feeding and discharge take place only over a limited period of time. Therefore a storage tank for raw leachate is necessary. As mentioned above raw leachate should be stored to equalize flow rate and quality for nearly every kind of treatment system. In general it is not feasible or allowed to discharge treated leachate in a short time into a water body. Also additional treatment steps need a more or less continuous feeding. In both cases an effluent storage tank is necessary.

Although conventional activated sludge and SBR plants consist of nearly the same treatment steps, there is a very important difference between both systems. In totally mixed (mixer or aeration) conventional activated sludge plants the concentrations (COD and nitrogen) in the reactor are the same as the effluent. At most SBR plants the leachate feed happens only over a short period of time. The resulting concentration ($C_{mixture}$) can be easily calculated respecting the tank volume (V) and feed quantity (V_{feed}):

$$C_{mixture} = [(V - V_{feed}) * C_{effluent} + V_{feed} * C_{influent}]/V$$

This can be of advantage for several leachate compounds if their elimination rate depends on the concentration in the reactor. For leachate treatment the SBR should be dimensioned for nitrification and denitrification. The reaction rate of these processes is only in a very low concentration range influenced by the substrate concentration.

On the other hand high concentrations of leachate compounds can result in inhibition or toxic effects. As already has been discussed earlier a problematic parameter in leachate is the ammonium nitrogen which, as ammonia, could inhibit both nitrification steps. The oxidation of nitrite to nitrate is primarily inhibited with an increase of nitrite. Laboratory-scale experiments (Dockhorn et al., 1997; Yalmaz and Öztürk, 2001) have shown that during nitrification very high ammonium concentration could be measured with a nitrite increase over a relative short phase. In this case maximum ammonium concentrations of 200–450 mg N/L were measured. If such values are present a pH regulation (pH = 7) should be used to prevent further shifting from ammonium to ammonia. To reduce high ammonium concentrations the feeding can be stretched over a longer period of the aeration phase.

Published nitrification rates from Robinson et al. (2003a) with 0.02–0.04 kg N/kg MLVSS*day at 10°C, Balslev et al. (2005) with 0.01 kg N/kg MLVSS*day, Strachan et al. (2007) with 0.006–0.040 kg N/kg MLSS*day, and from Robinson et al. (2007) with 0.021 kg N/kg MLVSS*day show nearly the same values as given in Box 10.3.5 for conventional activated sludge plants. Artemis (2010) described a conversion rate of 0.06 kg N/kg MLSS*day in an SBR plant in Germany. Using these data the design procedure as presented in Box 10.3.5 may be used to design the biological active SBR volume (V_{active}). In order to calculate the total SBR volume the phases of settling and discharge must be considered.

Example (as shown in Fig. 10.3.8): cycle = 24 h; estimated time of settling and discharge = 2 h; feeding during the aeration phase; 1 h postaeration to remove remaining external carbon, calculated volume V using the information in Box 10.3.5.

$$\text{Time for nitrification and denitrification} = 24 - 2 - 1 = 21\text{ h}$$

$$\text{Total volume of SBR tank: } V_{total} = V * 24/21\ (m^3)$$

$$\text{Nitrification phase} = 21\text{ h} * V_N/V;\ \text{Denitrification phase} = 21\text{ h} * V_{den}/V$$

Table 10.3.4 Performance data of activated sludge sequencing batch reactor systems at different landfill sites.

Parameter (mg/L)	Efford, 2003 (Robinson and Last, 2003)		Llanddulas, 2002 (Robinson and Last, 2003)		Arpley, 2001 (Robinson et al.t, 2003a)	
	Influent	Effluent	Influent	Effluent	Influent	Effluent
COD	942	462	3410	762	5990	1060
BOD_5	72	22	1520	9	688	<1
NH_4-N	820	1.59	965	1.2	1460	3.2
Nitrate-N	0.21	423	<0.3	668	1.9	238
Nitrite-N	0.04	0.56	<0.1	<0.1	<0.1	0.1

Efford, 2003 and Llanddulas, 2002, are cited by Robinson and Last(2003) while Arpley, 2001, is cited by Robinson et al., (2003a).

Performance data of some SBR plants are given in Table 10.3.4.

Fixed film reactors

An alternative technology for nitrification is fixed film reactors. These are tanks filled with a coarse solid material. This solid material should have sufficient voids for free flow of suspended solids and a large surface area where the microorganism may attach and build up a film. But it is difficult to fulfill both requirements. For treatment of easy degradable organic substances, the pores should be very coarse and the surface area is less important. But also with very large pores it is difficult to prevent blocking as a result of intensive biomass growth. The biomass production of nitrification is low, and the involved microorganism prefers fixation on a solid surface. Therefore in this case a large surface area is important and the pore size is not so relevant. The advantage of fixed film reactors is the fixation of microorganism on the surface of the solid material. The fixation increases the SRT (sludge age) to very high values. With increasing thickness the film may be physically eroded by the substrate flow. For these reasons fixed film reactors are mainly used for low biodegradable substances and nitrification. The preferred solid material is specifically designed plastic material with surface areas up to 500 m^2/m^3 (average 180–250 m^2/m^3) and relative large pores. But such material is relatively expensive and in many cases it is more economic to build larger activated sludge tanks than smaller tanks filled with this plastic material. For this reason most types of fixed film reactors with fixed or suspended solid material are seldom in use.

RBCs are also used for leachate treatment (Fig. 10.3.9). They consist of an axis with motor and carrier material for the biomass on this axis. Common material is flat or corrugated plastic disc with 3–5 m diameter; several of these discs may be placed in parallel. Alternative round cylinders built

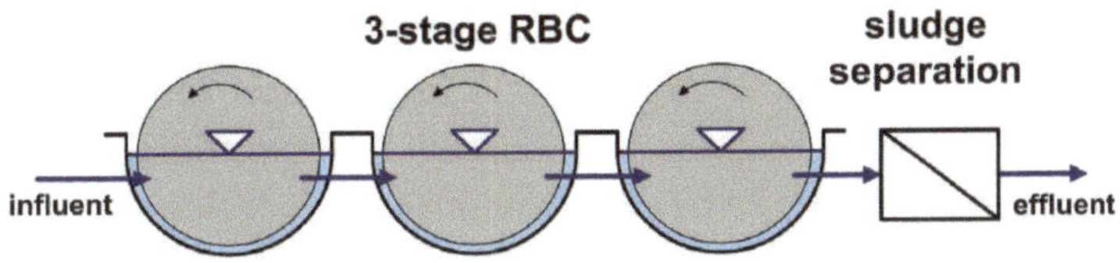

Figure 10.3.9 Three-stage rotating biological contactor (RBC).

from plastic material also with 3–5 m diameter are in use. Both systems are constructed with a large surface. Due to the distance between the disks, clogging can be avoided. Only 40%–50% of the RBC material is submersed, and the RBCs rotate with four to five rotations per minute. As a consequence the biomass on the RBC surface changes permanently between the substrate in the leachate and the air. For this reason, no external aeration is necessary. The only energy input is the low energy for the rotation of the RBCs.

The long SRT can only be used if the bacteria are not overgrown from other microorganism (e.g., rapidly growing heterotrophic bacteria). Nitrification bacteria can use this advantage only in the absence of higher concentrations of biodegradable COD.

Leachate with biodegradable organics concentrations higher than 4000–5000 mg COD/L may cause in RBCs a serious biomass accumulation problem. After some month of treating such a leachate in a large-scale pilot plant, the biomass on the disk became so heavy that the very stable axis of the first-stage RBC broke.

Ehrig (1983) reported data from laboratory experiments with a four-stage RBC for leachate nitrification. In contrast to activated sludge plants, over a wide range of loading rates, high nitrification rates had been measured. The reaction rate of nitrification increases with increasing pH value. But the toxicity effect of ammonia increases also. As a consequence the pH value should be near 7. During nitrification the alkalinity of leachate is reduced with decreasing pH values. Without a combination with a denitrification step, pH control by adding sodium hydroxide would be necessary. The addition of lime could lead to precipitates on the solids surface. To prevent short-circuiting, RBC should be designed with more than one stage in a sequence. Overloading results in increased nitrite values. To prevent overloading of the first stage with increasing nitrite concentrations not more than three stages should be realized. At such cascade systems an overloading of the first stage resulting in inhibition effects leads to a breakdown of the first stage with the result of overloading the next stage and so on. Effluent water from the last disc may be returned to dilute the influent, which is an option to reduce inhibiting concentrations of ammonia in the first stage.

The loading rate for an RBC may be in the range of 1–2 g N/m^2 day. If the concentrations of the solids, inorganics, that may precipitate, and biodegradable organics in the influent are low, the nitrogen load could be increased to 4–5 g N/m^2 day. RBC plants should be roofed because rainfall could wash off the biomass from the disks. As a consequence the temperature in the RBC reactor may be regulated.

Wittek and Fleischer (1985) presented data from a full-scale RBC (plastic material with a specific area of 220 m^2/m^3, total surface area of 28,500 m^2) at loading rates between 1 and 2 g N/m^2 day (Table 10.3.5). Knox (1992) published data from an RBC plant at Pitsea landfill (plastic material with a specific area of 180 m^2/m^3, total surface area of 30,000 m^2) (Table 10.3.5).

At some leachate treatment plants in Germany, RBCs are used after chemical oxidation to reduce the newly produced biodegradable organic substances. Biomass production from such pretreatment

Table 10.3.5 Rotating biological contactor (RBC) treatment data at Ottensen landfill (Witteck and Fleischer, 1985), at Pitsea landfill (Knox, 1992), and in two RBC plants after chemical oxidation with ozone (ATV, 1996)

Parameter (mg/L)	Influent	Effluent
OTTENSEN LANDFILL (WITTECK AND FLEISCHER, 1985)		
COD	481	278
BOD_5	71	10
$N\text{-}NH_4$	201	0.28
$N\text{-}NO_2$	0.05	0.02
$N\text{-}NO_3$	0.3	211
PITSEA LANDFILL (KNOX, 1992)		
$N\text{-}NH_4$	na	2.96
BOD	na	14
PERFORMANCE VALUES AT A LOADING RATE OF 3.1 GN/M² DAY		
BOD	27	23
TOC	303	264
$N\text{-}NH_4$	153	0.1
$N\text{-}NO_x$	34	105
$N\text{-}NO_2$	11	0.1

Table 10.3.5 Rotating biological contactor (RBC) treatment data at Ottensen landfill (Witteck and Fleischer, 1985), at Pitsea landfill (Knox, 1992), and in two RBC plants after chemical oxidation with ozone (ATV, 1996)—cont'd

Parameter (mg/L)	Influent	Effluent
RBC PLANTS AFTER CHEMICAL OXIDATION WITH OZONE (ATV, 1996)		
	Effluent Oxidation	*Effluent RBC*
PLANT 1		
COD (average)	130	70
COD (range)	50–200	18–150
BOD_5	20–40	<10
AOX (average)	0.16	0.09
AOX (range)	0.08–0.25	0.04–0.18
PLANT 2		
COD (average)	71	15
COD (range)	4–138	2–28

processes is too low and variable for activated sludge plants. But on an RBC, the attached biomass has a slower decay rate. As a consequence, under such circumstances, RBCs (and other fixed film reactors) can provide more stable operation conditions than activated sludge plants. Results from two plants are presented in Table 10.3.5. The extinct biomass washed out in relatively big flocks with good sedimentation properties. Only very small clarifiers are necessary. In several cases, cloth filter is used for sludge separation.

Until now the utilization of other fixed film reactors is rare. In Germany, one leachate treatment plant with plastic solid material (surface area 240 m^2/m^3) fixed in an aerated tank is in operation. With loading rates of 3.6–4.8 g N/m^2 day, full nitrification had been achieved (Spiess, 1995).

As mentioned above, denitrification with high carbon degradation does not make sense in RBCs and other fixed film reactors. The consequence could be an activated sludge plant for denitrification and a fixed film reactor for nitrification. Such a combination is more complex than a single system and can reduce the advantages of higher reaction rates. Table 10.3.6 shows influent and effluent data of three combinations of an activated sludge plant for denitrification and an RBC or submersed fixed film reactor for nitrification (ATV, 1996).

Table 10.3.6 Influent and effluent data of three combinations of an activated sludge plant for denitrification and a rotating biological contactor (RBC) or submersed fixed film reactor for nitrification (ATV, 1996)

Parameter	Units	Influent	Effluent
ACTIVATED SLUDGE PLANT WITH DENITRIFICATION AS A FIRST STEP AND A SUBMERSED FIXED FILM REACTOR AS A SECOND STEP			
Observation Phase 1			
COD	mg/L	1571	614
NH_4	mgN/L	633	0.81
Observation Phase 2			
COD	mg/L	1506	709
N-NH_4	mgN/L	579	0.26
N-NO_3	mgN/L	44	288
ACTIVATED SLUDGE PLANT FOR DENITRIFICATION WITH SETTLING TANK AND CLOTH FILTER, RBCS FOR NITRIFICATION WITH CLOTH FILTER			
COD	mg/L	668	551
		446–872	353–697
BOD_5	mg/L		43
		10–220	5–110
N-NH_4	mgN/L	180	1.8
		80–396	0.02–25.2
N-NO_3	mgN/L	53.5	240
		5.7–136.4	55.7–460
ACTIVATED SLUDGE PLANT FOR DENITRIFICATION WITH SETTLING TANK AND RBCS FOR NITRIFICATION WITH MICROSTRAINER			
BOD_5	mg/L	145–253	21–40
COD	mg/L	1543–2305	736–1145

Table 10.3.6 Influent and effluent data of three combinations of an activated sludge plant for denitrification and a rotating biological contactor (RBC) or submersed fixed film reactor for nitrification (ATV, 1996)—cont'd

Parameter	Units	Influent	Effluent
N-NH_4	mgN/L	230–659	0.1–48
N-NO_3	mgN/L	106–280	416–845

BIOLOGICAL COTREATMENT OF SEWAGE AND LANDFILL LEACHATE

Cotreatment of sewage and leachate is practiced successfully worldwide. In most cases, leachate is discharged into existing sewage treatment plants (STPs). This means it has to be investigated what are the free capacities and how much leachate at which concentration the STP can accept without exceeding the discharge limit values. This depends on the design values of the plant and its actual loading. Of course a leachate treatment, and not a simple dilution by the sewage, has to be achieved. In general, leachate is biologically well treatable and there are in general no toxic effects to be expected. There may be some elevated concentrations of some heavy metals (e.g., zinc in the acid phase). During biological treatment, most of the metals precipitates and remains in the sludge. In addition there may be some noneasy degradable trace organics in the leachate (see section "General Remarks") and it is due to the specific discharge regulation, if the leachate has to be pretreated or not. In specific cases the authorities may request a proof of the degradability of the specific compounds during cotreatment (e.g., by a laboratory test). In Germany, regulations for the concentrations of heavy metal and trace organic in the influent of an STP also exist to prevent only dilution of these compounds.

A very important point to consider is the treatment capacity of the STP. This capacity is often overestimated because the daily flow rate of leachate is normally relatively low compared to the amount of the incoming sewage. Only the available treatment capacity can be used for the elimination of biodegradable COD and nitrogen. In some cases it is possible to treat more leachate than calculated if the leachate is fed into the plant during times of low sewage loadings (e.g., nighttime). It should be considered that in some cases the nonbiodegradable COD and/or the adsorbable organic halogens (AOXs) can become very important for cotreatment if these parameters are considerably higher in the leachate than in the effluent of the STP. As a consequence these parameters could increase the effluent concentrations of the STP significantly.

The considerations for leachate addition to STPs must encounter all aspects as organic loading, nitrification, denitrification, alkalinity for pH stabilization, nutrient requirement, oxygen supply, etc. If some of these conditions cannot be met, the STP has either to be enlarged, the leachate has to be pretreated, or only part of the leachate can be cotreated. If pretreatment is selected, the designed plant should produce year-round the required effluent quality values. The treatment efficiency of

Parameter	
Test 1	
BOD_5 (mg/l)	5.800 - 6.500
COD (mg/l)	9.590 – 9.800
BOD_5/COD	0,6 – 0,66
Test 2	
BOD_5 (mg/l)	15.500 – 17.240
COD (mg/l)	20.100 – 25.350
BOD_5/COD	0,68 – 0,77

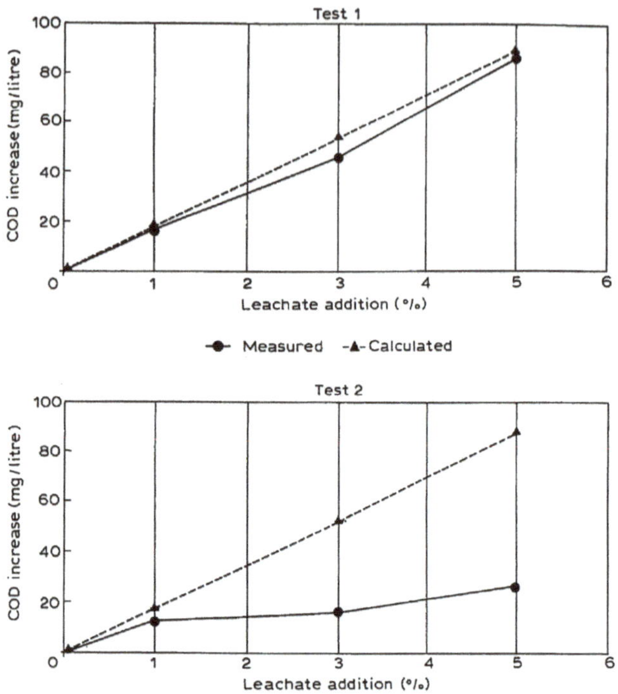

Figure 10.3.10 Cotreatment of sewage and leachate: calculated and measured effects of volumetric addition of leachate (range 0%—5% with respect to sewage flow) on the COD concentrations in the treated effluent (Knoch, 1974). *BOD*, biological oxygen demand; *COD*, chemical oxygen demand.

STPs may be partly reduced during cold periods. In this case the increased concentrations of the supplied leachate may lead to serious problems.

Fig. 10.3.10 shows the effect of cotreatment in laboratory-scale tests using two different kinds of leachates. For both tests, leachate from a landfill in the acetic phase had been used. The different results of the two tests are often confirmed by the operators reporting in some cases a COD increase as calculated, in other cases only a small or no increase in COD concentrations in the effluent. Doedens and Cord-Landwehr (1984) published data of cotreatment over 10 days from an STP with an organic load of 0.05 kg BOD_5/kg MLSS*day. The added leachate in a quantity of 3.2% with $BOD_5 = 80$ mg/L and COD = 1552 mg/L had no effect on the effluent values. From biological leachate treatment tests an increase of 30 mg COD/L was predicted. Particularly during short periods of leachate supply, adsorption effects on the sludge surface and/or precipitation may be an explanation of the different results.

Box 10.3.9 presents a rough calculation of potential leachate loads for cotreatment in an STP with a free treatment capacity of 10%. It should be considered that as an influent sewage quality German standard values had been used. The planning should be made with real sewage data. As presented in

Box 10.3.9 Calculation of leachate feed depending on the free capacity of a sewage treatment plant

Population equivalent of wastewater in Germany:
Q = 200 L/day; BOD_5 = 60 g/day;
COD = 120 g/day; nitrogen = 12 g/day.
For 1000 inhabitants:
Q = 200 m^3/day; BOD_5 = 60 kg/day;
COD = 120 kg/day;
estimated COD effluent = 40 mg/L;
nitrogen = 12 kg/day.
The following calculation estimated a free capacity of the domestic sewage plant in the range of 10%.
Possible leachate addition:
Example 1: acetic phase leachate
BOD_5 = 5000 mg/L; COD = 10,000 mg/L;
$COD_{biodegradable}$ = 9000 mg/L; BOD_5/COD with 0.5 in the same range as domestic wastewater;
nitrogen = 1000 mg/L.
Possible leachate feeding:
$BOD_5 = 0.1 * 60/5 = 1.2$ m^3/day;
nitrogen = $0.1 * 12/1 = 1.2$ m^3/day.
For 100 m^3/day of leachate a domestic sewage plant for >91,000 inhabitants is necessary.
Increase of COD effluent = $(200 * 40 + 1.2 * 1000)/201.2 = 46$ mg/L.

Example 2: methanogenic phase leachate:
BOD_5 = 300 mg/L; COD = 3000 mg/L;
$COD_{biodegradable}$ = 1500 mg/L;
nitrogen = 1000 mg/L.
Possible leachate feeding:
For estimation of organic load, $COD_{biodegradable}$ should be used.
$BOD_5 = 0.1 * 60/1.5 = 4$ m^3/day;
nitrogen = $0.1 * 12/1 = 1.2$ m^3/day.
Increase of COD effluent = $(200 * 40 + 1.2 * 1500)/201.2 = 49$ mg/L.
For nitrogen removal a treatment plant with nitrification and denitrification is necessary.
For domestic sewage a biodegradable COD for denitrification of $4.5 * 12 = 54$ kg COD/day is necessary.
COD effluent approximately 40 mg/L = 8 kg/day; estimated 60 mg/L with lower biodegradability and not suitable for denitrification = 12 kg/day;
COD for denitrification of leachate = 120 − 54 − 20 = 46 kg/day.
The remaining denitrification capacity is sufficient for 46/4.5 = 10.2 m^3/day of leachate.

Box 10.3.9 in the case an STP has a free capacity of 10% a leachate input of only approximately 0.6% of sewage inflow may be acceptable. The critical aspect is the additional ammonium load from the supplied leachate. For design and operation, in most cases, nitrification is the most relevant process in STPs. Using the abovementioned loading rates, the COD effluent concentrations increased by more than 20% when leachate from a landfill in the methanogenic stage has to be cotreated. In this example an increase of the AOX concentrations from <10% until 36% had been estimated (AOX sewage 0.05 mg/L, AOX leachate 0.5–3 mg/L).

References

Albers, H., 1985. Untersuchungen zur Behandlung von Sickerwässern in Belebungsanlagen und belüfteten Teichen: Investigation into leachate treatment with activated sludge plants and aerated lagoons, vol. 39. Veröffentlichungen des Instituts fur Stadtbauwesen TU Braunschweig, p. 228.

Artemis, 2010. Sickerwasserreinigung auf der Deponie "Weisser Weg" Chemnitz: Leachate treatment at the landfill "Weisser Weg" Chemnitz, vol. 18. Artemis GmbH-Info Brief, p. 2.

ATV (German Association for Water, Wastewater and Waste), 1996. Beschreibung von Sickerwasserbehandlungsanlagen (Description of leachate treatment plants). Korrespondenz Abwasser 43 (7), 1303.

Balslev, P., Pedersen, P.D., Eljassen, H., Fossum, T.R., 2005. Nitrogen removal from leachate by nitrification and recirculation. In: Proceedings Sardinia 2005, Tenth International Waste Management and Landfill Symposium, S. Margherita di Pula, Cagliari, Italy.

Blakey, N.C., Cossu, R., Maris, P.J., Mosey, 1992. Anaerobic lagoons and AUSB reactors: laboratory experiments. In: Christensen, Cassu, Stegmann (Eds.), Landfilling of Waste: Leachate. Elsevie Applied Science, ISBN 1-85166-733-4, p. 520 (Chapter 3.5).

Chang, L., 1993. Bemessung von einstufigen Belebungsanlagen zur Behandlung von Sickerwasser aus Hausmülldeponien (Dimension of one stage activated sludge plants for sanitary landfill leachate treatment). Korrespondenz Abwasser 40, 319–327. Germany.

Chang, L., 1998. Auslegung von einstufigen Belebungsanlagen zur Stickstoffelimination bei Sickerwässern aus Siedlungsabfalldeponien: Design of one stage activated sludge plants with nitrogen elimination for leachate from municipal solid waste landfills, vol. 62. Veröffentlichungen des Instituts für Siedlungswasserwirtschaft, University of Braunschweig.

Chang, L., 1999. Dimensioning of single-stage activated sludge treatment plants for nitrogen removal from landfill leachate. In: Christensen, T.H., Cossu, R., Stegmann, R. (Eds.), Proceedings Sardinia 99, Seventh International Landfill Symposium, vol. II. CISA, Environmental Sanitary Engineering Centre, Cagliari, Italy, pp. 207–214.

Chen, T., Esnault, D., Koenig, A., 1997. First year operation of the NENT landfill leachate treatment works in Hongkong. In: Christensen, T.H., Cossu, R., Stegmann, R. (Eds.), Proceedings Sardinia 97, Sixth International Landfill Symposium, vol. II. CISA, Environmental Sanitary Engineering Centre, Cagliari, Italy, pp. 219–228.

Cossu, R., 1981. Biological treatment of landfill leachate in a full scale plant. In: 5th European Sewage and Refuse Symposium EAS, Munich 22–28.6.1981, p. 605.

Denecke, M., Rekers, V., Walter, U., 2007. Einsparpotenziale bei der biologischen Reinigung von Deponiesickerwasser (New ways for the elimination of nitrogen from landfill sewage). Müll und Abfall (1), 4.

Dockhorn, T., Chang, L., Dichtl, N., 1997. Removal of nitrogen from landfill leachate using SBR-technology. In: Proceedings Sardinia 1997, Sixth International Landfill Symposium, S. Margherita di Pula, Cagliari, Italy.

Doedens, H., Cord-Landwehr, K., 1984. Sickerwasser-Kreislaufführung auf Deponien – neue Erkenntnisse und betriebliche Varianten (Leachate recirculation – new experiences and operational methods). Müll und Abfall 16, 68.

Ehrig, H.J., 1983. Biological oxidation of sanitary landfill leachates with high ammonium concentrations. In: International Seminar on Rotating Biological Discs, 6–8 October 1983, Fellbach, Germany.

Ehrig, H.J., Robinson, H.D., 2011. Chapter 10.11: Landfilling: leachate treatment. In: Christensen (Ed.), Solid Waste Technology & Management. Wiley, ISBN 978-1-4051-7517-3, p. 858.

Ehrig, H.J., Albers, H., 1987. Nitrifikation von Abwässern mit hohen Salz- und Stickstoffgehalten: Nitrification of wastewaters with high salt and nitrogen concentrations, vol. 42. Veröffentlichungen des Instituts fur Stadtbauwesen TU Braunschweig, p. 264.

Fisher, M., Fell, C., March 1999. Ammonium removal from high strength leachate using a sequencing batch reactor. Wastes Management 25.

Frascari, D., Bronzini, F., Giordano, G., Tedioli, G., Nocentini, M., 2004. Long-term characterization, lagoon treatment and migration potential of landfill leachate: a case study in an active Italian landfill. Chemosphere 54, 335.

Knoch, J., 1974. Gemeinsame biologische Reinigung von Müllsickerwasser und kommunalem Abwasser (Biological co-treatment of sewage and leachate). In: Proceedings: Reinigung von Sickerwasser aus Mülldeponien Kolloquien und Seminare im Hause Edelhoff, vol. 3.

Knox, K., 1992. Rotating biological contactors. In: Christensen, T.H., Cossu, R., Stegmann, R. (Eds.), Landfilling of Waste: Leachate. Elsevier Science Publishers, ISBN 1851667334, p. 211.

Maehlum, T., Haarstad, K., 1997. Leachate treatment in ponds and wetlands in cold climate. In: Christensen, T.H., Cossu, R., Stegmann, R. (Eds.), Proceedings Sardinia 97, Sixth International Landfill Symposium, vol. II. CISA, Environmental Sanitary Engineering Centre, Cagliari, Italy, p. 337.

Mennerich, A., 1988. Beitrag zur anaerob-aeroben Behandlung von Sickerwässern aus Hausmülldeponien: Contribution to the aerobic-aerobic treatment of sanitary landfill leachate, vol. 44. Veröffentlichungen des Instituts für Siedlungswasserwirtschaft, University of Braunschweig.

Mester, T., 1993. Sickerwasserreinigungsanlage der Deponie Wilsum – Belebungsanlage, Flockung, Aktivkohle: Leachate treatment plant - activated sludge plant, precipitation, activated carbon adsorption, vol. 8. Veröffentlichungen des Zentrums für Abfallforschung, TU Braunschweig, p. 263.

Pelkonen, M., Koto, M., Rintala, J., 1999. Biological nitrogen removal from landfill leachate: a pilot scale study. Waste Management and Research 17, 493.

Robinson, H., Carville, M., Walsh, T., 2003a. Advanced leachate treatment at Buckden landfill. Journal Environmental Engineering and Science 2, 255. Huntington, U.K.

Robinson, H.D., Knox, K., May 2001. Pollution Inventory Discharges to Sewer or Surface Waters from Landfill Leachates. Final Report Ref REGCON 70, prepared for the UK Environment Agency by Enviros and Knox Associates.

Robinson, H.D., Last, S.D., June 2003. On-site leachate treatment expands rapidly. Local Authority Waste and Environment.

Robinson, H.D., Chen, C.K., Formby, R.W., Carville, M.S., 1995. Treatment of leachates from Hong Kong Landfills with full nitrification and denitrification. In: Christensen, T.H., Cossu, R., Stegmann, R. (Eds.), Proceedings Sardinia 95, Fifth International Landfill Symposium, vol. I. CISA, Environmental Sanitary Engineering Centre, Cagliari, Italy, pp. 511–534.

Robinson, H.D., Farrow, S., Last, S., Jones, D., 2003b. Remediation of leachate problems at Arpley Landfill Site, Warrington, Cheshire. In: Paper Presented to Sardinia '93, the Ninth International Waste Management and Landfill Symposium, S. Margherita di Pula, Cagliari, Sardinia, 6–10 October 2003.

Robinson, H.D., van der Merve, W., Mitchell, C., Gombault, E., Novella, P., Carville, M.S., 2007. Treatment of leachate from a large hazardous waste landfill site in south Africa. In: Proceedings Sardinia 2007, Eleventh International Waste Management and Landfill Symposium, S. Margherita di Pula, Cagliari, Italy.

Spiess, A., 1995. Biologische Sickerwasserbehandlung in Festbettanlagen (Biological leachate treatment with fixed film reactors). In: ATV-Symposium "Behandlung von Deponiesickerwasser", 1–2.3.1995, Magdeburg, Germany.

Strachan, L.J., Robinson, H.D., Last, S.D., Payne, G., Wright, M., 2007. Development of leachate treatment at a large new tropical landfill site. In: Proceedings Sardinia 2007, Eleventh International Waste Management and Landfill Symposium, S. Margherita di Pula, Cagliari, Italy.

Walter, U., Rekers, V., Denecke, M., 2007. Deammonofikation von Deponiesickerwasser (De-ammonification of landfill leachate). Umweltmagazin 1, 14.

Wittek, J., Fleischer, B., March 1985. Erfahrungsbericht über die Sickerwasserbehandlung in Tauchtropfkörpern (Operation report of a rotating biological contactor treating landfill leachate). In: Sickerwasser aus Mülldeponien – Einflüsse und Behandlung. Fachtagung, vol. 39. University of Braunscheig, Germany (Veröffentlichungen des Instituts für Stadtbauwesen, p. 401.

Yalmaz, G., Öztürk, I., 2001. Biological ammonia removal from anaerobically pre-treated landfill leachate in sequencing batch reactors (SBR). Water Science and Technology 43 (3), 307.

Further Reading

Cossu, R., Stegmann, R., Andreattola, G., Cannas, P., 1989. Biological treatment. In: Chrisensen, Cossu, Stegmann (Eds.), Sanitary Landfilling: Process, Technology and Environmental Impact. Acadamic Press, ISBN 0-12-174255-5, p. 251.

Ehrig, H.-J., Ahnert, L., 1992. Co-treatment of leachate with sewage in chapter 3.1 leachate treatment. In: Christensen, T., Cossu, R., Stegmann, R. (Eds.), Landfilling of Waste: Leachate, p. 403.

Ehrig, H.-J., Hagedorn, S., 1998. Sickerwasserbehandlung, Belastungen, Verfahren, Entwicklung und Bewertung (Leachate treatment, pollution, processes, development and evaluation). In: Entwicklungstendezen in der Deponietechnik, 1. Hamburger Abfallwirtschaftstage 28–29 January 1998, Hamburger Berichte, Band 12. Economica Verlag, p. 261.

Ehrig, H.J., 1984. Treatment of sanitary landfill leachate: biological treatment. Waste Management & Research 2, 131.

Ehrig, H.J., June 23–28, 1985a. Biological treatment of sanitary landfill leachate with special aspects on the high ammonium concentrations. In: New Directions and Research in Waste Treatment and Residuals Management. University of British Columbia, Vancouver, Canada.

Ehrig, H.J., 1985b. Untersuchungen zur Behandlung von Sickerwässern in Festbettanlagen: Investigation on leachate treatment in fixed film reactors, vol. 39. Veröffentlichungen des Instituts für Stadtbauwesen, TU Braunschweig, p. 287.

Ehrig, H.J., 1986. Sickerwasserbehandlung auf Deponien –Überprüfung und Korrektur von Planungsmassnahmen in der Praxis (Leachate treatment on landfills- verification and correction of design in praxis). In: Fehlauu, Stief (Eds.), Abfallwirtschaft in Forschung und Praxis, vol. 15, p. 319.

Ehrig, H.J., 1992. Biologische Reinigung von Sickerwasser (Biological leachate treatment). In: Seminar at the Technical Academy Wuppertal, 8–9.10.1992.

Ehrig, H.J., 1998a. Biological processes. In: Symposium: Management and treatment of MSW landfill leachate, Venice, 2–4.12.1998.

Ehrig, H.J., 1998b. Co-treatment in domestic sewage facilities. In: Symposium: Management and Treatment of MSW Landfill Leachate, Venice, 2–4.12.1998.

Ehrig, H.J., 2001. Sickerwasser aus Abfallablagerungen (Leachate from landfills). In: ATV-Handbuch Industrieabwasser, Dienstleistungs- und Veredelungsindustrie, Verlag Ernst & Sohn (Eds.), ATV-DVWK (German Association for Water, Wastewater and Waste), ISBN 3-433-01468-X, pp. 347–380.

Heyer, K.-U., Stegmann, R., Ehrig, H.J., June 24–26, 1998a. Leachate Management: Leachate Generation, Collection, Treatment and Costs, Present and Future of MSW Landfills. Padua.

Heyer, K.-U., Stegmann, R., Ehrig, H.J., 1998b. Leachate treatment. Principles and options. In: Symposium: Management and Treatment of MSW Landfill Leachate, Venice, 2–4.12.1998.

Kayser, R., Ehrig, H.J., 1985. Grenzen und Möglichkeiten der Einleitung von Deponiesickerwässern in kommunale Kläranlagen (Limitations and possibilities of the discharge of leachate into sewage treatment plants). In: Proceedings: Sicherheit und Risiko wassertechnischerAnlagen, 18. Essener Tagung 1985, Gewässerschutz-Wasser-Abwasser, vol. 75.

Maehlum, T., 1995. Treatment of landfill leachate in on-site lagoons and constructed wetlands. Water Science and Technology 32 (3), 129.

Robinson, H.D., Knox, K., March 2003. Updating the Landfill Leachate Pollution Inventory Tool. R&D Technical Report No. PI-496/TR("), prepared for the UK Environment Agency by Enviros and Knox Associates.

Witz, U., Ehrig, H.J., 1993. Sustainable leachate treatment systems for landfill aftercare using nitrification and denitrification. In: Proceedings Sardinia 1993, Fourth International Landfill Symposium, S. Margherita di Pula, Cagliari, Italy.

10.4

PHYSICAL—CHEMICAL LEACHATE TREATMENT

Raffaello Cossu, Hans-Jürgen Ehrig and Aldo Muntoni

INTRODUCTION

Leachate pollution potential in MSW landfills is largely originated by a series of organic and inorganic compounds. Further to experience gained in the field of sewage treatment and the focus on high organic pollution, the biological treatment of leachate was initially implemented in many countries. However, the effluent concentrations that remained following biological treatment, featuring a low biochemical oxygen demand(BOD) and relatively high residual chemical oxygen demand (COD), resulted in the development of a wide range of additional chemical/physical treatment steps, or even in the total replacement of biological treatment. In some cases, the need for removal of specific inorganic compounds (salts, heavy metals etc.) may require targeted chemical/physical treatment. The majority of the physical—chemical processes described are applied in combination with biological treatment or other physical—chemical processes. The same processes can be successfully applied to the treatment of leachates from industrial waste landfilling.

This chapter illustrates the range of feasible physical—chemical processes suitable for application in leachate treatment. Processes are described, and design principles and calculations are illustrated. Performances of both pilot- and full-scale experiences are discussed.

PRECIPITATION, COAGULATION, AND FLOCCULATION

Principles and processes

Precipitation is the transformation of a dissolved chemical compound into a less soluble chemical compound. When the concentration of products in a chemical reaction exceeds specific solubility limits, solids are formed (precipitation). To allow them to settle, the solid particles need to come into contact with each other, particularly with the aim of neutralizing electrical surface charge (coagulation). During coagulation colloidal particles are destabilized to enhance agglomeration into larger particles and allow removal by gravity. Destabilization is obtained through the addition of chemical reagents (coagulants), thus enabling repulsive forces to be minimized through the neutralization of electrical charges present in colloidal particles. This occurs by means of bonding or adsorption mechanisms.

The above considerations relate mainly to nonhydrophilic colloidal particles, which are stabilized on the basis of negative electrical charges. The most common coagulant agents are Al and Fe^{3+} salts, characterized by multivalent ions with opposite charges. These salts display an acid behavior and, consequently, modify the physical–chemical characteristics (pH, alkalinity) of leachate.

In view of their ability to induce charge neutralization (cationic polyelectrolyte) and increase bridging between particles, polymeric organic compounds (polyelectrolyte) are also frequently used as coagulant agents, (Cossu et. al., 1992).

The agglomeration of destabilized colloidal particles is often enhanced by the addition of specific flocculating agents. Among these, activated silica or clay (inorganic flocculants) and organic flocculants may be used.

Reactions between precipitation/coagulation agents and compounds present in the water occur extremely rapidly, with a high risk of unwanted side effects. Highly intensive mixing with high energy input over a very short period of time is required. To produce readily settleable flocks, additional slow mixing with low energy input is applied. The use of exceedingly lengthy intensive mixing phases may result in new electrical charges and destruction of produced flocks.

The above-described process is also applied in the treatment of leachate, particularly to reduce large organic molecules such as fulvic and humic substances, both of which include a wide range of different organic compounds with a relatively high molecular weight. However, the latter is not a chemically uniform and clearly defined group. Analytical methods used for both groups include gravimetric analysis with precipitation steps at very low or very high pH values and photometry. Higher quantities of nonbiodegradable compounds can be precipitated at low and/or high pH values.

Precipitation is also used on a technical scale. Following addition of aluminum or iron salts different kinds of hydroxides are produced, which may build up very bulky flocks. Organic particles are incorporated into or adsorbed onto the surface of these flocks. Chemical reactions between metals and organics, including complex formation, cannot be excluded.

A common practice in wastewater treatment involves the utilization of lime for the precipitation of metals and phosphorous, with lime acting as a precipitation and flocculation agent. Lime has also been tested for use in the removal of organics from leachate, where it acts more along the lines of an adsorption and flocculation process.

Precipitation, coagulation, and flocculation processes are widely applied for the purpose of removing metals (particularly heavy metals) resulting from the formation of metal hydroxide or metal sulfide. These processes are also used in NH_4–N removal using MAP (magnesium-ammonium-phosphate precipitation) or phosphorus removal by formation of insoluble compounds with cationic metals, including Al or Fe coagulants.

Reduction of organic leachate pollution with precipitation, coagulation, and flocculation

Leachates deriving from landfills in the methanogenic phase contain significant amounts of humic and fulvic acids, therefore resulting in a relatively high COD. During aerobic biological treatment part of the organic content of leachate with low molecular weight is metabolized into organic compounds with a high molecular weight. Ferric and aluminum sulfates, or chlorides, can be used to remove a large portion of these organic compounds. These organic molecules are so large that it may not be

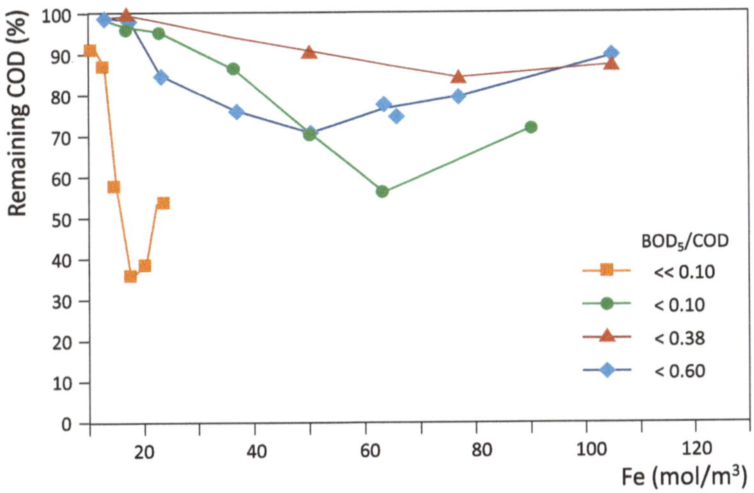

Figure 10.4.1 COD effluent (% of influent) after precipitation versus precipitation agent ($FeClSO_4$) based on BOD_5/COD ratios of four different leachates (1 mmol = 56 mg Fe) (Ehrig, 1989a,b).

appropriate to refer to them as being dissolved; the process involved is rather coagulation without precipitation.

Based on the relationship between BOD_5 and COD, a very rough description of the characteristics of organic compounds present in leachate is obtained. The BOD_5/COD ratio is an indicator of biodegradability (>0.5 good biodegradability; <0.1 very low biodegradability). The biodegradability of organic compounds decreases in line with an increase in high molecular weight organic compounds. As a consequence, the lower the BOD_5/COD ratio the more effective the application of precipitation and coagulation processes for the removal of organics from the majority of leachates should prove to be. Since a wide range of organic compounds may be present, leachate characterized by higher BOD_5/COD values may also contain molecules suitable for removal by means of these processes.

Fig. 10.4.1 shows the percentage of COD that remains subsequent to precipitation and coagulation used to treat leachate with different BOD_5/COD ratios using a ferric solution. The elimination of more than 50%–60% COD was only possible at very low BOD_5/COD ratios. Due to the biodegradability of a significant amount of organic compounds (see Chapters 10.2 and 10.3) leachate from landfills in the methanogenic stage (BOD_5/COD values less than 0.1) may also be successfully biologically treated before precipitation and coagulation. The effectiveness of precipitation and coagulation depends on pH value and—up to a certain limit—on the amount of added metal solution. The necessary pH range is a combination of increased insoluble organic compounds and sufficient hydroxide production depending on the amount of added Al and Fe. Fig. 10.4.2 (left side) shows the precipitation effect of pH using ferric solution. Optimum pH values for use of a ferric solution are between 4.5 and 5.0. The results obtained through the addition of aluminum salts are presented in Fig. 10.4.2 (right side),

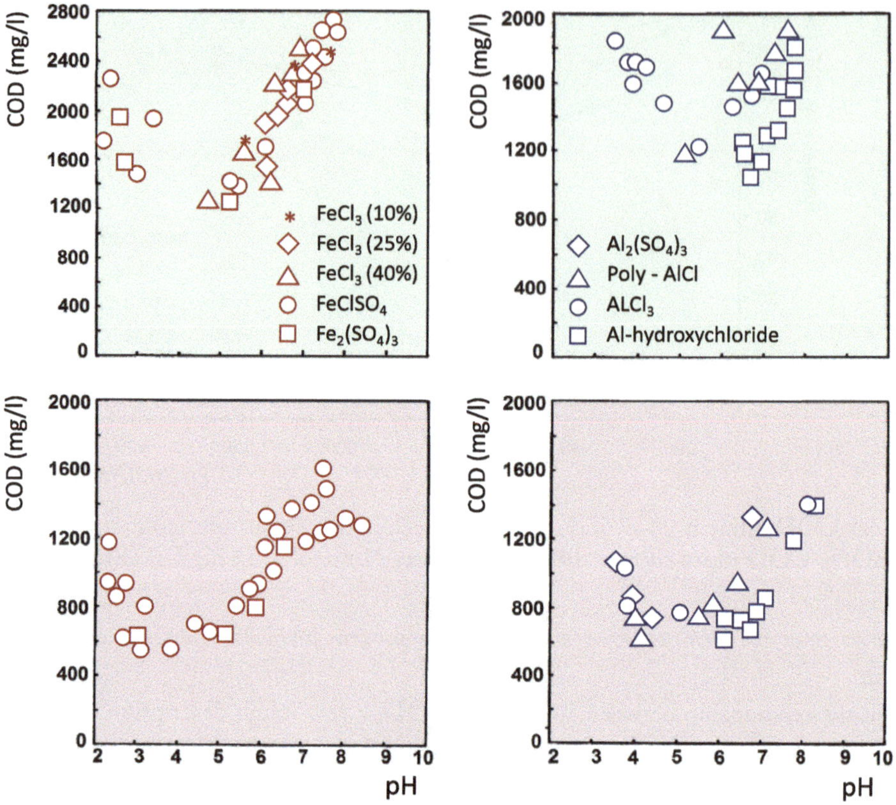

Figure 10.4.2 COD effluent values of two different leachates after precipitation with Fe (left side) and Al (right side) according to pH values (Ehrig, 1986).

highlighting an optimum pH value of between 5.0 and 5.5. However, the combination of precipitation agent and the unknown structure of the organic compounds present in leachate were intrinsic in identifying pH conditions for each leachate. In addition to pH value the optimum dose of precipitation agent should be determined. In the majority of cases, adjustment of pH with the precipitation agent is of scarce effectiveness, as the precipitation agent produces sludge. It is, however, possible to adjust pH without the production of sludge by using acid or lye. Fig. 10.4.3 shows pH-effect and influence of the added ferric solution on the reduction of organics. Precipitation effects and the appropriate dosage can be determined with relative ease using simple laboratory batch tests. The results obtained largely resemble those of a full-scale application.

BOD_5/COD ratio is not only of use in estimating COD removal rates due to precipitation; the type of organic compounds and their solubility depending on pH is an important, but largely unknown, parameter. Additional findings relating to the precipitation of organics are illustrated in Table 10.4.1, showing the wide variation of removal rates. In some cases, relatively high removal rates are detected at higher BOD_5/COD ratios. On comparing precipitation with Fe and Al, the removal rates obtained are

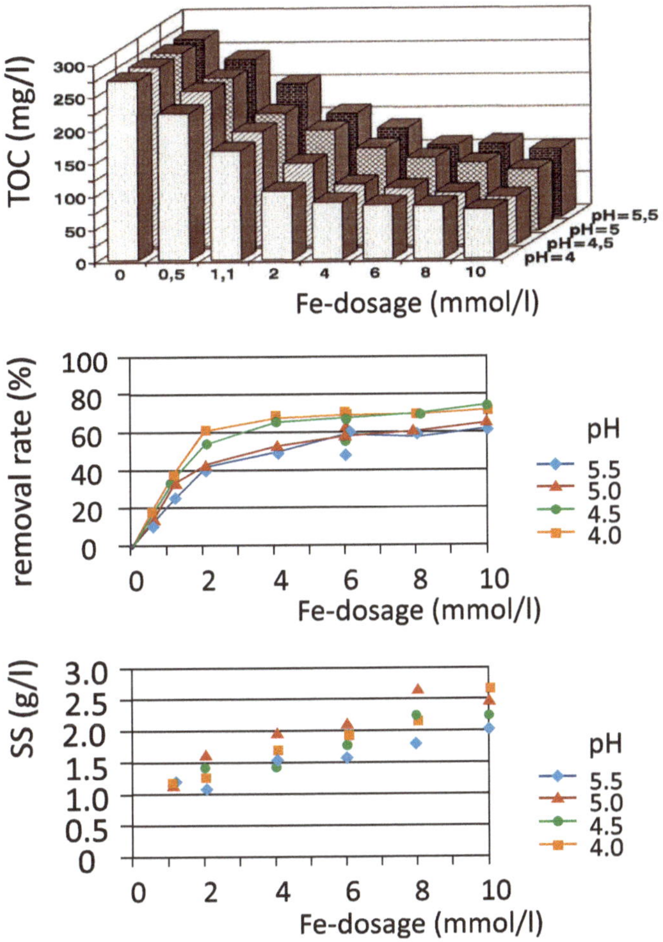

Figure 10.4.3 TOC effluent values and elimination rates (%), precipitation with Fe, dosage from 0.5 to 10 mmol/L (1 mmol = 56 g Fe) under controlled pH conditions (pH 4–5.5) (upper images). Lower image shows the associated production of suspended solids (g/l) (Ehrig, 1998).

often similar. However, in the majority of cases in which different removal rates had been detected, the lowest rate was achieved when using aluminum.

At variance with other processes of precipitation and coagulation the flocks produced are very robust. Studies undertaken with low energy input using a series of different mixing periods lasting more than 24 h revealed no destruction of flocks.

The production of sludge, mainly a metal hydroxide sludge containing coprecipitated organic and inorganic compounds, is a fundamental aspect. Fig. 10.4.3 (bottom graph) shows the production of solids obtained in these experiments. For a first estimation of sludge production, please refer to data presented in Table 10.4.2. The lower part of the table illustrates the results obtained from different

Table 10.4.1 Chemical oxygen demand elimination by precipitation processes with ferric and aluminum solution

References	COD (mg/L)	BOD_5 (mg/L)	BOD_5/COD	Precipitation Agent	COD Elimination	Notes
Amokrane et al. (1997)	4100	200	0.05	Fe 35 mL/L Al 35 mL/L	55% at pH 4.5–5 42% at pH 5.1–5.5	
Ahn et al. (2002)	2800–4200	1300–1950	0.46	1500 mg $FeCl_3$/L = 9.2 mmol/L	56% at pH 4.5–5	
Tatsi et al. (2003)	70,900	26,800	0.38	$FeCl_3$ $FeCl_3$ + polyelectrolyte $Al_2(SO_4)_3$	20%–30% Up to 26%–54% 20%–40%	pH Approximately 10 control by lime
Tatsi et al. (2003)	5350	1050	0.2	$FeCl_3$ $Al_2(SO_4)_3$	70%–80% 60%–70%	Partly stabilized leachate
Galvez et al. (2005)	24,675	8250	0.33	Fe and Al Fe and Al + polyelectrolyte	10%–18% 24%–26%	No pH adjustment
Mendez-Novelo (2005)	5740			50–300 mg/L Fe and Al	12.4% at pH 6 increasing to 42.1% at pH 1 (pH adjustment unknown)	
Aziz et al. (2007)	2098	377	0.15	$Al_2(SO_4)_3*18H_2O$ $FeCl_3$ $FeSO_4$	30% at pH 4 26.9% at pH 6 42.6% at pH 4 44.7% at pH 6 9.1% at pH 4 21% at pH 6 14.9% at pH 12	
Zgajnar Gotvajn et al. (2007)	1200	420	0.35	$Al_2(SO_4)_3$ 1000 mg/L $FeCl_3$1750 mg/L	~25% ~26%	
Loizidou et al. (1992)	4000–8810	650–1150	0.12–0.19	Alum 2–3.5 g/L	22%–24% at pH 6.5–7	
Kurniawan et al. (2006)	4100–5690			Ferric chloride Al salt	55%–80% 38%–42%	Leachate from hazardous waste landfill
Wu et al. (2004)	7500			Ferric chloride (900 mg/L)	60%	

Table 10.4.2 Chloride and sulfate increase during precipitation process

	Precipitation Agent		
	$FeClSO_4$	$Fe_2(SO_4)_3$	$FeCl_3$
Increase of Cl concentration in mg/L per mmol Fe/L	35.5		106.5
Increase of SO_4 concentration in mg/L per mmol Fe/L	96	144	
Sludge production	0.125 kg dry matter per mol Fe + inorganic and organic solids from leachate (approximately 0.5–1.5 kg/m^3)		
Possible solids content in thickener	100–150 kg dry solids/m^3		70–100 kg dry solids/m^3
Possible solids in belt filter press	22%–25%		14%–16% (muddy)
Possible solids in chamber filter press	28%–31%		~20–25% (pasty)

Sludge production and dewatering effects due to precipitation (Ehrig, 1987a).

dewatering processes. The table moreover shows how the use of iron chloride as precipitation aid resulted in the production of chemical sludge with the most unfavorable dewatering characteristics. The use of alum salts will yield sludge with very similar properties to those obtained using $FeCl_3$. As a consequence of higher optimum pH values and lower weight of aluminum salt, a lower production of sludge production is obtained.

The resulting sludge is acidic (pH 4.5–5.5), which may produce an important effect on sludge dewatering. pH value in the leachate effluent ranged from 4.5 to 5.0 (with Fe salt), or 5.0 to 5.5 (using Al salt). A pH value of $\geq$4.3 in the effluent can only be achieved by applying aeration. To prevent acidic effluents—e.g., in the presence of addition of excessively high amounts of Fe salts, thus resulting in pH $<$ 4.3—chemical neutralization should be performed using lye. The risk of eliciting an overdose of aluminum salts is much lower.

The addition of aluminum and iron salts will result in a considerable increase in salt content in the leachate effluent. The effect of addition of three different iron salts on effluent quality is presented in Table 10.4.2. If, in addition to the utilization of iron salts for precipitation, acids are also added for pH adjustment, the additional increase in salt content should be taken into account.

In some cases lime-induced precipitation was tested; the results are shown in Table 10.4.3. Overall, the removal rates achieved are relatively low, with pH value being the most important parameter in this process. Similar removal rates have also been obtained using sodium hydroxide.

Table 10.4.3 Chemical oxygen demand elimination by means of precipitation processes using lime

References	COD (mg/L)	BOD_5 (mg/L)	BOD_5/COD	Precipitation Agent	COD Elimination	
Salem et al. (2008)	3792	980	0.26	lime 0.5 −10 g/L	11.4% with 10 g/L ~8.5% with 5 g/L	
Tatsi et al. (2003)	5350	1050	0.2	$Ca(OH)_2$ (7 g/L)	30%–45% at pH 12	Partly stabilized leachate
Loizidou et al. (1992)	4000–8810	650–1150	0.12–0.19	lime 4 g/L	28% at pH 9–10	
Ehrig (1987a,b)	2170		<0.1	2 kg $Ca(OH)_2/m^3$	14% pH = 11 no increase with higher dosing	The same removal rate was observed with sodium hydroxide at pH = 11
	2000		<0.1	3 kg $Ca(OH)_2/m^3$ 7.5 kg $Ca(OH)_2/m^3$	45.5% pH = 11 49.5% pH = 11	

BOD, biochemical oxygen demand; *COD*, chemical oxygen demand.

The use of precipitation, coagulation, and flocculation in reducing the content of metals and suspended solids in leachate

The above-mentioned lime-based processes are frequently applied in the removal of heavy metals in the form of hydroxides (Thorton and Blanc, 1973; Ho et al., 1974; Keenan et al., 1983). Using a high dosage of lime for certain heavy metals a removal efficiency of 50%–70% was observed (Keenan et al., 1983). Favorable results in removing heavy metals were also obtained by treating "young" leachate (see Chapter 10.2) (Kurniawan et al., 2006) using a dose of 8 g/L lime and adjusting pH to 11.0 resulted in an acceptable removal of Cu^{2+}, Pb^{2+}, Fe^{2+}, Mn^{2+}, and Ni^{2+}. Lime, sodium hydroxide, or soda may all be used to achieve this kind of hydroxide precipitation. This is a commonly used process in the treatment of wastewaters with a high level of metal pollution. However, the solubility of the produced metal hydroxides is relatively high (often 2–10 mg/L). In leachate with a high organic content (particularly nonbiodegradable organics with higher molecular weight), inorganic compounds are often bound by complexation. Hydroxide precipitation of this kind of compounds is frequently unsuccessful. The use of sulfides as a precipitation agent results in production of metal sulfides with a significantly lower solubility and improved ability to precipitate metals bound through complexation. This increased ability may be achieved using organic sulfides, chemicals used to remove metals from wastewaters of flue gas scrubbers. The high toxicity of sulfides, however, requires the elimination of excess sulfide, e.g., by adding $FeCl_3$ to generate FeS.

The above-described precipitation processes are likewise applied to remove suspended solids (SS) from wastewater. The optimum pH value for SS removal is frequently in the range pH 6.5–7.5.

Li et al. (2010) conducted a series of experiments to investigate the application of coagulation/flocculation processes in combination with activated carbon adsorption in the removal of COD, SS, turbidity, and metals from leachate obtained from a landfill in the methanogenic stage. Several coagulants, including aluminum sulfate ($Al_2(SO_4)_3$), ferric chloride ($FeCl_3$), polyaluminum chloride and polyferric sulfate, were tested; among these, polyferric sulfate proved to be the most suitable in terms of removal efficiency of COD (70%), SS (93%), turbidity (97%), and toxicity reduction (74%); in addition, the lowest amount of sludge volume (32 mL) was produced. Using this coagulant the following optimum conditions were elaborated: pH 5.5–6.0; coagulant: 0.6 g Al^{3+}/L as $Al_2(SO_4)_3$, or polyaluminum chloride, 0.6 g Fe^{3+}/L as $FeCl_3$ and 0.3 g Fe^{3+}/L as polyferric sulfate.

Reduction of ammonium by precipitation

The removal of ammonium from wastewater can be achieved using magnesium and phosphate as a precipitation agent (Ehrig, 1998). Although the suitability of these chemical conditions have been acknowledged for decades, the high costs involved have to date prevented their practical application. However, recent developments have demonstrated how the chemicals used can be recycled with relative ease, highlighting the good quality of the remaining ammonium.

Ammonium can be precipitated as magnesium-ammonium-phosphate and the reaction described by the following equation:

$$Mg^{2+} + NH_4^{+} + PO_4^{3-} + 6H_2O \rightarrow MgNH_4PO_4{}^{*}6H_2O$$

Conversely to the above-described precipitation processes the end product of this reaction is presented in the form of crystals, a clean solid product. The pH value in the solution should be adjusted to 9. The advantage of this process is the low solubility of the final product and an almost quantitative precipitation rate. The influence of temperature or other parameters on the final result is very low. The crystals produced are of a substantially consistent size (10–30 μm) and feature a good settling characteristic. As a result of these conditions, secondary pollution with coprecipitated by-products is generally quite low.

Removal efficiencies of up to 90%–98% may be achieved by maintaining pH in the range of 7.5–9.0; these results have been presented in the literature with reference to NH_4–N concentrations of 2300–5600 mg/L in leachate (Li and Zhao, 2001; Ozturk et al., 2003; Calli et al., 2005). Additional results are presented in Table 10.4.4.

To date, the end product has been directly used as an expensive fertilizer. In addition to direct utilization of the product it could be heated to temperatures of up to 80–100°C, at which point the

Table 10.4.4 Effect of MAP precipitation on ammonium removal

References	NH_4 (mg N/L)			
	Influent	Effluent		
Li et al. (1999)	5618	210	pH 6.7	Stoichiometric addition of Mg and PO_4
		112/158	pH 6.7	10% overdosing of Mg or PO_4
		65	pH 8.6	Stoichiometric addition of Mg and PO_4; pH adjustment with NaOH
	5325	154	pH 9	$MgCl_2*6H_2O + Na_2HPO_4*12H_2O$ with highest increase of total dissolved solids from 14460 to 54500 mg/L
		1952	pH 9	$MgO + 85\% H_3PO_4$
		1398	pH 9	$Ca(H_2PO_4)_2*H_2O + MgSO_4*7H_2O$
Kabdash et al. (2000)	Raw leachate			
	2365	1770	pH 6.1	
		217–300	pH 6.5–8.9	87%–91% removal over the given pH interval COD elimination: Influent 13625 mg/L; effluent 2560–5880 mg/L
	Anaerobic pretreated leachate			
	2170	230–240	pH 7.5–8.1	COD influent 4560 mg/L; elimination 33%–39%
Kim et al. (2007)	800–2200	85–90 removal (%)	pH 9 adjusted	NH_4–N: Mg:PO_4–P = 1:1.2:1.2 (molar ratio) Mg^{2+} and PO_4^{3-} could added in any order but pH adjusting only after chemical addition (the removal rates of other order of pH adjusting are only in the range of 40%–70%)

magnesium-ammonium-phosphate fertilizer may be reverted to the original substances. The resulting ammonium and other components could then be recycled. Overall, the chemicals required in this process are rather costly. Moreover, the magnesium-ammonium-phosphate produced contains as little as approximately 8% ammonium, meaning that the remaining 92% is made up of added chemicals. A fundamental task will therefore be the recovery of these added substances. The total process represents an interesting alternative to other nitrogen removal processes including biological nitrogen removal. Biological treatment processes for nitrogen removal need energy (for nitrification) and chemicals, such as acetic acid or methanol (for postdenitrification), and fail to produce any usable final products (new processes as anammox reduces both—see Chapter 10.3). The application of the process of ammonia stripping will result in an end product suitable for use as fertilizer or directly as ammonium. From the point of view of the environment and resource savings, use of this type of treatment technology will become increasingly important.

Final remarks

Precipitation of "non"biologically degradable organic compounds (mostly humiclike and fulviclike acids) in leachate is a cost-effective advanced treatment process. The combination of leachate composition and precipitation agent determines elimination rate. Application of an additional precipitation step would not result in further reduction of organic compounds; to increase reduction rates alternative processes will need to be identified. Moreover, precipitation increases the concentrations of chloride and/or sulfate present in the leachate, resulting in the production of a significant amount of sludge.

From an environmental point of view, MAP-precipitation of ammonium and reuse of the fertilizer thus produced is a substantially sustainable process; however, from an economical perspective, at the current time this process is simply too expensive to warrant widespread application.

MEMBRANE PROCESSES

Principle of membrane processes

Membrane processes involve a kind of filtration or—more precisely—separation of different components in a solution and are characterized according to the dimension of the component to be separated. The bottom part of Fig. 10.4.4 shows the range of pore sizes for the different membrane technologies, while the upper part illustrates the range of particle sizes for different compounds present in leachate. This chapter will subsequently describe how membranes with finer pores are required to separate compounds with decreasing particle sizes [e.g., reverse osmosis (RO)] and will need to be operated at higher pressures; as a consequence the required energy input will increase drastically.

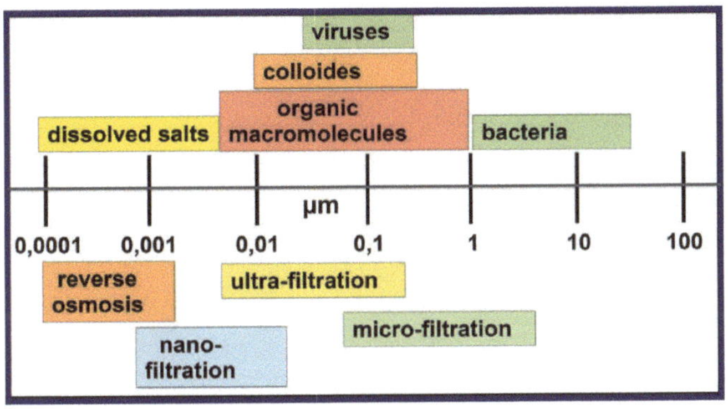

Figure 10.4.4 Dimension range of different water compounds (μm) (upper part) and permeation range of different membrane processes.

Microfiltration is frequently used in wastewater treatment to remove very small particles, such as sludges. Microfiltration is a full-blown filtration system and has no potential for advanced wastewater treatment. Although some of the other three systems (Fig. 10.4.4) are also referred to as "filtration", they are actually all-membrane processes. Ultrafiltration is commonly used to separate solids and particles from the effluent of biological treatment processes down to the size of viruses. Different types of ultrafiltration and technical application of these processes are presented in Chapter 10.3. Ultrafiltration may be effective in removing large macromolecules from leachate, although a scarce reduction of COD is generally achieved.

This chapter focuses on RO and nanofiltration processes. The fundamental difference between these membrane types relates to their permeability—nanofiltration is permeable for some monovalent compounds. If chloride concentrations in the effluent do not need to be removed due to a lower pressure requirement and less concentrated residuals, nanofiltration may be the less costly solution.

The upper section of Fig. 10.4.5 shows the principle of osmosis and RO. When two solutions are divided by a semipermeable membrane water from the least concentrated side permeates to the highest concentrated side until the concentration of both sides are equal. This results in increased pressure on the initially highest concentrated side. The principle of RO lies in the application of pressure on the

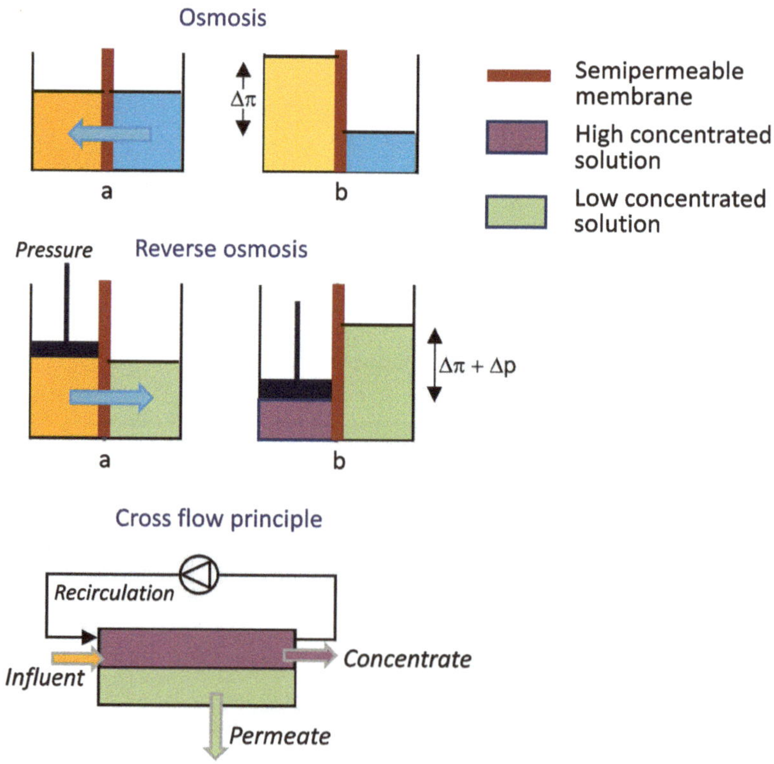

Figure 10.4.5 Description of osmosis principle with osmotic pressure $\Delta\pi$ and the principle of reverses osmosis with required pressure $\Delta\pi + \Delta p$. The principle of cross-flow in membrane modules is also graphically described, (Ehrig and Robinson, 2011).

side with a higher osmotic pressure, to enable the permeation of water through the semipermeable membrane. The more water permeates through the membrane the higher the concentration will be on the left side (Fig. 10.4.5), and a higher pressure will be necessary.

In the lower section of Fig. 10.4.5 the flow scheme is presented. In most membrane systems the flow direction is vertical to the membrane surface, and the filter needs to be continuously or periodically backwashed. If a vertical flow is applied, the membrane will be blocked immediately. To avoid blocking (fouling), cross-flow membrane systems are used in which the flow direction is parallel to the surface of the membrane. By recirculating the concentrate, the surface of the membrane will be cleaned to a high extent. The theoretical calculation of the permeate flow through a membrane is presented in Box 10.4.1. This calculation can be used for solutions containing only one compound. Calculation of

Box 10.4.1 Theoretical calculation of reverse osmosis systems

The following equations can be used to calculate osmotic pressure:

$$\pi = \Delta C \cdot R \cdot T \quad (1)$$

Therefore, it can be deduced that for constant temperature (T) the osmotic pressure π is directly proportional to the difference of molar concentrations (ΔC) of solutions across the membrane; consequently, for equal mass concentration, higher molar mass compounds involve lower osmotic pressure and, in turn, lower operating costs.

Referring to the scheme in Fig. 10.4.5, three parameters can be used to identify the performance of the process:

$$\text{Efficiency: } R = \frac{C_f - C_p}{C_f} \cdot 100 \quad (2)$$

$$\text{Salts transport: } ST = \frac{C_p}{C_f} \cdot 100 \quad (3)$$

$$\text{Volumetric concentration factor: } C_{fv} = \frac{Q_f}{Q_c} \cdot 100 \quad (4)$$

where C_f = feed (raw wastewater) concentration, C_p = permeate concentration, Q_f = raw wastewater flow rate, and Q_c = concentration flow rate.

According to the Merten model (Merten, 1966), solvent (permeate) and concentration flow can be described by the following equations (Weber and Holz, 1989):

$$\Phi_p = \frac{K_p \cdot a}{1} \cdot (\Delta P - \pi) = A \cdot (\Delta P - \pi) \quad (5)$$

$$\Phi_s = \frac{K_s \cdot a}{1} \cdot (C_f - C_p) = B \cdot (C_f - C_p) \quad (6)$$

and

$$\Phi_s = \Phi_p \cdot C_p \quad (7)$$

where K_p = permeability coefficient of membrane versus permeate, a = membrane surface area, 1 = membrane thickness, K_s = permeability coefficient of membrane versus salt, and A = membrane permeability to permeate.

According to Van't Hoff's law,

$$\pi = R \cdot T \cdot (C_f - C_p) = b \cdot (C_f - C_p) \quad (8)$$

Eq. (3) can be rearranged to:

$$\Phi_p = A \cdot [\Delta P - b \cdot (C_f - C_p)] \quad (9)$$

Since usually $C_f >> C_p$, Eqs. (9) and (6) can be rewritten

$$\Phi_p = A \cdot (\Delta P - b \cdot C_f) \quad (10)$$

$$\Phi_s = B \cdot C_f \quad (11)$$

Permeate flux depends on applied pressure, whereas solute flux depends only on concentration. If the process is studied with reference to the free-solute situation ($\Phi_o = A \cdot \Delta P$), the following equation (Weber and Holz, 1989) is valid.

the flow rate of wastewater with a wide variety of compounds—e.g., landfill leachate—is not a feasible option; consequently, tests using the specific membrane and substrate should always be carried out.

As a general rule, two types of membranes are used in the application of RO: cellulose Acetate (diacetate and triacetate) and aromatic Polyamide. Polysulfone amide membranes are used less frequently. The most important characteristics of cellulose acetate and aromatic polyamide membranes are reported in Table 10.4.5.

If leachate concentrate comes into contact with the semipermeable membrane, it may damage the membrane. The following substances are particularly problematic: acids, brine (potentially used for cleaning), organic solvents, etc. Solid materials such as small particles may also damage the membranes or clog flow channels and membranes. The two processes, scaling and fouling, represent one of the major risks in the performance and life span of these semipermeable membranes.

Scaling is the agglomeration of inorganic substances ($CaSO_4$, $CaCO_3$, $BaSO_4$, SiO_2, $Mg(OH)_2$, etc.) on the membrane surface. Fouling can be divided into organic and inorganic fouling. Colloids and biological substances may result in the growth of biofilms on the membranes, and metal hydroxides may likewise produce an inorganic film on membranes. Fouling of membranes entails lower efficiency, increased operating costs, and shortening of membrane life. In the case of treatment of landfill leachate,

Table 10.4.5 Characteristics of acetate and polyamide membranes

Material	Aromatic Polyamide	Cellulose Acetate
Configuration	Hollow fine-fiber modules	Tubular, spiral wound and hollow fine-fiber modules
Normal working pressure	28 bar	30–42 bar
Maximum back-pressure of treated water	3–5 bar	
Maximum operating temperature	45°C	30°C
Maximum storage temperature	40°C	30°C
pH acceptable	2–12	4–6.5
Hydrolysis	Unaffected	Highly sensitive
Bacterial attack	Unaffected	Highly sensitive
Chlorine	Highly sensitive	Unaffected
Operating life	3–5 years	2–3 years
Salt passage (NaCl)	5%–10%	5%–10%

Modified from Degremont (1979).

organic matter and SS are present and may result in the precipitation of salts, one of the main reasons underlying fouling. Prefiltration and appropriate pH adjustment are routinely applied to limit clogging by solids and precipitates, respectively. In "young" leachate, organic matter consists mainly of readily biodegradable volatile fatty acids. Biological pretreatment may therefore prove effective in preventing the biofouling of membranes. This biological stage may also be used to remove ammonium (NH_4) from leachate, possibly responsible—due to small molecular size—for the creation of a bottleneck when applying membrane technology.

Due to their chemical and physical stability Cellulose acetate membranes are widely used. In particular, resistance to chlorine is one of their primary strengths; this property is extremely relevant when using strong oxidants to reduce the risk of membrane fouling. A lower degree of fouling may be achieved in the presence of a smoother surface. Fouling may also be reduced by acetate end-groups determining a limited surface polarization, to ensure that acetate membranes are not electrically charged, contrary to polyamide membranes. On the other hand, acetate is sensitive to pH; particularly diacetate membrane should be operated in the optimum pH range of 4–6 to avoid material hydrolysis and a consequent dramatic deterioration of membrane performance.

Polyamide membranes enhance the production of high fluxes and therefore a lower pressure should be applied during operation, thus resulting in lower costs. These membranes are also more tolerant to pH (acceptable pH range: 2–12) and high temperatures (up to 45°C). On the other hand—as mentioned previously—they may be more prone to fouling.

Membranes are routinely arranged in modules, several kinds of which are in use (Table 10.4.6).

Table 10.4.6 Principle description of different membrane modules

	Tubular Module	Plate Module	Capillary Tube Module	Hollow Fiber Module	Spiral Wound Module
Diameter or feed channel dimension	5–25 mm	1–3 mm	0.5–5 mm	0.04–0.5 mm	0.25–1 mm
Packing density	<80–400 m^2/m^3	100–400 m^2/m^3	<1.000 m^2/m^3	<10.000 m^2/m^3	<1000 m^2/m^3
Solids in feed stream	+	+	–	–	–
Risk of blocking	–	–	+	++	++
In situ cleaning	++	+	+	+	+
Backwashing	–	–	+	+	–

+ more positive; – more negative

Cellulose acetate membranes are mainly used in tubular, spiral wound, and hollow fiber modules, whereas aromatic polyamide membranes are used in hollow fine-fiber modules. Whatever the configuration, a uniform distribution of inflow on the surface of the membrane should be ensured, to establish a consistent behavior throughout the treatment process.

Technical use of reverse osmosis and nanofiltration membranes

Fig. 10.4.6 shows the principle of tubular and capillary modules. An important difference between these two types of modules is represented by the diameter of membrane elements (see Table 10.4.6).

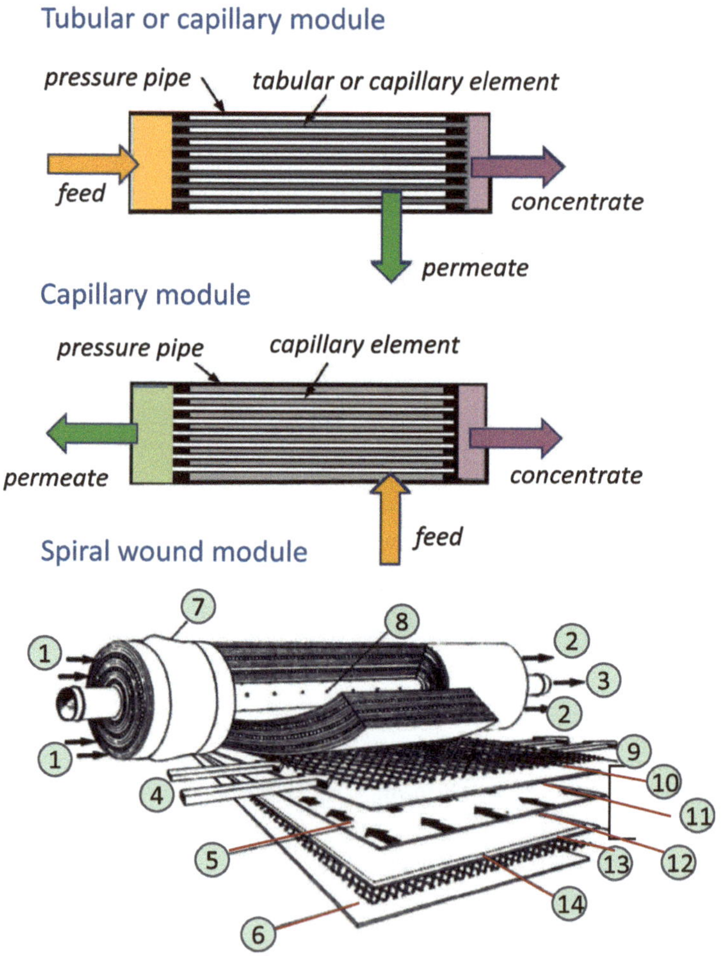

Figure 10.4.6 Schematical view of tubular and capillary modules (upper images) (Ehrig, 2018), and of a spiral wound module (lower image) (Degremont, 1979). LEGENDA: 1.Raw water, 2.Reject, 3.Permeate outlet, 4.Direction of raw water flow, 5. Direction of permeate flow, 6. Protective coating, 7.Seal between module and casing, 8.Perforated tube for collecting permeate, 9. Spacer, 10. Membrane, 11.Permeate collector, 12. Membrane, 13. Spacer, 14.Line of seam connecting the two membranes.

Owing to the larger diameter the stability of the tubular element is relatively low; the direction of flow is from the inside towards the outside of the element. The low diameter of capillary and hollow fiber elements results in a much higher stability and allows a flow in both directions.

The tubular module shows a lesser tendency towards fouling and cleansing is facilitated, although on the other hand it requires more space. In addition, this module features the highest investment and energy costs. Tubular modules are often used as a first treatment step aimed at retaining fine solids and as many compounds as possible in an effort to prevent fouling or scaling on the surface of more sensitive modules. With regard to leachate treatment, tubular modules are currently the most frequently used. However, these modules are characterized by low membrane packing density, which entails a significant number of membrane tubes and, in turn, the need for a large space (Li et al., 2009; Schiopu et al., 2012; Thörneby et al., 2003). Laboratory and full-scale investigations have also demonstrated a rather low permeate recovery rate. The main strength of the hollow fine-fiber module is the impressive membrane surface area per volume unit.

Although plate or disc modules were the first to be applied, due to the high amount of material required for construction, they are applied less and less frequently. Development of the disc-tube RO module—based on the principle of frame and plate filters by placing flat membranes between discs in a tube (Fig. 10.4.7)—has advanced quite significantly. This system has proved to be very efficient in treating leachate, is relatively easy to maintain, and the rotation of the disc modules limits the risks of fouling and surface polarization. Liu et al. (2008), Peters (2010), Renou et al. (2008), Schiopu et al.

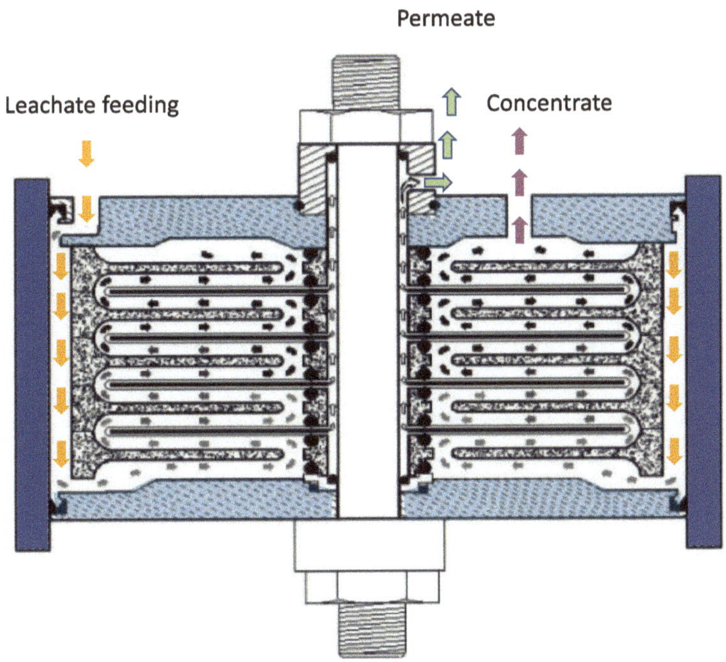

Figure 10.4.7 Disc-tube module.

(2012), Sir et al. (2012), and Ushikoshi et al. (2002) were among the first to extensively investigate the application of the disc-tube system for leachate treatment. Among others, the flexibility, maintainability, and reliability of this system have been positively highlighted. Moreover, as required, the membrane plates can be easily removed from the carrier plates and cleaned off-site.

In general, the spiral wound module is currently one of the most widely used modules in wastewater treatment; a scheme is shown in Fig. 10.4.6. One of the first RO systems in operation was a two-stage unit using tubular membrane modules in the first, and spiral-wounded membrane modules in the second stage; this system was first installed in 1986 at the wastewater treatment plant in Wijster (the Netherlands) (Sir et al., 2012).

However, over time, the use of different kinds of pretreatment processes reduces the flow rate (flux) through the membrane. Tubular membranes can be cleaned mechanically with specific balls or by means of chemical cleansing. All other modules should be exclusively cleaned by means of chemical treatment to remove deposits. This cleaning process should remove as much blocking material as possible, but should not damage the membrane material. In the majority of cases cleaning cannot restore either initial flux volumes or flux achieved after last cleaning (Fig. 10.4.8).

In practice, leachates with a similar composition may display a very different behavior during RO processes. For this reason, the carrying out of pilot-scale trials may be advised. During these tests flow rates, specific cleaning requirements for membranes, and, very importantly, the operational life of individual membranes should be determined. One crucial factor is represented by permeate recovery rate, i.e., the proportion of incoming leachate, which can be recovered as permeate; the recovery of up to 75%–95% may be possible. However, higher values will require use of high-pressure RO. The remaining proportion of liquid represents the volume of concentrate that will be produced. Operational

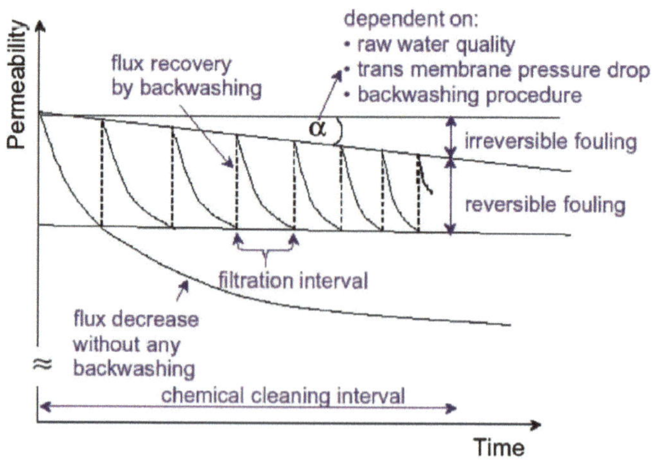

Figure 10.4.8 Principle of flux decrease without backwashing and cleaning. Flux recovery by backwashing and resulting elimination of reversible fouling and the remaining irreversible fouling (Wassertechnologie, 2013).

pressure normally reaches a maximum of 60 bar, although some experiments may be operated at a pressure of up to 200 bar. Following treatment at such high pressure may yield concentrate with a relatively low water content, which may be beneficial if further treatment of concentrate is planned.

Peters (1999) reported that flux rates depend on a series of different parameters, describing average flux rates of 13–15 L of permeate per square meter of membrane per hour. Additional flux rates are given in Tables 10.4.8 and 10.4.9.

Although in some cases the membranes feature a defined permeability—depending on pressure and other parameters—larger particles may pass through the membrane in small amounts. Monovalent ions (Na, Cl, K, etc.) can permeate in the range of 0.5%–5% and divalent ions (SO_4, Ca, Mg, etc.) may permeate through the membrane in the range of 0.5%–2%.

The performance of a leachate treatment plant (Logeman and Glas, 1989) showed that two-stage RO is capable of guaranteeing an efficiency exceeding 99% for COD, BOD, and TKN removal. High removal efficiency was also observed for heavy metals (>90%), with the exception of Cd and As for which approximately 70% efficiency was observed. Table 10.4.7 presents data from several two-stage RO plants. Additional data are shown in Table 10.4.8 (including biological pretreatment). Leachate treatment at the Venneberg-Lingen landfill, consisting of activated sludge oxidation with ammonia removal in the biological section and a subsequent two-stage RO plant is presented in Table 10.4.9 (Weber and Holz, 1989). RO performances were positively affected by biological pretreatment and it was observed that a pH of 6.5 and careful dosage of precipitation inhibitors were pivotal in improving ammonia removal and preventing sulfate scaling. Experimental data indicated very high efficiency of COD and adsorbable organic halides (AOX) removal, although it was not possible, even with two-stage RO, to obtain concentrations below 10 mg/L with ammonia from raw leachate.

In leachate from old landfills or landfills housing biologically pretreated waste, humic acids may represent the most prevalent organic compounds (from tens to hundreds mg/L). The presence of these high molecular weight compounds may have significant adverse effects on the process, as reported by Sir et al. (2012) who observed a decrease of 18% on average for permeate flux and 20% on average for rejection in the presence of 50 mg/L humic acids. Pretreatment using inorganic coagulants and bioflocculants (Zouboulis et al., 2004), adsorption and combination of adsorption and chemical oxidation (Fan et al., 2007) is effective in reducing the presence of humic compounds. A combined process of Fenton oxidation, submerged membrane bioreactor and RO was studied for old municipal landfill leachate treatment (Zhang et al., 2012). The RO section was found to benefit from the Fenton process in terms of reduction of membrane fouling.

Ammonium is a somewhat problematic parameter due to its ability to permeate in larger amounts through the membrane under neutral or alkaline conditions. Acidic pH values change the equilibrium of ammonium ion between gas and liquid phases, enabling NH_4^+ to form ammonium salts with anions (Li et al., 2009; Schiopu et al., 2012). Nevertheless, pretreatment by means of stripping or biological nitrification–denitrification has been proposed as more suitable than a two-stage RO plant (Ehrig, 1989b). Indeed, apart from the operational problem of pH control (i.e., NH_3–N removal efficiency

Table 10.4.7 Treatment results from different two-stage reverse osmosis plants: A, B, C landfill sites in Germany (Ehrig, 2001), Niemark and Rest of Germany and Suldoro and Lamego (Portugal) landfill sites (Gerdes and Nicolaou, 2006)

Plant			Influent		Effluent RO 1		Effluent RO 2	
			Range	∅	Range	∅	Range	∅
A	COD	mg/l	1.590 – 2.980	2.285	18 – 130	74	4 – 25	14,5
	NH_4-N	mg/l	900 – 1.800	1.350	50 – 220	135	4,4 – 8,8	6,6
	AOX	mg/l	1.500 – 1.900	1.700	–	–	0,002 – 0,021	0,0115
	BOD_5	mg/l	230 – 1.320	775	–	–	< 2 – 3	2,5
B	COD	mg/l	2.619 – 5.498	4.124	9 – 25	20	–	–
	NH_4-N	mg/l	398 – 794	577	1 – 18	8	–	–
	BOD_5	mg/l	70 – 185	122	< 1 – 3	1	–	–
C	COD	mg/l	1.301 – 1.707	1.504	–	–	5,2 – 16,8	11
	NH_4-N	mg/l	0,5 – 1,9	1,2	–	–	< 0,05 – 0,48	0,26
	AOX	mg/l	0,78 – 2,43	1,6	–	–	< 0,01 – 0,02	0,015

		Niemark Landfill (Germany)		Rastorf Landfill (Gemany)	
		Influent	**Effluent**	**Influent**	**Effluent**
COD	mg/l	1.024	15	2.500	22
BOD_5	mg/l	40	0,6	–	–
NH_4-N	mg/l	388	6,1	2.100	4
NO_3-N	mg/l	3,44	0,1	–	–
conductivity	µS/cm	8.310	48	18.100	78
pH	–	7,44	6,5	6,4	4,33
		Suldoro landfill (Portugal)		**Lamego Landfill (Portugal)**	
		Influent	**Effluent**	**Influent**	**Effluent**
COD	mg/l	17.780	28	17.029	23
BOD_5	mg/l	10.000	8	11.350	15
NH_4-N	mg/l	3.140	9	891	1
NO_3-N	mg/l	101	0,8	–	–
conductivity	µS/cm	20.000	80	15.400	18
pH	-	8,9	5,4	6,9	5,7

AOX, adsorbable organic halides; *BOD*, biochemical oxygen demand; *CD*, conductivity; *COD*, chemical oxygen demand; *in*, influent; *out*, effluent; *P*, parameter; *RO1*, first RO stage; *RO2*, second RO stage. All units in mg/L except pH and conductivity (µS/cm).

Table 10.4.8 Treatment results from different reverse osmosis and nanofiltration plants

		Robinson et al. (2007)		Ritz (1989)		
		Influent After Biological Treatment	Effluent (≥50% Recovery → flux 10–16 L/m²*h)	Influent	Effluent 1. Stage	Effluent 2. Stage
COD	mg/L	1341	<10			
BOD_5	mg/L			800	80	8
Cl	mg/L	4267	85	1900	380	20
SO_4	mg/L			210	20	1
NH_4–N	mg/L	47	0.1	15,00[a]	600[a]	120[a]
NO_3–N	mg/L	185	80			
AOX	mg/L			25	2	0.2
SS	mg/L	<5	<1			
Alkalinity (as $CaCO_3$)	mg/L	1750	7			
Conductivity	μS/cm	25,500	1200	15,000	1500	15

		Rautenbach and Melis (1991)				
		Two-Stage RO		BT + 1-Stage RO	BT + 1-Stage NF	BT + 2-Stage NF
		Influent	Effluent	Effluent	Effluent	Effluent
COD	mg/L	3900	<10	22	52	<20
NH_4–N	mg/L	1450	15	<10	<10	<10
AOX	mg/L	2.95	<0.01	0.023	0.55	<0.3
flux	L/m²*h		18	25	44	44

	Li (2013) Disc-Tube Module (See Fig. 10.4.7) (65 bar)				
	Conductivity (μS/cm)	COD (mg/L)	NH_4–N (mg/L)	Cl (mg/L)	Na (mg/L)
Influent	17.250	1797	366	2830	4180
Effluent	382	15	9.8	48.4	55.9

AOX, adsorbable organic halides; *BOD*, biochemical oxygen demand; *COD*, chemical oxygen demand; *NH_4 BT*, biological treatment; *NF*, nanofiltration; *RO*, reverse osmosis.

Table 10.4.9 Treatment results of a two-stage reverse osmosis plant and biological pretreatment (first stage cellulose acetate membrane = CA or thin film composite membrane = TFC; second stage only TFC membranes in spiral wound modules) (Weber et al., 1989)

			Influent	(1)	Flux (L/m²*h)	Effluent	
						Stage 1	Stage 2
COD	mg/L	Raw leachate	5.000	CA	15	300	<15
				TFC	13	175	<15
		(2)	2.000	CA	28	100	<15
				TFC	25	60	<15
		(3)	1.500	CA	31	75	<15
				TFC	27	40	<15
TKN	mg/L	Raw leachate	2.000	CA	15	400	45
				TFC	13	260	30
		(2)	1.700	CA	28	340	35
				TFC	25	235	25
		(3)	100	CA	31	20	<2
				TFC	27	10	<1
NH_4–N	mg/L	Raw leachate	1.800	CA	15	375	40
				TFC	13	250	25
		(2)	1.600	CA	28	325	30
				TFC	25	225	20
		(3)	<10	CA	31	<2	<1
				TFC	27	<1	<1

(Continued)

Table 10.4.9 Treatment results of a two-stage reverse osmosis plant and biological pretreatment (first stage cellulose acetate membrane = CA or thin film composite membrane = TFC; second stage only TFC membranes in spiral wound modules) (Weber et al., 1989)—cont'd

			Influent	(1)	Flux (L/m²*h)	Effluent	
						Stage 1	Stage 2
NO_3–N	mg/L	Raw leachate	0	CA	15	0	0
				TFC	13	0	0
		(2)	0	CA	28	0	0
				TFC	25	0	0
		(3)	400	CA	31	40	3
				TFC	27	20	2
AOX	μg/L	Raw leachate	4.000	CA	15	600	100
				TFC	13	400	75
		(2)	2.500	CA	28	400	75
				TFC	25	250	<50
		(3)	2.000	CA	31	300	<50
				TFC	27	200	<50

(1) Two different membrane types are used: *CA*, cellulose acetate; *TFC*, thin film composite.
(2) Biological treatment with activated sludge plant.
(3) Biological treatment with activated sludge plant and RBC.

is a function of pH), the presence of ammonia in concentrate flux complicates concentrate treatment, particularly when evaporation treatment is applied.

A significant advantage of RO is the reduction of very high percentages of all leachate compounds. Conversely, however, the total amount of leachate compounds is concentrated in the residue. Treatment of raw leachate produces approximately 10–15 kg dry solids per m^3. After pretreatment residues are reduced to approximately 5–10 kg dry solids per m^3, only a small part of which is of organic nature. All residues, however, are soluble and should be sealed to the environment.

Indeed, treatment of concentrate represents one of the most problematic aspects of RO. This flux was previously disposed of in landfill; this practice is however no longer acceptable as concentrate is considered a hazardous waste requiring special management. Furthermore, the returning of concentrate to the landfill does not result in an increased concentration of leachate, but rather to a never-ending recirculation of leachate pollution. An increase of leachate pollution through recirculation is completely unfeasible, particularly as this volume had been removed from the landfill shortly before. A series of technologies based on evaporation and drying processes have been developed to minimize the volume of concentrate to be dealt with.

In conclusion, the RO process has been increasingly applied to leachate treatment in view of its efficiency, modularity, and possibility of easy automatic control, and currently represents the core section of many in situ leachate treatment systems. The disadvantages of the process can be summarized as follows (Seyfried and Theilen, 1991):

- retention of small molecules (e.g., ammonia, small AOX molecules) is not fully satisfactory;
- high concentration of organics and precipitation of inorganics may result in fouling, biofouling, and scaling at the membrane surface;
- energy consumption is high as a consequence of high operational pressures (30–200 bar);
- residue production is relatively high (the amount of total solids of leachate);
- concentrate treatment by evaporation and drying is very expensive; and
- storage of solid residues is expensive.

As a treatment concept aimed at reducing the amount of residues and high energy input, a combination of nanofiltration with a secondary treatment step was developed. When applying nanofiltration, the permeability for monovalent compounds such as chlorides is very high. Nanofiltration is routinely operated at a pressure of 3–12 bar. Approximately 30%–75% of monovalent ions and 5%–30% of divalent ions permeate through the membrane. As a consequence, the water content of concentrate is higher than that yielded by reserve osmosis. To achieve the prescribed discharge values, chemical oxidation or activated carbon adsorption may be applied as a second treatment step to promote the removal of residual organic compounds from this concentrate; of course other process combinations are also available. More information relating to these combinations of processes is presented in Chapter 10.5. Results obtained with nanofiltration treatment are shown in Table 10.4.10.

Final remarks

RO systems are capable of removing high percentages of leachate compounds in one or two stages. The remaining residue is a highly viscous liquid concentrate containing most of the leachate compounds. The final discharge of this concentrate into landfill results in a never-ending circulation of concentrate. The evaporation of RO concentrate is a very expensive process and produces a highly soluble solid product. This end product is mostly characterized as a hazardous waste and—preferably after treatment—should be disposed of in special landfills designed to ensure the safety of the surrounding environment.

In some cases, nanofiltration may be applied following biological treatment to retain higher than monovalent leachate compounds. The concentrate may then be treated in activated carbon filters or by means of an oxidation step. As a result, the effectiveness of both processes is increased. Effluent emitted from these processes may be returned to the biological treatment system. It is estimated that some compounds may accumulate in this circulation stream.

Table 10.4.10 Treatment results from different nanofiltration plants

		Campos et al. (2013) (Flux 20 L/m²*h)		Cakmakci and Özyaka (2013) (Flux 28 L/m²*h)[a]		
		Influent	Effluent[b]	Influent	Effluent (10 bar)	Effluent (25 bar)
COD	mg/L	2437	433–460	23,900	350	212
TOC	mg/L	671	201–239	10,340	130	141
Cl	mg/L			3059	237	335
NH_4–N	mg/L	2595	782–824	754	154	106

		Trebouet et al. (2001) (20 bar)		
		Influent	Effluent[c] (Flux 14.3 L/m²*h)	Effluent[c] (Flux 13 L/m²*h)
SS	mg/L	130	0	–
COD	mg/L	500	130	100
TKN	mg/L	540	420	380
NH_4–N	mg/L	430	380	340
Na	mg/L	520	440	435
Ca	mg/L	140	90	61
Cl	mg/L	700	600	620

[a]Four membrane types tested; the type with the lowest effluents are presented.
[b]Results from experiments with 5, 6,7, and 8 bar.
[c]Two different membranes.

Nanofiltration, however, may at times represent a sufficiently efficient pretreatment step to warrant indirect discharge into the sewer for further treatment in a sewage treatment plant.

ADSORPTION

Basic principles of adsorption

Adsorption is the transfer of organic substances from a liquid phase onto the surface of a solid phase. Adsorption material should be characterized by a maximum surface area and a minimum volume. The efficiency of adsorption processes depends on the chemical and physical properties of the soluble substances and of the solid surface. A series of materials can be used in the adsorption process: typical adsorption materials include activated carbon, zeolithes, scavengers, activated alumina, lignite coke, and bentonite. However, in the treatment of organic substances present in leachate activated carbon alone has demonstrated a feasible relationship between adsorption capacity and process cost. Many different kinds of activated carbon, however, are available for a series of different uses.

Activated carbon

Activated carbon consists of carbon materials derived from various sources (coal, wood, peat etc.) after it has been subjected to pyrolysis. Activated carbon has a very large inner surface area (800–1200 m^2/g). As an example, 1 cm^3 of pulverized carbon has a weight of 0.4–0.8 g. By applying an average of 0.6 g/cm^3, it ensues that 1 cm^3 has an inner surface area of 480–720 m^2; this area represents the surface of a large number of pores with different sizes and structures. Pore size distribution is an important factor in the adsorption of organic compounds with different molecular size and structure, as the molecules must be able to penetrate into this pore system. In general, adsorption capacity increases in line with the increasing molecular weight of organic compounds but decreases in the presence of an excessively high molecular weight of organics. However, other molecular properties also exert a considerable influence on adsorption capacity. In the field of domestic wastewater treatment, or treatment of potable waters, activated carbon is frequently used to remove organic micropollutants and a series of nonbiodegradable organic substances. The adsorption capacity of many individual organic substances may indeed be estimated. Raw and pretreated wastewaters contain so many unknown organic components that it would prove impossible to obtain a theoretical calculation. In the case, however, of mixtures between a number of substances and most unknown substances, adsorption capacity should be calculated on the basis of results obtained in laboratory or pilot-scale experiments.

Activated carbon can be used in the form of either pulverized activated carbon (PAC) or granulated activated carbon (GAC). PAC has a particle size of less than 150 μm with a main fraction between 4 and 70 μm. GAC has a particle size of 0.2 mm–5 mm. For water treatment in filter columns a particle size of 0.8–2.5 or 3 mm is routinely applied. GAC carbon can be thermally reactivated, with less than 10% loss in mass during each reactivation. During this process, the adsorbed organic substances are oxidized by gasification and subsequently incinerated. PAC is cheaper, but cannot be reactivated. PAC must be dosed into a mixing tank, and subsequently separated by flocculation and settling. When drying the remaining sludge the particles agglomerate, and the cost of reactivating this material would simply be too expensive. In view of the fact that the carbon is loaded with organic substances present in the leachate, PAC should be incinerated.

Organic compounds will be adsorbed at the surface of the adsorbents; as a result the concentration of these compounds in leachate will decrease. Under steady state conditions the concentration of a substance on the inner surface area of the carbon is in equilibrium with the solution in the surrounding liquid. If the concentration in leachate is lower than the adsorbed material this will result in transport of the adsorbed substance back from the surface into the liquid. The effectiveness of the adsorption process can be described by the adsorption isotherm, which presents the mass of adsorbed material per mass of activated carbon (e.g., mg COD or TOC/g activated carbon) as a function of the equilibrium concentration in the leachate (for example: mg COD/L or mg TOC/L). The upper graph in Fig. 10.4.9 shows the results of an experimental series (with different activated carbon dosages) and resulting equilibrium concentration Ce (mg TOC/L) and carbon load X/M (mg TOC/g activated carbon). The values obtained can be described with a Freundlich-Isotherm (explained in Fig. 10.4.9). A real linear relationship in logarithmic scale is relatively rare. As a routine, nonlinear isotherms, as shown

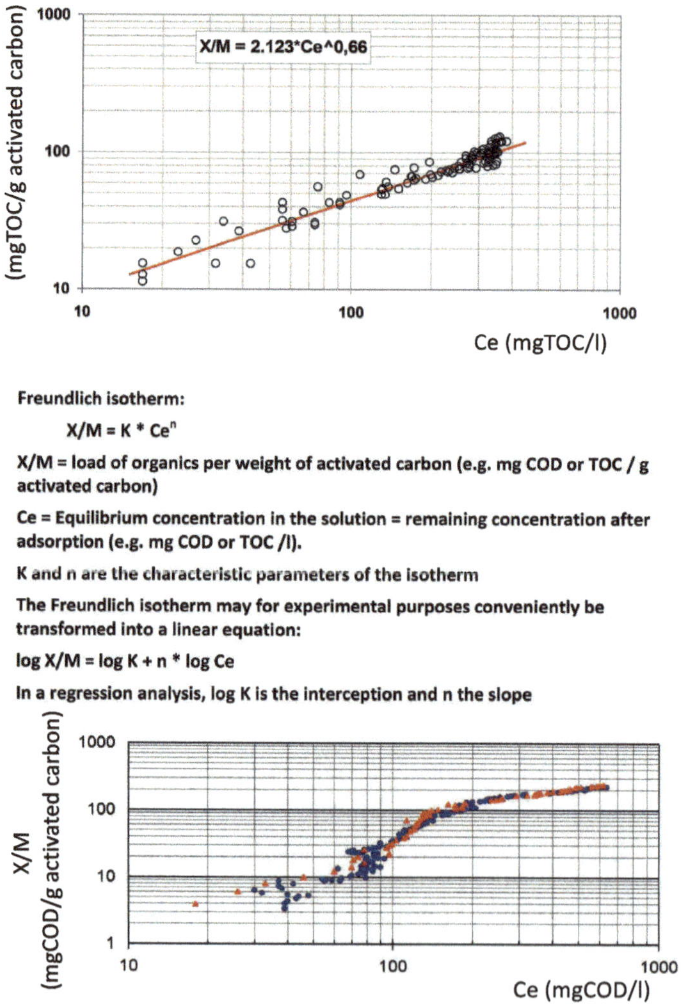

Figure 10.4.9 Relationship between effluent values Ce [COD or TOC (mg/L)] and activated carbon load X/M (mg COD or TOC/g activated carbon). A linear relationship in log–log scale can be described with Freundlich isotherm (Ehrig, 2018).

in Fig. 10.4.9, are detected (lower graph). The leachate in question contains substances or groups of substances characterized by highly diverse adsorption. The upper part of the isotherm shows a readily adsorbable group of compounds indicated by a low gradient; the steeper gradient represents a less adsorbable group. The lowest part is too diffuse to give any explanation.

Laboratory tests can be conducted to estimate the performance of a technical-scale plant. Mixing different amounts of carbon with leachate and measuring the concentration until steady state conditions are reached will describe the adsorption behavior of the specific PAC. PAC consumption may be estimated using a graph similar to Fig. 10.4.9. Similar tests with GAC will require at least 24–48 h to reach steady state conditions and may produce significant differences on real filter conditions. Although GAC has limited dimensions of 0.8–3 mm, the route of diffusion into the material is lengthy, and complete diffusion, i.e., reaching the end of the pores, cannot be achieved using short-term mixing tests. Filtration tests using GAC samples may represent a more appropriate solution. In the bottom graph in Fig. 10.4.9 (red and blue) the results of the operation of a technical-scale filter (Ø 2.5 m) and a laboratory-scale filter (Ø 3.5 cm) running in parallel are shown demonstrating how the same isotherm can be achieved for both systems.

The range of commercially available activated carbons feature different adsorption characteristics; only very vague descriptions are provided to assist selection for the specific application. In the majority of cases parallel tests using different types of carbon will produce largely diverse results. Use of another leachate with a similar composition may show a completely different behavior. In addition, quality of any specific type of carbon may vary over time, sometimes quite drastically.

Leachate composition exerts a significant effect on the adsorption capacity of AC. Different leachate concentrations may also be associated with a change in organic composition which could influence adsorption capacity; in some cases it may be necessary to resort to use of another type of carbon. Measured adsorption capacities are in the range of 100–250 g COD/kg carbon. Maximum loading rates of 420 g COD/kg carbon have been achieved.

Several COD compounds present in leachate originating from landfills in the acetic phase, such as volatile fatty acids, are only adsorbed to a very low extent. This low adsorption of several biodegradable leachate compounds may increase the costs of adsorption considerably. Biological pretreatment may be capable of increasing effectiveness of the process and lowering the costs. In some German landfills activated carbon adsorption is used as a first treatment step; these landfills, however, are in a highly stable methanogenic phase with a low content of biodegradable compounds but remaining high ammonium values. The use of activated carbon adsorption removes nonbiodegradable organic compounds. In some cases, the effluent from the carbon filter is fed into the domestic sewage treatment plant, where ammonium will be reduced. The last step is quite economic, thus implying that in the presence of biologically degradable organic compounds, the increase in carbon costs may be limited.

Organic compounds are characterized by a wide range of adsorbability. Indeed, previously adsorbed organics may as a result be replaced by organic compounds with a better adsorption behavior. This effect may result in a transient increase in effluent concentration (even exceeding influent concentration) of compounds that had been adsorbed first. Indeed, this effect was observed at one plant equipped with GAC filters: after a phase of AOX reduction in the filter effluent—as result of displacement by other COD compounds—values increased to exceed influent concentrations. AOX elimination subsequently stabilized at a lower level.

In contrast to studies from Ehrig (1987a,b) Foo and Hameed (2009) reported an effective removal of ammonium nitrogen from landfill leachate.

Application of different technologies for activated carbon adsorption

The two types of activated carbon (PAC and GAC) should be differentiated as handling in the treatment process varies considerably. PAC is dosed as a suspension in a mixing and reaction tank. GAC is used with a sand, or comparable type of, filter and leachate flows through the filter.

PAC is mainly stored in a carbon silo; PAC is then transported from the silo to a mixing tank to produce a carbon/water suspension. This suspension is pumped to the dosing point in the mixing and reaction tank. The carbon/water suspension is not stable and may lead to precipitation and clogging of pipes; for this reason, pipes should be short and cleaned on a routine basis. With regard to the mixing device, it should be taken into account that the carbon suspension is extremely abrasive. Mixing with a high amount of air has not proved problematic. PAC can either be removed with the activated sludge, or by means of a flocculation/precipitation process. COD removal efficiency of 95% was observed when leachate from the Thessaloniki landfill in Greece was treated using varying dosages (in the range 0.2–10.0 g/L) of pulverised activated carbon (PAC); the adsorption process could be well described by the Freundlich-isotherm (Diamadopoulos, 1994).

A filter system with three filter columns for adsorption filled with GAC is shown in Fig. 10.4.10. A minimum of two filters should be used for adsorption in GAC filters. The operation of GAC filters is considerably different from that of sand filters. Sand filters remove suspended particles and pressure drag increases in the presence of an increasing load. If a given limit of pressure drag is reached the filter will be backwashed and then operated as before. However, in carbon filters the inner surface of the carbon is loaded with the removed organics—backwashing of these organics is not effective. Removal of these organic compounds (reactivation) may only be achieved using a thermal gasification process. Due to the complex technology involved, this process takes place in central reactivation plants. Thermal reactivation destroys the adsorbed organic compounds and results in a carbon loss of 7%–10%. As mentioned previously, activated carbon features only slight loss of quality compared to the original GAC and may be used further. As a general rule, GAC delivery consists in a mixture of fresh and reactivated material; when the filter material is exchanged in one filter the second filter will be in operation, this ensuring continuity of the leachate treatment process.

If the effluent quality of a filter column system (Fig. 10.4.10) approaches effluent limit value, one of the filters should be emptied and filled with new GAC. Following reloading of the filter with fresh GAC, the filter with the lowest loading rate should be used for final treatment. This filter should be positioned as the last filter in the row. Operation of this type of filter system consisting of three columns is arranged

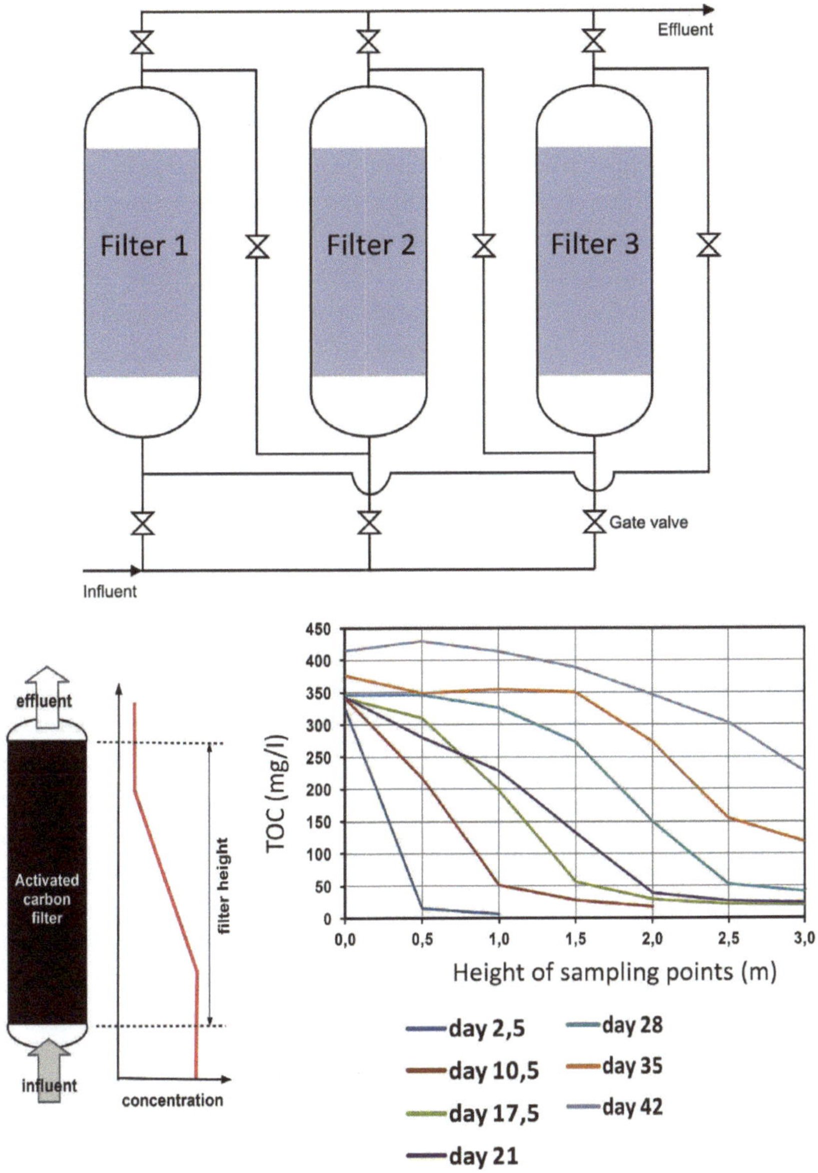

Figure 10.4.10 Activated carbon system consisting of 3 filters and connecting pipe system to operate the filter in every order (upper image). The lower image shows the principle concentration curve in the activated carbon filter (left side) and the measured moving concentration curve of an operating filter (right side) (explanation see Chapter 10.4) (Ehrig, 2018).

by changing the filter from (1-2-3) over (2-3-1) and (3-1-2) back to (1-2-3). To avoid the need to actually move filters, a pipe system is used to change the order. In general, the GAC delivery truck takes the loaded GAC back to the reactivation plant on the same trip. As a consequence, an additional filter or transfer tank may be required to enable the truck to unload the fresh GAC before loading the old GAC.

As shown in Fig. 10.4.9, equilibrium concentration in the solution should be effectively balanced with the load (X/M), irrespective of the kind of isotherms used (straight line or curve). Fig. 10.4.10 (lower image) shows on the left side a GAC filter with the main concentration of COD or TOC over the filter height. This concentration curve moves through the filter from the influent to effluent. The right side of Fig. 10.4.10 (lower image) shows the movement of a measured concentration curve through the filter from day 2.5 until day 42. The curve of GAC loading rate is similar to concentration curve. Using concentration curves for the calculation of the number of filter and/or the filter height is presented in Fig. 10.4.10 (lower image) (see Box 10.4.2).

The different means of utilization of PAC and GAC will significantly affect the carbon requirement of both systems, largely due to the different equilibrium concentrations of the systems. PAC is dosed in a mixing tank and equilibrium concentration corresponds to concentration in the tank and effluent concentration. Calculation of GAC filter column length (Box 10.4.2) highlights how the first filter will be emptied at a much higher equilibrium concentration (optimum: effluent concentration = influent concentration) than the following one. The X/M load of GAC filter systems is much higher than for PAC in mixing tanks; as a consequence the carbon requirement is lower (calculation see Box 10.4.2).

Fig. 10.4.11 shows influent and effluent data from three GAC filter plants. Both upper plants were equipped with 2 filters, whereas the lower plant had three filters and the fourth filter was temporarily used to exchange carbon. The vertical lines indicate carbon exchange, also illustrating succession with the change of color. The topmost graph highlights a significant difference between influent concentration and maximum effluent concentration of Filter 1. Such a large difference is an indicator that the adsorption capacity was only marginally used. The effluent limit of leachate (graph in the middle) was 400 mg COD/L. The difference between influent and effluent concentrations in Filter 1 was also considerable, but the variation of influent data was particularly extensive. Additional filters (upmost graph: ≥ 2 new filter; middle graph: ≥ 1 filter) had been installed in both cases. The bottom graph shows an attempt to achieve the closest correspondence between effluent and influent concentrations. It is very difficult to achieve the latter in the presence of three filters due to the very low effluent limit value of 100 mg COD/L. By using one additional filter GAC consumption could be further reduced. Overall, the installation of additional filters is usually less expensive compared to the increase in carbon consumption.

Other adsorption materials

The relatively low adsorption capacity for ammonia is due to the nonpolar surface of activated carbon. For this reason, several studies have focused on modifying the AC surface or producing composite adsorbents that may well interact with either polar or nonpolar compounds. Halim et al. (2010) compared performance in terms of COD and ammonia removal of zeolite, activated carbon and a composite material

Box 10.4.2 Calculation of activated carbon length and carbon requirement (Ehrig, 2018)

Calculation of the Number of Filter or Filter Length

The right side of Fig. 10.4.10 shows the actual passage of the concentration through a filter (height 3.5 m) from operating Day 2.5 until operating Day 42. It can be seen that the concentration slope becomes lower with time. With an effluent requirement of 70 mg TOC/L this value is nearly achieved at Day 28. However, at this time the adsorption capacity of carbon between ~1 and ~3 m of the filter height has only been partly used. A change of carbon over subsequent days wastes adsorption capacity. The optimum usage of GAC is reached if effluent values of the first filter are approximately similar to influent values and the entire concentration slope is moved into the next filter or filters. Using the concentration slope at Day 42 an additional length of at least 3–4 m is necessary to reach influent concentration over the total length of Filter 1. With such an experiment the optimum number of filters can be calculated. Using the graph of Fig. 10.4.10 the following filter system is necessary: filter 1 = 3m + 3–4m + a short security length of 0.5–1 m (as a consequence of changing filter order all filter must have the same length). Solution: 3 filter with 3 m length.

Calculation of Carbon Requirement of PAC in Mixing Tanks and GAC in Filters

For both the Freundlich-isotherm in Fig. 10.4.9 is used.
Filter: the total adsorption capacity of first filter is used.
The influent concentration is 310–410 mg TOC/L (average 350 mg/L)
The effluent limit = 70 mg TOC/L.
A calculated effluent of 50 mg TOC/L is used for safety reasons.

Mixing Tank

Equilibrium concentration = effluent concentration = 50 mg TOC/L

$$X/M = 2.123 \cdot 500.66$$
$$= 28 \text{ mg TOC/g activated carbon}$$

Carbon requirement =
(influent TOC − effluent TOC)/(X/M)
= (350 − 50)/28
= 10.7 g carbon per litre of leachate.

Filter System

Average estimated concentration over the filter length: 340 mg TOC/L

$$X/M = 2.123 \cdot 3400.66$$
$$= 99.5 \text{ mg TOC/g activated carbon}$$

Carbon requirement = (influent TOC − effluent TOC)/(X/M)
= (350 − 50)/99.5
= 3 g carbon per litre of leachate

containing 45.94% zeolite, 15.31% limestone, 4.38% activated carbon, and rice husk carbon, as well as 30% of ordinary Portland cement as a binder; the zeolite surface is hydrophilic, whereas the surface of carbon is hydrophobic with pore sizes in the nanometer range or above. The results showed that adsorption capacity of the composite adsorbent for ammonia nitrogen was higher than that of zeolite and activated carbon; the capacity was comparable to the adsorption capacity of activated carbon for COD.

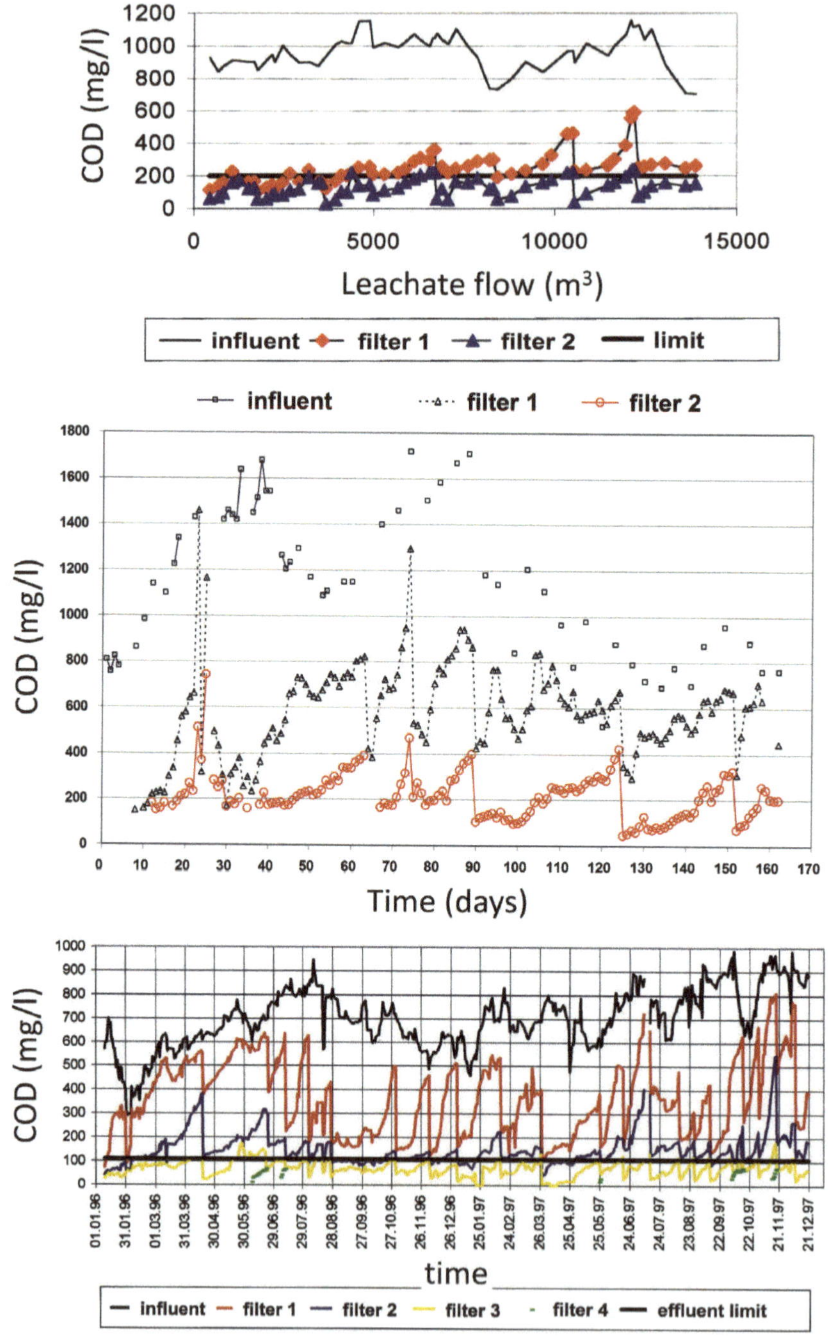

Figure 10.4.11 Influent and effluent data of three technical scales activated carbon filter plants (Filter 4 in the lower image was only in operation temporarily) (Ehrig, 2018).

Final remarks

PAC is significantly less effective than GAC, and in some cases PAC is used in combination with Fe-precipitation. Both processes complement one another and result in highly constant effluent values; these are also achieved when there is a considerable variation of flow rate and concentration. The use of activated carbon filters represents a highly effective advanced treatment process. The advantage of this process is the easy adaptation to changes in quantity and quality of the leachate. No dosage system is required to react on variation of flow rate or concentrations—only the duration of carbon exchange will be reduced or extended. Thereafter, the consumption of carbon is self-regulated to adapt to the requirement. The removed organic compounds are destroyed during carbon reactivation, and no further residues are produced.

CHEMICAL OXIDATION

Basic principles of oxidation

Pollutants (mainly organic compounds) can be oxidized into a series of less harmful compounds through addition of an oxidation agent. During chemical oxidation, one or more electrons transfer to the oxidant from the targeted pollutant, causing its destruction. The optimal end products of organic oxidation are water and carbon dioxide. A previously applied oxidation process focused on alkaline chlorination; however, in view of the complex organic matrix of leachate, high quantities of hazardous substances were produced. Ozone and hydrogen peroxide are the most widely agents in chemical oxidation at normal temperature and pressure. The Fenton process is also applied in some countries, with combinations of different agents and UV-light representing possible alternatives. In the past, a large number of combinations of oxidation processes have been investigated for use in leachate treatment. However, only a few types and combinations have yielded feasible treatment options for use in full-scale applications.

Chemical oxidation involves a similar oxidation process to biological oxidation, but is considerably stronger. However, the costs of chemical oxidation agents are considerably higher than the costs of biological treatment. It may therefore be helpful to implement a biological pretreatment step with the aim of producing a full biologically stabilized leachate. Many compounds present in leachate may affect oxidation efficiency, thus implying a need for oxidation tests to be carried out. A systematic and stable oxidation of ammoniacal-N is frequently difficult to achieve. Observations made at technical oxidation plants, however, have shown this type of oxidation occurs from time to time with increased consumption of oxidation agents.

Reactions produced by the oxidation of ozone and hydrogen peroxide are known in principle, although the extremely short reaction times hinder the exact description of the process using complex polluted leachate. Only a limited part of the organic compounds can be broken down into the end-products water and carbon dioxide. Organic residues can be divided in biodegradable and nonbiodegradable compounds. As a general rule leachate contains relatively high concentrations of chloride, and consequently, during the initial phase (short oxidation time) with low oxidation efficiency—when

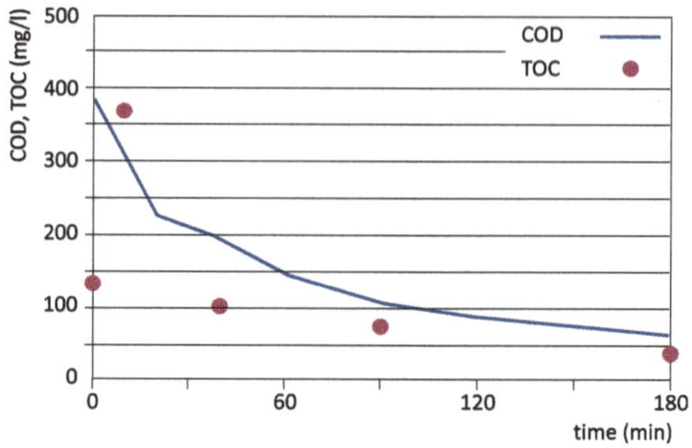

Figure 10.4.12 COD and TOC reduction during oxidation with ozone. A specific result of oxidation processes is decrease of the original COD/TOC-ratio (Ehrig, 2018).

in the presence of high concentrations of chloride—organic halogens are produced. During the second phase, organic substances are largely oxidized into organic fragments, many of which are biodegradable. Only during longer oxidation phases will a significant proportion of organic substances be oxidized into carbon dioxide. In many cases, COD is reduced following chemical oxidation, whereas BOD_5 is increased and at times exceeds the prescribed discharge limits. In these cases biological posttreatment is recommended.

Theoretically, the production of toxic organic substances during oxidation is a possibility. Fig. 10.4.12 shows the quality yielded following oxidation. The relationship between COD and TOC is reduced from approximately 3 to 1.5 as oxidation time increases. The impact of the treated leachate on the environment should be further investigated, particularly as the resulting organic compounds are not the same as those contained in the original leachate.

Oxidation processes with hydrogen peroxide

The standard concentration of hydrogen peroxide (H_2O_2) solution is 30%–50%, at which concentrations it is presented as a clear liquid. Industrially used H_2O_2 solution contains stabilizer to prevent dissociation in the short term. Higher concentrations may lead to spontaneous explosive dissociation. A stable H_2O_2 solution will be dissociated in a short time during contact with several materials; this should be taken into account when installing storage tanks and dosing systems. The environmental risk of hydrogen peroxide is relatively low due to it rapidly breaking down into O_2 and water when it comes into contact with other substances.

Hydrogen peroxide has been widely used as an oxidant to remove low concentrations of sulfide from leachates, by converting the sulfide to elemental sulfur. This is a highly specific reaction, and the required dosing rates can be predicted accurately on the basis of chemical stoichiometry. Previous experience has highlighted how complete oxidation of sulfides requires a contact time of approximately 30 minutes at a dose of 1.5–3.0 parts of H_2O_2 per part dissolved sulfide. A reduction of sulfide to 0.5 mg/L was observed (Robinson and Maris, 1979).

Contrary to sulfide oxidation the oxidation of organic components with hydrogen peroxide is relatively slow. The oxidation rate depends on specific conditions such as UV-light application, pH values, and leachate composition. Relatively high oxidation rates can be obtained by the addition of UV-light, and at pH values of approximately ≤ 4. The production of UV-light consumes large amounts of energy; the color and turbidity in leachates result in increased energy requirements due to a lower penetration of UV-light. Additionally, low pH values consume large amounts of acid and increase the salt content of the treated leachate. During the contact of hydrogen peroxide with organic polluted water the peroxide breaks down very rapidly to radicals, which are strong oxidation agents. These radicals react with the organic substances in water. Fig. 10.4.13 (upper image) shows the results of laboratory-scale experiments of leachate oxidation using hydrogen peroxide. After a reaction time of 6 h, 68% of TOC had been reduced. Fig. 10.4.14 (lower image) shows how elimination is strongly dependent on pH value. The influence of UV-light is very strong and is more important than reaction time. Fig. 10.4.14 (upper image) shows the influence of energy input (with low pressure UV lamps) on COD elimination (5 different leachates).

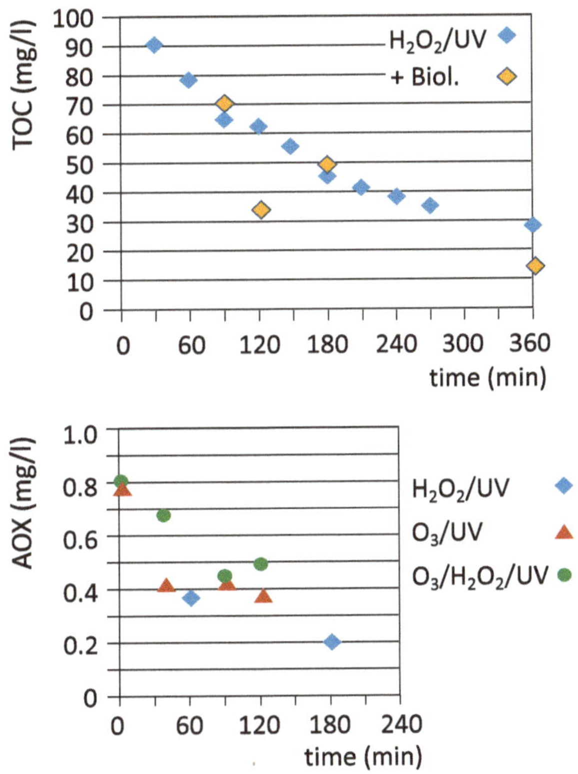

Figure 10.4.13 TOC values after oxidation with H_2O_2 enhanced by UV-light application. The *open rectangle* shows results when additional biological treatment was applied (upper graph). Oxidation effects using H_2O_2 and ozone on AOX reduction (lower graph) (Ehrig, 2018).

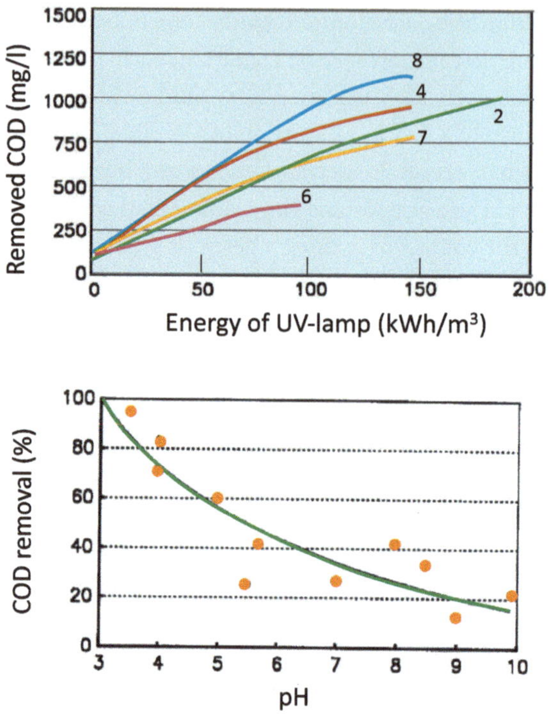

Figure 10.4.14 Effect of energy input on COD elimination of H_2O_2/UV-oxidation (5 different landfill leachates) (upper graph). Relive COD elimination with H_2O_2/UV-oxidation depending on pH value during oxidation (elimination at pH 3 = 100%) (Steensen, 1993).

The bottom part of Fig. 10.4.13 illustrates the results of AOX oxidation. Similar to AOX oxidation from groundwater remediation, H_2O_2 oxidation combined with UV is highly effective, although in comparison with ozone oxidation.

In the majority of cases the oxidized effluent is characterized by increased BOD_5 values. The open rectangles in Fig. 10.4.13 (upper image signed with "bio") show biologically treated samples after oxidation.

The Fenton reaction is a combination of hydrogen peroxide and Fe^{2+} salt used in the oxidation of organic compounds. During Fenton reaction, ferrous iron reacts under optimum pH with hydrogen peroxide ($Fe^{2+} + H_2O_2 \rightarrow Fe^{3+} + {\bullet}OH + OH^-$) and is oxidized to ferric iron, a hydroxyl radical and a hydroxyl anion (Deng, 2007; Umar et al., 2010; Cortez et al., 2011). Iron acts as catalyst by changing form between Fe^{2+} and Fe^{3+} and since the presence of H^+ ions is necessary for the decomposition of H_2O_2, initial pH is generally adjusted to acid values and may also be controlled continuously at the desired value by adding sulfuric acid or sodium hydroxide. Optimal pH values for landfill leachate treatment are reported to be within the range between 2.0 and 4.5 (Deng and Englehardt, 2006). With regard to the effects of main operating conditions, Zhang et al. (2005) performed an extensive study using batch reactors. The results obtained show that a short reaction time of 30 min at the observed optimal pH of 2.5; the efficacy of Fenton process was greater when both reagents were added in multiple steps rather than in a single step. Optimal H_2O_2 to Fe^{2+} molar ratio was 1.5, and given the favorable H_2O_2 to Fe^{2+} molar ratio, removal efficiency increased as dosage increased. Finally, the addition of the appropriate polymer (Percol 710 according to 0.5 mg/L) improved sludge settling characteristics, and COD removal efficiency increased slightly as temperature was increased from 13–15°C up to 35–37°C Sabour et al. (2011) described the oxidation of highly polluted leachate (COD 34,920 mg/L; BOD_5 16,840 mg/L) with the Fenton reaction. The optimum molar ratio H_2O_2/Fe^{2+} was 4 and optimum oxidation time 1–2 h. With increasing Fe addition ($H_2O_2/Fe^{2+} = 4$) COD elimination increased. However, in the presence of more than 10,000 mg Fe-precipitation of organic compounds may at times ensue. Table 10.4.11 shows the results of Fenton reaction cited from Sabour et al. (2011). Although COD concentrations in the leachates concerned are, in the majority of cases, much lower, a higher addition of Fe than that required for chemical precipitation is necessary. In addition to the assumed onset of precipitation effects, a marked increase in effluent salt concentration and high sludge production may be unavoidable.

Fenton reactor characteristics, operational problems, and treatment schemes have been discussed by Deng and Englehardt (2006). Commercial Fenton reactors are typically operated in batch mode

Table 10.4.11 Treatment from different oxidation experiments with Fenton reaction

Initial COD (mg/L)	Oxidation pH	H2O2/Fe^{2+} Molar Ratio	Fe^{2+} (mg/L)	Oxidation Time (min)	Coagulation pH	COD Removal Efficiency (%)
1100–1300	3.0	3.0	3351	120	8	55.0
1000	2.5	1.5	2792	30	7.5–8	61.3
2000	2.5	1.5	2792	30	7.5–8	49.4
3000	2.5	1.5	2792	30	7.5–8	37.5
1200–1500	3.5	3.1	875	180	3–6	72.0
1340–1660	6.0	1.1	300	10	N/A	70.0
1720	4.0	1.0	800	10	N/A	74.7
1500–2100	3.0	1.2	2000	N/A	N/A	51.9
10,540	3.0	19.8	830	120	8.5	55.0
10,915	3.2	23.2	700	120	N/A	50.0

Cited from Sabour et al. (2011).

according to the following phases: oxidation, neutralization, flocculation, and solid–liquid separation. The reactors are stirred and provided with pumps for the addition of acid and base Fenton reagents with pH sensor/controllers. Operational problems in the field may affect COD removal efficiency compared to the results obtained in laboratory tests. Foaming may be caused by the CO_2 produced from carbonate species at acidic pH, as well as by organic foaming agents in leachate, and result in the need for a considerably larger tank volume. Significant quantities of acid are required to adjust the pH of typical leachate, thus implying safety and corrosion issues. Hydroxyl radicals are not effective in oxidizing ammonia, with some organic compounds being recalcitrant to Fenton treatment, and total dissolved solids may be increased as a result of the low pH and Fenton reactions. Finally, further to treatment, the residual sludge should be appropriately disposed of. Deng et al. (2012) reported a significant reduction of COD removal rates through the Fenton reaction at chloride concentrations above 1250 mg Cl/L.

Overall, the total process involved should be closely monitored. The best results are obtained in batch processes, thus rendering integration in a continuous leachate treatment process difficult. The produced Fe^{3+} remains as residual sludge. To date, the majority of experiments have been carried out in laboratory or small pilot plants. Practical experiences with full-scale plants and a continuous operating mode are extremely rare. Similar to other leachate treatment processes, important issues were only observed at full-scale plants.

The addition of UV-light to the Fenton process could prove to be problematic due to the fact that during ozonization in the sole presence of leachate in the reactor, a virtually continuous cleaning of the lamps will be required.

Oxidation processes by means of ozone

The majority of full-scale chemical oxidation plants use ozone (O_3) in the oxidation of leachate components. Ozone is a gas characterized by an extremely rapid decay and should therefore be produced immediately prior to use. Ozone can be produced from the oxygen in the air or directly from pure oxygen. At most technical-scale plants ozone is produced in a corona-discharge generator. If air is used in ozone production, it should first be dried. Irrespective of the use of air or oxygen as gas source, only a part of the oxygen can be converted to ozone. Using air approximately 20–60 g ozone/m^3 gas and using oxygen up to 180 g ozone/m^3 gas can be produced. Energy consumption for the production of 1 kg ozone is 12–18 kW with air and 6–10 kW using oxygen. As a result of decreasing costs of pure oxygen and easier handling, in the majority of cases pure oxygen is used in technical-scale plants. Ozone reacts directly with organic pollutants similar to hydrogen peroxide during breakdown with radicals.

Ozone is a toxic gas and must be destroyed in the outlet air. Normally, the oxidized effluent contains ozone. In a closed system the oxygen can be used in biological treatment plants, whereas in other cases a stripping of ozone and subsequent destruction will be required.

Fig. 10.4.15 illustrates results obtain in the oxidation of leachate by means of ozone. Indeed, ozone was the sole gas used to oxidize leachate. Oxidation with ozone may be further improved by the addition

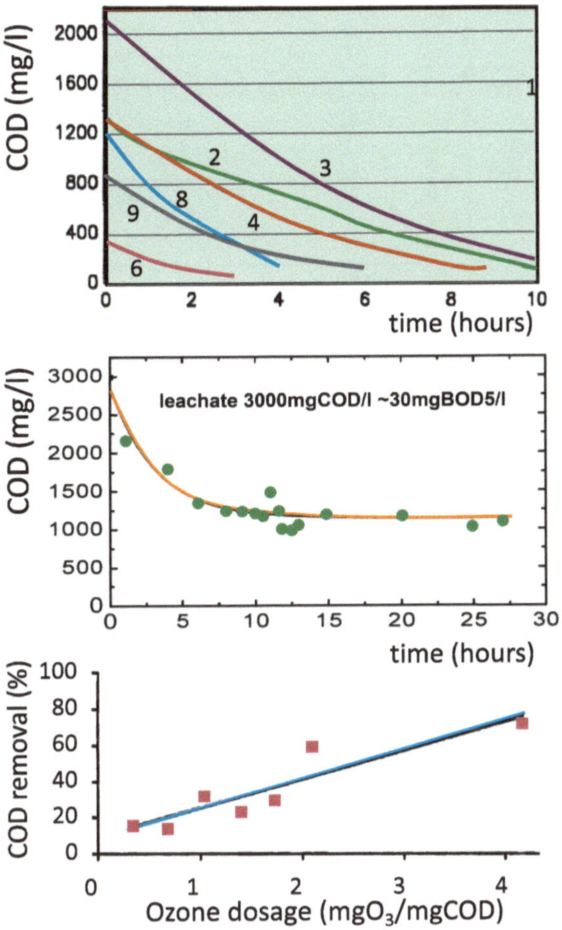

Figure 10.4.15 COD removal according to detention time or dosage of ozone (oxidation at original leachate pH values) (Steensen, 1993—six different leachates; Derco et al., 2002; Chaturapruek et al., 2005).

of UV-light or H_2O_2 (Fig. 10.4.16). The oxidation illustrated in both figures was operated at a pH value present in leachate. The organic pollution of all leachates shown was characterized by a low BOD_5/COD ratio. Additional removal rates achieved by means of ozone oxidation from Baig et al. (1996) are reported in Table 10.4.12. The addition of UV-light or H_2O_2 (Table 10.4.12) does not always, however, elicit an improvement. Overall, in many cases the enhancement effect has been found to be limited, and the potential benefits should be closely weighed up on the basis of the additional costs involved.

Fig. 10.4.16 shows a comparison between H_2O_2 and ozone oxidation of biologically treated leachate under actual pH conditions. The higher oxidation rate reported for ozone is significant. However, a reduction of pH during H_2O_2 oxidation will result in more comparable findings.

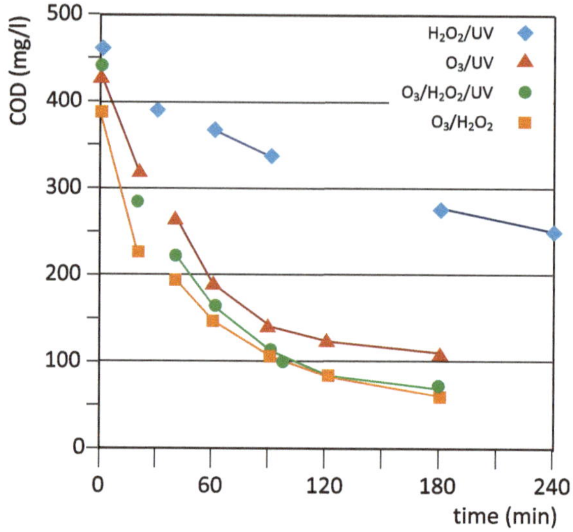

Figure 10.4.16 Comparison of oxidation with H_2O_2/UV with O_3/UV, O_3/H_2O_2 and O_3/UV/H_2O_2 at original pH value of the leachate (Ehrig, 2018).

Similar to oxidation performed by means of hydrogen peroxide (Fig. 10.4.13—upper image) and the Fenton reaction, oxidation conducted with ozone produces biologically degradable organic compounds, which can be more economically degraded by biological processes (Fig. 10.4.17). As a result of the large variation of biodegradable compounds in the effluent fixed film reactors enable a more stable operation. More natural systems such as reed beds may represent an alternative option for biological posttreatment. Robinson and Knox (2003) reported in great detail nearly 10 years of experience, including operational costs; at this plant dosing of ozone at rates of up to 150 mg/L were practised, following biological treatment in sequencing batch reactors (SBRs) and reed beds. It was reported that this plant was successful in removing up to 0.5 mg/L of isoproturon—but could only remove about 10% or 15% of a residual hard COD of about 400 mg/L (following degradation of ozone breakdown products in a final polishing reed bed). Presence of brominated compounds in intermediate stages of the treatment was noted; in this case ozone was generated using air, not oxygen.

Fig. 10.4.18 shows the combinations of oxidation with ozone and biological degradation. The thin lines show TOC reduction by oxidation with ozone combined with UV-light or H_2O_2. The marks show the TOC of oxidized samples and an additional biological treatment for each of the samples. It is evident how in all cases a further marked reduction of TOC is achieved. The most effective system is shown with the thick line. After 1.5 h of oxidation the leachate was biologically treated and, following stabilization, was then oxidized in an additional further step. By means of this procedure very low effluent values combined with low ozone requirement may be obtained. An increasing elimination effect, up to between three and five cycles has been reported by Steensen (1998). In these systems the intermediate biologically degradable products are removed from leachate biologically prior to

Table 10.4.12 Treatment results from ozone oxidation processes

Leachate Type	Initial COD (mg/L)	Oxidation Agent[a]	Ozone Dosage	COD Removal (%)
Young	881	O_3/H_2O_2 0.7	1.4	40
		O_3	1.4	43
Young MBR	2500	O_3/H_2O_2 0.4	2.8	94
		O_3	4.3	70
Biological	1250	O_3/UV	3	90
		O_3	3	75
–	1700	O_3/UV	1.5	94
		O_3	1.5	36
Biological	750	O_3/UV	0.7	70
		O_3		
Stabilized	2300	O_3/UV or O_3	1.5	60
		O_3/H_2O_2 0.3	1.5	92
		O_3	1.5	89
Stabilized	1400	O_3/catalyst	1.5	89
Biological		O_3	1.5	63

[a]H_2O_2 addition in g/g ozone.
Cited from Baig et al. (1996).

the subsequent chemical oxidation phase. This procedure was tested in a technical-scale plant with a continuous recirculation and a fixed film bioreactor, yielding largely similar results. Indeed, the consumption of expensive ozone can be reduced by means of this type of procedure.

Highly detailed calculations of chemical oxidation processes and additional biological posttreatment processes are described by Steensen (1998). Box 10.4.3 comprises a simplified abstract of these calculations.

In principle, chemical oxidation without addition of precipitable compounds (e.g., Fe) is a process devoid of residue production. However, at several ozone oxidation plants the production of calcium

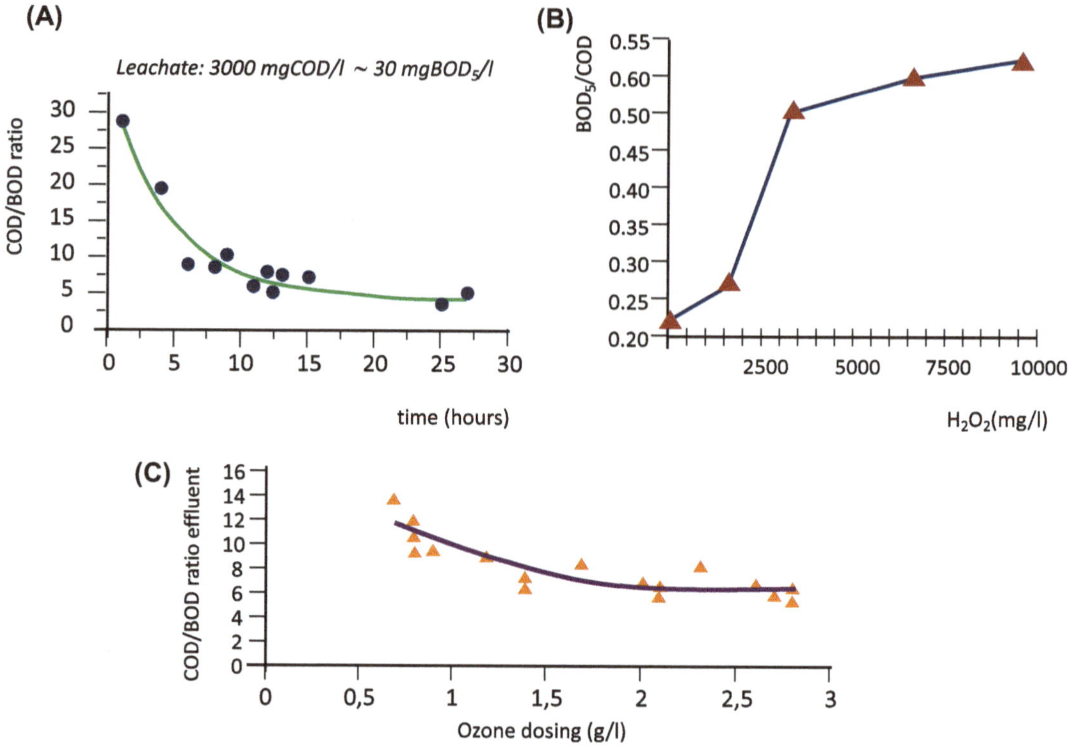

Figure 10.4.17 Increase of biodegradability (measured as BOD_5/COD or COD/BOD_5) during oxidation (Oxidation with ozone; upper right: oxidation with Fenton reaction) (Derco et al., 2002; Lopez et al., 2004; Geenens et al., 2000).

oxalate during oxidation has been observed with variable degrees of incrustation problems; this effect may significantly reduce the effectiveness of UV-light treatment. Up until now information on the production of residues from other oxidation processes has been lacking due to a paucity of practical experiences.

Final remarks

Chemical oxidation is typically applied as a posttreatment step as, compared with biological treatment processes, the oxidation of biodegradable organics and ammonium is a costly procedure. The oxidation of organic compounds present in leachate by means of hydrogen peroxide (H_2O_2) is, with the exception of a few specific compounds (e.g., some AOX components), only effective under low pH conditions, with pH adjustment increasing inorganic salt concentration significantly. The use of Fe^{2+} in the specific Fenton reaction has to date been used mainly in batch operations. The recommended addition of Fe^{2+} (Table 10.4.11: approximately 5 to 60 mmol Fe/L) is higher than that required for Fe-precipitation. As a consequence a very high quantity of sludge will be produced.

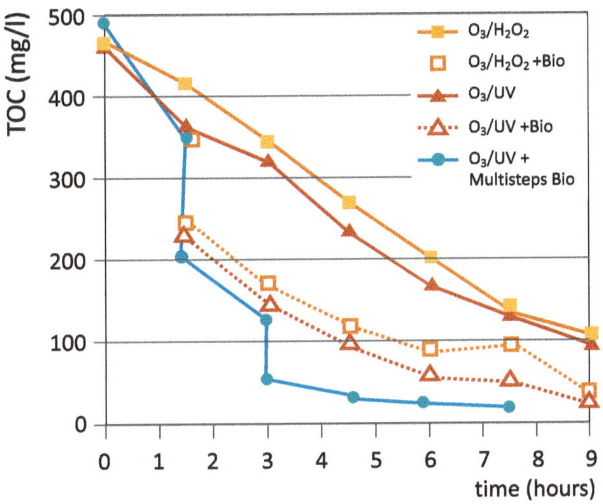

Figure 10.4.18 TOC elimination during oxidation processes. *Orange and red lines*: oxidation with O_3/H_2O_2 and O_3/UV; *dotted lines*: biological post-treatment after oxidation with O_3/H_2O_2 and O_3/UV ; *Blue line*: removal of sample after 1.5 h of oxidation—biological treatment and subsequent phase of oxidation with additional biological treatment using a fixed film reactor (five oxidation phases and four biological treatment phases) (Ehrig, 2018).

Box 10.4.3 Calculation of ozone requirement and reactor volume (Steensen, 1998)

Ozone Requirement

Theoretical ozone consumption is 3 kg of ozone per kg of COD.
In practice often in the area of 2.3–3 kg ozone per kg COD.
Ozone requirement in oxidation processes when only ozone is used:

$$fCOD = \frac{1.35 + 0.005 \cdot (COD)}{1 \cdot e^{-0.1} * [100 - (COD)]}$$

where f(COD) = ozone consumption in g ozone/g COD and η(COD) = COD reduction in %.

Activation of ozone by UV-light or hydrogen peroxide had only a limited effect.
The recirculation of leachate through oxidation and biological treatment is capable of reducing ozone consumption to between 1.5 and 2 kg per kg COD.
Estimation of possible COD degradation during biological posttreatment:
approximately 60% of COD oxidized.
Necessary reactor volume:

volumetric reaction rate: 0.15 kg/m^3 × h (only ozone used), and
volumetric reaction rate: 0.2 kg/m^3 × h (addition of H_2O_2).

Chemical oxidation by means of ozone represents an acknowledged procedure for use in full-scale posttreatment of leachate under existing pH conditions. One disadvantage is represented by the production of ozone in the plant with a limited variation of quantity. Ozone is a toxic gas and must be removed from exhaust gas and from the effluent.

An alternative oxidation process for leachate treatment is represented by the electrochemical oxidation. Lab experiences have been performed successfully, (Cossu et al. 1998). The effects of current density, pH, and chloride concentration on the removal of both chemical oxygen demand (COD) and ammonium nitrogen were investigated. Titanium coated with lead dioxide (PbO_2) or tin dioxide (SnO_2) was used as the anode. COD was removed from 1200 up to a value of 100 mg L^{-1}, and ammonia was totally eliminated. Results indicated that the organic load was removed by both direct and indirect oxidation (by chlorine or hypochlorite originating from oxidation of chlorides). The upgrade of the process to full scale application could be impaired by the high energy consumption and related costs.

With the exception of a small amount of calcium oxalate, oxidation is a residue free process. Contrary to other physical—chemical processes organic compounds present in leachate are converted mainly to water and carbon dioxide. However, the effluent may be characterized by varying amounts of organic compounds. These organic compounds are modified during the oxidation process and may thus assume totally different characteristics.

STRIPPING

Stripping comprises the mass transfer of dissolved gaseous or volatile organic compounds (VOC) from the water phase to the air phase. Henry's law describes the equilibrium between both phases. Typically dissolved compounds that can be removed from water by air stripping include VOCs and ammonia. The dissolved compounds must first move from the bulk liquid solution to the water—air interface and subsequently from the interface to the air, being transferred by molecular diffusion, which, in the majority of cases, is a very slow process. The transfer rate, however, may be markedly increased by turbulent conditions, an intensive transport of dissolved compounds to the water—air interface and a continuous surface exchange between water and air.

Air stripping is mainly carried out in bubble-aerated tanks or in stripping towers. Tanks or lagoons are intensively aerated and the compounds removed in the exhaust air. Stripping towers are filled with packing material and, in most cases the water flows in a downstream direction and the air upstream. A wide range of packing material is used, although an extensive free space and a large surface area should be ensured to allow high turbulence and intensive mixing between water and air.

Exhaust gas from the air stripping process contains removed compounds, which are frequently hazardous the environment. The use of aerated tanks or lagoons for air stripping may at times complicate treatment of the resulting exhaust air. Exhaust air from stripping towers and closed tanks, however, may be collected and treated.

When implementing leachate treatment, it may be beneficial to remove a gaseous compound from water before starting the treatment process. One such compound is represented by the methane originating from landfill gas. Indeed, when treatment is carried out in an enclosed building, and methane is stripped from the leachate in an uncontrolled manner, the combination of even low methane concentrations in the air may result in an explosive mixture. In the majority of cases methane is stripped naturally from the leachate during transportation from the landfill to the treatment plant. However, the emission of methane from manholes along the transport pipe may be problematic. One solution may be to implement controlled air stripping; moreover, to prevent odor problems linked to the smell of the compounds present in the exhaust gas could be incinerated.

Generally, the concentration of VOCs in leachate is considerably low due to the fact inside a landfill volatile compounds are an integrated part of landfill gas. In the presence, however, of higher concentrations in leachate, it may be necessary to remove VOCs as these may pose a serious health and/or environmental risk. VOCs in industrial wastewater are frequently removed by means of steam stripping in stripping towers. Here, the VOC-containing steam is condensed, and the VOCs are separated from the water and reused. Leachate containing VOCs is a highly complex mixture that varies over time and should be handled as hazardous waste. Significant removal of a series of trace organic components often present in landfill leachate can be achieved during air stripping treatment processes. Robinson and Knox (2001, 2003), has provided removal efficiencies for several volatile compounds in Table 10.4.13.

A crucial compound in leachate originating from a sanitary landfill is ammonium (NH_4), reaching concentrations of up to 5000 mg nitrogen per liter. Particularly in the case of high ammonium concentrations corresponding to thousands of milligrams per liter, air stripping may represent an interesting treatment alternative. To implement a stripping process ammonium (NH_4) should first be converted into ammonia (NH_3), with a chemical equilibrium dependent on temperature and pH value existing between the two (Fig. 10.4.19). In principle, ammonia can be stripped independently of temperature

Table 10.4.13 Stripping of volatile components from leachate (Robinson and Knox, 2001, 2003)

Parameter	LOD (μg/L)	Presence (%)	Median Value (μg/L)	% Removal
Ethylbenzene	10	15	10	40
Mecoprop (MCPP)	0.1	98	11	50
Naphthalene	0.1	70	0.46	40
Toluene	10	54	21	25
Xylenes	10	35	35	40

LOD, limit of detection achievable routinely in leachate samples. Presence (%) represents percent of samples in which compound was above the limit of detection.

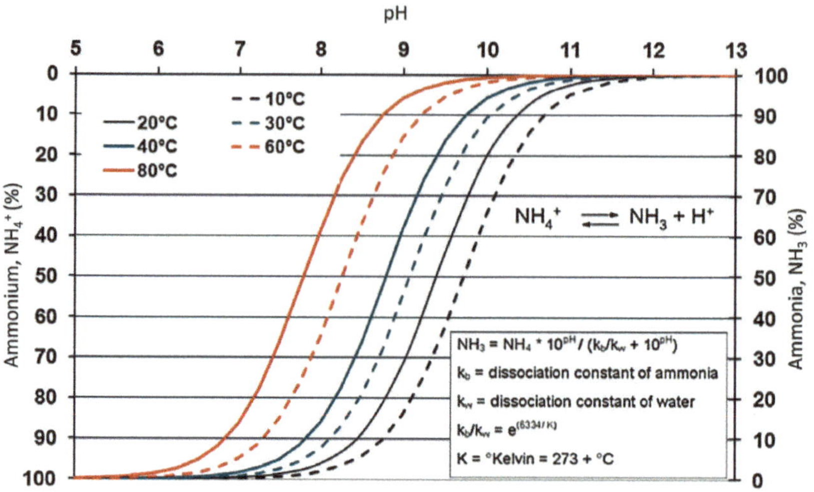

Figure 10.4.19 Relationship between ammonium and ammonia depending on pH and temperature.

and pH value to achieve extremely low effluent values, as with every air-correlated mass transfer of ammonia the equilibrium between ammonium and ammonia is newly adjusted. However, the efficiency of air stripping increases in line with increasing temperature and pH value. It is of the utmost importance that while performing air stripping a given relationship m^3 air/m^3 water reduces the compound to be stripped to a fixed percentage. Consequently, the higher the influent concentration the more economic the stripping pretreatment will be. Overall, air stripping of ammonium requires exceedingly high amounts of air. For example: at $pH = 11$ and 15°C for 90% removal approximately 21.000 m^3 air and for 99% removal approximately 230.000 m^3 air is required. Calculations for ammonia stripping from a single aeration tank are shown in Box 10.4.4.

A simple solution is represented by air stripping in aeration tanks. The equilibrium between concentrations of ammonia gas in water and air is reached within a short distance of gas bubbles through water (10–20 cm). As a consequence, the required depth for stripping tanks can be relatively low. In general, aeration tanks may be used in the presence of constant pH values and temperature. In laboratory-scale experiments a good correspondence can be achieved between theoretical calculations and tests, although in full-scale plants the differences may increase significantly (Ehrig, 1987a,b). Stoll (1979) and Keenan et al. (1984) described a plant using lime addition to increase pH values, stripping tank (950 m^3) and an activated sludge plant (2×75.7 m^3 aeration tank). At inflow rates of 80 m^3/d ammonium values were reduced from influent values of 890 mg N/L to effluent values from the stripping tank of 412 mg N/L, and further to effluent values from the activated sludge plant of <10 mg N/L. With higher inflow rates of 190–303 m^3/d ammonium concentrations amounted to 758, 350 and 75 mg N/L. The effectiveness of stripping achieved is low—and per m^3 even lower than ammonium

Box 10.4.4 Calculation procedure of ammonium stripping in a single aeration tank system

Calculation procedure for ammonium stripping with air: (single tank, constant pH value, constant temperature) air flow/water flow-ratio

$$\frac{V(\text{air})}{V(\text{water})} = \left(\frac{C_i}{C_e} - 1\right) \cdot a \cdot F$$

$C_{effluent}/C_{influent}$

$$\frac{C_e}{C_i} = \frac{1}{1 + \frac{V(\text{air})}{V(\text{water})} \cdot \frac{F}{\alpha}}$$

V_{air}/V_{water} = relationship between water and air flow; $C_i = NH_4{-}N$ influent concentration; $C_e = NH_4{-}N$ effluent concentration

$$F = \frac{10^{pH}}{\frac{Kb}{Kw} + 10^{pH}}$$

$$\frac{Kb}{Kw} = e^{6334 \cdot T}$$

ρ = density of ammonia

$$p = \frac{17,000}{22,413 \cdot \frac{T}{273,15}}$$

see Fig. 10.4.19

$$\alpha = H/\rho$$

H = solubility coefficient; $H = 3.564 \times 10^6 \times e^{(-0.0525 \times \theta)}$; θ = temperature in °C; T = temperature in K.

removal in the activated sludge plant. Damhaug and Jahren (1981) used a pilot-scale system of five closed tanks operated in a cascade for ammonium stripping. The air flows in counter current through the cascade. The supplied air was used in the aeration of all tanks one after the other, subsequently leaving the plant only from the first tank. To remove influent values of 42–200 mg NH_4–N/L down to 6–25 mg NH_4–N/L (pH: 9.5–12.5; temperature: 10–19°C) an air flow of 2.000–3.700 m^3/m^3 water was necessary.

Pilot tests conducted in Germany at a treatment plant consisting of an activated sludge stage followed by pH control (with lime and soda) and a stripping unit, produced satisfactory removal rates; pH was adjusted to 11 and the leachate/air ratio was established in the range of 1/3000–1/4000, this situation was considered optimum according to previous studies (Kettern, 1989).

To limit the toxicity of ammonia for aerobic microorganisms Hasar et al. (2009), proposed the use of an ammonia-stripping step prior to biological treatment. The leachate undergoing treatment was characterized by an ammonia content of up to 2150 mg/L. SS and part of the COD content were removed by coagulation–flocculation, and removal efficiencies of 90% and 92%, respectively, were achieved in an aerobic/anoxic membrane bioreactor for COD and total inorganic nitrogen by means of biooxidation and nitrification/denitrification. Finally, RO reduced COD effluent concentration from 450 mg/L to less than 4.0 mg/L. Stripping in a stirred (400 rpm) and aerated reactor allowed for a maximum ammonia removal rate of roughly 93% at pH 12 after 6 h. In the absence of forced air aeration, the application of stirring failed to produce pH optimization efficiencies exceeding 30%.

Air stripping was applied as a first step in a combined process consisting of air stripping, oxidation (Fenton treatment), SBR, and coagulation (Guo et al., 2010). The results showed that air stripping (pH 11 and aeration time 18 h) removed 96.6% of initial ammonia content (up to 1750 mg/L).

It is not an easy task to maintain constantly high pH values and temperatures throughout the detention time in a full-scale plant; as a consequence a reduced effectiveness and increased air requirement has been observed. If this system is not completely enclosed the exhaust air will transport the stripped ammonia into the atmosphere where it is diluted. Considering the high air requirement, the ammonia concentration of the air is extremely diluted, but it nonetheless sufficient to produce air pollution.

Stripping towers or stripping columns (Fig. 10.4.20-left side) and steam stripping columns represent alternative options for use in stripping procedures (Fig. 10.4.20-right side). Stripping columns operate over several levels with perforated plates or slated frames, or may be filled with packed material. The influx is distributed on the topmost level and flows down through the column, whereas air flows in a counter current direction.

Particularly in the presence of very high concentrations of ammonium in leachate, stripping processes are increasingly cost-effective Knox (2001) reported data for a pilot-scale leachate treatment plant with an air stripping tower in Hong Kong, treating a very strong methanogenic leachate (Table 10.4.14). This stripping tower pretreatment was combined with a subsequent biological treatment process to achieve the removal of organic substances. Eden (2001) described the realization of the stripping system for the Hong Kong landfills.

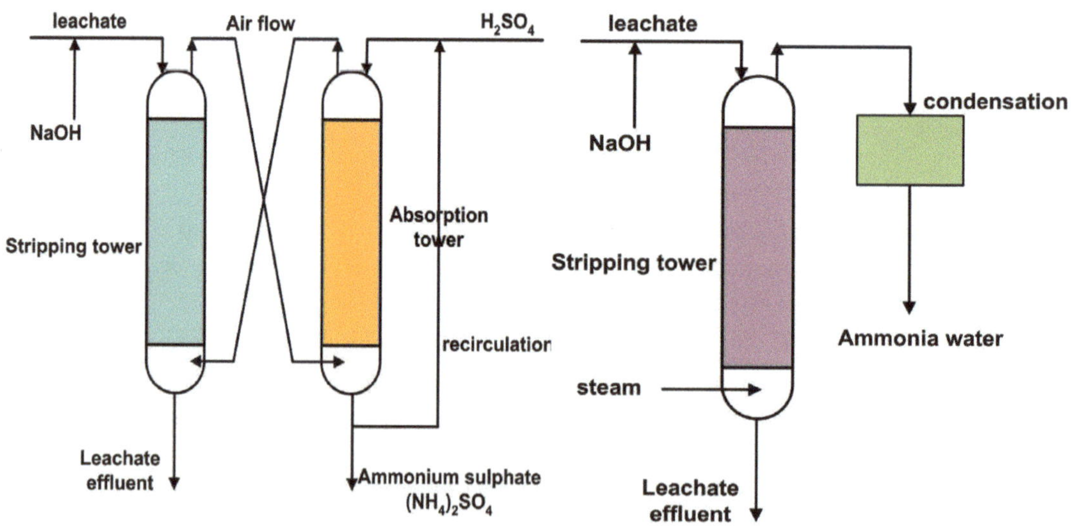

Figure 10.4.20 Schematic view of stripping tower (stripping column) and adsorption tower (left side) and steam stripping tower with steam condensation (right side).

Table 10.4.14 Technical and operation data of a pilot stripping tower (Knox, 2001)

Parameter	Value
Leachate flow rate	4.3 m^3/d
Temperature	70°C
Air flow rate	210 m^3/h
Air–liquid ratio	1167
Influent ammonia concentration	4000 to 4500 mgN/L
Effluent ammonia concentration	50–62 mg N/L

Stripping towers can be easily built as an enclosed system and provide for ammonia recovery. Through use of these enclosed systems environmental pollution produced by ammonia is prevented, and the natural resource ammonia can be reused. An example of this system is shown in Fig. 10.4.20—left side. Air used in the stripping process comes into contact with sulfuric acid in the second packing tower, and ammonia reacts with sulfuric acid to produce ammonium sulfate. If this product is not polluted by other leachate compounds it can be used directly as a fertilizer or in the chemical industry.

An alternative to air stripping is represented by steam stripping (Fig. 10.4.20—right). At the high steam temperatures the stripping process is more effective (see Fig. 10.4.19), although featuring a considerably higher energy burden. The steam loaded with ammonia is condensed and as a result ammonia water is produced. If ammonia concentrations are too low for technical utilization ammonia can be further concentrated.

It is theoretically possible to dimension stripping tanks and stripping columns with a specific degree of accuracy; however, when handling highly complex polluted waters such as leachate, practical tests are mandatory.

Final remarks

The process of air stripping in which exhaust air is released untreated directly into the environment is no longer considered acceptable. Air stripping or steam stripping with ammonium recovery may therefore represent an important process for ammonium recovery. The application of ammonium on agricultural land or its utilization in industrial processes may only be feasible if large enough quantities are produced and no other compounds pollute the ammonium. This type of system is appropriate for use at large landfills and with high ammonium concentrations (>3.000 mg N/L).

EVAPORATION

The aim of evaporation is the separation of solid compounds in the leachate from the liquid phase. A number of systems have been applied:

- direct evaporation of raw leachate,
- evaporation of the concentrate from RO treatment, and
- two-step evaporation of RO concentrate (first step: slurry production up to 10%–30% dry solids; second step: production of dry material with more than 90% dry solids).

A series of different technologies may be applied in an evaporation plant. Initial plant designs encountered serious problems with corrosion. The high costs linked to material problems, and high energy consumption lead to the development of several new evaporation processes. However, only a very few plants of each type were actually constructed and operated. As a consequence, the comparison of treated wastewater quality, residues and costs of the different plants is rather problematic.

Important features of evaporation processes include energy consumption, quality of solids produced, and the quality of treated leachate. Treated leachate quality is mainly affected by the presence of volatile components in the leachate. A critical parameter is ammonium, which can be transformed as ammonia and subsequently recovered by condensation.

Evaporation can be separated into direct and indirect processes. When applying direct processes, the exhaust gas or high temperature gas comes into direct contact with the leachate. Conversely, when indirect evaporation processes are applied the energy is transferred into the leachate using a heat exchanger. To reduce problems with corrosion, indirect processes are often operated at low evaporation temperatures and under vacuum.

The scheme of a generic evaporation plant is shown in Fig. 10.4.21. A simple evaporator system has a single evaporation chamber or "effect". An increase in the number of "effects" can boost the economy of the evaporator. A multiple "effect" system uses the vapor from the first "effect" as the steam source for each subsequent "effect". As the temperature decreases in each subsequent stage, evaporation continues as the pressure and boiling point are also reduced. The installation of each additional "effect" increases the energy efficiency of the system. The number of "effects" can be increased to the point where the capital cost of the next "effect" exceeds the savings in energy costs.

Residues from evaporation are constituted mainly by solids with variable water content. Should it not be possible to dispose of these as solid waste, an additional solidification step may be needed. Raw solids from evaporation processes are often characterized by a high solubility and consequently high environmental risk. As a result, in Germany underground storage (abandoned mine shafts) has been identified as a suitable disposal option for these solids. Solidification may be applied to reduce the solubility of these solid components and facilitate their disposal in landfills. Solid residues from evaporated leachate are typically in the range of 5–20 kg/m^3, depending on leachate strength. Evaporation further to biological treatment yields solids with a content of ammonium and nitrate, thus implying the possibility of exothermic reactions.

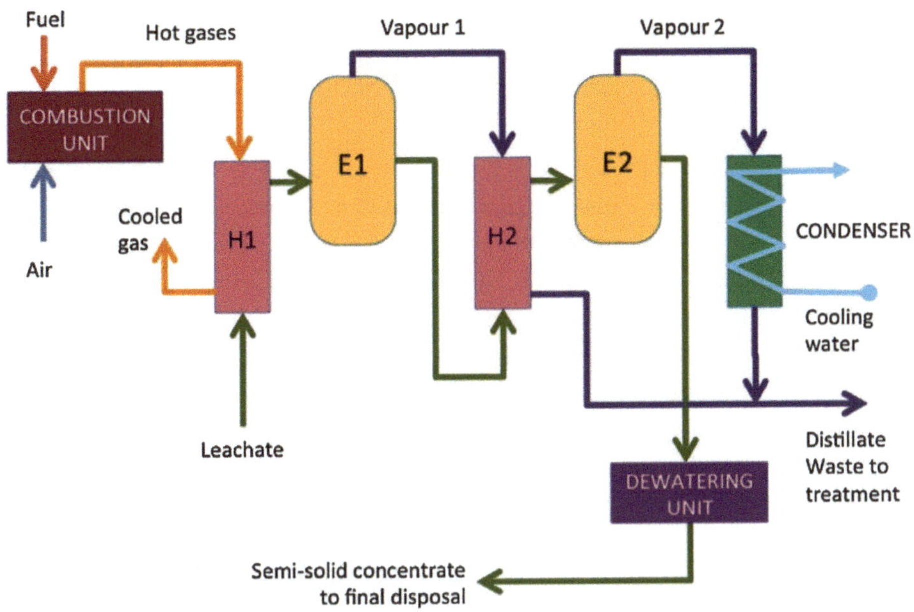

Figure 10.4.21 Scheme of a generic leachate evaporation treatment. Two evaporation effects are represented. *H*, Heat exchanger; *E*, Evaporation unit.

The gases produced during evaporation are condensed. This liquid may, however, be polluted by ammonium and volatile hydrocarbons. During the evaporation process, detergents and antifoaming agents are often required, and residues from these products may remain in the condensate or in the solid residues.

Ettala (1997) presented data from an experimental scale evaporation plant in Finland. The plant consisted of a sand filter, pH adjustment step to 4, a degasifier to remove carbon dioxide and an evaporator with a 1500 m^2 of heat transfer surface made of plastic. The plant operated with an absolute pressure between 15.8 and 17.6 kPa at temperatures of between 55 and 57.3°C. The energy consumption amounted to 13 kWh/m^3 of leachate treated. Nearly 18% of the leachate was recovered as concentrate and disposed back into the landfill. The COD was reduced from 227 to <30 mg/L and ammoniacal nitrogen from 120 to 0.1 mg/L.

In some RO plants the concentrates are evaporated and dried to produce a solid material, which is then stored underground as hazardous waste (German situation).

Final remarks

In line with the majority of currently enforced environmental regulations, evaporation and drying processes are essential in the treatment of RO concentrate. The highly polluted nature of these concentrates may imply a need to adopt special solutions for these processes. The direct evaporation of leachate has not been deemed a fully feasible option, although it may at times be taken into account in highly specific situations.

CONCLUSIONS

All the treatment processes investigated are associated with the presence in the leachate of organic substances, in particular volatile acids, small AOX molecules, but also ammonia.

Accordingly, several available methods may be better suited for use in the separation of organic substances with an extremely high molecular weight, such as humic acids (flocculation/precipitation, ultrafiltration), whereas others may be more appropriate for compounds with an average molecular weight such as fulvic compounds (RO, adsorption). Thus, with the exception of ammonia, salts and heavy metals, physical–chemical focuses mainly on the removal from leachate of organic substances that are largely refractory to biostabilization. No method, however, can guarantee a specific efficiency for compounds with a low molecular weight, mainly due to the issue of their polarity (adsorption) and dimensions (membrane processes, flocculation/precipitation). Therefore, to promote removal of readily degradable organic substances, physical–chemical treatment should be performed together with biological treatment; as an example, this is the case when treating leachate from young landfills. Moreover, the use of biological pretreatment is recommended in the separation of some types of organic compounds and contributes towards avoiding operational problems, i.e., biofouling in RO, and elimination of interference in adsorption processes. As a decrease in alkalinity facilitates a reduction in the amount of reactants and provides for better pH control in subsequent treatment steps, nitrification as a pretreatment will produce a positive influence on the processes of flocculation/precipitation and RO.

Ultimately, each of the above-described processes is characterized by both advantages and disadvantages. Although adsorption, membrane filtration, and chemical precipitation have to date been applied most frequently in the removal of recalcitrant organic compounds, it is evident that none of these technologies is universally applicable or fully effective (Kurniawan, 2006a).

It is generally acknowledged that a combination of physical–chemical treatment or physical–chemical and biological treatments is recommended to ensure effective treatment; indeed, the combination of different physical–chemical treatment processes may result in enhancement of the removal of recalcitrant organic compounds from stabilized leachate, whereas a combination of physical–chemical and biological treatment systems is aimed at effectively removing NH3–N and COD (Kurniawan, 2006a).

Nevertheless, it should be taken into account that leachate treatment today continues to remain a technological challenge due to both the complexity of the composition and variation of the leachate quality (both over time and from site to site). For this reason no general recommendations can be provided. Major criteria to be considered include initial leachate quality and future development, residue production and disposal, in addition to local discharge limit values.

Combinations of different biological and physical–chemical processes are summarized in Chapter 10.5.

References

Ahn, D.-H., Chung, Y-Ch, Chang, W.-S., 2002. Use of coagulant and zeolite to enhance the biological treatment efficiency of high ammonia leachate. Journal of Environmental Science and Health A37 (2), 163–173.

Amokrane, A., Comel, C., Veron, J., 1997. Landfill leachates pretreatment by coagulation–flocculation. Water Research 31, 2775–2782.

Aziz, H.A., Alias, S., Assari, F., Adlan, M.N., 2007. The use of alum, ferric chloride and ferrous sulphate as coagulants in removing suspended solids, colour and COD from semi-aerobic landfill leachate at controlled pH. Waste Management and Research 25, 556–565.

Baig, S., Thieblien, E., Zuliani, F., Jemy, R., Coste, C., 1996. Landfill leachate treatment: case studies. Water Research 30, 21.

Cakmakci, M., Özyaka, V.S., 2013. Aerobic composting leachate treatment by the combination of membrane processes. Waste Management and Research 31, 187–193.

Calli, B., Mertoglu, B., Inanc, B., 2005. Landfill leachate management in Istanbul: applications and alternatives. Chemosphere 59, 819–829.

Campos, J.C., Mauricio, R.G., Alves, M.C.M., de Castro, M.C., Dos Santos, M.F., 2013. Treatment of landfill leachate using nanofiltration and zeolite. In: Fourteenth International Waste Management and landfill Symposium, S. Margherita di Pula, Cagliari, Italy, 30 September–4 October 2013.

Chaturapruek, A., Visvanathan, C., Ahn, K.H., 2005. Ozonation of membrane bioreactor effluent for landfill leachate treatment. Environmental Technology 26, 65–73.

Cortez, S., Teixeira, P., Oliveira, R., Mota, M., 2011. Evaluation of Fenton and ozone-based advanced oxidation processes as mature landfill leachate pre-treatments. Journal of Environmental Management 92, 749–755.

Cossu, R., Serra, R., Muntoni, A., 1992. Physico-chemical treatment of leachate. In: Christensen, T.H., Cossu, R., Stegmann, R. (Eds.), Landfilling of Waste. Leachate, Elsevier Science Publishers, UK, pp. 265–304.

Cossu, R., Polcaro, A.M., Lavagnolo, M.C., Mascia, M., Palmas, S., Renoldi, F., 1998. Electrochemical treatment of landfill leachate: oxidation at Ti/PbO2 and Ti/SnO2 anodes. Environmental Science and Technology 32 (22), 3570–3573.

Damhaug, T., Jahren, P.E., 1981. Ammonia stripping from leachate by countercurrent aeration in shallow tanks. In: Proc. ISWA 1981, Munich, DE, 22–26 June 1981, pp. 71–73.

Deng, Y., Englehardt, J.D., 2006. Treatment of landfill leachate by the Fenton process. Water Research 40, 3683–3694.

Degremont, C., 1979. Water Treatment Handbook, fifth ed. International Standard Book. (Firmin-Didot S.A., Paris, France).

Deng, Y., 2007. Physical and oxidative removal of organics during Fenton treatment of mature municipal landfill leachate. Journal of Hazardous Materials 146, 334–340.

Deng, Y., Rosario-Muniz, E., Ma, X., 2012. Effects of inorganic anions on Fenton oxidation of organic species in landfill leachate. Waste Management and Research 30, 12–19.

Derco, J., Gulyasova, A., Hornak, M., 2002. Influence of ozonation on biodegradability of refractory organics in a landfill leachate. Chemical Papers 56, 41–44.

Diamadopoulos, E., 1994. Characterization and treatment of recirculation stabilized leachate. Water Research 28, 2439–2445.

Eden, R., 2001. Options for landfill leachate treatment in Hong Kong. In: Eighth International Waste Management and landfill symposium, S. Margherita di Pula, Cagliari, Italy, 1–5 October.

Ehrig, H.J., 1986. Flocculation as a Post-treatment Step for High-Strength Organic Waste Waters, Recycling in Chemical Water and Wastewater Treatment. Schiftenreihe ISWW, Karlsruhe, DE. Part 50, 91.

Ehrig, H.J., 1987a. Weitergehende Reinigung von Sickerwässern aus Abfalldeponien (Advanced Treatment of Landfill Leachate). Veröffentlichungen des Instituts für Stadtbauwesen 41 (Technical University Braunschweig, ISSN: 0341-5805).

Ehrig, H.J., 1987b. Leachate treatment: physico-chemical processes. In: Proc. Sardinia 87, 1st International Landfill Symposium, S.Margherita di Pula, IT, vol. 1.

Ehrig, H.J., 1989a. Leachate treatment overview. In: Proc. Sardinia 89, 2nd International Landfill Symposium, Porto Conte (IT), vol. 1.

Ehrig, H.J., 1989b. Physico-chemical treatment. In: Christensen, T.H., Cossu, R., Stegmann, R. (Eds.), Sanitary Landfilling: Processes, Technology and Environmental Impacts. Academic Press, Publishers, ISBN 0-12-174255-5, pp. 285–297.

Ehrig, H.J., 1998. Pyisical-chemical processes –strpping, precipitation, flocculation. In: Symposium: Management and Treatment of MSW Landfill Leachate, Venice, 2.-4.12.1998.

Ehrig, H.J., 2001. Sickerwasser aus Abfallablagerungen (Leachate from sanitary landfills). In: ATV-DVWK (Ed.), ATV-Handbuch Industrieabwasser, Dienstleistungs- und Veredelungsindustrie. Verlag Ernst & Sohn, ISBN 3-433-01468-X, pp. 347–380.

Ehrig, H.J., 2018. Leachate Management (in press).

Ehrig, H.J., Robinson, H., 2011. Landfilling: leachate treatment. In: Christensen, Th H. (Ed.), Solid Waste Technology & Management. Wiley, ISBN 978-1-4051-7517-3, pp. 858–897.

Ettala, M., 1997. Full-scale leachate treatment using new evaporation technology. In: Christensen, T.H., Cossu, R., Stegmann, R. (Eds.), Proceedings Sardinia 97, Sixth International Landfill Symposium, vol. II. CISA, Environmental Sanitary Engineering Centre, Cagliari, Italy, pp. 423–426.

Fan, H.J., Chiu, T., Yang, H.S., Chen, W.C., Furuya, E., 2007. Removal of humic acids, fulvic acids and non humic substance from landfill leachate. Journal of Environmental Engineering and Management 17 (5), 325–333.

Foo, K.Y., Hameed, B.H., 2009. An overview of landfill leachate treatment via activated carbon adsorption process. Journal of Hazardous Materials 171, 54–60.

Galvez, A., Zamorano, M., Ramos, A., Hontoria, E., 2005. Coagulation-flocculation pretreatment of a partially stabilized leachate from a Sanitary landfill site at Alhendin (Granada, Southern Spain). Jornal of Environmental Science and Health 40, 1741–1751.

Gerdes, C., Nicolaou, N., 2006. Reverse Osmosis in Leachate Treatment. Clarke Energy & Haase Energietechnik AG.

Geenens, D., Bixio, B., Thoeye, C., 2000. Combined ozone-activated sludge treatment of landfill leachate. Water Science and Technology 44, 359–365.

Guo, J., Abbas, A.A., Chen, Y., Liu, Z., Fang, F., Chen, P., 2010. Treatment of landfill leachate using a combined stripping, Fenton, SBR, and coagulation process. Journal of Hazardous Materials 178, 699–705.

Halim, A.A., Aziz, H.A., Johari, M.A.M., Ariffin, K.S., 2010. Comparison study of ammonia and COD adsorption on zeolite, activated carbon and composite materials in landfill leachate treatment. Desalination 262, 31–35.

Hasar, H., Unsala, S.A., Ipeka, U., Karatasa, S., Cınarc, O., Yamand, C., Kınacıe, C., 2009. Stripping/flocculation/membrane bioreactor/reverse osmosis treatment of municipal landfill leachate. Journal of Hazardous Materials 171, 309–317.

Ho, S., Boyle, W.C., Ham, R.K., 1974. Chemical treatment by coagulation and precipitation. Journal of Environmental Engineering-ASCE 99, 535–544.

Kabdash, I., Tünay, O., Öztürk, I., Yilmaz, S., Arikan, O., 2000. Ammonia removal from young landfill leachate by magnesium ammonium phosphate precipitation and air stripping. Water Science and Technology 41, 237–240.

Keenan, J.D., Steiner, R.L., Fungaroli, A.A., 1983. Chemical-physical leachate treatment. Journal of Environmental Engineering-ASCE 109, 1371–1384.

Keenan, J.D., Steiner, R.L., Fungaroli, A.A., 1984. Landfill leachate treatment. Water Pollution Control Federation 28.

Kettern, J.T., 1989. Problems of landfill leachate treatment and proposal of solutions. In: Proc. Sardinia 89, 2nd International Landfill Symposium, Porto Conte, it, vol. 2.

Kim, D., Ryu, H.-D., Kim, M.-S., Kim, J., Lee, S.-I., 2007. Enhancing struvite precipitation potential for ammonia removal in municipal landfill leachate. Journal of Hazardous Materials 146 (1–2), 81–85.

Knox, K., 2001. Development of a novel process for treatment of leachate with very high ammoniacal nitrogen concentrations. In: Christensen, T.H., Cossu, R., Stegmann, R. (Eds.), Proceedings Sardinia 2001, Eighth International Landfill Symposium, vol. II. CISA, Environmental Sanitary Engineering Centre, Cagliari, Italy, pp. 207–213.

Kurniawan, T.A., Lo, W., Chan, G.Y.S., 2006. Physico-chemical treatments for removal of recalcitrant contaminants from landfill leachate. Journal of Hazardous Materials 80–100.

Li, X.Z., Zhao, Q.L., Hao, X.D., 1999. Ammonium removal from landfill leachate by chemical precipitation. Waste Management 19, 409–415.

Li, F., Wichmann, K., Heine, W., 2009. Treatment of the methanogenic landfill leachate with thin open channel reverse osmosis membrane modules. Waste Management 29, 960–964.

Li, W., Hua, T., Zhou, Q., Zhang, S., Li, F., 2010. Treatment of stabilized landfill leachate by the combined process of coagulation/flocculation and powder activated carbon adsorption. Desalination 264, 56–62.

Li, F., 2013. Development of open channel membrane modules for landfill leachate treatment. In: Fourteenth International Waste Management and landfill Symposium, S. Margherita di Pula.

Li, X.Z., Zhao, Q.L., 2001. Efficiency of biological treatment affected by high strength of ammonium–nitrogen as pretreatment. Chemosphere 44, 37–43.

Liu, Y., Li, X., Wang, B., Liu, S., 2008. Performance of landfill leachate treatment system with disc-tube reverse osmosis units. Frontiers of Environmental Science and Engineering in China 2, 24–31.

Logeman, F.P., Glas, H., 1989. Using the reverse osmosis process for leach water treatment. Recycling International 3, 1965–1971.

Loizidou, M., Vithoulkas, N., Kapetanios, E., 1992. Physical chemical treatment of leachate from landfill. Journal of Environmental Science and Health A27 (4), 1059–1073.

Lopez, A., Pagano, M., Volpe, A., Di Pinto, A.C., 2004. Fenton's pre-treatment of mature landfill leachate. Chemosphere 54, 1005–1010.

Mendez-Novelo, R.I., Castillo-Borges, E.R., Sauri-Riancho, M.R., Quintal-Franco, C.A., Giacoman-Vallejos, G., 2005. Physico-chemical treatment of Merida landfill leachate for chemical oxygen demand reduction by coagulation. Waste Management and Research 23, 560–564.

Merten, U., 1966. Transport properties of osmotic membranes. In: Marten, U. (Ed.), Desalination by Reverse Osmosis. MIT Press, Cambridge, MA, USA.
Ozturk, I., Altinbas, M., Koyuncu, I., Arikan, O., Gomec-Yangin, C., 2003. Advanced physico-chemical treatment experiences on young municipal landfill leachates. Waste Management 23, 441–446.
Peters, T., 1999. Past and future of membrane filtration for the purification of landfill leachate. In: Christensen, T.H., Cossu, R., Stegmann, R. (Eds.), Proceedings Sardinia 99, Seventh International Landfill Symposium, vol. II. CISA, Environmental Sanitary Engineering Centre, Cagliari, Italy, pp. 335–344.
Peters, T., 2010. Membrane technology for water treatment. Chemical Engineering and Technology 33, 1233–1240.
Rautenbach, R., Melis, R., 1991. Was bringt die biologische Sickerwasser-Vorbehandlung? EntsorgungsPraxis 5, 244–250.
Renou, S., Givaudan, J.G., Poulain, S., Dirassouyan, F., Moulin, P., 2008. Landfill leachate treatment: review and opportunity. Journal of Hazardous Materials 150, 468–493.
Ritz, W.K., 1989. Deponiesickerwasser-reinigung. Schweizer Ingenieur und Architekt 107, 328–331.
Robinson, H., Maris, P.J., 1979. Leachate from Domestic Waste; Generation, Composition and Treatment: A Review. Technical report TR 108. Water Research Centre, Medmenham Lab., Marlow, UK.
Robinson, H.D., van der Merwe, W., Mitchell, C., Gombault, E., Novella, P., Carville, M.S., 2007. Treatment of leachate from a large hazardous waste landfill site in South Africa. In: Eleventh International Waste Management and Landfill Symposium, S. Margherita di Pula, Cagliari, Italy, 1–5 October 2007.
Robinson, H.D., Knox, K., May 2001. Pollution Inventory Discharges to Sewer or Surface Waters from Landfill Leachates. Final Report Ref REGCON 70, prepared for the UK Environment Agency by Enviros and Knox Associates, 19 pp. and Appendices.
Robinson, H.D., Knox, K., March 2003. Updating the Landfill Leachate Pollution Inventory Tool. R&D Technical Report No. PI-496/TR("), prepared for the UK Environment Agency by Enviros and Knox Associates, 56 pp.
Sabour, M.R., Lak, M.G., Rabbani, O., 2011. Evaluation of the main parameters affecting the Fenton oxidation process in municipal landfill leachate treatment. Waste Management and Research 29, 397–405.
Salem, z., Hamouri, K., Djemnaa, R., Allia, K., 2008. Evaluation of landfill leachate pollution and treatment. Desalination 220, 108–114.
Schiopu, A.M., Piuleac, G.C., Cojocaru, C., Apostol, I., Mămăligă, I., Gavrilescu, M., 2012. Reducing environmental risk of landfills: leachate treatment by reverse osmosis. Environmental Engineering and Management Journal 12 (11), 2319–2331.
Seyfried, C.F., Theilen, U., 1991. Leachate treatment by reverse osmosis and evaporation, effects of biological pre-treatment. In: Proc. Sardinia 91, 3rd International Landfill Symposium, S.Margherita di Pula, IT, 14–18 October 1991, vol. 1, pp. 919–928.
Sir, M., Podhola, M., Patocka, T., Honzajkova, Z., Kocurek, P., Kubal, M., Kuras, M., 2012. The effect of humic acids on the reverse osmosis treatment of hazardous landfill leachate. Journal of Hazardous Materials 207–208, 86–90.
Steensen, M., 1993. Chemische Naßoxidation zur weitergehnden Sickerwasserreinigung. Korrespondenz Abwasser 40, 308.
Steensen, M., 1998. Chemical oxidation and biological post-treatment for advanced leachate treatment", Institut für Siedlungswasserwirtschaft. Chemische Oxidation und biologische Nachreinigung zur weitergehneden Sickerwasserbehandlung, vol. 63. Technical University, Braunschweig. ISSN 0934–9731.
Stoll, B.J., 1979. Leachate treatment demonstration municipal solid waste. In: Land Disposal Proceedings 5th Annual Research Symposium, Orlando, p. 313. EPA-600/9-79-023a.
Tatsi, A.A., Zouboulis, A.I., Matis, K.A., Samaras, P., 2003. Coagulation-flocculation pretreatment of sanitary landfill leachates. Chemosphere 53, 737–744.
Thörneby, L., Hogland, W., Stenis, J., Mathiassan, L., Somogyi, P., 2003. Design of a reverse osmosis plant for leachate treatment aiming for safe disposal. Waste Management and Research 21 (5), 424–435.
Thorton, R.J., Blanc, F.C., 1973. Leachate treatment by coagulation and precipitation. Journal of Environmental Engineering-ASCE 99, 535–544.
Trebouet, D., Schlumpf, J.P., Jaouen, P., Quemeneur, F., 2001. Stabilized landfill leachate treatment by combined physicochemical-nanofiltration processes. Water Research 35, 2935–2942.
Umar, M., Aziz, H.A., Yusoff, M.S., 2010. Trends in the use of Fenton, electro-Fenton and photo-Fenton for the treatment of landfill leachate. Waste Management 30, 2113–2121.
Ushikoshi, K., Kobayashi, T., Uematsu, K., Toji, A., Kojima, A., Matsumoto, K., 2002. Leachate treatment by the reverse osmosis system. Desalination 150, 121–129.
Wassertechnologie TH Nürnberg, 2013. UVT: Aufbereitung von Wasser durch Membranverfahren.

Weber, B., Holz, F., 1989. Significance of the biological pretreatment of sanitary landfill leachate on the efficiency of the reverse osmosis process. In: Proc. Sardinia 89, 2nd International Landfill Symposium, Porto Conte, it, vol. 1, pp. 1–13.
Wu, J.J., Wu, C., Ma, H., Chang, C., 2004. Treatment of landfill leachate by ozone-based advanced oxidation processes. Chemosphere 54, 997–1003.
Zgajnar Gotvajn, A., Derco, J., Tisler, T., Zagorc-Koncan, J., 2007. Optimization of coagulation and flocculation processes for pretreatment of industrial landfill leachate. In: Eleventh International Waste Management and Landfill Symposium, s. Margherita di Pula, Cagliari, Italy, 1–5 October 2007.
Zhang, H., Choi, H.J., Huang, C., 2005. Optimization of Fenton process for the treatment of landfill leachate. Journal of Hazardous Materials 166–174.
Zhang, H., Wua, X., Li, X., 2012. Oxidation and coagulation removal of COD from landfill leachate by Fered–Fenton process. Chem. Eng. J 210, 188–194.
Zouboulis, A.I., Chai, X.L., Katsoyiannis, I.A., 2004. The application of bioflocculant for the removal of humic acids from stabilized landfill leachates. Journal of Environmental Management 70 (1), 35–41.

10.5

COMBINATION OF DIFFERENT MSW LEACHATE TREATMENT PROCESSES

Hans-Jürgen Ehrig and Rainer Stegmann

INTRODUCTION

Many leachate treatment plants operate with a combination of different processes. Environmental and/or economic boundary conditions and/or specific requirements can be the reason for such combinations. With the construction of large landfills in the second half of the 20th century the awareness and the knowledge of the high and relative complex leachate pollution investigations for appropriate treatment technology started. This search resulted in an adaption of different treatment processes for leachate treatment.

Leachate analysis shows a wide range of different compounds (see Chapter 10.2). The treatment of the organic substances is not without problems also because the chemical composition of the organic pollution (measured as COD or TOC) is mostly unknown. One classification is the differentiation between the biodegradable and nonbiodegradable organics. The problematic of this differentiation is described in Chapter 10.2. Using biological treatment only the biodegradable and nitrogen compounds can be removed. If the required COD effluent limit values cannot be met by means of biological treatment additional treatment steps will be necessary.

Unlike the publication of experiences from sewage treatment, experiences and results from full-scale biological and chemical physical leachate treatment plants are only scarcely published. It is also for this reason difficult to design a combination of different treatment processes. Different processes are available for the elimination of the same substances and there are great differences between treating raw or pretreated leachate. The execution of laboratory or better pilot scale tests is recommended to select the optimum treatment combination.

Treatment processes require energy, sorbents, and/or additives as, e.g., external organic carbon, flocculation agents, and activated carbon. Flocculation agents and activated carbon, for example, react very differently when treating different leachate qualities; this may even be the case if leachate quality data seem nearly to be the same. Several processes produce residues which must be further treated or discharged what adds to the cost.

Fig. 10.5.1 presents a selection of realized leachate treatment combinations.

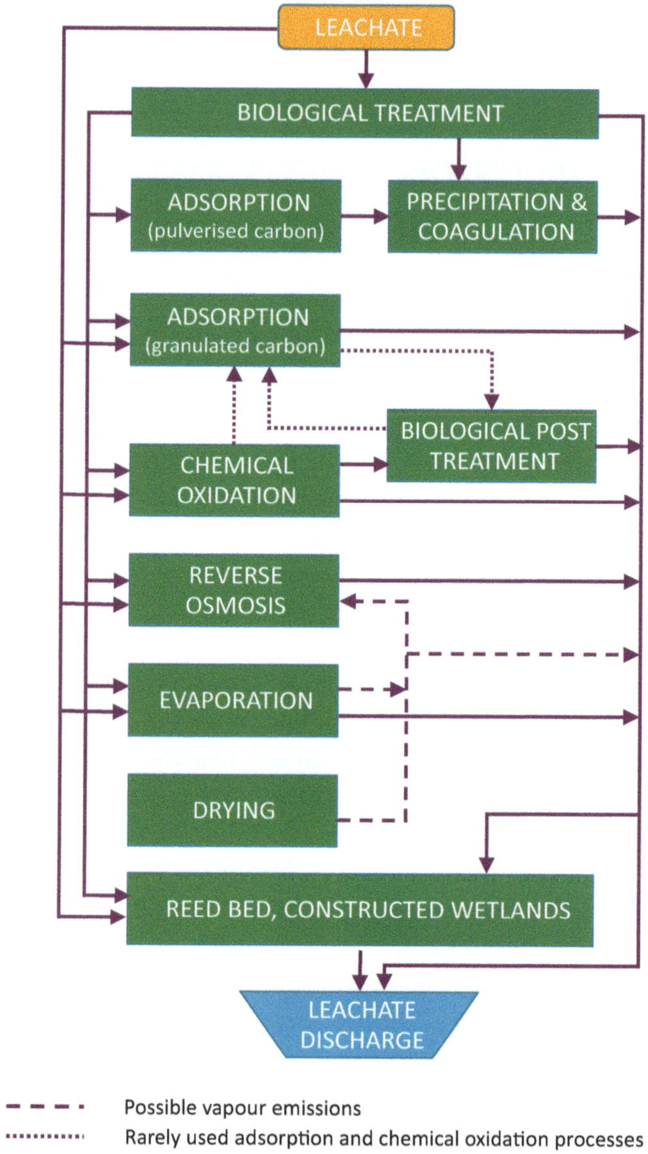

Figure 10.5.1 Different combinations of leachate treatment technologies used in full scale plants. Chemical oxidation generally is carried out by using ozone + UV or ozone + H_2O_2.

UPGRADING BIOLOGICAL TREATMENT PLANTS

In several cases biological treatment plants such as aerated lagoons or activated sludge plants can meet the required discharge limits. But due to the nature of the process aerated lagoons emit biomass with the effluent. Due to the existence of the clarifier in activated sludge plants this problem should be solved. But very often very light flocks of biomass may leave the plant via the effluent. If the temperature in the bioreactor decreases, the emissions may increase above given effluent requirements.

For an artificially aerated lagoon the installation of a natural separation section for biomass sedimentation and effluent polishing is recommended. In nonaerated lagoons excess biomass becomes a sediment and stabilizes at the bottom of the lagoon. In most cases the sludge has to be removed in a time interval of several years (mostly >10 years). Operating a completely aerated lagoon the produced nitrate can in general not be reduced. During warm weather periods the nitrate leads to algae blooms with secondary water pollution. During cold weather periods the ammonium concentration in the effluent increases. During temperature changes high nitrite concentrations may be measured in the effluent. Installing a constructed wetland or reed bed as a posttreatment it may be possible to reduce nitrogen compounds also during cold periods.

Although the suspended solids in the clarifier effluent of an activated sludge plant have a much lower concentration than in the effluent of aerated lagoons also in this case a polishing step may make sense. In Tables 10.5.1a and 10.5.1b results from activated sludge plants designed as a sequential batch reactor (SBR) with subsequent constructed reed beds are presented.

Robinson et al. (2003b) reported that suspended solids and colloids in the effluent may be further reduced by using an air flotation step between SBR and reed bed.

Table 10.5.1a Leachate treatment in a sequential batch reactor with subsequent constructed reed beds. Mean values at Efford Landfill (UK). From Robinson and Olufsen (2007).

Parameter	Raw Leachate	SBR Effluent	Final Effluent
COD (mg/L)	942	462	309
BOD_5 (mg/L)	72	22	3
NH_4 (mgN/L)	820	1.6	0.5
NO_3 (mgN/L)	0.21	423	384
NO_2 (mgN/L)	0.04	0.56	0.87
Cl (mg/L)	1502	1442	1507
PH	7.6	8.2	8.4
Alkalinity ($mgCaCO_3/L$)	3830	1229	1138

Table 10.5.1b Leachate treatment in a sequential batch reactor with subsequent constructed reed beds. Mean values at two different Landfills in South Africa (Robinson and Carville, 2007): A. Mariannhill (SBR approx. 78 m^3; reed bed 300 m^2; maximum flow rate 50m^3/day). B. Vissershok

Parameter	Raw Leachate		SBR Effluent		Final Effluent	
	A	B	A	B	A	B
Suspended solids (mg/L)	250	-	102	277	28	58
COD (mg/L)	760	4560	320	1470	435	1510
BOD_5 (mg/L)	170	550	90	28	8	10
NH_4 (mgN/L)	435	1554	1,9	0,2	<0,5	0,4
NO_3 (mgN/L)	<0,1	9,2	443	1461	471	1471
NO_2 (mgN/L)		0,4		<0,1		1
pH	7,5	8,18	7,1	7,39	6,7	7,61
Alkalinity (mg$CaCO_3$/L)	2422	9652	260	935	300	1133
Cl (mg/L)	1690	4626	1550	4545	1860	4455
Phosphate (mgP/L)	1,4	13	12	10,5	8,6	10,7
Sodium (mg/L)	590	2825	1113	3979	1469	3904
Magnesium (mg/L)	80	195	118	192	163	183
Calcium (mg/L)	105	198	118	123	170	115

ACTIVATED SLUDGE PLANTS WITH SUBSEQUENT PRECIPITATION AND ADSORPTION STEPS

As a first development in the early 1980s in Germany to further reduce nondegradable or difficult to degrade organics, precipitation with subsequent pulverized activated carbon (PAC) adsorption was practiced (Plant "Pohlsche Heide", Germany). Particularly for biologically treated leachate of landfills in the methanogenic phase precipitation of residual organics, e.g., using iron salts is a very effective and economic treatment step. To meet the required effluent standards a subsequent adsorption process using pulverized PAC was added which at that time was less costly than using granular activated carbon (GAC). This situation may be different in different countries. Treatment results of such plants are presented in Box 10.5.1.

Box 10.5.1 Leachate treatment in an activated sludge plant with subsequent post-treatment by precipitation and activated carbon adsorption (Albers and Mennerich, 1987; Mester, 1993)

	Landfill			
Plant Components (Volume in m³)	**Heisterholz (1985/87)**	**Heisterholz (1990/91)**	**Pohlsche Heide**	**Wilsum**
Denitrification tank	725	417	2 × 494	940
Nitrification tank		1670	2 × 1446	2 × 720
Clarifier 1	19	146	180	–
Fixed film reactor	–	460	88	–
Adsorption tank	36	80	81	50
Flocculation tank	36	30	10	2.5
Clarifier 2	60	277	267	–
Neutralization tank	131	48	21	n.d.
Clarifier 3	60	–	–	–
Filtration	–	–	–	Sand filter
Average flow rate (m³/d)	96	96	93	90

			Effluent	
Parameter		**Influent**	**Activated Sludge**	**After post-treatment**
LANDFILL HEISTERHOLZ (1985/87)				
BOD_5	mg/L	350	n.d.	7
COD	mg/L	1375	n.d.	52
AOX	mg/L	0.742	0.605	0.122
LANDFILL HEISTERHOLZ (1990/91)				
BOD_5	mg/L	91	n.d.	2.1
COD	mg/L	686	n.d.	51
NH_4	mgN/L	528	4.6	0.7
AOX	mg/L	1.65		0.23
LANDFILL POHLSCHE HEIDE (1990/91)				
BOD_5	mg/L	410	n.d.	3.5
COD	mg/L	1571	614	76
NH_4	mgN/L	633	0.81	0.69
AOX	mg/L	1.31	n.d.	0.17
LANDFILL POHLSCHE HEIDE (1994)				
BOD_5	mg/L	207	n.d.	1
COD	mg/L	1506	709	94

NH_4	mgN/L	579	0.26	0.09
NO_3	mgN/L	44	288	296
AOX	mg/L	1.45	n.d.	0.182
Cl	mg/L	1244	n.d.	1125
SO_4	mg/L	139	n.d.	746
LANDFILL WILSUM				
BOD_5	mg/L	n.d.	n.d.	3
COD	mg/L	896	401	73
NH_4	mgN/L	459	0.34	0.24
NO_2	mgN/L	0.12	0.04	0.84
NO_3	mgN/L	7.1	149	118
AOX	mg/L	1.04	0.48	0.15
Phosphate	mgP/L	3	n.d.	0.1
Cl	mg/L	1715	n.d.	1950
Copper	mg/L	0.58	n.d.	0.03
Nickel	mg/L	0.48	n.d.	0.15

n.d., no data available.

Due to easier handling at a later stage at "Pohlsche Heide" *the adsorption* plant was operated as filters with GAC. Fig. 10.5.2 shows the relative strong relationship between both processes (Plant Heisterholz). Using an optimized precipitation process with increased COD elimination the dosing of PAC can be reduced, which resulted in a significant reduction of the treatment costs. The dewatering of iron sludge is greatly improved as a mixture of GAC and iron sludge.

ACTIVATED SLUDGE PLANTS WITH ACTIVATED CARBON FILTERS

This combination is most commonly used for leachate treatment in Germany. Important reasons for this solution are the relatively low investment and low operation costs. In addition the required amount of carbon can be automatically adapted on the leachate quality. The higher costs of GAC compared to PAC can be partly compensated by higher carbon adsorption loading rates. Using GAC filters it is possible to reduce remaining organic and trace substances in the leachate. But it should be noted that trace substances are only adsorbed if the main portion of the organics are removed. In such cases a specific operation of two GAC filters in a row may be necessary. In the first filter the organics should

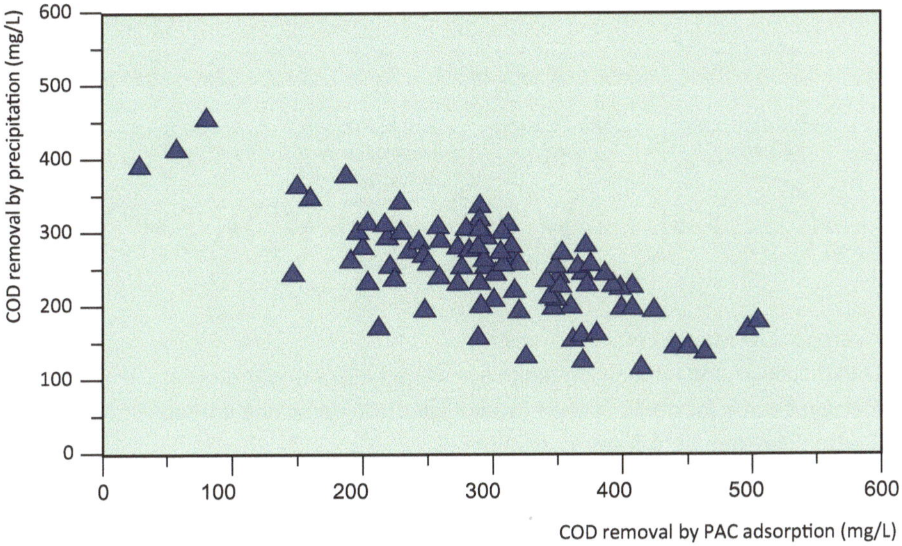

Figure 10.5.2 Relationship between COD removal by precipitation and pulverized activated carbon adsorption in a combined treatment system (Albers and Mennerich, 1987).

be reduced to the range of the required COD discharge limit values so that the in the subsequent filter trace substances can be adsorbed. Such a specific situation must be tested in laboratory or pilot scale tests. This is also necessary because a great number of different activated carbon qualities exist; the optimum quality can only be determined by performing tests.

To avoid early clogging of the pores in the AC the effluent of the activated sludge plant should be filtered before entering the GAC filters. Sand filter, fabric filter, and ultrafiltration are methods that are available for fine particle removal. The higher the removal rate of suspended solids upfront the higher will be the adsorption capacity of the GAC; as a consequence costs may be reduced. The ultrafiltration process reduces fine particles to the highest extent. Results from different treatment plants using the combination of activated sludge and GAC filters are presented in Box 10.5.2.

Some activated sludge plants in Germany with subsequent GAC filters are combined with nanofiltration as a third treatment step. The nanofiltration membranes are permeable for small molecules as chloride, but most inorganic and organic molecules are retained. The nanofiltration can be placed after the ultrafiltration of the activated sludge plant. The ultrafiltration is used to retain the suspended solids of the activated sludge plant. The permeate of ultrafiltration is fed in to the nanofiltration plant. The concentrate of the nanofiltration is fed to GAC filter, and its effluent is returned back to the activated sludge plant. The idea of this combination is to retain "all" organic substances and increase their concentrations. The consequence of higher organics (COD values) in the concentrate stream of the nanofiltration is an increase of adsorption capacity from 20% to 30% of GAC (by weight) to values up to

Box 10.5.2 Combination of an activated sludge plant with an activated carbon filter (ATV, 1996; Ehrig, 2001)

Plant layout (plant A) of the combination activated sludge plant and activated carbon filter.

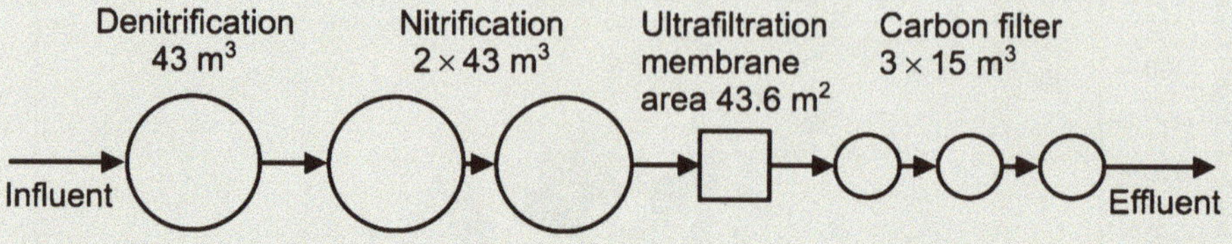

- Average flow rate: 80 m^3/day
- Denitrification and nitrification tanks are closed tanks under a pressure of 300 kPa
- Carbon source for denitrification: wastewater from biowaste composting
- Carbon consumption: 3 kg/m^3 leachate

Plant	Parameter	Unit	Influent	Effluent Biological Treatment	Effluent Activated Carbon Filter
A	BOD_5	(mg/L)	1500–6000	<5	<5
	COD	(mg/L)	1000–12000	500–800	<200
	AOX	(mg/L)	1.0–2.5	0.5–2.0	0.1–0.7
	NH_4	(mgN/L)	400–800	<10	<10
B	BOD_5	(mg/L)	n.d.	n.d.	n.d.
	COD	(mg/L)	3000–6000	500–900	<110
	AOX	(mg/L)	0.8–2.0	0.6–1.4	0.1–0.48
	NH_4	(mgN/L)	400–800	<10	<10
C	BOD_5	(mg/L)	280	<20	<20
	COD	(mg/L)	1700	900–1200	<400
	AOX	(mg/L)	1.1	0.5–0.9	0.1 –< 0.5
	NH_4	(mgN/L)	900	<10	<10
D	BOD_5	(mg/L)	90–210	<20	<20
	COD	(mg/L)	1100–1600	300–900	<200
	AOX	(mg/L)	0.8–1.7	0.9–1.2	<0.5
	NH_4	(mgN/L)	500–900	<10	<10
E	BOD_5	(mg/L)	50–150	<10	<10
	COD	(mg/L)	450–800	250–320	<200
	AOX	(mg/L)	n.d.	n.d.	n.d.
	NH_4	(mgN/L)	250–400	<50	<50
F	BOD_5	(mg/L)	500–800	<10	<10
	COD	(mg/L)	2000–5600	750–1200	<200
	AOX	(mg/L)	0.6–2.2	0.5–1.5	0.1–0.87
	NH4	(mgN/L)	600–1800	3–52	<20

n.d., no data available.

more than 50% of GAC (by weight). But some organic and most inorganic substances cannot be adsorbed. A small amount of these substances will be adsorbed at the surface of the suspended solids in the activated sludge plant. But the predominant part is enriched in the returned effluent of the GAC. To avoid too high accumulation in the activated sludge part of the effluent from the GAC has to be removed and further treated and/or disposed. Another idea is the treatment of the concentrate from the nanofiltration by chemical oxidation. But also in this case there may be a slight increase of the concentration of some compounds in the returned effluent from the GAC.

In general it is not useful to install GAC adsorption as first step before the activated sludge leachate treatment plant. But in the case of leachate cotreatment in a sewage treatment plant it may be an option if low organic concentrations of biodegradable organic substances (measured as BOD or BOD_5/COD $< 0,1$) in the leachate (leachate of the methanogenic phase) are required. With the preadsorption also a part of biodegradable organic substances are adsorbed with the consequence of higher GAC costs. But these higher costs may be compensated with lower treatment costs in the sewage treatment plant.

ACTIVATED SLUDGE PLANTS WITH OXIDATION PROCESS

Another option for the further reduction of organics and particularly trace organics in the leachate is the chemical oxidation. As agents, hydrogen peroxide and ozone and, in some cases, Fenton reagent are used. In most cases, in Germany, ozone in combination with UV is mostly used. In principle it is possible to oxidize organic substances to the end products CO_2 and H_2O. But in the reality some organics are difficult to oxidize and other substances can only be oxidized to lower molecular organic substances. The oxidation products may be biodegradable or nonbiodegradable. With increasing oxidation rate the production of biodegradable substances decreases. But since the biodegradation of organic substances with air is less expensive than oxidation with ozone several plants in Germany have an additional biological posttreatment step. Since the degradable BOD concentration is mostly lower than 100 mg/L in most cases the biological plant is a fixed film reactor (e.g., RBC, rotating biological contactor) (Box 10.5.3). Robinson et al. (2003a) presented another posttreatment process with a constructed reed bed (Box 10.5.3). Some plants in Germany have very short contact time in the oxidation reactor and the effluent is directly returned into the nitrification tank of the activated sludge plant. With such an operation the ozone requirement could be reduced up to 50% because biodegradable compound are immediately eliminated in the bioreactor and cannot consume any more ozone.

In a very few cases a partly oxidation can produce toxic trace compounds. This may be possible due a chemical reaction of oxidable organics and high chloride concentrations. In this case a nonsufficient oxidation could produce halogenated organic substances (AOX). In most cases this situation may be avoided by increasing the oxidation period. As a result lower effluent values than required may be produced and more ozone will be used resulting in a higher energy consumption.

In principle it is not possible to oxidize ammonium with ozone. But at most oxidation plants an occasional oxidation of a varying part of ammonium could be observed. Considering the high oxygen

Box 10.5.3 Activated sludge plant with a subsequent oxidation with ozone (ATV, 1996; Robinson et al., 2003a,b)

Plant layout (plant A) with activated sludge plant and oxidation with ozone.

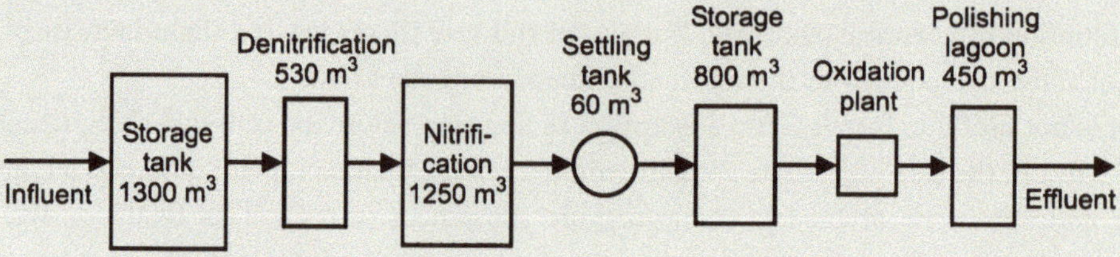

- Average flow rate: 55 m^3/day
- Acetic acid for denitrification: 3 L/m^3 leachate
- Energy consumption of total plant: 200 kW
- Ozone consumption 2.2-2.8 kg ozone/kg COD
- Energy consumption of oxidation stage: 170 kW with 40 kW for UV-light

Plant B: denitrification 950 m^3, nitrification 3700 m^3, clarifier, cloth filter, oxidation (3–12 kgO_3/h, 6 UV-reactors), RBC (4400 m^2 surface), cloth filter, flow rate 120 m^3/d.

Plant	Parameter	Unit	Influent	Effluent Biological Treatment	Effluent Oxidation With Ozone	Effluent Biological Posttreatment
A	COD	(mg/L)	320–5796	236–854	66–176	–
	NH_4	(mgN/L)	125–1350	<1.2	<0.9	–
B	COD	(mg/L)	1200–4000	300–1000	50–200	18–150
	NH_4	(mgN/L)	600–1900	0.1–9	n.d.	n.d.
	AOX	(mg/L)	1.0–3.8	0.5–2.0	0.08–0.25	0.04–0.18
	BOD_5	(mg/L)	n.d.	<20	20–40	<10

Plant C: Biological treatment with two parallel sequential batch reactor + reed bed (2000 m^2) + ozonation (maximum 150 mgO_3/L) + reed bed (500 m^2) (Robinson et al., 2003a).

Parameter (mg/L)	Influent	Effluent SBR	Effluent RB 1	Effluent Oxidation	Effluent RB 2
Suspended solids	96	26/14	9	54	4
COD	600	424/399	390	395	350
BOD_5	35	4/6	4	11	5
NH_4–N	405	1.6/<0.1	0.7	0.8	1
Mecoprop (μg/L)	~200	n.d.	n.d.	<0.05	<0.05
Isoproturon (μg/L)	~300	n.d.	n.d.	3	1.9

Plant D: Plant similar to plant B with following additions: filtration with sand filter and 3 activated carbon filters.

Parameter	Influent	Effluent Biological Treatment	Effluent Oxidation	Effluent RBC	Effluent Adsorption
COD	1045	563	71	15	27
NH_4–N	660	n.d.	n.d.	n.d.	0.1
AOX	1.41	n.d.	n.d.	n.d.	0.34

n.d., no data available.

requirement for the oxidation of ammonium this is a very expensive oxidation. The resulting fluctuation of the ozone requirement can lead to the high increase of effluent organics. As a consequence the activated sludge plant should nitrify almost all of the ammonium as stable as possible.

Remaining suspended solids in the effluent of the activated sludge plant can reduce the efficiency of the oxidation process and may result in problems (incrustation) in the oxidation reactor. These problems may increase if the oxidation process is intensified by using UV-light because the penetration depth of the UV-light will be further reduced. To prevent such problems, a prefiltration is necessary.

BIOLOGICAL TREATMENT WITH SUBSEQUENT REVERSE OSMOSIS

In a number of leachate treatment plants in Germany reverse osmosis is used as a posttreatment step after biological treatment. Using reverse osmosis almost all molecules of the compounds in the leachate may be retained. Under given conditions with lower total solids concentrations in the leachate the permeate flow is higher or the required pressure may be lower. For this reason in most cases the reverse osmosis plant consists of several modules. The membrane of the first module has a wider pore diameter to handle low concentrations of suspended solids. In Box 10.5.4 and 10.5.5 treatment plants and the corresponding results are presented.

Treating raw leachate with reverse osmosis produces a mixture of inorganic and reactive organic substances in the concentrate. If the influent of the reverse osmosis plant contains ammonium and nitrate, there is a potential risk of thermal reactions in the dried concentrate. The treatment/disposal of the concentrates is still a great problem.

Further information is presented in Chapter 10.4.

Box 10.5.4 Biological treatment with subsequent reverse osmosis (concentrate treatment not shown) (ATV, 1996)

Plant layout: biological treatment (activated sludge plant) for denitrification and rotating biological contactor (RBC for nitrification) and reverse osmosis.

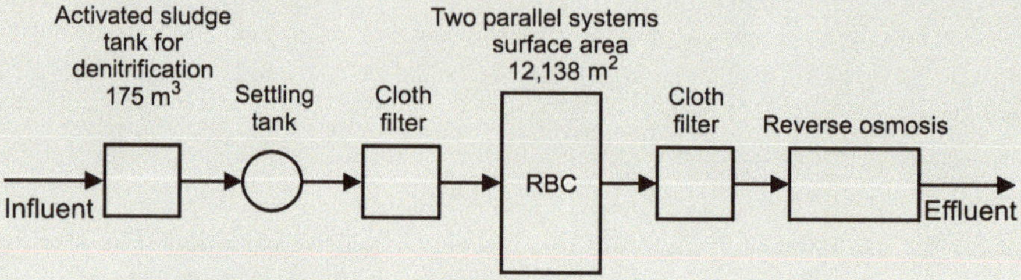

- Average leachate flow: 120 m^3/day
- Reverse osmosis: tubular modules with 280 m^2 membrane area
- Concentrate: 25%-30% of leachate (infiltrated back into landfill)
- Energy consumption: 13.5 kWh/m^3 leachate
- Sulphuric acid: 1 kg/m^3 leachate (pH-adjustment)
- Citric acid: 0.07 kg/m^3 leachate (membrane cleaning)
- Cleansing agent: 0.03 kg/m^3 leachate (membrane cleaning)

Parameter	Unit	Influent	Effluent Biological Treatment	Permeate Reverse Osmosis
COD	(mg/L)	446–872	353–697	5.3–27
BOD_5	(mg/L)	10–220	5–110	1–19
NH_4	(mgN/L)	80–396	0.02–25.2	0.03–10.1
AOX	(mg/L)	0.4–1.4	n.d.	<0.01–0.05

Box 10.5.5 Biological treatment with subsequent two-stage reverse osmosis (ATV, 1996; Ehrig and Robinson, 2011)

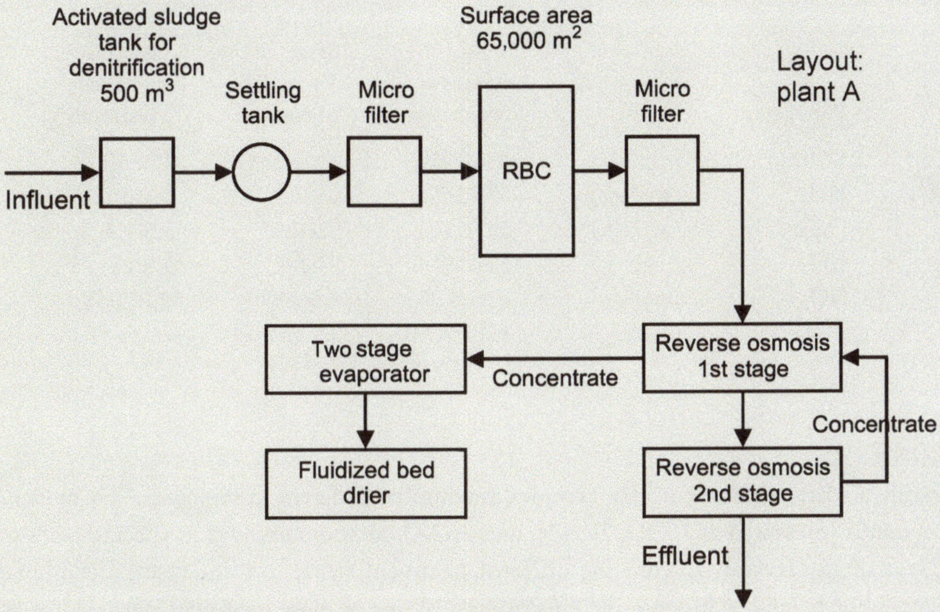

- Average flow rate: 65 m^3/day
- Reverse osmosis: 1st stage: tubular moduls with a membrane area of 425 m^2
- Reverse osmosis: 2nd stage: spiral wound moduls with a membrane area of 128 m^2
- Evaporator: 2 stage forced circulation evaporator
- Residuals: 0.7% of leachate influent

Plant	Parameter	Unit	Influent	Effluent Biological Treatment	Permeate Reverse Osmosis I	Permeate Reverse Osmosis II
A	COD	(mg/L)	1543–2305	736–1145	36–55	4.3–22
	NH_4	(mgN/L)	230–659	0.1–48	0.1–9	0.1–4.2
	BOD_5	(mg/L)	145–253	21–40	1.7–6.7	0.8–4.8

Plant	Parameter	Unit	Average/ Maximum	Effluent Biological Stage Average/ Maximum	Permeate Reverse Osmosis Average/ Maximum
B	COD	(mg/L)	3561/4590	1301/1755	15.8/45
	NH_4	(mgN/L)	833/1651	0.1/0.2	0.1/0.12
	Organic N	(mgN/L)	251/1103	82/125	5.3/9.4
	NO_3	(mgN/L)	1.3/2.9	84/140	9.8/21
	NO_2	(mgN/L)	0.15/5.25	1.2/30.9	0.12/0.79
	BOD_5	(mg/L)	1512/2860	23.7/64	2/5.3

CONCLUSIONS

Sanitary landfill leachate contains a very complex mixture of different compounds. To reduce the organic compounds (measured as COD, BOD_5, and AOX), ammonium, and in specific cases other parameters, it is often necessary to combine different treatment steps. In some cases, a combination of processes could be helpful to increase the effectiveness of one or more treatment steps and/or result in a reduction of the leachate treatment costs. There are several combinations of treatment processes described that are operated in technical scale; in addition results from these plants are presented.

References

Albers, H., Mennerich, A., 1987. Chemisch/physikalische Nachreinigung von Deponiesickerwasser in Minden-Heisterholz (Chemical/physical post-treatment of landfill leachate in Minden-Heisterholz). Müll u. Abfall, p. 326.

ATV work-group 7.2.26 "Landfill Leachate"(ATV = German Association for Water, Wastewater and Waste), 1996. 1st Supplement to the Work Report "Description of Leachate Treatment Plants", vol. 43. Korrespondenz Abwasser, p. 1303.

Ehrig, H.J., 2001. Sickerwasser aus Abfallablagerungen (Leachate from landfills). In: ATV-DVWK (Ed.), ATV-Handbuch Industrieabwasser, Dienstleistungs- und Veredelungsindustrie. Verlag Ernst & Sohn, ISBN 3-433-01468-X, pp. 347–380.

Ehrig, H.J., Robinson, H.D., 2011. Landfilling: leachate treatment. In: Christensen (Ed.), Solid Waste Technology & Management. Wiley, ISBN 978-1-4051-7517-3, p. 858 (Chapter 10.11).

Mester, T., 1993. Sickerwasserreinigungsanlage der Deponie Wilsum – Belebungsanlage, Flockung, Aktivkohle (Leachate treatment plant - activated sludge plant, precipitation, activated carbon adsorption), vol. 8. Veröffentlichungen des Zentrums für Abfallforschung, TU Braunschweig, p. 263.

Robinson, H., Carville, M., Walsh, T., 2003a. Advanced leachate treatment at Buckden landfill, Huntingdon, UK. Journal Environmental Engineering and Science 2, 255.

Robinson, H., Farrow, S., Last, S., Jones, D., December 2003b. Remediation of Leachate Problems at Arpley Landfill Site, Warrington, Cheshire, UK. CIWM Scientific & Technical Review, p. 18.

Robinson, H.D., Carville, M., 2007. The Design, Commissioning and Operation of Leachate Treatment Plants at Large Landfills in Tropical Regions. Paper presented to "Waste a global resource", the 2007 annual exhibition and conference of the UK Chartered Institution of Wastes Management, June 2007. Torbay, UK.

Robinson, H., Olufsen, J., 2007. Full biological treatment of landfill leachate; a detailed case study at Efford landfill in the New Forest, Hampshire, UK. In: Proceedings Sardinia 2007, Eleventh International Waste Management and Landfill Symposium. S. Margherita di Pula, Cagliari, Italy.

11 CHAPTER

LANDFILL COVERS

11.1

LANDFILL COVERS: PRINCIPLES AND DESIGN

Raffaello Cossu and Francesco Garbo

INTRODUCTION

Following deposition in a landfill, wastes should be covered to separate them from the surrounding environment.

A series of diverse reasons relating to life span of the landfill, typology of waste, landfill design concept, local climatic, economic, and social conditions may contribute toward the choice of cover.

Based on the above, four kinds of landfill covers may be envisaged:

- Daily cover
- Intermediate cover
- Temporary capping
- Permanent capping

The first two are prevalently associated with landfill operations and are aimed at reducing management problems on a daily and short-term basis.

Temporary capping and permanent capping are installed once the deposition of waste has been completed to address medium- and long-term issues related to postclosure care. These two types of cover provide a physical barrier, which represents the interface between the waste mass and the atmosphere (Chapter 7.1). Through this interface the exchange of water (precipitation, leachate, irrigation, etc.) and gas (landfill gas fugitive emissions, air intrusion) is controlled.

Water infiltration, leachate regulation, and air intrusion are important aspects in regulating landfill functionality according to the adopted concept (see Chapter 2.1) and, consequently, the management of leachate and gas emissions in the short-, medium-, and long-term. The latter all strongly impinge on technical and economic aspects related to aftercare and overall environmental protection targets.

In the following sections, the different typologies of cover and their required functions are illustrated. Materials and design criteria are discussed with specific reference to top barrier covers.

TYPOLOGIES OF COVER

Daily cover

Daily cover is an infralayer cover applied on a routine basis at an operational landfill site. The cover is referred to as a daily cover due to the fact that it is habitually placed on top of the open waste surface at the end of each operating day, although a different time frame may be adopted in relation to the type and quality of deposited waste.

The functions required for a daily cover in a municipal solid waste (MSW) landfill are as follows:

- Minimization of wind blown litter and dust
- Odor minimization
- Control of disease vectors (contact reduction for insects, flies, rodents, birds, and other animals)
- Fire prevention
- Control of scavenging
- Vehicle mobility and access to working place
- Optimization of the draining capability of waste

These requirements are typical for MSW landfills but may vary substantially according to other waste typologies and eventual pretreatments. However, optimal drainage, minimization of dust transport, and vehicle mobility are aspects of fundamental importance for all kinds of landfill.

Soil-based materials are commonly used as a daily cover. Quality and prescriptions may be detailed by regulations and guidelines, in terms of thickness (generally 10–15 cm), permeability ($k > 10^{-3}$ cm), and composition (no leaching, nondegradable, noncompressible, predominantly inorganic). Removable plastic tarps may also be conveniently used when permitted by climatic conditions and landfill morphology. Foams and sprays have also been used (Carson, 1992).

Selected and prequalified alternative waste materials have been proposed, including shredded tires, construction and demolition waste fractions, glass aggregates, foundry sand, bottom ash, dehydrated and stabilized sludge, etc. The quality of all materials should comply with existing regulatory prescriptions.

Critical aspects to be taken into account in the design of daily cover are occupancy of volume, interference with gas extraction wells, and poor draining quality of some materials. The customary practice of excavating and using silty or clay soil available on site should be avoided. Use of low-permeability materials may result in leachate ponding, odor emissions, and operational issues during stormy weather. Additionally, the use of these materials in a naturally aerated landfill would significantly reduce gas permeability of the waste mass.

Intermediate cover

When filling in a defined area is halted temporarily and the area remains exposed for a period of time, an intermediate cover might be needed.

The goals of intermediate covers can be summarized as follows:

- Control of water infiltration to minimize leachate generation
- Erosion control
- Control of fugitive gas emissions
- Prevention of leachate breakouts
- Seclusion from animals and scavengers
- Air intrusion prevention
- Fire prevention
- Control of littering and animals

Intermediate cover should provide service for months, before further landfilling in the area restarts. They need to be robust and, conversely to daily cover, should be made of impervious or low-permeability materials, either in terms of soil or of geomembranes. Removable structured tarps may be conveniently used as an alternative material. Intermediate cover should be regularly inspected and erosion phenomena controlled. If soil is used as a cover, and in line with functioning time, grass vegetation would contribute toward controlling erosion. When waste placement recommences, the cover should be completely removed to prevent leachate ponding.

Temporary capping

When a landfill has reached its capacity an intermediate cover may be applied while waiting for the primary waste to settle. An additional deposition of waste may take place until the final authorized or design profile is achieved. Subsequently, a temporary capping may be emplaced prior to installation of the permanent capping. The main role of the temporary capping is to perform as a normal top barrier until significant mechanical and biological stabilization of the waste mass has occurred.

The main features required of temporary capping are

- Functional to the adopted landfill design concept until waste is sufficiently stabilized
- Drainage and control (not prevention) of water infiltration
- Gas drainage
- Attenuation of fugitive gas emissions
- Contribution to leachate management (treatment, recirculation)
- Respect of aesthetics and surrounding landscape
- Provide support for vegetation growth
- Flexibility and adaptation to settling
- Erosion control

The first point is of particular relevance. Indeed, permanent capping is generally conceived by national regulations as a tool for use in sealing landfills to avoid or markedly minimize the generation of leachate. However, for specific types of degradable or leachable waste, this may prove detrimental, prolonging the long-term emissions of leachate, and failing to guarantee landfill sustainability (see Chapter 2.2). In addition, some sustainable landfill types such as semiaerobic bioreactors require air intrusion and circulation in the waste mass.

Materials suitable for use in the construction of temporary cappings may include

- Materials indicated previously for physical barriers (Chapter 7.1), such as low permeable soil, sand and gravel, geomembrane, geocomposite clay liner (GCL), and other geosynthetics.
- Materials fulfilling specific functions, such as methane oxidasing material, natural soil to support vegetation, low stem vegetation with high evapotranspiration (ET) effect (grass, bushes, etc.).

According to the planned working time, specific components for alternative temporary cover, such as solar geomembrane cover, may also be proposed. The latter system combines a normal geomembrane with a solar photovoltaic (PV) panel attached to the geomembrane allowing the generation of solar energy. When adopting impervious materials, the design of the temporary capping should be such to include, whenever necessary, devices for water and/or air immission (wells, dispersion underdrain, pipes, etc.).

Permanent capping

Once the waste mass has been mechanically stabilized (target prescribed by some landfill regulations) and significant contaminant mobilization (degradation, leaching) has occurred, a permanent capping could be installed.

Permanent capping will need to explete some of the functions requested of a temporary top cover, including

- Functional to the adopted landfill design concept
- Drainage and control (not prevention) of water infiltration
- Gas drainage
- Attenuation of fugitive gas emissions
- Contribution to leachate management (treatment, recirculation)
- Provide support for vegetation growth
- Erosion control

In addition, permanent capping will need to meet the following specific requirements:

- Functionality with respect to the landfill afteruse
- Leachate generation minimization
- Long-term durability and sustainable maintenance
- Vehicle mobility and access to gas and leachate management facilities
- Prevention of any gas emission (in combination with a gas extraction system)
- Simple and long-lasting monitoring network
- Permanent management of water runoff
- Resistance to erosion and subsidence phenomena

Permanent capping is responsible for controlling medium- and long-term emissions and consequently should be efficiently maintained until the designed final quality of landfilled waste and emissions is achieved (Final Store Quality; see Chapter 2.2).

Functions requested of each individual type of landfill cover are summarized in Table 11.1.1.

FUNCTIONS AND MATERIALS IN LANDFILL CAPPING

Mass transport at the atmosphere-landfilled waste interface

As highlighted earlier, landfill capping controls the exchange of mass along the waste—atmosphere interface (see Fig. 11.1.1). This exchange of mass mainly comprises water and air that enters the waste mass and gas emissions exiting the waste mass.

Mass exchange should be controlled for the following reasons:

- Water is needed to mobilize and reduce the contaminant potential of the waste mass (degradation, leaching).

Table 11.1.1 List of the required functions, indicative duration, and materials to be used in the four different typologies of landfill cover

Typology	Time Frame	Required Functions	Materials
Daily cover	Days, weeks	• Minimization of wind blown litter and dust • Odor minimization • Control of disease vectors (contact reduction for insects, flies, rodents, birds, and other animals) • Fire prevention • Control of scavenging • Vehicle mobility and access to working place • Optimization of the draining capability of waste	• Sand • No leaching, nondegradable alternatives: shredded tires, construction and demolition, glass aggregates, foundry sand, bottom ash, etc., highly permeable materials • Removable tarps (when climatic conditions and deposit morphology would allow their use) • Foams and sprays
Intermediate cover	months	• Control of water infiltration to minimize leachate generation • Erosion control • Control of fugitive gas emissions • Prevention of leachate break-outs • Seclusion from animals and scavengers • Air intrusion prevention • Fire prevention • Control of littering and animals	• Removable Tarps • Low-permeable soil
Temporary capping	years	• Functional to the adopted landfill design concept until waste is sufficiently stabilized • Drainage and control (not prevention) of water infiltration • Gas drainage • Attenuation of fugitive gas emissions • Mitigation of residual gas emissions • Contribution to leachate management (treatment, recirculation) • Respect of aesthetics and surrounding landscape • Provide support for vegetation growth • Flexibility and adaptation to settling • Erosion control	• Low-permeable soil • Sand and gravel • Geomembrane • Geocomposite clay liner (GCL) and other geosynthetics • Methane oxidasing material (compost, dried sludge, etc.) • Natural soil • Vegetative soil for low stem vegetation • Technical cover (e.g., solar geomembrane cover)

(Continued)

Table 11.1.1 List of the required functions, indicative duration, and materials to be used in the four different typologies of landfill cover—cont'd

Typology	Time Frame	Required Functions	Materials
Permanent capping	decades	• Coherent with adopted landfill design concept • Functionality with respect to the landfill afteruse • Water Drainage • Minimization of water infiltration • Attenuation of fugitive gas emissions • Mitigation of residual gas emissions • Gas drainage • Contribution to leachate management (treatment, recirculation) • Provide support for vegetation growth • Erosion control • Leachate generation minimization • Long-term durability and sustainable maintenance • Vehicle mobility and access to gas and leachate management facilities • Prevention of any gas emission • Simple and long-lasting monitoring network • Permanent management of water runoff • Resistance to erosion and subsidence phenomena	• Low-permeable soil • Sand and gravel • Geomembrane • GCL and other geosynthetics • Methane oxidasing material (compost, dried sludge, etc.) • Natural soil • Vegetative soil for low stem vegetation • Technical cover (e.g., solar geomembrane cover) • Materials and products to support specific landfill afteruse

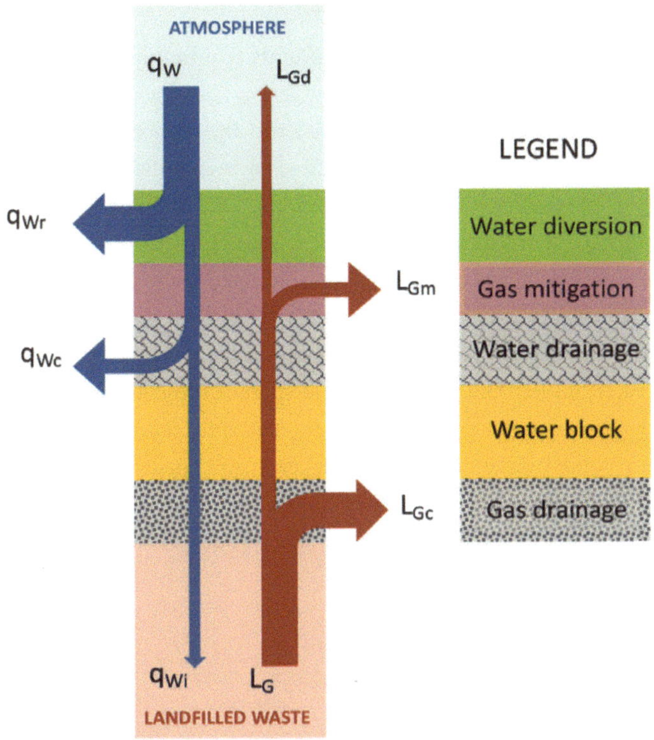

Figure 11.1.1 Role of the main possible functional units in controlling mass transfer of water and landfill gas (LFG) contaminants in a landfill capping. Possible intrusion of air or lateral escape of leachate is not represented. q_W = Water flow rate; q_{Wr} = flow rate of diverted water; q_{Wi} = flow rate of infiltrating water; L_G = residual load of contaminants in LFG; L_{Gc} = load of contaminants in the collected LFG; L_{Gm} = biodegraded load of contaminants in LFG; L_{Gd} = load of contaminants in dispersed LFG.

- Water generates leachate and controls the quality of the same; high volumes of leachate might create technical and economical operational problems.
- Oxygen is required by aerobic landfill models; conversely, air intrusion is detrimental for anaerobic degradation.
- Landfill gas fugitive emissions from degradable waste landfills are among the major sources of greenhouse gases and contribute significantly to global climate change (i.a. Bogner et al., 2007).
- Landfill gas produced under anaerobic conditions in an MSW landfill may represent a significant energy source only if the quality and quantity of the gas supports conversion from both a technical and economical viewpoint; if for any reason (mechanical biological pretreatment, aging of the waste and decay in the generation curve, air intrusion in the waste mass) the quality and quantity of gas is no longer effective for energy conversion, the residual gas could be treated in biomitigation systems integrated in the landfill capping (see Chapter 9.5).
- According to landfill morphology capping should also drain and contain possible lateral leachate outflows.

Capping structure and sequence of components

In general, landfill capping includes a series of components required in the control of mass exchange, as schematically represented in Fig. 11.1.2.

The design of the units (sequence and dimensions) and the choice of materials to be used are not a univocal, repetitive process but are rather strictly dictated by waste quality and properties, adopted landfill model, availability of materials, waste site conditions, and landfill management procedures. The units represented in Fig. 11.1.2 are those which could potentially be proposed, although not necessarily all present in the same capping combination. They would need to be carefully selected according to the specific requirements of each individual landfill project.

The required functions for temporary and permanent capping are linked to a proper choice of materials to assure an efficient service throughout the life span of the landfill and to prevent deterioration, cracking, and creeping phenomena (Walhström et al., 2009; Cossu et al., 2000).

Design therefore is a fundamental aspect to be taken into consideration in view of potential instability phenomena such as erosion, adverse weather conditions, waste settlement and general mechanical stability (see Chapters 5.1 and 5.2), frost/thawing, plant roots penetration, landfill morphology and tensile strains, etc.

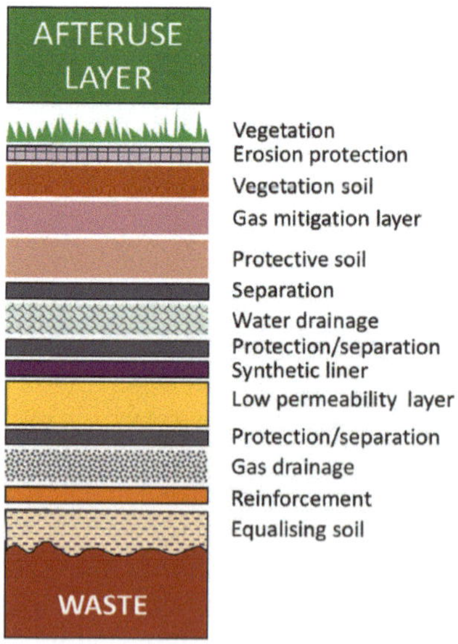

Figure 11.1.2 Different potential components and sequencing in the structure of a landfill capping. Sequencing and final design will include some of these based on waste quality, adopted landfill model, local conditions, availability of materials, required time of service, compliance with national regulations, afteruses, costs, etc.

Settlement phenomena may be uniformly distributed throughout the entire surface and may be manifested prior to realization of the top cover; however, situations may occur in which the system is damaged by localized settlements several years after landfill closure (Ivanova et al., 2008; Gourc et al., 2007).

In the design phase, consolidation and stability of the following structural units should be well assessed:

- foundations
- leachate collection and drainage system
- waste layers, including intermediate cover layers
- individual components of the top cover system

Due to the heterogeneities characterizing the waste body, settlement phenomena cannot be predicted with a high degree of certainty, although they can be distinguished between "total settlements" or "differentiated settlements" (Camp et al., 2005).

Settlements involving the whole waste body ("total settlements") depend on several processes: mechanical consolidation due to overloads, biochemical decomposition, and collapse of cavities. There is usually an initial, rapid primary consolidation followed by a secondary consolidation, which could last for several years. This kind of settlement is maximized if the waste is rich in organic, biodegradable compounds, proving the importance of degradation phenomena.

If settlements are expected to be relevant, the construction of the permanent cover should be delayed and the use of a temporary capping is strongly suggested.

Differentiated settlement represents the worst situation: the top cover system must be designed to support and pander such movements (Camp et al., 2005).

The structure of a landfill capping is made up of a sequence of different components which, starting from the top, can be listed as follows (see Fig. 11.1.2):

- Superficial layer (erosion control, water diversion-runoff, evaporation, ET—vegetation, gas mitigation)
- Protective, supportive layer
- Water drainage layer
- Protective layer
- Water block layer (lining, water infiltration control, gas flow control)
- Biogas drainage layer
- Waste-contact layer (reinforcement, equalization)

Specific additional elements such as geotextiles could be used to separate soils with different granulometry or to protect geomembranes.

The construction of the different elements in the capping structures involves the use of a series of different materials as summarized in Table 11.1.2.

Superficial layer

The superficial layer is in direct contact with the atmosphere and thus will need to be able to resist erosion (water and wind) and to divert water by optimizing the hydrological balance (runoff, ET,

Table 11.1.2 Materials that could be used in landfill cappings (Cossu et al., 2000; Andreas et al., 2005; Burkhardt et al., 2009)

Layer	Usable Materials
Superficial layer	• Vegetative soil • Low stem vegetation • Tall vegetation • Erosion control (geosynthetics or natural mats) • Gas mitigation materials (compost, dried sludge, etc.) • Cobbles • Paving materials • Technical products (e.g., solar geomembrane panel)
Protective layer	• Unclassified soil • Cobbles • Geosynthetics • Suitable residues from waste treatment
Water drainage layer	• Gravels or sand • Geonet or geocomposite • Suitable residues from waste treatment
Water block layer	• Compacted clay • Geomembrane • Composite barriers • Sand + gravel (capillary barrier)
Biogas drainage layer	• Gravels or sand • Residues from waste treatment
Waste-contact layer	• Soil for equalization • Reinforcement products (e.g., geogrids)

etc.). Consequently, the superficial layer may include erosion protection material (geosynthetics, mats); drainage and channeling of rainfall (see Chapter 8.2); vegetative soil to support vegetation growth, to store water for increased ET, to meet the plants needs, and to accommodate root development. The superficial layer may also be integrated into the structural materials for the biomitigation of gas.

Natural vegetative soil

Natural vegetative soil is used to support the growth of vegetation species and allow for storage of water. It is usually composed of local uncompacted soil, 0.15–1.50 m deep, which is often available in proximity of the landfill site at a reasonable cost. The material should display a good field capacity. The depth of the vegetative soil layer will depend on climatic conditions, type of vegetation (root length is important), and overall cover structure (temporary or permanent capping).

Vegetation

The role of vegetation in landfill capping is multiple:

- To improve erosion control;
- To maximize ET effects;
- To improve general aesthetics of the site;
- To create a synergy with "green afteruses" (parks, gardens, natural reserves, etc.).

Specific installation of plant species may provide an effective control of the hydrological balance without need for liners as exploited by Phytocapping covers (see Chapter 11.2).

The vegetation layer may also integrate phytotreatment units for leachate treatment as discussed further on.

Geosynthetic materials or mats/blankets

Geosynthetic products or mats/blankets made of natural materials (coconut fibers, jute, straw, etc.) may be efficiently applied to limit soil erosion, particularly along the slopes, and provide good resistance to scavenging animals. Some of these erosion control products are biodegradable and could be hydroseeded to promote vegetation on the slopes.

Cobbles

Cobbles are used as an alternative to natural soils in regions in which precipitation is insufficient to sustain stable vegetation growth, while ensuring protection against erosion. There are, however, some disadvantages:

- Locally available material is needed, and the layer must be thick enough to resist the action of wind and water
- Water infiltration into the waste mass is relevant
- Stability problems in case of significant slope angles

Paving materials

In case of specific uses of the landfill site (e.g., car parking) and when the waste body is mechanically stable, paving materials such as asphalt may be used (Burkhardt et al., 2009).

Maintenance problems and risks of unpredicted settlement phenomena should be carefully considered.

Gas mitigation materials

The gas mitigation layer may comprise materials capable of supporting the bacterial population responsible for the degradation of methane and other trace gases. Humic soil and waste materials such as compost, garden and stabilized food waste, straw, stabilized dewatered sludge, etc., could be employed (Chapter 9.5). The material should be spread over the entire landfill surface (biocover) or in openings within low-permeability covers (biowindows).

Drainage layer

The drainage layer permits the diversion of precipitation water infiltrating through the superficial layer. The drained water is then conveyed to collecting devices (pipes, manholes, ditches, etc.).

The drainage layer may be constructed using sand, gravel, or geodrain.

Water should be able to flow easily throughout the drainage layer; therefore, the drainage pipes should be well maintained to avoid obstruction. Conversely, the layer may become saturated and an excessive pore water pressure may develop, prejudicing stability of the top cover.

Sand and gravels

Sand and gravels are largely applied due to simplicity of installation, although at times limited by the availability and economic convenience of materials. The efficiency of the system is directly proportional to the hydraulic conductivity of the layer.

Geosynthetics

Geosynthetics in general may afford a simple alternative to natural materials if these are not locally available. Geodrains, geonets, and geotextiles, largely available on the market, are usually applied directly above the impermeable barrier. The main drawbacks are poor protection of the underlying barrier from thermal shocks and mechanical behavior on slopes, when tensile strains are present.

Residues from waste treatments

The utilization of incinerator residues (bottom and vitrified fly ashes), crushed construction debris, compost, etc. has been successfully tested in lab-scale and pilot-scale tests (Andreas et al., 2005, 2014; Tham et al., 2005). It is an interesting and promising option, although the potential release of contaminants should be evaluated carefully with leaching tests (Tham et al., 2005; Travar et al., 2005).

Protective layer

If present (the use of a protective layer is optional), this layer is intended to separate and protect the underlying low-permeability layer. In some cases, the superficial layer and the protective layer can be integrated into one single layer.

Unclassified soil

Unclassified soil is often used as it is widely available and is highly efficient, with the exception of specific cases in which specific requisites must be met.

Residues from waste treatments

As described previously, the use of residues from waste treatments (bottom and fly ashes, compost, etc.) is an interesting and promising option, although the potential release of contaminants should be evaluated.

Water block

The role of the water block layer is to minimize water infiltration into the waste body both directly, acting as an impermeable layer, and indirectly, causing the accumulation and drainage of water in the overlying layers, from which it will be removed by ET, runoff, or internal drainage. It also limits the upward flow of biogas. The most widely used materials in the realization of a low-permeability layer are compacted clay, geomembranes, and composite barriers. Capillary barriers may be of interest if capping needs to be permeable to gas.

Liner

Compacted clay

The mineral liner is composed of a layer of compacted clay, minimum 0.5 m deep, made up of several layers with a thickness ≤0.15–0.20 m and a hydraulic permeability of 1×10^{-9} m/s or lower.

In spite of its simplicity, compact clay liners may deteriorate over time due to several drawbacks (Miller and Mishra, 1994; Albrecht and Benson, 2001; Albright et al., 2006; Barbisan et al., 2013):

- Compaction may be difficult where the underlying layer is not structurally stable
- If not well protected, the layer may suffer due to dessication, freeze/thaw cycles, resulting in cracking
- Animal activity
- Plant roots penetration
- The system cannot be repaired after being damaged

Geosynthetic clay liners may represent a convenient alternative to compacted clay, featuring a series of advantages in the construction phase and promoting a saving of volume. However, the interface shear strength between the different layers of the top cover should be carefully considered and analyzed, particularly along significant slopes. Failures resulting from excessive gas or pore water pressure have been reported (e.g., Richardson et al., 2000).

Geomembrane

Conversely to the use of geomembrane in the bottom barrier, in capping systems there is no direct exposure to leachate, although the material may be exposed to high-temperature variations and to stresses generated by waste settlement. The geomembrane is anchored in a peripheral trench. HDPE and LDPE are the most widely proposed geomembranes for capping applications, with a thickness ranging from 1.8 to 2.5 mm.

This type of system is easily installed and the materials are readily available. They offer optimal resistance to humidity and waste settlement (Burkhardt et al., 2009).

Failures in the use of geomembrane landfill capping are related to perforations (mainly by drainage grains or root penetrations), aging, and potential slips with respect to the superimposed layers. Suitable geotextiles should always be used in the protection of geomembranes (see Chapter 7.1).

Low-permeable materials

Low-permeable materials may be applied in some types of capping in which water is permitted to infiltrate the landfill mass to sustain processes of contaminant mobilization (degradation, leaching) or when residual gas should be allowed to flow freely in the presence of biocovers with gas mitigation layers. Suitable materials include silty and clay soils with a permeability $k <$ around 10^{-7}–10^{-8} m/s.

Capillary barriers

Capillary barriers represent an effective alternative to other water block options, described in the previous paragraphs (Parent and Cabral, 2006).

Interest in this option is related to

- Good capacity to maintain efficiency after waste settlement phenomena
- Possibility to be installed on surfaces having high slopes up to 20 degrees
- Easy maintenance
- Good gas permeability
- Selective water infiltration (extremely important to adjust the biological degradation of waste)

A capillary barrier is based on exploitation of the capillary effect originating at the interface between two materials having a different grain size, under unsaturated conditions (Fig. 11.1.3). The barrier is composed of a layer of sand (capillary layer) overlying a gravel layer (capillary block). When the capillary retention force is stronger than force of gravity, water will accumulate in the sand until the hydraulic head is higher than capillary pressure and water will downflow into the gravel. However, if the two layers are inclined (5–15 degrees) water will fail to accumulate and will be diverted along the inclined surface (Barrès and Bonin, 1994; Kämpf and Montenegro, 1997; von Der Hude et al., 1999; Barth and Wohnlich, 1999; Weiß and Witzsche, 2005).

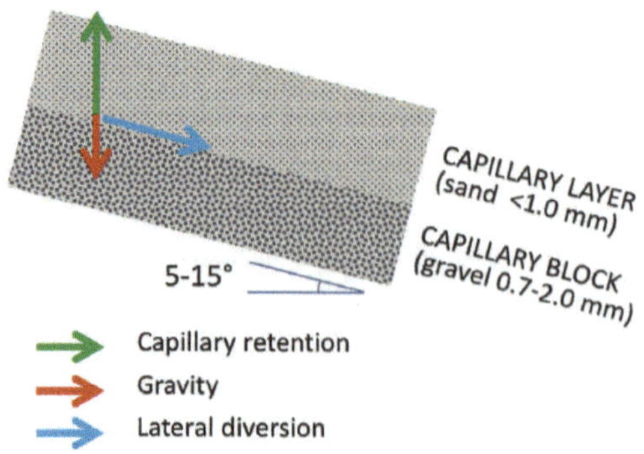

Figure 11.1.3 Components of a capillary barrier (slope between 5 degrees and 15 degrees).

When the entity of the hydraulic flux (as in case of intense rainfall), despite horizontal flow, generates a relevant water head, the capillary forces cannot retain water, which enters the waste mass (breakthrough). A layer of soil placed on top of the capillary barrier might allow a more correct functioning of the capillary barrier, acting as a storage volume for the water moving through the barrier (Suzuki et al., 2005). A certain number of breakthroughs per year may be acceptable if the adopted landfill model requires water for mobilization of contaminants.

In the long term, capillary barriers, when compared with traditional systems, may represent a more reliable water block because they are not affected by dessication issues and are less sensitive to settlement.

Biogas drainage layer

The drainage of biogas allows removal of the biogas produced by the anaerobic digestion of waste, being responsible for draining and delivering the gas to the pipes that convey it to combustion systems. Similar to the water drainage layer, gravels or sand (or residues from waste treatments) may be used. For the biogas drainage layer, the alternative use of geodrains is not recommended.

Waste-contact layer

Equalization

The surface of the waste deposit at the end of landfilling is generally rough and irregular. Permeable unclassified soil might be used to create a smooth surface for installion of the various layers that compose the final capping structure. Geotextile could be superimposed as a separation with the overlying gas drainage layer.

Reinforcement

Geogrids may represent a suitable material aimed at attenuating differential settlement. This is not normally the case in an MSW landfill where settlement phenomena are consistent. However, in the case of special waste landfills or in particular afteruse installations, adoption of geogrids may be required to control differential settlement (Caldwell et al., 2000; Rajesh and Viswanadham, 2012). The use of geogrids is strongly recommended when the top cover is part of a vertical expansion program and additional waste needs to be landfilled (Dayal et al., 1994).

MAIN LANDFILL CAPPING COMBINATIONS AND DESIGN CRITERIA

Sequencing of components and legal requirements

According to the different landfill requirements, three main forms of capping can be proposed:

- Sealing capping (no gas, no water exchange along the landfill—atmosphere interface)
- Low-permeable capping (no gas exchange, low water infiltration allowed)
- Gas permeable capping (permeable to gas, higher water infiltration allowed)

Table 11.1.3 Recommendations for "surface sealing" of landfills by the European Union (Council Directive, 1999/31/EC)

Top Cover Components	Landfill Category	
	Nonhazardous	Hazardous
Gas drainage layer	required	Not required
Artificial sealing liner	not required	required
Impermeable mineral layer	required	required
Drainage layer > 0.5 m	required	required
Top soil cover > 1 m	required	required

As already highlighted, the choice should be dictated by the quality of waste, the adopted landfill model, by the local situation, and by national regulations. The latter are in general oriented toward use of sealing or low-permeable capping. One exception is Australia where gas permeable capping based on the phytocapping concept is authorized (Chapter 11.2). In general, however, prescriptions for top final covers are more heterogeneous compared with the requirements for bottom barriers.

The European Landfill Directive (Council Directive, 1999/31/EC) leaves it up to the individual member states to legislate on the prescription of surface sealing "after a consideration of the potential hazards to the environment finds that the prevention of leachate formation is necessary." The recommendations for top covers are summarized in Table 11.1.3, according to the category of waste to be landfilled.

In Italy, legislation specifies the characteristics for each single layer, thus minimizing the possibility for designers to implement alternative solutions. In Germany, several differences are reported for top covers for use in Class I and Class II landfills, where only properties of the water drainage layer are specified. In the United Kingdom, prescriptions are provided as a kind of guideline, whereas in the United States the use of two layers is mandatory (superficial and low-permeability layer). Legal requirements for permanent capping in some countries are summarized in Table 11.1.4. In Italy, use of a simpler from of temporary capping is permitted until waste has significantly settled. According to the Sustainable Landfill Guidelines of the Lombardy Region (2014), the temporary cover should be installed once mechanical and biological stability has reached specific target values to be defined in the landfill design.

Table 11.1.4 Legal requirements for the realization of the final top cover in nonhazardous municipal solid waste landfills in Italy, Germany, the United Kingdom, the United States, Japan, and China

Top Cover Components	Italy	Germany		The United Kingdom	The United States	Japan	China
	Nonhazardous Waste Landfills	Landfills Class I[1]	Landfills Class II[1]	Nonhazardous Waste Landfills	Nonhazardous Waste Landfills	Nonhazardous Waste Landfills	Nonhazardous Waste Landfills
Superficial layer	$\geq$1 m	Mandatory	Mandatory	$\geq$1 m	Earthen material $\geq$6 inches ($\approx$0.15 m)	$\geq$0.5 m	Nutrient vegetation layer ($\geq$0.15 m) on the top of supporting soil layer ($\geq$0.45 m, $k \geq 10^{-6}$ m/s)
Water drainage layer	$\geq$0.5 m	$\geq$0.3 m, $k \geq 10^{-3}$ m/s	$\geq$0.3 m, $k \geq 10^{-3}$ m/s	m/s			$\geq$0.3 m, $k \geq 10^{-4}$ m/s
Low-permeability layer	Compacted mineral layer $\geq$0.5 m, $k \leq 10^{-8}$ m/s	Mandatory	Mandatory	Mandatory	Earthen material $\geq$ 18 inches ($\approx$0.45 m), k less than the underlying liner system, in any case $k \leq 10^{-7}$ m/s		**1.** Geomembrane on the top of compacted mineral layer, $\geq$ 0.45 m, $k \leq 10^{-7}$ m/s or **2.** Compacted mineral layer $\geq$0.90 m, $k \leq 10^{-9}$ m/s or **3.** HDPE or LDPE geomembrane, $\geq$ 1–1.5 mm, $k \leq 10^{-14}$ m/s, service life $\geq$30 years
Biogas drainage layer	$\geq$0.5 m		Mandatory in some case	Mandatory			$\geq$0.3 m (particle size 25–50 mm)
Equalization layer	Mandatory	Mandatory in some cases	Mandatory in some cases				
Reference	D. Lgs. 36/2003. Attuazione della direttiva 1999/31/CE relativa alle discariche di rifiuti	Deponieverordnung. DepV/2009		The Landfill (England and Wales) Regulations, 1559/2002	Code of Federal Regulations (Title 40)	Ministry of Environment, No.12,2017	Technical code for municipal solid waste sanitary landfill (GB 50869-2013 and GB 51220-2017)

[1]Landfill of class I and II are landfills for nonhazardous waste (e.g., MSW). Class II landfills contain a higher amount of biodegradable waste compared to Class I. Typically include WSW landfills.

Sealing capping

Sealing capping is prevalently addressed at preventing water infiltration into the waste mass, minimizing leachate generation, and promoting efficient collection of the generated gas, thus avoiding gas emissions.

A complete combination for reaching these targets is given in Fig. 11.1.4 (option A).

Equalizing and soil reinforcement will depend on the local situation. Coarse sand or granules are generally used for gas drainage in a 0.3–0.5 thick layer, topped by a 0.5 m layer of compacted clay with $k < 10^{-9}$ m/s. A geotextile separates the mineral liner from the gas drainage layer. An HDPE geomembrane of 1.5–2.0 mm is superimposed in direct contact with the surface of the clay liner. The geomembrane is protected by a thick high-density geotextile structured for this purpose. Above the geotextile, a layer of 16–32 mm gravel constitutes the water drainage layer (0.5 m thick). Protective/supporting soil is superimposed over the water drainage layer and a good quality geotextile emplaced to promote filtration. Soil should display good permeability properties with a thickness ranging from 0.5 to 1.0 mm, depending on the overlying vegetative layer (humic soil minimum 0.5 m) and the type of vegetation to be planted.

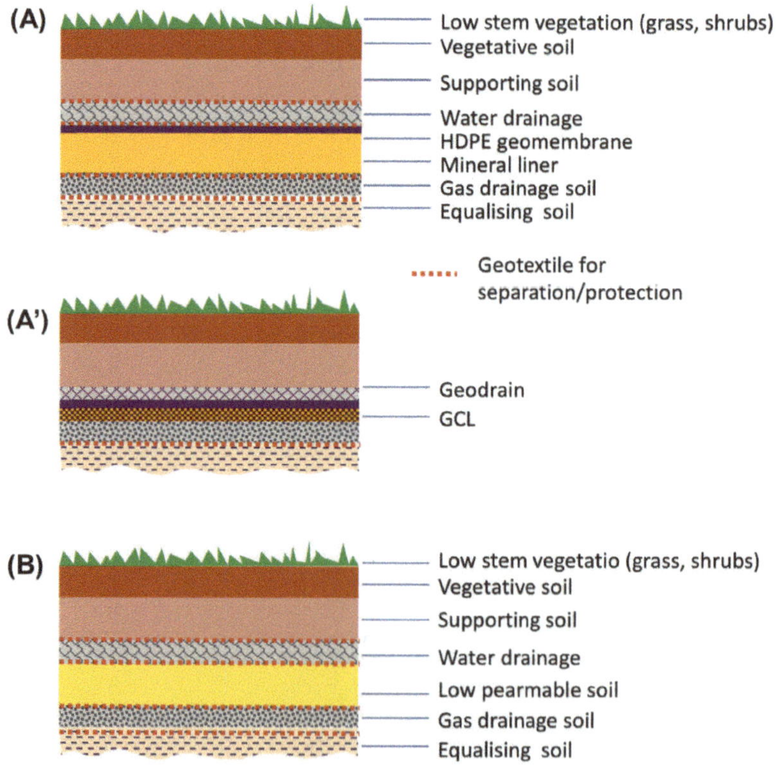

Figure 11.1.4 Structures of landfill capping with water infiltration block. A = Sealing capping with use of natural materials; A′ = sealing capping with use of geosynthetics; B = low-permeable capping.

The same level of efficiency for this combination of capping could be obtained by means of geosynthetics. A GCL could substitute the mineral liner and/or a geodrain could replace the gravel in the water drainage layer (option A′, Fig. 11.1.4), thus resulting in volume saving and facilitating construction along the slope. The design, as mentioned earlier, should take into account the interface shear strength between the different layers. A drawback of this option is that it may not comply with some national regulations prescribing a minimum thickness of 0.3–0.5 m for the mineral liner and the drainage layer.

If water flow is required for process purposes, the sealing capping should be engineered with facilities that allow an undercapping immission of water. For this purpose, gas extraction wells, if present, could be conveniently used. This type of capping requires careful maintenance, but if well maintained it may last for extremely long periods.

Low-permeable capping

This type of capping essentially features the same targets as sealing capping, but with a higher possibility of water leakage (option B, Fig. 11.1.4). The layer of mineral soil above the gas drainage layer features a minimum thickness of 0.5 m and a minimum permeability of 10^{-8} m/s. The superimposed layers are the same as those applied in the sealing combination. This combination meets the minimum requirements established by Italian and German landfill regulations (Table 11.1.4).

The main advantages of this solution are the lower cost compared with the sealing capping option, more favorable mechanical conditions along the slope, and controlled infiltration of water to better support the controlled mobilization of contaminants (degradation, leaching). Capping will be easier to maintain and will last longer. The main disadvantages are that fugitive emissions will be less effectively controlled.

Capillary capping

The structure of landfill capping based on the capillary barrier is given in Fig. 11.1.5 (option C). Equalization, reinforcement, and the protective and superficial layers are the same as those discussed previously for sealing capping. The efficiency of a capillary barrier is related to the choice and installation of the materials (Cossu et al., 2000; von Der Hude et al., 1999; Barth and Wohnlich, 1999). The granulometry of both capillary layer and capillary block is the most important parameter. A difference in granulometries should be ensured to produce a barrier effect, although at the same time should not be too large as this may result in interpenetration between the layers.

Small sand particles, with a granulometric distribution ranging from 0 to < 1.0 mm should be used to build the capillary layer; materials with a granulometry ranging from 0.7 to 2.0 mm, maximum 3.2 mm, should be used for the capillary block (Suzuki et al., 2005; Weiß and Witzsche, 2005; Cossu et al., 2000). Barrès and Bonin (1994) suggested a granulometry of 0.25 mm for sand.

With regard to thickness of the layers, the most frequent values are 0.30–0.60 m for the capillary layer and 0.15–0.30 m for the capillary block.

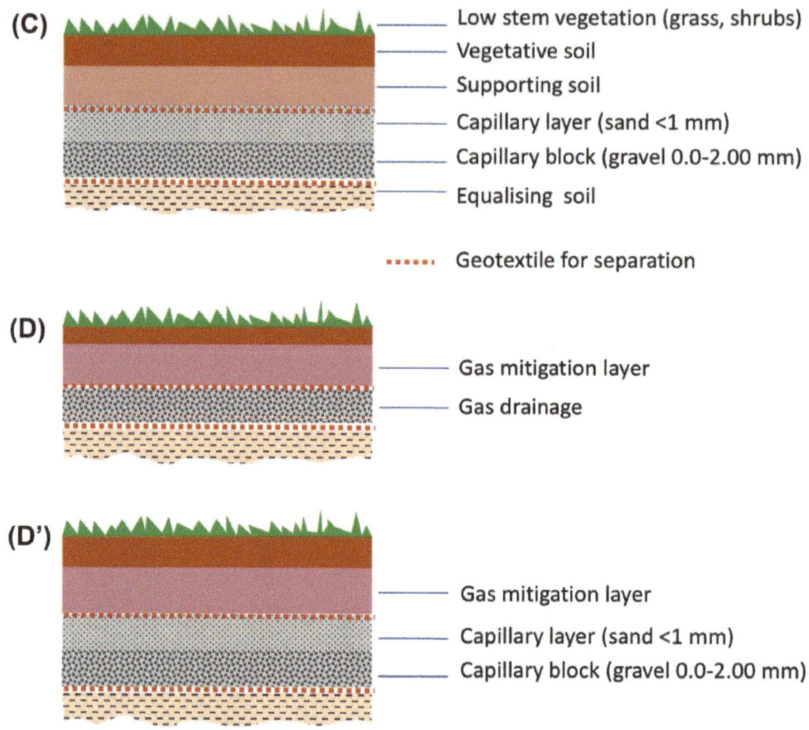

Figure 11.1.5 Structures of landfill capping permeable to gas. C = Capillary barrier; D = biocover; D′ = biocover with capillary barrier.

The slope of the cover system is essential for the optimal working of the barrier, as the presence of both flat and hollow areas limit the drainage of water retained at the interface between the two layers. The suggested slope ranges between 5 and 15 degrees (Suzuki et al., 2005).

The overall structure of a capillary capping is represented in Fig. 11.1.5.

In terms of performance of capillary barriers, the experience gained to date has derived mainly from pilot installations.

Barrès and Bonin (1994), using a capillary layer 30–40 cm thick with sand 0.25 mm, a capillary block 30 cm thick with 2–8 mm, observed an efficiency of approximately 90%, dropping rapidly to 80%, and lower when rainfall exceeded 10 mm/day.

A large-scale application has been implemented in the context of a European Project (LIFE96 ENV/D/000197) for a period of 6 years (1996–2002). The pilot project was developed on 2 ha of a sloping section of the landfill at Breinermoor, Lower Saxony in Germany. In the flat section of the test field, 23 mm breakthroughs in the year 2000 with an average rainfall of 800 mm were observed. It was concluded that "from a technical and financial point of view, the use of capillary barriers could prove to be an advantage to many federal states, regional authorities, and commercial landfill operators

Table 11.1.5 Total precipitation and water balance (from December 2002 to January 2004) in the pilot-scale test performed by Suzuki et al. (2005)

	Amount (mm)	Ratio (%)
Precipitation	2056	100
Evapotranspiration	695	34
Surface Runoff	508	25
Lateral drainage by soil	1	0
Lateral drainage by sand layer	668	32
Lateral drainage by gravel layer	136	7
Permeation	48	2

in Germany. In fact, capillary barriers could be used in any country where the climatic conditions resemble those of Germany" (von Der Hude et al., 1999).

Suzuki et al. (2005) performed a pilot-scale test in the Northeast of Japan, where the average annual precipitation is 1707 mm with around 400 mm snowfall in winter. The performance of a capillary barrier, made up of 0.30 m thick sand and 0.25 m thick gravel layers under 0.5 mm thick native soil layer was examined. The slope of the capillary barrier was 5%. Table 11.1.5 shows the measurements obtained over a 14-month period in each layer. Permeation to the bottom of the capillary barrier type was 48 mm (2%) of precipitation, demonstrating the effectiveness of capillary barriers to reduce water infiltration into the landfill body.

Weiß and Witzsche (2005) built a control plot (368 m^2) in a German landfill in which a capillary barrier made up of 0.40 m thick sand and 0.30 m thick gravel layers under 0.5–1.5 m thick native soil layer, was used. The experiment lasted for 24 months: the average permeation into the landfill body ranged between 0 and 7.5% of the precipitation (0.97% on average).

Biocover

The most simple biocover structure is represented by that described as option D in Fig. 11.1.5. Equalization, reinforcement, and gas drainage (in this case with the role of gas distribution layer) are the same as that described for the sealing cover, but with no liner and with a biomitigation layer (indicatively 0.8–1.0 m thick). For details on biocover processes and engineering see Chapter 9.5.

Phytocapping

Phytocapping represents a valuable soil—plant alternative to traditional low-permeability top covers. It relies on the presence of a porous substrate soil, in which stored precipitation is removed either by surface evaporation or plant transpiration, thus controlling water infiltration into the landfill body (Chapter 11.2). The process of water removal due to evaporation and transpiration is known as ET.

Phytocapping activity is based on an interaction between local climate, substrate (soil), and vegetation (Fig. 11.1.6), and therefore a site-specific design is required. The ideal phytocapping should be able to store a high amount of water, matching the needs of the growing vegetation. The selection of the plant species is based on compatibility with the substrate and climatic conditions.

Phytocappings can be combined with methane biological oxidation: the use of a porous substrate ensures the presence of aerobic conditions capable of promoting microbial methane oxidation (Huber-Humer et al., 2008). Recent studies have demonstrated that plant roots enhance microbial activity, significantly improving the methane oxidation potential (Bohn et al., 2011; Reichenauer et al., 2011; Venkatraman and Ashwath, 2009a,b,c).

From extensive studies carried out in Australia over a 4-year period at several landfill sites sited under different climatic conditions, with precipitation ranging between 500 and 1000 mm, the infiltration rate was maintained below 4% (see Chapter 11.2).

Alternative/integrative options

Some activities, such as solar energy production, cultivation of crops for the production of biofuels (energy crops), phytotreatment of wastewater, etc., require extensive use of land, which might compete with traditional uses (agriculture, animal grazing, etc.). Closed landfill may render available a huge surface for the noncompetitive undertaking of these kinds of activities due to the fact that traditional activities related to food production are often forbidden at closed landfill sites.

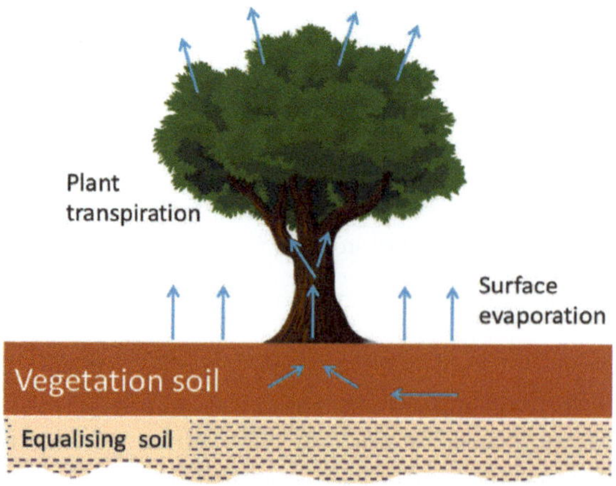

Figure 11.1.6 Example of a phytocapping cover.

Under specific conditions and restrictions, the abovementioned activities may be conveniently included in the landfill capping structures.

Phytotreatment/energy crops covers

Phytotreatment of wastewaters is a sustainable process with very low operational and maintenance costs, which is suited for the treatment of weak leachates from old landfills, or to polish leachates that have been partially pretreated by other processes. Several plant species have been tested at different scales by a series of authors and, in general, phytotreatment has displayed efficiency in the removal of the investigated contaminants. Plants treated with leachate often grew better than those irrigated with water (Lavagnolo et al., 2016). The use of plants for leachate treatment greatly increases ET compared with unvegetated sites, in which transpiration does not take place (Ettala, 1989). Phytoremediation of wastewater could be combined with the production of renewable energy, such as wood from short rotation coppice, bioethanol from lignocellulosic biomass, or biodiesel from oleaginous crops. In recent years, leachate phytotreatment using plants for biomass and/or biofuel production has been explored by the scientific community, resulting in excellent performances of pollutant removal (Garbo et al., 2017; Lavagnolo et al., 2016; Dimitriou and Aronsson, 2010; Duggan, 2005; Jones et al., 2006).

Energy crops might be grown on the top of closed landfills, not only for leachate phytotreatment purposes and for renewable energy generation but also to offer a pleasant view of the site, especially if plants with high aesthetic value are chosen (e.g., sunflowers) (Lavagnolo et al., 2016). The landfill leachate, collected and recirculated on the top of the landfill, can be phytotreated in an area with lower slope. At the same time, it is possible to save huge amounts of water that would had been necessary to irrigate these types of plants (Garbo et al., 2017).

Lavagnolo et al. (2016) performed a series of lab-scale tests in which three oleaginous plant species, *Helianthus annuus* (sunflower), *Glycine max* (soybean), and *Brassica napus* (rapeseed), grown both on sandy and clayey soil, were used to treat old landfill leachate at increasing dosages over time. Garbo et al. (2017) focused on the behavior of sunflowers grown in tanks filled with clayey soil under two different hydraulic regimes: vertical flow and horizontal subsuperficial flow. The effect of effluent recirculation was investigated too.

Significant removal efficiencies of the investigated contaminants (COD, nitrogen, phosphorous) were achieved in both studies, especially for sunflower and rapeseed growing on clayey soil, as reported in Table 11.1.6. It should be pointed out that nitrogen removal performances were excellent up until flowering point, followed by a sudden decrease, which may have been due to plant senescence. Generally, plants are known to favor the growth of microorganisms, which play a dominant role in the nitrification, by conveying oxygen to the rhizosphere (Akinbile et al., 2012). As senescence occurs, oxygen transfer to the soil by plants is reduced, thus decreasing the possibility of ammonia oxidation.

Garbo et al. (2017) also evaluated the ET capacity of phytotreatment units: an invariably significant finding was obtained under the different operating conditions, as reported in Table 11.1.7. In the majority of cases, more than 50% of influent volume was removed by ET processes, resulting in a limited volume of outflow.

Table 11.1.6 COD, nitrogen, and phosphorous removal efficiencies (RE %) in the experiment reported by Lavagnolo et al. (2016) and Garbo et al., 2017

		RE sunflower (%)		RE Soybean (%)		RE rapeseed (%)	
	Influent Concentration (mg/L)	Sandy soil	Clayey soil	Sandy soil	Clayey soil	Sandy soil	Clayey Soil
COD	55–676	>60	>80	>70	>80	>80	>80
N	27–363	>20 (>80[1])	>55 (>90[1])	>40 (>85[1])	>60 (>70[1])	>65 (>70[1])	>75 (>85[1])
P	0.5–6.7	>90	>90	>75	>65	>90	>90

[1]removal efficiency up until flowering point.

Table 11.1.7 Evapotranspiration (ET) in the experiment reported by Garbo et al. (2017). clayey soil was used

	Influent Volumes ($L/m^2/d$)	Average ET (%)
VF	4–12	68
HSSF	8–20	40
VF (with effluent recirculation)	3–12	77
HSSF (with effluent recirculation)	6–15	69

HSSF, horizontal subsurface flow; *VF*, vertical flow.

In the experiment reported by Lavagnolo et al. (2016), potential renewable energy generation was assessed. In particular, oil content amount and free fatty acids composition were analyzed and compared with values detected in sunflowers and soybeans grown in traditional ways, with leachate appearing to stimulate the accumulation of oil in the seeds.

Phytotreatment units can be installed on the landfill surface according to two different schemes as represented in Fig. 11.1.7. Phytotreatment with horizontal flow may be integrated in sealing capping structure, using soil or gravel as substrate for the plants. The effluent will be collected and discharged outside the landfill body (scheme A). Phytotreatment with horizontal flow might be more appropriate for installation in gas permeable capping and the treated water will infiltrate into the landfill body (scheme B).

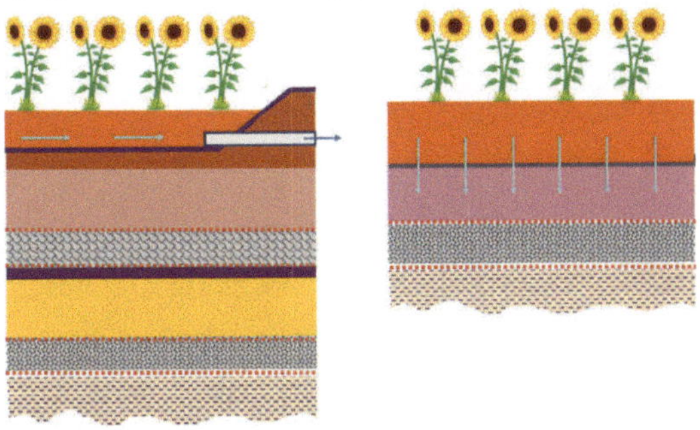

Figure 11.1.7 Different functional schemes for the use of phytotreatment in a landfill capping. A = Horizontal flow in a sealing capping, B = vertical flow in a biocover capping.

Solar energy recovery capping

The production of solar energy, including both concentrating solar power (CSP) and the PV technologies, may be implemented on top of closed landfills in the presence of a few specific favorable conditions:

- Significant land surface
- Adequate morphology and geographic position
- Climatic conditions (cloud cover, sun intensity, humidity, rainfall, snowfall, wind, etc.)
- Bearing capacity of the surface
- Mechanical stability of the landfill

With regard to the possible application of solar power production on landfills, two main categories of systems may be identified:

- Ground-mounted systems
- Surface-integrated systems

An important issue to be taken into account with ground-mounted systems is the footing support for sun capturing fixed or rotating devices. These vary (in terms of weight, foundation complexity, mechanical resistance, etc.) according to the solar power technologies used. PV technologies, the most widely used, use fixed or rotating panels, whereas CSP technologies use parabolic or linear mirrors and reflectors to concentrate and collect the solar energy.

In general, PV technologies are easier to install (Sampson, 2009). To this regard, PV systems use three different kinds of silicon cell materials in manufacturing the panels, each of which featuring different power energy efficiency, weight, and costs:

- Monocrystalline panels, black colored, are the more efficient and expensive
- Polycrystalline, blue colored, featuring a simpler production process, are less expensive but also less efficient and heavier.
- Amorphous silicon cell, noncrystalline and nonrigid, are used as thin, flexible film solar panels. Their efficiency is significantly lower than the other two materials.

Figure 11.1.8 Aerial surface view of the Hickory Ridge Landfill, Atlanta, Georgia, USA. *Courtesy of Google Earth.*

Monocrystalline and polycrystalline silicon cell panels are more rigid and need an adequate structure to prevent cracking phenomena. Indeed, installation of these panels over landfill surfaces should carefully assess differential settling and mechanical stability issues. As a consequence, despite the lower efficiency, amorphous film panels are more versatile for use in landfills.

The installation of solar energy technologies on top of closed landfills may provide significant support to energy needs during the aftercare phase. Their design life is estimated to be around 20 years (DeVita et al., 2017). One of the most significant case studies available relates to the installation of 7000 laminated panels, for a total surface of 4 ha, over the 19.4 ha geomembrane covered surface of the Hickory Ridge Landfill, Atlanta, Georgia, USA, to generate 1 MW of energy (DeVita et al., 2017). An areal view of the landfill site is given in Fig. 11.1.8.

CONCLUSIONS

Landfill cover represents an important structural work in landfill engineering, playing a fundamental role in controlling long-term emissions. The choice of the most appropriate top cover depends on waste quality, the adopted landfill concept, and landfill morphology, all of which influence the mechanical behavior of the deposit. Efficiency of landfill capping should be maintained for the time necessary to achieve environmentally sustainable conditions.

References

Akinbile, C.O., Yusoff, M.S., Zuki, A.Z.A., 2012. Landfill leachate treatment using sub-surface flow constructed wetland by *Cyperus haspan*. Waste Management 32, 1387–1393.

Albrecht, B.A., Benson, C.H., May 2001. Effect of desiccation on compacted natural clays. Journal of Geotechnical and Geoenvironmental Engineering 67–75.

Albright, W.H., Benson, C.H., Gee, G.W., Abichou, T., Tyler, S.W., Rock, S.A., 2006. Field performance of three compacted clay landfill covers. Vadose Zone Journal 5, 1157–1171.

Andreas, L., Herrmann, I., Lidström-Larsson, M., Lagerkvist, A., 2005. Physical properties of steel slag to be reused in a landfill cover. In: Sardinia 2005, Proceedings of the Tenth International Waste Management and Landfill Symposium. CISA Publisher.

Andreas, L., Diener, S., Lagerkvist, A., 2014. Steel slags in a landfill top cover—experiences from a full-scale experiment. Waste Management 34, 692–701.

Barbisan, E., Busana, S., Favaretti, M., 2013. Use of silt-clay mixtures in MSW landfills cover systems. In: Sardinia 2013, Proceedings of the Fourteenth International Waste Management and Landfill Symposium. CISA Publisher.

Barrès, M., Bonin, H., 1994. The Capillary Barrier for surface capping. In: Christensen, T.H., Cossu, R., Stegmann, R. (Eds.), Landfilling of Waste: Barriers. E & FN Spon, London, pp. 139–148. ISBN:0-419-15990-8.

Barth, C., Wohnlich, S., 1999. Proof of effectiveness of a capillary barrier as surface sealing of sanitary landfill. In: Sardinia 1999, Proceedings of the Seventh International Waste Management and Landfill Symposium, vol. III. CISA Publisher, pp. 389–392.

Bogner, J., Abdelra e Ahmed, M., Diaz, C., Faaij, A., Gao, Q., Hashimoto, S., Mareckova, K., Pipatti, R., Zhang, T., 2007. Waste management. In: Metz, B., Davidson, O.R., Bosch, P.R., Dave, R., Meyer, L.A. (Eds.), Climate Change Mitigation. Contribution of Working Group III to the Fourth Assessment Report of the Intergovernmental Panel on Climate Change. Cambridge University Press, Cambridge, United Kingdom and New York, NY, USA.

Bohn, S., Brunke, P., Gebert, J., Jager, J., 2011. Improving the aeration of critical fine-grained landfill top cover material by vegetation to increase the microbial methane oxidation efficiency. Waste Management 31, 854–863.

Burkhardt, G., Egloffstein, T.H., Humeniuk, A., 2009. Landfill liner systems – experiences in different countries and under different climatic conditions. In: Sardinia 2009, Proceedings of the Twelth International Landfill Symposium. CISA Publisher. ISBN:9788862650076.

Caldwell, J.A., Edward Kavazanjian, E., Abid Loan, A., Suji Somasundaram, S., 2000. Design and Construction of a Geogrid-Reinforced Cover for a Hazardous Waste Landfill. Accessed from Internet: http://www.infomine.com/library/publications/docs/caldwell2000.pdf.

Camp, S., Gourc, J.P., Ple, O., Villard, P., Rey, D., 2005. Landfill cap cover issue: improvement of the capability to sustain differential settlements. In: Sardinia 2005, Proceedings of the Tenth International Waste Management and Landfill Symposium. CISA Publisher.

Carson, D.A., 1992. Municipal solid waste landfill daily cover alternatives. Geotextiles and Geomembranes 11 (4–6), 629–635.

Cossu, R., Muntoni, A., Cappai, G., 2000. Tecniche di copertura finale delle discariche e confronto tra alternative. In: Seminario Internazionale La progettazione delle nuove discariche e la bonifica di quelle vecchie, Padova.

Council Directive 1999/31/EC on the Landfill of Waste. Accessed from Internet: https://www.google.it/search?q=Council+Directive+1999%2F31&ie=utf-8&oe=utf-8&client=firefox-b&gfe_rd=cr&dcr=0&ei=C3yVWuvINLTBXoDQksAF.

Dayal, U., Gardner, J., Chiado, E.D., 1994. Design consideration of a new liner system over an existing landfill. In: Christensen, T.H., Cossu, R., Stegmann, R. (Eds.), Landfilling of Waste: Barriers. E & FN Spon, London, pp. 169–182. ISBN:0-419-15990-8.

DeVita, J., Kanishka, Perera, Roberts, M., 2017. What is a solar energy land cover system? Waste. Accessed from Internet: www.hdrinc.com/sites/default/files/2017-06/InsideWaste-Solar-Energy-Landfill-Cover-System-Judy-DeVita.pdf.

Dimitriou, I., Aronsson, P., 2010. Landfill leachate treatment with willows and poplars – efficiency and plant response. Waste Management 30, 2136–2145.

Duggan, J., 2005. The potential for landfill leachate treatment using willows in the UK – a critical review. Resources, Conservation and Recycling 45, 97–113.

LIFE96 ENV/D/000197. Capillary Barrier for Landfill—Construction of a Capillary Barrier for the Surface Sealing System of a Landfill. Accessed from Internet: http://ec.europa.eu/environment/life/project/Projects/index.cfm?fuseaction=search.dspPage&n_proj_id=732.

Ettala, M.O., 1989. Vegetation and hydrological aspects in landfill management. In: Sardinia '89 Proceedings, Second International Landfill Symposium, pp. 1–12.

Garbo, F., Lavagnolo, M.C., Malagoli, M., Schiavon, M., Cossu, R., 2017. Different leachate phytotreatment systems using sunflowers. Waste Management 59, 267–275.

Gourc, J.P., Nousheen, A.K., Olivier, F., 2007. Long term settlement of domestic waste in landfill: ISPM method. In: Grenoble, 18ème Congrès Français de Mécanique.

Huber-Humer, M., Gebert, J., Hilger, H., 2008. Biotic systems to mitigate landfill methane emissions. Waste Management & Research: The Journal of the International Solid Wastes and Public Cleansing Association, ISWA 26 (1), 33–46.

Ivanova, L.K., Richards, D.J., Smallman, D.J., 2008. The long-term settlement of landfill waste. Proceedings of the Institution of Civil Engineers-Waste and Resource Management 161 (3), 121–133. Thomas Telford Ltd.

Jones, D.L., Williamson, K.L., Owen, A.G., 2006. Phytoremediation of landfill leachate. Waste Management 26, 825–837.

Kämpf, M., Montenegro, H., 1997. Assessing the effectiveness of capillary barriers. In: Sardinia 1997, Proceedings of the Sixth International Landfill Symposium, vol. III. CISA Publisher, pp. 421–428.

Lavagnolo, M.C., Malagoli, M., Garbo, F., Pivato, A., Cossu, R., 2016. Lab-scale phytotreatment of old landfill leachate using different energy crops. Waste Management 55, 265–275.

DGR Lombardia n. X/2461/2014 (Lombardy Region Decree X/2461). Linee guida per la progettazione e gestione sostenibile delle discariche. Accessed from Internet: http://download.acca.it/BibLus-net/OpereEdili/LineeGuida_Lombardia_Discariche.pdf.

Miller, C.J., Mishra, M., 1994. Failure mechanisms for clay cover liners. In: Christensen, T.H., Cossu, R., Stegmann, R. (Eds.), Landfilling of Waste: Barriers. E & FN Spon, London, pp. 149–156. ISBN:0-419-15990-8.

Parent, S.É., Cabral, A., 2006. Design of inclined covers with capillary barrier effect. Geotechnical and Geological Engineering 24 (3), 689–710.

Rajesh, S., Viswanadham, B.V.S., January 2012. Centrifuge modeling and instrumentation of geogrid-reinforced soil barriers of landfill covers. Journal of Geotechnical and Geoenvironmental Engineering, American Society of Civil Engineers (ASCE) 138 (1). https://doi.org/10.1061/(asce)gt.1943-5606.0000559.

Reichenauer, T.G., Watzinger, A., Riesing, J., Gerzabek, M.H., 2011. Impact of different plants on the gas profile of a landfill cover. Waste Management 31, 843–853.

Richardson, G.N., Thiel, R.S., Marr, W.A., 2000. Lessons Learned From the Failure of a GCL/Geomembrane Barrier on a Side Slope Landfill Cover. Accessed from Internet: http://www.geocomp.com/files/technical_papers/Lessons LearnedfromtheFailureofaGCL-GeomenbraneBarrieronaSideSlope1999_WEB.pdf.

Sampson, G., 2009. Solar power installations on closed landfills: technical and regulatory considerations. In: Research Report for US EPA Office of Solid Waste and Emergency Response, Office of Superfund Remediation and Technology Innovation. Accessed from Internet: https://clu-in.org/download/studentpapers/Solar-Power-Installations-on-Closed-Landfills-Sampson.pdf.

Suzuki, M., Imai, J., Saita, Y., Nishigaki, M., 2005. Substantiations and analysis for the capping with the capillary barriers. In: Sardinia 2005, Proceedings of the Tenth International Waste Management and Landfill Symposium. CISA Publisher.

Tham, G., Mellström, A., Sjöblom, R., Lagerkvist, A., Andreas, L., 2005. Utilization of secondary construction materials in a landfill cover system. In: Sardinia 2005, Proceedings of the Tenth International Waste Management and Landfill Symposium. CISA Publisher.

Travar, I., Andreas, L., Tham, G., Lagerkvist, A., 2005. Field test of landfill covers with secondary construction materials. In: Sardinia 2005, Proceedings of the Tenth International Waste Management and Landfill Symposium. CISA Publisher.

Venkatraman, K., Ashwath, N., 2009a. Can phytocapping technique reduce methane emission from municipal landfills? International Journal of Environmental Technology and Management 10, 44–55.

Venkatraman, K., Ashwath, N., 2009b. Phytocapping - can it reduce percolation of water into buried waste?. In: Sardinia 2009, Proceedings of the Twelth International Landfill Symposium. CISA Publisher. ISBN:9788862650076.

Venkatraman, K., Ashwath, N., 2009c. Phytocapping: importance of tree selection and soil thickness. Water, Air & Soil Pollution: Focus 9, 421–430.

von Der Hude, N., Melchior, S., Mockel, S., 1999. Construction of a capillary barrier in the cover of the Breinermoor landfill. In: Sardinia 1999, Seventh International Waste Management and Landfill Symposium, vol. III. CISA Publisher, pp. 393–402.

Walhström, M., Laine-Ylijoki, J., Vahanne, P., 2009. Materials for Construction of Top Cover in Landfills. Experience in the Nordic Countries. Norden, Copenhagen. ISBN:9789289318860.

Weiß, J., Witzsche, A., 2005. Results of the efficiency of the capillary barrier system at the landfill Grix Offenbach-Germany. In: Sardinia 2005, Proceedings of the Tenth International Waste Management and Landfill Symposium. CISA Publisher.

11.2

PHYTOCAPPING OF LANDFILLS

Melissa Salt, Sam T.S. Yuen, Nanjappa Ashwath, Jianlie Sun, Pia Benaud, Grant X. Zhu, Mark B. Jaksa, Hossein Ghadiri, Margaret Greenway and Andy B. Fourie

INTRODUCTION

Phytocapping presents a sustainable alternative to the conventional top barrier cover design (see Chapter 11.1). It relies on the capacity of a porous substrate (usually of locally available soil) to store water together with the natural processes of surface evaporation and plant transpiration to remove the stored water as a means of controlling water ingress into the landfill. They are thus often appropriately called evapotranspiration (ET) covers, soil—plant covers, store-and-release covers, or monolithic covers, as they rely on the capacity of the soil layer to "absorb" water and the plant community acting as biological "pump" to remove the stored water. The term phytocap is in predominant use in Australia due to its inclusion of phyto (the Latin prefix for plant), which emphasizes the importance of the plant-based element of the system.

In contrast to impermeable or low permeability barriers, the performance of phytocaps is expected to improve over time as the vegetation community matures. This advantage, together with the potential for phytocaps to enhance the ecological value of a site, gives phytocaps the potential for greater long-term performance and sustainability.

As the conventional barrier covers commonly include drainage layers aiming to reduce the hydraulic head acting on barriers to minimize percolation, their design is therefore inherently more complex and costly. The construction cost of phytocaps has been reported to be lower, typically at only 35%—72% of the conventional covers (Hauser, 2009).

Using phytocaps to oxidize methane and reduce greenhouse gas emissions offers another major advantage over conventional impermeable covers. It has been demonstrated that porous biotic cover systems can mitigate landfill gas (LFG) emissions by creating favorable aerobic environment to promote microbial methane oxidation in soil covers (see Chapter 9.5). The methane oxidation potential of phytocaps can be considered as a type of biotic cover where microbial activity is enhanced by plant roots. Recent laboratory-scale and field experiments have demonstrated that plant cover could significantly improve soil methane oxidation potential (e.g., Reichenauer et al., 2011; Bohn et al., 2011; Venkatraman and Ashwath, 2009).

Phytocap functionality relies on the inherent properties and interaction between the local climate, the substrate (soil), and the established plant community. Due to the reliance on local site characteristics, the design of phytocaps is necessarily specific to each landfill. When designing a phytocap, it is important to transfer the phytocap design methodology rather than a site-specific design. The ideal phytocap substrate is one of high water storage capacity with properties that promote sustained growth of the plant community. However, as the choice of substrate available locally is often limited, the thickness and quality of the soil can be manipulated to provide the required critical storage capacity during seasons when transpiration rates are low. Phytocap design considerations have been covered in detail by Hauser (2009) and Albright et al. (2010).

The selection of plant species relies on the adaptability of the chosen species to local climate, soil substrate, and their inherent strategies to alter transpiration rates in accordance with the moisture availability in the phytocap soil. Site assessment would involve defining broad climatic characteristics from historical data, characterizing the soils used for both hydraulic and plant growth parameters, and selecting suitable native plant species/communities. Another core selection criterion for plant selection is the inclusion of diverse group of plant species that would complement each other to ensure that the established species will survive and continue to regulate transpiration rates.

AUSTRALIAN ALTERNATIVE COVER ASSESSMENT PROGRAM

Australian Alternative Cover Assessment Program (A-ACAP) was a field and laboratory research program (2006—11) that was established to investigate phytocap alternatives to conventional landfill covers in the Australian context (Wong, 2007). The major goals were to demonstrate that phytocaps can perform to the satisfaction of regulators and to develop guidelines for their application, design, and construction. The guidelines aim to address the following: (1) control of percolation of water into the waste; (2) reduction of greenhouse gas emissions by methane oxidation; and (3) long-term sustainability of vegetative covers comprising a diverse range of native plant species.

The program established five full-scale test facilities across Australia to investigate the effects of a wide range of bioclimatic conditions. From tropical in the country's north to temperate in the south, the test facilities were located across all five mainland states of the country (Queensland, New South Wales, Victoria, South Australia, and Western Australia) representing an excellent bioclimatic diversity. Fig. 11.2.1 shows the test sections installed at each of the five sites and the climate for each location. The full-scale test facilities were constructed between 2007 and 2008.

Central to the project's experimental approach is the use of side-by-side comparisons of both conventional covers and candidate phytocaps (in Victoria, South Australia, and Queensland). Large-scale (10×20 m) lysimeters together with other instrumentation were used to assess their hydrological performance. As an important improvement to similar studies conducted in the past (Albright et al., 2006), all test facilities were placed directly on top of active landfills, and additional unlined test sections (i.e., without lysimeters) allowed the field experiment to investigate the methane oxidation potential of phytocaps in reducing landfill greenhouse emissions. The field program was supplemented by laboratory and glasshouse experiments to investigate the changes in soil structure over successive wetting and drying cycles, performance of native species, as well as LFG transport related to methane oxidation (Sun et al., 2011). A detailed description of the test cells, lysimeters, and field instrumentation is provided in Salt (2013).

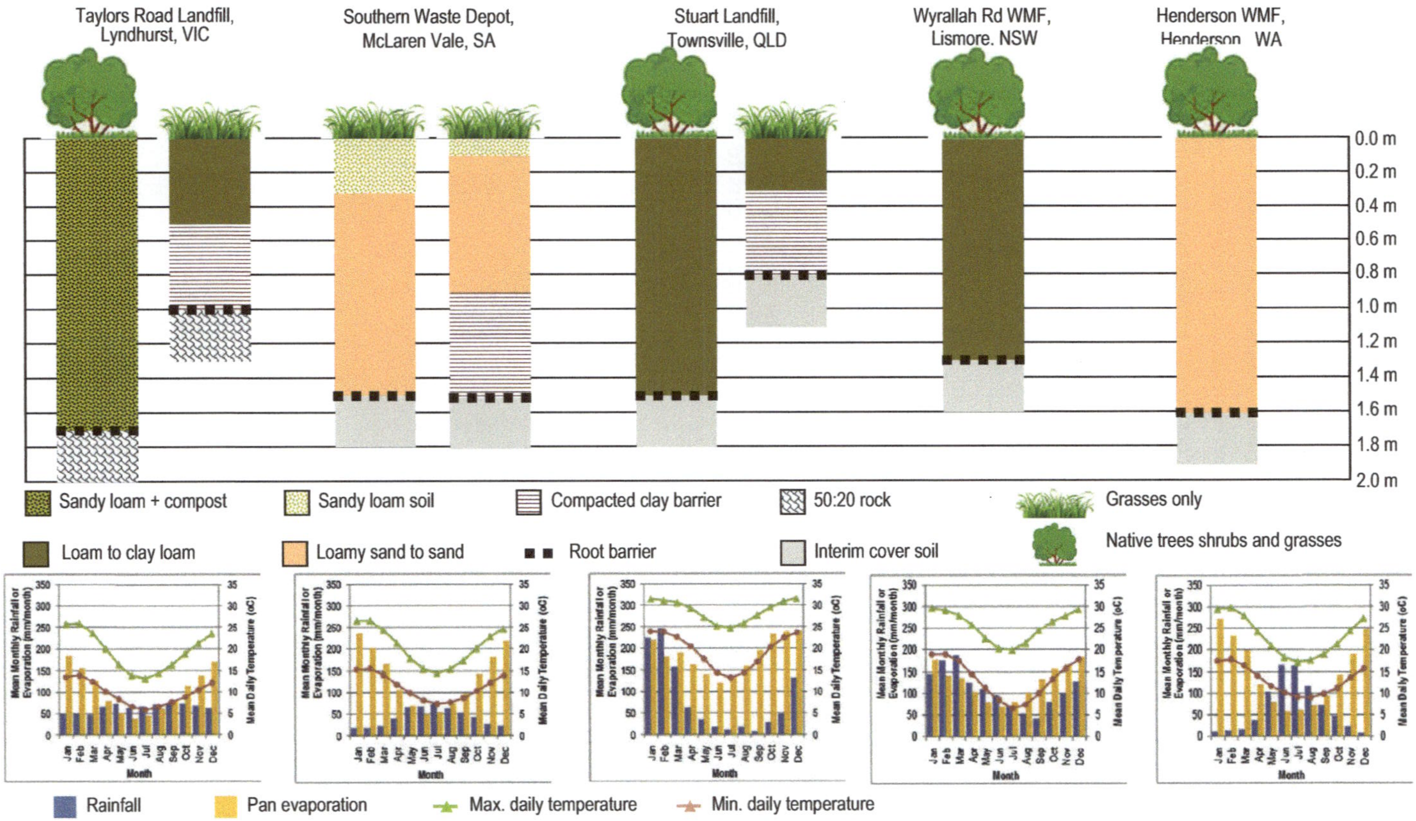

Figure 11.2.1 Schematic of capping profiles constructed and climate graphs at the five trial sites.

STUDY FINDINGS

Hydrological Performance—Phytocaps Versus Conventional Covers

The hydrological performance of the landfill covers has been measured for 3–6 years at all sites. Drainage was recorded at all sites over the monitoring period, as shown in Table 11.2.1, although in the case of the Southern Waste Depot trial site, irrigation was also applied to stress the cover systems. For most sites, the drainage from the phytocaps and conventional covers was <5% of precipitation received, including the Stuart Landfill that has a monsoonal tropical climate. The highest drainage was recorded at the Henderson Waste Recovery Park, with over 300 mm measured, or 15% of the rainfall received in the 3-year monitoring period.

The monitoring periods commenced either before or shortly after vegetation was established. As a result, the monitoring data include periods with little or no vegetation on the experimental plots. Focusing on the final complete year of monitoring showed that drainage for all sites was ≤4% of the precipitation received (Table 11.2.2). The precipitation received at most sites in the final year was above the long-term average. Despite this, the drainage was still low for the sites, including Stuart Landfill at Townsville that received 3 times the average rainfall in 2010. The conventional drainage was less than the phytocap drainage in 2010 for the southern trial sites (Taylors Road and Southern Waste Depot); but by 2012, the drainage from the conventional cover was more than the phytocap. At the northern sites, conventional cover drainage was more than the phytocap at the northern site (Stuart Landfill). For the Henderson Waste Recovery Park, the drainage appears to have decreased significantly, as the rainfall received at this site in 2010 was one of the lowest on record.

Table 11.2.1 Summary of total precipitation and drainage measured over the monitoring period

Site	Monitoring Period (dd/mm/yr)[1]	Precipitation (P)	Phytocap		Conventional Cover			
			Drainage		Lateral Flow[2]		Drainage	
		Mm	mm	%P	mm	%P	mm	%P
Taylors Rd Landfill	16/02/07 to 17/4/11	3783.4	78.9	2.1	46.0	1.2	21.4	0.6
Southern Waste Depot	05/07/07 to 09/02/13	3043.0	34.2	1.1	10.6	0.3	8.0	0.3
Stuart Landfill	01/01/08 to 07/08/11	7942.1	332.4	4.2	73.2	0.9	281.9	3.5
Wyrallah Rd WMF	01/01/08 to 31/07/11	747.3	1.4	0.2	Not constructed			
Henderson WMF	15/02/08 to 08/06/11	2294.2	336.7	14.7	Not constructed			

[1]Monitoring commenced prior to 1/1/08 for Stuart Landfill and Wyrallah Rd; however, data were incomplete.

[2]Later flow is the horizontal flow collected from the interface between the overlying soil and the compacted clay barrier.

Table 11.2.2 Long-term average annual precipitation and total precipitation and drainage measured during the final year of measurement

Site	Average Precipitation (mm/yr)	Final Year	Final Year Precipitation (P) (mm)	Phytocap Drainage (mm)	Phytocap Drainage (%P)	Conventional Cover Lateral Flow[1] (mm)	Conventional Cover Lateral Flow[1] (%P)	Conventional Cover Drainage (mm)	Conventional Cover Drainage (%P)
Taylors Rd Landfill	801	2010	1185.4	8.9	0.8	43.2	3.6	1.3	0.1
Southern Waste Depot	520	2010	623.6	10.7	1.7	1.0	0.1	0	0
		2012	605.6	3.6	0.6	2.9	0.5	6.5	1.1
Stuart Landfill	990	2010	2829.6	62.9	2.2	19.7	0.7	107.5	3.8
Wyrallah Rd WMF	1340	2010	1271.3	8.5	0.7	Not constructed			
Henderson WMF	790	2010	475.2	15.3	3.2	Not constructed			

[1]Monitoring at Southern Waste Depot continued following completion of the A-ACAP monitoring period; 2010 and 2012 data have been shown.

The hydraulic performance of the phytocaps at the five trial sites showed that drainage tends to decrease over time, whereas the drainage from the conventional covers tends to increase. The former appears to be related to the slower rate of canopy build up in the phytocaps, whereas the latter occurs due to drying and desiccation of the compacted clay. Some variability was noted for the drainage from the conventional covers, which was caused by the rate at which the wetting front moved through the soil prior to the clay wetting up.

The establishment of grasses in the phytocap appeared to result in decreased drainage occurring sooner, as observed at Taylors Road Landfill and Southern Waste Depot, but not at the Henderson Waste Recovery Park. This illustrates the importance of establishing a functionally diverse plant species both to regulate transpiration rates and to ensure long-term survival of the established vegetation. It also confirms studies by Zhang et al. (2001), which suggested that ET from grasslands was similar to treed communities in rainfall areas $\leq$ 500 mm/year, approximately.

It is apparent from these results, collected in the Australian context, that the performance of a phytocap and a conventional cap is related to the seasonality of precipitation more than the ratio between precipitation and ET that has been suggested in some previous studies (Hauser and Gimon, 2001). For example, although significantly less rainfall was received at the southern trial sites (Taylors Road Landfill, Southern Waste Depot, and Henderson Waste Recovery Park) compared with the northern sites (Lismore Waste Facility and Stuart Landfill), drainage represented a similar proportion of precipitation. This appears to be related to precipitation predominantly coinciding with cooler temperatures at the southern sites, whereas the precipitation coinciding with higher temperatures, and hence higher ET, in the northern sites. In addition, the impact on drainage through cracking of clay in the conventional covers is more pronounced with intense storm events, such as that occurs in tropical and subtropical environments of Queensland and New South Wales, respectively.

Plant Study

The performance of a wide range of native plant species selected for the five phytocap trial sites has been monitored throughout the duration of the study, with some sites continuing to be periodically photographed to the present. This variety of vegetation provided valuable information on the practices of growing native species on phytocapping systems, including species selection, planting and establishment techniques, weed control, and monitoring. The established plant communities have been subjected to seasonally drier and wetter conditions.

Plant survival and growth, in terms of species number and change in height and canopy cover, have been used as primary indicators of each species' performance and its tolerance to the landfill cover conditions to which they have been exposed. In addition, root depths, rainfall interception, stem flow, and soil nutrients were monitored at selected sites. Using the results of these studies, the A-ACAP program ascertained which of the species selected in the trials would have the best potential to perform well in future full-scale phytocap plantings under a wide range of agroclimatic conditions (WMAA, 2011).

Vegetation mortality was relatively high during plant establishment, with the exception of Southern Waste Depot where native grasses only were planted. This mortality was compensated by replanting and supplementary irrigation at the affected sites. Overall, plant growth (canopy cover) was slow for up to 1 year of establishment, and by second and third year, the vegetation structure had stabilized across all sites. As an illustration, Fig. 11.2.2 shows the survival of species at Stuart Landfill and Fig. 11.2.3 shows the growth of a selection of tree species from one of the trial sites (Wyrallah Road WMF), demonstrating an excellent survival and relatively slow growth in the first 12 months following planting.

During the monitoring period, vegetation at Stuart Landfill and Southern Waste Depot was placed under climatic stress. At Stuart Landfill, a Category 5 cyclone was experienced in February 2011, which caused only minimal damage due to the presence of a diverse range of plant species, and then by a bushfire in August 2011, which burnt the conventional cover plots but only singed the edges of the phytocap vegetation community. Unfortunately, the fire destroyed all the monitoring equipment, and data immediately prior and during the fire are not available. At the Southern Waste Depot trial site, native grasses were able to fully recover within 2 years after dieback in 2009, which was likely to be caused by a combination of a very dry summer and herbicide application. As indicated by the soil moisture content changes, the roots had returned to 700 mm following two winter periods, as shown in Fig. 11.2.4.

In addition, subsoil studies carried out at all sites revealed that plant roots had reached over 100 cm into the soil profile. More detailed studies were undertaken at Taylors Rd Landfill, Wyrallah Rd

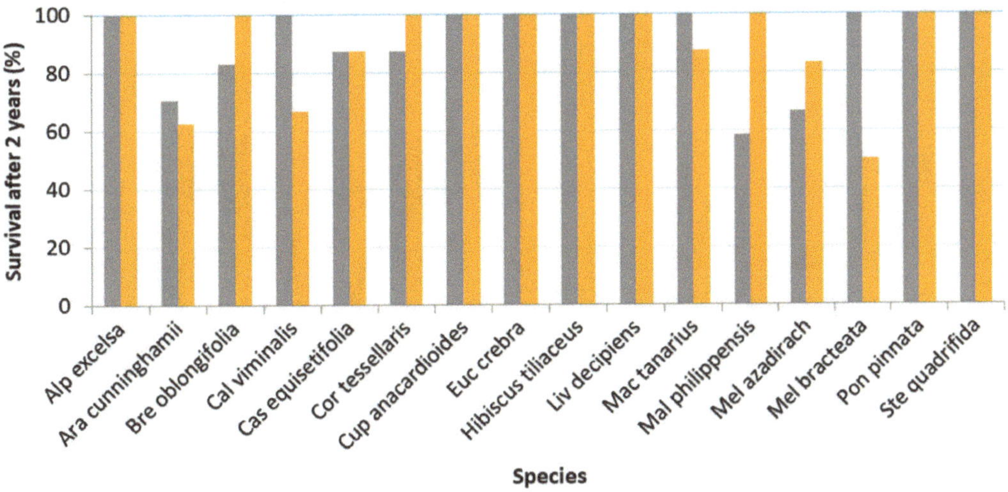

Figure 11.2.2 Survival percentage (at 2 years after planting) of various tree and shrub species established in phytocap control (first bar) and phytocap lysimeter (second bar). *Alp,* Alphitonia; *Ara,* Araucaria; *Bre,* Breynia; *Cal,* Callistemon; *Cas,* Casuarina; *Cor,* Corymbia; *Cup,* Cupaniopsis; *Euc,* Eucalyptus; *Liv,* Livistona; *Mac,* Macaranga; *Mal,* Mallotus; *Mel,* Melaleuca; *Pon,* Pongamia; *Ste,* Sterculia.

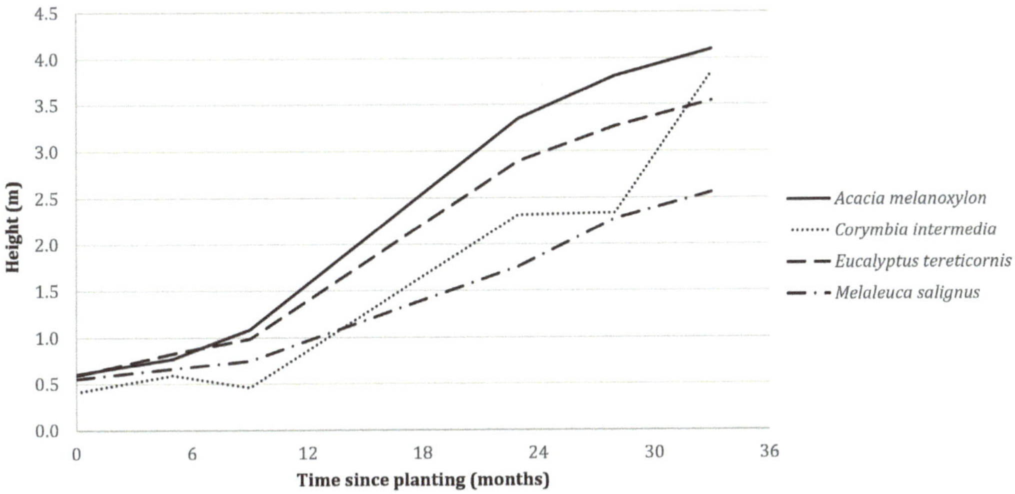

Figure 11.2.3 Mean height of selected species at Wyrallah Road WMF since April 2008.

WMF, and Henderson WMF; root density was found to be greater in the top half of the soil profile when compared with the lower half.

Toward the end of 3 years of establishment, the phytocapped sites were fully revegetated, plants showed up to 100% survival (Fig. 11.2.2) and produced up to 80% canopy cover, with the exception

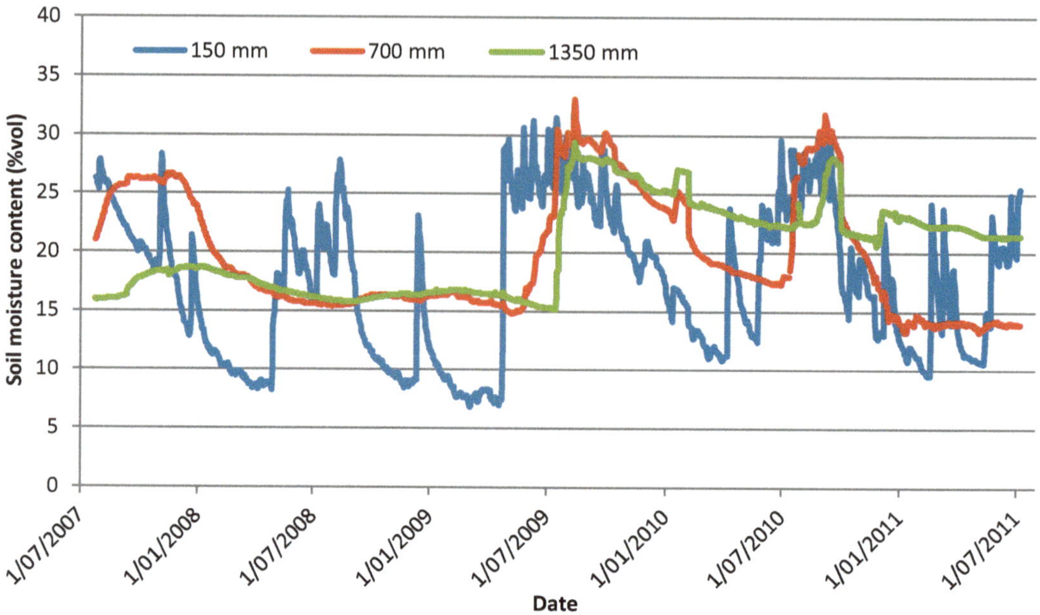

Figure 11.2.4 Average soil moisture content at southern waste depot. Rapid movement moisture to 700 and 1350 mm is noted after plant dieback in 2009, but moisture depletion in subsequent summers shows roots have grown back to 700 mm within 2 years.

of Henderson Waste Recovery Park where vegetation establishment has been slower, due to less favorable environmental and management conditions. Benaud (2010) provides details for mortality and growth for each of the species across sites, with data for Stuart Landfill reported by Ashwath (2010).

Methane Oxidation

A number of methods were deployed to estimate LFG oxidation and emissions as follows:

- Static flux chambers were employed to monitor methane emissions from phytocaps and conventional covers in the field trials.
- Soil gas probes were used to investigate soil gas profiles in both covers.
- A stable 11.2-carbon isotopic technique was also employed to quantify oxidation in chambers and probes.
- A glasshouse experiment was conducted under controlled conditions to support the field investigation.

Sun et al. (2011) provided a summary of their findings. The results indicate that vegetation can alter the soil physical properties to enhance oxygen availability, which will likely increase the methane oxidation capacity of the phytocaps. In addition, given soil moisture is an important factor for methane oxidation, the rates of methane oxidation and emission would be temporal and seasonal in a phytocap when considering the changing soil moisture due to transpiration in addition to surface evaporation.

Venkatraman and Ashwath (2009) found significant reduction in methane concentrations under certain trees compared with the other trees indicating that the presence of certain species can oxidize methane more efficiently than the others.

PHYTOCAP GUIDELINES AND AUSTRALIAN STATE-BASED REGULATIONS

Based on the research from the A-ACAP, the Waste Management Association of Australia (WMAA) released in 2011 the "Guidelines for the Assessment, Design, Construction and Maintenance of Phytocaps as Final Covers for Landfills" (WMAA, 2011).

The guidelines summarized the findings of the A-ACAP and provide guidance for regulators, site owners, and consultants on sustainable phytocapping on proceeding (see in Fig. 11.2.5).

New South Wales EPA's "Environmental Guidelines for Solid Waste Landfills" released in 2016 reference the WMAA Phytocap Guidelines for advice on phytocaps for landfills and provide a performance criterion for drainage from the phytocap of 5% of rainfall. Phytocaps have now been permitted at a number of sites including landfills at Lismore and Armidale. Further site owners are known to be investigating phytocapping options, particularly sites west of the Great Dividing Range.

The remaining states and territories have not released guidelines since the release of the WMAA Guidelines, though none preclude the use of phytocaps. EPA Victoria released their guidelines prior to completion of the A-ACAP project and has determined a stringent drainage requirement, which is equivalent to 75% of the anticipated seepage rate through a best practice base liner, and also requires lysimeter-based trials to be undertaken before approval is finalized. Regardless, phytocaps have been designed and/or constructed at a number of metropolitan and regional landfill sites, including Hanson's Wollert Landfill (which was a precursor to the A-ACAP trials), in Victoria, Queensland, and South Australia.

PHYTOCAPPING IN PRACTICE

Subsequent to the A-ACAP, a number of authors have been involved in further trials and full-scale construction of phytocaps. From this experience, the following recommendations may be made to assist in designing and constructing phytocaps.

Design

- Use a consultant experienced in phytocapping design and with knowledge of soil and plant science—this is a specialized science field and is required in addition to geotechnical and civil engineering;
- In addition to regulatory requirements, the design must consider the sensitivity of performance to the inputs, the availability of suitable soil materials and plant species, and the constructability and maintenance requirements of the constructed cap;
- Plant species selected for the phytocap need to consider the potentially limited rooting depth and the variability of the weather. For example, plants with dense canopy ("umbrella shaped," e.g., *Hibiscus tiliaceus*) are more susceptible to cyclones and strong winds than those having sparse canopy and smaller leaves (e.g., casuarinas, melaleucas, and pines);

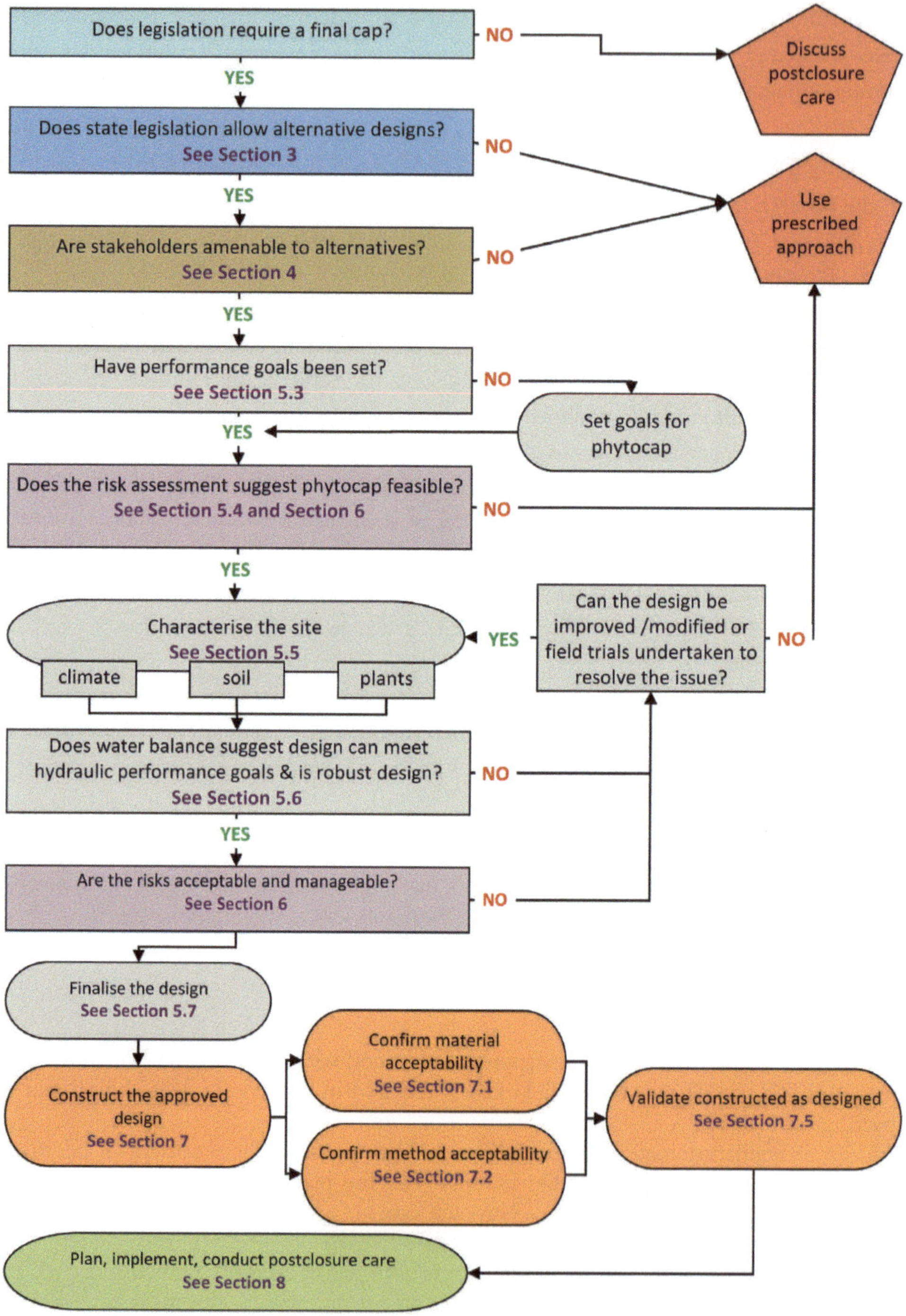

Figure 11.2.5 Decision tree for phytocaps (WMAA, 2011).

- Species and a mixture of species that have the ability to regulate their transpiration rates and complement each other in canopy growth must be considered. Plants facilitate this regulation via reduced transpiration rate, loss of leaves, survival under extreme desiccation due to production of lignotubers, resilient roots, or other structures that will help them regenerate on the onset of wet season.
- Plants that provide ecosystems services, such as the habitats for local fauna, possess conservation values (rare and threatened species), improve esthetic value of the site, and meet certain needs of the local community (e.g., bushwalking, picnics, bicycle ride) that should be included in the phytocaps.

Construction

- Prior to construction, consider constructing a trial pad to develop a methodology for placement that meets the specification. If constructed in advance of works, the trial pad may also be used to determine settlement so soil materials can be placed with little or no compaction but self-weight and rainfall assist in settlement. In trials conducted with a range of materials, the soil profile can be placed 10%–20% thicker than designed and settles to the designed thickness;
- Maximize lift thickness as far as practical;
- Supervise all works! "Corner cutting" by poorer quality earthwork contractors includes not ripping compacted haul road; planting tubestock without removing the plastic container; using valuable topsoil materials to construct stormwater contour banks; tube placing; not undertaking week control practices in a timely manner, which results in establishment of unwanted species and/or planted species being accidentally mowed or slashed;
- In designing the setout of trees and shrubs, ensure sufficient interrow space to allow for vehicles and maintenance equipment;
- Understand which "pest" is likely to be an issue at the site, e.g., kangaroos (which love eating Allocasuarina), rabbits (which ringbark anything), or various insect pests (which can reduce plant survival or health);
- Construction costs have been estimated to be 30%–70% lower for phytocapping compared with conventional covers but is highly dependent on the availability and haulage distance of soil materials;
- Plant vegetation at the best time of year to maximize competitiveness and survival;
- Rather than using irrigation (which is expensive), consider planting in contour rip-lines, mulching, and other techniques to minimize runoff and maximize water retention.

Maintenance

- Weed control is very important, particularly during plant establishment. Contract a specialist contractor to ensure weed control is appropriate and effective;
- Replace lost plants as soon as practical to limit invasion by weeds.

CONCLUSIONS

The results obtained from the A-ACAP study strongly suggest that phytocaps can provide a cost-effective and sustainable alternative when compared with traditional top barriers. The obvious advantages are their lower cost, utilizing available resources (i.e., use of local soils and native plants), high ecological site improvement, and greenhouse emission reduction potential.

The design of phytocaps is necessarily specific to each landfill and requires specialized skills in natural systems as the phytocap functionality relies on the inherent properties and interactions between the local climate, substrate, and the established plant community. Construction and maintenance of phytocaps requires less technical skills and engineering infrastructure; such covers could have the potential of making a significant improvement in the way developing countries are capping the waste disposal sites to address their technical and financial constraints (Yuen et al., 2010).

As a result of this Australian study, the WMAA has produced in 2011 a document entitled "Guidelines for the Assessment, Design, Construction and Maintenance of Phytocaps as Final Covers for Landfill" and phytocaps have now been permitted at a number of sites.

Acknowledgments

The A-ACAP program is co-funded by the Australian Research Council and the Waste Management Association of Australia and is conducted in collaboration with the University of Melbourne, the University of Adelaide, Griffith University, Central Queensland University, and the University of Western Australia.

References

Albright, W.H., Benson, C.H., Gee, G.W., Abichou, T., Tyler, S.W., Rock, S.A., 2006. Field performance of three compacted clay landfill covers. Vadose Zone Journal 5 (4), 1157–1171.

Albright, W.H., Benson, C.H., Waugh, W.J., 2010. Water Balance Covers for Waste Containment: Principles and Practice. ASCE Press, Reston, USA.

Ashwath, N., 2010. Stuart Landfill Phytocapping Monitoring Report to A-ACAP Committee. CQUniversity, Rockhampton.

Benaud, P., 2010. A Comparative Study of Vegetation Performance on Landfill Phytocovers in Australia. Honours thesis of B.Sc. Environment. Griffith University, Queensland, Australia.

Bohn, S., Brunke, P., Gebert, J., Jager, J., 2011. Improving the aeration of critical fine-grained landfill top cover material by vegetation to increase the microbial methane oxidation efficiency. Waste Management 31 (2011), 854–863.

Hauser, V.L., 2009. Evapotranspiration Covers for Landfills and Waste Sites. CRC Press, Roca Raton, USA.

Hauser, V.L., Gimon, D.M., 2001. Vegetated landfill covers and phytostabilization – the potential for evapotranspiration – based Remediation at Air Force bases. In: Air Force Center for Environmental Excellence Technology Transfer Division, Brooks, Texas, USA.

Reichenauer, T.G., Watzinger, A., Riesing, J., Gerzabek, M.H., 2011. Impact of different plants on the gas profile of a landfill cover. Waste Management 31 (2011), 843–853.

Salt, M., 2013. Water Balance and the Influence of Temporal Factors on Final Covers for Landfill Closure. Doctoral Thesis submitted to School of Civil, Environmental and Mining Engineering, University of Adelaide June 2013.

Sun, J., Yuen, S.T.S., Bogner, J., Chen, D., 2011. Investigation of phytocap methane oxidation and emission: a full scale field trial and a controlled glasshouse experiment. In: Proceedings of the Thirteenth International Waste Management and Landfill Symposium in Sardinia, Italy, 5–9 October 2011.

Venkatraman, K., Ashwath, N., 2009. Can phytocapping technique reduce methane emission from municipal landfills? International Journal of Environmental Technology and Management 10, 44–55.

WMAA, 2011. Guidelines for the assessment, design, construction and maintenance of phytocaps as final covers for landfills. In: Waste Management Association of Australia. October, 2011.

Wong, J.S.C., Yuen, S.T.S., Davey, R.B., Baker, A.J.M., Bogner, J., 2007. The Australian alternative covers assessment program (A-ACAP). In: Proceedings of the Eleventh International Waste Management and Landfill Symposium in Sardinia, Italy, October 2007.

Yuen, S.T.S., Michael, R.N., Salt, M., Jaksa, M.B., Sun, J., 2010. Phytocapping as a cost-effective cover option for waste disposal sites in developing countries. In: International Conference on Sustainable Built Environment Proceedings 2010 (ICSBE-2010), Kandy, Sri Lanka, 13-14 December 2010, Vol 1, pp. 145–210.

Zhang, L., Dawes, W.R., Walker, G.R., 2001. Response of mean annual evapotranspiration to vegetation changes at catchment scale. Water Resources Research 37, 701–708.

12 CHAPTER

LEACHATE RECIRCULATION

12.1

LEACHATE RECIRCULATION: HISTORY, OBJECTIVES, AND CONCEPTUAL DESIGN

Keith Knox, Richard P. Beaven and Raffaello Cossu

INTRODUCTION

Since the 1970s leachate recirculation has been practised at many landfills worldwide. It has been used for a wide range of purposes. Initially it was most often viewed as a cheap and easy means of temporary disposal of leachate, utilizing absorptive capacity of dry solid wastes. Even today, one of its most common uses is for the management of flow peaks, to regulate the volumes routed to treatment and disposal.

Stricter regulations evolved in many countries during the 1980s, 1990s, and 2000s that discouraged any additions of liquid to landfills and encouraged a "dry tomb" approach. This led to a decline in use of recirculation in some countries. For example, in the United States the dry tomb landfill concept was incorporated in the US EPA's Resource Conservation and Recovery Act (RCRA), Subtitle D (1976), and dominated landfill practices for the next 20–30 years. This legislation sought to prevent rainfall infiltration by the installation of impermeable top covers to prevent leachate generation and degradation of putrescibles in the waste mass. Under this approach the wastes were kept dry for as long as the constructed barriers maintained their integrity. A similar dry tomb approach was also an inevitable consequence of top cover and leachate minimization requirements in the EU Landfill Directive (1991/31/EC) from 1997 onward.

During the 1990s and 2000s, the dry tomb approach was increasingly recognized as being incompatible with sustainable development, because moisture is essential to complete the degradation of wastes and to flush pollutants from the landfills to progress toward final storage quality (FSQ) this can be considered as the point at which no further active management of leachate or gas, and no further monitoring, are needed (see Chapter 16.1). Consequently, recirculation began to be used in bioreactor landfills from the 1990s onward and is now in common use for a range of purposes, many related to acceleration of degradation. The landfill bioreactor concept is now recognized and accepted in many countries such as the United States, Australia, and several EU countries (e.g., France, Germany). In the United States, rainfall in some areas is too low to encourage optimal degradation rates and external sources of nonpotable water addition have been considered, e.g., fracking fluids.

This chapter will discuss the objectives of leachate recirculation, the key elements of a recirculation scheme, and the approach to the conceptual design of a scheme adoptable for a given case study (Environment Agency, 2009).

OBJECTIVES OF LEACHATE RECIRCULATION

Before describing the elements of a recirculation scheme, it is useful to consider the wide range of purposes for which leachate recirculation is used. The main ones are as follows:

1. Managing leachate flow rates
   - **a.** Reduction of net leachate volumes by utilizing the absorptive capacity of dry wastes.
   - **b.** Buffering seasonal and short-term peak flows by storing them as transient unsaturated flows percolating slowly downward through waste, to provide a more regulated flow to leachate treatment and disposal facilities.
2. Accelerating stabilization of degradable organic waste
   - **a.** Stimulating gas generation to increase landfill gas (LFG) generation and hence increase revenues from electricity export.
   - **b.** Accelerate progress toward landfill FSQ, also referred to as completion, and reduce long-term aftercare costs and liabilities.
   - **c.** Accelerating settlement to increase available airspace whereas the landfill is still operational.
   - **d.** Accelerate physical stability of waste prior to final restoration, and thereby accelerate progress toward FSQ/completion.
3. Managing leachate quality
   - **a.** Prevent peaks in chemical oxygen demand (COD) from acetogenic leachate reaching a leachate treatment plant (LTP) designed for methanogenic leachate, by recirculating high COD leachate from operational cells into methanogenic wastes in previously filled cells. This is also likely to maximize the opportunity for gas generation and capture, as an incidental benefit.
   - **b.** Seeding basal drainage layers in new cells, to accelerate the establishment of methanogenesis, to protect basal drainage layers from clogging from prolonged exposure to acetogenic leachate and to minimize COD in leachate collected from the new cell.
4. Water supply for irrigation of phytocovers
   - **a.** Use of relatively dilute leachates with sufficiently low salinity for plant tolerance and avoidance of damage to soil structure. This may help with disposal of leachate as well as providing nutrients to stimulate growth of biomass crops.
5. Control of dust
   - **a.** Probably limited to dilute leachates and/or operational areas, so that it does not lead to contaminated runoff reaching surface waters off site.
6. Controlling fires/reduce elevated temperatures
   - **a.** Leachate injection has been used to reduce the temperature of "hot spots" in wastes or to help extinguish subsurface landfill fires.
7. Raise temperature by injecting heated leachate to stimulate methanogens
   - **a.** May be most applicable in areas that experience extended periods of subzero surface temperatures. Has also been considered for temperate regions, to accelerate declining activity in the tail end of the gas curve.

8. Flushing of contaminants
   a. Removal of soluble contaminants from the waste mass, to accelerate progress toward FSQ/completion, and reduce long-term aftercare costs and liabilities.
   b. Denitrification of oxidized ammonia (nitrate) from leachate pretreatment plants. Would form part of a contaminant flushing strategy to accelerate the removal ammoniacal nitrogen without increasing the rate of leachate discharge off site.

ELEMENTS OF A RECIRCULATION SCHEME

The physical components of a recirculation system are shown schematically in the context of a landfill cross section in Fig. 12.1.1.

The schematic shows extracted leachate being reintroduced to the landfill via a range of methods, sometimes after pretreatment and sometimes mixed with other sources of water. It also indicates how the injected leachate can follow a variety of flow paths through the wastes, depending on the physical characteristics of the landfilled materials and the rate of injection (intensive injection over a small area will lead to more rapid channel short-circuiting than diffuse recirculation over a large area). In addition

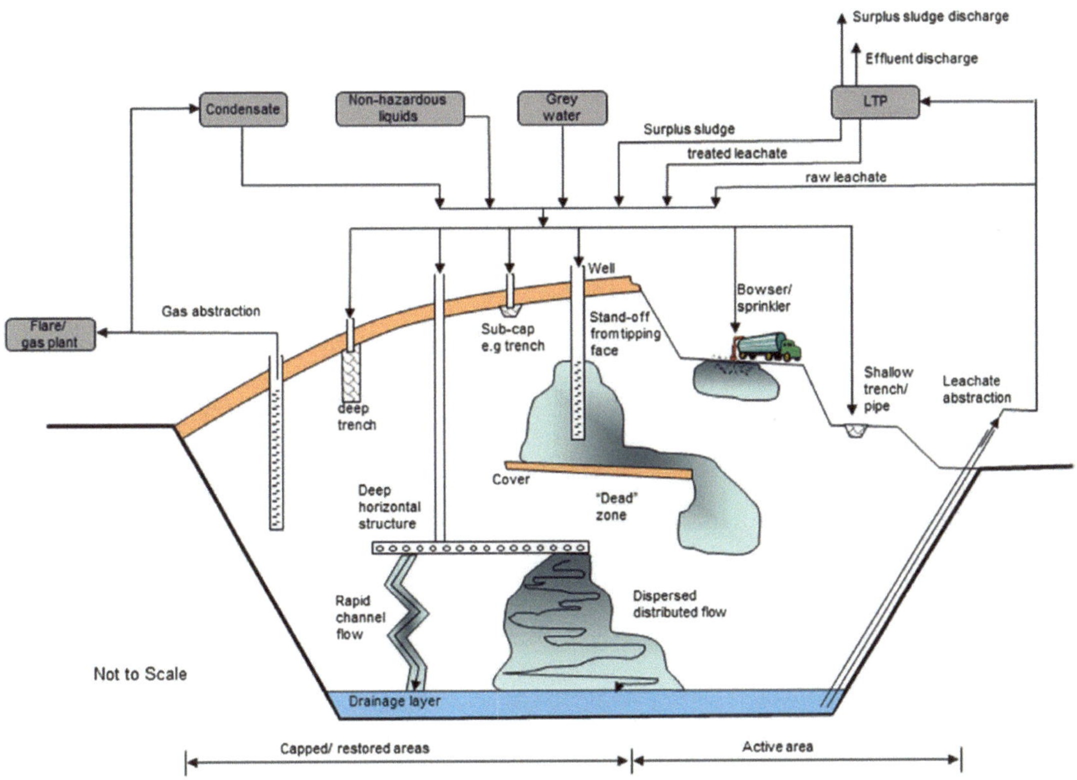

Figure 12.1.1 Schematic showing typical components of leachate recirculation systems. *LTP*, leachate treatment plant. (LANDSS, 2016).

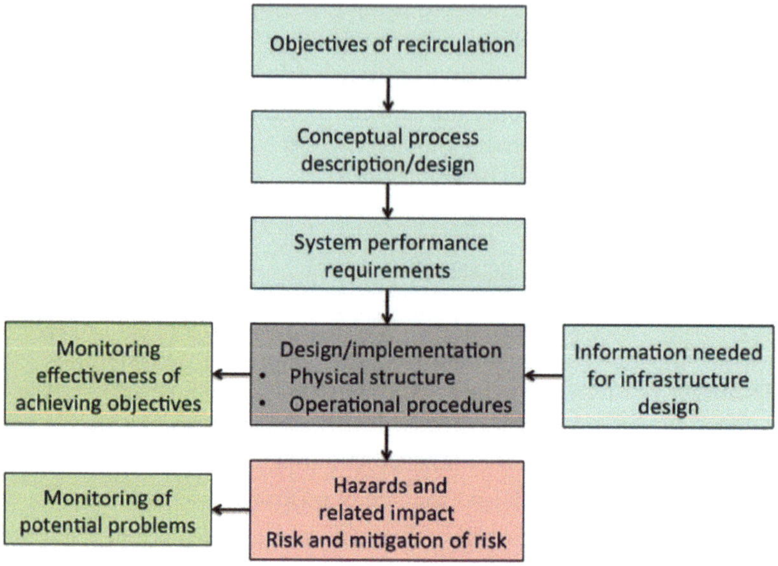

Figure 12.1.2 Conceptual framework for considering leachate recirculation. *Adapted from Environment Agency (2009).*

to using leachate extracted from basal collection systems, leachate is sometimes extracted from vertical gas wells, and sometimes intercepted at higher levels from, for example, seepage zones from perched leachate on an exposed waste slope.

For a well-designed and well-operated recirculation scheme, the physical components should be seen as constituting just one of three essential elements. These three elements are as follows:

- *Conceptual design*, leading to a clear understanding of what the infrastructure has to achieve, according to the objectives for which recirculation is being undertaken
- *Physical infrastructure*: engineering design; installation; operation; and maintenance
- *Monitoring* of performance and environmental impact

The relationship between these elements is shown within a conceptual framework in Fig. 12.1.2.

Physical infrastructure and monitoring are described in Chapter 12.2. The meaning and relevance of conceptual design are discussed below, with reference to some typical examples of the use of leachate recirculation.

CONCEPTUAL DESIGN

As in most engineering projects, a conceptual process description, developed at an early stage, is an important precursor for detailed design, construction, operation and control, and for deriving cost estimates.

The scale and design details of a leachate recirculation scheme will depend on the objectives of the scheme and the size and characteristics of the landfill. The quantities of leachate needed, or accommodated, in recirculation systems may vary by three orders of magnitude, depending on its purpose. At the lower end, quantities may be as little as 5 L per tonne of waste in place, while at the high end, they may exceed 3000 L per tonne of waste in place. This range has a direct impact on the choice of infrastructure and on its detailed design, to ensure that it introduces the correct volumes of leachate into the appropriate zones of the landfill, and to ensure that it remains fit for the whole design life during which recirculation will continue. Examples taken from the LANDSS (2016) decision support system are described below:

- for seasonal flow balancing
- for stimulation of gas generation

Seasonal flow management

Seasonal flow balancing is one of the most common reasons for recirculation. It uses the storage capacity in landfills, whether by absorption or as transient unsaturated flow, to balance peak flows in

Table 12.1.1 Estimate of scale of recirculation and system performance requirements for seasonal flow balancing at a UK landfill

	Value	Units
Winter surplus rainfall, in excess of discharge capacity	100–200	mm
Uncapped or poorly capped area generating surplus	5 50,000	ha m^2
• Volume of surplus to be recirculated	5,000–10,000	m^3
Period of winter surplus	100	days
• Average rate of recirculation required	50–100	m^3/day
Area on which recirculation is to be carried out	5	ha
• Average areal rate to be applied	10–20 1–2	m^3/ha.day mm/day
Peak to average ratio during winter surplus period	2:1	
• Areal rate required from infrastructure	20–40 2–4	m^3/ha.day mm/day

leachate production that will normally occur at wetter times of year. An example of a conceptual design calculation is shown in Table 12.1.1.

The table first calculates the volume of winter surplus to be accommodated, using the uncapped area of the landfill and the rainfall data. This may be the volume above some more or less constant treatment capacity within a LTP, or perhaps the volume in excess of a limitation on a discharge permit. In this example, the surplus volume ranges from 5000 to 10,000 m^3. The next variable is the period over which this surplus will arise—in this case 100 days. The volume is then divided by the period, to give the average rate at which leachate will have to be introduced to the landfill, in m^3/day. Next, the area over which recirculation is to be applied is specified, in this example 5 ha. This leads to the areal rate of application that is needed. Finally consideration is given to a peaking factor to reflect the fact that leachate flows may be episodic, and the system must accommodate flows significantly higher than the average. In this example a peak:average ratio of 2:1 has been used, but site-specific data should be used if available. It is this final value, incorporating a peaking factor that is used as the system performance requirement, to inform the selection of the type of infrastructure that is to be used, and its detailed engineering design. Infrastructure performance data are given in Chapter 12.2.

Table 12.1.1 is not the end of the conceptual design calculations: it is also necessary to consider the fate of the leachate after reintroduction into the waste mass. It is necessary to evaluate the absorptive capacity of dry wastes within the recirculation zone, and the capacity of the waste mass to accommodate a transient volume of slowly draining leachate.

Using the rainfall surplus values from Table 12.1.1, the recirculation quantity can be placed in the context of waste absorptive capacity (Table 12.1.2):

Quantities of 5–20 L/t may be within the absorptive capacity of much municipal solid waste (MSW) and similar dry solid wastes in some countries. However, once the waste is at field capacity, storage of winter peaks can only be accommodated as transient water, percolating slowly downward through the unsaturated waste mass. This condition would often be reached after a few winters or rainy

Table 12.1.2 Winter rainfall surplus expressed as water volume per tonne waste

	Value	Units
Winter surplus rainfall, in excess of discharge capacity, say	100–200	mm
Average waste depth, say	10–20	m
Average waste density, say	1	t/m^3
• Low end recirculation needed, per winter (100 mm/20 t)	5	L/t
• High end recirculation needed, per winter (200 mm/10 t)	20	L/t

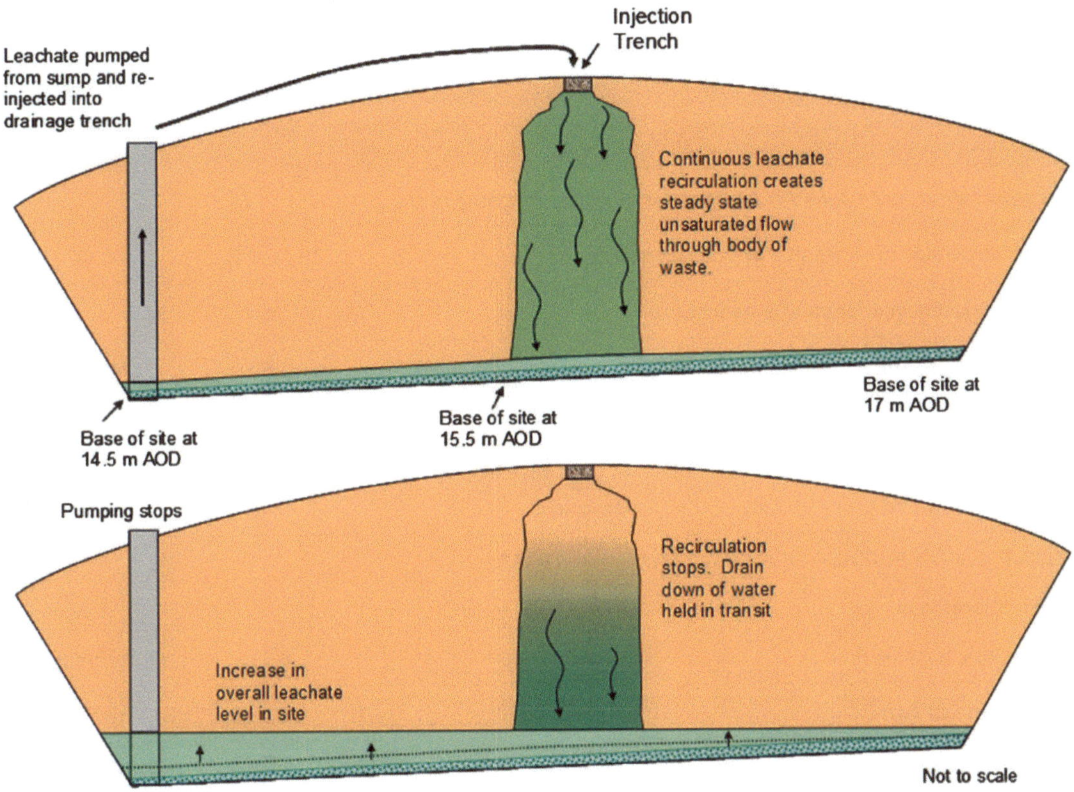

Figure 12.1.3 Illustration of drain-down study modeled by White et al. (2010, 2011).

seasons and is likely to be the starting condition in countries where MSW has a high initial moisture content.

The capacity of wastes at field capacity to accommodate recirculating leachate as transient flow has been investigated by White et al. (2010, 2011). They modeled field data from the drain down of leachate injected via a single injection trench, as illustrated in Fig. 12.1.3.

An example of a conceptual design calculation based on their work is shown in Table 12.1.3, taken from LANDSS (2016).

This worked example shows that the maximum recirculation would be achieved via three linear trenches running the length of an area of approximately 0.6 ha (100 m × 60 m). It shows that transient storage might accommodate 52 days of leachate generated in excess of the existing treatment and disposal capacity. If climate and rainfall data for the site show that this is not long enough, the calculation indicates that additional areas would be needed to accommodate transient leachate, or that alternative disposal or storage arrangements would be needed.

Table 12.1.3 Example calculation of transient storage of a winter flow surplus

Parameters	Value	Units
CHARACTERISTICS DERIVED FROM WHITE ET AL. MODELING		
Lateral dispersion to either side of injection trench	10 (5–15)	m
• Spacing required for parallel trenches	20	m
Transient storage of water within affected zone of waste	35 3.5	L/m^3 waste % v/v
Time for 95% drain down through 20 m waste	50	days
• Average vertical Darcy velocity	0.4	m/day
EXAMPLE APPLICATION AT A FIELD SCALE LANDFILL		
Problem Definition		
Winter peak leachate flow	85	m^3/day
Average annual flow	25	m^3/day
LTP winter flow capacity	45	m^3/day
• Winter daily surplus	40	m^3/day
CONCEPTUAL PROCESS EVALUATION		
Available width of area suitable for injection trenches	60	m
• Maximum number of trenches at 20 m spacing	3	
Available length of area suitable for injection trenches	100	m
Depth of waste	10	m
• Volume of waste "available" (100 m long × 60 m wide × 10 m depth)	60,000	m^3
Transient storage capacity at 3.5% v/v (3.5% × 60,000 m^3)	2,100	m^3
Number of days surplus that can be kept in transient storage (2100 m^3 ÷ 40 m^3/day)	52	days
Linear application rate in trenches (40 m^3/day ÷ (3 × 100 m))	133	L/m.day

LTP, leachate treatment plant.
LANDSS (2016). https://landss.soton.ac.uk/.

This type of conceptual calculation should be seen as part of a feasibility study: it allows the operator to judge whether it is worth proceeding to a more detailed design and costing stage.

Stimulation of Gas Generation

Numerous researchers during the last 30 years have shown a direct correlation between both moisture content and moisture movement, and gas generation rates (e.g., Rees and Grainger, 1982; Kasali and Senior, 1989; Burton et al., 2004; Klink and Ham, 1982).

Hard-and-fast rules are difficult to draw from the literature, but in general, exploitation of LFG in waste with a moisture content greater than ~40% w/w appears unlikely to be restricted by lack of moisture to any significant degree. At the other end of the scale, degradation rates in wastes with a moisture content of 20% w/w or less are likely to be considerably below their maximum and may be severely restricted.

In some regions (e.g. much of southeast Asia), most MSW contains excess moisture when landfilled, generating free leachate as soon as it is compressed. For these wastes, the addition of more water via recirculation is unlikely to stimulate useful additional degradation and may instead make leachate management and gas recovery more problematic.

In other regions (e.g. much of Europe and North America), MSW is typically dry of optimum. For such wastes, the addition of moisture via recirculation may result in a significant increase in degradation rates and gas generation.

The question arises as to how much water to add and over what proportion of the site to add it, to make a worthwhile difference to degradation rates and gas generation rates.

A useful case study is provided by an experiment at Yolo County, California, United States (Mehta et al., 2002; Augenstein et al., 2005a,b). The approach taken was to raise the moisture content of 90% of the solid waste by an average of 10%–20% w/w. In a ~10,000 t test cell, the median water content was raised from ~15% to ~35% w/w. This achieved more than a doubling of the methane yield compared with a control cell.

A generic example of applying this approach to a 1 million tonne landfill cell is shown in Table 12.1.4.

The first four rows of the table calculate the total volume of water addition needed. The next row stipulates the period over which water addition is to be achieved. It is obvious that the shorter the period chosen, the sooner degradation rates will increase, but the greater will be the burden on the performance of the infrastructure. Also, the volumes required are large, and the period stipulated may actually be dictated by the restricted availability of sufficient volumes of leachate or other water sources. Therefore, a compromise may often be needed. The final three rows calculate the absolute rate of water addition needed per year and per day (relevant to the evaluation of availability of suitable water sources) and the areal recirculation rates needed, which influence the choice of infrastructure.

Table 12.1.4 Scale of recirculation and system performance requirements to stimulate gas generation

Parameters	Value	Units
Proportion of waste mass to be targeted for wetting	90	%
Proposed moisture content increase in affected zone	10–20	% wet weight
Quantity of waste in targeted zone	10^6	t
• Water quantity needed	90,000–180,000	m^3
Period stipulated to achieve moisture content increase	2	years
• Average rate of recirculation or water addition needed	45,000–90,000 123–246	m^3/a m^3/day
Area of target zone of landfill	10	ha
Average areal recirculation rate required over target zone	12.3–24.6 1.23–2.46	m^3/ha.day mm/day

It is only by carrying out conceptual design calculations such as in Table 12.1.4 that a realistic assessment can be made of the prospects of stimulating gas generation to a measurable degree.

It has been the experience of the authors that few operators have taken this rational approach. Many schemes undertaken for the ostensible purpose of stimulating degradation are likely to have achieved no significant increase, because the volumes of water or leachate added have been far too low.

Once the feasibility of recirculation has been considered, by the way of a conceptual design, the project can proceed to consider the type of infrastructure to use, and its detailed engineering design, together with a risk assessment and development of a suitable monitoring program. These are discussed in Chapter 12.2.

References

Augenstein, D., Yazdani, R., Kieffer, J., Benemann, J., 2005a. Yolo County, California controlled landfill program: a summary of results since 1994. In: Proceedings Sardinia 2005, Tenth International Waste Management and Landfill Symposium, S. Margherita di Pula, Cagliari, Italy.

Augenstein, D., Yazdani, R., Kieffer, J., Sananikone, K., Benemann, J., March 2005b. Yolo County controlled landfill project status after 12 years of operation. In: Proceedings of 29th Annual SWANA Landfill Gas Symposium, St. Petersburg, Florida, pp. 27–30.

Burton, S.A.Q., Beaven, R.P., White, J.K., 2004. The effect of moisture content in controlling landfill gas production and its application to a model for landfill refuse decomposition. In: Proceedings Waste 2004 Conference, Stratford on Avon, September 28–30, 2004, pp. 333–342.

Environment Agency, 2009. A technical assessment of leachate recirculation. In: Final Report to the Environment Agency Under Research Contract P1–516/3b, by the University of Southampton and Knox Associates. Environment Agency, Bristol, ISBN 978-1-84911-147-8. Science Report SC030144/R6.

Kasali, G.B., Senior, E., 1989. Effects of temperature and moisture on the anaerobic-digestion of refuse. Journal of Chemical Technology and Biotechnology 44 (1), 31–41.

Klink, R.E., Ham, R.K., 1982. Effects of moisture movement on methane production in solid waste landfill samples. Resources and Conservation 8, 29–41.

LANDSS, 2016. https://landss.soton.ac.uk/.

Mehta, R., Barlaz, M.A., Yazdani, R., Augenstein, D., Bryars, M., Sinderson, L., 2002. Refuse decomposition in the presence and absence of leachate recirculation. Journal of Environmental Engineering-ASCE 128 (3), 228–236.

Rees, J.F., Grainger, J.M., 1982. Rubbish dump or fermenter – prospects for the control of refuse fermentation to methane in landfills. Process Biochemistry 17 (6), 41–44.

White, J.K., Beaven, R.P., Powrie, W., Knox, K., 2010. Leachate recirculation: useful hydraulic insights from a case study, using a simple 1-D model. In: Waste 2010, Waste & Resource Management: Putting Strategy Into Practice, Stratford-on-Avon, September 28–29, ISBN 0-9539301-5-7, pp. 69–79.

White, J.K., Beaven, R.P., Powrie, W., Knox, K., 2011. Leachate recirculation in a landfill: some insights obtained from the development of a simple 1-D model. Waste Management 31, 1210–1221.

12.2

LEACHATE RECIRCULATION: DESIGN, OPERATION, AND CONTROL

Richard P. Beaven and Keith Knox

INTRODUCTION

Components of leachate recirculation systems can conveniently be divided into elements that are installed/built into the landfill prior or during waste placement, and those elements that are retrofitted into existing waste. The majority of elements that are preinstalled will relate to leachate collection infrastructure (e.g., basal drainage systems) but operators have installed certain types of injection infrastructure during landfilling (e.g., horizontal pipe runs on top of a lift of refuse, and/or injection pads prior to overfilling). However, most leachate injection infrastructures tend to be retrofitted or installed after landfilling of a cell has completed and are described further in this chapter.

Leachate recirculation predominantly occurs through the zone of unsaturated wastes in a landfill, and the main effect of this is to change the amount of water held by the waste. As the unsaturated permeability of the waste changes with water content, the detailed mechanics of recirculation is rather complicated, especially if the role of landfill gas generation on leachate flow is included. Further details about unsaturated flow theory can be found in Chapter 6.2 on landfill hydraulics. One of the biggest unknowns relating to leachate recirculation is the extent to which preferential flow paths through waste control moisture movement.

Design of leachate recirculation systems should be linked to the objectives and concepts adopted in relation to the specific study case, as discussed in Chapter 12.1. It is not the purpose of this chapter to provide detailed design criteria, although some basic concepts are covered. The engineering design of the installation will be introduced, considering the type of infrastructure, and related operational strategies and procedures. Risk assessment, environmental impacts, and suitable monitoring program are also considered and discussed.

ENGINEERING DESIGN AND INSTALLATION

The different elements of leachate recirculation systems need to fulfill some fundamental design requirements that relate to structural strength, including resistance to landfill settlements and potentially elevated landfill temperatures, chemical durability, sufficient capacity to transmit the design flow rate of leachate, consideration of the clogging potential of leachate, and accessibility for remediation.

A fundamental design requirement of leachate injection systems relates to application rates. EA (2009) defined two different application rates.

- The first is the *infrastructure injection rate* that is related to a physical characteristic of the injection system. For surface application and injection into drainage pads, the area of application is chosen as the reference dimension. Application rates are normalized to liter per square meter per day (L/m^2 day). The reference dimension for horizontal injection pipes, trenches, and vertical wells can be related to liters per linear meter of pipe/trench/well screen per day (L/m/day). Occasionally, it is feasible to give infrastructure injection rates based on both area and length of injection system. This would normally be appropriate for closely spaced systems where even wetting between linear infrastructure is highly likely. An example of this would be band drains.
- The second measure is the *areal application rate.* This is the rate when averaged over the whole of the area assessed as being affected by the recirculation activity.

Having obtained a design infrastructure injection rate, a further distinction needs to be made between average design flows and flows necessary to achieve even distribution of flow, i.e., consideration of how the system is going to be operated.

Consider the mechanics of injecting leachate into a 10 m by 10 m injection pad, constructed out of tire chips to a final installed depth of 300 mm. The capacity of the pad to hold leachate can be calculated as 12 m^3 (assuming a tire chip porosity of 40%—Beaven et al., 2006). Suppose (through other calculations) that an average injection rate of 4 L/m^2 day (equivalent to an infiltration rate of ~1500 mm/year) was required. Hence the total average injection rate to the pad would be 0.4 m^3/day, equivalent to a flow rate of ~0.28 L/min.

If the waste underlying the injection pad has a hydraulic conductivity of 1×10^{-7} m/s and assuming a unit hydraulic gradient and, for simplicity, saturated vertical flow, Darcy's law indicates that an area of ~46 m^2 is required to allow this infiltration rate to occur. Consequently, if leachate was injected constantly to the tire pad at a rate of 0.28 L/min the tendency would be for the flow not to spread out over the whole injection pad, and there would be a risk that infiltration would be concentrated locally around the area of the injection point. In the above case the calculations indicate that approximately 50% of the pad area would be utilized, but if the waste hydraulic conductivity was on average 1×10^{-6} m/s, then less than 5% would be utilized.

A further constraint on operating the system at this low flow rate is the nature of the pump required to deliver flow at this low rate. For reliable operation at this flow rate, a dosing (e.g., piston operated) or peristaltic pump would typically be required. More usually the type of pumps used on landfills are centrifugal, which tend to operate at flow rates one or two orders of magnitude higher than the average design flow in this example. It is highly inefficient and unreliable to attempt to restrict flow from these types of pumps to the low flow required in this example.

One possible way to operate such a system would be to inject a pulse of leachate, say 8 m^3, into the injection pad over a matter of a few hours (this could be linked to existing pump capacity on site) and then allow the injected fluid to infiltrate the waste over a period of ~20 days (8/20 = 0.4 m^3/day).

This would initially occupy approximately two-thirds of the porosity of the tires, so the liquid would distribute itself evenly over the whole area. In reality, for a hydraulic conductivity of 1×10^{-7} m/s drainage would be completed within approximately 10 days (as infiltration is assumed to occur over the full 100 m^2 area) and there would be a period when there was no significant infiltration. The volume injected should ideally be similar to, but less than, the freely draining capacity of the injection pad so that the full areal extent of the pad is brought into action. However, pushing more water into the pad than it can hold will result in overpressurization of the system, the impact of which would need to be assessed. Adverse impacts could include the effect on waste stability and the wetting up of waste above the top of the injection pad. The potential problem with the latter point is that when leachate is pumped into the injection pad, gas will need to be displaced out. The permeability of waste to gas decreases with increasing water content, so wetting up the waste above the top of the pad may restrict routes for gas to escape.

INFRASTRUCTURE AND ITS PERFORMANCE

A wide variety of different leachate recirculation systems have been used on landfills, on varying scales and with varying degrees of success. Leachate recirculation systems contain two main elements: systems to collect leachate and possibly store leachate, and infrastructure necessary to allow reinjection into the landfill. Leachate collection systems are covered elsewhere in this publication, so this chapter concentrates solely on injection system infrastructure, which can be classified into seven broad classes below. Typical examples of injection rates into different types of infrastructures are provided in Table 12.2.1.

Low Pressure Surface Application

At its simplest level this category includes the use of a bowser/sprinkler bar to irrigate leachate at the tip face or on temporary capped areas, or the use of water cannons to spray leachate over exposed waste. Data on irrigation at the tipping face suggest that application rates between 10 and 40 m^3/day are achievable in practice (EA, 2009), which was equated to an application rate of between 10 and 40 L/m^2/day. Higher irrigation rates of 66 L/m^2/day have been reported using a water canon at a tipping face in the United States.

Other systems include open trenches, pits or unlined lagoons dug in the surface of the waste, and open-ended pipes laid on the waste surface. Application rates of 5–8 L/m^2/day were obtained in large-scale trials using surface infiltration ponds in the United States. Pits achieve very localized "point-injection", whereas irrigation/application at the tipping face achieves, by definition, a broader distribution.

The presence of open pits or lagoons of leachate at the surface can give the impression of poorly controlled activities and, depending on leachate quality, can lead to odors. Consequently, many operators have desisted from using this type of system. Similarly the use of cannons to spray

Table 12.2.1 Application rates for different types of injection infrastructure

System Type	Subset	Infrastructure Application Rate	Areal Application Rate (m^3/ha/day)	Note	Scale of Study	Reference
Low pressure surface applications	Infiltration lagoons	5.1–7.8 L/m^2 day	2.5		Field	Reinhart (1996) Townsend
	Irrigation at tipping face	10–40 L/m^2 day 34–66 L/m^2 day		From bowsers Large spray cannon	Field	EA (2009) Thiel (2005)
	Spray irrigation	No information		Reported use in the United Kingdom and United States		EA (2009), Morris et al., (2003)
	Maximum infiltration into waste with hydraulic conductivity of between 10^{-5} and 10^{-7} m/s	864 and 8.6 L/m^2 day		Assumes no low permeability cover, and takes no account of impact of gas generation	Theoretical	EA (2009)
Trenches and horizontal pipes	Tire filled trench	830 L/m day		Tire-filled trench 30 × 3 × 1 m	Field	EA (2009)
	Gravel filled trench	137 L/m day		Brogborough test cell 2. Gravel filled trench 20 m × 0.9 m × 1 m. Applied head = 0–3 m	Field	Mouchel Consulting Ltd. (2001)
	Deep (~5–7 m deep) tire-filled trench	205–274 L/m day		Liquid injection trenches – 130,000 m^3 in 1,300 m of trench in 1 month	Field	EA (2009)
	Tire chip–filled trenches.	600 L/m day short term 290–380 L/m day per meter of applied head 29 L/m day long term	4.7	Alachua County, Florida 1,738 linear meters of trenches: 11 trenches at 15 m lateral and 6 m vertical spacing. Applied head = 0–15 m	Field	Reinhart (1996); Townsend et al. (1995)

	Perforated pipes in gravel filled trenches	10 L/m day	1.1	Pecan Row Landfill, South Georgia 30 m spacing off 3 pressurized force mains	Field	Reinhart (1996)
	32 mm High Density PolyEthylene (HDPE) pipe	12–70 L/m day	11–13	Yolo County Project XL Bioreactor Horizontal pipes with 2.3 mm hole every 6 m. Spacing between 8 and 12 m	Field	Yazdani et al. (2006)
	75 mm OD PVC pipe in gravel-filled trenches.	10 L/m day	2.4–6.1	Busta 5 × 130 m long trenches at 25 m spacings	Field	Barina et al. (2001, 2003), Barina (2005)
	75 mm OD PVC pipe in gravel filled trenches.	8 L/m day	5.5	Drambon 5 × 70 m horizontal trenches	Field	Barina (2005), Barina et al. (2005)
	Perforated pipes in injection trenches	7 L/m day	9	Landgraaf Injection trenches 5 × 35 m at mid height of cell and 6 × ~60 m at 90 degrees orientation under cap	Field	Woelders et al. (2005)
	Modeled injection trenches	2000–8000 L/m day 200–800 L/m day		For K = 1×10^{-5} m/s For K = 1×10^{-6} m/s	Model	McCreanor and Reinhart (1999, 2000), McCreanor and Reinhart (1996)
	Modeled injection trenches	30–130 L/m day 3,000–13,000 L/m day		For K = 1×10^{-7} m/s and Head = 0–5 m For K = 1×10^{-5} m/s and Head = 0–5 m	Model	Haydar and Khire (2005)
Pads	50 × 50 × 2 m tire-filled pad	24 L/m^2 day		9000 m^3 over 5 months	Field	EA (2009)
	13 tire-filled pits at 8 m spacings	19 L/m^2 day	27	Yolo county test cells –8000 m^3 injected over 9 year period	Field	Augenstein et al. (2005)

(Continued)

Table 12.2.1 Application rates for different types of injection infrastructure—cont'd

System Type	Subset	Infrastructure Application Rate	Areal Application Rate (m^3/ha/day)	Note	Scale of Study	Reference
	60 × 9 × 0.15 blanket of crushed recycled glass	28 L/m^2 day (long term) 1440 L/m^2 day (short term)		3200 m^3 injected over 7-month period	Field and model	Haydar and Khire (2007)
	408 m^2 geocomposite drainage layer	21 L/m^2 day (long term) 705–1765 L/m^2 day (short term)	N/A	Short term tests (over 7 months) undertaken on a 12 m × 34 m geocomposite blanket fed by a central 12-m perforated pipe Applied head = 2–8 m	Field	Khire and Haydar (2007)
Wells	4 shallow wells 2 m into upper surface of waste	6875 L/m day		In operation for over a year. Drilled at 500 mm diameter. Applied head = 3 m	Field	EA (2009)
	18 injection wells at 30 m spacing	90 L/m day	~8.9	La Vergne—Some wells had split response zones	Field	Bureau et al. (2005)
	134 vertical wells in 45 clusters at 15 m spacing	50 L/m day (long term) 144–700 L/m day (steady state)	8.4	New River Regional Landfill, Florida Bioreactor Applied head = 1–15 m	Field	http://bioreactor.org/nrrl Jain et al. (2006)
	12 shallow pin wells at a spacing of 10–15 m	~4000–6000 L/m day (assumed) 110–160 L/m^2 day	11.7	Assumer 24 m of well screen (assume 2 m screen per pin). Injection rate 100–145 m^3/day A:900 m^2. Applied head <30 m	Field	Reinhart (1996)

	Modeled injection wells	22–89 L/m day		Saturated hydraulic conductivity of 1×10^{-5} m/s	Model	McCreanor and Reinhart (1999)
	Modeled injection wells	Very wide range predicted.		Saturated hydraulic conductivity of 10^{-5},10^{-6}, 10^{-7} m/s. Various injection heads modeled	Model	Khire and Mukherjee (2007)
Band drains	Band drains at 1 m centers on 40×40 m grid	11 L/m day 125 L/m^2 day	Not known	Each band taken to alternating depths of 5, 10, and 15 m ~ 18,000 m drain in 40×40 m grid. Q = 200 m^3/day over 45 days	Field	EA (2009)

Adapted from EA (2009).

leachate can lead to odors and aerosols and is considered by many to be too uncontrolled for a modern waste disposal operation.

A further application of surface irrigation is onto a vegetated top-cover or phyto-cap, which may be designed to reduce leachate volume through evapotranspiration and/or provide some in situ treatment of leachate quality. The most commonly used vegetation is different kinds of grasses and foliiferous trees, often within the family of Salix (e.g., willows), which may be harvested as a short-rotation tree crop. Improvement in leachate quality may occur through a number of processes: nitrogen can be converted by nitrification/denitrification; phosphorous can be adsorbed in the soil, as too can some organic material which in low doses can also be degraded by microorganisms; and metals can be oxidized and precipitated or be sorbed to the soil particles.

Optimum irrigation areas have a slope between 1:10 and 1:30. A steeper slope means an increased likelihood of surface runoff, which ought to be minimized. A flatter slope than 1:50 leads to the risk of standing leachate, which has an adverse impact especially for grass vegetation. Trees and bushes with a low sensitivity to high water content in the ground will function even on horizontal ground.

The possible hydraulic loading on an irrigation area is mainly dependent on the structure of the soil and the vegetation's tolerance of water saturation. With good conditions, large amounts of water could be added. In favorable years, there are examples of additions of up to 3340 mm (9 L/m^2/day) (Lagerkvist and Cossu, 2005).

Horizontal Trenches and Pipes

A wide variety of designs are included in this section, and include shallow (i.e., just below the cap) trenches (with or without perforated pipes) and more deep seated systems either involving "spiders" running off a central access shaft or subparallel horizontal pipes or trenches with vertical access points or side slope risers.

These types of system tend to be based around linear trenches (e.g., tire or rubble filled) or perforated pipes in a trench backfilled with drainage material. Some systems are designed to be horizontal and some to include a fall in the trenches. A typical design would involve HDPE (High Density Polyethylene) pipes with a diameter ranging from 32 to 150 mm, often installed in a stone or tire-filled trench.

Experience of injection rates into horizontal systems varies considerably (Table 12.2.1). Some operators (EA, 2009) reported difficulties maintaining hydraulic access to buried infrastructure of this type, leading to long-term performance issues.

However, there are many examples of successful injection into horizontal trenches and pipes. Short term injection rates of between 600 and 830 L per meter of trench per day have been reported, although this invariably drops during longer term applications (to nearer 7–70 L/m/day). One of the most successful long-term horizontal leachate injection schemes that has been reported was into a site where the use of traditional intermediate daily cover was replaced with green waste (Yazdani et al., 2006). An areal rate injection rate of 13 m^3/ha/day, at linear infrastructure rates of 12–70 L/m/day was achieved. The high areal rate at relatively modest linear rates was achieved by having a relatively close spacing of pipes (between 8 and 12 m).

More typically horizontal trenches and pipes have been installed in sites at spacings upwards of 20 m, although this does not mean that larger spacings are necessarily the most optimum, either in terms of performance or on a cost-effectiveness basis. Some work (Barina, 2005) has concluded that the mean zone of influence of a properly operating injection trench was possible of the order of 15 m laterally, whereas others using numerical modeling approaches (e.g., Haydar and Khire, 2005; White et al., 2011) concluded that wetting widths much over 10 m were difficult to achieve. However, more work is required on assessing the actual distribution of leachate injected into horizontal pipes (or all types of injecting infrastructure for that matter) and a better theoretical understanding of the role of waste anisotropy and heterogeneity is also important. Kumar et al. (2009) concluded that the spatial response of leachate recirculation to in situ water contents can be highly variable.

A variation on horizontal trenches and pipes are deep vertical trenches installed into existing deposited waste. One big advantage of deep trenches is that they cut through landfill heterogeneities, especially intermediate cover layers that have the potential to affect leachate recirculation. Trench depths of ~6 m by ~1.5 m wide are feasible with commonly available plant, but specialist excavators and safety equipment (e.g., cages) can lead to trenches in excess of 8 m being installed. Typically the bottom of the trench is infilled with drainage material, often used tires, and then the upper reaches of the trench loosely backfilled with the excavated waste material. Some operators have installed HDPE drainage pipe in the bottom of the trench. Although deep trenches cut in landfills containing municipal solid waste can be remarkably stable, the installation of deep trenches is a major undertaking. In addition to health and safety concerns, problems associate with odors and the ingress of air into the landfill need to be addressed. The latter is especially important if there is active landfill gas extraction occurring in the area to be trenched. However, deep trenches cut into one aged and methanogenic landfill in the United Kingdom did not result in any of these problems manifesting themselves. An example of long-term liquid injection into deep trenches is 75–100 m^3/linear meter/year. Assuming trenches are spaced at 50 m apart, then an areal rate would be up to 55 m^3/ha/day.

Pads and Drainage Blankets

A variation on subsurface trenches, subsurface pads are rectangular pits (or areas) filled with drainage material, often whole or shredded tires.

Two distinct types of design have been used. The first involves grids of small pads (typically between 1 and 10 m^2) filled with drainage material each with an individual leachate delivery pipes. Mixed results have been experienced with this type of system. Success stories include the use of 13 tire-filled pits at 8 m spacings used to inject 8000 m^3 leachate into a ~8000 tonne test cell over a 9-year period (Augenstein et al., 2005). The long-term infrastructure injection rate was approximately 19 L/m^2 day. However in contrast, EA (2009) reported that many UK operators had obtained poor performance from smaller multiple pads, especially in comparison to the second type of design that is single large pads of tires.

Single large pads of tires can potentially be any size, although one of the largest is a 2-m deep pad of used tires with dimensions of 50 × 50 m (EA, 2009). This was the as-constructed dimensions, and a considerable reduction in the depth of the tires would have occurred as the tires compressed following surcharging with new lifts of waste. The performance of this type of tire-filled pad is generally good (e.g., 24 L/m^2 day).

Other types of drainage material have been used in injection pads. Haydar and Khire (2007) investigated the use of a 60 m × 9 m "permeable blanket" consisting of 0.15 m thick layer of crushed glass aggregate. A volume of 3200 m^3 of leachate was injected over a period of 7 months, giving a long-term infrastructure injection rate of 28 L/m^2 day.

Wells

Vertical wells are one of the most commonly used types of infrastructure for leachate injection. This is because existing leachate abstraction/monitoring wells or gas wells can easily be adapted for leachate injection. A few operators have investigated the use of pin wells. These are percussion installed wells approximately 5–10 m deep typically at 20 m centers. It is also very easy to inject leachate under pressure into vertical wells, and injection pressures of over 30 m water head have been reported.

A vertical reduction in hydraulic conductivity with depth will occur in many landfills (Powrie and Beaven, 1999; Al-Thani et al., 2004; Jain et al., 2006) meaning that vertical wells with large screened horizons may deliver more leachate to higher levels than to lower.

Typically few operators have monitored flow rates into individual injection wells, so performance data on this type of system are scarce. However, where data are available (Table 12.2.1) areal recirculation rates of approximately 8 m^3/ha/day have been achieved which compares favorably with the rates achieved by trenches and horizontal pipework distribution system. When injection rates are related to well screen length, long-term infrastructure injection rates of between 50 and 90 L/m day have been achieved, with shorter term injection rates of as high as between 144 and 700 L/m day.

Injection into shallow wells can result in short circuiting to side slopes, gravel filled gas collection trenches or the surface itself (Jain et al., 2006). High pressure injection (~30 m water head) into a set of 12 pin wells achieved short term injection rates of between 4000 and 6000 L/m day and appeared to be a good way of increasing the water content of waste in advance of overtipping (Reinhart, 1996).

There has been some limited research (e.g., Skhiri et al., 2006) into the radius of influence of injection wells. For a leachate well discharging into the unsaturated waste a radius of influence of approximately 5–10 m was determined. This result can be supported by the modeling work of both McCreanor and Reinhart (1999) and Khire and Mukherjee (2007) (see Fig. 12.2.1).

Subsurface Band Drains (Constructed During Infilling)

Band drains (a form of geotextile drainage "socks") have been installed on at least one landfill (EA, 2009). The band drains were installed by percussion at 1-m centers on a 40 × 40 m grid, to alternating depths of 5, 10, and 15 m (Fig. 12.2.2). The top surface of the "socks" was then overlain by a

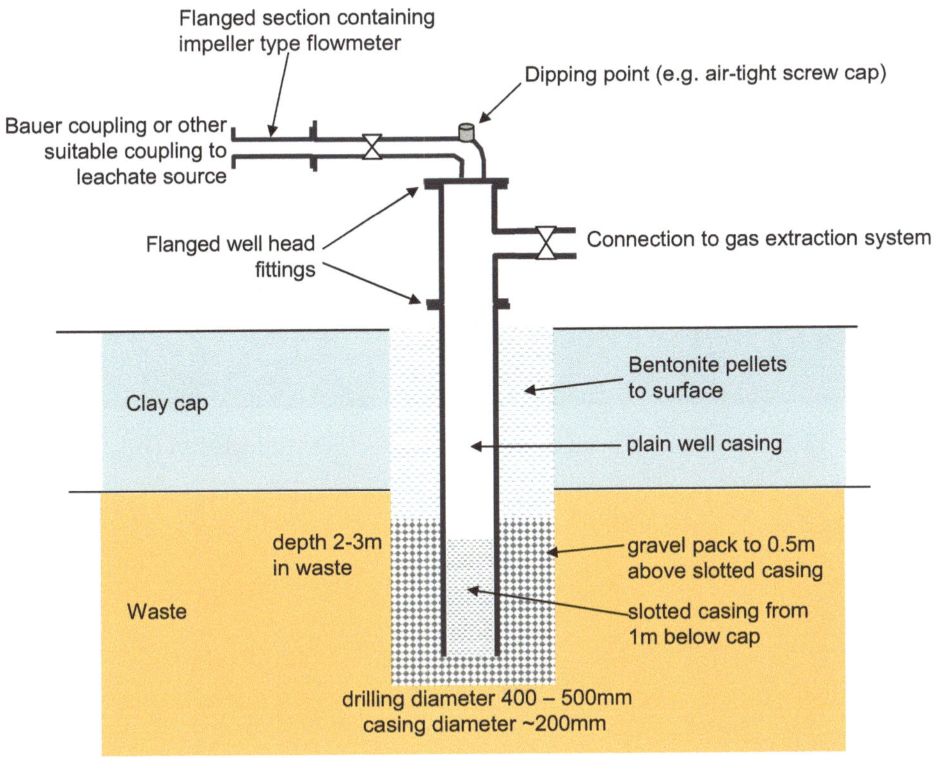

Figure 12.2.1 Example of injection well design for use in leachate recirculation (EA, 2009).

bed of drainage material to distribute injected leachate evenly over the whole area. This was accessed from the surface by a vertical pipe. The first of these systems was installed in 1998. During a trial an injection rate of 200 m^3/day was maintained into one system over a period of 45 days, giving an areal injection rate of 1250 m^3/ha/day. The close spacing of individual band drains means that many of the problems associated with the heterogeneity of the waste are effectively engineered out. The combination of vertical and horizontal injection structures also overcomes problems with layering. The system may be more prone to clogging than other types of infrastructure, so special attention should be given to the quality and need for any pretreatment of leachate injected (see Fig. 12.2.2).

OPERATIONAL AND ENVIRONMENTAL ISSUES

The key operational and environmental issues of leachate recirculation systems are identified in Table 12.2.2 and discussed in the following paragraphs.

Infrastructure Design

This topic is covered more thoroughly in the previous section, but in summary there is a large array of different infrastructure designs that can and have been used. There is a wide variation in reported

Figure 12.2.2 Example of band drain used for leachate recirculation (EA, 2009).

Table 12.2.2 Potential operational and environmental issues

Operational Issues	Environmental Issues
Infrastructure design and hydraulic performance	Odors and uncontrolled gas release
Clogging/reduction in performance of injection infrastructure	Adverse impact on leachate quality
Consider need for leachate pretreatment prior to reinjection	Adverse impact on leachate head
Flooding of gas wells	Effect on liner systems
Daily cover	Perching/surface outbreaks
Effects of settlement	Surface water contamination
Clogging of basal drainage layer	Short circuiting
Obtaining sufficient volumes to recirculate	Interaction of leachate and gas
Slope instability	

performance, but it is difficult to conclude on performance grounds that any particular type of infrastructure should be avoided in any future schemes. Although there are many examples of an infrastructure not working well, this could be balanced by reports of a similar type of infrastructure at another site performing adequately. A key factor influencing how well any injection system will work relates to the hydraulic properties (especially permeability) of the waste surrounding the infrastructure, and tests to characterize these properties may be advisable. However, it is also important to recognize that flow rates achieved in short term tests may not be sustainable in the long term.

There may be a need for leachate storage tanks or lagoons for flow balancing, providing at least 1–2 day's storage. Header tanks may also be required for controlling the pressure of injection into infrastructure.

Clogging and Reduction in Performance of Injection Infrastructure

Many operators have reported problems with clogging or poor hydraulic performance of various types of infrastructure, many exhibiting deterioration over time. However, few leachate flow and quality data are available, and in many instances it is not possible to differentiate between reduction in performance caused by changes in hydraulic properties of the waste/hydraulic gradient and those caused by clogging per se. However, clogging over a period of just a few weeks appears common when the injected leachate is acidogenic (i.e., high concentrations of organic acids). When the leachate is methanogenic, systems have operated for periods of up to several years without apparently clogging. These findings corresponds to research on clogging of leachate drainage layers by acidogenic and methanogenic leachates (e.g., Rowe, 2005; Beaven et al., 2013).

Need for Leachate Treatment Prior to Reinjection

To minimize the risks of clogging of reinjection infrastructure, abstracted leachate may be subjected to simple pretreatment to remove solids that are readily settleable or that may precipitate during recirculation. Pretreatment may comprise a short period of aeration (e.g., ~2 h), to strip off CO_2, raise the pH value, and encourage precipitation of calcium carbonate, followed by settling. The settled leachate could then be passed through a sand filter or similar, to ensure more complete solids removal. No documented examples quantifying the benefits of this type of pretreatment have been reported in the literature.

Before being recirculated to the landfill body leachate may also undergo other pretreatment to optimize the recirculation objectives. Leachate may be biologically oxidized to reduce the BOD/COD ratio to a value <0,1 to reduce odor problems where low pressure leachate surface addition is used (Doedens and Cord-Landwehr, 1989). Biological treatment could also be coupled with phytotreatment, installing vertical flow, open bottom, units directly on top of the landfill surface (Lavagnolo et al., 2011).

Flooding of Gas Wells

Localized flooding of gas wells as a consequence of leachate recirculation has commonly been reported. Increased leachate levels in gas wells have occurred from both low pressure surface applications and injection into deeper infrastructure. There will be a direct correlation between the risk of flooding and the distance to, and injection rate into, any injection infrastructure.

A further problem relating to the interaction of landfill gas is that the increased water content of waste that is a consequence of leachate recirculation reduces the permeability of the waste to gas and hence the ability to extract landfill gas. An alternating or rotating mode of operation (i.e., injecting leachate intermittently) between several injection zones is a method that has been used routinely by several operators, which often helps to overcome these problems.

Daily Cover

Low permeability cover has the potential to cause problems—it inhibits vertical flow, encourages horizontal movement, and creates bigger risk of flooding gas infrastructure and of causing lateral surface seepages. Some types of infrastructure have the ability to overcome the problems of daily cover (e.g., band drains), but in general it is a problem that can affect most types of recirculation systems. The use of green waste as daily cover was judged as a key ingredient to a successful leachate recirculation trial, which ran over 9 years at a long-term areal application rate of 27 m^3 ha day (Augenstein et al., 2005). When more conventional covers were used at the site, problems with leachate recirculation occurred.

As recirculation becomes more widespread, and as the recirculation flow rates increase (e.g., for flushing), the issue of cover will become more and more problematic. There is clearly a challenge to achieve the nuisance control objectives of daily cover, but using methods and materials that do not cause hydraulic barriers.

Effects of Settlement

There are many instances of failure of pipe work in horizontal pipes and radials, attributed to settlement. However, the cause of failure of pipe work at depth in landfill is difficult to establish with certainty. Settlement on most landfills will be large, and experience has shown it is likely to be locally increased in the zones around injection systems. Drainage systems will need to be designed to accommodate this. Some types of infrastructure may be better able to accommodate settlement than others, e.g., pads, band drains, and surface applications may accommodate settlement better than some horizontal pipe systems.

Clogging of Basal Drainage Layer

There is no direct evidence of any increased clogging of basal leachate drainage layers as a consequence of leachate recirculation. However, the mechanisms of clogging are now well understood and any increase in the flux of acidogenic leachate reaching a basal drainage layer is likely to increase the clogging potential, whereas any increase in the flux of methanogenic leachate will not, to any significant extent. Leachate recirculation is likely to result in the flushing of more organic acids from the waste, but whether these reach the drainage blanket will depend on flow paths, the rate of flow and the capacity of the landfill to microbiologically convert these acids to carbon dioxide and methane. It has been suggested that the practice of maintaining drainage layers in an unsaturated condition may be an important cause of the clogging of basal layers, and that maintaining a layer of waste saturated with methanogenic leachate may form a useful buffering role (Beaven et al., 2013).

Obtaining Sufficient Volumes to Recirculate

Some landfills have run out of leachate to recirculate when trying to increase gas generation. In some countries where it is allowed (e.g., the United States), importation of some nonhazardous liquids (e.g., sewage effluent) has been necessary to provide the moisture needed to stimulate gas production.

It should be straightforward, via a conceptual process design, to estimate the quantities of liquid needed for this purpose and to assess whether addition of external water is necessary. It could then be integrated with the site water balance and leachate management plan.

Slope Instability

The adverse effects of a raised pore pressure on slope stability in soils are well known in soil mechanics. As a potential hazard this must also be considered in landfills. Some instances of catastrophic slope failure have occurred, mainly (though not exclusively) at landfills in tropical locations, following extreme rainfall events. Slope failure tends to occur where pore pressures have become high and a significant slope exists.

The risk of slope failure as a result of leachate recirculation will be exacerbated in systems using high pressure injection. Recirculation in any case should not occur in areas close to any internal landfill slopes, and for above ground landfills an understanding of the impact of leachate recirculation on internal pore water pressures will be required.

Odors and Uncontrolled Gas Release

Odor problems resulting from recirculation via injection into open trenches or soakaways can occur, although they are likely to be strongly related to the quality of leachate being recirculated. Odor complaints are less likely when well-stabilized methanogenic leachate is recirculated, whereas the surface application of a high strength acidogenic leachate is likely to be problematic leachate recirculation at one site created a problem with hydrogen sulphide odors, attributed to recirculating through deposits of sulfate rich industrial waste.

Adverse Impact on Leachate Quality

A very small number of instances have been reported (EA, 2009) where operators thought there was an adverse impact (i.e., leachate concentrations increased). However, there do not seem to be data to support this perception. At least as many operators thought from their own experiences that there was no detectable impact on leachate quality. Some anecdotal evidence has noted short-lived flushes of acidogenic leachate reaching the site base on uncapped areas, following extreme rainfall events, and it is possible that the same could occur with high rates of recirculation.

Effect on Liner Systems

The EA (2009) review found no evidence to support a concern that leachate recirculation would increase the head on liner systems. Available data suggest the reverse is true, in that for a given volume of leachate within the landfill, recirculation will keep a significant amount of it "in transit" within the waste. This was also a conclusion from a US review (Benson et al., 2007). One UK operator provided a monitored example of recirculation lowering leachate heads at the base of the landfill, and some regulators have reported heads rising when recirculation is stopped. Knox and Shaw (2006) give an example in which leachate heads in the basal drainage system rose by 2.1 m in 48 days when the leachate was allowed to collect there, following interruption of recirculation.

A further indirect, but nevertheless important potential effect on liner systems relates to temperature. If leachate recirculation results in increased temperatures in the body of the landfill (as a result of increased biodegradation rates) and if elevated temperatures are transferred to a membrane liner then this could result in increased long-term degradation of the liner (Rowe, 2005). However, the temperature of the liner is regulated by the temperature in the external geological formations, so any increase in waste temperature will not be transmitted in full to the liner.

Perching/Surface Outbreaks

Perching or surface outbreaks of leachate as a result of leachate recirculation are potential problems, and a variety of different causes have been reported in EA (2009).

Injection into a pipe/trench system erroneously laid on a shallow gradient beneath a lapped synthetic cap led to a surface outbreak. Introduced leachate flowed to the lowest point, where localized heads built up under the cap. Similar systems laid accurately level along the contour under a clay cap resulted in fewer (if any) problems.

An outbreak of leachate as a spring line above an old restoration layer was reported at one landfill and a further leachate breakout 30 m from an injection trench at another.

Injection into subcap infrastructure where the receiving capacity was exceeded, led at one site to the overflowing of leachate at the surface. One operator solved this problem with manual controls, whereas other operators felt that more sophisticated control systems with fail/safe mechanisms were necessary to reduce risk of surface outbreaks etc. (e.g., from a burst pipe).

Avoiding surface breakouts requires consideration of several factors:

- The lateral zone of influence of most types of infrastructure appears to be typically on the order of 5–10 m, with a maximum of ~15 m. Therefore a lateral exclusion zone of 20 m or more from the edge of a slope should help to minimize the risk.
- The exclusion zone may have to be greater where pressure injection is being undertaken.
- Low permeability daily cover clearly increases the risk of surface breakouts. Consideration might therefore be given to using a permeable material particularly at the edges of slopes.

Surface Water Contamination

Pipelines carrying leachate across restored surfaces to injection points present a potential risk to surface water in the event of leakage. The potential for surface water contamination is a factor that should be considered explicitly in any recirculation scheme. It should be controllable through good system design and management.

Short Circuiting

There is limited evidence that short circuiting of leachate has been a problem in recirculation systems, although it is a difficult topic to assess. High pressure injection has been observed to cause hydrofracturing and consequent short-circuiting to the base of the cell at one site and to surface seeps in another.

The occurrence of short circuiting will be difficult to predict, as it will depend on the presence of preferential drainage paths and other unquantifiable inhomogeneities within the waste. However, its main effect will be to limit the effectiveness of the recirculation scheme, with the only major adverse environmental impact being the possible transmission of acidogenic leachate to basal drainage layers and an increased theoretical potential for clogging.

MONITORING OF RECIRCULATION SCHEMES AND PERFORMANCES

Monitoring will help the operator to assess the performance of their recirculation scheme against design objectives and also protect the environment, although some individual monitoring activities may fulfill both requirements. The need and purpose of monitoring can be separated into four functional groups as depicted in Fig. 12.2.3, and discussed further below.

In general the monitoring of existing leachate recirculation schemes on landfills has been inadequate and has certainly not been linked properly to design objectives. The exact requirements and frequency of monitoring will depend on the site, objectives, and scale of the scheme, and they should not be prescriptive. It is not appropriate or helpful for all of the listed monitoring to be done on all recirculation projects, as this would not contribute to the protection of the environment, nor necessarily improve performance. Rather, the lists (below) set out the areas that may need monitoring, and the reasons why certain types of monitoring may be required. By understanding the potential problems and risks relevant to a particular recirculation scheme, the operator and regulator will be able to develop a monitoring regime that is both targeted and proportionate.

Operational Performance of Recirculation Infrastructure

The purpose of the monitoring detailed under this section is purely to assess the ongoing performance of the recirculation scheme, and to ideally relate this to the original design objectives of the scheme (Table 12.2.3).

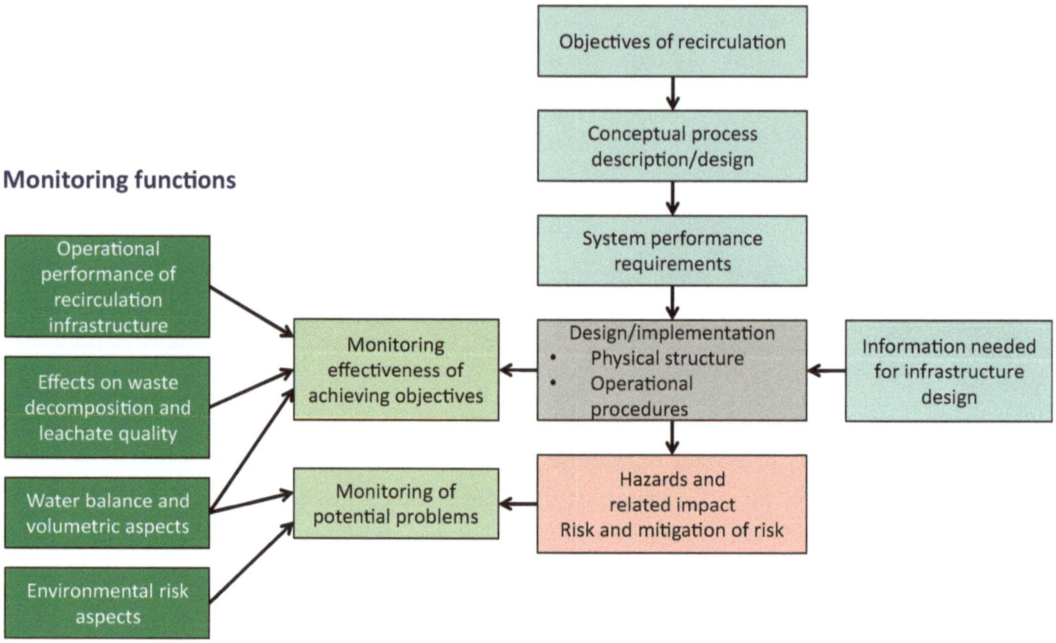

Figure 12.2.3 Monitoring issues in relation to the conceptual framework for evaluation of leachate recirculation. *Adapted from EA (2009).*

A fundamental requirement is to monitor the rate and volumes of fluid that are recirculated into different elements of the infrastructure over time. It is surprising the extent to which many recirculation schemes have previously been run with no monitoring and hence knowledge of these data. Depending on the design of the recirculation system monitoring at more than one point may be required. With knowledge of the infrastructure design, both infrastructure and areal application rates can be calculated. Maintaining up-to-date time series graphs of average rate and total volumes recirculated is a powerful management tool for assessing performance. It may also be useful to measure (and record on the graphs) the back pressure or head observed in the infrastructure systems during operation. Where a system is being operated intermittently, observations of the rate at which the back pressure declines when recirculation is turned off can provide valuable information on performance. During long-term operation, it is to be expected that flow rates will reduce and back pressure increase, a more rapid deterioration may be symptomatic of more serious failure, perhaps through clogging.

With some types of recirculation systems it may be useful to calculate the liquid to solid (LS) ratio, which is a measure increasingly used to assess flushing of wastes. The LS ratio has derived from waste acceptance criteria leaching tests where clean water is added to a known mass of waste and then leachate removed. Hence LS ratio (in its strict sense) should only be applied to the volume of clean water (e.g., rainfall) added and should not be related to the volume of *leachate* recirculated: its use in

Table 12.2.3 Monitoring of operational performance

Parameter	Derived Parameter	Comment
Rate (Q) and Volume (V) injected for: Leachate recirculation New water additions	Hydraulic infiltration rate Liquid Solid ratio	m^3/ha day for areal systems; L/linear metre/day for linear systems (e.g., trenches) Liters/dry kg. Requires estimate of volume/mass of waste affected by recirculation
Back pressure in injection infrastructure		Standing head (measured with dipmeter or transducer) or back pressure (measured with transducer or pressure gauge and then converted to head) within reinjection structures Rate of head decline in reinjection structures when flow is stopped
Duration and timing of on/off cycles		For intermittent schemes this will be important
Leachate levels in monitoring wells	Radius or distance of influence of recirculation	Without bespoke in situ water content probes the radius or distance of influence of a recirculation structure is difficult to establish. However, any changes in leachate levels in surrounding leachate or gas wells can provide useful information.
Visual/CCTV inspection of injection media and access pipes		Looking for accumulation of solids
Clogging potential of recirculated leachate		e.g., SS, VSS, Ca, alkalinity etc. Possibly measure before and after pretreatment such as aeration and filtration
Clogging/loss of performance of basal drainage layers		Not clear how to achieve this, but CCTV surveys may be of some help

Adapted from EA (2009).

Table 12.2.4 Monitoring of waste decomposition and leachate quality (EA, 2009)

Parameter	Comment
Gas temperature at well heads	Indicate local changes in reaction rates
Gas temperature at manifolds	Indicate overall changes in reaction rates
Leachate temperature in monitoring wells	
Leachate temperature at abstraction points	
Gas flow rates at well heads	
Gas flow rate from recirculation cell	
Settlement rate	Settlement gauges and/or site survey
Gas quality	CH_4, CO_2, H_2 etc
Leachate quality in monitoring points near recirculation zone	Look for flush of acetogenic leachate, or increased NH_4–N

this context will (and has) cause confusion. Consequently, in terms of monitoring, it is important that both the volume of leachate recirculated and the volume of clean water introduced to the system (perhaps calculated through water balance techniques) are recorded. Further details are provided under the third functional group in this list.

Effects on Waste Decomposition and Leachate Quality

The main issues related to the monitoring of waste decomposition and leachate quality are summarized in Table 12.2.4.

Evidence that leachate recirculation has influenced waste degradation can be obtained from a number of sources (see Chapter 12.1). Any increase in gas and leachate temperature might indicate increased biological activity, as too would increases in waste settlement rates. More direct evidence would come from increases in gas generation rates (e.g., m^3/tonne/year) in the recirculation zone before and after recirculation.

Changes in leachate quality (e.g., higher NH_4–N or BOD and COD) may indicate increased biological activity, although it may also simply represent increased flushing from the unsaturated zone.

Water Balance and Volumetric Aspect

For sites where the primary purpose of leachate recirculation is to provide seasonal buffering and storage capacity within the site, the functional monitoring summarized in Table 12.2.5 is likely to be important. It may also be beneficial for most schemes, whatever their purpose.

The basic requirement is to collect data that allow a proper water budget of a landfill cell to be undertaken. This includes all water (leachate) inputs and outputs, and seasonal change in storage that will relate to both leachate levels in the site, but also to the volume of leachate held in transit within the unsaturated zone (i.e., the volume that will drain when leachate recirculation is stopped).

It is likely that information to populate the following time series graphs will be required:

- leachate level and head within cells; calculated seasonal fluctuation in quantity of stored leachate (important to know geometry of cell base, to convert head changes to volume changes, also important to state assumptions regarding saturated storage coefficients);

Table 12.2.5 Monitoring of water balance

Parameter	Derived Parameter	Comment
Volume of leachate removed from leachate collection systems, wells, etc	Flow from individual landfill cells	
Flow to leachate treatment plant or to discharge off site		
Rate (Q) and Volume (V) injected for: Leachate recirculation New water additions	Volumes recirculated New water additions include imported water, effective rainfall and surface water ingress (from adjacent areas)	Record location of reintroduction
Imported water or sludge		
Rainfall Evapotranspiration	Effective rainfall	Can use nearby Met Office or amateur station data Use Met Office or estimated daily values
Leachate level	Change in water content of cell	Highly dependent on an understanding of absorptive capacity of waste and on drainable/total porosity
Depth of leachate in any storage tanks or lagoons	Volume of leachate held in storage	For sites that have large leachate storage capacity, (e.g., lagoons) this will be important

Adapted from EA (2009).

- flow from cells (daily and cumulative in year);
- recirculated flow (daily and cumulative in year);
- flow to leachate treatment or discharge (daily and cumulative in year); and
- estimated effective rainfall.

Environmental Risk Aspects

The range of possible monitoring tasks related to environmental risks from leachate recirculation is summarized in Table 12.2.6.

Table 12.2.6 Monitoring of environmental risk aspects

Parameter	Derived/Inferred Parameters	Comment
Visual inspection for surface breakouts Continuous EC measurements in main surface water collection ditches on site		Look for surface breakouts, overtopping of cell bunds etc Automatic monitoring solution
Visual inspection of transfer pipelines		Look for evidence of leaks that might affect surface water quality, especially where lines cross-restored surfaces
Visual inspection of pipelines around injection zones		Differential settlement around injection zones may lead to leaks in gas and leachate pipelines
Pressure changes in gas pipelines		May indicate blockage by leachate from recirculation
Pressure changes in leachate distribution pipelines		Increase may indicate blockage by leachate from recirculation, whereas a rapid decrease may indicate a (catastrophic) leak in distribution pipework
Water levels in monitoring wells	Pore pressures	Slope instability Compliance monitoring
CH_4 concentration at site boundary		Set threshold as in some environmental permits
Odor, H_2S at site boundary		Set H_2S threshold
LFG quality—CH_4, CO_2, N_2, O_2	Air ingress	Has leachate recirculation (e.g., through increased settlement) impacted on recovered LFG quality
Temperature on basal liner		Assess impact of recirculation on liner longevity

Adapted from EA (2009).

The nature of environmental risk monitoring undertaken will depend to some extent on the day-to-day presence of staff on site. The greatest environmental risks are likely to relate to surface breakouts of leachate resulting from either some type of failure of pipework or the appearance of leachate seeps at the surface, perhaps caused by localized perching or landfill inhomogeneity (e.g., daily intermediate cover). Routine visual inspection and monitoring of the recirculation system and surrounding areas by suitably experienced staff will be an appropriate level of monitoring in many cases. Automatic monitoring of water pressures in distribution pipework and/or electrical conductivity monitoring of surface water in collection ditches may provide rapid warning of leachate discharges in sensitive situations, with the ability for the recirculation system to be automatically shut down.

The risk of slope failure caused by increased pore water pressure in the landfill will also need to be considered at some sites. Where significant waste slopes exist near to leachate recirculation areas monitoring of leachate levels and/or pore water pressures are more likely to be required.

The operation of a leachate recirculation system should not impede the ability to capture and control the migration of landfill gas (and odors), so monitoring the performance of the LFG control system, which should be done anyway, will take account of recirculation activities.

If there is a concern that leachate recirculation will significantly increase waste biodegradation and increase landfill temperature, then monitoring of the temperature of the basal liner system may be required.

Although studies have indicated that during leachate recirculation the head on the basal liner is likely to decrease, increased monitoring of leachate levels may still be advisable, especially if leachate from other parts of the site have been introduced into an area. In particular monitoring, how leachate levels respond when leachate recirculation is switched off (or deliberately run in intermittent mode) may be required.

If recirculation is expected to increase waste degradation rates, or to access previously unwetted wastes, it may be appropriate to revisit the frequency and scope of ambient gas monitoring at the site perimeter for methane and H_2S.

CONCLUDING REMARKS

Recirculation of leachate is likely to continue to be an important activity at many landfills. Therefore, operators and regulators need to understand the need for conceptual design and for monitoring, and for monitoring to be focused and proportionate to the context in which recirculation is being used.

Greater attention must be paid to the conceptual design than has been the case up to now, to ensure that the engineering design and infrastructure meet the objectives of recirculation. Too many schemes have been implemented that have not been fit for their stated purpose. Leachate recirculation should be incorporated in operators' leachate management plans.

The exact requirements and frequency of monitoring will depend on the site, objectives, and scale of the scheme. An overall evaluation framework has been provided as a context for a series of checklist tables that can be used to develop a site-specific monitoring scheme. It must be emphasized that not all of the monitoring in these tables will be needed, or even helpful, for all recirculation schemes. They should be used as an aide memoir only and applied in a way that is proportionate and necessary for individual proposals.

References

Al-Thani, A.A., Beaven, R.P., White, J.K., 2004. Landfill Process Modelling Workshop. Modelling flow to leachate wells in landfills (in special issue: Landfill Process Modelling). Waste Management 24, (3), 271–276. https://doi.org/10.1016/j.wasman.2003.11.010. PMID:15016416.

Augenstein, D., Yazdani, R., Kieffer, J., Benemann, J., 2005. Yolo county, California controlled landfill program: a summary of results since 1994. In: Proceedings Sardinia 2005, Tenth International Waste Management & Landfill Symposium, S. Margherita di Pula, Cagliari, Italy.

Barina, G., March 2005. Effect of leachate recirculation on the stabilisation of organic matter in municipal solid waste. In: Proceedings of International Workshop Hydro-Physico-Mechanics (HPM) of Landfills. LIRIGM, Grenoble, p. 21.

Barina, G., Budka, A., Gisbert, T., Guyonnet, D., Puglierin, L., Cirino, N., 2001. Identification and assessment of leachate recirculation effects at real-scale landfill. In: Christensen, T.H., Cossu, R., Stegmann, R. (Eds.), Eighth International Waste Management and Landfill Symposium, Proceedings Sardinia, 2001, vol. II. CISA, S. Margherita di Pula, Cagliari, Italy, pp. 95–104.

Barina, G., Gisbert, T., Puglierin, L., Boin, B., Lefebvre, X., Gemignani, A., 2003. Integrated study of waste characteristics, effluents and waste mass evolution after 4 years of large scale recirculation. In: Christensen, T.H., Cossu, R., Stegmann, R. (Eds.), Ninth International Waste Management Symposium, Sardinia 2003, Cagliari, Italy.

Barina, G., Oberti, O., Budka, A., October 2005. Leachate recirculation by horizontal trenches: a new French experience. In: Cossu, R., Stegmann, R. (Eds.), Proceedings Sardinia 2005, Tenth International Waste Management and Landfill Symposium, S. Margherita di Pula, Cagliari, Italy. CISA, Italy, pp. 3–7.

Beaven, R.P., Powrie, W., Hudson, A.P., Parkes, D.J., 2006. Compressibility of tyres for use in landfill drainage systems. Waste and Resource Management 159 (WR4), 173–180. https://doi.org/10.1680/warm.2006.159.4.173.

Beaven, R.P., Hudson, A.P., Knox, K., Powrie, W., Robinson, J.P., 2013. Clogging of landfill tyre and aggregate drainage layers by methanogenic leachate and implications for practice. Waste Management 33 (2), 431–444. https://doi.org/10.1016/j.wasman.2012.10.021.

Benson, C.H., Barlaz, M.A., Lane, D.T., Rawe, J.M., 2007. Practice review of five bioreactor/recirculation landfills. Waste Management 27 (1), 13–29.

Bureau, N., Aran, C., Hebe, I., 2005. Bioreactor monitoring: assessment of performances. In: Rathje, E.M. (Ed.), Proceedings Geo-frontiers 2005, Austin, Texas, 24–26 January–geotechnical Special Publications Nos. 130–142. ASCE, Austin, Texas.

Doedens, H., Cord-Landwehr, K., 1989. Leachate recirculation. In: Christensen, T.H., Cossu, R., Stegmann, R. (Eds.), Sanitary Landfilling: Process, Technology and Environmental Impacts. Academic Press, pp. 231–250. ISBN:0-12-174255.

EA, 2009. A technical assessment of leachate recirculation. In: Final Report to the Environment Agency under Research Contract P1–516/3b, by the University of Southampton and Knox Associates. Environment Agency, Bristol, ISBN 978-1-84911-147-8. Science Report SC030144/R6.

Haydar, M.M., Khire, M.V., 2005. Leachate recirculation using horizontal trenches in bioreactor landfills. Journal of Geotechnical and Geoenvironmental Engineering 131 (7), 837–847.

Haydar, M.M., Khire, M.V., 2007. Leachate recirculation using permeable blankets in engineered landfills. Journal of Geotechnical and Geoenvironmental Engineering 133 (4), 360–371.

Jain, P., Powell, J., Townsend, T.G., Reinhart, D.R., 2006. Estimating the hydraulic conductivity of landfilled municipal solid waste using the borehole permeameter test. Journal of Environmental Engineering-ASCE 132 (6), 645–652.

Khire, M.V., Haydar, M.M., 2007. Leachate recirculation in bioreactor landfills using geocomposite drainage material. Journal of Geotechnical and Geoenvironmental Engineering 133 (2), 166–174.

Khire, M.V., Mukherjee, M., 2007. Leachate injection using vertical wells in bioreactor landfills. Waste Management 27 (9), 1233–1247.

Knox, K., Shaw, P.J., 2006. An Experimental Study of the Hydraulic Response of Landfills to Infiltration Events. Final Report to the Norlands Foundation. Entrust Project no. 222285.003.

Kumar, D., Jonnalagada, S., Jain, P., Gawande, N., Townsend, T., Reinhart, D., 2009. Field evaluation of resistivity sensors for in situ moisture measurement in a bioreactor landfill. Waste Management 29, 1547–1557. https://doi.org/10.1016/j.wasman.2008.10.020.

Lavagnolo, M.C., Cossu, R., Malagoli, M., Alibardi, L., 2011. Energy crops cover in landfills. In: Cossu, H., Kjeldsen, M., Reinhart, S. (Eds.), Proceedings Sardinia 2011-Thirteenth International Waste Management and Landfill Symposium. CISA Publisher, ISBN 978-88-6265-000-7, pp. 457–458.

Lagerkvist, A., Cossu, R., October 2005. Leachate recirculation: concepts and applications. In: Cossu, R., Stegmann, R. (Eds.), Proceedings Sardinia 2005, Tenth International Waste Management and Landfill Symposium, S. Margherita di Pula, Cagliari, Italy. CISA, Italy, pp. 3–7.

McCreanor, P.T., Reinhart, D.R., 1999. Hydrodynamic modeling of leachate recirculating landfills. Waste Management & Research 17 (6), 465–469.

McCreanor, P.T., Reinhart, D.R., 2000. Mathematical modeling of leachate routing in a leachate recirculating landfill. Water Research 34 (4), 1285–1295.

McCreanor, P.T., Reinhart, D.R., 1996. Hydrodynamic modeling of leachate recirculating landfills. Water Science and Technology 34 (7–8), 463–470.

Morris, J.W.F., Vasuki, N.C., Baker, J.A., Pendleton, C.H., 2003. Findings from long-term monitoring studies at MSW landfill facilities with leachate recirculation. Waste Management 23 (7), 653–666.

Mouchel Consulting Ltd, 2001. The Brogborough Test Cells: Final Report, Gas Monitoring and Hydraulics of Leachate Recirculation. Environment Agency, R&D. Technical Report: HOCO-231/1. Contract no: P1-352, 122 pp.

Powrie, W., Beaven, R.P., 1999. Hydraulic properties of household waste and implications for landfills. In: Proceedings of the Institution of Civil Engineers (Geotechnical Engineering), vol. 137 (4), pp. 235–247. https://doi.org/10.1680/geng.1997.137.4.235.

Reinhart, D.R., 1996. Full-scale experiences with leachate recirculating landfills: case studies. Waste Management & Research 14 (4), 347–365.

Rowe, R.K., 2005. Long-term performance of contaminant barrier systems. Geotechnique 55 (9), 631–678.

Skhiri, N., Guerbois, M., Hebe, I., 2006. Technical and environmental performances assessment for a bioreactor. In: Lagerkvist, A. (Ed.), ICLRS 06 – Abstract Proceedings of the 4th Intercontinental Landfill Research Symposium, Gallivare, Sweden, pp. 183–184.

Thiel, R., October 2005. Observed benefits and problems associated with leachate recirculation. In: Proceedings Sardinia 2005, S. Margherita di Pula, Cagliari, Italy, pp. 3–7 (CD only).

Townsend, T.G., Miller, W.L., Earle, J.F.K., 1995. Leachate recycle infiltration ponds. Journal of Environmental Engineering 121 (6), 465–471.

White, J.K., Beaven, R.P., Powrie, W., Knox, K., 2011. Leachate recirculation in a landfill: some insights obtained from the development of a simple 1-D model. Waste Management 31, (6), 1210–1221. https://doi.org/10.1016/j.wasman.2010.10.022.

Woelders, H., Luning, L., Van Velthoven, F., Hermkes, H., Oonk, H., October 2005. Dutch Sustainable Landfill Research program: four years experience with the bioreactor test cell Landgraaf. In: Cossu, R., Stegmann, R. (Eds.), Proceedings Sardinia 2005, Tenth International Waste Management and Landfill Symposium, S. Margherita di Pula, Cagliari, Italy. CISA, Italy, pp. 3–7.

Yazdani, R., Kieffer, J., Sananikone, K., Augenstein, D., 2006. Full Scale Bioreactor Landfill for Carbon Sequestration and Greenhouse Emission Control. Final Technical Progress Report. Planning and Public Works Department, D.O.E., Yolo County. Award Number DE-FC26–01NT41152. http://www.yolocounty.org/recycle/bioreactor.htm.

13 CHAPTER

PLANNING AND DESIGN

13.1

LANDFILL SITING AND PUBLIC INVOLVEMENT

Giovanni De Feo and Raffaello Cossu

INTRODUCTION

Landfill siting should be based on a multidisciplinary approach that incorporates natural, physical–social sciences, politics, and ethics. It is a fundamental part in the most general landfill planning and design approach (Raga et al., 2010).

A siting process can be broadly developed with a focus on three main phases (De Feo and De Gisi, 2014).

The first phase comprises a "feasibility analysis." The aim of this first phase is to identify "unsuitable areas" and "potentially suitable areas." Identification should be performed on the basis of "exclusion criteria" as defined by legislation and literature. If an area is affected by an exclusion criterion, it is deemed unacceptable and should be excluded from the selection process.

The second phase relates to spatial multicriteria analysis. This is aimed at identifying a list of sites through use of "preferential criteria" and "penalizing criteria." A preferential criterion indicates the presence of elements of suitability as well as advisability for the siting of the landfill. Conversely, a penalizing criterion indicates the presence of contraindications advocating the construction of a landfill only if particular care is taken during the design and construction phases. A penalizing criterion acts as a discriminant and does not exclude the siting of the landfill.

The third phase focuses on selection of the most suitable sites among those deemed potentially suitable.

A landfill siting process can be conveniently developed combining spatial techniques, such as geographical information systems (GIS) and multicriteria decision analysis (MCDA) (Sumathi et al., 2008; Tavares et al., 2011). The use of GIS allows an initial screening process to be performed to eliminate unsuitable areas, mainly working at the feasibility analysis level. Secondarily, the application of MCDA facilitates selection of the most suitable site.

An example of landfill siting procedure is proposed at the beginning of the chapter. Siting criteria and identification of alternatives are also discussed. Subsequently, an example of how to define a preliminary set of evaluation criteria and subcriteria to be used in assessing siting alternatives is provided. Finally, the methods employed to compare siting alternatives by means of MCDA techniques are described. The last paragraph of the chapter is focused on public involvement and landfill siting.

SITING PROCEDURE

Extensive reports are present in literature describing landfill siting procedures. In this chapter, the procedure proposed by De Feo and De Gisi (2014) is described: this procedure enhances the selection of a landfill site while simultaneously taking into consideration other types of waste management facilities, and ensuring that land suited for occupation by other types of waste management options is not overlooked. A case study relating to the siting of a hazardous waste landfill, undoubtedly a complicated situation, is also presented.

The proposed methodology for landfill siting consists in the following eight phases:

1. Definition of siting criteria on the basis of national and regional legislation for all facilities within the solid waste management system.
2. Mapping of potentially suitable areas for each specific type of solid waste facility considered.
3. Elaboration of a "land use map of potentially suitable areas" for the siting of all solid waste facilities.
4. Identification of siting alternatives with a punctual verification of exclusion criteria.
5. Definition of criteria to be applied in evaluation of siting alternatives (evaluation criteria).
6. Definition of an alternatives matrix.
7. Definition of criteria priorities by means of MCDA techniques, to obtain a ranking of alternatives.
8. Identification of the best site.

To this regard, phases 1–4 are grouped in section "Siting criteria and identification of alternatives," phases 5–6 in section "Evaluation criteria," and phases 7–8 in section "Comparison of siting alternatives with multicriteria decision analysis techniques."

SITING CRITERIA AND IDENTIFICATION OF ALTERNATIVES

Siting criteria for all facilities comprised in a solid waste management system should be defined at the beginning of the process. In this chapter, the following facilities are taken into consideration: hazardous waste landfills; nonhazardous waste landfills; inert waste landfills; waste-to-energy plants; mechanical and biological treatment plants; composting and anaerobic digestion plants.

Criteria sets for each single type of facility should be identified. As illustrated in Table 13.1.1, these can be defined on the basis of existing national and regional legislation, regulations, experiences and expertise and should cover natural, socioeconomic, technical, and environmental aspects.

The second phase is represented by the application of spatial multicriteria analysis aimed at identifying "unsuitable areas" and "potentially suitable areas" on the basis of "exclusion criteria" for all types of solid waste facilities considered. Thus, a suitability map is obtained for each type of plant.

Table 13.1.1 Example of siting criteria to be used in the mapping of potentially suitable areas (feasibility analysis) for all types of solid waste facility considered (the example is referred to Italian and regional legislations, Campania Region of Southern Italy) (De Feo and De Gisi, 2014)

No.	Criteria	Solid Waste Facilities[a]					
		F1	F2	F3	F4	F5	F6
1	Altimetry	***	***	***	***	***	***
2	First category seismicity_1	***	***				
3	First category seismicity_2				**	**	**
4	Second category seismicity	**					
5	"High"[b] and "very high"[b] landslide risk[c]	***	***	***	***	***	***
6	"Moderate"[b] and "low"[b] landslide risk[c]	**	**	**	**	**	**
7	"High"[b] and "very high"[b] hydraulic risk[c]	***	***	***	***	***	***
8	"Moderate"[b] and "low"[b] hydraulic risk[c]	**	**	**	**	**	**
9	Fluvial dynamics	***	***	***	***	***	***
10	Surface karstification	***	***	***	***	***	***
11	Wooded areas_1	***	***	***			
12	Wooded areas_2				**	**	**
13	Permanent pastures	***	***	***	***	***	***
14	Distance from urban areas	***	***	***	***	**	**
15	Distance from water sources	***	***	***	***	***	***
16	Groundwater vulnerability	***	***	***	***	***	***
17	Wastewater treatment plants efficiency				*	*	*
18	Distance from water bodies	***	***	***	***	***	***
19	Landscape	***	***	***	***	***	***

(Continued)

Table 13.1.1 Example of siting criteria to be used in the mapping of potentially suitable areas (feasibility analysis) for all types of solid waste facility considered (the example is referred to Italian and regional legislations, Campania Region of Southern Italy) (De Feo and De Gisi, 2014)—cont'd

No.	Criteria	Solid Waste Facilities[a]					
		F1	F2	F3	F4	F5	F6
20	Protected natural areas	***	***	***	***	***	***
21	Natura 2000 network areas	***	***	***	***	***	***
22	Faunal repopulation areas	**	**	**	**	**	**
23	Industrial areas				*	*	*
24	Distance from infrastructure	***	***	***	***	***	***
25	Disused industrial areas				*	*	*
26	Mining activity areas (quarries)	*	*	*	**	**	**
27	Contiguous industrial areas			*	*	*	*
28	Agricultural areas_1	***	***	***			
29	Agricultural areas_2				**	**	**
30	Archaeological areas	***	***	***	***	***	***
TOT_{EX}	Excluding criteria	18	18	17	15	14	14
TOT_{PE}	Penalizing criteria	4	3	3	7	8	8
TOT_{PR}	Preferential criteria	1	1	2	4	4	4
TOT	Total number	23	22	22	26	26	26

Type of criteria: *, preferential criteria; **, penalizing criteria; ***, exclusion criteria.

[a]Solid waste management facilities: *F1*, hazardous waste landfill; *F2*, nonhazardous waste landfill; *F3*, inert materials landfill; *F4*, waste-to-energy plant; *F5*, mechanical biological treatment plant (MBT); *F6*, composting and anaerobic digestion plant.

[b]The values assumed by the risk zoning descriptors (both for landslide risk and hydraulic risk) are directly furnished by the catchment plans of the following local authorities: National Authority of "Liri-Garigliano and Volturno" (http://www2.autoritadibacino.it/), Regional Authority of "Sarno" (http://www.autoritabacinosarno.it/), Regional Authority of "Nord-Occidentale della Campania" (http://www.autoritabacinonordoccidentale.campania.it/), Regional Authority of "Destra Sele" (http://www.autoritabacinodestrasele.it/), Regional Authority of "Puglia" (http://www.adb.puglia.it/public/news.php).

[c]Risk is defined as a measure of the probability and severity of an adverse effect to health, property, or the environment (Fell et al., 2008).

All other sources: Italian (Decreto Legislativo 3 aprile 2006, n. 152, "Norme in materia ambientale", Gazzetta Ufficiale n. 88 del 14 aprile 2006 - Supplemento Ordinario n. 96, in Italian) and European Union Legislation (Council Directive 1999/31/EC of April 26, 1999 on the landfill of waste, Official Journal L 182, July 16, 1999, 1–9).

The main aim of the third phase is to provide a "land use map of potentially suitable areas." As shown in Fig. 13.1.1, this map is obtained through georeferenced overlapping of the single suitability maps obtained in the previous phase. The novelty of the siting procedure proposed by De Feo and De Gisi (2014) is that it allows all types of solid waste facilities present within the solid waste management system to be taken into consideration simultaneously. The most important goal is to minimize space wastage and ensure the appropriate use of land.

The "land use map of potentially suitable areas" includes the following:

- Areas meeting exclusion factors for the nonsuitability of siting.
- Areas potentially suitable for the siting of inert waste landfills.
- Areas potentially suitable for the siting of nonhazardous waste landfills.
- Areas potentially suitable for the siting of hazardous waste landfills.
- Areas for the siting of solid waste technological plants.

The fourth phase of the procedure is the identification of siting alternatives with a punctual verification of exclusion criteria. Application of "preferential" and "penalizing" criteria to the macro areas suitable for siting of a waste landfill (inert, nonhazardous, or hazardous waste landfills) enhances identification of each single site, which can then be cordoned off. Areas characterized by preferential

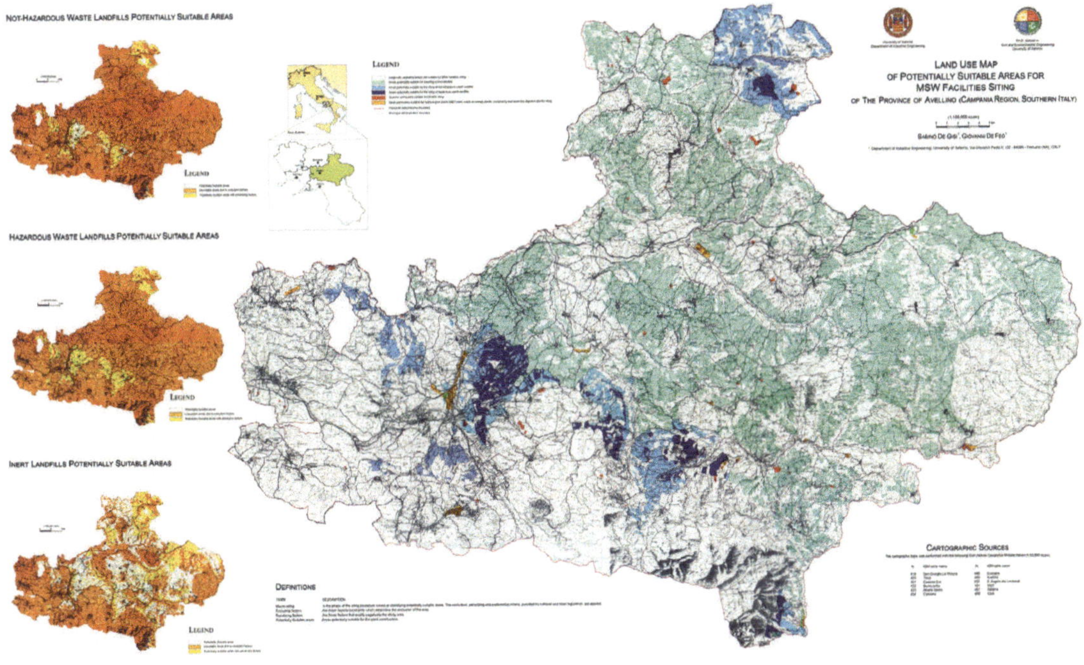

Figure 13.1.1 Land use map of potentially suitable areas (De Feo and De Gisi, 2014).

factors and/or a lower number of penalizing factors should be preferred. Once a single site has been identified, further verification is required to evaluate the exclusion criteria on a more detailed mapping scale (De Feo and De Gisi, 2014).

EVALUATION CRITERIA

The defining of a preliminary set of evaluation criteria and subcriteria to be used in evaluation of siting alternatives is the aim of the fifth phase of the siting procedure. For this purpose, the following seven macrocategories should be taken into account:

- population presence (C_1);
- groundwater risk (C_2);
- agricultural value (C_3);
- additional traffic on local roads (C_4);
- economy (C_5);
- protected areas (C_6);
- climate (C_7).

Naturally, in addition to the above, which includes the criteria listed in the procedure described by De Feo and De Gisi (2014), other criteria may also be adopted. Several other examples of evaluation criteria for landfill siting processes are described in literature (Kontos et al., 2005; Zamorano et al., 2008; Moeinaddini et al., 2010; Eskandari et al., 2012; Gorsevski et al., 2012; Korucu and Erdagi, 2012; Demesouka et al., 2013; Gbanie et al., 2013).

The criteria applied may be of a quantitative or qualitative nature; the significance of each criterion is indicated by the subcriteria defined. A numeric value is given to the qualitative criterion on the basis of the so-called "average value of intervals" defined by De Feo et al. (2014).

Resident population ($C_{1,1}$) and floating population ($C_{1,2}$) are representative of subcriteria for population presence (C_1). Moreover, people living at a distance of 0–500 m ($C_{1,1,1}$), 500–1000 m ($C_{1,1,2}$), or 1000–1500 m ($C_{1,1,3}$) could be taken into account for both $C_{1,1}$ and $C_{1,2}$. The value to be attributed to these subcriteria is equal to the sum of the resident population in the range 0–500, 500–1000, and 1000–1500 m from the site. When the resident population is unknown, an estimate could be obtained by identifying the number of houses available and assigning to each house a number equal to the average number of residents/home in that area. In the MCDA process, the subcriteria related to resident population needs to be minimized to identify the best site.

Groundwater risk (C_2) may be evaluated both in terms of vulnerability ($C_{2,1}$) and presence of wells and springs ($C_{2,2}$). Groundwater vulnerability can be assessed using the GOD method (Foster et al., 2002) in conjunction with the groundwater vulnerability map of the area concerned. GOD index, which is based on three parameters (Groundwater confinement, Overlying strata, Depth to groundwater), will range between values of 0 and 1. For instance, applying the "average value of intervals" defined by De Feo et al. (2014) the following numeric values could be attributed to $C_{2,1}$: very low = 0.125; low = 0.375; moderate = 0.625; high = 0.875. Groundwater vulnerability should be minimized to identify the best site with MCDA.

The presence of wells and springs over a 1000 m radius, for instance, may also be assessed taking into account the presence of thermomineral wells ($C_{2,2,1}$), drinking water wells ($C_{2,2,2}$), irrigation and industrial wells ($C_{2,2,3}$), and springs ($C_{2,2,4}$). These subcriteria should be minimized with MCDA.

Agricultural value (C_3) may be evaluated in terms of organic farms within a 1000 m radius ($C_{3,1}$), Farmhouses within a 1000 m radius ($C_{3,2}$) and land use ($C_{3,3}$). For instance, the following numeric values could be attributed to $C_{3,3}$: uncultivated = 0.083; shrubbery = 0.25; pasture = 0.416; arable = 0.583; heterogeneous agricultural areas = 0.75; permanent crops = 0.916. These subcriteria should be minimized with MCDA.

Traffic (C_4) can be evaluated in terms of interference of additional traffic with local roads. The value to be attributed to the criterion is calculated on the basis of the following classes: very low (0.1) when the position of the site from the main road only slightly interferes with local traffic; low (0.3): when the route to the nearest main road does not cross urban areas and the distance is less than 4 km; low-medium (0.5): when the route to the nearest main road does not cross urban areas and the distance is more than 4 km, or when the route requires marginal crossing of urban areas and the distance from the nearest main road is less than 4 km; medium (0.7): when the route requires marginal crossing of urban areas and the distance from the nearest main road is more than 4 km, or the distance is less than 4 km but there is an appreciable risk of interference with heavily trafficked roads; high (0.9): when the route interferes with the local road network. The criterion should be minimized with MCDA.

Economy (C_5) may be evaluated on the basis of distance from the main road ($C_{5,1}$) and accessibility ($C_{5,2}$). The value to be attributed to criterion $C_{5,1}$ is calculated as the distance between the center of the site and the nearest main road. This criterion should be minimized.

The value to be attributed to $C_{5,2}$ can be established by answering the question: "Is the site directly accessible?" A site should be considered accessible if no construction of a new access road is required. Two possible answers can be given: yes (0.750); no (0.250). This criterion should be maximized by MCDA.

Naturalistic value (C_6) may be evaluated according to protected areas. The value attributed to C_6 is provided by answering the question: "Are there any protected areas within a range of 0–2000 m from the site?" Possible answers are yes (0.750) or no (0.250). The criterion should be maximized by MCDA.

Climate (C_7) may be evaluated in terms of rainfall ($C_{7,1}$) and average wind velocity ($C_{7,2}$). The value to be attributed to criterion $C_{7,1}$ is based on the intensity of rain at the site location. This value should be measured in millimeters of rain per year. The criterion should be minimized by MCDA.

The value to be attributed to criterion $C_{7,2}$ may be calculated as average wind velocity at the site location (for instance, in m/s). The criterion is minimized by MCDA.

Evaluation criteria and subcriteria in the preliminary set are not necessarily present in the alternatives matrix. This occurs when the sum of values obtained by one or more criteria and/or subcriteria for all alternative sites considered is equal to zero.

Definition of the alternatives matrix represents the sixth phase in the procedure and involves attributing a value to the criteria for all considered alternative sites. The alternatives matrix contains as many rows as the number of possible sites identified, and as many columns as the number of effective criteria and subcriteria. With regard to the data acquired, in the present study raster data were used, although they can be developed in a GIS environment.

Table 13.1.2 illustrates an example of alternatives matrix for the siting of a hazardous waste landfill. Six subcriteria from the preliminary set of 20 evaluation criteria were not effective due to the lack of effective values (values assigned were equal to zero) for all selected alternatives: $C_{1,2,1}$, $C_{1,2,2}$, $C_{2,2,1}$, $C_{2,2,2}$, $C_{2,2,3}$, $C_{3,2}$.

COMPARISON OF SITING ALTERNATIVES WITH MULTICRITERIA DECISION ANALYSIS TECHNIQUES

A series of multicriteria decisionn analysis (MCDA)-based techniques can be used to compare siting alternatives. In this section, two different techniques for use in identifying the best alternative are described:

1. Analytic hierarchy process (AHP) as an MCDA technique (Saaty, 1996) coupled with the priority scale weighting (PSW) tool (De Feo and De Gisi, 2010) to evaluate criteria priorities (in short "AHP–PSW").
2. Simple additive weighting (SAW) (Ching-Lai and Kwangsun, 1981) as a multicriteria technique used to define alternatives priorities, coupled with the paired comparison technique (PCT) (Mondy and Noe, 2008) as a means of evaluating criteria priorities (in short "SAW–PCT").

Literature reports provide further examples of comparison of siting alternatives using MCDA techniques (Chang et al., 2008; Sumathi et al., 2008; Sharifi et al., 2009; Moeinaddini et al., 2010; Eskandari et al., 2012; Demesouka et al., 2013; Gbanie et al., 2013).

Definition of criteria priorities using the PSW tool and application of AHP to the alternatives matrix weighted with PSW tool

As shown in Fig. 13.1.2, the AHP method breaks down a complex multicriteria decisional issue into a hierarchy and is based on a pairwise comparison of the importance of different criteria and subcriteria (Saaty, 1977, 1996, 2001, 2005; Forman and Selly, 2001).

The AHP method is based on a pairwise comparison of the importance of different criteria. The fundamental scale for pairwise comparison (the so-called "Saaty scale") defines and explains values 1 thorough 9 to judgments in comparing pairs of like elements at each level of a hierarchy against criteria in the next highest level. In particular, value 1 ("equal importance") indicates an equal contribution of two criteria to the objective. Value 3 ("moderate importance) indicates that experience and judgment tend to slightly favor one criterion over another. Value 5 ("high importance) implies a strong preference for one criterion over another based on experience and judgment. Value 7 ("very high importance") indicates how one criterion is deemed highly preferable over another. Finally, value 9 ("extreme importance") indicates that evidence favoring one criterion over another is of the highest possible order of affirmation. Values 2, 4, 6, and 8 are used as a compromise between the above values (Saaty, 2001).

Table 13.1.2 Example of alternatives matrix for the siting of a hazardous waste landfill (De Feo and De Gisi, 2014)

Alternatives	Effective Evaluation Criteria and Subcriteria													
	$C_{1,1,1}$ (m)	$C_{1,1,2}$ (m)	$C_{1,1,3}$ (m)	$C_{1,2,3}$ (m)	$C_{2,1}$	$C_{2,2,4}$	$C_{3,1}$	$C_{3,3}$	C_4	$C_{5,1}$ (km)	$C_{5,2}$	C_6 (km)	$C_{7,1}$ (mm/year)	$C_{7,2}$ (m/s)
S_1	0	6	9	10	0.375	0	0	0.583	0.3	26.3	0.25	8.37	584.2	4.227
S_2	3	33	48	0	0.375	4	0	0.75	0.1	8	0.25	4.94	1542.2	4.252
S_3	3	54	60	15	0.625	5	0	0.916	0.1	27	0.25	3.13	879.8	1.727
S_4	3	12	42	120	0.625	0	1	0.583	0.3	54.4	0.75	0.78	1479.6	2.785

Criteria: C_1, population presence; C_2, groundwater risk; C_3, agricultural value; C_4, traffic; C_5, economy; C_6, protected areas; C_7, climate.
Subcriteria:

- Resident population ($C_{1,1}$); floating population ($C_{1,2}$); people at 0–500 m ($C_{1,1,1}$); people at 500–1000 m ($C_{1,1,2}$); people at 1000–1500 m ($C_{1,1,3}$, $C_{1,2,3}$).
- Vulnerability ($C_{2,1}$; very low = 0.125; low = 0.375; moderate = 0.625; high = 0.875); wells and springs in 1000 m radius ($C_{2,2}$); springs ($C_{2,2,4}$).
- Organic farms in 1000 m radius ($C_{3,1}$); land use ($C_{3,3}$; uncultivated = 0.083: shrubbery = 0.25; pasture = 0.416; arable = 0.583; heterogeneous agricultural areas = 0.75; permanent crops = 0.916);
- Distance from the main road ($C_{5,1}$, km); accessibility ($C_{5,2}$; a site was considered "accessible" when there was no need to widen the access road or to build a new road; accessible = 0.75, not accessible = 0.25);
- Rainfall ($C_{7,1}$, mm/year) and average wind velocity ($C_{7,2}$, m/s).

De Feo and De Gisi (2010) have developed a novel criteria weighting tool (CWT) aimed at defining the siting criteria weights vector for each decision-maker in line with the AHP theory: the "priority scale" (PSW) shown in Fig. 13.1.3. The term CWT refers specifically to the usefulness of the tool in identifying decision-maker priorities, and to shaping the scale of the same.

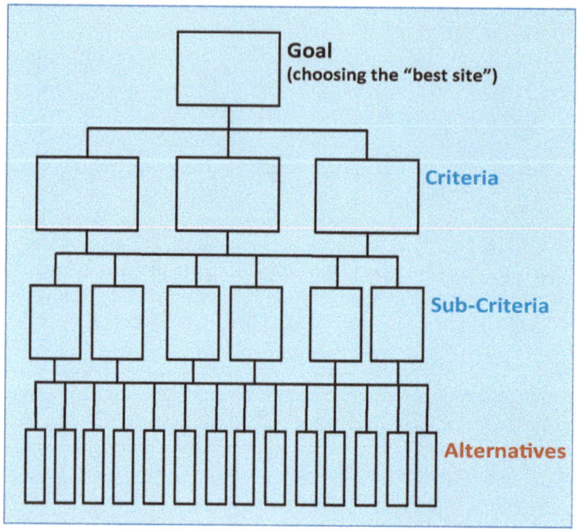

Figure 13.1.2 Framework of the analytic hierarchy process.

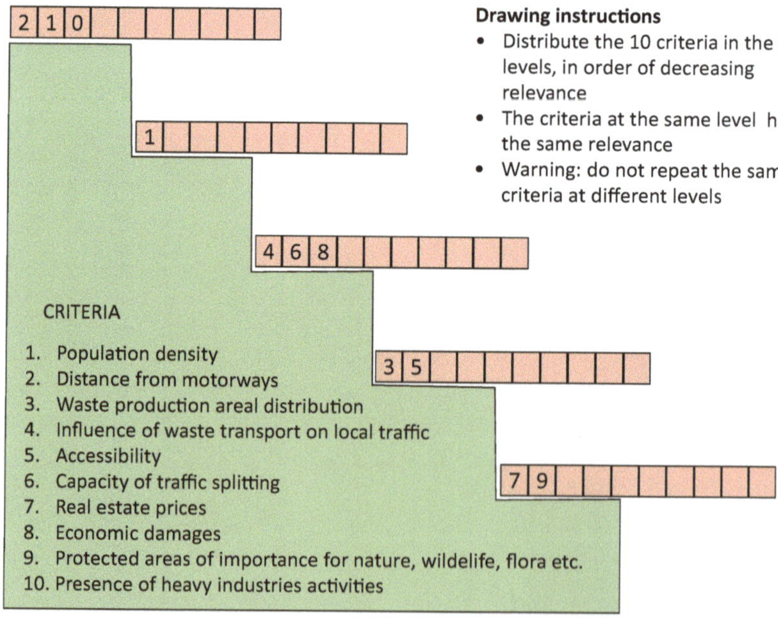

Figure 13.1.3 An example of the priority scale weighting tool (De Feo and De Gisi, 2010).

Criteria priorities should be defined both at subcriteria (local priorities) and at criteria levels (global priorities).

Following drafting of the priorities scales, these are subsequently transformed into the corresponding criteria weights vector according to the procedure described for the pairwise comparison in De Feo and De Gisi (2010):

- two criteria on the same level assume an equal importance ("equal");
- criteria on a higher level are more important than criteria on a lower level;
- a level of difference between two criteria corresponds to "moderate" in favor of the criteria on the higher level;
- two levels of difference between two criteria corresponds to "strong" in favor of the criteria on the higher level;
- three levels of difference between two criteria corresponds to "very strong" in favor of the criteria on the higher level;
- four levels of difference between two criteria corresponds to "extreme" in favor of the criteria on the higher level.

PSW adopts only values 1 (equal), 3 (moderate), 5 (strong), 7 (very strong), and 9 (extreme) of the Saaty scale (Saaty, 1996). Therefore, for the example reported in Fig. 13.1.3, it can be stated that criteria 2 is "moderately" more important than criteria 1; criteria 2 is "strongly" more important than criteria 4, 6, and 8; criteria 2 is more important than criteria 3 and 5 in a "very strong" way; criteria 2 is extremely more important than criteria 7 and 9; criteria 2 and 10 have the same importance; etc.

PSW graphically collects criteria priorities with the aim of preventing mistakes from being made, and to provide a comprehensive view of all criteria priorities. Each priority scale should subsequently be transformed into a weights vector using Expert Choice software (Expertchoice, 2004).

Fig. 13.1.4 shows an example to define criteria priorities to be used with AHP (as well as SAW) using the PSW tool.

All priorities can be defined by technical consultants; however, they may also be assigned by other means, for instance, by adopting a stakeholder involvement approach as explained in the section "Public involvement".

Table 13.1.3 shows local and global priorities vectors related to the example given in Fig. 13.1.4. For the first level of criteria, local and global criteria coincide. At each level, the sum of local priorities is equal to 1. For instance, local priorities for $C_{1,1}$ and $C_{1,2}$ are 0.833 and 0.167, respectively, while local priorities for $C_{1,1,1}$, $C_{1,1,2}$, and $C_{1,1,3}$ are 0.735, 0.207, and 0.058, respectively.

The global priority 0.422 for C_1 is divided into global values 0.352 for $C_{1,1}$ and 0.070 for $C_{1,2}$. The value 0.352 is obtained by multiplying 0.422 by the local value 0.833, while the value 0.070 is obtained by multiplying 0.422 by the local value 0.167.

Application of AHP to the alternatives matrix of Table 13.1.2 weighted with the calculated global priorities vector of Table 13.1.3 yielded the alternatives priorities reported in Table 13.1.4. The global

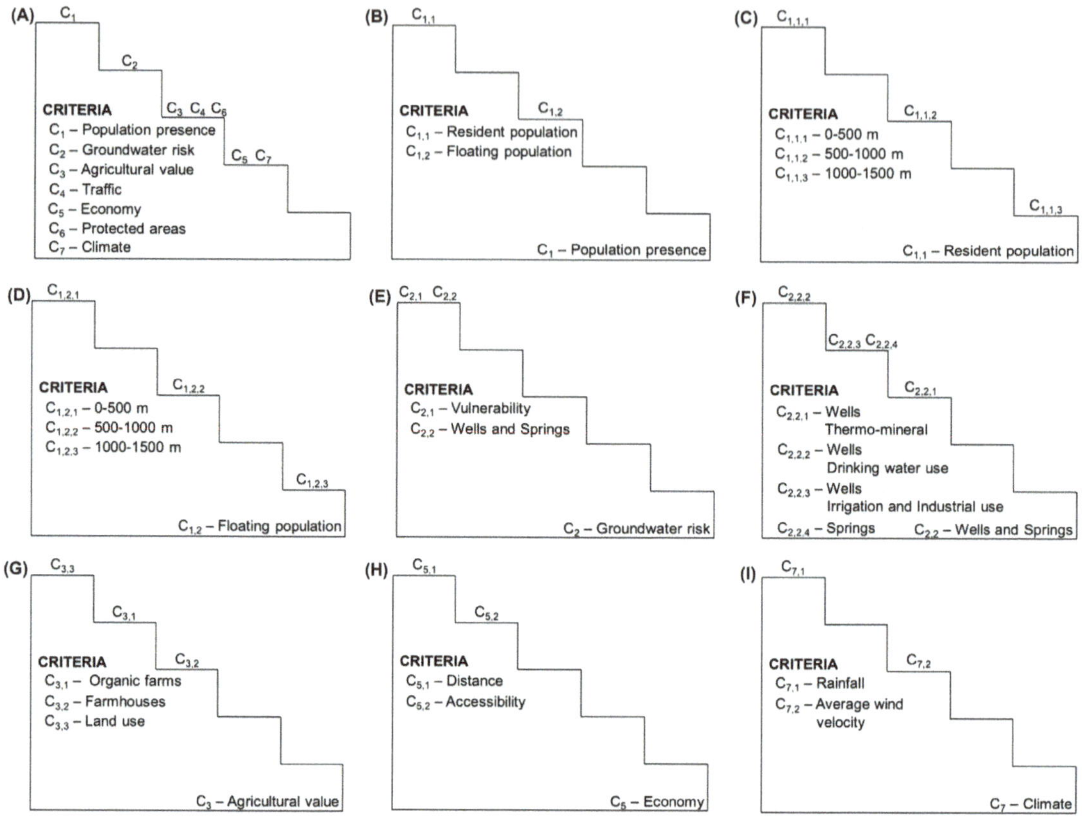

Figure 13.1.4 Example of priority scales for the definition of criteria and subcriteria priorities related to the siting of hazardous waste landfills. (A) global priorities among the seven criteria; (B) local priorities between the two subcriteria of C_1; (C) local priorities among the three subcriteria of $C_{1,1}$; (D) local priorities among the three subcriteria of $C_{1,2}$; (E) local priorities between the two subcriteria of C_2; (F) local priorities among the four subcriteria of $C_{2,2}$; (G) local priorities among the three subcriteria of C_3; (H) local priorities between the two subcriteria of C_5; (I) local priorities between the two subcriteria of C_7. *Adapted from De Feo and De Gisi (2014).*

priority of each single alternative is obtained by summing the priority obtained by the alternative for each single subcriterion.

Both the ranking obtained and single alternatives priorities may be normalized respect to the maximum priority obtained for S_1 (e.g., 0.467). In practice, normalization is obtained by multiplying each value in Table 13.1.4 by 1/0.467. The results obtained are reported in Table 13.1.5 and in Fig. 13.1.5.

The ranking list is S_1, S_2, S_3, and S_4. The difference between the priority of the first and second best alternatives was 55.4%. It was, therefore, evident that S_1 is the best site on which to place a hazardous waste landfill using the AHP–PSW approach.

Table 13.1.3 Local and global priorities of evaluation criteria based on the priority scale weighting method related to the example of Fig. 13.1.4

Criteria's Level			Priorities					
1st	2nd	3rd	Local			Global		
C_1			0.422			0.422		
	$C_{1,1}$			0.833			0.352	
		$C_{1,1,1}$			0.735			0.259
		$C_{1,1,2}$			0.207			0.073
		$C_{1,1,3}$			0.058			0.020
	$C_{1,2}$			0.167			0.070	
		$C_{1,2,3}$			1.000			0.070
C_2			0.221			0.221		
	$C_{2,1}$			0.500			0.111	
	$C_{2,2}$			0.500			0.111	
		$C_{2,2,4}$			1.000			0.111
C_3			0.093			0.093		
	$C_{3,1}$			0.250			0.023	
	$C_{3,3}$			0.750			0.070	
C_4			0.093			0.093		
C_5			0.038			0.038		
	$C_{5,1}$			0.750			0.029	
	$C_{5,2}$			0.250			0.010	
C_6			0.093			0.093		
C_7			0.038			0.038		
	$C_{7,1}$			0.833			0.032	
	$C_{7,2}$			0.167			0.006	

Table 13.1.4 Final priorities of landfill siting alternatives calculated by means of the analytic hierarchy process—priority scale weighting method

Alternative	$C_{1,1,1}$	$C_{1,1,2}$	$C_{1,1,3}$	$C_{1,2,3}$	$C_{2,1}$	$C_{2,2,4}$	$C_{3,1}$	$C_{3,3}$	C_4	$C_{5,1}$	$C_{5,2}$	C_6	$C_{7,1}$	$C_{7,2}$	Priority
S_1	0.137	0.038	0.011	0.034	0.059	0.059	0.012	0.037	0.000	0.009	0.005	0.049	0.017	0.000	0.467
S_2	0.000	0.017	0.003	0.037	0.059	0.012	0.012	0.019	0.049	0.015	0.005	0.027	0.000	0.000	0.255
S_3	0.000	0.000	0.000	0.033	0.001	0.000	0.012	0.000	0.049	0.009	0.005	0.015	0.012	0.003	0.139
S_4	0.000	0.034	0.004	0.000	0.001	0.059	0.000	0.037	0.000	0.000	0.000	0.000	0.001	0.002	0.138

Table 13.1.5 Final normalized priorities of landfill siting alternatives calculated by means of the analytic hierarchy process—priority scale weighting method

Alternative	$C_{1,1,1}$	$C_{1,1,2}$	$C_{1,1,3}$	$C_{1,2,3}$	$C_{2,1}$	$C_{2,2,4}$	$C_{3,1}$	$C_{3,3}$	C_4	$C_{5,1}$	$C_{5,2}$	C_6	$C_{7,1}$	$C_{7,2}$	Priority
S_1	0.293	0.081	0.024	0.073	0.126	0.126	0.026	0.079	0.000	0.019	0.011	0.105	0.036	0.000	1.000
S_2	0.000	0.036	0.006	0.079	0.126	0.026	0.026	0.041	0.105	0.032	0.011	0.058	0.000	0.000	0.546
S_3	0.000	0.000	0.000	0.071	0.002	0.000	0.026	0.000	0.105	0.019	0.011	0.032	0.026	0.006	0.298
S_4	0.000	0.073	0.009	0.000	0.002	0.126	0.000	0.079	0.000	0.000	0.000	0.000	0.002	0.004	0.296

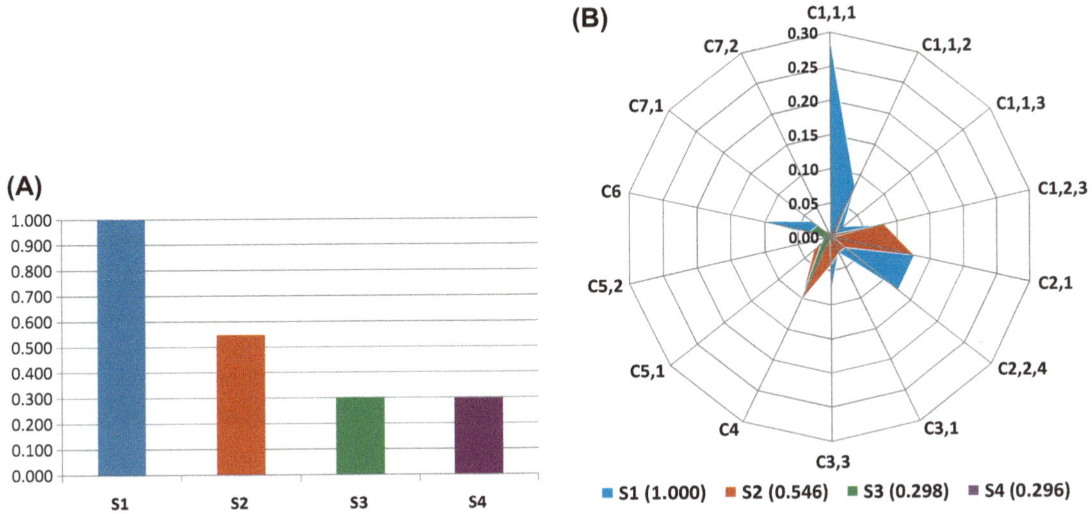

Figure 13.1.5 Analytic hierarchy process—priority scale weighting method. Global normalized alternatives rankings in the developed example related to the siting of hazardous waste landfills (A), and contribution of each evaluation criteria to the global alternatives ranking (normalized values are reported in brackets) and normalized values obtained for each criterion (B). *Adapted from De Feo and De Gisi (2014).*

Definition of criteria priorities by means of PCT and application of the SAW method to the alternatives matrix weighted with PCT method

The priority scale tool may be used as a starting point in the performing of pairwise comparisons using the PCT method as defined by Mondy and Noe (2008).

With the PCT technique, paired comparison between criteria C_i and C_j are coded as follows: C_i is given a priority 1 if it is preferred over C_j; C_i has priority 0 if C_j is preferred over C_i; both C_i and C_j have priority 0.5 if they are of equal importance. Moreover, to avoid assigning a 0 priority, each criterion is considered preferable over a dummy criterion. The priority of each criterion is calculated dividing the sum of its preferences by the sum of preferences of all criteria (De Feo et al., 2013).

Fig. 13.1.6 illustrates application of the PCT technique for the attribution of local criteria priorities to comparisons of priority scales seen in Fig. 13.1.4. Global criteria priorities should be calculated as described for the AHP technique in the previous section. For instance, the global weight for criterion $C_{1,1,1}$ is obtained by multiplying the global weight of C_1 (0.250) by the local weight of $C_{1,1}$ (0.667), and, finally, by the local weight of $C_{1,1,1}$ (0.500).

Table 13.1.6 illustrates the local and global priorities for evaluation criteria using the PCT method related to the example given in Fig. 13.1.4.

	C1	C2	C3	C4	C5	C6	C7	Cd	Sum	Weigth
C1		1	1	1	1	1	1	1	7	**0.250**
C2	0		1	1	1	1	1	1	6	**0.214**
C3	0	0		0.5	1	0.5	1	1	4	**0.143**
C4	0	0	0.5		1	0.5	1	1	4	**0.143**
C5	0	0	0	0		0	0.5	1	1.5	**0.054**
C6	0	0	0.5	0.5	1		1	1	4	**0.143**
C7	0	0	0	0	0.5	0		1	1.5	**0.054**
Cd	0	0	0	0	0	0	0		0	0.000
									28	1.000

Population presence (C1)

	C1,1	C1,2	Cd	Sum	Weigth
C1,1		1	1	2.000	**0.667**
C1,2	0		1	1.000	**0.333**
Cd	0	0		0.000	0.000
				3.000	1.000

Resident population (C1,1)

	C1,1,1	C1,1,2	C1,1,3	Cd	Sum	Weigth
C1,1,1		1	1	1	[illegible]	**0.500**
C1,1,2	0		1	1	2	**0.333**
C1,1,3	0	0		1	1	**0.167**
Cd	0	0	0		0	0.000
					6	1.000

Floating population (C1,2)

	C1,2,3	Cd	Sum	Weigth
C1,2,3		1	1	**1.000**
Cd	0		0	0.000
			1	1.000

Groundwater risk (C2)

	C2,1	C2,2	Cd	Sum	Weigth
C2,1		0.5	1	1.500	**0.500**
C2,2	0.5		1	1.500	**0.500**
Cd	0			0.000	0.000
				3.000	1.000

Wells and Springs in 1000m radius (C2,2)

	C2,2,4	Cd	Sum	Weigth
C2,2,4		1	1.000	**1.000**
Cd	0		0.000	0.000
			1.000	1.000

Global weight for $C_{1,1,1}$

→ **0.250×0.667×0.500=0.083**

Agricultural value (C3)

	C3,1	C3,3	Cd	Sum	Weigth
C3,1		0	1	1	**0.333**
C3,3	1		1	2	**0.667**
Cd	0	0		0	0.000
				3	1.000

Economy (C5)

	C5,1	C5,2	Cd	Sum	Weigth
C5,1		1	1	2.000	**0.667**
C5,2	0		1	1.000	**0.333**
Cd	0	0		0.000	0.000
				3.000	1.000

Climate (C7)

	C7,1	C7,2	Cd	Sum	Weigth
C7,1		1	1	2.000	**0.667**
C7,2	0		1	1.000	**0.333**
Cd	0	0		0.000	0.000
				3.000	1.000

Figure 13.1.6 Example of application of the paired comparison technique for the attribution of criteria priorities to comparisons of priority scales present in Fig. 13.1.4.

The first step in the application of the simple additive weighting technique is to assign a numeric value to each qualitative value. A numerical value can be attributed to each qualitative value, according to a technique termed by De Feo et al. (2014) the "average value of intervals." By means of this technique, a numerical value is assigned to each qualitative value by applying Eq. (1):

$$V_i = \frac{(i-1)}{n} + \left(\frac{1}{n}\right)\left(\frac{1}{2}\right), \quad 1 \le i \le n, \ \ 0 < V_i < 1 \tag{1}$$

where V_i is the numeric value that corresponds to the qualitative value of the considered criterion occupying position "i" in the increasing ranking of qualitative values for the criterion, and "n" is the total number of qualitative values of the considered criteria.

For instance, as shown earlier, criterion $C_{5,2}$ can be evaluated by answering the question: "Is the site directly accessible?" A site is considered accessible when no construction of a new access road is required. The two possible replies are of course "no" or "yes", and the criterion should be maximized by applying Eq. (1) in which $n = 2$, 0.25 is assigned to "no" and 0.75 to "yes".

Similarly, values attributed to $C_{2,1}$ (very low = 0.125; low = 0.375; moderate = 0.625; high = 0.875) derived from Eq. (1), with $n = 4$.

Table 13.1.6 Local and global priorities for the evaluation criteria using the paired comparison technique related to the example given in Fig. 13.1.4

Criteria's Level			Priorities					
1st	2nd	3rd	Local			Global		
C_1			0.250			0.250		
	$C_{1,1}$			0.667			0.167	
		$C_{1,1,1}$			0.500			0.083
		$C_{1,1,2}$			0.333			0.056
		$C_{1,1,3}$			0.167			0.028
	$C_{1,2}$			0.333			0.083	
		$C_{1,2,3}$			1.000			0.083
C_2			0.214			0.214		
	$C_{2,1}$			0.500			0.107	
	$C_{2,2}$			0.500			0.107	
		$C_{2,2,4}$			1.000			0.107
C_3			0.143			0.143		
	$C_{3,1}$			0.333			0.048	
	$C_{3,3}$			0.667			0.095	
C_4			0.143			0.143		
C_5			0.054			0.054		
	$C_{5,1}$			0.667			0.036	
	$C_{5,2}$			0.333			0.018	
C_6			0.143			0.143		
C_7			0.054			0.054		
	$C_{7,1}$			0.667			0.036	
	$C_{7,2}$			0.333			0.018	

The next step is represented by normalization of the alternative matrix, rendered necessary as each criterion uses a distinct evaluation technique in measuring alternatives performance. Consequently, the aim of the normalization phase is to facilitate comparison between the performances of different criteria. The result of normalization is indicated by a value in the range [0—1].

Indicated with x_{ij} the value assumed for the i-alternative by the j-criterion, the x_{ij} normalized value should be evaluated using the relations $x_{ij}/Max(x_j)$ and $Min(x_j)/x_{ij}$, respectively, if the criteria needs to be maximized or minimized (De Feo et al., 2008). The single criterion should be maximized if it represents a benefit, or minimized if it represents a cost (De Feo et al., 2014).

Table 13.1.7 shows the normalized version of the example given in Table 13.1.2 of alternatives matrix related to the considered siting of hazardous waste landfills.

The global alternatives priority for each single alternative is obtained by multiplying each single value in Table 13.1.7 by the corresponding global criterion priority (see Table 13.1.6), and summing the values in each row, thus resulting as illustrated in Table 13.1.8.

Both the ranking obtained and each single alternatives priorities may be normalized respect to the maximum priority obtained for S_1 (e.g., 0.818). In practice, normalization is achieved by multiplying each value in Table 13.1.4 by 1/0.818. The results obtained are reported in Table 13.1.9 and in Fig. 13.1.7.

The ranking list is S_1, S_2, S_3, and S_4. The difference between the priorities yielded for the first and second best alternatives was 55.4%. Thus, S_1 was clearly the best site on which to place a hazardous waste landfill using the AHP—PSW approach.

PUBLIC INVOLVEMENT

Landfills have long been considered an undesirable facility and are commonly perceived as a threat to public health, quality of life, and natural ecosystems (Kikuchi and Gerardo, 2009). Accordingly, landfill sitings are burdened by increasing public opposition (Lober, 1995).

Social opposition to landfill construction is extensively described by a series of purpose-created acronyms, including NIMBY (not in my backyard), NIMTO (not in my term of office), LULU (locally unacceptable land use), BANANA (build absolutely nothing anywhere near anything), and others (Eiselt and Marianov, 2015).

The landfill siting process will be undermined if public opinion is not taken seriously into account.

Eiselt and Marianov (2015) suggest several steps that decision-makers can take to ensure a successful landfill siting process:

1. Obtain an agreement among all stakeholders that the status quo is untenable,
2. Guarantee well-defined and closely monitored safety standards, and ensure that appropriate compensation is offered to those adversely affected,
3. Invite municipalities and counties to offer appropriate sites, and
4. Choose a site in an open process.

Table 13.1.7 Normalized version of the example given in Table 13.1.2 of alternatives matrix related to the siting of hazardous waste landfills

Alternatives	Criteria													
	$C_{1,1,1}$	$C_{1,1,2}$	$C_{1,1,3}$	$C_{1,2,3}$	$C_{2,1}$	$C_{2,2,4}$	$C_{3,1}$	$C_{3,3}$	C_4	$C_{5,1}$	$C_{5,2}$	C_6	$C_{7,1}$	$C_{7,2}$
S_1	1.00	1.00	1.00	0.92	1.00	1.00	1.00	1.00	0.00	0.61	1.00	1.00	1.00	0.01
S_2	0.00	0.44	0.24	1.00	1.00	0.20	1.00	0.50	1.00	1.00	1.00	0.55	0.00	0.00
S_3	0.00	0.00	0.00	0.88	0.00	0.00	1.00	0.00	1.00	0.59	1.00	0.31	0.69	1.00
S_4	0.00	0.88	0.35	0.00	0.00	1.00	0.00	1.00	0.00	0.00	0.00	0.00	0.07	0.58

Table 13.1.8 Final priorities of the landfill siting alternatives calculated by means of the analytic hierarchy process—priority scale weighting method

Alternative	$C_{1,1,1}$	$C_{1,1,2}$	$C_{1,1,3}$	$C_{1,2,3}$	$C_{2,1}$	$C_{2,2,4}$	$C_{3,1}$	$C_{3,3}$	C_4	$C_{5,1}$	$C_{5,2}$	C_6	$C_{7,1}$	$C_{7,2}$	Priority
S_1	0.08	0.06	0.03	0.08	0.11	0.11	0.05	0.10	0.00	0.02	0.02	0.14	0.04	0.00	0.818
S_2	0.00	0.02	0.01	0.08	0.11	0.02	0.05	0.05	0.14	0.04	0.02	0.08	0.00	0.00	0.613
S_3	0.00	0.00	0.00	0.07	0.00	0.00	0.05	0.00	0.14	0.02	0.02	0.04	0.02	0.02	0.389
S_4	0.00	0.05	0.01	0.00	0.00	0.11	0.00	0.10	0.00	0.00	0.00	0.00	0.00	0.01	0.274

Table 13.1.9 Final normalized priorities of landfill siting alternatives calculated using the simple additive weighting–paired comparison technique method

Alternative	$C_{1,1,1}$	$C_{1,1,2}$	$C_{1,1,3}$	$C_{1,2,3}$	$C_{2,1}$	$C_{2,2,4}$	$C_{3,1}$	$C_{3,3}$	C_4	$C_{5,1}$	$C_{5,2}$	C_6	$C_{7,1}$	$C_{7,2}$	Priority
S_1	0.10	0.07	0.03	0.09	0.13	0.13	0.06	0.12	0.00	0.03	0.02	0.17	0.04	0.00	1.000
S_2	0.00	0.03	0.01	0.10	0.13	0.03	0.06	0.06	0.17	0.04	0.02	0.10	0.00	0.00	0.748
S_3	0.00	0.00	0.00	0.09	0.00	0.00	0.06	0.00	0.17	0.03	0.02	0.05	0.03	0.02	0.475
S_4	0.00	0.06	0.01	0.00	0.00	0.13	0.00	0.12	0.00	0.00	0.00	0.00	0.00	0.01	0.334

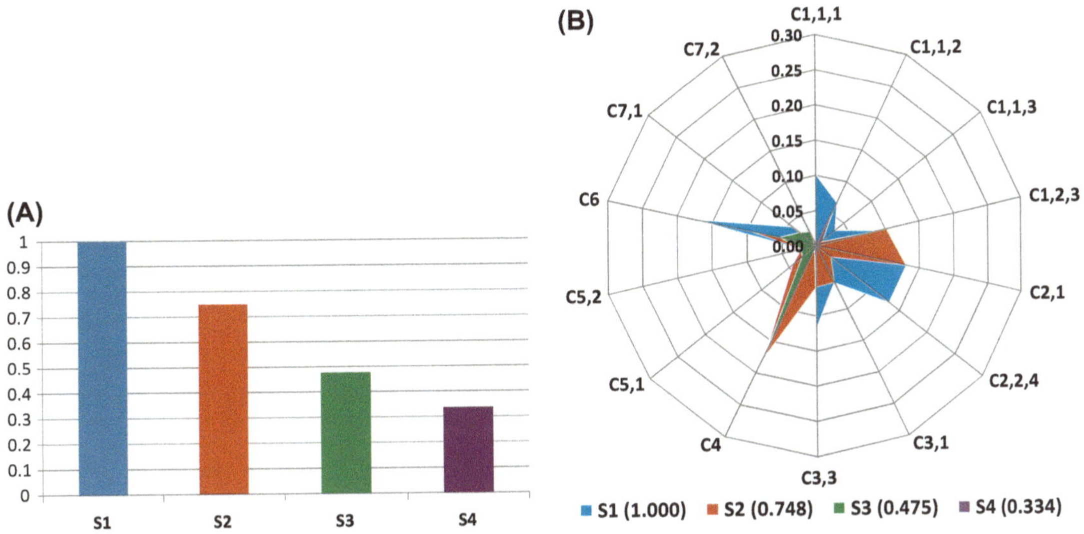

Figure 13.1.7 Simple additive weighting—paired comparison technique method. Normalized global alternatives rankings in the developed example related to the siting of hazardous waste landfills (A), and contribution of each evaluation criteria to global alternatives ranking (normalized values are reported in brackets) and normalized values obtained for each criterion (B). *Adapted from De Feo and De Gisi (2014).*

Public opinion can be involved in assigning weights to the landfill siting criteria. Two fundamental pieces of information are involved: the "who" and "how" weights assigned to the criteria. Indeed, the means of assigning (and treating) weights is a crucial step in correctly and effectively interpreting the will of decision-makers, and avoiding the unwanted onset of the NIMBY syndrome.

Bobbio (2002) described a deliberative democratic process in the Province of Torino (6829 km^2), Italy, in which local communities were involved in a decision concerning the siting of an incinerator and an MSW landfill. The process lasted 17 months and concluded with a mutual decision. The author focused on identifying the best alternative by considering 14 criteria for the incinerator and 13 criteria for the landfill. Weights were assigned by 43 members of a special commission (23 representatives of local administrations, 17 representatives of citizen councils, and 4 representatives of MSW management enterprises). Electre III (Roy, 1990) was used as an MCDA technique. The commission initially agreed on the method to be applied in assigning weights to avoid strategic behavior oriented favoring a specific criterion (e.g., a weight of 90% for a single criterion). Moreover, the commission decided that each criterion should be assigned a minimum weight based on the two following rules: (1) the most important criteria could not be assigned an evaluation more than 20-fold higher than that given to the less important one, (2) the maximum percentage weight for a single criterion was 50%. No group aggregation methods were adopted as these may likely have resulted in leveling of the stronger preferences and may have altered the participants' points of view. To ensure a transparent process for the expressed preferences, participant evaluations were not aggregated, and each was submitted to the MCDA process. This resulted in the formation of a series of rankings equal to the number of participants expressing criteria weights (14 for the landfill and 27 for the incinerator). Each decision-maker was separately ranked during the final meeting of the commission. The next step was to compare all rankings and reach a final decision (De Feo and De Gisi, 2010).

The priority scale (PSW) described and used in the previous sections, and defined by De Feo and De Gisi (2010), represents a highly innovative CWT for use by stakeholders in ranking MSW facility sites; moreover, it provides an easy use for technical and nontechnical decision-makers in collecting noncontradictory criteria preferences. The PSW is likewise suitable for extensive use by stakeholders in ranking MSW plants (or other kinds of alternatives) while using the AHP or other MCDA techniques, both with and without group aggregation methods.

It is of the utmost importance that the opinions of future generations should be given due consideration when siting a landfill in an area of historic unpopularity. De Feo and Williams (2013) reported on the current views and awareness of young people (university students) on the siting of landfills in the Campania Region in Southern Italy. This region has been affected by a series of major municipal solid waste management issues since the mid-1990s, leading to significant public disorder and subsequent media coverage. 56.4% of students were opposed to the construction of a landfill in their city, with only 25.6% agreeing to the siting of a landfill if their town was identified as the best alternative. The majority of respondents (58.9%) opined that technicians, politicians, and citizens should all be part of the decision process when siting a landfill. Hence, while the majority of students interviewed were contrary to the siting of an MSW facility in their town, the majority agreed that a siting process should actively involve all stakeholders (Al-Yaqout et al., 2002; Rahardyan et al., 2004). University faculties with the highest "a priori opposition" to the siting of a landfill in their town (shortly "no, a priori") were education, law, and pharmacy, while students in engineering, political, and economics were favorable to hosting a landfill if their town was identified as the best alternative (shortly "yes, if"). Students of technical faculties were slightly more favorable than their peers in humanities faculties. Students from Faculty of Engineering attend courses on pollution, environmental quality, and waste management in their second and fourth years, and students in Environmental Science study waste management. It was, therefore, to be expected that students with knowledge of MSW facilities would be more inclined to accept their presence locally. It is an acknowledged fact that fears of the risks involved increase in the presence of a limited knowledge of MSW facilities (Rahardyan et al., 2004). A decrease in the percentage of "no, a priori" answers and an increase in the percentage of "yes, if" among students in higher academic years is of no significance.

Landfill siting in developing countries was discussed by Owusua et al. (2012), who presented a case study from Ghana. The issue was compounded by weak town governance that resulted in communities failing to take part in any meaningful way in the decision to locate and operate a landfill in their town.

However, siting of LULUs such as landfills in a peri-urban area undergoing rapid change from one land use (in this case, stone quarry) to another (i.e., residential land use) requires adequate consultation and active participation of community members as well as conflict resolution mechanisms. The main causes of conflict were as follows: decisions relating to the siting and operation of the landfill; subsequent management issues relating to improper maintenance of the site (odor and noise); littering of the community by waste trucks; the view that chiefs would be the sole beneficiaries of royalties paid by Greater Accra Metropolitan Area.

Glucker et al. (2013) discussed the issue of public participation in environmental impact assessment, underlining the fact that whereas from a democratic point of view, it is desirable that the populations affected by a proposed project should give their consent for the project, one should be aware of the far-reaching policy implications of such an approach (O'Faircheallaigh, 2010). For instance, if affected citizens were allowed to decide on the siting of a toxic waste landfill, the outcome would most likely be policy deadlock (Leroy and Van Tatenhove, 2001; O'Faircheallaigh, 2010).

References

Al-Yaqout, A.F., Koushki, P.A., Hamoda, M.F., 2002. Public opinion and siting solid waste landfills in Kuwait. Resources, Conservation and Recycling 35, 215–227.

Bobbio, L., 2002. Smaltimento dei rifiuti e democrazia deliberativa (Waste Disposal and Deliberative Democracy). Working Paper 1. Department of Political Science, University of Torino, Torino, Italy. Available online at: http://www.dsp.unito.it/download/wpn1.pdf (in Italian).

Chang, N.B., Parvathinathan, G., Breeden, J.B., 2008. Combining GIS with fuzzy multicriteria decision-making for landfill siting in a fast-growing urban region. Journal of Environmental Management 87 (1), 139–153.

Ching-Lai, H., Kwangsun, Y., 1981. Multiple Attribute Decision Making. Springer Verlag, Berlin-Heidelberg, New York.

De Feo, G., Cerrato, F., Siano, P., Torretta, V., 2014. Definition of a multi-criteria, web-based approach to managing the illegal dumping of solid waste in Italian villages. Environmental Technology 35 (1), 104–114.

De Feo, G., De Gisi, S., 2014. Using MCDA and GIS for hazardous waste landfill siting considering land scarcity for waste disposal. Waste Management 34 (11), 2225–2238.

De Feo, G., De Gisi, S., 2010. Using an innovative criteria weighting tool for stakeholders involvement to rank MSW facility sites with the AHP. Waste Management 30, 2370–2382.

De Feo, G., De Gisi, S., Galasso, M., 2008. Definition of a practical multi-criteria procedure for selecting the best coagulant in a chemically assisted primary sedimentation process for the treatment of urban wastewater. Desalination 230 (1), 229–238.

De Feo, G., Galasso, M., Landi, R., Donnarumma, A., De Gisi, S., 2013. A comparison of the efficacy of organic and mixed-organic polymers with polyaluminium chloride in chemically assisted primary sedimentation (CAPS). Environmental Technology 34 (10), 1297–1305.

De Feo, G., Williams, I.D., 2013. Siting landfills and incinerators in areas of historic unpopularity: Surveying the views of the next generation. Waste Management 33, 2798–2810.

Demesouka, O.E., Vavatsikos, A.P., Anagnostopoulos, K.P., 2013. Suitability analysis for siting MSW landfills and its multicriteria spatial decision support system: method, implementation and case study. Waste Management 33 (5), 1190–1206.

Eiselt, H.A., Marianov, V., 2015. Location modeling for municipal solid waste facilities. Computes and Operations Research 62, 305–315.

Eskandari, M., Homaee, M., Mahmodi, S., 2012. An integrated multi criteria approach for landfill siting in a conflicting environmental, economical and socio-cultural area. Waste Management 32 (8), 1528–1538.

Expert choice, 2004. Quick Start Guide and Tutorials. Expert Choice Inc., Arlington, VA, USA.

Fell, R., Corominas, J., Bonnard, C., Cascini, L., Leroi, E., Savage, W.Z., 2008. Guidelines for landslide susceptibility, hazard and risk zoning for land-use planning. Engineering Geology 102 (3), 99–111. https://doi.org/10.1016/j. enggeo.2008.03.014.

Forman, E.H., Selly, M.A., 2001. Decision by Objectives. How to Convince Others That You Are Right. World Scientific Publishing Co. Pte. Ltd., London.
Foster, S., Hirata, R., Gomes, D., D'Elia, M., Paris, M., 2002. Groundwater Quality Protection: A Guide for Water Utilities, Municipal Authorities, and Environment Agencies. World Bank Publication.
Gbanie, S.P., Tengbe, P.B., Momoh, J.S., Medo, J., Kabba, V.T.S., 2013. Modelling landfill location using geographic information systems (GIS) and multi-criteria decision analysis (MCDA): case study Bo, Southern Sierra Leone. Applied Geography 36, 3–12.
Glucker, A.N., Driessen, P.P.J., Kolhoff, A., Runhaar, H.A.C., 2013. Public participation in environmental impact assessment: why, who and how? Environmental Impact Assessment 43, 104–111.
Gorsevski, P.V., Donevska, K.R., Mitrovski, C.D., Frizado, J.P., 2012. Integrating multi-criteria evaluation techniques with geographic information systems for landfill site selection: a case study using ordered weighted average. Waste Management 32 (2), 287–296.
Kikuchi, R., Gerardo, R., 2009. More than a decade of conflict between hazardous waste management and public resistance: a case study of NIMBY syndrome in Souselas (Portugal). Journal of Hazardous Materials 172, 1681–1685.
Kontos, T.D., Komilis, D.P., Halvadakis, C.P., 2005. Siting MSW landfills with a spatial multiple criteria analysis methodology. Waste Management 25 (8), 818–832.
Korucu, M.K., Erdagi, B., 2012. A criticism of applications with multi-criteria decision analysis that are used for the site selection for the disposal of municipal solid wastes. Waste Management 32 (12), 2315–2323.
Leroy, P., Van Tatenhove, J.P., 2001. Environment and participation - the shifting significance of a double concept. In: Driessen, P.P., Glasbergen, P. (Eds.), Greening Society. The Paradigm Shift in Dutch Environmental Politics. Kluwer Academic Publishers, Dordrecht.
Lober, D.J., 1995. Why not here: the importance of context, process and outcome on public attitudes towards the siting of waste facilities. Society and Nature Resources 9, 375–394.
Moeinaddini, M., Khorasani, N., Danehkar, A., Darvishsefat, A.A., Zienalyan, M., 2010. Siting MSW landfill using weighted linear combination and analytical hierarchy process (AHP) methodology in GIS environment (case study: Karaj). Waste Management 30 (5), 912–920.
Mondy, W., Noe, R., 2008. Human Resource Management. Prentice-Hall, USA.
O'Faircheallaigh, C., 2010. Public participation and environmental impact assessment: purposes, implications, and lessons for public policy making. Environmental Impact Assessment Review 30, 19–27.
Owusua, G., Oteng-Ababio, M., Afutu-Kotey, R.L., 2012. Conflicts and governance of landfills in a developing country city, Accra. Landscape and Urban Planning 104, 105–113.
Rahardyan, B., Matsuto, T., Kakuta, Y., Tanaka, N., 2004. Residents's concerns and attitudes towards solid waste management facilities. Waste Management 24, 437–451.
Raga, R., Cossu, R., Lagerkvist, A., 2010. Landfilling: planning, siting and design, Chapter 10.12. In: Christensen, T.H. (Ed.), Solid Waste Technology & Management, vols. 1 and 2. John Wiley & Sons, ISBN 978-1-4051-7517-3, pp. 898–913.
Roy, B., 1990. The outranking approach and the foundations of ELECTRE methods. In: Bana e Costa, C.A. (Ed.), Readings in Multiple Criteria Decision Aid. Springer, pp. 155–183.
Saaty, T.L., 1977. The Analytic Hierarchy Process. McGraw-Hill, New York.
Saaty, T.L., 1996. The Analytic Hierarchy Process. McGraw Hill, New York, USA.
Saaty, T.L., 2001. Decision Making for Leaders. The Analytic Hierarchy Process for Decision in a Complex World. New Edition RWS Publications, Pittsburgh, USA.
Saaty, T.L., 2005. Theory and Applications of the Theory of the Analytic Network Processes. Decision Making with Benefits, Opportunities, Costs, and Risks. RWS Publications, Pittsburgh, USA.
Sharifi, M., Hadidi, M., Vessali, E., Mosstafakhani, P., Taheri, K., Shahoie, S., 2009. Mehran Khodamoradpour, integrating multi-criteria decision analysis for a GIS-based hazardous waste landfill sitting in Kurdistan Province, western Iran. Waste Management 29 (10), 2740–2758.
Sumathi, V.R., Natesan, U., Sarkar, C., 2008. GIS-based approach for optimized siting of municipal solid waste landfill. Waste Management 28, 2146–2160.
Tavares, G., Zsigraiova, Z., Semiao, V., 2011. Multi-criteria GIS-based siting of an incineration plant for municipal solid waste. Waste Management 31, 1960–1972.
Zamorano, M., Molero, E., Hurtado, Á., Grindlay, A., Ramos, Á., 2008. Evaluation of a municipal landfill site in Southern Spain with GIS-aided methodology. Journal of Hazardous Materials 160 (2–3), 473–481.

13.2

LANDFILL PLANNING AND DESIGN

Raffaello Cossu and Rainer Stegmann

INTRODUCTION

The planning and design of a landfill is no longer a purely technical task, but in recent years it has assumed an increasingly social dimension. This is due to the fact that awareness of environmental problems relating to traditional landfilling has not been appropriately addressed by politicians and has not been adequately supported by scientifically updated regulations aimed at controlling nuisances and long-term emission of contaminants from landfills. In the meantime, the growing public sensitivity to environmental issues, together with an optimistic political view of waste recycling and zero waste management, has rendered landfilling, in the eyes of the public, an obsolete and hazardous technology. As a consequence, public acceptance of new landfills is extremely low.

As highlighted in the Introduction to this book (Chapter 1.1), taking into account that landfilling is an unavoidable phase required to close the material loop in any modern waste management strategy, the planning and design of landfills should consequently be developed according to the following principles:

- Projects should be developed in compliance with the principle of environmental sustainability, with the monitoring and control of short-, medium-, and long-term emissions and nuisances being expected to reach an environmental equilibrium within a period not exceeding one generation (30 years approximately).
- Environmental equilibrium is defined on the basis of Final Storage Quality (FSQ), which should be established by regulations combining tables of values of the most significant parameters and site-specific risk assessments. FSQ should be viewed as a design target to be achieved through the project, operational management, and postclosure phase (aftercare), in the same way as the achieving of effluent quality targets in a wastewater treatment facility.
- The design should be developed to meet the FSQ target by adopting a multibarrier approach and controlling the different terms of the mass balance of potential contaminants (Chapter 2.2). Waste pretreatment options, operational management, and on-site and off-site aftercare activities should be considered.
- Methodologies and technologies to be used to achieve environmental sustainability must be described, substantiated in detail, sized, computed, and cost estimated.
- Landfills should be designed preferably to receive homogeneous fractions of waste to facilitate possible subsequent landfill mining and recovery of resources.

- With regard to space planning and land use, landfilling should not represent the final destination of a given area, but rather a tool for obtaining new functional destinations, planned in compliance with existing official plans, in agreement with local Authorities, in the interest of the local population.
- Works and infrastructures required to finalize the intended destination of the landfill site must be specifically planned from the outset and coherently realized throughout the various construction stages of the landfill.
- Landfills should be built and operated in accordance with the highest standards in terms of materials, construction techniques, safety measures, monitoring, and quality control assessment.
- Decision and planning of landfills should be transparent and proactively discussed with the local communities. These people should be continuously informed and/or involved, e.g., by implementing a roundtable including all stakeholders. This kind of "landfill neighborhood" should also be maintained throughout the operational phase of the landfill.

A landfill should be constructed, controlled, operated, and maintained as a normal civil work installation of public interest (buildings, roads, airports, waste treatment facilities) and meet the same high technical standards. This implies a need for experienced engineers to execute the planning, construction, and operation of the landfill. Only well-defined, high-quality materials should be used; indeed, it is no longer feasible to advocate use of cheap, easily available construction wastes, whereas high-quality recycled waste materials that meet established requirements are appropriate. The argument that some materials or constructions are too costly and should be substituted by cheaper materials or replaced is applied more frequently in the case of landfills than for municipal solid waste (MSW) incinerators. In addition, landfills should provide for intensive monitoring aimed at documenting operations and emission control. It is of course inherent that all procedures should be undertaken in full compliance with relevant national laws and regulations.

The aim of this chapter is to illustrate the logical process to be applied in the technical development of a landfill and highlight any additional considerations. The information provided herein is not intended to be fully detailed and should be viewed as indicative. The majority of the information provided originates from the "field" experience of the Authors.

LAWS AND REGULATIONS IN THE EUROPEAN UNION AND UNITED STATES

Landfilling in the European Union is regulated by EU Council Directive 1999/31/EC of April 26, 1999, on the landfilling of waste (EU Council Directive, 1999).

In the US MSW landfills fall under several federal regulations, including the RCRA Subtitle D landfill regulations found in 40 CFR 258 (US Government, 2012; Townsend et al., 2015).

Different regulations and standards for landfills in China are presented in Chapter 15.1 (e.g., China Technical Code, 2013).

Based on the EU framework as the minimum standard, the different member states establish their own detailed regulations. Specific additional requirements may be enacted by the regional and/or local administration. In planning a landfill not only the waste act but also numerous other laws and regulations focused on environmental protection, surface and groundwater, air quality, etc., should be taken into account and complied with. Prior to the planning stage, the consultants and planning bodies should draw up an overview of specific requirements that will need to be adhered to. As an example, in many cases the height of the landfill is locally regulated to not exceed that of any surrounding mountains, trees, or dikes.

To respond to the protests of the people, frequently manifested against landfill construction, some national regulations (e.g., in Germany and Italy) establish a need for the population to be involved. Each single question and complaint raised by individual members of the community is discussed and answers from the authorities presented to an authorized group comprising the different stakeholders. If the answers are not satisfactory—e.g., due to the specific situation—although complying with the prescribed regulations, design changes may be requested by the court. In some cases, the entire project has been halted. This procedure prolongs the planning process (and in many cases also costs) significantly. A planning period of up to 10 years in many cases is not unrealistic (see Chapter 1.3).

LAND PLANNING PHASES

A landfill can be classified as a civil work installation if construction is extended over a lengthy period of time. Construction works are of course required initially (waste pretreatment, bottom barriers, infrastructures, emission treatment plants, etc.). Construction work is also further required during landfill operation and aftercare (leachate and gas extraction and collection facilities, in situ treatments, top barriers, etc.), as well as at the time of implementation of the final destination of the landfill site subsequent to completion of the aftercare phase.

The time frame involved from identifying the need for a landfill to solve a waste management issue up until landfill completion and subsequent transformation into a site to be returned to the community for public use may be extremely lengthy, at times spanning more than one generation.

The different steps to be undertaken between the start and end points considered are summarized in Fig. 13.2.1. Compliance and administrative commitments and the main specific activities to be carried out at each individual step are indicated.

The sequence of the above steps may be subdivided roughly into four main sections:

I. Preliminary phase
II. Detailed and executive design
III. Afteruse and delivery of the landfill area back to the people
IV. Landfill life (realization, management, aftercare, and final release)

The laws and regulations implemented by each individual country define the specific requirements and the official documents to be elaborated. It should be mentioned, however, that these regulations do not always include the latest developments and may in some cases be questionable.

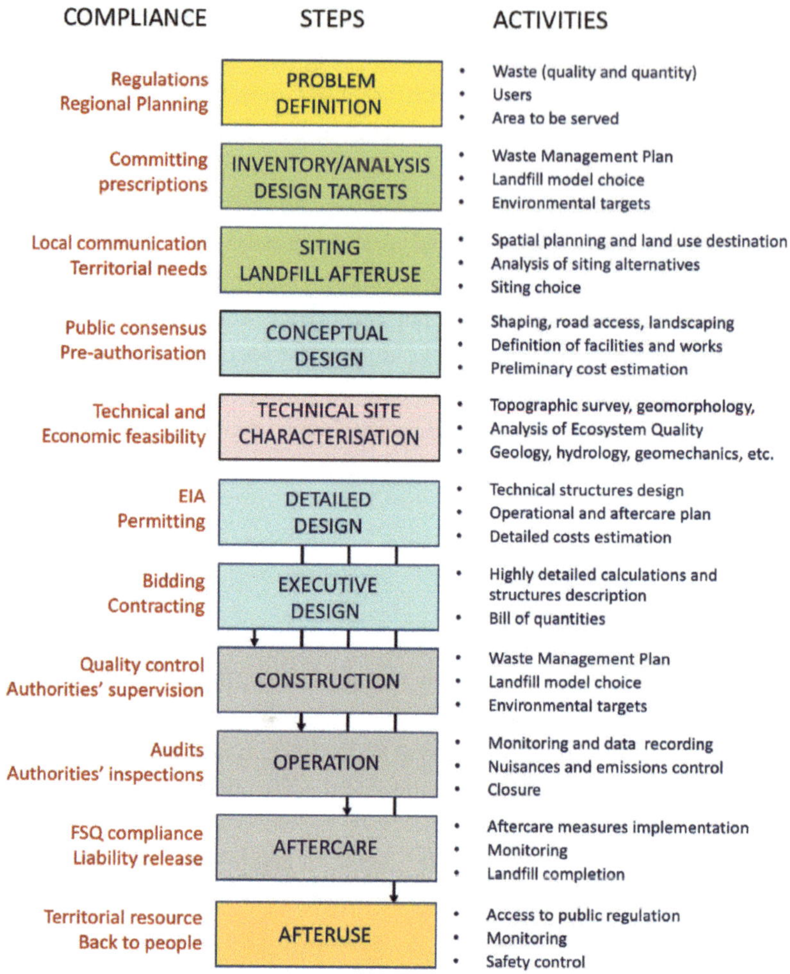

Figure 13.2.1 Scheme of different steps in landfill planning, design, and realization. Compliance administrative steps and main specific activities to be carried out at each individual phase are indicated. *EIA*, Environmental Impact Assessment; *FSQ*, Final Storage Quality.

The preliminary phase comprises definition of the problem, an inventory and analysis of existing data and waste management plans, definition of the environmental and other design targets, definition of the input waste quality, choice of the most appropriate landfill model, critical study and evaluation of potential suitable sites, choice of the site, and landfill afteruse in the frame of any existing local spatial planning. The final step of the preliminary phase is the conceptual design representing the tool with which to obtain first stage authorization and public consensus and to assess technical and economic feasibility of the landfill.

The detailed and executive design phase of landfill planning is the phase in which all technical decisions are made, structures and infrastructures of the landfill are dimensioned, calculations are developed, materials are chosen, control of emissions in the short-, medium-, and long term are decided and designed, and plans for construction, operation, and aftercare of the landfill are produced.

The third phase is represented by the life span of the landfill. The following steps can be identified: construction of all necessary structures, operation, closure, and aftercare up until the landfill is released by the competent Authority. During this phase, all aspects comprised in the detailed design should be implemented, including the measures for aftercare management.

The last phase commences when the landfill site has been released from any residual liability and the site can be given back—back to the people, back to the hydrological cycle, "Back to the Earth" (see Box 2.2-1).

Design adjustments might be necessary throughout the different phases.

PRELIMINARY PLANNING

Waste categories and volumes

Problem definition

A prerequisite in designing a landfill is knowledge of the region, city, or county or industry to be served. How large is the population, how will it develop in the future? What is the productive process in the given industry? What kinds of waste will need to be landfilled (see Chapter 1.2)? Are there any seasonal influences on waste production and quality as, e.g., by tourism? What are the main peculiarities of the area to be served (traffic, access roads, specific restrictions, etc.)? With this background information and further local details, a first preliminary estimation of the amount of waste in the different categories to be disposed of over the life span of the landfill can be made.

Inventory, analysis, and design targets

An inventory of all significant available data, inputs from waste management plans, and performance of existing facilities should be analyzed. Is any kind of pretreatment necessary to meet quality requirements for the waste to be landfilled (see Chapters 4.1, 4.2, 4.3)? Will pretreatment be carried out at the landfill site? What kind of waste collection system(s) are in place in the communities served and what are their plans for the future?

Targets aimed at guaranteeing environmental sustainability should be set and accordingly the most appropriate waste treatment process and landfill models should be defined.

Depending on the operational period of the landfill, the total waste disposal volume and waste input per year can be estimated. It should also be decided whether additional space for future landfill expansion should be identified. In addition, space needed for potential treatment and storage facilities may be roughly estimated.

Landfill siting and planning objectives

Prior to embarking on the conceptual design of a landfill, it is important to carry out an inventory and collect documentation of the existing ecosystem (fauna, flora, surface water body, etc.) of the wider area where the landfill is to be sited. Documentation of the groundwater situation, depth, flow velocity, and direction as well as groundwater quality, is of particular importance. As a minimum requirement wells should exist or be constructed, particularly in the direction of the flow at both edges of the landfill, and further downstream. This is likewise of importance in the case of lawsuits filed following the detection of groundwater pollution, and the landfill is blamed for being the source of pollution. In many cases, a preexisting groundwater contamination may have originated from agriculture, industry, or other sources.

Due to the decade-long operations of landfills and even longer aftercare times, planning should—as far as possible—respect future new potential developments and regulations. This may be achieved, for example, by ring-fencing space around the facilities and on the landfill itself, installing higher capacities for the provision of electricity, etc., and providing additional connection points.

Landfill siting is a fundamental first step in the realization of a landfill. This procedure is described in Chapter 13.1 and lays the foundation for the acceptance and quality—but also for estimating the economics—of the entire project. As already mentioned, during this process any initial existing contamination (e.g., groundwater pollution, odors from other facilities) should be monitored and documented.

Before undertaking any further planning steps the project objectives should be defined. The kind of facilities to be built at any stages of the landfill life should be defined as well. These may include an MSW pretreatment plant, a composting and/or anaerobic digestion plant for biological degradable waste, storage facilities for household hazardous waste, acceptance area for separated collected waste compounds, waste sorting plant, community center for citizens (where they can deposit their household waste and recycling waste in different containers), biological polluted soil treatment plant, in situ treatment, etc.

Due to synergistic effects, the operation of additional plants on the landfill area may be advantageous. This approach may contribute toward rendering landfill operations more economical (e.g., though a more efficient employment of personnel on-site). This may be particularly important if the waste volumes to be landfilled are reduced. Landfills may also act as a final sink for residues resulting from the operation of treatment plants on-site. The inclusion of other plants on the site is highly situation specific and difficult to generalize. This chapter focused on the planning of landfills and related emission control units, this type of situation is beyond the scope of the paper.

Separate storage areas, as an example for waste that may be recycled in the future, should be taken into consideration.

Conceptual and preliminary design

In addition to ground- and surface water protection other objectives for emission control should be taken into account during the planning stage:

- The existing infrastructure should be investigated to ensure there is access to drinking water, electricity (including potential to feed in electricity generated from landfill gas (LFG) utilization), heat, telephone, Internet, roads, etc. Infrastructure installations should be planned based on the existing situation.
- *Gas emission control*: In addition to the required LFG control by active extraction, additional measures aimed at reducing the emission of gases may be required (e.g., methane oxidation layer on top of the landfill).
- *Odor emission control*: Due to specific waste composition (also potential uncontrolled digestion in the collection bins, trucks, etc.) and climatic situation special measures (although these are quite limited) should be implemented (e.g., immediate cover of the incoming waste, in situ aeration).
- *Noise reduction*: Potential measures may comprise the selection of adequate machinery, noise protection walls, etc.
- *Habitat preservation*: Specific local animal and/or plant protection measures may be planned.
- *Landfill reclamation*: The time when the landfill will be ready for reclamation should be established as part of the planning procedure. In addition, the final shape and potential afteruse should be planned. To shorten the time required until the landfill can be reclaimed specific operational measures may be implemented (e.g., landfill aeration (see Chapter 16.2)).
- *Energy recovery*: The following points should be discussed:
  - prediction of gas production with time (see Chapter 9.1),
  - identification of potential users for LFG in form of electricity, heat, or natural gas quality, on or near the landfill site,
  - if required, provision of access to the public grid or natural gas system,
  - calculation of expected revenues from energy sales,
  - the possibility of installing photovoltaic and/or wind power plants on the landfill site (after closure and biological stabilization) or on the landfill area.
- *Control of Leachate flow*: Planning of operational and/or technical measures for the reduction of high leachate production rates, which may result from high permanent or seasonal precipitation rates and/or high moisture content of incoming waste.
  - In the case of high precipitation measures may focus on diversion of surface water from the landfill surface, e.g., by means of mobile plastic covers. What kind of technology should be adopted and what kind of surface water storage capacities are required?
  - In the case of high moisture content of the incoming MSW (e.g., in Asia) pretreatment (e.g., partial thermal or biological drying) and/or specific landfill operation technologies (e.g., intermediate layers of coarse materials (Stegmann, 2017)) may be considered (see Chapters 10.1, 10.2 and 11.1).
- *Leachate treatment aspects*: The following points should be clarified:
  - Definition of the receiving system for treated leachate (sewage treatment facility, natural water bodies, etc.); which leachate discharge volumes and/or limit values have to be met? What constructive consequences may there be (kind of leachate treatment, leachate pipes, reservoirs, transfer station for transport to sewage treatment plant, etc.)?

- Where will the produced residual sludge or concentrate from leachate treatment be treated or disposed? What kind of constructions may be required to realize this (dewatering system, thermal drying, etc.)?
- Is co-treatment in a sewage treatment plant an option and what kinds of constructive measures will be necessary (transfer station for transport by truck to sewage treatment plant, transport pipes, etc.).

- *Landfill operation* (see also Chapter 15.1): The different operation activities and their goals should be defined in advance, e.g., if rapid waste stabilization in the landfill body is aimed at, how and where will the different waste categories be landfilled, will daily cover be provided, etc.? What kind of technical installations will be required (e.g., storage for soil, etc.).
- *Landfill morphology*: Landfills can be constructed according to different morphologies as represented in Fig. 13.2.2.

  As far as possible, landfills should be built above ground level (mound landfill, option B and also D). In this case, leachate will be able to flow out from the landfill by gravity and will not need to be pumped off for decades from deep shafts, as would be the case for below ground level landfills (pit landfill, option A, C, and E). In addition, for options B and D, cleaning and maintenance of drainage pipes will be easier. In the majority of cases the distance between the landfill bottom and groundwater level is higher, contact between the landfill surface and the ground is reduced (in the case of mound landfill), and there is no contact at all between slopes and below ground levels. The landfill is also more visible, which is of importance in view of the long-term life span of the landfill and the risk of overlooking its presence. Additionally, landfill mining can be carried out with greater ease. An example of the organization of a mound landfill is given in Fig. 13.2.3.

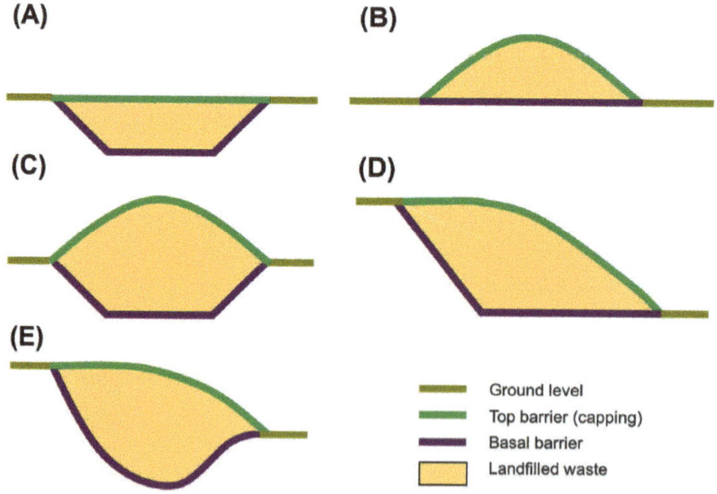

Figure 13.2.2 Scheme of different landfill morphologies. (A) Below grand level, (B) Above the ground level, (C) Below and above ground level, (D) On a slope, (E) On a valley.

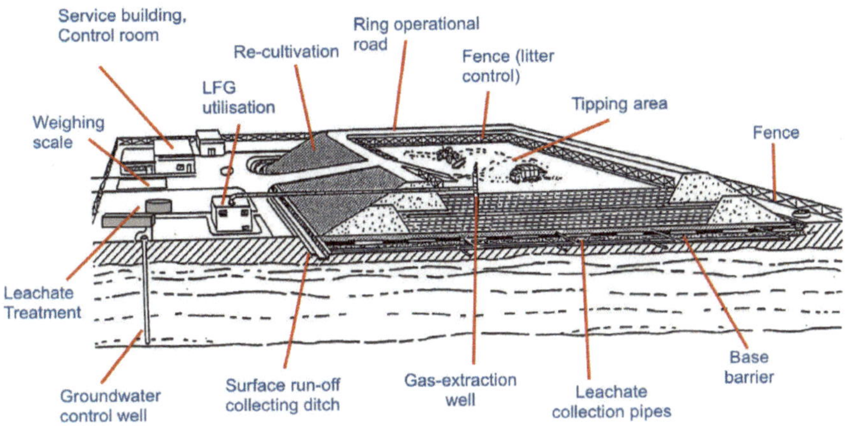

Figure 13.2.3 Scheme of a landfill mount. *LFG*, landfill gas. *Modified after Bilitewski et al. (1997).*

If it proves impossible to construct the landfill above ground level, another acceptable morphology would be option D, a landfill situated on a slope (Fig. 13.2.2). In this case, control of the water regime is fundamental in avoiding problems of mechanical instability. The landfill should also be well drained (including the sides) and rainfall and clean drain water well diverted.

As a general rule, landfills should never be built up to fill a valley; this may run the risk of generating uncontrolled waterways and contaminating piped waterways existing underneath the landfill and receiving superficial water bodies.

For both slope and pit landfills, the base barrier system should adhere to the same high standards used for mound landfill.

When a landfill is constructed in a pit, the above described drainage and lining should be installed at all surfaces between soil and waste using the same standard of construction as planned for the bottom liner. In addition, a specific design should be applied for leachate extraction.

DETAILED AND EXECUTIVE DESIGN

Once the site has been chosen, a highly accurate technical survey should be carried out during the design stage to characterize in detail the following aspects:

- Topographic survey,
- Climatological regime
- Geological structure
- Geomorphological processes analysis
- Pedological characteristics of the soil
- Groundwater and surface water hydrology
- Geomechanical characterization of the site

A detailed design may then be developed using the results of the technical site survey as fundamental data input. The detailed design should take into consideration the definition and computing of all landfill facilities to be constructed during the landfill life span, to provide an Environmental Impact Assessment study and a series of plans to regulate operations, aftercare, and afteruse (financial plan, monitoring plan, operation plan, safety plan, etc.). A detailed cost estimation should be produced. The detailed design will be presented to the authority to obtain the necessary permits and licenses. Once these have been granted, the executive design will develop in greater detail all aspects relating to construction works subject to a bid procedure previously. The executive design should also contain an accurate bill of quantities, technical prescriptions for materials and construction methodologies, and the structure of the bidding documents.

Space for waste landfilling

The size of the landfilling area will be closely dependent on the specific situation, the available space, the amount and kind of incoming waste, the time period of operation, and the height and topography of the final landfill. Prediction of the amount and quality of incoming waste over several decades is a very difficult task. Growth of the population and the commercial sector, together with the impact of future political decisions, will also need to be borne in mind.

For economical reasons, landfill operations should continue at least while financial depreciation of the investment is ongoing (e.g., machines 6–8 years, constructions 20–30 years).

Additional space is needed for a ditch and service road that surrounds the landfill (Fig. 13.2.4).

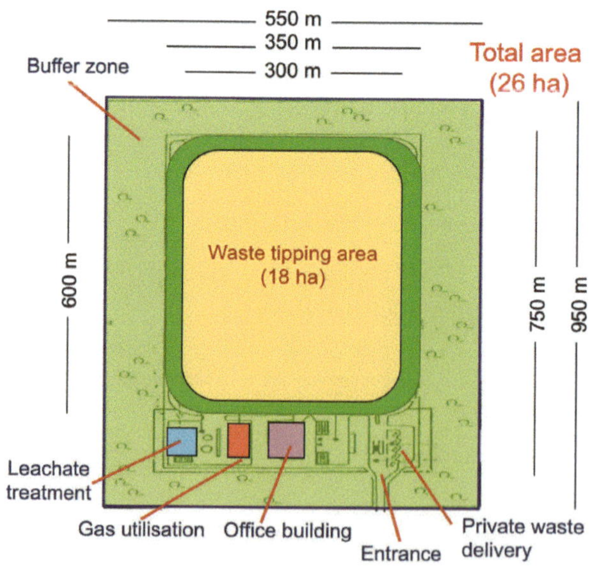

Figure 13.2.4 Example of a ground plan of a landfill. The buffer zone between waste disposal and outside area might be planted for secluding the view from outside, reducing noise and odors.

An area around the landfill that creates a buffer zone between the landfill itself and the surrounding outside area (inside the fence) should be planned. At an early stage trees should be planted in the buffer zone to create a visual barrier. This buffer zone should not be used to house facilities and will provide a series of advantages including noise control, litter control, aesthetics, etc. (Fig. 13.2.4).

Planning of the various landfill components and facilities

Once the location and size of the landfill have been defined, the landfill can then be designed in detail. The following landfill components should be envisaged:

- A series of measures for the preparation of the underground foundations and—in the case of construction of a pit landfill—the sides of the landfill area should be planned to provide stability and minimize the potential for settling and deformation. To prepare the underground for installation of the base liners, any existing small ditches or ponds will need to be diverted or eliminated, vegetation, including roots, should be removed and any existing constructions eliminated. Removed soil should be stored and used for landfill covering, only if appropriate (Chapter 11.1). The amount of soil to be moved or exchanged should be calculated. An exchange of soil materials may be necessary when the existing soil is not suitable to meet the specific requirements. Alternatively, the existing soil may need to be artificially compacted.
- Selection and planning of the type of bottom liner system, consisting of the drain system (see Chapters 8.1, 8.2) and the liner system (see Chapters 7.1, 7.2, 7.3) to be adopted:

  In the majority of cases the type of bottom liner system is prescribed by national or regional regulations. Indeed, a combination liner should be selected that consists of two different materials, such as high-density polyethylene and a low permeable mineral liner such as clay. Of course, the prescribed quality of the mineral liner material (as e.g., the permeability) should be guaranteed, implying in many cases a need to pretreat (i.e., homogenize or upgrade) the natural clay material. The clay/claylike material should be applied in thin layers (see Chapter 7.1).

  The drainage layer should be made up of an appropriately thick layer of inert coarse gravel (see Chapter 8.1). In planning the bottom liner system, calculations of the amounts of the required materials should be made and availability guaranteed. Any substitution of these materials by cheaper products that fail to meet the required standards should be avoided. The bottom liner system should be constructed to last for an indefinite period of time and is the main component responsible for the environmental impact of the landfill. Mistakes that are made cannot be corrected. Design of the slopes to be covered by the bottom liner should be guaranteed, and if the baseliner presents with local defects or even when the liner system is fully functional, water should be able to flow by gravity into the drainage pipes and exit the landfill (see Chapter 8.1).
- In the installation of a surface liner system the final shape of the landfill should be designed in such a way that it allows for adequate construction and functionality (e.g., transport of surface and drainage water out of the landfill). Planning of the final surface liner should adhere to the same rules described for the bottom liner. However, the design of the drain system on top of the surface liner, having the main function of avoiding water accumulation in the cover soil to avoid roots entering the impounded water zone from dying off, may differ. A layer of coarse material should be placed underneath the surface liner to act as a kind of "gas drainage" (see Chapter 11.1). To avoid damage of the liner system, in many cases the landfill is recultivated using plants with short roots (e.g., grass, small bushes). Another option may be use of a commercial asphalt

liner (see Chapter 11.1). Once again the availability of the quality and volumes of materials required should be verified.

- The surface water control system should be planned to avoid ponding on the landfill surface, as well as soil erosion in the slope area. An on-site surface water collection system may be necessary when high precipitation rates are expected and when only parts of the landfills are intermediately lined (intermediate cover with low permeable materials or plastic liners); surface water runoff should be monitored once the final surface liner system has been implemented (see also Chapter 8.1).

  In the case of a mound landfill, a ditch surrounding the entire landfill with an adequate gradient and a receiving pond should be planned and sited. This pond may also be used for fire fighting purposes and/or water supply for dust control. Dependent on the specific situation, gutters for surface water collection and transport of surface water on the landfill surface or on the slopes may be required. If the landfill has been constructed on a mountain slope, the situation becomes increasingly complex. In this case, rainwater flowing from the mountain will need to be diverted from the landfill. If landfills are operated as a pit landfill surface water control becomes increasingly complex. While the best option would be to prevent surface water from the surrounding area entering the pit and to divert surface water from the waste surface (daily cover soil or intermediate cover), this is extremely difficult to achieve in practice. In an attempt to solve the situation, surface water may be directed to the drainage layer located at the slopes, collected at the bottom of the landfill, and pumped out. It is, however, important to avoid a mixing of surface water with leachate.
- The gas extraction system, including the distribution stations, piping, and blower station should be designed in detail. In addition, the location of the gas treatment system (flare, treatment, and/or utilization unit) should be defined and its capacities calculated based on the predicted-time dependent—gas volumes (LFG prediction model (see Chapter 9.1)). When planning the gas extraction and utilization system, the possibility of future in situ aeration should also be taken into consideration (see Chapter 16.2).
- The leachate treatment system, including basins, buildings, pipes, pumps, etc., should be planned in detail. This is particularly complex in view of the fact that leachate quality and volume will vary over time (see Chapters 10.1 and 10.2). Once the location of the plant has been identified, all pipes and pumps should be dimensioned and their local installation planned. If necessary, a sealed pond for intermediate leachate storage (treated or untreated) may have to be considered. To avoid odors the installation of a simple aeration system could be envisaged. In the case of co-treatment in an existing sewage treatment plant the availability of sufficient treatment capacities should be investigated and compliance with discharge limits ensured.

  Due to the variation of leachate volumes and quality with progressing development (size and volume) and age of the landfill, leachate treatment should be designed to be as flexible as possible (see Chapters 10.3–10.5). If a biological plant is planned as a first step, after a certain period of time this plant may no longer be required and may be used for other purposes.

Operational infrastructures

In addition to the landfilling area the infrastructure on the entire landfill area should be planned.

- Effective security fencing should be erected around the landfill. This should serve to prevent vandalism and theft, prevent people from entering the landfill, and for safety and insurance reasons. Signs indicating the danger of entering the landfill should be posted. CCTV (Closed Circuit Television) systems may also be installed.

- The entrance should be secured by an automatic gate with video surveillance. An information board including opening hours should be located near the gate. A lifting barrier for traffic control should be installed in front of the weighing station.
- An electronic scale with automatic registration of the weight of waste trucks and automatic accounting is an essential part of the weighing station. By these means data relating to volume and quality of the incoming waste will be collected. These data represent the basis for the future development and planning of the landfill and the entire waste management system of the region. As an alternative, mechanical scales may be used instead of electronic scales. Due to the amount of the waste to be landfilled, several scales may have to be constructed. With the exception of very small landfills, in general two scales—one for the incoming and one for the outgoing trucks—will be required. When calculating the number of scales it should be taken into account that numerous trucks will confer waste at peak times, as waste collections are largely commenced at the same time. In addition, the build up of traffic congestion should be avoided due to the lack of space for queuing vehicles.
- In the entrance area a center for the private delivery of source segregated different kinds of household or commercial waste should be installed. The conferred waste consists larger or greater volumes and/or kinds of waste components that may not be disposed of in household bins (e.g., small amounts of rubble, garden and park waste, large packaging material, wood, etc.). Furthermore, electrical and electronic wastes, used clothes, etc., may be accepted at this point. Conferment of small amounts of hazardous wastes from households (e.g., residues of paints, detergents, solvents, poison, etc.) may also be permitted. Frequently, this type of plant consists in a series of easily accessible containers. To avoid contact with rainfall and littering these containers may be placed under a roof.
- A multipurpose building containing office rooms, a conference room (also for visitors), small kitchen, bathrooms with showers and changing room, as well as a small laboratory and a storage room for samples, equipment, etc., should be planned. In the past, these buildings were often simply constructed (at times mere containers) and poorly maintained. Even in view of budgetary constraints relating to total costs of building a landfill and its life span, there is no reason justifying the lack of provision of a high-quality, well-designed house. The quality of the building will give visitors their first impression of the site, demonstrating how landfills are high quality technical constructions in need of good care and maintenance (an example in Fig. 13.2.5).
- If access to the local sewer system is lacking, a sewage treatment plant for wastewaters (not leachate) produced in the office building and other facilities should be provided.
- A parking lot for visitors and employees should be sufficiently sized and be located adjacent to the office building. All roads on the landfill outside the deposition area should be paved. The traffic on the site has to be regulated by adequate signs. There should be a special lane for the compactor. As far as possible temporary roads on the landfill site should be planned.
- Areas around the office building should be planted with trees, bushes, and flowers to create a positive image and provide good working conditions. This will of course need to be appropriately planned, with maintenance works being included in costings.
- A building containing a workshop and garage for the vehicles on-site (compactor, street cleaning machine, tank vehicle for dust control, etc. (see Chapter 15.1)) should be planned.

Figure 13.2.5 Entrance, weighing scale, and service building at Minden landfill site, Germany.

The following additional infrastructures may be envisaged:

- storage area for daily cover
- wheel washing station
- tanks/ponds for intermediate storage of raw or treated leachate
- recycling plant (waste segregation)
- MSW pretreatment plant (e.g., mechanical biological treatment (MBT) plant) (Chapters 4.1–4.3)
- composting/anaerobic digestion plant for separately collected kitchen and yard and/or green waste
- biological soil treatment plant
- inspection area for incoming waste
- installation of an intermediate storage facility for bulky and hazardous waste from households
- devices for measuring at the gate the radioactivity level of the waste to be landfilled (if justified and requested by the Authorities)
- fire–safety systems
- climatological station
- fuels and lubricant storage area

As indicated previously, operating a treatment plant on a landfill area may result in synergistic effects with regard to the treatment of polluted water, biogas utilization, etc.; additionally, residues originating from the plant may be landfilled on-site.

Fig. 13.2.6 presents an aerial view of a landfill site consisting of a landfill, an MBT plant, and a composting facility for biological waste treatment.

Figure 13.2.6 Aerial view of a landfill with an MBT plant and biological treatment facility for separately collected organic waste in Bassum, Germany.

Plans and reports

Plans and reports, as indicated earlier, should be included in the design Project documentation, in conformity with national regulations.

In Italy, national regulations (Legislative Decree No. 36/03) require the following plans and documents:

- Operational management plan
- Aftercare management plan
- Financial plan
- Environmental recovery plan
- Monitoring and control plan
- Annual report to be compiled by the Landfill manager.

The annual report should comprise all results obtained from the monitoring and maintenance required by law of the authorized party or otherwise prescribed (e.g., entry checks on waste being delivered, price charged, waste volumes delivered and remaining, air quality, topographical surveys, maintenance of safeguards, problems, etc.).

CONSTRUCTION, OPERATION, AND AFTERCARE

Construction

During construction, a need to adjust the detailed design may arise from unexpected issues (springs, unforeseen geological anomalies, etc.) and changes in the local situation or requirements.

Construction works should be subjected to technical supervision and rigorous quality checks with regard to materials used and construction techniques applied. As an example, dirty sand or gravel may subsequently result in clogging issues. All clay liners placed on the bottom and slopes of the landfill should comply with the Proctor parameters.

The Technical Committee encharged with testing the quality of the structure and the performance of the service to be provided should commence activities in line with onset of landfill construction. For instance, in a multilayered physical barrier each individual component should be tested immediately after construction/installation.

Manufacts and materials emplaced should not remain exposed for lengthy periods due to the possibility of drainage layers clogging due to vegetation growth, whereas geotextiles may be degraded by ultraviolet solar radiation, oxidation, etc. The presence of animals may also result in damages.

Operation

Landfill operations, as mentioned earlier, may be regarded as ongoing construction work, and a similar rigor in the checking and testing of materials and works should be applied.

All important tasks scheduled to be undertaken during the operational phase (see Chapter 15.1) should already be computed, described, and detailed in the Operational Plan, which is an integral part of the detailed design project. Depositing position and time, related typologies of waste should be mapped by using a computer based tool (GIS).

It is mandatory that a rigorous Quality Control and quality assessment procedure based on the items described below should be set up during operations.

- *Qualified and well-trained staff* with precise definition of roles. Multiple roles may potentially be assigned to a single manager, depending on the size and potential of the landfill. The Lombardia Region Guidelines for landfills (2014) establish a need for the following:
  - Technical Manager, to assume technical and legal responsibilities in managing the landfill.
  - Waste Delivery Manager, to ensure administrative compliance of the individual waste deliveries and the weighing thereof; responsible for the filling in of loading and unloading records.
  - Site Manager, present at the site throughout opening hours and is responsible for ensuring the application of operating standards.
  - Quality Manager.
  - Eco-management and environmental audit manager (if required).
- *Internal auditing.* An internal Quality control should be aimed at checking periodically procedures, observance of all safety measures, measure of operational parameters such as density of landfilled waste, leachate levels within the waste mass, referencing of the location of the daily waste tipping, etc.).
- *External quality assessment programme.* An experienced third party should verify periodically that the Operation Plan is fully complied with and that all facilities are performing at their best. Diffuse emissions should be measured and environmental situations (groundwater quality, air quality, etc.) kept under control.

Furthermore, an annual report should be compiled according to the operational plan, specifically indicating whether the planning forecasts for the various parameters leading to FSQ are accurate or not, and providing details of any corrective measures.

If leachate is recirculated, the annual report must include the following information:

- quantity of leachate recirculated and sent for disposal compared to the estimated levels indicated in the application for authorization;

- assessment of the effects of recirculation on the quantities and characteristics of biogas, characteristics of the leachate, waste settlement;
- estimated water balance by assessing rainfall infiltration, waste dampness, quantity of leachate produced by the degradation of waste and recirculation, evaporation, and leachate sent for disposal to verify the absence of significant infiltrations of leachate into the subsoil and efficacy of the collection and transportation system.

Aftercare

During this phase, it should be verified whether waste settlement has decreased sufficiently to not compromise the overall future closure plans for the landfill, as per the express indications of the approved project. It should be kept in mind that during an eventual in situ aeration an enhanced settling may occur.

Indeed, according to some Guidelines (Lombardia Region, 2014) waste settlement is deemed negligible when the percentage of settlement after 1 year does not exceed the previously measured 3%. This may be different if in situ aeration is adopted.

All parameters that define FSQ should be measured on a yearly basis and compared with Tables of values defined in the Detailed Design documentation (see Chapter 16.1).

AFTERUSE

As already indicated, the final shape and envisaged future utilization of the landfill, once completed and released, should be designed at project start-up. No use of the landfill should be implemented prior to the achieving of desired environmental targets (FSQ). The numerous possibilities for reutilization of a landfill include a public park, commercial use, education center, golf course (an example in Fig. 13.2.7),

Figure 13.2.7 An example of landfill afteruse: a golf course in Los Angeles, United States.

cultivation of energy plants, natural habitat, animal grazing, photovoltaic facilities, and windmill for electricity production.

A prerequisite for utilization is of course a well-operated landfill with a top liner system. However, also under these conditions the monitoring of residual gas volumes that will continue to be produced, and even leachate, will be required; although the amount of gas production is very low, gas may accumulate over time in specific areas. Therefore, all installations on the landfill should be equipped with gas control facilities and safety and monitoring systems.

Detailed aspects of landfill afteruse, presentation, and discussion of case studies are illustrated in Chapter 16.3.

FINAL REMARKS

Landfills should be designed using a multibarrier system for the specific purpose of achieving environmental sustainability. They should be planned in detail to include all components and facilities, establishing the final shape and afteruse destination. Although input data and operational mode may vary over time, this should have been envisaged and all aspects of planning should be appropriately completed.

Mistakes made during the planning stage may prove to be irreversible and result in high maintenance and repair costs, as well as potential environmental damage. Unfortunately, still today mistakes encountered in landfill construction and management are frequently caused in an attempt to reduce costs. However, the reduction of costs during the stages of construction and operations may ultimately result in even greater and at times economically unsustainable expenses during aftercare.

References

Bilitewski, B., Härdtle, G., Marek, K., Weissbach, A., Boeddicker, H., 1997. Waste Management. Springer-Verlag, Berlin, Heidelberg, Germany. ISBN 978-3-540-59210-5.

China Technical Code, 2013. Technical Code for Municipal Solid Waste Sanitary Landfill (GB 50869—52013).

EU Council Directive, 1999. Council Directive 1999/31/EC of 26 April 1999 on the Landfilling of Waste.

Regione Lombardia, October 10, 2014. Guidelines for design and sustainable management of landfilling (Linee guida per la progettazione e gestione sostenibile delle discariche). Bollettino Ufficiale (41).

Stegmann, R., 2017. Discussion of landfill problems and proposals for solutions. In: Christensen, Cossu, R., Stegmann, R. (Eds.), Proceedings SARDINIA 2017 Symposium. CISA Publisher, Cagliari.

Townsend, T.G., Powell, J., Jain, P., Xu, Q., Thabet, T., Reinhart, D. (Eds.), 2015. Sustainable Sustainable Practices for Landfill Design and Operation. Springer Verlag. ISBN:978—1493926619.

US Government, 2012. US Code of Federal Regulations. Title 40 Part 258. Criteria for Municipal Solid Waste Landfills.

13.3

LANDFILLING IN DEVELOPING COUNTRIES

Maria Cristina Lavagnolo

GENERAL ISSUES RELATED TO MUNICIPAL SOLID WASTE MANAGEMENT IN DEVELOPING COUNTRIES

The country classifications established by a series of international institutions (World Bank, United Nations, International Monetary Fund) and the mission of these institutions will determine the definition of a status of developing country (DC). However, there is a lack of a general agreement over how to define a poor person and hence a poor country. The author of this chapter, in referring to a DC, prefers to imply a country in which the majority of the population is poor, with production intended for self-consumption far exceeding that sold on the open market. The presence of a limited market demand is strictly linked to the scarce availability of infrastructures, political stability, and education. However, as all DCs feature an individual set of characteristics, the situations described below will not necessarily apply to all DCs or even to an entire nation.

To understand the relevance of the issue of municipal solid waste (MSW) management in DCs, it should be underlined that population growth in DCs is not only higher than in industrialized countries (Hoornweg and Bhada-Tata, 2012) but is also mainly concentrated in urban areas. When there is increasing urbanization in a country, its economic wealth also increases. This is also reflected by a higher consumption of goods and more services, which results in the generation of more waste (Table 13.3.1).

Table 13.3.1 Average waste generation per capita by income level (modified from Hoornweg and Bhata-Tata, 2012)

Income Level	Waste Generation (kg/capita/day)
High	2.1
Middle - Upper middle	0.79 - 1.2
Lower	0.60

Figure 13.3.1 Yaoundé, Cameroon : typical situation in a poor neighborhood (A) and combination of solid waste management and water pollution (B). *Picture by Maria Cristina Lavagnolo.*

Solid waste is a by-product of a consumer-based economic lifestyle, which is actually more concentrated in urban areas. It is indeed a source of high concern for the municipalities because MSW management in low income countries usually falls under the competency of the local government. However, municipalities frequently lack budgetary and technical capacities to deal with waste management.

Of even greater concern are the nefarious environmental and health issues related to littering, insufficient quantity and quality of drinking water, rapidly worsening air quality, degradation of the urban environment, impurity of the streets (Yhdego, 1995), river pollution, and climate change (Ayuba et al., 2013). Moreover, any uncontrolled dump site will constitute a potential health hazard due to scavengers, groundwater pollution, spread of infectious diseases, highly toxic smoke from continuously smoldering fires, and foul odors from decomposing refuse (Ogwueleka, 2009). Impacts on the environment due to air, water, and soil pollution and associated health risks may also ultimately affect the economy (Fig. 13.3.1).

Clearly, therefore, the introduction of new technologies in countries encumbered by the above-cited conditions should take into account the evident constraints and challenges to be faced.

TECHNOLOGY TRANSFER LIMITATION

To transfer without distinction waste concepts, technologies and models from the industrialized world to DCs is a significant mistake due to the wide variance of conditions in which industrialized and DCs are called on to work to solve MSW management (UNEP, 2005). The mistake is not related solely to the shifting of a model that is not relevant to such a situation rather than applying outdated methods or obsolete technologies. A sound MSW management program for DCs should always be based on new and advanced concepts, internationally conceived and experienced as well as effective, well adapted to the specific situation, easy to operate, and well-proven technologies.

Proper management and technologies specifically focused on ensuring efficacy at least in a medium-term perspective must take into account the limitations of the existing situation.

Constraints, typically recognizable as barriers for a valid technology transfer, are determined by a series of different reasons:

- low education at different levels: universities often have little competence or are not equipped to prepare engineers and qualified technicians (shortage of specific courses and research laboratories); insufficient specifically driven courses for operators (management and maintenance); notably absent education on environmental issues in the primary and secondary schools (still today people grow up with no proper awareness of environmental pollution and MSW practises);
- political instability with failure of long-term MSW management actions;
- MSW management is not always a high priority for local and national policy makers and planners (Marshall and Farahbakhsh, 2013);
- a scarce awareness among administrations of the basic needs of the population and the lack of willingness to promote proper actions result in weak and ineffective public sector services. According to Ogwueleka (2009), local authorities in DCs spend 77%–95% of their revenue on collection and the balance on disposal, although only collecting approximately 50%–70% of MSW;
- weak institutional structures and pervasive corruption: regulatory and legislative frameworks have been established in many countries, but effective MSW management requires clear roles and the legal responsibilities of institutions and government bodies to be clearly defined to avoid controversies, ineffectiveness, and inaction. Institutional strengthening and capacity building becomes a major driver for MSW management in DCs (Marshall and Farahbakhsh, 2013);
- inappropriate international funding and loans that support the project just for a short period of time; this kind of technology transfer may at times produce an adverse effect, blocking appropriate concepts for many years. In the majority of cases, the international financial institutions cover the investment costs but do not cover operations for a long enough period to warrant successful transfer of the project to the local authorities. In addition, the implemented technologies are frequently not properly selected and sized to meet local needs (Coffey and Coad, 2010);
- DCs often seek to implement technologies of the highest standards, which they are subsequently not in a position to operate due to missing spare parts for repair and/or the lack of well-trained personal. The main aim should be focused on providing robust and well-proven technologies suited for management by local people, thus providing an opportunity for further education and jobs.

To design and build a sustainable MSW management system, the different interests and needs of the stakeholders (public and private) need to be met. The system should be resilient to political interferences, be flexible to further developments (e.g., market, technology, social), and should promote an increasing awareness of environmental issues. It is fundamental not only to identify the best technology but rather to find custom-made technologies in line with social, cultural, economical, and local requirements.

TECHNICAL CONSTRAINTS

The accumulations of waste on the road, open dumps, and open burning are widespread practices applied in dealing with garbage. The latter situations are those that most frequently need to be urgently addressed through implementation of a proper waste management concept.

In addition to social, cultural, and economical conditions, several other issues drive the selection of the most appropriate solutions for waste management in DCs.

As far as possible recent and reliable information should be collected on MSW quality and production, population size, and growth rate, middle- and long-term urban development (expansion of the residential area; land use planning for commercial or industrial area). This may prove difficult as frequently this type of information is not readily available, is not exhaustive, and is disorganized or scattered among different stakeholders (NGOs, WHO, public and private companies, local and national administrations, universities). Rapid unplanned population growth in cities, mainly concentrated in slums and poor neighborhoods, markedly impact on this situation. In most cases, waste collection does not cover all urban areas and frequently neglects rural areas. Moreover, the various streams of valuable materials are managed by the informal sector, thus resulting in a lack of information in the official data.

The putrescible organic content of MSW is generally higher in DCs than in industrialized countries (Table 13.3.2). Important differences in waste composition and quantities depend also on the impact of agriculture, climate, and its seasonal variations. In a recent project implemented in Bissau (Guinea-Bissau)—due to the relevant lifestyle—analysis of waste composition showed a significant variation among the different city districts. In the downtown areas a much higher percentage of glass bottles was detected, whereas the fines <2 cm—due to road sweeping and home combustions for cooking and water heating—amounted to around 66%–67% of the total waste (Lavagnolo et al., 2016).

In DCs it is especially important to distinguish between rural and urban areas. In rural areas waste can be handled to a high degree locally (e.g., home composting for the organic fraction) and the rest may be reused and/or centrally collected and transported at appropriate times to central plants in the next city/larger development.

Table 13.3.2 Composition of waste by income level (modified from Hoornweg and Bhada-Tata, 2012)

Income Level	Fractions (%)					
	Putrescibles	Paper	Plastics	Glass	Metal	Others
Low	64	5	8	3	3	17
Upper - Lower middle	54 - 59	14 - 9	11 - 12	5 - 3	3 - 2	13 - 15
High	28	31	11	7	6	17

Lack of infrastructures such as no roads or paved roads for transportation limit the siting of waste management facilities and increase the costs for construction materials particularly when these are not locally available. Insufficient efficiency in the energy supply (poor technologies, intermittent shortage of fuel, etc.) may lead to prefer simplest and low technology options.

Many low income countries are located in dry climate areas with a scarcity of accessible water or in areas characterized by high and seasonal rainfall (Munawar and Fellner, 2013). Particularly for the design and operation of open composting plants and landfills (leachate and gas emissions, kinetics of degradation processes, moisture saturation, contaminants diffusion control, etc.), the climatic aspects are of severe influence.

LANDFILL MODELS FOR DEVELOPING COUNTRIES

Landfill sites may be classified according to their environmental safety level. The *open dump* approach is the primitive stage of landfill characterized by uncontrolled disposal of waste and limited or no measures in place to control operations; this of course includes no control of the environmental and social effects of landfills. A *Controlled landfill* is often the first stage in a country's efforts to upgrade landfills. The first step and challenge in upgrading open dumps to sanitary landfills involves reducing nuisances such as odors, dust, vermin, and birds. The landfill therefore operates with some form of inspection and recording of incoming wastes, practices extensive compaction of waste, and controls the tipping front and the application of soil cover. Subsequently, *Engineered landfills* embody further attempts to minimize environmental impacts. Finally, *Sanitary landfills* incorporate a full set of measures to include adequate siting, avoidance of nuisances and vermin, control and utilization of landfill gas (LFG), collection and treatment of leachate, application of a daily soil cover on waste, and implementation of plans for closure and aftercare long after waste has ceased being deposited to the site.

Accordingly, the concept of sustainable landfill should be implemented. Cossu (2007) defined a sustainable landfill as a system that should reach an acceptable equilibrium with the environment within one generation (30–40 years). In other terms, a sustainable landfill is the one that within a limited period of time has reached a state at which the wastes no longer pose a potential threat to the environment (Huber-Humer and Lechner, 2013). Different landfill concepts are presented in Chapter 2.1.

From a technical point of view, Sukholthaman and Shirahada (2015) have suggested small-scale, low capital cost, energy-efficient landfills that are controlled by the local community. To ensure the sustainability of an engineered landfill in DCs, Allen (2002) highlighted similar suggestions: low cost solutions in terms of development, operation, and maintenance; simple easily implemented technologies; and a maximum utilization of natural resources and the properties of in situ materials. In particular the financial sustainability of the project is strongly affected by the use of artificial lining systems, the single most costly feature of modern landfills. Generally speaking, the few examples of engineered sanitary landfills in the larger cities in DCs have operational problems such as the cost of the daily cover (up to 50% of the operating costs), the managing of the LFG, and the leachate control and treatment (Oakley and Jimenez, 2012).

In particular, the semiaerobic landfill seems to meet the specific needs of low income countries as it can be managed in a simpler way with less energetic demand (no biogas extraction and leachate outflow by gravity) and produces a shorter long-term impact due to the aerobic degradation of organics.

Technical Suggestions

Taken for granted the international recommendations for landfill construction (barriers, slopes, cover) are widely discussed in the other book chapters, several technical suggestions specifically addressed for DCs can be resumed from different experiences worldwide (Table 13.3.3) (UNEP, 2005; Munawar and Fellner, 2013; Lavagnolo et al., 2016; Matsufuji, 2016).

Table 13.3.3 Overview of the suggestions and of alternative materials to be applied in landfill construction and management in developing countries

	Factors to Be Evaluated	Technical/Economical Constraints	Solutions
Site selection	**1.** topography; availability of local materials; transport distance (costs); accessible roads; water and energy supply available **2.** religious and cultural sites; ownership and land prices	**1.** no machine and staff for on-site survey; no proper analytical laboratories **2.** external technicians are often not aware of the local situation	**1.** refer to a company involved in road or building construction **2.** participatory action involving local community
Climate condition	temperature; rainfall	erosion; water saturation of waste layer; leachate management	maximum slope 25% and surface water control in case of high rainfall and berms for high landfill
Bottom liner	**1.** HDPE liner **2.** clay layer	**1.** availability and costs **2.** hydraulic conductivity $>10^{-9}$ m/s	**1.** bituminous asphalt or a single baseliner with 0.4–1 m clay liner permeability $\leq 10^{-9}$ m/s **2.** 20–30 cm clay reprocessing
Drainage and cover materials	**1.** bottom and intermediate drainage **2.** top cover drainage **3.** daily cover	**1.** availability and costs **2.** availability and costs **3.** availability and costs	**1.** C&D blocks or crashed stones or scrap tires **2.** appropriately shaped glass or properly selected C&D waste materials **3.** raw compost or properly selected C&D waste materials

TABLE 13.3.3 Overview of the suggestions and of alternative materials to be applied in landfill construction and management in developing countries—cont'd

	Factors to Be Evaluated	Technical/Economical Constraints	Solutions
Biogas drainage and collection (anaerobic)	**1.** biogas well **2.** extraction and reutilization system	**1.** availability and costs **2.** costs and energy demand	**1.** bamboo or concrete pipes combined with: crushed stones or C&D blocks or waste tires **2.** local use with mobile station for industrial use or flaring or methane oxidation layer
Venting system (semiaerobic)	vertical pipes	availability and costs	bamboo or concrete pipes combined with: crushed stones or C&D blocks or waste tires
Leachate drainage, collection, and treatment	**1.** leachate collection and transport **2.** leachate pipes for transportation **3.** treatment	**1.** energy demand (cost and availability) **2.** availability and costs **3.** cost and energy demand	**1.** large area collection system using crushed stones, or similar material, drain by gravity **2.** (handmade) concrete or bamboo or any plastics pipes (if possible HDPE) **3.** properly designed ponds and phytotreatment; wind-driven aerators (TORNADO and ECO-FAN)

ECO-FAN, Eco-Friendly Aeration Nippon; *HDPE*, high-density polyethylene.

Siting

Site selection criteria are the same for all landfills worldwide. Siting should take into account the overall costs of landfilling and its potential environmental impact (topography, hydrology, hydrogeology, accessible roads, availability of local materials, transport costs) and the essential availability of energy and water supplies. An additional important factor to be taken into account, not only in DCs, is the involvement of the local people. Sustainability of landfill construction and operation depends very much on the initial acceptance of the local community. This criterion is potentially difficult to evaluate as external technicians are not allowed to participate at local community level. To this regard, it is crucial to refer to the chiefs of the villages involved, also with a view to avoiding misunderstanding over the location of specific religious and cultural sites and identification of the ownership of the land, including the related prices.

Specific characterization of the geological and hydrogeological properties of the site may be difficult to achieve due to the lack of proper laboratories with adequate equipment. It may therefore be advisable to locally present foreign companies with experience in the construction of roads and buildings; in general they are willing to collaborate.

Slope

Landfill slopes should be carefully designed to avoid the possibility of landslides in DCs, particularly when located in tropical areas. This phenomenon is largely caused by water saturation of waste layers. To minimize this risk, Munawar and Fellner (2013) suggested that the incline of the slope should not exceed 25%. In the case of very large landfills (height exceeding 30 m), the maximum slope angle should be reduced, or at least one berm should be constructed.

Bottom Liner and Drainage Systems

In most affluent countries a composite lining system consisting of a clay layer (about 0.40–1 m thick) and a 2 mm high-density polyethylene (HDPE) foil is applied. If HDPE foil is not affordable, a single baseliner consisting of compacted clay—especially if locally available—may be used. When the hydraulic conductivity of the existing clay below the planned landfill is higher than 10^{-9} m/s, the existing clay may, for example, be improved by the on-site reworking of the first 20–30 cm of virginal clay with a heavy compactor and/or the addition of additives. In other cases a man-made base sealing is needed.

To reduce operational costs and allow long-term drainage of leachate, the latter should be able to flow by gravity—i.e., without pumping—out of the landfill. The drainage and collection system should be designed in such a way that leachate drains from the landfill mound in a short time (this is particularly recommended in tropical climate with high rainfall rates) (Fig. 13.3.2).

Leachate Management

The quality and quantity of leachate obtained from landfills in both DCs and other countries are comparable (see Chapter 10.2). In the majority of DCs waste quality is characterized by a high content of readily degradable organics (Table 13.3.2), which is reflected in leachate quality (organics, ammonium, and easy soluble salts). The concentration of pollutants may increase during dry seasons (less dilution of leachate by rainwater) and decrease in the wet season.

Leachate treatment methods should address this situation by providing an adequate storage volume. Leachate treatment plants should be both effective and low cost. The quality of treated leachate should be monitored, as a minimum requirement, on the basis of COD and ammonia measurements. During dry seasons leachate may be recirculated back to the top of the landfill to reduce the volume by evaporation and to enhance the biodegradation processes. In this case, electrical energy will be required for

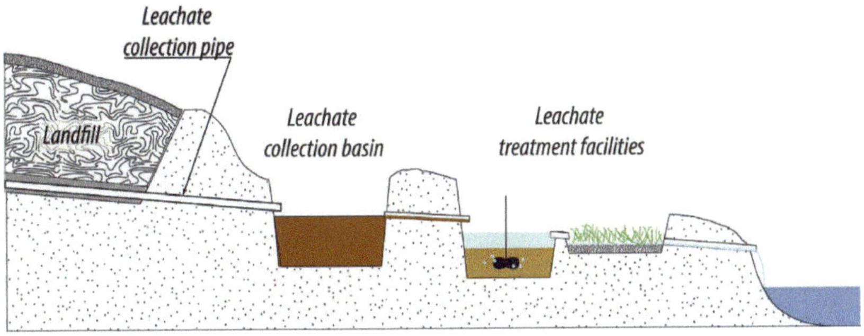

Figure 13.3.2 Example of leachate gravity flow (edit by Grossule, 2018).

pumping, which adds to the cost. The latter, however, may in some cases be compensated due to the need to treat lower amounts of leachate. Recirculation should be executed in a controlled manner, preferably when the landfill is closed (Chapters 12.1 and 12.2).

Leachate is treated mainly in multistage ponds in the following sequence: anaerobic, aerobic with artificial aeration, and refining (maturation pond often combined with constructed wetland as a final step). Ponds must be carefully sized to properly retain incoming leachate volume and to allow an adequate treatment time; this is particularly important during the heavy rain season (tropical climate) (Chapter 10.3). The following formula may be used to calculate the volume for each single pond:

$$V = r \times P \times A \times HRT \times 10^{-3}$$

where: V, leachate pond volume (m^3); r, ratio between rainfall and leachate production (0.4–0.7); P, maximum daily rainfall at the site (mm/day); A, landfill surface area (m^2); HRT = retention time in the specific pond (day).

Munawar and Fellner (2013) suggested P as the maximum precipitation rate for a representative storm over the period of HRT; Matsufuji, 2016, suggested P as the average of the maximum daily volumes in the reference year (TRY = test reference year) removing the extreme cases (2013); Lavagnolo et al. (2016) suggested for P the average of the daily precipitation volume in the reference year.

Landfill Gas Management

Methane production in low-compacted landfills or dumps may be reduced due to (partly) aerobic processes and fires that may occur inside a landfill. LFG should be strictly controlled, i.e., the gas should be extracted and either be flared or preferably used for energy production (Chapter 9.4), also with the aim of reducing greenhouse gas (GHG) emissions. The following options for LFG treatment may be considered:

- gas extraction and utilization for local electricity generation or extraction and pumping to a nearby industrial plant (if present) (Chapters 9.3 and 9.4). These solutions, however, require a relatively high technological input and high capital investment. Financial support can be applied under the framework of clean development mechanism (CDM);
- extraction and flaring without energy recovery, this option cuts the investment costs without generating revenue and does not require a particularly high degree of technology;
- methane oxidation layer to convert methane into CO_2 on the landfill surface to reduce the environmental impact (Chapter 9.5).

It is always difficult to adjust blowers and gas utilization plants to varying/changing gas production rates. For this reason—if relevant—instead of one big engine, two or more smaller engines providing a similar total capacity should be considered for installation. In some cases the location of the gas engine and flare could be set up to enable the gas flow to be easily switched from one system to the other. This may be a way of dealing with low gas production rates produced in dry seasons that require flaring.

The number, size, and radius of influence of the gas extraction wells should be carefully investigated and designed. In most cases 2–3 test wells should be installed and test runs should be carried out (Chapter 9.3). Biogas extraction wells can be installed/drilled once a landfill sector is closed; by these means the investment costs may be recouped from the operational costs.

A major issue in LFG utilization in DCs is represented by the low electricity costs that often render a project unprofitable or even produce a deficit. On the other hand—when the village has no access to the public grid—generators driven by diesel engines often produce electricity; in these cases, LFG may be used as a substitute for diesel fuel. The latter may be an economic solution as the cost of diesel is relatively high and only marginally subsidized, if at all.

Alternative Materials for Landfill Construction and Management

The use of alternative solutions is suggested when proper materials for landfill construction are not available or costs cannot be covered.

Bottom Liner

The potential substitution of a synthetic bottom liner, as discussed in the previous paragraph, should be carefully evaluated. If the local subsoil consists in a layer of several meters of homogeneous low permeable clay and the groundwater level is several meters below the surface, a substitute may be used. A 2 mm HDPE liner may be replaced by a bituminous asphalt liner or by an artificially emplaced clay or bentonite liner.

Where neither HDPE nor proper clay liners are available, the other components of the barrier system (particularly the drainage system with sufficient slopes) should be strengthened to minimize the potential impacts; it may even be advisable to reconsider the entire waste management system (collection, treatments, and final disposal).

Drainage System and Covers

Handmade concrete pipes may be used as a substitute for HDPE leachate drains and collection pipes. The openings in the drain pipes may be fabricated locally using simple steel molds, with diameters of about 25 mm. Other possible materials for pipes that may be used are bamboo (Fig. 13.3.3A) and plastic (other than HDPE) as well as any other suitable locally available materials.

Materials that are available locally can substitute for drainage material, thus potentially resulting in a reduction of transportation costs. Instead of sand and gravel, C&D blocks, crashed stones, and scrap tires may be used for the bottom drainage layer. C&D waste should also have very low sulfate content to avoid H2S production (toxic, very odorous) and corrosion. Moreover, to avoid carbonate precipitation (danger of clogging drain pipes), C&D waste should feature a low lime content (Chapter 8.1). Among the different materials used, scrap tires could represent an alternative solution. In terms of their permeability several studies have demonstrated the suitability of coarse shredded tires as landfill drainage media (Fig. 13.3.3B) (Beaven et al., 2013).

Kaushik et al. (2015) reported that shredded/chipped tires (capable of yielding a relatively free drainage) maintained a high hydraulic conductivity under high vertical overburden pressure levels over relatively long periods of time even for high strength MSW landfill leachate. Leachate analysis highlighted a significant removal of suspended solids and a significant release of iron from the shredded tires. In a review paper, Das and Singh (2013) stated that scrap tires do not seem to be a hazardous waste and may not produce significant adverse effects on water quality. On the contrary, shredded tires showed a significant sorption capacity for organic compounds, which may be an interesting property for a reactive barrier in landfills.

Properly shaped shredded waste glass may also be used for top cover drainage.

Nonutilized compost and shredded C&D waste can be used as an alternative daily cover material.

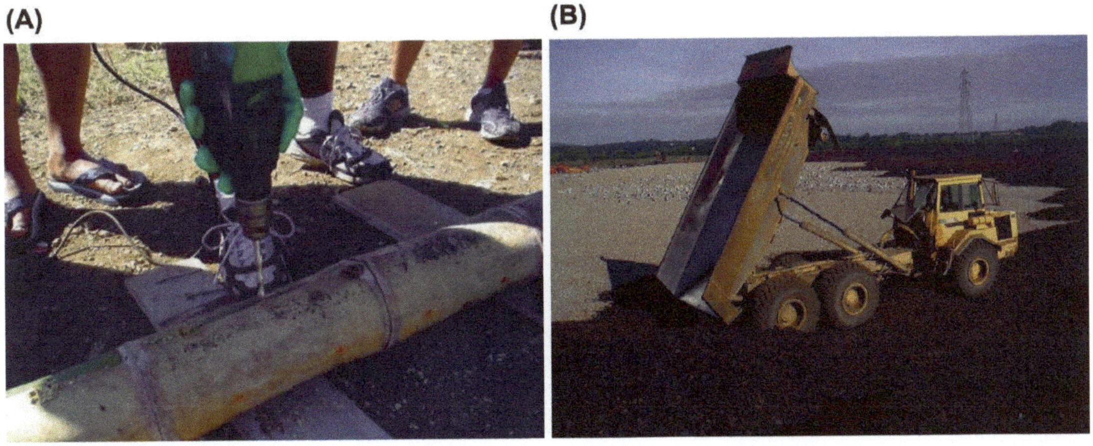

Figure 13.3.3 Alternative use of materials such as (A) bamboo and (B) tire for bottom drainage system (courtesy of (A) prof. Matsufuji from SPREP-JICA, 2010; (B) prof. Beaven, 2013).

Biogas Wells and Venting System

Biogas extraction wells (in case of anaerobic landfill, Fig. 13.3.4A) or venting systems (in case of semiaerobic landfill, Fig. 13.3.4C) may be built using both bamboo and concrete pipes combined with crushed stones, C&D blocks, or waste tires.

The specific system illustrated in Fig. 13.3.4A is a biogas extraction unit built using a wooden frame, wire mesh, and crushed stone. An LFG flare is then designed by sealing the top of the collection unit and inserting a 50 mm metal pipe into the stone; a metallic can is welded to the top of the pipe and perforations are made on the side of the can, as shown in Fig. 13.3.4B (UNEP, 2005). Using this kind of system the upper 4–5 m below the final landfill surface should be sealed to capture the gas without outside air intrusion (Chapter 9.3). If no sealing is applied, the wells will promote enhanced gas emission into the atmosphere, which should be avoided at all costs. Accordingly, the proposal illustrated in Fig. 13.3.4A is not an appropriate solution.

Leachate Treatment

All types of landfill generate leachate due to infiltration of rainfall and percolation of water through the waste, which will subsequently require treatment to reach the required standards. These standards are often not provided for by local legislation but can be identified, depending on the specific on-site situation, to avoid or at least to minimize the pollution of soils, rivers, and groundwater. To date, the use of ponds seems to constitute one of the most suitable solutions when on-site technicians are present and no materials and energy supplies are available. Under certain conditions organic compounds can be aerobically degraded in ponds by means of natural aeration, as long as the ponds are shallow (depth 0.5–1.0 m) and COD is not too high, i.e., ponds will need to be adapted ad hoc to the specific situation.

(A) (B) (C)

Figure 13.3.4 Alternative materials for (A) biogas well; (B) gas flare in conventional anaerobic landfill (UNEP, 2005); and (C) gas-venting system in semiaerobic landfill (*courtesy of prof. Matsufuji, 2016*).

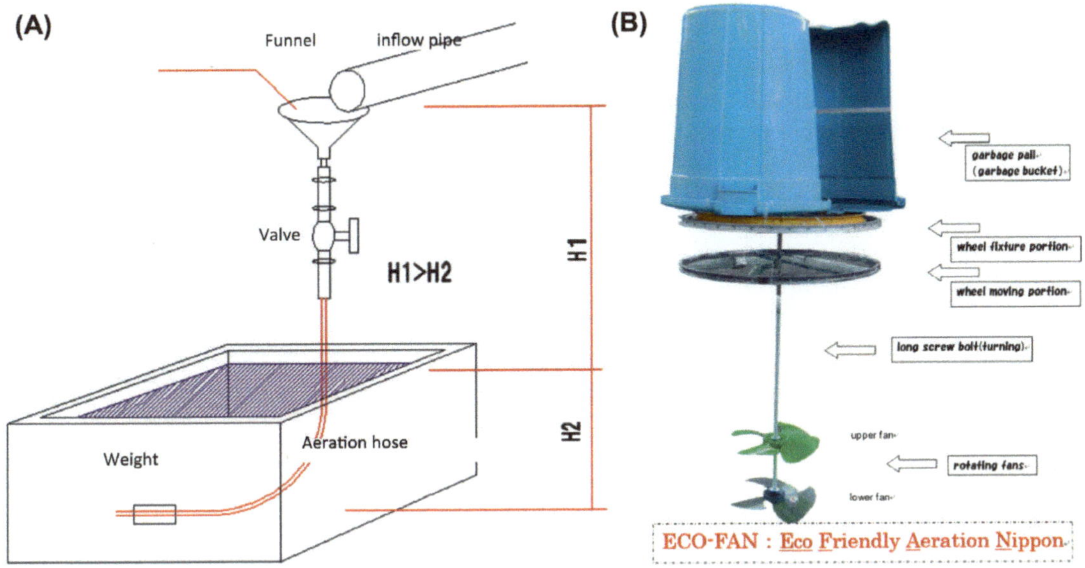

Figure 13.3.5 Low cost leachate treatment devices: (A) the TORNADO aeration system; (B) the ECO-FAN wind mixer. *Courtesy of Prof. Matsufuji, 2017.*

Matsufuji et al. (2013) purpose-designed several specific devices to enhance the on-site treatment of landfill leachate in ponds (Fig. 13.3.5). Should artificial aeration be required due to a pond depth of >1.0 m, systems such as *Tornado* and Eco-Friendly Aeration Nippon (*ECO-FAN*) may be used; these are simple physical aeration and stirring devices used to enrich oxygen content and to mix the leachate in the pond. Using the so-called *Tornado* (Fig. 13.3.5A), leachate enters the lagoon by means of a circuitous pathway, entrapping air during the flow and thus transferring oxygen into the lagoon. The *ECO-FAN* (Fig. 13.3.5B) is a kind of wind-driven mixing device built with recovered materials including wheels from old bikes and purpose-molded garbage bins, which is installed on the surface of the lagoon. By these means the biodegradation of organics will be enhanced and—under certain conditions—will allow simultaneous nitrification/denitrification. Other wind-driven and physical aerators are used in Asia to aerate fish ponds. Ponds should under no circumstances exceed a depth of 1.5–2.0 m.

Table 13.3.3 provides suggestions for alternative materials that may be used if the "normally required" materials are not available. These alternative materials should be used with care and only following investigation into their suitability for the specific situation.

A GLANCE AT THE WORLD OF DEVELOPING COUNTRIES

Various studies published over the last decade have confirmed that open dumping (Fig. 13.3.6) is the most common practice for the disposal of urban solid wastes in the less developed regions of Africa, Asia, and Latin America (Achankeng, 2004; Pugh and Caine, 2000; Dedinec et al., 2014;

Figure 13.3.6 Open dumps in developing countries: Bissau, Guinea-Bissau (A); Battambang, Cambodia (B). *(A) Picture by Maria Cristina Lavagnolo; (B) Courtesy of Prof. Werner Bidlingmaier.*

Guerrero et al., 2013; Oakley and Jimenez, 2012; Ofori-Boateng et al., 2013; Oteng-Ababio et al., 2013; Otieno and Taiwo, 2007; Thi et al., 2015). Where no official disposal sites are present, communities often carry out the illegal disposal of waste on the land, in rivers, lakes, oceans, drainage channels, and roadsides.

Waste Atlas 2013 Report revealed that globally almost 40% of the waste generated is not disposed in an environmentally sound way, often in open dump sites. Mavropoulos et al. (2015) listed the worldwide 50 largest active dump sites (Table 13.3.4).

Table 13.3.4 Characteristics of the 50 largest dump sites listed per continent (courtesy of CISA publisher, in Mavropoulos et al., 2015)

	Number of Dump Sites Among the 50 Biggest	Waste in Place (million tonne)	Total Size (ha)	Number of People Living And/Or Working on Sites	Number of People Living in a Radius of 10 km (million)
Africa	18	76–88	500	22,000	29.8
Asia	17	109–155	835	16,000	23.8
Europe	2	7–10	40	20	570,000
Latin America	8	33–48	490	8,000	5.6
Caribbean	5	33–48	325	6,700	3.8

An open dump is defined as an open space on which rubbish is dumped and left uncontrolled with (SPREP-JICA, 2010) the following:

- no soil cover,
- no leachate collection/control,
- no drainage,
- no compaction,
- poor access to the tipping area, especially in a wet season,
- open to scavenging, and
- uncontrolled open burning.

As indicated previously, open dumping of waste is an environmentally and socially dangerous method of waste disposal (Mavropoulos et al., 2015; Ogwueleka, 2009). The number and size of open dumps is raising alarmingly due to higher waste generation rates in growing—and already overpopulated—cities. There is a growing awareness of citizens and politicians in DCs of this dangerous situation, although there is still considerable room for improvement.

Case studies of disposal methods in DCs are reported in Table 13.3.5. According to the description of the technologies and installations used, a categorization of landfills is proposed (open dump, controlled landfill, engineered landfill, sanitary landfill, and sustainable landfill). Specifically, the

Table 13.3.5 Overview of case studies in developing countries, with proposed labeling

Case Study	Disposal Technologies	Costs	Source
AFRICA			
Cape Town (South Africa)	*SANITARY LANDFILL* Engineered liner; odor control, dust mitigation, and pest control measures; secure entrances and perimeter fencing measures; surface and groundwater monitoring; landfill gas management (capture and energy production/flaring), air quality monitoring.	~24.1 US$/ tonne of waste	Engledow (2007) and UNFCCC/CCNUCC (2014a,b)
Durban (South Africa)	*SANITARY LANDFILL* Multibarrier bottom liner system; Sequencing Batch Reactor unit and lined reed bed for leachate treatment.		Johannessen and Boyer (1999), Moodley (2013), Pass and Wilson (2005), and Parkin et al. (2006)
Mauritius	*SANITARY LANDFILL* Composite liners: compacted clay plus HDPE 1.5 mm geo membrane; inefficient drainage system; geo membrane top cover; leachate buffer lagoons; LFG wells and pneumatic pumps; LFG-to-energy engines/flares.	83 US$/tonne of waste	Knox and Kowlesser (2011) and Beerachee (2012)

(Continued)

TABLE 13.3.5 Overview of case studies in developing countries, with proposed labeling—cont'd

Case Study	Disposal Technologies	Costs	Source
Dares Salaam (Tanzania)	*OPEN DUMP* No cover material; limited compaction; spreading of wastes by a bulldozer; waste pickers; open fires.	0.69 US$/ tonne of waste	Breeze (2012)
Tanga (Tanzania)	*CONTROLLED LANDFILL* Sand quarry with a permeability of 10^{-5}–10^{-9} m/s; no bottom liner, "dilute and disperse" method; groundwater level rises to the ground surface; LF near to existing dwellings (30 m); in contrast to the plan of settlements expansion of the LF.		Mato (1999)
Nairobi (Kenya)	*OPEN DUMP* Uncontrolled MSW dump sites without any environmental protection.	11–17 US$/ tonne of waste	Henry et al. (2006)
Mombasa (Kenya)	*OPEN DUMP* No top cover; presence of scavengers; surface water and groundwater pollution due to the proximity of the sea; flooded roads; no leachate treatment; no efforts in dealing with pest control, dust control, and methane gas control.	7.67 million US$ (French donation)	Tan (2012)
Kampala (Uganda)	*CONTROLLED LANDFILL* Weighbridge; crawler trucks; waste is sometimes sprayed with insecticide to kill off flies before it is covered with soil; mechanical aeration plus wetland to treat leachate.		Komakech et al. (2014) and Mwiganga and Kansiime (2005)
Yaoundé (Cameroon)	*SANITARY LANDFILL* Excavations 15 m deep; 15 cm of soil as top cover (compacted on-site materials + geo membrane + sand possibly artificial protective drains in Geotextile); additional compensation layer using excavated on-site material; LFG extraction and collection system; three-stage leachate treatment ponds; leachate and rainwater are drained separately.		UNFCCCC/CCNUCC (2014a,b) and Achankeng (2004)
Bamenda (Cameroon)	*OPEN DUMP* No natural or synthetic lining, no soil cover, no leachate or gas collection system, or measures against odors, mosquitoes, and dust; no fence; open fires; a bulldozer; or a front-end loader is used to push the waste in the river valley.		Achankeng (2004)

TABLE 13.3.5 Overview of case studies in developing countries, with proposed labeling—cont'd

Case Study	Disposal Technologies	Costs	Source
Fed. Cap. Territory Abuja (Nigeria)	*CONTROLLED LANDFILL* Multimembrane barrier (clay—sandy soil mixed with bentonite); fence, an administrative building; dilapidated access road inaccessible during rainy season.		Ayuba et al. (2013)
Ibadan (Nigeria)	*OPEN DUMP* Prefeasibility study in 2010 for an LFG-to-electricity projects; installation of liners and a leachate control system.		CPE (2010)
Kumasi Metropolitan Area (Ghana)	*CONTROLLED LANDFILL* Well-lined base, leachate collection, and treatment ponds; security fence; geology is covered by lateritic soil consisting mainly of sand- to silt-sized quartz particles in a matrix of clay.		Owusu-Sekyere et al. (2013)
Freetown (Sierra Leone)	*OPEN DUMP* Poor operational practices; no environmental safeguards.		Gogra et al. (2010)
ASIA			
Weifang City (China)	*SUSTAINABLE LANDFILL* Semiaerobic; liner system, leachate collection pipes, leachate pond, leachate treatment plant, gas ventilation pipes, runoff drainage, site office, and laboratory facility.		FMG, Environment Bureau (2010)
Shanghai (China)	*ENGINEERED LANDFILL* Bottom natural barrier, 17 m deep with a permeability coefficient of 10^{-9} m/s; leachate collection; aerated lagoons as treatment method; passive gas-venting system, four wells per ha: 1 m in diameter, perforated pipes surrounded by stone fill.		Johannessen and Boyer (1999)
Beijing (China)	*ENGINEERED LANDFILL* Natural clay deposits with a bentonite liner; leachate collection and treatment with aerated lagoons; passive gas-venting system, four wells per ha: 1 m in diameter perforated pipe surrounded by stone fill.		Johannessen and Boyer (1999)
Phnom Penh (Cambodia)	*CONTROLLED LANDFILL* Waste compaction; lining and leachate drainage system with frequent outbreaks; no soil cover and no landfill gas management.	US$ 0.75/ tonne of waste	Sang-Arun and Heng (2011) and Mongtoeun et al. (2014)
Chonburi (Thailand)	*CONTROLLED LANDFILL* Composite HDPE/clay liner and daily cover with		Pugh and Caine (2000)

(*Continued*)

TABLE 13.3.5 Overview of case studies in developing countries, with proposed labeling—cont'd

Case Study	Disposal Technologies	Costs	Source
	in situ excavated material; no specific compaction machinery; density achieved: 0.82–0.95 t/m^3.		
Sarawak (Malaysia)	*ENGINEERED LANDFILL* Intensive geographical and geotechnical research; TRISOPLAST bottom liner; top drainage system with gravel and perforated HDPE pipes; compaction up to about 1.2 t/m^3; welded HDPE liners covering the top of the waste cells; gas collection system consisting of horizontal and vertical pipes; LFG flaring.	US$ 7.89/ tonne of waste	Damiecki and Bilong (2009)
Malaysia	*SUSTAINABLE LANDFILL* Semiaerobic landfill; site selection: presence of a natural clay soil with low permeability.	US$ 8.89/ tonne of waste	Chong et al. (2005)
Metro Manila (Philippines)	*CONTROLLED LANDFILL* Artificial liners of polyethylene (2.5 mm HDPE liner) at the bottom; leachate treatment in aerated and facultative ponds; leachate recirculation		Johannessen and Boyer (1999)
Delhi (India)	Landfill mining.		Ranjan et al. (2014)
Ahmedabad (India)	*ENGINEERED LANDFILL* Security cabin, entrance gate, weighbridge, platform, administrative building with laboratory, parking shed, workers' toilet and drinking room, electrical substation and panel room, compound wall and wire fencing, underground water and bore well, truck parking area and truck wash system, test or piezometer bore and wells around landfill.	US$ 2.44/ tonne of waste	(Kedar, 2015)
Amman (Jordan)	*CONTROLLED LANDFILL* Lining system: • Needle-punched Bentofix Geosynthetic Clay Liner with Thermal Lock; • HDPE Carbofol 2 mm thick (smooth for the base and structured for the slopes); • Secutex protection nonwoven; • Gravel leachate collection system with 30 cm of thickness.		Tadros (2007)

TABLE 13.3.5 Overview of case studies in developing countries, with proposed labeling—cont'd

Case Study	Disposal Technologies	Costs	Source
Upolu (Samoa)	*SUSTAINABLE LANDFILL* Upgrade from an open dump to a semiaerobic landfill (Fukuoka method); all season access road; rolling and the compaction of the floor (made by old rubbish), gatehouse, office, recycling areas, leachate treatment system with simple cost-effective technology.	US$ 400,000	Blake and Chapman (2017) and SPREP-JICA (2010)
LATIN AMERICA			
Rio Grande City (Brazil)	*SANITARY LANDFILL* Reverse osmosis plant.		Zanon et al. (2013)
Santiago (Chile)	*SANITARY LANDFILL* Composite liners of a low-permeability clay covered directly with a polyethylene liner; successful recovery and utilization of landfill gas.		Johannessen and Boyer (1999)
Guayaquil (Ecuador)	*SANITARY LANDFILL* Electronic and manual scales; a net of leachate drainage canals are placed and LFG are treated by incineration; The impermeability of the LF was ensured by the presence of 10 m of natural clay.		Von Buchwald (2005)
Villanueva (Honduras)	*ENGINEERED LANDFILL* Semimechanized trench method; filled trench is covered with 1 m of excavated soil; natural barrier with clay topsoil (depth = 10 m) and groundwater depths of 8–10 m; negligible leachate generation: leachate collection was not practiced and natural attenuation for leachate management; rock chimneys were used for methane ventilation.	US$ 4.60/ tonne of waste	Oakley and Jimenez (2012)

HDPE, high-density polyethylene; *LFG*, landfill gas; *MSW*, municipal solid waste; LF, landfill

semiaerobic landfill has been classified as a sustainable landfill. As many of the landfill types proposed for use as a sanitary landfill lack a proper maintenance program, some might, over time, become downgraded to a controlled landfill or an open dump. The main challenge illustrated in Table 13.3.5 is the cost estimation section. The large variation in estimated costs and the lack of information relating to disposal costs in some cities in DCs have resulted in a scarce availability of verified sources. Many cities do not know the total costs involved in solid waste collection and disposal services. Solid waste management (SWM) systems should address all incurred costs from the planning and design stage to capital service, operation and monitoring, and safe disposal of wastes.

CLEAN DEVELOPMENT MECHANISM PROJECTS

The CDM allows emission-reduction projects in DCs to earn certified emission-reduction (CER) credits, each equivalent to 1 t of CO_2. These CERs can be traded and sold and used by industrialized countries to meet a part of their emission-reduction targets under the Kyoto Protocol framework. Therefore, DCs should focus their attention on the CDM project as a means of funding biogas extraction and utilization projects.

In 2011 the percentage of GHG emissions deriving from waste management in Africa (8% of total emissions) was more than three times higher than those in the United States; the main reason for this situation is the high CO_2 equivalence of CH_4 (factor of about 25) (Couth and Trois, 2011, 2012). In Brazil, China, India, and Mexico the percentage of emissions of CH_4 from MSW disposal sites were, respectively, 2.4, 1.2, 1.1, and 8.1 (Hoornweg and Bhada Tata, 2012). The upgrading of landfill management practices provides an opportunity to reduce GHG emissions; this is the aim of LFG CDM projects. Some experiences underline how international marketplace developers should better grasp both the challenges and opportunities of these projects (Lee et al., 2007).

A fundamental problem in expanding methane recovery efforts in DCs is the lack of reliable data. Most estimates of methane emissions from landfills are based on a "top–down" approach, in which the quantities and types of decomposable waste deposited are estimated and multiplied by estimated methane generation rates (Ayub et al., 2011). UNFCCC (United Nation Framework Convention on Climate Change) statistics demonstrate that several projects in DCs show—compared to the predicted values—lower amounts of gas extraction rates. These discrepancies are also related to the uncertain estimations of local conditions used in the models. The principal factors affecting production and collection of LFG are as follows:

- loss of degradable carbon because of fire and aerobic conditions (low compaction, over sucking of LFG extraction systems);
- reduction of the radius of influence of the gas wells due to elevated water tables, sludge deposits, etc.;
- adverse conditions for gas production due to dry waste and high or low temperatures, low or high pH, etc.

These factors (alone or in combination) are capable of eliciting a reduction in the production of LFG amounting to—in comparison with the predicted values—almost 100%. These influencing factors should also be considered very carefully when planning JI (Joint Implementation)/CDM projects (Rettenberger, 2011). In addition, the relationship between calculated LFG production and actual gas extraction rate is strongly associated with the design and operation of the LFG system and amounts to ±50%.

Tawfig et al. (2009) explained how an LFG extraction and combustion CDM project in Sudan was shifted to a composting CDM project. The latter was dictated on the grounds of lower capital investment needs and a higher emission-reduction potential that was more sustainable and more economically viable. Moreover, with regard to the local situation, no existing facilities, which could have used the energy produced if the LFG project had been realized, were present in the vicinity of the landfill. Furthermore, in the arid country of Sudan there is a market demand for the compost produced. Thus, an integrated view of all waste management options may generate win—win solutions.

CDM projects focused on waste minimization, and avoidance measures could be even more effective and efficient, although they remain a utopia due to the prevalent view worldwide of mass production and mass construction. In addition, educational and sensitizing projects require lengthy periods to manifest the desired effects. The dramatic public health and sanitary situation observed throughout the majority of DCs should be addressed as a matter of priority and adequate solutions identified. Accordingly, CDM projects set up to provide investments in the collection and landfilling of waste should be implemented. However, in contrast with conventional international policies, new concepts of sustainable landfilling including bioreactor and semiaerobic landfills should be supported. The latter method is capable of reducing the production of GHGs by as much as 40% compared with an anaerobic landfill (SPREP-JICA, 2010). Of course, the use of bioreactor landfills providing for LFG collection and thermal treatment from the start of landfill operations may result in even higher reduction rates.

CONCLUSIONS

In view of the highly specific social, economical, and technical situations in DCs, landfilling is deemed in the majority of cases to be the most appropriate solid waste treatment solution. An adequate landfill design, construction materials, leachate, and biogas treatment systems should all be carefully selected to ensure compliance with appropriate environmental standards. Politicians and administrators should implement effective legislation and technical regulations for the purpose of guiding companies and technicians in the proper design and management of an environmentally sound landfill. In this context it should be kept in mind that the "perfect may be the enemy of the good," meaning that the highest standard may not always represent the best solution due to an excessive cost, an overcomplex technology, etc.

To be deemed sustainable an SWM system should above all rely on local human resources; indeed, every effort should be undertaken to raise the environmental awareness of the population by means of both educational school programs and household education. The contribution of affluent international communities, organizations, and universities would prove to be extremely beneficial in this process. This importance of setting up similar educational projects is on a par with the importance of CDM projects.

References

Achankeng, E., 2004. Sustainability in Municipal Solid Waste Management in Bamenda and Yaoundé, Cameroon. Thesis submitted for the Degree of Doctor of Philosophy (PhD) at School of Social Sciences, Department of Geographical and Environmental Studies (GES), University of Adelaide.

Allen, A.R., 2002. Low cost natural solutions to landfill management for developing nations. In: Kocasoy, G., Atabarut, T., Nuhuoðlu, Ý. (Eds.), Appropriate Environmental and Solid Waste Management and Technologies for Developing Countries.

Ayub, S., Khan, A.H., 2011. Landfill practice in India: A review. J. Chem. Pharm. Res. 3 (4), 270–279.
Ayuba, K.A., Latifah, A.M., Abdullah, H.S., Sulaiman, W., Nur, A., 2013. Current status of municipal solid waste management practise in FCT Abuja. Research Journal of Environmental and Earth Sciences 5 (6), 295–304.
Beaven, R.P., 2013. Clogging of leachate collection systems. Lesson at the University of Southampton.
Beaven, R.P., Hudson, A.P., Knox, K., Powrie, W., Robinson, J.P., 2013. Clogging of landfill tyre and aggregate drainage layers by methanogenic leachate and implication for practice. Waste Management 33 (2), 431–444.
Beerachee, B., 05 September 2012. Overview of Wastes Management in Mauritius. Solid Waste Management Division, Ministry of Local Government and Outer Islands, Mauritius, Seoul, Korea.
Blake, E., Chapman, B.. Meeting the Challenge – Landfill in Samoa (visited in 2017). www.wasteminz.org.nz.
Breeze, R., 2012. Municipal Solid Waste Management in Dar Es Salaam Draft Baseline Analysis. A World Bank Report.
Chong, T.L., Matsufuji, Y., Hassan, M.N., 2005. Implementation of the semi-aerobic landfill system (Fukuoka method) in developing countries: a Malaysia cost analysis. Waste Management 25 (7), 702–711.
Coffey, M., Coad, A., 2010. Collection of Municipal Solid Waste in Developing Countries. UN-HABITAT, Malta.
Cossu, R., 2007. Landfill sustainability – concepts and models. In: Proceedings Sardinia 2007, Eleventh International Waste Management and Landfill Symposium. S. Margherita di Pula, Cagliari, Italy: October 1–5, 2007.
Couth, R., Trois, C., 2011. Waste management activities and carbon emissions in Africa. Waste Management (New York, N.Y.) 31 (1), 131–137.
Couth, R., Trois, C., 2012. Sustainable waste management in Africa through CDM projects. Waste Management 32 (11), 2115–2125.
CPE-Centre for People and Environment (CPE), 2010. Landfill Recovery and Use in Nigeria (Pre-feasibility Studies of Using Landfill Gas-to-Electricity). Final Report.
Damiecki, R., Bilong, J., 2009. Landfill gas management at a sanitary landfill in extreme tropical conditions, Kuching, Sarawak, Malaysia. In: Proceedings Sardinia 2009, Twelfth International Waste Management and Landfill Symposium. S. Margherita di Pula, Cagliari, Italy: October 5–9, 2009.
Das, Tapas, Singh, Baleshwar, 2013. Benefits and Impacts of scrap tyres used inn Geotechnical Engineering. Journal of Environmental Research And Development 7 (3), 1262–1271.
Dedinec, A., Markovska, N., Ristovski, I., Velevski, G., Verica, G., Obradovic, G.T., Zdraveva, P., 2014. Economic and environmental evaluation of climate change mitigation measures in the waste sector of developing countries. Journal of Cleaner Production 88, 234–241.
Engledow, S.-A., 2007. Integrated Analysis Solid Waste Baseline Report. Sustainability Institute, Stellenbosch.
Fukuoka Municipal Government (FMG), Environment Bureau, 2010. The Fukuoka Method, Semi-aerobic Landfill Type. Unpublished report.
Gogra, A.B., Jun, Y., Victor, T.S.K., Edward, H.S., Gyula, Z., Solomon, P.G., Tamba, S.B., 2010. A situational analysis of waste management in freetown. Chemical Technology 6 (5), 124–135.
Guerrero, L.A., Maas, G., Hogland, W., 2013. Solid waste management challenges for cities in developing countries. Waste Management 33, 220–232.
Henry, R.K., Yongsheng, Z., Jun, D., 2006. Municipal solid waste management challenges in developing countries – Kenyan case study. Waste Management 26 (1), 92–100.
Hoornweg, D., Bhada-Tata, P., 2012. A Global Review of Solid Waste Management. World Bank Urban Development Series Knowledge Papers.
Huber-Humer, M., Lechner, P., 2013. Future visions: sustainable landfilling or a sustainable society without landfills? In: Cossu, R., van der Sloot, H. (Eds.), Sustainable Landfilling. CISA Publisher, Padova, pp. 45–62. ISBN 978-88-6265-005-2.
Johannessen, L.M., Boyer, G., 1999. Observations of Solid Waste Landfills in Developing Countries: Africa, Asia, and Latin America. Urban Development Division – Waste Management Anchor Team – The World Bank.
Kaushik, M.K., Kumar, A., Bansal, A., 2015. Performance assessment of gravel—tire chips mixes as drainage layer material using real active MSW landfill leachate. Geotech Geol Eng 33 (4), 1081–1098.
Kedar, R.J., 2015. Sustainable Solid Waste Management for Ahmedabad. MSc Thesis, TU Delft, India.
Knox, K., Kowlesser, P., 2011. A case study of leachate management at tropical landfills: Mare Chicose, Mauritius. In: Proceedings Mauritius 2011, 1st International Conference on Waste Management in Developing Countries.
Komakech, A.J., Banadda, N.E., Kinobe, J.R., Kasisira, L., Sundberg, C., Gebresenbet, G., Vinnerås, B., 2014. Characterization of municipal waste in Kampala, Uganda. Journal of the Air & Waste Management Association 64 (3), 340–348.

Lavagnolo, M.C., Pieretti, G., Boscaro, F., Lambertini, M., 2016. Studio di Prefattibilità: Identificazione del Sito e della Tipologia di Discarica per la Citta' di Bissau – Guinea Bissau.
Lee, C., Bogner, J., Toefy, A., 2007. Landfill gas CDM projects: current trends and future opportunities for developing countries. In: Proceedings Sardinia 2007, Eleventh International Waste Management and Landfill Symposium. S. Margherita di Pula, Cagliari, Italy: October 1–5, 2007.
Marshall, R.E., Farahbakhsh, K., 2013. Systems approaches to integrated solid waste management in developing countries. Waste Management (New York, N.Y.) 33 (4), 988–1003.
Mato, R.R.A.M., 1999. Environmental implications involving the establishment of sanitary landfills in five municipalities in Tanzania: the case of Tanga municipality. Resources, Conservation and Recycling 25 (1), 1–16.
Matsufuji, Y., 2016. A Road to Sanitary Landfill, vols. 4 and 5. (NPO)SWAN-Fukuoka, Fukuoka Environmental Foundation.
Matsufuji, Y., Tachifuji, A., Yoshimura, Y., 2013. On-site simple leachate treatment in landfill by stirring device eco-fun using natural energy. In: Proceedings of the XIII International Waste Management and Landfill Symposium. CISA Publisher, Cagliari.
Mavropoulos, A., Velis, C., Mavropoulou, N., Tsakona, M., 2015. Identification of the World's 50 biggest dumpsites. In: Proceedings Sardinia 2015, Fifteenth International Waste Management and Landfill Symposium. S. Magharita di Pula, Cagliari, Italy; October 5–9, 2015.
Mongtoeun, Y., Takeshi, F., Sour, S., 2014. A study of commercial solid waste generation and composition in Phnom Penh City, Cambodia. Journal of Natural Sciences Research 4 (13), 49–54.
Moodley, L., 2013. An overview of landfill operations. In: 4th Waste Management Officers Khoro, Durban, KwaZulu-Natal, October 14–16, 2013.
Munawar, E., Fellner, J., 2013. Guidelines for Design and Operation of Municipal Solid Waste Landfills in Tropical Climates. ISWA.
Mwiganga, M., Kansiime, F., 2005. The impact of Mpererwe landfill in Kampala – Uganda, on the surrounding environment. Physics and Chemistry of the Earth 30 (11–16 SPEC. ISS.), 744–750.
Oakley, S.M., Jimenez, R., 2012. Sustainable sanitary landfills for neglected small cities in developing countries: the semi-mechanized trench method from Villanueva, Honduras. Waste Management 32 (12), 2535–2551.
Ofori-Boateng, C., Lee, K.T., Mensah, M., 2013. The prospects of electricity generation from municipal solid waste (MSW) in Ghana: a better waste management option. Fuel Processing Technology 110, 94–102.
Ogwueleka, T. Ch, 2009. Municipal solid waste characteristics and management in Nigeria. Iranian Journal of Environmental Health Science & Engineering 6 (3), 173–180.
Oteng-Ababio, M., Melara Arguello, J.E., Gabbay, O., 2013. Solid waste management in African cities: sorting the facts from the fads in Accra, Ghana. Habitat International 39, 96–104.
Otieno, F.A.O., Taiwo, O., 2007. Current state of urban solid waste management in some cities of Africa. In: Proceedings Sardinia 2007, Eleventh International Waste Management and Landfill Symposium. S. Margherita di Pula, Cagliari, Italy: October 1–5, 2007.
Owusu-Sekyere, E., Osumanu, I.K., Yaro, J.A., 2013. Dompoase landfill in the Kumasi Metropolitan area of Ghana: a 'blessing' or a 'curse'? International Journal of Current Trends in Research 2, 87–95.
Parkin, J., Strachan, L., Bowers, A., Wright, M., Winn, R., 2006. Extreme landfill engineering: developing and managing South Africa's busiest and largest landfill facilities. In: Proceedings Wastecon 2006, International Congress, IWMSA.
Pass, J., Wilson, C., 2005. Two slides that damaged parts of a lining at Bisasar Road landfill, Durban. In: Landfill 2005 Conference Proceedings.
Pugh, M., Caine, M., 2000. The development of low cost landfilling techniques for emerging countries. In: Intercontinental Landfill Research Symposium. Lulea University of Technology. Luleå, Sweden: December 11–13, 2000.
Ranjan, M.R., Ramanathan, A.L., Tripathi, A., Jha, P.K., 2014. Landfill mining: a case study from Ghazipur landfill area of Delhi. International Journal of Environmental Sciences 4 (5), 919–925.
Rettenberger, G., 2011. Recovery of landfill gas under the conditions of landfills in developing countries. In: Proceedings Mauritius 2011, 1st International Conference on Waste Management in Developing Countries and Transient Economies, Mauritius, Africa; September 5–8, 2011.
Sang-Arun, J., Heng, C.K., 2011. A guide for Technology Selection and Implementation of Urban Organic Waste Utilisation Projects in Cambodia. IGES Poilicy Report 2011-06.
SPREP-JICA, 2010. A Practical Guide to Landfill Management in Pacific Island Countries and Territories: Inland-Based Waste Disposal, vol. 1. Apia, Samoa.

Sukholthaman, P., Shirahada, K., June 2015. Technological challenges for effective development towards sustainable waste management in developing countries: case study of Bangkok, Thailand. Technology in Society 43.
Tadros, Z., 2007. Site selection and management of Solid Waste Disposal Site. Case study, Ghabawi Landfill/ Jordan. In: Proceedings Sardinia 2007, Eleventh International Waste Management and Landfill Symposium. S. Margherita di Pula, Cagliari, Italy. October 1-5, 2007.
Tan, Y.J., 2012. The Management of Residential Solid Waste in Mombasa, Kenya. Independent Study Project (ISP) Collection. Paper 1388.
Tawfig M., Couth R., Pearson G., Strachan L. (2009). Development of sustainable waste disposal in Sudan, in Proceedings Sardinia 2009, Twelfth International Waste Management and Landfill Symposium. S. Margherita di Pula, Cagliari, Italy: October 5-9, 2009
Thi, N.B.D., Kumar, G., Lin, C.-Y., 2015. An overview of food waste management in developing countries: current status and future perspective. Journal of Environmental Management 157, 220–229.
UNEP, 2005. Solid Waste Management. ISBN: 92-807-2676-5.
UNFCCC/CCNUCC, 2014a. Component Project Activities Design Document (CPA-DD) – City of Cape Town Landfill Gas Extraction and Utilisation Programme.
UNFCCC/CCNUCC, 2014b. Project Design Document (PDD) for CDM Project Activities – Nkolfoulouu Landfill Gas Recovery Project.
Von Buchwald, F., 2005. The operation experiences in the sanitary landfill "Las Iguanas" Guayaquil – Ecuador. In: Proceedings Sardinia 2005, Tenth International Waste Management and Landfill Symposium. S. Margherita di Pula, Cagliari, Italy. October 3-7, 2005.
Waste Atlas 2013 Report, 2013, ISSN 2241-2484, ed. D-Waste.
Yhdego, M., 1995. Urban solid waste management in Tanzania issues, concepts and challenges. Resources, Conservation and Recycling 14 (1), 1–10.
Zanon, T., Ferrari, A., Lima, L., Costa, D., 2013. Implantation of leachate treatment by reverse osmosis and influence of controlled infiltration of retentate on biogas generation at a sanitary landfill – a case study. In: Proceedings Sardinia 2013, Fourteenth International Waste Management and Landfill Symposium. S. Margherita di Pula, Cagliari. September 3-4 October 2013.

Further Reading

Ayub, L., Widjaya, E.R., 2006. Environmental and economic analysis of waste disposal option for traditional markets in Indonesia. Waste Management 26 (10), 1180–1191.
Cossu, R., 2009a. From triangles to cycles. Waste Management (New York, N.Y.) 29 (12), 2915–2917.
Cossu, R., 2009b. Driving forces in national waste management strategies. Waste Management 29 (11), 2797–2798.
Cossu, R., Febbraio 2012. Anche le discariche possono essere sostenibili (*in Italian*). Ambiente Rischio Comunicazione 2.
Visvanathan, C., Trankler, J., Basnayake, B.F.A., Chiemchaisri, C., Kurian, J., Gonming, Z., 2013. Landfill management in Asia – possible approaches to sustainable solutions. In: Cossu, R., van der Sloot, H. (Eds.), Sustainable Landfilling. CISA Publisher, Padova, pp. 84–92.
Yukihito, Y., Matsufuji, Y., 2016. A Road to Sanitary Landfill, vol. 4. (NPO)SWAN-Fukuoka, Fukuoka Environmental Foundation.

14 CHAPTER

SUSTAINABLE LANDFILLING

14.1

LANDFILLING OF MECHANICALLY BIOLOGICALLY PRETREATED WASTE

Rainer Stegmann, Kai-Uwe Heyer and Karsten Hupe

INTRODUCTION

The overall strategy of waste management in Europe is toward more material and energy recovery from waste and less landfilling. The European Union (EU) has, via the Landfill Directive, set a target for reducing the biologically degradable waste fraction going to landfills (see Chapter 4.1). In Austria, target values for the mechanical biological pretreatment (MBT) of municipal solid waste (MSW) as a prerequisite for landfilling are already in power from the beginning of 2005. This is also the case in Germany where through the implementation of the Landfill Directive in June 1, 2005 only thermally or mechanically biologically pretreated MSW are allowed to be landfilled (Anonymous, 2000). The disposal of MBT waste has to meet the requirements for a proper disposal described in the Abfallablagerungsverordnung (Anonymous, 2001) (German Waste Disposal Act, particularly Appendix 3).

German limit values for waste going to landfills are presented in Table 14.1.1.

Table 14.1.1 Selection of limit values for the quality of waste going to landfills (Anonymous, 2000)

Parameter	Target Value
Respiration activity[a] (RA_4)	≤5 mg O_2/g dry matter
Gas formation potential[b] (GF_{21})	≤20 Nl/kg dry matter
TOC_{eluate}[c]	≤250 mg/L
TOC_{solid}	≤18 Mass-%
Gross calorific value	≤6000 kJ/kg

[a]By means of bacteria in 4 days dependent on the amount of substrate available (indirect method to measure the biodegradable fraction of the waste sample).
[b]Gas formation potential (GF_{21}): Gas formation of the waste sample in 21 days.
[c]TOC in the eluate produced in an elution test (1:10 solid/liquid ratio and 24 hours shaking).

GENERAL ASPECTS OF MECHANICAL BIOLOGICAL PRETREATMENT WASTE LANDFILLING

The incoming mechanically and biologically waste material heats up during transportation so that during the unloading of the material at the landfill temperatures around 50°C may be measured. Often water vapors escape from the waste. This phenomenon shows that this material is still biologically active. The existing easy degradable organic fractions in the MBT material may be partly produced during biological pretreatment from more difficult degradable organics.

For a better long-term emission control, landfills in general should be built as a mound. To guarantee the mechanical stability of the landfill beyond others the water content and pore water pressures of the landfilled MBT material should be controlled and monitored. In addition, a landfill with intermediate layers of coarse material is recommended (Stegmann and Heyer, 2001). Aerobic conditions should be permanently achieved within the landfill body by means of suitable constructional and operational measures. Should anaerobic conditions restore, a rise in leachate concentrations and methane formation—however low—can be expected.

To avoid problems during the emplacement, MBT material may be pressed to bales and landfilled using a forklift loader (Stegmann, 2005).

To describe the emission behavior of biologically pretreated residual waste, landfill simulation experiments in the laboratory have been carried out (Stegmann, 1997). By choosing appropriate milieu conditions for the test, an enhanced biological degradation and elution process of the waste will be achieved. By these means, the emission quality and potential—represented by gas quality and production as well as leachate concentrations and loads—of the waste samples can be estimated within reasonable periods of time. Table 14.1.2 presents a summary of results from investigations using landfill simulation reactors (LSRs), which show that the emissions regarding organics and nitrogen as well as the gas production can be reduced by approximately 90%.

Table 14.1.2 Effects of mechanical biological pre-treatment (MBT) on landfill emissions (results from laboratory scale tests) (Stegmann and Heyer, 2001)

Emission	Raw Waste	MBT-Waste	Reduction
COD (mg/kg dry matter)	25.000–40.000	1.000–3.000	90%
Total nitrogen (mg/kg dry matter)	1.500–3.000	150–300	90%
Gas production GB21 (Nl/kg dry matter)	150–200	0–20	90%

Since the gas formation potential is low, an active gas extraction may be considered; in any case the residual amounts of produced gas need to be biologically oxidized in the top area before leaving the landfill.

Today in Germany there are already 10 years of experiences available regarding the operation of MBT-waste landfills, but there are not too many results from field investigations available.

After MBT, also the landfilling characteristic of the waste is significantly different from raw MSW. Due to the separation of the high calorific value fraction refuse derived fuel (RDF) from the raw waste, MBT waste has a lower content of structural material. Due to the low amount of macropores the water and gas permeability are lower and water saturation of the landfilled material is reached at lower water contents.

DESIGN AND OPERATION OF MBT LANDFILL

Investigations of Mechanical Properties at Test Fields and in the Laboratory

The results from investigations at test fields and laboratory analysis may not be representative and cannot be transferred directly to actual landfilling; in addition, due to the kind of mechanical (dependent also on the quality of the produced RDF) and the kind of biological treatment [aerobic or anaerobic (wet or dry AD process)] the quality of the MBT material may vary significantly.

Eitner and Tiedt (2005) summarized investigations and experiences with the landfilling of MBT waste as follows:

Shear strength (f: 27–36°) and cohesion (C:11–62 kN/m^2) of MBT waste is similar to raw waste, but the tensile strength is nearly zero.

- It can be expected that slopes of 1:3.5–1:3 are probably sufficiently stable ($f = 35°$, $C > 15$ KN/m^2). This should be confirmed by waste mechanical calculations.
- Densities in the surface area—dependent on the material quality—can be expected between 1.0–1.5 t/m^3 wet weight (wwt) resp. 0.7–1.0 t/m^3 dry weight (dwt).
- Compaction may be difficult to achieve due to different quality and water content of the to-be-landfilled MBT waste as well as a kind of “mattress” effect.

Experiences From Landfill Operation

Due to the homogeneity of the MBT material and a relatively high achievable density more even and lower settling rates can be expected. As a consequence a higher load may weigh on the bottom of the landfill and drainage pipes. Due to a low content in coarse material (e.g., plastic pieces, wood) and the resulting lower reinforcement of the material the sheer strength and friction are reduced (Bräcker, 2010).

Based on present experiences the following procedures for the landfilling of an MBT material may be applied:

- To achieve a low permeability the waste may be highly compacted in thin layers, (30–50 cm) by means of about 3–4 compaction runs (Figs. 14.1.1 and 14.1.2). Different kinds of machines for emplacement, compaction, and surface flattening can be used (compactor, roller, sheep’s foot roller).

Figure 14.1.1 Distribution and compaction of mechanical biological pretreatment material (landfill Minden-Lübbecke).

Figure 14.1.2 Surface layer of a mechanical biological pretreatment landfill.

- Due to the more homogeneous character of the waste material as well its lower pore volume and higher disposal density, the permeability of the emplaced MBT material is between kf: 10−5 to 10−8 m/s; the great range reflects influences of the quality of the raw waste the kind of pretreatment (e.g., sieve size for RDF separation, kind of biological treatment) the initial emplacement density and the depth of the landfill (load of the overlaying waste mass). When gas fills part of the pores the density may further increase up to values of kf = 10-10 m/s.

- Maximum densities of 18 kN/m^3 may be reached (Bräcker, 2010), which are in practical operation significantly lower (10−15 kN/m^3) (Entenmann and Wendt, 2005).
- Settling rates in the area of in total 2.5% of the final height can be expected.
- The working area should, in general, be small but can be extended when the waste has a higher incoming moisture content than recommended. In this case the pore water can decrease due to consolidation on the surface.
- The Proctor water content should be between 25%−40% (wwt). No landfilling should be practiced during (intensive) rainfall. If the moisture content on the surface is too high the emplacement will become difficult because the machines and trucks will sink into the waste. If this is the case the incoming waste has to be intermittently stored and landfilled when it is dry weather again. During rain the working phase may be covered with a tarpaulin (reusable plastic foils that should be rolled and unrolled). By these means also the leachate production will be reduced. Areas that are not in operation and not yet cultivated should be intermittently lined with plastic foil (Stegmann and Heyer, 2002).
- Precipitation of low intensity may be completely absorbed by the waste; if the upper waste layer is water saturated and the precipitation rate is higher than the infiltration, water may accumulate on the surface. If the surface is quite flat, significant surface water runoff can be expected. The amount of runoff depends on the intensity of the rain and the degree of the slope (e.g., at high precipitation rate and a slope of 1:2 about 2/3 of the precipitation became surface water runoff (observation at a test area) (Bräcker, 2010).
- Therefore, the planning and operation of a surface water collection system is an important task. As already mentioned the plastic membranes used for temporary cover should be reused as far as possible after resuming operation of those areas. A collection (e.g., in a pond) and treatment of the surface water that had contact with the waste surface may be necessary where a significant pollution reduction can be achieved by removing particles by means of settling (Ziehmann et al., 2002).
- During dry periods the surface may become dry, and dust may be produced.
- Owing to the pretreatment, there will be no food for birds and mammals.
- When the landfill is closed the installation of a final surface cover/liner system will be necessary. An often-used system is the combination liner consisting of a clay liner and plastic membrane as well as a soil layer on top. Other systems, e.g., capillary barrier, may be encountered (see Chapter 11.1).
- The stability of the landfill mound needs the highest attention. When the waste is water-saturated, pore water over pressure can build up, e.g., due to the weight of the overlaying waste. In addition, the produced gas (even at low rates) can be compressed in the pores and it adds to the positive pressure. This situation results in lower friction values of the landfilled waste, which means less stability. As a consequence pore water pressures should be measured also during landfill operation. With lower friction values the incline of the slope of the MBT landfill mounds has to be adapted. In any way it should be $<1:3$. Since the waste material is more homogeneous and therefore more similar to soil waste mechanical calculation should be used to determine the safe physical stability of the landfill. But it should be kept in mind that the MBT material can be quite different depending on the kind of mechanical and biological pretreatment. Also due to this situation geotechnical investigations of the material to be used should be obligatory.
- In addition, it should be the intention to reduce the pore water pressure. As a measure the installation of intermediate layers of about 50 cm height consisting of course material ("no" carbonate

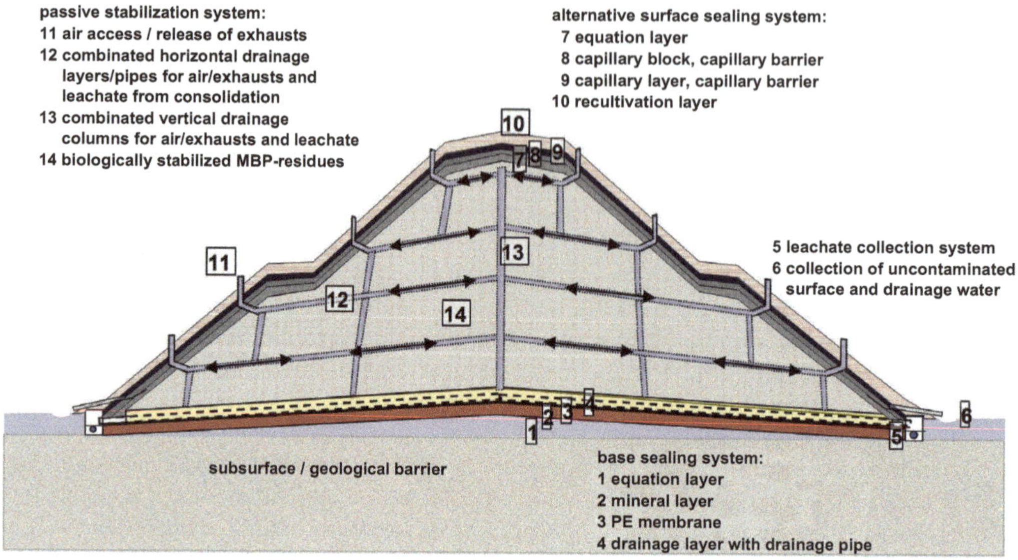

Figure 14.1.3 Concept of a mechanical biological pretreatment landfill (Stegmann and Heyer, 2001).

and sulfur content) should be encountered. These should be constructed about every 5 m with a slope to the outside of the landfill (Fig. 14.1.3). By these means water over pressure resulting from water and gas can be reduced and the total friction of the landfill will increase. This system could also support natural aeration of the landfill once forced air in situ aeration has been terminated. This system is used at several landfills in Germany.

Emission Control

MBT landfills still produce leachate and gas. Regarding long-term total gas production rates, it can be expected that still about 10–40 Nm^3/t wwt. (10%–20% of the total gas potential of the raw MSW) will be produced. After a short active gas production period (due to a kind of aerobic breakdown of heavy degradable compounds during the biological pretreatment) the remaining organic substances are not easily degradable; as a consequence low gas production rates over long periods of time can be expected.

Leachate production will still occur where it is difficult to predict leachate production rates. These will be dependent on the specific climatic conditions as well as the waste mechanics of the emplaced material and the landfill operation. Leachate production rates in the average of about 5% of the annual precipitation rate are predicted by Bräcker (2010). These values will depend very much on the surface water control of precipitation, the MBT quality and the kind and time period of landfill operation. The authors have no access to actual monitored leachate production rates; this is also due to the fact that in Germany in most cases MBT material is landfilled on the top of existing landfills.

The leachate quality (see also Chapter 10.2) will be similar to the leachate quality from existing landfills in the stable methanogenic phase. The nitrogen content may be lower. To meet the German discharge standards for leachate discharge into natural waters treatment will still be required. Leachate concentrations from a study by Robinson et al. in 2005 are presented in Table 14.1.3. These data are difficult to judge, since the material parameters (achieved respiration rate: AT4, and gas production potential, GB21) as well as the kind of landfill operation are not known. In the view of the authors, these concentrations seem to be on the higher end of what can be expected, if the German requirements for MBT material are reached. In this case COD values in the range of 500–2000 mg/L, BOD_5 values of <50 mg/L, and NH_4–N values of <500 mg/L seem to be more realistic. Of course, anaerobically pretreated waste may result to somehow different leachate quality data than aerobically treated material; this is also due to the different amounts of water used during the different processes (elution factor). In addition the disposal height with time and the achieved density are of influence. In the first years of landfill operation the concentrations may be somewhat higher than with increasing age. Since no acidic phase is expected the concentration curves dependent with time are quite flat. A better understanding of leachate quality evolution along the time and the influence of the pretreatment typology will occur when more data from field experiences will be available.

Table 14.1.3 Range of concentrations of leachate parameters from MBT landfills (study by Robinson et al. in 2005).

Leachate Composition	Low–Medium-General Range
pH value (-)	7.5–8.5
Conductivity (μS/cm)	10,000–20,000
COD (mg/L)	1000–5000
BOD_5 (mg/L)	20–200
TOC (mg/L)	500–2000
Chloride (mg/L)	4000–8000
Sulfate (as SO_4) (mg/L)	1000–5000
Phosphate (as P) (mg/L)	1.0–15
Alkalinity (as $CaCO_3$) (mg/L)	2000–6000
Ammonia-N (mg/L)	50–1000
Kjeldahl-N (mg/L)	100–1300

The leachate should be treated if possible together with the municipal wastewater in sewage treatment plants or it has to be treated separately. Owing to the low content in biological degradable substances, activated carbon adsorption and membrane separation processes are suitable for a treatment to achieve, e.g., the German target values for the discharge of treated leachate (Anonymous, 1996). For further information on leachate treatment refer to Chapters 10.3 and 10.4.

FINAL REMARKS

Landfilling of MBT waste is quite different from landfilling of raw MSW. This implies modified landfill design and operation. Important aspect to consider is the mechanical stability of the landfill mound. MBT waste has a high sorption capacity; pore water overpressures have to be avoided. As a consequence the water content of the incoming MBT material has to be strictly controlled.

If the landfill mound stability cannot be guaranteed, dams built of soil could be constructed at the landfill foot. There is still leachate produced and has to be treated. The quality is similar to leachate from an MSW landfill in the stable methanogenic phase. About 10%–20% of the total gas production is still produced and has either be captured and treated or oxidized in the specially designed landfill surface. There are advantages regarding landfill operation as less birds and rats, but during dry periods dust emissions may happen.

References

Anonymous, 1996. Allgemeine Rahmen- Verwaltungsvorschrift über Mindestanforderungen an das Einleiten von Abwässer in Gewässer. Rahmen-Abwasser VwV-GMBI, Anhang 51: Ablagerung von Sielungsabfällen. www.bmu.de.

Anonymous, 2000. Verordnung über die umweltverträgliche Ablagerung von Siedlungsabfällen und über biologische Behandlungsanlagen. www.bmu.de.

Anonymous, 2001. 30.Verordnung zur Durchführung des Bundesemissionsschutzgesetzes (Verordnung über Anlagen zur biologischen Behandlung von Abfällen – 30. BImSchV) vom 31.01.2001. www.bmu.de.

Bräcker, 2010. Deponietechnik für mechanism-biologisch behandelte Abfälle, Abfallwirtschaftsfakten 4.3, Staatliches Gewerbeaufsichtsamt Hildesheim, Zentrale UnterstützungsstelleAbfall, Gentechnik und Gerätesicherheit (ZUS AGG).

Eitner, R., Tiedt, M., 2005. Konzepte zum Deponiebetrieb mit MBA-Output. In: Gallenkemper, Bidlingmaier, Doedens, Stegmann (Eds.), 9. Münsteraner Abfalltage, Münsteraner Schriften zur Abfallwirtschaft, Band 8. Labor für Abfallwirtschaft der Fachhochschule Münster.

Entenmann, W., Wendt, P., 2005. Einbauversuche mit AbfAblV-konformen MBA-Output. In: Gallenkemper, Bidlingmaier, Doedens, Stegmann (Eds.), 9. Münsteraner Abfalltage, Münsteraner Schriften zur Abfallwirtschaft, Band 8, Labor für Abfallwirtschaft der Fachhochschule Münster.

Robinson, H.D., Knox, K., Bone, B.D., Picken, A., 2005. Leachate quality from landfilled MBT-waste. Waste Management 25.

Stegmann, R., 1997. Description of a laboratory scale method to investigate anaerobic degradation processes taking place in landfills. In: Christensen, Cossu, Stegmann (Eds.), Sardinia 97, Proceedings, CISA Publisher, Cagliari.

Stegmann, R., Heyer, K.-U., 2001. Landfill Concept for mechanical-biologically treated residual waste. In: Christensen, Cossu, Stegmann (Eds.), SARDINIA 2001, 8th International Landfill Symposium, Proceedings. CISA Publisher, Italy.

Stegmann, R., Heyer, K.-U., 2002. Konzept für eine nachsorgearme MBV-Deponie. In: Stegmann, Rettenberger, Bidlingmaier, Ehrig (Eds.), Deponietechnik 2002, Hamburger Berichte 18. Verlag Abfall aktuell, Stuttgart.

Stegmann, R., 2005. Landfilling of mechanically biologically pre-treated waste. In: Christensen, Cossu, Stegmann (Eds.), SARDINIA 2001, 10th International Landfill Symposium, Proceedings, CISA, via Marengo 34, 09123 Cagliari, Italy.

Ziehmann, G., Münnich, K., Fricke, K., 2002. Einbau von MBV-Materialien. In: Stegmann, Rettenberger, Bidlingmaier, Ehrig (Eds.), Deponietechnik 2002, Hamburger Berichte 18. Verlag Abfall aktuell, Stuttgart.

14.2

SEMIAEROBIC LANDFILLING

Yasushi Matsufuji, Ayako Tanaka and Raffaello Cossu

INTRODUCTION

The implementation of aerobic processes in landfills may be achieved by means of either natural ventilation or forced aeration (Chapter 14.1). The natural ventilation method is dependent on a specific landfill design aimed at promoting the advective flow of air, due to differences in temperature between the landfilled waste mass and external ambient. The semiaerobic landfill system has been extensively studied in Japan, since the early 1970s, at the University of Fukuoka (Matsufuji et al., 1978; Hanashima et al., 1981), hence known as the "Fukuoka method." From laboratory- and pilot-scale research studies, the system was subsequently applied to full-scale procedures with a series of applications both in Japan and abroad (China, Malaysia, Samoa). Semi-aerobic principle has been applied also in some European landfill (Cossu, 2015).

Semiaerobic landfill systems, being based on a mechanism of natural air ventilation, promote, without recourse to highly technological solutions, the presence of oxygen in the waste mass, thus accelerating waste stabilization. Although semiaerobic landfilling is characterized by a lower stabilization efficiency compared with aerobic systems, it represents a valuable solution for use in achieving a form of sustainable landfilling, acting as a compromise between the technological complexity and high costs of aerobic systems, and the need to reduce the long-term impacts generated by anaerobic systems.

Accordingly, the Fukuoka method is a cost-effective, low-technology landfill system suited to meeting sustainability requirements, particularly in developing countries, where financial constraints and limited technical capability are frequently the main cause of inadequate waste disposal.

This chapter illustrates the principles and theories underlying semiaerobic landfilling, providing findings of scientific studies conducted to investigate mass balance of elements and quality of emissions [landfill gas (LFG) and leachate], design guidelines, calculation models, and codes of practice for effective operating procedures. Examples of dimensioning and case studies are presented.

PROCESSES

Principles

Semiaerobic-type landfills are based on the simple physical principle described graphically in Fig. 14.2.1. Air flows into waste layers by means of a natural advection process in the presence of a

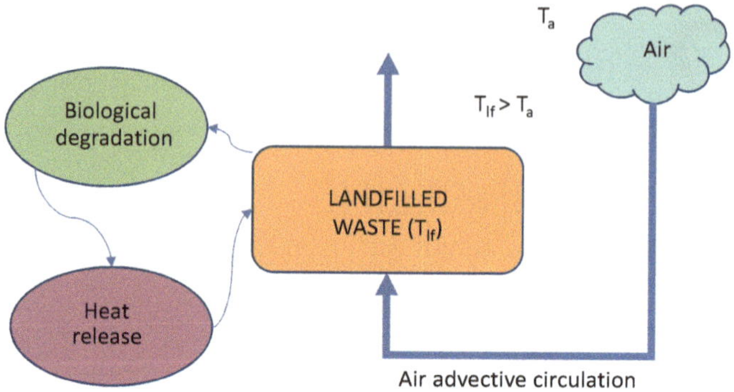

Figure 14.2.1 Principle of air advective flow in semiaerobic landfill (T_{lf} and T_a = Temperatures, respectively, in landfill and in external ambient).

difference in temperature between landfill layers (T_{lf}) and external ambient (T_a). This difference is established when the landfilled waste contains significant quantities of putrescible organics and exothermic degradation reactions occur. Under aerobic conditions the heat released can increase waste mass temperature by 50–70°C.

The theory of natural ventilation in a semiaerobic landfill is explained by the fluid dynamics theories of gaseous flow and diffusion through the porous media. The main parameter driving natural ventilation is the difference in density in gas mixtures as a consequence of temperature differences. Density differences produce the buoyant force that allows the warm gas mixture to flow upward throughout the waste layers, thus promoting aerobic degradation, resulting in a significant decrease in methane generation and concentrations of organic substances [biological oxygen demand (BOD) and chemical oxygen demand (COD)] in leachate. From a structural point of view, a semiaerobic landfill system consists of a network of horizontal pipes installed at the bottom of landfill sectors and vertical venting pipes erected at specific intersections of the horizontal pipes (Fig. 14.2.2).

This network of pipelines promotes the circulation of air within the waste mass, thus accelerating waste stabilization and producing positive effects on the quality and quantity of emissions. Organic compounds are degraded more effectively under aerobic conditions rather than under anaerobic ones, and ammonia is oxidized, generating leachate, which is easier and more economical to manage. The generation of CH_4 and H_2S is significantly reduced, contributing toward the prevention of global warming (in a semiaerobic process the proportion of CO_2 and CH_4 is approximately 4:1, much higher than the ratio 1:1 in an anaerobic landfill) (Matsufuji et al., 1997).

Due to the impossibility in semiaerobic systems of achieving aerobic conditions throughout the entire waste mass, both aerobic and anaerobic areas continue to coexist, enhancing denitrification of oxidized nitrogen compounds with subsequent release of N_2 gas.

The mechanisms and concepts used in the semiaerobic system are illustrated in Fig. 14.2.3 (Chong et al., 2005).

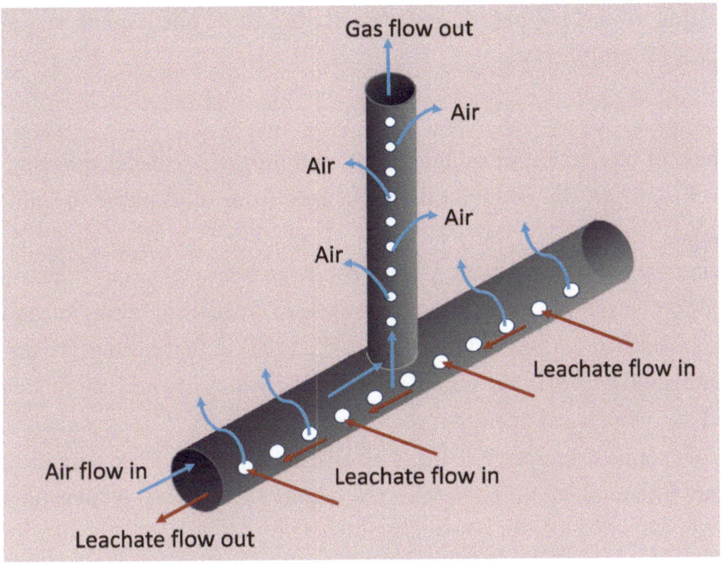

Figure 14.2.2 Conceptual sketch of horizontal collection and vertical gas venting pipes in a semiaerobic landfill system. *Modified from Chong et al. (2005).*

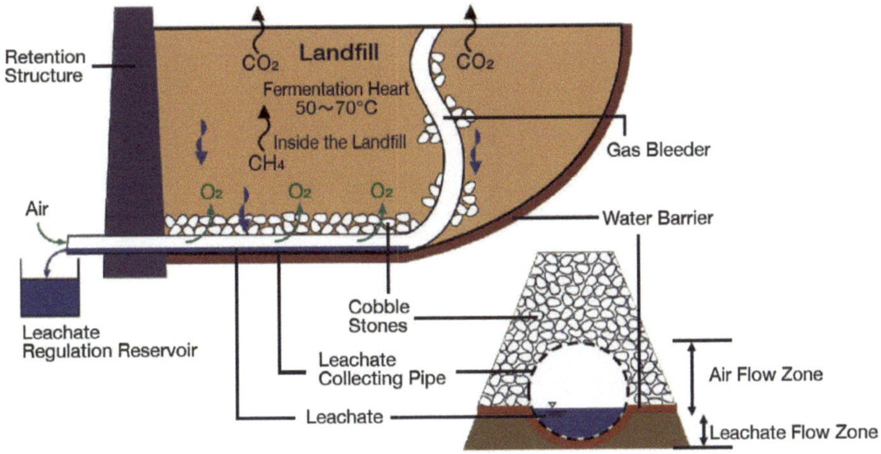

Figure 14.2.3 The semiaerobic landfill mechanism—the Fukuoka method (Zhang, 2012).

The perforated horizontal pipes not only collect and quickly drain off leachate generated in the waste layers but also introduce air into the waste layers through the advection process.

The semiaerobic processes described above can be improved by recirculating leachate back into the landfill, with the advantage of further promoting the denitrification processes mentioned above, and potentially contributing toward trapping heavy metals. Both these phenomena may significantly reduce the long-term impact of landfills.

Mass balance

In landfills, the grade of conversion of mobile organics during waste decomposition is effectively represented by a mass balance of the compounds transferred from solid phase to liquid phase (leachate) and gas phase (biogas).

Extended experimental studies conducted using two large lysimeters, one operating under semiaerobic and the other under anaerobic conditions, have yielded interesting data on the cumulative loss of weight in the form of leachate and gas (Matsufuji et al., 1997). The findings are graphically presented in Fig. 14.2.4. These graphs display the cumulative production of gas and cumulative leaching of contaminants observed after 4 years of operation for the two lysimeters. The organic content in leachate is expressed in terms of volatile solids (at 600°C). The original organic matter present in the lysimeter was 139.6 kg dry weight. The total amount of converted organics (total loss) was virtually identical in both

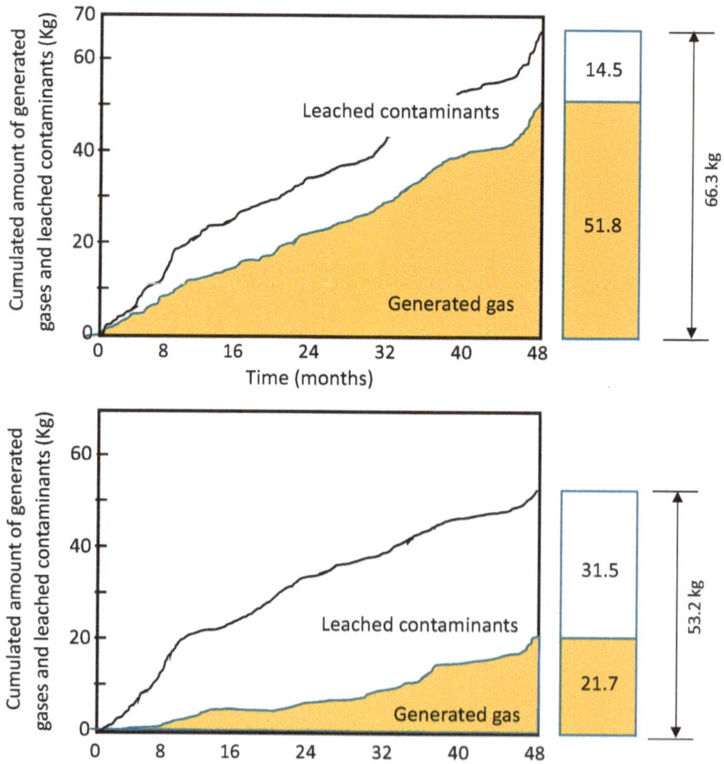

Figure 14.2.4 Cumulative amount of generated gases and leaching contaminants from lysimeters (Shimaoka et al., 2000).

lysimeters: 66.3 kg for the semiaerobic lysimeter and 53.2 kg for the anaerobic one. However, compared with aerobic processes, in the semiaerobic process a higher quantity of organic substances is converted into gas rather than leachate. The ratio of generated gases/leached organic contaminants obtained in these experiments yielded values of 8/2 for semiaerobic and 4/6 for anaerobic landfills (Matsufuji et al., 1997).

EMISSIONS

Increased waste stabilization achievable with a semiaerobic landfill is reflected in the quality of the emissions [leachate, LFG] and quantity of produced gas.

Quality of leachate

The positive influence of semiaerobic landfilling on leachate quality is illustrated in the graphs in Fig. 14.2.5, which show the fluctuations in BOD concentration over time in landfills operating under anaerobic, semiaerobic, and aerated conditions.

Similar results obtained from three sanitary landfill sites featuring different types of landfill and waste compositions are presented in Fig. 14.2.6, covering a period of 10 years from start-up of landfilling.

This figure clearly highlights how BOD concentration peaked (1000 mg/L) at A, B, and C immediately after landfill start-up. After 1 year of landfilling in a semiaerobic landfill (A, B), BOD concentration dropped below 100 mg/L. However, in semiaerobic landfill B (noncombustible waste), BOD concentration first dropped, and then rose again; the concentration failed to drop below 100 mg/L for at least 3 years following termination of waste conferment (Yanase et al., 1983).

The concentration of NH_4^+-N peaked a little later than BOD levels. With regard to combustible wastes, the peak occurred 3 years after landfill start-up and dropped more gradually than BOD (Fig. 14.2.7). Therefore, the stabilization of leachate quality in a semiaerobic or aerobic landfill occurs more rapidly compared with an anaerobic landfill (Yanase et al., 1983).

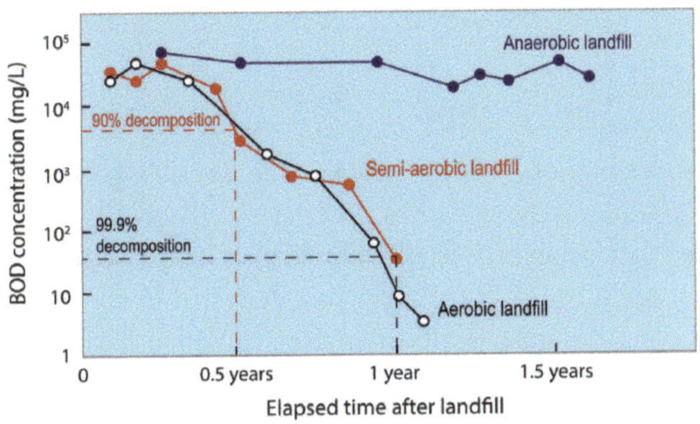

Figure 14.2.5 Relation between leachate quality and landfill type. *BOD*, biological oxygen demand.

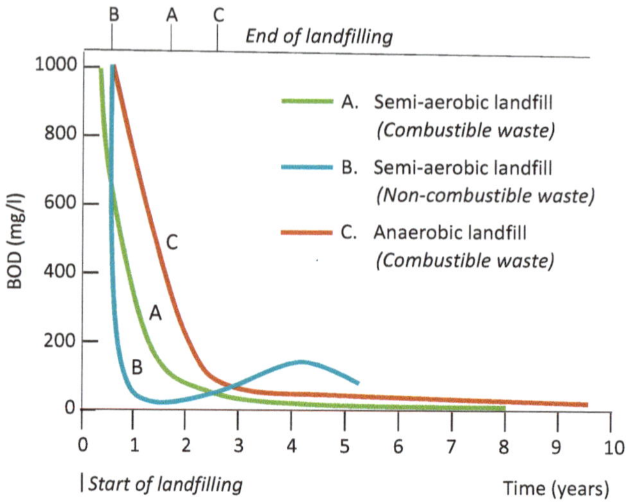

Figure 14.2.6 Fluctuation of BOD according to landfill type (Yanase et al., 1983). *BOD*, biological oxygen demand.

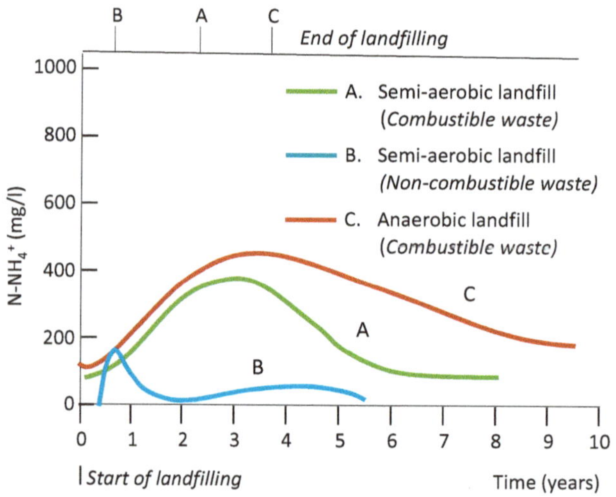

Figure 14.2.7 Fluctuation of NH_4^+-N according to landfill type (Yanase et al., 1983).

Fig. 14.2.8 shows the relationship between COD values and water level in a leachate collection pipe of a semiaerobic sanitary landfill sited at the previously described location A. When leachate volume increased to a level exceeding the upper limits of leachate collection pipe capacity, air supply to waste layers of the landfill via the pipes was shut off, rendering the landfill anaerobic. Concurrently, COD level increased, although as climatic conditions improved and water level dropped, the landfill returned to semiaerobic conditions and COD concentration fell accordingly (Yanase et al., 1983).

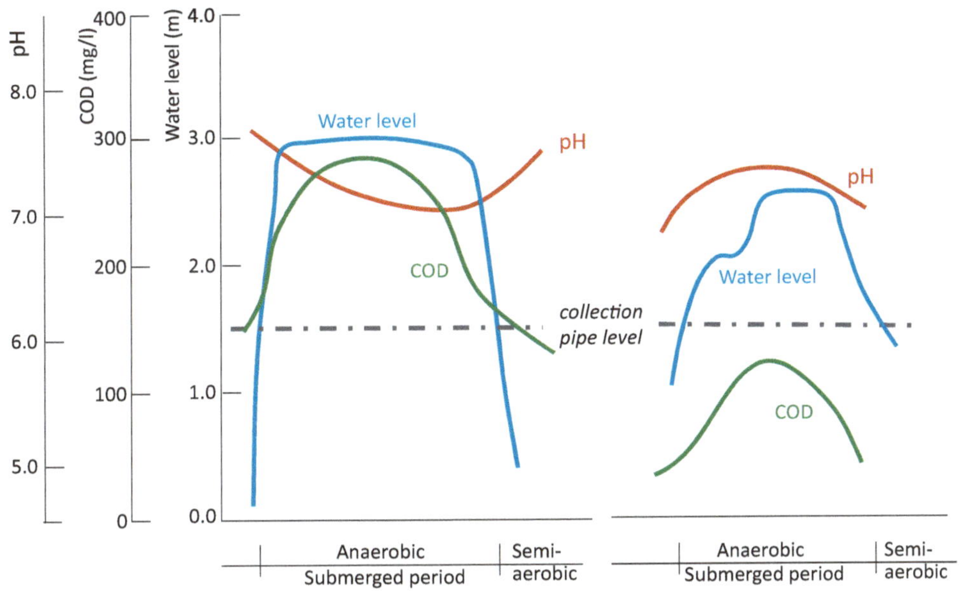

Figure14.2.8 Fluctuation of leachate quality according to the amount of trapped leachate in the landfill.

Furthermore, the thickness of landfilled waste layer, size of the catchment area, climatic conditions, and in particular, ambient temperature and rainfall, each significantly affect leachate quality. However, quantitative data on these effects are not available to date.

Levels of BOD, COD, and T-N concentration in a semiaerobic landfill vary over time, as shown in Fig. 14.2.9 (Matsufuji et al., 2005).

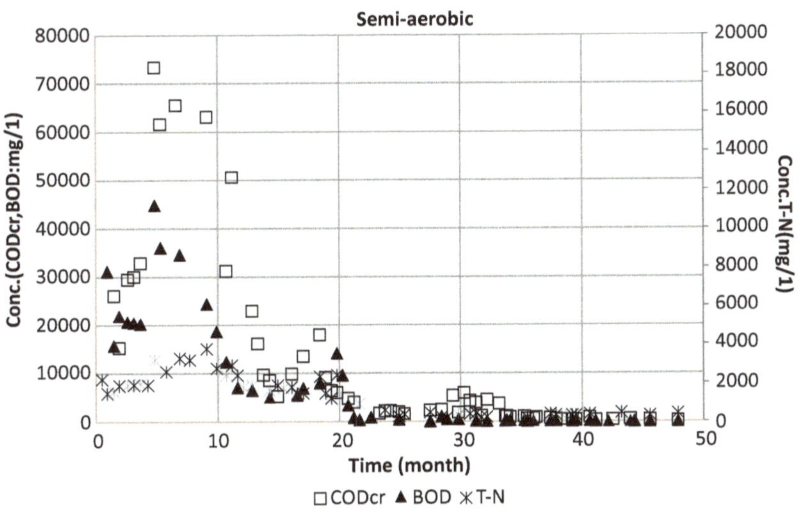

Figure 14.2.9 Fluctuation of BOD, COD, and T-N (Analyzed data by large lysimeter experiment). *BOD*, biological oxygen demand; *COD*, chemical oxygen demand.

Results obtained in China in lysimeter studies using a series of wastes confirm results obtained in Japan, as shown in Fig. 14.2.10 (He et al., 2011).

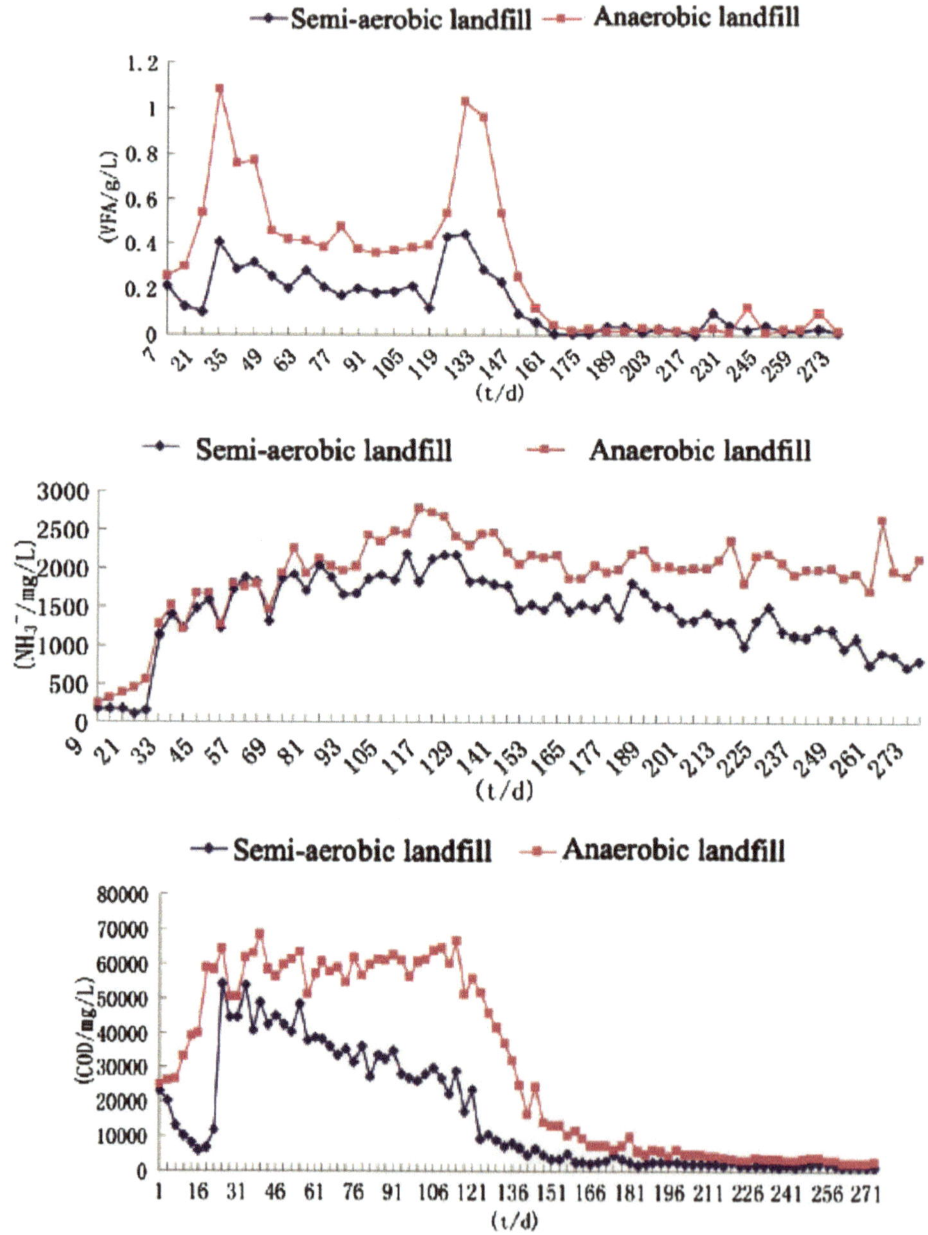

Figure 14.2.10 Changing curve of VFA, COD, and NH_3-N over time for two lysimeters reproducing a semiaerobic and an anaerobic landfills type (He et al., 2011). *COD*, chemical oxygen demand; *VFA*, volatile fatty acid.

On analyzing the figures, the following observations may be made:

- No acid accumulation was detected in semiaerobic leachate.
- Ammonia nitrogen concentrations in semiaerobic leachate were lower than those in anaerobic leachate.
- COD concentrations in semiaerobic leachate were lower and rate of degradation was higher than that in the anaerobic system.

Gas quality

A significant difference in volumetric gas composition was observed between anaerobic and semiaerobic landfills. Fig. 14.2.11 illustrates the results obtained in extended lysimeter studies. Under semiaerobic conditions, methane concentrations were constantly 10%–20% lower than those under anaerobic conditions. Some authors estimate the reduction in methane concentration to be around 54% (Blake et al., 2008); conversely, CO_2 was higher, as a result of natural air advection. In view of the importance of global warming, methane reduction produced under semiaerobic conditions represents an important feature of the system.

Gas composition from two Japanese landfills filled with a series of different wastes is reported in Table 14.2.1.

In a semiaerobic landfill, as mentioned previously, natural ventilation is not sufficient to promote homogeneous aerobic conditions within the waste mass, with the consequent presence of several anaerobic zones. Fig. 14.2.12 shows oxygen concentration contours of waste layers filled with incombustibles in an experiment using three large semiaerobic lysimeters (3 m width, 0.5 m depth, and 4 m height) without venting pipes, with only vertical pipes, and with vertical and horizontal pipes. In the absence of a venting pipe, it can be clearly observed how air circulation is substantially confined to the bottom and surface areas; on the contrary, in the presence of gas venting pipes, air inflow is better distributed throughout the waste layers (Park et al., 1994).

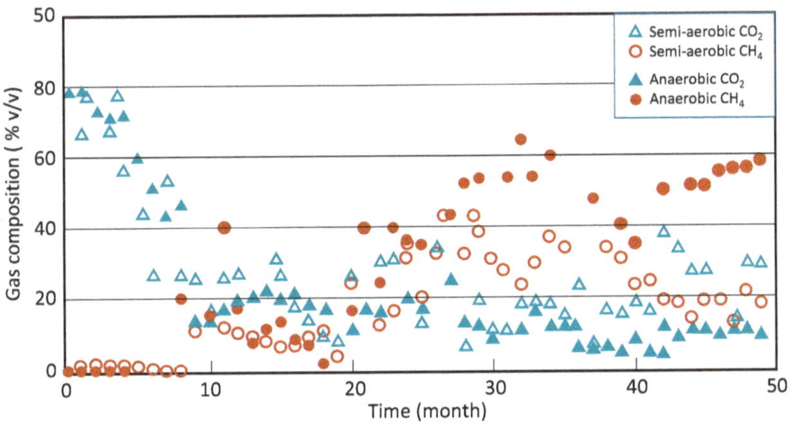

Figure 14.2.11 Difference of gas quality between semiaerobic and aerobic landfills.

Table 14.2.1 Relation between methane concentration of landfill gas and the waste composition

Site Name	Landfill Type	Landfilled Area (ha)	Landfilled Wastes (t)	Waste Composition (%)		Methane in Landfill Gas (%)		
						10 years After Close	15 years After Close	25 years After Close
New Kamata	Semiaerobic	7.3	259,000	Combustibles	59.5	1.8–60.0 (Avg. 32.5)	0.0–50.0 (Avg. 26.3)	0.0–20.0 (Avg. 6.3)
				Noncombustibles	24.1			
				Incineration ash	16.4			
Busetsu A	Semiaerobic	1.7	82,800	Combustibles	6.4	6.5–31.0 (Avg. 20.0)	0.0–20.0 (Avg. 9.6)	0.0–13.0 (Avg. 4.8)
				Noncombustibles	68.3			
				Incineration ash	25.3			

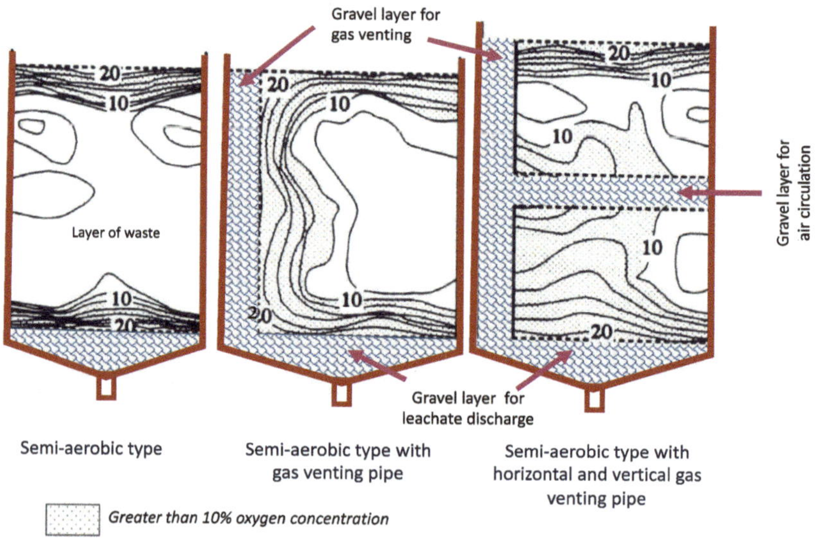

Figure 14.2.12 Oxygen concentration contours of waste layers.

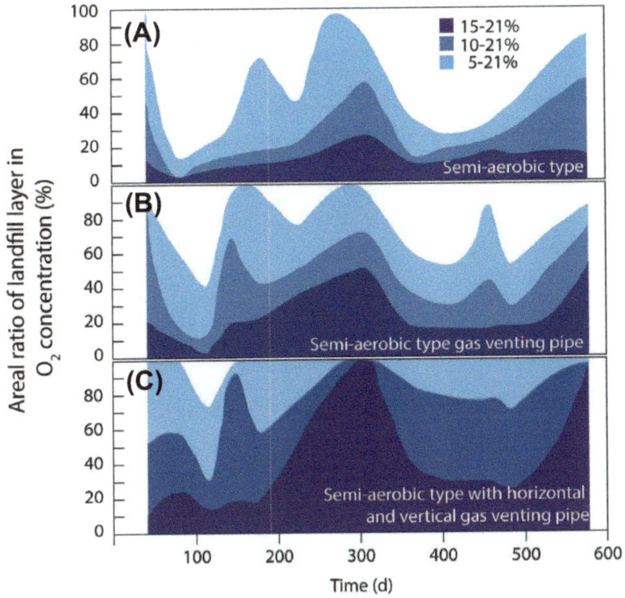

Figure 14.2.13 Change in oxygen concentration contours of waste layers over time.

Over the winter months, ambient temperatures decrease and consequently, as biodegradation and heat generation processes progress, the temperature gradient between waste mass and the surrounding ambient rises, thus contributing toward an extended presence of oxygen-rich zones (Fig. 14.2.13) (Park et al., 1994).

DESIGN GUIDELINES

General structure

A carefully conceived engineering design is mandatory to establish semiaerobic conditions in landfill. The general structure of a semiaerobic landfill, as anticipated earlier and described by Fig. 14.2.14 (Matsufuji, 2004), consists in a network of horizontal pipes (laid in graded stones and covered by gravel, installed at the bottom of the landfill) and vertical pipes (surrounded by gravel, rising from the bottom to the landfill surface) (Matsufuji and Hanashima, 1991; Matsufuji, 2004). The use of crushed rock around leachate pipes provides mechanical protection for the horizontal pipes, promoting the inflow of leachate and outflow of air (Zhang, 2012).

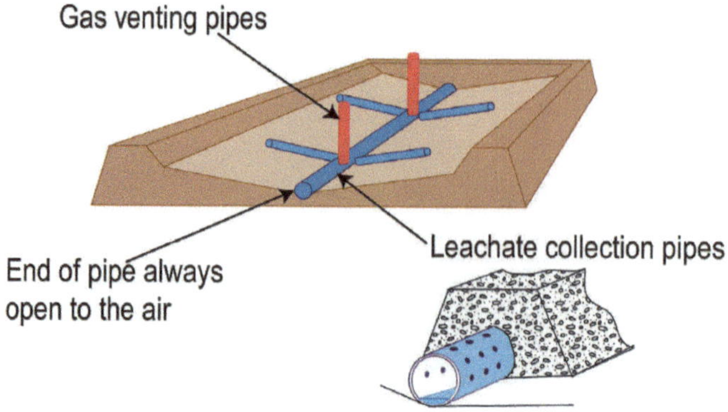

Figure 14.2.14 Leachate collection and gas venting pipes system (Zhang, 2012).

Leachate collection pipes

Leachate collection pipes may be placed in situ according to a series of different layouts in line with site morphology and shaping of the landfill sectors, as described in Fig. 14.2.15 (Matsufuji, 2007).

The leachate collection system consists of a central pipe (main collection pipe) with branch pipes on either side, laid on a slope to enhance leachate collection (2% main pipes, 1% branch pipes). Examples of leachate collection pipeline layouts in full-scale landfills are given in Fig. 14.2.16 (Matsufuji, 2007).

The main collection pipe terminates in an open leachate retention pond or collection well, in which water level is maintained below the outlet of the collection pipe to allow fresh air to flow into the pipe.

To facilitate leachate outflow and air inflow into the waste layers, all collection pipes are perforated with a large number of holes with an approximate diameter of 2.5 cm. The distribution of holes along and around the collection pipes is described in Fig. 14.2.17 (Matsufuji, 2007).

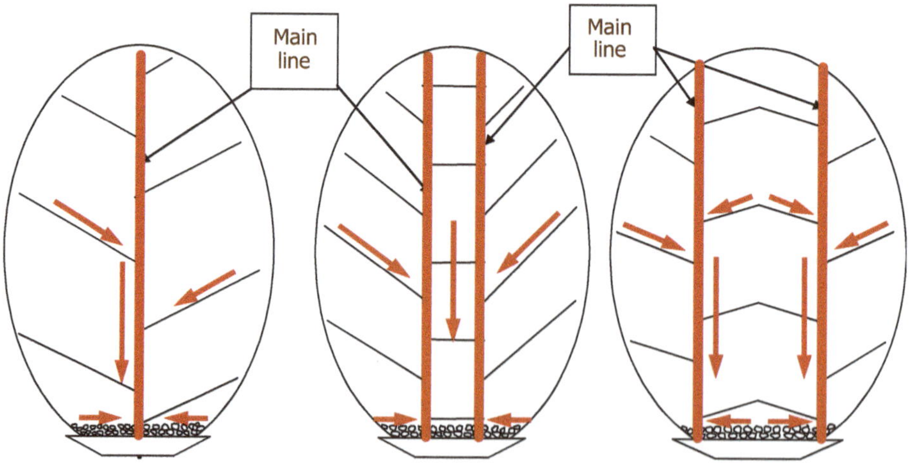

Figure 14.2.15 Placement pattern for leachate collection discharge pipes.

Figure 14.2.16 Placement example for leachate collection discharge pipes at Fukuoka city Busetsugaura and Nakata Landfills.

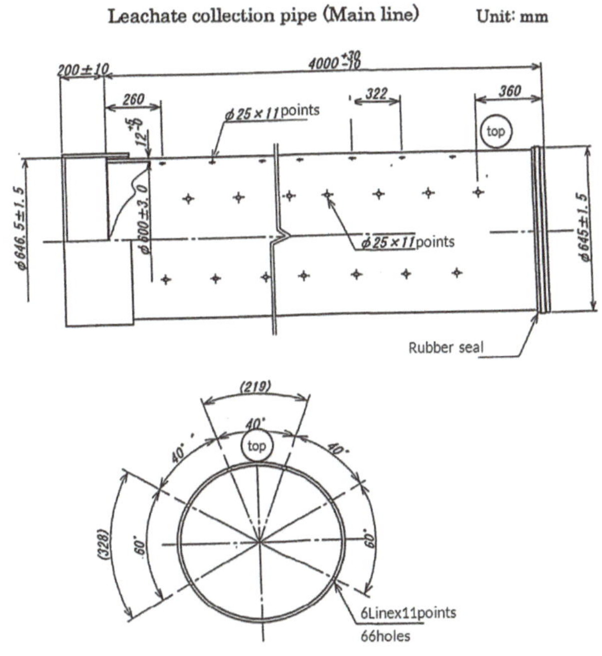

Figure 14.2.17 Example of perforations in leachate collection pipes.

As the leachate collection pipes are situated at the bottom of the landfill, maintenance control after landfill start-up may be problematic. Ponding of leachate and sludge/sediments inside the pipe should be avoided, as this may severely reduce air inflow section inside the horizontal pipes. These issues, however, cannot easily be accounted for and quantified during the design stage. Therefore, as a rule of

Figure 14.2.18 Samples of leachate collection pipes: (A) cement, (B) HDPE, and (C) Bamboo.

thumb dictated by experience, a venting pipe diameter of more than 600 mm should be used. To obtain good mechanical resistance, the use of appropriate materials such as fiberglass reinforced plastics should be ensured. Other materials, including perforated concrete or synthetic polymer pipes (commonly HDPE), may be used, although characterized by a lower resistance to pressure. In some cases, particularly in developing countries, alternative materials such as bamboo or scrap tires have been used (Fig. 14.2.18) (Matsufuji, 2004).

When using lower resistance materials, the diameter of the main collection pipe might be reduced to 450 mm or less, and the frequency of branch transverse pipes increased.

The graded rocks used to protect pipes from clogging and damage during operation are in the size range 10–25 cm in diameter. The layer of cover material above the pipe reaches a thickness of more than 50 cm. The containment barrier is similar to that employed for a normal landfill (Fig. 14.2.19) (Matsufuji, 2007).

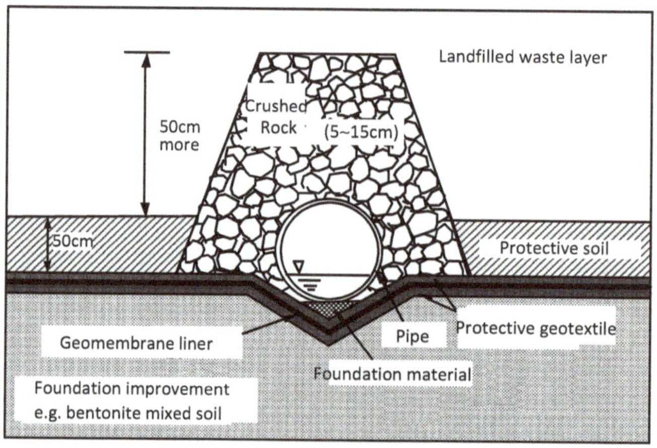

Figure 14.2.19 Placement examples for leachate collection pipes.

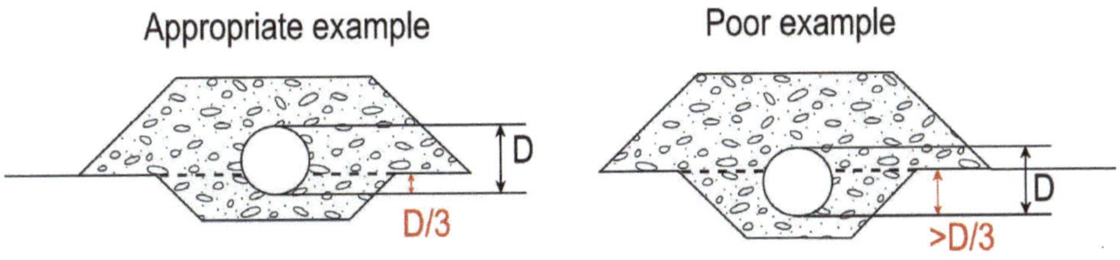

Figure 14.2.20 Correct and incorrect installation of the main collection pipe in semiaerobic systems.

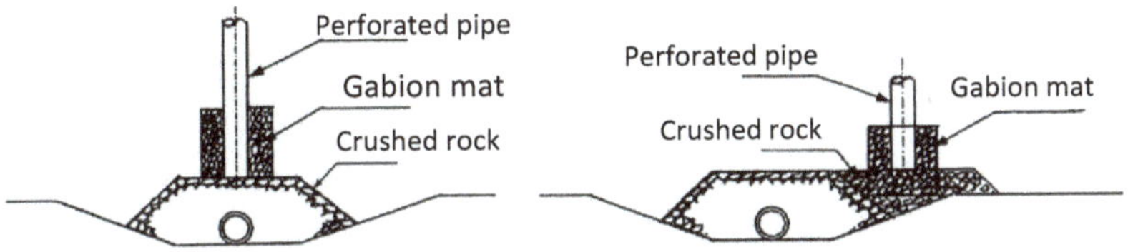

Figure 14.2.21 Structure of gas venting pipe.

Leachate collection pipelines promote both natural air circulation and leachate drainage; to guarantee air circulation at all times, two-thirds of the pipe section must be free of leachate. This can be achieved by defining the configuration of collection pipes and cover material layer at the design stage, having less than one-third of the diameter of pipes installed below the landfill bottom level (Fig. 14.2.20) (Matsufuji, 2007).

Gas venting pipes

Gas venting pipes are expected to fulfill four main tasks:

- Expand aerobic area, consequently shortening or preventing CH_4 generation
- Release gases generated from the landfill layers
- Allow a rapid discharge of leachate
- Promote the recirculation of leachate

The gas venting facility consists of horizontal and vertical/inclined pipes, of about 30 cm diameter, connected to the leachate collection system as illustrated in the configuration represented in Fig. 14.2.21 (Matsufuji, 2007). Crushed stones used in Japan in vertical–horizontal connections have an indicative size of 15 cm.

Vertical gas ventilation pipes are installed on the top of a connection pit at each intersection of the main collection pipe with the branch pipes and at the end of each branch pipe. The gas venting pipes are enclosed in protective perforated casing (simple low-cost structure, such as perforated oil drums, can be conveniently used) and filled with graded rocks (Fig. 14.2.22) (Matsufuji, 2007).

Figure 14.2.22 Examples of gas venting pipes with high- and low-cost structure (SPREP-JICA, 2010).

Figure 14.2.23 Different types of vertical gas venting facility (SPREP-JICA, 2010).

Connection pits should be fitted with a strong cover to support the weight of the gas venting columns (pipes/gravel). Gas venting columns should be mounted at least 30 cm higher than ground level to prevent runoff water from filling the column.

The use of filter materials around the gas venting pipes is intended to extend the aerobic zone: used/waste materials can substitute for more expensive materials and are capable of efficiently performing the same function, as described by the examples in Fig. 14.2.23 (Matsufuji, 2007). Gravel (10–15 cm) may be used as filter material in the gas venting column.

To obtain a consistent diffusion of gas, venting pipes should be installed every 20–40 m on flat landfill sites (Fig. 14.2.24), depending on landfill depth (Matsufuji, 2007).

Indicative combinations between landfill depth and distance between gas venting columns are given in Table 14.2.2.

Figure 14.2.24 Example of gas venting pipe installation.

Table 14.2.2 The indicative optimal combinations between the landfill depth and the distance between the gas venting columns in semiaerobic systems

Planned Landfill Depth (m)	Distance Between the Gas Venting Columns (m)
10	40
10–20	30
20	20

In the presence of wastes with a high putrescible content, additional air flow pipes should be located at 10–20 m intervals over the landfill height to limit presence of anaerobic zones within the waste mass.

An optimal balance between vertical venting pipes and additional air supply pipes should be established to minimize future operational issues.

CALCULATION MODELS

Equations for gas flow in waste layers, as schematized in Fig. 14.2.25, are expressed by the following equations (Hanashima et al., 1981):

$$J_x = -\frac{\kappa_x}{\mu}\frac{\delta p}{\delta x} \tag{1}$$

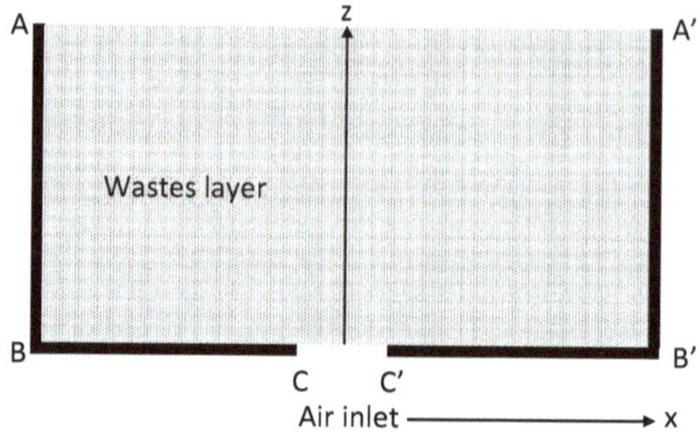

Figure 14.2.25 Cross-sectional view of waste layers and location of air inlet.

$$J_z = -\frac{\kappa_z}{\mu}\left[\frac{\delta p}{\delta z} + \rho g\right] \tag{2}$$

$$\frac{\delta}{\delta x}(J_x) + \frac{\delta}{\delta z}(J_x) \tag{3}$$

where p is pressure (N/m^2); J_x and J_z are superficial gas fluxes defined by ρv_x & ρv_z (kg/m^2 s); ρ is density of gas mixture (kg/m^3); v_x and v_z are velocity of flowing gas mixture (m/s); κ_x and κ_z are air permeability of solid wastes layer (kg/m); μ is viscosity (kg/m s); and g is gravity (m/s^2).

The following assumptions may be made with regard to gas flow:

1. flow of gas mixture in the waste layers is a quasi steady-state, temperature-dependent fluid flow;

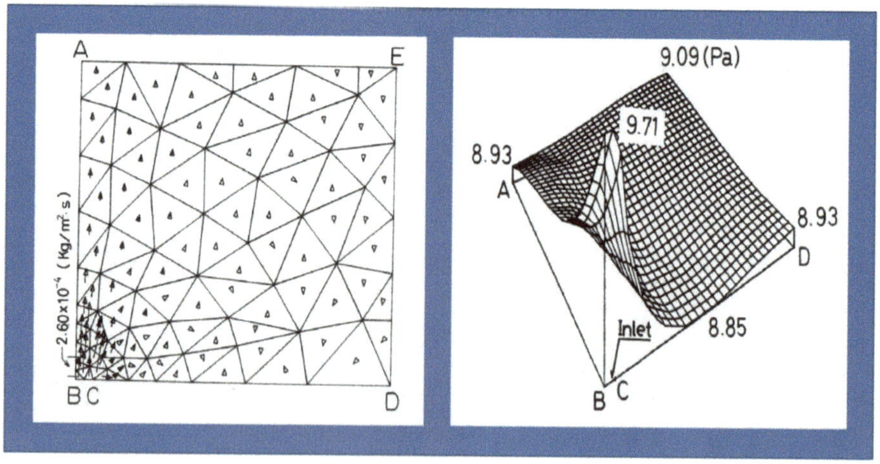

Figure 14.2.26 Total pressure potential (A) and element flux of gas mixture (B).

2. flow adheres to Darcy's law, in which permeability coefficients are constant;
3. gas mixture viscosity is constant;
4. apparent rate of gas production can be neglected.

The calculated total pressure potential and plotted flux vector (J_x, J_z) for each element is illustrated in Fig. 14.2.26, in which a flow region from BC to AE can be observed in the right-half parabolic area. Since the region is confined above the inlet, oxygen concentration is expected to be high in the region and low in the adjacent area. Aerobic biodegradation promotes production of thermal energy in the confined region rather than at the side of the inlet.

CODE OF PRACTICE FOR OPERATION

Degradation processes in a semiaerobic landfill should be continuously monitored to verify actual aerobic conditions.

The main control parameters are as follows:

- Temperature. A decrease in temperature indicates the presence of decreased air infiltration or localized oxygen consumption by putrescible organics.
- Gas composition. A significant increase in methane concentration may indicate the onset of anaerobic degradation.
- Leachate composition. Increase in BOD and ammonium nitrogen and decrease in pH are indicative of prevailing anaerobic conditions.

Leachate collection and gas venting pipes play a crucial role in a semiaerobic system; for this reason, issues such as breakage or blockage of the pipes that may affect their functioning should be avoided. Particular care should therefore be taken during landfill operation to maintain functionality of the pipelines. Likewise, handling of daily cover materials should be carried out with caution to prevent fine particles from clogging the filter layers surrounding leachate collection pipes, as had evidently occurred in the cases represented in Fig. 14.2.27.

Figure 14.2.27 Example of leachate collection pipe and gas venting pipe clogged with cover material and small particles of wastes.

Figure 14.2.28 Good example of leachate collection pipe opening.

The layout of leachate collection pipes should be simple, with pipes being easily inspectable, and (as highlighted previously) ensuring the pipe extremity remains open to the atmosphere. During the design stage, ease of maintenance for collection pipes should be taken into account (Fig. 14.2.28).

Leachate collection pipes installed on the bottom of the landfill should be of an appropriate size and constructed using appropriate materials, to resist the static weight of waste layers, taking into account the final planned height of the landfill. However, at start-up of landfilling, when waste thickness is less than 3 m, pipes may be directly influenced by the kinetic force of transport vehicles and heavy operational machinery (compactors, etc.). In particular, bottom collection systems may be exposed to the adverse pressure of heavy machinery, resulting in damage to the pipes. To protect the system therefore during the early stages of landfilling, an iron plate may be conveniently used to better distribute machinery pressure. The abovementioned situations are graphically described in Fig. 14.2.29.

CASE STUDIES

Semiaerobic landfills in Japan

The demonstrated improvement obtained in leachate quality following application of the semiaerobic landfill type, not only in laboratory-scale and extended lysimeter studies but also in new full-scale landfills, has led to the majority of local governments in Japan adopting the semiaerobic method for the purpose of addressing environmental problems linked to leachate management. Japanese technical guidelines have long indicated semiaerobic landfills as the standard landfill model. Fig. 14.2.30 shows examples of landfill sites applying the semiaerobic landfill type. As can be seen, leachate collection systems and the crushed stones that cover and protect the pipes on these sites are extremely large.

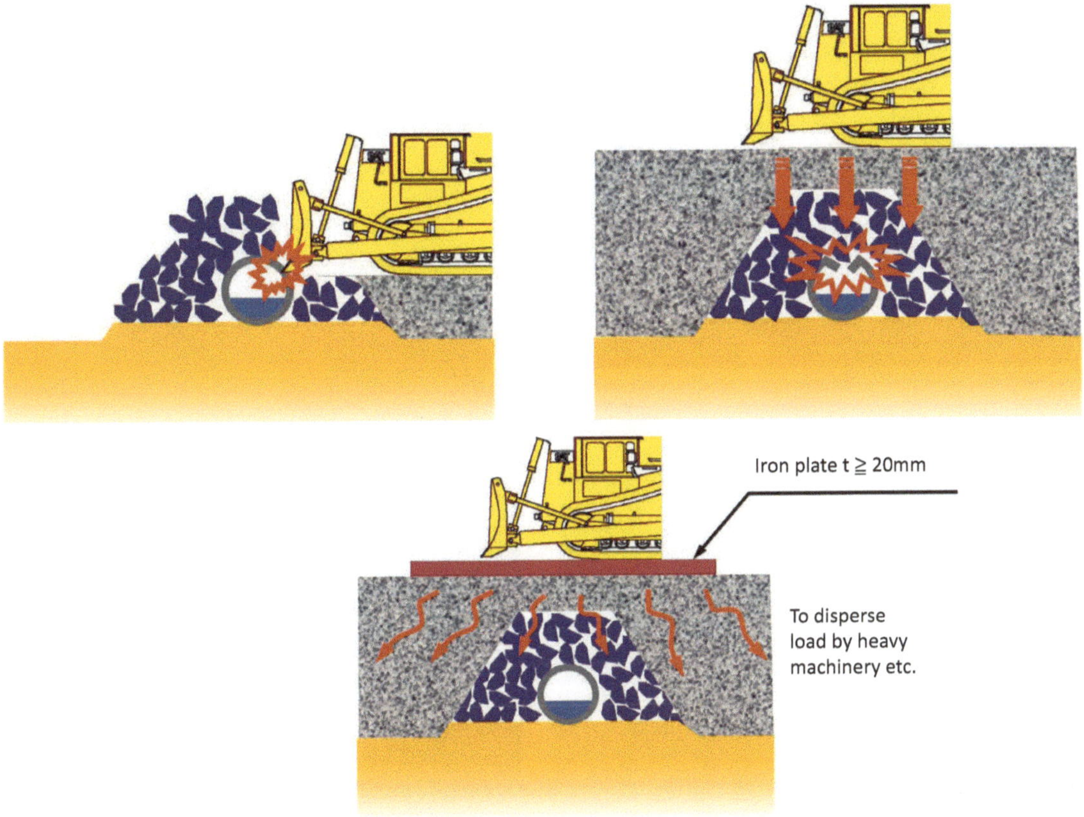

Figure 14.2.29 Protection of leachate collection pipe during landfilling operation.

Semiaerobic landfills in other countries

As mentioned in the introduction, in addition to Japan, the semiaerobic model has been applied successfully in several other countries, including China, Iran, Malaysia, Pakistan (Fig. 14.2.31), Samoa, and Vietnam (Fig. 14.2.32).

In Europe a semiaerobic landfill sector has been constructed at the Scala Erre landfill site in Sassari, Sardinia (Cossu, 2015). This landfill is operated following the PAF landfill concept, which combines the Fukuoka method with biological waste pretreatment and flushing (Cossu et al., 2003).

The model, graphically described in Fig. 14.2.33, provides for a combined mechanical biological pretreatment (P) with in situ natural aeration (A), and open covers to allow flushing (F). Using this landfill model, final BOD and ammoniacal nitrogen concentrations close to those required by Italian law for the drainage of wastewaters in superficial bodies of water have been obtained (Cossu et al., 2003).

In some of these cases (again as mentioned previously), a series of locally available alternative materials such as bamboo, scrap tires, plastic waste, and oil drums have been conveniently used to construct the semiaerobic structure.

Figure 14.2.30 Examples of landfill applying semiaerobic landfill type in Japan.

Figure 14.2.31 View of the structure of pipeline network in a semiaerobic landfill in Pakistan.

Figure 14.2.32 View of different construction phases in a semiaerobic landfill in Vietnam.

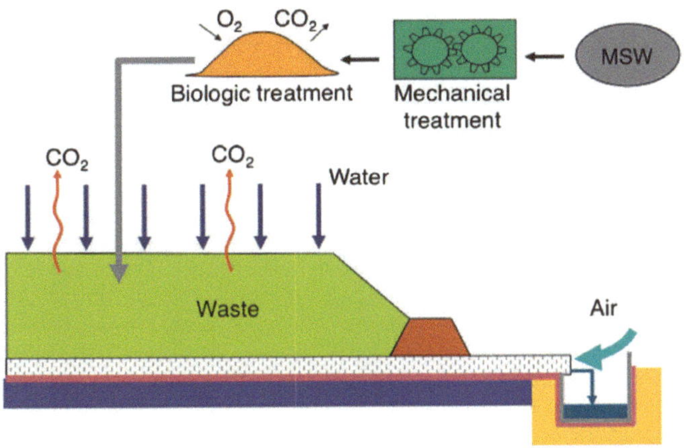

Figure 14.2.33 Scheme of PAF (Mechanical biological pretreatment combined with aeration by natural convection and natural waste flushing) integrated model (Cossu et al., 2003). *MSW*, municipal solid waste.

In developing countries, landfill sites are often managed under anaerobic conditions as open dumps. Since 2008, the Japanese Ministry of the Environment (JMOE) has promoted a cobenefit Clean Development Mechanism (CDM) program. The main aim of the cobenefit CDM for landfills is not only to reduce GHG emissions but also to improve the management of old landfills and leachate using the semiaerobic landfill method.

The JMOE finances 50% of the initial construction costs in the potential cobenefit CDM scheme, on condition that 50%–100% of the generated Certified Emission Reductions are transferred, free of charge, from the Country receiving the financial support to the Japanese national registry.

The Pulau Burung landfill site in Malaysia represents the first example of this cobenefit CDM project.

References

Blake E., Bruce Chapman B., 2008. Meeting the challenge — landfill in Samoa. WasteMINZ Conference 2008, accessed the 20th of October 2018. From www.wasteminz.org.nz/pubs/meeting-the-challenge-landfill-in-samoa/.

Cossu, R., Raga, R., Rossetti, D., 2003. The PAF model: an integrated approach for landfill sustainability. Waste Management 23, 37–44. Elsevier Ltd. ISSN: 0956-053X.

Cossu, R., 2015. Design Project of the Scala Erre Landfill for the Municipality of Sassari (Personal Comunication).

Chong, T.L., Matsufuji, Y., Hassan, M.N., 2005. Implementation of the semi-aerobic landfill system (Fukuoka method) in developing countries: a Malaysia cost analysis. Waste Management 25, 702–711.

He, P., Yang, N., Gu, H., Zhang, H., Shao, L., 2011. N_2O and NH_3 emissions from a bioreactor landfill operated under limited aerobic degradation conditions. Journal of Environmental Sciences 23, 1011–1019.

Hanashima, M., Yamasaki, K., Kuroki, T., Onishi, K., 1981. Heat and gas flow analysis in semiaerobic landfill. Journal of the Environmental Engineering Division ASCE 107 (1), 1–9.

Matsufuji, Y., Hanashima, M., 1991. Sanitary landfill design and operation in Japan. In: 5th International Symposium 91 Solid Waste Management Technology, pp. 275–325.

Matsufuji, Y., Hanashima, M., Nagano, S., 1978. Waste and landfill type. In: Proceedings of 33th Annual Conference of Japan Society of Civil Engineers, vol. 33, pp. 133–134.

Matsufuji, Y., Tanaka, A., Hanashima, M., 1997. Biodegradation process of municipal soild waste by semiaerobic landfill type. In: Proceeding of the First Korean – Japan Joint International Symposium. Korea Society of Waste Management, pp. 87–94.

Matsufuji, Y., Tachifuji, A., Matsugu, H., 2005. Biodegradation process of municipal solid waste by semiaerobic landfill type. In: Proceeding of 10th International Waste Management and Landfill Symposium. CISA Publisher, No.C, pp. 113–120.

Matsufuji, Y., 2004. A Road to Sanitary Landfill, 2. Ministry of Environment, Japan.

Matsufuji, Y., 2007. Point of Attention in Application of "Fukuoka Method" (Semiaerobic Landfill Technology). Japan International Cooperation Agency (in Japanese).

Park, S., Kusuda, T., Shimaoka, T., Matsufuji, Y., Hanashima, M., 1994. Effect of air permeability in solid waste landfill layer on stabilization of leachate quality. Journal of Japan Society of Waste Management Experts 5 (5), 19–28.

Shimaoka, T., Matsufuji, Y., Hanashima, M., 2000. Mechanism of Self-Stabilization of Semi-Aerobic Landfill. Proceedings of the 5th Annual Landfill Symposium, Solid Waste Association of North America, 171–186.

SPREP-JICA, 2010. A Practical Guide to Landfill Management in Pacific Island Countries and Territories Volume 1 – Inland-based waste disposal. Apia, Samoa. https://wedocs.unep.org/bitstream/handle/20.500.11822/8957/-A%20practical%20guide%20to%20landfill%20management%20in%20Pacific%20Island%20countries%20and%20territories:%20Volume%201%20-%20Inland-based%20waste%20disposal%20(2nd%20edition)-2010Landfill.Guidelines_2010_3.pdf?sequence=3&isAllowed=y. (accessed on the 22nd October 2018).

Yanase, R., Hanashima, M., Yamasaki, K., Matsufuji, Y., Nagano, S., 1983. Field study of semi-aerobic landfill type. In: Proceeding of the 34th Annual Meeting of Japan Council for Solid Waste Management, pp. 115–120 (in Japanese).

Zhang Chen, Santha A/P Chenayah@Ramu, Fauziah Shahul Hamid, 2012. Environmental Evaluation of Sanitary Landfills Establishment: Malaysian Case Studies. Jornal of Advanced Materials Research vol. 599, pp. 224–228.

Further Reading

Hanashima, M., 1981. Study on leachate control by accelerated landfill stabilization. Verhandlungen des Internationalen Verein Limnologie 21, 1101–1104.

Matsufuji, Y., Tachifuji, A., 1997. Greenhouse gases emission by landfill type. Journal of Japan Society of Waste Management Experts 8 (6), 438–446.

14.3

LANDFILL BIOREACTORS

Raffaello Cossu and Valentina Grossule

INTRODUCTION

In landfill, the control of degradable organics and related environmental impacts may be achieved by a series of options:

- Avoidance of putrescible fraction in landfilled waste (separate collection);
- Prestabilization of waste prior to landfilling (thermal treatment, mechanical biological treatment, see Chapters 4.1 and 4.2);
- In situ treatment to enhance biodegradation during landfilling and/or the aftercare phase;
- A combination of the above listed options.

In recent years, the term "landfill bioreactor" has been coined specifically to indicate a landfill in which different in situ measures are undertaken to enhance biological degradation and increase removal of ammonia and recalcitrant organics. These measures may include leachate recirculation, introduction of water, and natural or forced aeration, as graphically represented in Fig. 14.3.1.

Improved control of biochemical kinetics, moisture content, and redox conditions may result in a significant shortening of the aftercare phase and consequently, in a quicker achieving of environmentally sustainable targets (Chapter 2.2).

Advantages gained by enhancing biodegradation processes in landfill bioreactors (depending on the different bioreactor models) may include the following:

- Shortening of the aftercare phase
- Reduction of environmental impacts
- Reduction of aftercare costs
- Reduction of landfill owner's environmental liability
- Improvement of leachate quality
- Faster and time-specific biogas production
- Faster mechanical stabilization of waste mass
- Increased waste settlement and density
- Less postclosure operations required
- Removal of long-lasting compounds from traditional landfills (ammonia, recalcitrant organics, etc.)

Disadvantages (in line with type of bioreactor model) may include the following:

- Increased capital and/or management costs
- Higher complexity of construction

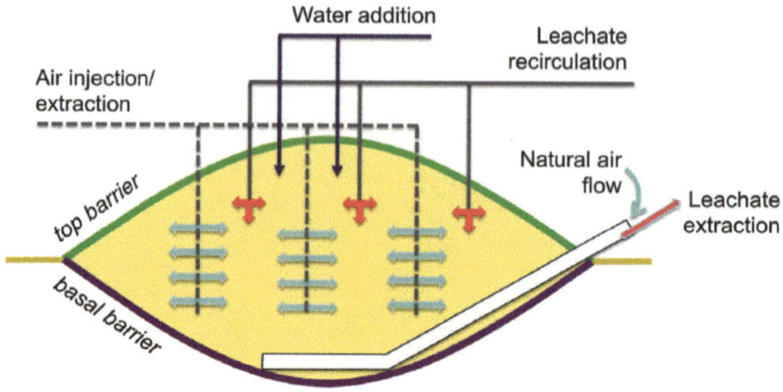

Figure 14.3.1 Scheme of the potential measures that may be adopted to enhance the biodegradation and removal of ammonia and recalcitrant organics in landfill bioreactors. Some potential theoretical measures (such as pH adjustment, sludge, or liquid waste addition) have not been mentioned.

- Energy for aeration
- More complex management and monitoring program
- Specific disadvantages may arise with leachate recirculation such as:
  - Increased odor generation
  - Risk of mechanical instability (particularly on slopes) during the operational phase due to higher moisture content
  - Higher production of leachate

According to type of bioreactor, different methane generation yields and waste stabilization rates may be achieved.

The choice of a specific bioreactor landfill is regulated by the objectives to be pursued (energy recovery landfill gas, waste stabilization, sustainability targets, etc.) and by economic issues (balance between capital and operational costs and long-term savings).

The way that each individual in situ measure affects the processes within the landfill is schematically represented in Fig. 14.3.2.

This chapter presents a series of different technologies capable of enhancing biodegradation in landfills with particular focus on hybrid bioreactors. Experiences, both on lab, pilot, and full scale, are described and outputs are discussed.

LANDFILL BIOREACTORS TYPOLOGIES

According to biodegradation conditions, landfill bioreactors can be subdivided into four main typologies: anaerobic, aerobic (aerated), semiaerobic, and hybrid, which consists in a sequence of aerobic and anaerobic conditions (Cossu et al., 2017).

The waste to be landfilled may either undergo biological pretreatment or not, to the extent of pretreatment being functional to the environmental sustainability targets.

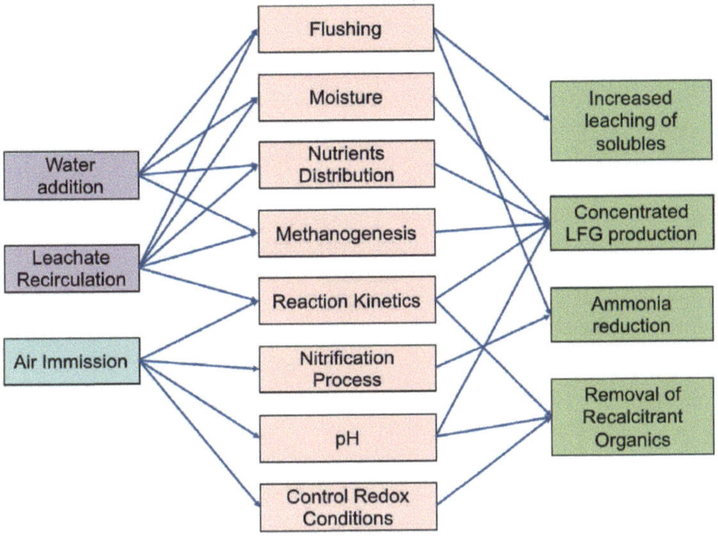

Figure 14.3.2 Graphical representation of how the most important measures that may be adopted in a landfill bioreactor can influence processes and consequent quality of emissions.

The main features of the four typologies of landfill bioreactors are graphically represented in Fig. 14.3.3.

Anaerobic Landfilling

In anaerobic landfilling, enhancement of the biodegradation process is obtained by increasing the leachate recirculation rate (Chapters 12.1 and 12.2). When the concept of landfill bioreactor was originally launched, this was the most popular model. It has been applied since the 80s in a series of landfills in the United States, Canada, Australia, and in Europe, mainly in France (i.e., Reinhart and Townsend, 1998; Pattison and Yuen, 2007; Rivière et al., 2011). This enhancement is reflected in leachate quality, with lower biochemical oxygen demand (BOD) and chemical oxygen demand (COD), when compared with traditional landfilling, and in an enhanced methane generation rate. Under anaerobic conditions ammonia is not significantly reduced. Some positive effects might be detected in relation with the flushing effect of leachate recirculation.

In some countries the regulators are reluctant to permit leachate recirculation due to concerns over an increased potential of environmental impacts and costs (Pattison and Yuen, 2007).

To overcome the problem of ammonia accumulation in anaerobic conditions, an alternative solution consists in external aerobic pretreatment of leachate prior to recirculation in the anaerobic bioreactor, to facilitate onset of both nitrification and denitrification processes and thus remove nitrogen compounds (De Abreu et al., 2005; Price et al., 2003; Berge et al., 2005; Zhong et al., 2009). The system aims to ensure the recovery of energy caused by methane production by maintaining the whole landfill in anaerobic conditions. To remediate nitrogen pollution in leachate, the latter is aerobically treated to nitrify n

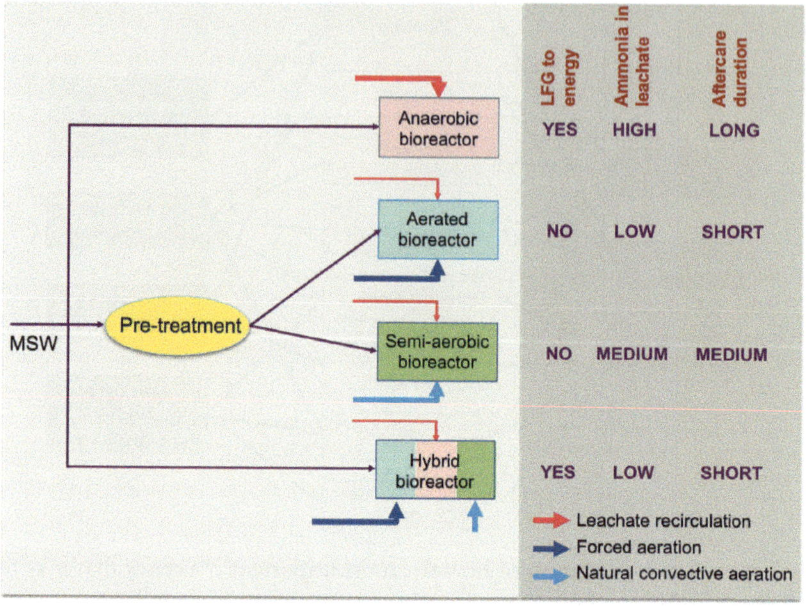

Figure 14.3.3 Scheme and general features of the different landfill bioreactor typologies.

and subsequently reinjected into the landfill to denitrify the produced nitrates. This system has also been patented in the United States (US639895, 2002) as a facultative landfill and has been tested in lab and full scale with positive results (Price et al., 2003).

Aerated Landfilling

Aerobic landfilling can be performed under forced or natural conditions: under forced conditions air is injected into the waste mass using several kinds of blower devices. In aerated landfilling, the waste should undergo preliminary pretreatment to reduce the biodegradable content and prevent a rapid increase in temperature.

A series of different technologies and strategies have been developed for use in situ aeration, including high-pressure aeration, low-pressure aeration, and active aeration with or without off-gas extraction (Ritzkowski and Stegmann, 2012; Chapter 16.2, this book).

Amongst others, nitrogen removal is one of the most significant benefits to be gained from an aerobic system (Berge et al., 2006; Shao et al., 2008). Although air injection might theoretically inhibit the denitrification process, complete aerobization of the waste mass is never achieved in a full-scale landfill, and anaerobic areas will still be present inside the landfill and both processes can take place simultaneously, even under scarcely biodegradable conditions (Berge et al., 2006; Shao et al., 2008; Ritzkowski and Stegmann, 2003, 2005; Ritzkowski, 2011; Giannis et al., 2008).

Leachate recirculation represents an additional in situ leachate treatment tool aimed at improving stabilization performance. In particular, increased frequency of leachate recirculation accelerates the stabilization rate of waste, although an excessive recirculation may result in saturation, ponding, and acidic conditions (Šan and Onay, 2001). Slezak et al. (2015) observed that the higher the recirculation rate, the higher and faster the reduction of carbon and nitrogen parameters in leachate, although O_2 diffusion was limited, thus leading to lower waste stabilization.

Kim et al. (2011) investigated the fate of metals in aerobic and anaerobic landfill bioreactors. With the exception of the initial acidic phase, the mobility of heavy metals is reduced under aerobic conditions due to high pH and positive redox conditions, affecting solubility and sorption properties of metals retained in waste by sorption, carbonate precipitation, and hydroxide precipitation (Borglin et al., 2004; Giannis et al., 2008).

Aeration rates and mode exert a varying influence on degradation performance: Slezak et al. (2010) compared stabilization performance in four aerobic lysimeters with different aeration rates, obtaining similar changes in leachate parameters, demonstrating that above the minimum aeration requirements, increased rates do not yield any additional benefit. Intermittent aeration has been demonstrated to be much more effective than continuous aeration (Cossu et al., 2015; Morello et al., 2017).

Nowadays, forced aeration is mainly applied in the remediation of old anaerobic landfills, rather than representing merely a design option for landfill management (Chapter 16.2). Moreover, aeration of landfill mass is a fundamental pretreatment for landfill mining procedures (Raga and Cossu, 2014; Ritzkowski and Stegmann, 2012).

Depending on the concentration of putrescibles in the waste (pretreated or non-pretreated waste), high temperatures may develop and waste moisture could be negatively affected (waste could dry out!). The process is energy consuming and could prove to be complex and expensive. Application of this process to a normal active landfill may create operational problems. The method is proposed and widely applied with a view to shortening aftercare time and in the remediation of old landfills (Chapter 16.2).

Semiaerobic Landfilling

The process involving natural aeration of landfills was initially proposed in Japan, in early seventies, by Hanashima at Fukuoka University (see Chapter 14.2). This system, referred to as semiaerobic landfilling or the "Fukuoka method," aims to achieve aerobization of waste mass with a proper engineering design in which ambient air naturally flows into the waste body through the leachate collection pipes, moved by the temperature gradient present inside and outside the landfill (Chapter 14.2). The semiaerobic system has been widely applied in Japan and, to a lesser extent, in many other countries. Through aerobic landfilling, BOD and COD concentrations decrease significantly; likewise, Total Kjeldhal Nitrogen (TKN) is also decreased due to the nitrification of ammonia. Methane production is low, with the exception of the initial phases. Leachate recirculation in aerobic landfills may be undertaken, although waste saturation should be carefully monitored to prevent potential obstacles in the circulation of air and extraction of the exhausted gas.

A combination of both mechanical biological pretreatement, flushing, and semiaerobic landfilling (PAF) has been suggested by Cossu et al. (2003). The results obtained with this combination have been very positive in terms of stabilization of the biomass and reduction of BOD, COD, and TKN in leachate.

Hybrid Landfilling

Hybrid landfill bioreactors are conceptually based on the principle of combining both aerobic and anaerobic processes by alternating over time aeration of the waste mass. The main purpose is to achieve the benefits yielded by the two systems (Cossu et al., 2017):

- Significant enhancement of biodegradation and shortening of aftercare time
- Leachate with lower concentrations of BOD and COD
- Nitrification of ammonia and denitrification
- Degradation of recalcitrant compounds such as lignin and aromatic substances
- Better control of pH during acidic enzymatic hydrolysis by early oxidation of the most putrescible substances.
- Quicker and more concentrated production of methane
- Enhancement of biogas generation

The immission of air could be implemented by means of either natural or forced aeration.

The sequencing of aerated and non-aerated phases, aeration mode (continuous or intermittent), and the duration of these characterizes the different schemes of hybrid bioreactors that have been proposed as summarized in Fig. 14.3.4.

Aerobic–Anaerobic Sequencing

The mechanical biological pretreatment of waste prior to anaerobic landfilling may be regarded as a kind of hybrid reactor, with off-site forced aeration followed by in situ anaerobic reactions. MBP is aimed at achieving rapid stabilization of the landfilled waste, although the resulting production of landfill gas may not be sufficient for energetic exploitation.

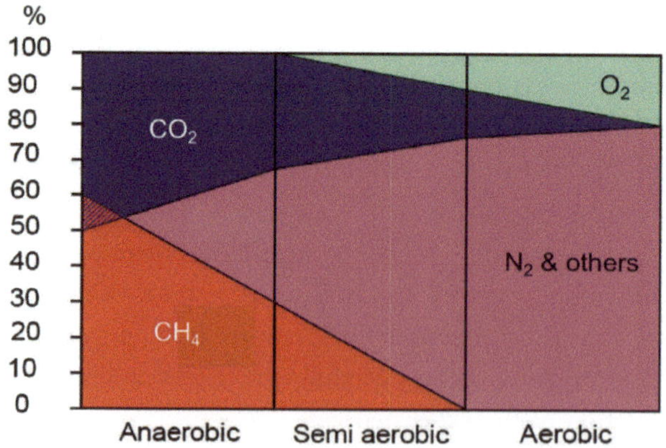

Figure 14.3.4 Composition of landfill gases in anaerobic, semiaerobic, and aerobic bioreactors (Grossule et al., 2018).

Aerobic–anaerobic sequencing may be applied completely in situ. However, in this case, preliminary aeration of the waste mass, rather than stabilizing the waste, would be mainly focused on accelerating the initial acidogenic phase, achieving optimum pH and VFA conditions for methanogenesis much earlier (Mali Sandip et al., 2012; Xu et al., 2014; Morello et al., 2017). Cossu et al. (2015) tested aerobic–anaerobic hybrid bioreactors with continuous and intermittent aeration up until an optimum pH and VFA concentration for methanogenesis was achieved. Both aeration modes were beneficial in accelerating waste stabilization and acidogenic phase, however, intermittent aeration up until optimum pH values was more efficient in enhancing stabilization kinetics and methane generation.

Anaerobic–Aerobic Sequencing

An example of double stage AN-AE is provided by application of the aeration technology to the remediation of old landfills, (Ritzkowski and Stegmann, 2012, Chapter 16.2). The lengthy duration of the anaerobic process over the life span of the landfill is followed by forced aeration. The same technology may also be applied to modern anaerobic landfills to shorten the aftercare phase, once the biogas generation has dropped to a level no longer viable for energy exploitation. Lab scale tests have shown that residual recalcitrant organics left over from the anaerobic phase are degraded under aerobic conditions (Long et al., 2009).

Aerobic–Anaerobic–Aerobic Sequencing

Three step sequencing promotes the best removal efficiency for ammonia (Repetti et al., 2013; Cossu et al., 2015; Morello et al., 2017).

A scheme proposed by Repetti et al. (2013), known as S.An.A, includes a first semiaerobic step followed by an anaerobic phase and subsequently forced aeration. Column tests were performed and the results obtained were quite encouraging (Morello et al., 2017).

This scheme may require the preliminary treatment of waste to reduce putrescibles content.

Other schemes could include the following:

- Preliminary forced aeration to remove readily degradable organics and enhance biogas generation
- Anaerobic phase for massive production of biogas
- Aerobic phase (semiaerobic or forced aeration) shorten the aftercare duration

Compared Performance

The composition of gas generated by the various bioreactor alternatives is represented in Fig. 14.3.5. Under anaerobic conditions, during the methanogenic phase, composition ranges typically around 60% CH_4 and 40% CO_2 (by vol.), with consequent CH_4/CO_2 ratio values in the range of 1.5. Values from lab-scale tests are consistent with full-scale landfill reactor interstitial gas concentration during stable methanogenic phase (Raga and Cossu, 2014; Ritzkowski and Stegmann, 2007). Under semiaerobic conditions, the CH_4/CO_2 ratio is around 0.48 (Jeong et al., 2015) with methane ranging between 0 and 30%, CO_2 from 10% to 30%, oxygen 0%–10%, with the remainder largely constituted by N2.

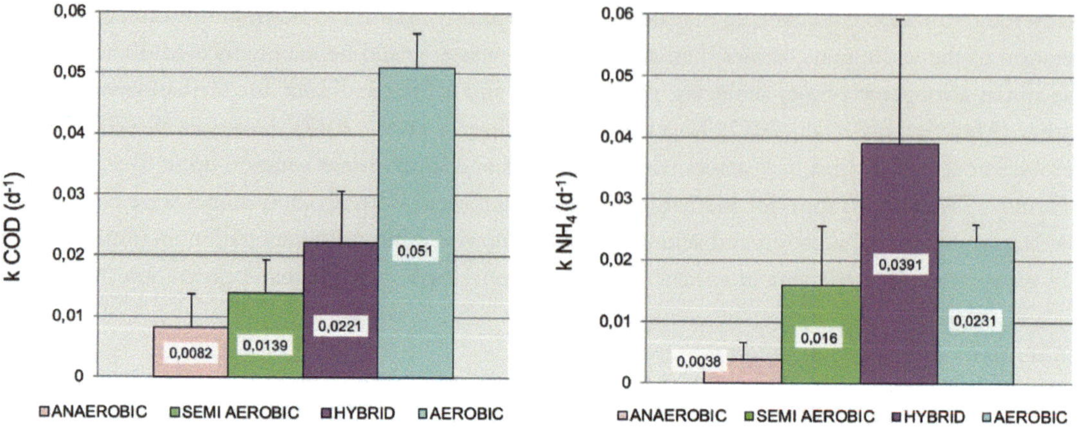

Figure 14.3.5 Mean values and associated standard deviations of ammonia and chemical oxygen demand (COD) removal kinetics associated with the individual landfill bioreactor typologies (Grossule et al., 2018).

Grossule et al. (2018) analyzed the results of several studies and experiences of enhanced degradation carried out using the various bioreactor landfill technologies. Assuming an overall first-order kinetics for the conversion (biodegradation, flushing, volatilization, etc.) of COD and $N-NH_4$ in leachate, they calculated the kinetics constant k and standard deviations for both parameters in relation to the different technologies (Fig. 14.3.5).

Despite the variety of operational conditions adopted in the studies (recirculation rate, waste composition, L/S ratio, air injection and experimental period, scale of the testing), all methods significantly reached similar conclusions, with values ranging around the mean value. The benefits yielded by aerobic conditions are evident in the maximization of COD removal with 0.051 day^{-1} average COD removal kinetic. The performance of hybrid and semi-aerobic bioreactors was situated midway between the anaerobic and aerobic reactors, displaying 0.0221 day^{-1} and 0.0139 day^{-1} average k values. On the contrary, for ammonia the best results were obtained with hybrid reactors ($k_{N\text{-}NH4} = 0.0391$ day^{-1}). The higher variability of values around the mean, however, renders these results particularly reliable as they were strongly influenced by the specific hybrid reactor application.

By calculating mean COD and ammonia removal kinetics, it is easy to predict and compare the required stabilization time for each bioreactor typology as representative average behavior. Considering the same starting point for COD and ammonia concentrations in reference to time (T) required under aerobic conditions to obtain 95% contaminant removal (Fig. 14.3.6), the time required to achieve the same rate of COD removal under hybrid, semiaerobic, and anaerobic conditions is increased by 2.3, 3.7 and 6.2-fold, respectively. In the case of ammonia removal, the time is reduced by 0.7-fold under hybrid conditions, whereas it increases by 1.7 and 3.7-fold under semiaerobic and anaerobic conditions, respectively.

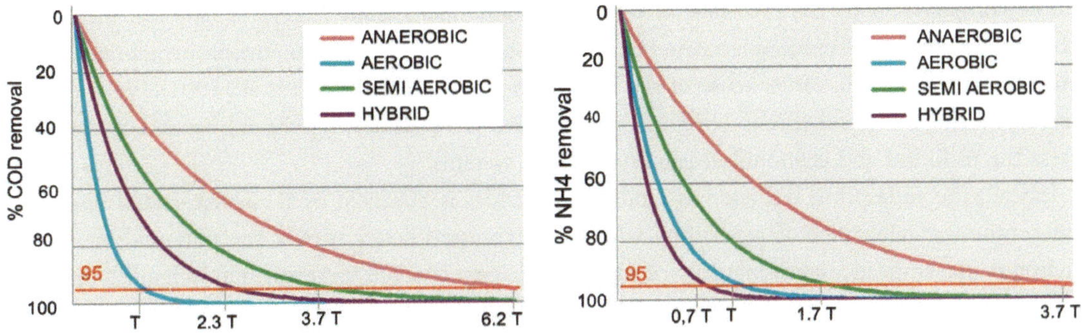

Figure 14.3.6 Variation of % chemical oxygen demand (COD) and % ammonia removal over time according to the mean removal kinetics in Fig. 14.3.5 versus time for the different bioreactor models. T is the assumed reference time corresponding to the time required to achieve 95% removal under aerobic (aerated) conditions. For both graphs, a common starting concentration was assumed for all models (Grossule et al., 2018).

To achieve a 95% removal of COD with an anaerobic landfill bioreactor, a period 6.2-fold higher than under aerobic conditions is required.

With regards to ammonia, hybrid reactors have displayed better performances than aerated bioreactors, achieving a 95% removal rate in a time below the reference time.

FINAL REMARKS

Landfill bioreactors, by implementing different in situ measures for the enhancement of biodegradation of putrescible waste, such as leachate recirculation and air immission, may elicit a significant shortening of the aftercare phase. Anaerobic bioreactors produce an improvement in methane generation rate but exert no relevant impact on ammonia accumulation, with degradation kinetics remaining slow. Aerated reactors increase ammonia and COD removal kinetics up to 10-fold those obtained under anaerobic conditions; they therefore represent an effective alternative to traditional anaerobic processes, although the need for forced ventilation systems, the complexity of operation and management, and high energy consumption associated with high operational and capital costs often fail to render aerated landfill a technically and economically feasible option. Semiaerobic landfill achieves a performance rate situated midway between that obtained by anaerobic and aerobic reactors, although with lower operational costs. Naturally, recourse to aerobic and semiaerobic landfills does not provide the possibility of energetic exploitation of landfill gas.

Hybrid bioreactors, operated under various combinations of aerobic and anaerobic conditions, may achieve energy recovery, fast waste stabilization, and significant removal of ammonia by means of a combination of nitrification–denitrification processes.

In view of the rigorous operational and construction requirements of bioreactor landfills, the required capital and operating costs are higher than those of traditional landfills. However, these increased costs will be compensated by the provision of future economic gains from bioreactor landfills, including a shorter aftercare period, reduced leachate treatment costs, reduced long-term environmental risks, longer active life of the landfill, earlier reuse of the land (Berge et al., 2009; Read et al., 2001).

Despite the successful results obtained at lab scale, full-scale testing should be undertaken in to assess the technical and economic feasibility of these systems.

Indeed, the widespread diffusion of landfill bioreactors is hindered by the above-stated high costs and continuing reluctance of regulators, who have expressed concern over the potential short-term environmental impacts, scarce maturity of the technologies, and the technical skill of landfill operators in implementing advanced technologies (Pattison and Yuen, 2007). Studies carried out to date have demonstrated that benefits achieved in lab-scale tests are much higher than those obtained in full-scale application, largely due to the reproduction of optimum homogeneous conditions in the laboratory (Ritzkowski and Stegmann, 2012; Kylefors et al., 2003). However, assessment of the general behavior of each type of bioreactor at lab scale will facilitate identification of the most appropriate bioreactor options to be used on a large scale, in line with the objectives established and in situ conditions.

References

Berge, N.D., Reinhart, D.R., Townsend, T.G., 2005. The fate of nitrogen in bioreactor landfills. Critical Reviews in Environmental Science and Technology 35, 365–399. https://doi.org/10.1080/10643380590945003.

Berge, N.D., Reinhart, D.R., Dietz, J., Townsend, T., 2006. In situ ammonia removal in bioreactor landfill leachate. Waste Management 26, 334–343. https://doi.org/10.1016/j.wasman.2005.11.003.

Berge, N.D., Reinhart, D.R., Batarseh, E.S., 2009. An assessment of bioreactor landfill costs and benefits. Waste Management 29, 1558–1567. https://doi.org/10.1016/j.wasman.2008.12.010.

Borglin, S.E., Hazen, T.C., Oldenburg, C.M., Zawislanski, P.T., 2004. Comparison of aerobic and anaerobic biotreatment of municipal solid waste. Journal of the Air & Waste Management Association. https://doi.org/10.1080/10473289.2004.10470951.

Cossu, R., Raga, R., Rossetti, D., 2003. The PAF model: an integrated approach for landfill sustainability. Waste Management 23, 37–44. https://doi.org/10.1016/S0956-053X(02)00147-2.

Cossu, R., Morello, L., Dongbei, Y., 2015. Hybrid bioreactor landfill: a review of the possible systems. In: Proceedings of Sardinia 2015. Fifteenth International Waste Management and Landfill Symposium, S. Margherita di Pula, Cagliari, Italy; 5–9 October 11.

Cossu, R., Lavagnolo, M.C., Morello, L., Pivato, A., Raga, R., Yue D., 2017. Role of semi-aerobic phase in hybrid landfilling: the S.AN.A method. In: Proceedings of Sardinia. Sixteenth International Waste Management and Landfill Symposium. S. Margherita di Pula, Cagliari, Italy; 2-6.

De Abreu, R., La Motta, E., McManis, K., 2005. Facultative landfill bioreactors (FLB): results of a pilot-scale study. Waste Containment Remediat 1–13.

Giannis, A., Makripodis, G., Simantiraki, F., Somara, M., Gidarakos, E., 2008. Monitoring operational and leachate characteristics of an aerobic simulated landfill bioreactor. Waste Management 28, 1346–1354. https://doi.org/10.1016/j.wasman.2007.06.024.

Grossule, V., Morello, L., Cossu, R., Lavagnolo, M.C., 2018. Bioreactor Landfills: comparison and kinetics of the different systems. Detritus, 03, 100–113. https://doi.org/10.31025/2611-4135/2018.13703.

Jeong, S., Nam, A., Yi, S.M., Kim, J.Y., 2015. Field assessment of semi-aerobic condition and the methane correction factor for the semi-aerobic landfills provided by IPCC guidelines. Waste Management 36, 197–203. https://doi.org/10.1016/j.wasman.2014.10.020.

Kim, H., Jang, Y.C., Townsend, T., 2011. The behavior and long-term fate of metals in simulated landfill bioreactors under aerobic and anaerobic conditions. Journal of Hazardous Materials 194, 369–377. https://doi.org/10.1016/j.jhazmat.2011.07.119.

Kylefors, K., Andreas, L., Lagerkvist, A., 2003. A comparison of small-scale, pilot scale and large -scale tests for predicting leaching behaviour of landfilled wastes. Waste Management 23, 45–59.
Long, Y., Long, Y.Y., Liu, H.C., Shen, D.S., 2009. Degradation of refuse in hybrid bioreactor landfill. Biomedical and Environmental Sciences: BES 22, 303–310. https://doi.org/10.1016/S0895-3988(09)60060-X.
Mali Sandip, T., Khare Kanchan, C., Biradar Ashok, H., 2012. Enhancement of methane production and bio-stabilisation of municipal solid waste in anaerobic bioreactor landfill. Bioresource Technology 110, 10–17. https://doi.org/10.1016/j.biortech.2011.12.027.
Morello, L., Raga, R., Lavagnolo, M.C., Pivato, A., Ali, M., Yue, D., Cossu, R., 2017. The S.An.A.® concept: semi-aerobic, Anaerobic, Aerated bioreactor landfill. Waste Management. https://doi.org/10.1016/j.wasman.2017.05.006.
Pattison, E.C., Yuen, S.T.S., 2007. Obstacles to progression of landfill bioreactor technology in Australia. In: Proceedings of International Solid Waste Association (ISWA) World Congress 2007, Amsterdam, 24–27 September 2007.
Price, G.A., Barlaz, M.A., Hater, G.R., 2003. Nitrogen management in bioreactor landfills. Waste Management 23, 675–688. https://doi.org/10.1016/S0956-053X(03)00104-1.
Raga, R., Cossu, R., 2014. Landfill aeration in the framework of a reclamation project in Northern Italy. Waste Management 34, 683–691. https://doi.org/10.1016/j.wasman.2013.12.011.
Read, A.D., Hudgins, M., Phillips, P., 2001. Aerobic landfill test cells and their implications for sustainable waste disposal. Geographical Journal 167, 235–247. https://doi.org/10.1111/1475-4959.00021.
Reinhart, D.R., Townsend, T.G., 1998. Landfill Bioreactor Design and Operation. Lewis Publishers, Boca Raton, FL, ISBN 1-56670-259-3, p. 189.
Repetti, R., Testolin, G., Cossu, R., Raga, R., 2013. Discarica S.AN.A?: Un sistema innovativo per la sostenibilità ambientale delle discariche. In: Proceedings Sardinia 2013. Fourteenth International Waste Management and Landfill Symposium, S. Margherita di Pula, Cagliari, Italy, 30 September–4 October 2013.
Ritzkowski, M., 2011. How does landfill aeration impact on leachate composition ?. In: Proceedings Sardinia 2011, Thirteenth International Waste Management and Landfill Symposium S. Margherita di Pula, Cagliari, Italy; 3–7 October 2011.
Ritzkowski, M., Stegmann, R., 2003. Emission behaviour of aerated landfills : results of laboratory scale investigations. In: Proceedings Sardinia 2003, Ninth International Waste Management and Landfill Symposium. S. Margherita di Pula, Cagliari, Italy; 6–10 October 2003.
Ritzkowski, M., Stegmann, R., 2005. Mechanisms affecting the leachate quality in the course of landfill in situ aeration. In: Proceedings Sardinia 2005, Tenth International Waste Management and Landfill Symposium. S. Margherita di Pula, Cagliari, Italy; 3–7 October 2005.
Ritzkowski, M., Stegmann, R., 2007. Controlling greenhouse gas emissions through landfill in situ aeration. International Journal of Greenhouse Gas Control 1, 281–288. https://doi.org/10.1016/S1750-5836(07)00029-1.
Ritzkowski, M., Stegmann, R., 2012. Landfill aeration worldwide: concepts, indications and findings. Waste Management 32, 1411–1419. https://doi.org/10.1016/j.wasman.2012.02.020.
Rivière, D., Vigneron, V., Baffaleuf, N., Le Lay, G., Guerbois, M., Rivero, J.C., 2011. Diagnostic of a French bioreactor landfill after 8.5 years of degradation. In: Proceedings Sardinia 2011, Thirteenth International Waste Management and Landfill S Ymposium. CISA Publisher, ISBN 978-88-6265-000-7.
Šan, I., Onay, T.T., 2001. Impact of various leachate recirculation regimes on municipal solid waste degradation. Journal of Hazardous Materials 87, 259–271. https://doi.org/10.1016/S0304-3894(01)00290-4.
Shao, L.M., He, P.J., Li, G.J., 2008. In situ nitrogen removal from leachate by bioreactor landfill with limited aeration. Waste Management 28, 1000–1007. https://doi.org/10.1016/j.wasman.2007.02.028.
Slezak, R., Krzystek, L., Ledakowicz, S., 2010. Simulation of aerobic landfill in laboratory scale lysimeters - effect of aeration rate. Chemical Papers 64, 223–229. https://doi.org/10.2478/s11696-009-0113-8.
Slezak, R., Krzystek, L., Ledakowicz, S., 2015. Degradation of municipal solid waste in simulated landfill bioreactors under aerobic conditions. Waste Management 43. https://doi.org/10.1016/j.wasman.2015.06.017.
US6398958 B1, 2002. Hater, G. R., Green, R. B., Harris, J. M. Patent Facultative Landfill Bioreactor. Waste Management Inc.
Xu, Q., Jin, X., Ma, Z., Tao, H., Ko, J.H., 2014. Methane production in simulated hybrid bioreactor landfill. Bioresource Technology 168, 92–96. https://doi.org/10.1016/j.biortech.2014.03.036.
Zhong, Q., Li, D., Tao, Y., Wang, X., He, X., Zhang, J., Zhang, J., Guo, W., Wang, L., 2009. Nitrogen removal from landfill leachate via ex situ nitrification and sequential in situ denitrification. Waste Management 29, 1347–1353. https://doi.org/10.1016/j.wasman.2008.10.014.

15 CHAPTER

LANDFILL OPERATION

15.1

LANDFILL OPERATION

Nicole D. Berge, Eyad S. Batarseh, Debra R. Reinhart, Rainer Stegmann,
Stephanie C. Bolyard and Wenjing Lu

INTRODUCTION

A successful landfill requires sound operation and monitoring in addition to good siting and design. The technical operation of a landfill requires the integration of several necessary components, such as equipment, waste filling sequences and placement methods, waste compaction, and daily cover placement. Additionally, aspects such as temporary road placement, safety and security, waste input control, and storm water management must be addressed and/or implemented. The technical operation of efficiently placing the waste in the landfill is accompanied by environmental procedures controlling noise, odors, fire, litter, dust, and vectors at the site.

Landfills have to be monitored for their performance with respect to leachate, air, groundwater, liner, and settlement. The results have also to be supplied to the authorities and should be accessible to the public.

The organization of a landfill has to cover the following topics:

- definition of waste admitted to disposal,
- entrance control on landfills for municipal solid waste,
- handling of special types of waste,
- biological pretreatment of municipal solid waste (where relevant),
- techniques of waste disposal,
- general improvement of landfill conditions, and
- documentation and recording of landfill activities.

All site-specific regulations have to be compiled with in a “Basic Operation Manual”.

WASTE ADMITTED TO DISPOSAL

For each landfill a list of wastes has to be defined giving information about the types of waste admitted to disposal. This list has to enclose the following:

- the key number of the waste according to a unique waste classification system,
- a short description, and
- characteristic parameters.

In general on landfills for municipal solid waste household waste, commercial waste (similar to household waste), and the nonhazardous waste, which is accepted by the local administration and the landfill operator, can be disposed of.

The nonmunicipal waste, which is admitted to disposal, has to be defined and listed. In addition, a list of industrial and hazardous wastes should exist, which names the waste not admitted to disposal on the municipal landfill site. The types of waste listed there should be in accordance with national or international waste classification waste systems.

Entrance control

The type of waste delivered has to correspond with the list of waste admitted to disposal. This list must be available on the landfill site and must be part of the operation manual. Delivered waste that is doubtful due to the yardmen should be unloaded on a paved small place near the entrance area for further check (sampling etc.).

The waste should be checked at the entrance as well as at the dumping place (to detect hidden components). The landfill workers are to carry mirrors with long sticks and ladders (to check the open vans and containers) and walkie-talkies for the communication between the entrance and the dumping place.

At the entrance the waste is to register and to record. According to the EC-Directive on the Landfill of Waste the following reception procedures are to be respected by the operator:

- checking of the transport papers (company, vehicle, s) in case of municipal waste and in addition checking of the waste documentation in case of nonmunicipal waste,
- determination of weight or volume of the waste,
- check of the waste by its properties (odor, color, consistency, components, in many cases also for radioactivity etc.) at the entrance and at the point of deposit to verify the conformity with the description provided in the documentation of the waste generator,
- keeping a register of the quantities and characteristics of waste deposited, and
- exhibit a bill.

Handling of special types of waste

The following recommendations are given for the handling of special types of waste on smaller landfills for municipal solid waste:

Recyclable materials

Recyclable materials should be rejected from disposal, if there are any possibilities for their recycling in the respective region. A separate area for temporary storage of recyclable materials and containers should be provided at the entrance area. The term "recyclable waste" includes separately delivered paper/plastic as well as green waste/biowaste, but also tires etc.

Bulky waste

The disposal of specific types of bulky wastes (such as refrigerators or iron materials) should be organized in a way that dumping on landfills is not necessary.

Debris or construction waste

The disposal of debris or construction waste on municipal landfills should be reduced to a minimum. Debris should only be accepted for maintenance of operation roads or for covering of dumping areas. If debris cannot be used as recycling material, special landfills for inert materials are to be established.

Sludge

The disposal of sewage sludge may cause problems. Often the contamination of the sludge has as a consequence that the acceptance criteria are not fulfilled. In addition, layers of spread sludge can cause problems for the stability of the landfill (risk of sliding etc.). A practical rule is that sludge should not exceed 10% of the total weight of waste disposed of.

Industrial waste and other nonhazardous waste of nonmunicipal origin

Industrial waste and nonhazardous waste of nonmunicipal origin (mining waste, slag etc.) have to be rejected from disposal on municipal landfills. There is just one exception for the generator who has the allowance from the local administration and from the operator, which have to consider waste properties and landfill capacities.

Hazardous waste

According to the definitions of the EC-Directive on the Landfill of Waste, the disposal of hazardous wastes is not to be allowed on municipal landfills (landfill category 2).

TECHNICAL OPERATION

Technical operation of a landfill typically involves moving equipment, waste filling sequences and placement methods, waste compaction, daily cover placement, temporary roads, safety and security, waste input control, and storm water management.

Equipment

Machinery to move, place, spread, and compact the waste and daily and final cover soils is necessary. Service or support equipment, such as excavators, service and water trucks, and grinders, is also necessary to maintain a landfill.

Fig. 15.1.1 presents pictures of some landfill equipment, including a crawler tractor, compactor, excavator, and wheel loader. Of these, the crawler tractor is excellent at spreading both waste and cover soil. Additionally, it is able to compact both materials to a certain extent, achieving waste densities between 475 and 725 kg/m^3 (Tchobanoglous et al., 1993). Densities as well as shredding effects are—compared to compactors—comparably low. Another attribute of the crawler tractor but also for wheel loader is its ability to excavate, which may be necessary in instances in which a cover soil source is a borrow pit on site or must be excavated at a nearby off-site location. Crawler tractors as well as wheel loaders are also able to prepare the landfill site and build temporary roads.

Figure 15.1.1 Different landfill operation equipment.

Compactors designed specifically for landfill operations are almost universally used and preferred (Worrell and Vesilind, 2012). They are heavy and equipped with knobbed steel wheels, which aid in effectively shredding, spreading, and compacting large amounts of waste. Densities between 725 and 950 kg/m^3 are often achieved (Worrell and Vesilind, 2012). Compactors, however, are poor at excavating and hauling materials. Thus, at a site in which cover soil must be excavated from on-site borrow pits or hauled long distances, additional equipment is necessary.

Pans and scrapers have been used to excavate and place daily cover soils at landfills. These machines are specially designed to scrape up cohesive dirt that can be removed in layers, haul it to the landfill, and then run down the compacted slope, effectively placing dirt evenly over the waste. For longer distance transport hydraulic excavators and trucks have been shown to move dirt faster and at lower cost than scrapers (Taylor, 2000). The articulated (or artic) dump truck has become the preferred hauling vehicle at many US landfills (Bader, 1998).

Since in Germany and other countries in Europe soil is in general not excavated to produce landfill volume and daily cover material (different from, e.g., the United States), machinery used for these purposes are not necessary. Different from operation in many countries, in Germany no or only little amounts of daily cover material is used. Also due to this reason mainly the compactor as the universal equipment is used. Crawler tractors may be on site for additional tasks (e.g., construction of dams and roads on site).

The exact type, number, and size of equipment needed are highly dependent on individual landfill requirements. Fig. 15.1.1 depicts several of the more common equipment types, e.g., in the United States. Just to give a rough estimate, a medium-sized compactor may over an 8-h day have an annual capacity of 20,000 t of waste.

Filling sequences and waste placement

After the site has been prepared (i.e., placement of liner), the landfill is ready to receive waste. There are many different waste placement methods. The three most common methods are the excavated cell method, the area method, and the canyon/depression method.

- When filling via the excavated cell method, cells are excavated, lined with synthetic and/or low permeability clay liners to prohibit subsurface movement of leachate and gases, and then filled with waste. This method of landfilling is best suited in areas in which the water table is not near the surface and where the excavated soil may function as cover material (standard method in the United States and other countries).
- The area method differs from the excavated method in that there is no excavation; the waste is placed above grade. This method is preferred, e.g., in the United States when the terrain is unsuitable for the excavation method or when the groundwater table is near the surface. A liner and leachate collection system is constructed above grade. Cover material can either be hauled to the landfill from an off-site source or excavated on site (Tchobanoglous et al., 1993). The area method is, e.g., mandatory in Europe to ensure natural drainage of leachate to ditches also after closure of the landfill.
- The canyon/depression method involves the placement of waste into an existing canyon, ravine, borrow pit, or quarry. This technique allows the geometry of the depression to dictate filling and compaction procedures (Tchobanoglous et al., 1993). This method needs ambitious surface and groundwater control and at least leachate has to be pumped out of the landfill for unlimited periods of time. Due to the fact that a liner may not be completely tight and may not last forever, leachate pumping may also be at a later point necessary when the landfill is covered with an impermeable liner on top. This method should be rethought since it may cause severe long term problems.

Waste is generally placed in lifts, or thin layers of waste. The working face must be large enough to allow for unloading of several vehicles simultaneously. Typically, 4–6 m is required per vehicle (Worrell and Veselind, 2012). Ramke (2001) mentions a working phase area for a medium size landfill of about 2500 m^2; large landfills may have more than one working phase.

The first layer placed consists of preselected waste, with sharp and heavy objects removed, and is placed directly above the leachate collection system (Fig. 15.1.2). It must be placed and compacted in a manner in which all equipment is kept from damaging the leachate collection system. This layer functions to protect the leachate collection system and is often called the operational layer

Figure 15.1.2 Installation of a composite bottom liner system (German landfill); left side: placement of a geomembrane directly on top of a clay liner; right side: gravel layer placed on top of a geomembrane (with a geotextile in between).

(Worrell and Veselind, 2012). In some experiences in Germany composted preselected waste has been placed as a first lift (approx. 1.5–2.0 m) in order to protect mechanically the drain system and acts as a biological filter to reduce organic degradable leachate compounds (Stegmann and Spendlin, 1989). Subsequent filling continues in thin highly compacted lifts.

If a horizontal gas collection system shall be installed after placement of several lifts, construction may begin as part of an active gas extraction plan during landfilling (see also Chapter 9.3).

Compaction

As waste is placed in the landfill, it is compacted to allow for placement of a maximum quantity of waste. But even as important as the use of less volume are the following effects: avoidance/reduction of vermin (no holes/caves, low waste surface), fires (low air intrusion), odors (low waste surface), littering (highly compacted waste surface), etc. In addition less/no cover soil may be used.

In addition to compaction, other factors such as cell geometry and depth impact efficiency of waste placement have to be considered. Compaction increases waste density and can be thought of in terms of a compaction ratio (CR). The CR is defined as (Tchobanoglous et al., 1993):

$$CR = V_i / V_f \qquad (1)$$

where V_i, volume of waste prior to compaction; V_f, volume of waste after compaction.

The CR is a function of several factors, including slope, moisture content, height of waste layer, waste composition, number of passes a compactor makes over the waste and compaction vehicle. Slope of the landfill affects the CR; flatter slopes generally result in higher CRs than steeper slopes because they are easier to compact by the landfill equipment. Up to certain percentage higher moisture contents result in higher CRs than drier waste. Additionally, as lift thickness increases, the degree of compaction attainable decreases. As one would expect the more passes a compacting device makes over the waste and the heavier the compactor, the higher the CR and thus waste density. This is also due to an increasing shredding effect.

Fig. 15.1.3 presents some techniques of waste disposal; in any way the waste should be—after shredding—compacted in thin layers at a slight slope, tipping over the edge is not acceptable and represents dumping.

Typical specific weights of landfilled waste (no cover soil) should at least have a density of 800 kg/m^3. Compacted mineral waste as waste incineration bottom ash may have densities of more than 1000 kg/m^3. Specific weight of the waste placed in the landfill is often determined using landfill tipping records and surveys. In many cases landfills contain 15%–20% by volume of cover soil. The density of the cover soil is approximately 2600 kg/m^3. Thus, when calculating the in-place specific weight of an entire landfill, both the soil and waste-specific weights should be accounted for. A typical waste disposal area is presented in Fig. 15.1.4.

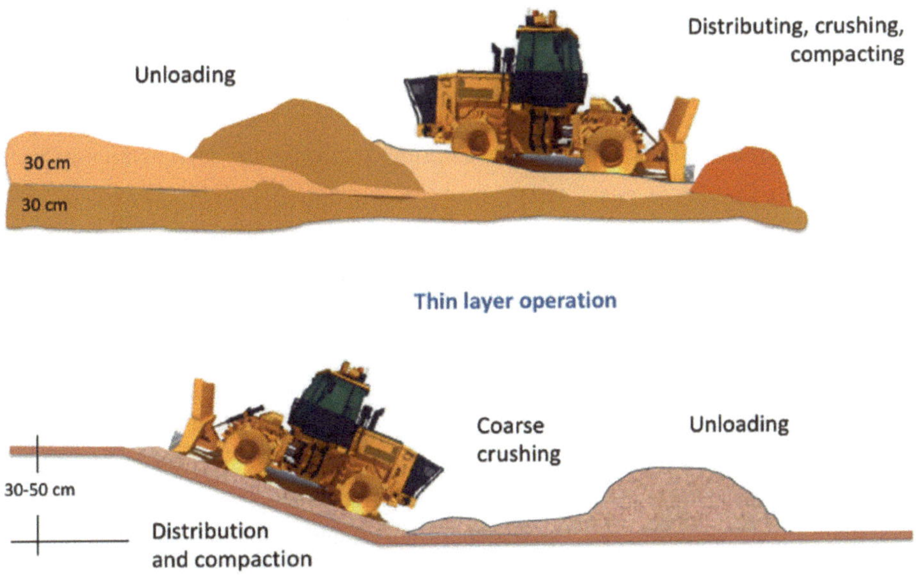

Figure 15.1.3 Thin layer waste emplacement and compaction (Ramke, 2001).

Figure 15.1.4 Typical waste working phase at a landfill site.

In Europe, for example, the waste is generally spread in 30–50 (70) cm lifts and shredded by several passes with the compactor. The waste should be emplaced in about 50 cm lifts and compacted by 6–10 passes of the compactor. By these means, densities of up to 1.2 t wwt/m^3 may be reached (Ramke, 2001)

Daily cover

After the last load of waste is placed in the landfill for the day, it is in many places required that a layer of soil (10–25 cm) or alternative cover material be placed on the waste. This cover functions:

- to control the amount of water infiltrating into the waste and thus reduces the early generation of leachate,
- to minimize the amount of litter, dust, and air pollution,
- to prevent air intrusion, and
- to control the burrowing and emerging of small animals and insects, such as rodents and flies.

On the other hand 15%–20% of the landfill is filled with soil or other inert material.

The utilization of daily cover is prescribed in many countries. In those countries where it is not prescribed using high amounts of daily cover may be rethought. Since the aforementioned benefits of daily cover may be to a high degree also achieved with a thorough high compaction by means of a heavy weight compactor. In this case, only low amounts of or even no daily cover soil may be applied. This kind of operation is practiced, e.g., in Germany. If landfill sections are not in use for a longer period of time intermediate cover will be put in place.

Soil is generally used as daily cover material. But not all soil qualities are appropriate to act as a daily cover as, e.g., cohesive soils with a significant content of clay etc. Using these kinds of soil may result in severe operational problems during rain. When soil is unavailable or not economical for use as daily cover material, alternative cover materials may be used (Tchobanoglous et al., 1993). Compost and mulch have been used quite successfully as daily cover material. Additionally, tarps, construction and

Figure 15.1.5 Use of geosynthetic alternative daily cover.

demolition wastes, and agricultural residues have been used as alternative cover materials. Removable geosynthetic tarpaulins or spray-on foam covers (seldom used) do not consume valuable airspace that becomes available for additional waste placement. To ensure adequate vertical moisture movement, selection of permeable or removable daily cover is essential (Fig. 15.1.5).

Temporary roads and fences

In general, temporary access roads are necessary to allow collection vehicles and other landfill equipment to continue filling of the landfill as the working face changes location. Landfill employees using equipment on-site usually complete road construction. As the working face is moved, road location is changed accordingly. Often materials received at the landfill can be used to construct the road, such as concrete rubble. Access roads that are used for long periods may also be paved or build of prefabricated (removable) concrete plates.

SAFETY AND SECURITY

The health and safety of both the landfill employees and the general public is a critical factor to consider when operating a landfill. Potential hazards include contact with hazardous materials (pathogens, toxic air pollutants, asbestos), high accident risk, allergies due to environmental hazards (gas, dust, litter), and excessive noise and side effects caused by excessive vibration due to landfill equipment operation (Otineo and Magagula, 1999; Wilhelm, 1989). Necessary protective equipment should be supplied to employees. Additionally, access to landfills is generally restricted. A health and safety plan must be developed and prominently displayed (see Chapter 15.2).

A fence to avoid prohibited people entering the landfill should surround each landfill. This is also necessary in regard to safety and—in case of accidents—associated liability.

Storm water management

Because infiltration of all rainfall, storm water runoff, intermittent streams, and artesian springs add to the total amount of leachate generated and thus the amount that must be managed, an effective storm water management plan is necessary for landfills. In addition, improper storm water control can result in erosion of cover material. The most effective control measure is the use of a properly designed cover (or daily cover) with an appropriate slope, generally 3%–5% (Taylor, 2000). Runoff can then be effectively diverted off of the top of the landfill via concrete lined trapezoidal ditches. If operating the landfill as a bioreactor, it can be advantageous to design slopes to capture and retain storm water, allowing for it to infiltrate into the waste and act as a moisture source. Of course necessity of these measures is very much dependent on the total amount and the pattern of precipitation over the year. The run-off should be collected in ditches, that are constructed around the landfill, the water quality has to be checked and in case of noncompliance with target values be treated.

Noise control

Noise may be excessive at landfill sites because of routine movement of collection trucks and the loud engines of the landfill equipment, causing a nuisance to both landfill employees and any residences located near the site. Thus, a noise control program should be considered dependending on the specific situation (location, amount and kinds of vehicles etc.). Landfill employees should be equipped with hearing protection devices. To prevent noise from reaching adjacent residences, several control methods may be used including tree planting to create a buffer zone to mitigate noise (which should exist also for other reasons like litter and dust control, reduction of the visibility of the landfill), maintaining equipment properly, regulation of hours, and ensuring the working face (the loudest area of the landfill) is located at a suitable distance from nearby residences (Farquhar et al., 1995). Research by Nalamothu et al. (2008) shows that setbacks of 1.9 km are necessary to minimize noise disturbances.

Odor control

Odors exist due to putrescible wastes, disposed sludge, landfill gases, and leachate seeps but also when organic rich waste is—especially in hot climate—unloaded by the collection truck. Good overall landfill operation, high compaction, the placement of daily cover, and even the immediate covering of materials that have an offensive odor, aid in controlling odor (Farquhar et al., 1995). Employing an effective gas control system will also aid in mitigating odor. Collected gas can be deodorized via thermal oxidation, wet gas scrubbing, activated carbon filtration, and biofiltration (Frechen, 1989; Feinbaum, 2000). In some cases more frequent waste collection may also help to reduce odors in the incoming waste. In extreme cases chemicals can be employed to attenuate odor including ozone and mixtures containing plant oils and surfactants (O'Maley, 2003). Odor-neutralizing chemicals are often dispensed in a perimeter misting system. These measures are taken only in a few cases. Bad odors at landfills in countries located in moderate climate zones are in most cases a result of nonoptimal operation.

Temperature and fire control

In situ landfill conditions are generally anaerobic. But although—compared to aerobic processes—there is only a low energy production from anaerobic processes, as biological activity increases, internal temperatures also rise. Waste has high insulating properties, thus heat generally accumulates. In situ temperatures as high as 65°C have been observed (see Chapter 3.3). This has to be encountered when choosing the appropriate material for construction inside the landfill as, e.g., pipes, liners, and monitoring equipment.

When methane concentrations of 5%–15% are present in air, it is explosive and thus can initiate internal fires (see Chapter 9.4) (Tchobanoglous et al., 1993). Generally, this is a special issue on low compacted landfill where methane emits through the surface of the landfill and air can reach into the landfill by diffusion but mainly by wind (especially at slopes) and atmospheric pressure changes. Potential ignition sources may be pieces of glass (magnifying effect), open fires on the landfill, chemical processes in the landfill, etc. This situation can be especially observed at many dumps and low compacted landfills in developing countries. Therefore high compaction is also due to this reason imperative. However, in active landfills in which the gas extraction system is not properly maintained, oxygen intrusion may also occur, resulting in explosive mixtures. A special situation exists when landfills are aerated. But inside the landfill there exist not such ignition potentials, except for potential chemical processes, which are not well defined. Experiences show that there may be a low ignition potential when the temperatures in the landfill body rise to more than 60–70°C. Therefore efforts should be made to keep the temperatures down to max, 50°C. Chapters 3.3, 9.3, 9.4, and 16.2 address this issue in more detail.

Litter control

Litter must also be controlled and is the main complaint of nearby residents. Litter results from delivery of uncovered loads, wind, and operational practices. The most effective method may be unloading waste on a minimal surface area and immediate high compaction; in addition operating the working face so as to minimize wind disturbance (creating it in the opposite direction of the wind). In addition common control method is the placement of portable screens near the working face. Other mitigation methods may include more frequent covering of waste on windy days, requiring all delivered loads to be covered, manual collecting litter as necessary. Whenever permanent or mobile fences are used, frequent cleaning of the fences is needed to minimize their vulnerability to being damaged from strong winds (Fig. 15.1.6). To avoid long distance transport of litter outside, the landfill buffer zones around the landfills planted with trees are helpful. It has to be kept in mind that in general the landfill operator is responsible for litter outside the landfill and has to collect it.

Dust control

Unloading of waste, earthmoving and compaction activities, certain cover soil qualities, wind, and traffic on roadways may be reasons for dust problems. Excessive dust can cause health problems

Figure 15.1.6 Mobile net for litter control.

Figure 15.1.7 Water tanker used to control dust at a landfill.

for landfill employees (allergic reactions, lung problems, etc.), increases in equipment maintenance costs and frequency and nuisances for nearby residents. The most common control measure is to spray access roads with water (see Fig. 15.1.7). Other methods to control dust include using dust-free roads (i.e., asphalt), enforcing speed limits via speed bumps, using vegetation to reduce wind speed at ground level and selecting an appropriate time to move soils to minimize the amount of dust (Farquhar et al., 1995).

Small animal and insect control

Small animals and insects such as birds, rodents, and flies can cause problems at landfills and thus must be controlled. Seagulls and other kinds of birds frequent landfills and may cause problems by picking up components of the waste and dropping them in nearby fields or surface waters (Farquhar et al., 1995). In addition they may be a danger when the landfills are close to an airport (their distance rules for landfill siting near airports). Birds are particularity difficult to control. Methods that have been used include loud noises or recordings of birds in distress, netting systems, shooting, dogs, birds of prey (e.g., hawks) firing devices and chemical poisons. A landfill site in the United Kingdom not too far away from an airport has been experimenting with a netting system to totally enclose the filled landfill to protect it from birds (Terry and Willett, 1999). This technique has been shown to be quite effective; however, general upkeep of the netting systems can become costly.

Rodents are also problematic. They are sometimes brought to the landfill in waste loads and will remain if a food source and shelter are present. For this reason the most effective way to control rodents is high compaction and operation as discussed earlier. Larger animals such as pigs, donkeys, deer, etc., which are common on dumps can also be dangerous and may disrupt the cover if there is any. Site-specific control measures should be implemented if disruption by larger animals is expected. In general, they should be kept away from landfills.

Insects, such as flies and mosquitoes, are also a problem. They spread disease and are usually brought in with the waste load. High compaction and covering the waste will help eliminate the food source. Extermination by using pesticides has also been used, though measures should be taken to ensure the health of the employees is not compromised (Farquhar et al., 1995).

LANDFILL MONITORING

It is important to record as exact as possible the kind and amount of waste that is landfilled daily and in which landfill section waste it is disposed. This is important for liability reasons (producer product reliability, potential remediation, or even mining). Monitoring during both pre-and post-closure periods is necessary. Regulations generally require monitoring of leachate (quantity, quality, leakage), local air quality, groundwater quality both upstream and downstream of the landfill, and landfill performance parameters, such as head on the liner and cap integrity. The EU Landfill Directive in addition requires that meteorological data (precipitation, air temperature, and wind) are monitored (or obtained from a nearby weather station) to provide data for water balance calculations.

In addition, as landfill design transitions to bioreactor technology, monitoring of in situ parameters such as temperature and moisture content is necessary for adequate system operation and process control. Bioreactor landfills are operated to accelerate stabilization of the waste by controlling the chemical and biological processes occurring within the landfill to have the ability to respond to changes in these conditions through adjustment of operating parameters.

Leachate and groundwater monitoring

To minimize adverse environmental impacts from leachate, it is important to closely monitor leachate quality, quantity, leakage through the liner, and outbreaks on side slopes. Landfill operators are generally responsible for ensuring leachate is managed effectively and the leachate collection and removal system is maintained throughout the design period of the landfill.

Leachate quality analyses are necessary to ensure leachate water quality standards are not violated. All analyses must be conducted on leachate sampled from a location at which the leachate is representative of the in situ landfill conditions, such as in the leachate collection and removal system, and not changed or altered as a result of sampling, storage, and transportation (see also Chapters 10.2 and 10.3). Leachate leakage is also critical to monitor. In landfills with a liner system consisting of two separate liners with a drain in between the leak detection system generally consists of a highly permeable material. Leachate is collected from this system and in cases of monitoring leachate it can be concluded that the primary liner system has failed (Koerner, 2003). In most cases composite liners are in place where two liners are directly attached to each other. By these means in case of a local leakage in the upper liner (mainly a geomembrane) the leachate gets into contact with the second liner (mainly mineral liner) only locally, in the other case the leachate may be distributed over the entire surface of the second liner.

Additionally, leachate leakage monitoring can be conducted in the zone between the landfill liner and the groundwater table by direct monitoring (placing lysimeters under the liner) or indirect monitoring (instruments placed near or under the liner system) that detect changes in moisture content or chemical concentration (salinity in most cases) (Bagchi and Bagchi, 1994). These systems are rarely used. The question is always what to do if failures of liners placed underneath landfills have been identified. Repairing those under this condition is a great problem.

Monitoring of leachate seeps is also important. Leachate outbreaks may occur on side slopes because of local waste heterogeneities containing impermeable objects, impermeable cover soil placement, or leachate built up. Conducting routine visual inspections of side slopes is the most common detection method. If seeps are found, measures should be taken to install drains, slope cover material towards the inside of the landfill, reduce leachate injection volumes and leachate levels in the landfill, or prevent infiltration of storm water.

Monitoring groundwater around landfill sites is necessary to aid in detecting leachate leakage of contaminants (see Chapter 18.2).

Air and gaseous emission monitoring

Monitoring of both air/gas quality and migration is important at and around landfill sites to protect the environment and to ensure public health and safety is not compromised. Escape and migration of gaseous contaminants from landfills can cause several negative environmental impacts (see Chapters 18.1 and 15.2). Air and gas monitoring at landfills generally include monitoring of ambient air quality, extracted gases from the landfill, gases present in the vadose zone and any off gases from a treatment or energy recovery facility.

Ambient air quality is monitored to detect concentrations and migration of harmful gases, such as methane and hydrogen sulfide. Locations around the landfill can be equipped with explosive gas meters, hydrogen sulfide meters, and sample collection equipment for samples to be taken at different locations around the site. These measures are in most cases executed when there are specific problems at a site. The ambient air quality in buildings located on-site must be measured routinely. Global positioning systems (GPS) can be used to aid in developing contour maps depicting plume concentrations of gases measured around the landfill site, allowing determination of gas migration patterns to be more accurate (Curro, 2000). Gas samples can be collected in either an evacuated canister, syringe, or air collection bag and stored for subsequent analysis by gas chromatography 1 or can be analyzed using a portable infrared gas meter.

Settlement monitoring

Settlement is an important parameter to monitor also because of its impact on landfill cap integrity. The cap functions to ensure long-term integrity and to support post closure uses of the landfill site (see Chapter 16.3).

Settlement is a result of waste decomposition and overburden pressures exerted on the landfill due to added moisture, waste, or cover soil. Settlement rarely occurs uniformly in the landfill, thus cracks and breaks in the cap may occur if there is excessive differential settlement. The extent of settlement is dependent on compaction, characteristics of the waste, amount of decomposition, amount of water in the landfill and height of the landfill. Generally, the majority of settlement occurs during the intensive bio-degradation phase of a landfill but stays over periods of several decades. Settlement can be measured using a variety of techniques. The most common include settlement plates, GPS surveying and flyover aerial photography. Routine surveys of the landfill surface using the GPS technology is becoming more common and can be used to accurately determine the amount of settlement. Depending on the number of survey points used, both differential and average settlements can be measured with great accuracy. Additionally, as the waste settles an air space may be recovered and subsequently used to add additional waste to the landfill. By using in situ aeration in the aftercare phase the time period when settling occurs may be reduced (see Chapter 19.2).

Temperature monitoring

Temperature monitoring of in situ landfill conditions in conventional landfills is conducted primarily to aid in preventing or detecting excessive heat generation. Knowledge of in situ temperature profiles can provide landfill operators an indication of what areas of the landfill may be reaching combustible temperatures so prevention measures can be implemented. In addition, temperature monitoring can detect locations of potential and actual subsurface fires. Because waste degradation involves biochemical reactions producing significant amounts of heat, temperature can be used as an indirect measure of waste degradation; higher temperatures indicating an area containing more rapidly degrading waste. The temperature within a landfill may be determined by a balance between heat production during the biological degradation of organic waste fractions and the loss of heat to the surrounding soil and atmosphere (Rees, 1980). The microbial processes are capable of significant heat generation, particularly—up to a certain limit—at higher moisture conditions.

Temperature is a relatively simple and inexpensive parameter to monitor. Common temperature measuring devices used in landfills include thermocouples and thermistors. Thermocouples are pairs of dissimilar metal wires joined at one end. A temperature difference between the paired wires generates a net thermoelectric voltage. Thermistors measure temperature as a function of the change in electrical resistance of materials as a result of temperature changes. Thermocouples tend to be more rugged and less expensive than thermistors, however they are slightly less accurate. The cost of measuring the temperatures are associated with the emplacement of the devices in the landfill, except they are installed in existing wells, boreholes, etc.

Moisture management

Moisture content and water flux control is the most critical parameter for successful bioreactor landfill operation (see Chapter 2.4). Moisture control is usually achieved by the recirculation of (in many cases treated) leachate, although additional liquids, such as groundwater, wastewater, biosolids, and stormwater, are sometimes required to achieve optimal moisture content. Liquid is added to a landfill using vertical injection wells, permeable blankets, spraying as waste is placed, and horizontal trenches. In many cases these devices can be constructed during waste placement or underneath the final cover, allowing for liquid injection to occur while filling and for a more uniform liquid distribution system after final landfill capacity is attained. Subsequent lifts and gas extraction systems are continually added until final design grade has been met, after which a final cover is constructed above the waste. Vertical gas wells may be installed at that time for long-term gas extraction.

Water addition during operation will not be necessary in areas with all year long high precipitation rates. Beside the water content water movement through the waste is necessary to transport degradation products out of the landfill.

Calculation of a water balance by monitoring moisture addition volume and leachate generation rates provides a means of determining the average moisture content of the landfilled waste. Spatial distribution of in situ moisture can be determined by removal of solid waste samples and using traditional gravimetric techniques or by placing sensors inside the landfill to detect the spatial distribution of water. Quite often, using a drill rig and auger to remove a solid waste sample is costly and difficult to accomplish. Waste heterogeneities make drilling in some locations difficult. Additionally, as the waste sample is removed, water loss may occur and result in inaccurate moisture content measurement. In bioreactor landfills, the weight of added moisture also serves to increase the in-place specific weight.

Placing water content sensors in situ is advantageous, as destructive sampling is not required and measurement can be continuous. In situ moisture content devices available for use in landfills were developed primarily for use in the agricultural industry for irrigation purposes. Landfills, by their nature, pose severe challenges to the application of these devices. The most commonly used devices in landfills include TDR (Davis and Annan, 1977; Topp et al., 1980), neutron probes (Evett, 1998), electrical resistance (McCann et al., 1992), capacitance probes (Bell et al., 1987), and time domain transmissometry. A more complete description and discussion can be found elsewhere (Thomas, 2001).

Monitoring sustainable landfill performance

To ensure that the objectives of a sustainable landfill are met, evaluation of system performance is necessary. Townsend et al. (2015) recommended the following parameters for control of sustainable landfills: liquid addition flow rate, cumulative liquids added, liquid injection pressure, leachate composition, air (if added) and gas flow rate, gas pressure, gas composition, and moisture content.

LANDFILL OPERATION IN ASIA: EXAMPLE CHINA

Most information and experiences about landfill operation that are presented are derived from experiences and practice in USA and Europe. But landfilling in tropical countries, where high rainfall rate might be expected, or in countries where high moisture content waste—often rich in organics—has to be landfilled is very challenging and the operation has to be adapted to that situation. As an example, a short report about landfill operation in China is presented.

MSW COMPOSITION

In China, MSW is generally divided into organics, paper, glass, metal, plastic, textile, bamboo, and dust. Table 15.1.1 displays the respective composition of municipal solid waste (MSW) in some big cities in China. In general, waste composition in China is dominated by a high organic and moisture

Table 15.1.1 Waste composition at different Chinese cities (% wet weight)

City (Year)	Organic Waste	Paper	Glass	Metal	Plastic	Textile	Bamboo	Dust
Beijing (2010)	63.40	11.10	1.60	0.30	12.70	2.50	1.80	6.60
Shanghai (2009)	66.70	4.46	2.72	0.27	19.98	1.80	1.21	2.77
Hangzhou (2009)	57.0	15.0	8.0	3.0	3.0	2.0	2.0	4.0
Yangzhou (2009)	57.81	10.49	3.05	0.45	16.08	3.78	2.10	6.84
Xizang (2009)	72.0	6.0	–	1.0	12.0	7.0	–	–
Shenzhen (2008)	51.10	8.40	3.00	1.10	14.70	6.90	5.90	8.90
Nanjing (2008)	70.59	8.32	1.45	0.08	14.18	3.05	1.04	1.29

content, since the amount of the kitchen waste in urban solid waste makes up the highest proportion of approximate 60%. It is attributed to the different diet custom compared with western countries, say, more fresh vegetables and fruit as well as more cooking. It could be predicted that kitchen waste will continue to occupy a high ratio of the waste in China in the future. The high moisture content of MSW and poor management of kitchen waste are usually related to increased leachate production/percolation, with the potential of causing in landfill areas serious sanitary and environmental problems such as unpleasant odors, high leachate concentrations, leachate build up, reduced physical landfill stability etc., as well as an increased potential of groundwater contamination. In contrast, the quantity of paper, plastics, and glass is in many cases relatively small because most of the recyclable paper and glass are collected—often by scavengers—before entering the MSW management system.

The composition of rural solid waste (RSW) is quite different. Generally, it is dominated by ash and organic waste, while the proportions of recyclable compounds in RSW are highly less than that of MSW. However, with the urbanization and rapid economic development of rural China, the composition of RSW is becoming more and more similar to MSW.

Municipal sewage sludge often goes to landfills or is dumped without any effective treatment. About 30 million tonnes of sewage sludge was produced in 2010; however, until 2014, more than 50% of sewage sludge was not safely disposed. The high water content of the sewage sludge (80% moisture content) could not only lead to build up physical barriers in the landfill body where water may accumulate, but also might influence the aerobic and anaerobic waste degradation process.

Another type of waste that needs attention is restaurant waste (RW). It is estimated that about 30 million tonnes of RW are generated every year in China. In the future the RW must be collected separately and it will be forbidden that it is mixed with MSW or treated in MSW treatment facilities. Today the separate collection of RW is rather low, i.e., most of this waste goes to landfill.

Regulations and standards for sanitary landfilling

China has established various regulations and standards on landfill construction, operation, management, and pollution control etc. Significant regulations and standards are listed as the following:

- Technical Code for Municipal Solid Waste Sanitary Landfill (GB 50869-2013),
- Standard for Pollution Control on the Landfill Site of Municipal Solid Waste (GB 16889-2008),
- Technical Requirement for Environmental Monitor on Sanitary Landfill Site of Domestic Refuse (GB/T 18772-2008),
- Technical Specification for Operation and Maintenance of Municipal Solid Waste Sanitary Landfill (CJJ 93-2011),
- Emission Standards for Odour Pollutants (GB14554-93),
- Integrated Wastewater Discharge Standard (GB 8978-88),
- Standard for Assessment on Municipal Solid Waste Landfill (CJJJ/T 107-2005),
- Technical Code for Geotechnical Engineering of Municipal Solid Waste Sanitary Landfill (CJJ 176-2012),

- Technical Specification for Operation and Maintenance of Landfill Gas Collection Treatment and Utilization Projects (CJJ 175-2012),
- Leachate Treatment Project Technical Specification of Municipal Solid Waste Landfill (HJ 564-2010), and
- Technical Code for Municipal Solid Waste Sanitary Landfill Closure (CJJ 112-2007).

Despite of the Technical Specification for the Operation and Maintenance of Municipal Solid Waste Sanitary Landfills (CJJ 93-2011), many cities such as Beijing, Shanghai, etc. established in addition local regulations for the management of sanitary landfills. Scavengers are no longer allowed to enter into the landfill or pick up recyclable materials on the working face. Sprinklers move around the landfill site and spray liquid deodorant—especially in summer—into the air to reduce olfactory defense to surrounding communities. In addition disinfection is provided regularly to kill insects and small animals. Fences and nets are set up to avoid littering. Most sanitary landfills in China are under standardized management currently.

Sanitary landfill siting

There are mainly two types of landfills in China: built on flat lands as a mound or below ground in valleys. The type of landfill usually depends on the topography of the area. In most northern cities such as Beijing and Zhengzhou, landfills are built above ground and could reach heights up to 50 m. In southern cities like Suzhou and Shenzhen, landfills are built directly in valleys (Fig. 15.1.8).

The very important step in the process of constructing a landfill is siting. In recent years, landfill siting has become the biggest problem that government is facing when a new landfill is planned. People usually show strong opposition to solid waste treatment facilities, including transfer stations, landfills and incineration plants. The "not in my backyard effect" is becoming more and more obvious in China, and conflicts often happen in the process of such facilities siting.

Figure 15.1.8 Two typical landfills in flat land (northern China) and in a valley (southern China).

Sanitary landfill operation

MSW is transferred to landfills by different kind of garbage vehicles. The waste truck enters the weighbridge at the entrance of the landfill and the total weight is recorded. The weighbridge system is usually connected with the Smart Environmental and Sanitation Management Platform, which serves for the local City Environmental & Sanitation Bureau; relevant information of the garbage vehicle is uploaded to the platform immediately as basic data collection.

A sanitary landfill is often divided into several working units or cells. Some landfills divide units for different kinds of waste, while some landfills divide the working unit only for management convenience. The incoming waste is unloaded on the working face. A bulldozer then distributes the waste in layers less than 60 cm; after that, the waste is continuously compacted by compactors until a final density no less than 600 kg/m^3 is reached (Fig. 15.1.9). Several layers of compacted waste are piled up until a height of is about 2–4 m, but no more than 6 m is reached.

The different kinds of cover of a landfill include daily cover, intermediate cover and final cover. It is required to cover the compacted waste on the working face (i.e., the open cell) daily with 20–25 cm soil or HDPE/LDPE membranes no less than 0.5 mm thick (Fig. 15.1.10). When a unit is filled up with waste, it is covered by no less than 30 cm of clay or HDPE/LDPE membranes no less than 0.5 mm thick, which is called intermediate cover. When the landfill reached its final height, a final cover has to be constructed.

The management and disposal of leachate have long been a problem that many Chinese landfills face. Large amounts of leachate are generated also due to the high moisture content of MSW, and it varies significantly dependent on different seasons and the existing climate. Leachate is collected at the bottom of the landfill and transported to the surface by using a pump. Generally, leachate is stored in a leachate-conditioning pool for short periods (Fig. 15.1.11), from where it flows to the leachate treatment unit. Most leachate conditioning pools are open to the air, while in some landfills they are covered especially to prevent odor emissions.

Figure 15.1.9 Waste dumping, bulldozing, and compacting on the working face.

Figure 15.1.10 Daily and intermediate cover of a landfill in China.

Figure 15.1.11 Covered and open leachate-conditioning pools.

At some landfills the leachate is recirculated to reduce the amount of leachate and adjust the water quality. Often a surface spray system is used. This technology is widely used especially at landfills that have a low leachate treatment capacity since it is cost-effective and flexible. However, it could also bring many problems, such as odor emissions, surface leachate breakouts, slope instability, etc.

References

Bader, C.D., 1998. The articulated dump truck: workhorse of the landfill. MSW Management 56–61.

Bagchi, A., Bagchi, A., 1994. Design, Construction, and Monitoring of Landfills, second ed. J. Wiley, New York.

Bell, J.P., Dean, T.J., Hodnett, M.G., 1987. Soil moisture measurement by an improved capacitance technique, part II. Field techniques, evaluation and calibration. Journal of Hydrology 93, (1–2), 79–90.

Curro, J., 2000. Using the global positioning system in landfill gas system monitoring. MSW Management.

Davis, J.L., Annan, A.P., 1977. Electromagnetic detection of soil moisture: progress report I. Canadian Journal of Remote Sensing 3, (1), 76–86.

Evett, S.R., 1998. Some aspects of time domain reflectometry, neutron scattering, and capacitance methods for soil water content measurement. In: Comparison of Soil Water Measurement Using the Neutron Scattering, Time Domain Reflectometry and capacitance methods, IAEA, pp. 5–49.

Farquhar, G.J., Rovers, F.A., McBean, E.A., 1995. Solid Waste Landfill Engineering and Design. PTR Prentice Hall, United States of America.

Feinbaum, R., 2000. Odor control, going the extra mile. MSW Management.

Frechen, F.B., 1989. Odour emissions and controls. In: Christensen, T.H., Cossu, R., Stegmann, R. (Eds.), Sanitary Landfilling: Process, Technology and Environmental Impact. Academic Press, London, UK, pp. 509–519.

Koerner, R., 2003. In: Linear system design and materials, USEPA Workshop on bioreactor landfills. Arlington, VA (CD-ROM).

McCann, I.R., Kincaid, D.C., Wang, D., 1992. Operational characteristics of the watermark model 200 soil water potential sensor for irrigation management. Applied Engineering in Agriculture 5, 603–609.

Nalamothu, R.C., Reinhart, D.R., Wayson, R.L., Martin, A., Rogoff, M., 2008. Urban infilling impacts on Florida solid waste facilities. MSW Management 60–71.

O'Maley, P.G., 2003. Controlling odor at transfer stations and MRFs: there's more than one way to reduce complaints. MSW Management 44–52.

Otineo, F., Magagula, C.S., 1999. Management strategies of odour problems at landfill sites. In: Proceeding Sardinia 99, Seventh International Landfill Symposium, CISA Publisher, Cagliari, Italy, pp. 589–596.

Ramke, H.-G., 2001. Appropriate design and operation of sanitary landfills. In: International Conference on Sustainable Economic Development and Sound Resource Management in Central Asia, October 3–5, 2001, Tashkent, Uzbekistan.

Rees, J.F., 1980. Optimisation of methane production and refuse decomposition in landfills by temperature control. Journal of Chemical Technology and Biotechnology 30 (1), 458–465.

Stegmann, R., Spendlin, H.H., 1989. Enhancement of degradation, German experiences. In: Christensen, C., Stegmann (Eds.), Sanitary Landfilling: Process, Technology and Environmental Impact. Academic Press, ISBN 0121742555.

Taylor, M.A., 2000. Landfill final cover storm drainage. MSW Management p. 7. https://foresternetwork.com/weekly/msw-management-weekly/landfill-management/landfill-final-cover-stormwater-drainage/ (accessed on the 23rd October 2018).

Tchobanoglous, G., Theisen, H., Vigil, S., 1993. Integrated Solid Waste Management: Engineering Principles and Management Issues. McGraw-Hill.

Terry, R.J., Willett, D.B., 1999. Bird control at landfill sites: total enclosure netting systems. In: Proceeding Sardinia 99, Seventh International Landfill Symposium. CISA, Cagliari, Italy, pp. 581–587.

Thomas, P.A., 2001. The Testing and Evaluation of a Prototype Sensor for the Measurement of Moisture Content in Bioreactor Landfills. University of Central Florida.

Topp, G.C., Davis, J.L., Annan, A.P., 1980. Electromagnetic determination of soil water content: measurements in coaxial transmission lines. Water Resources Research 16 (3), 574.

Townsend, T., Xu, Q., Jain, P., Powell, J., Tolaymat, T., Reinhart, D., 2015. Sustainable Practices for Landfill Design and Operation, Waste Management Principles and Practice. Springer Science, New York, NY.

Wilhelm, V., 1989. Occupational safety at landfills. In: Christensen, C., Stegmann (Eds.), Sanitary Landfilling: Process, Technology and Environmental Impact. Academic Press, UK, ISBN 0121742555, pp. 509–519.

Worrell, W.A., Vesilind, P.A., 2012. Solid Waste Engineering, second ed. Cengage Learning. ISBN: 978-1439062159.

15.2

OCCUPATIONAL HEALTH AND SAFETY

Volkmar Wilhelm

INTRODUCTION

The potential hazards on the environment and local community caused by emissions from a landfill usually get much focus, while the occupational hazards on those actually working at the landfill often are overlooked.

Actually, workers at landfills and waste treatment plants are, however, subject to a variety of hazards and stress. There are now many occupational health and safety directives in the EU, which must be respected also at landfills and waste treatment plants. Goal of these regulations is to protect the workers on their job from injury or disease.

The chapter provides an overview about occupational health and safety regulations and about occupational hazards and adverse effects of employment at landfills as well as the measures necessary to deal with them.

General Remarks on Occupational Safety

The term "occupational safety" refers here to all of those measures that need to be taken to protect staff from dangers and risks caused by their working conditions. This includes, for example, measures to ensure safety from fire, explosion, injury by machines, health risks, as well as such questions as humane working procedures and working hours.

The implementation of occupational safety measures should be a self-evident responsibility, since protecting oneself and others from possible harm is one of the basic ethical and moral foundations of our society. On the other hand, the implementation of occupational safety measures also makes economic sense, since unsafe working conditions and dangerous conduct at working places can lead to serious economic loss.

Clearly though, ethical and moral precepts and economic interests do not always have enough of a motivating force to impel people to act.

Experience has shown that only legal regulations with detailed procedures lead to the actual implementation of occupational safety measures in individual cases. Taking this into consideration a number of national and international regulations have been introduced to ensure safety at work.

The numerous rules that currently regulate occupational safety are, however, minimum requirements and do not by any means take into account every individual case, so that more extensive measures may be necessary. For this reason and to make working places safe and free from health hazards, the dangers and health risks must first be determined. Once they are known, relevant safety measures must then be implemented. In this context, a clear hierarchy or sequence of safety measures can be observed. The primary task is to determine those technical and organizational measures that will completely eliminate present hazards to people and health ("direct safety measures"). The complete elimination of hazards cannot be achieved, however, in many cases (e.g., it is not possible to prevent the development of biogas in landfills) so that secondary steps must be taken to make sure people are not directly involved in hazardous situations. This can be achieved by putting either time or space between the individual and the potential risk ("indirect safety measures").

An example of an indirect safety measure would be a protective shield on a machine or tool. If there are still remaining hazards, a last step would be the use of "directive safety measures," such as the wearing of personal protective clothing, or the use of respirators, furthermore, the measuring of the concentration of hazardous substances, the employment of qualified training staff, the placing of warning signs, and insistence on the observation of regulations and instructions. These measures are, however, hardly effective since they are not automatic but require the cooperation of those involved. Frequently they are simply ignored, for example, when staff enters gas-filled underground shafts without respirators or measuring instruments.

Hazards and Adverse Working Conditions at Landfills

The hazards and health risks posed to personnel by the working conditions at landfills can be of very different kinds and must be determined individually and characterized from case to case. Frequent hazards and impositions, for example, are as follows:

- Contact with hazardous substances (e.g., biogas, acids, bases, silicon dust, infectious substances)
- Accident risk through collision with heavy vehicles
- Dangerous technical equipment (e.g., danger points on pumps, defective hydraulic pipes on scrapers, defective electrical and soldering equipment)
- Vibration caused by heavy vehicles
- Accidental contact with sharp or pointed objects
- Impositions placed on personnel by dust, wet, cold, heat, smells, and insects

How these hazards and harsh working conditions at landfills can be eliminated or reduced will be discussed later in the chapter.

LEGAL FRAMEWORK IN THE EU AND IN GERMANY

The European Community is committed to the goal, to harmonize working conditions in the member states, in particular the measures for occupational safety and health. The working conditions and the working environment should have the same standards everywhere and are continuously developed. To achieve this, the EU has adopted a variety of EU directives on health and safety. The legal bases for the adoption of these guidelines are articles 114 and 153 of the Treaty on the Functioning of the European

Union (TFEU). The member states must implement EU directives into national law (https://osha.europa.eu/de/legislation/directives).

The guidelines referred to in article 114 of the TFEU are the reduction of trade barriers. They harmonize the safety requirements for equipment as well as chemical substances as a priority. They are also called internal market directives. The following directives have been adopted so far:

- Asbestos
- General product safety
- Electrical equipment designed for use within certain voltage limits
- Safety of toys
- The protection of workers against hazards by noise at work
- Simple pressure vessels
- Personal protective equipment
- Machines
- Classification, packaging, and labeling of dangerous substances
- Equipment and protective systems intended for use in potentially explosive atmospheres
- Building products
- Lifts
- Gas appliances
- Electromagnetic compatibility
- Aerosol
- Pressure equipment

In the national implementation of these policies, the standards must be strictly observed, not to impede the free exchange of goods.

The directives referred to in article 153 of the TFEU lay down minimum standards for the occupational safety and health in Europe. In contrast to the directives by article 114 of the TFEU, the member states are allowed to maintain or establish more stringent measures. The European Framework Directive on Safety and Health at Work (Directive 89/391 EEC) adopted in 1989 was a substantial milestone in improving safety and health at work. It guarantees minimum safety and health requirements throughout Europe.

The following directives were adopted on the basis of the framework directive:

1. Workplaces
2. Use of work equipment
3. Using personal protective equipment
4. Handling heavy loads
5. Work with visual display units
6. Exposure to carcinogenic agents
7. Exposure to biological agents
8. Construction
9. Safety and health labeling
10. Protection of pregnant workers

11. Extraction of minerals through drilling
12. Above ground or underground mineral-extracting industries
13. Onboard fishing vessels
14. Exposure to chemical agents
15. Improvement of the health and safety of workers who may be endangered by explosive atmospheres
16. Hazard arising from physical agents (vibration)
17. Hazard from physical agents (noise)
18. Hazard from physical agents (electromagnetic fields)
19. Exposure to artificial optical radiation

Many of those listed directives must be respected during operation of landfills and waste treatment plants.

The framework directive introduces as a key element the principle of risk assessment and defines its main elements, e.g., hazard identification, worker participation, introduction of adequate measures with the priority of eliminating risk at source, documentation, and periodical reassessment of workplace hazards.

The following risk types assume it:

1. Mechanical hazards
2. Electrical hazards
3. Risks from dangerous substances
4. Risks from biological agents
5. Fire and explosion hazards
6. Hazards of cold and hot media
7. Hazards of bad climatic conditions
8. Hazards due to insufficient lighting
9. Hazards of noise
10. Hazards due to vibration
11. Hazards of radiation
12. Recording of information, handling of control actuators
13. Exposures to physical stress
14. Hazards of mental stress

Risk assessments on landfills and waste treatment plants revealed that there are hazards that match almost all the abovementioned types.

In Germany, the occupational safety and health framework provided by the EU is further concretized through a variety of government regulations and technical rules (BAuA Website, 2016) Examples include the technical regulations for safety (TRBS) (e.g., TRBS 1112 maintenance, TRBS 1201 examination of work equipment and requiring monitoring equipment, TRBS 2121 danger caused by crash, TRBS 2152 hazardous explosive atmosphere), the technical rules for hazardous substances, TRGS (e.g., TRGS 400 risk assessment for activities involving hazardous substances, TRGS 519 asbestos: demolition, renovation, or maintenance work, TRGS 900 occupational exposure limits), and the technical rules for biological agents (TRBA) (e.g., the TRBA 214 waste treatment systems including sorting waste, TRBA 400 guidance on risk assessment, and for the information of employees in activities with biological agents, TRBA 500 basic measures when working with biological agents).

In addition to public regulations, there are also the rules and regulations of the Berufsgenossenschaften (BGs). The BGs are the professional associations and represent the statutory accident insurance. The set of rules and regulations of the BGs includes accident prevention regulations, principles, rules, and information (DGUV Website, 2016). From this set of rules, the BG Rule 127 "landfills," the BGI 893 "risk assessment for biological agents at work on landfill sites," and the sample collection "explosion protection measures when working on and in landfills" concern the landfill sector in particular.

In BG Rule 127 "landfills" are regulated particularly in the following topics:

- Construction and equipment, e.g.,
  - Operation buildings, shafts, tunnels, roads, work, and rescue equipment
  - Personal protective equipment
  - Requirements for landfill gas systems
- Operating, e.g.,
  - Traffic regulations, operating instructions, laying down protective measures, drainage facilities, flame fittings
  - Regulations for work on the landfill gas system, for work in shafts, tunnels, and underground structures
  - Regulations for the storage of asbestos-containing waste
- First aid and rescue
- Medical care
- Fire protection
- Tests

The BGI 893 "guidance risk assessment for biological agents at work on landfills" serves as an aid to the identification and assessment of risks in activities with biological agents on landfills as well as to the selection of appropriate measures.

The sample collection "explosion protection measures when working on and in landfills" gives assistance to the assessment of explosion hazards in landfills for the responsible entrepreneurs and the insured, as well as specialists for occupational safety, safety officer, staff councils, and planners.

HAZARDS AND ADVERSE EFFECTS OF LANDFILL GAS

Biogas Constituents

The term biogas is used to refer to the gaseous products of the microbial decomposition of organic waste deposited in landfills, as well as waste substances that have vaporized. As a rule, biogas consists of about 55%–60% methane (CH_4) and 40%–45% CO_2. Methane is a nontoxic, colorless, odorless, energy-rich gas, lighter than air. In concentrations of 5%–15% volume in air, explosive mixtures form. It has a slightly narcotic effect; however, only in concentrations of more than 50% it is noticeably anesthetic. Carbon dioxide is a colorless, caustic gas, with a slightly acidic taste and smell, 1.5 times heavier than air. High concentrations of CO_2 are toxic even when enough oxygen is available for breathing. Air containing 4%–5% CO_2 leads to unconsciousness, and concentrations of over 9% cause death within minutes.

In addition to the main constituents of methane and carbon dioxide, biogas can also contain a variety of trace substances. Table 15.2.1 gives an overview of the range of values of trace gases in biogas and the corresponding occupational exposure limit values (OELs), called in German AGW ("Arbeitsplatzgrenzwert").

The AGW value list (TRGS 900) is revised every year by the German Research Foundation.

OELs are the average permissible values for an 8-h work shift. If these limits are observed, then there is usually no risk to health. Excesses of short duration are permissible.

The OELs do not apply to mixtures. A procedure for assessing the adverse effects of gas mixtures on the basis of a scientific analysis of the interaction of substances is not yet available. The system employed in the United States is not strictly scientific. It is based on the following procedure: If there is no positive evidence suggesting that the effects of the individual constituents of a gas mixture are independent of each other, then it is assumed that their effects are cumulative. For each substance, the ratio between its concentration in the air and its limiting value is calculated. If the sum of these ratios is greater than 1, then the limiting value for the mixture has been exceeded; if the sum is 1 or less than 1, then the limiting value has not been exceeded.

$$\frac{c1}{\mathrm{AGW1}} + \frac{c2}{\mathrm{AGW2}} + \frac{c3}{\mathrm{AGW3}} + \ldots \leq 1$$

In spite of the criticisms laid against this method it can be extremely helpful, if the formula is not used to assess the interaction of substances but as a guideline for implementing practical safety measures.

Various trace gases are toxic (for example, dichloromethane and hydrogen sulfide), or they are potentially carcinogenic (for example, benzole and vinyl chloride). For substances, are potentially carcinogenic, it is impossible to work out OELs values since cancer only becomes apparent after many years or even decades. Long-term exposure leads potentially to a large accumulation of mutations. Whether cell damage is repaired or not and, if so, to what extent, is unknown at present. Safe levels of exposure cannot be determined on the basis of epidemiological statistics as there are neither adequate analytical data on the intake of substances over sufficient periods of time exist, nor sufficiently large numbers of exposed persons and control groups are available. In Germany the Hazardous Substances Ordinance regulates in article 10 special protective measures for activities involving hazardous substances that are carcinogenic, mutagenic, and toxic to reproduction. The TRGS 900 contains a directory of these substances.

Properties of Biogas and The Resulting Hazards

Asphyxiation

Although biogas is, as a rule, somewhat lighter than air, escaping biogas is capable of displacing air and collecting in shafts, rooms, or cavities. If the oxygen content of air decreases from an average 21 vol.-%. to below 17 vol.-%., then there is a threat of asphyxiation. With respect to biogas this means that when biogas stands in a ratio of 1:4 to air, asphyxiation is possible.

Table 15.2.1 Trace substances in biogas and their occupational exposure limit values (OELs), expressed in mg/m^3 measured in air free conditions

Individual Gas	Concentration in Landfill Gas	OELs
Propane	1.4–13	1800
Butane	0.3–23	2400
Pentane	0–12	3000
Cyclohexane	0.03–11	700
Hexane	3–18	180
Heptane	3–8	2100
Octane	0.05–75	2400
Isopropylbenzol	0–32	100
Ethylbenzol	0.5–236	88
Toluene	0.2–615	190
Xylol	0–376	440
Dichloromethane	0–6	260
Trichloromethane	0–2	2.5
1,1,1-Trichloroethane	0.5–4	1100
Dichloroethylene	0–294	8
Trichloroethylene	0–182	269
Tetrachloroethene	0.1–142	138
Chlorobenzol	0–0.2	47
Hydrogen sulfide	0–600	7.1
Tetrachloroethane	0–0.6	7
Benzene	0.03–7	
Chloroethene	0–264	

Toxic Effects and Health Risks

The toxic properties of biogas derive mainly from the presence of carbon dioxide and hydrogen sulfide. Concentrations of more than 9% carbon dioxide can cause death within minutes. As biogas has a CO_2 content of between 10% and 50%, it loses its toxicity when it is mixed with air to a ratio of 1:4. The amount of hydrogen sulfide present in biogas is usually below the toxic level of 800 ppm. However, higher values do at times occur, especially at landfills for hazardous wastes (up to 17 000 ppm). As long as the gas can escape over the whole landfill area and is sufficiently mixed with air, there is no immediate danger. In places where concentrated discharges of gas are possible, for example, from substantial fissures in the landfill cover or from vertical gas discharge shafts, the limits permitted at places of work may be exceeded.

There are no reliable data on the long-term effects on health of exposure to biogas. The toxic properties of a gas mixture such as biogas cannot be determined. The physiological effects of the individual substances are known and corresponding limits have been laid down. When mixed, however, the toxic properties of the individual substances may decrease or increase. Toxicologists refuse to set limits for mixtures. Personnel employed at landfills frequently point to cases of colleagues dying of carcinomas and believe this to result from long-term exposure to biogas. Such claims have not yet been scientifically proved.

Fire and Explosion Hazards

Methane, an important constituent of biogas, can, when mixed with air, cause fires, explosions, and fulminations. Within the upper and lower limits of explosion (between 5% and 15% vol. methane), methane can be detonated. Gas mixtures below and above the limits of explosion are not explosive but quantities $\sim$15% represent a fire hazard. When methane has been mixed with air to the usual ratio over the surface of the landfill, then there is no immediate danger. Biogas can collect in shafts, trenches, and cavities to form an explosive atmosphere, which can then be detonated by gas burn-offs, biogas compressors, open fires or lights, or hot surfaces such as car or truck exhausts.

There have been many reports of explosions or fulminations at landfills, which have, however, rarely been considered in the relevant literature. In one case, explosive biogas was detonated at a distance of 150 m from a gas discharge shaft and then began to burn in the discharge shaft itself.

Safety Measures

To eliminate or reduce the risks created by biogas, the following safety measures must be taken:

1. The waste delivered to landfills should, as far as possible, have had all hazardous substances removed; these should be collected, for example, at mobile or stationary collection points and then disposed off at special dumps or processing plants for dangerous materials. In this way substances that may vaporize and are responsible for the dangerous trace gases in biogas can be reduced.
2. Biogas must not have harmful effects on the health of the personnel at landfills. This can be achieved during the construction of a landfill by the installation of horizontal gas extraction equipment and by making sure that vertical vent pipes are properly sealed and discharged gas is

prevented from collecting and concentrating. If these measures are not practicable, then respirators using an independent air supply should be worn during work in the vicinity of escaping gas.

3. Buildings that have to be entered by personnel should be constructed in such a way that no dangerous concentrations of biogas can form (e.g., by observing the principle that all underground buildings should have the necessary technical ventilation devices installed).
4. Areas in which there may be dangerous concentrations of biogas (for example, shafts and enclosed leachate sumps) should only be entered by personnel after it has been ascertained by means of portable measuring instruments that the atmosphere does not contain dangerous concentrations of gas. When warning devices detect dangerous concentrations, then a safe atmosphere must be created by the deployment of ventilators. Ventilators may be regarded as sufficient if they fulfill the following technical requirements: (1) At the lowest point in shafts and underground gullies, an air current of 600 m^3/h per m^2 of shaft cross section is produced; (2) In other constructions such as pump stations, the air should be renewed at a rate of six to eight times an hour.

Personnel must present themselves regularly for medical checks. This aspect is dealt with in greater detail in a following section.

HEALTH HAZARDS FROM BIOLOGICAL AGENTS

There is a variety of microorganisms in the household waste that can cause infections, sensitizing effects, or toxic effects on workers. Infectious agents may occur especially at injection, animal carcasses, diapers, and in the leachate (Fig. 15.2.1). Typical infection pathogens in the waste are coliforms, salmonella, and hepatitis viruses.

Sensitizing and toxic effects are induced especially from fungi and bacteria (in particular actinomycetes). The occurrence of fungi and bacteria is expected particularly in organic waste. The sensitizing and toxic effects manifest themselves mainly in the form of respiratory disease. The microorganisms can get in the human body by airborne dust particles or by direct contact with the waste.

The transfer of microorganisms in indoor areas such as cabins and operation buildings is typical for the landfill operation (Fig. 15.2.2). There, the microorganisms can quickly multiply as a function of humidity and temperature.

Animals can also cause hazards. Sometimes rats or mice are found in landfills and waste treatment plants, especially where the wastes are a source of food for these animals. In addition to the rodents, health hazards can be assumed by birds, such as pigeons. The threats are not tied to the direct contact with the animals. The infectious agents are partially released with the excrement of animals. An infection can occur through direct contact with the skin or the ingestion through the respiratory system when cleaning. An example of this is the Hantavirus that can be found in the feces of the vole and that can lead to renal failure.

Depending on the transmission path, protective clothing, protective gloves, eye protection, and appropriate respiratory protection must be available. All activities associated with a direct contact with biological agents, personal protective equipment must be used, depending on the result of risk assessment.

Direct dealing with biological substances can consist, for example, in sampling, quality control, and temperature measurements. Special requirements for the selection of personal protective equipment are

Figure 15.2.1 Hospital waste disposed jointly with household waste.

Figure 15.2.2 Transmission of microorganisms into the cabin.

necessary, for example, if there is a risk of infectious agents. This can be when cleaning activities are conducted in areas where rodents or birds have stayed. In these cases, particle filtering half mask FFP3 is necessary.

In these cases the gloves need to be protected against biological risks. Corresponding protective gloves are also necessary if animal carcasses must be eliminated.

HEALTH HAZARDS FROM ASBESTOS

One of the most important risks to health is the inhalation of hazardous dust. Such dust may contain, for example, asbestos fibers or silicotic particles. Long-period exposure to even inert dust particles may be detrimental to the health of personnel (through deposition in the lungs).

The disposal of asbestos-containing materials in landfills is particularly important. In doing so, asbestos fibers can be released and the workers can inhale or swallow them. Fibers with a length of less than 5 μm and a diameter of maximum 3 μm can reach the alveoli of the lungs and can trigger asbestosis (asbestos hypertension disease). Another disease caused by asbestos is pleural mesothelioma. Particularly dangerous in this context are weakly bound asbestos products that have a fiber content of 60% and more. These include, for example, asbestos, asbestos dust, asbestos-containing fire protection boards, asbestos textiles, and filter materials.

Because of the harmful effects of asbestos fibers, in Germany there has been a ban on the use of asbestos and asbestos-containing materials since 1993 in accordance with the Ordinance on Hazardous Substances. This prohibition applies since 2005 at the European level.

In accordance with the Ordinance on Hazardous Substances, measures with the highest level of protection are necessary for activities with asbestos exposure, e.g., demolition, renovation, and maintenance work, dealing with mineral resources and landfilling.

Most asbestos-containing materials to be stored in landfills come from this demolition, renovation, and maintenance work. The occupational safety and health measures are regulated in Germany in the TRGS 519. This includes the disposal of the waste.

Further information on the disposal of asbestos waste is represented in the BGR 127 "landfills" and cited below:

- Asbestos-containing waste is in agreement with the competent authority to treat, pack, and cover that no asbestos dust can get in the environment.
- Weakly bound asbestos waste may only be deposited when it is bound with hydraulic binders. Fine-grained or woven waste in small quantities can be deposited without binding, if it is covered immediately. Waste with tightly bound asbestos fibers, which are not immediately installed, is to be kept damp to avoid dust development.
- Asbestos-containing waste should not be discarded from vehicles or tilted.
- Asbestos-containing waste may only be crossed by vehicles and equipment if a sufficiently thick layer of a suitable inert material, such as earth material has been applied previously.
- Workers have to use protective suits, gloves with cuff, shoe covers, and filtering devices with particle filters of class P 2 in conjunction with half masks.

- Asbestos-containing waste must be stored in a separate part of a landfill section or in a separate section of the landfill.
- Landfill parts containing asbestos waste must be marked in a map of the landfill.
- An industrial vacuum cleaner of the H class is to provide for the cleaning of contaminated clothing and contaminated equipment.
- Appliances must be equipped with pressure cabins.
- For the unloading and installation of waste, spraying devices for water are to be kept.

During demolition, renovation, and maintenance work, waste already at the construction site must be treated in such a way that during transport recording, during transport and at the subsequent disposal on landfills, no asbestos fibers are released.

Appropriate accumulation bins for plate-shaped waste are, e.g., as follows:

- Well lockable plastic woven bags (big bags, record bags) for storing plates or coarse waste (Figs. 15.2.3 and 15.2.4)
- Tear-resistant PE swarf, to pack plates stack
- Tear-resistant PE plastic bags for storing small fragments, fasteners, protective clothing, breathing protection filters, contaminated materials, etc.

Weakly bound asbestos or dusts from filtration systems must be solidified with a binder and then dust packed before deposition at the landfill. The solidification must be carried out in a closed system.

SAFETY MEASURES FOR WORKING IN SHAFTS AND UNDERGROUND BUILDINGS

Entering shafts, tunnels, and underground structures and working at these places are linked to a variety of hazards and stress. Therefore these buildings should be designed in such a way that they must not be used for regular inspection and maintenance.

Figure 15.2.3 Asbestos waste packed in big bags.

Figure 15.2.4 Record bags for flat and corrugated boards.

This can be achieved, if, for example,

- leachate can be removed through pipes;
- the amount of leachate can be determined with inductive measuring devices (Fig. 15.2.5);
- nozzles, measuring devices, or television cameras can be introduced from above in bow-shaped leachate pipes.

Figure 15.2.5 Inductive flow meter in a bypass to the leachate pipe.

Most important hazards when driving shafts and underground structures are as follows:

- Open manhole (crushing of hands and feet, slipping, and falling)
- Fall into open buildings
- Crash when going down
- Hazardous atmosphere (fire, explosion, and health hazards)
- Electric current
- Microorganisms
- Drowning in dammed up water

In addition to these threats the psychological pressure that may be related with the wearing of full-face masks or working in very deep and narrow structures should also be taken into account.

Protective Measures Against Hazardous Atmosphere

Ventilation must be ensured before and during work, so that hazardous explosive atmosphere, lack of oxygen, gases, or vapors in harmful concentrations cannot occur at workplaces in shafts and underground structures.

To ensure sufficient ventilation, concentrations of the relevant gases and vapors with acoustic and optical signal indication have to be measured continuously. Sufficient ventilation occurs when gases or vapors are so diluted that following air units are available:

- Methane <0.5 vol.-%
- Oxygen >20 vol.-%
- Carbon dioxide <0.5 vol.-%
- Hydrogen sulfide <10 mL/m^3

Effective ventilation can normally be ensured only by blowing air. The suction point of the ventilation unit should be in accordance with the direction of wind 5 m from the shaft and within approximately 1.50 m height above the landfill body to avoid that gases from the surface area can be sucked in.

For shafts and not ventilated buildings, the minimum airflow volume must be 10 m^3/min per m^2 of the cross section. If a gas influx also occurs above the end of the pipe, the ventilation system must be designed for a minimum airflow volume of 30 m^3/min per m^2 of cross section.

After switching on the ventilation, shafts and underground constructions can be entered safely, when the air has changed at least five times, and measures have demonstrated compliance with the permissible concentrations of oxygen, toxic gases, and flammable gases.

In special cases, e.g., when the influx of landfill gas occurs in large quantities, these measures are not sufficient. The investigation of an explosion accident, which occurred by the flushing of a clogged leachate pipe, proved that the formation of hazardous explosive atmospheres could not be prevented by the currently known safety measures.

The leakage of landfill gas in shafts and underground constructions must be avoided. This can be achieved, e.g., by opening the pipes individually and keeping these openings as small as possible. For the camera inspection or flushing leachate pipes in galleries, mobile ventilation devices are required in addition to the fixed ventilation devices for the safe disposal of the invading landfill gas.

To check the atmosphere the used stationary or mobile gas detectors must satisfy the following requirements:

- Explosion proof and continuous measuring
- Measuring cells for the components of oxygen, methane, carbon dioxide, and hydrogen sulfide
- Tested for function even at high concentrations of carbon dioxide, oxygen, humidity, and catalyst poisons

Safe measurements can be carried out, for example, by lowering acoustic display measuring instruments or by using sampling pipes.

Prevention of Ignition Hazards

As long as a potentially explosive atmosphere cannot be ruled out with any certainty, in shafts and underground structures,

- working with ignition hazards is not allowed and
- tools with a danger of ignition cannot be used.

Danger of ignition occurs when friction, impact, and grinding operations (e.g., handheld machines, using spark sweeping tools), fire work (welding), electrostatic discharge, as well as friction and shock processes between aluminum and rusty steel are carried out. Electrical equipment must have the type approval for use in potentially explosive atmospheres.

Driving with nozzles and television cameras in rechargeable, nonconductive pipes, where landfill gas may be present, is only allowed after performing additional explosion protection measures. An additional explosion protection measure is a continuous blowing ventilation of the pipe before the rinsing process and use of an explosion-proof TV camera.

Protective Measures Against Electric Shock Hazards

For the use of electrical equipment in shafts and underground buildings, protection against increased electrical hazards shall be taken.

Increased electrical hazards can be avoided, by using

- electrical tools with protective extra-low safety voltage or protective isolation,
- hand lamps, except such match for protective low voltage, at least the protection class IP X 5,
- flexible connecting cables of type HO7RN-F or AO7RN-F or an equivalent type.

These measures are to be taken regardless of the measures against explosion.

Safeguarding of Workers

Driving shafts and underground buildings on landfills, as well as working in these buildings, is associated with specific risks. Therefore, it is essential to secure the persons working in those buildings by measures that allow rapid exits and the prevention of crashes, for example, due to corroded crampons.

The following points should be noted in any case:

- There must be at least a second person present to secure. The workers must be in constant sight or call connection. A sight connection is preferable for the safety of people. Television cameras for the sight connection must be made explosion proof. If the sight connection is not guaranteed, an explosion-proof radio with emergency call button is required.
- Each employee rising up has to be secured to enable a fast and safe rescue at any time with a full body harness and safety rope with absorber.
- Rescue lifting devices must be secured to sufficiently viable anchorage devices vertically above the entry point. Anchor points are, for example:
  - tripod secured against shifting and disassembling sliding feet,
  - crane cantilever arm attached on a vehicle, or
  - in manhole openings wearing items.
- Rescue lifting equipment must be operated by a second person at the entry point.
- Entering shafts with depths >5 m is only allowed with silo drive-in facilities (Fig. 15.2.6).

To protect the workers adequately against the event of a blowback from the effects of hot and toxic combustion gases, they must wear, when working in shafts and underground buildings, where gas accesses can occur, breathing apparatus with full-face masks and compressed air cylinders.

Equipment for Work and Rescue

Work and rescue operations must have at least the following equipment:

- ventilation equipment in explosion-proof execution;
- portable explosion-proof and continuously measuring multiple gas detector or separate devices for methane, carbon dioxide, hydrogen sulfide and oxygen;
- rescue lifting device with safety rope, safety harness, and absorber;

Figure 15.2.6 Silo drive-in facility.

Figure 15.2.7 Equipment for work and rescue.

- tripod or an equivalent anchor device for fixing the rescue lifting device;
- explosion-protected hand lamp;
- free portable breathing apparatus acting independently of the surrounding atmosphere;
- stretcher.

  Figure 15.2.7 shows equipment for work and rescue.

Organizational Protection Measures

The necessary measures for health and safety have to be explained in a manual. Works that are connected to specific risks must be regulated with permits and released. This includes working with ignition hazards in hazardous areas, the opening of gas pipes, such as landfill gas pipelines and leachate lines. Workers are to be instructed on hand of these operating instructions. The employees have to train the use of the rescue equipment and facilities at the beginning of the work and thereafter at least twice per year.

MEDICAL CARE

In the previous sections it has been shown that the health of landfill personnel is put at risk by a number of hazards. Whether or not the health of landfill personnel is adversely affected depends especially on the substances deposited.

Therefore, employees must be informed about the health risks and burdens before starting the activities. This includes also the information about preventive measures and possible vaccinations.

In Germany the law states that if hazardous substances exceed statutory limits or if particular jobs are carried out under certain conditions, then medical examinations are necessary both before employment is taken up and thereafter according to regular intervals. Medical examinations are required by law when respirators are worn and in case of contact with the following substances, which are frequently present at landfills: asbestos, benzene, fluorine and its inorganic compounds, silicogenic dust, sandblasting materials, vinyl chloride, toluene, xylol, trichloroethylene, and tetrachloroethene.

Owing to biological hazards, occupational medical checkups must be offered to workers on landfill sites and in waste treatment and sorting systems in many cases.

If an examining doctor finds any doubts on the state of health of an employee, the latter must cease work until safety measures have been taken to eliminate harmful effects or the permissible limit is no longer exceeded.

CONCLUDING REMARKS

The staff of landfills and waste treatment plants is exposed to a variety of risks and burdens. The occupational safety and health framework in the EU oblige the employer to set appropriate measures according to the risk assessment, carry them out, and review their effectiveness on a regular basis. In Germany, BGR 127 “landfill” delivers valuable information for the organization of an effective occupational health and safety system on landfills.

References

BAuA (Bundesanstalt für Arbeitsschutz und Arbeitsmedizin- Federal Institute for Occupational Safety, Health) Webpage, 2016. http://www.baua.de/en/Topics-from-A-to-Z/Topics-from-A-to-Z.html.

DGUV (Deutschen Gesetzlichen Unfallversicherung) Webpage, 2016. http://www.dguv.de/content/media_publications/index.jsp.

European Agency for Safety and Health at Work Webpage, 2016. https://osha.europa.eu/de/legislation/directives.

CHAPTER 16

AFTERCARE

16.1

AFTERCARE COMPLETION: FINAL STORAGE QUALITY ASSESSMENT

Raffaello Cossu and Alberto Pivato

INTRODUCTION

The long-term emission behavior and sustainability of landfills is a fundamental issue that drives all debate on landfilling. Although no conventional definition of sustainable landfill has been established, there is a general consensus throughout the scientific community that a landfill may be considered sustainable if emissions do not significantly modify the quality of the surrounding environmental compartments: air, water, and soil (Hjelmar and Hansen, 2005; Stegmann et al., 2003).

This concept is strictly linked to the concept of final storage quality (FSQ). The term "final storage" was coined in the mid-eighties by Baccini, Henseler, and others from the Swiss working group on landfills (Belevi and Baccini, 1989). It refers to the quality reached by emissions and waste at a specific point in time, when all active control measures can be safely removed. To guarantee sustainability, this equilibrium needs to be reached within the span of one generation, commonly taken to be approximately 30 years, so as not to "compromise the ability of future generations to meet their own needs" (Brundtland, 1987).

Current European landfill legislation (European Landfill Directive, 1999/31/CE) prescribes that aftercare should continue for at least 30 years after closure of a landfill; during this period, monitoring and maintenance activities should be covered by a specific financial provision.

There is an increasing recognition that the legislative, time-based approach does not represent the best criterion for assessment of the endpoint of landfill aftercare (Fourie and Morris, 2003). Indeed, the stabilization degree of landfilled waste depends not only on time but also on waste characteristics, operational techniques, and site-specific conditions. To achieve FSQ over a sustainable period of time, the kind and amount of landfill emissions to be deemed acceptable for the environment should be established.

How to put this principle into practice represents one of the main research issues to be addressed by those involved in the field of landfilling (Heyer et al., 2005a; Knox et al., 2005). Several approaches have been proposed: the first provides for the application of table reference values for biogas, leachate, and residual waste, whereas the second takes into account the specific site conditions of each landfill through the application of risk analysis (Pivato, 2003).

The aim of the chapter is to provide a reliable methodology for use in defining landfill stability and aftercare completion based on a combination of a minimum set of table values with a set of specific target values to be estimated on the basis of risk analysis, taking into account the local site conditions.

Additionally, this chapter aims to underline how achievement of the sustainability target is not something that "happens" but rather the result of a specifically detailed planning and design in which a combination of pretreatment, in situ treatment, and aftercare treatment should be applied to control the accumulation of mobile contaminants within the landfill body.

ASSESSMENT OF LONG-TERM EMISSIONS

Emission Potential of Landfills

As discussed in Chapter 2.2, the accumulation of residual mobile contaminants in a landfill represents the emission potential in terms of leachate and gas generated. The emission potential may vary throughout the life span of the landfill according to trends that depend on waste quality and the processes/technologies adopted in treatment of the same.

Various types of barriers (see Chapter 2.1) are implemented to control landfill emissions. The duration of physical barriers (lining and drainage) implemented to prevent uncontrolled emissions is generally inferior to the duration of long-term emissions.

To achieve sustainability targets, a combination of efforts (see Table 2.2.4) is required to control the trend of emissions over time (Fig. 16.1.1).

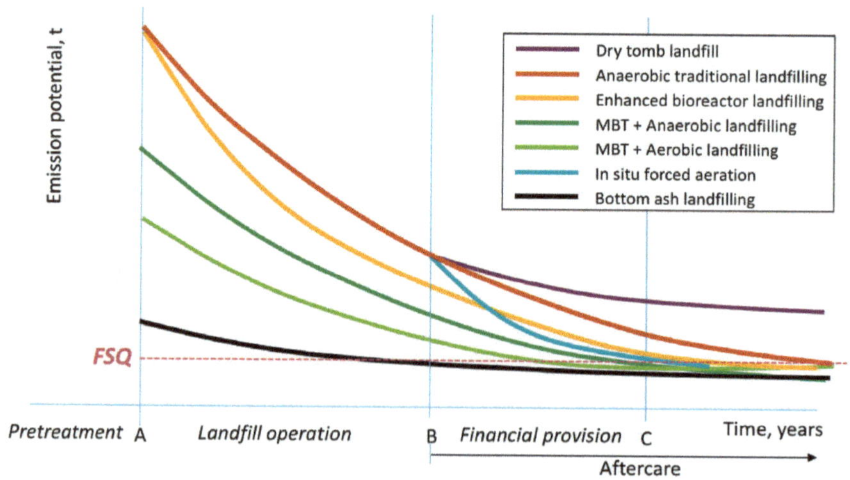

Figure 16.1.1 Evolution of contaminants emission potential in landfills according to the application of different kinds of waste treatment. *A*, beginning of landfilling operation; *B*, landfill closure at the end of landfill operation; *C*, end of the financial provision time set by regulations; *FSQ*, Final Storage Quality; *MBT*, mechanical–biological treatment.

Emission Parameters

The emission potential of the landfill body during the aftercare period may be monitored by means of qualitative and quantitative analysis of biogas, leachate, and deposited waste. The parameters commonly tested and used as indicators of emission quality are listed in the following paragraph: some may be more significant in evaluating the degree of landfill stabilization and verifying achievement of FSQ.

Biogas

- *Biogas composition (%)*: The concentrations of individual gas components, such as CH_4, CO_2 and O_2, can be easily measured as volumetric percentages. The percentages depend on biodegradation processes (anaerobic, aerobic, semiaerobic) and stages reached by the process.
- *Total gas production (Nm^3LFG/h)*: This may be determined by means of extraction tests, in the presence of a Landfill Gas (LFG) extraction system, or using a standardized LFG prediction model. The total amount of LFG extracted from the landfill, irrespective of size, should not exceed the minimum volume that can be treated in a high-quality standard flare or in a gas utilization plant (Heyer et al., 2005a).
- *Areal gas production ($NlLFG/m^2*h$)*: This parameter may be used to assess surface gas emission (when LFG extraction is insufficient or absent). To estimate whether surface emission is acceptable, areal gas production is compared with CH_4 oxidation potential of the top cover. Different investigations show methane oxidation rates, dependent on particular characteristics of the cover system, in the range of 0.34–5.6 Nl CH_4/m^2h (Heyer et al., 2005a). Accordingly, LFG emission through the surface should not exceed these values.

To evaluate the degree of stabilization of landfilled waste, the composition of biogas may not be of fundamental importance; indeed, biogas produced by anaerobic degradation of a stabilized waste has the same composition as that produced from raw waste (Vettorazzi, 2005). Conversely, it may be of greater significance to evaluate the fraction of total organic carbon converted to CO_2 and CH_4.

Moreover, it may also be of interest to assess the $(CH_4 + CO_2)/N_2$ ratio: the latter may prove to be of use in characterizing biogas quality both in anaerobic and aerobic conditions. The ratio is calculated according to N_2 in view of the fact that, conversely from O_2, the former is not implicated in biological degradation of organic matter. This parameter is clearly significant as a stabilization index only in the case of old landfills, at the point in which biogas production decreases and air begins to infiltrate into the landfill body.

Leachate

- *Chemical Oxygen Demand, COD (mgO_2/L)*: This parameter measures the total amount of organic substance (biologically degradable and nondegradable) present in leachate. Values obtained are strictly linked to the characteristics of landfilled waste (unprocessed MSW or MBP waste), biodegradation process (anaerobic or aerobic), and the process phase. Even after the completion of aftercare COD might display high values (in the range of hundreds of mg O_2/L) due to the presence of humic acids produced by the stabilization of degradable organics.

- *Biological Oxygen Demand, BOD_5 (mgO_2/L)*: This parameter measures the amount of biologically degradable organic substance in leachate.
- *BOD_5/COD*: This parameter provides information on waste stability and leachate biodegradability. BOD_5/COD ratio usually decreases with leachate aging and following mechanical–biological pretreatment.
- *Total Organic Carbon, TOC (mg/L)*: This measures the total amount of organic substance present in leachate; it is usually used in mass balances due to the possibility of taking measurements from both liquid and solid matter.
- *Total Kjeldhal Nitrogen, TKN (mg/L)*: This measures total nitrogen present in leachate expressed as organic nitrogen and ammonium.
- *N–NH_4+ (mgN/L)*: Under anaerobic conditions ammonia cannot be oxidized and remains in the leachate.
- *N–NOx (mgN/L)*: Oxidized forms of nitrogen are usually present only in leachate from aerobic or semiaerobic landfills.

The more significant parameters are represented by BOD_5/COD ratio and ammonium concentration. The first provides information on the degree of stabilization of the leached waste, whereas the second, although not strictly a stabilization index, is of considerable importance in view of the fact that ammonia is largely involved in the long-term impact of an anaerobic landfill.

Deposited Waste

- *Respiration index, RI_4* (mgO_2/gTS): This parameter measures the amount of biologically degradable substances still present in landfilled waste.
- *Fermentation index, GB_{21} (Nl/kgTS)*: By means of an anaerobic fermentation test, this index calculates the amount of biogas produced by a sample of waste. Accordingly, it may be considered, on a par with respiration index, an indicator of the quantity of biologically degradable substances present in the deposited waste.
- *TOC (%)*: This parameter measures total organic carbon in a waste sample.
- *Eluate quality*: By means of an elution test the potential leachate emission of a sample of waste can be characterized. Several parameters can be analyzed: BOD_5, COD, BOD_5/COD, TOC, TKN, N–NH_4^+, N–NOx.

The more significant of the above parameters are RI_4 and GB_{21}, providing information on the degree of stabilization of the deposited waste.

DEFINITION OF LANDFILL QUALITY PARAMETERS

The application of a methodology to evaluate the achievement of FSQ implies the creation of a proper terminology. The following definitions are all linked to the concept of bottom values, i.e., values attainable through application of specific technologies or particular combinations of the latter.

- *CSQVt (Current Storage Quality Value at t years)*
  The parameters values are measured at a certain time t after closure of a landfill.

- *BATV (Best Available Technology Value)*
  These are values achievable following application of the best available techniques method. They represent the most effective and advanced stage in the development of activities and their methods of operation indicate the practical suitability of particular techniques in reducing emissions into the environment. They include techniques developed on a scale that allows implementation under economically and technically viable conditions, taking into account costs and advantages, whether or not the techniques are used or produced inside the country, as long as they are reasonably achievable to landfill operators.
- *BTV (Best Technology Value)*
  These are values achievable through application of the most effective technology with the aim of establishing a high level of environmental protection without any cost restriction, as in the case of the best available technology.
- *LTV (Legislation Threshold Value)*
  These are reference table values, which must be fulfilled at any given time, as, established by law.
- *FSQV (Final Storage Quality Value)*
  These are values of site-specific parameters, which define a quality of the landfill matrices (solid, liquid, gas) in equilibrium with the environment without creating any kind of hazard.
- *RQV (Rock Quality Value)*
  This value indicates a theoretical condition in which there is a substantial absence of any contaminant emissions.

Values attainable for the most significant parameters by means of application of the best available techniques/technologies are presented in Table 16.1.1, on the basis of data originating from different experiences. The table also includes a proposal of threshold values set by some national regulations.

In the first column, typical values (CSQVt) for old sanitary landfills are illustrated (Bonelli, 2003). Values show emission quality achieved by small landfills of the approximate age of 20–30 years operated in a traditional way (biogas capture, leachate collection, sealing surface, no water infiltration, or leachate recirculation). These data are compared with data from old open dumps, illustrated in the second column (Bonelli, 2003). It can be observed that emissions from open dumps achieve lower values than those observed for traditional landfills. This is the result of uncontrolled conditions: open burning, water infiltration, no compaction, air circulation etc., all of which promotes the mobilization of contaminants (biological degradation, leaching) and stabilization of the waste mass.

Subsequently, data relating to a series of operational techniques are illustrated: the third column shows a summary of data from different European anaerobic landfills of varying ages in which waste subjected to mechanical–biological treatment was deposited (Bone et al., 2003; Bonelli, 2003).

The fourth column illustrates a series of data from bioreactor landfills: a bioreactor landfill is a sanitary landfill in which the extent of organic waste decomposition, conversion rates, and process effectiveness are increased by means of specific management activities and operational modifications. The main feature of a bioreactor landfill is not only the addition of liquid, commonly leachate but also stormwater, wastewater, and wastewater treatment plant sludges. Various configurations of bioreactor landfill are encountered: aerobic, anaerobic, and hybrid. The latter accelerates waste degradation by employing a sequential aerobic–anaerobic treatment; the result is the earlier onset of

Table 16.1.1 Summary of the Current Storage Quality Value (CSQV), the lowest values Attainable through the application of the Best Technologies (BTV), Law Threshold Values in national regulations (LTV) and Proposed Threshold Values (TV)

	CSQV Old Traditional Landfills	CSQV Old Open Dump	BTV MBT	BTV Bioreactor Landfill	BTV Aerobic Landfill	BTV Thermal Treatment	LTV Discharge Values	TV Proposed
BIOGAS								
CH_4 (%)	38–61			1–50	2–8			
CO_2 (%)	30–41			3–37	2–15			
CO_2/CH_4	0.67–0.78							
O_2 (%)	0–1.6			0–20	5–20			
LFG (Nm^3 CH_4/h)				4.45	10			<25
LFG (Nl $CH_4/m^{2*}h$)								<0.5
LEACHATE								
COD (mg/L)	478–8356	38–29.4	228–4670	400–4125	1000–4020	20–40	160^a–200^b	
BOD_5 (mg/L)	45–2720	4–13	3–202	30–100	50–260	5	20^b–40^a	
BOD_5/COD	0.06–0.29	0.044–0.1	0.003–0.08		0.03–0.1			0.1
TOC (mg/L)		21–148	78–1480			10.3		
TKN (mg N/L)	183–3232	<5–94		129–200	1449	0–2	$<70^b$	

$N-NH_4^+$ (mg L/L)	175–3080		1.8–1130	18–1000	100–1400	0.2–1	<15[a]	10–50
$N-NO_x$ (mg L/L)	<0.1–6.2		<1–16.1	0.1–195	3.7		<20[a]	
DEPOSITED WASTE								
Solid								
RI_4 (mgO_2/mg TS)	1.97–6.1		7.32	3.95	0.55			<2–2.5
GB_{21} (Nl CH_4/kg TS)	0.59–15.6		15.7	47	7.1			<5–10
Elaute								
COD (mg/L)	114–1577		481.50	70	219	80–250		
BOD_5 (mg/L)	17–176		40.83	14	15–34	40–140		
BOD_5/COD	0.08–0.26		0.09	0.11	0.13			
TKN (mg L/L)	25–109		30.00	6.5	26	<1.5		
$N-NH_4^+$ (mg L/L)	22–40			3.3	17–55	<5		
$N-NO_x$ (mg N/L)					0.3	<0.7		

The values come from different laboratory and full-scale experiences and are not necessarily related to each other.
MBT, mechanical–biological treatment.
[a]Table 3 of Appendix 5 to Part 3 of D.Lgs. 152/2006, Italy.
[b]AbwV-Abwasserverordnung (1997)—51st Appendix, Germany.

methanogenesis compared with aerobic landfill. In particular, the data illustrated were derived from analysis of several hybrid bioreactor landfills in which leachate recirculation had been applied (Green et al., 2005; Magnani et al., 2005). Furthermore, data from PAF (process combining Pretreatment, natural Aeration and Flushing) treatment on a laboratory scale lysimeters have been included (Cossu et al., 2003).

The fifth column provides the results of analysis of several landfills operated under aerobic conditions, both by means of forced (Cossu, 2004; Ritzkowski et al., 2006; Ritzkowski and Stegmann, 2007; Heyer et al., 2005b) and natural aeration (Hanashima, 1999; Chong et al., 2005).

The sixth column shows a summary of data from European landfills for MSW incinerator residues, such as bottom ash and fly ash (Robinson et al., 2004a,b).

The wide variety of ranges observed for the various technologies should be underlined. The main reason is that collected data are often not homogeneous from a temporal point of view. In fact, both the age of the analyzed landfills and duration of treatment may vary considerably. Moreover, analysis of the data revealed how promising technologies, such as in situ aeration and bioreactor landfills, did not achieve the expected results The latter may have been due to the fact that data were obtained from young landfills, 3–4 years old, (particularly for bioreactor landfills), or from landfills in which aeration treatment had not reached completion. Indeed, an underestimated cause may be represented by the high leachate hydraulic level within the waste mass. This would make air circulation and gas extraction difficult. However long-term monitoring would lead to more satisfactory results.

Discharge values established by Italian and German regulations are illustrated in the seventh column. Leachate discharge values reported are derived from regulations relating to urban wastewater treatment and prevention of water pollution; no specific current regulations define target values for leachate, biogas, and residual waste at the end of the aftercare period. To achieve FSQ it is mandatory that migration of leachate into the groundwater is not allowed to produce any "significant" changes to groundwater quality, where "significant" should be interpreted as a level that society agrees to tolerate in the context of sustainable development. In this way, fixed discharge values may be of use in evaluating completion attainment, although analysis of Table 16.1.1 reveals how the latter cannot be achieved in the absence of secondary treatment, particularly with regard to COD and ammonia concentrations.

A proposal of LTVs is included in the last column. The total amount of LFG extracted from the landfill, irrespective of size, should not exceed 25 m^3 CH_4/h, i.e., the minimum volume that can be treated in a high quality standard flare (Stegmann et al., 2006). The proposed value of 0.5 lN CH_4/m^2h for areal LFG production is compatible with methane oxidation in top-cover systems (Chapters 9.5 and 11.1).

Proposed values for leachate quality are limited to BOD/COD ratio and ammonium concentration; the values are derived from the work of Knox et al. (2005) and from the Guidelines of the Regione Lombardia (2014).

Proposed values for RI_4 and GB_{21} respect target values suggested in previous studies of Stegmann et al. (2006) and, again, in the Guidelines of the Regione Lombardia (2014).

Analysis of the waste quality attained following application of best technologies reveals the achievement of lower values.

THE ASSESSMENT PROCEDURE

A methodology based on table values and risk analysis is hereby proposed for landfills.

The procedure is based on two sequential steps: during the first step, it is necessary to verify whether residual landfill emissions respect reference table values for a selected set of parameters; the second step consists in a risk analysis aimed at evaluating whether emissions are compatible with the specific quality of local environmental media. Risk analysis implementation allows LTV to be set at scarcely restrictive levels.

The most significant parameters, suited to evaluating emission quality and FSQ achievement, should be selected.

The procedure structure is iterative: a point of analysis and compliance with LTV is set for each of the polluting emission sources, biogas, leachate, and residual waste. The evaluation procedure

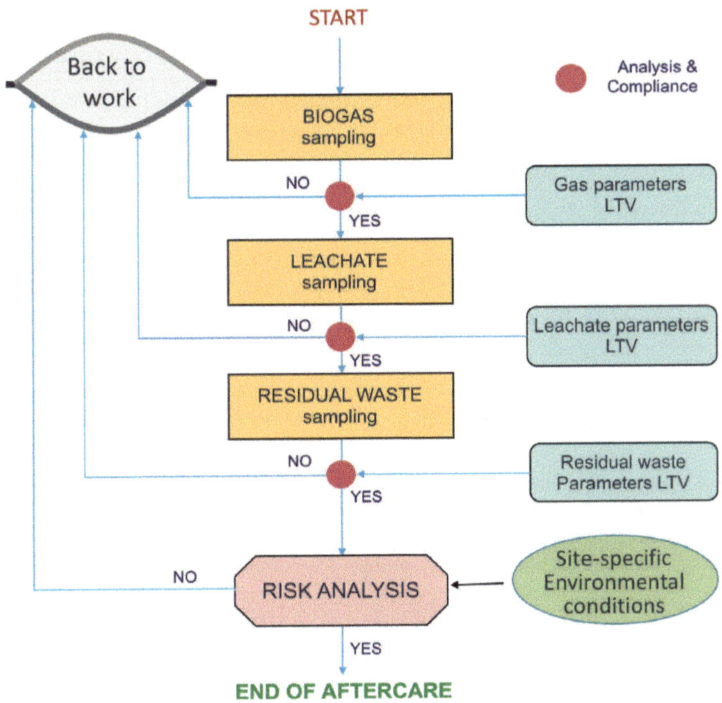

Figure 16.1.2 Scheme of the procedure proposed to assess Final Storage Quality of a landfill.

commences with biogas analysis, due to the fact that biogas generation is shorter lasting compared with other emissions (Höring et al., 1999).

If biogas parameters comply with target values, the evaluation continues with leachate and residual waste; should the target values not be met, it will then be necessary to return to the landfill and proceed further with landfill operations. To ensure LTV compliance, it may prove necessary to modify the operational technique, potentially implementing in situ enhancing measures. The same procedure is then applied to the other points of analysis and compliance. When LTV are met for all emission sources, risk analysis may proceed to evaluate whether emissions are compatible with the specific quality of the local environmental media and verify attainment of completion.

The procedure should continue for the entire duration over which a landfill represents a possible source of risk for both the environment and humans.

A schematic representation of the procedure applied to assess FSQ of a landfill is shown in Fig. 16.1.2.

CONCLUSIONS

Current legislative approaches establish the length of the aftercare period while failing to emphasize the fact that commonly practiced management techniques are not capable of guaranteeing the adequacy of the period to reach an equilibrium point with the environment.

The proposed methodology combines the use of table values and risk analysis. The definition of target values is linked to the long-term observation of development of landfill emissions. Nevertheless, the majority of available information refers to traditional sanitary landfills, while there is a scarcity of long-term observations for modern landfills operated through application of new technologies (pretreatment, in situ treatment, etc.; see Chapters 4 and 14).

Compliance with reference table values may be determined through use of currently available tests to assess the extent of waste stabilization.

Following this first phase, risk analysis can be applied. Once an acceptable risk value has been established, the best techniques to be applied to achieve FSQ and the amount of time needed can be established.

References

AbwV, 1997. Verordnung über Anforderungen an das Einleiten von Abwasser in Gewässer (Abwasserverordnung - AbwV). https://www.gesetze-im-internet.de/abwv/BJNR056610997.html.

Belevi, H., Baccini, P., 1989. Long-term behavior of municipal solid waste landfills. Waste Management & Research: The Journal of the International Solid Wastes and Public Cleansing Association, ISWA 7, 43–56. https://doi.org/10.1016/0734-242X(89)90007-4.

Bone, B.D., Knox, K., Picken, A., Robinson, H.D., 2003. The effect of mechanical and biological pre-treatment on landfill leachate quality. In: Proceedings Sardinia 2003. Ninth International Waste Management and Landfill Symposium. CISA Publisher.

Bonelli, G., 2003. EVAPASSOLD – Evaluation and Preliminary Assessment of Old Deposits. LIFE99 ENV/A/000390. http://ec.europa.eu/environment/life/project/Projects/index.cfm?fuseaction=search.dspPage&n_proj_id=1566&docType=pdf.

Brundtland, G.H., 1987. Our common future: report of the World Commission on environment and development. United Nations Comm 4 (300). https://doi.org/10.1080/07488008808408783.

Chong, T.L., Matsufuji, Y., Hassan, M.N., 2005. Implementation of the semi-aerobic landfill system (Fukuoka method) in developing countries: a Malaysia cost analysis. Waste Management 25, 702–711. https://doi.org/10.1016/j.wasman.2005.01.008.

Cossu, R., 2004. Attraversamento delle discariche di via Caruso da parte della linea ferroviaria ad alta capacità. Condizionamento ambientale dei rifiuti: modulo sperimentale (In Italian: In situ waste treatment of waste at the Via Caruso Landfill in Moden, in view of the excavation for building the fast train track line). Final report. IMAGE Department, University of Padova, Italy.

Cossu, R., Raga, R., Rossetti, D., 2003. The PAF model: an integrated approach for landfill sustainability. Waste Management 23 (1), 37–44.

Decreto Legislativo 3 aprile, 2006. n. 152 (D. Lgs. 152/2006) Norme in Materia Ambientale (In Italian). http://www.gazzettaufficiale.it/atto/serie_generale/caricaDettaglioAtto/originario?atto.dataPubblicazioneGazzetta=2006-04-14&atto.codiceRedazionale=006G0171.

Fourie, A.B., Morris, J.W.F., 2003. The irrelevance of time as a criterion for aftercare provision. In: Proceedings Sardinia 2003. Ninth International Waste Management and Landfill Symposium. CISA Publisher.

Green, R.B., Hater, G.R., Goldsmith, C.D., Kremer, F., Tolaymat, T., 2005. Commercial-scale aerobic-anaerobic bioreactor landfill operations. In: Proceedings Sardinia 2005. Tenth International Waste Management and Landfill Symposium. CISA Publisher.

Hanashima, M., 1999. Pollution control and stabilization process by semi-aerobic landfill type: the Fukuoka methods. In: Proceedings Sardinia 1999. Seventh International Waste Management and Landfill Symposium. CISA Publisher.

Heyer, K.U., Hupe, K., Stegmann, R., 2005a. Landfill aftercare-Scope for actions, duration, costs and quantitative criteria for the completion. In: Proceedings Sardinia 2005. Tenth International Waste Management and Landfill Symposium. CISA Publisher.

Heyer, K.U., Hupe, K., Ritzkowski, M., Stegmann, R., 2005b. Pollutant release and pollutant reduction – impact of the aeration of landfills. Waste Management 25 (4), 353–359. https://doi.org/10.1016/J.WASMAN.2005.02.007.

Hjelmar, O., Hansen, J.B., 2005. Sustainable landfill: the role of final storage quality. In: Proceedings Sardinia 2005. Tenth International Waste Management and Landfill Symposium. CISA Publisher.

Höring, K., Kruempelbeck, I., Ehrig, H.J., 1999. Long-term emissions behaviour of mechanical-biological pre-treated municipal solid waste. In: Proceedings Sardinia 1999. Seventh International Waste Management and Landfill Symposium. CISA Publisher.

Knox, K., Braithwaite, P., Caine, M., Croft, B., 2005. Brogborough landfill test cells: the final chapter. A study of landfill completion in relation to Final Storage Quality (FSQ) criteria. In: Proceedings Sardinia 2005. Tenth International Waste Management and Landfill Symposium. CISA Publisher.

Laner, D., Crest, M., Scharff, H., Morris, J.W.F., Barlaz, M.A., 2012. A review of approaches for the long-term management of municipal solid waste landfills. Waste Management 32, 498–512. https://doi.org/10.1016/j.wasman.2011.11.010.

Magnani, M., Magnani, P., Pinamonte, S., 2005. Aerobic-anaerobic treatment of MSW organic fraction in landfill: a bridge to bioreactor technology. In: Proceedings Sardinia 2005. Tenth International Waste Management and Landfill Symposium. CISA Publisher.

Pivato, A., 2003. An overview of the fundamentals of risk assessment applied to the aftercare landfill impact. In: Proceedings Sardinia 2003. Ninth International Waste Management and Landfill Symposium. CISA Publisher.

Regione Lombardia, 2014. Linee guida per la progettazione e gestione sostenibile delle discariche (Guidelines for design and sustainable management of landfilling). Bollettino Ufficiale, n.41, 10 ottobre 2014. https://www.insic.it/Tutela-ambientale/Notizie/Lombardia-nuove-Linee-guida-per-la-progettazione-e-gestione-/c8c6b67b–48f2-4490-9105-32adc96906e8.

Ritzkowski, M., Stegmann, R., 2007. Controlling greenhouse gas emissions through landfill in situ aeration. International Journal of Greenhouse Gas Control 1, 281–288. https://doi.org/10.1016/S1750-5836(07)00029-1.

Ritzkowski, M., Heyer, K.U., Stegmann, R., 2006. Fundamental processes and implications during in situ aeration of old landfills. Waste Management 26, 356–372. https://doi.org/10.1016/j.wasman.2005.11.009.

Robinson, H.D., Knox, K., Bone, B.D., 2004a. Environment Agency Science Report P1–494/SR1. Improved Definition of Leachate Source Term from Landfills. Phase 1: Review of Data from European Landfills. https://www.gov.uk/government/uploads/system/uploads/attachment_data/file/290387/scho0904bigd-e-e.pdf.

Robinson, H.D., Knox, K., Formby, R., Bone, B.D., 2004b. Environment Agency Science Report P1–494/SR2. Testing of Residues from Incineration of Municipal Solid Waste. https://www.gov.uk/government/uploads/system/uploads/attachment_data/file/290379/scho0105bijb-e-e.pdf.

Stegmann, R., Heyer, K.U., Hupe, K., Ritzkowski, M., 2003. Discussion of criteria for the completion of landfill aftercare. In: Proceedings Sardinia 2003. Ninth International Waste Management and Landfill Symposium. CISA Publisher.

Stegmann, R., Heyer, K.-U., Hupe, K., Willand, A., 2006. Deponienachsorge – Handlungsoptionen, Dauer, Kosten und quantitative kriterien für die Entlassung aus der Nachsorge. Dessau, Umweltbundesamt, p. 172 cited in Laner et al., 2012.

Vettorazzi, G., 2005. Bilancio di massa del carbonio e dell'azoto in alcuni modelli di discarica per la valutazione della sostenibilità. Thesis work. IMAGE Department. University of Padua.

16.2

LANDFILL IN SITU AERATION

Marco Ritzkowski and Rainer Stegmann

INTRODUCTION

Solid waste disposal sites, ranging from open dumps to sanitary landfills as such cannot be considered a sustainable option for waste management. During waste deposition, closure and postclosure emissions are generated that do have a negative impact on the environment and potentially represent a threat to human health.

According to the Council Directive 1999/31/EG (EU, 1999) and the German Landfill Ordinance (BMU, 2009) landfill gas (LFG) has to be captured and preferably used for energy production during waste deposition, landfill closure and aftercare. Due to the progression in waste bio-stabilization LFG generation rates are gradually decreasing and gas treatment becomes economically unattractive after a period of some decades (see also Chapter 9.4). At this time and in the long term, methane is continuously produced and eventually released into the environment. These residual LFG emissions account for approximately 10% to 15% of the total LFG generation potential of the municipal solid waste (MSW) (Fig. 16.2.1). Against this background, landfill aeration represents only one part in a sequence of measures, starting with waste deposition and ending with the termination of aftercare. Aeration effectively stimulates processes that are capable to transfer a landfill from the postclosure phase, where gas utilization is no longer profitable and/or feasible, to a long-term phase where active provisions for emission control are no longer needed. At that time, the final top cover can be installed, and a basis for multiple options of landfill after use has been established. Moreover, the German Landfill Ordinance prohibits landfilling of untreated MSW since June 2005.

Worldwide, LFG contributes approximately 25%–27% (67–90 million tonnes per year) to the overall anthropogenic methane emissions (IPCC, 2013). Furthermore, for the vast majority of developing countries no legal requirements toward landfill emission control exist, thus open up a significant potential for future climate projects. Through the application of LFG collection and treatment in combination with landfill aeration, a controlled, fast, and sustainable reduction of these emissions could be achieved.

BASIC PROCESSES DURING LANDFILL AERATION

During controlled air supply into a landfill, two fundamental reactions are observed. Firstly, the biologically available organic compounds (e.g., hydrocarbons, partly none or only hardly degradable under anaerobic

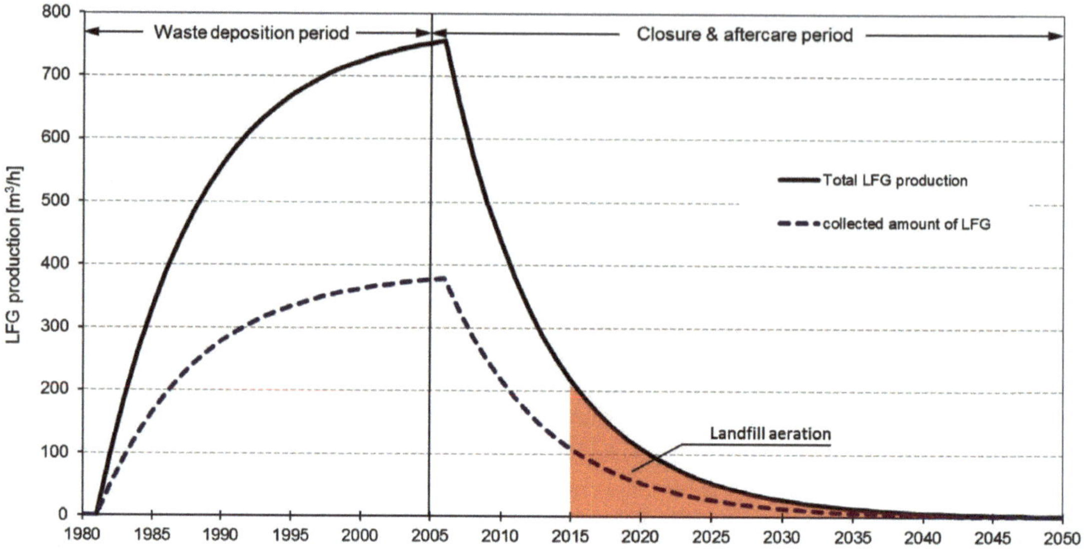

Figure 16.2.1 Residual landfill gas (LFG) potential (red area) that can be avoided by landfill aeration.

conditions) are decomposed into oxidized compounds such as carbon dioxide and water. In addition—compared to anaerobic conditions—significant higher energy is produced (about factor 8–10). As a result a temperature increase in the landfill can be observed. Moreover, humiclike organic substances are produced (Leikam et al., 1998; Prantl et al., 2004). Thus, landfill aeration is associated with decomposition and stabilization of the waste organic fraction including nitrogen containing compounds. Processes to occur can be distinguished between the three phases such as gases, solids, and liquids.

Changes in the gas phase

During air injection, existing LFG becomes diluted due to the supply of oxygen and nitrogen. Microbial methane oxidation may lead to a further reduction in methane concentration whereas aerobic metabolism [respiration, carbon dioxide (CO_2) formation] will result in further increasing carbon dioxide concentrations. The intensity of the biological decomposition processes is reflected by the prevailing oxygen concentration in the off-gas. In consequence, O_2 concentrations can be considered as an indicator for the operating condition of the aerated landfill: Air supply rates should be adjusted in a way that oxygen does not become a limiting factor for the microbial activity, but at the same time oxygen excess should be limited for economic reasons. To maximize the decomposition processes, a constant and sufficient air supply is mandatory. By this means both macro and micropores can be implied in the aeration process.

Changes in the solid waste material

The proceeding biostabilization during aeration can be assessed by analyzing solid waste samples. During aeration, the volatile solid content as well as the total organic carbon (TOC) content will be reduced in consequence of microbial metabolic processes. At the same time the moisture content is reduced due to the release of water vapor from the heated waste mass and water discharge through moisture saturated off-gas. Under the impact of increasing temperatures, nitrogen mineralization processes (ammonification) are enhanced leading to reductions in the N-content of solid waste material. On the other hand, nitrification will be inhibited at temperatures >45°C, and part of the mobilized nitrogen remains in the waste mass since N-uptake into the microbial biomass as well as incorporation into humiclike compounds will occur.

For determination of the stabilization process, solid waste samples are subjected to biological tests under both aerobic and anaerobic conditions. The respiration index (RI_4) will provide information on the remaining oxygen demand (for complete stabilization) whereas the biogas formation test (e.g., GP 21) serves to estimate the residual LFG generation potential of the stabilized waste.

Changes in the liquid phase

Analyzing landfill leachate is a suitable means to gain evidence on the impact of aeration. Decreasing concentrations of organic pollutants [biological oxygen demand (BOD_5), chemical oxygen demand(COD)] are indicators of the proceeding biostabilization process of the waste organic fraction. The major share of organic carbon contained in the waste mass is mobilized and discharged with the off-gas (with CO_2 as the predominating compound). Consequently, the amount of organic carbon dissolved in the liquid phase is decreasing since the metabolism pathway through volatile fatty acids and alcohol (predominating under anaerobic conditions) is no longer controlling the waste decomposition. Reductions in ammonium−nitrogen (NH_4−N) concentrations on the other hand are mainly due to nitrification processes inside the aerated waste mass. This additional transformation pathway, in combination with potential denitrification in the course of leachate percolation through anoxic or anaerobic zones, is promoting the improvement in leachate quality. However, unlike the accelerated discharge of organic carbon under aerobic conditions the nitrogen mineralization rates (i.e., ammonification) are only increasing due to higher temperature levels. Long-term reductions in leachate nitrogen loads are therefore mainly attributed to redistribution and chemical structuring processes, leading, e.g., to the formation of stable humiclike macromolecules.

Subsequent to the wide decrease of the organic carbon compounds as well as the oxidation of ammonium−nitrogen reduced sulfur compounds are oxidized. This reaction leads to a dissolution of sulfate ions in the leachate, associated by a temporarily moderate increase in heavy metal concentrations. As the sulfate oxidation is a widely reversible process, an interruption of the aeration and the consequent change toward anaerobic conditions leads to the reformation of sulfur in unsoluble metal-sulfide compounds. Moreover, under aerobic conditions metals become immobilized through reaction with carbonates, etc., as well as complexation with the humiclike macromolecules.

ENGINEERING ASPECTS OF LANDFILL AERATION

Air injection

Air is actively injected into the landfilled waste mass either through vertical gas wells or horizontal piping systems (Fig. 16.2.2). Typically, the applied positive pressure differences are in a range of 20–80 mbar and do not exceed a maximum of 0.3 bars. Passive aeration concepts on the other hand follow the basic approach of air venting (i.e., the introduction of ambient air into the landfill through its surface or eventually through open gas wells) driven by a negative pressure induced inside the landfill body. Inside the waste mass the air is further distributed by means of convection and diffusion processes. Depending on the rate and duration of aeration, a gradual aerobisation of the landfill occurs. By applying a combined air supply and off-gas extraction operation, the aeration effect and the air distribution can be optimized.

Off-gas extraction and treatment

The off-gas is collected by means of a gas collection system, operated in parallel to the air injection devices. Alternatively, the injected air may pass through the landfilled waste and leave the landfill after its flow through a special designed biocover. The latter has to provide a high methane oxidation potential to minimize unintended emissions of greenhouse gases. If the off-gas is actively extracted from the landfill, methane oxidation is realized either by means of regenerative thermal oxidation (RTO) or, in case of very small methane fluxes, combinations of bioscrubbers and biofilters. The operation of the RTO is based on the exothermic oxidation of residual methane present in the off-gas under high temperatures (>1000°C) in a ceramic bed. The bed is electrically heated prior to the operation; no additional energy supply is required to maintain optimal operation temperature, thus ensuring low operating costs and minimized secondary CO_2 emissions.

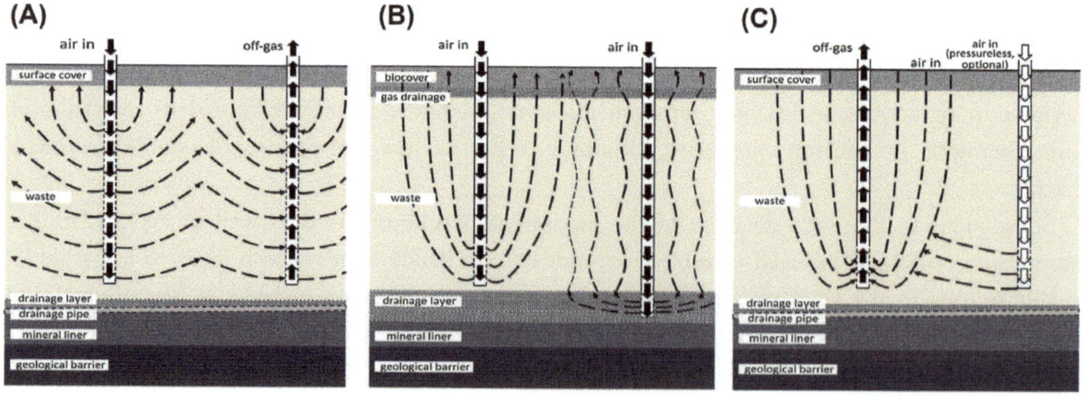

Figure 16.2.2 Schematic outlines of different aeration concepts (A) active aeration with parallel off-gas extraction; (B) active aeration w/o off-gas extraction; (C) passive aeration/air venting (Ritzkowski and Stegmann, 2012).

System operation

During aeration, a gas well system leads air into the landfilled waste. Simultaneously, the slightly contaminated off-gas is collected in a controlled manner by means of additional gas wells operated in parallel. Each gas well is connected to a distribution station by means of separate main headers. At each header, the pipes are connected to the distribution system for either aeration to the system or for the collection of off-gas. Furthermore, each connection includes armatures (valves, ball, stopcocks, etc.) for monitoring gas flux and composition. The installation of the separate main headers may take place on the existing surface cover with a continuous slope toward the gas well heads. A schematic description of the in situ aeration installations is given in Fig. 16.2.3.

To collect the off-gas, negative pressures are adjusted at the gas wells, leading to continuous gas extraction within the area of influence of the specific gas well. By operating the system as described, uncontrolled LFG emissions via the landfill surface and gas migration via the soil vapor into the neighboring subsoil can be completely avoided. The aeration distribution network is connected to a compressing unit for aeration by the main supply duct. The gas collection system for the collection of off-gas is provided with a condensate separator and connected to a blower unit for the collection of waste gas by the main extraction pipe.

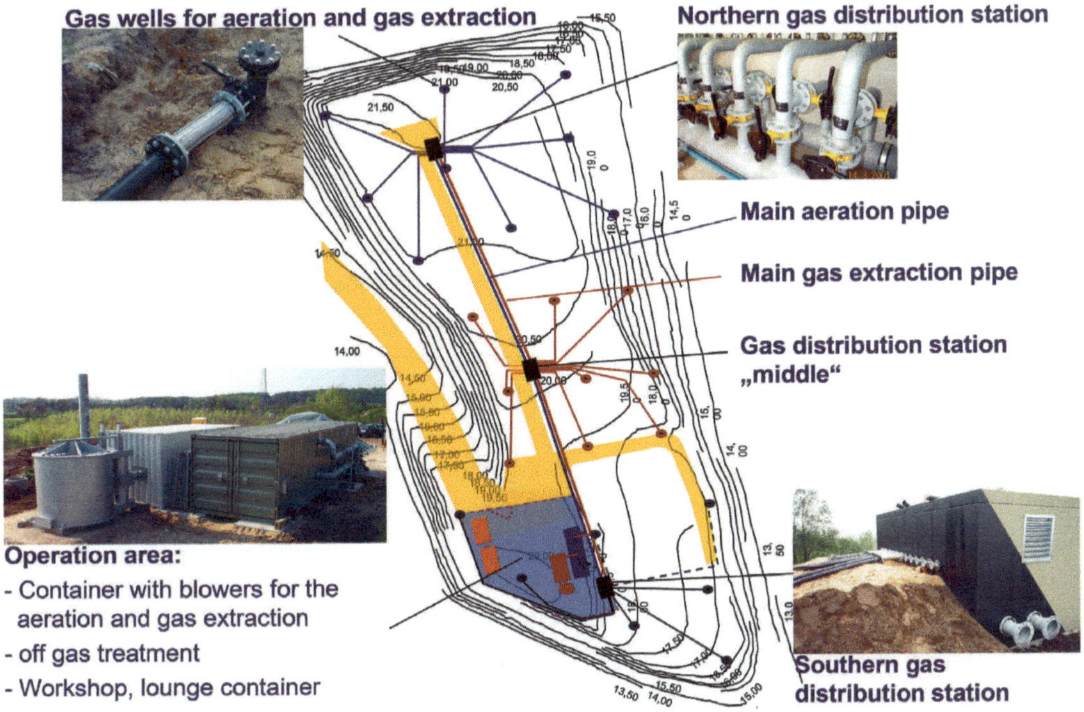

Figure 16.2.3 System components and design plan for landfill aeration according to the low-pressure concept.

The design and operation of the aeration and off-gas collection devices are chosen in such a way that, under normal automated operation, no explosive conditions are to be expected at the off-gas side and, at the same time, a high utilization level of the added oxygen is achieved (meaning that oxygen concentration is low in the off-gas).

MONITORING CONCEPT

On-site monitoring

During landfill aeration on-site monitoring serves operational and environmental safety, contributes to understand the progression in waste stabilization and sets the basis for a cost-benefit analysis.

Since methane is a flammable gas and LFG might potentially exhibit an explosive composition, online monitoring of the off-gas compounds (particularly O_2 and CH_4 concentrations) is mandatory for all aeration projects to ensure both operational and labor safety (see Chapter 9). Moreover, temperature measurements in the extracted off-gas as well as inside the aerated waste mass (eventually amended by periodical measurements of CO concentrations) provide evidence about the operation at noncritical temperatures. As a rule, waste temperatures should not exceed 55°C over longer time periods.

Environmental safety is mainly related to diffuse emissions that may occur either in the gas or liquid phase. Measurements at regular intervals on the landfill surface by means of flame ionization detector are recommended to detect any potential methane emission and to optimize plant operation. Quality inspections of the groundwater (upstream and downstream) are a useful means for both efficiency control of the aeration and reinsurance that any potential impact remains at an environmental acceptable level.

The progression of the biostabilization process can be reliably followed by means of a carbon balance. Under aerated conditions, the major component in the extracted off-gas is carbon dioxide whereas relatively smaller concentrations of methane occur. In comparison with anaerobic landfill conditions the conversion rates of organic carbon are significantly increased thus allowing a faster reduction of the remaining landfill emission potential. Once the amount of biodegradable organic carbon (BOC) contained in the waste mass is known, it might serve as a target value for the complete biostabilization of a landfill. One option to determine the amount of waste biodegradable organic compounds is a long-term test in lab scale lysimeter. More details are provided in the chapter on off-site monitoring. Landfill settlement caused by the aeration process should be monitored as an additional indicator of the progression in biological stabilization. However, as a major share of the determined height reduction is caused by subsidence and/or consolidation, mainly depending on the former landfill operation technique (e.g., intensity of compaction, landfill load, waste characteristic), trends should be identified by biannual recording and plotting of the time depending settling behavior.

Active landfill aeration demands additional energy supply for operating blowers, eventually thermal off-gas treatment devices as well as monitoring equipment. For cost-benefit analysis, these energy amounts have to be considered and converted into secondary emissions. On-site monitoring should therefore include permanent recording of the energy consumption, at least of the total aeration plant, better, for each individual appliance.

Off-site monitoring

Off-site monitoring activities are meaningful amendments to on-site measurements. To conduct these activities, solid waste samples have to be taken from the aerated landfill during different stages: before, during, and after the completion of in situ aeration. Samples and sampling locations have to be selected in a way ensuring maximum representativeness, knowing that due to the complexity and severe heterogeneity of landfills this is difficult to achieve.

The solid waste samples should at least be analyzed for the following parameter: total solids (TS), TOC, RI_4 and gas formation potential (GP_{21}).

The TS content is a meaningful indicator whether the intended biostabilization process will be eventually inhibited by either too low or too high moisture content. If the average moisture content is less than 25% (w/v, related to TS) an inhibition effect of the biological decomposition can be expected. In this case, aeration may be amended by in situ moistening measures. Too high moisture contents may be an indicator for potential leachate accumulation that can limit the air distribution inside the aerated waste mass.

In addition, the determination of the biological stability (RI_4 and/or GP_{21}) in combination with the TOC content facilitates the preliminary assessment of the wastes' stabilization level. In comparison with fresh MSW, a RI_4 value in the range of 5 mgO_2/gTS and a GP_{21} value in the range of 20 $l_N/kgTS$ can be associated with a decomposition rate of up to 90%, thus indicating a suitable starting point for aeration. Based on experiences from former aeration projects, typical TOC concentrations are in the range of 8%—12% (wwt, related to TS) at the beginning of in situ aeration.

Complementary long-term investigations in laboratory scale landfill simulation reactors may be conducted to aid in determining a more precise estimation of the actual and future emission behavior of the waste components. Under optimal boundary conditions (e.g., temperature, moisture content, leachate recirculation, etc.) the potential of organic carbon to be discharged, both via gas and liquid phase is quantified. Details about the investigation procedure, waste characteristic, and comprehensive results from different aeration projects have been published elsewhere (Ritzkowski, 2005; Ritzkowski et al., 2006).

The main indications derived from this off-site monitoring measure are as follows:

- assessment of the residual LFG production potential;
- assessment of the total amount of biodegradable organic carbon;
- assessment of the net oxygen demand required for wide stabilization of the landfill;
- estimation of the necessary time frame for the active aeration measure; and
- prognosis of the leachate quality development during and after aeration.

These meaningful results support the execution of the aeration in full scale and provide the landfill operator with additional and authoritative information in connection with the ongoing project.

IMPACT OF AERATION ON LANDFILL EMISSION BEHAVIOR

Reduction of the organic content of the waste

As an example the degradation of the organic content under aerobic conditions (monitored) and anaerobic conditions [predicted by using a gas prediction model (see Chapter 9.1)] of the German landfill Milmersdorf is presented in Fig. 16.2.4. It becomes obvious that the organic carbon degradation is by a factor faster than under anaerobic conditions. Using a relatively short half-life (8 years) in the model or a longer one (12 years) results in factors of 5 to >7 that the organic carbon reduction can be enhanced.

Reduction of greenhouse gases emissions from landfills

Climate protection is a global task and the reduction of greenhouse gases (GHG) represents one of the major challenges of today. Methane from former waste disposal sites as well as sanitary landfills contributes significantly to the global anthropogenic methane emissions. At the same time, these particular emissions can be relatively simply reduced by the installation and operation of LFG collection and utilization systems. Landfill aeration represents a reasonable amendment to these measures and might be even considered an alternative under certain boundary conditions. During the aeration, methane production is gradually reduced, and the methane generation potential (i.e., the amount of biodegradable organic compounds) is sustainably minimized. In combination with a suitable off-gas treatment (thermal oxidation) GHG emissions can be reduced by more than 90% in comparison with an anaerobic landfill. Therefore, landfill aeration is an effective contributor toward climate protection.

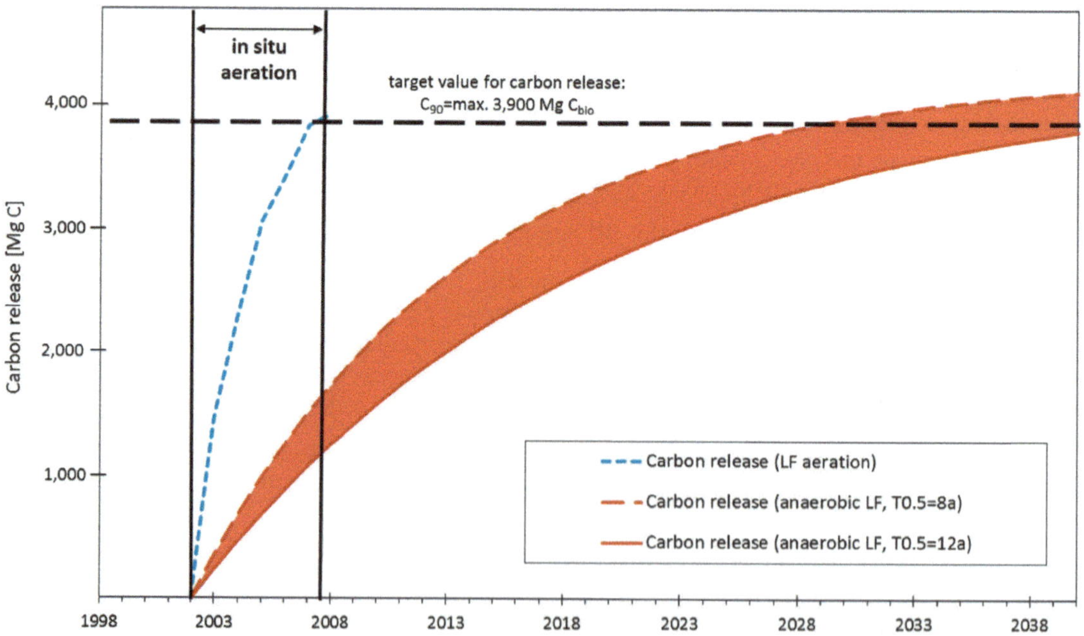

Figure 16.2.4 Carbon release from aerated (measured) and anaerobic (estimated) landfill (example Milmersdorf landfill, Brandenburg, Germany).

Situation in Western Europe

A significant number of sanitary landfills have been closed between 1995 and 2005. Many of these landfills exhibit a volume of several million cubic meter of deposited MSW. For conservativeness an average waste mass of 1 million tonnes per landfill is applied for the following calculations.

It is assumed that the residual 10%–15% of the total LFG generation potential are to be reduced through landfill aeration. Based on 200 kg BOC per tonne MSW a theoretical LFG generation rate of 20–30 m^3/t MSW or 7–11 kg CH_4/tonne MSW can be expected. The aeration of one average landfill (1 M tonnes MSW) would therefore contribute to a reduction of 7000–11,000 t methane corresponding to 147,000–231,000 t $CO_{2,e}$ respectively (calculation based on a 21 times stronger GHG potential of CH_4 in relation to CO_2 following the IPCC, Intergovernmental Panel on Climate Change guidance). Secondary emissions from the operation of the aeration equipment (including a thermal off-gas treatment for the complete avoidance of methane releases) are taken into account by application of 10% attenuation. Potential nitrous oxide emissions (mainly N_2O) are considered for the calculation through application of the IPCC default value of 0.027 kg N_2O/tonne MSW (dry).

During a 7-year crediting period an average annual avoidance of 13,000–24,000 t $CO_{2,e}$ per landfill (aerated) might be realized.

In Germany in situ aeration is declared as a climate gas reduction measure and therefore, financially supported by the government (investment 50% up to 450,000 €, in addition preinvestigations) (Anon, 2015).

Actual monitoring at in situ aerated landfills show high CO_2 avoidance rates as reported in Table 16.2.1.

Table 16.2.1 Greenhouse gas savings by different German landfill aeration projects

Landfill Location	Avoided Methane Emissions (In Terms of $CO_{2,e}$) by Landfill Aeration
Old Kuhstedt landfill, county Rotenburg (Wuemme) Net aeration period: 5.5 years (project completed)	25,000 Mg $CO_{2,e}$ Landfill volume: 240,000 m^3
Old Milmersdorf landfill, county Uckermark Net aeration period: 5 years (project completed)	45,000 Mg $CO_{2,e}$ Landfill volume: 580,000 m^3
Old Doerentrup landfill, county Lippe Net aeration period: 6 years (project ongoing)	50,000 Mg $CO_{2,e}$ Landfill volume: 1,100,000 m^3
Old Suepplingen landfill, county Helmstedt net aeration period: 5 years (project ongoing)	40,000 Mg $CO_{2,e}$ Landfill volume: 450,000 m^3

Situation in economic developing countries

By July 2010, a total number of 475 projects had been registered at the United Nations Climate Secretariat (UNFCCC) in the area "waste treatment and disposal" (UNFCCC, 2010). Based on the figures for projects under German participation, the intended emission reductions are ranging from 13,000 to 1 million tonnes $CO_{2,e}$ per year at an annual average value of about 390,000 t (based on the calculations in the project design documents).

For an exemplary landfill (deposited waste mass: 1 million tonnes) the residual BOC content is assumed to be 60 kg/t MSW deposited. The theoretical methane generation potential is estimated to 22 kg methane per tonne MSW. If this potential would be reduced by 90% in the course of landfill aeration, 462,000 t of $CO_{2,e}$ could be avoided (calculation based on a 21 times stronger GHG potential of CH_4 in relation to CO_2 following the IPCC guidance). Under consideration of secondary emissions (energy supply plus the formation of nitrous oxides during aeration) and provided that no thermal off-gas treatment is applied, a net reduction of approximately 305,000 t $CO_{2,e}$ could be realized. Based on a 7-year crediting period this value corresponds to an annual average of approximately 43,500 t $CO_{2,e}$.

Changes in leachate quality

With regard to the intended enhancement of leachate quality, data are limited and available results (from full scale projects) differ significantly. The general possibility to reduce leachate pollution in a sustainable way has been manifold demonstrated in lab scale tests. If these positive effects become immediately noticeable in the field or not, mainly depends on the specific boundary conditions (e.g., water level inside the landfill, type of aeration system, temperature regime). In the long term, however, the achieved biostabilization of the solid waste material will always promote the reduction in contaminate loads (organics and nitrogen) to be discharged with the leachate. In the following section, a brief overview on available data regarding both, full- and laboratory-scale investigations is provided.

Results from full-scale aeration projects

Data on full-scale results have been reported for Italian, Austrian, Swiss, and German landfill sites. Cossu et al. (2007a) reported on an aerated landfill near the city of Legnago (northern Italy). During the first 14 months no clear indications regarding the effects of aeration on leachate quality could be gained. However, an overall increase of ammonia concentrations was observed. For another Italian landfill, Cossu et al. (2007b) observed a slight increase of nitrate concentrations while ammonia showed variable concentrations with an increasing tendency during the first 6–7 months. At the same time the easily degradable organic pollution (expressed as BOD_5) and the total volatile acids (TVA) decreased at both sites by 30%–50% of the initial concentrations measured prior to the start of aeration.

Prantl et al. (2006) reported on the results of the Austrian Innovative Technologies for Remediation of Landfills and Contaminated Sites (INTERLAND) project. Based on monitoring measurements at the aerated landfill Mannersdorf the leachate quality could be improved. During an aeration period of 12 and 18 months respectively, both the ammonia and the COD concentrations have been reduced by 60%.

Bachofner et al. (2010) summarized the results of the first landfill aeration project in Switzerland. The impacts of the first 24 months of aeration are evaluated by means of a groundwater quality monitoring in the direct downstream of the landfill. For ammonia nitrogen and chloride concentrations an increasing tendency has been observed whereas dissolved organic carbon concentrations remained at an unchanged level. For the investigated heavy metals (As, Pb, Cd, Cr, Cu, Ni, Zn) no significant changes in comparison to the initial situation have been observed.

For an aerated landfill in northern Germany, Heyer et al. (2005) reported on a considerable decrease in the ammonia—nitrogen concentration measured in a groundwater monitoring well near the landfill (down gradient) over a period of 24 months. Further changes in the consistency of the aquifer had been detected regarding the pH value and the redox potential.

Results from lab scale aeration tests

The results from lab scale experiments (simulated aerated landfills) conclusively show an efficient conversion of the biodegradable organic compounds in the waste during aeration. In particular, a fast and significant decrease in both COD and BOD_5 was observed. The concentrations of ammonium—nitrogen (NH_4—N) are noticeably reduced under aerated conditions.

In comparison, under anaerobic landfill conditions only slight decreases in NH_4—N concentrations are observed over long periods of time. Causal are ammonification rates that normally exceed the microbial ammonium uptake necessary for bacteria growth. A reduction in NH_4—N concentrations under anaerobic conditions is therefore closely related to the leachate flow and volume (dilution and wash-out effect).

Under aerated conditions NH_4—N concentrations are significantly reduced. There are indications of nitrification (oxidation of NH_4^+ via nitrite to nitrate) and simultaneous denitrification processes, whereas the latter are assumed to occur mainly in the process water storage container. Ammonia volatilization (and release as NH_3 via the gas phase) can be widely excluded due to the prevailing neutral/slightly alkaline pH values.

Regarding organic compounds aeration leads to an efficient reduction in particular of the easily degradable substances (BOD_5). COD concentrations (that include the moderate and hardly degradable organics) were also significantly reduced, whereas the anaerobically operated controls showed only a slight reduction. In the longer term, concentrations continuously decreased under both, aerobic and anaerobic conditions, suggesting that the influence of aeration decreases. At this stage reduction of COD is more a result of leachate discharge whereas at the beginning of aeration biological conversion processes dominated organics removal.

COMPLETION OF THE AERATION PROCESS

Landfill aeration is a temporary activity that aims at biostabilization to the greatest possible extent. It has to be pointed out that zero-emissions or complete stabilization cannot and will not be achieved by means of the aeration approach that is based on the promotion of biological decomposition processes. Accordingly, the desired state of biostabilization has to be defined by certain criteria representing an endpoint for the active aeration process. These criteria for the completion of aeration refer to both the directly and indirectly affected subjects of protection. They are determined during the ongoing aeration activity (i.e., "online" criteria) as well as through sampling and laboratory investigations (i.e., "laboratory criteria). Moreover, observations describing the progress of biostabilization in a more qualitative way (i.e., long-term trends; "supplementation" criteria) are supplementing the quantitative criteria. While the main online criterion (i.e., BOC conversion rate) is case specific and may vary for each aerated landfill, other criteria are associated with predefined target values or observation trends.

"Online" criteria

Online criteria are the most important indicators for the progress of the biological stabilization process. They are determined without sampling of solid or liquid material and therefore considered as cost-effective and contemporary.

Residual methane generation rate

LFG extraction tests are to be performed toward the final phase of aeration. Through comparison with the initial LFG production rates determined prior to aeration, the progressing stabilization can be evaluated. It is advisable to conduct the LFG extraction test following an intended down time period of the aeration plant of up to 3 weeks. Further information regarding the procedure of LFG extraction tests are published elsewhere (LUA NRW, 2004). The accordant limit value for completion of landfill aeration is a residual methane generation rate (RMGR) of <0.5 L/(m^2 h), corresponding to an LFG production rate of approximately 1 L/(m^2 h). The application of this low value ensures the feasibility of a wide biological methane oxidation in a qualified landfill top cover, regardless of seasonal climatic conditions.

Reduction in the biodegradable organic carbon conversion rate

The amount of BOC in a landfill determines the potential quantity and quality of the main emissions. Therefore, the wide and controlled reduction of this compound is of particular importance. Unlike the procedure for the LFG extraction test, waste samples have to be taken prior to the start of aeration and tested in the laboratory. The samples are analyzed by means of long-term tests (either in lysimeters or by means of standardized gas formation tests) for assessment of the total amount of BOC under anaerobic conditions. The methodology for the lysimeter approach has been described in detail elsewhere

(Ritzkowski and Stegmann, 2007). The setup for gas formation tests is described in the German Landfill Ordinance (BMU, 2009).

Under consideration of the optimized conditions for lab tests in comparison with the significant inhomogeneity of landfills in full scale, the determined TOC decomposition rate is attenuated by 10%. Through this means a specific C90 value for a given landfill is established. In the course of landfill aeration, off-gas volumes and compounds are measured continuously, and the carbon discharge is calculated. Following compliance with the C90 value the residual LFG emissions are assumed to be at a very moderate level, allowing for the wide biological oxidation within the methane oxidation layer of a qualified top cover. The final verification is made by means of an LFG extraction test (RMGR value). At the same time significantly reduced concentrations of organic leachate pollutants are to be expected due to the wide reduction of organic waste compounds.

Laboratory criteria

By means of biological tests under both aerobic and anaerobic conditions, the stabilization process of an aerated landfill can be determined. Unlike the procedure for the C90 value, target values for these criteria are not related to specific landfills but can be universally applied. However, measuring these criteria requires waste sampling from the aerated landfill at regular intervals, associated with a noticeable time and labor effort.

Gas formation potential

Generally landfill aeration follows the same approach as the one often applied for mechanical biological waste treatment (MBT). However, it has to be taken into consideration that the MBT process deals with residual MSW showing a relatively high amount of BOC. Landfill aeration, on the other hand, starts after a wide prestabilization of these compounds under anaerobic conditions inside the landfill. Consequently, the existing limit value for MBT material (German Landfill Ordinance, BMU, 2009) is reduced by a minimum of 75% for application to biostabilized landfills. The proposed target value for GP_{21} is between <1.0 and 5.0 L/kg TS. At a 90% reduction of the BOC in the aerated landfill (compliance with the C90 value) the criterion for the GP_{21} is assumed to be met as well.

Respiration activity

In addition to GP_{21} the respiration activity under aerobic conditions (RI_4) can be applied as a further completion criterion. The respiration test enables an estimation of the amount of residual organic material to become decomposed under aerated conditions. In this respect the RI_4 values represent a suitable amendment to the determination of anaerobic test values. As for the completion of landfill aeration, the proposed target value is between <0.5 and 2.5 mgO_2/g TS. This value is normally kept once the residual BOC has been reduced by 90%.

Supplementation criteria

As supplementation to "online" and "laboratory" criteria, additional aspects like "settling rate" and "waste temperature" may be applied. Both criteria can be measured, either continuously or on a regular basis without severe financial and time efforts. However, unlike the other criteria no target values are defined. Despite, long-term trends are analyzed and qualitatively described.

An improvement of the leachate quality might be applied as a supplementation criterion as well. For landfills showing a base sealing and leachate collection system the quality can be determined directly, whereas for unsealed landfills waste samples can be taken for laboratory investigations (leaching tests). For the latter case, long-term results of groundwater monitoring in the landfills vicinity can be applied as well.

The criteria and proposed target values that can be proposed for completion of landfill aeration are reported in Table 16.2.2.

Table 16.2.2 Criteria and proposed target values for the completion of landfill aeration

Criteria	Definition	Proposed Value or Dimension
Biodegradable organic carbon (BOC) conversion rate (Carbon balance)	A significant reduction in the BOC is associated with both minimized emissions and small residual emission potential. Determination through permanent off-gas analysis (CO_2 and CH_4); target value based on landfill simulation reactor (LSR) tests.	$C_{discharge,\ aerated\ LF} \geq 90\%$ BOC [Mg C]
Landfill gas (LFG) generation	Verification of minimized LFG generation rates by means of an LFG extraction test (to be conducted after a period of 3 weeks without active aeration). LFG expressed as CH_4.	CH_4-formation ≤ 0.5 [l/m^2 h]
LFG formation potential (GP_{21})	Verification of minimized LFG generation potential (laboratory test using waste material taken from the aerated landfill).	GP_{21}: $\leq$1.0–5.0 (l/kg DM)
Respiration index (RI_4)	Verification of a wide reduction in BOC under aerobic conditions (laboratory test using waste material taken from the aerated landfill).	RI_4: $\leq$0.5–2.5 (mg O_2/g DM)
LF settlements	Verification of widely completed BOC conversion and warranty of mechanical LF stability (allowing for the installation of a final surface cover).	Significant reduction (trend)
LF temperatures	Verification of widely completed BOC conversion. Online measurement inside the LF at different depth and areas.	Average temperature in a range comparable to the situation before aeration.

LF, landfills.

ECONOMIC ASPECTS RELATED TO LANDFILL AERATION

For the low-pressure aeration concept (parallel aeration and off-gas extraction and/or treatment) costs are indicated in a range from 0.45 € up to 7 € per tonne of waste material to be aerated (Heyer et al., 2007; van Vossen et al., 2009; Huber-Humer, 2011). These costs include the investment and operation over a period of up to 8 years and are related to landfill aeration projects in Europe. In Germany cost of 1–3 €/t (wwt) of landfilled waste (dependent also on the size of a landfill) can be expected. For aerobic landfills in the United States, cost estimates have been reported in a range of US $3–5 per tonne of waste (Read et al., 2001). These costs include piping requirements, air and leachate injection equipment, gas monitoring, and operation. A more recent assessment of qualitative bioreactor costs and benefits is presented by Berge et al. (2009). The presented model results indicate that aerobic bioreactor landfills are less costly than traditional anaerobic landfills.

The differences in cost estimations (cost ranges) are mainly associated with specific local conditions: small landfills without existing infrastructure (e.g., gas wells, blowers, etc.) exhibit higher investment costs in comparison to larger landfills with gas extraction systems already in place. Furthermore, the complexity of the monitoring program has to be considered when determining the overall costs.

The costs for landfill aeration aiming at an accelerated and wide biological waste stabilization (European approach) might be offset by considerable cost savings during both landfill closure and aftercare. These potential cost savings are mainly related to the required provision of capital for:

- Replacement of a cost-intensive standard surface cover by an alternative long-life surface cover, adjusted to the minimized emission potential of the biostabilised wastes (lower investment costs and particularly lower maintenance costs, especially relevant in Germany).
- Lower costs in connection with groundwater decontamination (in particular at old landfills without a base liner and no leachate collection system in place) due to reduced leachate pollutants.
- Lower operating costs resulting from reduced leachate treatment complexity and shorter treatment durations in landfills with a base liner and a leachate collection system.
- Reduction of the aftercare period (earlier release from aftercare) by at least several decades potentially associated with an earlier revitalization (for possible transformation to a green space) and afteruse (might be of importance particularly in congested urban areas).

When balancing these potential cost savings with the costs for aeration, cost reductions are mainly expected in the medium and long term. Calculations show that landfill aeration implies a total cost reduction potential of at least 10%–25% during landfill closure and aftercare in comparison with the costs associated with anaerobic landfills (Heyer et al., 2007).

References

Anon, 2015. Aktionsprogramm Klimaschutz 2020. www.bmub.bund.de/N51378.

Bachofner, A., Meier, W., Düring, A., 2010. Aerobisation of landfills in Switzerland – discoveries from operation. In: Lorber, K.E., Adam, J., Aldrian, A., Arnberger, A., Bezama, A., Kreindl, G., Müller, P., Sager, D., Sarc, R., Wruss, K. (Eds.), DepoTech 2010, ISBN 978-3-200-02018-4, p. 887.

Berge, N.D., Reinhart, D.R., Batarseh, E.S., 2009. An assessment of bioreactor landfill costs and benefits. Waste Management 29, 1558–1567.

BMU, 2009. Verordnung über Deponien und Langzeitlager (Deponieverordnung – DepV) (Ordinance on Landfill Sites and Long-Term Storage Facilities) vom April 27, 2009 (BGBl. I S. 900), zuletzt geändert durch Art. 5 Abs. 28 Gesetz vom Februar 24, 2012 (BGBl. I S. 212).
Cossu, R., Raga, R., Rosetti, D., Cestaro, S., 2007a. Case study of application of the in situ aeration on an old landfill: results and perspectives. In: Cossu, Diaz, Stegmann (Eds.), Proceedings of SARDINIA 2007-Eleventh International Waste Management and Landfill Symposium. CISA Publisher, pp. 239–240. Session B14.
Cossu, R., Raga, R., Rosetti, D., Cestaro, S., Zane, M., 2007b. Preliminary tests and full scale applications of in situ aeration. In: Cossu, Diaz, Stegmann (Eds.), Proceedings of SARDINIA 2007-Eleventh International Waste Management and Landfill Symposium. CISA Publisher, pp. 885–886. Session B14.
EU, July 7, 1999. Council directive 1999/31/EC of 26 April 1999 on the landfill of waste. Official Journal of the European Communities.
Heyer, K.-U., Hupe, K., Koop, A., Stegmann, R., 2007. Aerobic in situ stabilisation of landfills: long term experience and new developments. In: Cossu, Diaz, Stegmann (Eds.), Proceedings of SARDINIA 2007-Eleventh International Waste Management and Landfill Symposium. CISA Publisher, pp. 241–242. Session B14.
Heyer, K.-U., Hupe, K., Ritzkowski, M., Stegmann, A., 2005. Pollutant release and pollutant reduction – impact of the aeration of landfills. Waste Management 25, 353–359.
Huber-Humer, M., 2011. Grundlagen der in-situ-Aerobisierung. In: Dokumentation des 3. ÖVA-Technologie Workshops 04–2011, April 28/29, 2011. Innsbruck.
IPCC, 2013. Climate Change 2013: The Physical Science Basis. Final Draft Report of the Working Group I Contribution to the IPCC 5th Assessment Report.
Landesumweltamt Nordrhein-Westfalen – LUA NRW, 2004. Materialien Nr. 65 – Arbeitshilfe Deponiegas. LUA NRW (Hrsg.). ISSN: 0948–079X.
Leikam, K., Heyer, K.-U., Stegmann, R., 1998. Einflussnahme auf das Deponieverhalten durch In-Situ Stabilisierung, 1. Hamburger Abfallwirtschaftstage Januar 28–29, 1998, Hamburger Berichte, Band 12, Hrsg.. In: Stegmann, R., Rettenberger, G. (Eds.), Entwicklungs-tendenzen in der Deponietechnik. Economica Verlag Bonn, pp. 83–104.
Prantl, R., Tesar, M., Huber-Humer, M., Heiss-Ziegler, C., 2006. Technical Guideline "In Situ Aeration of Old Deposits" (Technischer Leifaden In-situ Aerobisierung von Altablagerungen). Outcome of the INTERLAND Project. http://interland.arcs.ac.at.
Prantl, R., Tesar, M., Huber-Humer, M., Heiß-Ziegler, C., Lechner, P., 2004. Veränderungen der organischen Substanz im Zuge der In-situ Aerobisierung von Altablagerungen. Altlasten Spektrum 6, 336–344.
Read, A.D., Hudgins, M., Phillips, P., 2001. Perpetual landfilling through aeration of the waste mass; lessons from test cells in Georgia (USA). Waste Management 21, 617–629.
Ritzkowski, M., 2005. Beschleunigte aerobe in-situ Stabilisierung von Altdeponien (Dissertationsschrift). Verlag Abfall aktuell, Stuttgart, ISBN 3-9810064-1-0.
Ritzkowski, M., Heyer, K.-U., Stegmann, R., 2006. Fundamental processes and implications during in situ aeration of old landfills. Waste Management 26 (4), 356–372.
Ritzkowski, M., Stegmann, R., 2007. Controlling greenhouse gas emissions through landfill in situ aeration. International Journal of Greenhouse Gas Control I 281–288.
Ritzkowski, M., Stegmann, R., 2012. Landfill aeration worldwide: concepts, indications and findings. Waste Management 32 (7), 1411–1419.
UNFCCC, 2010. Website of the United Nations Framework Convention on Climate Change, Clean Development Mechanism (CDM). http://cdm.unfccc.int/Projects/projsearch.html.
van Vossen, W., Heyer, K.-U., Woelders, H., 2009. Feasibility study of sustainable emission reduction at the existing landfills Wieringermeer and Kragge in the Netherlands: processes in the waste body and design and costs of enhancing measures (infiltration/aeration). In: Cossu, Diaz, Stegmann (Eds.), Proceedings SARDINIA 2009, Twelfth International Waste Management and Landfill Symposium. CISA Publisher, pp. 145–146. Session B11.

Further Reading

Eurostat, 2015. In: Waste Statistics – Waste Treatment 2012. http://ec.europa.eu/eurostat/statistics-explained/index.php/Waste_statistics.
Gourlay, K.A., 1992. World of Waste – Dilemmas of Industrial Development. Zed Books, London.
UN Statistics Division, 2009. In: Environmental Indicators – Waste. http://unstats.un.org/unsd/environment/wastetreatment.htm.

16.3

AFTERUSE OF LANDFILLS

Anna Artuso, Elena Cossu and Rainer Stegmann

GENERAL ASPECTS

Municipal solid waste management is consistently linked to a strong interrelationship between technology, the environment, the terrain, and society. In the past, although in the presence of a dearth of disposal technologies, this interrelationship was concealed by the small scale of interventions, low population density, limited mobility of the population, and by a limited (at times absent) environmental awareness. Subsequent economic and demographic growth has raised the visibility of land use and environmental issues deriving from waste management. Means of transport, technology, and the Internet have all promoted an increased mobility and enhanced the possibility of a widespread contact throughout a region.

The locality no longer provides residual spaces to be used for the purpose of waste disposal but dictates how waste management systems should be comprised as part of an overall land use plan, linking the diverse areas of the local area in a process ultimately identifying the intended use of the areas within the context of an organized community. Accordingly, the disposal of waste evolves from an inexpert task into a process and tool intended to foster new forms of land use. The planning and design underpinning the management and realization of a waste disposal facility should represent a planning activity capable of contributing to local land use planning and fully integrating with other urban structures (intended as the locations on which the attention of the community is focused), rather than being seen merely as the consequence of exclusions and concealments or limitations dictated by a state of emergency. Moreover, planning and design should strive toward obtaining consent at the time of decision-making, and in no case later than siting of the works. Consent should never be applied retrospectively in the presence of an organized opposition carried out by the populations concerned (see Chapter 4.1 for further details).

In terms of quality of the local terrain, it is no longer a sustainable practice to earmark extensive areas to be occupied by landfill, although the importance placed on this issue may vary according to the reference context concerned. In some countries, land reclamation is a firmly established condition aimed at guaranteeing quality of life under threat from an uncontrolled industrial development and population growth necessitating use of the land in question and avoidance of misuse (Fig. 16.3.1).

In this regard, a landfill represents potential value in line with the context in which it is located. Functional reclamation of the Hong Kong landfills (Sai Tso Wan, Jordan Valley, Sheung Wan,

Figure 16.3.1 View from the Monongahela River to the administrative district in Pittsburg, USA, 1974. *Photo by John F. Alexandrowicz from Alba (2015a).*

etc.) is a concrete example of this concept (Fig. 16.3.2). The latter landfills, all sited on areas of virgin land, were identified as valuable land to be used in creating leisure facilities and green spaces that were inevitably lacking throughout the territory.

Functional reclamation of the majority of landfills, when envisaged, consists mainly in the siting of revegetation works on the final cover with the aim of mitigating impact, although these works are frequently limited to a mere restyling that rarely leads to an effective functional reuse of the area.

However, numerous cases of functional reclamation of old landfills have been described worldwide, attesting the real possibility of undertaking works for the good of the community. Thanks to a cultural approach that tends to view this type of work favorably; and due to a general lack of space, old landfills in Spain, Japan, and China (for example) are frequently used for the purpose of creating urban green spaces with leisure and sport facilities (Box 16.3.1).

Figure 16.3.2 Jordan Valley, Hong Kong. Formerly, a landfill site with landfill gas and leachate, it was restored and converted into a popular recreational park. It is currently a big green lung in Hong Kong becoming a popular recreational facility for the public. Jordan Valley Park was opened to the public in August 2010. The Park (6.3 ha) features a radio-controlled model car racing circuit, horticultural education center, community garden, children's play areas, elderly exercise corner, jogging track, etc. *Photo by James Tsz Fung Wong (2015), from Waste to Photo, Sardinia, 2015.*

Box 16.3.1 Moerenuma Park, an example of landfill afteruse in Sapporo

Moerenuma Park is unique in its typology; it is the biggest park in the world, entirely, built on a landfill, realized in Sapporo, following a master plan by the Japanese–American sculptor, Isamu Noguchi.

Moerenuma is a comprehensive park that is intended to be the base of the "Circular Greenbelt Concept" that combines the green spaces of the city of Sapporo within a loop. Construction commenced in 1982, and the park had its grand opening in 2005 (Fig. 1).

Sculptor Isamu Noguchi created the basic design based on the concept of "the whole being a single sculpture."

In Fig. 2, the site plan and the main structures of the park are represented.

Figure 1 Aerial view of the park.

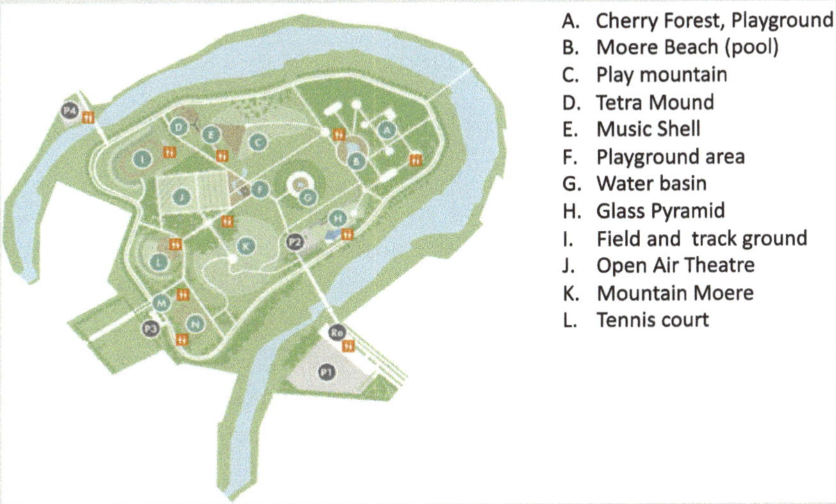

Figure 2 Site plan and the main structures of the park.

Box 16.3.1 *(cont'd)*

The site is designed in a meticulous way. The fountain and hills form many geometric shapes in the expansive grounds, facilities for play equipment are arranged in an orderly manner, and the landscape can be enjoyed as a fusion of nature and art. Visitors may enjoy cross-country skiing and sledding across the snow-covered landscape during winter. The park is conceived to be used for the whole year, with attractions for all seasons.

Serving as the symbol of Moerenuma Park, the Glass Pyramid is the central facility of the park that is considered to be one complete sculpture. This facility accommodates a gallery introducing sculptor Isamu Noguchi with visuals and books, a multipurpose space that serves as a place for citizens' cultural activities, restaurants, shops, etc. (Figs. 3 and 4).

The park is equipped with a green zone that includes seven Play Equipment Areas, which are interconnected by paths.

A shallow pond from Isamu Noguchi's planar form is installed in the middle of the cone-shaped site surrounded by promenades.

Noguchi drew up plans for the park in November 1988, but died, aged 84, just 1 month later.
It took 17 years to complete the final project.

Edited info by Arcoplan Associates, Padova.

Figure 3 The path leading to the top of Play Mountain.

Figure 4 Visitors in winter time.

Options resulting in a renewed use of areas are required to undergo a decisional/design process aimed at assessing the most appropriate final choice in terms of land reclamation, impact on the landscape, environmental sustainability, and community consensus. These options are considerably limited by the fact that landfills are conceived to represent the final solution for a specific area, with reclamation not being taken into account during the important design stage. The limitations to be addressed during this process are linked mainly to management of the waste volumes concerned. Indeed, aboveground waste void space is developed to allow the deposition of as much waste as possible, resulting in the formation of masses that significantly limit reuse options. Old landfills, initially developed with a specific intended use, although not strictly viewed as pollutants, constitute a potential "polluting" presence on the area that frequently deface the landscape and functionality of the area (for example, the typical troncopyramidal shape, etc.).

OPTIONS FOR LANDFILL AFTERUSE

The potential afteruse of landfills and main options can be summarized as follows:

- "Natural" biotope, with or without significant maintenance
- Public park (with or without installations, infrastructure facilities, buildings)
  - Recreational—sport area—thematic parks
  - Botanic garden
  - Golf course
- Grassland
- Agricultural use
  - Grazing land for sheep and cattle
  - Plants for food production
  - Energy crops
- Energy use
  - Photovoltaic installations
- Commercial use
  - Waste treatment facilities (e.g., composting, anaerobic digestion plants, recycling center, etc.)
  - Diverse commercial plants (storage facilities, handicraft businesses, small production business (mostly with office building))
  - Open air research centers
- Housing
  - Houses for permanent living should in general not be built on a landfill. Even if the potential risks have been overcome by means of adequate measures, there will always remain the psychological stress living on a landfill.
  - Concern may arise from, e.g., rumors regarding emissions arising from potential toxic compounds deposited in a landfill. Even if analytical results show no contamination in the atmosphere, often this concern may remain.
  - Offices could be built only if the highest safety precautions are in place (see also Chapter 15.2).

To avoid risks to the environment and to people staying on or adjacent to a landfill, it should only be considered for afteruse if the landfill has a low emission potential and all necessary safety measures have been implemented.

Afteruse of landfills can be manifold. There is a distinction regarding whether the landfill will be open to the public or not. In case the landfill will be used without public access, it may make a difference if food products are grown on the landfill surface or if animals graze on it.

Examples for landfill afteruse without public access are the construction of windmills and photovoltaic plants on the surface. If public access should be allowed, the safety measures should be more stringent. Examples for this kind of utilization may range from recreation areas to commercial use, up to housing. In any case for each kind of utilization, it has to be ensured that there are no direct (e.g., explosions, suffocation) or indirect (e.g., food contamination) risks.

Some examples of potential afteruse options are given in Fig. 16.3.3. Different kinds of utilization can be applied to the same landfill.

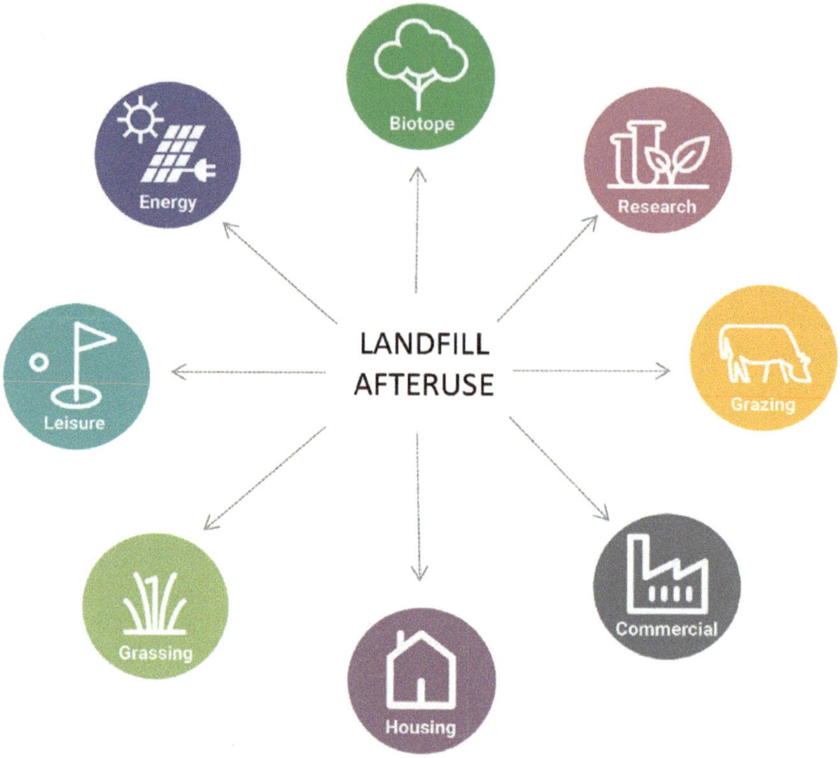

Figure 16.3.3 Graphical representation of the main categories of possible landfill afteruses.

METHODOLOGICAL APPROACH

Generally speaking, designing a final solution for the area and studying the morphology and configuration the site will possess once waste deposition has been completed, implies an inherent need to identify the relationship that the new development will establish with the surrounding natural and cultural landscape, thus determining the success of future relationships. Therefore, the forms assumed by the environment, the history of the area, the locations, the presence of places of interest throughout the area, should all be reviewed in detail with a view to defining the relationship of the facility with the local context, what functions it may fulfill, and—in case of designing, e.g., a park—what species of plants should be used to cover the site, while at the same time envisaging use of the waste mass as a "plastic" element (an architectural term for rich, three-dimensional or sculptural presence of an architectural element or building) for the reconstruction of a coherent and redeveloped landscape.

Potential risks to be considered in case of landfill utilization

Before planning a specific afteruse project, potential risks that may derive from the landfill should be known. Even if landfills are closed for several decades, there may still be risks that are associated mainly with gas emissions, mechanical landfill stability, and settling. Even if the gas production is very low and restricted to certain areas of the site, landfill gas may accumulate over time (perhaps years) especially in constructions built on, into, or adjacent to landfills. These may be in houses, manholes, shafts, pipes, cable ducts, etc.

Dangers that can emanate from landfill gas emissions may be many:

- Inhalation of toxic trace compounds may cause health effects. In some cases health or well-being of people living close to or on a landfill has been influenced because people were afraid of these toxic traces although they could not have been detected in the living environment of the people, i.e., the psychological effect should be taken seriously.
- There is a great risk for explosions (Chapters 9.3 and 9.4) when explosive mixtures are present in the accumulated gas and a spark is caused, e.g., by switching a light on or open fires. There are many examples of casualties in this regard. At a landfill transformed into a park children entered into a manhole as part of the surface water collection system and ignited a small fire that caused an explosion.
- In shafts and manholes gas may accumulate and substitute oxygen. If people enter this area, they may suffocate. This happened in the past more often than explosions.

Adequate planning and supervision can avoid these potential problems.

Risks associated with the mechanical stability (Chapter 5.2) are as follows:

- Nonhomogeneous settling may occur also at old landfills and can have different effects: landfill surface liners may be affected in their functionality (depressed areas where water can accumulate, fissures, cracks in clay or bentonite liners, as well as holes and tears in plastic liners).
- If the mechanical landfill stability is not secured, e.g., due to water accumulation, sludge disposal, steep slopes, and/or low compaction landfill sliding may occur.
- Because the waste underlying a landfill surface is in general not as stable as soil, another potential danger may be the collapsing of constructions on the surface (e.g., windmills).
- Trees and other deep rooting plants should be avoided because they may influence the integrity of the drain system and the surface liner.

Prerequisite for landfill afteruse

To avoid the potential problems described above, landfills have to be constructed and operated respecting regulations and the state of the art. In addition, the landfill should be biologically stable with a low emission potential (Chapter 16.1). Uncontrolled leachate and gas emissions have to be avoided; residual gas emissions have to be captured and treated or the landfill has to be transferred into and kept in the aerobic stage (Chapter 16.2).

The surface cover system should be engineered in order to be primarily functional to technical requirements according to the different landfill concepts and technologies (Chapter 11.1).

Construction of buildings above the landfill should not be in direct contact with the landfill surface so that air can circulate underneath and infrastructure facilities as pipes, manholes, etc., should be ventilated if necessary.

The standard a landfill should meet may be also dependent on the kind of afteruse. As a consequence, different kinds of landfill utilization may take place at different degrees of stability reached (Chapter 16.1). In each case a risk assessment should be made for the different options.

Monitoring program

When a closed landfill is considered for utilization, or is already being used, an extended monitoring program will be necessary. Potential gas emissions through the landfill surface can be detected, e.g., by means of a flame ionization detector; in all areas where gases may accumulate or where preferential pathways can be expected gas monitoring devices may be used to measure the gas quality (in general CH_4, CO_2). Also inside constructions and buildings the gas quality has to be supervised. In addition to CH_4 and CO_2, H_2S should also be measured and from time to time trace components have to be analyzed. In areas where gas may accumulate also the O_2 content should be measured. Gas control monitoring should also be done in the near neighborhood of a landfill.

Settlement rates have to be monitored. Also the functionality of the leachate collection and treatment system has to be secured and water levels have to be ascertained; eventual uncontrolled leachate emissions have to be detected. Also ground water control monitoring regarding levels and quality is necessary. Of course the prescribed landfill monitoring program in the aftercare phase has to be executed.

Assessment of the surrounding area

The following paragraphs focus mainly on landscaping, integration into the surrounding area, and public perception. In addition, several examples will be presented. In line with the most elementary principles of land use planning, the assessment of the surrounding area undeniably represents the most important step to be undertaken prior to drawing up of the project and irrespective of the type of intervention to be carried out (Fig. 16.3.5, Scenarios A, B, or C).

In the same way as for any other type of area, the renaturalization of a landfill, whether it be in existence or still at the design stage, is largely linked to an understanding of the area and the territorial structures present to enhance the integration of the area and the establishment of necessary bonds with the preexisting urban and natural structures.

The complexity of the landscape and the wealth of elements characterizing the area should be carefully analyzed in the light of the predominant factors present; the stratification of the area should be investigated and the history of the same derived to enable a coherent future development of the area to be hypothesized.

Once the main objectives of the project have been established, the methodological process should include an accurate assessment of the local area.

Analysis of the landscape and terrain

Assessments of the terrain and land use are fundamental with a view to analyzing the predominant features of the area concerned. In particular, a detailed analysis of the agricultural context and the dominant environmental factors is mandatory in identifying the lines to which the new landform design should adhere and to establish the specific land use of the restored site.

This type of analysis is based on the identification of the environmental components that will be involved in the project. The environmental system is thus split into elementary components. In the same way as in routine land use planning, this type of analysis is conducted by means of overlay mapping of the natural system in which the major components (elevations, waterways, etc.) have been broken down.

Analysis of the residential/infrastructural system

This assessment possibly represents the most important stage of the preliminary analyses and is fundamental in ascertaining the size of the catchment area of potential users, also in view of the vicinity/distance from the main towns and cities that will need to be taken into account. A thorough knowledge of the history of the area and of the existing facilities is mandatory with a view to identifying a potential lack of infrastructure and determining future relationships.

Analysis of the productive system

The existence and articulation of resources (underpinning the specific local characteristics) should be highlighted: identification of the latter will contribute toward revealing the magnitude of the environmental values and pinpointing of those areas displaying signs of incompatibility with potential transformation works. Subsequently, any existing restrictions should be examined.

Analysis of prevalent activities—this should be carried out to assess the possible links to the project and to identify potential involvement of the local manufacturing industry.

In general, an activity is subjected to elaboration of the geography of environmental values that correspond to the geography of compatibility of use of the territory.

The concept of an environmental system that summarizes these aspects should be analyzed according to the division of the three main areas affected—population, activities, and locations—which, through their links with the spaces, society, and the economy and the relationships each has with the history of the area will provide a detailed picture of the system concerned.

PROJECT REQUISITES

Downstream of the outcomes of the preliminary territorial analysis, the planning and design should be developed with the aim of fostering a harmonious insertion of the intervention in the local context, in respect of the morphology of the surrounding landscape, and recreating the essential forms and lines of spatial continuity. A project for reuse of a landfill should restore the terrain to its natural moreover providing an added value and meeting the requirements of the local community by bridging a potential gap in the lack of infrastructures. It should furthermore provide a series of educational, social, cultural, and leisure facilities to attract potential users.

The design concept should moreover promote a positive outcome and economies of scale, bearing in mind the economic feasibility of the transformation, and should strive toward creating an "exportable model" for use as a reference project in similar contexts.

Participation and consensus during the design and planning stage

The success of a project will depend largely on the level of consensus received from the local community who will benefit from the presence of a quality-controlled environment, from local manufacturers and consumers. Indeed, an environmental reclamation project, in the same way as any other type of policy impinging on the environment, implies a conflict of interest between preoccupations relating to quality that underpin the project and the need for economic growth and consumer use. Intense dialog and political debate may attenuate these contrasts: the population, businesses, and representative pressure groups should be kept informed of the choices made, the impact of these, and of associated economical, social, and political values to overcome the scepticism of the public with regard to the objectivity of the information provided and the motivations underlying the initiatives. This is particularly important when doubt is shed on the credibility of the institutions in the light of past experiences.

In a context involving a series of different, at times conflicting, interests, public involvement should be focused mainly on establishing a situation of trust between the administration, the businesses, and the end users, in identifying any significant issues for the community, and in raising consensus by limiting occasions for conflict and improving the decision-making process.

The acquiring of consensus may be obtained by means of awareness programs developed ad hoc for each type of user highlighting the objectives to be achieved and the specific reference context. A series of methods are available for use in raising awareness. Experimental studies have shown, for example, how meetings and seminars are particularly beneficial in establishing and fostering relationships of mutual trust and credibility. Two possible involvement processes are illustrated in Fig. 16.3.4. In

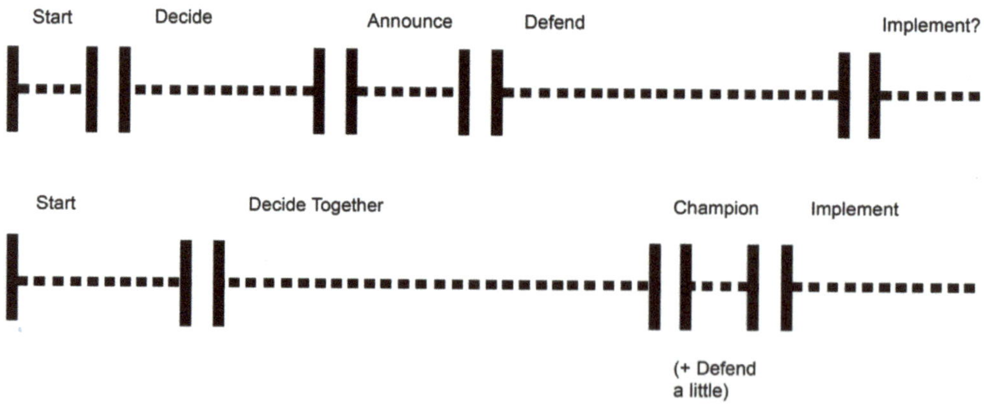

Figure 16.3.4 Involvement processes: diagram 1—Decide—Announce—Defend (common process)—and diagram 2—Engage—Design—Deliver (recommended approach). *Source: (Bishop, 2015).*

the approach "Engage—Design—Deliver," some key representatives of the wider community are brought in at the very start to discuss and agree the scope of the project and the involvement process with the technical team. This takes longer than the start in the old model. The team, the core group, and many others then engage in the process "Decide Together." Then, instead of having to announce the proposals to an unsuspecting public, not only are (most of) the public already aware but some will have contributed directly to those proposals. Instead of objecting they will often support, even "champion" what many by now consider to be "their" project. Some defending is, however, always still needed.

AFTERUSE OF LANDFILLS: POTENTIAL SCENARIOS OF INTERVENTION

When considering the reuse of an existing landfill or hypothesizing the functional use of a landfill yet to be constructed, the main features to bear in mind are environmental conditions, safety, technologies used to mitigate impact, the impact on land use and infrastructure exerted by the facility within the context in which it has been sited, and landscaping in view of the value of transformation elicited on the existing landscape by siting of the landfill on the area.

Functional reuse of a landfill may be envisaged in a series of different situations as follows:

A. Existing landfills requiring works of environmental reclamation or recovery
B. Existing modern landfills either undergoing construction or operational (or in which extension works are required)
C. Landfills still to be designed (details relating to siting, waste volumes to be deposited, etc., may not be known).

Accordingly, three different scenarios are described below.

Scenario A/old landfills: when intended use is dictated by the shape

Reclamation of an old landfill is characterized from the outset by a predominantly negative view of the intervention: to reclaim, i.e., to undertake the recovery of an environmentally compromised situation and reestablish links with the surrounding area. From a morphological viewpoint, old landfills were invariably constructed without any consideration for the landscape or urban planning, with the sole constraint of complying with the height limits defined by the competent Authorities at the time of authorization. In an attempt to maximize the capacity of the facility, landfills were frequently developed without paying due attention to aboveground waste volumes, thus resulting in the formation of improbable-shaped masses that interrupted the natural lines of the landscape. A recurrent skyline was represented by a truncated pyramidal formation that, although designed according to laws enforced at the time (in terms of maximum height, slopes, etc.…), left an invasive and permanent anthropic mark on the landscape. The resulting land conformation significantly limited the choice of intervention, with the preexisting landfill constituting a highly visible element with which the restoration project would be required to interact (Fig. 16.3.5, Scenario A).

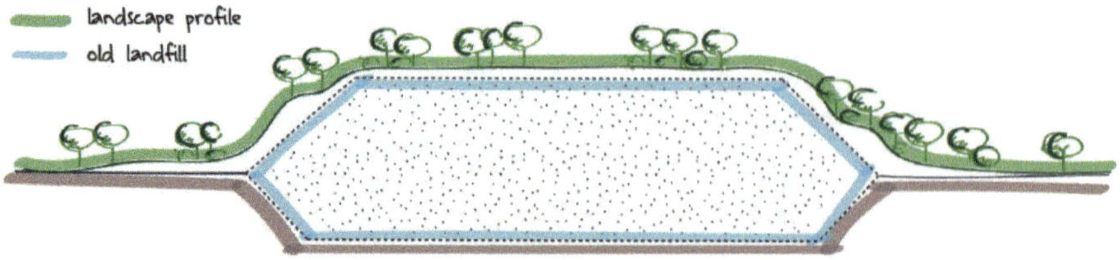

SCENARIO A / Old landfills: when intended use is dictated by the shape

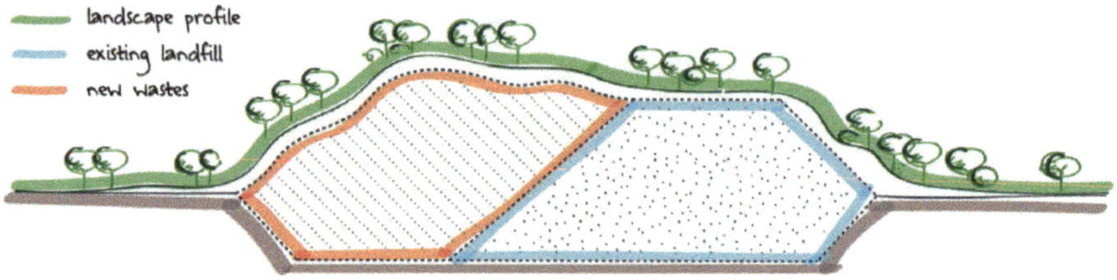

SCENARIO B / Intervention on operational landfills: the use of waste as a plastic material

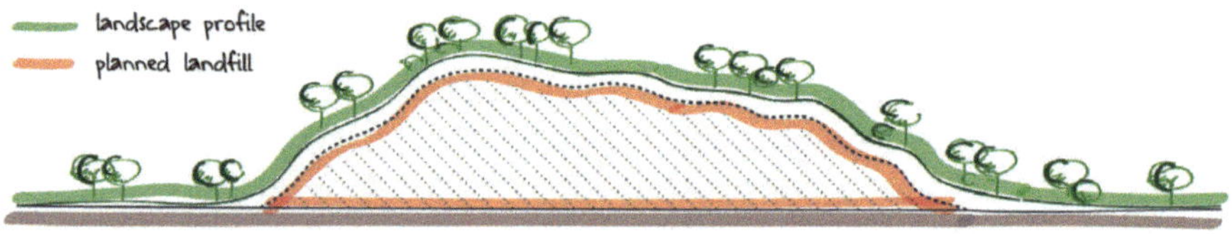

SCENARIO C / Planned landfills: the project should be developed according to the intended use

Figure 16.3.5 Scenarios of intervention: Scenario A diagram of the typical truncated pyramidal shape of old landfills, Scenario B diagram of a possible extension in existing landfills, and Scenario C diagram of a future landfill: the wastes are used as a plastic material to shape the land.

A good functional rehabilitation project should therefore assign an appropriate role to the preexisting landfill, without attempting to mitigate or conceal the same, but rather seeing it as a means of interpreting the entire project, as a continuing testimony to the previous life of the site.

A relevant example that clarifies this concept is provided by the restoration of the Hiriya landfill in Tel Aviv, an impressive 60 m high "mountain" of waste distributed over an area of more than 450,000 m^2 (Fig. 16.3.6). The landfill restoration project provided for transformation of the area into one of the largest urban parks in the world, thus changing the aspect and features of the "mountain" of waste from a negative element of the landscape into a symbol of ecological renovation. The volume of the old landfill stands at the center of the project like a large evocative totem, in clear view of visitors in all corners of the park. Taking into account the preexisting facility, the transformation of the Hiriya landfill was designed as a sort of enormous theme park focusing on waste recycling and evolved into an innovative center for research on recycling technologies that also hosts educational activities.

Figure 16.3.6 Hiriya Landfill, Tel Aviv. *Courtesy of Studio Latz + Partner.*

As a general rule, a functional restoration and landscaping project will need to address the issue of the complex management of a landfill during the postfilling stage. It will need to take into account the presence of the visible impact of some infrastructure (i.e., leachate drainage and/or treatment plants, biogas extraction plants, etc.), of subsidence and settling of the waste, of the existing slopes and topography in general, of water management and layering of the final cover. All these elements, and potential safety problems, if not assessed previously with a view to final recovery, will further limit any possible form of reuse.

The landscaping and agronomic management of the site should be carefully investigated during the planning stage. The area should be planted to enhance a correct balance between surface runoff, infiltration, and evapotranspiration, thus promoting a correct control of infiltration water (see Chapter 11.1).

Indigenous trees and shrubs are to be preferred, having roots that are suited to the substrate and do not interfere with the surface liner and leafy coverage to enhance runoff and evapotranspiration. Herbaceous species should be fast establishing, hardy, and with a low degree of flammability.

In the context of Scenario A, Fig. 16.3.5, possible outcomes are represented by nature parks or green spaces with leisure facilities (theme parks, motocross tracks, cycle paths, golf courses, model plane fields, and sport and leisure facilities in general). Other more complex and structural constructions are hard to achieve.

It is, however, possible to achieve a good outcome in the functional reclamation and landscaping of old landfills denoted by high quality architectural features, as attested by a wide series of projects undertaken throughout the world.

Reclamation of the Barcelona landfill, situated in the Vall d'en Joan in the natural Garraf Park, is still today deemed a reference project of excellence for this type of intervention. The project for restoration of the landfill was drawn up by the Spanish architects Enric Batlle and Joan Roig, and in one sole intervention was aimed at achieving three major aspects: solving a complex technical problem, creating a new community area, and providing a new landscape (Figs. 16.3.7 and 16.3.8).

The complex technical issues deriving from the closure and final covering of the landfill underpinned the rationale behind the working hypothesis: the organization of the stepped consolidation

Figure 16.3.7 Land view of the El Garraf Massif, Barcelona. Project for a community park on the Vall d'en Joan landfill designed by the architects Enric Battle and Joan Roig (Sànchez Fabra, 2015).

Figure 16.3.8 Aerial view of the El Garraf Massif, Barcelona. Project for a community park on the Vall d'en Joan landfill drawn up by the architects Enric Battle and Joan Roig (Sànchez Fabra, 2015).

terraces, the containment banks and road access marked the geometry of the landfill and determined the placing of the pipelines required by the gas treatment plant to generate electricity, for the drainage system and the transfer of leachate. The third objective, the development of a new landscape, was influenced by a desire to merge the old landfill with the Natural Garraf Park. Naturally, the morphology of the site today differs vastly compared to the original features. However, the Garraf Park comprises cultivated valleys that have been modified by means of agricultural techniques (Fig. 16.3.9) purpose adapted to the local geography using a system of terraces, drainage, and cultivation to meet the technical requirements of closure and final covering of the landfill.

The final outcome saw the construction of eleven stepped terraces planted with native drought-tolerant species compatible with integration of the landscape. Moreover, an irrigation system was set up throughout the area to facilitate watering. An underground drainage system was devised to separate the pollutant liquids and recirculate the water to irrigate the park. The landfill also provided biogas used in the production of electricity.

Some wastes were left on the top of the landfill in large steel cages to remind visitors of the origin of the site (Fig. 16.3.10).

Figure 16.3.9 Constructive and vegetation systems, inherited from traditional agriculture (Sànchez Fabra, 2015).

Figure 16.3.10 Detail of plant walls, built of nondegradable products of the old landfill (Sànchez Fabra, 2015).

A similar strategy was adopted by the architect Israel Alba who curated the transformation of the Valdemingómez landfill in Madrid (Fig. 16.3.11). The project for the recovery and transformation of the Valdemingómez landfill in Madrid required the application of complex environmental engineering processes, as well as new architectural strategies. Today, it is a place that can be incorporated, with full guarantees, into the city structure, as long as it is viewed as a monumental public space. It is

Figure 16.3.11 Valdemingómez landfill, Madrid. Project by Israel Alba.

recovered ground, capable of becoming a new, free metropolitan space that can respond to the current and future needs of society, especially if it remains as such over time.

The architectural project undertaken involved the proposal of new strategies to create an area that would remain open, flexible, and dynamic throughout time, in a search for equilibrium between city and nature. The Valdemingómez landfill constitutes an example of a proposed model of continuity between the forest and the surrounding area; a pseudobotanical garden with indigenous species seeking integration into the Parque Regional del Sureste (Southeast Regional Park). It was transformed into a free, public area with pedestrian paths and bicycle lanes, along with woods and wetlands, which have helped to create small, localized ecosystems. The life of both the nature and the city can be observed within.

The project for the general reclamation of the old "Fresh Kills" landfill in New York is one of the most widely known. Firstly, as it is the largest existing landfill worldwide, since closure of the landfill in 2001, the value and significance of this urban site of more than 890 ha have changed considerably. This artificial landscape located to the east of Staten Island is viewed today as an extraordinary resource for the fast-growing population of New York with its pressing need for green spaces. Only 45% of the Fresh Kills site has actually been used as a landfill. The remaining area features the extensive swamps that have characterized the New York archipelago since its origin. The wide variety of natural environments present forms a habitat for numerous indigenous and migratory animal species. The aim of the project was to transform Fresh Kills into a 21st-century urban park (Fig. 16.3.12), while maintaining both the large dimensions and the essential underlying character of the area. The buildings and the activities are all confined to specific areas, thus leaving the rest of the site as open and natural as possible.

The project will be completed over a 30-year period, with the first challenging 10-year stage comprising work on the northern and southern sections of the park. The implementation strategy is based on a series of flexible and incremental steps aimed at ensuring a balanced execution of works to close the landfill, start up the site management processes, and transform the area into a public park (Fig. 16.3.13).

Figure 16.3.12 Illustrative aerial view of Fresh Kills Park (Alba, 2015a,b).

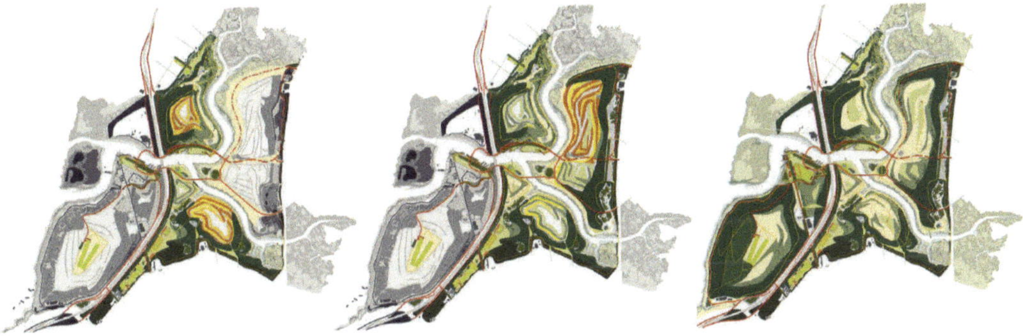

Figure 16.3.13 Chronological evolution of the landscaping of the Fresh Kills landfill, New York. *Graphics: Studio Arcoplan.*

Scenario B/intervention on operational landfills: the use of waste as a plastic material

Reclamation works performed on a landfill will have a far greater chance of success when studied and developed during the stage of active management of the landfill (during the construction stage, during waste deposition, or at the time of a potential extension). Indeed, the employing of new wastes as a plastic material with which to mold the shape extends the pool of potential uses assigned to the project and provides greater freedom in reinventing the final configuration of the area (Fig. 16.3.5, Scenario B), in addition to facilitating important budgetary savings in terms of material to be delivered to modify the structure of the site.

The possibility of intervening during the operational phase of the landfill rather than during the postmanagement phase not only enhances the possibility of modeling the banking of new wastes in line with the project design (Fig. 16.3.14) but also enables operations aimed at the functional recovery and landscaping of the area to be commenced when the landfill is still operational (e.g., in the previously established closed sectors). In other words, when carrying work as described in Scenario B, Fig. 16.3.5, an immediate functional use of the areas is obtained with a progressive involvement of other sectors.

Figure 16.3.14 Cross section of the landfill. The wastes are used as a plastic material to shape the land. *Graphics: Studio Arcoplan.*

The main difficulties encountered in the planning and design are due to coexistence of the ongoing operations of waste deposition and remediation of the landscape, thus requiring careful planning throughout. The establishing of a timeline representing the chronological development of the project is fundamental and should define the different operations to be carried out in line with the state of waste deposition in the landfill sectors (Fig. 16.3.15) according to a series of future short-, medium-, and long-term scenarios developed over a period of no less than 30 years.

The distribution of vegetation should be implemented in full compliance with the original existing landscape and should be completed taking into account not only the technical aspects described above but also the functional reclamation of the area to be achieved once waste placement operations have terminated and the relative sectors have been covered. Plant distribution throughout the area therefore should be envisaged in synergy with the operations and functions provided for by the reclamation project (Fig. 16.3.16).

Furthermore, the undertaking of works at this stage enables the planning of more complex operations that foresee a marked local involvement, thus promoting a widespread consequent positive impact. Accordingly, a project for valorization of an area should be developed in line with the specific aim of creating important synergies with the locally present manufacturing and entrepreneurial concerns.

Scenario C/planned landfills: the project should be developed according to the intended use

When designing a project on the basis not solely of the designated use of the site as a landfill, but rather in view of the ultimate use established, wastes may be employed from the start of the project to shape and model the final morphology of the landscape. From the outset therefore the structure of the project is dependent on the final use of the area. Accordingly, in line with the parameters established by the project, the deposited wastes may be used today as a shaping material for the purpose of creating the area for tomorrow (Fig. 16.3.5, Scenario C).

The location, the form, and the volume of wastes to be banked, the architectural materials, and language of the buildings destined to house the facilities, together with all other features of the project, will need to establish operational synergies to enable this dual function: the first temporary function associated with operations of disposal and treatment of the wastes, and a second more permanent aspect linked to the future intended use determined during the stage of landfill design (Fig. 16.3.17).

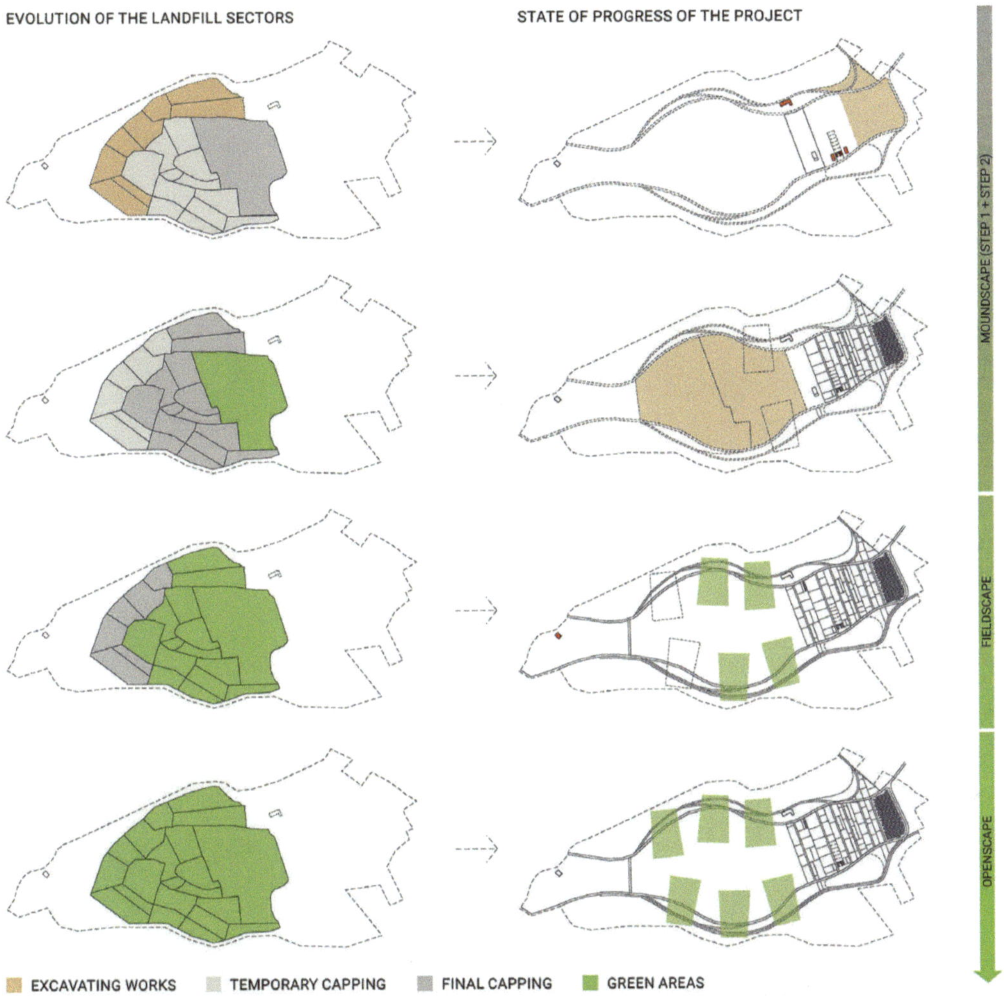

Figure 16.3.15 Example of a timeline: state of progress of the project with regard to evolution of the landfill sectors. The chronological study of the project establishes a link between the banking of soils excavated from the new areas undergoing development and the reclamation works being realized. *Graphics: Studio Arcoplan.*

The planning and design phase for a new landfill therefore provides all the necessary conditions to envisage a coherent contextual change in line with the needs of the locality.

The construction of a new landfill according to a project aligned with the surrounding landscape and urban spaces is dependent on a fully integrated project strategy and a multidisciplinary approach that can only be achieved by relying on a team of designers with competence in numerous sectors (environmental engineering, geology, agronomy, landscape architecture, etc.).

Figure 16.3.16 Example of landscape design based on the intended use provided for in the general project. *Graphics: Studio Arcoplan.*

Figure 16.3.17 Aerial view of the future use of a landfill site. *Graphics: Studio Arcoplan, Padova.*

The above principles are, however, deemed extremely innovative when compared with common practice as they dictate a change in the terms of the project; we are no longer the designers of mere landfills, but rather of community spaces.

To make provisions for the project as a whole, the planning of the intended future use of the area should be articulated on the basis of the three subsequent levels of in-depth technical analysis provided for by current law, i.e., a preliminary design, a definitive project, and an executive project. The study should be subjected to detailed analysis to provide for a detailed budgetary control from the outset. This will allow sufficient sums to be accrued during the operational phase of the landfill to effectively finance the carrying out of all works.

These important concepts, which should drive the planning and design of new landfills and shift the focus of the designer, unfortunately often fail to be taken into account.

CONCLUSIONS

A sanitary landfill, intended as the last link in a circular economy, should be seen as an endeavor that is developed throughout the duration of the operational phase. Completion of construction will only be achieved once the global project has been terminated. Accordingly, the operational phase of a landfill should be seen as an extension of the building of the concern.

In a vision of a circular economy, landfills occupy an unavoidable rather than a disputable position and should be viewed as a necessary facility in the same way as all other services a community relies on. Indeed, it is the overwhelming need for landfills to legitimize construction of the same and likewise justifies the need to ensure they remain in the vicinity of the community they serve.

An environmentally sustainable landfill should assume programmatic significance through a planned intervention that contributes toward the overall design of the locality and meets the requirements of the two life phases of the landfill itself: the operational phase (temporary use as a landfill) and postoperational phase (final use created on the basis of the deposited wastes). Therefore, a precise time horizon and specific technical regulations should be established in line with the final intended use of the site as determined in the initial project.

Municipal waste sanitary landfills should therefore be seen as an integral part of local area planning with a functional rather than a "disposable" intended use. It should thus no longer be construed merely in the light of its intended use, but as a process heralding the development of a "new form of intended use."

References

Alba, I., 2015a. Los paysajes del desecho. Reactivactiòn de los lugares del deterioro. Doctoral Thesis. Universidad Politécnica de Madrid, p. 46.

Alba, I., 2015b. A recovered landfill in the construction of a metropolis: Valdemingómez forest park, over time. In: Cossu, R., He, P., Kjeldsen, P., Matsufuji, Y., Reinhart, D., Stegmann, R. (Eds.), Proceedings Sardinia 2015, 15th International Waste Management and Landfill Symposium. CISA Publisher. ISBN:9788862650212.

Bishop, J., 2015. Working with community: faster, cheaper, better! In: Cossu, R., He, P., Kjeldsen, P., Matsufuji, Y., Reinhart, D., Stegmann, R. (Eds.), Proceedings Sardinia 2015, 15th International Waste Management and Landfill Symposium. CISA Publisher. ISBN:9788862650212.

Sànchez Fabra, I., 2015. Waste landfill environmental restoration at the Garraf's Natural Park, Barcelona. In: Cossu, R., He, P., Kjeldsen, P., Matsufuji, Y., Reinhart, D., Stegmann, R. (Eds.), Proceedings Sardinia 2015, 15th International Waste Management and Landfill Symposium. CISA Publisher. ISBN:9788862650212.

Wong, J.T.F., 2015. Waste to photo 2015. In: Photo Contest at Sardinia 2015 15th International Waste Management and Landfill Symposium, 5–9 October 2015, Cagliari, Italy.

Further Reading

Cosgrove, D., 1990. Social Formation and Symbolic Landscape, University of Wisconsin Press, ISBN 978-0-299-15514-5.

LAE, Landscape Architecture Europe, 2006. Fieldwork. Birkhauser LAE Foundation. ISBN 978-3-7643-7508-9.

Maciocco, G., Sanna, G., Serreli, S., 2011. The Urban Potential of External Territories. Franco Angeli, Milano. ISBN 978-88-568-3739-1.

McLeod, V., 2008. Detail in Contemporary Landscape Architecture. Laurence King Publishing Ltd. ISBN 978-1-7806-7023-2.

17 CHAPTER

ENVIRONMENTAL IMPACT

17.1

ENVIRONMENTAL IMPACTS ASSESSMENT

Raffaello Cossu, Alberto Pivato and Alberto Barausse

INTRODUCTION

Environmental impact assessment (EIA) is a procedure aimed at ensuring that the environmental implications of decisions are fully considered before the decisions are made.

The term "environmental impact" is used to define the alteration of the environment with respect to a "reference" state caused by human activities carried out to implement a program or carry out work. The term "reference state" is used to indicate the state of the environment, whether this be a natural environment or an environment previously modified by man, before the alteration. In this context, the concept of environment includes the complex set of physical, social, cultural, and esthetic factors affecting the individuals and the communities they belong to.

The procedure of EIA aims at evaluating, in the sense of identifying and quantifying, the relationship between the proposed work and the environment in which it is to be implemented. This task is carried out by taking information of various kinds (environmental, social, economic, regulatory, etc.) into account to make an informed judgment on the environmental sustainability of the task. Numerous methods have been proposed for the assessment of environmental impact, many of which emerged during the first half of the 1970s. In the present chapter a specific approach for landfills, based on the works of Cossu et al. (1986) and Andreottola et al. (1989) is presented and commented.

ENVIRONMENTAL IMPACT ASSESSMENT METHODOLOGIES

The methodologies put forward to date for use in the EIA of landfilling are based on the reformulation of broad methodologies, summarized in the following sections (Moore 1973; US-EPA, 1974; US-EPA, 1998; Canter, 1999).

"Overlapping map" method

This method consists in the overlapping of maps describing all the considered elements of impact onto maps of various themes (socioeconomical, morphological, hydraulic, landscape, etc.) to highlight the areas of minimum and/or maximum impact. This method has been applied mainly to studies relating to the siting of infrastructures, roads, motorways, oil and gas pipelines, etc.

Lists of questions and checklists

The list of questions is a series of queries focusing on the different environmental problems that may be encountered. These lists are necessarily compiled by involving as many experts, corporations, citizens, and other stakeholders as possible, and focus on aspects which will subsequently represent the object of the study. The control list, or checklist, represents an evolution of the list of questions; it attempts to provide a systematic means of identifying the activities and elements of impact that may influence both the environment and the environmental categories present.

Networks

The method introduces a falling sequence of cause/effect conditions: in this way it facilitates the identification of primary, secondary, and tertiary impacts, thereby enabling the assessment of cumulated impact as long as the probability of the event happening (the occurrence of impact), its degree of importance and its size are taken into consideration. On closer examination, the latter two factors represent an analogue of the two numbers examined in Leopold's matrix (Leopold, 1971). This method, also known as "impact tree," was applied by Sorensen (Sorensen, 1971) to assess the impact generated by construction of a road.

Correlation matrices

The matrices may be considered as two-dimensional control lists illustrating on one dimension the individual characteristics of a project (proposed activities, elements of impact, etc.), while on the other listing the environmental categories that may potentially be affected by the project. The effects or potential impacts are therefore identified by intersecting the two control lists. The differences between the various proposed types of matrix can largely be observed in the variety, number, and specificity of control lists, as well as in the system of evaluation of the identified impact. With regard to evaluation, this ranges from mere identification of the potential impact to qualitative (weak, moderate, strong, etc.) or quantitative evaluation, which may be either absolute or relative: the evaluation is generally conducted in relation to the outcome of the impact (positive or negative). Numerical evaluation is frequently subject to criticism as it may suggest the introduction of a criterion of objective judgment that is actually hard to achieve, also given that environmental impacts typically affect multiple environmental categories and stakeholders, each with their own requirements for the status of the environment. In Table 17.1.1, examples of different scale of evaluations are reported.

Modeling methods

This class groups different methodologies, all featuring the use of mathematical models to quantify environmental impact. In the case of landfills, important methodologies employing modeling techniques include life cycle impact assessment (LCIA), risk assessment (RA) and environmental noise impact assessment procedures. In some cases, the results of these methodologies are reported in the form of correlation matrices.

Table 17.1.1 Example of different scales commonly applied for environmental impact evaluation using the matrix approach

Class of Impact	Numerical Scale	Symbol Scale	Chromatic Scale
Extremely positive value	+2	☺☺	Dark green
Positive value	+1	☺	Light green
Neutral value	0	😐	White
Negative value	−1	☹	Orange
Extremely negative value	−2	☹☹	Red

THE LANDFILL MATRIX APPROACH

The approach is based on the works of Cossu et al. (1986) and Andreottola et al. (1989) who used the matrix approach with chromatic tonalities to facilitate the understanding of the final results of the EIA performed for sanitary plants.

The methodology

The methodology is based on a "decisional procedure" consisting in a series of steps common to other standard environmental assessments (LCA, Multicriteria analysis, RA, etc.).

All steps should be followed to successfully apply the operative module described below to sanitary landfills.

Scope

The scope of the study should be defined. Typical objectives include the following:

- To assess different scenarios to define the feasibility of alternative solutions, such as: a single project on a single site; more than one project but related to only one site; one project at different sites; more than one project at more than one site; a project that has already been carried out.
- To identify and quantify the direct and indirect significant effects of the landfill on the following factors (environmental categories): population and human health; biodiversity; land, soil, water, air and climate; material assets, cultural heritage and the landscape; interaction between the factors previously mentioned.
- To investigate the feasibility, effectiveness, and implications of the proposed mitigation or compensation measures.
- To identify, predict, and evaluate the residual environmental impacts (i.e., after all practicable mitigations) and cumulative effects expected to arise during the entire service life cycle of the landfill.

- To design and specify the environmental monitoring and audit requirements.
- To select the most sustainable alternative.

Definition of temporal and spatial scale

The temporal scale of the assessment should cover all life stages of the service life cycle of the landfill (Yang et al., 2014), particularly as important impacts may arise both in the construction and aftercare phases (See Fig. 17.1.1):

1. Construction phase
   a. Site operation; for example, excavation and backfilling of soil and stone.
   b. Construction of the main parts of landfill body including groundwater drainage, barrier layer, bottom liner, leachate, and LFG (landfill gas) collection.
   c. Construction of other facilities in the landfill site, such as monitoring wells, on-site roads, and official buildings.
2. Operation phase
   a. Operation of the landfill; for example, the transport of waste from the outside to the landfill, the placement and compaction of waste, intermediate soil covers, top cover.
   b. Leachate and LFG management.
3. Aftercare Phase
   The operations planned for this phase consist in monitoring and maintenance activities, mainly: cap maintenance and monitoring; leachate recirculation operation and maintenance (where leachate is present); leachate collection system operation and maintenance; landfill gas collection system maintenance and monitoring; landfill gas migration control and monitoring; groundwater and surface water monitoring; and security and ground stability maintenance. The choice of the duration of the "Aftercare Phase" to be considered in the analysis is important since landfills can generate emissions to air or groundwater for decades and therefore this choice, to some extent subjective, may influence the result of the assessment.

Each phase may be characterized by an "ordinary" or "extraordinary" function. The latter term refers to the occurrence of exceptional events (earthquakes, hurricanes, etc.). Despite the precautionary principles and principles of the European Directive 2014/52/EU, extraordinary conditions are rarely considered in EIA and are not implemented in the current approach.

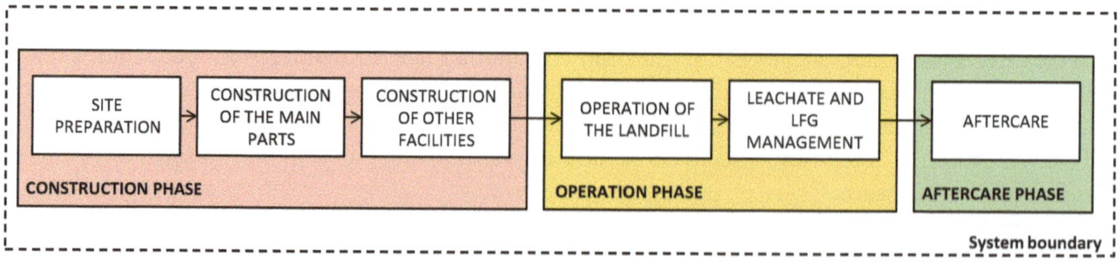

Figure 17.1.1 Service life cycle of the landfill. *LFG*, landfill gas.

Spatial scale ranges from global, regional to local scales (see Table 17.1.2). The definition of spatial scale is based on spatial (area) gradations (reduction) of environmental impacts from the source (landfill) of impacts. Analysis is performed by means of technical analysis, mathematical modeling, or on the grounds of expert judgment.

"Local" distances are taken as ending within 1 km of the landfill; the term "global" is considered relevant for distances more than 1000 km from the point of application. "Regional" is an area that falls between "global" and "local" scales and is therefore likely to be governed by intermediate processes between those governing the former areas.

Impact evaluation

Impact evaluation is the prediction, in quantitative terms as far as possible, of the significant direct and indirect effects of the landfill on environmental categories. Estimation is achieved by means of the following steps:

- *Identification.* The procedure through which all, or at least the most relevant, impact elements of landfills are defined. Impact elements can be quantified using one or more indicators. To plan the best measures to reduce environmental impacts, causes must be linked with impacted environmental categories. The environmental categories may be defined as those components of the environment influenced by the effects generated by the elements of impact. These include not only the biophysical components of the environment (land, soil, water, air, climate, biodiversity, and ecosystem functioning) but also those more directly related to human activity (human health, material assets, cultural heritage, and landscape) (see Fig. 17.1.3). A general (although not exhaustive) list of impact elements is reported in Table 17.1.3.
- *Normalization.* Normalization is a procedure intended to express the previously quantified impact indicators in a way to allow their mutual comparison (e.g., comparing groundwater contamination and methane emission into the air). This procedure transforms the impact values into dimensionless values (devoid of a unit of measurement) using one or more logic-mathematical functions. The most common method applied implies the normalization of the potential impact to an "acceptable-value" represented by a regulation limit or other recognized values by international organizations.

Table 17.1.2 Spatial scale for environmental assessment of landfills

Scale	Criteria
Local	Impact is registered at the distance 1 km from the source (landfill)
Regional	Impact is registered at the distance from 1 to 1000 km from the source (landfill)
Global	Impact is registered at the distance exceeding 1000 km from the source (landfill)

Table 17.1.3 Example of checklist of the impact elements for the matrix for landfill environmental impact assessments (see Also Andreottola et al., 1989; Pivato, 2011; Pivato et al., 2013; Ranzato et al., 2012; Palmiotto et al., 2014; Yang et al., 2014)

Impact Elements	Indicators to Quantify the Impact Element	Causes of the Impact Element	Environmental Categories Affected by the Impact Element	Mitigation and Compensation Measure to Reduce the Impact Element	Scale Analysis	
					Temporal Scale	Spatial Scale
Emission of greenhouses gases	• Emission of methane, carbon dioxide, steam, etc.	• Point emissions from flares and engines • Diffusive emission from landfill surface • Breakdown of biogas treatment plant • Intermediate cover/daily cover operation • Failure of the top cover system • Leaking gas wells	• air • Climate	• Air quality monitoring (to inform subsequent measures) • Waste pretreatment • Waste compacting degree optimization • Cover soil operation optimization (frequency, quality and quantity of materials, etc.) • Top cover typology • In situ aeration, biocover • Job opportunities	• Operation phase • Aftercare phase	• Global
Landfill pollutant gas emissions	• Emission of nonmethane organic compound (NMOCs) which include volatile organic compounds (VOCs), CFC, H_2S and other hazardous air pollutants (HAPs)	• Point emissions from flares and engines • Diffusive emission from landfill surface • Intermediate cover/daily cover operation • Failure of the top cover system	• air • Population and human health • Biodiversity	• Air quality monitoring (to inform subsequent measures) • Waste pretreatment • Waste compacting degree optimization • Cover soil operation optimization (frequency, quality and quantity of materials, etc.) • Top cover typology • Water spraying • Job opportunities	• Operation phase • Aftercare phase	• Local • Regional

Emission of dust and particles	• Emission of particulate (Kg PM_{10}/d) • Concentration of particulate (μg PM_{10}/m^3)	• Supply of inert material • Transport of waste (traffic) • Intermediate cover/daily cover operation • Top cover system • Waste unloading	• Air • Population and human health • Biodiversity	• Waste compacting degree optimization • Cover soil operation optimization (frequency, quality and quantity of materials, etc.) • Top cover typology • Alternative traffic routes • Waste disposal and spreading time optimization • Job opportunities	• Construction phase • Operation phase	• Local
Annoying odor emissions	• Odor concentration, expressed in odor units per cubic meter (ouE/m^3), e.g., determined through dynamic olfactometry • Public interview	• Type of waste • No intermediate cover/daily cover operation • No final top cover • Bad function leachate treatment plant or of gas collection system • Low compaction • Waste unloading	• Air • Population and human health	• Air quality monitoring (to inform subsequent measures) • Waste pretreatment • Waste disposal and spreading time optimization • Waste compacting degree optimization • Cover soil operation optimization (frequency, quality and quantity of materials, etc.) • Top cover typology • Improvement of the construction and operation of the gas extraction system • Job opportunities	• Operation phase • Aftercare phase	• Local • Regional
Noise	Noise level monitoring (dB)	• Installation and start of construction • Transport of material for construction • Transport of waste • Waste compaction	• Population and human health • Biodiversity	• Effective exhaust mufflers for work vehicles • Waste transport, disposal and spreading time optimization • physical noise barriers as soil berms • Job opportunities	• Construction phase • Operation phase	• Local

(Continued)

Table 17.1.3 Example of checklist of the impact elements for the matrix for landfill environmental impact assessments (see Also Andreottola et al., 1989; Pivato, 2011; Pivato et al., 2013; Ranzato et al., 2012; Palmiotto et al., 2014; Yang et al., 2014)—cont'd

Impact Elements	Indicators to Quantify the Impact Element	Causes of the Impact Element	Environmental Categories Affected by the Impact Element	Mitigation and Compensation Measure to Reduce the Impact Element	Scale Analysis	
					Temporal Scale	Spatial Scale
Uncontrolled leachate emission into the environment	• Volume of leachate (m^3 of leachate/d) • Concentration of leachate-originated pollutants in nearby groundwaters and surface waters (mg of contaminanton/m^3) • Leachate toxicity measured using whole Effluent Toxicity (WET)	The cause of leachate leakage are most related to the reliability of liners, and consequently of their failure: • Bad geomembrane seams and/or clay compaction • Installation damage • Failure to safeguard liner in operation • Pipes penetrating liner • Clogging of the leachate collection and removal system • Geotechnical failure • Breach by vertical pipes. • Elevated water tables in landfill, seeping out of slopes or overflow (pits)	• Surface waters and groundwater • Population and human health • Biodiversity	• Water quality monitoring (to inform subsequent measures) • Waste pretreatment • Waste compacting degree optimization • Cleaning drain pipes, water extraction from landfills, surface liner, groundwater encapsulation and lowering groundwater table underneath landfill, control of external water penetration into landfill body • Job opportunities	• Operation phase • Aftercare phase	• Local • Regional
Stability of ground/ waste	Altimetric profile measures	• Excavations • Type of waste • Waste compaction underlying soil compaction • Inadequate subsoil (compressible) • Final top cover • Landfill leachate level above bottom liner system	• Soil • Population and human health	• Monitoring and adjusting of landfill slope • Waste compacting degree optimization • Surface run-on and runoff controls • Landfill surveilling • Job opportunities	• Construction phase • Operation phase • Aftercare phase	• Local

Occupation of area and volume, landscape impact	• Surface occupation (m^2) • Land vegetation coverage index • Landfill elevation (m)	• Installation and start of construction • Height of landfill	• Material assets • Cultural heritage • Landscape • Land and soil	• Green compensation measures such as green belts, representing both a windbreak, an odor barrier and a buffer zone, and wide wooded zones acting as core natural areas • Improve the connectivity of important nearby ecological units and greenways • Job opportunities	• Construction phase • Operation phase • Aftercare phase	• Local • Regional
Traffic	Number of cars and trucks per day	• Installation and start of construction • Transport of waste	• Air • Population and human health • Biodiversity	• Alternatives routes • Waste transport, disposal and spreading time optimization • Job opportunities	• Construction phase • Operation phase	• Local • Regional
Consume of material and energy for construction and operation	• Megajoules of energy or liters of diesel • Materials used for the construction of the landfill body and other facilities and for operations (tonne of sand, clay, gravel, etc.)	• Installation and start of construction	• Material assets	• Development and utilization of renewable (e.g., wind, solar) energy (including locally produced biogas) • Utilization of recycled materials as construction materials • Job opportunities	• Construction phase • Operation phase	• Local • Regional
Adverse health and hygiene effects	• Incidence of cancer and chronic diseases, mortality studies, increase of undesirable pest species (e.g., rats)	Most of the causes indicated for the above impact elements can directly or indirectly determine adverse health effects, in particular emissions to air, soil and water, noise and traffic.	• Population and human health	• Air and water quality monitoring • Noise monitoring • Waste pretreatment • High compaction • Pest species monitoring and control	• Construction phase • Operation phase • Aftercare phase	• Local • Regional

(*Continued*)

Table 17.1.3 Example of checklist of the impact elements for the matrix for landfill environmental impact assessments (see Also Andreottola et al., 1989; Pivato, 2011; Pivato et al., 2013; Ranzato et al., 2012; Palmiotto et al., 2014; Yang et al., 2014)—cont'd

Impact Elements	Indicators to Quantify the Impact Element	Causes of the Impact Element	Environmental Categories Affected by the Impact Element	Mitigation and Compensation Measure to Reduce the Impact Element	Scale Analysis	
					Temporal Scale	Spatial Scale
Adverse effects on biodiversity in the surrounding environment	• Variations in population densities of special or protected species	Most of the causes indicated for the above impact elements can directly or indirectly determine adverse effects on biodiversity, in particular emissions to air, soil and water, noise and traffic.	• Biodiversity	• Improve the connectivity of important nearby ecological units and greenways • Waste pretreatment • Air and water quality monitoring and Noise monitoring seperately	• Construction phase • Operation phase • Aftercare phase	• Local • Regional

- *Weighting*. To achieve a more reasonable and reliable prediction of impacts, it is recommended that the relative importance of each assessment indicator is taken into account. This can be accomplished by assigning weights to the assessment indicators based on the importance attached to them by decision makers. Although the process of weighting is inherently subjective and reflect societal and stakeholder preferences, methods such as the analytic hierarchy process (AHP), which has been utilized in a wide range of decision making areas, can at least provide a viable and rigorous procedure for determining such weights (Saaty, 1980; Wang et al., 2010).
- *Representation*. Considering the difficulty frequently encountered in quantifying the entity of interactions between the various control lists present in each matrix (as described below) and communicating this synthetically and effectively, a chromatic representation, including five colors reflecting positive and negative influences and a neutral (no influence) condition, can be employed to describe the latter in a qualitative form. Two different chromatic scales can be used, corresponding to positive or negative influences and including a total of four assessment levels (expressed by different tonalities) plus a neutral value. The four chromatic tonalities correspond to extremely positive, positive, negative, and extremely negative values; the neutral value can be represented using a neutral color (white) (see Table 17.1.1). The chromatic representation of impact allows decision makers to immediately identify in a visual manner the most critical elements of impact to be assigned priority in the case of mitigation and compensation actions.

Evaluation of mitigations and compensations

It is virtually impossible to achieve a true zero impact. Accordingly, mitigation and compensation constitute the key steps in EIA. Mitigation may be defined as the measures put in place to avoid, reduce, and, if possible, remedy significant adverse effects of a plan or project, as envisaged, for example, in the European Directive 85/337/EEC. Mitigation actions can target the sources of impact or, alternatively, reduce the exposure that receptors undergo, and be carried out either during a plan or project or following completion. When effective mitigation is impossible, compensation may be provided. Compensation is defined as the replacement of lost (because of adverse environmental impacts) environmental functions and values with equivalent ones, for example, by reconstructing a new habitat if one is destroyed during the creation of a project.

Summing up, the evaluation of mitigations and compensation is aimed at:

- identifying and assessing measures to avoid, reduce, or remedy the impacts determined previously;
- assessing the effectiveness of the above mitigation/compensation measures; and
- defining the residual environmental impacts, which are the net impacts remaining even if the mitigation and compensation measures are in place.

Evaluation of residual environmental impacts

The residual environmental impacts refer to the net environmental impacts remaining after mitigation and compensation, taking into account the background environmental conditions and the impacts from existing, committed, and planned projects.

Final decision

The final decision presents the best alternative, identified as the decision having the lowest residual negative impacts or least overall environmental impact. Viewed from a sustainability perspective, the most acceptable alternative is one that meets international, national, and regional environmental standards while being socially and economically acceptable.

Application of the method to sanitary landfills

The operative approach for sanitary landfills is represented by a sequence of four matrix schemes that correlate the interactions between the causes of impact, elements of impact, environmental categories, and mitigation/compensation measures over the service life of a landfill.

Fig. 17.1.2 shows a block scheme of the method based on the matrices commented in detail in the next paragraphs. The iteration represented in the scheme means that the number of evaluations of residual impact will be equal to the product of the number of projects (N) for the number of sites (M) deemed suitable. All these assessments will subsequently be the object of cross-analysis prior to final decision. The whole scheme will be repeated for each of the life stages of the service's life cycle of the landfill.

Matrix of the causes and elements of impact (Matrix A)

The first type of matrix identifies the activities, processes, and features related to plant construction, functioning, and existence underlying the onset of elements of impact (i.e., processes and factors capable of affecting the state of environmental categories, see Table 17.1.3). Since the method should cover all three different phases of a given project (construction, operation, and aftercare phases), three corresponding matrices focusing on the various elements of impact for each phase should be envisaged. The importance of the causes with regard to determination of a specific element of impact is assessed by means of a series of chromatic tonalities.

The first matrix permits the clarification of environmental impact of a project or plant, thus allowing the need for improvement to be identified. In the case of more alternative projects to be assessed, this matrix represents a useful tool in establishing a sound background for decision making.

Matrix of potential impacts (Matrix B)

This matrix presents as control lists the elements of impact, already defined in matrix "A" and environmental categories (see Table 17.1.3).

From the intersection of the two lists, potential impacts on the environment arising from landfill construction, operation, and aftercare can be singled out, and the "B" matrix can therefore be used to provide a global, synthetic overview of all potential issues related to the landfill.

Indeed, should it be necessary to assess the impact of a particular landfill on various sites, a number of different "B" matrices would be carried out and compared to formulate a judgment as to the most suitable site. Clearly, in this case the diversity of the various "B" matrices should be aligned to the

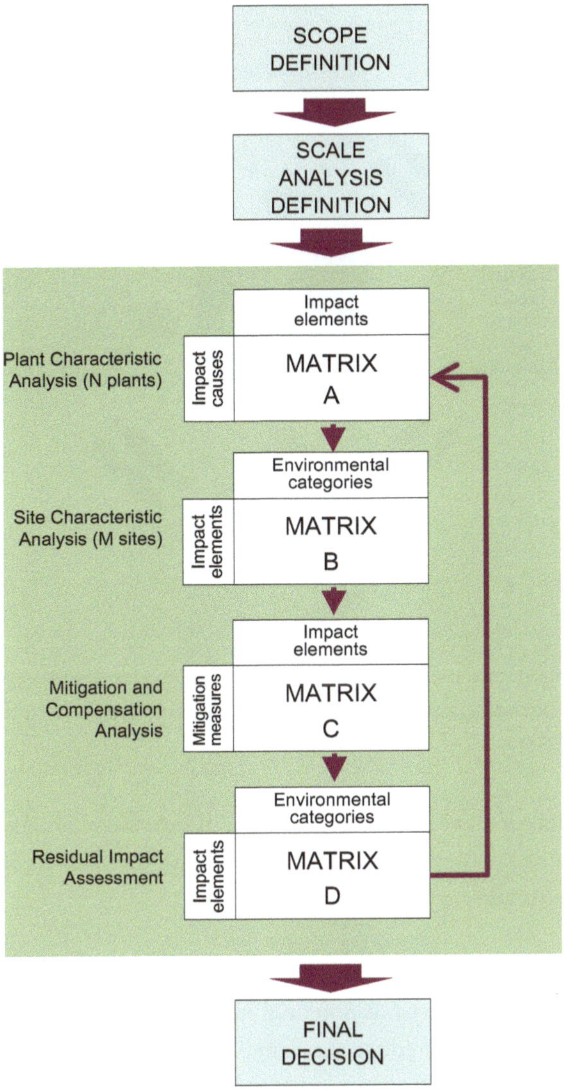

Figure 17.1.2 Scheme of the environmental impact assessment methodology through chromatic matrices.

different characteristics of the sites, while the "A" matrix remains unchanged since it depends on the intrinsic characteristics of the landfill.

Vice versa, if it were necessary to assess the impact of different types of landfill on a single site, the resulting various "B" matrices would be influenced by the corresponding variations in "A" matrices.

In the case of a decision dependent on the results of consideration of more than one project based on more than one site, the number of "B" matrices taken into account would increase considerably, as they

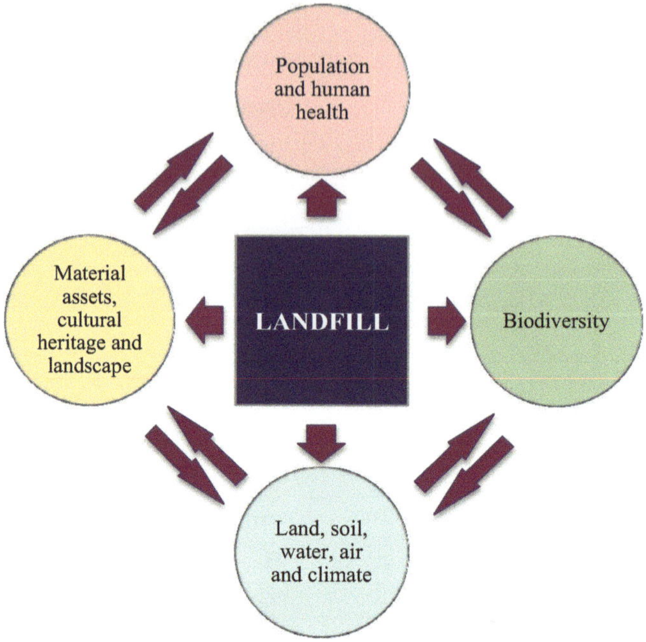

Figure 17.1.3 The principal environmental categories (factors) affected directly or indirectly by a landfill. The interactions between the categories should be explicitly evaluated to carry out a rigorous environmental impact assessment.

are equal to the combination of all possible cases. Finally, in the case of an already existing single installation, the "B" matrix identifies potential impacts, highlighting those issues towards which monitoring efforts and improvements should be focused.

Matrix of mitigations and compensations (Matrix C)

The third matrix considers, on the basis of potential negative impact identified in matrix "B", all the actions to be adopted to eliminate or reduce the negative impacts to acceptable environmental levels, or at least to limit these impacts. When implemented, these actions should impinge on the causes largely contributing to the onset of negative elements of impact; for this reason, identification of the causes should be based on careful evaluation of the "A" matrix. To this regard, two important points should be highlighted. Firstly, the obtaining of a clear picture of all possible interactions is mandatory, as any single element of impact may be influenced by more than one introduced improvement. Moreover, it should not be overlooked that a given same measure may produce simultaneously positive effects on some elements and negative side effects on others. As an example, in the field of sanitary landfilling, a collection system and an on-site treatment process aimed at limiting the impact caused by the presence and uncontrolled emissions of leachate may also produce negative effects such as annoying odors generated in various parts of the treatment plant. Therefore, the overall efficacy of the measures should be assessed using the two different chromatic scales (negative and positive) defined previously.

Matrix of residual impacts (Matrix D)

On the basis of the selected mitigation and compensation measures and their efficacy, evaluated using the "C" matrix, the method provides a fourth matrix to assess residual impacts remaining once the operations of compensation and mitigation have been set up to counteract the various elements of impact.

This matrix is analogous to the "B" matrix used for assessment of potential impact although, unlike the latter, examination of the "D" matrix allows us to express a definitive judgment on the degree of compatibility of a landfill with regard to its impact on the surrounding environment. Moreover, the comparison of "B" and "D" matrices allows one to visually appreciate the efficacy of compensation and mitigation measures.

FINAL REMARKS

The assessment scheme based on the use of chromatic matrices appears to represent a convenient tool for use in assessing the environmental impact of a landfill. The application of this method allows for a careful and systematic identification of all elements of impact to be taken into account, providing a rational organization of the results of the study, and a synthetic and effective graphical representation. The proposed method supports the identification of a solution to the problem of EIA at all levels, particularly as it is clearly comprehensible by both technicians and nonexperts such as decision makers.

The assessment scheme illustrated, moreover, presents the advantage of explicitly taking all phases of the life cycle of the landfill into account, some of which are often overlooked but are nonetheless important.

In conclusion, this scheme is proposed for use at various levels. A public administration official who has no specific technical knowledge in the field of waste disposal, but who is called upon to express an opinion on an installation (taking environmental, social, economic and political implications into account), gains an immediate and easy understanding of the potential and residual impact on the environment caused by the installation through the "B" and "D" matrices. However, if the reader of the EIA matrices is an expert in the field, by examining the entire series of matrices, he or she will be in a position to determine all the interactions with their corresponding weights, which have led to the final result. Another advantage, of fundamental importance given the impact that public concerns and protests, at times fueled by misinformation, may have on landfill construction, is the possibility afforded by this tool to present the results of the impact study to the public in a simple and easily understandable way.

References

Andreottola, G., Cossu, R., Serra, R., 1989. A method for the assessment of environmental impact of sanitary landfills. In: Thomas H. Christensen, Raffaello Cossum, Rainer Stegmann (Eds.), Sanitary Landfilling: Process, Technology and Environmental Impact, ISBN 0-12-174255-5.

Canter, L.W., 1999. Environmental impact assessment. In: David, H.F.L., Lipták, B.G. (Eds.), Environmental Engineers' Handbook, second ed. CRC Press.

Cossu, et al., 1986. Studio di Impatto Ambientale dell'impianto di depurazione di Pero (Ctip, Roma, Cogefar, Milano).

Leopold, L., 1971. A Procedure for Evaluating Environmental Impacts (US Geological Survey Circular 645/1971). Washington.

Moore, J.L., et al., 1973. A Methodology for Evaluating Manufacturing Environmental Impact Statements for Delaware's Coastal Zone (Report to the State of Delaware). Battelle Memorial Institute, Columbus, Ohio.

Palmiotto, M., Fattore, E., Paiano, V., Celeste, G., Colombo, A., Davili, E., 2014. Influence of a municipal solid waste landfill in the surrounding environment: toxicological risk and odor nuisance effects. Environmental International 68, 16–24.

Pivato, A., 2011. Landfill liner failure: an open question for landfill risk analysis. Journal of Environmental Protection 2 (3).

Pivato, A., Vanin, S., Palmeri, L., Barausse, A., Mangione, G., Rasera, M., Gianluca, M., 2013. Biopotentiality as an index of environmental compensation for composting plants. Waste Management 33, 1607–1615.

Ranzato, L., Barausse, A., Mantovani, A., Pittarello, A., Benzo, M., Palmeri, L., 2012. A comparison of methods for the assessment of odor impacts on air quality: field inspection (VDI 3940) and the air dispersion model CALPUFF. Atmospheric Environment 61, 570–579.

Saaty, T.L., 1980. The Analytic Hierarchy Process. McGraw-Hill Book Co., New York.

Sorensen, J.C., 1971. A Framework for Identification and Control of Resources Degradation and Conflict in the Multiple Use of the Coastal Zone. University of California, Berkley.

U.S. EPA, April 1974. A Review of Environmental Impact Assessment Methodologies. U.S. Environmental Protection Agency, Washington, DC. EPA-600/5-74-002.

U.S. EPA, 1998. Principles of Environmental Impact Assessment Review: Appendix B: Environmental Impact Assessment Methodologies. U.S. Environmental Protection Agency, Washington, DC.

Wang, H.S., Che, Z.H., Wu, C., 2010. Using analytic hierarchy process and particle swarm optimization algorithm for evaluating product plans. Expert Systems with Applications 37, 1023–1034.

Yang, N., Damgaard, A., Lü, F., Shao, L.M., Brogaard, L.K., He, P.J., 2014. Environmental impact assessment on the construction and operation of municipal solid waste sanitary landfills in developing countries: China case study. Waste Management 34, 929–937.

17.2

LIFE CYCLE ASSESSMENT OF LANDFILLING

Davide Tonini, Simone Manfredi, Ioannis Bakas, Line Kai-Sørensen Brogaard and Anders Damgaard

INTRODUCTION

Landfills are expected to provide handling of the waste with the lowest possible load on the environment and risk to human health. To compare the environmental performance of different landfill (and waste treatment) technologies and to provide sound scientific support for decision- and policy-making processes, life cycle assessment (LCA) is increasingly being applied.

This chapter illustrates the fundamental methodological aspects and crucial assumptions involved in LCA modeling of landfilling scenarios. The importance of these methodological aspects and assumptions are illustrated with two case studies. The first involves a comparison among different landfilling scenarios and the second highlights the contribution of a number of individual waste fractions to the total potential environmental impact from landfilling of mixed municipal solid waste (MSW).

RELEVANT METHODOLOGICAL ASPECTS

LCA is a standardized methodology (ISO, 2006a,b) commonly used for evaluating waste and bioenergy systems. LCA allows for a holistic and systematic assessment of both direct and indirect impacts, as well as resources consumption. The LCA case studies presented in this section were performed using consequential LCA. The term consequential, as opposed to attributional, refers to the methodological approach used when carrying out the LCA. The choice between attributional and consequential LCA is determined by the decision context of the LCA, as exemplified in the International Reference Life Cycle Data System (ILCD) handbook of the European Commission (2010). Consequential LCA is intended to highlight the environmental consequences of a decision (for example deciding between alternative A, B, C, etc.).

LCA consists of four phases: (1) goal and scope definition, (2) inventory analysis, (3) impact assessment, and (4) interpretation. The first phase mainly includes specification of the aim of the LCA and definition of system boundaries and functional unit (the unit that qualitatively and quantitatively describes the service provided by the system under assessment, e.g., landfilling of 1 tonne of MSW). This phase also includes the specification of the temporal, geographical, and technological scope considered. In the second phase, all relevant direct and indirect emissions associated with upstream and downstream processes are collected and listed based on the functional unit. In the third phase, the emissions are characterized and aggregated conformingly with the considered impact categories. In the fourth phase, the results of the impact assessment are interpreted and discussed in the perspective of the goal and scope defined in the first phase. This chapter focuses on the application of LCA methodology to the assessment of selected landfilling scenarios. Background theory and details of the LCA methodology can be found elsewhere (e.g., ISO, 2006a,b; European Commission, 2010, 2011).

Functional unit

The typical service provided by a given waste management system is to collect and treat the waste, including eventual final disposal. Conformingly, the functional unit is typically defined as the management of a certain amount of waste with a defined material composition or the waste generation of a region within a specific time frame (Laurent et al., 2014). Furthermore, the temporal (time period for which the LCA results will be valid for, e.g., until 2015, 2050, etc., and for how long emissions are tracked), geographical (region of interest), and technological scope (level of technology assumed, which influences the choice of temporal scope) must be defined along with the functional unit (e.g., European Commission, 2010). The temporal, geographical, and technological scope inherently affects the choices and assumptions about waste composition, technologies, and marginal energy and products substituted by energy and materials produced by the waste management system.

System boundary

In product LCA all inputs and outputs to and from the system under assessment are typically followed from "cradle to grave." However, in LCA of waste management the inputs of waste are typically followed from the point when these products are discarded (conformingly with the definition of waste): the waste is therefore considered as a zero-burden input to the system, and all the upstream activities required to produce the product before becoming waste are not included in the LCA (Clift et al., 2000). This is correct unless one of the systems in the comparison involves waste minimization strategies: in this particular case, the upstream activities associated with the generation of the product later becoming waste have to be included, and the modeling becomes more complex (Cleary, 2010; Gentil et al., 2011). Failing to do this would lead to an inconsistent and unfair comparison.

Time perspective and long-term emissions

LCA adopts a fairness principle, according to which emissions, and subsequently their impacts, should be time integrated regardless of which point in time or part of the life cycle they occur (Guinée and Heijungs, 1993). The application of this principle by most life cycle impact assessment (LCIA) methodologies creates issues related to the realistic representation of environmental impacts within the LCA context, when the emissions causing the impacts occur for a long time frame but in low concentrations.

These issues typically appear in LCA applied to solid waste management systems and in landfill processes in particular. Landfilling of waste leads to a release of pollutants to different environmental compartments (e.g., greenhouse gases from waste degradation to air and heavy metals in landfill leachate to soil, fresh-, or groundwater) for very long time frames. Heavy metal emissions may take place over many tens of thousands of years but in very low concentrations (Doka and Hischier, 2005; Astrup et al., 2006). The accurate prediction of the long-term emissions[1] development is hindered by uncertainty that is related mainly to the many parameters shaping landfill chemistry and whose influence on pollutants mobility may change over time. Moreover, besides the fact that obviously no field measurements exist on long-term emissions, the validation of prediction models is very difficult.

The time integration principle of LCA creates problems to the long-term emissions impact assessment, since all organic waste is bound to degrade and the heavy metal content of waste is bound to leach over a virtually infinite time period. In the case of metals, this causes bias to the overall impact assessment since toxicity-related impacts (to which metals are the main contributors) might dominate the assessment.

Many researchers propose a separation of time periods to overcome this problem: heavy metal emissions are divided into short- and long-term in the inventory phase of LCA and only the former or both are included in the impact assessment (Hischier et al., 2010). Most often, the time frame separating short- and long-term emissions is arbitrarily set to 100 years.

If only short-term emissions are included, then the impact estimation is more credible since there is reliable information on the development of heavy metal emissions in such a short time frame, usually set to 100 years. On the other hand, a significant proportion of the potential impact is ignored, as a small fraction of the metal content would have leached within 100 years from waste deposition (Hauschild et al., 2008).

Therefore, it is often proposed to use both approaches, namely present the results for both with and without long-term emissions so that a complete representation of impacts is achieved and more informed decisions can be made on the basis of LCA.

Biogenic carbon sequestration

Biogenic carbon is released during all biological and thermal conversion processes involved in waste management systems. When considering the carbon released by food waste or paper (i.e. biomass with typically short rotation time), the impact of this carbon on global warming (GW) can be reasonably assumed to be the same as that of fossil carbon. However, the biogenic carbon derives from atmospheric CO_2 previously uptaken by growing plants. For this reason, it is common practice in LCA to assign a GW potential of zero to the biogenic carbon dioxide emitted, as the emission is counterbalanced by the uptake. However, this practice may induce inconsistencies and errors when the landfill is assessed: when a selected time horizon is considered (e.g., 100 years) a share of the biogenic carbon will stay undegraded in the landfill body after the selected time horizon. Thus, the landfill becomes a carbon sink. To credit it for this, Christensen et al. (2009a,b) suggest setting the Global Warming Potential (GWP) of the biogenic carbon stored in the landfill after 100 years equal to −3.67 kg CO_2-eq. per kg

[1]Long-term emissions are defined in LCA as all emissions occurring after 100 years.

C bound (3.67 = 44 kg CO_2/12 kg C). Summarizing, the GWP assigned to biogenic and fossil carbon would be:

- Fossil C emitted as CO_2: 1 kg CO_2-eq. per kg CO_2 emitted
- Fossil C bound in landfill: 0 kg CO_2-eq. per kg C bound
- Biogenic C emitted as CO_2: 0 kg CO_2-eq. per kg CO_2 emitted
- Biogenic C bound in landfill: −3.67 kg CO_2-eq. per kg C bound

An alternative is to model the uptake of biogenic carbon in the growing plants and trees (later becoming waste) and to assign a GWP of 1 to the CO_2 biogenic (same as for CO_2 fossil). This is often applied in bioenergy studies (e.g., Tonini et al., 2012). In respect to this, it should be borne in mind that recent studies suggest even more advanced dynamic approaches (time-dependent GWP) when dealing with wood waste (i.e. long-rotation biomass). The reader is referred to the state-of-the-art literature for this.

The amount of carbon remaining sequestered is directly linked to the amount of biogenic carbon entering the landfill subtracted the amount that leaves the landfill again through emissions of gas and leachate. The amount of biogenic carbon entering the landfill is calculated based on the composition of the waste, thus the composition of the waste is an important part of describing the functional unit. The composition of the waste entering a landfill, together with information on the degradability of the different waste fractions is essential to calculate how much methane and carbon dioxide is formed within the timeframe of the study.

Modeling capital goods used in waste landfilling

Landfills are large constructions compared to other waste management technologies. The materials needed for the construction of the capital goods are quantified in studies by Bagchi (1994), Ecobalance (1999), and Christensen (2011). Only a few studies have addressed the environmental impacts from the capital goods of landfills (Brogaard et al., 2013; Frischknecht et al., 2007; Ecobalance, 1999; Ménard et al., 2004).

Capital goods define the materials and goods used for construction. Embedded in the capital goods is as well the energy consumption for construction of the landfill. The life cycle of capital goods includes production of materials and goods, construction of landfill, maintenance and decommissioning. The latter is only relevant for the goods above ground like buildings, fence, and infrastructural items since the capital goods underground will stay with the waste.

The amount of materials used to build a landfill depends on its design. The top cover can be constructed as a plastic liner with layers of soils or only of layers of different compacted soil types. The choice of liner system also depends on the design but most landfills are constructed with a high density polyethylene geomembrane and under that a compacted layer of clay. It is important to know the landfill design beforehand to include capital goods in the LCA of waste landfilling. The landfill design depends on the type of waste to landfill and the landscape where the landfill is being built (see sections about the different types of landfills).

Different construction techniques are used for building a landfill in a valley than a hill type landfill. All landfills are constructed differently due to site, waste, and climate, and therefore it is recommended to sketch the design of the landfill with all layers and materials used. Thereafter it is possible to calculate the amounts of materials used and the energy consumption needed.

It is important to estimate the lifetime of the landfill and thereby need for maintenance. The lifetime can vary depending on the size of the landfill and with that the filling time. In the case study investigated by Brogaard et al. (2013), a landfill with a capacity of 3,500,000 t, a filling time of 10 years and aftercare period of 30 years were assumed. The capital goods above ground will be used during the filling time and aftercare period, and the maintenance of these can be relevant for the LCA. From Brogaard et al. (2013), some examples are presented of capital goods to include in the sketching and planning of an LCA of waste landfill:

- Preparing earthwork
- Membrane system
- Leachate system
- Collection and inspection wells
- Pipes for leachate transport and wells
- Pumping station
- Leachate tank
- Systems for surface water collection
- Top cover and gas collection equipment
- Gas utilization equipment (or piping systems for transporting the gas for utilization)
- Fence and gates
- Monitoring wells
- Reception area—weighbridge, offices, workshop, and garages.

Energy is used for each of the points mentioned in the form of diesel to excavate soil, to transport the materials to site or around at the site. Machinery should be included in the LCA as use of diesel per hour but also related to the amount of goods/materials transported/moved per hour. The share of life for the machinery used for constructing the landfill should be included in the LCA. Machineries needed to construct a landfill could be the following (Brogaard et al., 2013):

- Bulldozer
- Compactor (for soil and for waste)
- Dumper
- Excavator
- Front loader
- Truck/Lorry.

The main environmental impacts are caused by the transport of the large amounts of materials needed (clay, gravel). Other impacts derive from producing the materials that are not recovered at the end of the life cycle, like the liner system (Brogaard et al., 2013).

MODELING THE LANDFILL SYSTEM

A general structure of the landfill system is presented in Fig. 17.2.1. The external box defines the boundary of the environmental impact assessment, i.e., all the processes within and across the box

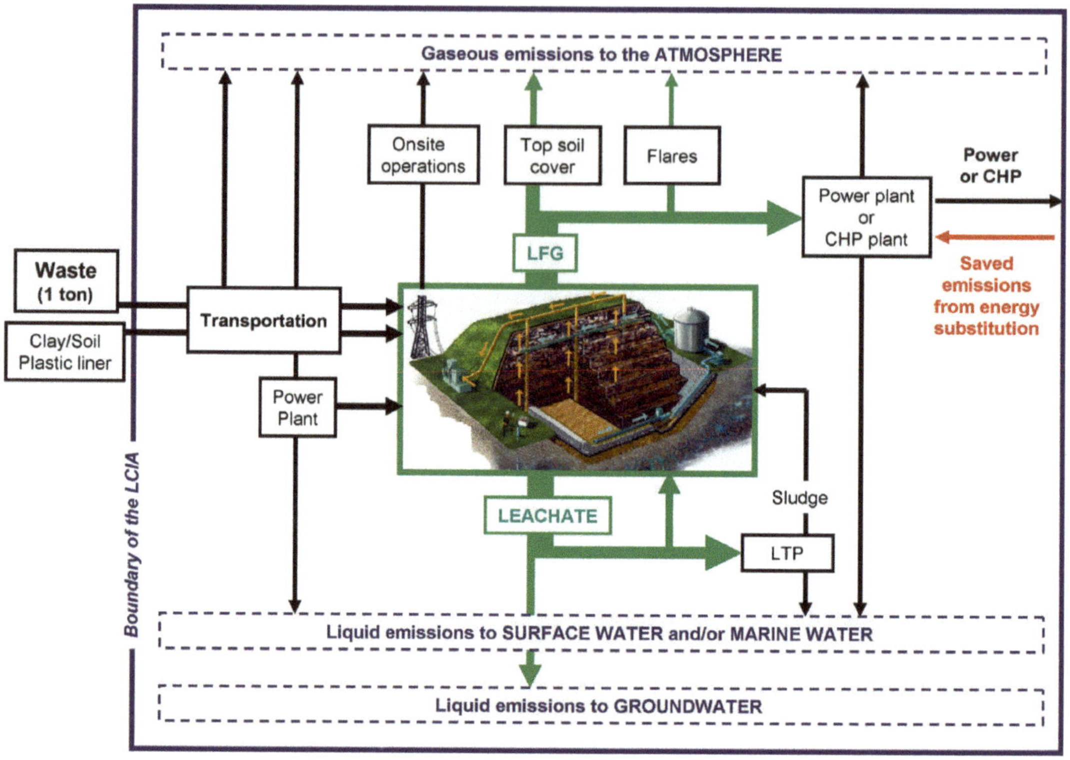

Figure 17.2.1 General structure of the landfilling technologies and boundary of the assessment (Manfredi and Christensen, 2009). *LCIA*, life cycle impact assessment; *LFG*, landfill gas; *LTP*, leachate treatment plant; *CHP*, combined heat and power.

should, in principle, be considered in the LCA. However, to simplify the modeling, selected processes (e.g., production of capital goods) may be disregarded provided that their contribution to the environmental impact is negligible compared with the emissions occurring during the landfill operation.

Clay and soil are used to construct soil covers and impermeable liners, and the amounts needed are specified as mass per tonne of wet waste disposed. The emissions related to the fuel consumed for transport of the construction materials and for onsite operations (trucks, vehicles) are included in the assessment. The emissions associated with the electricity consumption are also accounted for. If the landfill gas produced from the waste degradation is collected and converted into electricity, corresponding amount of electricity produced from the marginal electricity source (e.g., fossil fuels or the average country mix) is assumed to be substituted and the associated emissions avoided. This is in accordance with the principle of consequential LCA (see e.g., ISO, 2006a,b; Wenzel et al., 1997; Finnveden et al., 2009). Leachate and gas emissions from the landfill should be modeled on the type of landfill technology considered in the assessment. As an example, gas and leachate modeling details for open dump, conventional landfill, bioreactor landfill, and semiaerobic landfill based on a dedicated

Table 17.2.1 Technical and environmental parameters for landfill gas (LFG) and leachate generation, collection and management in selected landfill technologies during four time periods (Manfredi and Christensen, 2009)

Open Dump	Time Period 1	Time Period 2	Time Period 3	Time Period 4
LFG generated (% of LFG potential)	2 years; 2%	3 years; 8%	35 years; 70%	60 years; 16%
LFG collected	None	None	None	None
Leachate generated (mm/year)	2 years; 500 mm/year	8 years; 500 mm/year	30 years; 450 mm/year	60 years; 400 mm/year
Leachate collected	None	None	None	None
CONVENTIONAL TECHNOLOGIES				
LFG generated (% of LFG potential)	2 years; 2%	3 years; 8%	35 years; 70%	60 years; 16%
LFG collected (% of generated)	2 years; none	3 years; 90%	35 years; 90%	60 years; none
LFG management	None	Flare/energy recovery	Flare/energy recovery	None
Leachate generated (mm/year)	2 years; 500 mm/year	8 years; 250 mm/year	30 years; 200 mm/year	60 years; 180 mm/year
Leachate collected (% of generated)	20 years; 95%	20 years; 70%	30 years; none	30 years; none
BIOREACTOR TECHNOLOGIES				
LFG generated (% of LFG potential)	2 years; 2%	8 years; 72%	5 years; 8%	85 years; 16%
LFG collected (% of generated)	2 years; none	8 years; 95%	5 years; 90%	85 years; none
LFG management	None	Energy recovery	Energy recovery	None
Leachate generated (mm/year)	2 years; 500 mm/year	8 years; 250 mm/year (recirculation)	30 years; 200 mm/year	60 years; 180 mm/year

(Continued)

Table 17.2.1 Technical and environmental parameters for landfill gas (LFG) and leachate generation, collection and management in selected landfill technologies during four time periods (Manfredi and Christensen, 2009)—cont'd

Open Dump	Time Period 1	Time Period 2	Time Period 3	Time Period 4
Leachate collected (% of generated)	20 years; 95%	20 years; 70%	30 years; none	30 years; none
SEMIAEROBIC TECHNOLOGY				
LFG generated (% of LFG potential)	2 years; 2%	6 years; 50%	7 years; 20% (aerobic)	85 years; 10%
LFG collecte (% of generated)	2 years; none	6 years; 95%	7 years; none	85 years; none
LFG management	None	Energy recovery	None	None
Leachate generated (mm/year)	2 years; 500 mm/year	6 years; 250 mm/year (recirculation)	7 years; 200 mm/year	85 years; 180 mm/year
Leachate collected (% of generated)	20 years; 95%	20 years; 70%	30 years; none	30 years; none

study (Manfredi and Christensen, 2009) are reported in Table 17.2.1. A brief definition of the landfill technologies (as they are intended in this section) follows.

Open dump

The open dump is defined as a landfill where no measures for gas and leachate control are implemented. The gas and leachate generated during the landfill operation are assumed to be emitted entirely to atmosphere and groundwater.

Conventional landfill

Conventional landfills are typically defined as technologies that do not implement measures to affect waste degradation. However, technical measures to collect and manage gas and leachate are implemented, including bottom liner, leachate collection and treatment system, top soil cover, and gas collection and treatment system (flaring or power/heat generation).

Bioreactor landfill

A bioreactor landfill is typically defined as an engineered landfill where waste degradation is enhanced to achieve faster stabilization of the waste. This is generally accomplished through leachate recirculation (here defined as standard bioreactor) or recirculation of leachate along with considerable amounts of water (here defined as flushing bioreactor). In the latter, the flushing rate typically ranges from 1 to 5 m^3 of total liquid per tonne of waste (Blakey et al., 1997; Hupe et al., 2003). Measures to reduce ammonia concentration are commonly used in flushing bioreactors: leachate is nitrified before being recirculated in the waste mass and the produced NO_3 are later reduced to N_2 thanks to the anaerobic conditions established in the landfill body.

Semiaerobic landfill

This technology combines anaerobic and aerobic degradation to achieve a rapid waste stabilization (Matsufuji et al., 2005). During the first 5–10 years the degradation is anaerobic enhanced by leachate recirculation. During this period it is assumed that the CH_4 generation is too low (due to the relatively shallow landfill) to justify a system of collection and recovery. The following period is characterized by aerobic conditions obtained by inflowing air from the bottom of the landfill (chimney effect, Hanashima, 1999). Compared with anaerobic degradation, the aerobic conditions allow for a faster waste stabilization, higher landfill gas production, and lower leachate generation.

Low-organic landfill

This is typically defined as a landfill receiving pretreated waste characterized by a certain degree of biological stabilization (i.e., with a methane potential generally lower than 20 Nm^3 CH_4 per tonne wet waste). This is at least 4–5 times lower than the observed methane potential of mixed municipal solid waste (80–100 Nm^3 CH_4 per tonne wet waste).

Impact of assumptions and choices

There are a number of critical issues and choices taken in an LCA for waste landfilling, as these choices and issues are crucial for the results. It is thus critical to make sure that these choices reflect the reality they are used in (e.g., type of landfill, gas recovery efficiency, methane oxidation rate in the top cover, local conditions for heat and electricity substitution, etc.), and that a thorough sensitivity assessment is carried out to ensure the robustness of the result (Clavreul et al., 2012). Some of the most important choices that should be considered are described in the following.

Landfill type

As described, there is a large difference in the type landfills that can be designed. As a waste LCA is often a prospective study, it is important to assess what a change in landfill technology will have for the overall impact. An example could be a conventional landfill that is converted into a bioreactor landfill to ensure faster degradation.

Collection rates

The main impact of a landfill is often due to the release of landfill gas and leachate (Damgaard et al., 2011), though the capital goods related to construction and decommissioning also plays a role (Brogaard et al., 2013; Frischknecht et al., 2007; Ménard et al., 2004). It is therefore important to collect the leachate and gas produced in the landfill. The operation of a landfill varies considerably, and this impacts the collection of gas and leachate. This explains why it is important to consider the variation of these parameters and make sure that the full variation is considered within the modeling. A recent study by NCSU (2013) found that a review of landfill gas collection efficiencies ranged from 14% to 99.6% with a suggested average of 75% for a final cover. This highlights the need to consider how the landfill will be constructed and maintained. The same applies to leachate collection, where the collection efficiencies are typically higher, but also rely on the design of the collection system and the maintenance over time, as discussed elsewhere in this book.

Energy recovery

The way of utilizing the collected landfill gas also plays an important role in the overall environmental performance. If the collected gas is used to generate electricity or heat in an engine or boiler, it will be able to displace energy that, otherwise, should have been produced by other means. This thus credits the system by subtracting the avoided energy production. Damgaard et al. (2011) found that the most important fact with regards to landfill gas management is whether or not the gas is collected, while whether or not it is used for energy generation or flared has a lower environmental significance. But this may be of importance when comparing different types of landfills (Fig. 17.2.3).

Oxidation

Some of the landfill gas that is not collected will be oxidized before it reaches the surface of the landfill. The amount of oxidation will lower the amount of greenhouse gas generation due to the release of methane and other greenhouse gases. The oxidation is related to the flux through the top cover as well as the type of cover. Thus it is important to consider the type, operation, and geographical location of the landfill. A review by Chanton et al. (2009) found that the oxidation rate ranged from 10% to 89%, with averages of 27% and 54% at 30° and 55°N latitude, respectively. The choice of oxidation rate can have a determining impact on the impact of a landfill at low gas collection rates, and it's therefore important to understand how these two parameters are linked (i.e., a higher gas flux through the cover will mean a lower oxidation rate).

Concentrations of substances

The concentration of certain compounds in the landfill gas and leachate also plays a large role. Leachate composition is directly related to the type of landfill, the age of landfill, and the type of material deposited in the landfill. The problem is that when doing a predictive LCA the intent is to forecast what will happen in the future, but availability of data is very limited in this respect. So the modeling is based on data from current landfills. This raises the issue that field studies on landfills only properly cover the past couple of decades, while the attempt is to model 100 years or more. There will therefore be a large degree of uncertainty with regards to the concentration of substances in landfill gas and leachate. A review by Kjeldsen et al. (2002) found that differences of several orders of magnitude exist on reported concentrations for substance in landfill leachate. It is therefore important to consider these ranges when modeling the landfill and not only relying on a single study with average values.

LIFE CYCLE ASSESSMENT CASE STUDIES

Case study I: comparison among landfilling scenarios

This section illustrates the results of an LCA case study (Manfredi and Christensen, 2009) comparing six landfilling scenarios involving open dump, conventional (with flaring and with energy recovery), bioreactor (standard and flushing), and semiaerobic landfill.

Key modeling assumptions

The functional unit used in the assessment was "landfilling of 1 t of wet household waste in a 10 m deep landfill, considering a time horizon of 100 years." The waste composition as well as the chemical composition (including methane potential of the material fractions) was assumed based on Riber et al. (2009). The environmental assessment was performed conformingly with the LCA method EDIP 1997 (Wenzel et al., 1997). The environmental category considered were GW, photochemical ozone formation (POF), ozone depletion (OD), acidification (AC), nutrient enrichment (NE), human toxicity via soil (HTs), via water (HTw) and via air (HTa), and ecotoxicity in soil (ETs) and in water (ETwc). To this, a specific category representing the potential impact on groundwater was considered (spoiled groundwater resource, SGR). The related environmental impact was calculated by dividing the amount of each compound leached by the accepted concentration limit in drinking water (World Health Organization, 2006), thereby calculating the volume of groundwater that may be contaminated up to the limit values.

An overview of the modeling details for gas and leachate production and collection at the individual landfill technologies considered is presented in Table 17.2.1. For flushing bioreactor it was assumed partial ammonia removal through leachate nitrification/onsite denitrification. The savings associated with the amount of biogenic carbon left in the landfill body after the 100-year time horizon were not considered in this study. The consequences of this assumption are discussed in the key results section. The modeling was facilitated by the LCA-dedicated software EASEWASTE (Kirkeby et al., 2007).

Key results

Emissions of landfill gas (LFG) and leachate were the most important contributors to the environmental impacts in the assessed environmental categories. This is illustrated in Fig. 17.2.2 for the case of the conventional landfill with energy recovery. However, utilization of the LFG for energy production (in the conventional landfill with energy production, and standard and flushing bioreactor) brought to remarkable savings (Fig. 17.2.2) thanks to offsetting of energy production from fossil fuels (i.e., avoided emissions from fossil fuel production and combustion).

As expected, the comparison of the landfilling scenarios (Fig. 17.2.3) highlighted open dump as the worst scenario in most of the categories due to massive leachate and LFG emissions. This, however,

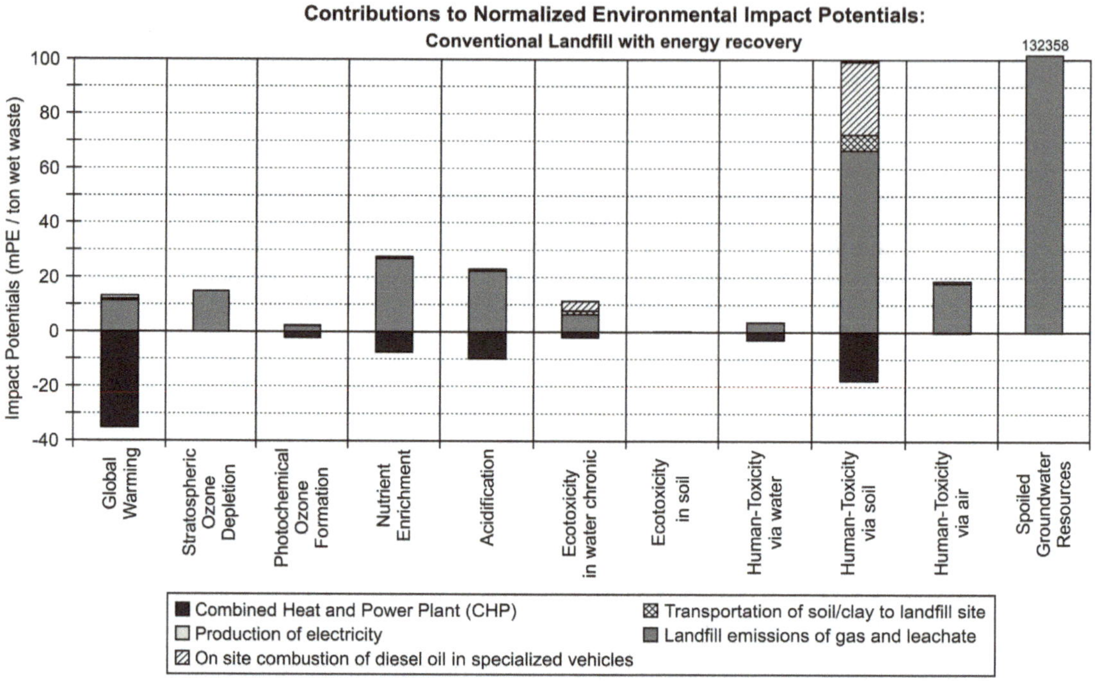

Figure 17.2.2 Life cycle assessment results of the conventional landfill with energy recovery, as normalized impact potentials (mPE/tonne wet waste). Standard and toxicity-related environmental impact categories; time horizon 0–100 years (Manfredi and Christensen, 2009).

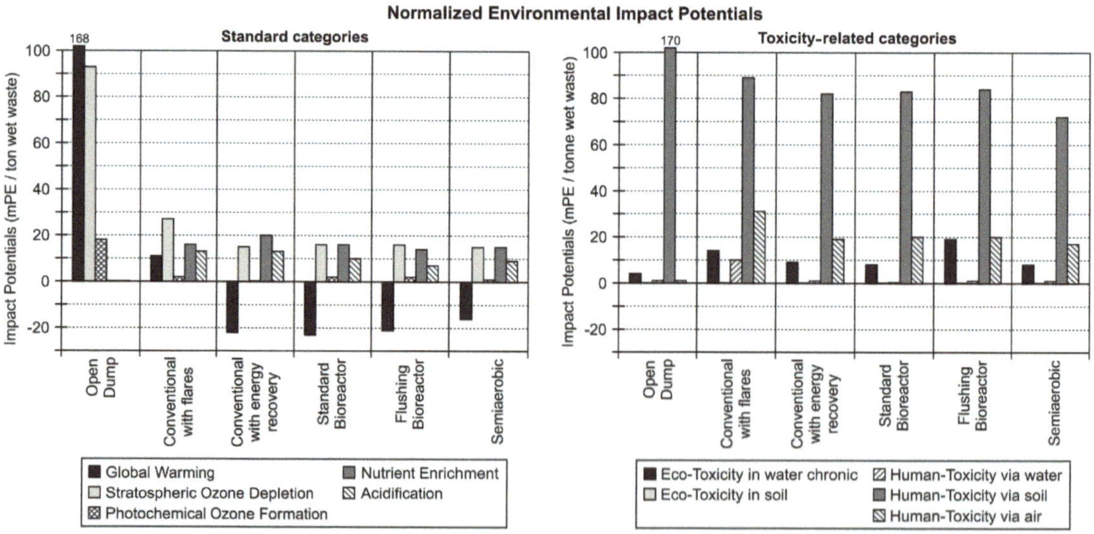

Figure 17.2.3 Life cycle assessment results of six landfilling technologies, as normalized impact potentials. Standard and toxicity-related environmental impact categories; time horizon 0–100 years (Manfredi and Christensen, 2009).

was not true for the toxic categories ETs, HTw, and HTa where open dump achieved lower impacts due to the absence of emissions from electricity generation and onsite operations. On most standard and toxic environmental categories the performance of conventional landfill with energy recovery, bioreactors, and semiaerobic landfill was comparable. This primarily highlights the importance of the gas collection and treatment/utilization system. The flushing bioreactor showed by far the best performance on SGR thanks to the ammonia removal through leachate nitrification/onsite denitrification.

Manfredi and Christensen (2009) also highlighted how leachate recirculation in bioreactors increased the rate of LFG production and, as a consequence, the fugitive gas emissions during the first 15 years of operation compared with conventional landfill. However, at the same time, most of the CH_4 potential could be exploited for energy production within the first 15 years, thereby optimizing collection and utilization. In addition, after this initial period the CH_4 emissions became low, therefore inducing low impacts on GW. In contrast, LFG emissions (and associated GW impacts) from conventional landfill and open dump were considerable throughout the 100-years period.

As aforementioned, the biogenic C left after the considered time horizon was not accounted for. This affects the overall magnitude of the GW savings which would increase correspondingly. However, the overall carbon degradation over 100 years (which is represented by the share of the total methane potential liberated over 100 years) was similar for all the landfilling scenarios, inducing comparable carbon sequestration across the six scenarios. Therefore, the relative ranking of the scenarios on GW was not affected. Notice that the same assumption would induce GW results completely inconsistent if the comparison was between landfilling and thermal technologies such as incineration. In this case, in fact, carbon sequestration would only occur in landfilling and not accounting for it would make the comparison mathematically inconsistent (Christensen et al., 2009a,b).

Case study II: environmental impact associated with individual waste fractions

Decades of experience with landfilling of mixed waste resulted in a consistent scientific framework available for environmental assessment of landfills. However, little is known about the contribution from the individual waste material fractions to the overall impact potentials from landfilling of the mixed waste. Typically, data derived from empirical models are used for LCA studies focusing on individual waste fractions: these are based on accelerated degradation tests on the individual materials (e.g., Belevi and Baccini, 1989; Barlaz 1998, 2006). However, as already mentioned, inconsistencies in assumptions about the GHG benefits associated with carbon sequestration and energy recovery/substitution in the landfill may lead to an unfair comparison with other options such as recycling and incineration.

Manfredi et al. (2010) performed an LCA study where the major focus was the quantification of the environmental impacts associated with the individual waste material fractions. A wrap-up of key assumptions and results follows.

Table 17.2.2 Waste fractions and subfractions included in the study, mass (wet basis) distribution in the mixed waste (%), degradability factors D (%), and methane potential M (Nm^3 CH_4/tonne wet fraction)

Fraction	Waste Subfractions	Mass (%)	D (%)	M
Organics	Vegetable waste, animal waste, kitchen tissues	32.3	86.5	117
Paper	Magazines, advertisements, books and phonebooks, office paper, other clean paper, paper and carton containers, cardboard, milk carton and alike, dirty paper, dirty cardboard, carton with alu-foil	33.2	40	115
Plastic	Soft plastic, hard plastic, plastic bottles, nonrecyclable plastic	6.9	1	0
Metals	Al containers, Al trays/foil, metal containers, metal foil, other metal	2.3	50	0
Glass	Clear glass, green glass, brown glass, other glass	9.0	50	0
Other combustible waste	Yard waste, animals, diapers and tampons, cotton stick, other cotton, wood, textiles, shoes leather, rubber, office articles, cigarette buts, vacuum cleaner bags, other combustible	12.9	27	22
Other noncombustible waste	Soil, stones and gravel, residuals, ceramics, cat gravel, batteries, other noncombustibles	3.4	1	0

What does it mean 50% decomposition factor for glass and metals? Degradability refers to the decomposition and mineralization of the materials in a waste matrix, which is usually a prerequisite for an emission of this material (Doka, 2007).
The Table is taken from Manfredi et al. (2010).

Key modeling assumptions

The functional unit used in the assessment was "landfilling of 1 t of wet household waste in a 10-m deep landfill, considering a time horizon of 100 years." The goal was to quantify the contribution of each individual fraction to the total impact. The fractions considered were: "organics," "paper," "plastic," "metals," "glass," "other combustible waste," and "other noncombustible waste" (Table 17.2.2). The LFG was assumed collected and utilized for energy recovery substituting corresponding energy produced from fossil fuels. The environmental emissions (both gas and leachate) for the individual fractions were calculated with an empirical model. This is reported in Manfredi et al. (2011) and makes use of the results from Belevi and Baccini (1989), Barlaz (2006), and Doka (2007) about long-term degradation of MSW landfills. As input the model required the chemical composition of the mixed waste and the degradability factors for the individual fractions. This refers to the decomposition and mineralization of the materials in a waste matrix, which is usually a prerequisite for an emission of this material (Doka, 2007; see Table 17.2.2). The waste composition as well as the chemical composition

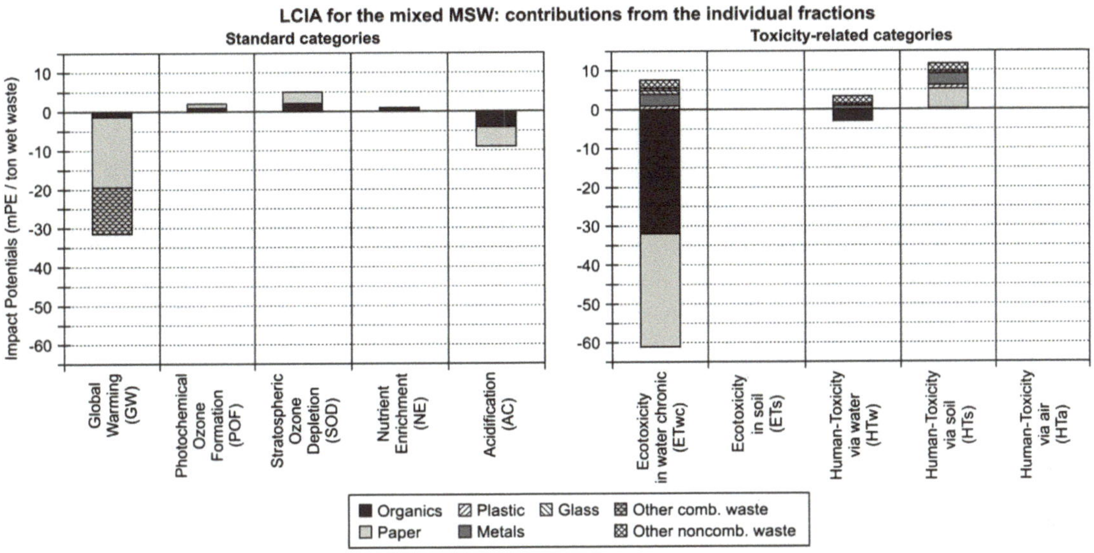

Figure 17.2.4 Life cycle assessment results for 1 t of mixed municipal solid waste (MSW) with focus on the relative contributions from the individual waste fractions (Manfredi et al., 2010). *LCIA*, life cycle impact assessment.

(including methane potential of the material fractions) was assumed based on Riber et al. (2009). The environmental assessment was performed conformingly with the LCA method EDIP 1997 (Wenzel et al., 1997). The environmental category considered were GW; POF; OD; AC; NE; HTs, HTw, and HTa; and ETs and ETwc.

Key results

The results (Fig. 17.2.4) showed that the impact potentials estimated for the standard categories were principally caused by "organics" and "paper." The remaining fractions caused most of the impacts for the toxic categories. "Organics," "paper," and "other combustible waste" were the main contributors to the GW savings. This was due to (1) high CH_4 potential and degradability for "organics": this induced high production of LFG and therefore high energy recovery; (2) biogenic carbon left undegraded after 100 years for "paper": this also has a high CH_4 potential but a relatively low degradability (here 40%); therefore, the most remarkable contribution to GW was attributed to the biogenic carbon stored rather than to use of LFG generated by its degradation; and (3) biogenic carbon left undegraded after 100 years for "other combustible waste." (2) and (3) highlight the importance of whether carbon storage is included or not, when compared to case study I. On HTw and HTs, the main responsibility for the impacts was leaching of metals from paper, metals, and plastic. This was also the case for ETwc, where, however, the benefits associated with avoided energy production from fossil fuel exceeded the loads associated with leaching, determining an overall saving.

Landfilling compared with other technologies: overview of recent life cycle assessment studies

A number of LCA studies have evaluated the environmental performance of landfilling compared with alternative waste management systems. The goal of this section is to provide an overview of recent selected LCA studies on the subject. Crucial assumptions and uncertainties are also highlighted.

Manfredi et al. (2011) evaluated the most environmentally sound option among landfilling, incineration, recycling for individual waste fractions. In general, recycling was the best option for metals and plastic. For organic waste, incineration and landfilling with energy recovery were comparable (except for the potential impact on the groundwater resource which was considerably higher for landfilling); recycling (in the form of direct composting and further use-on-land) showed higher savings on the toxicity categories but lower on GW and on the remaining standard categories due to fugitive emissions and energy consumptions. This performance would considerably improve if the organic waste was exploited for biogas production and energy recovery prior to final composting and use-on-land, as demonstrated by other studies (Montejo et al., 2013; Boldrin et al., 2011). For paper, recycling and incineration offered comparable environmental performances across the different categories. Landfilling resulted in remarkable savings only on GW thanks to the credits derived from biogenic carbon left undegraded after 100 years. These results are mostly in line with those of WRAP (2006) where a comprehensive review of LCA studies published up to 2005 was performed. However, the focus here was mainly on recycling, and crucial LCA assumptions and choices for landfilling were not thoroughly addressed.

Christensen et al. (2009a,b) analyzed the GW potential of 40 waste management scenarios involving recycling, landfilling, incineration, mechanical–biological treatment (MBT), and different combinations of those. The authors concluded that a number of different waste management schemes may lead to significant GHG savings, including those involving landfilling provided that the carbon left undegraded after the selected time horizon is credited with GHG savings. However, the same study showed that the GHG benefits associated with scenarios involving thermal energy recovery from the waste (e.g., incineration and MBT) were generally larger than those from landfilling. This conclusion, however, was largely affected by the choice of the energy fuel substituted by the energy produced in the waste system. For example, Tonini et al. (2013) highlighted how bioreactor landfills may be the most favorable strategy in the situation where the electricity substituted is based on natural gas and no-heat recovery is considered (e.g., no-heat demand).

References

Astrup, T., Mosbæk, H., Christensen, T.H., 2006. Assessment of long-term leaching from waste incineration air-pollution-control residues. Waste Management 26, 803–814.

Barlaz, M.A., 2006. Forest products decomposition in municipal solid waste landfills. Waste Management 26, 321–333.

Bagchi, A., 1994. Design, Construction and Monitoring of Landfills, second ed. Wisconsin Department of Natural Resources, John Wiley & Sons, Ltd., USA.

Barlaz, M.A., 1998. Carbon storage during biodegradation of municipal solid waste components in laboratory-scale landfills. Global Biogeochemical Cycles 12, 373–380.

Blakey, N.C., Bradshaw, K., Reynolds, P., Knox, K., 1997. Bio-reactor landfill — a field trial of accelerated waste stabilisation. In: Proceedings Sardinia 1997, Sixth International Waste Management and Landfill Symposium. S. Margherita di Pula, Cagliari, Italy, vol. 1, pp. 375–386.
Belevi, H., Baccini, P., 1989. Long-term behavior of municipal solid waste landfills. Waste Management & Research 7, 43–56.
Boldrin, A., Neidel, T.L., Damgaard, A., Bhander, G.S., Møller, J., Christensen, T.H., 2011. Modelling of environmental impacts from biological treatment of organic municipal waste in EASEWASTE. Waste Management 31, 619–630.
Brogaard, L.K., Stentsøe, S., Willumsen, H.C., Christensen, T.H., 2013. Quantifying capital goods for waste landfilling. Waste Management & Research 31 (6), 585–598.
Chanton, J.P., Powelson, D.K., Green, R.B., 2009. Methane oxidation in landfill cover soils, is a 10% default value reasonable? Journal of Environmental Quality 38, 654–663.
Christensen, T.H., 2011. In: Christensen, T.H. (Ed.), Solid Waste Technology & Management, vol. 1 & 2. John Wiley & Sons, Ltd., Chichester, United Kingdom.
Christensen, T.H., Simion, F., Tonini, D., Møller, J., 2009a. Global warming factors modelled for 40 generic municipal waste management scenarios. Waste Management & Research 27, 871–884.
Christensen, T.H., Gentil, E., Boldrin, A., Larsen, A.W., Weidema, B.P., Hauschild, M., 2009b. C balance, carbon dioxide emissions and global warming potentials in LCA-modelling of waste management systems. Waste Management & Research 27, 707–715.
Cleary, J., 2010. The incorporation of waste prevention activities into life cycle assessments of municipal solid waste management systems: methodological issues. The International Journal of Life Cycle Assessment 15, 579–589.
Clift, R., Doig, A., Finnveden, G., 2000. The application of life cycle assessment to integrated solid waste management — part 1-methodology. Process Safety and Environmental Protection 78, 279–287.
Clavreul, J., Guyonnet, D., Christensen, T.H., 2012. Quantifying uncertainty in LCA-modelling of waste management systems. Waste Management 32, 2482–2495.
Damgaard, A., Manfredi, S., Merrild, H., Stensøe, S., Christensen, T.H., 2011. LCA and economic evaluation of landfill leachate and gas technologies. Waste Management 31 (7), 1532–1541.
Doka, G., December 2007. Life Cycle Inventories of Waste Treatment Services. Ecoinvent Report No. 13. Swiss Centre for Life Cycle Inventories, Dübendorf.
Doka, G., Hischier, R., 2005. Waste treatment and assessment of long-term emissions. The International Journal of Life Cycle Assessment 10 (1), 77–84.
Ecobalance, 1999. Life Cycle Inventory of a Modern Municipal Solid Waste Landfill. Prepared for: Environmental Research and Education Foundation by Ecobalance Inc. 7101 Wisconsin Ave, Suite 700 Bethesda, MD 20814, USA.
European Commission, 2011. Supporting Environmentally Sound Decisions for Waste Management. A Technical Guide to LCT and LCA for Waste Experts and LCA Practitioners. Joint Research Centre, Institute for Environment and Sustainability. Available at: http://lct.jrc.ec.europa.eu/assessment/publications.
European Commission, March 2010. International Reference Life Cycle Data System (ILCD) Handbook — General Guide for Life Cycle Assessment — Detailed Guidance. EUR 24708 EN, first ed. Publications Office of the European Union, Luxembourg.
Finnveden, G.A., Hauschild, M.Z., Ekvall, T., Guinee, J.B.J., Heijungs, R., Hellweg, S., Koehler, A., Pennington, D., Suh, S., 2009. Recent developments in life cycle assessment. Journal of Environmental Management 91, 1–21.
Frischknecht, R., Althaus, H.-J., Bauer, C., Doka, G., Heck, T., Jungbluth, N., Kellenberger, D., Nemecek, T., 2007. The environmental relevance of capital goods in life cycle assessments of products and services. The International Journal of Life Cycle Assessment 12 (Special Issue 1), 7–17.
Gentil, E.C., Gallo, D., Christensen, T.H., 2011. Environmental evaluation of municipal waste prevention. Waste Management 31, 2371–2379.
Guinée, J., Heijungs, R., 1993. A proposal for the classification of toxic substances within the framework of life cycle assessment of products. Chemosphere 26 (10), 1925–1944.
Hanashima, M., 1999. Pollution control and stabilization processes by semi-aerobic landfill type: the Fukuoka method. In: Proceedings Sardinia 1999, Seventh Waste Management and Landfill Symposium. S. Margherita di Pula, Cagliari, Italy, pp. 313–326.
Hauschild, M., Olsen, S.I., Hansen, E., Schmidt, A., 2008. Gone…but not away—addressing the problem of long-term impacts from landfills in LCA. The International Journal of Life Cycle Assessment 13, 547–554.

Hischier, R., Weidema, B., Althaus, H.-J., Bauer, C., Doka, G., Dones, R., Frischknecht, R., Hellweg, S., Humbert, S., Jungbluth, N., Köllner, T., Loerincik, Y., Margni, M., Nemecek, T., 2010. Implementation of Life Cycle Impact Assessment Methods. Ecoinvent Report No. 3, v2.2. Swiss Centre for Life Cycle Inventories, Dübendorf.

Hupe, K., Heyer, K., Stegmann, R., 2003. Water infiltration for enhanced in situ stabilization. In: Proceedings Sardinia 2003, Ninth Waste Management and Landfill Symposium. S. Margherita di Pula, Cagliari, Italy.

ISO 14040, July 01, 2006a. Environmental Management-Life Cycle Assessment-Principles and Framework, second ed.

ISO 14040, July 01, 2006a. Environmental Management-Life Cycle Assessment-Requirements and Guidelines, first ed.

Kjeldsen, P., Barlaz, M., Rooker, A.P., Baun, A., Ledin, A., Christensen, T.H., 2002. Present and long-term composition of MSW landfill leachate: a review. Critical Reviews in Environmental Science and Technology 32 (4), 297–336.

Kirkeby, J.T., Birgisdottir, H., Bhander, G.S., Hauschild, M., Christensen, T.H., 2007. Modelling of environmental impacts of solid waste landfilling within the life-cycle analysis program EASEWASTE. Waste Management 27, 961–970.

Laurent, A., Clavreul, J., Bernstad, A., Bakas, I., Niero, M., Gentil, E., Christensen, T.H., Hauschild, M.Z., 2014. Review of LCA studies of solid waste management systems—part II: methodological guidance for a better practice. Waste Management 34, 589–606.

Manfredi, S., Tonini, D., Christensen, T.H., 2011. Environmental assessment of different management options for individual waste fractions by means of life-cycle assessment modelling. Resources Conservation and Recycling 55, 995–1004.

Manfredi, S., Tonini, D., Christensen, T.H., 2010. Contribution of individual waste fractions to the environmental impacts from landfilling of municipal solid waste. Waste Management 30, 433–440.

Manfredi, S., Christensen, T.H., 2009. Environmental assessment of solid waste landfilling technologies by means of LCA-modeling. Waste Management 29, 32–43.

Matsufuji, Y., Tachifuji, A., Matsugu, H., 2005. Biodegradation process of municipal solid waste by semiaerobic landfill type. In: Proceedings Sardinia 2005, Tenth Waste International Waste Management and Landfill Symposium S Margherita di Pula, Cagliari, Italy.

Ménard, J.-F., Lesage, P., Deschênes, L., Samson, R., 2004. Comparative life cycle assessment of two landfill technologies for the treatment of municipal solid waste. The International Journal of Life Cycle Assessment 9 (6), 371–378.

Montejo, C., Tonini, D., Márquez, M.C., Astrup, T., 2013. Mechanical-biological treatment: performance and potentials. An LCA of 8 MBT plants including waste characterization. Journal of Environmental Management 128, 661–673.

NCSU, 2013. Critical Analysis of Literature on Landfill Gas Collection Efficiency. Report by Morton Barlaz. North Carolina State University, Raleigh, USA.

Riber, C., Petersen, C., Christensen, T.H., 2009. Chemical composition of material fractions in Danish household waste. Waste Management 29, 1251–1257.

Tonini, D., Hamelin, L., Wenzel, H., Astrup, T., 2012. Bioenergy production from solid biomass: a consequential LCA of 12 bioenergy scenarios. Environmental Science & Technology 46 (24), 13521–13530.

Tonini, D., Sanchez, M.V., Astrup, T., 2013. Material resources, energy, and nutrient recovery from waste: are waste refineries the solution for the future? Environmental Science & Technology 47, 8962–8969.

World Health Organization (WHO), 2006. Guidelines for Drinking-Water Quality – First Addendum to Third Edition, vol. 1. WHO Press, 20 Avenue Appia, 1211 Geneva 27, Switzerland.

Wenzel, H., Hauschild, M.Z., Alting, L., 1997. Environmental Assessment of Products. Volume 1 Methodology, Tools and Case Studies in Product Development.

Waste & Resource Action Program – WRAP, 2006. Environmental Benefits of Recycling. An International Review of Life Cycle Comparisons for Key Materials in the UK Recycling Sector. Available at: http://www.wrap.org.uk/content/environmental-benefits-recycling.

Further Reading

European Commission, 2013a. Communication from the Commission to the European Parliament and the Council: Building the Single Market for Green Products – Facilitating Better Information on the Environmental Performance of Products and Organisations. Available online at: http://eur-lex.europa.eu/LexUriServ/LexUriServ.do?uri=COM:2013:0196:FIN: EN:PDF.

European Commission, 2013b. Product Environmental Footprint (PEF) Guide. Available at: http://ec.europa.eu/environment/eussd/smgp/product_footprint.htm.

18 CHAPTER

UNCONTROLLED EMISSIONS

18.1

LFG MIGRATION

Peter Kjeldsen

INTRODUCTION

Historically, landfills have received large quantities of organic wastes which in most countries still is the case. The organic waste will undergo anaerobic degradation (Chapter 3.1) that leads to production of biogas mainly consisting of methane (CH_4) and carbon dioxide (CO_2).

Landfill gas (LFG) is produced at all landfills, receiving municipal solid waste. Beside CO_2 and CH_4 it contains trace gases (hydrogen sulfide, volatile organic compounds (VOCs) (such as aromatic hydrocarbons), chlorinated and fluorinated compounds, etc. (see Chapter 9.2). The produced LFG should be extracted and utilized at almost all landfills Chapter 9.3), but the extraction is never 100% effective. Worldwide the LFG is still not extracted and utilized at many landfills; gas management by methane oxidation processes may be an option (Chapter 9.4). In both cases the risk of off-site gas migration through surrounding soil strata exists.

The potential effects of migrating and emitting LFG are illustrated in Fig. 18.1.1. They relate to different scales. As shown in the figure the effects may be local (less than 1 km distance), regional (1−10 km), and global. In addition, the effects may be also highly different in nature:

Explosion and fire hazards. LFG is explosive mainly due to its high content of CH_4. The explosive limit range for methane in air lies between 5% and 15%. LFG entering buildings through cracks in foundations, leaking pipes, etc., may result in explosive mixtures where a spark from an electric engine, switching on the light, etc., may initiate an explosion (see Chapters 9.4. and 15.2).

Damage to vegetation. Many cases of damages to vegetation in the vicinity of landfills are reported in the literature. Gendebien et al. (1992) describe 31 different cases from UK, USA, Germany, Canada, and Japan. The main reason for damages to vegetation from LFG is asphyxia by removal of oxygen in the root zone. This removal can be either due to a displacement of the oxygen by LFG or by oxidation of methane. High concentrations of carbon dioxide (>20%) are also toxic to plants (Rettenberger, 1985) and some trace compounds (hydrogen sulfide, haloorganic compounds, etc.) are toxic as well.

Toxic compounds in air. Carbon dioxide displaces oxygen in the respiratory system and causes indisposition even at low concentrations. As compared to the threshold limit value (TVL) of 0.5% or 5000 ppm, LFG has high concentrations of carbon dioxide (Hoather and Wright, 1989). This means that in general a 10 times higher dilution factor is needed not to exceed the TVL value for CO_2 in comparison to what is needed not to exceed the lower explosion limit of 5% methane. Methane

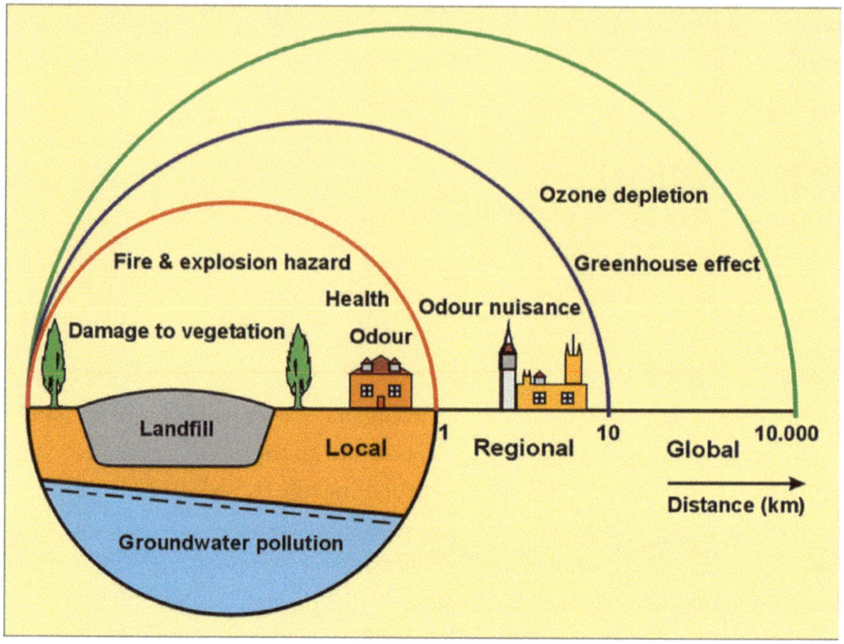

Figure 18.1.1 The different scales of landfill gas (LFG) effects. *After Kjeldsen (1996).*

oxidation occurring in the affected soil will produce CO_2, which could lead to increased levels of CO_2 in the affected soil. Besides CO_2, the most critical compounds have been benzene and vinyl chloride due to their carcinogenic nature. Several studies, especially of older origin, have found potential health effect caused by these compounds (Kjeldsen, 1996).

Odor nuisances. The main LFG compounds giving rise to odor problems are hydrogen sulfide and mercaptans, the latter originating from anaerobic breakdown of organic waste. In the initial acid phase, also volatile fatty acid can lead to odor nuisances. Odor problems have been recognized at several landfills (Dincer et al., 2006; Duan et al., 2014).

Greenhouse effect. CH_4 is one of the most important greenhouse gases , with a 100-year global warming potential 28 times that of carbon dioxide (CO_2) because of its stronger molar absorption coefficient for infrared radiation and its longer residence time in the atmosphere (Scheutz et al., 2009). The digital value of 28 does not include CO_2 from CH_4 oxidation or climate-carbon feedbacks (IPCC, 2013). In 2011, landfills contributed to 4907 and 3052 Gg CH_4, accounting for 17.5% and 19.6% of anthropogenic CH_4 emissions in the United States and Europe, respectively (EEA, 2013; US EPA, 2013).

Ozone depletion. The LFG content of volatile chlorinated and fluorinated hydrocarbons is a source to ozone depletion. Owing to their chemical stability especially the fluorinated hydrocarbons (the "freons") reach the stratosphere, where the chlorine atom is separated, and the radical causes ozone to break down (Rowland and Molina, 1975). A main waste source to emission of "freons" is insulation boards consisting of polyurethane and polystyrene rigid foam. The release of the "freons" from insulation foam is very slow if the foam is not crushed down to small particles (Kjeldsen and Scheutz, 2003).

Groundwater pollution. LFG migrating in the surrounding unsaturated zone is exposed to infiltrating water. Some of the compounds in the gas are highly water-soluble and can be leached and transported with the infiltrating water to the groundwater zone. There are a few examples reported where LFG has been the source for groundwater contamination (Kerfoot et al., 2004; Kerfoot, 1994).

CONCEPTUAL FRAMEWORK FOR LANDFILL GAS MIGRATION AND EMISSION

The conceptual framework of LFG migration and emission is given in Fig. 18.1.2 (after Kjeldsen, 1996). The figure shows the LFG transport routes eventually leading to emission of the gas to the surrounding atmosphere together with the basic processes and the factors affecting these processes.

The basic processes are divided into transport processes and sinks (or sources) as follows :

- Transport processes
  - diffusion,
  - advection, and
  - dilution/dispersion.

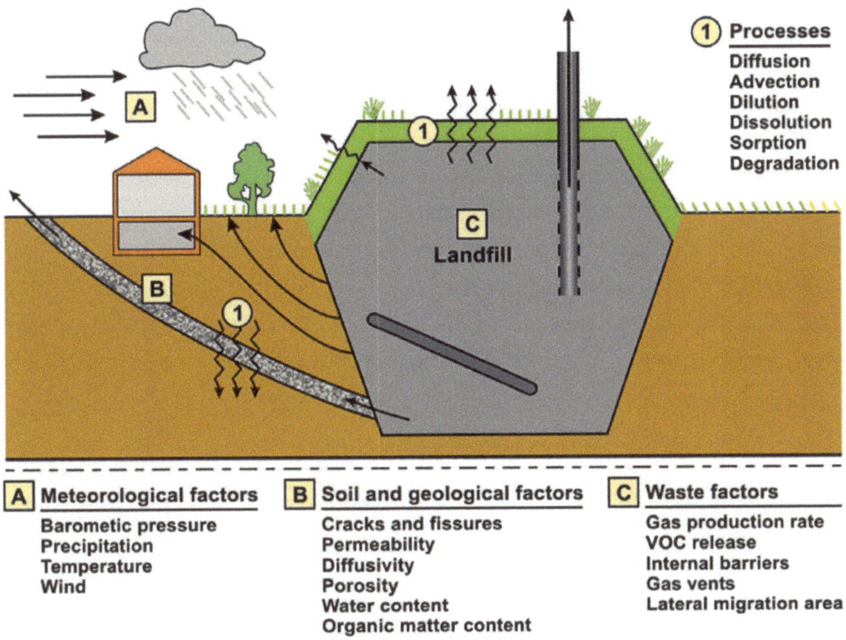

Figure 18.1.2 The conceptual framework of landfill gas (LFG) migration. The processes governing the migration are shown together with the factors affecting the governing processes. *VOC*, volatile organic compounds. *After Kjeldsen (1996).*

- Sink processes
  - dissolution of LFG constituents in water,
  - sorption to soil particles, and
  - oxidation (and other degradation processes).

The factors affecting the processes can be divided into three classes as shown on Fig. 18.1.2:

- meteorological conditions,
- soil and geological conditions, and
- waste and landfill conditions.

The transport and the sink processes together with the most important governing factors will be described in the following sections.

TRANSPORT PROCESSES AND GOVERNING FACTORS

Advection

Advective flux of a gas constituent through a porous medium is caused by a pressure gradient and is described by Darcy's law:

$$J_a = -C \cdot \frac{k_a}{\mu} \cdot \frac{\partial p}{\partial x} \tag{1}$$

where J_a is the advective flux (g/m^2 s), C is the gas concentration of the constituent (g/m^3), k_a is the gas permeability of the soil (m^2), μ is the viscosity of the gas (N s/m^2), and $\partial p/\partial x$ is the pressure gradient (Pa/m). Examples of parameters are given in Kjeldsen (1996).

The parameters that govern the advective flux the most are the gas permeability and the pressure gradient. The gas permeability varies orders of magnitude between different soil types and is also highly dependent on the water content of the soil (Kjeldsen, 1996). The water content is highly dependent on the precipitation, which can vary a lot both on short term and seasonal term.

The pressure gradient can be built up by the LFG generated in the interior of the landfill or by sudden changes in the barometric pressure. If the advective transport is occurring in cracks or animal burrow, even very small pressure differences can drive significant gas fluxes. Also wind can create spatial pressure differences that can increase advectively driven gas migration (Poulsen, 2005).

Diffusion

Diffusional fluxes are caused by variation in gas concentrations within the porous medium. The diffusional flux can be described by Fick's law:

$$J_d = -\varepsilon_a \cdot D \cdot \frac{\partial C}{\partial x} \tag{2}$$

where J_d is the diffusional flux (g/m^2 s), ε_a is the gas-filled porosity (fraction), D is the diffusion coefficient in the soil (m^2/s), and $\partial C/\partial x$ is the concentration gradient (g/m^4).

The diffusional flux is often driven by the presence of large concentration differences; for instance, methane concentration from 50% to close to zero over a cover soil, but the diffusion coefficient is dependent of the total porosity of the soil, and especially on the water content. At high-water content the diffusion coefficient approaches zero [further details are given in Kjeldsen (1996)]. Thus, meteorological factors are—as for the advective transport—important for the diffusional transport.

Diffusional fluxes have, in many years, been regarded as the most important transport process for gas migration and emission. However, through the later years it has been recognized that the advective processes can drive much higher fluxes of gas away from the waste especially if permeable soils have been used or poorly covered areas exist. Also the presence of cracks or man-made pathways, such as drain pipes, sewers may route large fluxes driven by advective forces (Kjeldsen, 1996).

Dilution/gas dispersion

During advective-controlled gas flow through soil layers, pressure fluctuation introduced by wind may lead to enhanced mixing of LFG with atmospheric air due to dispersion processes (Poulsen and Sharma, 2011). This may be important for the microbial oxidation of methane in the soil layers. Gas dispersion in soil has generally not been studied in great detail. However, the dispersion process is expected to be of major importance for the transport of LFG in unsaturated soil. Even a higher dispersion can be expected as compared to dispersion in the groundwater zone, since a larger variability in gas permeabilities exists due to highly variable water content (directly affecting the gas permeability).

SINK AND SOURCE PROCESSES

The migration of LFG is also governed by different chemical and microbial reaction that can occur during the transport processes in the surrounding soil layers.

Gas to water transfer

When LFG migrates, it may stay in direct contact with the infiltrating water and water in the pores. Under these circumstances a transfer of the gas constituents into the water phase is taking place. At equilibrium it is controlled by the Henry's law:

$$C_g = H \cdot C_w \tag{3}$$

where C_g is the gas phase concentration (mg/L), C_w is the water phase concentration (mg/L), and H is the Henry's Law constant (dimensionless). Under nonequilibrium conditions, kinetic factors control the transfer process. In such a case, the move of conditions from equilibrium determines the magnitude of the gas/water phase transfer. The transfer process can be effective in each direction—a sudden decrease in gas concentration can "retransfer" compounds back into the gas phase from the water phase.

The significance of the transfer process for a certain gas constituent is determined by the magnitude of the Henry's Law constant. For methane and carbon dioxide the Henry's law constant is 29 and 1.1, respectively. This means in general that a higher proportion of the carbon dioxide in migrating LFG tends to be "absorbed" by the water content in the soil, than is the case for methane. The Henry's law coefficients for many of the trace organics in LFG are low as well, which means that these compounds also have a (low) tendency to be transferred to infiltrating water.

Sorption of pollutants to soil

LFG components can be sorbed onto the soil particles, mainly by the organic matter content of the soil. The sorption process—together with the gas to water transfer process—leads to a retarded transport of the gas constituents during the migration process. Simple modeling approaches (confer Kjeldsen, 1996) show that methane is only slightly retarded, while carbon dioxide is slightly more retarded solely due to its high water solubility. VOCs such as TCE and o-xylene exhibit significant retardation.

Ward et al. (1992) studied the behavior of trace organics in migrating LFG at a landfill site in the United Kingdom and found that the less lipophilic compounds (i.e., lower-sorption capacity) vinyl chloride and dichlorofluoromethane, peak furthest away from the landfill (>50 m) compared to chloroethane, dichloromethane (30 m), and 1,2-dichloroethane (20 m).

Degradation and formation of gas constituents

Methane can undergo oxidation when oxygen is present and can be an important process to control methane emissions from landfills by implementation of biomitigation systems relying on methane oxidation —as described in Chapter 9.5. The methane oxidation process can also be a very important process for the migration of methane away from landfills. During the migration process, the LFG can be mixed with soil and gas containing considerable amounts of oxygen due to gas dispersion processes as described previously. If the migration occurs through conduits such as cracks and fissures or man-made facilities such as sewers or other buried installations the mixing with soil gas may be significantly reduced leading to anaerobic conditions, where methane will not be degraded. If gas migration is happening in deep geological sandy layers the methane oxidation capacity of the soil may be constrained by less optimal conditions for the methane-oxidizing bacteria (for instance, low nutrient content).

As already discussed in Chapter 9.5 many of the VOCs contained in LFG can be degraded in soils—especially at the presence of oxygen. The abovementioned considerations in respect to oxidation of methane are also valid for the oxidation of VOCs during the migration process.

The oxidation of methane and VOCs is leading to the formation of CO_2 as degradation product of the processes. This is especially important for methane oxidation due to the much higher concentration level in the migrating LFG. Thus the LFG migration may lead to elevated concentrations of CO_2 in the soil gas, which can have adverse effect (suffocation) at intrusion of the gas into buildings or manholes.

The degradation processes is depending on several factors. The soil temperature is important since the degradation occurs faster at elevated soil temperatures. The degradation may also depend on the nutrient and water content of the soil. For transport in very coarse and dry soil strata, the environmental conditions may not be in favor of degradation processes.

CONCEPTUAL MODEL AND MIGRATION RISKS

As described in the introduction, several risks related to emission and migration of LFG exist. The conceptual model and the governing processes and factors show that the prediction of risks especially related to migrating LFG is very complex due to high spatial and temporal variability in governing factors. This is the case especially for risk related to worse-case scenarios such as the risk for odor nuisances and explosion due to intruding LFG. Here, special meteorological conditions such as sudden decrease in barometric pressure and gas permeability of the top soil layer due to heavy rain can enhance the migration of LFG with sudden high methane concentrations in nearby buildings due to soil gas intrusion. For other gas constituents the risks are more of a chronic nature. For instance high concentrations of CO_2 (even further elevated by the oxidation of methane) in the soil gas may constantly be a source to elevated CO_2 level in buildings, which as described previously give rise to nuisances to people staying in the houses. The same is the case for carcinogenic gas constituents such as benzene and vinyl chloride, where the effects are of chronic nature.

EXPERIENCES WITH MIGRATING GAS AT LANDFILLS

Several cases of incidents related to migration of LFG have been reported. In a review from the United Kingdom (Gregson, 2000) covering the period 1970—95, 63 cases are listed. The cases cover incidents where elevated methane concentrations were measured in buildings and facilities, over smaller explosions in manholes and sewers, to larger explosions in houses leading to severe injuries or death. Kjeldsen and Fischer (1995) reported from Skellingsted landfill where a gas explosion accident occurred in a house close to the landfill around March 21, 1991. Fig. 18.1.3 shows the barometric pressure and the precipitation for March 1991. Heavy rain and a substantial drop in barometric pressure are seen the days before the accident. The heavy rain filled the pores of the upper soil layer and decreased the vertical gas transport. The soil below the house was not sealed by water. The barometric pumping enhanced the horizontal migration of LFG, and the only place where the gas could escape was through the dry soil layers below the house and through sewer and water ducts into the house. The unfavorable combination of high precipitation and barometric pressure drop has also been reported as the main factor at an LFG explosion close to a landfill in Wisconsin (Franklin and Vetter, 1987) and at Loscoe landfill, where a bungalow was completely destroyed by an LFG explosion (Williams and Aitkenhead, 1991).

Overall, most of the reported incidents of significant migration of LFG away from the landfill have occurred when the landfill has been constructed in gravel or limestone quarries (Gandolla et al., 2002; Franzidis et al., 2008; Kjeldsen and Fischer, 1995)—in most cases without liners for prevention of the migration of the generated LFG directly out into the adjacent soil layers. Several of the reported cases did not have gas extraction systems installed at the site.

METHODS TO CONTROL GAS MIGRATION

Several measures to prevent the migration of LFG have been reported. The most widely used method is implementation of side liners (HPDE or alike) to prevent gas migration. Establishment of side liners

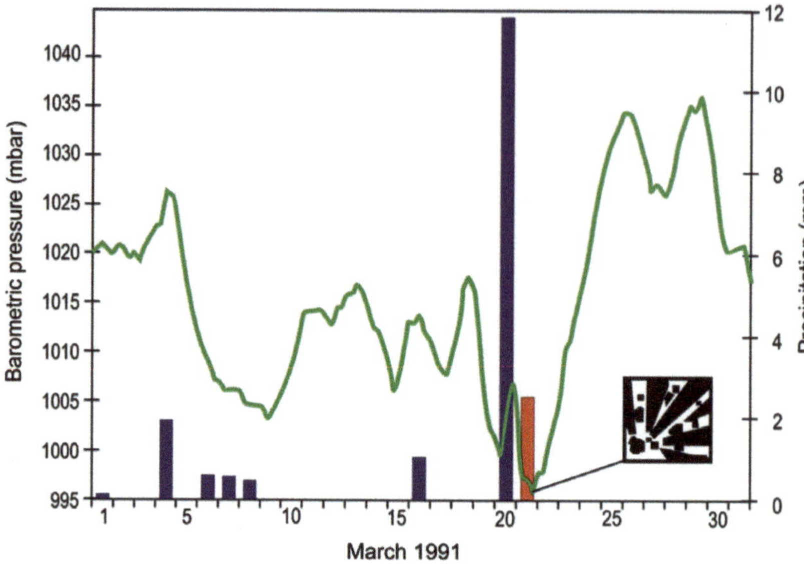

Figure 18.1.3 Daily precipitation (columns) and barometric pressure (line) for March 1991 as measured at a meteorological station close to Skellingsted landfill, Denmark. An explosion accident happened on March 21 as indicated. *After Kjeldsen and Fischer (1995).*

on steep slopes can be a difficult task, and there have been reported examples that such liners have been imperfect leading to significant migration. The side liner is obviously to be constructed prior to the disposal of waste in the quarry. For landfills already established without side liners, passive impermeable barriers have been constructed outside the border of the landfill to prevent migration. Such impermeable barriers can be established as low-permeable clay barriers constructed by pouring bentonite slurry into an excavated trench, by steel sheet piling, or by installation of a vertical liner (of HDPE or similar material) into an excavated trench.

As an alternative to a low-permeable barrier, an intercepting trench filled with high-permeable materials (such as pebbles or gravel) can be used. The concept here is to form a low-pressure volume below the surface to encourage LFG to flow toward the barrier. The venting barrier may also be constructed by driving vent notes into the ground (Wilson and Shuttleworth, 2002). To reduce the emissions of methane or odoros/toxic compounds into the atmosphere from the venting barrier, the barrier may be constructed with a layer of a biologically active and coarse material (such as compost) on top of the intercepting trench. The additional compost trench is able to oxidize the gas constituents that significantly reduce the emission. The concept of the technology is shown in Fig. 18.1.4. Such an intercepting biotrench was successfully established as a pilot trail at an Australian landfill (Dever, 2009).

The migration may also be avoided by improved gas extraction from the landfilled waste. In many cases the migration problem has partly been a result of the lack of a LFG extraction system in place. Establishment of an active gas extraction system will reduce the pressure inside the landfill, and by that

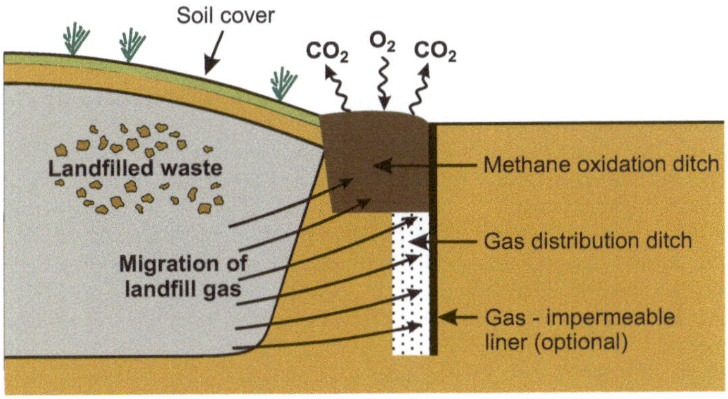

Figure 18.1.4 Conceptual sketch of an intercepting biotrench for avoiding landfill gas (LFG) migration to the surrounding area and simultaneously avoiding emission of methane to the atmosphere. *After Dever (2009).*

reduce the advectively driven gas migration into the surroundings. The quality and quantity of the extracted gas may be low especially for older landfills. In such cases the extracted low-quality LFG can be led to a compost-based biofilter for reducing the methane content and content of odorous constituent prior to release to the atmosphere (see Chapter 9.5 for details).

Finally, problems with migration of LFG have been remediated by establishment of soil-venting systems, consisting of either passive or active (the latter by use of gas pumps) venting gas wells installed outside the waste along the perimeter of the landfill. In many cases the gas wells cannot be directly vented to the atmosphere not to introduce a source to odor problem for the neighbors and/or due to regulation on the direct release of the greenhouse gas, methane, to the atmosphere. Instead the collected soil pore gas has, in some cases, been led to an activated carbon filter (to remove odorous/toxic compounds) or a biofilter to remove methane and the odorous/toxic compounds simultaneously.

References

Dever, S.A., March 30, 2009. Passive Drainage and Biofiltration of Landfill Gas: "Behaviour and Performance in a Temperate Climate" (Dissertation 1-448). School of Civil and Environmental Engineering, University of New South Wales, Sydney, Australia.

Dincer, F., Odabasi, M., Muezzinoglu, A., 2006. Chemical characterization of odorous gases at a landfill site by gas chromatography-mass spectrometry. Journal of Chromatography A 1122, 222–229.

Duan, Z., Lu, W., Li, D., Wang, H., 2014. Temporal variation of trace compounds emission on the working surface at a landfill in Beijing, China. Atmospheric Environment 88, 230–238.

EEA, May 2013. Annual European Union Greenhouse Gas Inventory 1990–2011 and Inventory Report. European Environment Agency. Technical Report No. 8/2013. http://www.eea.europa.eu/publications/european-union-greenhousegas-inventory-2013.

Franklin, J.D., Vetter, R.J., May 1987. Emergency response to landfill gas migration. Public Works 84–85.

Franzidis, J.-P., Heroux, M., Naslev, M., Guy, C., 2008. Lateral migration and offsite surface emission of landfill gas at city of Montreal landfill site. Waste Management & Research 26, 121–131.

Gandolla, M., Acaia, C., Fischer, C., 2002. LFG Migration in the Subsoil: Experiences of Control and Remediation. Rev./2002. Econs SA – Environmental Engineering, Bioggio, Switzerland.

Gendebien, A., Pauwels, M., Constant, M., Ledrut-Damanet, M.-J., Nyns, E.-J., Willumsen, H.-C., Butson, J., Fabry, R., Ferrero, G.-L., 1992. LFG – from Environment to Energy. Commision of the European Communities, Directorate-General Telecommunications, Information Industries and Innovation, Luxembourg.
Gregson, E.M., 2000. Review of Landfill Gas: Incidents and Guidance. TD5/030. Health and Safety Executive. http://www.hse.gov.uk/foi/internalops/td_din/other-gas/td5_030.htm.
Hoather, H.A., Wright, P.A., 1989. Landfill gas: site licensing and risk assessment. In: Alston, Y.R., Richards, G.E. (Eds.), Landfill Gas and Anaerobic Digestion of Solid Wastes, Proceedings Conference Chester, UK. Harwell Laboratories, UK.
IPCC, 2013. Contribution of Working Group I to the Fifth Assessment Report of the Intergovernmental Panel on Climate Change. In: Stocker, T.F., Qin, D.H., Plattner, G.K., Tignor, M.M.B., Allen, S.K., Boschung, J., Nauels, A., Xia, Y., Bex, V., Midgley, P.M. (Eds.), Climate Change 2013: The Physical Science Basis. Cambridge University Press, Cambridge, United Kingdom and New York, NY, USA.
Kerfoot, H.B., Baker, J.A., Burt, D.M., 2004. Geochemical changes in ground water due to landfill gas effects. Ground Water Monitoring and Remediation 24 (1), 60–65.
Kerfoot, H.B., 1994. Landfill gas effects on groundwater samples at a municipal solid-waste facility. Journal of the Air & Waste Management Association 44 (11), 1293–1298.
Kjeldsen, P., 1996. Landfill gas migration in soil. In: Christensen, T.H., Cossu, R., Stegmann, R. (Eds.), Landfilling of Waste: Biogas. E & FN Spon, London, GB.
Kjeldsen, P., Fischer, E.V., 1995. Landfill gas migration – field investigations at Skellingsted landfill, Denmark. Waste Management & Research 13, 467–484.
Kjeldsen, P., Scheutz, C., 2003. Short and long term releases of fluorocarbons from disposal of polyurethane foam waste. Environmental Science & Technology 37, 5071–5079.
Poulsen, T.G., 2005. Impact of wind turbulence on landfill gas emissions. In: Proc. Sardinia'05. Tenth International Waste Management and Landfill Symposium, 3–7 October 2005, CISA, Environmental Sanitary Engineering Centre, Cagliari, Italy.
Poulsen, T.G., Sharma, P., 2011. Apparent porous media gas dispersion in response to rapid pressure fluctuations. Soil Science 176 (12), 635–641.
Rettenberger, G., 1985. Gasförmige Emissionen aus Abfalldeponien im Hinblick auf Umweltrelevante Schadstoffe, in German. In: Umwelteinflüsse von Abfalldeponien und Sondermüllbeseitigung. Proc. Mülltechnisches Seminar und Fachgespräch Sondermüll, November 1984, München, Germany, pp. 40–56.
Rowland, F.S., Molina, M.J., 1975. Chlorofluoromethanes in the environment. Reviews of Geophysics 13 (1), 1–35.
Scheutz, C., Kjeldsen, P., Gentil, E.C., 2009. Greenhouse gases, radiative forcing, global warming potential and waste management – an introduction. Waste Management & Research 27 (8), 716–723.
US EPA, April 2013. Inventory of U.S. Greenhouse Gas Emissions and Sinks: 1990–2011. US Environmental Protection Agency, EPA 430-R-13-001. http://www.epa.gov/climatechange/Downloads/ghgemissions/US-GHG-Inventory-2013-Main-Text.pdf.
Ward, R.S., Williams, G.M., Hills, C.C., 1992. Changes in Gas Composition During Migration from the Foxhall Landfill, Suffolk. Technical Report WE/92/22. Fluid Processes Group, British Geological Survey, Nottingham, UK.
Williams, G.M., Aitkenhead, N., 1991. Lessons from Loscoe: the uncontrolled migration of landfill gas. Quarterly Journal of Engineering Geology 24, 191–207.
Wilson, S.A., Shuttleworth, A., January 2002. Design and performance of a passive gas migration barrier. Ground Engineering 34–37.

18.2

GROUNDWATER QUALITY MONITORING IN LANDFILL AREAS

Giovanni Pietro Beretta and Raffaello Cossu

INTRODUCTION

Groundwater contamination by leachate represents a major concern for environmental contamination originating from landfilling. For this reason, barriers are installed in landfills according to the multi-barrier concept (see Chapter 2.1).

To control the efficiency of barriers, a groundwater quality assessment and information system should be planned at the time of landfill design and later implemented during the different phases of the landfill life span (construction, operation, aftercare) as described in Fig. 18.2.1. The core of the information system is a monitoring network for groundwater quality analysis upstream and

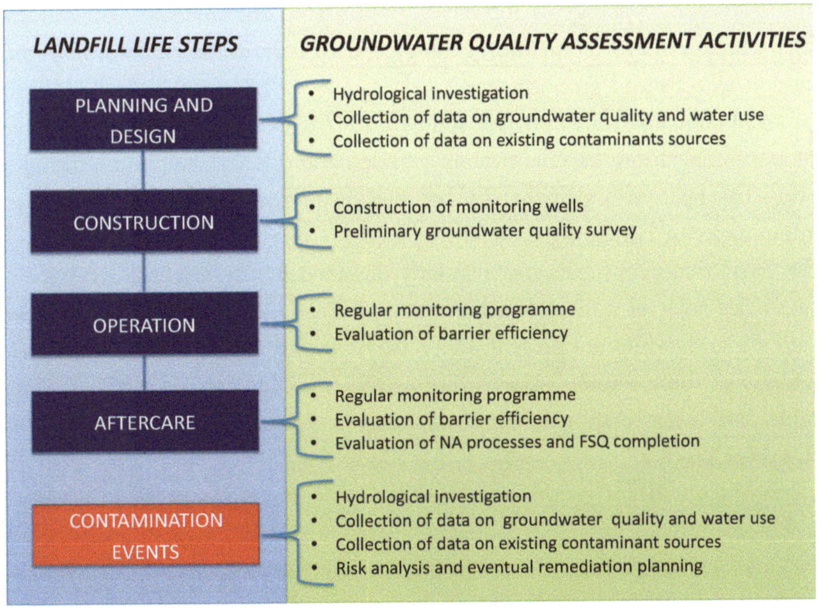

Figure 18.2.1 Groundwater assessment activities to be carried out during the various life steps of a landfill.

downstream of the landfilling deposit area and a database to be continuously updated with regard to hydrogeological data, water use in the area, groundwater quality external to the landfill area, and other existing potential contaminating sources.

The goal of groundwater monitoring is to acquire data that can be used to minimize the risks of groundwater pollution by leachate from landfilled waste throughout each phase of the facility life.

A monitoring network project should be aimed at clarifying the following issues:

- efficiency of leachate collection and extraction systems;
- need for investigation to prevent groundwater pollution;
- data for verification of risk analysis;
- need for measures to reduce or clean up groundwater pollution;
- landfill closure and residual potential risks for the environment or for the human health.

Despite the adoption of measures to prevent the uncontrolled emission of leachate from landfill, diffuse and/or concentrated emissions may occur, either on a short- or long-term basis.

This may be manifested particularly with landfills in which technical measures are merely represented by physical barriers (liners, drainage systems). The main critical aspect of these types of landfill is the life span of barrier efficiency, which is generally shorter than the duration of the environmental impacts. Liners may leak and drainage may clog (Rollin et al., 1994, Chapter 7.1). Indeed, high levels of leachate within the waste mass, as a consequence of drainage clogging, may increase (according to the Darcy law) the speed of the uncontrolled leachate leakage.

The consequent impact on groundwater quality will depend on the quality of waste, quality of leachate, landfill operation, typology of top cover, unsaturated zone beneath the landfill, aquifer and groundwater characteristics (hydrodynamics, geology, red-ox conditions), extension of diffusion plume, and the attenuation in soils and groundwater.

Contamination phenomena for municipal solid waste (MSW) landfills are typically associated with the presence of ammonia, chlorides, chemical oxygen demand (COD), sulphates, iron, and some heavy metals. Attention has been also paid in the last decade to the increasing presence of xenobiotics, including Contaminants of Emerging Concern (CECs).

Many of the most frequent substances regularly detected in the leachate are widely produced by several other activities such as agriculture, livestock farming, industrial production, and housing. In some areas where these activities are present, these activities may serve to mask the impacts solely originated by landfills. For these reasons, monitoring activities should rely on a robust investigation program and reliable data interpretation to avoid mistakes in overlooking other potential contamination sources (see Box 18.2.1).

Leachate contaminants in the groundwater environment undergo a series of attenuation processes which need to be carefully considered, studied, and implemented in a risk analysis scheme. Natural attenuation (NA) should be a strategic asset in considering the long-term sustainability of landfills.

This chapter introduces the potential contaminants in landfill leachate, describing the criteria for site conceptual modeling and illustrating the main features of the groundwater monitoring network design.

The transport of solutes in groundwater and the management of groundwater quality data are discussed. Finally the chapter illustrates the use of natural (radioactive, chemical, and microbiological) tracers to evaluate contamination issues around a landfill site.

SOURCE OF CONTAMINANTS

With regard to a typical landfill receiving MSW and special urban waste including commercial, sewage sludge, etc., the potential contaminants detected in the leachate (see Chapter 10.2) may be grouped into the following categories:

- Natural organic substances: proteins, fats, hydrocarbons and their degradation products (volatile fatty acids, humic soluble matter, etc.), generally cumulatively measured as COD, biochemical oxygen demand (BOD), or total organic carbon (TOC).
- Inorganic macrocomponents: calcium, magnesium, sodium, manganese, ammonium, iron, manganese, chloride, sulfate, carbonate, etc.
- Heavy metals and metalloids: cadmium, zinc, chromium, copper, arsenic, cadmium, etc.
- Xenobiotic organic compounds (XOCs): BTEX, C3-benzenes, bicyclic compounds, naphthalene, chlorinated aliphatics, phenols, pesticides, phthalates, etc.

Natural organic substances

The content of dissolved organic matter in leachate is generally described by a series of parameters, including TOC (Total Organic Carbon), COD, and BOD.

These parameters are comprehensive of a large variety of organics (Chapter 10.2). A low BOD/COD ratio in leachates indicates the presence of a low concentration of volatile fatty acids and relatively higher amounts of fulvic- and humic-like compounds (Kjeldsen et al., 2002).

Inorganic macrocomponents

The concentrations of some inorganic macrocomponents may vary significantly following pH variation that occurs throughout the anaerobic degradation process (Chapter 10.2).

Ammonia is released from waste mainly through the decomposition of proteins, and nowadays it represents the most relevant component of concern for long-term landfill emissions. Concentrations may exceed 3000 mg/L.

In an anaerobic environment, there is no mechanism to promote the biological degradation of ammonia (see Chapter 3.1). Ammonia concentration may only be reduced by the flushing effect of infiltrating water.

Heavy metals and metalloids

The content of heavy metals in MSW leachate varies widely between landfills, although as a general rule, metals concentration is low and of scarce concern in relation to environmental impacts from MSW landfilling (Chapter 10.2).

Mechanisms such as sorption and precipitation, and pH variations during the anaerobic process, play a significant role in the immobilization of metals and subsequent low concentrations in leachate.

Under anaerobic condition, the passage from Fe^{3+} (insoluble) to Fe^{2+} (soluble) occurs, as well in the case of Mn^{4+} to Mn^{2+}; these metals are generally present in the aquifer solid matrix.

Processes such as complexation between metals and organic and inorganic compounds and sorption to colloids are capable of mobilizing heavy metals, and therefore increased concentrations may be detected in the mobile aqueous phase (Kjeldsen et al., 2002).

Xenobiotic organic compounds

XOCs invariably display very wide concentration ranges, reflecting differences in landfill technologies, waste composition, and waste age.

Along the last decades, the typical heterogeneity of MSW becomes more and more complex. Further, to the traditional xenobiotics, an increasing number of Contaminants of Emerging Corncern (CECs) can be found in modern life products, such as plastics, paper, pharmaceuticals, personal care items, etc. Chemicals are added for different reasons (performance, production, and distribution, etc.), but then they can be leached out once the products become waste and they are landfilled or recycled (Pivnenko et al., 2016 a,b). Indeed, more than 200 individual compounds or classes of compounds may be identified in screening procedures for XOCs (Kjeldsen et al., 2002).

The most commonly found compounds are monoaromatic hydrocarbons (benzene, toluene, ethylbenzene, xylenes), together with halogenated hydrocarbons such as tetrachloroethylene and trichloroethylene. Other significantly observed groups of pollutants include phenols, cresols, chlorophenols, pesticides (ametryn, atrazine, chlorpropham, dichlobenil, hexazinon, etc.), phthalates, aromatic sulfonates, phosphonates, and others. Concentrations of single XOCs were detected in a range from <0.1 to 2220 μ/L (Baun et al., 2004).

From an investigation carried out in 2011 across the United States, a total of 129 out of 202 analyzed CECs were detected in one or more samples over 19 different landfills (Masoner et al., 2014). The distribution of concentrations found in 50% or more samples is reported in Fig. 18.2.2.

Among these substances BPA (a plastics component), DET (present in insects-repellent products), and nicotine derivates were the most frequently detected substances.

Several of the XOCs found in leachate further to be potentially toxic, carcinogenic, teratogenic, mutagenic, etc., they can be persistent and bioaccumulated.

SITE CONCEPTUAL MODEL

The first phase for design of the groundwater monitoring network in a landfill site relates to characterization of the conceptual model.

From a hydrogeological point of view, a conceptual model is a simplified representation of the physical hydrogeological system in relation to landfill site, interpreted as a source of contaminants.

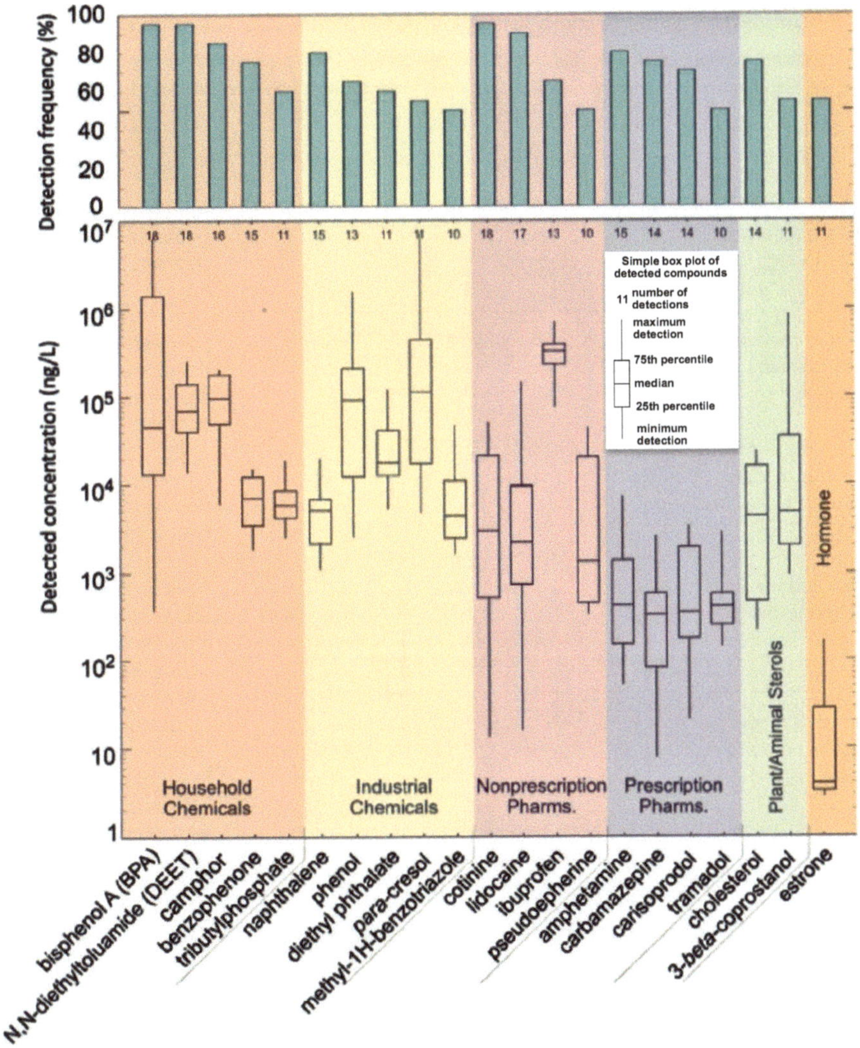

Figure 18.2.2 Distribution of concentrations for the main Contaminants of Emerging Concern detected in 19 different landfills across the United States. *Masoner et al., 2014.*

The pathway from the source to the receptor terms identifies the resultant potential risks to water resources.

The geological media is highly variable, heterogeneous, and anisotropic; this media plays a primary role in controlling groundwater flow and contaminant transport.

Five general geologic type settings can be defined as described in Table 18.2.1 (NRC, 2005; Sale and Newell, 2011).

Table 18.2.1 Summary of geologic type settings (NCR, 2005; Sale and Newell, 2011)

Geologic setting	Permeability (m^2)	Hydraulic conductivity (m/s)	Porosity (%)	Properties
Type I: Granular media with mild heterogeneity and moderate to high permeability	$k > 10^{-14}$	$K > 10^{-7}$	5–40	Uniform material; relatively uncommon (e.g., sand or gravel deposits)
Type II: Granular media with low heterogeneity and low permeability	$k < 10^{-14}$	$K < 10^{-7}$	5–40	Low permeability materials with no secondary permeability features, i.e., fractures (e.g., clay deposit)
Type III: Granular media with moderate to high heterogeneity	$k > 10^{-14}$	$K > 10^{-7}$	5–40	Large spatial variations in permeability (e.g., deltaic, fluvial, and glacial deposits)
Type IV: Fractured media with low matrix porosity	$k < 10^{-17}$	$K < 10^{-10}$	<1 (both fractures and matrix)	Little void space exists in unfractured matrix; transmissive features due to fractures (e.g., crystalline rock)
Type V: Fractured media with high matrix porosity	$k < 10^{-17}$	$K < 10^{-10}$	<1 (fractures) 1–40 (matrix)	Large void spaces exist in unfractured matrix; transmissive features due to fractures (e.g., limestone, sandstone, and clays with secondary permeability features)

In these geological environments (type), the existing information should enhance the identification of the preliminary conceptual model, capable of indicating the preferential pathway for contamination and subsequently validating this model through the application of geological, geophysical, hydrogeological, and hydrochemical field investigations.

Preliminary conceptual model

A preliminary conceptual model is performed on the basis of existing data and inspection; the geological and hydrogeological conditions of the site area are described.

In this feasibility study, a geological map and geological cross sections are acquired or rebuilt through boring logs.

A hydrology study should be executed to identify climatology (temperature, precipitation, and evapotranspiration) and surface water in relation to flooding and groundwater.

The structure and properties of aquifer, aquitard, and aquiclude of the subsoil are examined, in particular for the uppermost aquifer, which is the most vulnerable to pollution.

The uppermost aquifer is defined by "the geologic units positioned close to the surface, that can store and subsequently yield significant volumes of groundwater and it is confined to the bottom by a unit that, due to its low permeability, represents the boundary level or substrate of the aquifer" (National Water Well Association, 1986).

Groundwater quantity and quality, groundwater movement, significance, and use of groundwater resources are examined.

Geological hazards, such as floods, earthquakes, landslides, should be evaluated with regard to the area surrounding the landfill.

Literature review and field reconnaissance of geologic, hydrologic, hydrogeological, and hydrochemical features of the site should be included in a first report.

The potential receptors for contamination are (U.K. Environment Agency, 2008)

- groundwater beneath the site;
- groundwater at the downgradient boundary;
- water extraction;
- potential water resources;
- biodiversity (ecological value);
- amenity.

The result of preliminary conceptual model facilitates preparation of an investigation plan for the area surrounding the landfill, taking into account the characteristics of the following general geological types.

Geologic type I—This type, formed by eolian sands, is relatively rare for landfill siting and preferential flow is not clear, but it is influenced by local factors.

Geologic type II—The lacustrine sediments feature a low permeability and flow limited to sandy horizons with limited thickness. In general, this type may be assimilated to stagnant groundwater.

Geologic type III—The structure of fluvial sediments determines preferential flows in the horizons of gravel and sand, while featuring a slow diffusion and storage in the low-permeable horizons.

Geologic type IV—This type relates largely to karst rocks, in which permeability varies in 10 orders of magnitude. Groundwater flow is also significant due to the presence of wide-open caves and fractures.

Geologic type V—In this rocky type, in practice, groundwater flow only occurs in the fractures forming the preferential paths.

Other geologic types include glacial deposits; the glacial till is unsorted sediment, and groundwater flow rate is generally reduced due to low hydraulic conductivity; the glacial deposits may also contain lacustrine and fluvial deposits.

Many regulations require a preventive measurement of the piezometric levels over time to establish a safety level to avoid the groundwater inflow through the bottom of the landfill.

The objective of this zone of safety (unsaturated zone) is avoiding the eventual flooding of part of the waste by an unforeseen increase of the groundwater table, following extraordinary aquifer recharge events.

Such event causes (1) a significant increase in the amount of leachate to be collected and treated, (2) a reduction of the efficiency and the stability of the geological barrier (with effects on the overall efficiency of the plant), and (3) changes in the chemical—physical and biological reactions that take place in the wastes.

In absence of a significant historical series of piezometric level data and in presence of a short distance between the landfill bottom and the groundwater table, it is possible to (1) install in piezometers a level data logger to ensure the updates knowledge of groundwater levels and (2) plan a groundwater level stabilization in case of extreme recharge events with active (drainage wells) or passive interventions (trenches, engineering barriers).

Validation of the conceptual model

A validation of preliminary conceptual model is performed by means of field investigations and interpretation of the same to formulate a definitive conceptual model for design of the groundwater monitoring network.

The goal can be achieved by defining the following elements in three phases:

- hydrogeological characterization of the site (subsurface structure definition and determination of hydraulic conductivity values for each significant horizon);
- experimental resolution of groundwater flow, including the three-dimensional (3D) scale, as well through the realization of cluster wells (piezometers measuring groundwater levels at different depth);
- experimental resolution of the problems related to transport of contaminants (groundwater quality monitoring wells) and subsequently to the definition of target zones.

The definitive conceptual model may be further improved during the management of groundwater monitoring data.

Phase 1

The geological reconstruction of landfill areas is performed by means of boring for stratigraphic and geotechnical purposes.

The number of suggested boreholes is illustrated in Table 18.2.2; this number may be reduced in relation to geological features of the area derived from the preliminary conceptual model (homogeneous surface geology, only soil or rock in explored subsoil, little influence of tectonics, etc.) and use of geophysical methods.

Efforts to achieve stratigraphic reconstruction will provide a log of soil and/or rock (with recognized field and laboratory procedures of standard classification) and geological cross-section execution.

Indirect methods of investigation (geophysical) are applied to enhance direct methods of investigation aimed at identifying potential pathways.

Geophysical investigations complete the knowledge of lateral and vertical distribution of lithotypes identified by drilling the boreholes, thus allowing their reduction.

Table 18.2.2 Number of borings in relation to landfill surface (Waste Management of North America, 1991 modified)

Surface (ha)	Number of geothecnical boring	Number of deep stratigraphic boring recommended
<4	3	1
4–20	6	2
20–40	10	4
40–80	15	5
>80	18 + 1 boring/each additional 4 ha	6 + 1 boring/each additional 16 ha

Seismic methods and low-frequency methods explore the geological structure of the subsurface (bedrock, fractures, and faults, etc.).

Geoelectrical methods (Schlumberger or Wenner array, dipole-dipole method, etc.) characterize the distribution in the subsurface of different permeability units (coarse, sand, clay, etc.) and defects in geomembranes inside the landfill.

Furthermore, the presence of contaminants in the subsurface may be assessed using the induced polarization method, electrical tomography method, and electromagnetic method; the latter is also applied in study of the continuity of clay barriers in landfill.

In the context of geotechnical characterization, soil and rock samples are obtained during drilling to permit analysis of some parameters: grain size, bulk and matrix density, porosity, permeability, and Rock Quality Designation (RQD). Other parameters to be determined include cation-exchange capacity (CEC) and fraction of organic carbon (foc) in fine sands, silts, clays, and peat.

If the area is suspected as being contaminated, soil samples should also be obtained for chemical analysis.

For the hydrogeological characterization of soil during the drilling phase, permeability tests are carried out using constant water-level methods or rising or failing water-level methods (slug test or Lefranc test).

Pumping tests for hydrogeological characterization are more challenging from a technical and economical point of view, but they provide more significant parameter values, as they involve a greater volume of aquifer.

With secondary permeability, due to presence of discontinuities, the Packer test technique for hydraulic conductivity measurement is better indicated.

The results obtained in hydrogeological characterization define the presence, geometry, and properties of aquifers, aquitards, and aquicludes in the area.

Phase 2

The borehole can be completed as a piezometer according to the methods suggested in the subsequent Paragraph.

The measure of groundwater levels in the piezometer, with a precision of 1 cm, should allow 3D reconstruction of groundwater flow.

Groundwater flow direction and hydraulic gradient are commonly reconstructed using contour piezometric maps, although the reconstruction of cross sections may also be required to identify the horizontal and vertical components of flow and recharge/discharge zones.

The construction of groundwater flownet (intersection of equipotential and flow lines) is also fundamental in establishing the location of the well upgradient and downgradient of the landfill.

The groundwater flownet must be reconstructed in 3D (map and section) to recognize the zones target of any contamination caused by the landfill, as shown in Fig. 18.2.3.

For this reason, an exact design of the depth of the boreholes and of the position of the piezometer screens are way important.

As illustrated in Table 18.2 and in absence of hystorical series of the piezometric data, groudwater level measurements should be collected over a period of at least 1 year to detect groundwater level fluctuations in relation to seasonal variations and other natural or man-made reasons.

Boreholes not devolved for use as piezometers should be decommissioned by means of cementation.

The results of Phases 1 and 2 will also define the type of aquifer—unconfined, semiconfined, or confined—as well as the monitoring strategies to be applied to assess landfill impact.

A perched aquifer with low flow may also be present at times, which may direct groundwater and possible pollutants to the underlying unconfined aquifer.

On a site featuring a flat groundwater surface and with low hydraulic conductivity aquifer, as is the case for the majority of old landfills, a change in the direction of flow lines (the downgradient becomes upgradient and vice versa) may be manifested; the arrangement of monitoring wells should take this variable situation into account at the time of interpretation of results.

Phase 3

In the third phase, the distribution of permeability in the subsoil and flownet will contribute toward identifying the target zones defined as "areas in which groundwater flow if preferential and in which the observation of possible release of pollutants from landfill is more likely."

The location of the possible supplementary piezometers (to detail groundwater flow) and the monitoring wells should be appropriately designed and implemented.

The depth and screen elevation of these monitoring wells should be included in the target zones.

Drawings and sketches of a baseline for the quality of groundwater should be prepared in this phase.

A series of water samples should be obtained and analysis performed by selecting parameters on the basis of the general characteristics previously acquired in implementation of the preliminary conceptual model, relating to the type of waste to be disposed of and expected landfill leachate.

These surveys should be conducted on a quarterly basis for a period of at least 1 year.

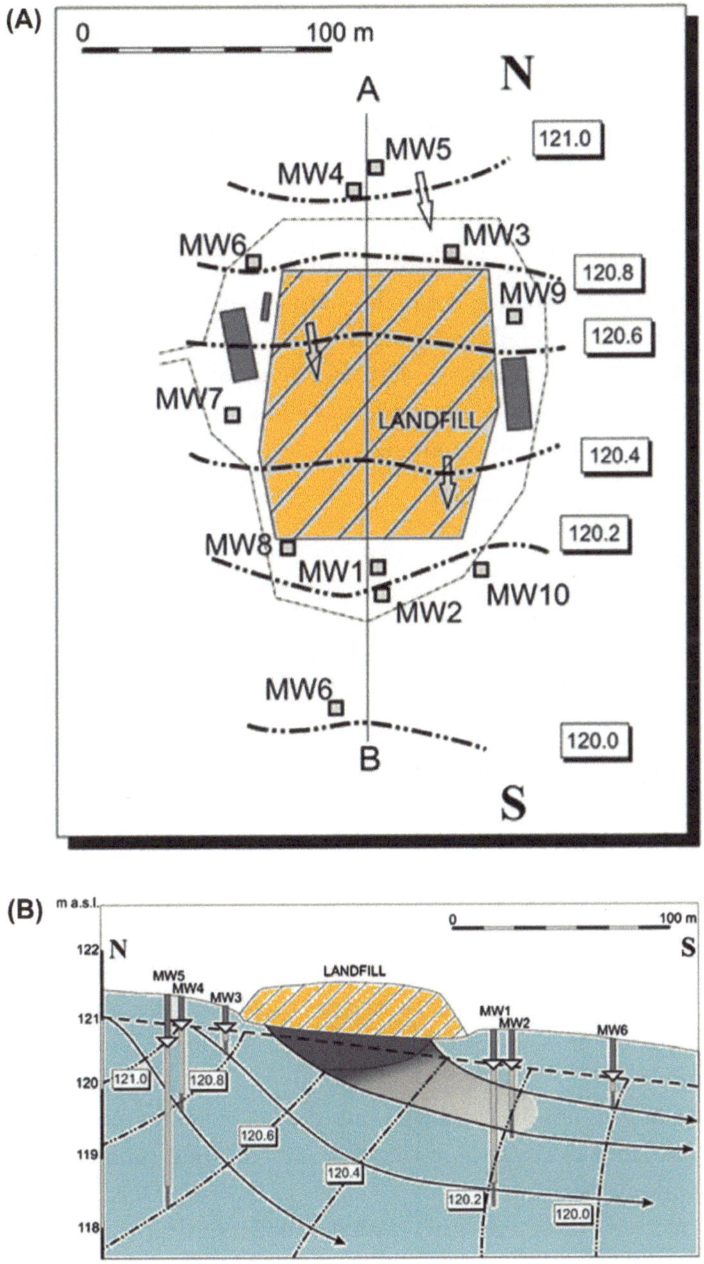

Figure 18.2.3 (A) Isopiezometric map (m a.s.l.) of groundwater in landfill area. (B). Groundwater flownet in landfill area.

In the absence of previous hydrochemical data, a large spectrum of parameters should be measured.

Evaluation of a background value should be obtained by means of the procedure described in Section 4.2.

A groundwater monitoring plan should be developed for upgradient and downgradient wells to provide details on the type of parameters and frequency of sampling and measurement.

GROUNDWATER MONITORING NETWORK DESIGN AND MANAGEMENT

The purpose of groundwater monitoring in the area surrounding the landfill is to obtain a water sample representative of hydrochemical conditions in the subsoil.

Monitoring results are used to decide the need for any actions to be taken with regard to groundwater: ordinary operating conditions or extraordinary conditions in the case of contamination.

Three elements should be taken into consideration:

- design of monitoring wells arrangement;
- monitoring well design and installation;
- groundwater sampling.

Design of monitoring wells arrangement

A landfill groundwater monitoring network may be designed based on two options.

- Single line of monitoring wells

A single line of monitoring wells around the landfill is intended to measure any releases of leachate into the groundwater. The monitoring wells should be located on the sides of the landfill (no closer than 10–20 m from waste perimeter).

- Double line of monitoring wells

A double line of monitoring wells is designed to measure any release of leachate into the groundwater and NA of solutes in the aquifer. A double line contains a buffer zone in which potential effects of leachate are eliminated or minimized; the buffer zone should be dimensioned using a geometric criterion of 30–100 m wide (as proposed by Government of Ontario, 2012) or studied using groundwater flow and transport models.

The existence of a buffer zone also facilitates remediation works in the case of contamination.

The choice of approach is linked to local factors, hydrogeological conditions, and landfill structure (elevated, in the pit or on slope).

To ascertain groundwater flows and water quality at different depths in the same location, monitoring wells may be installed either in a single borehole (cluster monitoring wells) or in a series of different boreholes (single monitoring well).

The number of monitoring wells in the same location is in relation to preferential flows originating from the landfill identified by the conceptual model.

In general,

- a divergent pattern of flow lines requires a greater number of monitoring wells than a converging pattern;
- an important vertical component of hydraulic gradient requires a greater number of monitoring wells compared with a prevailing horizontal component of hydraulic gradient.

The location of the monitoring wells is of similar importance, and different criteria may be applied.

The first criterion is a geometric layout of wells, with a spacing of 20—50 m, not based however on hydrogeological characterization; this may represent the initially adopted method to be improved during the course of work.

The second criterion is based on monitoring efficiency defined as the ratio between the area liable to produce a detected contamination and total landfill area.

Optimization of a monitoring network is achieved by means of subsequent approximations. The efficiency of the system may be enhanced using the method proposed by Wilson et al. (1992): generation of virtual contamination sources in the landfill and effective observation of the corresponding plumes at the monitoring points.

The third criterion, finalized toward the optimization of groundwater monitoring networks, is based on sensitivity, capital, and Operation and Maintenance costs (de Cleen et al., 1995).

Monitoring well design and installation

The importance of proper design and installation of monitoring wells is frequently largely underestimated.

However, an incorrect implementation of monitoring wells may determine false analytical results for groundwater quality and impact subsequent decisions relating to possible pollution by landfill leachate release.

Accordingly, appropriate technical and economic resources should be ring-fenced for design and installation.

The provisions laid down in standard methods or guidelines, such as in Nielsen (1991) or the ASTM (1990) and NGWA (1986) guidelines, should be adhered to as summarized below.

The construction of monitoring wells may be developed in two phases:

(A) during field investigations to implement the definite conceptual model and to identify preferential flow paths;
(B) during construction of the landfill and environmental monitoring facilities, according to the findings obtained by the conceptual model.

As an example, the location, number, depth, and screen of monitoring wells may be easier to identify during Phase A.

Drilling

Drilling is performed according to one of two main techniques: fluid circulation and direct push. The diameter is a function of the diameter of the casing that will need to be introduced into the borehole.

The choice of drilling system is related to the geological features and depth and should be sufficient to allow reconstruction of the stratigraphy of the subsoil and the sampling of soil and rocks.

Other criteria are related to local conditions such as health and safety, accessibility and noise, fluid disposal facilities, water availability, costs, etc.

A report with boring log together with casing, screen, filter pack, annular seal, and surface protection features should be compiled (monograph of the well) and brought to the site for inspection during sampling.

Casing

Monitoring well casing should be capable of providing resistance to strain induced by geologic materials crossed in the borehole, to corrosion (chemical resistance and durability). and to reactions impacting the groundwater and geological materials.

The three most common materials used in the casing are carbon steel, stainless steel, and polyvinyl chloride

The hydrochemical features identified during field investigations should guide the selection of casing materials.

The diameter of monitoring wells is an additional parameter to be taken into account, and this will vary in line with use of the well (water level measurement, water sampling, or both). A minimum diameter of 3 inch is recommended, but pumps with diameter less than 2 are also available.

Screen and filter pack

The monitoring well needs to capture water from one aquifer and should be constructed using the same material as the casing.

Well screens allowing the ingress of water into the monitoring well are located along the preferential pathways of the groundwater flow originating from the landfill area.

The main type of screen is slotted pipe and continuous-slot screen with diameter depending on casing diameter; an extended open area is available for the continuous-slot screen.

The velocity of water ingress in the screen should not cause turbulence or allow access of sediments into the well.

In the naturally developed well, a permeable material collapses around the screen and fine materials are removed together with water during pumping.

In the artificially developed well, a chemically inert granular media is installed in the annular space available using a gravity method, tremie pipe, or reverse circulation and backwashing.

A series of methods are available for use in sizing the filter pack in relation to aquifer grain size (Johnson Division, 1966). Due to the reduced diameter of the borehole, filter pack installation is rather complex; in some cases, an artificial filter pack is inserted together with the screen, or the screen is protected by means of geotextile.

Annular seal

The annular seal in the monitoring well is aimed at preventing vertical groundwater movements along the borehole: from the surface and to subsoil and between different permeable horizons, in an attempt to restore the natural stratigraphy.

Commonly used low-permeability materials include neat cement (mixture of Portland cement, aggregate, or sand and water) or bentonite (granular, pelletized, powdered, etc.)

Following mixing, the neat cement is pumped into the annulus; dry bentonite is emplaced into the subsoil (pellets, granular) or mixed with water on the surface (powdered).

Portland cement should be applied in the absence of chemicals to avoid cross contamination of groundwater sampling.

Surface completion

The monitoring well is completed with surface seals and protective casing (above ground) or cement or metal manholes (below ground).

An identification code and reference points for piezometric level measurements and elevation above sea level are required to identify the head of the monitoring well in the field.

Particular care should be taken in the field in the identification of cluster monitoring wells (two or three casings in a single borehole).

Development

The aim of the development is the removal of fine materials and drilling fluids from the space immediately surrounding the monitoring well screen.

In general, this operation is carried out using compressed air or submersible pumps up until clear water is withdrawn.

Water sampling cannot be carried out immediately, but only a few days subsequent to this procedure.

Unsaturated zone monitoring

The groundwater monitoring in the unsaturated zone is a less widespread specialized activity than monitoring the saturated zone.

In the installation of the samplers, it is of the utmost importance not to set up preferential flow paths and to ensure continuity between the soil, backfill material, and sampler device.

Water flow is generally measured with tensiometer, recording the matrix potential at different depth.

Pore fluid sampling is difficult depending on the saturation degree of unsaturated soils, and it is possible to sample water at the pressure higher than the melting point.

The choice of type of sampling device is dependent of various local condition: lithology, depth, morphology of landfill, temperature and rainfall regime, type of substances researched (hydrophilitic or hydrophobic), etc.

The following is a list of types of device (ASTM, 1992; Nielsen, 1991):

- Suction samplers: vacuum lysimeter, pressure-vacuum lysimeter, high pressure-vacuum lysimeter, suction lysimeter, filter tip samplers.
- Free drainage lysimeter: pan lysimeter, glass block lysimeter, trough lysimeter, vacuum through lysimeter, caisson lysimeter, wicking soil-pore liquid sampler, sand-filled funnel sampler.

There are many limits to the significance of measurement results.

The interpretation of the measures carried out in the unsaturated zone serves for an initial assessment of the conditions of infiltration of water and dissolved substances which must then be deepened with the monitoring of the saturated zone.

Groundwater measuring and sampling

During monitoring the data gathered should be reliable and fit for purpose to allow identification of the potential impact of landfill on groundwater.

A specific procedure should be included in the monitoring Plan based on Quality Assurance (QA) and Quality Control protocols as indicated in the following paragraph "Measurement of hydrochemical parameters".

The plan should contain certification of monitoring infrastructure, details of sampling methods, laboratory analysis, data management, interpretation and reporting, and procedures applied to assess the adequacy and appropriateness of monitoring activities (U.K. Environment Agency, 2008).

Field measures and laboratory analysis should be carried out using updated standardized methods.

Groundwater measures should be performed to define the flow system in the proximity of the landfill.

Groundwater level should be periodically measured (monthly frequency) inside the piezometer of the monitoring network during the design stage and throughout the early years of landfill life.

Daily measurements with level logger in at least one monitoring well are recommended to enable association of groundwater levels with rainfall, river levels, etc. No groundwater samples should be obtained from this monitoring well to not dislodge the logger; measurement relating to pH, Eh and electrical conductivity of water may also be combined.

Although rarely, it may at times be necessary to measure the vertical flow in the well using vertical flow meter equipment.

Groundwater samples are collected for the purpose of analyzing inorganic and organic compounds and to define landfill impact.

Static sampling performed by means of bailer equipment is undertaken when the presence of a free phase of nonaqueous phase liquids (NAPL) in groundwater is suspected.

The latter suspicion should be based on the organoleptic characteristics of groundwater (color, smell) or very high concentrations in solution of light (LNAPL) or heavy (DNAPL) compounds.

Dynamic sampling is performed using traditional or alternative new methods.

The traditional method consists in a submergible pump for the extraction of groundwater.

A sample is taken by applying one or more of the following criteria:

- water volumes extracted from the well (generally 3–6 times of well volume);
- stabilization of some hydrochemical indicators measured in situ (temperature, pH, Eh, dissolved O_2, electrical conductivity);
- analysis of sets of samples (corresponding to the extraction of 1,2,4 times the volume of the well);
- hydrogeological parameters of the aquifer.

The alternative new methods are particularly suited for use in aquifers with low flow rate.

The traditional sampling of groundwater may result in considerable turbidity, determining an overestimation in the concentration of some compounds, such as metals and hydrophobic organic substances, with filtration of the sample consequently resulting in an underestimation of the level of contamination; these problems may be reduced through use of low-flow purging (US Environmental Protection Agency, 1996a).

The aim is to extract the water in such a way as to minimize the stress imparted to the subsoil system, identified by the rate of lowering of piezometric level.

A peristaltic pump with a range of 0.1–0.5 L/minute withdrawal is used; these values, however, are dependent on the hydrogeological conditions of site characteristics, and coarse-grained aquifers have even sampled at a rate of 1 L/minute.

Significant differences in the concentration of organic compounds in groundwater are manifested according to the sampling method used.

An additional element is represented by the need to filter samples, due to the fact that a turbid sample is not representative of groundwater and therefore treatment undertaken may affect the original characteristics of the sample.

Table 18.2.3 shows how statistically significant differences in heavy metal concentrations may be observed when measured on filtered or unfiltered samples (iron, potassium, manganese, zinc); this aspect should indeed be taken into consideration in the discussion of survey results.

Sample containers are identified on the basis of substances to be examined:

- metals: polyethylene or fluorocarbon resin with polypropylene cap;
- organic compounds: glass bottles with fluorocarbon resin-lined cap.

Sampling equipment should be decontaminated if a single portable submergible pump is used for all monitoring wells.

An appropriate decontamination procedure for pumps and rubber collection tubes used to sample inorganic compounds is as follows (Nielsen, 1991):

- wash with tap water and soapy water;
- rinse with dilute acid solution (10% nitric or hydrochloric);
- rinse with distilled or deionized water.

Table 18.2.3 Parameters to undergo routine analysis and observation frequencies in groundwater monitoring networks in landfill survey

Landfill life	Monitoring goals and activities	Measures and analysis
Design and construction	Groundwater knowledge to support the planning and approval process Groundwater quality before waste disposal	
Quantity	Parameter	Groundwater level
	Frequency	Monthly
Quality	Parameter	T, pH, electrical conductivity Eh, O_2, Ca^{2+}, Mg^{2+}, Na^+, K^+, HCO_3^-, Cl^-, SO_4^{2-}, NH_4^+, NO_3^-, NO_2^-, CN^-, TOC, DOC, As, Si, metals, BTEX, TPH, phenols, chlorinated solvents, pesticides, PAH
	Frequency	Quarterly
Management	Determine potential impacts of waste disposal on groundwater In existing landfills define the requirements of a revised monitoring plan	
Quantity	Parameter	Groundwater level
	Frequency	Monthly
Quality	Parameters	T, pH, electrical conductivity Eh, O_2, Cl^-, SO_4^{2-}, Na^+, NH_4^+, NO_3^-, NO_2^-, TOC, DOC, As, Metals, TPH, phenols, chlorinated solvents
	Frequency	Quarterly
Closure	Using data of the regular monitoring program for evaluation of transition to postclosure phase.	
Quantity	Parameter	Groundwater level
	Frequency	Quarterly
Quality	Parameter	T, pH, electrical conductivity Eh, O_2, Ca^{2+}, Mg^{2+}, Na^+, K^+, HCO_3^-, Cl^-, SO_4^{2-}, NH_4^+, NO_3^-, NO_2^-, TOC, DOC, Si, As, metals, BTEX, TPH, phenols, chlorinated solvents, pesticides, PAH
	Frequency	Quarterly

TABLE 18.2.3 Parameters to undergo routine analysis and observation frequencies in groundwater monitoring networks in landfill survey—cont'd

Landfill life	Monitoring goals and activities	Measures and analysis
Postclosure	The monitoring program may be varied, depending on risks to groundwater and downgradient receptors For old landfill understanding of the groundwater in the surrounding area.	
Quantity	Parameter	Groundwater level
	Frequency	Six monthly
Quality	Parameters	T, pH, electrical conductivity Eh, O_2, Ca^{2+}, Mg^{2+}, Na^+, K^+, HCO_3^-, Cl^-, SO_4^{2-}, NH_4^+, NO_3^-, NO_2^-, CN^-, TOC, DOC, Si, As, Metals, BTEX, TPH, phenols, chlorinated solvents, pesticides, PAH
	Frequency	Six monthly (3 years) and then yearly

And for the sampling of organics, it is as follows:

- initial wash with water (tap, distilled, or deionized) and laboratory-grade detergent;
- solvent rinse (acetone, hexane, isopropyl alcohol, methanol alone or in some combination);
- rinse with distilled or deionized water.

To further minimize the risk of cross contamination, the sampling collection may be commenced at the less contaminated monitoring wells in the field, and the rubber tube may be replaced at each sampling.

Blank samples are also inserted amongst the test samples; field blanks are used to evaluate the effectiveness of decontamination procedures and trip blanks to evaluate the impact on groundwater quality of shipping and handling procedures.

Another equally important aspect concerns the preservation, handling, and shipping of samples: some parameters do not remain chemically stable and some preserving agent should be used such as HNO_3, HCl, cool at 4°C, etc. and described by guidelines why in U.S. Environmental Protection Agency (1996b).

All samples should be collected, labeled, and recorded in chain of custody and shipped (at a constant temperature of 4°C) to the analytical laboratory to comply with a holding period ranging from several days to weeks, as described by Nielsen (1991).

Measurement of hydrochemical parameters

Two issues should first be addressed: the type of elements and compounds and the frequency of measurements should be distinguished.

Elements and compounds should be selected according to the following criteria:

- a baseline parameter that characterizes the groundwater (over 99% of solute);
- an indicator parameter that flags changes in hydrochemical parameters compared with baseline values;
- specific parameters related to waste and landfill leachate (expected or measured);
- typical hydrochemical parameters of the area identified during characterization of the conceptual model;
- parameters that influence the local use of groundwater;
- parameters already exceeding limit values (before landfill design and construction);
- new parameters identified during subsequent analysis.

The list of parameters to be assessed may vary in line with age of the landfill and according to the stage of design, construction, operation, closure, and postclosure.

In situ measurements are performed for temperature, electrical conductivity, pH, Eh, and dissolved oxygen in groundwater.

The frequency of measurements may also vary in line with landfill age and particularly with the hydrological and groundwater regime identified during characterization of the conceptual model.

Taking into account these numerous and complex aspects, Table 18.2.3 shows the recommended parameters and frequency of measurement, which should, however, be evaluated as described above.

During monitoring, these frequencies may be reduced or increased, and specific substances or compounds may be added to or removed from the analytical set.

Therefore, the initial groundwater monitoring plan (static plan) may vary throughout the life span of the landfill (dynamic plan).

The criteria underlying choice of the frequency of measurements and sampling are based prevalently on the hydrological regime of the area or the risk of impact on a potential receptor, such as a well, at x (m) distance from landfill.

With regard to a conservative species (for example, chlorides, nitrates, etc.), groundwater travel time t (s) from landfill to well may be calculated by means of the acknowledged relationships:

$$u = k_i/n$$

$$t = x/ua$$

where u (m/s) is groundwater seepage velocity, i hydraulic gradient, and n the effective porosity.

The US Environment Agency (2008) suggests a monitoring frequency as reported in Table 18.2.4.

Table 18.2.4 Groundwater monitoring: example of minimum survey frequencies based on travel time (U.K. Environment Agency, 2008)

Travel time to receptor (months)	Minimum recommended frequency
>24	Six monthly
12÷24	Quarterly
6÷−12	Monthly
<6	Result of risk assessment

Quality assurance project plan in monitoring systems

The results of monitoring surveys are expressed either as a spatial distribution of concentrations or as time-related evolution of the reference parameters.

Measures are implemented by the plant administrator and the controlling Board to ensure the yielding of reliable, comparable results. It is therefore fundamental that a QA Project Plan (Table 18.2.5) should be defined according to the legal, administrative, and economic consequences the observed measures will produce on subsequent decisions.

Quality targets are considered the most important part of a QA Project Plan, being expressed by the following quantitative (a) and qualitative (b) elements:

- Method detection limit (a): the smallest concentration of an element that can be reliably measured;
- Accuracy (a): the difference between a measured and actual parameter; it enhances the occurrence of bias in the measurement system;
- Precision (a): the variability of results that defines the scatter between two or more consecutive measurements of a single characteristic of the same sample;
- Completeness (a): the amount of acceptable data produced by a data measurement system compared with total amount of data that would be generated if the measurement system is operated without error (efficiency of the measure system);
- Representativeness (b): the degree of reliability under which the measured values represent the entire observed system;
- Comparability (b): the confidence with which the data to be generated can be compared with others of the same type.

Technical and operative assessment of all the above measures relating to the choice, realization, and management of the monitoring points (in addition to the qualitative–quantitative measures carried out on groundwater) should be considered in the preparation of a QA Project Plan aimed at defining operative conditions and performance requirements capable of providing accurate data.

Table 18.2.5 Quality assurance (QA) levels

Elements of QA project plan	Level			
	1	2a	2b	3
Title page and table of contents	X	X	X	X
Project description	X	X	X	X
Project organization	X	X		
QA level determination	X	X	X	X
Sampling procedure	X	X	X	X
Sample custody	X			
Analytical procedures	X	X	X	X
Calibration frequency	X			
Data quality objectives	X	X	X	
Internal QA checks	X	X	X	
Performance/system audits	X	X	X	
Preventive maintenance	X			
Data assessment procedures	X	X	X	
Corrective action	X	X	X	
QA report management	X	X	X	

1, enforcement, regulatory, and policy decisions; 2a, "high visibility" to public scrutiny; 2b, method development, feasibility studies, and preliminary assessments; 3, qualitative screening (Plumb, 1987).

The Plan should be prepared during the initial stage and the approvals obtained from the board to plan the actions and to evaluate, over time, the quality of data generated.

In selecting data quality targets, a numeric value should be assigned to each objective, to highlight the difference between the best data and those required for the solution of the problem, to produce discussion of each individual point, taking into account that quality objectives may vary over time.

As a general rule, to define the acceptability limits of decisional errors in terms of required performance of the monitoring system, a set of elaborations is drawn up to assist determination of the consequences of such incorrect decisions.

Data are applied in the procedure adopted to select the hypothesis related to environmental conditions, i.e., the null hypothesis (Ho) and the alternative hypothesis (Ha): a decisional error is deemed to have occurred when:

- false positive F (+): the null hypothesis is rejected although true;
- false negative F (−): the null hypothesis is not rejected although false.

To minimize decisional errors, the number of samples and degree of accuracy of determinations should be defined in the monitoring plan.

Report on groundwater monitoring results

The periodic monitoring report provides information on the results of quantitative and qualitative status of groundwater with a minimum yearly frequency or in correspondence of specific events.

The report, complete with maps, graphs, and tables, provides results and analysis of recent sampling events and compares these with the historical analysis. The contents of the report should comprise

- measurements, samples, and chemical analyses undertaken;
- trend in recorded values;
- evidence that trigger levels and/or limit values have been exceeded;
- actions taken to reduce potential pollution;
- possible changes to the monitoring program.

The report should include validation or modifications of the conceptual model and identification of risks for downgradient receptors (wells, rivers, lake, etc.).

In some cases, the Public Authorities may require the landfill owner or operator to provide an independent verification of reports.

The report should be delivered at prespecified times and incidents of pollution and noncompliance with groundwater requirements notified immediately to the Public Authorities.

TRANSPORT OF SOLUTE IN GROUNDWATER AND NATURAL ATTENUATION

A landfill monitoring network may detect a contaminating plume that originates from leachate.

Once reference values are exceeded, according to the above mentioned procedure, tests will need to be carried out to ascertain whether a reduction in contaminant concentration may be produced via dispersion, absorption, and/or biodegradation (natural or intrinsic attenuation following chemical–physical or biological reactions) or whether more specific remediation measures are required.

The term "monitored natural attenuation" (MNA) in a specific controlled and monitored site refers to the reliance on NA processes to achieve site-specific remedial objectives within a reasonable time frame compared with other more conventional methods. NA reduces mass, toxicity, mobility, volume, and concentration of the contaminants avoiding any intervention and protecting the human health and the environment.

Transport of solute in groundwater

Knowledge of the physical laws regulating the transport of solutes is essential in the evaluation of hydrochemical data recorded by the groundwater monitoring system.

Considering a groundwater flow along the x axis of a Cartesian coordinate system, the source in the origin of the axis, the simplified formula of solute transport is

$$D_x\frac{\delta^2 C}{\delta x^2} + D_y\frac{\delta^2 C}{\delta y^2} + D_z\frac{\delta^2 C}{\delta z^2} - u\frac{\delta C}{\delta x} = R\frac{\delta C}{\delta t} + \lambda C$$

where C is the concentration (mg/L), u the seepage velocity (m/s) in x direction, $D_x = D_l$, $D_y = D_t$ and D_z (m^2/s) the dispersion coefficient defined, respectively, by the adjective longitudinal, transversal, and vertical, t (s) the time, R the retardation factor, and λ (1/s) the first-order rate constant of degradation.

$$u = i/n$$

$$D_l = \alpha_l u + D_m$$

$$D_t = \alpha_t u + D_m$$

$$D_z = \alpha_z u + D_m$$

Transport parameters for solutes in groundwater are k the hydraulic conductivity (m/s), i the hydraulic gradient, n the effective porosity, α_l, α_t, α_z the horizontal, transversal and vertical dispersivity (m), and D_m the diffusion coefficient (m^2/s) which features a very low value in groundwater (about $10^{-10}\ m^2/s$).

From a practical point of view, dispersivity values can be calculated if a reference distance from the site L (m) is known (Gelhar and Axness, 1981; Xu, and Eckstein, 1995, 1997) by means of the following equations:

$$\alpha_l = 0.83(\log L)^{2.414}$$

$$\alpha_t = 0.1 \div 0.33\alpha_l$$

$$\alpha_z = 0.0.25 \div 0.0.056\alpha_l$$

The retardation factor, R, that expresses the rate between the velocities of groundwater and transported elements and which depends on adsorption phenomena is given by the following expression:

$$R = 1 + \frac{1-n}{n}\rho_s Kd$$

where ρ_s is the volumetric mass (kg/m^3) of the solid phase and n effective porosity.

K_d is the distribution coefficient defined by the following relation (for linear adsorption):

$$K_d = \frac{S}{C}$$

where S represents the concentration in the solid (mg/kg) and C in the liquid (mg/L) phase of an element and depends on the chemical and physical conditions of inorganic elements (for example, pH).

Organic compounds are adsorbed by organic particles contained inside or distributed on the surface of the solid medium and their behavior may be estimated by the parameter Kow (n-octanole-water partition coefficient) (Karickhoff, Brown, 1979):

$$K_d = K_{oc} f_{oc}$$

$$\log K_{oc} = -0.21 + \log K_{ow}$$

K_{oc} is the organic Carbon partition coefficient and f_{oc} the soil fraction of organic carbon.

The partial derivatives equation describing solute transport can be solved by analytical methods (for homogeneous and isotropic aquifer) or numerical methods (finite differences, finite elements, method of characteristic, random walk) by means of available software (free, commercial, or scientific).

Contaminant attenuation

Several fundamental parameters should be applied in the study of groundwater environmental fate in the landfill site.

- Solubility is one of the most important factors that determine the transport of pollutants in the subsurface. It represents the ability of a substance to be dissolved in water and is indicated by the ratio between the mass of solute and the volume of water.
- Fluid density, also defined in relation to water, is an important parameter as it determines the behavior of fluids in the soil with respect to water. Poorly soluble organic compounds (NAPL) may be lighter (LNAPL) or heavier (DNAPL) than water and, if a free phase is present, may float or sink in the aquifer.
- The volatility of a compound may be considered in relation to vapor pressure and Henry's constant; in the scientific literature, both values are acknowledged, and the compounds are divided into volatile organic compounds and semivolatile organic compounds.
- Dilution by water is a process that decreases the concentrations of a substance, and attenuation of the conservative pollutants (chlorides, nitrates, etc.) is only due to this process.
- Adsorption, according to the linear, Langmuir, or Freundlich isotherms, expresses the process in which the solute adheres to the solid surface of subsoil; this process affects the retardation factor previously shown.

- Ion exchange determines, particularly in mineral clays, the exchange of one ion with another ion. Cation exchange is measured and defined as cationic exchange capacity (CEC).
- Precipitation involves a change of phase of a dissolved chemical substance, which is deposited in the solution in relation to the achievement of the solubility limit. The main parameters that affect precipitation are pH and Eh, and a widely used phase diagram may be derived from PHREEQC software (Parkhurts and Appelo, 2013).
- Hydrolysis is a reaction that involves the presence of organic compounds and water or water ions. Several compounds are resistant to hydrolysis (benzene, phenols, PAH-Polycyclic Aromatic Hydrocarbons, pesticides, etc.), whereas others are susceptible to hydrolysis (amides, carbamates, carboxylic acid esters, etc.).
- Oxidation—reduction (ox—redox) of an atom involving loss or acquisition of electrons and the subsequent reactions are mediated by microorganisms. These reactions are of importance for Fe and Mn, as an example, in which the reduced forms (Fe^{2+} and Mn^{2+}) are mobile in groundwater compared with the oxidized form (Fe^{3+}) and Mn^{4+}) conversely, the oxidized form of Cr is mobile in groundwater (Cr^{6+}) while the reduced form is only slightly soluble (Cr^{3+})
- Complexation implies the formation of an ion product through the reaction of cation (central atom) or anion (ligand) with some molecules. The central atom is represented by a large number of atomic metals and the ligand by a series of species dissolved in groundwater such as halogens, sulphates, carbonates, etc.
- Biodegradation is a process in which microorganisms make use of several substances and convert them into other simpler substances in an aerobic or anaerobic environment, as outlined below.

Organic compounds

Although high concentrations of contaminants may be observed in concomitance of the source, these concentrations are not able to evolve as an equilibrium condition is established between the masses of transported contaminants and microorganism biodegradation in the soil.

This acknowledged condition, manifested by the stationary nature of the plume, is further emphasized by an additional series of events, including progressive depletion in dissolved oxygen and reduction of the amounts of nitrates, iron, and sulphates detected by the monitoring system.

In the initial phase, aerobic biodegradation takes place: this is the most significant event as biodegradation velocity is high, and high energy levels are produced by reaction.

Subsequently, once oxygen becomes depleted, anaerobic biodegradation may occur, thanks to other electron acceptors (in the order: nitrates, iron, sulphates, carbon dioxide).

The monitoring of electron acceptors in various physical parameters and in final products is crucial in demonstrating the onset of intrinsic biodegradation.

The "utilization factor" is defined to assess "biodegradation capacity" on the basis of stoichiometric equations of associated reactions; it is given by the mass of the electron acceptor required to degrade a given mass of dissolved hydrocarbon (Wiedemeier et al., 1994). By dividing the background groundwater concentration of an electron acceptor by the utilization factor (Table 18.2.6), an evaluation of the aquifer NA ability may be obtained.

Table 18.2.6 Characteristics of the natural biodegradation reaction of organic compounds (Wiedemeier et al., 1994)

Electron acceptor	Type of reaction	Metabolic by-product	Eh (mV) (pH = 7)	Utilization factor	Reaction preference
Oxygen	Aerobic	CO_2	+820	3.14	Most
Nitrate	Anaerobic	N_2 and CO_2	+740	4.9	↓
Iron (3+) (solid)	Anaerobic	Iron (+2) (dissolved)	−50	21.8	↓
Sulfate	Anaerobic	H_2S	−220	4.6	↓
Carbon dioxide	Anaerobic	CH_4	−240	2.17	Least

The organic compounds undergo degradation in an aerobic and anaerobic environment, and this transformation is mediated either by microorganisms (biotic) or by chemical reactions (abiotic).

Table 18.2.7 summarizes a series of degradation reactions for some organics compounds widely diffused in groundwater, while Tables 18.2.8 and 18.2.9 define hydrochemical parameters applied to define aerobic or anaerobic conditions.

As an example, microorganisms implicated in the anaerobic degradation of chlorinated solvents are *Dehalococcoides ethenogenes strain 195* (PCE to ethylene) and *Dehalobacter* sp. (PCE to cis-DCE) (U.S. Environmental Protection Agency, 2006).

A further process underlying the onset of biodegradation is cometabolism.

The microorganisms that metabolize an organic substance diluted in groundwater, such as a carbon source, produce nontoxic terminal products (CO, CO_2, H_2O) and enzymes that react with other organic substances.

It has been demonstrated, for example, that the degradation of toluene, propane, and butane (in an oxidized environment) also supports the cometabolism of chlorinated solvents such as trichloroethene.

Nitrogen organic compounds in groundwater are routinely present in an MSW landfill and may also be detected in industrial waste landfills.

The mineralization of organic nitrogen that originates from proteins, nucleic acids, and urea takes place in two phases: ammonification and, if conditions allow, nitrification.

Ammonification leads to the formation of NH_3 by a scarcely specific process involving a series of microorganisms, including *Bacillus*, *Clostridium*, *Flavobacter*, *Pseudomonas*, etc..; as a general rule, this stage is almost always achieved.

Table 18.2.7 Biologic and abiotic degradation mechanisms for various anthropogenic organic compounds

Compound	Degradation mechanism
PCE	Reductive dechlorination
TCE	Reductive dechlorination, cometabolism
DCE	Reductive dechlorination, direct biological oxidation
VC	Reductive dechlorination, direct biological oxidation
TCA	Reductive dechlorination, hydrolysis, dehydrohalogenation
1,2-DCA	Reductive dechlorination, direct biological oxidation
Chloroethane	Hydrolysis
Carbon tetrachloride	Reductive dechlorination, cometabolism, abiotic
Chloroform	Reductive dechlorination, cometabolism
Methylene chloride	Direct biological oxidation
Chlorobenzenes	Direct biological oxidation, reductive dechlorination, cometabolism
Benzene	Direct biological oxidation
Toluene	Direct biological oxidation
Ethylbenzene	Direct biological oxidation
Xylenes	Direct biological oxidation
1,2-Dibromoethane	Reductive dehalogenation, hydrolysis, direct biological oxidation

DCA, dichloroethane; *DCE*, dichloroethene; *PCE*, tetrachloroethene; *TCA*, trichloroethane; *TCE*, trichloroethene; *VC*, chloroethene (vinyl chloride) (U.S. Environmental Protection Agency, 1998).

The landfill groundwater monitoring system should first assess the area surrounding the facility and subsequently locate the presence of NH_4^+ dissolved in groundwater.

Nitrification occurs under highly specific conditions that result in the formation of nitrite and nitrate.

Nitrosation occurs, thanks to *Nitrosomonas*, a microorganism activated under strictly aerobic conditions and featuring scarcely specific needs; optimum pH is between 6.8 and 9, although higher values

Table 18.2.8 Analytical parameters and weighting for use in preliminary screening for anaerobic biodegradation processes

Analysis	Concentration in Most Contaminated Zone	Interpretation	Value
Oxygen[a]	<0.5 mg/L	Tolerated, suppresses the reductive pathway at higher concentrations	3
Oxygen[a]	>5 mg/L	Not tolerated; however, VC may be oxidized aerobically	-3
Nitrate[a]	<1 mg/L	At higher concentrations may compete with reductive pathway	2
Iron II[a]	>1 mg/L	Reductive pathway possible; VC may be oxidized under Fe(III)-reducing conditions	3
Sulfate[a]	<20 mg/L	At higher concentrations may compete with reductive pathway	2
Sulfide[a]	>1 mg/L	Reductive pathway possible	3
Methane[a]	<0.5 mg/L >0.5 mg/L	VC oxidizes Ultimate reductive daughter product, VC Accumulates	0 3
Oxidation Reduction Potential[a] (ORP) against Ag/AgCl electrode	<50 mV <-100mV	Reductive pathway possible Reductive pathway likely	1 2
pH[a]	5 < pH < 9 5 > pH >9	Optimal range for reductive pathway Outside optimal range for reductive pathway	0 -2
TOC	> 20 mg/L	Carbon and energy source; drives dechlorination; can be natural or anthropogenic	2
Temperature[a]	> 20 °C	At T >20 °C biochemical process is accelerated	1
Carbon Dioxide	>2x background	Ultimate oxidative daughter product	1
Alkalinity	>2x background	Results from interaction between CO_2 and aquifer minerals	1
Chloride[a]	>2x background	Daughter product of organic chlorine	2
Hydrogen	>1 nM	Reductive pathway possible, VC may accumulate	3

(*Continued*)

TABLE 18.2.8 Analytical parameters and weighting for use in preliminary screening for anaerobic biodegradation processes—cont'd

Analysis	Concentration in Most Contaminated Zone	Interpretation	Value
Hydrogen	<1 nM	VC oxidized	0
Volatile Fatty Acids	>0.1 mg/L	Intermediates resulting from biodegradation of more complex compounds; carbon and energy source	
BTEX[a]	>0.1 mg/L	Carbon and energy source; drives dechlorination	2
Tetrachloroethene		Material released	0
Trichloroethene[a]		Material released Daughter product of PCE	0 2
DCE[a]		Material released Daughter product of TCE If cis is $>80\%$ of total DCE it is likely a daughter product 1,1-DCE can be chemical reaction product of TCA	0 2
VC[a]		Material released Daughter product of DCE	0 2
1,1,1-Trichloroethane[a]		Material released	0
DCA		Daughter product of TCA under reducing conditions	2
Carbon Tetrachloride		Material released Daughter product of DCE	0 2
Chloroethane[a]		Daughter product of DCA or VC under reducing conditions	2
Ethene/Ethane	>0.01mg/L >0.1 mg/L	Daughter product of VC/ethene	2 3
Chloroform		Material released Daughter product of Carbon Tetrachloride	0 2
Dichloromethane		Material released Daughter product of Chloroform	0 2

[a]Required analysis. a/ points awarded only if it can be shown that the compound is a daughter product (i.e., not a constituent of the source NAPL) (U.S. Environmental Protection Agency, 1998).

DCA, dichloroethane; *DCE*, dichloroethene; *PCE*, tetrachloroethene; *TCA*, trichloroethane; *TCE*, trichloroethene; *VC*, chloroethene (vinyl chloride).

Table 18.2.9 Interpretation of points awarded during screening step of Table 18.2.6–18.2.7—biodegradation with reductive dechlorination (U.S. Environmental Protection Agency, 1998)

Score	Interpretation
0–5	Inadequate evidence for anaerobic biodegradation of chlorinated organics
6–14	Limited evidence for anaerobic biodegradation of chlorinated organics
15–20	Adequate evidence for anaerobic biodegradation of chlorinated organics
>20	Strong evidence for anaerobic biodegradation of chlorinated organics

are also tolerated; the microorganisms are active at low temperatures and are characterized by strong concentrations of NH_4^+.

Nitrification occurs, thanks to *Nitrobacter*, a strictly aerobic organism which carries out reactions under specific conditions: enzymatic activity is inhibited at pH lower than 5 and higher than 9, is inactive at temperatures below 5 °C and above 40 °C, and is inhibited by high concentrations of NH_4^+.

Monitoring wells located at a significant distance from the landfill are capable of recording the presence of nitrates in the facility; nitrites are metastable and have a very short life.

In general, natural conditions very rarely lead to the denitrification of groundwater, and, therefore, nitrates may be considered as "conservative" compounds.

Inorganic compounds

A survey of NA may involve inorganic compounds (metals and nonmetals), occurring in leachate and in groundwater, that may be subject to a series of reactions (precipitation, adsorption, etc.).

Issues related to radioactive decay should be discussed as a separate entity; however, for all other forms of attenuation, the type of reaction responsible for the attenuation should be ascertained as numerous reactions are stable (i.e., precipitation), while others are reversible (adsorption, organic compound partition, etc.), chemical, and physical conditions (as pH and Eh), which may affect the mobility of compounds should also be investigated.

As reported in Table 18.2.10, the main attenuation process related to a landfill barrier system is influenced by factors such as percolation velocity, degree of saturation in the unsaturated zone, organic matter, type of clay, soil texture, and stratigraphy.

Inorganic compounds are persistent in the soil as they do not degrade, are scarcely mobile, are not bioavailable, or are devoid of harmful forms.

Table 18.2.10 Attenuation mechanisms of landfill leachate constituents

Element or compound	Attenuation mechanism	Mobility in clayey environment
Al	P	Low
As	P, A	Moderate
B	A, P	High
Ba	A, S, P	Low
Be	P, S	Low
Ca	P, S	High
Cd	P, A	Moderate
Cl^-	D	High
CN^-	A	High
Cr	P, S, A	Low (Cr^{3+})–high (Cr^{6+})
Cu	A, S, P	Low
F^-	S	High
Fe	P, S, A	Moderate–high
Hg	A, P	High
K	A, S	Moderate
Mg	S, P	Moderate
Mn	P, S	High
Na	S	Low-High
NH_4^+	S, B	Moderate
Ni	A, P	Moderate
NO_3^-	B, D	High

TABLE 18.2.10 Attenuation mechanisms of landfill leachate constituents—cont'd

Element or compound	Attenuation mechanism	Mobility in clayey environment
Pb	A, S, P	Low
Se	A, S	Moderate
Si	P	Moderate
SO_4^{2-}	S, D	High
Zn	S, A, P	Moderate

A, adsorption; *B*, biological uptake; *D*, diluition; *P*, precipitation; *S*, cation and anion exchange (Bagchi, 1994).

The oxidized form of Cr (Cr^{6+}) is more mobile in groundwater than the reduced form; the same is also partly true for nickel. Iron and manganese are only mobile in a redox environment.

GROUNDWATER QUALITY DATA MANAGEMENT

Data obtained via the monitoring system are used to check the status of groundwater quality in the proximity of the plant: a database should therefore be compiled and updated using appropriate software.

The goal of data management is to identify whether the landfill is impinging on groundwater quality; an upgradient well represents background conditions, and a downgradient well represents background conditions together with a potential impact on the facilities due to leachate release.

To achieve this, goal processes facilitating data representations in space and time and statistical evaluation should be implemented.

Statistical data processing

Preliminary data processing should be carried out to estimate basic statistical parameters (mean, median, standard deviation) and to verify data distribution (normal, Poisson, log-normal) with a view to the subsequent application of parametric or nonparametric tests.

Statistical analysis should be undertaken to cover a homogeneous sample of hydrochemical data.

In the presence of uncertainty as to the homogeneity of the sample (for example, water from a stratified aquifer), pretreatment may be carried out using various methods: groundwater classification with synthetic methods such as Schoeller or Piper diagrams, separation of subsamples, etc.

Other aspects of processing relating to the evaluation of outliers include nondetects (distribution censored), seasonality, and trends in measured values.

These problems can be solved with the help of statistical software and by means of specific methodologies described in detail in several texts (i.a.Gibbons, 1994; Gilbert, 1987; Naval Facilities Engineering Command, 2003; U.S. Environmental Protection Agency, 2009).

Data value representation

Hydrochemical maps

Hydrochemical data may be used to produce discrete or continuous maps.

Discrete mapping reports measured values of the piezometric levels or chemical parameters at the monitoring points.

If a sufficient number of data points are available, continuous mapping may be derived from the interpolation of data using deterministic methods or, preferably, with geostatistical methods to represent lines of equal concentration (contours) of parameters of interest; a significant reconstruction will likewise depend on the homogeneity of hydrochemical composition according to depth.

A series of monitoring points outside the landfill area should be established to map pollutant distribution in groundwater and to aid recognition of a potential plume.

Time series analysis

Representation of the chemical composition varies over time and facilitates evaluation of the evolution of groundwater conditions.

Four types of processing may be possible:

- changes over time of one or more parameters;
- changes over time of one or more parameters with precipitation;
- changes over time of one or more parameters with leachate head;
- changes over time of one or more parameters with piezometric levels.

Graphs may be further processed using regressions to highlight a trend over time (stationary, increase, or decrease in value).

Should a progressive decreasing in concentration occur, a constant of degradation may be obtained by plotting concentration (log scale) versus time (arithmetic scale); the slope of linear interpolation represents the first-order rate constant of degradation λ (day^{-1}).

A seasonal trend may be influenced by precipitation or groundwater levels.

If further analysis is needed, a statistical approach such as the Mann-Kendall test (Connor et al., 2012) may be applied.

Background values

An appropriate statistical analysis of background groundwater values should be identified based on the age of the landfill.

Indeed, numerous chemicals may already be present in the subsoil in relation to geochemical conditions (As, Fe, ammonia, etc.) or nonpoint sources (nitrates, pesticides, chlorinated solvents, etc.).

Data measured by the groundwater monitoring network should be subjected to pretreatment to remove outliers and highlight any potential trends. Dixon's method should be used to remove any outliers.

Table 18.2.11 Strategies of background values evaluation of groundwater quality

New landfill	Existing or old landfill
Upgradient and downgradient monitoring wells can be used to define background groundwater quality	Only monitoring wells unaffected by the landfill can be used to define the background groundwater quality
Background values evaluation are made via intrawell comparison	Background value evaluationS are made via interwell comparison.
Upper tolerance limit (UTL) or Shewhart-CUSUM control chart limits are used to set decision values of background	UPL is used to set decision values of background
Multiple downgradient wells are compared with background UTL or individual monitoring well is compared with its control chart limits	Multiple downgradient wells compared with upgradient UPL is necessary to verify exceedance with two sets of independent samples.

Modified from State of Idaho (2014).

To evaluate the trends of seasonality variations in concentration, the Kruskal-Wallis test should be applied, whereas to assess an increase or decrease over time of secular trend reporting data, the Mann-Kendall test or Sen's slope method should be applied.

Once the outliers and trend have been ascertained, two different strategies may be applied (Table 18.2.11).

The approach suggested above is derived from the State of Idaho (2014) and calculations can be performed using ProUCL software (US Environmental Protection Agency, 2013a,b).

New landfill

For a new landfill, the intrawell approach may be applied in wells in which groundwater quality has not yet been impacted by leachate and represents background values of the site.

Applicability of the interwell method is indicated for a previously defined conceptual model: hydrogeological conditions (upgradient and downgradient wells) and hydrogeochemical conditions (sampling producing statistically independent data).

A specific exceedance (coverage) may be tolerated to minimize false negatives.

The procedure consists in the calculation of upper tolerance limit (UTL) that defines a range into which a proportion of the population (coverage of 95%) will fall for a proportion (95%) of the time.

In a parametric approach using the Shapiro–Wilk test, the value is calculated by means of the following formula:

$$\mathrm{UTL} = \mathrm{m} + \mathrm{Ks}$$

Table 18.2.12 Value of parameter K (Gibbons, 1994)

Sample size	95% Coverage	Sample size	95% coverage
8	3.188	16	2.523
9	3.032	17	2.486
10	2.911	18	2.453
11	2.815	19	2.423
12	2.736	20	2.396
13	2.670	25	2.292
14	2.614	30	2.220
15	2.566	35	2.166

where m is the arithmetic mean, s the standard deviation, and K (Table 18.2.12) a constant depending on the proportions used (Gibbons, 1994).

A nonparametric approach for data corrected for seasonal effects may be applied to calculate sample sizes (N) required to achieve the desired coverage at varying confidence levels. UTL represents the highest value using N seasonally adjusted background sample values (Table 18.2.3).

According to Conover (1999), "q" is the proportion of the population covered by the tolerance interval. The value of sample size in Table 18.2.13 is required to ensure that the probability of at least q of all future samples is greater than or equal to 1α.

Use of the intrawell method may be preferable in situations in which high heterogeneous hydrogeological media and a small number of upgradient wells are present.

In the presence of insufficient data for statistical analysis, three simple methods may be applied in calculation of the Alternative Concentration Limit (ACL) (a temporary value) from available data:

- ACL1 = the largest of the 12 most recent data values collected
- ACL2 = m + 1.65s
- ACL3 = median + 1.65 IQR (in which IQR = the interquartile range).

Existing and old landfill

The interwell approach may also be applied to an existing or old landfill (State of Idaho, 2014). In a parametric approach, future measurements in downgradient wells are compared with an upper

Table 18.2.13 Sample Size for the Upper Tolerance Limit (Conover, 1999)

1α	q = 0.500	0.700	0.750	0.800	0.850	0.900	0.950	0.975	0.980	0.990
0.700	2	4	5	6	8	12	24	48	60	120
0.750	2	4	5	7	9	14	28	55	69	138
0.800	3	5	6	8	10	16	32	64	80	161
0.850	3	6	7	9	12	19	37	75	94	189
0.900	4	7	9	11	15	22	45	91	144	230
0.950	5	9	11	14	19	29	59	119	149	299

prediction limit (UPL) based on background groundwater quality data from upgradient wells (possibly pooled), according to the following formula:

$$\mathrm{UPL} = \mathrm{m} + \mathrm{Cs}$$

where the parameter C (Table 18.2.14) depends on the number of measures per well per year, in relation to years, wells, and inverse number of contaminants of concern (COCs):

d = number of measures/well/year × years x number of wells × number of COCs.

According to the nonparametric approach, the UPL value is set equal to the maximum value from N independent background samples required to achieve a specified confidence level for a specified number of future comparisons.

Data provided in Table 18.2.15 summarize confidence levels for various background sample sizes and future comparisons and the same verification.

Threshold value assessment

The proposal of thresholds (trigger levels) of attention and possible intervention is a complex matter that is highly dependent on local conditions such as hydrogeological structure, quality and use of water, previous conditions of groundwater, type of substance examined, presence of limit values in groundwater quality, data availability, requests from Public Authorities, etc.

A general indication is provided hereafter that should be compared and adapted with the local situations.

In the ordinary and extraordinary management of groundwater, threshold values of interventions should be established on the basis of three criteria: (A) statistical assessment of measured values, (B) maximum limit values allowed in groundwater, and (C) risk analysis.

If a limit value for poor groundwater quality has been exceeded, criterion (A) alone will apply; conversely, if the limit value has not been exceeded, criteria (A), (B), and (C) may all be applied.

Table 18.2.14 Value of parameter C at $\alpha = 0.05$ for a verification protocol where one or both verification resamples must confirm initial exceedance (State of Idaho, 2014)

	d = Number of future comparisons				
N	10	20	30	40	50
4	2.02	2.42	2.65	2.82	2.94
8	1.37	1.61	1.75	1.84	1.92
12	1.21	1.42	1.54	1.62	1.68
16	1.14	1.33	1.44	1.52	1.58
20	1.10	1.28	1.39	1.46	1.51
24	1.08	1.25	1.35	1.42	1.47
36	1.03	1.20	1.29	1.36	1.41
48	1.01	1.17	1.27	1.33	1.38

Table 18.2.15 Confidence levels where exceedance is verified when one or both verification resamples also exceed the limit

	d = Number of future comparisons							
N	10	20	30	40	50	60	80	100
4	0.5585	0.4393	0.3759	0.3347	0.3050	0.2822	0.2491	0.2257
8	0.7616	0.6522	0.5836	0.5348	0.4976	0.4678	0.4225	0.3890
12	0.8538	0.7676	0.7072	0.6613	0.6246	0.5942	0.5463	0.5095
16	0.9023	0.8356	0.7852	0.7449	0.7115	0.6831	0.6368	0.6001
20	0.9305	0.8785	0.8369	0.8024	0.7729	0.7473	0.7044	0.6695
25	0.9516	0.9126	0.8798	0.8516	0.8268	0.8047	0.7668	0.7350
35	0.9729	0.9492	0.9279	0.9087	0.8912	0.8751	0.8463	0.8211
50	0.9858	0.9727	0.9604	0.9488	0.9379	0.9275	0.9083	0.8908

In Gibbons (1994) (State of Idhao, 2014).

For the B criterion, during landfill design and construction hydrochemical parameters were measured, and the mean (m) and the standard deviation values (s) for the initial statistical sample may therefore be defined.

If the upgradient monitoring wells do not record significant changes in concentration, attention should be focused on the downgradient monitoring wells.

- *Threshold 1: (m+2s)*—This situation requires no further action as deemed to fit with initial measured values. Under ordinary conditions and in the presence of a meaningful number of samples, the normal distribution of values will contain 95% of all individual measurements.
- *Threshold 2: (m+3s)*—This situation may be a signal of change in natural concentrations but may also indicate a worsening of water quality due to leachate release. Sampling and water analysis should be repeated and the Plan of possible interventions updated. A check of the reasons underlying the possible release, including accidental spill, high leachate head, and/or barrier failure, should be established.
- *Threshold 3 (overcoming a limit value):* This situation may signal a worsening of water quality due to the release of leachate and subsequent noncompliance of detected concentrations of compounds with legal limits. Sampling and water analysis should be immediately repeated (to avoid false positives) and, should the concentrations be confirmed, the first interventions provided for by the Plan should be implemented. Monitoring should be intensified and the effects of interventions verified.

With regard to the (C) criterion, the risks that a landfill may present to groundwater and human health will determine the level of controls required, which may be developed in line with the three levels of risk of growing complexity.

Trigger levels (or threshold levels) may be established to minimize or prevent the impact on receptors of episodes of groundwater pollution due to release of leachate (use of [1] simple initial screening, [2] analytical methods, and [3] deterministic or probabilistic numerical methods).

USE OF NATURAL TRACERS IN GROUNDWATER MONITORING—INVESTIGATION AROUND LANDFILL SITES

An increasingly complicated and controversial issue is represented by the plan to be adopted to investigate contamination in areas surrounding the landfill site. Inappropriate landfilling may indeed constitute a significant source of contaminants for groundwater. However, several other sources of contaminants may be present in the same areas, including agriculture, farming, industries, services, old dumps, and urban settlements. In some cases, the groundwater "symptoms" and "anamnesis" are clear and "diagnosis" is univocal.

However, frequently source attribution is unclear or controversial and "diagnostic" errors can easily be made (see Box 18.2.1).

In these cases, the use of natural tracers already present in the leachates or originating from degradation phenomena of the same (microbial flora, nitrates, etc.) may provide a significant contribution.

A tracer by definition is any substance or tool unequivocally linked to the contaminants transported by water and capable of providing significant information on the potential source of contamination.

Box 18.2.1 Groundwater contamination from landfill leachate: when appearances are deceiving!

Contamination phenomena for muncipal solid waste landfills are almost exclusively associated with the presence of ammonia, chlorides, chemical oxygen demand (COD), sulphates, iron, and some heavy metals. When the above substances are detected in groundwater downstream of a landfill by environmental protection authorities, no further analyses are carried out frequently , and the landfill is automatically the cause of contamination. Even if this is not the case! Below are a few (young and old) examples of landfills, located in Northern Italy, that have found themselves at the heart of an unjustified technical and public frenzy.

Case A: In a small stream running along an MSW landfill, periodic fish mortalities were registered downstream (Fig. 1), resulting in protests by the population, anxiety for the town mayor and subsequent closure of the landfill. Early one morning, a maintenance worker randomly bumped into a farmer discharging blood collected at a nearby slaughterhouse into the stream near the landfill. The farmer thought he was doing nothing wrong: "We even feed animals with the blood!". It's too bad that blood has a biochemical oxygen demand (BOD) of 200,000 mg/L and entails such a high consumption of oxygen that fish in the creek were completely deprived of it!

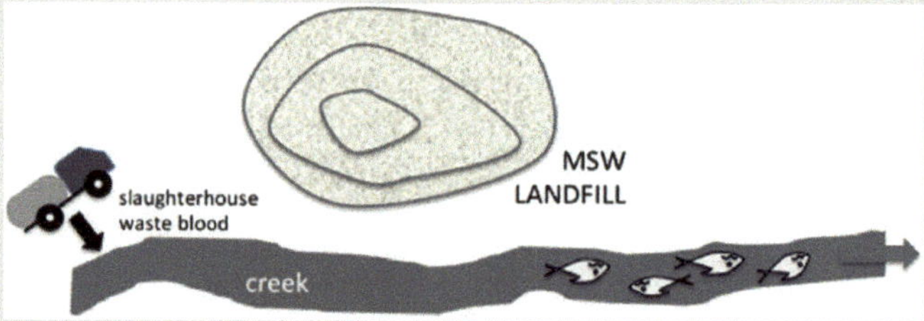

Figure 1 Fish kill downstream of a landfill (1980, S. Liberale, Marcon, Venice).

Case B: In a groundwater monitoring well at the foot of a municipal solid waste landfill, located along a river entering the sea approximately 1 km from the landfill, trivalent chromium concentrations were found to exceed legal limits (Fig. 2). No traces of chromium were, however, detected in the well upstream the landfill. Natural conclusion is that the landfill was to be blamed for the contamination. Widespread panic, anxiety for the town mayor, and subsequent closure of the landfill. The true source has been found to be the river discharging into the sea chromium-rich wastewaters from an area of tannery industries. The chromium was then transported upstream as results of the saltwater inclusion. The guilty party responsible for the contamination was the sea! Or rather the tannery industries located 80 km from the landfill site!

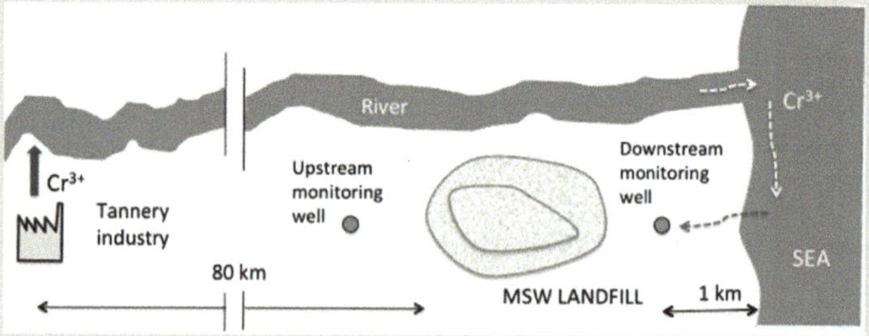

Figure 2 Chromium contamination of groundwater downstream a landfill, as a consequence of salt intrusion.

Case C: Ammonia concentrations (approx.130 mg/L) were detected in a monitoring well located 10 m. close to the border of two million cubic meters landfill site (Fig. 3). Widespread panic, alerts by local authorities, judicial intervention, closure, and repossession of the landfill. A court-appointed expert declared a leak in the landfill lining system to be the cause of the contamination. An old small open dump, which had been inactive for 30 years, was located 50 m from the landfill. A number of technical experts, in an attempt to defend the landfill management authority, blamed this smaller landfill. More careful analyses revealed that the leachate from neither the large nor the small dumpsite was the main responsible for the contamination. Indeed whereas the ammonia and chloride ratio in leachate was 1:1, the same ratio in the groundwater was 10:1. Chlorides would have been emitted together with ammonia if landfills were to be the main pollution source. Indeed, the groundwater upstream the landfill proved to be contaminated by 50 mg/L of nitrates.

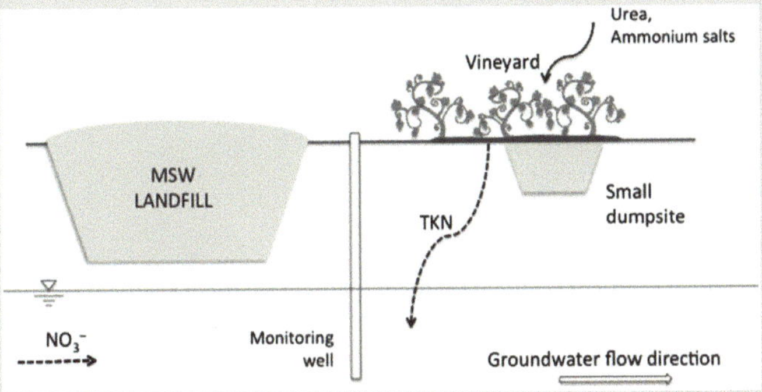

Figure 3 Groundwater contamination by ammoniacal nitrogen in the surroundings of a landfill, as a consequence of the presence of vineyards.

Nobody had paid attention to this detail or to the fact that a vineyard had been planted on the old open dump and that the landfill was located in one of the most important wine areas in Italy. Vineyards are fertilized with urea, ammonium salts....but not chlorides! The upstream presence of nitrates was just the consequence of the oxidation in the groundwater of ammoniacal nitrogen emitted from the vineyards located several kilometers from the landfill site.

Case D: In a monitoring well located 2.5 km upstream of an MSW landfill, the local environmental protection agency detected high concentrations of trichloromethane (Fig. 4). With no further analyses the officer on duty ruled: "The landfill is to be blamed!". Notice to the town mayor, intervention order to proceed immediately with remediation. A more careful analysis revealed that trichloromethane was also found in the immediate surroundings of the landfill, albeit in much lower concentrations, and that said substance was never detected in the landfill leachate. Chloroform is an industrial product widely used for several purposes (production of refrigerants and plastics, solvent, dry cleaning agent, fumigant, component in fire extinguishers, etc.). The area in which the landfill is located is an industrial area where several factories use the substance. Nobody had even thought of them!

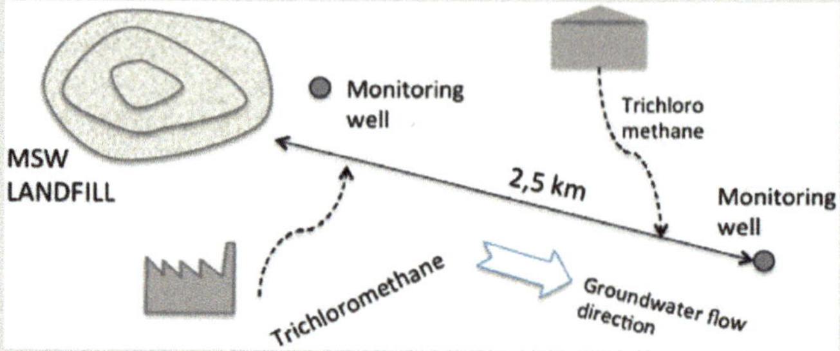

Figure 4 Contamination of a groundwater downstream a landfill due primarily to emissions from industrial sources.

In conclusion, traditional landfilling, particularly when mismanaged, may represent the cause of significant groundwater contamination. However, the murderer is not necessarily always the butler!*Raffaello Cossu, Editorial - adapted from Waste Management (2013), 33, 1793-94.*

The characteristics of a good tracer are

- possible interactions with environmental matrices and behavior in groundwater should be well known and understood;
- reasonably conservative;
- methodologies for detection (sampling, analysis) should be simple, reliable, available, and affordable.

Potential natural tracers suitable for use in monitoring groundwater quality around landfill sites may be grouped into three main categories:

- Chemical
- Isotopic
- Biological

Chemical tracers

The traditional chemical parameters characterizing leachate quality (COD, BOD, ammonia, nitrates, heavy metals, etc.) may also be present in several other kinds of wastewaters. However, ratios among these parameters may be representative and source specific.

Additionally, a huge variety of xenobiotic substances (see Chapter 18.2) may indeed be present in products employed on a daily basis (plastics, pharmaceuticals, packed food, paper, etc.). Some of these substances do not undergo degradation or any other attenuation processes, and consequently their persistence could be used to trace a potential landfill-derived groundwater contamination.

Nitrogen–chloride ratio

Nitrogen in leachate is characterized by the TKN parameter (Total Kjeldhal Nitrogen, which includes ammonia and organic nitrogen). TKN and chloride are invariably present in MSW leachate in concentrations typically ranging between 500 and 3000 mg/L. Concentration values depend on a series of factors, including quality of the waste, age of the landfill, and liquid/solid ratio. The ratio of these compounds, however, is quite stable and is comprised in the range of 1–1.5 (see Chapter 10.2). Waste pretreatment and in situ treatment may of course alter this ratio.

Emission of these substances from landfills in the long term will follow a similar trend, maintaining a more or less constant ratio (Chapter 10.2).

In groundwater, despite the different processes involving ammonia conversion and attenuation (nitrification–denitrification, volatilization), the N/Cl ratio will tend to remain significant and representative of leachate presence.

Total N/Cl^- ratio has been demonstrated to be a valid tool in studying plumes and detecting leachate contaminations in groundwater (see Box 18.2.1), also in view of its lack of association with individual nitrogen forms.

The parameter cannot be used when there is seawater intrusion into freshwater aquifers.

Specific xenobiotics

Amongst the xenobiotics investigated as being of potential use in tracing leachate contamination in groundwater, the following may represent promising markers:

- *Propyphenazone:* An active component of several antiinflammatory, analgesic, and antipyretic drugs. It is a persistent compound and relevant concentrations have been detected in groundwater contaminated by MSW leachate. Propyphenazone has been indicated as an effective marker compound (Heim, 2004).
- *NBBSA (N-butylbenzenesulphonamide):* Used commercially as a plasticizer and commonly found in liquid form in landfill leachates at μg/l range (Eggen et al., 2010). NBBSA is persistent in groundwater environment, being resistant to biodegradation and NA (Eggen et al., 2010; Filho et al., 2003).
- *DEET (N–N-diethyltoluamide)*: This substance is widely used as a pesticide in insect repellents. It is significantly present in leachate (Masoner et al., 2014), does not present composition alteration in leakage water, and is found in contaminated groundwater (Heim et al., 2004).
- *Chlorofluorocarbons:* Despite an international ban, CFCs are still present in domestic and commercial waste. They have being indicated as useful tracers by Foley et al. (2012).

Isotope tracers

Isotope investigation methods are based on the study of variations of specific element isotopes. The analysis is an address to identify the sources of contamination, the fate of contaminants, and in support of the definition of background values. Landfill leachates present significantly different isotopic composition compared with that of meteoric waters and background values.

Isotopes of hydrogen and oxygen

The natural isotopes of Hydrogen and Oxygen are employed increasingly to evaluate the impact of landfill on groundwater and NA of pollutants.

Hydrogen stable isotopes are ^{1}H and ^{2}H with relative abundance of 99.985% and 0.015%, respectively.

Oxygen stable isotopes are ^{16}O, ^{17}O, and ^{18}O with relative abundance of 99.76%, 0.04%, and 0.20%, respectively.

Isotope analysis of hydrogen and oxygen involves measurement of the relative abundance, with ratios being expressed according the SMOW (Standard Mean Ocean Water) international standard.

The stable hydrogen isotopes undergo significant fractionation due to several landfill processes and enrichment of Deuterium (^{2}H), and ^{18}O allows these isotopes to be used as tracers of potential pollution; the initial methanogenesis phase is characterized by a leachate enrichment of ^{2}H (Hacley et al., 1996; Rank et al., 1996).

Tritium (^{3}H) is the radioactive isotope of hydrogen, featuring a decay time of 12.3 years. It is the only radioactive isotope displaying a behavior identical to that of water. Thus, retardation, sorption, and chemical processes exert virtually no impact on Tritium.

The presence of tritium is expressed as Tritium Units (TU), where one TU represents a $^{3}H/^{1}H$ ratio of 1018. One TU is equivalent to water activity in Tritium of 0.118 Bq/l or 3.221 pCi/L and matches one Tritium atom per 1018 hydrogen atoms; therefore 1 Bq/l = 8.475 TU and 1 pCi/L = 0.311 TU.

Environmental tritium occurs by means of precipitation through the interaction of cosmic radiation with atmospheric gases and from anthropogenic nuclear testing and reactors. Generally, values obtained for ^{3}H in precipitation are in the range of 4–14 TU, with similar values of tritium activities measured in groundwater subjected to rapid renewal. On the other hand, extremely high values of ^{3}H, almost 100-fold the ^{3}H values obtained from current precipitations, have been measured in MSW landfill leachates (Nigro et al., 2017).

The isotope is often used in biomedical research to trace materials, with other uses including self-luminous devices, including exit signs, wristwatches, military-style compasses, shooting ranges, key chains, etc.; the material can be disposed of in MSW landfills, with the major source appearing to be gaseous tritium lighting devices.

Tritium concentrations detected in leachate from several US landfills corresponded to 33,800 pCi/L, with peak values of 304,000 pCi/L (Mutch and Mahony, 2008; Association of State and Territorial Solid Waste Management Officials, 2009); the finding of elevated values in groundwater have long been acknowledged (Egboka et al., 1983).

In Italy, tritium concentrations in leachate of 400–1700 TU have been detected in some landfills and 16–260 TU in downgradient groundwater (Tazioli, 2011).

Isotopes of carbon

Stable isotopes of carbon are ^{12}C and ^{13}C and have a diffusion of 98.98% and 1.11%, respectively.

The contents of isotope $\delta^{13}C$ ‰ vary depending of the origin: 0/2 carbonates and marine shells, −4/−12 carbonates and continental water, −7 atmospheric CO_2, −20/−30 organic matter and soil CO_2.

In 2008, the US Environmental Protection Agency suggested a procedure according to which an initial isotopic signature represents the measure of the source (leachate for landfill in this case), prior to eventual transformations and the evaluation of reactions in groundwater.

These measures are essential when leachate is represented by a mixture of various organic and inorganic substances deriving from the chemical–physical and biological reactions of waste.

In the traditional monitoring approach, the plume is expected for the reduction of contaminant concentration in downstream groundwater caused by dilution and dispersion. If the reduction is also caused by biodegradation or abiotic transformation, the study of stable isotopes is necessary. The study by means Compound-Specific Isotope Analysis identifies if it is possible to apply the monitoring natural attenuation (MNA) as remediation measure.

Microbial process, responsible of decreasing contaminant concentration downgradient of a landfill, uses the light isotope more easily, causing an enrichment in the heavy one with respect to the starting point.

The molecules of organic compounds made up of the lighter ^{12}C isotope tends to be processed more rapidly than molecules of the heavy ^{13}C isotope; the shift displayed during a reaction will be of $^{13}C-^{12}C$.

If $f = C_t/C_0$, in which C is the concentration at time t and at initial time 0,

$$f = e^{\left(\delta^{13}\text{Cgroundwater}-\delta^{13}\text{Csorce}\right)/\varepsilon}$$

where the parameter ε is the stable isotope enrichment factor:

$$\varepsilon = (\alpha - 1)1000$$

where α is given by

$$\alpha = R_a - R_b$$

R_a and R_b are the stable isotope ratio ($^{13}C/^{12}C$) of the compound at time zero (t0) and at a later point (t).

Prior to and following the reaction, the isotopic ratio may shift from negative to positive values.

Not only an enrichment of ^{13}C manifested due to degradation of organic carbon but aslo an enrichment of leachate in ^{2}H compared to local precipitation is also recorded.

Carbon isotopes may also be used in source characterization or differentiation, qualitative proof of biodegradation or abiotic transformation, and quantification of biodegradation or abiotic transformation processes (U.S. Environmental Protection Agency, 2008).

Applications have also been extended to chlorinated solvents, aromatic petroleum hydrocarbons, and fuel oxygenates.

The radiogenic isotope of carbon is ^{14}C, having a decay time of 5730 years (β particles) and 1.2×10^{-10} abundance in nature.

With regard to leachate, an enrichment of ^{14}C with values up to 150 pMC in CH_4 and DIC (dissolved inorganic carbon) is observed compared with values obtained in groundwater surrounding the landfill; these values of ^{14}C are explained by past nuclear explosions in the atmosphere (Hackley et al., 1996).

Chloride isotopes

Isotopes of chloride may also be used in specific cases to differentiate leachate release from other sources (Coleman and Joseph, 2001).

This technique of investigation is applied as in the case of carbon isotopes (^{12}C and ^{13}C). The chlorine isotopes (^{35}Cl and ^{37}Cl) identify progressive biotic or abiotic dechlorination of halogenated compounds.

Isotopes of nitrogen

Nitrates are commonly present in groundwater mainly having as a potential source of agricultural activities or landfilling. Nitrate and other nitrogen compounds (urea, ammonium) are important and widely used components of synthetic fertilizers. The reduced form of nitrogen, both in fertilizers and leachate, is converted to nitrates under aerobic conditions. Stable isotopes (nitrogen with oxygen) may be convenient to trace the origin of nitrates in ecosystems (Clark and Fritz, 1997; Kendall et al., 2007; Michalsky et al., 2015). The $\delta^{15}N$ values for commercial nitrate-based fertilizers ranged around 2‰, similar to nitrogen in the air (Aravena et al., 1993; Michalsky et al., 2015). Values obtained for organic nitrogen in soil are higher (3‰–8‰) and much higher (10‰–20‰) for human and animal waste nitrate (Freyer and Aly, 1974; Kreitler, 1975; Gormly and Spalding, 1979; Heaton, 1986).

Additional information is provided by the oxygen isotope $\delta^{18}O$, which tracks the oxygen in nitrate molecules: atmospheric oxygen nitrates from fertilizers and water oxygen in nitrates from nitrification processes.

Combined use of $\delta^{15}N$ values and $\delta^{18}O$ values may allow precision tracing of the origin of nitrates in the concomitant presence of agricultural activities and landfilling in the same areas. This has been clearly observed in a study carried out in the period 2014–15 in Apulia Region, South of Italy.

In the study area, there was concern for the presence of nitrates, exceeding the limit values for drinking water. In the area are present both intense agriculture activities and a landfill, organized in sectors of different ages (Cossu et al., 2017). The results (Fig. 18.2.4) showed that nitrate content of all groundwater samples was very well correlated to fertilizer effects. Nitrates originating from ammoniacal nitrogen in leachate would have behaved similarly to the nitrates originated from manure or septic system effluents.

Biological Tracers

The biological conversion of degradable organics in groundwater implicates microbial communities that vary according to the prevailing geochemical conditions produced by different contaminants sources.

The microbiological method assesses and quantifies microbial activity for identification the contaminant biodegradation.

The BACTRAP is an in situ microcosm system in which a tube filled with activated carbon (^{13}C), and microcosm is exposed to groundwater. The microcosm is extracted and analyzed after exposure for 6 weeks, and the biodegradation is evaluated.

The fluorescence in situ hybridization technique (FISH and CARD/FISH) is a molecular method for microscopic identification, visualization, and quantification of communities of interest.

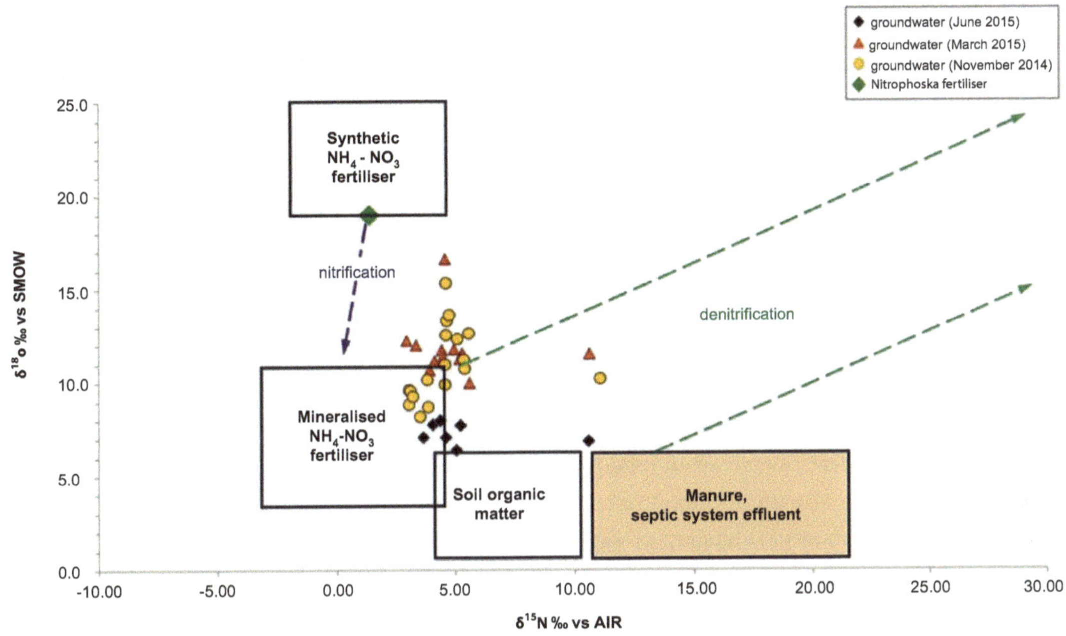

Figure 18.2.4 Isotope investigation carried out in Apulia Region, Italy, in an area where either agriculture activities or municipal solid waste landfill are present. The stable isotope composition of dissolved nitrates in groundwater and most popular fertilizer in the area (Nitrophoska) is represented. The rectangles highlight the main types of nitrate sources (Cossu et al., 2017).

The quantitative real-time polymerase chain reaction (qPCR) assesses the microbial dechlorination activity by nucleic acid analysis as it determines gene and transcript number in groundwater.

Research in this field is on a continual increase and the results obtained to date are promising. Incorporating biological data into the groundwater monitoring network appears to represent an instrumental trend capable of modifying the traditional standard for groundwater investigation (i.a. Lu et al., 2012; Mauser et al., 2010).

FINAL REMARKS

Groundwater quality assessment activities have to be planned at the time of landfill design and later implemented during the different phases of the landfill life span (construction, operation, aftercare).

A monitoring network for groundwater quality analysis should be organized at the landfill site on the base of conceptual model.

From a hydrogeological point of view, a conceptual model is a simplified representation of the physical hydrogeological system in relation to landfill site, as a potential source of contaminants.

The increasing huge variety of contaminants, including CECs, should be considered for updating the traditional data set of leachate parameters.

The results of monitoring surveys will be expressed either as a spatial distribution of concentrations or as time-related evolution of the reference parameters.

On the base of the results compliance with operational and environmental sustainability, targets should be assessed.

In case of failures of the landfill barrier system, an uncontrolled emission of contaminants may occur. Leachate contaminants in the groundwater environment undergo a series of attenuation processes which need to be carefully considered, studied, and implemented in a risk analysis scheme. MNA should be a strategic asset in considering the long-term sustainability of landfills (Christensen et al., 1993).

Many of the most frequent substances regularly detected in the leachate are widely produced by several other activities such as agriculture, livestock farming, industrial production, and housing. In some areas where these activities are present, these activities may serve to mask the impacts solely originated by landfills. For these reasons monitoring activities should rely on a robust investigation program and reliable data interpretation to avoid mistakes in overlooking other potential contamination sources. In these cases, the use of natural tracers (chemical parameters specific ratio, isotopes of different elements, microbial communities) already present in the leachates or originating from degradation phenomena of the same may provide a significant contribution.

References

Aravena, R., Evans, M.L., Cherry, J.A., 1993. Stable isotopes of oxygen and nitrogen in source identification of nitrate from septic systems. Ground Water 31, 180–186 [cit. in Cossu et al., 2017]. https://doi.org/10.1111/j.1745-6584.1993.tb01809.x.

Association of State and Territorial Solid Waste Management Officials, 2009. Lack of Tritium Exit Signs Control and Contamination of Landfill Leachate. Federal Facilities Research Center's Radiation Focus Group, Washington, D.C.

ASTM-American Society for Testing Material, 1990. Standard Practice for Design and Installation of Ground Water Monitoring Wells in Aquifers. ASTM D 5092-90, Philadelphia, PA.

ASTM-American Society for Testing Material, 1992. Guide for Pore-liquid Sampling from the Vadose Zone. ASTM D-4696, Philadelphia, Pennsylvania.

Bagchi, A., 1994. Design, Construction and Monitoring of Landfills. John Wiley & Sons, Inc, New York.

Baun, A., Ledin, A., Reitzel, L.A., Bjerg, P.L., Christensen, T.H., 2004. Xenobiotic organic compounds in leachates from ten Danish MSW landfills—chemical analysis and toxicity tests. Water Research 38 (18), 3845–3858.

Christensen, T.H., Bierg, P.L., Rügge, K., Albrechtsten, H.J., Heron, G., Pedersen, J.K., Foverskov, A., Skov, B., Würst, B., Refstrup, M., 1993. Attenuation of organic leachate pollutants in groundwater. In: Proceedings Sardinia 93, Fourth International Landfill Symposium, 11–15 October, S.Margherita di Pula, Cagliari.

Clark, I., Fritz, P., 1997. Environmental Isotopes in Hydrogeology. Lewis Publishers, Boca Raton. FL. ISBN 1566702496.

Coleman, M.L., Joseph, J.B., 2001. Stable isotope differentiation of chloride in leachates and groundwaters. In: Proceedings Sardinia 2001, 1–5 October, vol. III, pp. 665–673 (S.Margherita di Pula, Cagliari, Italy).

Connor, J.A., Farhat, S.K., Vanderford, M., 2012. GSI Mann-Kendall Toolkit for Constituent Trend Analysis. GSI Environmental Inc.

Conover, W.L., 1999. Practical Nonparametric Statistics, third ed. John Wiley & Sons, New York, NY.

Cossu, R., Zuffianò, L.E., Limoni, P.P., De Giorgio, G., Pizzardini, P., Miano, T., Mondelli, D., Garavaglia, R., Carella, C., Polemio, M., 2017. A multi-methodological approach to clarify the landfill role in the case of nitrate groundwater quality degradation (Southern Italy). In: Proceedings Sardinia 2017, Sixtheen Waste Management and Landfilling Symposium. CISA Publisher.

De Cleen, M.P.T.M., Schurink, E., Driessen, J.H.A., 1995. Systematic design of a monitoring network. In: Proceedings Sardinia 95, Fifth International Landfill Symposium, 2–6 October, S.Margherita di Pula, Cagliari.

Egboka, B.C.E., Cherry, J.A., Farvalden, R.N., Frind, E.O., 1983. Migration of contaminants in groundwater at a landfill: a case study. Journal of Hydrology 63, 51–80.

Eggen, T., Moeder, M., Arukwe, A., 2010. Municipal landfill leachates: a significant source for new and emerging pollutants. The Science of the Total Environment 408 (21), 5147–5157.
Filho, I., von Mühlen, C., Schlosser, P., Caramaõ, E.B., 2003. Identification of some plasticizers compounds in landfill leachate. Chemosphere 50, 657–663.
Foley, A.E., Atkinson, T.C., Zhao, Y., 2012. Chlorofluorocarbons as tracers of landfill leachate in surface and groundwater. Quarterly Journal of Engineering Geology and Hydrogeology 45, 61–70.
Freyer, H.D., Aly, A.I.M., 1974. Nitrogen 15 variations in fertilizer nitrogen. Journal of Environmental Quality 3, 405–406 [cit. in Cossu et al., 2017].
Gelhar, L.W., Axness, C.L., 1981. Stocastic Analysis of Macro-dispersion in Three-dimensionally Heterogeneous Aquifer. Geophysical Research Center, Hydrology Research Program, Rep. No. 118. New Mexico Inst. of Mining and Technology, Socorro, New Mexico.
Gibbons, R.D., 1994. Statistical Methods for Groundwater Monitoring. John Wiley & Sons, New York, NY.
Gilbert, R.O., 1987. Statistical Methods for Environmental Pollution Monitoring. Van Nostrand Reinhold, New York, NY.
Gormly, J.R., Spalding, R.F., 1979. Sources and concentrations of nitrate-nitrogen in ground water of the central Platte region, Nebraskaa. Ground Water 17, 291–301 [cit. in Cossu et al., 2017]. https://doi.org/10.1111/j.1745-6584.1979.tb03323.x.
Government of Ontario, 2012. Landfill standards: a guideline on the regulatory and approval requirements for new or expanding landfilling sites PIBS 7792e. Queen's Printer for Ontario.
Hackley, K.C., Liu, C.L., Coleman, D.D., 1996. Environmental isotope characteristics of landfill leachates and gases. Ground Water 34 (5), 827–836.
Heaton, T.H., 1986. Isotopic studies of nitrogen pollution in the hydrosphere and atmosphere: a review. Chemical Geology: Isotope Geoscience Section 59, 87–102 [cit. in Cossu et al., 2017]. https://doi.org/10.1016/0168-9622(86)90059-X.
Heim, S., Schwarzbauer, J., Littke, R., 2004. Monitoring of waste deposit derived groundwater contamination with organic tracers. Environmental Chemistry Letters 2, 21–25.
Johnson Division, 1966. Ground Water and Wells: A Reference Book for the Water-well Industry. Johnson Division, UOP Incorporated.
Karickhoff, S.W., Brown, D.S., 1979. Determination of Octanol/Water Distribution Coefficients, Water Solubilities,and S Diment/water Partition Coefficients for Hydrophobic Organic Pollutants. EPA-600/4-79-032, Athens.
Kendall, C., Elliott, E.M., Wankel, S.D., 2007. Tracing anthropogenic inputs of nitrogen to ecosystems. In: Michener, R., Lajtha, K. (Eds.), Stable Isotope in Ecology and Environmental Science, second ed. Blackwell Publishing Inc., Oxford UK, pp. 375–449.
Kjeldsen, P., Barlaz, M.A., Rooker, A.P., Baun, A., Ledin, A., Christensen, T.H., 2002. Present and long-term composition of MSW landfill leachate: a review. Critical Reviews in Environmental Science and Technology 32 (4), 297–336.
Kreitler, C.W., 1975. Determining the Source of Nitrate in Groundwater by Nitrogen Isotope Studies. (Report of Investigations No. 83). Bureau of Economic Geology, University of Texas at Austin, USA [cit. in Cossu et al., 2017].
Lu, Z., He, Z., Parisi, V.A., Kang, S., Deng, Y., Van Nostrand, J.D., Masoner, J.R., Cozzarelli, I.M., Suflita, J.M., Zhou, J., June 5, 2012. GeoChip-based analysis of microbial functional gene diversity in a landfill leachate-contaminated aquifer. Environmental Science & Technology 46 (11), 5824–5833. https://doi.org/10.1021/es300478j [Epub 2012 May 23].
Masoner, J.R., Kolpin, D.W., Furlong, E.T., Cozzarelli, I.M., Gray, J.L., Schwab, E.A., 2014. Contaminants of emerging concern in fresh leachate from landfills in the conterminous United States. Environmental Science: Processes & Impacts, Royal Society of Chemistry 10, 2–48. https://doi.org/10.1039/C4EM00124A.
Mauser, P.J., Rizzo, D.M., Druschel, G.K., Morales, S.E., Hayden, N., O'Grady, P., Stevens, L., 2010. Enhanced detection of groundwater contamination from a leaking waste disposal site by microbial community profiles. Water Research 46. https://doi.org/10.1029/2010WR009459. W12506.
Michalsky, G., Kolanowski, M., Rihaa, K.M., 2015. Oxygen and nitrogen isotopic composition of nitrate in commercial fertilizers, nitric acid, and reagent salts. Isotopes in Environmental and Health Studies. https://doi.org/10.1080/10256016.2015.1054821.
Mutch, R.D., Mahony, J.D., 2008. A study of tritium in municipal solid waste leachate and gas. Fusion Science and Technology 54 (1), 305–310.
National Water Well Association, 1986. RCRA Ground Water Monitoring Technical Enforcement Guidance Document (TEGD. NWWA, September, Dublin, Ohio.
Naval Facilities Engineering Command, 2003. Guidance for Environmental Background Analysis – Volume III Groundwater. Battelle Memorial Institute, Earth Tech, Inc., NewFields, Inc., Washington D.C.
NGWA-National Groundwater Association, 1986. Guidelines on Monitoring Wells.
Nielsen, D.M., 1991. Practical Handbook of Ground-water Monitoring. Lewis Publishers, Chelsea, Michigan. ISBN 978–0873711241.

Nigro, A., Sappa, G., Barbieri, M., 2017. Application of boron and tritium isotopes for tracing landfill contamination in groundwater. Journal of Geochemical Exploration 172, 101–108.

NRC, 2005. Contaminants in the Subsurface: Source Zone Assessment and Remediation. National Academies Press, Washington, DC, USA.

Parkhurts, D.L., Appelo, C.A.J., 2013. PHREQC version 3 – a computer program speciation, batch reaction, one-dimensional transport, and inverse geochemical calculations. U.S.Geological Survey Techniques and Methods. Book 6, Chap. A13, 497 pp.

Pivnenko, K., Eriksen, M.K., Martín-Fernández, J.A., Eriksson, E., Astrup, T.F., 2016a. Recycling of plastic waste: presence of phthalates in plastics from households and industry. Waste Management 54, 44–52.

Pivnenko, K., Olsson, M.E., Götze, R., Eriksson, E., Astrup, T.F., 2016b. Quantification of chemical contaminants in the paper and board fractions of municipal
solid waste. Waste Management 51, 43–54.

Plumb, 1987. Quality Assurance.

Rank, D., Papesch, W., Rajner, V., 1996. Environmental isotopes study at a research landfill (Breitenau, Lower Austria). In: Proceedings of a symposium on the Isotopes in water resources management (I.A.E.A, 1995), pp. 379–381.

Rollin, A.L., Mlynarek, J., Lafleur, J., Zanescu, A., 1994. Performance changes in aged in-situ HDPE geomembrane. In: Christensen, T.H., Cossu, R., Stegmann, R. (Eds.), Landfilling of Waste. Barriers, E & FN Spon, London, pp. 431–443. ISBN 0-419-15990-8/.

Sale, T., Newell, C., 2011. A Guide for Selecting Remedies for Subsurface Releases of Chlorinated Solvents ESTCP Project ER-200530. Department of Defense Environmental Security Technology Certification Program (ESTCP), Colorado State University GSI Environmental, Inc.

State of Idaho, 2014. Statistical Guidance for Determining Background Ground Water Quality and Degradation. Department of Environmental Quality, 1410 N. Hilton Boise, ID 83706, March.

Tazioli, A., 2011. Landfill investigation using tritium and isotopes as pollution tracers. AQUA Mundi 1, 83–91 n.1.

U.K. Environment Agency, 2008. Guidance on Monitoring of Landfill Leachate, Groundwater and Surface Water. LFTGN02.

U.S. Environmental Protection Agency, 1996a. Low-flow (Minimal Drawdown) Ground-water Sampling Procedures. By Robert W. Puls and Michael J. Barcelona. Office of Solid Waste and Emergency Response, Office of Research and Development. EPA/540/S-95/504.

U.S. Environmental Protection Agency, 1996b. Soil Screening Guidance: Technical Background Document. User's Guide. Office of Solid Waste and Emergency Response, Washington DC.

U.S. Environmental Protection Agency, 1998. Technical Protocol for Evaluating Natural Attenuation of Chlorinated Solvents in Ground Water. Office of Research and Development. EPA/600/R-98/128.

U.S. Environmental Protection Agency, 2006. Evaluation of the Role of Dehalococcoides Organisms in the Natural Attenuation of Chlorinated Ethylenes in Ground Water. National Risk Management Research Laboratory Office of Research and Development, Cincinnati, Ohio.

U.S. Environmental Protection Agency, 2008. A Guide for Assessing Biodegradation and Source Identification of Organic Ground Water Contaminants Using Compound Specific Isotope Analysis (CSIA) EPA 600/R-08/148. Office of Research and Development, National Risk Management Research Laboratory, Ada, Oklahoma.

U.S.Environmental Protection Agency, 2009. Statistical Analysis of Ground-water Monitoring Data at RCRA Facilities. Unified Guidance, Washington, DC. EPA. EPA 530/R-09–007.

U.S. Environmental Protection Agency, 2013a. ProUCL 5.0 Technical Guide. http://Www.Epa.Gov/Osp/Hstl/Tsc/Software.Htm.

U.S. Environmental Protection Agency, 2013b. ProUCL 5.0 Software and User Guide. http://www.epa.gov/osp/hstl/tsc/software.htm.

Waste Management of North America, 1991. Site Assessment Manual.

Wiedemeier, T.H., Wilson, J.T., Miller, R.N., Kampbell, D.H., 1994. United States guidelines for successfully supporting intrinsic remediation with an example form Hill Air Force Base. In: Proceedings of The National Ground Water Association Petroleum Hydrocarbon Conference, November 1994.

Wilson, C.R., Eimberger, C.M., Jackson, R.L., Mercer, R.B., 1992. Design of ground-water monitoring network using the Monitoring Efficiency Model (MEMO). Ground Water 30 (6).

Xu, M., Eckstein, Y., 1995. Use of weighted least-squares method in evaluation on the relationship between dispersivity and field scale. Ground Water 33 (6).

Xu, M., Eckstein, Y., 1997. Statistical analysis of the relation between dispersivity and other physical properties of porous media. Hydrogeology Journal 5 (4) (Verlag Heinz Heise, Hannover).

CHAPTER 19

REMEDIATION

19.1

INVESTIGATIONS ON CLOSED LANDFILLS

Raffaello Cossu and Roberto Raga

INTRODUCTION

Investigation on closed landfills may be carried out for different reasons and with different objectives, the most frequent being

- evaluation of current and potential landfill emissions, preliminary to the design of remediation works;
- evaluation of the potential cause of aquifer contamination or biogas migration in the landfill surroundings;
- assessment and enhancement of the efficiency of biogas and leachate collection networks;
- estimation of the duration of aftercare phase and definition of measures to accelerate landfill completion;
- collection of information for designing technical interventions (vertical expansion, rehabilitation measures, site afteruse);
- monitoring the results of on-going in situ remediation projects;
- evaluating the feasibility of landfill mining.

The characterization of closed landfills involves both landfill body and deposited waste and may provide the following:

- definition of landfill boundaries;
- assessment of leachate and biogas emissions;
- assessment of landfill mechanical stability;
- evaluation of hydraulic conditions (leachate levels, waste hydraulic properties, etc.);
- assessment of waste fractional composition for a possible recovery of resources;
- physical and biochemical characterization of deposited waste, evaluation of the emission potential;
- assessment of groundwater quality.

When remediation actions are needed, these could relate to either old landfills or modern landfills, built in accordance with current regulations. The reasons for remediation may include problems arising from inappropriate design, poor construction, poor operational management, counter-productive permitting prescriptions, inappropriate technical requirements set by landfill regulations, aging of materials, clogging of drainage systems, settling, etc (Cossu et al., 2011).

A general scheme for an investigation plan of closed landfills is described graphically in Fig. 19.1.1.

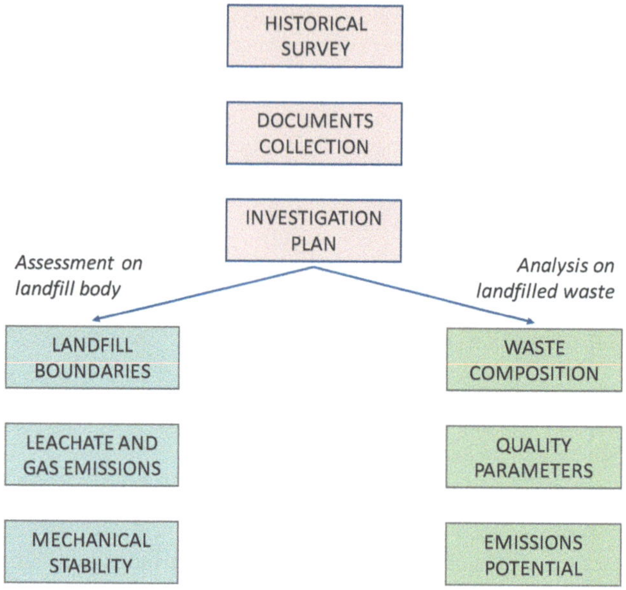

Figure 19.1.1 Scheme of a general procedure for investigating closed landfills. Investigations on the surrounding environment (groundwater quality, gas migration) are not included. See Chapters 9.5 and 18.2.

The first step of any investigation program should be addressed to the collection of historical information on the site and technical documentation (permissions, projects, etc.) whenever available. In the case of old non-engineered landfills, this may prove problematic as even basic information (areal extension, volume and depth of the waste deposit) may be missing.

Documents to be gathered and analyzed may comprise old maps and photos, reports of geological surveys, possible former quarrying activities, territorial studies, information on waste delivery (even newspaper articles if official records are not available) and any other study regarding the landfill area.

Subsequent steps in the investigation program may relate to different kinds of methodologies and techniques (geophysical surveys, geotechnical and hydraulic tests, sampling and analyses of waste and emissions, etc.) to be discussed in detail in the following sections.

GEOPHYSICAL SURVEYS

Geophysical surveys are convenient methodologies for use in the preliminary investigation of closed landfills as they do not involve penetration (at times problematic in engineered landfills), provide global information, are quick, and provide indications on where to conduct further expensive and intrusive field investigations.

Geophysical methods may be applied to achieve the following

- definition of location, geometry, and lateral extent of old dumps;
- determination of geology and pattern of the bedrock;

- detection of the presence of significant buried objects;
- mapping of the lateral and vertical distribution of relevant parameters characterizing waste deposit;
- monitoring leachate and biogas pockets, migration pathways in the subsurface, and pollution plumes in groundwater.

A short introduction of the geophysical methods most commonly applied in landfill investigations is given and some case studies briefly described.

Electrical resistivity tomography (ERT) is carried out by injecting electrical current into the subsoil by means of a set of electrodes and subsequently measuring electrical resistivity, which depends on porosity, moisture content, leachate quality, and temperature (Imhoff et al., 2007).

ERT has been used in the framework of landfill characterization projects and has provided evidence of anomalies of resistivity values beneath the base of the landfill, probably associated to leachate migration into the aquifer. Further confirmation of the development of a leachate plume in the aquifer below the landfill was provided by hydrochemical data (Bichet et al., 2016; Casado et al., 2015).

Electromagnetic (EM) methods are based on the application of an alternating primary magnetic field, which gives rise to a secondary magnetic field, directly proportional to the electrical conductivity of a material. EM survey methods are widely used in near-surface geology surveys, mapping variations in the electrical conductivity caused by changes in soil structure, nature, and physical characteristics (i.e., conductivity, water content) (McNeill, 1990). The contribution for landfill investigation consists in the fine resolution of zones of high conductivity within the ground, which may be related to the increase of solutes in groundwater, possibly due to leachate intrusion (Di Maio et al., 2018). Highly conductive zones in the landfill body are normally associated to leachate accumulation in waste layers (Cossu et al., 2009).

Induced polarization (IP) provides information related to the presence and relative abundance of highly conductive particles in the landfill body by means of the evolution of chargeability values. A signal is generated by means of different electric current pulses and is detected either during or after current circulation, to provide the well-known parameter chargeability, associated with the presence of a conductive medium. The research of buried metal objects underground and of highly conductive (leachate saturated) zones appears to be a natural field of application of the IP method, especially if integrated with other electric and/or EM methods (Cossu et al., 2001).

Seismic surveys are carried out by monitoring acoustic waves propagation underground. Sources of acoustic energy include a sledge hammer impacting a metallic plate, weight drops, and a number of different types of explosive sources. Wave velocity depends on the density and elastic properties of the material through which they travel. When the waves reach an interface where layer velocities change significantly, a portion of the energy is reflected back to the surface, and the remainder is transmitted into the lower layer. Receivers (geophones) record the incoming refracted and reflected waves. Primary application of the seismic methods is the determination of the subsurface stratigraphy and of anomalous structural and physical conditions. Therefore, the contribution to the study of old landfills consists essentially in the definition of landfill boundaries (Cossu et al., 2001).

A multimethodological geophysical approach may provide more reliable results; geophysical surveys might serve as a preliminary assessment, based on which the sampling and specific characterization might be carried out (Cossu et al., 2001; Di Maio et al., 2018).

An example of multimethodological geophysical approach and integration of geophysical survey with other test methods is reported, where EM, IP, and Seismic methods were used for the preliminary characterization of an old nonengineered landfill in the framework of a landfill mining project in Northern Italy. Following to the outcomes of the geophysical survey, waste and leachate sampling and analysis were carried out (Cossu et al., 2001). In particular, probably for the first time and following proper evaluation, explosives were used on a landfill site for seismic surveys, despite the presence of methane in the subsoil. In fact, due to strong absorption of acoustic energy in the waste body, a very powerful source of acoustic energy was needed. Fig. 19.1.2 provides an overview of the results. At a glance, the first three pictures exhibit very similar geometrical features. High conductivity value distribution (100–130 mmhos/m) characterizes the upper layers of the landfill (Fig. 19.1.2A), locally disturbed by small–medium, even more conductive (>130 mmhos/m) areas. These values are typical of material partly saturated with a conductive solute; the higher values being probably associated to perched leachate zones. The deeper part of the pseudo cross section appears much less conductive (30–70 mmhos/m) than the upper part and may be associated to the natural deposits of clay, silts, and sandy silts below the landfill body. The chargeability section derived from IP data (Fig. 19.1.2B) is in good agreement with EM results; the areas characterized by high conductivity values correspond to zones of high chargeability. Fig. 19.1.2C reports five seismic velocity intervals, the lower ones (0–350 m/s) associable to the waste mass. The transition from low to high velocity values delineates very well the change from waste mass to natural soil. The high seismic velocity value characterizing natural soil beneath the waste is compatible with the low conductivity and low chargeability areas shown in Fig. 19.1.2A and 2B, respectively. In Fig. 19.1.2D, measured leachate tables and natural clay depth at the bottom of the landfill are shown. The measurements were performed by means of purpose-built drillings for waste and leachate sampling and piezometer installation, in 3A, 3C, 3E, and 2A (Fig. 19.1.2D), located along the alignments for the geophysical tests. The results of the intrusive tests proved that previous seismic survey successfully identified landfill boundaries. Measurements in piezometers showed the presence of a leachate table 3–4 m from landfill bottom. However, highly conductive and chargeable zones were identified at lower depths in the landfill also, probably due to the presence of perched leachate. This was confirmed by the characterization of waste samples, which often looked saturated with leachate even when extracted from shallow parts of the landfill above leachate table.

GEOTECHNICAL AND HYDRAULIC TESTS

Information on waste geotechnical and hydraulic properties such as emplacement density, mechanical stability, hydraulic conductivity, and expected settlements are necessary for proper landfill design and in

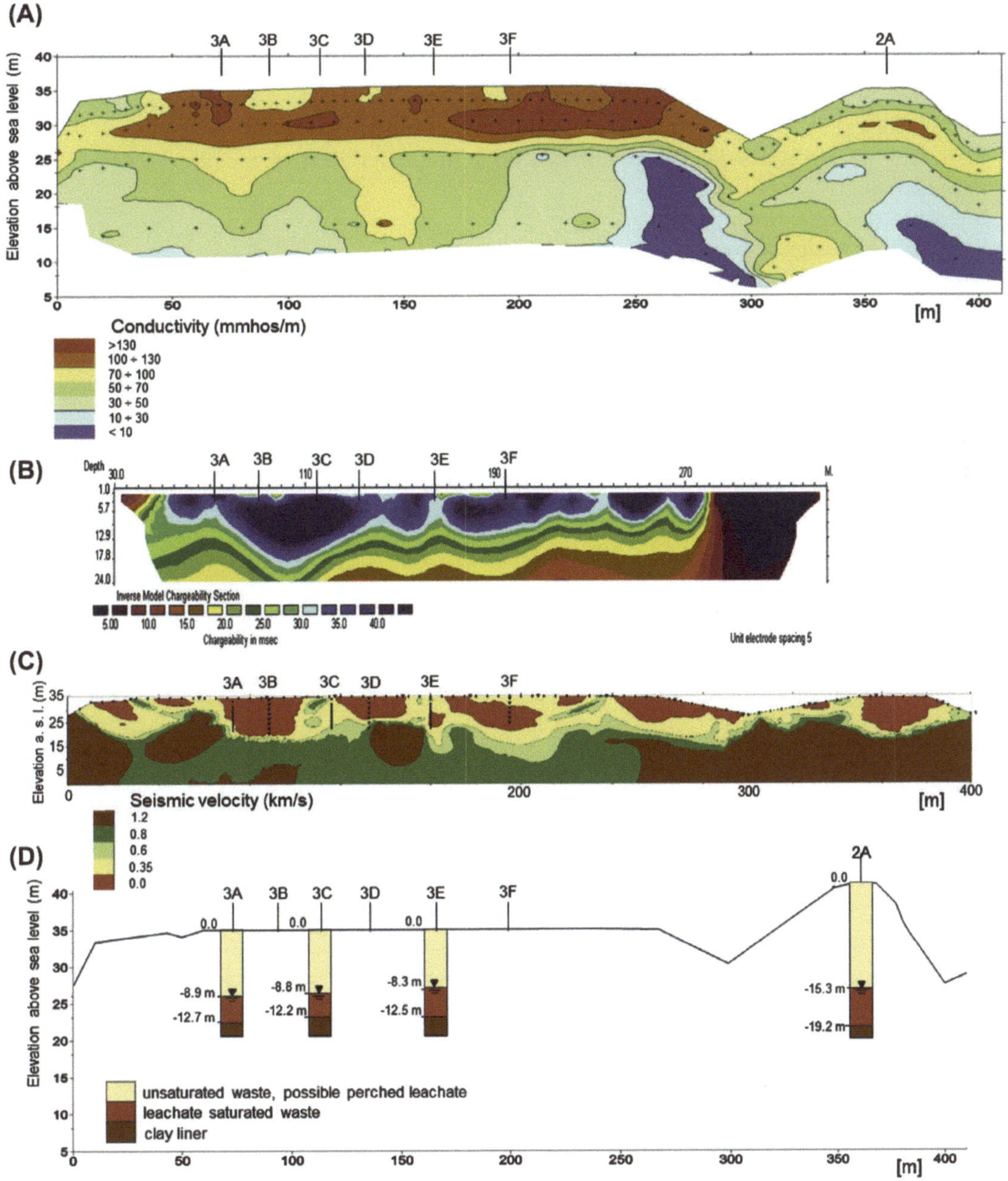

Figure 19.1.2 Comparison of results from three different geophysical methods: (A) Electromagnetic mapping; (B) Induced polarization; (C) Seismic refraction; (D) intrusive survey. (Cossu et al., 2001).

all other phases of landfill life, including landfill remediation, to properly address specific situations such as the risk of landslides and the undesired accumulation of leachate in the landfill body.

Mechanical stability concerns may arise in the case of

- landfill slopes;
- high pore water pressure due to uncontrolled water infiltration or inappropriate leachate recirculation;
- interstitial gas pressure in waste;
- inappropriate sludge disposal;
- high leachate level due to reduced efficiency of leachate extraction systems;
- landfill vertical extension;
- excavation activities during landfill mining.

High compaction and consequent low values of hydraulic conductivity may cause difficulties in leachate extraction and render leachate recirculation unfeasible. Difficulties in leachate extraction and the build up of significant leachate tables not only affect landfill mechanical stability but also result in an increased risk of uncontrolled leachate emissions. Another consequence of high leachate levels is the limited efficiency of biogas extraction systems, causing increased pressure and consequent biogas uncontrolled emissions. This is an important issue also when in situ aeration is practiced. For these reasons, reliable data on leachate hydraulic conductivity are necessary to design measures to prevent or to remediate such situations.

Tests used for geotechnical and hydraulic waste characterization are based on the principles of conventional soil mechanics. However, due to the wide variability of waste composition, models available in soil mechanics are not always applicable to waste (Powrie et al., 1999). As an example, soil particles are assumed to be incompressible in soil mechanics theories; however, results of test carried out in a large scale compression cell showed significant changes in waste particles density as the vertical stress increases (Beaven and Powrie, 1995).

Compaction, compressibility, hydraulic conductivity, pore water and gas pressure, and shear strength are typically considered for waste mechanical characterization. Compaction is determined by means of standard Proctor test where leachate can be used instead of water; the evolution of waste dry density at increasing values of moisture content is determined. Oedometers are used for the determination of primary and secondary compression, the latter mainly due to waste biodegradation. Triaxial tests are used for the evaluation of shear strength (Powrie et al., 1999; Reddy et al., 2009).

Standard lab-scale equipment used for soil geotechnical characterization may be too small for heterogeneous material such as Municipal Solid Waste (MSW); for this reason, tailor-made solutions should be studied, and specific equipment is designed and used to obtain reliable results (Stoltz et al., 2010). In the framework of old landfill characterization, a shear box might be purpose designed to take into account specific waste characteristics (Cossu et al., 2009). Inside a tailor-made direct shear box (Fig. 19.1.3), specimens were located side to side, then directly charged by means of steel slabs to provide vertical stress. The shear action was obtained by means of a hydraulic jack; the lateral force and the displacement were automatically recorded.

Figure 19.1.3 Set up of test with direct loading (left) and displacement of the upper half shear box at the end of the test (right) (Cossu et al., 2009).

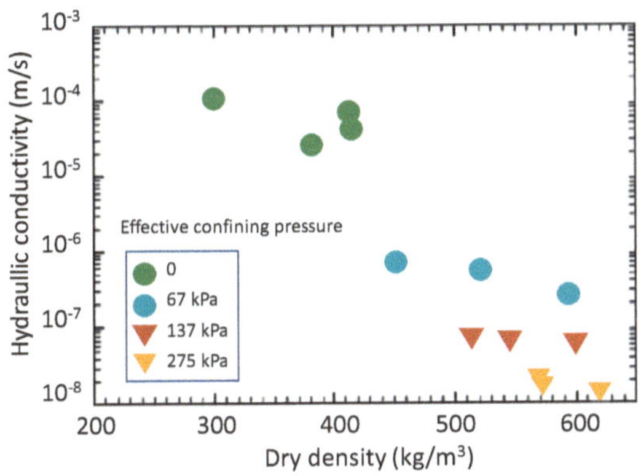

Figure 19.1.4 Relationship between hydraulic conductivity and dry density of fresh shredded MSW under different confining pressures (Reddy et al., 2009).

Hydraulic conductivity may be assessed at lab scale in permeameters or by means of triaxial tests; in Fig. 19.1.4, an example of relationship between hydraulic conductivity and dry density, obtained by means of triaxial equipment under different confining pressures for shredded MSW, is reported (Reddy et al., 2009). Experiments in large cells are used for specific investigations on waste mechanics. The results of tests for the assessment of the influence of biogas accumulation on waste hydraulic conductivity in a large scale (2 m diameter) compression cell (Fig. 19.1.5) at different pore water pressures are reported by Hudson et al. (2004) and Powrie et al. (2008). Biochemical, hydrological, and mechanical behavior of landfilled MSW was investigated at lab scale by means of large experiment experimental cells of approximately 180 m^3 (Liang-Tong Zhan et al., 2017). The results showed that the reduction

Figure 19.1.5 Large compression cell used by University of Southampton for mechanical tests on MSW samples from an old landfill (*Courtesy Richard Beaven*)

of waste hydraulic conductivity is caused not only by increased density but also by biogas pockets present below the leachate table.

However, in the case of landfilled waste, better insight into the hydraulic characteristics of landfill may be obtained by means of in situ tests; constant rate pumping tests and step-drawdown tests are commonly used (Cossu et al., 1997).

In general, due to the heterogeneity of landfilled waste, field measurements are considered more reliable for the assessment of most landfill parameters (Matasovic et al., 2011). In the framework of an investigation project on an MSW landfill, comprehensive in situ and laboratory tests were carried out for the evaluation of the influence of MSW shear strength on a waste-sliding movement in the landfill. The in situ shear tests were carried out by means of a tailor-made apparatus, consisting of two steel rings (internal diameter 800 mm and height 220 mm) pushed into undisturbed waste. Vertical stress was applied; consequently, shear stress was applied through a hydraulic jack, and related horizontal load and displacement were recorded (Mazzuccato et al., 1999).

SOLID WASTE CHARACTERIZATION

Tests and parameters to be analyzed for the characterization of solid waste are chosen based on the purpose of the investigations and on the landfill type. In the case of characterization prior to landfill mining for instance, the assessment of waste fractional composition is the main issue (see Chapter 19.3) to enable the design of reuse options of the different fractions. However, for very old landfills, where no information about disposal activities are available, getting a reliable assessment of waste fractional composition might be challenging, despite the extraction and analyses of a significant number of samples.

Several strategies are available (EPA, 2002) to decide the number and location of samples. Sampling at different depths offers the possibility of collecting waste of different ages. However, irrespective of the sampling strategy chosen and the number of samples, it is not possible to prevent the risk of missing small deposits of special waste that might have been disposed of in specific situations and that might need tailor-made treatment before and/or after excavation.

Figs. 19.1.6 and 19.1.7 illustrate a series of different sampling alternatives. Core drilling (Fig. 19.1.6) is carried out by means of the same equipment used for soil sampling for traditional

Figure 19.1.6 Core drilling in a closed landfill (left) and sample collection (right).

Figure 19.1.7 Large diameter auger drilling (left) and the extracted waste sample (right).

geotechnical laboratory tests, the only difference being that, contrary to soil sampling, in the case of landfilled waste, the use of water to extract the sample should be avoided, as this would result in elution of elements and compounds from the sample, and therefore, the results of further characterization tests would lose relevance. For this reason, the use of pressurized air alone is recommended for sample extraction. The use of core drilling enables samples related to specific depths in the landfill to be extracted and the results of characterization to be correlated with the expected age of waste. However, the drawback is the limited size of the sample, in comparison to those obtained by means of large diameter drilling (Fig. 19.1.7). The latter is therefore recommended particularly when waste fractional composition and grain size distribution analyses are planned, for the assessment of possible treatment options and destinations of the excavated material in a landfill mining project.

Examples of results of fractional composition analyses from municipal solid waste landfill in view of possible landfill mining and resource recovery are reported in Chapter 19.3.

Assessment of the biological stability of deposited waste is the main task if the investigation is aimed at analyzing contamination processes potentially related to the landfill or at monitoring landfill aftercare and/or in situ remediation processes.

Common tests used for this purpose involve the measurement of oxygen consumption (respiration index) in respiration tests and biogas production in fermentation tests of waste samples, as well as the characterization of leaching tests eluates.

Respiration index is measured as oxygen consumption for biological degradation processes of waste samples in laboratory scale reactors. Several approaches are available, a common test is described in German regulation (Anonymous, 2001), for the determination of AT_4 measured as mg of O_2 per g of dry matter (mgO_2/gDM), indicating the oxygen cumulative consumption in 4 days in small reactors in the Sapromat apparatus, at a controlled temperature of 20 °C. Another possibility is the use of larger lab-scale reactors with continuous air flow through the waste (Andreottola et al., 2005), providing the so-called dynamic respiration index ($mgO_2/kgSV$ h).

The fermentation test, providing cumulative biogas production over a period of 21 days (as NL of biogas per kg of total solids, NL/kg DM) may be carried out on the basis of current German regulations using a procedure set up by the German Institute for Standardization (DIN, 1985). Fermentation tests are carried out at 35°C; air tight glass bottles are filled with 50–100 g of sample, with the addition of microbial inoculum and water. Results of the assessment of respiration index and of biogas production in fermentation tests from waste samples are very well correlated, showing that both parameters are good indicators of waste biological stability and of the related expected biogas emission potential (Cossu and Raga, 2008).

The evaluation of BOD_5 to COD ratio may be carried out on leaching test eluates at significantly lower costs than respiration tests and at shorter duration than fermentation tests. Moreover, contrary to respiration and biogas production tests, the information given by the BOD_5 to COD ratio is still meaningful in the presence of toxic or inhibiting substances (Cossu et al., 2012). The results obtained are well correlated with those of other waste stability tests.

Leaching tests are also used in the characterization of the potential of constituent release from a landfill site. Eluates are analyzed by inductively coupled plasma optical emission spectrometry for major, minor, and trace elements such as Al, As, B, Ba, Ca, Cd, Co, Cr, Cu, Fe, K, Li, Mg, Mn, Mo, Na, Ni, P, Pb, S, Sb, Se, Si, Sn, Sr, Ti, V, Zn; dissolved organic carbon and TOC (total organic carbon) can be detected by means of a TOC analyzer; Cl, F, ammonium, and sulfate are normally analyzed by ion chromatography (van der Sloot, 2017).

Sample preparation affects the results of waste characterization. Shredding of waste samples to <10 mm prior to measurement of the respiration index and of biogas production potential has been suggested by German regulations (Anonymous, 2001). Another common approach is to conduct characterization tests on the undersieve 20 mm (Mönkäre et al., 2016; Andreottola et al., 2005; Leikam and Stegmann, 1997). Redon et al. (2005) report the results of an investigation carried out on waste samples to evaluate the influence of sample preparation on respiration index and biogas production in fermentation tests. The sample was divided into two parts: first one was sieved, and the fraction <10 mm was used for characterization; the other part was shredded and then sieved to obtain a <10 mm fraction for characterization. Higher values of the respiration index of the shredded sample were measured. This was due not only to sample heterogeneity but also mainly to the possible increase of availability of organic matter to biological degradation caused by shredding. Similar results had previously been obtained in other tests (Binner et al., 1997). Both approaches yield reliable information, provided that a clear description of the procedure for sample preparation is available.

EMISSION ANALYSIS

The evaluation of residual biogas emission potential from old landfills has been carried out using gas production models (Scharff and Jacobs, 2006).

However, models require the input of waste composition data, which might be lacking in the case of old landfills. An experimental approach foresees waste sampling from different locations within the landfill and the measurement of biogas emission potential in bioreactors. Large samples are needed to limit the effects of waste heterogeneity on test results and the use of diggers is a practical solution. However, in the case of deep deposits, large diameter drilling is recommended (Fig. 19.1.7). Sample size in lab-scale reactor tests varies considerably, from approximately 20 kg up to 200 kg (Scheutz et al., 2011; Cossu et al., 2016). Reactors are sealed, flushed with nitrogen to remove oxygen, and then run at a controlled temperature to simulate anaerobic landfill conditions, and potential emissions with leachate are also assessed. Most tests are carried out with leachate extraction and recirculation (Stegmann and Spendlin, 1989; Francois et al., 2007), weekly leachate sampling and analyses (mainly for pH, alkalinity, conductivity, TOC, COD, BOD_5, $SO_4^=$, Cl^-, ammonia nitrogen, volatile acids, heavy metals), continuous biogas collection and analysis (mainly for CO_2, CH_4, N_2, O_2, H_2S).

Characterization of biogas in the landfill body may be carried out through the installation of gas probes with screens at different depths. These devices are also used for the evaluation of biogas

Figure 19.1.8 Measuring biogas emissions from landfill surface by means of a flux chamber and a mobile flame ionization detector analyzer. The possible presence of a hot spot of methane emissions through the soil is generally indicated by the absence of grass.

composition profiles at different depths of biocovers intended for methane oxidation and limit uncontrolled emissions into the atmosphere.

Uncontrolled biogas emission through the landfill surface may occur when the biogas extraction system is missing or damaged (especially in old landfills), or malfunctioning due to lack of maintenance or bad design. A typical example is the accumulation of condense water and consequent clogging of biogas extraction pipes, resulting in increased internal pressure and uncontrolled biogas release through landfill surface. These emissions can be monitored by means of flux chambers made from a top of a drum and pushed into the ground for a couple of centimeters; the perimeter is then sealed with bentonite (Fig. 19.1.8). Gas composition is ascertained by using a mobile flame ionization detector analyzer.

The top of the chamber is equipped with sampling ports for the monitoring of gas composition, pressure, and temperature. The calculated methane emission is routinely expressed as gCH_4/m^2 h. If the biogas extraction system is present on the landfill and well maintained, uncontrolled methane emissions through the surface may be limited by means of proper adjustment of flow rates from the single biogas extraction wells.

Leachate accumulation in the landfill body is generally caused by drainage clogging, gas entrapped in pores, low waste permeability (particularly in the deeper landfill layers), and ruptures in leachate extraction wells. Accumulation may cause difficulties in biogas circulation and extraction. Consequently, the leachate hydraulic level may increase, thus enhancing the risk for leachate leakage through the bottom barrier. Assessment of leachate stored in the landfill body involves assessment of the piezometric level under static landfill conditions. This is habitually performed by the insertion of electric probes into biogas extraction wells in the landfill. However, biogas wells in old landfills may be damaged or even missing; in these cases, the core drills made for waste sampling will need to be equipped with piezometers, which will then also be available for pumping tests (Cossu et al., 1997).

Leachate sampling for characterization should be carried out to measure, as a minimum requirement, pH, alkalinity, conductivity, TOC, COD, BOD_5, $SO_4^{=}$, Cl^-, ammonia nitrogen, volatile acids, and heavy metals. Several other compounds may be present in landfill leachate such as xenobiotic organic compounds present in relatively low concentrations (see Chapter 18.2), from household and industrial chemicals (aromatic hydrocarbons, phenols, solvents among others), depending on the kind of waste disposed of.

In very old landfills, waste might display a good biological stability in terms of respiration index, BOD_5, etc., but leachate may still present very high concentrations of organic compounds non degradable under anaerobic conditions (i.e. ammonia) and soluble salts (i.e. chloride), making the landfill a potential source of contamination for a very long time.

The major concerns caused by leachate accumulation in landfills are related to the potential formation of leachate contamination plumes. In this case, comprehensive groundwater investigation should be carried out (see Chapter 18.2).

CONCLUSIONS

Landfill investigation projects may involve the application of several techniques for the comprehensive assessment of the landfill body and the deposited waste. The investigation plan should be designed based on the specific objectives and on the available information from previous reports or tests in such a way that the outcomes enable the design of site-specific solutions.

Should the results of the investigation reveal that landfill emissions are not managed in an appropriate manner, the design of mitigation measures should be undertaken, which may involve the enhancement of biogas extraction system, installation of methane oxidation surface covers, and enhancement of leachate extraction system.

If the investigation proves that aquifer contamination has been caused by a landfill under specific environmental situations, emergency measures such as the installation of hydraulic barriers may be undertaken.

In the absence of evidence of groundwater contamination caused by the landfill, and although the investigation may demonstrate a remaining significant emission potential for the landfill to be viewed as a potential source of contamination in the area, financial constraints may represent a limiting factor for further projects on the site. However, specific measures should be taken, aimed at enhancing waste stabilization by means of aerobization and enhanced leachate extraction, shortening the aftercare phase with the goal of achieving landfill completion within a reasonable time frame (Ritzkowski et al., 2006, 2016, Chapter 16.2).

References

Andreottola, G., Dallago, L., Ragazzi, M., 2005. Dynamic respirometric tests for assessing the biological activity of waste. In: Proceedings Sardinia 2005, Tenth International Waste Management and Landfill Symposium. CISA, Italy.

Anonymous, 2001. . Ordinance on Environmentally Compatible Storage of Waste from Human Settlements and on Biological Waste Treatment Facilities, Berlin, Germany 20.02.2001.

Beaven, R.P., Powrie, W., 1995. . Hydrogeological and geotechnical properties of refuse using a large scale compression cell. In: Proceedings Sardinia 95, Fifth International Landfill Symposium, Vol II. CISA Publisher, Italy, pp. 745–760.

Bichet, V., Grisey, E., Aleya, L., 2016. Spatial characterization of leachate plume using electrical resistivity tomography in a landfill composed of old and new cells (Belfort, France). Engineering Geology 211, 61–73.

Binner, E., Lechner, P., Widerin, M., Zach, A., 1997. Laboratory test methods characterizing the biological reactivity of waste. In: Proceedings Sardinia 1997, Sixth International Waste Management and Landfill Symposium, Vol. I. CISA, Italy, pp. 485–494.

Casado, I., Mahjoub, H., Lovera, R., Fernández, J., Casas, A., 2015. Use of electrical tomography methods to determinate the extension and main migration routes of uncontrolled landfill leachates in fractured areas. Science of Total Environment 506–507, 546–553.

Cossu, R., Raga, R., 2008. Test methods for assessing the biological stability of biodegradable waste. Waste Management 28, 381–388.

Cossu, R., Frongia, G., Muntoni, A., Nobile, A., Raga, R., 1997. Use of pumping tests for the assessment of leachate flow regime, waste hydraulic parameters and well efficiency. In: Proceedings Sardinia 1997, Sixth International Waste Management and Landfill Symposium, Vol II. CISA Publisher, Italy, pp. 53–62.

Cossu, R., Di Maio, R., Raga, R., 2001. An integrated geophysical survey of an old landfill site with use of explosive as energy source for the seismic methods. In: Proceedings Sardinia 2001, Eighth International Waste Management and Landfill Symposium. CISA Publisher, Italy.

Cossu, R., Del Greco, O., Godio, A., Oggeri, C., 2009. Integrated investigation of MSW for a landfill mining project: geophysical, geotechnical and hydraulic surveys. In: Proceedings Sardinia 2009, Twelfth International Waste Management and Landfill Symposium. CISA Publisher, Italy.

Cossu, R., Carrubba, P., Conte, M., Raga, R., Salieri, V., 2011. When wrong design is the main problem in landfilling. In: Proceedings Sardinia 2011, Thirteenth International Waste Management and Landfill Symposium. CISA Publisher, Italy.

Cossu, R., Lai, T., Sandon, A., 2012. Standardization of BOD5/COD ratio as a biological stability index for MSW. Waste Management 32, 1503–1508.

Cossu, R., Morello, L., Raga, R., Cerminara, G., 2016. Biogas production enhancement using semi-aerobic pre-aeration in a hybrid bioreactor landfill. Waste Management 55, 83–92.

Di Maio, R., Fais, S., Ligas, P., Piegari, E., Raga, R., Cossu, R., 2018. 3D geophysical imaging for site-specific characterization plan of an old landfill. Waste Management 76, 629–642.

DIN, 1985. DIN 38414 Part 8. German Standardised Procedures for Testing of Water, Wastewater and Sludge; Sludge and Sediments (Group S); Determination of Fermentation Behaviour (S 8). Beuth Verlag GmbG, Berlin.

EPA, 2002. Guidance on Choosing a Sampling Design for Environmental Data Collection. EPA/240/R-02/005. December 2002.

Francois, V., Feuillade, G., Matejka, G., Lagier, T., Skhiri, N., 2007. Leachate recirculation effects on waste degradation: study on columns. Waste Management 27, 1259–1272.

Hudson, A.P., White, J.K., Beaven, R.P., Powrie, W., 2004. Modelling the compression behaviour of landfilled domestic waste. Waste Management 24, 259–269.

Imhoff, P.T., Reinhart, D.R., Englund, M., Guérin, R., Gawande, N., Han, B., Jonnalagadda, S., Townsend, T.G., Yazdani, R., 2007. Review of state of the art methods for measuring water in landfills. Waste Management 27, 729–745.

Leikam, K., Stegmann, R., 1997. Mechanical–biological pre-treatment of municipal solid waste and the landfill behaviour of pretreated waste. In: Proceedings Sardinia 1997, Sixth International Landfill Symposium, Vol. I. CISA, Italy, pp. 463–474.

Matasovic, N., El-Sherbiny, R., Kavazanjian Jr., E., 2011. In situ measurements of MSW properties. Geotechnical characterization, field measurement and laboratory testing of municipal solid waste. In: Zekkos, Dimitrios (Ed.), ASCE Geotechnical Spec. Publication No. 209, pp. 153–193.

Mazzucato, N., Simonini, P., Colombo, S., 1999. Analysis of block slide in a MSW landfill. Proceedings Sardinia 1999. In: Seventh International Waste Management and Landfill Symposium, III. CISA Publisher, Italy, pp. 537–544.

McNeill, J.D., 1990. Use of electromagnetic methods for groundwater studies. In: Ward, S.H. (Ed.), Geotechnical and Environmental Geophysics, Vol. 1. Review and Tutorial. SEG, Tulsa.

Mönkäre, T.J., Palmroth, M.R.T., Rintala, J.A., 2016. Characterization of fine fraction mined from two Finnish landfills. Waste Management 47, 34–39.

Powrie, W., Beaven, R.P., Harkness, R.M., 1999. Applicability of soil mechanics principles to household waste. In: Proceedings Sardinia 1999, Seventh International Waste Management and Landfill Symposium, Vol III. CISA Publisher, Italy, pp. 429–436.

Powrie, W., Beaven, R., Hudson, A., 2008. The influence of landfill gas on the hydraulic conductivity of waste. In: GeoCongress 2008: Geotechnics of Waste Management and Remediation. ASCE, New Orleans, Louisiana, USA, pp. 264–271.

Reddy, K.R., Hettiarachchi, H., Parakalla, N.S., Gangathulasi, J., Bogner, J.E., 2009. Geotechnical properties of fresh municipal solid waste at Orchard Hills Landfill, USA. Waste Management 29, 952–959.

Redon, E., Lornage, R., Lagier, T., Hebe, I., 2005. Measurement and comparison of different stability parameters during a mechanical–biological pre-treatment before landfilling. In: Proceedings Sardinia 2005, Tenth International Waste Management and Landfill Symposium. CISA Publisher, Italy.

Ritzkowski, M., Heyer, K.-U., Stegmann, R., 2006. Fundamental processes and implications during in situ aeration of old landfills. Waste Management 26, 356–372.

Ritzkowski, M., Walker, B., Kuchta, K., Raga, R., Stegmann, R., 2016. Aeration of the Teuftal landfill: field scale concept and lab scale simulation. Waste Management 55, 99–107.

Scharff, H., Jacobs, J., 2006. Applying guidance for methane emission estimation for landfills. Waste Management 26, 417–429.

Scheutz, C., Fredenslund, A.M., Nedenskov, J., Samuelsson, J., Kjeldsen, P., 2011. Gas production, composition and emission at a modern disposal site receiving waste with a low-organic content. Waste Management 31, 946–955.

Stegmann, R., Spendlin, H.H., 1989. Enhancement of degradation: German experiences. In: Christensen, Cossu, Stegmann (Eds.), Sanitary Landfilling: Process, Technology and Environmental Impact. Academic Press, pp. 61–82. ISBN 0121742555.

Stoltz, G., Gourc, J.P., Oxarango, L., 2010. Characterisation of the physic-mechanical parameters of MSW. Waste Management 30, 1439–1449.

van der Sloot, H.A., Kosson, D.S., van Zomeren, A., 2017. Leaching, geochemical modelling and field verification of a municipal solid waste and a predominantly non-degradable waste landfill. Waste Management 63, 74–95.

Zhan, Liang-Tong, Xu, Hui, Chen, Yun-Min, Lan, Ji-Wu, Lin, Wei-An, Xu, Xiao-Bing, He, Pin-Jing, 2017. Biochemical, hydrological and mechanical behaviors of high food waste content MSW landfill: Liquid-gas interactions observed from a large-scale experiment. Waste Management 68, 307–318.

19.2

PRINCIPLES OF LANDFILL REMEDIATION

Raffaello Cossu

INTRODUCTION

It is well known that major environmental problems in landfilling are associated with long-term uncontrolled emissions of leachate and biogas (Chapter 2.2). Accordingly, an effective multibarrier system aimed at controlling landfill emissions should be installed (Chapter 2.1).

A graphical representation of a multibarrier landfill is given in Fig. 19.2.1. Quality of waste, top cover, waste degradation, drainage systems, artificial mineral and synthetic liners, geological barrier, and the aquifer itself are all barriers which may efficiently protect the sensitive environment in a landfill surrounding.

When one or more of these barriers is missing or failing, a load of contaminants associated with the uncontrolled emissions of leachate and biogas is released into the environment, creating problems or risks which may require the undertaking of remedial actions.

Traditionally, landfill remediation is an issue that addresses mainly old landfill sites. This is due to the fact that in the past the knowledge and awareness of environmental problems associated with the interment of waste in the ground was somewhat scarce, and consequently barrier systems either were not present or were poorly effective.

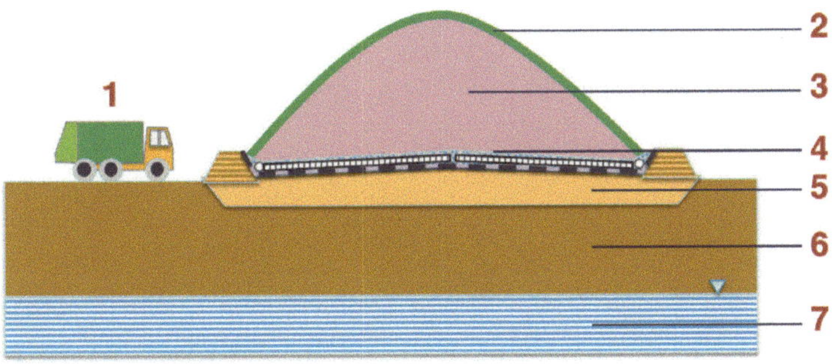

Figure 19.2.1 Graphical description of the landfill multibarrier system. *1*, quality of deposited waste (pretreatment); *2*, top cover system; *3*, waste in situ transformation (degradation, adsorption, complexation, solubilization, etc.); *4*, drainage system and leachate level control; *5*, artificial liner system (mineral and geosynthetic); *6*, geological barrier; *7*, aquifer.

However, more and more frequently nowadays, remedial works are required to deal with environmental problems arising from the operation of regularly authorized "modern" landfills, prevalently due to the implementation of artificial lining systems.

Indeed, the development of synthetic lining technologies and an excessive trust in their efficacy has led to the construction of landfills even in sensitive vulnerable areas. Even when applying optimum quality control procedures, the duration of physical barriers is limited and they do not last forever. Synthetic and mineral liners are subject to leakage (Farquhar, 1989; Rollin et al., 1994) and granular drainage to clogging (Brune et al., 1993).

The main criteria for landfill remediation should be based on constructing or rebuilding the missing or failing barriers. Related technologies and expected results will be discussed in the following sections.

UNCONTROLLED EMISSIONS FROM LANDFILLS

Fig. 19.2.2 illustrates a qualitative assessment of the emission potential and uncontrolled emissions patterns for an open dump and a traditional landfill at different phases.

An open dump or a nonengineered landfill presents the highest release of leachate and gas during the operational phase due to the lack of physical barriers (neither top nor bottom liner, no drainage, no

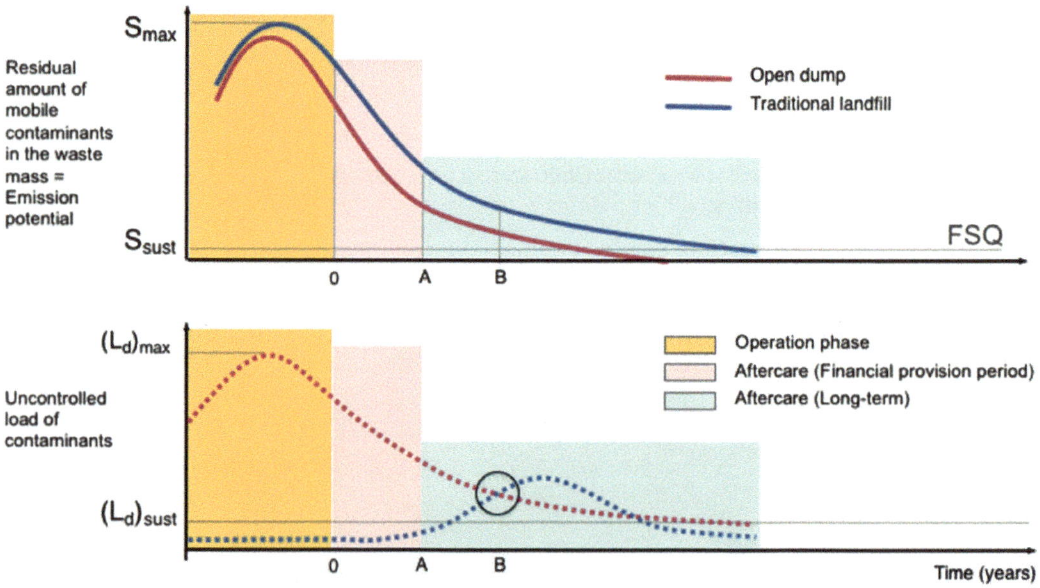

Figure 19.2.2 Trend versus time of the emission potential and load of contaminant associated to uncontrolled landfill emissions at different life phases for an open dump and a traditional landfill. *0*, landfill closure and start of aftercare; *A*, end of aftercare covered by financial provision; *B*, point where remediation is needed.

gas extraction systems, etc.); subsequently, however, in the aftercare phase several decades after closure, the residual emission potential may be very low.

On the contrary, contained landfills produce very limited uncontrolled emissions during the operational and active aftercare phases, in coincidence with the period of maximum efficiency of the physical barriers. However, as the barrier lifetime expires and liners fail, water infiltrates, gas is generated, and leachate leakage occurs. These uncontrolled emissions may be even higher than in open dumps, as the emission potential of the waste was preserved ("mummification effect") and the waste degradation/attenuation barrier was very low (Cossu, 2005). Results from an experimental investigation on small old rural landfills (lined and unlined) clearly support these considerations (Allgaier et al., 2001).

Sustainable landfills should be designed and managed to maintain uncontrolled emissions below levels defined based on the Final Storage Quality concept (Chapter 16.1), even once aging impinges on the efficiency of physical barriers (Chapter 2.2).

REMEDIATION OPTIONS

According to the multibarrier concept (Fig. 19.2.1), the main driving criteria for landfill remediation should be to build, rebuild, repair, or integrate the missing or failing barriers.

The actions undertaken should take into account the results of a preliminary investigation on the landfill site (Chapter 19.1), a site-specific risk analysis, and any other related issues (social aspects, costs, land planning, etc.).

The range of remediation options is graphically represented in Fig. 19.2.3.

The "No action" scenario may be considered in cases in which waste stability has been ascertained and residual potential emissions are negligible. This status will need to be agreed and certified by authorities according to the existing regulations. In other cases, where a leachate contamination plume is detected in the aquifer, it is still possible that no action may be required as the natural attenuation processes may be capable of addressing the contamination effects (Chapter 18.2).

Numerous authors have discussed the feasibility of remediation of landfill leachate plumes through natural attenuation (Christensen et al., 2000; Baun et al., 2003; Longe and Enekwechi, 2007; Yin et al., 2016). However, Christensen et al. (2000) highlighted how the long-time scale expected for remediation may represent a major constraint. Moreover, the large size of landfills, uncertainty over local hydrogeology, possibility of the presence of several plumes, and the variety of pollutants involved

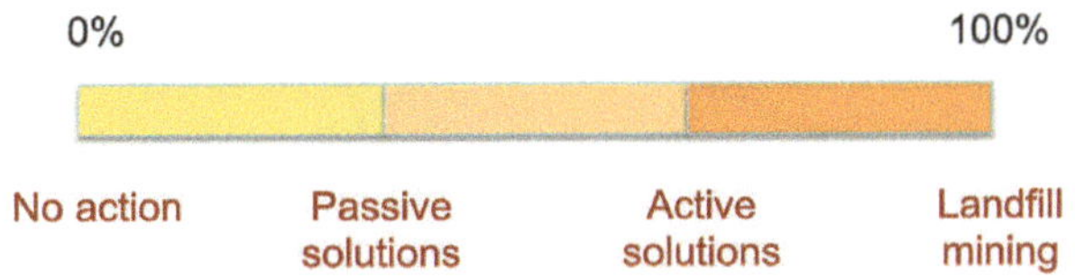

Figure 19.2.3 Scheme of the choice range between different remediation options according to increased efficacy, complexity, and costs.

might not allow prediction to be made and conclusions to be drawn as to the effectiveness of natural attenuation processes.

"Passive solutions" (sealed capping, waste encapsulation, leachate drainage) are aimed at preventing water infiltration and leachate accumulation within the waste mass (Fig. 19.2.4). By adopting these solutions leachate generation could be prevented and mechanical stability controlled.

Passive solutions adhering to the abovementioned criteria are aimed at building or repairing physical barriers, by providing lining and/or drainage systems. They should be adopted when the presence of biodegradable organics and soluble substances in the waste mass is low. Conversely, the limitations of a system based on physical barriers alone may once again be manifested and the related environmental impacts will be only delayed, as well described graphically in Fig. 19.2.5A.

The problem could be dealt with by providing for continuous maintenance of the capping and drainage systems and by keeping the pumping stations working constantly. For these reasons passive solutions cannot be classified as a sustainable solution.

"Active solutions" are intended to act as a reactive barrier to reduce the emission potential at the source with the aim of achieving quality of the residual waste and of emissions that does not exceed Final Storage Quality targets (Chapter 2.1).

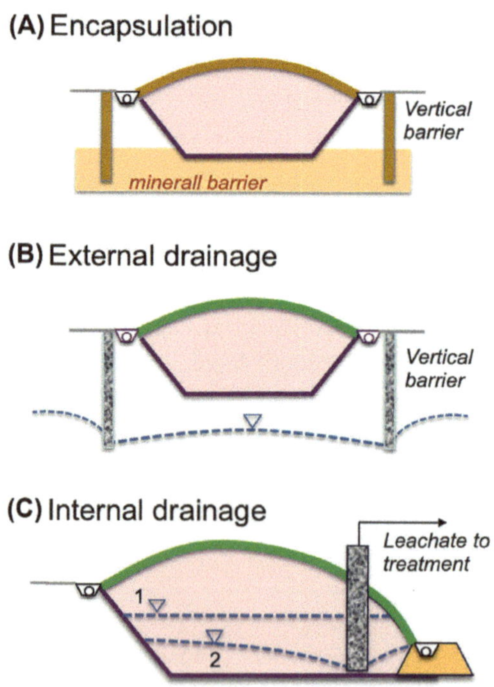

Figure 19.2.4 Possible passive measures for landfill remediation.

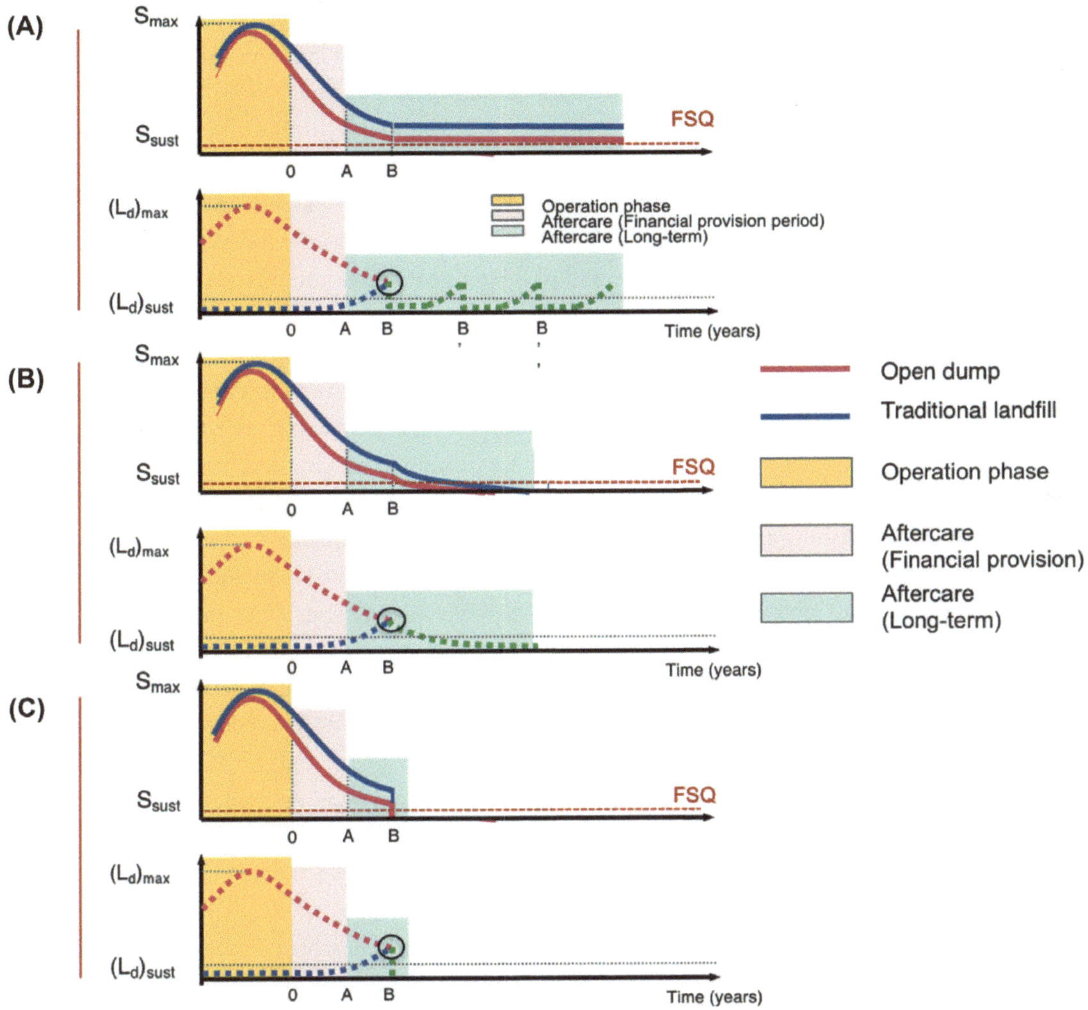

Figure 19.2.5 Effects on long-term uncontrolled emissions patterns according to different remediation actions performed on two kinds of problematic landfills (old landfill and contained landfill with liner failures). *A*, remediation with passive solutions; *B*, remediation with active solutions; *C*, landfill mining; *FSQ*, Final Storage Quality; $(L_d)_{max}$, maximum load of dispersed contaminants; $(L_d)_{sust}$, sustainable load of dispersed contaminants; S_{max}, maximum emission potential; S_{sust}, sustainable emission.

The active measures for remediation of contaminating landfills, according to the quality of landfilled waste, may include

- water addition to implement a controlled flushing of soluble contaminants or to control waste moisture;
- in situ forced aeration to stabilize degradable waste and/or convert ammoniacal nitrogen.

Water addition may only be adopted when the landfill features safe hydrogeological conditions.

The in situ aeration of landfills (Fig. 19.2.5B) is aimed at modifying landfill conditions from anaerobic to aerobic by injecting air into the waste mass (Hudgins and Harper, 1999; Cossu et al., 2001; Ritzkowski et al., 2016; Raga et al., 2015, Chapter 16.2, this book). Air injection is generally carried out by means of perforated wells, and oxygen is distributed by advective and diffusive transport (Chapter 16.2). The extracted oxidized gas is treated before being discharged into the environment.

In situ aeration, with an operation time of 7—10 years, produces waste stabilization and the reduction of biodegradable compounds in leachate and of biogas generation potential. It should be adopted whenever environmental problems from a landfill are associated with a significant residual presence of biodegradables in the waste mass.

Air migration and diffusion into the landfill body may also be obtained with the semiaerobic landfill method, where a natural advective flow of air is established by a difference of temperature between the waste mass and external ambient. (Chapter 14.2). For application of the semiaerobic landfill technology to the remediation of old landfills, a venting system (well and pipes) should be installed after assessing the possibility of establishing the necessary temperature gradient (Gagliardi et al., 2016).

In situ aeration is not intended to produce any beneficial effects on the attenuation of mobile nondegradable inorganics (i.e., chlorides, heavy metals). However, a synergic effect may be obtained by extracting leachate from the landfill during in situ aeration, by means of devices installed in the individual aeration wells. This will prove beneficial not only in the removal of mobile nondegradable inorganics from the landfill body but also in lowering leachate hydraulic levels and favoring air circulation throughout the unsaturated waste layers.

The long-term emission pattern in a landfill after remediation by in situ aeration is indicated in Fig. 19.2.5B. Once the waste is stabilized, the contaminant load remains stably low.

"Landfill mining" (LFM) is the radical solution adopted to remove the source of contaminants, representing what is referred to as a waste barrier (Chapter 2.1).

Deposited wastes are excavated, the resulting material processed, reusable fractions are obtained, and unrecoverables are relandfilled under controlled sustainable conditions (Chapter 19.3). There are several reasons for applying LFM, the main being the need to rapidly remove the source of groundwater contamination. This is of course the most expensive solution. To subsidize the cost for LFM, some other synergic interests could be perceived:

- recovery of volumes for new landfilling
- recovery of urban land or industrial settings
- recovery of useful recyclable resources

The long-term emission pattern in a landfill after remediation by LFM is indicated in Fig. 19.2.5C.

CONCLUSIONS

The guiding principles for the remediation of landfills manifesting environmental problems related to release of contaminants into the environment should be the same as those adopted in the design of new landfills: control of long-term emissions (sustainability) and multibarrier approach. The main remediation criteria, therefore, should be aimed at building, rebuilding, repairing, and integrating missing or failing barriers within a multibarrier landfill system, reaching an acceptable equilibrium with the environment. The adoption of passive remediation technologies (sealing, encapsulation, ...), if not supported by other active barriers, only postpones the environmental issues, in contrast therefore with the environmental principle of sustainability.

References

Allgaier, G., Ritzkowski, M., Stegmann, R., 2001. Investigation of small old deposits in rural areas. In: Christensen, T.H., Cossu, R., Stegmann, R. (Eds.), Proceedings Sardinia 2001, Eighth International Waste Management and Landfill Symposium, Vol. IV. CISA Publisher, Italy, pp. 465–472.

Baun, A., Reitzel, L.A., Ledin, A., Christensen, T.H., Bjerg, P.L., 2003. Natural attenuation of xenobiotic organic compounds in a landfill leachate plume (Vejen, Denmark). Journal of Contaminant Hydrology 65, 269–291.

Brune, M., Ramke, H.G., Collins, H.J., Hanert, H.H., 1993. Incrustation processes in drainage systems of sanitary landfills. In: Christensen, T.H., Cossu, R., Stegmann, R. (Eds.), Proceedings Sardinia 91, Third International Landfill Symposium, Vol. II. CISA Publisher, Italy, pp. 999–1035.

Christensen, T.H., Bjerg, P.L., Kieldsen, P., 2000. Natural Attenuation: a feasible approach to remediation of groundwater pollution at landfills? Groundwater Monitoring and Remediation 20, 22–126.

Cossu, R., Lavagnolo, M.C., Raga, R., 2001. In situ stabilisation of old landfills. In: Christensen, T.H., Cossu, R., Stegmann, R. (Eds.), Proceedings Sardinia 2001, Eighth International Waste Management and Landfill Symposium, Vol. IV. CISA Publisher, Italy, pp. 521–530.

Cossu, R., 2005. The sustainable landfilling concept. In: Cossu, R., Stegmann, R. (Eds.), Proceedings Sardinia 2005, Tenth International Waste Management and Landfill Symposium. CISA Publisher, Italy.

Farquhar, G.J., 1989. Overview of landfill liners using natural material. In: Proceedings Sardinia 89. Second International Landfill Symposium. CIPA srl, Milan (Paper IX.).

Gagliardi, C., Raga, R., Hirata, O., Matsufuji, Y., 2016. Aftercare method by semi-aerobic landfill type for old landfill with incombustible waste. In: Proceedings of 27th Annual Conference of Japan Society of Material Cycles and Waste Management, Wakayama, Japan, 27–29 Sep. 2016.

Hudgins, M., Harper, S., 1999. Operational characteristics of two aerobic landfill systems. In: Christensen, T.H., Cossu, R., Stegmann, R. (Eds.), Proceedings Sardinia 99. Seventh International Waste Management and Landfill Symposium, Vol. I. CISA Publisher, Italy, pp. 327–333.

Longe, E.O., Enekwechi, L.O., 2007. Investigation on potential groundwater impacts and influence of local hydrogeology on natural attenuation of leachate at a municipal landfill. International Journal of Environmental Science and Technology 4, 133–140.

Raga, R., Cossu, R., Heerenklage, J., Pivato, A., Ritzkowski, M., 2015. Landfill aeration for emission control before and during landfill mining. Waste Management 46, 420–429.

Ritzkowski, M., Walker, B., Kuchta, K., Raga, R., Stegmann, R., 2016. Aeration of the Teuftal landfill: field scale concept and lab scale simulation. Waste Management 55, 99–107.

Rollin, A.L., Mlynarek, J., Lafleur, J., Zanescu, A., 1994. Performance changes in aged in-situ HDPE geomembrane. In: Christensen, T.H., Cossu, R., Stegmann, R. (Eds.), Landfilling of Waste: Barriers. E&FN Spon, pp. 431–443. ISBN: 0-419-15990-8.

Yin, K., Tong, H., Giannis, A., Chang, V.W.C., Wang, J.-Y., 2016. Insights for transformation of contaminants in leachate at a tropical landfill dominated by natural attenuation. Waste Management 53, 105–115.

19.3

LANDFILL MINING

Roberto Raga and Raffaello Cossu

INTRODUCTION

Landfill mining (LFM) consists in the excavation and treatment of waste from landfills. The first case study reported in scientific literature dates back to 1953 in Israel and was meant to obtain fertilizers from the deposited waste (Savage et al., 1993). Subsequently, the interest in LFM grew in the 1990s, and several pilot- and full-scale applications have been reported so far (Cossu et al., 1995; Hogland et al., 2004; Jain et al., 2013; Krook et al., 2012; Rettenberger, 1995 among others).

The following motivations and objectives may trigger the application of LFM:

- removal of a (potential) source of contamination for the surrounding environment;
- elimination of technical and administrative issues related to landfill aftercare;
- energy and material recovery from deposited waste;
- creation of new volumes available for waste disposal in new landfill sectors;
- generation of valuable land for town expansion.

LFM projects should be properly planned and built on a sound technical, economical, and financial basis. In fact, reaching the abovementioned objectives should generate the resources to cover the costs of LFM activities.

For assessing the feasibility of LFM, a number of information have to be acquired such as waste composition, landfill environmental conditions, need for landfill conditioning before excavation, possible destination of excavated waste fractions, and the expected costs and revenues (Fig. 19.3.1).

PRELIMINARY TESTS FOR WASTE AND LANDFILL CHARACTERIZATION

The characterization of the deposited material involves the assessment of waste composition and its biochemical and physical characteristics, relevant to the design of LFM process and in view of material and energy recovery (Cossu et al., 1996; Krook et al., 2012; Masi et al., 2014; Prechthai et al., 2008).

The evaluation of landfill environmental conditions is necessary too, and it is aimed at preventing technical difficulties that can arise during the excavation due to high leachate levels and/or significant biogas emissions that could result in waste mechanical instability and accumulation of biogas in the excavation area, hampering the operations and causing safety concerns (Raga and Cossu, 2014; Raga et al., 2015).

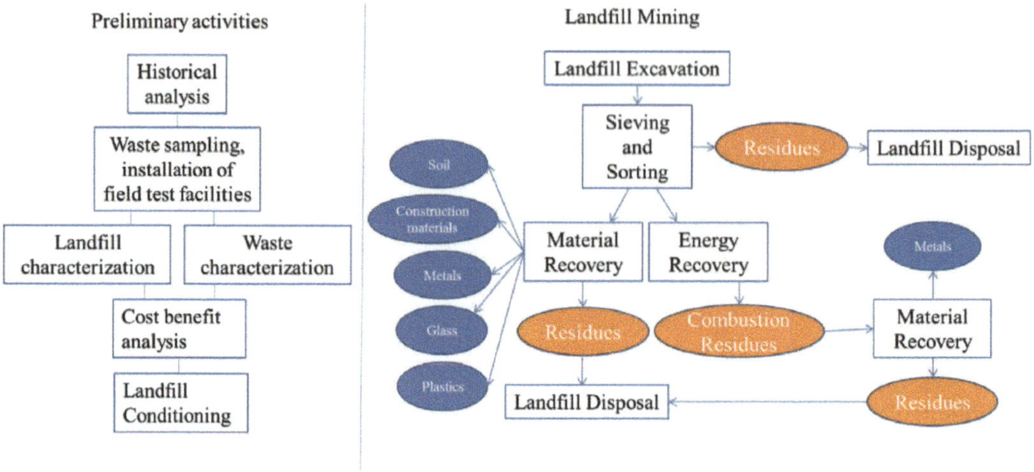

Figure 19.3.1 Landfill mining concept.

For these reasons, a series of ad hoc analyses is needed preliminary to LFM to gain detailed information on the specific aspects given below:

- Morphologic and geomechanic characteristics of the landfill: Especially in the case of old uncontrolled landfill, where no records of disposal activities are available, geophysical investigations may provide an evaluation of the thickness and extent of deposited waste, as well as of the presence of leachate saturated layers. Moreover, based on the results of geophysical survey, geotechnical investigations are carried out to provide detailed information on the stratigraphy of the site and on the physical characteristics of deposited material (i.e., humidity, density) and of the expected mechanical stability of waste during excavation (Cossu et al., 2009).
- Waste characterization: Fractional composition and grain-size distribution analysis on bulk waste samples are carried out in view of resource recovery from excavated waste. The flow chart of the activities carried out for sorting and characterization of the waste excavated during the first LFM pilot project in Italy (Cossu et al., 1995) is reported in Fig. 19.3.2. Data on waste characterization from the time of deposition are rarely available and reliable; moreover, a variation of waste composition over time is expected due to biodegradation processes in the landfill body (Pretchai et al., 2008). For these reasons, specific campaigns are necessary for sampling waste from different locations in the landfill body. Also, the quality of the selected fractions has to be investigated in view of the evaluation of suitability for treatment and recycling. Potential residual emissions should be measured and possible stabilization processes considered for those fractions destined to further landfill disposal.
- Landfill environmental conditions should be defined in terms of leachate production and accumulation, as well as biogas production potential and current emissions. This is needed for planning proper landfill pretreatment before excavation, for the selection of excavation technology and of site-specific measures for the prevention of environmental pollution and of hazardous conditions for workers.

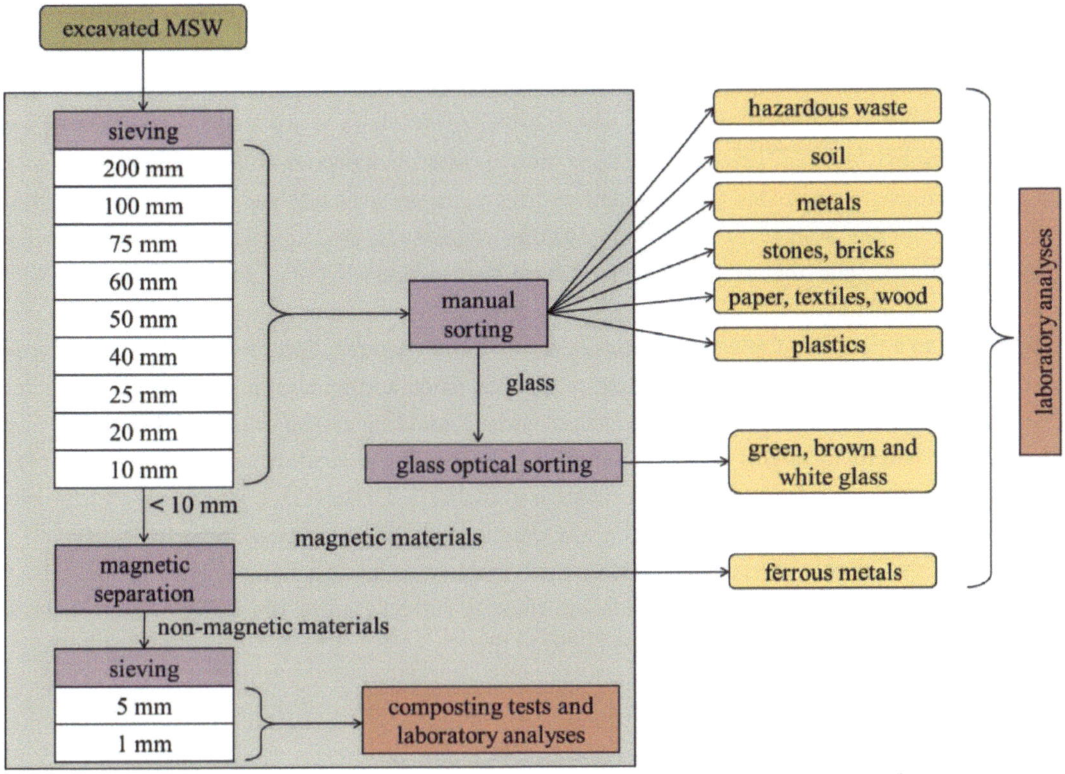

Figure 19.3.2 Flow chart of the activities carried out for sorting and characterization of excavated waste in Simbirizzi landfill, Italy. *Cossu et al. (1995); modified.*

RECOVERY OF RESOURCES

The potential for resource recovery from LFM depends on

- the waste fractional composition in the landfill;
- the quality of the different materials;
- the characteristics and efficiency of the technologies used for separation and further treatment of the different material fluxes;
- the available destinations for the recovered fractions.

Concerning the material composition, municipal solid waste (MSW) landfills operated before the introduction of separate collection programs should be more suitable for LFM. However, in many cases the limited efficiency of diversion practices causes the disposal in new MSW landfills of significant quantities of reusable and recoverable materials (Wagner and Raymond, 2015).

According to Jones et al. (2013), LFM should be "*embedded in a broad resource management perspective,*" where a significant role is played by resource recovery (including energy from biogas before excavation) from the landfilled waste. This is one of the key issues of the enhanced landfill mining

(ELFM), defined as "*the safe conditioning, excavation and integrated valorization of landfilled waste streams as both materials and energy, using innovative transformation technologies and respecting the most stringent social and ecological criteria.*" Based on this concept, both old and new landfills are considered as "*temporary storage place, awaiting future valorization*" (Geysen et al., 2009); after waste excavation and treatment, the nonrecyclable fractions are deposited in purpose-built landfill sectors, to be stored until proper technologies for treatment and recycling are available and further LFM becomes profitable. Temporary storage of waste before future valorization is currently an issue in numerous European countries, and the related challenges have been addressed (Wagner and Bilitewski, 2009; Passamani et al., 2016).

According to a number of research projects, a very common composition of old MSW landfills consists of 20%–50% of a fraction containing plastics, paper, wood and textiles (potentially combustibles), 40%–60% soillike materials, 10%–20% inorganic matter (including construction and demolition waste and glass), and a limited amount of metals (Zhou et al., 2014; Quaghebeur et al., 2013). An example of results of composition analyses of excavated waste is available in Table 19.3.1, where it is clearly visible that a harmonization is still necessary in the way data from LFM projects are reported.

Quality of the waste material after a long time within the landfill body is expected to be very different from at the time of deposition, and, according to several studies, obtaining marketable materials might be challenging and often not feasible from the economic and technical point of view (Krook et al., 2012).

According to Zhou et al. (2014), due to long-term physical and biochemical processes, landfilled plastics proved to have a lower content of carbon and sulfur but higher amount of oxygen, silicon, and aluminum compared with new plastics. This can be due to impurities (mainly, small particles of soil and sand, whose main components are SiO_2, Al_2O_3) associated to landfilled plastics waste and that are very difficult and expensive to clean. For this reason, the economic feasibility of recycling of plastics from LFM or its valorization through chemical treatment by pyrolysis, gasification, and hydrogenation is questionable, and in most cases, direct thermal treatment or the production of residue-derived fuels are the applied options.

Metals in MSW landfills are present in low concentrations compared with other fractions. Based on data from 60 LFM projects, Ratcliffe et al. (2012) report a mean concentration of metals of 2.0%. The characterization of Remo landfill in Belgium, in view of a huge LFM project, provided values for the concentration of metals in different sectors of the landfill ranging from 3% to 6% (Quaghebeur et al., 2013). Although metal recovery is expected to become increasingly attractive due to the global demand and the raw materials expected future costs, technical and economic feasibility might be questionable when metals are bound within articles and the need to apply specific treatment such as shredding, grinding, or other processing arises. Few full-scale LFM projects for metals recovery have been completed without the need of specific funding. An LFM research project was carried out on a landfill for MSW combustion residues and the economic feasibility of metal recovery ascertained (Wagner and Raymond, 2015). Based on the results of the study, according to the authors, metal recovery from landfilled MSW might be more cost effective after thermal treatment.

Table 19.3.1 Fractional composition of landfilled municipal solid waste after different years from deposition (all values as %)

Fractions	References							
	A	B	C	D	E	F	G	H
Plastics	9.4	5.9	23	17		12.9	10.6	13.9
Paper	4.7	2.8		7.5		3.7	0.15	4.6
Paper and textiles			6					
Plastics and cellulosic materials					0.3			
Composite materials	1.1	0.8						
Wood	10.1	5.4		7		1.4	2.4	
Textiles	2.5	1.6		6.8		3.6	1.5	2.6
Metals	2.5	3.3	4	2.8	2	1.1	0.41	5.1
Soil	9.8		11					
Glass	1.6		12		8.9		0.6	8.2
Stones	7.7		13		18.2		8.3	13.2
Glass and stones		3.5		1.3		14.9		
Aggregates				10				
Unidentified	3.9			3.8	0.1		1	
Burnt material					0.2			
Fines (10–20 mm)	14							
Fines (<10 mm)	32		31	44	70.4		75	52.4
Fines (<40 mm)		76.7						
Fines (<20 mm)						62.5		
Deposition period	1988–2005	1980–85	1972–1980	n.a.	1950–1980	1970–1983	1989–2004	1980–2010

n.a.: not Available.

References: A. Wanka et al. (2017), B. Obermeier and Saure (1995), C. Cossu et al. (1995), D. Jones et al. (2013), E. Masi et al. (2014), F. Raga and Cossu (2014), G. Zhou et al. (2015), H. Rong et al. (2017).

Depending on the destinations chosen for the excavated material, sorting can be designed to produce fluxes combining different material fractions. Commonly, a flux destined to energy recovery in a thermal treatment plant can be composed of paper, cardboard, plastics, wood, and textiles (Quaghebeur et al., 2013). These components can be found especially in the coarse fractions and can be separated from other waste as oversieve (mesh size of 40 or 50 mm). This fraction is characterized by a high calorific value, which varies between 7 ÷ 8 MJ/kg (Hogland et al., 1995) and 11 MJ/kg (Obermeier and Saure, 1995), enabling energy use without the need for auxiliary fuels. In other experiments, values up to 20 MJ/kg were obtained (Cossu et al., 1995; Rettenberger, 1995; Quaghebeur et al., 2013; Wanka et al., 2017). A viable option for this fraction is the production of solid recovered fuel (SRF), defined in compliance with the European standard EN 15359 and produced from nonhazardous waste. Depending on National legislation, the main requirements for SRF are related to its calorific value and chlorine and mercury content, often causing the need for further processing (i.e., biodrying) before utilization (Passamani et al., 2016).

The fine fraction, generally separated as undersieve 10–20 mm, contains higher levels of organic substances, and it was sometimes considered as possible additive to compost, although this option is normally hindered by high-quality legislation requirement for agriculture. In a case study in Thailand (Prechthai et al., 2008), the fine fraction from LFM (undersieve 25 mm) was successfully tested as additive in soils for growing energy crops.

Especially in the case where codisposal with industrial waste occurred in the landfill, the fines may contain high concentrations of heavy metals (Cu, Cr, Ni, and Zn) and provide opportunities for metal extraction and recovery. The development of proper technologies for resource recovery in such cases remains one of the main challenges for further development of LFM (Quaghebeur et al., 2013).

A very common use for the fine fraction from LFM is the daily cover in new landfill sectors in the same site or, based on its quality and the local requirements, outside the landfill to build embankments and dams or to fill land depressions.

Residual fraction from separation and treatment of the excavated materials will always be present and make an additional flux destined to final disposal in new landfill sectors.

PRETREATMENT BEFORE EXCAVATION

Landfill conditioning by means of in situ aeration can be necessary for waste pretreatment prior to LFM to reduce the effects connected to biogas uncontrolled emissions from the waste mass during the excavation (Cossu et al., 2003). Among the first examples of LFM in Europe, the case of Burghof landfill in Germany (Rettenberger, 1995) involved the application of an odor stabilization technique, consisting in air injection and gas extraction for 2 weeks before starting the excavation, carried out by means of 3.5 m probes pressed into the waste, at a distance of 5–6 m from each other. A similar approach has been adopted in mining Sharjah's landfill in the United Arab Emirates, the main difference being that in this case the aerobic conditions were maintained for a longer period of time (6 weeks) before the excavation (Goeschl and Rudland, 2007). Compared to landfill aeration within the scope of postclosure care and its completion (Ritzkowski and Stegmann, 2013), in the case of landfill conditioning before LFM, the aeration time can be shorter as the goal is not the abatement of the emission potential but the control biogas formation and migration in the excavation area (Ritzkowski and Stegmann, 2012). However, a prolonged aeration can also result in increased biological stability of the organic fraction, which would prove beneficial in case the final disposal in new landfill sectors is foreseen for (a fraction of) the excavated waste. As a matter of fact, despite the current development of innovative concepts such the ELFM, a very common purpose of LFM is still to relocate the waste from unlined landfill sectors to new lined ones (Jain et al., 2013), with the additional gain of new landfill volume.

In the many cases, where the material below the waste acts as a (natural or constructed) barrier to leachate migration and leachate extraction systems are either missing or not effective, relevant leachate tables are expected to be present, limiting the efficiency of aeration and hampering proper waste excavation (Fig. 19.3.3) and further processing. In such cases, each aeration well should be equipped with pneumatic ejectors (or other kinds of pumps) to enable leachate extraction.

Air injection/extraction coupled with leachate extraction is one of the main features of the Airflow technology, used for landfill conditioning in the framework of a reclamation project proposed for an old site in northern Italy, involving LFM with material and energy recovery and the construction of new landfill sectors in the same area for the nonrecoverable fraction of the excavated waste and for the future waste disposal needs of the region. In this case, an in situ aeration period of 6 months was considered adequate to reach a significant biological stabilization of the deposited waste before excavation, also in view of the further landfill disposal (Raga and Cossu, 2014).

The Airflow technology was applied to the old landfill of Modena (Italy), for waste conditioning before and during waste excavation for the construction of a 19 m deep trench through the landfill, for the high-velocity railway line connecting Milan to Bologna that is now crossing the landfill, laying

Figure 19.3.3 Waste excavation without preliminary leachate extraction.

on the natural soil at its bottom. The very specific challenges posed by this project were related to the expected leachate and biogas emissions into the excavation area from the adjacent and still active landfill sectors. The application of the Airflow technology preliminary to LFM enabled to minimize the leachate table in the landfill and to reach a significant enhancement of the biological stability not only of the waste to be mined but also of those contiguous to the excavation area, thus limiting the emission potential in view of mining activities. The Airflow plant was run during LFM also, and the flow rates were adjusted to maintain a depression on the slopes to prevent any emission into the trench (Raga et al., 2015).

Active aeration is among the recommended measures to prevent odorous emissions before the excavation of temporary biodegradable waste storage facilities also, where anaerobic conditions have developed due both to waste quality and storage conditions (Wagner and Bilitewski, 2009; Passamani et al., 2016). Conversely, in the case of very well-stabilized old waste deposits, the use of the in situ aeration before LFM can be unnecessary, as recently documented by Jain et al. (2013), who report that uncontrolled gas emissions and odor were not an issue during landfill excavation and the processing of the excavated material.

EXCAVATION, SEPARATION, AND TREATMENT

Various technologies can be employed for excavation, separation, and treatment, depending on the quality and nature of the landfilled waste. Age and depth of the waste, degree of degradation, and moisture content can influence the success of material recovery, therefore the equipment needs to be designed for the particular landfill under consideration (Ford et al., 2013; Jain et al., 2014).

Waste is firstly extracted using excavators and/or front-end loaders and then stored in stockpiles for further processing. Preliminary in situ aeration with leachate extraction is recommended to provide for proper working conditions and avoid leachate or biogas emissions in the excavation area. Fig. 19.3.4 shows waste excavation in the framework of the full-scale LFM project at the old MSW landfill of Modena, in Italy. Previously, the waste had undergone biological stabilization through landfill aeration involving leachate extraction from aeration wells by means of the Airflow technology (Raga et al., 2015).

Landfilled material is usually mixed and compacted; therefore, size reduction is necessary before screening and separation of recyclables. For this purpose, shredders and grinders are employed, considering that waste can have an abrasive potential which could damage the mechanical elements of the equipment. Low-speed shredders and multiple shaft machines are generally the most suitable devices for shredding operation of landfilled waste, although conditions of the excavated waste and its composition can affect separation efficiency (Ford et al., 2013). Waste is then separated by size using screens. Between the several types of screens those usually employed in LFM projects are trommel screens or vibrating screens. In particular, trommel screens are typically used for separating soil and fine organic material from other waste. They can be featured at different stages of the LFM process, depending on type of waste and requirements of the recovered material (Zhou et al., 2014; Ford et al., 2013).

Figure 19.3.4 Waste excavation in the framework of the full-scale landfill mining project at the old municipal solid waste landfill of Modena, Italy. The beneficial effects of preliminary leachate extraction are visible (Raga et al., 2015).

After being screened, waste is likely to go through air technologies, where light material (primary plastics) is separated from heavy fraction. Metal separation is the following step with the use of drum magnets or overband magnets to remove ferrous material. This should be a straightforward operation, with high removal efficiency, as it is already broadly applied in waste management industry. However, recovered metals from MSW landfills generally contain a lot of impurities, therefore other treatments could be necessary before their effective recycling.

To remove nonferrous metals, typically aluminum and copper, waste management plants employ eddy current separators, but the effectiveness of this technology is likely to be lower for compacted waste from landfills (Ford et al., 2013). After all the recyclables are separated from excavated waste, the unrecoverable material is relocated to a new sustainable landfill, in most cases built on the same site (Jain et al., 2014).

In some cases, coarse materials might be separated from the other components to undergo washing and conditioning treatment before being dispatched for recycling elsewhere (Rettenberger, 1995). Based on a similar concept, Wanka et al. (2017) carried out a wet mechanical treatment of the fraction 10–60 mm to end up with the separation of three main fluxes of different densities: a so-called dense inert fraction with a high content of glass and stones, a dense fine fraction, and a light fraction. The positive effect of the wet treatment was the removal of dirt particles form the surface, increasing the quality of all separated fraction in view of possible recycling. However, due to a calorific value up to 20 MJ/kg, the recommended final destination for the light fraction was thermal treatment for energy recovery.

Hydromechanical treatment tests were carried out by Bilitewski et al. (1995) to treat the fine fraction in two phases: in the first washing phase, heavy inert material is separated from the floating fraction (wood and plastic); in the second phase, the suspension obtained is separated into a fine inert fraction and into a sludge, where organic substances are more concentrated.

Encouraging results have been obtained in some experiences (among which Cossu et al., 1995) in the separation of glass from other inert materials via prewashing and a combined process of breaking and smashing, separating the resulting fraction via a fine-meshed screen.

COSTS AND FUNDING

The application of a cost—benefit analysis is of utmost importance for the assessment of the economic feasibility of LFM. Actually, despite the potential offered by LFM and the increasing interest aroused in the last 2 decades, the number of full-scale projects has been unexpectedly low so far, mainly due to the difficulties in proving the economic feasibility of the many candidate cases.

The need for building facilities of public interest may be a very strong driver (and money is no longer a limiting factor), as happened in the previously mentioned application of LFM to the old landfill of Modena (Italy) to enable the construction of the trench for the railway line (Raga et al., 2015).

LFM projects can be implemented by private enterprises, provided that the revenues for the company outweigh the costs. This is likely to occur when the LFM activities generate valuable land for town expansion or new volumes for the extension of landfill service life (Goeschl and Rudland, 2007; van der Zee et al., 2004; Raga and Cossu, 2014).

The creation of economy of scale may foster the application of LFM; the generation of additional waste streams to feed a waste thermal treatment plant may be needed in areas with strong population (and waste production) fluctuation over the year (Cossu et al., 2009).

In other cases, LFM profitability for private companies can be discussed and it depends on very site-specific conditions.

Wagner and Raymond (2015) demonstrated that ashfills can be profitably mined for metals without financial support from government. Except for the former and other few cases of monolandfills for industrial waste (Zanetti and Godio, 2006), LFM activities have not been carried out with focus on resource recovery so far, and the very limited research effort for the enhancement of the recycling potential for the excavated waste has led to discouraging results. For these reasons, the application of LFM currently does not seem realistic if the only motivation is material recovery; further research for innovative separation and transformation technologies is needed (Krook et al., 2012; Jones et al., 2013).

A full-scale application of life cycle inventory and impact evaluation showed that the reduction of greenhouse gas emissions is not likely to foster the mining of an MSW landfill and that a main driver could be the revenues associated to the solving of an environmental issue instead (Jain et al., 2014). The trigger for LFM can be the removal of the landfill as a (potential) source of groundwater and surface water contamination. This would grant a dramatic reduction of the aftercare costs for the landfill and the avoidance of the unpredictable cost for the possible long-term groundwater remediation. However, it is likely that public funding cannot be avoided in this case (van der Zee et al., 2004; Hull et al., 2005) unless a profitable enough use of the recovered land is foreseen.

According to the United States Environmental Protection Agency (EPA, 1997), most economic benefits associated with LFM are indirect and fall on a broad community, particularly on those living close to the landfill in question. To take them into account, social issues should be properly assessed (by assigning them a monetary value) and integrated in a methodological framework to support private and public actors in the evaluation process of LFM feasibility (van Passel et al., 2013) to avoid the risk of overlooking those cases in which LFM would guarantee large social benefits without any significant private revenues. In such cases, however, a fundamental role should be played by appropriate government policies, support schemes, and incentives (Jones et al., 2013). It is a common position among the scientific community that a comprehensive assessment of LFM projects should take into account not only private costs and revenues but also social and environmental issues. This should be achieved through the application of a cost—benefit analysis which includes the monetary evaluation of those costs and benefits (generally called externalities) affecting somebody (i.e., citizens living in the area, the society) other than the people engaged in the LFM project and which are not reflected in prices.

Some externalities commonly associated to LFM are

- reduction of environmental footprint (air, soil, surface water, and groundwater pollution);
- lower import dependency for energy and materials;
- nature restoration and creation of recreational areas;
- benefits for the citizens from the urban development in the recovered area;
- noise, smells, dust, and increased traffic during excavation.

All of them should be properly encompassed in the assessment procedures of an LFM project. To this regard, the implementation of a methodological framework is needed; some results of the assessment of private profitability and costs and benefits to society of LFM projects have been recently reported (van Passel et al., 2013; Marella and Raga, 2014).

CONCLUSIONS

LFM has been applied worldwide since a few decades to address site-specific issues, mainly related to the removal of obstacles to more profitable land uses or to enable the construction of bigger landfill sectors in the same area.

Unless specific incentives or public funding are available, the application of LFM to other situations where the main motivation is the removal of a potential source of contamination or the recovery of resources has been limited to very few cases so far due to the difficulties in reaching the economic feasibility of the project.

Currently, the concept of ELFM is opening encouraging scenarios for the full exploitation of LFM potential, and it is boosting the development of innovative approaches with the goal of achieving the economic sustainability of LFM activities focused on resource recovery. However, although several research projects have been investigating the potential of LFM for resource recovery for more than 20 years now, according to numerous authors, the development of cost effective technologies is still a challenge.

References

Bilitewski, B., Conrad, H., Grischek, H., 1995. Transportation and reconstruction of old landfill sites in east Germany. In: Proceedings Sardinia 95, Fifth International Landfill Symposium. CISA Publisher, Cagliari, Italy, pp. 807–818.

Cossu, R., Motzo, G.M., Laudadio, M., 1995. Preliminary study for a landfill mining project in Sardinia. In: Proceedings Sardinia 95, Fifth International Landfill Symposium. CISA Publisher, Cagliari, Italy, pp. 841–850.

Cossu, R., Hogland, W., Salerni, E., 1996. Landfill mining in Europe and the USA. In: ISWA Year Book, pp. 107–114.

Cossu, R., Raga, R., Rossetti, D., 2003. Full scale application of in situ aerobic stabilization of old landfills. In: Proceedings Sardinia 2003 – Ninth International Waste Management and Landfill Symposium. CISA Publisher, Cagliari, Italy.

Cossu, R., Del Greco, O., Godio, A., Oggeri, C., 2009. Integrated investigation of MSW for a landfill mining project: geophysical geotechnical and hydraulic surveys. In: Proceedings Sardinia 2009, Twelfth International Waste Management and Landfill Symposium. CISA Publisher, Cagliari, Italy.

EPA, 1997. Landfill Reclamation. EPA 530-F-97–9001. United States Environmental Protection Agency.

Ford, S., Warren, K., Lorton, C., Smithers, R., Read, A., Hudgins, M., 2013. Feasibility and Viability of Landfill Mining and Reclamation in Scotland. Zero Waste Scotland. www.zerowastescotland.org.uk.

Geysen, D., Jones, P.T., Van Acker, K., Van Passel, S., Craps, M., Eyckmans, J., Vrancken, K., Laenen, B., Laevers, P., 2009. Enhanced Landfill Mining – a future perspective for landfilling. In: Proceedings Sardinia 2009, Twelfth International Waste Management and Landfill Symposium. CISA Publisher, Cagliari, Italy.

Goeschl, R., Rudland, D., 2007. Full remediation of Sharjah's old landfill. In: Proceedings Sardinia 2007, Eleventh International Waste Management and Landfill Symposium. CISA Publisher, Cagliari, Italy.

Hogland, W., Marques, M., Nimmermark, S., 2004. Landfill mining and waste characterization: a strategy for remediation of contaminated areas. Journal of Material Cycles and Waste Management 6, 119–124.

Hogland, K.H.W., Jagodzinski, K., Meijer, J.E., 1995. Landfill mining tests in Sweden. In: Proceedings Sardinia 95, Fifth International Landfill Symposium. CISA Publisher, Cagliari, Italy, pp. 841–850.

Hull, R.M., Krogmann, U., Strom, P.F., 2005. Composition and characteristics of excavated materials from a New Jersey landfill. Journal of Environmental Engineering 131, 478–490.

Jain, P., Townsend, T.G., Johnson, P., 2013. Case study of landfill reclamation at a Florida landfill site. Waste Management 33, 109–116.

Jain, P., Powell, J.T., Smith, J.L., Townsend, T.G., Tolaymat, T., 2014. Life-cycle inventory and impact evaluation of mining municipal solid waste landfills. Environmental Science & Technology 48, 2920–2927.

Jones, P.T., Geysen, D., Tielemans, Y., Van Passel, S., Pontikes, Y., Blanpain, B., Quaghebeur, M., Hoekstra, N., 2013. Enhanced Landfill Mining in view of multiple resource recovery: a critical review. Journal of Cleaner Production 55, 45–55.

Krook, J., Svensson, N., Eklund, M., 2012. Landfill mining: a critical review of two decades of research. Waste Management 32, 513–520.

Marella, G., Raga, R., 2014. Use of the Contingent Valuation Method in the assessment of a landfill mining project. Waste Management 34, 1199–1205.

Masi, S., Caniani, D., Grieco, E., Lioi, D.S., Mancini, I.M., 2014. Assessment of the possible reuse of MSW coming from landfill mining of old open dumpsites. Waste Management 34, 702–710.

Obermeier, T., Saure, T., 1995. Landfill reconstruction, biological treatment of landfill waste. In: Proceedings Sardinia 95, Fifth International Landfill Symposium. CISA Publisher, Cagliari, Italy, pp. 841–850.

Passamani, G., Ragazzi, M., Torretta, V., 2016. Potential SRF generation from a closed landfill in Northern Italy. Waste Management 47, 157–163.

Prechthai, T., Padmasri, M., Visvanathan, C., 2008. Quality assessment of mined MSW from an open dumpsite for recycling potential. Resources. Conservation and Recycling 53, 70–78.

Quaghebeur, M., Laenen, B., Geysen, D., Nielsen, P., Pontikes, Y., Van Gerven, T., Spooren, J., 2013. Characterization of landfilled materials: screening of the enhanced landfill mining potential. Journal of Cleaner Production 55, 72–83.

Raga, R., Cossu, R., 2014. Landfill aeration in the framework of a reclamation project in Northern Italy. Waste Management 34, 683–691.

Raga, R., Cossu, R., Heerenklage, J., Pivato, A., Ritzkowski, M., 2015. Landfill aeration for emission control before and during landfill mining. Waste Management 46, 420–429.

Ratcliffe, A., Prent, O.J., van Vossen, W., 2012. Feasibility of material recovery from landfills (MFL) in the European Union. In: Proceedings, the ISWA World Waste Congress 2012, Florence, Italy, 17–19 September 2012.

Rettenberger, G., 1995. Results from a landfill mining demonstration project. In: Proceedings Sardinia 95, Fifth International Landfill Symposium. CISA Publisher, Cagliari, Italy, pp. 827–840.

Ritzkowski, M., Stegmann, R., 2012. Landfill aeration worldwide: concepts, indications and findings. Waste Management 32, 1411–1419.

Ritzkowski, M., Stegmann, R., 2013. Landfill aeration within the scope of post-closure care and its completion. Waste Management 33, 2074–2082.

Rong, L., Zhang, C., Jin, D., Dai, Z., 2017. Assessment of the potential utilization of municipal solid waste from a closed irregular landfill. Journal of Cleaner Production 142, 413–419.

Savage, G.M., Golueke, C.G., von Stein, E.L., 1993. Landfill mining: past and present. Biocycle 34, 58–61.

van der Zee, D.J., Achterkamp, M.C., de Visser, B.J., 2004. Assessing the market opportunities of landfill mining. Waste Management 24, 795–804.

van Passel, S., Dubois, M., Eyckmans, J., de Gheldere, S., Ang, F., Jones, P.T., Van Acker, K., 2013. The economics of enhanced landfill mining: private and societal performance drivers. Journal of Cleaner Production 55, 92–102.

Wagner, J., Bilitewski, B., 2009. The temporary storage of municipal solid waste - recommendations for a safe operation of interim storage facilities. Waste Management 29, 1693–1701.

Wagner, T.P., Raymond, T., 2015. Landfill mining: case study of a successful metals recovery project. Waste Management 45, 448–457.

Wanka, S., Münnich, K., Fricke, K., 2017. Landfill mining - wet mechanical treatment of fine MSW with a wet jigger. Waste Management 59, 316–323.

Zanetti, M., Godio, A., 2006. Recovery of foundry sands and iron fractions from an industrial waste landfill. Resources, Conservation and Recycling 48, 396–411.

Zhou, C., Fang, W., Xu, W., Cao, A., Wang, R., 2014. Characteristics and the recovery potential of plastic wastes obtained from landfill mining. Journal of Cleaner Production 80, 80–86.

Zhou, C., Gong, Z., Hu, J., Cao, A., Liang, H., 2015. A cost-benefit analysis of landfill mining and material recycling in China. Waste Management 35, 191–198.

20 CHAPTER

SPECIAL WASTE LANDFILLING

20.1

LANDFILLING OF DIFFERENT KINDS OF WASTE: LEACHING BEHAVIOR

Hans A. van der Sloot, David S. Kosson and Andre van Zomeren

INTRODUCTION

As discussed in Chapter 1.2, several kinds of waste can be landfilled. A single waste type can be deposited in dedicated areas (monofills) or different wastes can be deposited together, such as in municipal solid waste (MSW) landfills. The quality (physical and chemical properties) of landfilled waste is a primary factor in reducing the risk from potential release of constituents of potential concern (COPCs) to the environment. And, in particular, understanding the leaching behavior of the deposited waste is essential as part of any landfill design procedure (Mathlener et al., 2006).

What Is Leaching? In an environmental context, leaching is the release or dissolution of constituents from a solid material into contacting water. Dissolved constituents have the potential to contaminate surface water, groundwater, or soils. The release of inorganic constituents from a material or matrix is controlled by a combination of physical and chemical processes that depend on the properties of the solid material, the environmental exposure conditions or scenario, and the specific COPCs. The release of organic constituents is further complicated by microbial biological processes that may degrade, transform, or solubilize organic matter, as well as influence inorganic constituents through reduction—oxidation (redox) processes. In case of landfill, the release of substances affecting soil, surface water, and groundwater is the main concern, which is generally controlled by setting limits on the properties of waste that may be disposed (i.e., waste acceptance criteria). Leaching frequently controls other biological, physical—chemical processes relevant to human health and the environment (e.g., bioavailability, plant uptake, ecotoxicity); however, a description of how leaching influences these secondary processes is beyond the scope of this document.

Why Perform Leaching Tests? Leaching tests are used to *measure* the potential for environmental release of COPCs under controlled conditions in a laboratory setting. The results of leaching tests are then used to *estimate* leaching under field conditions through interpretation of intrinsic leaching behavior in the context of the anticipated field conditions or field scenario. "Source term" refers to an estimate of constituent release from a material that is then used in fate and transport modeling, for example, for setting limit values based on compliance with groundwater quality objectives (Hjelmar et al., 2001).

From the perspective of environmental protection, the goal of leaching assessment is to answer the question, “What is the potential for toxic constituent release from this material by leaching and, therefore, the risk associated with the selected management option or range of options?” Environmentally sound waste management also involves the waste generator and the waste management facility. Each stakeholder has their own interests, liabilities, and perspectives that modify the overarching question in the context of regulatory requirements (Kosson et al., 2002):

- From the waste generator’s perspective: “Which waste management options are acceptable for this waste?”
- From the waste management facility’s perspective: “Which wastes are suitable for disposal in different types of disposal facilities?” and “What treatment methods can be applied to reduce constituent mobility before disposal?”

Over the past decade, leaching tests for wastes have been developed to address long-term release characteristics of COPCs (van der Sloot et al., 1997; Kosson et al., 2002). This is a departure from the more traditional regulatory approach, where limit values have been applied on a pass/fail basis using simplistic single-point tests. Leaching characterization has concentrated on an understanding of the mechanisms of contaminant release and release controlling parameters for inorganic and organic constituents in waste (the so-called “comprehensive characterization” testing regime). In Europe, leaching tests have been developed through a variety of research initiatives (Van der Sloot et al., 1997) and have been standardized through work carried out under the control of CEN and ISO. In the United States (US), the US Environmental Protection Agency (US EPA) initially focused on the development of methods to simulate codisposal in an MSW landfill (e.g., Method 1311 Toxicity Characteristic Leaching Procedure, TCLP) for use as a discriminator between nonhazardous and hazardous solid wastes based on the plausible mismanagement scenario of codisposing hazardous waste in a municipal waste landfill. More recently, US EPA has standardized and validated more robust methods for waste characterization and assessment of disposal and beneficial use applications (Garrabrants et al., 2012a,b; US EPA, 2014a,b).

Waste acceptance criteria such as specified in the European Union landfill directive (EU LFD; Annex II, 2003) are based on test results from individual waste materials (EN12457-2, 2002; PrEN 14405, 2016). However, it is unclear whether these results have any significance with regard to the final behavior and release from a complete landfill where different kinds of wastes are landfilled together. Key release controlling factors are the individual waste dissolved organic carbon (DOC) levels and mobile inorganic (e.g., chloride and sulfate) and water-soluble organic contaminant levels. In the US, landfilling of waste is regulated primarily under the Resource Conservation and Recovery Act as well as additional federal and state regulations. Landfilling of materials is controlled based on origin, treatment standards and TCLP as a leaching toxicity test.

The development of acceptance criteria or other forms of decision thresholds based on leaching data needs to consider (1) the scenario resulting in leaching, including water contact mode, frequency, and quantity; (2) the COPC transport, including dilution and attenuation from the source (e.g., landfill) to the hypothetical receptor (human or ecological) or the regulatory point of compliance (e.g., property boundary); and (3) the human or ecological toxicity or standards associated with exposure to individual COPCs. Thus, leaching test results in conjunction with hydrologic information provide the basis for estimating COPC release from a landfill (i.e., a source term) that then must be evaluated in conjunction with local or regional COPC transport, fate, and exposure information or dilution and attenuation factors to be compared to decision criteria. This approach was taken in the development of EU landfill criteria (Hjelmar et al., 2001), Soil Protection Criteria in The Netherlands (SQD, 2007), and for evaluation of disposal of coal combustion residues in the US (US EPA, 2014a).

The long-term leaching behavior COPCs from a landfill is strongly affected by the macrochemistry (e.g., pH, redox, particulate, and DOC, primary constituents in the solid and liquid phases), which is currently largely ignored in monitoring of leachate quality. A major challenge is to develop means to predict the long-term leachate quality based on geochemical reactive transport modeling, taking into account the waste geochemistry as well as preferential flow aspects, which occur in landfills. The desired outcome is disposal practice implemented through more robust landfill waste acceptance and design criteria that will maintain contaminant release below acceptable levels during the long term. Assessing the long-term leaching behavior of a landfill is particularly relevant for determining aftercare needs.

LEACHING PROCESSES AND ENVIRONMENTAL LEACHING ASSESSMENT

Leaching is defined as the release of constituents from a solid material to the aqueous phase when contacted with water. Release to the aqueous phase can be determined by constituent liquid–solid partitioning (LSP) (including consideration of solubility, adsorption of the solid phase, available content for leaching, aqueous complexation, etc.), the physical properties of the material that limit mass transport, the degree to which equilibrium is achieved, and the properties of the contacting liquid. The solid materials of interest may be soils or sediments (with or without known contamination), wastes (from municipal, industrial, construction, or nuclear material processing), treated wastes or waste forms (e.g., cement-stabilized wastes, vitrified wastes, or products from a range of physical/chemical/thermal treatment processes), secondary materials under consideration for beneficial use (e.g., slags, flue gas desulfurization gypsum, coal fly ash), or construction materials. The contacting water may be from percolation through porous materials, flow around porous or nonporous (or fractured) monolithic materials, or from condensation processes. The material may be water saturated or unsaturated. The source and fate of the water (and any leached constituents) may include precipitation, runoff, groundwater, surface water, or collected leachate.

It is important to recognize that leaching of COPCs most often is strongly influenced by the leaching of other major and trace constituents in the material being evaluated and the constituents present initially in the contacting water, the general chemical state (e.g., pH, oxidation–reduction potential, and ionic strength) of the leachant in contact with the solid, and the physical characteristics of the material that impact water contact. All of the above factors influence the LSP of COPCs and the rate and extent to which equilibrium between the solid and liquid phases is approached.

For percolation scenarios the liquid-to-solid ratio (L/S in L/kg) can be related to a time scale based on the net infiltration rate, the density, and the height of the application/disposal (Hjelmar, 1990; Sanchez and Kosson, 2005). In covered landfill scenarios L/S = 1 L/kg already may cover a period of 1000 years (Van der Sloot et al., 1997; US EPA, 2014a).

The goal of environmental leaching assessment is to provide an estimate of constituent leaching potential for materials under possible management scenarios that is not only as accurate as practical or needed but also does not underestimate release of COPCs. The intended use of assessments may be to evaluate the environmental safety of specific management options for a class of materials (e.g., disposal scenarios for MSW), evaluate a specific disposal scenarios for a material (e.g., stabilized waste monofill, embankment, or structural fill), establish classes or performance characteristics of materials that may be acceptable for disposal in a given landfill type, compare the effectiveness of treatment processes for specific waste types (such as may be needed to ensure compliance with regulatory criteria), delisting of materials categorized as hazardous wastes based on the material's origin, and evaluate End-of-Waste[1] conditions for materials with potential for beneficial use or to determine remediation goals for contaminated soils or sediments. The constituents identified as COPCs are specific to the material being evaluated, with specific COPCs usually considered because of their inherent human or aquatic toxicity (e.g., arsenic, mercury).

STANDARDIZATION OF LEACHING TESTS

The release of soluble constituents on contact with water is regarded as a main mechanism of contaminant release, which results in a potential risk to the environment during the reuse or disposal of waste materials. However, all of the relevant aspects of leaching behavior cannot be addressed in one standard. In different CEN technical committees, leaching tests have been standardized to characterize release behavior of constituents, which can be leached from granular solid waste. This has been prompted by new requirements in relation to the environmental compatibility of materials for disposal, recycling, and utilization, which will be part of future European legislation. The European Landfill Directive (CEC, 1999) and the Annex with criteria (CEC, 2003) is one of the first EU rulings that require the use of leaching tests. In the US, leaching test methods for inorganic constituents have been standardized, undergone interlaboratory validation (Garrabrants et al., 2012a,b), and are included in the US EPA Test Methods for Evaluating Solid Waste (SW846).[2]

The leaching test methods developed in EU and US are presented in Table 20.1.1 by type and standard reference. Documentation supporting the development and use of the EU test methods is available in literature (van der Sloot, 2002; van der Sloot and van Zomeren, 2012; Hjelmar et al., 2012). The set of US EPA methods are referred to as the Leaching Environmental Assessment (LEAF) methods. The

[1]"End of Waste" is a terminology defined in European Waste Framework Directive 2008/98EC that specifies that once a material meets certain criteria, it is no longer considered a waste under waste management regulations.

[2]https://www.epa.gov/hw-sw846/validated-test-methods-recommended-waste-testing.

Table 20.1.1 Leaching test types corresponding across different fields and jurisdictions (European Union and United States)

	CEN/TC 345 Soil ISO/TC190 Soil	CEN/TC 292 Waste	CEN/TC 292 WG8	CEN/TC 351+60 Product TC's
	Matrix			
Test	Soil, Sediments, Compost, and Sludge	Waste	Mining Waste	Construction Products
pH dependence test	ISO/TS21268-4	EN14429	EN14429	EN14429[c]
		EN14497	EN14497	
	EPA 1313[a]	EPA 1313	EPA 1313	EPA 1313
Percolation test	ISO/TS21268-3	PrEN14405	PrEN14405	FprCENTS 16637-3
		NEN7373		NEN7373
	EPA 1314[a]	EPA 1314	EPA 1314	EPA 1314
Monolith test		EN15863		FprCENTS 16637-2
		NEN7375		NEN7375
	EPA 1315[a]	EPA 1315	EPA 1315	EPA 1315
Compacted granular test		NEN7347		FprCENTS 16637-2
	EPA 1315	EPA 1315	EPA 1315	EPA 1315
Redox capacity		CEN/TS 16660[b]		CEN/TS 16660[b]
Acid rock drainage			EN15875	
Reactive surfaces	ISO 12782 Parts 1-5	EN-ISO 12782 Parts 1–5		

CPR stands for EU Construction Products Regulation. The percolation test (PrEN 14405) was used as the source term for the criteria development in the EU Landfill Directive (Annex II).
[a]EPA methods included in SW846.
[b]CEN/TS16660 (2015) based on NEN 7348.
[c]Not yet adopted in CEN/TC 351 (very relevant for CPR).

characterization approach embodied in LEAF is fundamentally different than a simulation test-based approach because it focuses on measuring intrinsic material-specific leaching behaviors.

The primary parameters assessed using the LEAF tests control the release of COPCs from solid materials over a broad range of environmental conditions with applicability of the leaching data to specific disposal or use conditions (Kosson et al., 2002). The framework consists of four laboratory leaching methods, data management tools, and leaching assessment approaches (see also EN 12920, 2006). Coordinated development of LEAF has occurred between research laboratories in the US and the EU. The methods are designed to measure fundamental leaching parameters including the following:

- LSP as a function of eluate (endpoint) pH;
- LSP as a function of L/S under percolation (column flow) or batch extraction testing;
- Mass transfer rates for leaching from monolithic or compacted granular materials.
- Additional parameters for assessing special conditions, such as redox status, acid generation potential, and parameters for geochemical reaction and transport modeling.

Method 1313 and Method 1316 are parallel batch procedures intended to characterize the LSP at conditions approaching equilibrium as a function of final extract pH and L/S, respectively. Method 1314 and Method 1315 are test methods intended to measure the rate of constituent release under percolation or diffusive/dissolution mass transfer conditions, respectively (Box 20.1.1).

The test parameters and values specified in these methods have been described in a background information document on the LEAF leaching methods with fully validated methods available including comparative assessment with corresponding EU methods (Garrabrants et al., 2012a,b). Many of the EU methods have only minor deviations in test structure (e.g., the number of test fractions taken) or in test parameters (e.g., specified targets or time durations) from their LEAF counterparts. This means that the results obtained are equivalent as demonstrated in parallel validation (Garrabrants et al., 2012a,b).

Leaching tests are tools typically used to estimate the environmental impact associated with disposal or utilization of materials and wastes on the land (e.g., soils, sediments, industrial wastes, demolition debris, etc.). However, the results of leaching tests are often used without sufficient consideration of the type of data available or the applicability of the available data for the chosen disposal or utilization scenario. Furthermore, results of leaching assessments based on testing and interpretive models only provide a source term as one part of an evaluation of environmental safety. Additional factors in assessment or criteria development for use and disposal include (1) definition and application of appropriate fate and transport models and (2) establishment of risk-informed constituent concentration thresholds at defined points of compliance.

A frequently made false assumption is that the leaching test results represent the leachate that would occur in the field. No single leaching test can encompass the *range of conditions* that a material or waste may be subjected to over the duration of use or disposal. To address these limitations, a more robust science-based approach is needed for environmental leaching assessment that considers the impact on leaching of the chemical and physical characteristics of the tested materials and the range of

Box 20.1.1 Description of two different leaching test methods

pH dependence (EPA 1313, EN14429, ISO/TS21268-4)

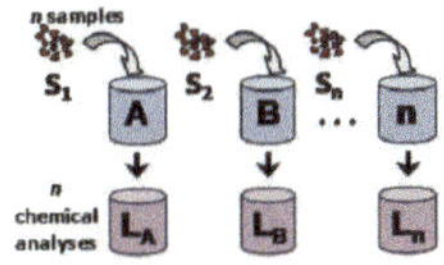

Equilibrium Leaching Test

- Parallel batch as function of pH

Test Specifications

- 8 specified target pH values plus natural condition
- Size-reduced material
- L/S = 10 mL/g-dry
- Dilute HNO_3 or KOH (NaOH)
- Contact time based on particle size
  - 18-72 hours
- Reported Data
  - Equivalents of acid/base added
  - Eluate pH and conductivity
  - Eluate constituent concentrations

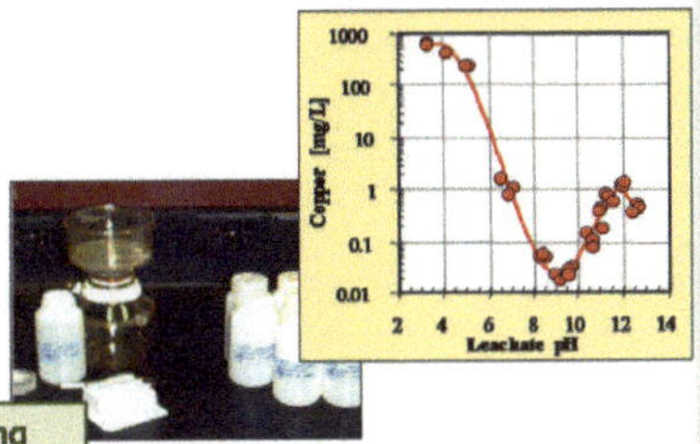

Titration Curve and Liquid-solid Partitioning (LSP) Curve as Function of Eluate pH

Up-flow column (EPA 1314, PrEN14405, ISO/TS21268-3)

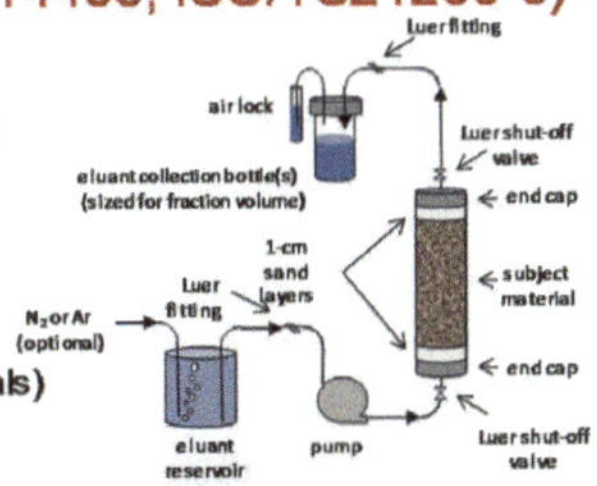

Equilibrium Leaching Test

- Percolation through loosely-packed material

Test Specifications

- 5-cm diameter x 30-cm high glass column
- Size-reduced material
- DI water or 1 mM $CaCl_2$ (clays, organic materials)
- Upward flow to minimize channeling
- Collect leachate at cumulative L/S
  - 0.2, 0.5, 1, 1.5, 2, 4.5, 5, 9.5, 10 mL/g-dry
- Reported Data
  - Eluate volume collected
  - Eluate pH and conductivity
  - Eluate constituent concentrations

Liquid-solid Partitioning (LSP) Curve as Function of L/S; Estimate of Pore Water Concentration

environmental conditions likely to be encountered during disposal and utilization. Even when used as a screening test, LEAF methods provide release estimates that are more accurate and reliable (i.e., less conservative or less of an overestimate) and robust (able to consider multiple or evolving physical–chemical conditions) than are obtainable using any single-point leaching test. Testing is considered to be more accurate because of the tailoring to the range of potential environmental conditions and intrinsic leaching characteristics of materials inherent in the design of LEAF.

The broad range of potential uses of environmental leaching assessment implies that there is a need for a graded or tiered approach that provides for flexible, scenario-based assessments and allows tailoring of the needed testing and information based on the type of intended use of the assessment and available prior or related information. The hierarchy in testing is presented for granular materials in Fig. 20.1.1. At the characterization level, the pH dependence test (Method 1313, SW846, 2009 or EN14429, 2015) and the percolation test (Method 1314, SW846, 2009 or PrEN 14405, 2016) provide already a broadly applicable assessment of a material's behavior under a variety of exposure conditions.

Materials from the same category (source) will generally provide the same basic leaching characteristic (illustrated here with the bandwidth of release behavior by a confidence interval), which allows it to

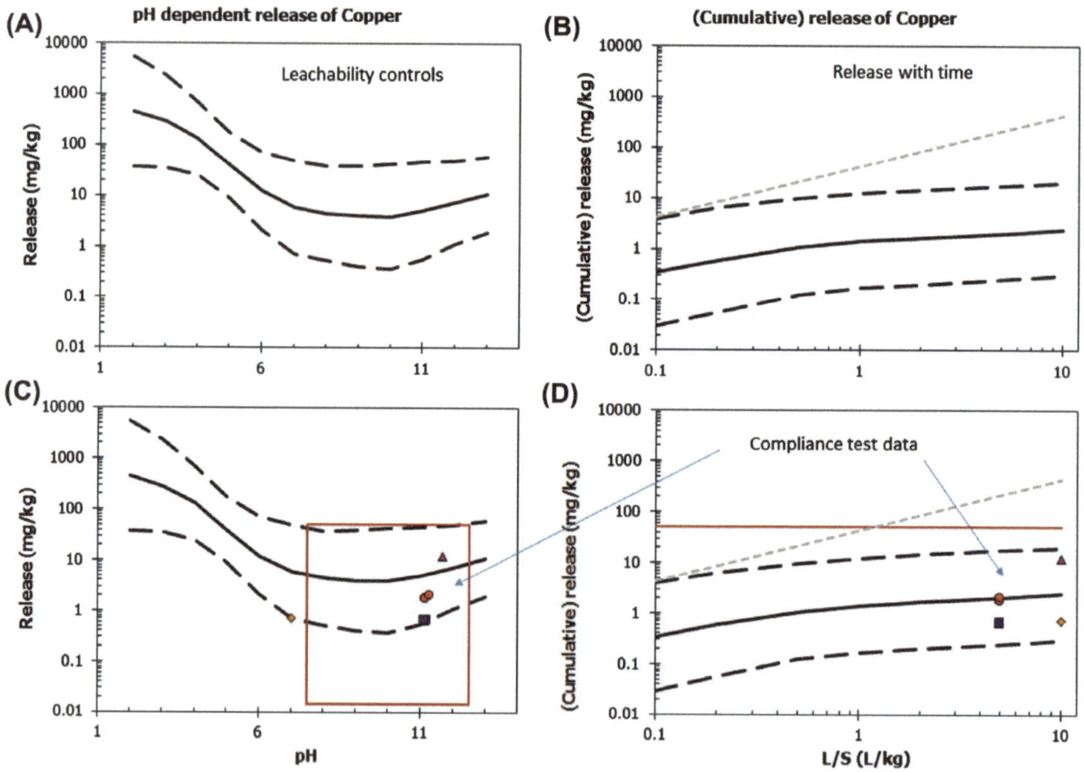

Figure 20.1.1 Leaching test relationships. Graphs (A) and (B) represent the characteristic release behavior of the material, whereas (C) and (D) show to what extent single-step test data (EN 12457-2, 2002) show compliance with the material characteristic. The confidence limits (*broken lines*) reflect the 95% confidence limits for the material as basis of reference for compliance testing (L/S = 10 batch test or cumulative column test results). The *dotted line* in graph (B) and (D) reflects solubility-limited release. The *red square* and *red line* reflect the relevant pH domain and the European Union Landfill Directive criteria. *L/S*, liquid-to-solid ratio.

be used as reference in compliance testing to decide whether one is still dealing with the same basic properties. Significant deviations from the basic character of the material would need to be verified by more extended testing as it is an indication of a change in material properties that affect the leaching behavior. The frequency of testing and the parameters to be tested depends on the degree of compliance with set regulations (e.g., the EU landfill Directive). A bonus-malus principle can be applied to stepwise decrease the level of testing to a low frequency, until an excursion occurs. Then the frequency is temporarily increased until compliance is again established.

Several physical–chemical factors such as preferential flow and reducing conditions can impact leaching and observed leachate conditions under field conditions but are not readily reproduced through routine laboratory testing (i.e., using standardized commercial test methods). As a consequence, while the LEAF test methods directly incorporate many of the important factors that impact leaching, additional factors such as reducing conditions and preferential flow are considered through chemical speciation and reactive mass transport simulations.

Developments in Leaching Relevant to Landfilling of Waste

A significant shift has occurred in landfill science in last decades (Landfill Symposium, Sardinia, 1997, 1999, 2001; 2003, 2005, 2007, 2009, and 2011). The emphasis on isolation technologies, which cannot guarantee long-term environmental compatibility, has shifted toward an understanding of processes in the landfill and subsequent control of these processes by acceptance of waste, separation of wastes, and pretreatment of waste prior to landfill. As has been already shown (Van der Sloot et al., 1997; Kosson et al., 2014), the leaching behavior of many elements is to a large extent solubility limited, which means that in terms of concentration there will be a relatively small difference between the result of a leaching test ($L/S = 10$) and the leachate composition ($L/S < 0.5$). In Fig. 20.1.2 this is illustrated for Zn. Solubility limiting phases for MSW are $ZnCO_3$, $Zn(OH)_2$, hydrozincite, hydrated ironoxide (HFO)-bound Zn, and dissolved organic carbon (DOC)-bound Zn. For PNW (predominantly nondegradable waste), these are willemite, calcium zincate, HFO-bound Zn, and DOC-bound Zn. Municipal solid waste incinerator (MSWI) bottom ash is the fully inorganic residue from MSW, which in terms of leaching corresponds well with PNW.

The overall composition of a landfill leads to integrated leaching chemistry, which often controls the leaching from individual wastes. For example, Fig. 20.1.3 illustrates Cu leaching using (1) EN 12457-2 from individual wastes, (2) the leaching of Cu in a 1:10 mixture of each specific waste with the overall waste mixture, and (3) pH dependence leaching of the overall waste mixture.

The overall waste mixture consists of all materials blended in the proportion each waste was delivered to the pilot cell of the PNW landfill (van Zomeren et al., 2005; Van der Sloot et al., 2016). In each 1:10 mixture, the pH is dominated by the overall PNW mixture. In addition, the leaching release may decrease by an order of magnitude due to the interaction with the overall mixture. Mixing effects are not addressed in current regulations that define waste acceptance criteria, but rather individual wastes are evaluated and then either accepted or rejected based on meeting the limit values for acceptance in a given landfill type. Obviously, when much larger proportions of a given waste with strong deviating

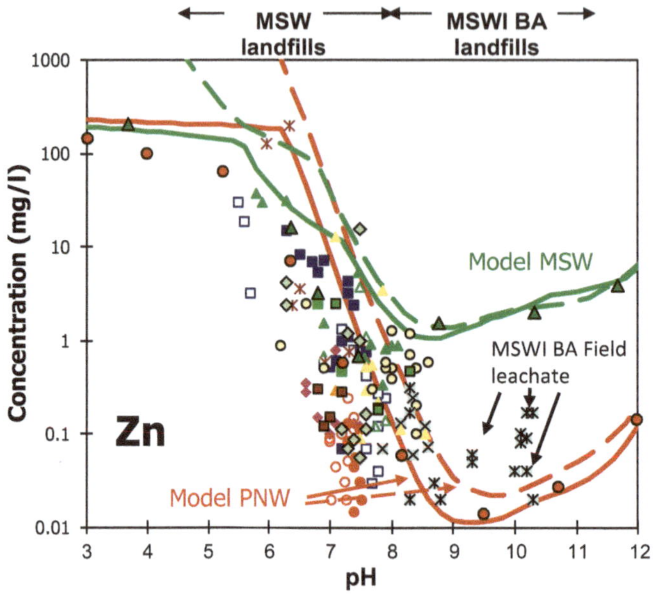

Figure 20.1.2 Leachate concentrations of Zn from a wide range of municipal and mixed waste landfills in comparison with leachate from a monofill of MSWI bottom ash (Meima, 1997), model results for MSW (measurements—*green triangles*) and for PNW (measurements—*red dots*) at L/S = 10 (*solid line*) and at L/S = 0.3 (*broken line*). *L/S*, liquid-to-solid ratio; *MSW*, municipal solid waste; *PNW*, predominantly nondegradable waste.

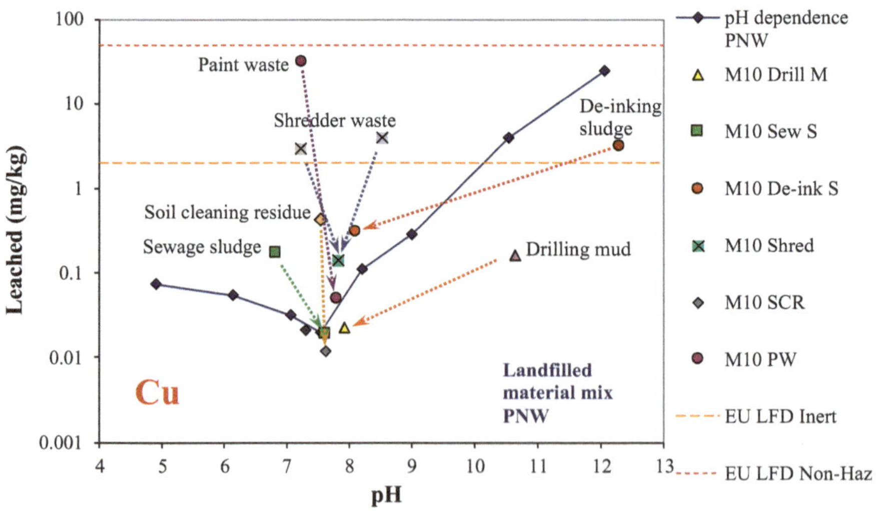

Figure 20.1.3 Comparison of leaching of Cu from individual wastes with 1:10 mixtures (indicated by dotted lines and arrows to illustrate change) of each waste with the overall predominantly nondegradable waste (PNW) mixture compared with the leaching behavior of PNW as a function of pH.

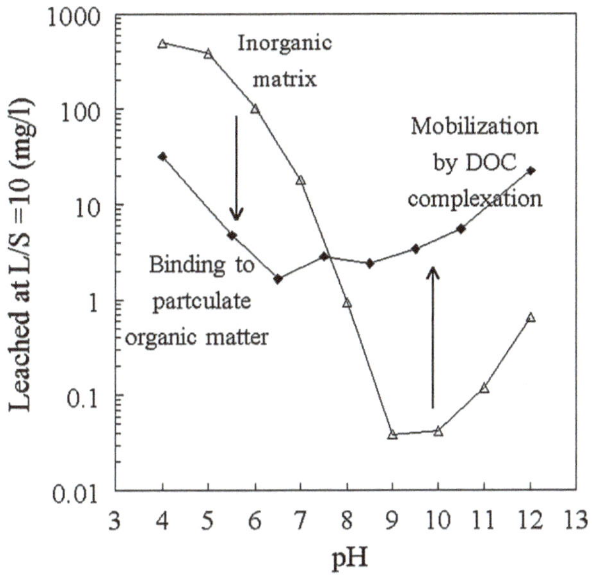

Figure 20.1.4 Role of organic matter in zinc binding and mobilization. *DOC*, dissolved organic carbon.

conditions are delivered the conditions of the entire cell and leachate characteristics may change more significantly.

Organic matter, which provides both binding sites and generates DOC through abiotic and biotic degradations, has a significant effect on metal mobilization, even under reducing conditions (Van der Sloot et al., 2016). In Fig. 20.1.4, this is schematically illustrated.

The binding and the mobilization is illustrated by the pH dependence leaching curve for Zn in bioreactor residue (Fig. 20.1.4). In the low pH domain, binding to particulate organic matter is dominant, whereas at pH > 7 at mobilization through complexation with DOC is important. In contrast, the curve for MSWI bottom ash reflects the leaching behavior of largely inorganic Zn. These types of leaching behavior have been observed in sewage sludge, compost, natural soil, and contaminated soil for all metals. Depending on the level of contaminants present and the amount of particulate organic carbon and DOC, the binding on the low pH end will be more or less pronounced as sorption sites may become saturated.

Although by far not as much work has been done on organic COPCs as on inorganic COPCs, organic matter also plays a major role as binding medium and mobilizing medium for low solubility organic COPCs, such as polyaromatic hydrocarbons (PAH) and polychlorinated biphenyl (PCB) (Grathwohl and Susset, 2009).

Chemical Speciation of Various Wastes

In the context of an EU-DGXII M&T project, a wide range of materials have been subjected to the characterization leaching tests (Van der Sloot et al., 1997, 2008; Van der Sloot, 2002). The range of materials covered is very broad and covers the spectrum of materials that in one form or another may end up in a landfill. The information gained from this project is directly relevant for the understanding of the behavior of materials mixed together in landfills. Generic properties such as acid–base neutralization capacity obtained from the pH dependence test, salt loading, and potential for DOC generation are very important in this respect. In this context it must be realized that one material with extreme properties can have a detrimental impact on a large amount of otherwise relatively harmless material.

A chemical speciation–based model of liquid–solid equilibrium partitioning as a function of pH can be developed based on pH-dependent leaching test results and previously reported speciation information (Van der Sloot et al., 2016). This has been accomplished previously for several cases using LeachXS, which includes the ORCHESTRA modeling framework (Meeussen, 2003; Kosson et al., 2014; Van der Sloot et al., 2016). Aqueous speciation reactions and selected mineral precipitates were taken from the MINTEQA2 database. Ion adsorption onto organic matter was calculated with the NICA-Donnan model (Kinniburgh et al., 1999), with the generic adsorption reactions for all elements as published by Milne et al. (Milne et al., 2001, 2003). Adsorption of ions onto iron and aluminum oxides was modeled according to the generalized two-layer model of Dzombak and Morel (1990).

A comparison of Cu leaching between a wide range of materials studied is given in Fig. 20.1.5.

In the graphs, the modeled partitioning between mineral phases, hydrated iron oxide sorption, and distribution between dissolved and particulate matter is provided. The measured data points of the pH dependence test are inserted for reference (red dots). The dark blue line between the light green (DOC) and dark green (particulate organic matter, POM) fields reflect the dissolved Cu (free and DOC-bound Cu) obtained from geochemical modeling (van der Sloot et al., 2007). For these simulations, the geochemical speciation model was calibrated simultaneously for a large number of elements (e.g., >20) based on a range of pH dependence test results and related literature. The following materials were studied: crushed concrete, MSWI fly ash, coal fly ash, green compost, MSW compost, drinking water sludge, sulphidic mining waste, PNW (mix), contaminated soil, and thermally treated soil. Many of these materials have in one form or another been landfilled.

The square denotes the most common pH domain for the material considered (vertical separation); the lower line of the box indicates analytical detection and the upper bound reflects the regulatory limit for disposal in a nonhazardous waste landfill (EU landfill Directive, 2002).

In the figures, the role of DOC has been shown to illustrate its relevance for Cu LSP. The partitioning is based on chemical speciation modeling (Dijkstra et al., 2004, 2006a,b) using LeachXS-Orchestra (Van der Sloot et al., 2008; Meeussen, 2003). The modeling parameters needed for sorptive sites (hydrated iron oxide, particulate and dissolved organic matter) can be quantified using protocols standardized in ISO/TC190 (ISO/DIS12782, 2010).

For compost, soil, contaminated soil, sewage sludge, and sediments materials with a high and low contamination levels have been selected. For the materials containing significant amounts of organic matter (compost, sewage sludge, soil, contaminated soil), high mobilization is noted at $pH > 8$. This is attributed to Cu association with DOC, for which it is known to have a high affinity. As

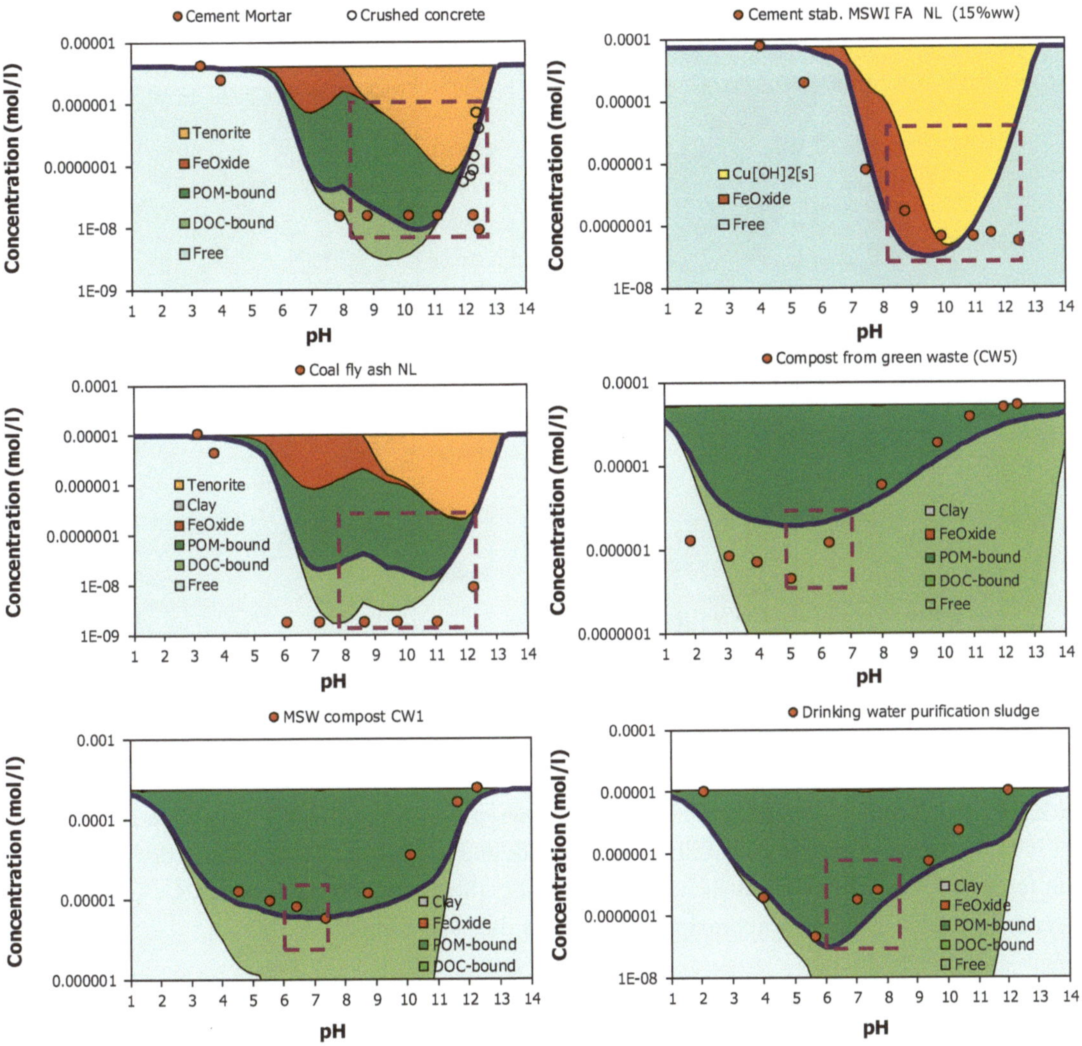

Figure 20.1.5 Leaching behavior of Cu from different materials using the pH dependence test. The box (*dashed lines*) indicates the most relevant pH domain for the respective materials. *MSW*, municipal solid waste; *MSWI*, municipal solid waste incinerator.

pH increases, the mobilization of DOC increases (van Zomeren and Comans, 2007). In construction products the leachability of Cu is generally low; this, therefore, also applies to demolition debris. From these observations similarities and common characteristics of element and material behavior can be identified. Similar comparisons are available for other elements as well. Solubility and sorption control observed leaching, and thus in landfills these same factors will be of importance. A key factor that controls solubility and sorption is pH. Because leachability may vary over orders of magnitude with highest

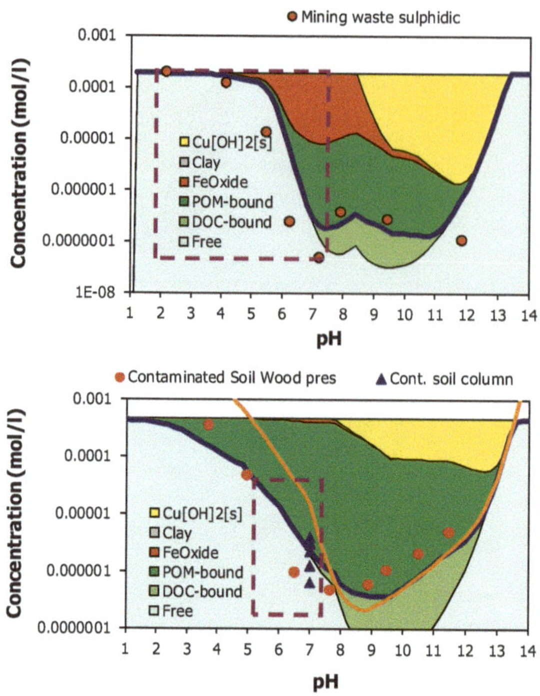

Figure 20.1.5 cont'd

leachability for metals at low pH and highest leachability for oxyanions at mildly alkaline pH, a neutral pH is an optimal condition for landfill for both metals and oxyanions. Total composition is only relevant for truly soluble components such as Na, Cl, and highly water-soluble organic COPCs, where the measured available content is approximately equal to the total content.

CONCLUSIONS

New developments in waste management call for new approaches in assessing the release characteristics of new formulations, treatment products, and new materials to the environment. The new approaches in assessing release in their turn provide new options for waste management. In landfilling of waste soluble salts (e.g., derived from flue gas cleaning wastes) and mobile water-soluble organics need to be controlled more strictly as their impact will be sure to take place over time. Dealing with a limited number of wastes enriched in these components prior to landfilling is easier than solving matters in leachate treatment at a full-scale facility. For both metals and oxyanions, a neutral pH is the optimal condition for retention of these substances in most scenarios, except in case of a monofill of stabilized waste. DOC mobilizes a variety of components: several major elements, metals, and organic COPCs. This process is already important at neutral pH. Therefore, minimization of DOC generation will directly reduce leachate concentrations of a wide range of constituents. Joint disposal of alkaline waste with waste containing water-insoluble organic COPCs should be avoided as DOC is mobilized by hydrolysis and thus carries otherwise immobile organic COPCs into the leachate.

In general there is a good agreement between the leaching behavior of the constituted waste mixture and the leachate as obtained from the full-scale pilot experiment. This indicates that in spite of an apparently very heterogeneous mix of materials, leaching is governed by well-defined solubility controls for many constituents. The key aspect is to identify which waste streams or external stresses can disrupt this balance to such an extent as to result in unacceptable leaching conditions.

Key controlling factors for the PNW disposal are controls on the individual waste DOC levels (as shown with geochemical modeling), mobile inorganic (e.g., Cl, sulfate), and water-soluble organic contaminant levels. The behavior of organic COPCs shows a very consistent behavior as well, in a similar data representation as used for inorganic constituents. In view of excessive DOC and colloid mobilization, a percolation test would seem more appropriate to assess organic contaminant release from a landfill than a single-step batch leaching test ($L/S = 10$).

The main differences in release between the laboratory, lysimeters, and the pilot experiment are caused by the different redox potentials measured at the different scales of testing. The different redox potentials lead to relatively large deviations in the release of Cu, Fe, and Mn on the different test scales.

Geochemical release modeling on column leaching test data showed that the model generally simulates the release of Ca, SO_4^{2-}, Mg, and K, rather well. Cu, Cr, and Pb show the right trend, but the simulated concentrations are higher than those measured. In case of Cu and Cr this is most likely linked to the predicted level of DOC (or dissolved humic acid, DHA; DHA is a fraction of DOC). This is still an uncertain factor and needs further work. This implies that the main chemical processes leading to release are understood, but improvement on the level is still needed. The results indicate that preferential flow aspects are important for the long-term prediction of the release of mobile elements from a landfill. The model simulates the release of soluble salts well. This modeling approach is comprehensive in that it deals with all major, minor, and trace elements simultaneously and their partitioning between dissolved (free and DOC bound) and solid phases (minerals, Fe-oxide, Al-oxide, and particulate organic matter). However, a few COPCs show deviations in concentration of over one order of magnitude and these deviations need to be sorted out in future research work.

Impact evaluation based on criteria derived from impact assessment is feasible; however, the assessment of individual wastes going to landfill does not consider the mutual waste interaction, as individual wastes may exceed criteria, but for the ultimate waste mix in a landfill the release behavior seems much more consistent than one would expect based on the heterogeneity of the wastes delivered to a landfill. This aspect was clearly identified in excavation of an MSW bioreactor pilot (Van der Sloot et al., 2016).

In view of the observed relationship of the controlling leaching chemistry imposed by the overall waste mixture, a good estimate of the release behavior of a landfill (cell) can be obtained by applying the pH dependence test and the percolation test on a composite sample from mixing core samples, while keeping individual samples for batch testing at own pH (endpoint pH for extraction with deionized water at L/S 10 mL/g) to get an impression of the spatial and depth variation. The integrated information on leaching as a function of pH and L/S (which is related to time) provides insight into the degree of degradation, the presence of hot spots, and the extent of solubility control for COPCs. Comparing the release over the applicable range of pH and L/S with limit values as set, for instance, in the EU Landfill Directive allows an evaluation on the performance of the landfill body as a whole. This can be particularly relevant in decisions on releasing a site from aftercare.

References

CEC, 1999. Council Directive 1999/31/EC of 26 April 1999 on the landfill of waste. Official Journal of the European Communities.

CEC (2003). Council Decision 2003/33/EC of 19 December 2002 establishing criteria and procedures for the acceptance of waste at landfills pursuant to Article 16 of and Annex II to Directive 199/31/EC.

CEN/TS16660, 2015. Characterization of Waste — Leaching Behaviour Test — Determination of the Reducing Character and the Reducing Capacity. CEN, Brussels.

Dijkstra, J.J., Meeussen, J.C.L., Comans, R.N.J., 2004. Leaching of heavy metals from contaminated soils: an experimental and modelling study. Environmental Science & Technology 38, 4390—4395.

Dijkstra, J.J., Van der Sloot, H.A., Comans, R.N.J., 2006a. The leaching of major and trace elements from MSWI bottom ash as a function of pH and time. Applied Geochemistry 21, 335—351.

Dijkstra, J.J., Van der Sloot, H.A., Comans, R.N.J., 2006b. A consistent geochemical modeling approach for the leaching and reactive transport of major and trace elements in MSWI bottom ash. Applied Geochemistry 23 (6), 1544—1562.

Dzombak, D.A., Morel, F.M.M., 1990. Surface Complexation Modelling: Hydrous Ferric Oxide. John Wiley & Sons, Inc., New York.

EN 14429, 2015. Characterization of Waste – Leaching Behaviour Tests – Influence of pH on Leaching with Initial Acid/Base Addition. CEN/TC292. CEN, Brussels.

EN-12457-2, 2002. Characterisation of Waste – Leaching – Compliance Test for Leaching of Granular Waste Materials and Sludges – Part 2: One Stage Batch Test at a Liquid to Solid Ratio of 10 l/kg for Materials with Particle Size below 4 mm (Without or with Size Reduction). CEN, Brussels.

EN 12920, 2006. Characterisation of Waste – Methodology Guideline for the Determination of the Leaching Behaviour of Waste under Specified Conditions. CEN, Brussels.

EU Landfill Directive, 2002. Annex II of Council Directive 1999/31/EC of 26 April 1999 on the landfill of waste. Official Journal of the European Communities.

Garrabrants, A.C., Kosson, D.S., Stefanski, L., DeLapp, R., Seignette, P.F.A.B., van der Sloot, H.A., Kariher, P., Baldwin, M., 2012a. Interlaboratory Validation of the Leaching Environmental Assessment (LEAF) Method 1313 and Method 1316, EPA-600/R-12/623. US Environmental Protection Agency, Washington, DC.

Garrabrants, A.C., Kosson, D.S., DeLapp, R., Kariher, P., Seignette, P.F.A.B., van der Sloot, H.A., Stefanski, L., Baldwin, M., 2012b. Interlaboratory Validation of the Leaching Environmental Assessment (LEAF) Method 1314 and Method 1315, EPA-600/R-12/624. US Environmental Protection Agency, Washington, DC.

Grathwohl, P., Susset, B., 2009. Comparison of percolation to batch and sequential leaching tests: theory and data. Waste Management 29 (10), 2681—2688.

Hjelmar, O., 1990. Leachate from land disposal of coal fly ash. Waste Management 8, 429—446.

Hjelmar, O., van der Sloot, H.A., Guyonnet, D., Rietra, R.P.J.J., Brun, A., Hall, D., 2001. Development of acceptance criteria for landfilling of waste: an approach based on impact modelling and scenario calculations. In: CISA – Environmental Sanitary Engineering Centre, Cagliari, Italy, S. Margherita di Pula, pp. 711—721.

Hjelmar, O., Hyks, J., Wahlström, M., Laine-Ylijoki, J., van Zomeren, A., Comans, R.N.J., Kalbe, U., Schoknecht, U., Krüger, O., Grathwohl, P., Wendel, T., Abdelghafour, M., Méhu, J., Schiopu, N., Lupsea, M., 2012. Robustness Validation of TS-2 and TS-3 Developed by CEN/TC351/WG1 to Assess Release from Products to Soil, Surface Water and Groundwater. NEN, Delft, The Netherlands.

ISO/DIS 12782 parts 1 – 5, 2010. Soil quality – parameters for geochemical modelling of leaching and speciation of constituents in soils and soil materials – Part 1: extraction of amorphous iron (hydr)oxides with ascorbic acid; Part 2: extraction of crystalline iron (hydr)oxides with dithionite; Part 3: extraction of aluminium (hydr)oxides with ammonium oxalate – oxalic acid; Part 4: extraction of humic substances from solid samples. Part 5: extraction of humic substances from aqueous samples, ISO, Geneve, Switzerland.

Kinniburgh, D.G., van Riemsdijk, W.H., Koopal, L.K., Borkovec, M., Benedetti, M.F., Avena, M.J., 1999. Ion binding to natural organic matter: competition, heterogeneity, stoichiometry and thermodynamic consistency. Colloids and Surfaces A: Physicochemical and Engineering Aspects 151, 147–166.
Kosson, D.S., van der Sloot, H.A., Sanchez, F., Garrabrants, A.C., 2002. An integrated framework for evaluating leaching in waste management and utilization of secondary materials. Environmental Engineering Science 19 (3), 159–204.
Kosson, D.S., van der Sloot, H.A., Garrabrants, A.C., Seignette, P.F.A.B., 2014. Leaching Test Relationships, Laboratory-to-Field Comparisons and Recommendations for Leaching Evaluation Using the Leaching Environmental Assessment Framework (LEAF). EPA-600/R-14/061. U.S. Environmental Protection Agency, Washington, DC.
Mathlener, R.A., Heimovaara, T., Oonk, H., Luning, L., van der Sloot, H.A., van Zomeren, A., 2006. Opening the Black Box, Process-Based Design Criteria to Eliminate Aftercare of Landfills. Dutch Sustainable Landfill Foundation, Den Bosch, The Netherlands, pp. 1–78. ISBN-10: 90-73573-30-0.
Meeussen, J.C.L., 2003. ORCHESTRA: an object-oriented framework for implementing chemical equilibrium models. Environmental Science & Technology 37, 1175–1182.
Meima, J.A., 1997. Geochemical Modelling and Identification of Leaching Processes in MSWI Bottom Ash. Thesis University of Utrecht.
Milne, C.J., Kinniburgh, D.G., Tipping, E., 2001. Generic NICA-Donnan model parameters for proton binding by humic substances. Environmental Science & Technology 35, 2049–2059.
Milne, C.J., Kinniburgh, D.G., van Riemsdijk, W.H., Tipping, E., 2003. Generic NICA-Donnan model parameters for metal-ion binding by humic substances. Environmental Science & Technology 37, 958–971.
PrEN 14405, 2016. Characterization of Waste – Leaching Behaviour Tests – Up-Flow Percolation Test (Under Specified Conditions). CEN/TC292. CEN, Brussels.
Sanchez, F., Kosson, D.S., 2005. Probabilistic approach for estimating the release of contaminants under field management scenarios. Waste Management 25, 463–472.
Sardinia Symposia Proceedings, International Landfill Symposium, S. Margherita di Pula, Cagliari, 1997, 1999, 2001, 2003, 2005, 2007, 2009, 2011.
SQD, December 20, 2007. Soil quality decree. Staatscourant (247), 67–90.
TCLP – EPA Test Method 1311-Toxicity Characteristic Leaching Procedure, SW846.
U.S. EPA, 2014a. Human and Ecological Risk Assessment of Coal Combustion Residuals. U.S. Environmental Protection Agency, Office of Solid Waste and Emergency Response, Office of Resource Conservation and Recovery, Washington, DC.
U.S. EPA, 2014b. Coal Combustion Residual Beneficial Use Evaluation: Fly Ash Concrete and FGD Gypsum Wallboard. U.S. Environmental Protection Agency, Office of Solid Waste and Emergency Response, Office of Resource Conservation and Recovery, Washington, DC.
Van der Sloot, H.A., 2002. Harmonisation of leaching/extraction procedures for sludge, compost, soil and sediment analyses. In: Quevauviller, P. (Ed.), Methodologies for Soil and Sediment Fractionation Studies. Royal Society of Chemistry, pp. 142–170.
Van der Sloot, H.A., Seignettte, P.F.A.B., Meeussen, J.C.L., Hjelmar, O., Kosson, D.S., 2008. A database, speciation modelling and decision support tool for soil, sludge, sediments, wastes and construction products: LeachXS™- Orchestra. In: Second International Symposium on Energy from Biomass and Waste, Venice, November 17–20, 2008.
Van der Sloot, H.A., Heasman, L., Quevauviller, P., 1997. Harmonization of Leaching/Extraction Tests. In: Studies in Environmental Science, vol. 70. Elsevier Science, Amsterdam, 292 pp.
Van der Sloot, H.A., van, Zomeren, 2012. Characterisation Leaching Tests and Associated Geochemical Speciation Modelling to Assess Long Term Release Behaviour from Extractive Wastes, Mine Water and the Environment. Mine Water Environ 31, 92–103.
van der Sloot, H.A., van Zomeren, A., Dijkstra, J.J., Meeussen, J.C.L., Comans, R.N.J., Scharff, H., 2007. Prediction of the leaching behaviour of waste mixtures by chemical speciation modelling. In: Haarstrick, A., Reichel, T. (Eds.), Landfill Modelling, CISA – Environmental Sanitary Engineering Centre, Cagliari, Italy.
Van der Sloot, H.A., Kosson, D.S., van Zomeren, A., 2017. Leaching, geochemical modelling and field verification of a municipal solid waste and a predominantly non-degradable waste landfill. Waste Management 63, 24–95.
van Zomeren, A., van der Sloot, H.A., Meeussen, J.C.L., Jacobs, J., Scharff, H., 2005. Prediction of the long-term leaching behaviour of a sustainable landfill containing predominantly inorganic waste. In: Proceedings Sardinia 2005, Tenth International Waste Management and Landfill Symposium S. Margherita di Pula, Cagliari, Italy; October 3–7, 2005. © 2005 by CISA, Environmental Sanitary Engineering Centre, Italy.
van Zomeren, A., Comans, R.N.J., 2007. Measurement of humic and fulvic acid concentrations and dissolution properties by a rapid batch procedure. Environmental Science & Technology 41, 6755–6761.

20.2

INCINERATOR RESIDUE MONOFILLS

Ole Hjelmar

INTRODUCTION

The various types of residues produced by municipal solid waste incinerators (MSWIs) and their main characteristics and environmental properties have been presented and discussed in Chapter 4.2. That chapter also discusses various treatment techniques that can be applied to recover valuable materials [e.g., ferromagnetic and nonmagnetic metals for bottom ash (BA)] and to improve the properties of the residues with respect to utilization or, when that is not possible, final disposal (landfilling). Finally, Chapter 4.2 also discusses some disposal strategies for incinerator residues and some of the challenges involved in optimizing the sustainability of incinerator residue management.

This chapter picks up on the discussion of landfill strategies for MSWI residue landfills with particular focus on the duration of the aftercare period, i.e., the period after closure of the landfill during which operation and maintenance of active environmental protection systems and control measures will still be required to ensure adequate environmental protection until final storage quality (FSQ) is attained (see Chapter 4.2). Since this period has often been underestimated both in regulation and in practise, the consequences of extended aftercare periods are of great concern to both landfill operators and regulators (see e.g., Beaven et al., 2014). Modern landfills are designed and constructed in a way that emissions can be avoided to a large extent during the operational phase and the aftercare period (van Zomeren and Hjelmar, 2015). However, in the longer term, continued release and potential emission of harmful substances with the leachate is generally considered the main environmental issue, whereas any gas production within the landfill is likely to have long subsided. In addition, with the rather common practice of landfill capping, the emission potential of the landfill may, if the capping is effective, remains largely unchanged after closure, and as a result potential problems with future emissions will be shifted to coming generations (and in principle "eternal" aftercare will be required). For this reason, it is important to find methods to assess the duration of the aftercare period under various conditions and to shorten this period by applying different techniques aiming to reduce the emission potential (e.g., leachate recirculation, flushing and/or aeration). For MSWI residue landfills, this implies that the source term will decrease as leaching progresses. Any assessment of the duration of aftercare period (and of whether or not FSQ has been attained) will therefore have to be site-specific. To allow such an assessment and to enable the development of strategies to reduce the emissions with time and possibly accelerate the process, it is necessary to have a reliable description of the source term as a function of time.

Unfortunately, very few useful and extended monitoring (time series) data on the corresponding composition and quantity of leachate from MSWI residue monofills are available. This is primarily due to a general lack of focus on the problems associated with aftercare duration in the past and a lack of understanding of the required quality of the data needed to address these issues. In addition, a certain degree of aversion to exposure and genuine inaccessibility of data probably also play a role. This chapter presents some of the few data that are available and further discusses some of the challenges involved in developing appropriate source terms for risk assessments at MSWI monofills.

LANDFILL OPTIONS AND PRACTICES FOR MSWI RESIDUES

Within the European Union (EU), fly ash (FA), and air pollution control (APC) residues (here meaning acid gas cleaning residues, often mixed with FA) from MSWIs are considered hazardous waste (CEC, 2000, 2014) and must therefore, if landfilled, since 2004 (CEC, 1999, 2003) be placed in landfills designated to receive hazardous waste. They normally need to be treated to meet the EU waste acceptance criteria for hazardous waste landfills (CEC, 2003). It is actually a requirement within the EU (CEC, 2003) that all waste, both hazardous and nonhazardous, must be treated prior to landfilling, although "treatment" in general has been interpreted rather flexibly.

MSWI BA is, on the other hand, within the EU mostly classified as nonhazardous waste and should therefore, if landfilled, be placed in landfills for nonhazardous waste (CEC, 1999, 2000, 2003, 2014). In some countries, MSWI BA is used as a covering material in landfills for other waste. The minor MSWI residue streams (grate siftings, boiler ash, and economizer ash) are generally landfilled (or utilized) alongside or mixed in with the BA or the FA/APC residues.

In most European countries, APC residues/FA and BA have been collected and landfilled separately, at least after 2000, whereas it has been customary to mix and landfill APC residues and BA together in the same cells in other parts of the world (e.g., the United States of America and Singapore). In Europe, the APC residues from the dry and semidry processes and FA are often placed in monofills after treatment. A common treatment prior to landfilling is stabilizing the APC residues with pozzolans (e.g., cement) and various additives to a relatively dry paste, which is then placed in a monofill and allowed to settle and harden. Although removal of the large content of calcium chloride and other soluble salts by washing prior to stabilization would probably save cement and lead to a product with better environmental properties (Hjelmar, 2014), this is generally not done. Residues from wet scrubbers (sludge from treatment of the scrubbing water) have often been mixed with FA prior to landfilling. In Northern Europe, large amounts of APC residues have been either deposited underground in old German salt mines, underground limestone mines in the United Kingdom, or they have been shipped to the Norwegian island of Langøya, where they have been deposited in a large limestone quarry after being neutralized by mixing with waste sulfuric acid from a nearby titanium oxide production plant (dry and semidry APC residues are normally alkaline due to the content of excess lime). Although there are a number of strategic long-term advantages associated with placing the MSWI BA in monofills when it cannot be otherwise utilized (e.g., Hjelmar, 1996), landfills or landfill cells containing only MSWI BA appear to be relatively rare, even in Europe.

LANDFILL STRATEGY AND RISK ASSESSMENT

For MSWI residue monofills, the most appropriate strategy to achieve FSQ within an acceptable time-frame is likely to be containment and collection of leachate (possibly supplemented by recirculation of leachate or addition of water to the surface) followed by controlled (restricted) contaminant release (Hjelmar, 1996). The first part requires active environmental protection systems, whereas the latter may require only passive systems. Table 20.2.1 shows three different definitions of FSQ, and the adoption of the controlled contaminant release strategy as the endpoint will require acceptance of definition C in the table. Most likely, only a strategy based on definition C in the table can ensure that FSQ is attained within a foreseeable future.

Any assessment at any given time of whether or not a given MSWI residue landfill (monofill) meets the requirements of FSQ or, if it does not, any estimate of the time remaining until it may meet the requirements will obviously have to be site-specific and accounts both for conditions related to the waste, the leachate, the landfill, and to the surrounding environment and its vulnerability, see Fig. 20.2.1.

The assessment to determine if the impact on downstream groundwater or surface water at a certain distance downstream of the landfill (often referred to as the "point of compliance") is likely to remain

Table 20.2.1 Different definitions of "final storage quality" (where the aftercare of a landfill can be safely stopped) in terms of quality and quantity of leachate.

Type	Character of the Contamination Potential	Potential Release of Contaminating Substances to Soil and Water	Necessary Stability of Passive Environmental Protection Systems
A	Total and remaining equilibrium between potentially contaminating substances in the landfilled waste and outside the landfill (same concentration in the leachate and the groundwater)	No gradient for transport of contaminating substances between the landfill and the surroundings	Not important
B	The level of concentration of potentially contaminating substances in the leachate is (and will remain) acceptable in the surroundings	Release/seepage of leachate to the surrounding groundwater or surface water will not lead to unacceptable impacts, regardless of the amount of leachate produced	Less important
C	The flux of potentially contaminating substances is (and will remain) acceptable in the surroundings	Release/seepage of leachate to the surrounding groundwater or surface water will not lead to unacceptable impacts if the flux does not exceed a given (site-specific) value	Long-term stability of passive measures to reduce infiltration may be required

Adopted from Hjelmar et al. (2013).

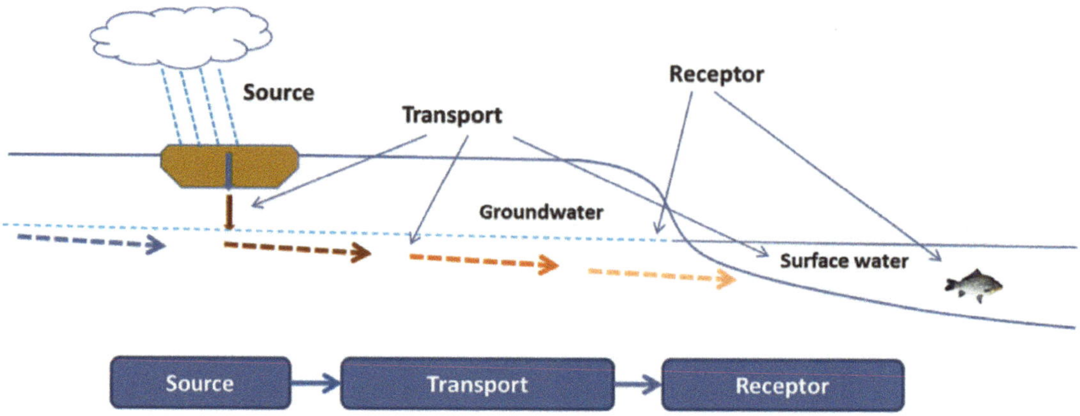

Figure 20.2.1 Cross-section illustrating the site-specific impact assessment required for a landfill to determine if and possibly when the aftercare may be discontinued. *After Hjelmar et al. (2015).*

acceptable in terms of meeting water quality standards requires a reliable source term (as a function of time) and reliable (mathematical/numeric) models to describe the transport of the substances emitted by the source term through the unsaturated zone, the aquifer and, if relevant, the surface water body to the receptor, taking into account the attenuation and dilution occurring on the way. Both the source term and the transport models require site-specific input. The methodology is further described, e.g., in Hjelmar (2012) and Hjelmar et al. (2013). When carrying out this assessment, it is important also to assess the source term situation occurring after collection and removal of leachate has been stopped, and a possible new hydraulic equilibrium has been established, particularly if this gives rise to a "bathtub effect" (see e.g., Hjelmar and Henriksen, 2015).

Whereas the transport modeling and the subsequent assessment of the impact at the receptor to a large extent is "state-of-the-art"—provided sufficient information on the geology and hydrogeology of the area surrounding the monofill is available or can be procured, and groundwater and/or surface water quality criteria relevant for the downstream receptor exist—the source term—and particularly its development over time—is in many respects still partly uncharted territory full of challenges. Often a substantial amount of the information required for the transport calculations is obtained already during the planning and permission application stage for the monofill, and in most cases it can be supplemented throughout the lifetime of site.

THE SOURCE TERM

For a given MSWI residue monofill, the total amount of a given substance, i, released or removed with the leachate per unit time, i.e., the flux, $M_i(t)$, as a function of time may be expressed as:

$$M_i(t) = C_i(t) \times Q(t),$$

where C_i (t) is the concentration of substance i in the leachate as a function of time, and Q(t) is the amount of leachate produced per unit time as a function of time.

The leachate production rate, Q(t), depends on a number of factors such as climate, top cover, landfill design, and operation, and may change with time, particularly when a landfill passes from the operation phase to the aftercare phase (and over a longer period of time with changes in the climate), but for MSWI monofills situated above groundwater and with no contact to surface water under steady state conditions (in hydraulic equilibrium), the mean value of Q(t) will depend mainly on the average local rate of infiltration of precipitation, which can be expected to be fairly constant over longer periods of time for the entire landfill, although the internal flow of infiltrated precipitation through the landfilled residues can be rather unevenly distributed. Depending on the information available, Q(t) can be estimated and predicted on the basis of leachate removal monitoring data or water balance calculations. To calculate the accumulated liquid to solid ratio, L/S, which may be used in the prediction of C(t), see below, both L and S should be known or estimated. L = Q(t) accumulated from 0 to t then represents the total amount of leachate produced from a given cell/site at a given time, t, and S is the total dry mass of MSWI residues in the cell/site. The relationship between the time since the occurrence of the first leachate, t (e.g., in years), and L/S (m^3/t or L/kg) may be described by the following equation (Hjelmar, 1990):

$$t = (L/S) \times d \times H/I$$

where d is the average dry bulk density of the landfilled MSWI residues (t/m^3), H is the average height of the monofill (m), and I is the average annual net rate of infiltration of precipitation into the monofill (m^3/m^2/year).

The leachate concentration of a given substance, i, as a function of time, $C_i(t)$, or as a function of L/S, $C_i(L/S)$, is generally not readily available for a given MSWI residue monofill. For a given substance and a given monofill, it depends on the chemical and physical properties of the landfilled residues, in particular on the leaching properties, and on the hydrogeochemical conditions (pH, redox potential, contact with CO_2 or O_2) within the landfilled residues (and their change over time as a result of internal processes or external impacts). Most MSWI residues are highly alkaline when landfilled, and the external influences which they may be subjected to after landfilling will invariably lower the pH in contact with water over time (sometimes a very long time). It is generally assumed that the concentrations of most/all substances released will decrease as a function of time or L/S due to depletion, but some substances, particularly some of those that are solubility controlled, may initially show increasing concentrations before they start to decrease.

Data that may contribute to the description of the composition of the leachate from an MSWI residue monofill as a function of L/S or time, i.e., C(t), roughly fall in three categories:

1. Data from percolation and pH dependence leaching tests carried out on the residues, or on similar materials, prior to landfilling in the monofill. Appropriate column tests would be prEN 14405 (Europe) and EPA 1314 (USA), and appropriate pH dependence tests would be EN 14997 or EN 14429 (Europe), and EPA 1313 (USA).

2. Data (time series) from routine monitoring of the quality and quantity of leachate from the monofill in question or from similar monofills.
3. Data from dedicated studies carried out on the monofill in question aimed at obtaining information on leachate quality, infiltration of precipitation, the hydraulic, physical, and chemical conditions within the monofill and the leaching properties of the landfilled residues at the time of the study.

It is interesting to note that the (local) equilibrium-based percolation tests that describe the release of substances from the residues as a function of L/S most likely also can be used to describe the release of substances from monofills, e.g., with stabilized MSWI APC residues. A Nordic study (Hjelmar et al., 2006) has shown that: (1) in Europe, cement-stabilized APC residues are generally landfilled as a paste that remains partially penetrable by infiltrating precipitation, and (2) due to the slow vertical rate of movement of infiltrating water compared with the rate of diffusion out of the paste matrix for most substances, the composition of the leachate at the bottom of such a site is likely to be at local equilibrium with the residues there, and the development of the leachate composition as a function of L/S can be reasonably well described by a percolation leaching test performed on the crushed monolithic material.

In all cases the information gained should preferably be combined from more than one of the above sources, and further combined with information on history, design, construction, operation, etc. Predictions of the time when FSQ may be attained should, for example, not rely solely on leachate monitoring data, but should include some theoretical assumptions and should include the potential consequences of the establishment of new hydraulic equilibria resulting from the discontinuation of leachate removal. The consequences of future changes in pH on the release of selected substances may be assessed on the basis of the results of pH dependence leaching tests (see e.g., Hjelmar and Hansen, 2005). Hydrogeochemical modeling can further support predictions of the effect of changes in pH, redox potential, and carbonation on the release of a number of substances. Such models can also be used to show the influence of organic material, expressed as dissolved organic carbon (DOC), on the leaching of various substances such as, e.g., Cu (see e.g., van Zomeren and Comans, 2004). Compared with leachate from, e.g., municipal solid waste, leachate from MSWI residue monofills typically exhibits relatively low concentrations of organics with DOC values from 1 to 500 mg/L (Hjelmar, 1996). DOC is generally somewhat higher in leachate from MSWI BA than in leachate from FA/APC residues, probably mainly due to the content of humic and fulvic acids as well as hydrophilic organic substances. MSWI FA and APC residues generally have a higher content of potentially hazardous organic substances than MSWI BA, but seldom high enough to contribute to the classification of these as hazardous waste. The concentrations of these potentially hazardous organic substances (e.g., PCDDs/PCDFs, PCBs, etc.) are normally very low (often below the limit of detection) in leachate from MSWI residues due to their very low water solubility.

In Table 20.2.2 some examples of maximum concentration levels of a number of substances observed in leaching tests performed on different MSWI residues are shown.

The source term for landfills, in particular monofills for predominantly mineral waste such as MSWI residues has, in some cases, been estimated and modeled as an exponentially decreasing

Table 20.2.2 Maximum concentration levels observed in eluates from leaching tests on MSWI residues

Concentration Level	Unit	Bottom Ash	Fly Ash and Dry/Semidry Acid Gas Cleaning Residues	Wet Scrubber Sludge + Fly Ash
>100	g/L		Cl^-, Ca	Cl^-
10–100	g/L	Cl^-	Na, K, Pb	Na, K
1–10	g/L	SO_4^{2-}, Na, K, Ca	Zn	SO_4^{2-}, Ca
100–1000	mg/L	DOC	DOC, SO_4^{2-}	
10–100	mg/L	Cu		
1–10	mg/L	Mo, Pb	Cu, Cd, Cr, Mo	DOC, Mo
100–1000	μg/L	Mn, Zn, As	As	
10–100	μg/L	Cd, Ni, Se, Cr		As, Cr, Zn
1–10	μg/L	Hg, Sn		Pb
<1	μg/L		Hg	Cd, Cu, Hg

DOC, dissolved organic carbon.
After Hjelmar (1996).

function of the concentration, C_i, of L/S or time (Hjelmar et al., 2001; Robinson et al., 2004; Hjelmar, 2012):

$$C_i(L/S) = C_{i0} \times e^{-\kappa \times (L/S)}$$

where C_{i0} is the initial (generally also the highest) concentration of substance i in the leachate; and κ (kappa) is substance-specific (and waste-specific) constant (in kg/L) that describes the rate of decrease of the concentration of substance i as a function of L/S. κ may be estimated for a given substance and a given type of waste from laboratory leaching data (percolation test) or from time series of leachate composition data that can be related to the L/S.

Unfortunately, longer-term time series of data describing the composition of leachate from MSWI residue monofills are rare, let alone such data that can be related to the hydraulic conditions, i.e., the volume of leachate produced over time which can then be used to calculate the L/S values corresponding to the composition data available. The main issues rendering most existing leachate monitoring data less useful in estimating source terms that can be used as predictive tools for future impacts on groundwater and surface water are as follows:

- lack of information on the water balance (in many cases both leachate composition and leachate production data only exist for the entire landfill, not for single cell monofills, thus not allowing the calculation of L/S associated with the leachate composition data), and
- in some cases only a limited number of substances, and not necessarily the "environmentally critical" substances, are analyzed.

Landfill operators should be strongly encouraged to ensure that routine monitoring during the entire lifetime of the landfill includes the collection of all the information necessary to develop reliable source terms for predictions of future impacts on the surrounding environment. This should include separate monitoring of individual landfill cells, and it may save both time and efforts in assessing when the aftercare can be ended.

Although sharing of information in the long run will benefit all stakeholders involved in the prediction of future leachate emissions from landfills, it should not be overlooked that not all landfill operators necessarily are interested in having their leachate monitoring data published. In some countries such data are on public record and hence in principle available, while the concerns in other cases can be overcome by anonymizing the data prior to publication. Furthermore, leachate monitoring data reported to the authorities may in principle be available, but they are often fragmented and hidden in written reports and pdfs and hence in practice often almost inaccessible. It is highly recommended to set up databases (with rigorous data quality control) at national, European, or global level to ensure that properly collected leachate monitoring data are made generally available.

In the following discussion a few examples of longer-term time series of the leachate development at monofills for MSWI residues are presented.

MONITORING DATA ON LEACHATE FROM MSWI RESIDUE LANDFILLS

Fig. 20.2.2 shows the development of pH and the concentrations of a number of substances in the leachate produced at two relatively shallow (thickness approximately 5 m) monofill cells for MSWI APC residues at the landfill site of AV Miljø near Copenhagen in Denmark (Wellendorph, 2016). The cells were uncapped throughout the periods shown. Unfortunately, the amount of leachate produced was not accounted for. One of the two data sets shows the annual average concentrations of some substances in the leachate over a period of 12 years (1989–2000) from a cell containing residue from the semidry MSWI acid gas cleaning process without prior removal of particulates (i.e., the FA is part of the residue). The residue was untreated except for some moistening to avoid dust problems. The other data set shows the annual average concentrations of the same substances in the leachate over a period of 8 years (1993–2000) from a cell containing a mixture of FA from electrostatic precipitators and sludge from scrubbers at an MSWI. It is interesting to see that the pH of the leachate from the semidry residue decreases over a few years, probably due to carbonation, decreases from approximately 10 to a level between 7 and 8, which is the same level maintained throughout the period by the leachate

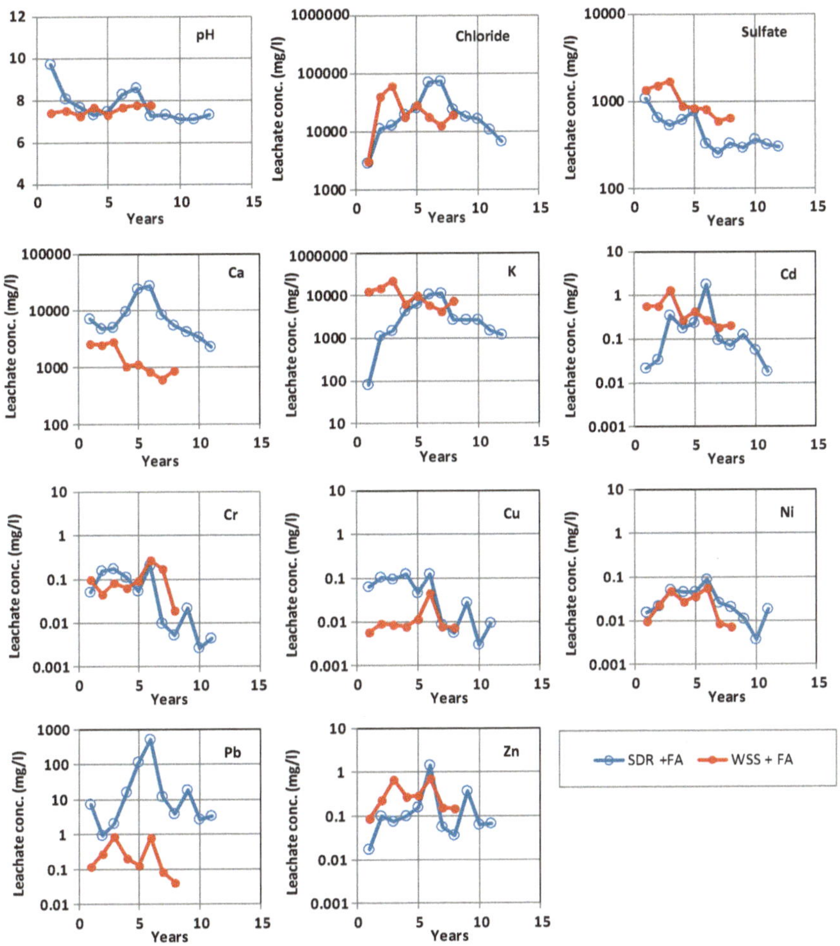

Figure 20.2.2 Development of the composition of leachate over 12 and 8 years from two Danish monofills for semidry municipal solid waste (MSW) incinerator (MSW air pollution control residue containing fly ash (SDR + FA) and MSWI wet scrubber sludge mixed with fly ash (WSS + FA).

from the wet scrubber residue/FA. The initial increase in concentration that is seen for many of the substances over the first 5–6 years is probably caused by changes in the hydraulic conditions and by the fact that the cells may still have been in operation during this period. During the last 5 years for the semidry residue and the last 3 years for the wet scrubber residue/FA—and for substances for longer—the general trend for all the substances measured is decreasing concentration level. The monitoring ended in 2000, because the APC residues were dug out and moved to another site.

Fig. 20.2.3 shows the development in pH and the concentration of a number of substances in leachate from a landfill containing approximately 10,000 m^3 of MSWI bottom and FA (approximately

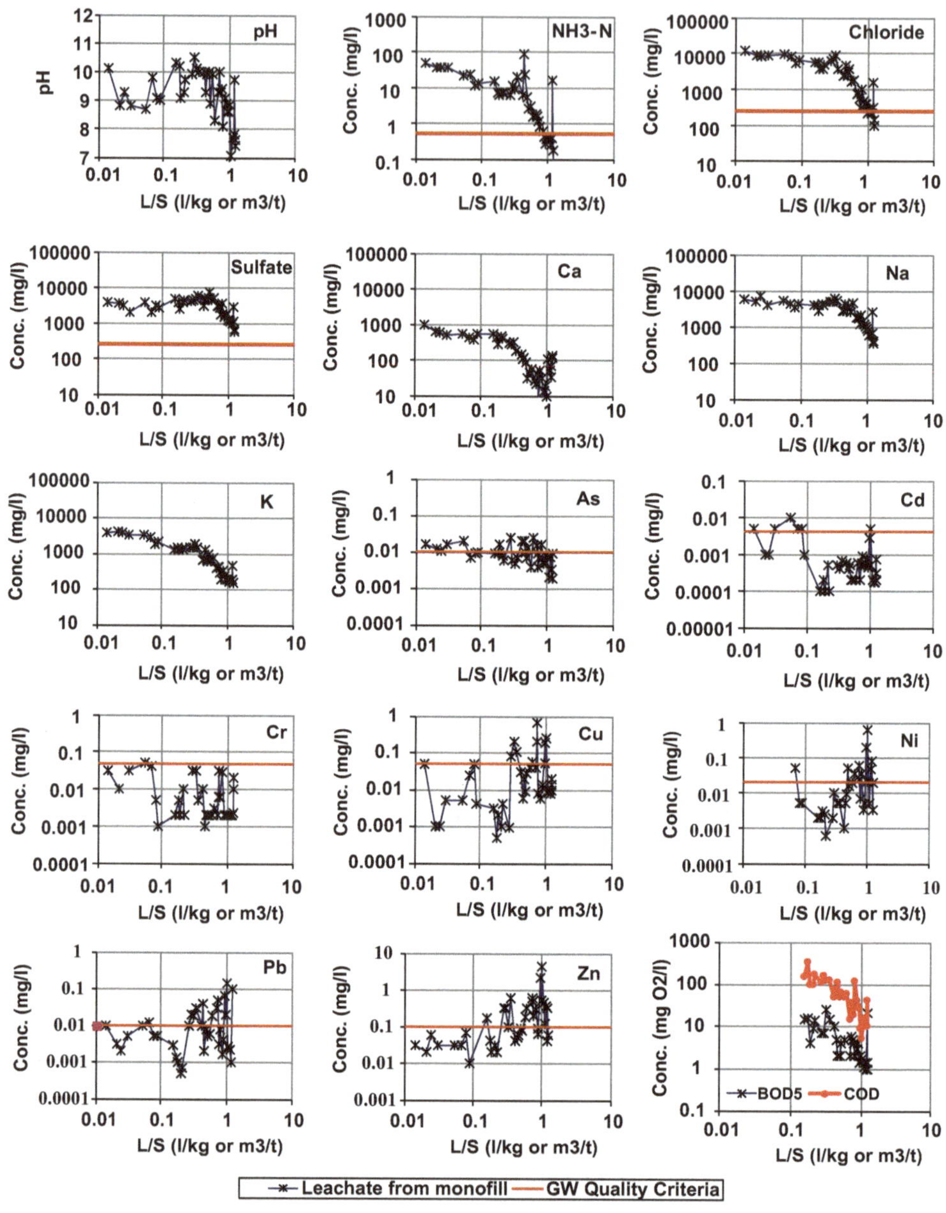

Figure 20.2.3 Development of the composition of leachate from a Danish MSWI bottom and fly ash monofill over a period of 37 years. Leachate composition is shown as a function of the liquid/solid ratio L/S.

85% BA and 15% FA) over a period of 37 years. The site, "Vestskoven," which is located in the Greater Copenhagen area in Denmark, has been well controlled and the quality and quantity of the leachate have been monitored at least once or twice a year throughout the 37 years. This means that it has been possible to calculate the accumulated L/S corresponding to each sample of the leachate that has been analyzed chemically, and hence it is possible to plot the composition of the leachate as a function of L/S. With an annual rate of infiltration corresponding to approximately 33% of the precipitation through the cover which simply consists of topsoil with natural vegetation (grass and bushes), an overall L/S value of 1.23 m^3/t has been reached during the 37 years. More details on the design of the site can be found in Hjelmar and Hansen (2005). It should be mentioned that the leachate samples are filtered through a 0.45 μm filter immediately after sampling at the site.

From Fig. 20.2.3, it can be seen that pH decreases from approximately 10 to a level between 7 and 8 over the period. The horizontal line shown in some of the graphs are examples of groundwater quality criteria. If the concentration of a substance in the leachate is lower than the groundwater quality criterion, then no further dilution is required for that substance if the leachate is discharged into the aquifer. However, there may still be a so-called "bathtub" effect resulting in higher concentrations in the leachate if the collection of leachate is stopped, and a new hydraulic equilibrium is established. In general, and for all the salts, the concentration level is seen to decrease with increasing L/S (and time). For some of the trace elements, in particular those that are pH-sensitive such as Ni, Pb, and Zn, the concentration level may show a slightly increasing trend over the latest 15–20 years. However, the levels are still low, and the uncertainty caused by the sampling, the chemical analytical procedures and the fluctuations of pH as a result of exposure to atmospheric air in the leachate collection well is substantial. The shapes of some of the graphs underline the warning given in the section "The source term" not to rely solely on monitoring data when predicting the future development of leachate composition: Totally different results may be reached, depending on which part of a given curve is used or is available.

More leachate quality data from MSWI residues can be found in Chapter 10.2.

References

Beaven, R.P., Knox, K., Gronow, J.R., Hjelmar, O., Greedy, D., Scharff, H., 2014. A new economic instrument for financing accelerated landfill aftercare. Waste Management 34 (7), 1191–1198.

CEC, 1999. Council Directive 1999/31/EC of April 1999 on the Landfill of Waste.

CEC, 2000. Commission Decision 2000/532/EC of 3 May 2000 Replacing Decision 94/3/EC Establishing a List of Wastes Pursuant to Article 1(a) of Council Directive 75/442/EEC on Waste and Council Decision 94/904/EC Establishing a List of Hazardous Waste Pursuant to Article 1(4) of Council Directive 91/689/EEC on Hazardous Waste.

CEC, 2003. Council Decision 2003/33/EC of 19 December 2002 Establishing Criteria and Procedures for the Acceptance of Waste at Landfills Pursuant to Article 16 of and Annex II to Directive 199/31/EC.

CEC, 2014. Commission Decision 2014/955/EU of 18 December 2014 Amending Decision 2000/532/EC on the List of Waste Pursuant to Directive 2008/98/EC of the European Parliament and of the Council.

Hjelmar, O., 1990. Leachate from land disposal of coal fly ash. Waste Management and Research 8, 429–449.

Hjelmar, O., 1996. Disposal strategies for municipal solid waste incineration residues. Journal of Hazardous Materials 47, 345–368.

Hjelmar, 2012. Description of the Methodology Used to Set the Leaching Criteria for Acceptance of Waste at Landfills for Inert Waste, Landfills for Non-hazardous Waste Accepting Stable, Non-reactive Waste and Landfills for Hazardous Waste Listed in Council Decision 2003/33/EC (Note for the Committee for the Adaptation to Scientific and Technical Progress of EC Legislation on Waste, Subgroup on Landfill, EU Commission.).

Hjelmar, O., 2014. Washing of residues from municipal solid waste incinerators. In: Cossu, van der Sloot (Eds.), Sustainable Landfilling. CISA, Padova, Italy, ISBN 978-886265-005-2.

Hjelmar, O., Hansen, J.B., October 3–7, 2005. Sustainable landfill: the role of final storage quality. In: Proceedings of 10th International Waste Management and Landfill Symposium, Cagliari, Sardinia, Italy.

Hjelmar, O., Henriksen, H.J., Aabling, J., Hansen, J.G., 2013. When is the aftercare period over for a landfill?. In: Proceedings Sardinia 2013, Fourteenth International Waste Management and Landfill Symposium, S. Margherita di Pula, Cagliari, Italy, September 30 to October 4, 2013.

Hjelmar, O., Henriksen, H.J., 2015. Completion of the Aftercare Period at Landfills (Miljøprojekt nr. 1726). Miljøministeriet, Miljøstyrelsen, Copenhagen, Denmark (in Danish, English summary).

Hjelmar, O., Holm, J., Gudbjerg, J., Bendz, D., Suér, P., Rosqvist, H., Wahlström, M., Laine-Ylijoki, J., 2006. Development of Criteria for Acceptance of Monolithic Waste at Landfills (TemaNord 2006:555). Nordic Council of Ministers, Copenhagen.

Hjelmar, O., Rosendal, R.M., Nielse, S., Nygaard, S.D., October 5–9, 2015. Assessing the aftercare duration at landfill: Theoretical considerations and practical issues. Presented at Sardinia 2015. In: S. Margherita di Pula. Cagliari, Italy.

Hjelmar, O., van der Sloot, H.A., Guyonnet, D., Rietra, R.P.J.J., Brun, A., Hall, D., 2001. Development of acceptance criteria for landfilling of waste: an approach based on impact modelling and scenario calculations. In: Christensen, T.H., Cossu, R., Stegmann, R. (Eds.), Sardinia 2001, Proceedings of the Eighth International Waste Management and Landfill Symposium, S. Margharita di Pula, Cagliari, CISA, vol. III. CISA, pp. 712–721.

Robinson, H.D., Knox, K., Bone, B.D., 2004. Improved definition of leachate source term from landfills. Phase 1: Review of data from European landfills (Science Report P1–494/SR1). Environment Agency, Bristol, United Kingdom.

van Zomeren, A., Comans, R.N.J., 2004. Contribution of natural organic matter to copper leaching from municipal solid waste incinerator bottom ash. Environ. Sci. Technol. 38, 3927–3932.

van Zomeren, A., Hjelmar, O., October 5–9, 2015. Assessment of long-term emissions from landfills as part of the landfill aftercare completion procedure. In: Proceedings Sardinia 2015, Fifteenth International Waste Management and Landfill Symposium, S. Margherita di Pula, Cagliari, Italy.

Wellendorph, P.P., 2016. Monitoring data on former landfill cells at AV Miljø, Avedøreholmen, 97, 2650 (Hvidovre, Denmark, containing MSWI APC residues (personal communication by Per Palle Wellendorph, Landfill site manager)).

21 CHAPTER

ECONOMIC ASPECTS

21.1

CAPITAL AND OPERATIONAL COSTS

Kai-Uwe Heyer

INTRODUCTION

For a landfill to be built and operated, the operators have to make sure that they follow certain steps. In most parts of the world, there are regulations that govern where a landfill can be placed and how it can operate. Very often sanitary landfilling is local government's responsibility. Before a city or other authority can build a landfill, a site selection and an environmental impact study must be done. Once the environmental impact study has been completed, permits must be obtained from the local, state, and federal governments. In addition, money will have to be raised from taxes, fees, or municipal bonds to build and operate the landfill.

The basic cost elements for the construction, operation, and aftercare of a landfill may be due to the specific regulations:

- planning and design, permissions:
  - feasibility study, economic study, siting
  - spatial needs depending on the planned capacity
  - permission procedures: cost of approval process, legal documentation, taxes, licenses
- landfill construction (capital expenditure):
  - land purchase/cost of land (with amortization)
  - geological barrier and bottom sealing system
  - leachate drainage and collection system, leachate treatment facility
  - gas collection system, gas treatment/utilization facility
  - infrastructure such as water and power supply, sewage installations, roads on the landfill, access to public road, weighbridge, wheel washing, office and other buildings, fence, laboratory, workshop/garage, entrance area
  - tree plantation, supporting areas
  - investment machinery (compactor, bulldozer, trucks, cars)
- landfill operation:
  - energy demand (and production), electricity, fuel
  - staff, human resources, management
  - equipment such as computers, telephone, etc.
  - maintenance costs
  - waste emplacement and compaction
  - daily cover

    - leachate collection, treatment, and disposal of residues
    - surface water management
    - gas collection and treatment/utilization
    - emission control during operation: noise, dust, fires
    - emission control and monitoring costs (leachate, groundwater, gas, settlements, etc.)
- landfill closure and aftercare
    - temporary cover (if necessary)
    - in situ stabilization/remediation methods (investments and operation costs): infiltration, aeration, landfill mining
    - final cover (investments, maintenance)
    - gradually reducing
        - leachate collection and treatment (reinvestments, maintenance, operation)
        - gas collection and treatment (reinvestments, maintenance, operation)
        - monitoring costs
        - regular inspections (as part of monitoring and maintenance)
    - final plannings, reports, and expertises
    - landfill after-use options for financial benefits such as commercial use, recycling facilities, material deposits, agricultural use such as Miscanthus grass or managed woodland

In the following, relevant cost elements will be addressed. Their range and some examples of the total costs for landfill construction, operation, and aftercare will be compiled.

ECONOMIC ISSUES IN THE EUROPEAN LANDFILL DIRECTIVE

The European Landfill Directive (Article 8(a)(iv)) states (CEC, 1999): "*...adequate provisions, by way of a financial security or any other equivalent, on the basis of modalities to be decided by Member States, has been or will be made by the applicant prior to the commencement of disposal operations to ensure that the obligations (including after-care provisions) arising under the permit issued under the provisions of this Directive are discharged and that the closure procedures required by Article 13 are followed. This security or its equivalent shall be kept as long as required by maintenance and after-care operation of the site in accordance with Article 13(d).*"

In addition the European Landfill Directive (Art.10) states: "*Member States shall take measures to ensure that all of the costs involved in the setting up and operation of a landfill site, including as far as possible the cost of the financial security or its equivalent referred to in Article 8(a)(iv), and the estimated costs of the closure and after-care of the site for a period of at least 30 years shall be covered by the price to be charged by the operator for the disposal of any type of waste in that site.*"

Moreover the European Landfill Directive states in Article 13(d): "*... for as long as the competent authority considers that a landfill is likely to cause a hazard to the environment ..., the operator of the site shall be responsible for monitoring and analysing landfill gas and leachate and groundwater regime in the vicinity of the site ....*" Worldwide landfill regulations often contain similar articles as is briefly reviewed by Scharff et al. (2011).

CALCULATION METHODS AND FINANCING OPTIONS

Calculation Methods

The unit costs per tonne of deposited waste (expressed as €/tonne or $/tonne) are mainly affected by the technical standard and operation costs of a landfill site, yearly input and the total capacity. The last two aspects together effectively determine the period over which waste is accepted, and thereby, the depreciation period for capital. Total capacity determines the quantity of material which can be used as the basis for effectively generating a fund to support aftercare expenditure.

For financial calculations, interest rates and depreciation periods, that reflect the technical life of the components being costed, should be obeyed. Sometimes investments are written off much quicker than necessary (reflecting required rates of return on capital). Depreciation periods of 6–8 years e.g., for gas engines are not unusual even if those capital items show a considerable working life (15–20 years or so).

Financing Options, Gate Fee, Landfill Tax

The gate fee paid represents a unit (usually per tonne) payment to the service provider to generate a stream of revenue. The payment is often made by the local authority if municipal solid waste is collected and delivered by them.

Additional public subsidies may support the construction and operation of the landfill.

A landfill tax or levy is a form of tax that is applied in some countries to increase the cost of landfill. The tax is typically levied in units of currency per unit of weight or volume. The tax is in addition to the overall cost of landfill and forms a proportion of the gate fee. A tax or fee may be imposed on landfills as a means of raising general revenues, to generate funds for inspection programs or long-term mitigation of environmental impacts related to disposal, or as a means of inhibiting disposal by raising the cost in comparison to preferable alternatives, in the same manner as an excise or "sin tax" (Wikipedia, 2014).

COST UNITS/COST ELEMENTS

Development Costs: Planning and Design, Permissions

Development costs comprise:

- feasibility study, economic study, land search
- spatial needs depending on the planned capacity
- site selection and **environmental impact study**
- detailed plannings and design of the landfill
- permission procedures: cost of approval process, legal documentation, taxes, licenses

A bill of quantities for all works and cost estimates of pay items for each component of the design has to be compiled. All construction quantities and costs shall be estimated to an accuracy of plus or minus 10%. The methods of payment per item (lump sum, unit cost) shall be defined, which would be most appropriate to enable and facilitate cost and quality control. Taxes which are anticipated for each pay item, such as value added taxes and customs duties have to be taken into account.

The development costs can require up to 10% of the total investment costs for landfill construction.

Landfill Construction

The main cost elements for landfill construction (capital expenditure) are:

- acquisition costs: land purchase/cost of land (with amortization)
  The costs for the purchase of needed land, and the way in which this is accounted for, can vary enormously, especially for landfills where site acquisition may reflect strategic investments.
  Within a country, as well as across countries, acquisition costs are difficult to specify in any formulaic manner. In some cases, the site may be acquired outright for a fee, in others, a royalty may be paid, or the site may be leased. It is difficult to generalize about the costs of acquisition and much depends on the landowner in determining these costs (Eunomia, n.d.).
- geological barrier and bottom sealing system with leachate drainage and collection system
  Capital expenditure and development costs are affected by legal regulations in terms of the requirement for liners, as well as the geology of the site, and the site's proximity to sensitive aquifers, etc. The detailed approach to regulation in this respect will determine whether the approach allows reduced expenditure (on liners, etc.) where a low risk is illustrated.
  A bottom sealing system under European regulations and requirements for
  - preparation/upgrading of the geological barrier
  - mineral sealing layer
  - geomembrane (such as HDPE-High Density PolyEthylene, 2.5 mm)
  - drainage layer with leachate collection system
  - additional protection layers/geotextiles

  can cause costs between 40–100 €/m^2.
- leachate treatment facility (see landfill operation)
- gas collection system, gas treatment/utilization facility
  Landfill gas (LFG) recovery requires investments in both extraction and utilization systems. The extraction system consists of the collection system (in the waste) and a suction system (of pumps, valves, etc.). The LFG collection system utilizes either vertical wells (placed after infilling of waste) or horizontal drains (placed during infilling of waste). On average, the two systems require the same level of investment; however, horizontal drains tend to fall at the higher end of the cost range. For an average 20-m-deep landfill, the investment in the collection system may range from €15,000 to €30,000 per hectare.
  The LFG extraction system consists of vacuum pumps, monitoring equipment, and control systems. The investment depends largely on the sophistication of the monitoring and control system and on the volume of LFG to be extracted. For an average 20-m-deep landfill, the LFG suction system requires investments ranging from €10,000 to €20,000 per hectare. Many developing countries may not need equipment of the highest level of sophistication, and the investment range for these countries may therefore be lower.
  Utilization of LFG is most commonly achieved through the production of electric power. This is the most dependable and applicable method for utilization of LFG in lower- and middle-income countries. However, conversion of LFG to fuel by separation of methane is of growing importance. So the level of investment for energy recovery depends primarily on the type of power to be generated and the distance over which the power can be delivered.
- infrastructure:
  - water and power supply, sewage installations to a remote site can be costly
  - roads on the landfill, long access roads from a public road to a remote site can be costly

- weighbridge, wheel cleaner, transfer station, office and other buildings, fence, laboratory, workshop/garage, entrance area
- tree plantation, supporting areas
- investment machinery (compactor, bulldozer, trucks, cars) depending on the size of the landfill, the type of waste and the annual waste acceptance

Landfill Operation

Leachate Treatment and Disposal of Residues

Plants currently in operation for the treatment of leachate often consist of several treatment methods to meet the limiting concentrations for the effluent (See Chapter 10.5, Figure 10.5.1). Leachate from landfills in the methanogenic stage may also be treated by a single step (e.g. multistep reverse osmosis).

The costs of leachate treatment in industrial countries can vary roughly between 10 and 50 €/m^3 sometimes up to 70 €/m^3 leachate. This has several reasons:

- the same treatment procedures may be realized in totally different ways e.g., treatment facilities may be installed in cheap containers or in expensive buildings
- the technical equipment can be very simple or very sophisticated, e.g., for online measurements of the leachate components
- the total capacity and the utilization coefficient of the treatment plant; a small capacity and a low utilization coefficient means high costs per m^3 of treated leachate
- a growing competition between companies who are producing treatment plants leads to lower prices
- changing prices for energy and chemicals such as oxygen or active carbon
- cut backs in the budgets of landfill operators for landfill operation (reduced waste quantities for disposal, dropping prices for waste, growing competition between landfill operators etc.)

In 2012 seven German treatment plants with capacities from 10,000 up to 80,000 m^3/a were investigated (Heyer et al., 2012). The total costs for investment and operation varied between 30 and 50 €/m^3 leachate. The biggest treatment plant with a capacity of almost 250,000 m^3/a required costs of only 10–20 €/m^3 leachate.

Gas Utilization, Energy production, Costs, and Benefits

The economics of LFG recovery depends on several factors. Two of the most important economic factors are the possibility of selling the produced energy and the determination of price. Thereby some revenues may be generated from the sale of energy from LFG depending on the regime governing energy sales. As a further option of LFG recovery, environmental benefits (e.g., reduction of greenhouse gases, replacement of fossil fuel) may be included.

In some cases, operators may contract out the management of landfill gas for energy recovery.

The annual operational and maintenance costs for an enhanced bioreactor landfill are higher than those of standard managed landfills. The additional costs reflect the operation of pumps to circulate leachate and the cost of extracting and utilizing LFG in larger plants, which have higher annual yields of LFG. The additional annual operation costs for an enhanced bioreactor landfill may therefore be from 40% to 60% higher than operation and management costs of conventional landfill gas project. This increase in cost may be compensated by higher annual energy revenues.

Revenues from LFG recovery will depend significantly on the type of energy produced, price fluctuations, and on country-specific regulations regarding the subsidy of renewable energy sources (to promote CO_2 reduction). Moreover any combustion of LFG will reduce emissions of methane, the powerful greenhouse gas. Utilizing LFG in controlled combustion for the purpose of producing energy and thereby displacing fossil fuel (and abating carbon emissions) is an added global environmental benefit.

Landfill Closure and Aftercare

The regime regarding aftercare potentially becomes more and more important in the wake of legal regulations like the EU Landfill Directive, and the fact that old big municipal solid waste (MSW) landfills are phased out and require a long-term aftercare. This requires adequate financial provisions to be made by the operator to cover the costs of aftercare.

Basic and optional measures in the closure period that determine the costs may be:

- optional installation of a temporary surface cover
  A temporary surface cover can be installed after the end of waste deposition period to control and reduce the climatic leachate generation (Fig. 21.1.1). The temporary surface cover should be applied as long as there are still significant settlements of the waste body that could damage a final surface cap. After the end of the significant settlements the temporary surface cover should be replaced by the final surface liner. Temporary covers may consist of soil with and without a drainage and/or an additional sealing layer. Costs for the installation can range between 10–40 €/m^2. On several German landfills, average costs of 18–28 €/m^2 became necessary for soil covers or by the installation of a 1.5-mm HDPE-geomembrane (Heyer et al., 2012).

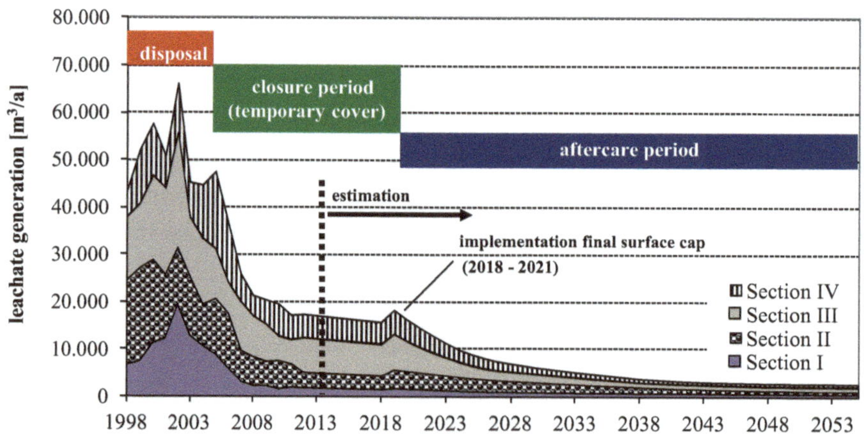

Figure 21.1.1 Estimation of future leachate generation based on monitored values respecting the application of a surface liner (German conditions).

- continuation of the leachate collection and treatment
  Leachate will continue to be produced for many years, although the quantity should be minimized by the optional temporary cover and the final surface cap that has to be properly maintained to limit and control water infiltration into the waste. Costs are necessary for the leachate control systems that collect, store, treat, and discharge leachate and have to be maintained during the aftercare period. Cost calculations especially for leachate treatment require an estimation of the future leachate quality and generation based on the monitoring results of the operation period. Fig. 21.1.1 shows an estimation of leachate generation in the closure and aftercare period of a German landfill divided in four sections.
- implementation or continuation of the landfill gas collection and treatment/utilization
  Landfill gas will continue to be generated by biological degradation of the organic waste fraction. Where a landfill has a gas collection system such as horizontal or vertical gas wells and a gas flare or a gas utilization plant on site, this will need to be maintained, and probably continuously staffed, during the closure and aftercare period as long as the gas emissions and their impacts or risks are still significant (many years to several decades).
- infiltration measures for the optimization of biological degradation processes in the landfill body
- aerobic in situ stabilization (aeration) for an accelerated reduction of biodegradable organic waste components for an improvement of the long-term behavior
  Infiltration measures and especially landfill aeration have been investigated, developed, and carried out on municipal waste landfills for more than 20 years. They reduce the emission potential and improve the long-term landfill behavior. Costs for an infiltration and/or aeration may range between 0.5 and 3 €/m^3 of landfill volume. The technical and financial demand requires about 2%–10% of the total costs for landfill closure and aftercare. On the other hand, the biologically stabilized landfill offers cost saving potentials of 15%–25% of the total costs for landfill closure and aftercare and an earlier release from aftercare of several decades (Heyer et al., 2009; see also Chapter 16.2).
- Installation of the (final) surface cap
  According to the legal requirements and regulations different surface sealing systems for similar nonhazardous waste landfills are possible. When the regulated surface sealing construction is known, a calculation of the cost is straightforward. Moreover the maintenance costs and the expected functional lifetime of the surface sealing respectively the necessary aftercare period is essential for cost calculations.
  A combination surface sealing is often composed of (from bottom to top) a support layer, a gas drainage layer, a mineral liner, a 2.5-mm HDPE-geomembrane, rainwater drainage, and a recultivation layer of a height of at least 1 m (see also Chapter 11.1). Assuming that regulations require a combination surface liner on an MSW landfill and that a gas extraction is already in place, the cost for capping can be calculated to €40–60 per m^2, sometimes up to 80 €/m^2.

Further measures that may affect the costs in the aftercare period:

- final plannings, reports, and expertises
- monitoring measures for a control of the landfill behavior
- optional: groundwater remediation measures
- optional: landfill mining (reconversion of landfill sites)
- landfill afteruse

CASE STUDIES, REFERENCE EXAMPLES

Reference Examples for Landfill Costs EU

Landfill costs can typically be disaggregated into the following components:

- Acquisition costs
- Capital expenditure (landfill construction) and development costs
- Operating costs
- Restoration
- Aftercare costs

As an example a breakdown for a new extension of an existing landfill site in the United Kingdom is shown in Table 21.1.1. The annual waste input amounts to 175,000 tonne over a period of 10 years (Eunomia, n.d.).

The costs which have been collected by Eunomia for 15 European countries are shown in Table 21.1.2. These are not costs in all cases. In some cases, only gate fees were available. There is

Table 21.1.1 Estimated landfill costs for a new extension of an existing landfill site in the United Kingdom (Eunomia, n.d.)

	Investment Costs (€)	Yearly Depreciation over 10 years (€)	Specific Costs (€/tonne)
CAPITAL COSTS			
Site assessment	320,000.00	45,560.80	0.26
Acquisition	1,600,000.00	227,804.00	1.30
Landfill construction and development	14,088,729.60	2,005,918.14	11.46
Landfill closure/restoration	960,000.00	136,682.40	0.78
Landfill aftercare	4,924,582.40	701,149.74	4.01
Total	21,893,312.00	3,117,115.09	17.81
OPERATING COSTS			
Operation		1,920,000.00	10.97
Energy recovery from landfill gas[a]			ca. 1.00
Total costs		5,037,115.09	27.78

[a]The energy recovery is taken on by a third party, which pays a royalty to the operator of approximately €1 per tonne.

Table 21.1.2 Comparative costs of landfilling in different EU member states

Country	Operational Cost (€/tonne)	Total Cost (Excl. Tax) (€/tonne)	Gate Fees (Excl. Tax) (€/tonne)	Landfill Tax (€/tonne)	Total Costs (Incl. Tax)[a] (€/tonne)
Austria		67		43	110
Belgium (Fl) Belgium (Wa)	No data No data	45	47.5 —[c]	52–55	100 45
Denmark	No data		44	50	94
Finland	4		37–46	15	52–61
France	3–8	31–85		9	40–94
Germany	7.3	20–51	35–220		30–51
Greece	1.5–15				9–30
Ireland	13		35–78	19	60–95[b]
Italy	13	52		Varies[c]	70–75
Luxembourg	35–43	123–147			123–147[d]
Netherlands	No data		43–100	64	107–164
Poland	No data		6–15		6–15
Spain	No data	25–35	6–40		25–35
Sweden			20–60	30.6	50.6–90.6[b]
United Kingdom	3–8	28	8–35	19.2	40–48[e]

[a]Where only gate fees are available, values are based on estimated average gate fees.
[b]Estimation is based on the assumption that the entire landfill tax is respected.
[c]Varies by region (and sometimes, degree of source separation in municipality or level of pretreatment).
[d]The costs reflect landfilling including cost for mechanical biological pretreatment.
[e]The costs are estimated for new landfills of different size and capacities. Older landfills (still operating) may be less costly (associated with aftercare and other needs).
Adapted from Eunomia (n.d.).

enormous variation in the cost also due to additional landfill taxes. The table shows that total costs are higher for small landfills so the size has a significant effect. The picture is likely to be quite dynamic in ensuing years, especially for those countries where costs have been, historically, at low levels.

Reference Examples in Australia

Estimations of the costs of landfilling in Australia (private companies) were compiled by the BDA Group (2009) (Table 21.1.3). They comprise the full cost of landfill disposal related to the landfill size. This reflects the impact of small, medium, and large landfills on costs and the corresponding waste disposal in tonnes per annum.

Case Study in Brazil

A study made on Brazilian sanitary landfills economic aspects was conducted in 2009 (Machline et al., 2011). The 40 landfills analyzed are privately owned. They receive both municipal and nonhazardous industrial wastes. These 40 landfills adhere strictly to national technical and legal standards in the construction, operation, and future remediation phases. To take into account the effects of economy of

Table 21.1.3 Cost estimations of landfilling in Australia (private sector) (A $[a] per tonne) (BDA Group, 2009))

Type of Cost	Small[b] ($/tonne)	Medium[c] ($/tonne)	Large[d] ($/tonne)
Land purchase	5	3	2
Approvals/site development	10	6	4
Best practice liner	13	8	5
Leachate collection	6	4	3
Gas recovery	6	4	3
Amenity management	1	1	1
Operations	34	20	14
Capping and remediation	10	6	4
Postclosure maintenance	15	9	6
Total costs	100	60	40

[a]1 A $ ≈ 0.7 € ≈ 1.1 US $.
[b]waste acceptance < 10,000 tonnes/year.
[c]waste acceptance 10,000–100,00 tonnes/year.
[d]waste acceptance > 100,000 tonnes/year.

scale, the landfills were divided into large, medium size and small categories, corresponding to 2000, 800, and 100 tonnes daily receiving capacity. The landfill lifecycle was divided in five periods:

- feasibility studies: 1-year duration
- construction and development: 1-year duration
- operation: 20 years
- closure: a month
- remediation and land recovery: 20 years

The total lifecycle considered was 42 years. The following costs focus on the larger landfills (adapted from Machline et al., 2011):

- Feasibility studies:
  economic study, land search, land purchase (965,314 m^2 × €1.2 per m^2), legal documentation, taxes, licenses: €1.56 Mio.
- Construction:
  - infrastructure, detailed construction plans, area enclosures, access roads, internal roads, water and sewage installations, electricity, power, telephone, facilities
  - disposal cells construction, soil removal, liner system (56,632 m^2 × €21 per m^2), leachate treatment system, surface water drainage, trees plantation, supporting areas, management, taxes, licenses: €7.0 Mio.
- Operation:
  disposal cells (continuation), liner system (509,686 m^2 × €21 per m^2), waste disposal, drainage, flare system, green areas plantation, and maintenance, environment monitoring, labor, management, taxes: 20-year costs: €177.5 Mio.
- Closure:
  covering works, with special clay. Cost: €2.5 Mio.
- Remediation and land recovery:
  leachate transportation and treatment in a sewage treatment plant, green areas maintenance, environment monitoring, labor. 20-year costs: €13.9 Mio.

The total cost per tonne of landfilling of waste amounted to about €14. Similar data collected for medium size and small landfills resulted in total costs of €15.6 and €27.6 per tonne.

Costs Related to Technical Standards

Furthermore costs are determined by the requirements and technical standards like in:

- highly industrialized countries (such as Germany, United Kingdom, France, United States, etc.)
- emerging countries at the stage of economic takeoff (India, Brazil, etc.)
- developing countries

Examples (ISWA, 2004):

- France: 60–80 €/tonne (including insurances, landfill taxes, final closure provisions, initial investment, investments during operation, etc.)
- United States, Delaware: 22 €/tonne for landfill operation only
- Private sector operated landfills, over 3000 tonnes/day in developing countries: 13–20 €/tonne

Research sponsored by the World Bank, estimated the cost of disposal, per tonne, to be in the range between €4 and €15 (cost for land not included). These studies have been carried out in developed, developing, and less developed countries (Cointreau, 2008). Results from newer calculations and experiences show that nowadays the costs have significantly increased.

Costs for Landfill Closure and Aftercare

Due to new regulations in Germany in 2005, landfilling of waste deposition of untreated municipal solid waste on landfills ended in Germany. Many landfills were closed and are now in the closure or aftercare period. In 2012 several German landfills were investigated concerning their costs for closure and aftercare (Heyer et al., 2012). In Fig. 21.1.2 the results of the evaluation related to the landfill volume are compiled under the following boundary conditions:

- All landfills (A–F) were operated by public authorities.
- The landfill bodies have a volume between 1.3 and 5.7 Mio. m^3. They cover an area of 6–23 ha and have a maximum deposition height of 20–70 m.
- Landfill A is already covered with a surface cap.
- The costs are related to the year 2014 and market prices without future inflation and value added taxes.
- The emission behavior of all landfills shows that a release from aftercare will probably not possible after 30 years. Therefore site-specific cost estimations for a longer aftercare period over 40–70 years were conducted.

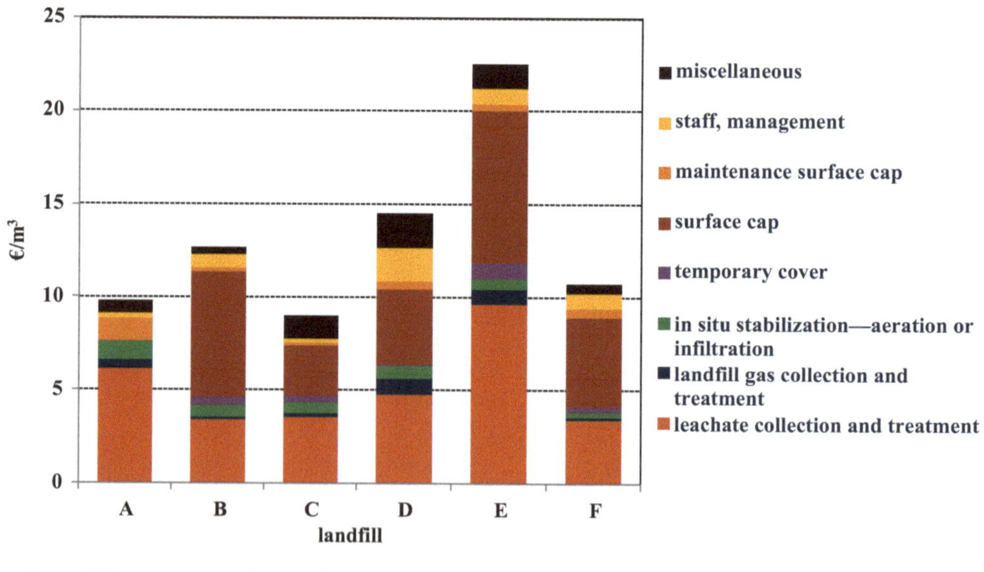

Figure 21.1.2 Estimation of costs for closure and aftercare for several German landfills (A–F) (Heyer et al., 2012).

The estimations show:

- Leachate collection and treatment is one of the most cost-demanding aftercare measures. It may require 21%—47% of the total costs for all closure and aftercare measures. That corresponds to 3.4—9.6 €/m^3 landfill volume.
- Costs for landfill gas collection and treatment/utilization are significantly lower compared to the leachate emissions and correspond to 0.2—0.9 €/m^3.
- In situ stabilization by aeration is for almost all municipal landfills an option to improve the emission behavior and to reduce costs in the aftercare period (reduced investments for the surface cap, reduced costs for leachate treatment, landfill gas treatment, monitoring, earlier release from aftercare, etc.). Costs for aeration of the landfills A—F amount to 0.5—1.0 €/m^3.
- The temporary cover requires costs of 18—28 €/m^2 respectively 0.1—0.9 €/m^3 landfill volume.
- The application of the final surface cap was calculated to 50—70 €/m^2 with a cost saving potential after a successful aeration of the landfill (specific German situation). This corresponds to 2.8—8.2 €/m^3 landfill volume and 15%—41% of the total costs for closure and aftercare. Maintenance costs were estimated to 5%—15% (0.2—0.5 €/m^3) of the investment costs. They are regarded as very uncertain and may be higher according to damages caused by settlements or nonappropriate afteruse of the surface.
- Costs for staff and management were estimated to 0.2—1.8 €/m^3 respectively 3%—9% of the total costs.
- The costs for further aftercare actions such as monitoring, insurances, maintenance of infrastructure, reports, etc., are in a range of 0.4—1.8 €/m^3.

According to these calculations and estimations the total costs for an aftercare period of 30 years were determined to be 9—22 €/m^3 with an average value of 14 €/m^3 (without future inflation). If a longer aftercare period of 40—70 years was taken into account the total costs rose to 12—27 €/m^3.

In a further German study, typical capping and aftercare costs of €10—12 per m^3 of waste were calculated for MSW landfills with an impermeable surface sealing and with an assumption of leachate emissions and aftercare for a period of 30 years (Eitner, 2010).

Comparable calculations in the Netherlands led to lower average ranges for the different closure and aftercare measures (Scharff and van Zomeren, 2011):

- Surface cap: 2.00—5.00 €/m^3 of waste
- Leachate treatment: 1.40—2.10 €/m^3 of waste
- Landfill gas control: 0.40—0.50 €/m^3 of waste
- Monitoring: 0.40—0.80 €/m^3 of waste
- Maintenance: 0.50—0.90 €/m^3 of waste
- Management: 0.30—0.70 €/m^3 of waste

The total financial demand for capping and aftercare in the Netherlands was calculated to be between 5 and 10 € per m^3 of waste.

CONCLUSIONS

General Comments

- In many countries, landfilling is still the cheapest waste management option.
- In the absence of other waste treatment facilities, higher cost for landfilling that are above actual cost may be charged to cross finance other kind of waste treatment such as recycling. This has to be in concurrence with the respective regulations in place.
- Cost estimates for a variety of positions may have significantly different degrees of accuracy. Hence the simple addition of all the estimates to arrive at a total cost may not be purposeful. Sensitivity analysis could be undertaken to determine whether changes in assumptions greatly affect the outcomes when a new landfill is planned.
- The calculation of additional potential cost for effects that are difficult to predict may be respected as e.g.,
  - Decreasing value of land, i.e., costs for the impact of a landfill on surrounding areas because of their proximity to the landfill
  - Groundwater contamination and remediation. The possible degradation may impose an intergenerational cost (Thom and MacGreggor, 2009).
- Quantitative criteria for the release of a landfill from aftercare should be developed. A combination of general and site-specific criteria shall consider the emission behavior and technical demands, but also general legal and economic conditions. These criteria provide the basis for the decision when a landfill may be released from aftercare within reasonable periods of time, preserve the collective good, and offer a better and more reliable estimation of the closure and aftercare expenditures (Heyer et al., 2007).

Remarks for Industrialized Countries

- As the EU regulations for landfilling become more stringent, landfills in Europe tend to increase in size. By these means, landfilling may become more economic.
- In industrialized countries, costs for landfilling may further increase over the next decade also due to the fact that old landfills are completed. This is likely to lead to smaller numbers of larger landfills in those countries where landfill is still important.
- Site reclamation and landfill mining may become an option especially when the occupied land has a high value. Due to the limitations of the availability of raw materials, the exploitation of resources from landfills and old dumps may be an option if the revenues from the recovered materials are high enough and there are significantly amounts of recyclable materials in the landfill. In addition the landfilling of the nonrecycled materials has to be respected. However, in the year 2015, raw material prices are still too low to make the recovery of materials from the landfills economic. In most cases the aftercare measures are still cheaper and more rewarding than the dismantling and the exploitation of the landfill. Due to relative low amounts of recyclable materials and their in general low quality in most German MSW landfills, raw material extraction is not worthwhile in 2015. The decision has to be made site specific on the basis of intense investigations.

Remarks for Developing Countries

- In most developing countries still open dumping is practiced. There is an as urgent need to remediate and close existing open dumpsites and implement sanitary landfills. Due to the higher cost of sanitary landfilling, little progress toward controlled landfill is made. To improve the situation adequate regulatory framework has to be implemented and enforced. Financial support e.g., by banks, international organizations, and companies may in many cases be the precondition to make this transition from open dumping to sanitary landfills.

References

BDA Group, July 13, 2009. The Full Cost of Landfill Disposal in Australia. Prepared for the Department of the Environment, Water, Heritage and the Arts. http://www.environment.gov.au/settlements/waste/publications/pubs/landfill-cost.pdf.

CEC, 1999. Council Directive 1999/31/EC of 26 April 1999 on the landfill of waste. Official Journal of the European Communities.

Cointreau, S., 2008. Landfill ER revenues versus landfill costs. http://www.worldbank.org/solidwaste.

Eitner, R., January 1998. R-29 Hamburger Berichte. In: Entwicklungstendezen in der Deponietechnik, 1. Hamburger Abfallwirtschaftstage, pp. 245–260. Economica Verlag Bonn.

Eunomia, n.d. Costs for Municipal Waste Management in the EU. (Final Report to Directorate General Environment, European Commission. Eunomia Research & Consulting, on behalf of Ecotec Research & Consulting). http://ec.europa.eu/environment/waste/studies/pdf/eucostwaste.pdf.

Heyer, K.-U., Hupe, K., Stegmann, R., Willand, A., 2007. Landfill aftercare – options for action, duration, costs and quantitative criteria regarding the release from aftercare. In: Proceedings Sardinia 2007, Eleventh International Waste Management and Landfill Symposium S. Margherita di Pula, Cagliari, Italy; October 1–5, 2007. CISA Publisher, Italy.

Heyer, K.-U., Hupe, K., Koop, A., Stegmann, R., 2009. Aerobic in situ stabilization - a new part in the German landfill ordinance. In: Proceedings Sardinia 2009, Twelth International Waste Management and Landfill Symposium S. Margherita di Pula, Cagliari, Italy; October 5–9, 2009. CISA Publisher, Italy.

Heyer, K.-U., Hupe, K., Biesterfeld, J., Stegmann, R., 2012. Kosten der Stilllegung und Nachsorge. In: Stegmann, R., Rettenberger, G., Kuchta, K., Fricke, K., Heyer, K.-U. (Eds.), Deponietechnik 2012, vol. 37. Verlag Abfall aktuell, Stuttgart, pp. 313–332. Hamburger Berichte.

ISWA, 2004. Data Provided by the International Solid Waste Association. Mentioned in: www.worldbank.org/INTURBANDEVELOPMENT/Resources.

Machline, C., del Bel, D., Caselani, C., Sandroni, P., Machline, V., 2011. Economics of sanitary landfills in Brazil. In: Proceedings Sardinia 2011, Thirteenth International Waste Management and Landfill Symposium S. Margherita di Pula, Cagliari, Italy; October 3–7, 2011. CISA Publisher, Italy.

Scharff, H., van Zomeren, A., 2011. Landfill end point criteria for groundwater impact. In: Proceedings Sardinia 2011, Thirteenth International Waste Management and Landfill Symposium S. Margherita di Pula, Cagliari, Italy; October 3–7, 2011. CISA Publisher, Italy.

Scharff, H., van Zomeren, A., van der Sloot, H.A., 2011. Landfill sustainability and aftercare completion criteria. Waste Management & Research: the Journal of the International Solid Wastes and Public Cleansing Association, ISWA 29, 30–40.

Thom, M.J., MacGreggor, F., 2009. The real costs of landfill design and operations. In: Proceedings Sardinia 2009, Twelfth International Waste Management and Landfill Symposium S. Margherita di Pula, Cagliari, Italy; October 5–9, 2009. CISA Publisher, Italy.

Wikipedia, 2014. Landfill Tax. http://en.wikipedia.org/wiki/Landfill_tax.

21.2

LANDFILL EXTERNALITIES

Ofira Ayalon

INTRODUCTION

The impact of waste management practices on the environment has been widely recognized. Since the 1990s, it has contributed to better siting and planning of facilities, upgraded engineering, and better infrastructure and management of all waste treatment technologies (landfilling, waste to energy, recycling, etc.). The generation and management of municipal waste and associated externalities have been addressed as one of the key targets in environmental policies (e.g., EC, 2000; Ayalon et al., 2000; Baddeley, 2008; Hogg et al., 2008).

In general, environmental burden associated with landfilling of waste (summarized in Table 21.2.1) can be divided into direct emissions to air, water, and land, as well as disamenities and noxious effects of the waste facility.

THE PRICE TAG OF THE ENVIRONMENT

Valuation methods of externalities

Externalities are costs (or benefits) arising from an economic activity that affects someone other than those engaged in it. These costs are not fully reflected in prices or direct costs of the activity (Economist, 2012) and are therefore nonmarketed goods. External costs refer to any loss of human well-being associated with those activities, and they include air pollution, land use and contamination, deterioration of water quality and quantity, noise, odor, visual intrusion, and so on (Bickel and Rainer, 2005; Maibach et al., 2008). On various levels, these impacts may cause climate changes, health effects, damage to crops and buildings, as well as disamenities (nuisances).

In recent decades, environmental economists have developed a range of valuation methods to quantify these changes in well-being. With the exception of some methods that assign approximate values to externalities (based on expert judgment and assessment), most of the methods rely on economic welfare theory that aims to maximize individual and social welfare through optimal resource allocation (EC, 2000). The methods are based on the values that individuals ascribe to their preferences regarding an environmental good with no observable price in the market.

The core of evaluation for externalities is the dose—response function (DRF), which is a production function and impact pathway method. In fact it provides reasonable basic data for various valuation techniques associated with the economic damage—cost approach. DRF measures the correlation

Table 21.2.1 Possible environmental impacts of waste disposal in landfills

Category	Examples of Externalities	Source	Major Impacts
Land use	Area of the landfill and affected area	The existence of the landfill	Impact on natural resources; land degradation
Resource depletion			Exhaustion of raw materials and need to produce new products
Emissions	Methane (CH_4)	Landfill gas (LFG)	Climate change
	Carbon dioxide (CO_2)	LFG	Climate change
	Local air pollutants, e.g., sulfur dioxide (SO_2), nitrogen oxides (NO_x), particulates (PM_{10}), nonmethane volatile organic compounds	LFG	Human and occupational health; damage to buildings, crops, forests, and water; ozone depletion; photochemical oxidant formation; acidification; radiation
	Emissions to soil and water	Leachate	Human and ecosystem toxicity
Disamenities	Odors, dust, wind-blown litter, illness vectors, vermin (e.g., rodents, flies and birds), noise, visual intrusion, risk of fires and explosions	Proximity to landfill site. Odors and fire/ explosion (mainly from methane emissions)	Nuisance

Sources: Eshet et al. (2005); El-Fadel et al. (1997); Nahman (2011).

between a unit concentration of a pollutant (dose) and its impact on an affected receptor (population, crops, buildings, water, etc.) based on scientific data (Tellus Institute, 1992; Rabl et al., 1998). Moreover, the methods can be divided into two main clusters: direct methods, which include the individuals' stated preferences, and indirect methods, which include the revealed preferences (RPs).

Direct valuation methods: stated preference methods

Stated preference (SP) methods are direct methods enabling estimations of economic values for a wide range of intangible commodities based on individuals' preferences, including contingent valuation method (CVM) and choice modeling. SP methods are mainly used in waste management to assess individuals' willingness to pay (WTP) for a specific degree of improvement in their welfare level, which was harmed by the level of air, water, or land pollution, as well as disamenity impacts related to different waste management options (i.e., Shechter, 1995).

Contingent valuation method

This is a marked creation method in which people are asked directly through a survey to state their WTP for a benefit or to avoid a cost, or conversely, to state their willingness to accept (WTA) compensation regarding a specific hypothetical policy. Econometric techniques are then applied to the survey results to derive the average bid value, i.e., the average WTP (Pearce and Howarth, 2000; Ayalon et al., 2006).

Choice modeling method

The method is a conjoint analysis that represents a broad term of survey methods asking individuals to rank, rate, or choose alternatives rather than explicitly expressing a WTP or WTA. The techniques (based on market research tools) rely on the idea that any "good" can be described in terms of its attributes and their levels. Changing attribute levels will essentially result in the production of a different "good," and the techniques focus on the value of such changes in attributes. A baseline status quo alternative is usually included to help establish the other alternatives in relation to the respondent's actual experience. It is generally assumed that choice modeling approaches are preferred over CVM in contexts where it is important to value individual attributes (Pearce and Howarth, 2000; Hanley et al., 2001).

Indirect valuation methods: revealed preference methods

RP methods are indirect methods inferring preferences and implicit value for externalities from actual market observations. Preferences are revealed indirectly when individuals purchase marketed goods and services that are supposed to eliminate or reduce the associated environmental problem (Shechter, 1995; ODPM, 2003).

Hedonic price method

This surrogate market method relies on the notion that people derive utility from various physical and environmental attributes of a house. The value that individuals place on an environmental attribute (e.g., air quality near a landfill or incinerator) is obtained from differences in housing prices at various distances from the sites (Shechter, 1995; Tellus Institute, 1992; Segeron, 2001; Hite et al., 2001; Nahman, 2011).

Averting behavior method

The method assumes that marketed goods can act as substitutes for environmental goods because households spend money to offset environmental impacts. For instance, expenditures such as installation of water filters or air conditioners in the vicinity of a landfill can be used to value clean water or air and can thus represent the value of the damage to groundwater caused by landfill leachate or air emissions (ODPM, 2003; Shechter, 1995).

Cost of illness method

This method estimates external costs through changes in private and public expenditures on medical commodities and earnings lost due to days not worked resulting from various impacts related to noxious facilities (Shechter, 1995; Navrud, 2001).

Health production function

Health production function (HPF) is a popular way of calculating the relative effects that a pollutant has on health. The measurement widely used is the mortality rate. The function depicts health as a good or an output that is a function of various inputs including the environmental factor. This approach assumes that increases in the purchases of health-care goods and services decrease the purchases of other goods and services, when controlling for household income constraints (Pearce and Howarth, 2000). In fact, HPF methods comprise elements both from averting behavior method and from cost of illness methods, and they are regarded as a more comprehensive approach.

Travel cost method

This method is mostly used for monetary valuation of recreational sites. This approach is based on the fact that a trip to a recreational site usually requires the individual to incur costs in terms of travel, entry fees, on-site expenditures, and time. These costs are used as a proxy for the value of the site land use and for changes in its quality (Pearce and Howarth, 2000; ODPM, 2003; Shechter, 1995). This approach has been used by Ayalon et al. (2006) to assess the benefits of the Hiria (Tel Aviv, Israel) landfill rehabilitation.

Experts' assessment of damage costs

Due to valuation limitations, environmental damages are often valued according to the knowledge, experience, and mainly the intuition and judgment of professionals in particular fields, such as engineers, who estimate, ex ante or ex post, the costs of repairing, restoring, or replacing a damaged asset, or of abating impacts. Various techniques that often refer to the term "expert assessment" are addressed below.

Control cost method

The method, which considers abatement, avoidance, and remediation costs, purports to infer the value that society attributes to pollutants, from the costs of implementing regulations that society imposes on itself to abate pollution (avoiding damage). The critics argue that the highest required cost for abatement of a specific pollutant should be taken as the minimum value that society places on removing it (Tellus Institute, 1992; EC, 2000).

Cleanup cost method

This method has an engineering approach; it assumes that once the damage resulting from pollution is done, the costs of rehabilitation to achieve the predamage situation will appear as a (minimum) proxy economic value of the damage. For example, cleanup cost to reduce leachate from a landfill will represent the damage cost of leachate (EC, 2000).

Replacement cost method

This method uses the cost of replacing or restoring a damaged asset to its original state as the measure of the benefit of restoration and thus as the cost of damage. The technique implies that complete replacement is, in fact, feasible (Pearce and Howarth, 2000).

Court decision assessment

Court decision assessment is an alternative indirect method presented in a disamenities–valuation study by Beloff et al. (2000). This method is based on observation of actual expenditure involved in citizens' lawsuits against noxious facilities. These costs include, inter alia, various government or municipality costs engaged in the investigation of a complaint, costs of the activities of citizens groups, environmental lawyers' fees, and advocacy costs of the companies.

Ancillary Methods

Value of statistical life

Value of statistical life (VSL) is not actually a valuation method for environmental damages in the usual sense; nevertheless, we refer to the VSL method because the assessment of external costs associated with human health in the framework of some studies is based on VSL, and it serves frequently as a basis for estimations in several valuation methods. The obvious moral objections to valuing human life are avoided by valuing small risks to life and normalizing to unit risk—"value of statistical life" (VSL), namely, the valuation of a loss of life. This can be performed either via CVM (summarizing information about people's WTP for small reductions in mortality risks/how much income they are ready to forgo to reduce the probability of a premature death) or by observation of actual payments, such as expenditure on safety devices and wage premium in risky occupations (Eyre, 1998). This information enables estimations of how much society is ready to spend to reduce a statistically determined number of premature deaths, and economists use the VSL in CBA to express the social costs of a life lost (EC, 2000).

Benefit transfer

Benefit transfer (BT) is an econometric tool for transferring existing estimates of nonmarket values (benefit or damage) from one site (study site/primary study) to a new different study (policy site/secondary study), potentially through most of the direct and indirect valuation methods. The aim of the BT process is mainly to save time and money that would be required to conduct new studies for new policy procedures. BT offers an efficient, low cost, and a fast way to extend the economic valuation of externalities and is, therefore, a popular tool (Brisson and Pearce, 1995; Brouwer and Spaninks, 1999; Eshet et al., 2007).

THE PRICE TAG OF ENVIRONMENTAL ASPECTS OF LANDFILLS

To quantify the environmental impacts of different waste management practices, a price tag can be attached to every activity and decisions can be more balanced.

Landfill externalities can be expressed in two ways:

- assessment of the overall disamenities from a landfill;
- aggregation of specific emissions of contaminants originated from landfilled waste.

Assessment of the overall disamenities from a landfill

The externalities are expressed as the total external costs associated with the overall disamenities from the landfill (the costs can further be divided by the total amount of landfilled waste to obtain external costs per tonne of waste). The main methodology used to assess these disamenities is the Hedonic Property Price Methodology. Brisson and Pearce (1995) aggregated several studies conducted in the 1970–90s and suggested the following equation:

$$\text{Change in Hedonic Property Price (\%)} = 12.79 - 3.76\ \text{D}.$$
$$(\text{D} = \text{distance from the site in miles.})$$

Therefore, a property is not affected by a landfill if it is situated more than 3.76 miles from the landfill site.

EMC (1996) suggested the following equation to estimate the WTP in Israel to move away from a landfill site:

$$\text{Average WTP (\$)} = 15{,}400\$ - 1730\$\cdot\text{D}.$$
$$(\text{D} = \text{distance from the site in km.})$$

Therefore, a property in Israel is not affected by a landfill if it is situated more than 8.9 km from the landfill site.

This aggregated "formula" reflects the NIMBY (Not In My BackYard) syndrome, which prevailed in the 1970s and even as recently as the 1990s, when landfills were improperly managed and were often a major nuisance to the neighboring communities. In the countries in which putrescible waste is still being landfilled and therefore still imposes an environmental threat (Australia—BDA Group, 2009; Nigeria—Arimah, 1996; South Africa—Du Preez and Lottering, 2009; Nahman, 2011), the literature still discusses these methods.

In developed countries, however, these methods are no longer discussed in relation to environmental threats, but when assessing the benefits of closure and rehabilitation of landfills (e.g., Ayalon et al., 2006; Kinnaman, 2009).

We may conclude that current state-of-the-art landfills are planned and located to minimize disamenities.

Aggregation of specific emissions of contaminants

The method is based on the aggregation of the specific emissions of contaminants from each tonne of landfilled waste, attaching a monetary valuation for each pollutant and calculating the total externality produced by each tonne of landfilled waste.

As mentioned before, various methods exist for putting a price tag on these environmental impacts. Eshet et al. (2005) reviewed studies conducted in the early 1990s and summarized the external costs associated with landfills. The results are presented in Table 21.2.2. Given the variability of the landfill operation and the variability of the assessment valuations of the externalities, wide variations can be seen among the different studies.

Table 21.2.2 gives rise to several comments and remarks:

- The results of the externality assessment are created by estimating the amount of pollutant emitted per tonne of waste, which is then multiplied by the evaluated monetary price of the pollutant. The result yields a finite number, whereas the actual costs are, at best, a good estimation.
- The study produced significant differences in values, which can reach two orders of magnitude. This may be due to the different assessment methods and different discount rates.
- The above results are relevant to specific studies and specific landfill sites. Nonetheless, the final costs do not take the geographic and climatic conditions of the landfill site into transparent consideration. In addition, the degree of landfill planning and management is not specified. Therefore, one cannot attach a price tag to every activity or engineering process conducted at the site.
- The main contributor to landfill externalities is GHG emissions (EC, 2000; El-Fadel et al., 1997; Baddeley, 2008; Hogg et al., 2008; BDA Group, 2009). Therefore, landfilling of biodegradable organic material is the main cause of landfill externalities.
- The externalities presented serve as a good approximation for setting the rate of a landfill tax (LFT), to reflect the environmental burden of low standard landfills.

As in the case of the landfill disamenity assessment, the amount of literature on aggregating costs of pollutants to assess landfill externalities is also decreasing. The most recent literature found is from the mid-2000s. This decrease has main reasons; the first is the "internalization" of the external costs, by better engineering and management of the waste and landfills and the second is that these externalities are reflected in the LFT. Therefore, the more recent literature focuses on the effects of the LFT.

LANDFILL TAX

The imposition of a LFT on waste disposed off in landfills is a means of internalizing the externalities into the price of landfilling. The major reason for implementing LFTs across EU countries is that each country has to comply with the 1999 EU Landfill Directive (99/31/EC). Article 5 (2) of the Landfill Directive implies that:

- by 2006 (2010),[1] biodegradable municipal waste (BMW) going to landfills must be reduced to 75% of the total amount (by weight) of BMW produced in 1995;
- by 2009 (2013), BMW going to landfills must be reduced to 50% of the total amount (by weight) of BMW produced in 1995;

[1]The years in parentheses indicate the targets for EU Member States that have been granted a 4-year delay due to relatively lower performance in the baseline year of 1995, i.e., >80% landfill rate in 1995.

Table 21.2.2 Valuation results (USD/tonne waste, 2003 value) of emissions from landfill disposal

Pollutant Study (year)/Method	CO_2	CH_4	Other (Conventional)	Transport Rural/Urban	Leachate	Energy Recovery	Total[b] Estimate
Schall (1992) [a]/ N.A.							2.42–14.14
CSERGE et al. (1993) [a]/ BT of temperature impacts (impact pathway); cleanup cost; insurance risks and monitoring costs.	0.128–2.03	0.72–8.64		0.08/1.904	0–1.44	0.72–3.07	0.91–14.64
Powell and Brisson (1994) [a]/ BT from existing studies that used DRF and SP methods. Cleanup cost for leachate + risk of accidents and monitoring costs.	0.512–0.736	2.176–3.776		0.144/ 0.608	0–1.43	1.29–1.79	1.6–6.4
Enosh (1996)/ BT from studies mainly used DRF and abatement cost methods.					0.01–0.7		6.5
EMC (1996)/ BT from European studies mainly used DRF and abatement cost methods (for Carbon Oxide (CO)).					0		3 (with energy recovery)
Miranda and Hale (1997)/ Damage cost functions (BT from several studies).					0–0.98		2.42–13.16
EC (2000) [a]/ BT of temperature impact assessment.	1–23		0–0.2		0–2	0–10	6–44
Eunomia (2002)/ Mainly based on existing studies; cleanup cost for leachate.	3.05–3.21	4.49–4.87	0.03–0.08		0–2.03		7.51–10.19
Dijkgraaf and Vollebergh (2003) [c]/ Abatement cost (assumed to reflect minimum WTP).	5.85				0	4.76	21

BT, benefit transfer; *DRF*, dose–response function; *SP*, stated preference; *WTP*, willingness to pay.
[a]The range refers to with (high) and without (low) energy recovery.
[b]Each of the estimates is a sum of different components and not necessarily the sum of the values that are shown in the far right-hand column.
[c]Landfill with energy recovery including calculation of land use—17.88 USD/tonne waste.

- by 2016 (2020), BMW going to landfills must be reduced to 35% of the total amount (by weight) of BMW produced in 1995.

Landfilling used to be considered the cheapest method of waste disposal. This was the case when the facility was actually a dump, or a poor standard landfill, with no means of minimizing potential harm to the environment. The main goals of imposing an LFT are to reflect the overall environmental impacts of landfill along with creating a policy mechanism to encourage more sustainable waste practices. Twenty European countries have introduced a tax on waste sent to landfills, with some variation regarding the waste types covered by the LFT (Fischer et al., 2012).

Table 21.2.3 presents the LFT for municipal solid waste (MSW) in several countries and the anticipated increase in the tax in the near future.

Fischer et al. (2012) showed that in 2009–10, the total revenue generated in Europe from the LFT was approximately €2.1 billion. They also showed that in some countries, the revenues have declined, and conclude that the tax has a positive effect in diverting waste away from landfills toward other and environmentally better waste management methods.

Fig. 21.2.1 illustrates the effect of LFT on the annual amount (per capita) of landfilled MSW.

It can be seen that, since 2000, the annual amount of landfilled waste per capita is decreasing, although the impact of the LFT on the amount of landfilled waste is different. In countries where the LFT is relatively low, more waste is generally landfilled. Sweden and Belgium, with a moderate LFT, were landfilling relatively small amounts of waste even before the LFT was imposed; nevertheless, the impact of the tax is clear. The increasing LFT in the UK stimulated a decrease in landfilled waste (468 kg per capita in 2000 and 255 kg in 2010). It is important to note that neither Germany nor Switzerland have imposed an LFT, but rather use a ban on untreated MSW landfilling.

THE LANDFILL TAX PARADOX

Alongside the decrease in actual externalities, a constant increase in LFT can be seen (Table 21.2.3). As mentioned above, the LFT is an economic policy measure to reflect the external costs of pollution and disamenities associated with landfilling waste. However, the pollution and the nuisance are determined by several factors, which include the following:

- Composition of the waste stream. Since putrescible waste is the source of GHG emissions, the lower the organic content of the waste, the fewer GHG emissions and leachates are produced.
- Physical characteristics and infrastructures at the landfill site. These include the size and the daily amount of landfilled waste, spatial location (annual rainfall and temperatures, topography, and so on), age of the site, level of pollution abatement in the facility (lining and leachate collection and treatment, gas collection efficiency, flaring of the gas or energy recovery, the recovery system efficiency and the pollution displaced), etc.

Therefore, if the site is well designed and managed, the environmental externalities are actually internalized into the capital costs of the construction of the site as well as into the operation and maintenance of the landfill.

Table 21.2.3 Landfill tax in several countries (2012)

Country	Landfill Tax (€/tonne)[a]	Remarks
Austria	Landfill tax since January 1, 2006: €87	Depending on composition of waste and standard of the landfill; ban since January 1, 2008 for wastes with TOC >5%
Belgium, Flanders	Household waste: €50	Adjustment of prices to annual consumer price index
Czech Republic	€19	
Denmark	Landfill tax since January 1, 2010: €63	Since January 1, 1997 ban on landfilling of waste suitable for incineration
Estonia	€14.5	Environmental tax will rise by 20% per year until 2015 (€30 in 2015)
Finland	Since January 1, 2013: €50/tonne	Landfill ban in preparation: ban for all organic (biodegradable + plastics) waste (TOC < 10%)
France	Landfill tax since 2011: • "Nonauthorized" landfills: €70 • "Authorized" landfills: €20 • "Authorized + ISO 14001" landfills: €17 • Landfills with energy recovery at a minimum of 75% level: €11	
Germany	No landfill tax	Landfill ban for untreated municipal solid waste (MSW) since June 1, 2005
Hungary	€25	Ban since 2015 for non–pretreated waste

Ireland	As of September 2011: €50	Plan to increase the tax to €65 in July 2012 and to €75 in 2013
Italy	€10–25 MSW, depending on region	
Israel	€10	Expected to increase to €22 in 2015
Lithuania	€22	
Netherlands	High tax €107.49—combustible waste Low tax €16.79—noncombustible waste	
Norway	€ 59/t waste > 10% TOC € 34/t < 10% TOC	Landfill ban for all waste >10% TOC introduced in July 1, 2009
Portugal	€3.5	Taxes are updated every year
Spain	€25.36–€40	
Sweden	€43	As of 2005, ban on organic waste
Switzerland	No landfill tax	Landfill ban for MSW since 2005
UK	£64 (€72.62)	Rising by £8 (€9.08) per annum until at least 2014–15

[a]Different VAT rates are collected according to each country's tax system. The LFT was imposed in each country in different years and hence, the rate of the tax increase is different.
Based on: CEWEP website, MoEP website.

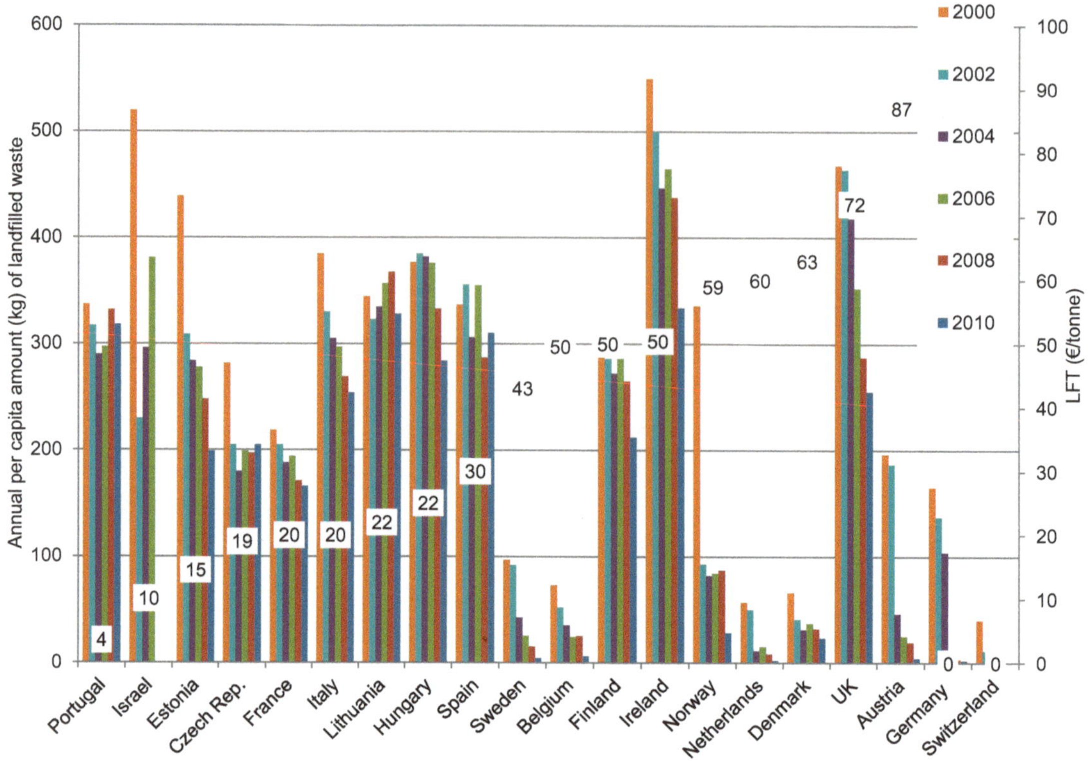

Figure 21.2.1 Landfill tax (LFT) (2012) and per capita amount of landfilled municipal solid waste in 2000–10 in selected countries.

Furthermore, if GHG emissions are avoided through landfill gas (LFG) recovery, not only the major contributor of externalities is reduced but also pollution from conventional power plants is displaced, so external costs are, therefore, converted into a benefit.

Due to better siting, planning, infrastructure, and engineering as well as strict regulations regarding the type of waste entering the landfills, the externalities are significantly decreased.

The other paradox of the LFT is the uses of the revenues. Fischer et al. (2012) showed that in some countries (Denmark, the Netherlands, Norway, and Sweden), 100% of the revenues go directly to the general state budget, whereas in other countries, the revenues are used both for the state budget as well as for the cleaning up of contaminated sites, waste management initiatives, and environmental initiatives. It is noteworthy that the LFT in Israel is deposited in a separate and dedicated account of the Maintenance of Cleanliness Fund to be used for the development, establishment, promotion, and greater efficiency of alternative means to landfilling, which have a lower impact on the environment as well (MoEP, Website, 2012). The impacts of the tax show that it is a strong driver in diverting waste away from landfills, yet given the uses of the tax, where the revenues are used directly to promote environmental issues in general, and improved waste practices in particular, the effect of the LFT is more significant in improving the environmental performance of waste management.

Looking at the LFT in the various countries (Table 21.2.3), the tax is far higher than the evaluated costs of the externalities (Table 21.2.2). In addition, the improved diversion of BMW from landfills and the upgraded infrastructure in landfills should, in fact, decrease the tax level.

Therefore, one may argue that the justified rationale for imposing a LFT was indeed a way to reflect environmental externalities, although today, the LFT is an economic means of reducing environmental burdens of waste as well as a means of enriching the treasury budget.

THE ROLE OF ASSESSING EXTERNALITIES IN WASTE MANAGEMENT

In the past, when most of the MSW was landfilled in rather poor standard sites, the public opposition and the NIMBY syndrome pushed policy makers to take various regulatory measures to reduce landfilling and promote other, more sustainable, waste practices. Assessment of externalities arising from landfills was the key to calculating the rate of the LFT, to reflect the effects of these externalities. To reduce the environmental burden of landfilled waste, organic material is continuously diverted from landfills and engineering infrastructure in landfills is improving.

Nevertheless, one must bear in mind that all other waste management technologies have an environmental impact as well. The term "zero pollution" does not exist as long as our society produces waste. The mission of environmental economists, waste management practitioners, and policy makers is to assess the externalities associated with each technology used, both transparently and objectively. Each alternative assessed should cover all aspects of possible air, water, and land pollution, along the entire waste cycle, including transportation, sorting, processing, and treatment of residuals. The evaluation can be used further to incorporate comprehensive economic incentives in waste management.

CONCLUSIVE REMARKS

The amount of landfilled waste has reduced in recent decades. Current landfilled waste streams contain less BMW, which is the major contributor to the environmental burden caused by landfills.

Apart from complying with the EU Landfill Directive and imposing an LFT, other policy instruments have helped to reduce landfilling; the introduction of separate collection schemes for biodegradable waste, the EU Packaging Directive (Directive 94/62/EC ,as amended by Directive 2004/12/EC), which has helped initiate the diversion from landfills of packages in general and biodegradable packages (paper, cardboard, and wood) in particular.

Environmental economics enables assessment and putting a monetary tag on the costs associated with landfilling of waste. Therefore, the initial aim and level of LFT was based on the level of externalities associated with landfills. In recent years, along with the decrease of the actual externalities, the

LFT has increased, no longer only to reflect the externalities, but to promote more sustainable waste technologies.

In addition, one must bear in mind that a broader scope of internalizing external costs must be used for all waste management practices (landfilling, incineration, recycling, anaerobic digestion, and composting), since they all carry a certain degree of emissions and environmental pollution. Therefore, all different strategies for waste solid treatment should be based on a life cycle perspective, and the price tag should be attached to all stages and all technologies.

LIST OF ABBREVIATIONS

BMW Biodegradable municipal waste
BT Benefit transfer
CBA Cost–benefit analysis
LFT Landfill tax
MBT Mechanical–biological treatment
RP Revealed preferences
SP Stated preferences
VSL Value of statistical life
WTA Willingness to accept
WTP Willingness to pay

References

Arimah, B.C., 1996. Willingness to pay for improved environmental sanitation in a Nigerian city. Journal of Environmental Management 48 (2), 127–138.

Ayalon, O., Avnimelech, Y., Shechter, M., 2000. Alternative MSW treatment options to reduce global greenhouse gases emissions – the Israeli example. Waste Management and Research 18, 538–544.

Ayalon, O., Becker, N., Shani, E., 2006. Economic aspects of the rehabilitation of the Hiriya landfill. Waste Management 26 (11), 1313–1323.

Baddeley, A., 2008. 'Biostabilisation' of Wastes: Making the Case for a Differential Rate of Landfill Tax, 16 pp. www.organics-recycling.org.uk/dmdocuments/Eunomia_Jan_ 2008_report.pdf.

BDA Group, 2009. The Full Cost of Landfill Disposal in Australia. BDA Group, Canberra, Australia. www.environment.gov.au/.../landfill-cost.doc.

Beloff, B.R., Beaver, E.R., Massin, H., 2000. Assessing societal costs associated with environmental impacts. Environmental Quality Management 10 (2), 67–82.

Bickel, P., Rainer, F., 2005. Externalities of Energy. Methodology 2005 Update. Institut fur Energiewirtschaft un Rationelle Energieanwendung - IER, Universitat Stuttgart, Germany. http://ec.europa.eu/research/energy/pdf/kina_en.pdf.

Brisson, I., Pearce, D., 1995. Benefits Transfer for Disamenities From Waste Disposal (CSERGE Working Paper WM 95-06). ISSN: 0967-8875, 14 pp.

Brouwer, R., Spaninks, F.A., 1999. The validity of environmental benefits transfer: further empirical testing. Environmental and Resource Economics 14 (1), 95–117.

CEWEP Website, 2012. http://www.cewep.eu/information/data/landfill/index.html.

CSERGE, Warren Spring Laboratory, EFTEL, 1993. Externalities From Landfill and Incineration (Report to the Department of Environment). HMSO, London, 111 pp.

Dijkgraaf, E., Vollebergh, H., 2003. Burn or Bury? A Social Comparison of Final Waste Disposal Methods, SIEV. Erasmus University Rotterdam and OCFEB, Rotterdam. papers.ssrn.com/sol3/papers.cfm?abstract_id=425281.

Du Preez, M., Lottering, T., 2009. Determining the negative effect on house values of proximity to a landfill site by means of an application of the hedonic pricing method. South African Journal of Economic and Management Sciences 12 (2), 256–262.

Economist, 2012. Economics A-Z Terms. http://www.economist.com/economics-a-to-z.

El-Fadel, M., Findikakis, A.N., Leckie, J.O., 1997. Environmental impacts of solid waste landfilling. Journal of Environmental Management 50, 1–25.
EMC, 1996. Waste Disposal Management, Analyzing Direct and Indirect Costs. Final report (70 pp.). Israel (in Hebrew).
Enosh, L.T.D., 1996. Solid Waste Management Externalities. Final report (pp. 140). Israel (in Hebrew).
Eshet, T., Ayalon, O., Shechter, M., 2005. A critical review of economic valuation studies of externalities from incineration and landfilling. Waste Management and Research 23 (6), 487–504.
Eshet, T., Baron, M.G., Shechter, M., 2007. Exploring benefit transfer: disamenities of waste transfer stations. Environmental and Resource Economics 37 (3), 521–547.
EU Landfill Directive, 1999. http://eur-lex.europa.eu/LexUriServ/LexUriServ.do?uri=CELEX:31999L0031:EN:NOT.
EU Packaging Directive, 1994 (As amended in 2004). http://eur-lex.europa.eu/LexUriServ/LexUriServ.do?uri=CELEX:01994L0062-20050405:EN: NOT.
Eunomia, 2002. . Economic Analysis of Option for Managing Biodegradable Municipal Waste (Final report and appendices). ec.europa.eu/environment/waste/.../pdf/econanalysis_finalreport.pdf.
European Commission (EC), DG Environment, 2000. A Study on the Economic Valuation of Environmental Externalities From Landfill Disposal and Incineration of Waste (Final main report). http://ec.europa.eu/environment/waste/studies/econ_eva_landfill.htm.
Eyre, N., 1998. Study on Energy and the Environment, Environmental Impacts of Energy (Paper prepared by Eyre Energy Environment for Royal Commission Environmental Pollution, UK), p. 46.
Fischer, C., Lehner, M., McKinnon, D.L., 2012. Overview of the Use of Landfill Taxes in Europe. European Topic Centre on Sustainable Consumption and Production, 96 pp. scp.eionet.europa.eu/publications/WP2012_1/wp/WP2012_1 (A.
Hanley, N., Mourato, S., Wright, R., 2001. Choice modeling approaches: a superior alternative for environmental valuation. Journal of Economics Surveys 15 (3), 435–462.
Hite, D., Chern, W., Hitzhusen, F., Randall, A., 2001. Journal of Real Estate Finance and Economics 22 (2–3), 185–202.
Hogg, D., Baddeley, A., Gibbs, A., North, J., Curry, R., Maguire, C., 2008. Greenhouse Gas Balances of Waste Management Scenarios (Report for the Greater London Authority). Eunomia Research & Consulting Ltd., 94 pp. http://static.london.gov.uk/mayor/environment/waste/docs/greenhousegas/greenhousegasbalances.pdf
Kinnaman, T.C., 2009. A landfill closure and housing values. Contemporary Economic Policy 27 (3), 380–389.
Maibach, M., Schreyer, C., Sutter, D., van Essen, H.P., Boon, B.H., Smokers, R., et al., 2008. Handbook on the Estimation of External Costs in the Transport Sector (Report produced within the study Internalisation Measures and Policies for All external Cost of Transport (IMPACT), Version 1.1, Commissioned by the European Commission DG TREN Delft, CE, The Netherlands). www.europadecentraal.nl/documents/dossiers/Transport/2008_01_15_handbook_external_cost_en.pdf.
Miranda, M.L., Hale, B., 1997. Waste not, want not: the private and social costs of waste-to-energy production. Energy Policy 25, 587–600.
MoEP Website, 2012. http://www.sviva.gov.il/subjectsEnv/Waste/Laws/Pages/Landfill-Levy.aspx
Nahman, A., 2011. Pricing landfill externalities: emissions and disamenity costs in Cape Town, South Africa. Waste Management 31 (9–10), 2046–2056.
Navrud, S., 2001. Valuing health impacts from air pollution in Europe. Environmental and Resource Economics 20 (4), 305–329.
ODPM (Office of Deputy Prime Minister), 2003. Valuing the External Benefits of Undeveloped Land: A Review of the Economic Literature. www.communities.gov.uk/documents/.../pdf/158136.pdf.
Pearce, D.W., Howarth, A., 2000. Technical Report on Methodology: Cost Benefit Analysis and Policy Responses (Report 481505020 for RIVM, EFTEC, NTUA and IIASA), p. 72. http://www.pbl.nl/en/publications/2000/Technical_Report_on_Methodology__Cost_Benefit_Analysis_and_Policy_Responses.
Powell, J.C., Brisson, I., 1994. The Assessment of Social Costs and Benefits of Waste Disposal (Working paper). ISSN: 0967-8875. www.cserge.uea.ac.uk/sites/default/files/wm_1994_06.pdf.
Rabl, A., Spadaro, J.V., Desaigues, B., 1998. Usefulness of Damage Cost Estimates Despite Uncertainties: The Example of Regulations for Incinerators (Final Report 1 Annex 7: COSTS), p. 14. arirabl.org/Publications_files/Rabl+98incineratorsCBA.pdf.
Segeron, K., 2001. Real estate and the environment: an introduction. Journal of Real Estate Finance and Economics 22 (2–3), 135–139.

Schall, J., 1992. Does the Solid Waste Management Hierarchy Make Sense? A Technical, Economic and Environmental Justification for the Priority of Source Reduction and Recycling. In: Yale Working Papers on Solid Waste Policy. http://www.yale.edu/pswp/#schall.

Shechter, M., 1995. Valuing the environment. In: Folmer, H., Gabel, H.L. (Eds.), Principles of Environmental and Resource Economics. A Guide for Students and Decision-Makers, Chapter 3. Edward Elgar Pub., pp. 73–103

Tellus Institute, 1992. Tellus Packaging Study – Assessing the Impacts of Production and Disposal of Packaging and Public Policy Measures, 1. Tellus Institute.

21.3

CLEAN DEVELOPMENT MECHANISM (CDM) IN LANDFILLING

Marco Ritzkowski

INTRODUCTION

Global warming is the result of a change in the atmospheric balance caused by anthropogenic emissions of greenhouse gases (GHGs). Among anthropogenic GHGs, methane has the second greatest effect on the climate after carbon dioxide (IPCC, 1996). Methane (CH_4) emissions from anthropogenic sources account for 50%–65% of all methane emissions worldwide (IPCC, 2013). Their overall contribution to global warming is significant because, over a 100-year time frame, methane is estimated to be 25 times more effective at trapping heat in the atmosphere than carbon dioxide.

Since the beginning of the Industrial Revolution (end of 18th century) the atmospheric methane concentration has increased by approximately 15% in total and almost tripled during the last 150 years (Keppler et al., 2006). According to latest estimations, a total of 67–90 million tonnes of CH_4 is emitted from landfills, old waste deposits, and further waste handling activities worldwide, accounting for up to 19% of the global anthropogenic CH_4 emissions. This ranks landfills second after ruminant livestock (87–94 million tonnes of CH_4 per year) (IPCC, 2013; Lelieveld et al., 1998).

In 1992, several countries joined an international treaty, the United Nations Framework Convention on Climate Change (UNFCCC), to cooperatively consider what they could do to limit average global temperature increases and the resulting climate change. To operationalize the Convention, the Kyoto Protocol was adopted in 1997 and entered into force in 2005. It commits industrialized countries to stabilize or reduce greenhouse gas emissions based on the principles of the Convention. By this commitment of the set targets, emission reductions got an economic value. To help countries meet their emission targets, and to encourage the private sector and developing countries (DCs) to contribute to emission-reduction efforts, negotiators of the Protocol included three market-based mechanisms—emissions trading, joint implementation, and the clean development mechanism (CDM).

The CDM allows emission-reduction projects in DCs to earn certified emission-reduction (CER) credits, each equivalent to 1 tonne of CO_2 ($CO_{2,e}$). These CERs can be traded and sold and used by industrialized countries to meet a part of their emission-reduction targets under the Kyoto Protocol. In previous years, many local governments in DCs experience a serious deficit in meeting their revenue need from the tax base. Waste management activities are therefore often underfunded or not realized at all. Against this background many stakeholders analyze possibilities in obtaining different funding for waste management improvement, where CDM is one of them. In the sectorial scope "Waste handling and disposal," landfilling is favorable for CDM projects as many DCs are widely depending on dumping solid waste. Climate projects are also a potential source of electricity generation, as methane capturing and utilization not only reduces the GHG emissions but may also contribute toward the generation of renewable energy.

CLEAN DEVELOPMENT MECHANISM METHODOLOGIES RELATED TO LANDFILLING

CDM methodologies are linked to a total of 15 sectorial scopes whereat landfilling is considered under the sectorial scope "Waste handling and disposal." With approximately 755 project activities (11% of all registered CDM projects), "Waste handling and disposal" represents the second largest scope behind "Energy industries" that covers approximately 75% of all registered CDM projects (Fig. 21.3.1).

Landfill-related projects account for 35% of all projects under the scope "Waste handling and disposal." A total of 213 projects (86%) are considered as large-scale activities (i.e., the estimated annual amount of emission reductions is greater than 60,000 tonnes $CO_{2,e}$), whereas 37 projects have been registered as small-scale (SSC) activities, each associated to specific approved methodologies. Table 21.3.1 provides an overview on the approved methodologies for landfill projects in 2013.

Large-scale projects basically refer to the approved consolidated methodology ACM0001 "Flaring or use of landfill gas" that may be supplemented by additional methodologies in case of electricity feed into the grid. For the latter cases both, large-scale and SSC methodologies for "Grid-connected electricity

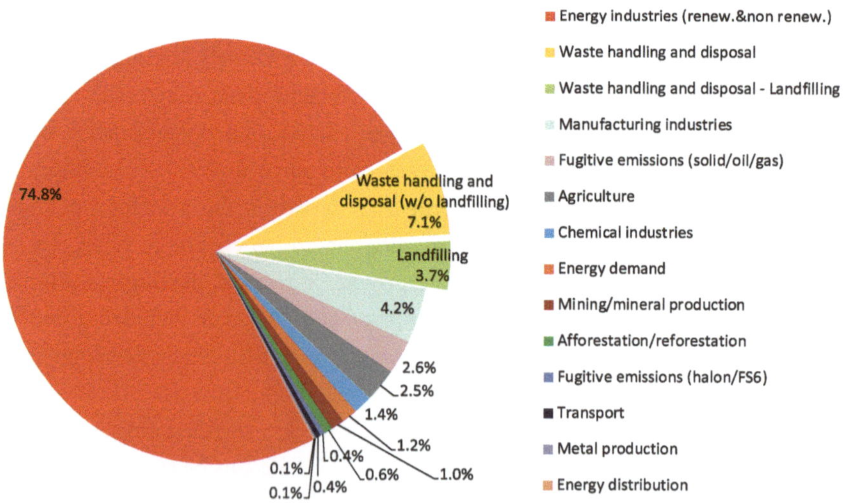

Figure 21.3.1 Distribution of registered clean development mechanism projects by sectoral scopes (in percentage of total project activities) in July 2013 (UNFCCC, 2013).

Table 21.3.1 Approved methodologies for large and small scale clean development mechanism landfill projects

	Large Scale Projects			Small Scale Projects	
Main methodologies	ACM0001			AMS-III.G.	
	AM0083	AM0093		AMS-III.AF.	AMS-III.AX.
Complementary methodologies	ACM0002	ACM0022	AMS-I.D.	AMS-I.D.	AMS-I.C.

generation from renewable sources" (ACM0002 or AMS-I.D.) may be applied, depending on the amount of electrical energy generation (less than 15 MW or above). In 2013, two landfill gas (LFG) projects included the anaerobic fermentation of organic waste (according to ACM0022 "Alternative waste treatment processes") in an approach where LFG and biogas together were utilized for energy generation.

SSC projects basically refer to the approved SSC methodology AMS-III.G. "Landfill methane recovery." Approximately two-thirds of all approved LFG SSC projects intending to feed electricity into the grid, based on the modalities laid out in the SSC methodology AMS-I.D. At the same time, only 8% of the approved SSC projects producing both, electrical and thermal energy (cogeneration) for users' supply to displace fossil fuel (AMS-I.C. "Thermal energy production with or without electricity"). In 2013, only one approved project activity was intended to solely utilize generated electricity for captive use (AMS-I.F.).

Two additional methodologies for large-scale projects in the scope of landfilling have been approved in 2009 and 2011, respectively. The methodology AM0083 "Avoidance of landfill gas emissions by in situ aeration of landfills" comprises guidance for landfill aeration by means of air venting (overdrawing) or low pressure air injection with the objective of avoiding anaerobic degradation processes inside the waste mass. Furthermore, the methodology AM0093 "Avoidance of landfill gas emissions by passive aeration of landfills" refers to semiaerobic landfills where the waste is treated aerobically on-site by means of passive aeration with the objective of avoiding or reducing anaerobic degradation processes inside the waste mass. Both methodologies may be applied under the condition that the waste mass to be treated is contained in closed landfills or closed landfill cells.

In connection with SSC projects, two additional methodologies have been approved by the UNFCCC Meth Panel in 2009 and 2011, respectively. The methodology AMS-III.AF. "Avoidance of methane emissions through excavating and composting of partially decayed municipal solid waste" comprises guidance on methane emissions reduction through landfill excavation and aerobic treatment (composting) of the excavated waste material. Prior to waste excavation, a preaeration phase (either by high or low pressure air injection) is foreseen in the methodology. The second methodology, AMS-III.AX. "Methane oxidation layer (MOL) for solid waste disposal sites" intents to reduce methane emissions from landfills through biological oxidation processes to occur in a constructed MOL on top of the deposited waste. This methodology is only applicable at landfills exhibiting low LFG production rates (<4 L CH_4/m^2 h) and no active gas extraction system in place.

CLEAN DEVELOPMENT MECHANISM PROJECTS RELATED TO LANDFILLING

Statistical data

In July 2013, 213 large scale and 37 SSC CDM landfill projects have been registered at the UNFCCC. As for the large-scale projects, the total amount of estimated emissions reduction according to the project design documents (PDDs) accounted to 41.5 million tonnes $CO_{2,e}$ at an average amount of 181,500 tonnes $CO_{2,e}$ per project. Approximately 53% of all large-scale projects requested issuance of CERs by July 2013, resulting in a total amount of issued CERs of almost 13 million tonnes $CO_{2,e}$ (115,000 tonnes $CO_{2,e}$ per project on average). Based on these figures, the average degree of performance for large-scale CDM landfill projects is calculated to approximately 50%. Furthermore, a significant margin for the individual project performance can be observed: While 5% of the projects exceed the calculated amount of CERs (maximum performance: 146%), some of the projects showing underperformance have been issued less than 10% of the amount of emission reductions calculated ex ante in the PDD. The reasons for these discrepancies are manifold and involve, beside others, model implied overestimations, differences between actual and assumed site conditions, and operational problems. More details about CDM project's performance are provided in the paragraph "Performance of CDM landfill projects" and "Pros and cons of CDM landfill projects—discussion."

During project registration the duration of the crediting period has to be selected. Project participants may either opt for a fixed 10 years crediting period or a 7 years period which may be renewed at most two times, provided that, for each renewal, the original project baseline is still valid or has been updated taking account of new data where applicable. The majority (almost 2/3) of large-scale CDM landfill projects are registered for the renewable 7-years term whereas only 37% selected the fixed 10-years crediting period. It can be assumed that project participants opting for the fixed crediting period are either expecting a shortfall in CERs in the long term or intent to avoid the necessary redetermination of the original project baseline by the designated operational entity.

With regard to the spatial distribution of large-scale CDM landfill projects (Fig. 21.3.2), it becomes apparent that the majority of activities is carried out in Asia (43%) followed by South and North America (35% and 15%, respectively) as well as Africa (7%).

From Fig. 21.3.2 it becomes apparent that comparatively larger projects are registered in South America (also in Africa, but to a significantly lesser extent) since the amount of calculated emission

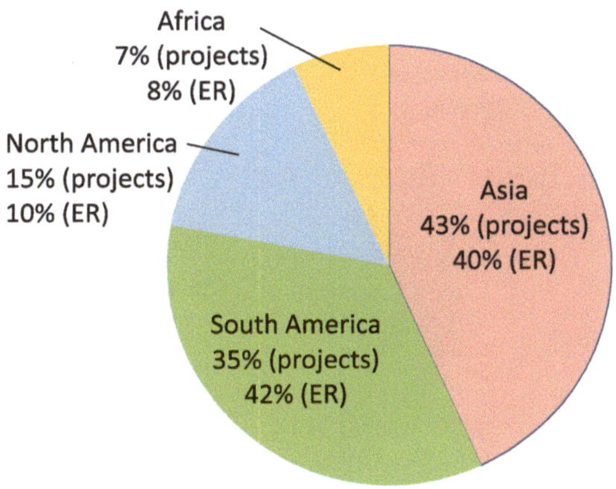

Figure 21.3.2 Spatial distribution of large-scale clean development mechanism landfill projects according to continents (July 2013), both by relative amount of projects and proportion of estimated emission reductions (ER) (UNFCCC, 2013).

reduction is exceeding the proportion of projects by numbers. Accordingly, the estimated amount of emission reductions per project is 221 kilotonnes $CO_{2,e}$ and 213 kilotonnes $CO_{2,e}$ for South America and Africa, respectively, whereas projects in Asia and North America are expected to generate 169 kilotonnes $CO_{2,e}$ and 122 kilotonnes $CO_{2,e}$, respectively.

As for the 37 registered SSC CDM landfill projects, the total amount of estimated emissions reduction according to the PDDs accounts to 1 million tonnes $CO_{2,e}$ at an average amount of 30,000 tonnes $CO_{2,e}$ per project (data refer to July 2013). However, until October 2013 only approximately 22% of all registered SSC projects requested issuance of CERs, resulting in a total amount of issued CERs of approximately 300 kilotonnes $CO_{2,e}$ (37,500 tonnes $CO_{2,e}$ per project on average and 20,000 kilotonnes $CO_{2,e}$ per project and year). Based on the data of these projects the average degree of performance for SSC CDM landfill projects is calculated to approximately 75%, thus significantly exceeding the performance of large-scale activities. The majority (76%) of SSC CDM landfill projects select a crediting period of 10 years probably due to the limited expectations of significant long-term emission reductions from comparably small sites.

With regard to the spatial distribution of SSC CDM landfill projects (Fig. 21.3.3) it becomes apparent that, as for the large-scale projects, the majority of activities is carried out in Asia (62%). Only 22% and 11% of all registered projects are located in South America and Africa, respectively, whereas minorities of 3% each are implemented in North America and Europe.

In contrast to the situation with large-scale projects, relatively larger project activities are conducted in Asia. Here, 71% of the total calculated emission reductions are to be generated by only 62% of

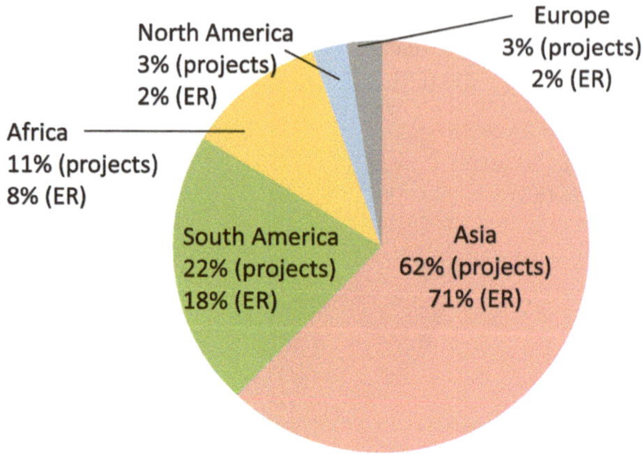

Figure 21.3.3 Spatial distribution of small scale clean development mechanism landfill projects according to continents (July 2013), both by relative amount of projects and proportion of estimated emission reductions (UNFCCC, 2013).

the total amount of projects. On project level the estimated amount of emission reductions is 33.5 kilotonnes $CO_{2,e}$ for Asia, 25 kilotonnes $CO_{2,e}$ for South and North America, respectively, 21.5 kilotonnes $CO_{2,e}$ for Africa, and 13.5 kilotonnes $CO_{2,e}$ for Europe.

Performance of clean development mechanism landfill projects

The proportion of registered large-scale CDM landfill projects that have requested issuance and being issued CERs amounts to approximately 50% (July 2013). A predominant proportion of these projects show a significant underperformance when comparing the estimated amount of emission reductions (based on the calculations made in the PDD) with the actual certified amount of emission reductions. On average, the performance is at 50% with a median value of 45.4%. The specific project performance is very variable, ranging from 1.7% to 146.5%.

On the lower extreme (i.e., <10% performance), the main reason for the observed underperformance is in connection with operational problems at the beginning of project activities. As a rule CERs have been issued only for a period of a few months, thus not adequately reflecting the future project performance over the complete crediting period of 7 or 10 years. This category covers approximately 4% of all projects, which have been issued CERs. In contrast, projects which are operational over several years and have been issued CERs over the same period often show performances in a range of 11%–30%. This category covers approximately 30% of all projects that have been issued CERs. Approximately 32% of all projects that have been issued CERs show a performance between 31% and 60% whereas another 29% reported a performance quota between 61% and 100%. Fig. 21.3.4 summarizes the performances observed for a total of 97 registered large-scale CDM landfill projects that have been issued CERs by July 2013.

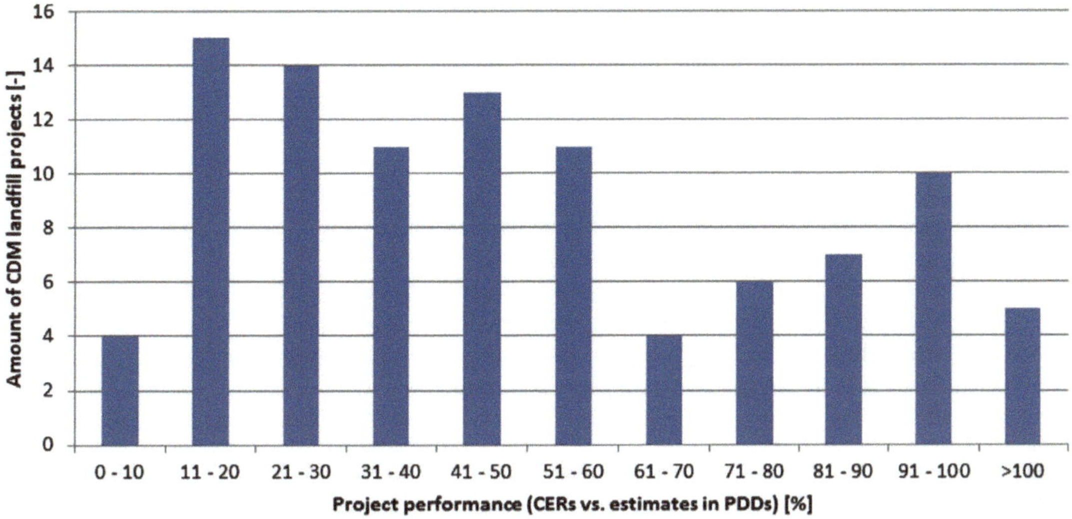

Figure 21.3.4 Performances [certified emission-reduction (CERs) versus estimates in project design documents (PDDs)] by categories observed for a total of 97 registered large-scale clean development mechanism (CDM) landfill projects which have been issued CERs by July 2013.

To identify applicable reasons for the observed underperformances is difficult since there is no obligation to justify negative differences from estimated values in registered PDDs. However, based on the information provided with the PDDs it is assumed that one or more of the following reasons are causal for the observed severe underperformance:

- relatively old waste deposited in the landfill (up to 50 years), thus biological decomposition of organic material has already taken place to a significant extent prior to project implementation;
- no or insufficient waste compaction (densities of 0.5–0.6 mg/m^3, based on wet waste) associated with enhanced aerobic decomposition and reduced methane formation potential;
- landfill fires may contribute to a reduction of degradable organics;
- poor (permeable) landfill top cover not allowing for LFG extraction at full capacity (lower methane concentrations due to air infiltration);
- operational problems, such as defect measuring devices (e.g., flow meters, data logger, etc.);
- delay in the installation of the projected extraction, flaring, and/or electricity generation capacity;
- change in waste composition between the time of establishment of the PDD and project activity implementation; and
- insufficient modeling results due to incorrect parameter estimation.

On the upper extreme, 5% of the projects show an overperformance by up to 147%. For these cases, project participants are obliged to provide an explanation for the increased amount of emission reductions included in the periodic monitoring report. Based on these reports the following reasons have been identified the most common in case of overperformance:

- higher CH_4 concentrations in the collected LFG in comparison with the assumption made in the PDD;
- increased amount of captured LFG due to a larger landfill area covered by the gas extraction system;
- increased amount of captured LFG due to systematic monitoring at individual gas wells and accordant optimization of the gas extraction system;
- increased gas extraction efficiency due to reduced leachate levels in both gas wells and landfill (installation of pumps for leachate discharge in the vertical gas wells);
- increased availability (in terms of operation time) of the LFG extraction plant;
- increased flare efficiency (upgraded flaring system); and
- higher availability of gas engines (in comparison with the assumption made in the PDD).

Moreover, there is a correlation between the project performance and the estimated amount of CERs as per the PDD. Whereas projects with a low performance (<30%) are relatively smaller (estimated annual amount of CERs averages 217 kilotonnes $CO_{2,e}$), projects with a medium performance (30%–60%) are relatively larger (estimated annual amount of CERs averages 243 kilotonnes $CO_{2,e}$), and projects with performance rates >60% show an average estimated annual amount of CERs of 251 kilotonnes $CO_{2,e}$.

PROS AND CONS OF CLEAN DEVELOPMENT MECHANISM LANDFILL PROJECTS—DISCUSSION

Landfill related subjects represent almost 4% of all registered CDM projects in 2013. During the first (2005–07) and second (2008–12) commitment period, 26 large scale and 6 SSC projects have annually been registered on average. In October 2013, the total number of CDM landfill projects amounted to 250.

Since landfills are some of the major contributors of anthropogenic GHG emissions (mainly through the diffuse emission of methane), projects aiming at emission reductions can be considered ecologically worthwhile. The overall intention of CDM landfill projects is to reduce GHG emissions through LFG capture and treatment, whereas the latter may be realized either through flaring or energy recovery. Revenues from the project activities can be gained by selling CERs on the accordant market. If project activities involve the generation of electrical and/or thermal energy to be fed into the grid, the amount of issued CERs can be increased since fossil fuels for energy generation are replaced.

However, it has to be discussed whether CDM projects in landfilling can provide the right incentives for a sustainable waste management in general and toward sustainable landfills in particular. First of all, significant amounts of LFG emissions are only generated if solid waste of high organic content is landfilled. Experiences from, e.g., Europe show that this strategy is associated with significant problems, since long-term emissions occur whereat the generation of polluted leachate is of particular importance. In consequence, landfills require long-term aftercare associated with relevant costs.

In addition, leachate emissions are capable to pollute both soil and water in the landfill vicinity. The European approach (set out in the Council Directive, 1999/31/EC) is therefore to reduce landfilling of organic waste and stimulating alternative treatment option such as composting, anaerobic fermentation, incineration, or mechanical biological treatment. In general, CDM projects for LFG capture and treatment do not contribute into this direction but rather provide a motivation for a continuation of organic waste deposition. Nevertheless, the recently approved methodology for the avoidance of LFG emissions by in situ aeration of landfills represents a suitable approach toward the creation of sustainable landfills. During the aeration process, LFG emissions are widely avoided, and CERs can be generated in an amount comparable to the one generated by LFG capture and treatment. Since aeration contributes toward an accelerated biological waste stabilization, residual emissions (both LFG generation and polluted leachate) are significantly reduced after the project has been completed. At that time the landfill shows a low-emission potential thus can be considered essentially more sustainable than before the project activity (see Chapter 16.2).

Another issue in the discussion about current and future CDM landfill projects is the question whether the invested money could produce a greater environmental benefit if spent outside the CDM framework, e.g., in awareness creation campaigns, introduction of source separated waste collection, increased recycling activities, etc. In this connection, it must be stated that in the absence of CDM landfill projects, less incentives for improved landfill standards would exist, and the technology transfer into DCs would not take place in this dimension. In addition, the potential for GHG emission reductions from (particularly) large landfill sites is significant and inevitable in the framework of global efforts for climate change mitigation. Beside the landfill sector the CDM sets incentives for the implementation of alternative waste treatment processes through the approval of the consolidated baseline and monitoring methodology ACM0022 in September 2012. This methodology regulates the application of different waste treatment options such as composting, anaerobic digestion, as well as thermal or mechanical treatment to reduce the disposal of fresh organic waste in solid waste disposal sites (SWDSs). Thus, it represents a necessary amendment to the LFG capture and flaring approaches since LFG emissions can be avoided and not only mitigated. By the end of 2013, a total of six projects had been registered under the accordant methodology. All of them are intended to avoid GHG emissions from landfills through the installation and operation of thermal treatment (incineration, gasification, thermal recovery of RDF) whereat one project includes waste composting. Based on the assessments made within the PDDs, 71,500 tonnes $CO_{2,e}$ are expected to be annually avoided on average per project. However, as of December 2013 no CERs have been issued for these project activities.

Another improvement for increasing the efficiency of CDM projects could be achieved if the extracted gas is not flared but used for energy production.

One of the most important criticisms about CDM landfill projects regards the significant underperformance in terms of CER generation.To attract more investors, increase the number of project activities, and to reduce the financial uncertainty it would be essential to provide a more realistic beforehand estimation of the targeted CERs. The latter is done by means of a first order decay (FOD) model, capable to calculate the amount of methane generated from a landfill using Eq. (1). The model differentiates between different types of waste with respectively different decay rates and different fractions of degradable organic carbon that have been deposited in the landfill.

$$BE_{CH_4,SWDS,y} = \varphi \cdot (1-f) \cdot GWP_{CH_4} \cdot (1-OX) \cdot \frac{16}{12} F \cdot DOC_f \cdot MCF \cdot \sum_{x=1}^{y} \sum_{j} W_{j,x} \cdot DOC_j \cdot e^{-k_j(y-x)} \cdot (1-e^{-k_j}) \tag{1}$$

where:

$BE_{CH_4,SWDS,y}$, Methane emissions avoided during the year y from preventing waste disposal at the SWDS during the period from the start of the project activity to the end of the year y (tonne $CO_{2,e}$)
φ, Default model correction factor to account for model uncertainties (0.9)
F, Fraction of methane captured at the SWDS and flared, combusted, or used in another manner
GWP_{CH_4}, Global warming potential of methane, valid for the relevant commitment period (tonne CO_2/tonne CH_4)
OX, Oxidation factor (reflecting the amount of methane from the landfill that is oxidized in the soil or other material covering the waste)
F, Fraction of methane in the SWDS gas (volume fraction) (0.5)
DOC_f, Fraction of degradable organic carbon (DOC) that can decompose
MCF, Methane correction factor
$W_{j,x}$, Amount of organic waste type j prevented from disposal in the SWDS in the year x (tonnes)
DOC_j, Fraction of degradable organic carbon (by weight) in the waste type j
k_j, Decay rate for the waste type j
j, Waste type category (index)
x, Year during the crediting period: x runs from the first year of the first crediting period ($x = 1$) to the year y for which avoided emissions are calculated ($x = y$)
y, Year for which methane emissions are calculated.

Based on the specific reasons mentioned for project underperformance within the monitoring reports, incorrect parameter estimation for the ex ante calculation of emission reductions is the dominating aspect. This can be demonstrated by means of several examples:

Methane oxidation factor

The methane oxidation factor is reflecting the amount of methane generated from a landfill that is oxidized in the top cover. Provided that the landfill is covered with oxidizing material such as soil or compost, the calculated methane emissions (as of the FOD approach) are to be attenuated by 10%. However, this quota does not adequately reflect the actual CH_4 oxidation potential of a top cover. Oxidation rates between 0.5 and 13 L $CH_4/(m^2\ h)$ have been found even for simple top covers (Röwer et al., 2011) whereby up to 1300 m^3 methane could be oxidized hourly at a 10 ha landfill. In consequence, calculation results tend to overestimate the actual LFG emissions.

Methane correction factor

The methane correction factor accounts for the fact that unmanaged landfills produce less methane from a given amount of waste than managed landfills, because a larger fraction of waste decomposes aerobically in the top layers of unmanaged landfills. In addition, for shallow landfills or sites with a high water table lower-methane emissions are considered. What is missing in these considerations is the fact that at unmanaged landfills the waste is usually not sufficiently compacted. The lack of compaction leads to significantly longer aerobic degradation processes and, in addition, lower performance of the gas extraction systems once the project activity is carried out. In consequence, calculation results tend to overestimate the actual LFG emissions.

Waste characteristic (amount of biodegradable organic carbon)

The amount of biodegradable organic carbon is calculated on the basis of table values. No validation of the calculation results by means of waste sampling and analysis is foreseen in the methodology. Impacts on the specific biodegradability such as landfill fires are not considered, thus calculation results tend to overestimate the actual LFG emissions.

Moisture content and distribution

The FOD model approach assumes the landfill to be a well-mixed system with optimum microbial kinetics. This is certainly not the case, and the attenuation made for general model uncertainties (reduction by 10%) is not sufficient to balance this false estimation. In particular, waste moisture content and water flow inside the waste mass are not adequately considered. Dry zones (i.e., moisture content less than 20%) do not contribute toward LFG generation, and an insufficient water flow will not allow for the necessary contact between substrate and microbial biomass. In both cases, actual LFG generation rates will be lower than estimated by means of the model approach, and potential emission reductions will be overestimated.

In summary, through consideration of the above mentioned aspects a more realistic ex ante assessment of the CERs to be generated in the projects framework seems to be possible.

Finally, the question about the relevance of GHG emission reductions in the framework of CDM landfill projects in comparison with the overall emissions from SWDS has to be discussed. According to latest estimations made with the Fifth Assessment Report Climate Change 2013 by the Intergovernmental Panel on Climate Change (IPCC) a total of 67–90 million tonnes of CH_4 are annually emitted from landfills, old waste deposits, and other waste treatment activities worldwide. Under consideration of the 25 times greater impact of methane toward heat trapping in the atmosphere (in comparison with CO_2), these emissions are equivalent to approximately 2 billion tonnes of $CO_{2,e}$ on average per year. At the same time estimated emission reductions from registered CDM projects amount to approximately 42.5 million tonnes of $CO_{2,e}$ (both large and SSC project activities). Under consideration of the specific performance rates of these projects, a total annual amount of 21.5 million - tonnes of $CO_{2,e}$ emission reductions can be expected in 2013 and following years. Consequently, at present time CDM landfill projects exhibit a GHG emission reduction potential of approximately 1% of the global GHG emissions from landfills and waste. Even if the estimations made with the

PDDs will be met, the quota will raise only up to 2% at its best. Large-scale projects are contributing more than 97% of these reductions.

Owing to the increase in population on global scale in connection with economic growth in many DCs and emerging economies, the amount of landfilled municipal solid waste (MSW) is expected to further increase over the next years. CDM projects as an instrument for GHG emission mitigation alone will most likely not be sufficient for an effective attenuation of the rising emissions, and certainly not for a significant emission reduction. At the same time, the potential for emission reductions (ERs) at landfills is very high since these anthropogenic GHG emissions are the largest emission category that is not connected to food production. Since the price for CERs is generally fluctuating and has decreased to a very low value in 2013, it is questionable whether the CDM provides the right incentives that are necessary. Fixed revenues for electricity sales, guaranteed by the national governments, could be a more efficient way to increase the LFG collection and utilization quota and reduced GHG emissions from landfills. This incentive has to come along with accordant adaptation in the national legislations. In this connection Germany provides a good example: Since the early 1990s the waste-related legislation has been modified, and organizational measures have been taken to reduce the amount of landfilled MSW. These measures comprised the source-separated collection and utilization of organic waste from households and commercial activities, increased collection quota for recyclables (e.g., glass, paper,cardboard, metals, and plastics) as well as the separate collection of packaging waste. Since 2005 the landfilling of MSW without pretreatment is ineligible. In parallel to these optimizations in the waste management system, technical installations for LFG capture and utilization are to be installed at MSW landfills on the basis of the national legislation. In consequence, methane emissions from German landfills could be reduced from 1.7 million tonnes in 1990 to 0.4 million tonnes in 2009. On average 27 million tonnes $CO_{2,e}$ could be avoided annually. In Germany also in situ aeration of closed old landfills is financially subsidized as a climate gas reduction measure by the NKI initiative of the German Government (Stegmann et al., 2015).

References

Council Directive, July 17, 1999. Council Directive 1999/31/EC of 26 April 1999 on the landfill of waste. Official Journal of the European Communities.

IPCC, 1996. Climate Change 1995: The Science of Climate Change. Intergovernmental Panel on Climate Change (IPCC). Cambridge University Press, Cambridge, United Kingdom.

IPCC, 2013. Climate Change 2013: The Physical Science Basis. Final Draft Report of the Working Group I Contribution to the IPCC 5th Assessment Report.

Keppler, F., Hamilton, J.T.G., Brass, M., Röckmann, T., January 12, 2006. Methane emissions from terrestrial plants under aerobic conditions. Nature 439.

Lelieveld, J., Crutzen, P.J., Dentener, F.J., 1998. Changing concentration, lifetime and climate forcing of atmospheric methane. Tellus 50B, 128–150.

Röwer, I.U., Geck, C., Gebert, J., Pfeiffer, E.-M., 2011. Spatial variability of soil gas concentration and methane oxidation capacity in landfill covers. Waste Management 31 (5), 926–934.

Stegmann, R., Heyer, K.U., Hupe, K., 2015. Landfill aeration in the course of the German climate protection initiative. In: Proceedings of the 15. International Waste Management and Landfill Symposium. CISA Publisher, ISBN 9788862650212.

UNFCCC, 2013. Website of the United Nations Framework Convention on Climate Change. Clean Development Mechanism (CDM). http://cdm.unfccc.int/Projects/projsearch.html.

INDEX

Note: Page numbers followed by "f" indicate figures, "t" indicate tables and "b" indicate boxes.

A

B

C

D

E

F

G

H

I

J

K

L

M

N

O

P

Q

R

S

T

U

V

W

X

Z

Kolibri 360 GmbH
Ferdinand-Jühlke-Straße 7
99095 Erfurt, Deutschland
produktsicherheit@kolibri360.de